T0392563

Materials Nanoarchitectonics

Materials Nanoarchitectonics
From Integrated Molecular Systems to Advanced Devices

Edited by

Katsuhiko Ariga

Research Center for Materials Nanoarchitectonics (MANA), National Institute for Materials Science (NIMS), Namiki, Tsukuba, Japan
Graduate School of Frontier Sciences, The University of Tokyo, Kashiwanoha, Kashiwa, Japan

Omar Azzaroni

Instituto de Investigaciones Fisicoquímicas Teóricas y Aplicadas (INIFTA)—Departamento de Química—Facultad de Ciencias Exactas—Universidad Nacional de La Plata—CONICET, La Plata, Argentina

ELSEVIER

Elsevier
Radarweg 29, PO Box 211, 1000 AE Amsterdam, Netherlands
The Boulevard, Langford Lane, Kidlington, Oxford OX5 1GB, United Kingdom
50 Hampshire Street, 5th Floor, Cambridge, MA 02139, United States

MATLAB® is a trademark of The MathWorks, Inc. and is used with permission. The MathWorks does not warrant the accuracy of the text or exercises in this book. This book's use or discussion of MATLAB® software or related products does not constitute endorsement or sponsorship by The MathWorks of a particular pedagogical approach or particular use of the MATLAB® software.

Notices

Knowledge and best practice in this field are constantly changing. As new research and experience broaden our understanding, changes in research methods, professional practices, or medical treatment may become necessary.

Practitioners and researchers must always rely on their own experience and knowledge in evaluating and using any information, methods, compounds, or experiments described herein. In using such information or methods they should be mindful of their own safety and the safety of others, including parties for whom they have a professional responsibility.

To the fullest extent of the law, neither the Publisher nor the authors, contributors, or editors, assume any liability for any injury and/or damage to persons or property as a matter of products liability, negligence or otherwise, or from any use or operation of any methods, products, instructions, or ideas contained in the material herein.

ISBN: 978-0-323-99472-9

For Information on all Elsevier publications
visit our website at https://www.elsevier.com/books-and-journals

Publisher: Matthew Deans
Acquisitions Editor: Ana Claudia A. Garcia
Editorial Project Manager: Andrea Gallego Ortiz
Production Project Manager: Prem Kumar Kaliamoorthi
Cover Designer: Greg Harris

Typeset by MPS Limited, Chennai, India

Working together
to grow libraries in
developing countries

www.elsevier.com • www.bookaid.org

Dedication

This book is dedicated to Dante and Ivo.

Contents

List of contributors

Juan A. Allegretto
Instituto de Investigaciones Fisicoquímicas Teóricas y Aplicadas (INIFTA)—Departamento de Química—Facultad de Ciencias Exactas—Universidad Nacional de La Plata—CONICET, La Plata, Argentina

Paula C. Angelomé
Chemistry Department & Nanoscience and Nanotechnology Institute, CAC, CNEA, CONICET, Buenos Aires, Argentina

Melina Arcidiácono
Instituto de Investigaciones Fisicoquímicas Teóricas y Aplicadas (INIFTA)—Departamento de Química—Facultad de Ciencias Exactas—Universidad Nacional de La Plata—CONICET, La Plata, Argentina

Katsuhiko Ariga
Research Center for Materials Nanoarchitectonics (MANA), National Institute for Materials Science (NIMS), Namiki, Tsukuba, Japan; Graduate School of Frontier Sciences, The University of Tokyo, Kashiwanoha, Kashiwa, Japan

Omar Azzaroni
Instituto de Investigaciones Fisicoquímicas Teóricas y Aplicadas (INIFTA)—Departamento de Química—Facultad de Ciencias Exactas—Universidad Nacional de La Plata—CONICET, La Plata, Argentina

Michal Bodik
Institute of Physics, Slovak Academy of Sciences, Bratislava, Slovakia; Centre for Advanced Materials Application, Bratislava, Slovakia

Vanina M. Cayón
Instituto de Investigaciones Fisicoquímicas Teóricas y Aplicadas (INIFTA)—Departamento de Química—Facultad de Ciencias Exactas—Universidad Nacional de La Plata—CONICET, La Plata, Argentina; GSI Helmholtzzentrum für Schwerionenforschung, Darmstadt, Germany

Cintia Belen Contreras
Instituto de Nanosistemas, Escuela de Bio y Nanotecnologías, Universidad Nacional de San Martín, San Martín, Buenos Aires, Argentina

M. Lorena Cortez
Instituto de Investigaciones Fisicoquímicas Teóricas y Aplicadas (INIFTA)—Departamento de Química—Facultad de Ciencias Exactas—Universidad Nacional de La Plata—CONICET, La Plata, Argentina

Priscila Destro
Institute of Chemistry, University of Campinas (UNICAMP), Campinas, SP, Brazil

Nidhi C. Dubey
Department of Molecular Medicine, Jamia Hamdard, New Delhi, India

Tanna E.R. Fiuza
Institute of Chemistry, University of Campinas (UNICAMP), Campinas, SP, Brazil; Brazilian Nanotechnology National Laboratory (LNNano), Brazilian Center for Research in Energy and Materials (CNPEM), Campinas, SP, Brazil

M. Cecilia Fuertes
Chemistry Department & Nanoscience and Nanotechnology Institute, CAC, CNEA, CONICET, Buenos Aires, Argentina

Divya Gaur
Department of Materials Science & Engineering, Indian Institute of Technology Delhi, New Delhi, India

Danielle S. Gonçalves
Institute of Chemistry, University of Campinas (UNICAMP), Campinas, SP, Brazil; Karlsruhe Institute of Technology (KIT), Institute of Catalysis Research & Technology (IKFT), Baden, Württemberg, Germany

Kuniharu Ijiro
Research Institute for Electronic Science, Hokkaido University, Sapporo, Japan

Rizwan Khan
Department of Electrical Engineering, Kwangwoon University, Seoul, Republic of Korea

Tathiana M. Kokumai
Institute of Chemistry, University of Campinas (UNICAMP), Campinas, SP, Brazil

Makoto Komiyama
Research Center for Advanced Science and Technology (RCAST), The University of Tokyo, Komaba, Meguro, Tokyo, Japan.

Gregorio Laucirica
Instituto de Investigaciones Fisicoquímicas Teóricas y Aplicadas (INIFTA)—Departamento de Química—Facultad de Ciencias Exactas—Universidad Nacional de La Plata—CONICET, La Plata, Argentina

Jun-Ge Liang
Engineering Research Center of IoT Technology Applications (Ministry of Education), Department of Electronic Engineering, Jiangnan University, Wuxi, P.R. China; School of Electronic Science and Engineering, Collaborative Innovation Center of Advanced Microstructures, Nanjing University, Nanjing, P.R. China

Hiromitsu Maeda
Department of Applied Chemistry, College of Life Sciences, Ritsumeikan University, Kusatsu, Shiga, Japan

Waldemar A. Marmisollé
Instituto de Investigaciones Fisicoquímicas Teóricas y Aplicadas (INIFTA)—Departamento de Química—Facultad de Ciencias Exactas—Universidad Nacional de La Plata—CONICET, La Plata, Argentina

Ana Paula Mártire
Instituto de Investigaciones Fisicoquímicas Teóricas y Aplicadas (INIFTA)—Departamento de Química—Facultad de Ciencias Exactas—Universidad Nacional de La Plata—CONICET, La Plata, Argentina

Tsuyoshi Minami
Institute of Industrial Science, The University of Tokyo, Tokyo, Japan

Hideyuki Mitomo
Research Institute for Electronic Science, Hokkaido University, Sapporo, Japan

Yuta Nishina
Research Core for Interdisciplinary Sciences, Okayama University, Okayama, Japan

Lijia Pan
School of Electronic Science and Engineering, Collaborative Innovation Center of Advanced Microstructures, Nanjing University, Nanjing, P.R. China

Miguel Ángel Pasquale
Instituto de Investigaciones Fisicoquímicas Teóricas y Aplicadas (INIFTA)—Departamento de Química—Facultad de Ciencias Exactas—Universidad Nacional de La Plata—CONICET, La Plata, Argentina

Ramakrishnan Prakash
Daegu Gyeongbuk Institute of Science & Technology (DGIST), Daegu, The Republic of Korea

Matías Rafti
Instituto de Investigaciones Fisicoquímicas Teóricas y Aplicadas (INIFTA)—Departamento de Química—Facultad de Ciencias Exactas—Universidad Nacional de La Plata—CONICET, La Plata, Argentina

Maria K. Ramos
Department of Chemistry, Federal University of Paraná (UFPR), Curitiba, Paraná, Brazil

Karen A. Resende
Institute of Chemistry, University of Campinas (UNICAMP), Campinas, SP, Brazil

Yui Sasaki
Institute of Industrial Science, The University of Tokyo, Tokyo, Japan

Sangaraju Shanmugam
Daegu Gyeongbuk Institute of Science & Technology (DGIST), Daegu, The Republic of Korea

Peter Siffalovic
Institute of Physics, Slovak Academy of Sciences, Bratislava, Slovakia; Centre for Advanced Materials Application, Bratislava, Slovakia

Galo J.A.A. Soler-Illia
Instituto de Nanosistemas, Escuela de Bio y Nanotecnologías, Universidad Nacional de San Martín, San Martín, Buenos Aires, Argentina

María Eugenia Toimil-Molares
GSI Helmholtzzentrum für Schwerionenforschung, Darmstadt, Germany

Yamili Toum Terrones
Instituto de Investigaciones Fisicoquímicas Teóricas y Aplicadas (INIFTA)—Departamento de Química—Facultad de Ciencias Exactas—Universidad Nacional de La Plata—CONICET, La Plata, Argentina

Christina Trautmann
GSI Helmholtzzentrum für Schwerionenforschung, Darmstadt, Germany; Technische Universität Darmstadt, Materialwissenschaft, Darmstadt, Germany

Bijay P. Tripathi
Department of Materials Science & Engineering, Indian Institute of Technology Delhi, New Delhi, India

Ianina L. Violi
Nanosystems Institute, University of San Martín, Buenos Aires, Argentina

Kazuhisa Yamasumi
Department of Applied Chemistry, College of Life Sciences, Ritsumeikan University, Kusatsu, Shiga, Japan

Youfeng Yue
Research Institute for Advanced Electronics and Photonics, National Institute of Advanced Industrial Science and Technology (AIST), Tsukuba, Japan; PRESTO, Japan Science and Technology Agency (JST), Kawaguchi, Japan

Daniela Zanchet
Institute of Chemistry, University of Campinas (UNICAMP), Campinas, SP, Brazil

Aldo J.G. Zarbin
Department of Chemistry, Federal University of Paraná (UFPR), Curitiba, Paraná, Brazil

Preface

Nanoarchitectonics as the method for everything in materials science

There are many social demands, such as energy conversion and storage, environmental remediation and sensing, and various medical applications. The development of new functional materials is an important means by which science and technology can meet these demands. To achieve functions beyond those of conventional materials, it is essential to control the shape and internal structure of the material, as well as to improve the materials' intrinsic properties. This requires the introduction of advanced technologies for structural control at the atomic/molecular and nanoscale, in addition to conventional methods of organic synthesis and materials' preparation. In other words, materials fabrication based on nanotechnological knowledge and techniques is highly desirable.

It is widely recognized that nanotechnology plays a central role in the development of materials having nanoscale structures. However, nanotechnology is not a discipline dedicated to material synthesis, as its main focus is on developing a better understanding of new phenomena and physical principles at the nanoscale. For the construction of functional materials from nanoscale unit structures, contributions from fields other than nanotechnology, such as supramolecular chemistry, material processing, and biotechnology are essential. Therefore the creation of functional materials from nanoscale units should be based on a new concept that combines the various research fields mentioned earlier with the concept of nanotechnology. This concept is "nanoarchitectonics."

Richard Feynman founded nanotechnology in the mid-20th century and Masakazu Aono proposed nanoarchitectonics as a postnanotechnology concept at the beginning of the 21st century. Nanoarchitectonics based on nanotechnology integrates organic chemistry, inorganic chemistry, polymer chemistry, supramolecular chemistry, coordination chemistry, other materials chemistry, biology-related chemistry, and microfabrication technology to create functional material systems from nanounits (atoms, molecules, and nanomaterials). This is a "unified methodology" to integrate the fields.

In nanoarchitectonics, functional materials are assembled by combining various elemental technologies. For example, functional structures are created by selecting and combining appropriate elements from atomic and molecular manipulation, chemical and physical material transformation, self-assembly/self-organization, orientation and stacking by external fields and forces, microfabrication techniques, and biological processes. The following are some of the most important aspects of materials development. Materials are originally composed of atoms and molecules. Therefore nanoarchitectonics, in which functional materials are constructed from nanounits such as atoms and molecules, can be a creation method for all materials. By analogy with the Theory of Everything in physics, nanoarchitectonics could be called the Method of Everything in materials science. Therefore the term "nanoarchitectonics" is widely used regardless of material systems, functions, and applications. This book provides an overview of nanoarchitectonics, illustrates the creation of functional structures with it, and discusses the possibilities.

Kasuhiko Ariga
Omar Azzaroni

Nanoarchitectonics: a land of opportunities

Omar Azzaroni[1] and Katsuhiko Ariga[2,3]

[1]*Instituto de Investigaciones Fisicoquímicas Teóricas y Aplicadas (INIFTA)—Departamento de Química—Facultad de Ciencias Exactas—Universidad Nacional de La Plata—CONICET, La Plata, Argentina* [2]*Research Center for Materials Nanoarchitectonics (MANA), National Institute for Materials Science (NIMS), Namiki, Tsukuba, Japan* [3]*Graduate School of Frontier Sciences, The University of Tokyo, Kashiwanoha, Kashiwa, Japan*

1.1 Bottom-up creation of functional materials and devices

It is a truism that scientific and technological progress is of great importance in accelerating the social and economic development of our societies. In this sense, to achieve sustainable growth and economic diversification, reduce unemployment, and support rising living standards, we have to solve various problems related to environmental, energy, and health issues. These demands are settled by a wide range of scientific efforts including organic and polymer syntheses to create new molecules and polymers [1], materials processing to fabricate functional structures [2], energy/materials conversion and catalysis [3], facile sensing and analysis [4], energy generation and storage with advanced solar cells [5], fuel cells [6], capacitors [7], and batteries [8], and biological and biomedical treatments [9]. To accomplish low-cost, environment-friendly, and less-emissive systems, these scientific and technical processes have to pursue high efficiency and specificity.

Developments in science and technology also have initiated one distinct discipline, nanotechnology, that has enabled us to observe, analyze, and fabricate nanoscopic systems and materials [10]. The development of observation and manipulation techniques at the nanoscale, which extends to the molecular and even atomic levels, has led to the discovery and understanding of nanoscale phenomena [11]. In addition, these scientific and technological advances have promoted understanding and advancements in related fields. The development of microfabrication techniques has also stimulated the advancement of device technologies [12]. In supramolecular chemistry, research on specific molecular recognition [13] and self-assembly [14] has been actively pursued, leading to a broad range of functional materials. In biological chemistry, the importance of nanostructures in functional systems has also been revealed with the advancement of analytical technology [15]. To promote these parallel developments in science and technology in a more rational manner, they should be treated as a unified concept. In particular, we should emphasize the fact that nanotechnology has played the role of a game changer in the observation and elucidation of nanostructures [16]. In order to harness these developments for the production of functional materials, a postnanotechnology concept is necessary. This role has been assigned to an emerging concept: *nanoarchitectonics* (Fig. 1.1) [17,18].

Materials Nanoarchitectonics. DOI: https://doi.org/10.1016/B978-0-323-99472-9.00019-5

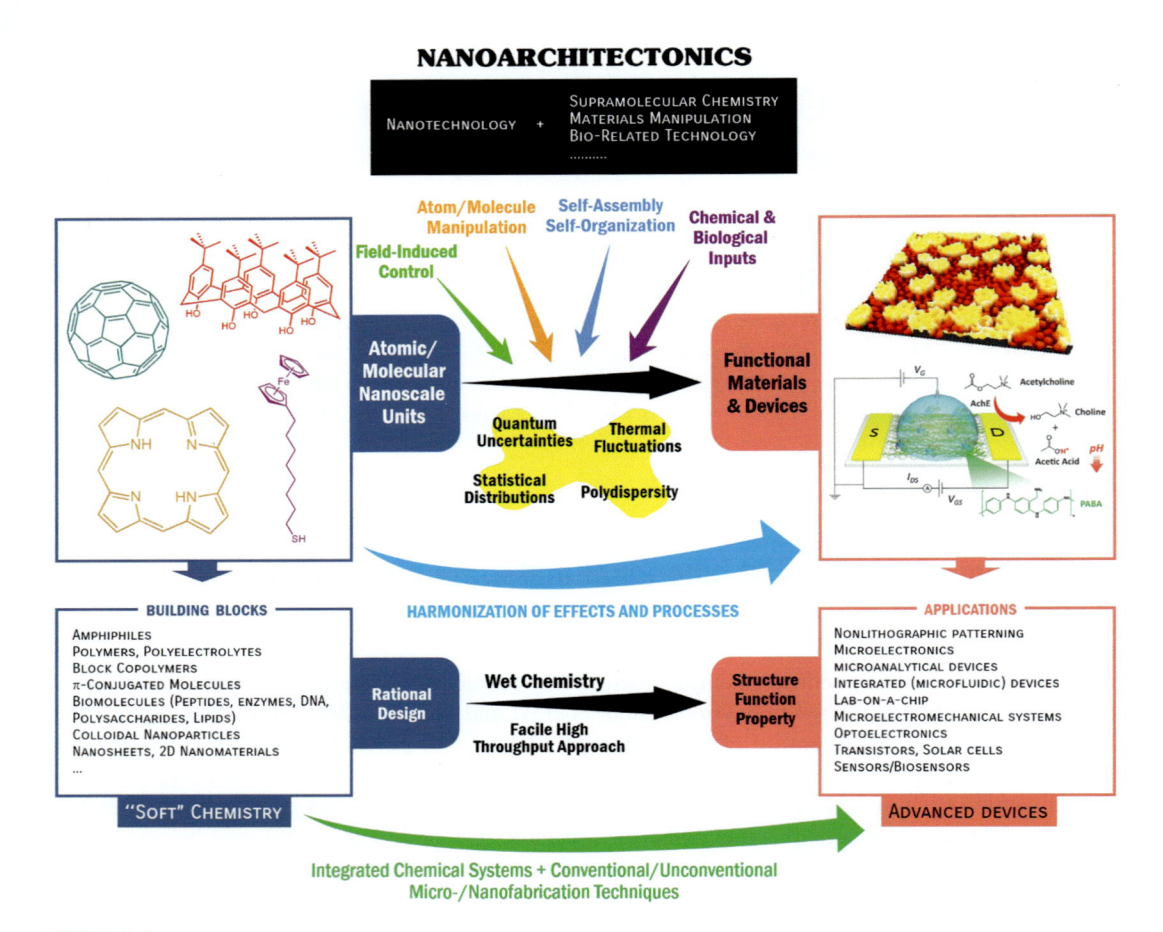

FIGURE 1.1

The nanoarchitectonics approach to produce the functional materials and devices upon structural construction from nanounits through the combination of nanotechnology concepts and the other research fields, such as organic chemistry, supramolecular chemistry, materials science, and bio-related technology.

Just as Richard Feynman paved the way for nanotechnology [19], the concept of nanoarchitectonics was proposed by Masakazu Aono [20−22]. Nanoarchitectonics aims to establish a discipline in which functional materials are fabricated from nanoscale components, such as atoms, molecules, and nanomaterials, using various techniques [23]. Therefore, nanoarchitectonics opens routes to form a more unified paradigm by integrating nanotechnology with organic chemistry, supramolecular chemistry, material chemistry, microfabrication technology, and biotechnology [24].

We can design functional materials through manipulation of the nanoscale, like carpenters construct houses and buildings. However, it is necessary to differentiate two key terms: nanotechnology and nanoarchitectonics. Nanotechnology has become a popular word these days. Even nonscientists may frequently use this word in daily conversation. Nanotechnology (and/or nanoscience) deals with nanoscale and nano-related objects. On the basis of scale, it can be regarded as an advanced form of microtechnology, which can be activated by the considerable developments of microfabrications. However, nanotechnology and microtechnology are somewhat different. Most of the phenomena at the microscale (10^{-6} m range and the related scales) can be estimated from macroscopic phenomena of the same materials. In contrast, various unexpected properties become apparent for materials at the nanoscale (10^{-9} m range and the related scales). The latter features can be seen in various quantum phenomena. Let us reflect for a moment on properties of nanocarbon materials. Fullerene, carbon nanotubes, and graphene are zero-, one-, and two-dimensional objects, respectively, where some dimensions are reduced to the nanoscale. Their properties cannot be deduced from the properties observed on macroscopic carbon materials such as carbon ash and diamonds. The unexpected phenomena found for nanocarbons actually create a novel scientific field.

As we move further into the 21st century, we are realizing that a paradigm shift from technology to architectonics in nanoscale science and technology is necessary. At this point, the reader may ask, what is the fundamental meaning of nanoarchitectonics? To answer this question, we need to compare material design and fabrication at different scales. In the macroscopic scale (visible-scale worlds), our craft hobbies, carpentry work, and building construction can be done according to design drawings and blueprints. We can create and fabricate materials and structures with 100% probability if we just obey the appropriate design. We can easily expect the fabrication results from the design. This principle can be applicable to fabrications in microscopic scales also. Fabrication in the microscopic invisible scale, the so-called microfabrication, can be done with advanced technologies such as photolithography. These fabrication processes within invisible scales can be also done exactly based on predetermined structural design. The fabrication of microscale objects exactly reflects their design drawings in microfabrication techniques. We can basically assign and expect structures and properties of the fabricated objects from their predesigned drawings at the microscopic scale.

However, when the scale of the systems is reduced to the nanoscale, unexpected disturbances and fluctuations have significant influence on the fabrication of materials and, consequently, systems become partially uncontrollable. Therefore, these fabrication processes are not always done decisively according to their predesigned drawings. Materials in nanoscale regions cannot fundamentally exclude the influence of thermal/statistical fluctuations and mutual interactions. These uncontrollable factors are inevitably included between component atoms, molecules, and materials.

Functional material systems are supposed to be nanoarchitected from these nanoscale units upon the selection and combination of various unit processes, including atomic/molecular manipulation, chemical/material conversion, self-assembly/self-organization, field-assisted arrangement, nano/microfabrication, and biochemical/biological treatment [25]. Since the basic concept of nanoarchitectonics is very general, it can be widely applied to fundamental subjects such as material synthesis [26] and structure control [27], as well as to applied fields, such as energy, the environment [28], catalysts [29], sensors [30], and devices [31].

In addition, the concept of nanoarchitectonics has been extended to various bio-based fields. The nanoarchitectonics concept is utilized in a wide range of bio-related areas such as the behavior of biomolecules [32], the construction of bioactive material systems [33], biomimetic approaches [34], artificial enzymes [35], and biomedical applications [36]. This is because the synthesis of functional materials by nanoarchitectonics is similar in some respects to the construction of functional systems in living organisms. Biological systems consist of the rational organization of constituent molecules. Their structures have highly asymmetric and hierarchical features that allow for chained functional coordination, signal amplification, and vector-like energy and signal flow [37]. Such a hierarchical structure is difficult to achieve in artificial systems based on the simple equilibrium process of self-assembly. In contrast, the process of nanoarchitectonics is based on the premise of combining several different processes, which makes it easier to obtain a hierarchical structure.

As discussed above, architecture at the nanoscale has different characteristics from architecture at the microscale and macroscale [38]. Underpinning the foundation of nanoarchitectonics are phenomena at the nanoscale level. Indeed, we have already mentioned that architecture at large scales is less susceptible to disturbances and, on the contrary, phenomena at the nanoscale level are not always precise and exact. There are a number of uncertainties that are latent in them. Due to thermal fluctuations, stochastic distributions, quantum effects, etc., nanoscale phenomena often contain ambiguities. These effects and contributions are difficult to control but cannot be ignored. In the nanoarchitectonics processes, molecular and materials organizations might be influenced by nonunified sets of statistical distributions in addition to complicated mutual interactions, so that the cause and effect cannot be represented by simple input and output signals. A simple summation of individual actions might not decide the final output for the constructed functional system. Therefore, when preparing nanoarchitected systems, it is necessary to consider the fine structures and compositions of nanosized components and any intercomponent interactions. Harmonization of multiple actions and interactions rather than their summation is a crucial factor in the nanoarchitectonics processes. Intelligent selection including possible errors and fluctuations is possible in the production of functional materials by the nanoarchitectonics concept. That is why, at the nanoscale, it is better to think of the various effects as being *harmonized under uncertainty*, rather than being added together. In fact, these characteristics are similar to those observed in biological systems. The functional molecules in living organisms are exposed to thermal fluctuations and work to harmonize their functions while undergoing uncontrollable perturbations [39].

From an historical perspective, much of the interest in nanoarchitectonics originated from its importance to create new functional nanoarchitectures through the synergetic combination of chemical nanofabrication, self-organization, and field-controlled organization, exploiting chemistry as a key enabler to rationally design molecular building blocks entirely from scratch. While the field of "nanoarchitectonics" has been very active over the last decade, the concept has been almost exclusively circumscribed to the design and construction of nanoarchitected materials. However, the view of nanoarchitectonics has changed at a rapid pace: it is now clear that nanoarchitectonics can also be considered as a concept capable of enabling practical applications that, in turn, might lead to real-world devices. Hence, for several years now, nanoarchitectonics has been no longer circumscribed to the realm of "materials design" but has begun to enter the domain of technology and engineering as well. This tendency was accelerated by the transdisciplinary integration of life sciences, physical sciences, and engineering as a strategy to tackle complex challenges and achieve new and innovative solutions to a broad variety of societal problems.

In these days, materials nanoarchitectonics has spread to many scientific and technological fields, including molecular machines and nanocars [40], amphiphilic assembly [41,42], control of molecular inclusion [43,44], nano- and micropatterning [45−49], organic electrochemical transistors [50−52], nanostructured responsive surfaces [53−58], supramolecular recognition [59,60], mechanofunctional materials [61], thin-film fabrications and functions [62−71], graphene fabrication [72], nanocarbon assembly [73], hybrid mesoporous materials [74−78], fabrication of hybrid materials and integrated systems [79−81], functional colloidal materials [82−87], catalysis [88−90], electrocatalysis [91−94], photocatalytic removal of pollutants [95,96], proton exchange membranes [97−99], optoelectronic devices [100,101], supercapacitors [102,103], batteries [104,105], biosensors [106−114], heterojunctions for photonic functions [115], nanofluidic devices [116−123], soft materials with conflicting properties [124,125], assembly of biomolecules on surfaces [126−131], DNA and enzymes [32,132,133], nucleic acid delivery [134], drug delivery [135−137], cell adhesion control [138−142], nanomedicine [143], biological applications of inorganic materials [144], bioimaging [145,146], and dynamic functions from molecules to structures of macroscopic size [38,147,148].

The aim in this introductory chapter was to reflect on the origin of nanoarchitectonics by first attempting to describe its evolution and then looking at the many faces of this cross-disciplinary research field. The field is quickly evolving and is now intricately interfacing with many different scientific disciplines, from chemistry to physics, to materials science, to engineering, and to biology.

In this sense, we can and must say that practitioners in nanoarchitectonics are coming from very different scientific disciplines. The fundamental of this increasingly important research area is unarguably about how to design/create/build/assemble new materials and devices with optimized, or even, unexpected functions and properties (Fig. 1.2). For this reason, chemists are playing a significant role since the bedrock of materials nanoarchitectonics is certainly about how to assemble atoms, molecules, and nanomaterials into nanostructures of desired coordination environment, sizes, and shapes. To obtain structurally well-defined nanosized architectures, the approach from the viewpoint of chemistry is decisive. For example, the design and synthesis of the component molecules for molecular electronics is strongly dependent on the relationship between the shapes and the functional groups of the molecules and nanoscale quantum phenomena.

Without hesitation, we should also emphasize the predominant role of the wet chemical route as a very effective technique to develop a variety of nanoarchitected materials and nanomaterial-based devices on a large scale. The method offers an easy route to achieve simple as well as innovative, efficient, and low-cost complex assemblies, materials, and devices. In this sense, a remarkable example is the development of chemical routes leading to high-throughput solution processing of large-scale graphene for device applications [149]. Nowadays, many researchers have shown how to standardize nanomaterials production to guarantee stability and reproducibility, while increasing yield. Eliminating the need for multimillion-dollar facilities, many research groups have consistently demonstrated that by using readily available and relatively affordable pieces of benchtop equipment, the production of nanomaterials and nanomaterial-based devices in a controlled, adjustable, and low-cost manner is totally feasible [150,151].

Wet chemical synthesis is one of the most popular and comprehensive techniques that have attracted researchers in nanoarchitectonics from different areas—let us call them *"nanoarchitects"* (Fig. 1.2). Over the years, this community of nanoarchitects has shown a persistent interest in developing new nanoarchitected materials and incorporating/integrating them into device fabrication. Indeed, due to ease of use and reproducibility, many laboratories have extensively exploited chemical solution synthesis and self-assembly as an appealing strategy for obtaining quick and reliable results in different steps of device fabrication.

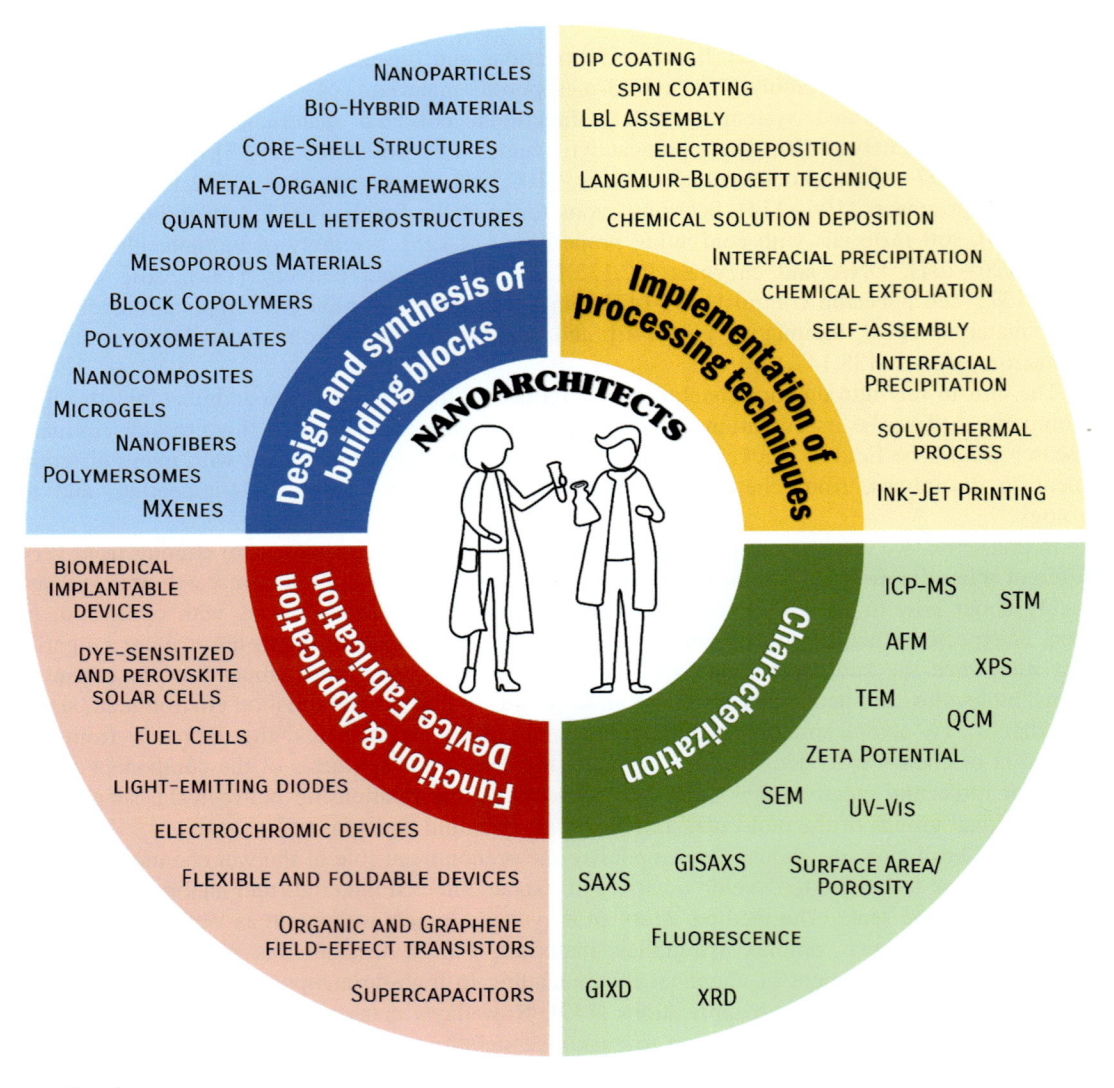

FIGURE 1.2

Today's nanoarchitects at work. Schematic illustration of the four steps involved in the fabrication of nanoarchitected materials and devices. These steps are (clockwise from top left): (1) design and synthesis of appropriate building blocks; (2) implementation of adequate processing conditions; (3) structural, morphological, chemical, and/or physical characterization of the built nanosystems; and (4) fabrication or integration into devices through different strategies that may include: atomic-/molecular-level manipulation, self-assembly, field-regulated materials arrangement, patterning, and so on. The acronyms of the characterization techniques refer to: inductively coupled plasma mass spectrometry (ICP-MS), transmission electron microscopy (TEM), dynamic light scattering (DLS), scanning tunneling microscopy (STM), X-ray photoelectron spectroscopy (XPS), quartz crystal microbalance (QCM), scanning electron microscopy (SEM), UV—Vis spectroscopy, grazing-incidence small-angle X-ray scattering (GISAXS), small-angle X-ray scattering (SAXS), grazing-incidence X-ray diffraction (GIXD), X-ray diffraction (XRD).

In this regard, a nonnegligible merit of physicists and engineers has been their willingness to reach beyond the boundaries in which they have been traditionally confined and to move toward the exploration and application of nanoarchitectonics concepts and notions to create novel devices.

While the present book was not intended to provide a comprehensive picture of the current state of the art of nanoarchitectonics, we do hope that the overview presented here captures the current excitement of this vibrant area of research.

References

[1] G. Povie, Y. Segawa, T. Nishihara, Y. Miyauchi, K. Itami, Science 356 (2017) 172.
[2] W. Chaikittisilp, N.L. Torad, C. Li, M. Imura, N. Suzuki, S. Ishihara, et al., Chem. Eur. J. 20 (2014) 4217.
[3] K.K.R. Datta, B.V.S. Reddy, K. Ariga, A. Vinu, Angew. Chem. Int. Ed. 49 (2010) 5961.
[4] Q. Ji, I. Honma, S.-M. Paek, M. Akada, J.P. Hill, A. Vinu, et al., Angew. Chem. Int. Ed. 49 (2010) 9737.
[5] W.S. Yang, B.-W. Park, E.H. Jung, N.J. Jeon, Y.C. Kim, D.U. Lee, et al., Science 356 (2017) 1376.
[6] C. Santoro, C. Arbizzani, B. Erable, I. Ieropoulos, J. Power Sources 356 (2017) 225.
[7] D. Sheberla, J.C. Bachman, J.S. Elias, C.-J. Sun, Y. Shao-Horn, M. Dinca, Nat. Mater. 16 (2017) 220.
[8] D. Lin, Y. Liu, Y. Cui, Nat. Nanotechnol. 12 (2017) 194.
[9] Q. Zou, M. Abbas, L. Zhao, S. Li, G. Shen, X. Yan, J. Am. Chem. Soc. 139 (2017) 1921.
[10] (a) M. Nishizawa, Bull. Chem. Soc. Jpn. 91 (2018) 1141.
    (b) Q. Hu, H. Li, L. Wang, H. Gu, C. Fan, Chem. Rev. 119 (2019) 6459.
[11] K. Ariga, T. Mori, J.P. Hill, Adv. Mater. 24 (2012) 158−176.
[12] (a) A.-G. Niculescu, C. Chircov, A.C. Bîrca, A.M. Grumezescu, Int. J. Mol. Sci. 22 (2021) 2011.
    (b) H.M. Yamamoto, Bull. Chem. Soc. Jpn. 94 (2021) 2505−2539.
    (c) J.J. Kim, Y. Wang, H. Wang, S. Lee, T. Yokota, T. Someya, Adv. Funct. Mater. 31 (2021) 2009602.
[13] (a) A. Ikeda, S. Shinkai, Chem. Rev. 97 (1997) 1713−1734.
    (b) K. Ariga, T. Kunitake, Acc. Chem. Res. 31 (1998) 371−378.
    (c) Y. Sasaki, X. Lyu, W. Tang, H. Wu, et al., Bull. Chem. Soc. Jpn. 94 (2021) 2613−2622.
[14] K. Ariga, M. Nishikawa, T. Mori, J. Takeya, L.K. Shrestha, J.P. Hill, Sci. Technol. Adv. Mater. 20 (2019) 51−95.
[15] Y. Sugimoto, P. Pou, M. Abe, P. Jelinek, R. Pérez, S. Morita, et al., Nature 446 (2007) 64−67.
[16] T. Shimizu, D. Lungerich, J. Stuckner, M. Murayama, K. Harano, E. Nakamura, Bull. Chem. Soc. Jpn. 93 (2020) 1079−1085.
[17] K. Ariga, Nanoscale Horiz. 6 (2021) 364−378.
[18] O. Azzaroni, K. Ariga, Mol. Syst. Des. Eng. 4 (2019) 9−10.
[19] R.P. Feynman, Calif. Inst. Technol. J. Eng. Sci. 4 (1960) 23−36.
[20] K. Ariga, Q. Ji, J.P. Hill, Y. Bando, M. Aono, NPG Asia Mater. 4 (2012) e17.
[21] This terminology was first proposed by Dr Masakazu Aono at the 1st International Symposium on Nanoarchitectonics Using Suprainteractions (NASI-1) at Tsukuba in 2000.
[22] P.S. Weiss, A conversation with Dr. Masakazu Aono: leader in atomic scale control and nanomanipulation, ACS Nano 1 (2007) 379.
[23] K. Ariga, Q. Ji, N. Nakanishi, J.P. Hill, M. Aono, Mater. Horiz. 2 (2015) 406−413.
[24] K. Ariga, K. Minami, M. Ebara, J. Nakanishi, Polym. J. 48 (2016) 371−389.
[25] K. Ariga, J. Li, J. Fei, Q. Ji, J.P. Hill, Adv. Mater. 28 (2016) 1251−1286.
[26] C. Tirayaphanitchkul, K. Imwiset, M. Ogawa, Bull. Chem. Soc. Jpn. 94 (2021) 678−693.

[27] Y. Sang, M. Liu, Mol. Syst. Des. Eng. 4 (2019) 11−28.

[28] N. Boukhalfa, M. Darder, M. Boutahala, P. Aranda, E. Ruiz-Hitzky, Bull. Chem. Soc. Jpn. 94 (2021) 122−132.

[29] N. Kumari, A. Kumar, V. Krishnan, J. Inorg. Organomet. Polym. 31 (2021) 1954−1966.

[30] M. Komiyama, T. Mori, K. Ariga, Bull. Chem. Soc. Jpn. 91 (2018) 1075−1111.

[31] K. Terabe, T. Tsuchiya, T. Tsuruoka, Adv. Electron. Mater. 8 (2021) 2100645.

[32] M.B. Avinash, T. Govindaraju, Nanoscale 6 (2014) 13348.

[33] D.B. Momekova, V.E. Gugleva, P.D. Petrov, ACS Omega 6 (2021) 33265−33273.

[34] S. Dutta, J. Kim, P.-H. Hsieh, Y.-S. Hsu, Y.V. Kaneti, F.-K. Shieh, et al., Small Methods 3 (2019) 1900213.

[35] M. Komiyama, K. Ariga, Mol. Catal. 475 (2019) 110492.

[36] S. Eom, G. Choi, G., H. Nakamura, J.-H. Choy, Bull. Chem. Soc. Jpn. 93 (2020) 1−12.

[37] K. Simons, D. Toomre, Nat. Rev. Mol. Cell Biol. 1 (2000) 31−39.

[38] M. Aono, K. Ariga, Adv. Mater. 28 (2016) 989−992.

[39] T. Yanadida, Y. Ishii, Proc. Jpn. Acad. Ser. B 93 (2017) 51−63.

[40] Y. Shirai, K. Minami, W. Nakanishi, Y. Yonamine, C. Joachim, K. Ariga, Jpn. J. Appl. Phys. 55 (2016) 1102A2.

[41] M. Ramanathan, L.K. Shrestha, T. Mori, Q. Ji, J.P. Hill, K. Ariga, Phys. Chem. Chem. Phys. 15 (2013) 10580.

[42] L.K. Shrestha, K.M. Strzelczyk, R.G. Shrestha, K. Ichikawa, K. Aramaki, J.P. Hill, et al., Nanotechnology 26 (2015) 204002.

[43] L. Zerkoune, A. Angelova, S. Lesieur, Nanomaterials 4 (2014) 741.

[44] K. Ariga, M. Naito, Q. Ji, D. Payra, Cryst. Eng. Comm. 18 (2016) 4890.

[45] O. Azzaroni, M.H. Fonticelli, G. Benítez, P.L. Schilardi, R. Gago, I. Caretti, et al., Adv. Mater. 16 (2004) 405−409.

[46] O. Azzaroni, P.L. Schilardi, R.C. Salvarezza, Appl. Phys. Lett. 80 (2002) 1061−1063.

[47] P.L. Schilardi, O. Azzaroni, R.C. Salvarezza, Langmuir 17 (2001) 2748−2752.

[48] O. Azzaroni, S.E. Moya, A.A. Brown, Z. Zheng, E. Donath, W.T.S. Huck, Adv. Funct. Mater. 16 (2006) 1037−1042.

[49] M.A. Auger, P.L. Schilardi, I. Caretti, O. Sánchez, G. Benítez, J.M. Albella, et al., Small 1 (2005) 300−309.

[50] G.E. Fenoy, J. Scotto, J.A. Allegretto, E. Piccinini, A.L. Cantillo, W. Knoll, et al., ACS Appl. Electro. Mater. 4 (2022) 5953−5962.

[51] G.E. Fenoy, C. von Bilderling, W. Knoll, O. Azzaroni, W.A. Marmisollé, Adv. Electron. Mater. 7 (2021) 2100059.

[52] G.E. Fenoy, R. Hasler, F. Quartinello, W.A. Marmisollé, C. Lorenz, O. Azzaroni, et al., JACS Au 2 (2022) 2778−2790.

[53] B. Sanz, C. von Bilderling, J.S. Tuninetti, L. Pietrasanta, C. Mijangos, G.S. Longo, et al., Soft Matter 13 (2017) 2453−2464.

[54] S.E. Herrera, M.L. Agazzi, M.L. Cortez, W.A. Marmisollé, C. von Bilderling, O. Azzaroni, Macromol. Chem. Phys. 220 (2019) 1900094.

[55] M.L. Agazzi, S.E. Herrera, M.L. Cortez, W.A. Marmisollé, C. von Bilderling, L.ía I. Pietrasanta, et al., Soft Matter 15 (2019) 1640−1650.

[56] E. Piccinini, M. Ceolín, F. Battaglini, O. Azzaroni, Chem. Chem 85 (2020) 1616−1622.

[57] J.A. Allegretto, A. Iborra, J.M. Giussi, C. von Bilderling, M. Ceolín, S. Moya, et al., Chem. Eur. J. 26 (2020) 12388−12396.

[58] G.E. Fenoy, J.M. Giussi, C. von Bilderling, E.M. Maza, L.I. Pietrasanta, W. Knoll, et al., J. Colloid Interf. Sci. 518 (2018) 92−101.

[59] M. Pandeeswar, H. Khare, S. Ramakumar, T. Govindaraju, Chem. Commun. 51 (2015) 8315.

[60] L. Zhang, T. Wang, Z. Shen, M. Liu, Adv. Mater. 28 (2016) 1044.

[61] O. Azzaroni, B. Trappmann, P. van Rijn, F. Zhou, B. Kong, W.T.S. Huck, Angew. Chem. Int. Ed. 118 (2006) 7600–7603.

[62] K. Ariga, M.V. Lee, T. Mori, X.-Y. Yu, J.P. Hill, Adv. Colloid Interface Sci. 154 (2010) 20.

[63] Lorenzo, W.A. Marmisollé, E.M. Maza, M. Ceolín, O. Azzaroni, Phys. Chem. Chem. Phys. 20 (2018) 7570–7578.

[64] E. Piccinini, J.S. Tuninetti, J.I. Otamendi, S.E. Moya, M. Ceolín, F. Battaglini, et al., Phys. Chem. Chem. Phys. 20 (2018) 9298–9308.

[65] M.L. Cortez, D. Pallarola, M. Ceolín, O. Azzaroni, F. Battaglini, Chem. Comm. 48 (2012) 10868–10870.

[66] J. Irigoyen, S.E. Moya, J.J. Iturri, I. Llarena, O. Azzaroni, E. Donath, Langmuir 25 (2009) 3374–3380.

[67] J. Cui, O. Azzaroni, A. del Campo, Macromol. Rapid Commun. 32 (2011) 1699–1703.

[68] W.A. Marmisollé, J. Irigoyen, D. Gregurec, S. Moya, O. Azzaroni, Adv. Funct. Mater. 25 (2015) 4144–4152.

[69] M.L. Cortez, A. Lorenzo, W.A. Marmisollé, C. von Bilderling, E. Maza, L. Pietrasanta, et al., Soft Matter 14 (2018) 1939–1952.

[70] M.L. Cortez, W. Marmisollé, D. Pallarola, L.I. Pietrasanta, D.H. Murgida, M. Ceolín, et al., Chem. Eur. J. 20 (2014) 13366–13374.

[71] G.E. Fenoy, J. Scotto, J. Azcárate, M. Rafti, W.A. Marmisollé, O. Azzaroni, ACS Appl. Energy Mater. 1 (2018) 5428–5436.

[72] H. Pan, S. Zhu, L. Mao, J. Inorg. Organomet. Polym. Mater. 25 (2015) 179.

[73] W. Nakanishi, K. Minami, L.K. Shrestha, Q. Ji, J.P. Hill, K. Ariga, Nano Today 9 (2014) 378.

[74] Andrieu-Brunsen, S. Micoureau, M. Tagliazucchi, I. Szleifer, O. Azzaroni, G.J.A.A. Soler-Illia, Chem. Mater. 27 (2015) 808–821.

[75] S. Schmidt, S. Alberti, P. Vana, G.J.A.A. Soler-Illia, O. Azzaroni, Chem. Eur. J. 23 (2017) 14500–14506.

[76] A. Calvo, M.C. Fuertes, B. Yameen, F.J. Williams, O. Azzaroni, G.J.A.A. Soler-Illia, Langmuir 26 (2010) 5559–5567.

[77] Brunsen, J. Cui, M. Ceolín, A. del Campo, G.J.A.A. Soler-Illia, O. Azzaroni, Chem. Comm. 48 (2012) 1422–1424.

[78] Calvo, B. Yameen, F.J. Williams, O. Azzaroni, G.J.A.A. Soler-Illia, Chem. Comm. (2009) 2553–2555.

[79] T. Govindaraju, N.B. Avinash, Nanoscale 4 (2012) 6102.

[80] R. Rajendran, L.K. Shrestha, K. Minami, M. Subramanian, R. Jayavel, K. Ariga, J. Mater. Chem. A2 (2014) 18480.

[81] K.L. Wang, K. Galatsis, R. Ostroumov, A. Khitun, Z. Zhao, S. Han, Proc. IEEE 96 (2008) 212.

[82] S.E. Herrera, M.L. Agazzi, M.L. Cortez, W.A. Marmisollé, M. Tagliazucchi, O. Azzaroni, Phys. Chem. Chem. Phys. 22 (2020) 7440–7450.

[83] R.E. Giménez, E. Piccinini, O. Azzaroni, M. Rafti, ACS Omega 4 (2020) 842–848.

[84] G.M. Segovia, J.S. Tuninetti, S. Moya, A.S. Picco, M.R. Ceolín, O. Azzaroni, et al., Mater. Today Chem. 8 (2018) 29–35.

[85] G.M. Segovia, J.S. Tuninetti, O. Azzaroni, M. Rafti, ACS Appl. Nano Mater. 3 (2020) 11266–11273.

[86] L.L. Coria-Oriundo, M.L. Cortez, O. Azzaroni, F. Battaglini, Soft Matter 17 (2021) 5240–5247.

[87] S.E. Herrera, M.L. Agazzi, M.L. Cortez, W.A. Marmisollé, M. Tagliazucchi, O. Azzaroni, Chem. Comm. 55 (2019) 14653–14656.

[88] K. Ariga, S. Ishihara, H. Abe, Cryst. Eng. Comm. 18 (2017) 6770.

[89]   M. Rafti, A. Brunsen, M.C. Fuertes, O. Azzaroni, G.J.A.A. Soler-Illia, ACS Appl. Mater. Interfaces 5 (2013) 8833−8840.

[90]   H. Abe, J. Liu, K. Ariga, Mater. Today 19 (2016) 12.

[91]   F.C.D. López, A.I. Bertoni, J.A. Allegretto, W.A. Marmisollé, O. Azzaroni, M. Rafti, et al., Chem. Cat. Chem 14 (2022) e202201015.

[92]   G.E. Fenoy, E. Maza, E. Zelaya, W.A. Marmisollé, O. Azzaroni, Appl. Surf. Sci. 416 (2017) 24−32.

[93]   A.P. Mártire, G.M. Segovia, O. Azzaroni, M. Rafti, W. Marmisollé, Mol. Syst. Des. Eng. 4 (2019) 893−900.

[94]   G.E. Fenoy, M. Rafti, W.A. Marmisollé, O. Azzaroni, Mater. Adv. 2 (2021) 7731−7740.

[95]   C.-M. Puscasu, E.M. Seftel, M. Mertens, P. Cool, G. Carja, J. Inorg. Organomet. Polym. Mater. 25 (2015) 259.

[96]   C.-M. Puscasu, G. Carja, C. Zaharia, Int. J. Mater. Prod. Technol. 51 (2015) 228.

[97]   B. Yameen, A. Kaltbeitzel, G. Glasser, A. Langner, F. Muller, U. Gösele, et al., ACS Appl. Mater. Interf. 2 (2010) 279−287.

[98]   Yameen, A. Kaltbeitzel, A. Langner, H. Duran, F. Müller, U. Gosele, et al., J. Am. Chem. Soc. 130 (2008) 13140−13144.

[99]   Yameen, A. Kaltbeitzel, A. Langer, F. Müller, U. Gösele, W. Knoll, et al., Angew. Chem. Int. Ed. 48 (2009) 3124−3128.

[100] M. Pandeeswar, T. Govindaraju, J. Inorg. Organomet. Polym. Mater. 25 (2015) 293.

[101] X. Chen, P. Li, H. Tong, T. Kako, J. Ye, Sci. Technol. Adv. Mater. 12 (2011) 044604.

[102] G.E. Fenoy, B. Van der Schueren, J. Scotto, F. Boulmedais, M.R. Ceolín, S. Bégin-Colin, et al., Electrochim. Acta 283 (2018) 1178−1187.

[103] L.D. Sappia, B.S. Pascual, O. Azzaroni, W. Marmisolle, ACS Appl. Energy Mater. 4 (2021) 9283−9293.

[104] K. Takada, N. Ohta, Y. Tateyama, J. Inorg. Organomet. Polym.Mater. 25 (2015) 205.

[105] K. Takada, Langmuir 29 (2013) 7538.

[106] T. Berninger, C. Bliem, E. Piccinini, O. Azzaroni, W. Knoll, Biosens. Bioelectron. 115 (2018) 104−110.

[107] G.E. Fenoy, W.A. Marmisollé, W. Knoll, O. Azzaroni, Sens. Diagnostics 1 (2022) 139−148.

[108] L.D. Sappia, J.S. Tuninetti, M. Ceolín, W. Knoll, M. Rafti, O. Azzaroni, Glob. Chall. 4 (2020) 1900076.

[109] Piccinini, S. Alberti, G.S. Longo, T. Berninger, J. Breu, J. Dostalek, et al., J. Phys. Chem. C. 122 (2018) 10181−10188.

[110] J. Scotto, E. Piccinini, C. Von Bilderling, L.L. Coria-Oriundo, F. Battaglini, W. Knoll, et al., Appl. Surf. Sci. 525 (2020) 146440.

[111] L.D. Sappia, E. Piccinini, C. von Binderling, W. Knoll, W. Marmisollé, O. Azzaroni, Mater. Sci. Engineering: C. 109 (2020) 110575.

[112] S. Alberti, E. Piccinini, P.G. Ramirez, G.S. Longo, M. Ceolín, O. Azzaroni, Nanoscale 13 (2021) 19098−19108.

[113] Piccinini, J.A. Allegretto, J. Scotto, A.L. Cantillo, G.E. Fenoy, W.A. Marmisollé, et al., ACS Appl. Mater. Interfaces 13 (2021) 43696−43707.

[114] Piccinini, G.E. Fenoy, A.L. Cantillo, J.A. Allegretto, J. Scotto, J.M. Piccinini, et al., Azzaroni, Adv. Mater. Interf. 9 (2022) 2102526.

[115] Y.J. Li, Y. Yan, Y.S. Zhao, J. Yao, Adv. Mater. 28 (2016) 1319.

[116] Laucirica, J.A. Allegretto, M.F. Wagner, M.E. Toimil-Molares, C. Trautmann, M. Rafti, et al., Adv. Mater. 34 (2022) 2207339.

[117] Pérez-Mitta, W.A. Marmisollé, C. Trautmann, M.E. Toimil-Molares, O. Azzaroni, Adv. Mater. 29 (2017) 1700972.

[118] G. Pérez-Mitta, L. Burr, J.S. Tuninetti, C. Trautmann, M.E. Toimil-Molares, O. Azzaroni, Nanoscale 8 (2016) 1470−1478.

[119] G. Laucirica, M.E. Toimil-Molares, C. Trautmann, W. Marmisolle, O. Azzaroni, ACS Appl. Mater. Interfaces 12 (2020) 28148−28157.

[120] G. Laucirica, Y.T. Terrones, V. Cayón, M.L. Cortez, M.E. Toimil-Molares, C. Trautmann, et al., TrAC. Trends Anal. Chem. 144 (2021) 116425.

[121] G. Laucirica, W.A. Marmisollé, M.E. Toimil-Molares, C. Trautmann, O. Azzaroni, ACS Appl. Mater. Interfaces 11 (2019) 30001−30009.

[122] A.S. Peinetti, R.J. Lake, W. Cong, L. Cooper, Y. Wu, Y. Ma, et al., Sci. Adv. 7 (2021) eabh2848.

[123] G. Laucirica, M.E. Toimil-Molares, C. Trautmann, W. Marmisollé, O. Azzaroni, Chem. Sci. 12 (2021) 12874−12910.

[124] X. Meng, M. Wang, L. Heng, L. Jiang, Adv. Mater. 30 (2018) 1706634.

[125] E. Maza, C. von Bilderling, M.L. Cortez, G. Díaz, M. Bianchi, L.I. Pietrasanta, et al., Langmuir 34 (2018) 3711−3719.

[126] D. Pallarola, N. Queralto, F. Battaglini, O. Azzaroni, Phys. Chem. Chem. Phys. 12 (2010) 8072−8084.

[127] D. Pallarola, C. von Bildering, L.I. Pietrasanta, N. Queralto, W. Knoll, F. Battaglini, et al., Phys. Chem. Chem. Phys. 14 (2012) 11027−11039.

[128] D. Pallarola, N. Queralto, W. Knoll, M. Ceolin, O. Azzaroni, F. Battaglini, Langmuir 26 (2010) 13684−13696.

[129] D. Pallarola, N. Queralto, W. Knoll, O. Azzaroni, F. Battaglini, Chem. Eur. J. 16 (2010) 13970−13975.

[130] M. Mir, M. Álvarez, O. Azzaroni, L. Tiefenauer, W. Knoll, Anal. Chem. 80 (2008) 6554−6559.

[131] L.D. Sappia, E. Piccinini, W. Marmisollé, N. Santilli, E. Maza, S. Moya, et al., Adv. Mater. Interf. 4 (2017) 1700502.

[132] S. Howorka, Langmuir 29 (2013) 7344.

[133] K. Ariga, Q. Ji, T. Mori, M. Naito, Y. Yamauchi, H. Abe, et al., Chem. Soc. Rev. 42 (2013) 6322.

[134] M.R. Molla, P.A. Levkin, P.A. Adv. Mater. 28 (2016) 1159.

[135] M.L. Agazzi, S.E. Herrera, M.L. Cortez, W.A. Marmisollé, O. Azzaroni, Colloids Surf. B: Biointerfaces 190 (2020) 110895.

[136] K. Ariga, K. Kawakami, M. Ebara, Y. Kotsuchibashi, Q. Ji, J.P. Hill, N. J. Chem. 38 (2014) 5149.

[137] M.L. Agazzi, S.E. Herrera, M.L. Cortez, W.A. Marmisollé, M. Tagliazucchi, O. Azzaroni, Chem. Eur. J. 26 (2020) 2456−2463.

[138] N.E. Muzzio, M.A. Pasquale, E. Diamanti, D. Gregurec, M.M. Moro, O. Azzaroni, et al., Mater. Sci. Engineering: C. 80 (2017) 677−687.

[139] N.E. Muzzio, M.A. Pasquale, D. Gregurec, E. Diamanti, M. Kosutic, O. Azzaroni, et al., Macromol. Biosci. 16 (2016) 482−495.

[140] N.E. Muzzio, M.A. Pasquale, W.A. Marmisollé, C. von Bilderling, M.L. Cortez, L.I. Pietrasanta, et al., Biomaterials, Science 6 (2018) 2230−2247.

[141] N.E. Muzzio, D. Gregurec, E. Diamanti, J. Irigoyen, M.A. Pasquale, O. Azzaroni, et al., Adv. Mater. Interf. 4 (2017) 1600126.

[142] E. Psarra, U. Konig, Y. Ueda, C. Bellmann, A. Janke, E. Bittrich, et al., P. ACS Appl. Mater. Interfaces 7 (2015) 12516.

[143] P. Kujawa, F.M. Winnik, Langmuir 29 (2013) 7354.

[144] K. Ariga, Q. Ji, M.J. McShane, Y.M. Lvov, A. Vinu, J.P. Hill, Chem. Mater. 24 (2012) 728.

[145] Komatsu, M. Akamatsu, Q. Ji, J.P. Hill, K. Ariga, Disp. Imaging 2 (2015) 3.

[146] A.P. Pandey, N.M. Girase, M.D. Patil, P.O. Patil, D.A. Patil, P.K. Deshmukh, J. Nanosci. Nanotechnol. 14 (2014) 828.

[147] Ariga, T. Mori, J.P. Hill, J.P. Langmuir 29 (2013) 8459.

[148] O. Azzaroni, G. Andreasen, B. Blum, R.C. Salvarezza, A.J. Arvia, J. Phys. Chem B 104 (2000) 1395−1398.

[149] S. Gilje, S. Han, M. Wang, K.L. Wang, R.B. Kaner, Nano Lett. 7 (2007) 3394−3398.

[150] R. Molinaro, M. Evangelopoulos, J.R. Hoffman, C. Corbo, F. Taraballi, J.O. Martinez, et al., Adv. Mater. 30 (2018) 1702749.

[151] V.C. Tung, M.J. Allen, Y. Yang, R.B. Kaner, Nat. Nanotech. 4 (2019) 25−29.

Nitrogen functionalities assisted nanoporous carbon materials for supercapacitor studies

2

Ramakrishnan Prakash and Sangaraju Shanmugam

Daegu Gyeongbuk Institute of Science & Technology (DGIST), Daegu, The Republic of Korea

2.1 Introduction

Supercapacitors (SCs) are electrochemical energy storage devices that combine the high power delivery capability of conventional capacitors and the high energy storage ability of conventional batteries [1,2]. SCs have found several attractive applications, from automobiles to heavy-duty machinery, as an instantaneous power supply for motors [2]. From a conceptual point of view, the cell construction of SCs is similar to that of batteries, but the cell chemistry is not, with a two-electrode system and a separator with electrolyte impregnation [3,4]. Unambiguously, SCs' characteristics, like superior power density (1500 W/kg), very long cycle life ($>$ 1,000,000 cycles), lower thermochemical heat dissipation, simple mode of operation, and convenient integration into electronics, have given them advantages over the conventional batteries for several commercial applications, specifically those that requires fast charging and discharging and long-term reliability [5,6]. On the other hand, the conventional SCs and their derivatives, such as asymmetric, symmetric, and hybrid, are technically far behind the battery's storage capability (200−250 Wh/kg), as shown in the Ragone plot in Fig. 2.1A [7,8]. A Ragone plot is defined as a straightforward representation of the specific energy and power densities of an energy device, calculated based on the combined mass of the electrode and electrolyte.

In principle, the conventional SCs supported by carbon-based electrode materials can achieve the charge storage process by adsorbing ions at the biased electrode/electrolyte interface, thereby creating an electric double-layer process. The capacitance realized by this process is called electric double-layer capacitance (EDLC). The overall EDLC characteristic is often limited by the electrolyte stability window (ESW) of the electrolyte; for instance, 1.23 V in the case of an aqueous electrolyte.

Hence, the energy density (E.D) of SCs is usually determined by the EDLC capability of the material and the potential acquired by the working electrodes, $E.D = \frac{1}{2} CV^2$. The power density (P.D) of SCs is governed by the overall resistance of the device or equivalent series resistance (Rs). The P.D of SCs can be calculated by $P = V^2/4Rs$. Generally, the factor Rs is affected by the following: poor electronic conductive behavior of electrodes, electrolyte resistance, interfacial resistance between the electrodes and the current collectors, separator permeability of ions motion, and incompatible ion size to the pores of active materials.

The commercially available organic electrolyte-supported SCs suffer from limited E.D ($<$ 10 Wh/kg) due to sluggish mass diffusion and restricted EDLC formation [9]. Thus it is obvious from our

Materials Nanoarchitectonics. DOI: https://doi.org/10.1016/B978-0-323-99472-9.00002-X

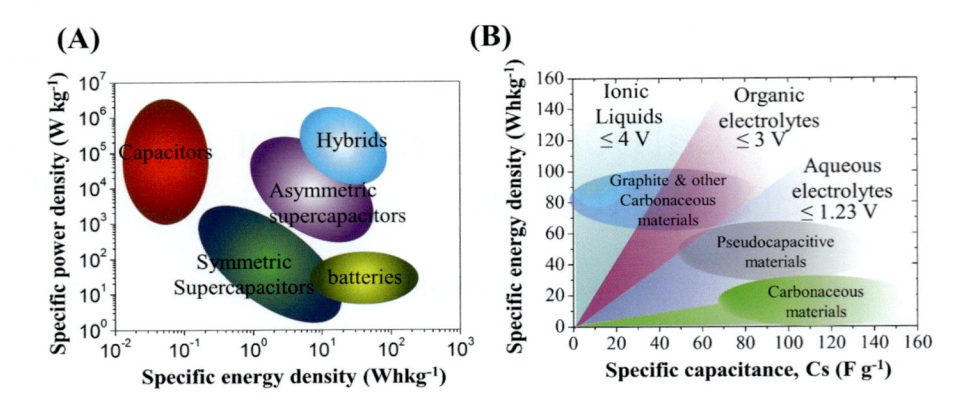

FIGURE 2.1

(A) The Ragone plot and (B) E.D versus Cs of various carbon materials under different electrolyte medium.

viewpoint that to increase the E.D of the SCs, one should either have high-capacity carbon electrodes or high-voltage stable electrolytes, or otherwise a combination of both. Regarding the improvement in carbon electrode capacity, a multitude of approaches like engineering surface area, tuning pore features, enhancing electrical conductivity, and polarizing electrode surface have been attempted in research. On the other hand, as perceived from Fig. 2.1B, the ESWs of various electrolytes on different EDLC materials have been investigated with the aim of improving the E.D. In addition to the above, hybrid SCs endow faradic pseudocapacitance in the carbon materials to bridge the gap between batteries and SCs. Unfortunately, such upgraded SCs show limited redox kinetics which results in undesirable cycling and power performance characteristics. Based on the above understanding, the major hurdle in SCs development is to unveil suitable carbon and electrolyte materials of interest to enhance the overall energy and power densities without compromising longevity.

2.2 Effect of morphology, pore structure, and heteroatoms functionalization on capacitive behavior

For many years, porous carbons have been the most popular choice of SCs electrode material, due to the overwhelming theoretical and experimental evidence [10−13]. In general, porous carbons offer customizable superior properties such as high surface engineering for double-layer formation, tuning of the desired pore texture for electrolyte ions permeation, functionalizing heteroatoms for electrode surface polarization, and wettability toward capacitance enhancement.

2.2.1 Controlled morphology influences capacitive behavior

The porous carbons allow versatile synthesis approaches using various carbon source precursors, such as biomass, phenolic resins, hypercrosslinked polymers, metal−organic frameworks, covalent organic frameworks, poly(ionic) liquids, etc. [14,15]. Generally, transforming the carbonized precursor to high-surface-area porous carbon involves a subsequent activation process: physical activation using oxidative atmosphere (CO_2, O_2, and gas mixtures) and chemical activation using corrosive chemicals (H_3PO_4, KOH, NaOH, etc.) [16]. Although the above two approaches yield high-surface-area porous carbon materials, the entire area is not necessarily electrochemically accessible, that is, the interior area is restricted to double layer formation. Thus controlling the morphology of porous carbon has been taken seriously into consideration. Accordingly, several reports have shown promising exploration of nanocarbon materials' surface topology and interior texture. The most used carbon materials of interest are three-dimensional (3D) carbon nanospheres, two-dimensional (2D) carbon nanosheets/nanoribbons, and one-dimensional (1D) carbon nanofibers/nanorods, which have been synthesized using a modified traditional approach to yield various textural morphologies [17]. For instance, Wang et al. developed 2,6-diaminopyridine-derived multichamber carbon microspheres of 250−3091 nm using the dual-surfactant-assisted assembly approach [18]; Yu et al. used dopamine as a carbon precursor and MgAl-layered double hydroxide as a 2D structure-directing template to develop carbon nanosheets with a thickness of ∼99 nm and they achieved a maximum E.D of 94 Wh/kg for an ionic liquid supported SC device [19]; Bo et al. reported a high-surface-area carbon nanofiber (1000 m^2/g) using self-assembled chitin fibers in a urea/NaOH solution and achieved a high E.D value of 58.7 Wh/kg in organic media [20].

Also, in our studies, we developed several advancements in nanocarbon morphologies such as carbon nanorods (CNR), hollow-shaped carbon nanorods (HCNR), arch-shaped carbon nanorods (ACNR), multinanochannel carbon nanorods (MCNR), and mesoporous carbon nanoform (MCNR) structures [21−23], as shown in Fig. 2.2A−E. Each morphology presents its own significance in terms of textural advancement toward capacitance improvement. The BET surface areas of these carbon materials were found to be in the following order: CNR (481.1 m^2/g) <HCNR (556.9 m^2/g) <ACNR (619.3 m^2/g) <MCNR (840.8 m^2/g) <MCNF (1564.0 m^2/g). All these carbon materials are turbostratic in nature. Conceptual understanding: Considering the CNR material, most of the charge accumulation is limited to the exterior surface, whereas the interior texture is hindered due to the inaccessible pore size and volume. Thus in the pursuit of exploring the interior texture, the HCNR, ACNR, and MCNR nanocarbons were developed systematically using a simple or coaxial electrospinning-assisted pyrolysis process; and the MCNF nanocarbon was developed using a metal-chelate complex-assisted pyrolysis process. The maximum specific capacitance (Cs) of CNR, HCNR, ACNR, and MCNR was found to be 152, 220, 232, and 340 F/g, respectively, in symmetric cell configuration at 1 M H_2SO_4 electrolyte, and the observed Cs values corroborate the surface area trend. The corresponding E.D values of CNR, HCNR, ACNR, and MCNR were found to be 5.0, 7.5, 8.4, and 11.2 Wh/kg.

2.2.2 Pore structure influences capacitive behavior

According to the International Union of Pure and Applied Chemistry (IUPAC), the pores are classified by their diameter into the following: micropores (<2 nm), mesopores (2−50 nm), and hierarchical porous structure (>50 nm) [24]. Notably, carbon materials with micropores, meso-/macropores, and hierarchical (micro-, meso-, and macropores) porous structures are correspondingly favorable to controlled diffusion, reduced interfacial resistance, and rapid ion transmission, respectively. In terms of high volumetric capacitance, interconnected hierarchical pore architecture is highly desirable as it

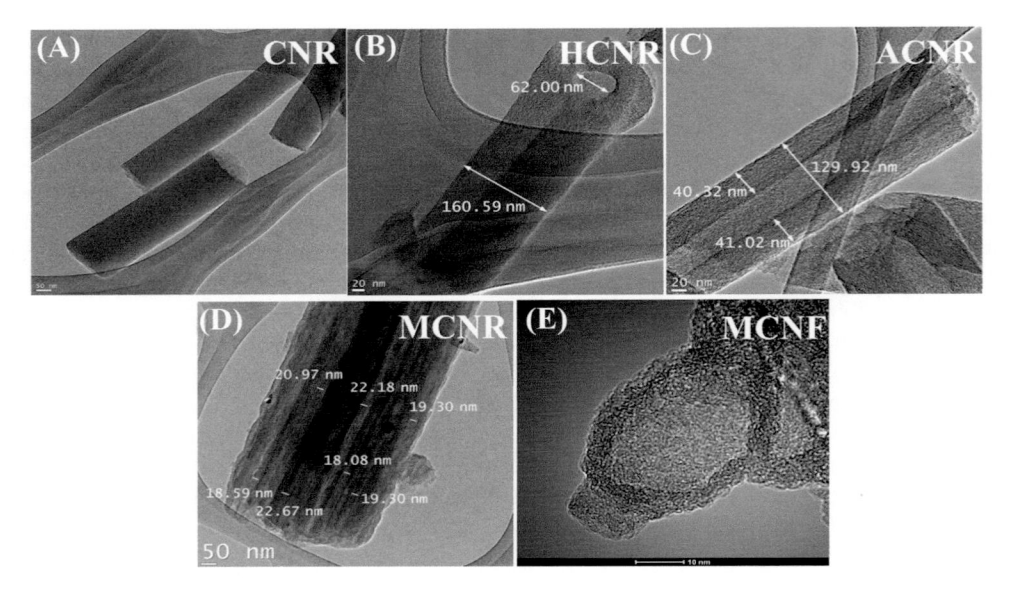

FIGURE 2.2

FE-SEM images of CNR (A), HCNR (B), ACNR (C), MCNR (D), and MCNF (E).

Parts (A—C) are reproduced with the permission from Ref. [21], Copyright 2015 Royal Society of Chemistry; part (D) is reproduced with the permission from Ref. [23], Copyright 2016 American Chemical Society; part (E) is reproduced with the permission from Ref. [22], Copyright 2016 Elsevier.

permits feasible ion-diffusion pathways within the geometry of the pores. It is noted that the commercially available activated carbons (AC) possess pores in the range of microns (<0.5 nm) and thus hinder the use of organic/ionic liquid electrolytes in which the size of the electrolyte ions are mismatched with the pore size of ACs [9]. Therefore, tailoring pore size is of the utmost importance for enhanced surface accessibility for maximized double layer formation. Reflecting on this point, to date, several nanoengineered and regulated pore size carbon materials have been developed using direct/modified hard and soft templates, template-free methods, and architectures derived from reticular chemistry [25].

Martin et al. investigated carbide-derived carbons of tunable pore size texture suitability in different electrolyte media, such as aqueous, organic, and ionic liquid electrolytes, and suggested high specific surface area and micropore volume led to higher capacitance in the mentioned electrolyte [11]. Young et al. attempted to determine the threshold of pore size for the electrolyte ions to penetrate a wide range of pores by experimenting on 13 porous carbon samples prepared by various methods. Their results indicated that variation of specific capacities was observed as the function of the effective surface area to the critical pore size (0—5 nm) [25].

Similarly, we also emphasized the importance of pore size for electrolyte ions accessibility in nanocarbons like HCNR, ACNR, MCNR, and MCNF. Specifically, the MCNR nanocarbons exhibited substantial mesopores ($\sim$3.79 nm) and subnanopores ($\sim$0.64 nm); this pore size variation is highly desirable for high-power rate performance in SCs application. The symmetric MCNF cell achieved excellent capacitance retention of 95.6% after 50,000 charge/discharge cycles at a 2 A/g current density in an aqueous electrolyte. The postmorphological analysis of the cycled MCNR electrode revealed the retention of multichannel carbon structures. For the consideration of nonaqueous-based ionic liquid electrolytes (ILs), the micropore/subnanopore textures are of least concern as they restrict the diffusion of the ions at high currents, owing to the pore-size limit following the relation $\tau = l^2/d$; where l is the ion transport length and d is the ion transport diffusion coefficient [26]. Thus we developed unimodal pore-size distribution (~ 3.9 nm) in MCNF nanocarbons, and it achieved an excellent cycle life retention (92.2%) at a high current density (20 A/g) after 15,000 cycles under the EMIMBF$_4$ electrolyte [22]. Furthermore, studies have shown that organic solvents-diluted ILs improved the electrosorption ability near the micropores region due to the partial desolvation of cations. Thus the effective formation of EDLC under the ILs can be realized either with the optimization of nanocarbons pore size or with the alteration of the solvated ILs ions.

2.2.3 Heteroatoms functionalization on capacitive behavior

It has been widely accepted that modifying the carbon surface with pseudocapacitive materials like transition metal oxides and electrochemically conductive polymers can facilitate faradic reaction to enhance the SCs device capacitance but at the cost of its cycle life. Alternatively, doping heteroatoms (N, O, S, B, and P) on the carbon materials have significantly improved the performance of SCs [27]. The heteroatom-doped carbon offers several advantages, such as polarizing the electrode surface to facilitate ion adsorption ability, promoting a lone pair of electrons on the carbon framework to enhance the intrinsic conductivity, and endowing carbon materials with redox ability [28]. Many recent studies have shown the synergistic effect of multiheteroatom-doped carbon materials on electrochemical performance. For instance, some experimental and theoretical studies have reported that excess single-/multiheteroatom doping on carbon materials could decrease the intrinsic conductivity behavior and hinder the pore-size texture development [29−31]. Therefore, considerable efforts have been made to find a suitable fabrication method to achieve the synergy between heteroatom functionalization and pore architecture on carbon materials.

In our studies, the N-content (wt.%) of the abovementioned nanocarbons (CNR, HCNF, ACNR, and MCNR) was reported as 9.78, 8.22, 8.70, and 9.55 wt.%, respectively. And their respective electrical conductivity follows the trend of CNR (5.1 S/cm) > HCNR (4.2 S/cm) >ACNF (4.5 S/cm) >MCNR (3.5 S/cm). However, the observed Cs of these nanocarbons followed only the surface area trend, as mentioned in the above section. It is noted that despite the CNR inheriting the high N-content and the electrical conductivity, it resulted in poor Cs. Thus the correlation of N-content with electrical conductivity or solely an electrical conductivity parameter should not directly account for the Cs observation. This attributed to the following: (1) in general, the N-functionalities on the carbon framework exist in various chemical forms (pyridinic-, pyrrolic-, quaternary-, pyridonic-, and pyridine N-oxide), and many studies have shown only quaternary-N or graphitic-N contributions to conductive character, for example, pyridinic-N accounts for a redox character; (2) theoretical studies have reported that the excess of N-content in pyridinic (14 wt.%) and quaternary (25.8 wt.%) N-oxide increases the bandgap and results in poor conductivity; and (iii) the electrical conductivity mainly depends on the nature of the sp^2 hybridized carbon framework in general [32−34].

2.3 **Electrolyte influence and its limitation on overall performance**

Recent studies have suggested that the scientific communities have extended their scope of energy storage enhancement to electrolyte aspects. As understood from the E.D and P.D formulae, voltage (V) is one of the common elements of device performance-deciding factors. This means that the electrolytes with higher ESW can increase the cell voltage of SCs, which further enhances the E.D and P.D. However, the higher operating voltage does not necessarily imply only benefits, but comes with the cost of electrolyte degradation and starvation issues. There is a limit to ESW in different electrolytes; nonaqueous (organic and ionic) electrolytes can yield up to 4 V, but aqueous electrolytes are limited to 1.23 V [35,36].

Thus, in addition to the high-capacity electrodes, the electrolyte also plays a crucial role in determining the overall performance of SCs. For instance, most nonaqueous electrolytes are claimed to exhibit low Cs values compared to the aqueous electrolytes. In addition, the Cs can also be influenced by other major properties of the electrolyte used, such as ionic conductivity, ESW, ion mobility, thermal stability, and operation temperature range [37].

Both aqueous and nonaqueous-based SCs are finding a promising marketplace; the former have been preferred for large installations due to their high safety, low-cost manufacturing, and low internal resistance, and the latter are known primarily for their use in high-energy storage applications. The variation of anion and cation sizes of these electrolytes is the bottleneck challenge to the effective formation of EDLC in relation to the surface area of the electrode. As noted in Table 2.1, the cation and anion ionic radii of the nonaqueous SCs are relatively bigger than the aqueous-based. So, the EDLC electrode should have the optimum adsorption sites or pore size adaptation for the electrolyte ions of interest, in accord with the relation, $C = \varepsilon A/d$, where ε is the electrolyte dielectric constant, A is the surface area of the electrode, and d is the distance between ions and the adsorption site. Hence, meticulous pore size design is required to obtain stable and maximal EDLC characteristics.

Table 2.1 Some of the commonly used electrolytes in supercapacitors.

Electrolyte types	Cation (nm)	Anion (nm)
Aqueous electrolyte		
H_2SO_4	[a]H^+ (0.26)	SO_4^{2-} (0.533)
KOH	[a]K^+ (0.36)	OH^- (0.300)
Na_2SO_4	[a]Na^+ (0.36)	SO_4^{2-} (0.533)
Nonaqueous (organic electrolyte)		
TEABF4	TEA^+ (0.686)	BF_4^- (0.458)
$TEMABF_4$	$TEMA^+$ (0.654)	BF_4^- (0.458)
TEATFSI	TEA^+ (0.680)	$TFSI^-$ (0.650)
Ionic liquids (solvent-free or molten salts)		
$EMImBF_4$	EMIm + (0.676)	BF_4^- (0.515)
DMPImTFSI	DMPIm + (0.773)	$TFSI^-$ (0.838)
BMPyFAP	BMPy + (0.771)	FAP^- (1.042)

2.3.1 Aqueous supercapacitors

The aqueous SCs can be classified into acid (H_2SO_4), base (KOH), and neutral (Na_2SO_4) electrolytes [35]. These electrolytes satisfy the rudimentary electrolyte criteria of SCs like high ionic conductivity and nonflammability. From the industrial aspect, the convenient cell assembly in air, noncorrosiveness, and low-cost manufacturing have gained the upper hand over the organic electrolytes [38]. However, to reiterate the fact, the ESW of aqueous-based SCs is far lower than that of organic-based SCs. Thus, according to the E.D relation, such low-potential limiting SCs yield moderate E.D values. Two possible approaches can address the shortcomings of the aqueous SCs. The first approach involves introducing redox-active materials to the electrode material. However, this approach mimics the battery-type electrode but has its demerits like poor reversibility over a period and large ESR values. The second approach involves adding redox mediators or additives to the aqueous electrolyte, termed here as redox-active electrolyte (RAE). The latter approach involves the facilitation of electron transfer from the electrolyte additives to the electrode surface during the charge—discharge process.

Notably, the electrode material of this approach is often any EDLC material as they possess high surface area characteristics. Besides, the chemical affinity of such EDLC materials majorly influences the resulting faradic capacitance. The redox mediators in SCs are of different types: inorganic compounds ($AgNO_3$, KBr, KI, $CuSO_4$, $FeSO_4$, and others), organic compounds (HQ, PPD, PNA, and others), and often a blending of two different redox materials (Na_2MoO_4 and KI) in aqueous media [39–42]. These aqueous-based RA have been demonstrated to be an effective approach to enhance the SCs' performance.

2.3.2 Inorganic redox-active electrolyte supercapacitors system

Among them, the KI-based RAE has been widely explored. For instance, Wang et al. claimed that for KI the redox couple (IO_x^-/I^-) in KOH solution had attained a maximum specific E.D of 7.1 Wh/kg and 14,000 cycles with 93% retention; notably, the system achieved a cell voltage of 1.6 V by suppressing the hydrogen evolution reaction [42]. Similarly, several other inorganic RA-supported SC systems like $Ce_2(SO_4)_3$ (Ce^{3+}/Ce^{4+}), KBr (Br^-/Br^{3-}), $VOSO_4$ (VO^{2+}/VO_2^+), $K_3Fe(CN)_6$ ($Fe(CN)_6^{3-}/Fe(CN)_6^{4-}$), and KSeCN ($SCN^-/SeCN)_2$) showed significant capacitance improvement [40,43,44]. In the advancement of RA-based SCs, the mixing of two different redox additives is considered the most effective approach to enhance the E.D of aqueous RAE SCs. The Na_2MO_4 and KI complex redox additives supported by aqueous H_2SO_4 RAE influence the concurrent redox reaction on the synergy basis as follows [45,46]:

$$Mo(VI)O_4^{2-} + 2H^+ \leftrightarrow H_2Mo(VI)O_4$$

$$H_2Mo(VI)O_4 + 2H^+ + 2e^- \leftrightarrow H_2Mo(VI)O_3 + H_2O$$

$$H_2Mo(VI)O_4 + H^+ + e^- \leftrightarrow HMo(V)O_3 + H_2O$$

This complex RAE system obtained 17.3 times capacity enhancement and achieved a maximum E.D of 65.3 Wh/kg. On the negative side, although some of the inorganic RA exhibited significant signs of potential improvement it did not to a substantial extent boost the E.D. In most cases, the reduction of the cycle life was observed more likely after adding these RA due to the irreversible growth of RA species on the electrode surface.

2.3.3 Organic redox-active electrolyte supercapacitors system

Considering the structural diversity and the material cost production, the organic compounds RAE system is a highly viable alternative. This has attracted a number of researchers, resulting in a plethora of organic RAE articles. Most organic redox additives are made of small molecules with a large HOMO−LUMO energy gap, thus resulting in a low electronic conductive behavior [47]. Such poor electronic organic redox compounds directly adversely influence the power density of SCs. In addition, the standard redox potentials of these additives should be within the ESW of the electrolytes and compatible with the choice of working electrodes. Further, the complete solubility and toxicity of these compounds should also be considered.

One of the major breakthroughs in electroactive organic compounds is hydroquinone (HQ). It has gained enormous attention owing to its two-electron transfer reaction preference: ketone functional groups are reduced to hydroxyls [48]. Several HQ-supported H_2SO_4 RAE with activated carbon//activated carbon supercapacitors (AC//AC SCs) were examined, and the results show significant improvement in E.D [49]. For instance, in our work, we reported a nitrogen-functionalized multichannel carbon nanorod (N-MCNR) that delivered a maximum E.D of 11.2 Wh/kg and remarkable long-term cycle stability of 50,000 cycles with 92.6% capacity retention. Utilizing the superior physicochemical properties of N-MCNR, such as 18−22 nm nanochannels, 840 m^2/g surface area, and 9.55 wt.% N-content, which motivated to subject in RAE system of HQ.

In this study, the optimization of HQ level is highly regarded because any high level of HQ results in surface meiotic formation or irreversible growth of HQ species accumulating at the pores, which directly affects the overall cell performance. The HQ optimization was performed in a three-electrode system HQ of different molarity (0.1, 0.3, and 0.5 M) in 1 M H_2SO_4. The HQ in 1 M H_2SO_4 of different molarities undergoes oxidation of HQ to p-quinone (Q) and the reduction of Q to HQ, which is a highly reversible redox reaction with two-electron transfer processes, as shown in Fig. 2.3A. The study shows an increase in Csp was observed with the increase in the HQ concentration from 0.1 to 0.3 M. However, the Csp value was decreased with a further increase in the HQ concentration of 0.5 M owing to the possible formation of HQ crystal aggregates on the N-MCNR electrode, which limited the full utilization of HQ. The understanding of the preliminary experiment guided the construction of an asymmetric two-electrode cell with 0.3 M HQ as an RAE SC cell. Also, for comparison, an RAE-free N-MCNR cell was designed.

Fig. 2.3B presents the GCD profiles at a low discharge current density of 0.25 A/g of an HQ-assisted cell (Blue trace, 1468 F/g) and HQ-free cell (black trace, 340 F/g). As the voltage profile of the HQ-assisted cell seems to be nonlinear, the C_{sp} was calculated with the utmost care by considering the area under the discharge curve profile. It is worth mentioning that for practical applications coulombic efficiency is highly warranted and critical as well. In this study, the RAE-SC device exhibits 70%−75% coulombic efficiencies at lower current densities (0.25−5 A/g); the coulombic efficiencies improved further by 85% to 90% at high current densities (10−35 A/g) as the GCD profiles approach near symmetry, as shown in Fig. 2.3C.

To further evaluate the practical applicability of HQ-supported RAE cell, a Ragone plot was generated, as shown in Fig. 2.3D. The HQ-assisted N-MCNR cell produced a high gravimetric E.D

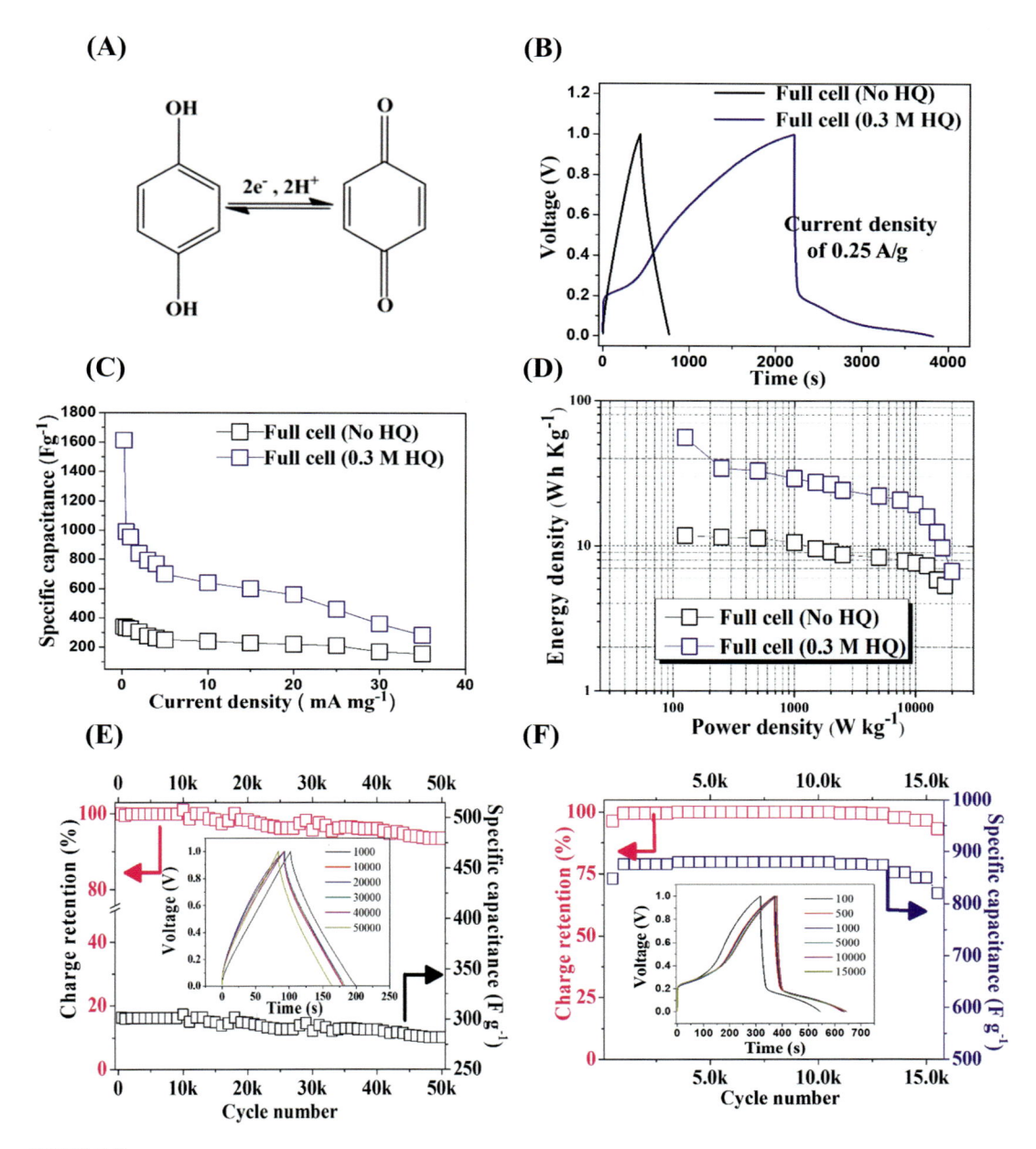

FIGURE 2.3

Redox process of HQ (A) and MCNR full cell electrochemical performance: galvanostatic charge-discharge profile (B), Cs as the function of current densities (C), E.D versus P.D plot (D), and cycle test profile with no HQ (E) and with HQ (F).

Part (E) is reproduced with the permission from P. Ramakrishnan, S. Shanmugam, Nitrogen-doped porous multi-nano-channel nanocarbons for use in high-performance supercapacitor applications, ACS Sustain. Chem. Eng. 4 (2016) 2439–2448, Copyright 2016 American Chemical Society.

of 50.2 Wh/kg (1613 seconds), with electrode mass loading of ~ 4.5 mg, which is 4.7 times higher than that measured for HQ-free N-MCNR cell. Even at a high gravimetric P.D of 14.7 kW/kg (1.1 seconds), the HQ-assisted cell delivered an E.D of 6.1 Wh/kg. Thus this device is suitable for both high-E. D and high-P.D applications. On the other hand, an HQ-free full cell reached an E. D of 11.2 Wh/kg at a P.D of 118.7 W/kg (340 seconds) and retained an E.D of 4.0 Wh/kg at a maximum P.D of 12.9 kW/kg (1.4 seconds). Thus this device can be used exclusively for high-power-rate applications.

The key advantage of SCs over other energy storage devices is their long-term cycling stability. Both devices were examined to demonstrate the long cycle life. The HQ-free N-MCNR cell (Fig. 2.3E) delivered 95.6% capacity retention at the end of 50,000 cycles. On the other hand, the HQ-assisted device (Fig. 2.3F) was tested for up to 15,000 cycles at a current density of 2 A/g. An initial increment in C_{sp} was observed up to 500 cycles, which indicated that the HQ redox reactions inside the pores were incomplete. At the end of the 15,000 cycles, it retained 94.7% charge retention. It is noted that the RAE cell was demonstrated to have fewer cycle numbers than the HQ-free cell, which corresponds to its continuous electrolyte starvation.

2.3.4 Nonaqueous supercapacitors

In general, the nonaqueous SCs can be subclassified into organic, ionic liquid, and mixtures. They are usually composed of a conductive salt (tetraethylammonium tetrafluoroborate (TEABF$_4$) or lithium perchlorate (LiClO$_4$)) dissolved in an organic electrolyte (acetonitrile (ACN) or propylene carbonate (PC)). This type of SCs can provide a maximum ESW value of 2.5–2.7 V. As the E.D relation equates to the square of the cell voltage, such a wide potential range electrolyte can largely benefit the E.D value.

Like the RAE-aqueous SC system, there are several organic additives, such as phenylenediamines (PPD), ortho-phenylenediamine (OPD), 1,10-phenanthroline (phen), ethyl viologen (EV), benzyl viologen (BV), π-extended viologen (π-EV), and decamethylferrocene (DmFc), which can be incorporated in the selected organic solvent to enhance the E.D further [50–54]. Some of the organic additives experience poor solubility issues with a selected organic solvent which results in the formation of organic moieties. In addition, most organic solvents are prone to flammability due to their high vapor pressure and flammable characteristics, particularly in high-temperature conditions. Thus ionic liquids (ILs) of nonvolatility, high thermal stability, and excellent electrochemical stability have been considered as alternatives.

In general, ILs are classified as aprotic, protic, and zwitterion type. Among them, both aprotic and protic type find applications in SCs. Specifically, the aprotic ILs (AILs) of SCs application are further classified into net ILs for EDLC, quasisolid-state ILs, and all-solid-state ILs, based on their composition. In general ILs for EDLC, imidazolium ILs ([EMI][BF$_4$], [EMI][FSI], and [EMI] [TFSI]) and pyrrolidinium ILs ([Pyr][TFSI], [Pyr$_{14}$][TFSI], and [Pyr$_{14}$][FSI]), have gained importance due to their appreciable ionic conductivities and wide ESW. Studies on different imidazolium and pyrrolidinium ILs have shown that critical cation size is necessary for acquiring ionic conductivity as well as influencing the formation of the double layer [55].

Also, studies on the correlation between the pore size of the electrode and the ion size of ILs suggest that when the pore size is close to the size of the ions the double-layer formation is limited; when the pore size is wider than that of the size of the ions, the solvent polarity, conductivity, and viscosity have an insignificant effect and hence there is facile double layer formation [56,57]. Moreover, examining different high surface area carbon materials corroborates that open pore size is highly desirable for enhancing the P.D.

We reported in our study that the nitrogen-doped mesoporous carbon nanofoam (N-MCNF) electrode possessing a surface area of 1564 m^2/g and a mesopore size of ~3.9 nm in symmetric cell configuration exhibited E.D of 63.4 Wh/kg and P. D of 35.9 kW/kg under [EMIM][BF$_4$] at 25 °C. The study involved the optimization of textural properties of the N-MCNF electrode at different synthesis conditions.

The E.D evaluation of the best-performed N-MCNF material at low (0°C), moderate (60°C), and high (90°C) temperatures corresponded to 31.9, 83.4, and 93.3 Wh/kg, respectively, as shown in Fig. 2.4A, B. The variation of E.D is correlated with the differences in ionic conductivities at different temperatures: ionic conductivities at 0, 60, and 90°C were found to be 0.89, 12.69, and 28.04 mS/cm. Further, the symmetric N-MCNF cell was demonstrated in an energy harvesting application using a polycrystalline silicon solar panel (186 mA at 3 V). The experiment involved charging the cell for 60 seconds with the circuit on state and allowed to discharge using the galvanostatic discharge technique assisted by potentiostat equipment, as schematically represented in Fig. 2.4C–F.

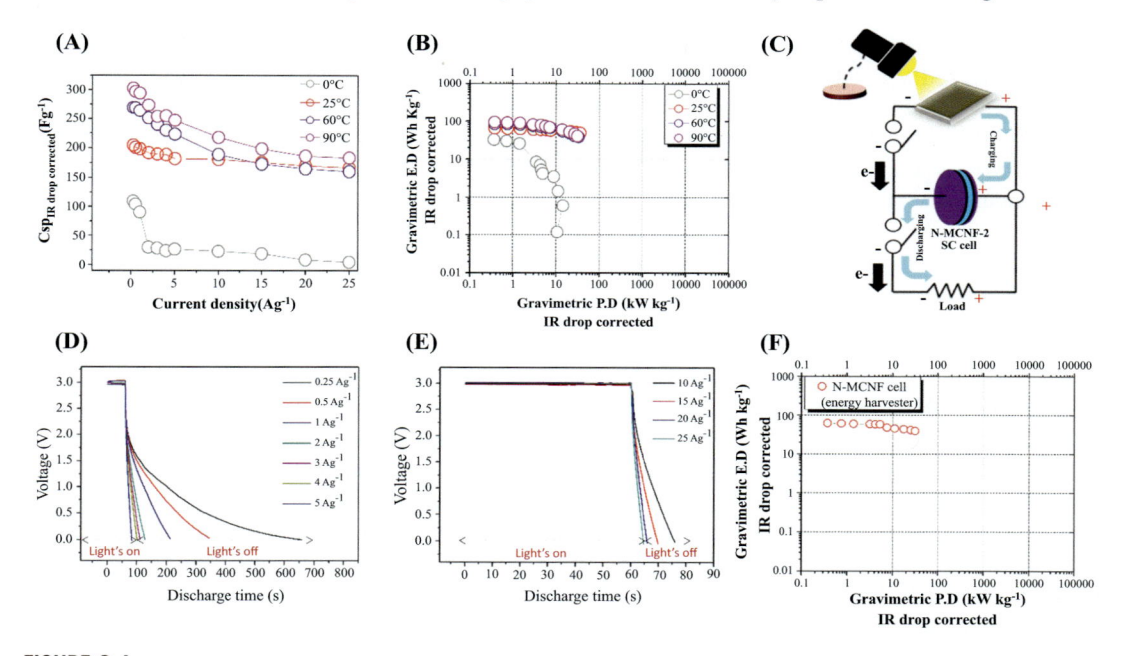

FIGURE 2.4

MCNF full cell electrochemical performance: (A) Cs as the function of different current densities at different temperature conditions, (B) E.D versus P.D plot evaluated at different temperature conditions, (C) schematic representation of energy harvester set up, evaluation of MCNF energy harvester at low (D) and high current densities € for the determination of E.D versus P.D (F).

Parts (A–F) are adopted with the permission from P. Ramakrishnan, S. Shanmugam, Nitrogen-doped carbon nanofoam derived from amino acid chelate complex for supercapacitor applications, J. Power Sources. 316 (2016) 60–71, Copyright 2016 Elsevier.

The SC N-MCNF cell was able to discharge at different current densities (0.25 to 25 A/g). Hence, the N-MCNF as an energy harvester is able to hold an E.D of 62.4 Wh/kg at room temperature.

One of the major drawbacks of ILs is that few of them maintain a solid form or viscous property at room temperature. In addition, ILs possess large organic cations (imidazolium, pyrrolidinium, and ammonium) with a suitable combination of anions, thus leading to high viscosity and resulting in low ionic conductivity. The standard LIBs electrolyte with the mixture of alkyl carbonate (EC/DMC) with 1 mol/L of $LiPF_6$ delivers an ionic conductivity of 12 mS/cm at RT [58]. Hence, new strategies like ILs mixtures, organic solvents mixed with aprotic and protic ILs, and redox-active assisted ILs were adopted in several works. Among them, the most promising approach is the ILs mixtures, as they showed significant improvement in conductivity and performance over a wide voltage window. For instance, $[Pyr][NO_3]$ + $[Pyr][TFSI]$ binary protic ILs mixture exhibited significant ionic conductivity with increasing concentration of $[Pyr][NO_3]$, and the $[EMIm][TFSI]$ + $[PMPyrr][TFSI]$ delivered a high thermal decomposition of 440 °C, a low melting temperature of −70°C, and ESW of 4.4 V versus Pt [59].

The other types of AILs, such as quasisolid and solid-state type SCs, are mainly applicable for flexible design approaches; such concepts offer electrolyte leakage-free systems. This approach mainly involves embedding polymer matrix in ILs, which is commonly referred to as ion gel. Several ion gel electrolytes have been reported using the following polymers: PVdF-HFP, PHEMA-co-PEGDMGA, PVA, P4VPh, and PUA. To highlight a few, ion gel composed of 25 wt.% PUA and 75 wt.% [EMI] [TFSI] delivered ionic conductivity of ~0.7 mS/cm and achieved E.D of 93.9 Wh/kg at ESW of 4 V [60]. The gel polymer Pyr-based PILTFSI matrix with ionic liquids ([MBIM][FSI] or [DPI][TFSI]) as a solid-state electrolyte SCs attained ESW of 3 V. This gel electrolyte achieved an ionic conductivity of 3.2 mS/cm (PIL-MBIMFSI) and 0.5 mS/cm (PIL-DPITFSI) [61].

2.4 Concluding remarks

In the pursuit of achieving high Cs with improved E.D, many studies, including the studies highlighted in this chapter, have adopted facile synthesis strategies to develop high-surface-area nanocarbons with controlled parameters, such as morphologies, pore size, pore texture, N-heteroatom functionalization, and metal−carbon composite. In addition, through exploiting such engineered nanocarbons within the wide ESW, ILs have gained importance due to their chemical inertness and wide operating temperature. Having said that, rudimentary mechanistic understanding via experimental/theoretical studies, simplified self-templating heteroatom functionalization, high-yield green synthesis approach, formulation of sale and solvent mixture in ILs, electrochemical standardization of ILs at wide temperature range, and improved carbon/electrolyte interface studies are still to be accomplished in the upcoming years.

References

[1] A. González, E. Goikolea, J.A. Barrena, R. Mysyk, Review on supercapacitors: technologies and materials, Renew. Sustain. Energy Rev. 58 (2016) 1189−1206. Available from: https://doi.org/10.1016/j.rser.2015.12.249.

[2] F. Naseri, S. Karimi, E. Farjah, E. Schaltz, Supercapacitor management system: a comprehensive review of modeling, estimation, balancing, and protection techniques, Renew. Sustain. Energy Rev. 155 (2022) 111913. Available from: https://doi.org/10.1016/j.rser.2021.111913.

[3] L. Kouchachvili, W. Yaïci, E. Entchev, Hybrid battery/supercapacitor energy storage system for the electric vehicles, J. Power Sources 374 (2018) 237−248. Available from: https://doi.org/10.1016/j.jpowsour.2017.11.040.

[4] X. Li, B. Wei, Supercapacitors based on nanostructured carbon, Nano Energy 2 (2013) 159−173. Available from: https://doi.org/10.1016/j.nanoen.2012.09.008.

[5] K.K. Patel, T. Singhal, V. Pandey, T.P. Sumangala, M.S. Sreekanth, Evolution and recent developments of high performance electrode material for supercapacitors: a review, J. Energy Storage 44 (2021) 103366. Available from: https://doi.org/10.1016/j.est.2021.103366.

[6] C. Wang, M. Muni, V. Strauss, A. Borenstein, X. Chang, A. Huang, et al., Graphene's role in emerging trends of capacitive energy storage, Small 17 (2021) 1−19. Available from: https://doi.org/10.1002/smll.202006875.

[7] M. Bigdeloo, E. Kowsari, A. Ehsani, A. Chinnappan, S. Ramakrishna, R. AliAkbari, Review on innovative sustainable nanomaterials to enhance the performance of supercapacitors, J. Energy Storage 37 (2021) 102474. Available from: https://doi.org/10.1016/j.est.2021.102474.

[8] X. Zhu, Recent advances of transition metal oxides and chalcogenides in pseudo-capacitors and hybrid capacitors: a review of structures, synthetic strategies, and mechanism studies, J. Energy Storage 49 (2022) 104148. Available from: https://doi.org/10.1016/j.est.2022.104148.

[9] L. Miao, Z. Song, D. Zhu, L. Li, L. Gan, M. Liu, Recent advances in carbon-based supercapacitors, Mater. Adv. 1 (2020) 945−966. Available from: https://doi.org/10.1039/d0ma00384k.

[10] N. Jäckel, P. Simon, Y. Gogotsi, V. Presser, Increase in capacitance by subnanometer pores in carbon, ACS Energy Lett. 1 (2016) 1262−1265. Available from: https://doi.org/10.1021/acsenergylett.6b00516.

[11] M. Oschatz, S. Boukhalfa, W. Nickel, J.P. Hofmann, C. Fischer, G. Yushin, et al., Carbide-derived carbon aerogels with tunable pore structure as versatile electrode material in high power supercapacitors, Carbon N. Y. 113 (2017) 283−291. Available from: https://doi.org/10.1016/j.carbon.2016.11.050.

[12] S. Kondrat, C.R. Pérez, V. Presser, Y. Gogotsi, A.A. Kornyshev, Effect of pore size and its dispersity on the energy storage in nanoporous supercapacitors, Energy Environ. Sci. 5 (2012) 6474−6479. Available from: https://doi.org/10.1039/c2ee03092f.

[13] D.E. Jiang, Z. Jin, D. Henderson, J. Wu, Solvent effect on the pore-size dependence of an organic electrolyte supercapacitor, J. Phys. Chem. Lett. 3 (2012) 1727−1731. Available from: https://doi.org/10.1021/jz3004624.

[14] F. Xu, D. Wu, R. Fu, B. Wei, Design and preparation of porous carbons from conjugated polymer precursors, Mater. Today. 20 (2017) 629−656. Available from: https://doi.org/10.1016/j.mattod.2017.04.026.

[15] B.N. Bhadra, A. Vinu, C. Serre, S.H. Jhung, MOF-derived carbonaceous materials enriched with nitrogen: preparation and applications in adsorption and catalysis, Mater. Today. 25 (2019) 88−111. Available from: https://doi.org/10.1016/j.mattod.2018.10.016.

[16] S. Dutta, A. Bhaumik, K.C.-W. Wu, Hierarchically porous carbon derived from polymers and biomass: effect of interconnected pores on energy applications, Energy Environ. Sci. 7 (2014) 3574−3592. Available from: https://doi.org/10.1039/C4EE01075B.

[17] Z. Yu, L. Tetard, L. Zhai, J. Thomas, Supercapacitor electrode materials: nanostructures from 0 to 3 dimensions, Energy Environ. Sci. 8 (2015) 702−730. Available from: https://doi.org/10.1039/c4ee03229b.

[18] T. Wang, Y. Sun, L. Zhang, K. Li, Y. Yi, S. Song, et al., Space-confined polymerization: controlled fabrication of nitrogen-doped polymer and carbon microspheres with refined hierarchical architectures, Adv. Mater. 31 (2019) 1−8. Available from: https://doi.org/10.1002/adma.201807876.

[19] J. Yu, C. Yu, W. Guo, Z. Wang, S. Li, J. Chang, et al., Decoupling and correlating the ion transport by engineering 2D carbon nanosheets for enhanced charge storage, Nano Energy 64 (2019) 103921. Available from: https://doi.org/10.1016/j.nanoen.2019.103921.

[20] B. Duan, X. Gao, X. Yao, Y. Fang, L. Huang, J. Zhou, et al., Unique elastic N-doped carbon nanofibrous microspheres with hierarchical porosity derived from renewable chitin for high rate supercapacitors, Nano Energy 27 (2016) 482−491. Available from: https://doi.org/10.1016/j.nanoen.2016.07.034.

[21] P. Ramakrishnan, S.-G. Park, S. Shanmugam, Three-dimensional hierarchical nitrogen-doped arch and hollow nanocarbons: morphological influences on supercapacitor applications, J. Mater. Chem. A 3 (2015) 16242−16250. Available from: https://doi.org/10.1039/C5TA03384E.

[22] P. Ramakrishnan, S. Shanmugam, Nitrogen-doped carbon nanofoam derived from amino acid chelate complex for supercapacitor applications, J. Power Sources 316 (2016) 60−71. Available from: https://doi.org/10.1016/j.jpowsour.2016.03.061.

[23] P. Ramakrishnan, S. Shanmugam, Nitrogen-doped porous multi-nano-channel nanocarbons for use in high-performance supercapacitor applications, ACS Sustain. Chem. Eng. 4 (2016) 2439−2448. Available from: https://doi.org/10.1021/acssuschemeng.6b00289.

[24] K.S.W. Sing, Reporting physisorption data for gas/solid systems with special reference to the determination of surface area and porosity (Recommendations 1984), Pure Appl. Chem. 57 (1985) 603−619. Available from: https://doi.org/10.1351/pac198557040603.

[25] C. Young, J. Lin, J. Wang, B. Ding, X. Zhang, S.M. Alshehri, et al., Significant effect of pore sizes on energy storage in nanoporous carbon supercapacitors, Chem. - A Eur. J. 24 (2018) 6127−6132. Available from: https://doi.org/10.1002/chem.201705465.

[26] J. Ye, P. Simon, Y. Zhu, Designing ionic channels in novel carbons for electrochemical energy storage, Natl. Sci. Rev. 7 (2020) 191−201. Available from: https://doi.org/10.1093/nsr/nwz140.

[27] Z. Li, J. Lin, B. Li, C. Yu, H. Wang, Q. Li, Construction of heteroatom-doped and three-dimensional graphene materials for the applications in supercapacitors: a review, J. Energy Storage 44 (2021) 103437. Available from: https://doi.org/10.1016/j.est.2021.103437.

[28] N.A. Rashidi, S. Yusup, Recent methodological trends in nitrogen−functionalized activated carbon production towards the gravimetric capacitance: a mini review, J. Energy Storage 32 (2020) 101757. Available from: https://doi.org/10.1016/j.est.2020.101757.

[29] J.D. Wiggins-Camacho, K.J. Stevenson, Effect of nitrogen concentration on capacitance, density of states, electronic conductivity, and morphology of N-doped carbon nanotube electrodes, J. Phys. Chem. C. 113 (2009) 19082−19090. Available from: https://doi.org/10.1021/jp907160v.

[30] K. Eklund, A.J. Karttunen, Effect of the dopant configuration on the electronic transport properties of nitrogen-doped carbon nanotubes, Nanomaterials 12 (2022). Available from: https://doi.org/10.3390/nano12020199.

[31] L. Hu, J. Hou, Y. Ma, H. Li, T. Zhai, Multi-heteroatom self-doped porous carbon derived from swim bladders for large capacitance supercapacitors, J. Mater. Chem. A 4 (2016) 15006−15014. Available from: https://doi.org/10.1039/C6TA06337C.

[32] G. Lota, K. Fic, E. Frackowiak, Carbon nanotubes and their composites in electrochemical applications, Energy Environ. Sci. 4 (2011) 1592−1605. Available from: https://doi.org/10.1039/C0EE00470G.

[33] F.M. Hassan, V. Chabot, J. Li, B.K. Kim, L. Ricardez-Sandoval, A. Yu, Pyrrolic-structure enriched nitrogen doped graphene for highly efficient next generation supercapacitors, J. Mater. Chem. A 1 (2013) 2904−2912. Available from: https://doi.org/10.1039/c2ta01064j.

[34] E. Paek, A.J. Pak, K.E. Kweon, G.S. Hwang, On the origin of the enhanced supercapacitor performance of nitrogen-doped graphene, J. Phys. Chem. C. 117 (2013) 5610−5616. Available from: https://doi.org/10.1021/jp312490q.

[35] C. Zhao, W. Zheng, A review for aqueous electrochemical supercapacitors, Front. Energy Res. 3 (2015) 1−11. Available from: https://doi.org/10.3389/fenrg.2015.00023.

[36] A.G. Pandolfo, A.F. Hollenkamp, Carbon properties and their role in supercapacitors, J. Power Sources 157 (2006) 11−27. Available from: https://doi.org/10.1016/j.jpowsour.2006.02.065.

[37] B. Pal, S. Yang, S. Ramesh, V. Thangadurai, R. Jose, Electrolyte selection for supercapacitive devices: a critical review, Nanoscale Adv. 1 (2019) 3807−3835. Available from: https://doi.org/10.1039/c9na00374f.

[38] Py Hung, H. Zhang, H. Lin, Q. Guo, Kt Lau, B. Jia, Specializing liquid electrolytes and carbon-based materials in EDLCs for low-temperature applications, J. Energy Chem. 68 (2022) 580−602. Available from: https://doi.org/10.1016/j.jechem.2021.12.012.

[39] I. Tanahashi, Capacitance enhancement of activated carbon fiber cloth electrodes in electrochemical capacitors with a mixed aqueous solution of H[sub 2]{SO}[sub 4] and {AgNO}[sub 3], Electrochem. Solid-State Lett. 8 (2005) A627. Available from: https://doi.org/10.1149/1.2087187.

[40] X. Tang, Y.H. Lui, B. Chen, S. Hu, Functionalized carbon nanotube based hybrid electrochemical capacitors using neutral bromide redox-active electrolyte for enhancing energy density, J. Power Sources 352 (2017) 118−126. Available from: https://doi.org/10.1016/j.jpowsour.2017.03.094.

[41] Q. Li, K. Li, C. Sun, Y. Li, An investigation of Cu^{2+} and Fe^{2+} ions as active materials for electrochemical redox supercapacitors, J. Electroanal. Chem. 611 (2007) 43−50. Available from: https://doi.org/10.1016/j.jelechem.2007.07.022.

[42] X. Wang, R.S. Chandrabose, S.-E. Chun, T. Zhang, B. Evanko, Z. Jian, et al., High energy density aqueous electrochemical capacitors with a KI-KOH electrolyte, ACS Appl. Mater. Interfaces. 7 (2015) 19978−19985. Available from: https://doi.org/10.1021/acsami.5b04677.

[43] P. Díaz, Z. González, R. Santamaría, M. Granda, R. Menéndez, C. Blanco, Enhanced energy density of carbon-based supercapacitors using Cerium (III) sulphate as inorganic redox electrolyte, Electrochim. Acta. 168 (2015) 277−284. Available from: https://doi.org/10.1016/j.electacta.2015.03.187.

[44] P. Bujewska, B. Gorska, K. Fic, Redox activity of selenocyanate anion in electrochemical capacitor application, Synth. Met. 253 (2019) 62−72. Available from: https://doi.org/10.1016/j.synthmet.2019.04.024.

[45] V. Moutarlier, M.P. Gigandet, J. Pagetti, L. Ricq, Molybdate/sulfuric acid anodising of 2024-aluminium alloy: influence of inhibitor concentration on film growth and on corrosion resistance, Surf. Coat. Technol. 173 (2003) 87−95. Available from: https://doi.org/10.1016/S0257-8972(03)00511-5.

[46] W.S. Li, L.P. Tian, Q.M. Huang, H. Li, H.Y. Chen, X.P. Lian, Catalytic oxidation of methanol on molybdate-modified platinum electrode in sulfuric acid solution, J. Power Sources 104 (2002) 281−288. Available from: https://doi.org/10.1016/S0378-7753(01)00961-2.

[47] T. Xiong, W.S.V. Lee, L. Chen, T.L. Tan, X. Huang, J. Xue, Indole-based conjugated macromolecules as a redox-mediated electrolyte for an ultrahigh power supercapacitor, Energy Environ. Sci. 10 (2017) 2441−2449. Available from: https://doi.org/10.1039/C7EE02584J.

[48] S.P. Ega, P. Srinivasan, Quinone materials for supercapacitor: current status, approaches, and future directions, J. Energy Storage 47 (2022) 103700. Available from: https://doi.org/10.1016/j.est.2021.103700.

[49] K. Nasrin, S. Gokulnath, M. Karnan, K. Subramani, M. Sathish, Redox-additives in aqueous, non-aqueous, and all-solid-state electrolytes for carbon-based supercapacitor: a mini-review, Energy Fuels 35 (2021) 6465−6482. Available from: https://doi.org/10.1021/acs.energyfuels.1c00341.

[50] H. Yu, J. Wu, L. Fan, S. Hao, J. Lin, M. Huang, An efficient redox-mediated organic electrolyte for high-energy supercapacitor, J. Power Sources 248 (2014) 1123−1126. Available from: https://doi.org/10.1016/j.jpowsour.2013.10.040.

[51] J. Fang, X. Zhang, X. Miao, Y. Liu, S. Chen, Y. Chen, et al., A phenylenediamine-mediated organic electrolyte for high performance graphene-hydrogel based supercapacitors, Electrochim. Acta. 273 (2018) 495−501. Available from: https://doi.org/10.1016/j.electacta.2018.04.009.

[52] A. Borenstein, S. Hershkovitz, A. Oz, S. Luski, Y. Tsur, D. Aurbach, Use of 1,10-phenanthroline as an additive for high-performance supercapacitors, J. Phys. Chem. C. 119 (2015) 12165−12173. Available from: https://doi.org/10.1021/acs.jpcc.5b02335.

[53] Y. Niu, J. Niu, Y. Ma, L. Zhi, Rational design of viologen redox additives for high-performance supercapacitors with organic electrolytes, Sci. China Mater. 64 (2021) 329−338. Available from: https://doi.org/10.1007/s40843-020-1418-9.

[54] J. Park, B. Kim, Y.-E. Yoo, H. Chung, W. Kim, Energy-density enhancement of carbon-nanotube-based supercapacitors with redox couple in organic electrolyte, ACS Appl. Mater. Interfaces. 6 (2014) 19499−19503. Available from: https://doi.org/10.1021/am506258s.

[55] V. Lockett, R. Sedev, J. Ralston, M. Horne, T. Rodopoulos, Differential capacitance of the electrical double layer in imidazolium-based ionic liquids: influence of potential, cation size, and temperature, J. Phys. Chem. C. 112 (2008) 7486−7495. Available from: https://doi.org/10.1021/jp7100732.

[56] Z.G. Cambaz, G.N. Yushin, Y. Gogotsi, K.L. Vyshnyakova, L.N. Pereselentseva, Formation of carbide-derived carbon on β-silicon carbide whiskers, J. Am. Ceram. Soc. 89 (2006) 509−514. Available from: https://doi.org/10.1111/j.1551-2916.2005.00780.x.

[57] C. Largeot, C. Portet, J. Chmiola, P.L. Taberna, Y. Gogotsi, P. Simon, Relation between the ion size and pore size for an electric double-layer capacitor, J. Am. Chem. Soc. 130 (2008) 2730−2731. Available from: https://doi.org/10.1021/ja7106178.

[58] R. Newell, J. Faure-Vincent, B. Iliev, T. Schubert, D. Aradilla, A new high performance ionic liquid mixture electrolyte for large temperature range supercapacitor applications (−70 °C to 80 °C) operating at 3.5V cell voltage, Electrochim. Acta. 267 (2018) 15−19. Available from: https://doi.org/10.1016/j.electacta.2018.02.067.

[59] L. Timperman, A. Vigeant, M. Anouti, Eutectic mixture of protic ionic liquids as an electrolyte for activated carbon-based supercapacitors, Electrochim. Acta. 155 (2015) 164−173. Available from: https://doi.org/10.1016/j.electacta.2014.12.130.

[60] J. Han, Y. Choi, J. Lee, S. Pyo, S. Jo, J. Yoo, UV curable ionogel for all-solid-state supercapacitor, Chem. Eng. J. 416 (2021) 129089. Available from: https://doi.org/10.1016/j.cej.2021.129089.

[61] E. Kim, J. Han, S. Ryu, Y. Choi, J. Yoo, Ionic liquid electrolytes for electrochemical energy storage devices, Mater. (Basel) 14 (2021). Available from: https://doi.org/10.3390/ma14144000.

Membrane nanoarchitectonics: advanced nanoporous membranes for osmotic power generation

Gregorio Laucirica[1], Yamili Toum Terrones[1], María Eugenia Toimil-Molares[2], Christina Trautmann[2,3], Waldemar A. Marmisollé[1] and Omar Azzaroni[1]

[1]*Instituto de Investigaciones Fisicoquímicas Teóricas y Aplicadas (INIFTA)—Departamento de Química—Facultad de Ciencias Exactas—Universidad Nacional de La Plata—CONICET, La Plata, Argentina* [2]*GSI Helmholtzzentrum für Schwerionenforschung, Darmstadt, Germany* [3]*Technische Universität Darmstadt, Materialwissenschaft, Darmstadt, Germany*

3.1 Introduction

During the last decades, the scientific community has devoted great efforts in the search for new ways to produce sustainable energy [1]. Given this scenario, the possibility of converting the osmotic energy (also referred to as salinity gradient energy [SGE] or blue energy) involved when two solutions of different concentrations are mixed into electrical energy is very attractive, due not only to its high availability but also to its relatively low environmental impact [2−4]. Up to now, different methods have been proposed to tackle this aim, such as reverse electrodialysis (RED), pressure retarded osmosis (PRO), and capacitive mixing [5,6].

In particular, the RED membrane-based method has attracted considerable attention due to its high efficiency and power [7−9]. This method involves the creation of a salinity battery by stacking ion-selective membranes that separate reservoirs filled with solutions of different salt concentrations (Fig. 3.1). RED systems have been typically constructed by using ion-selective membranes with subnanoporosity which provides a high selectivity but, as a downside, their high resistance limits the power generated. While there have been great advances in this field in the last years, the bottleneck of this technology (and also for PRO technology) remains in the limited achievable output power density, the low efficiency, and the high cost linked to membrane production [10−12]. In this context, the generation of new routes for the creation of advanced nanoporous membranes has been recently proposed as an alternative for achieving the balance between high selectivity and low resistance [13,14].

Throughout this chapter, the state-of-the-art of osmotic energy generation via the RED method by employing advanced nanoporous membranes is discussed. To provide an integral overview, the chapter structure consists of the fundamental concepts, fabrication techniques, recent advances, current limitations and challenges of this topic, and, finally, conclusions.

Materials Nanoarchitectonics. DOI: https://doi.org/10.1016/B978-0-323-99472-9.00021-3

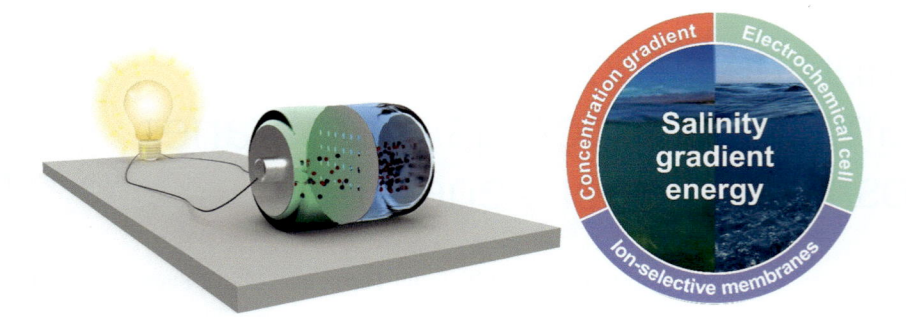

FIGURE 3.1

Scheme illustrating the main fundamentals of salinity gradient energy harvesting. The disposition of a concentration gradient separated by an ion-selective membrane in an electrochemical cell enables the conversion of Gibbs free energy of mixing into electrical energy.

3.2 Fundamental concepts

3.2.1 Reverse electrodialysis

One of the first antecedents of the generation of energy via RED was published by R.E. Pattle in 1954 [15]. In that article, the author reported the creation of a salinity battery by stacking 47 anion and 47 cation-selective membranes that separated fresh- and saltwater. In the following years, the advances in this field were limited due to the difficulties in creating efficient membranes [16−18]. However, in the last decades, the advent of new technologies and the knowledge gained in materials science promoted a considerable evolution in the synthesis of membranes and, consequently, in the creation of osmotic energy generators, such as RED systems.

The basis of this energy generation method relies on the conversion of the Gibbs free energy of mixing (ΔG_m) released in the mixing process of two electrolyte solutions with different activities [2,19]

$$\Delta G_m = G_m - (G_c + G_d) \tag{3.1}$$

where G is, in all the cases, the Gibbs free energy and the subscripts c, d, and m refer to the concentrated, diluted, and after-mixed solutions, respectively. After some thermodynamic considerations, Eq. (3.1) can be rewritten as follows [5,6,11]

$$\Delta G_m = \sum_i \left[C_{i,m} V_m RT \ln x_{i,m} \gamma_{i,m} - (C_{i,c} V_c RT \ln x_{i,c} \gamma_{i,c} + C_{i,d} V_d RT \ln x_{i,d} \gamma_{i,d}) \right] \tag{3.2}$$

where C_i, V, R, T, x_i, and γ_i correspond to the i-compound concentration, solution volume, universal gas constant, temperature, i-compound mole fraction, and i-compound activity coefficient respectively.

With Eq. (3.2) in hand, it is possible to estimate the theoretical energy involved in a mixing process. For instance, a mixing process between 1 m^3 of river water (0.01 M NaCl) and 1 m^3 of seawater (0.5 M NaCl) would involve energy around -1.5 MJ (a negative sign indicates released energy) [19].

However, in a common mixed process of two effluents with different salinity content, the phenomenon is irreversible and the released energy is rapidly dissipated [20]. For this reason, the main goal of a blue energy method is to enable the mixing in a controlled way, and thus harvest and convert the involved energy.

In the case of the RED method, this objective is tackled by intercalating ion-selective membranes between two solutions of different activities with a two-electrode arrangement (Fig. 3.2A). Under these conditions, the combination of a concentration asymmetry and membrane permselectivity promotes an ion flux across the membrane that generates a membrane potential (E_m) [21−23]

$$E_m = \varphi_{\text{diff}} - (\varphi_D'' - \varphi_D')$$

(3.3)

where φ_D'' and φ_D' are the Donnan potential at both sides of a given ion-selective membrane and φ_{diff} is the potential that arises due to the differences in the ion transport of cations and anions inside of the membrane (Fig. 3.2B). Moreover, E_m can be also directly related to different experimental values by the derivation of (3) with the Teorell−Meyer−Sievers theory [23,24]. In its simplest expression for a single membrane (half-cell design Fig. 3.2A), this procedure gives rise to the following equation [11,12]

$$E_m = 2(t_+ - 1)\frac{RT}{nF}\ln\frac{a_H}{a_L}$$

(3.4)

with a_H, a_L, n, and t_+ being the activity in the highly concentrated reservoir, activity in the diluted reservoir, valence number, and the cation transference number, respectively. For the case of stacked ion-selective membranes (full-cell design Fig. 3.2C), this equation is converted into (ideal conditions):

$$\text{OCV} = N \times E_m$$

(3.5)

where OCV and N correspond to the open-circuit voltage and the number of stacked membranes, respectively.

Conceptually, these equations denote that an increment in the driving force, the concentration gradient (a_H/a_L), generates an increment in the membrane potential. In addition, the equation shows that, to materialize the increment in E_m, it is also important to fulfill one of the most important requirements of RED: the membrane must present a good ion selectivity (ideally, $t_+ \rightarrow 1$ or $t_+ \rightarrow 0$).

In addition to the membrane selectivity, there are other important parameters to understand the energy capabilities of a given RED system, such as the power (P). If it is considered the case of the system connected to an external load resistance (R_{load}), an expression for the output power can be written as follows:

$$P = \frac{E_m^2 R_{\text{load}}}{(R_{\text{load}} + R_{\text{stack}})^2}$$

(3.6)

where R_{stack} corresponds to the cell resistance and contains several contributions such as membrane, solution, diffusion boundary layers, and electrode reactions. In particular, diffusion boundary layers are created due to the differences in the transference numbers in the bulk compared with the membrane. Consequently, a diminution (increment) in the electrolyte concentration is produced at the proximity of the membrane in the highly (lowly) concentrated reservoir (Fig. 3.2B) [11]. This fact, also called ion concentration polarization, generates a diminution in the effective gradient

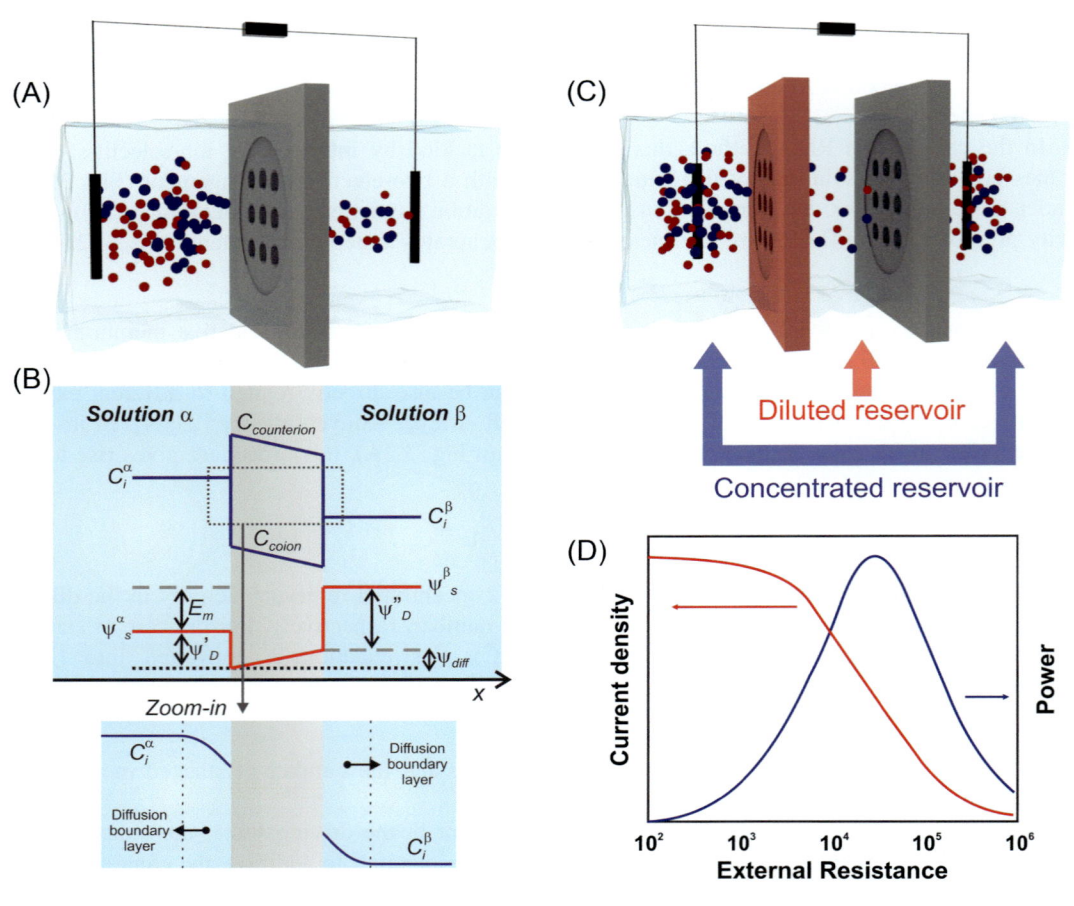

FIGURE 3.2

(A) Half-cell setup. Such system consists of a nanoporous ion-selective membrane that separates two solution of different activities. The energy involved is converted into electrical energy by means of a two-electrode arrangement. (B) Scheme of the different potentials involved in the region around the membrane when exposed to the concentration gradient. In the zoom-in, the creation of the diffusion boundary layers is schematized. (C) Full-cell setup. Such a system consists of the intercalation of nanoporous anion-selective and cation-selective membranes that separate the two solution of different activities. (D) Typical response of the current density and power versus resistance (i-R and P-R curves).

Scheme (B) was adapted from G. Laucirica, M.E. Toimil-Molares, C. Trautmann, W. Marmisollé, O. Azzaroni, Chem. Sci. 12 (2021) 12874 with permissions from the Royal Society of Chemistry.

(the driving force of osmotic power generation) and, concomitantly, vanishes the output power. Undoubtedly, the development of strategies to mitigate this effect involves one of the most crucial challenges in the development of RED systems. Considering the previous equation, a typical P

versus. R_{load} curve for a RED system is shown in Fig. 3.2D. It can be derived that the curve maximum, that is, the maximum power (P_{max}), is achieved when $R_{\text{load}} = R_{\text{stack}}$:

$$P_{\text{max}} = \frac{E_m^2}{4R_{\text{stack}}} \tag{3.7}$$

From a conceptual perspective, this relation denotes that, while membrane selectivity is a parameter of paramount relevance, it is not enough to achieve a good performance in terms of energy. In particular, obtaining an adequate balance between selectivity (represented by E_m) and R_{stack} is of paramount relevance. These concepts will be addressed throughout the chapter.

For comparative purposes, most of the works report the performance in terms of the maximum power density (PD_{max}):

$$\text{PD}_{\text{max}} = \frac{P_{\text{max}}}{A} \tag{3.8}$$

where A corresponds to the area and its definition is not a trivial issue. For the case of single-pore systems, the area is typically defined as the pore cross-section area, whereas, in the case of upscaled membranes, this magnitude corresponds to the exposed area of the membrane (working or testing area) [11,12,25]. Several discussions can arise behind this parameter (see current limitations and challenges section). In principle, the area definition for single-pore systems leads to excessively high (and unrealistic) PD_{max} values. For this reason, in some cases, the scalability of single-pore systems is estimated by extrapolating the P_{max} of an individual pore/channel to the theoretically achievable PD_{max} for an upscaled membrane with a certain pore-density. However, as we will see, beyond a certain limit of pore-density, PD_{max} largely deviates (negative deviation) from the extrapolated single-pore behavior [26−28]. The implications of the area in the osmotic energy conversion will be addressed in more detail in the following sections.

3.2.2 Ion selectivity in nanoporous membranes

When a membrane with charged surfaces is immersed in an aqueous electrolyte solution, the electrical potential arising from the surface promotes an ion redistribution in the proximity of the surface (Fig. 3.3A) [29,30]. This phenomenon gives rise to the electrical double-layer formation where counterions (ions with opposite charge polarity to the surface) will be enriched in the proximity of the surface, whereas coions (ions with the same charge polarity to the surface) will be depleted. The Debye length λ_D is typically referred to as the characteristic length of the electrical double-layer and denotes the effective range of electrostatic interactions in the solution [31]

$$\lambda_D = \sqrt{\frac{\varepsilon \varepsilon_0 k_B T}{2C_i e^2}} \tag{3.9}$$

where k_B is the Boltzmann constant, e is the electron charge, and ε_0 and ε are the free-space and relative permittivity, respectively.

Typically, a good ion selectivity is achieved for nanochannels with charged surfaces if the channel dimension is in the order of λ_D (Fig. 3.3A). Taking into consideration the previous equation, it is usually possible to increase the selectivity with smaller channels and lower electrolyte concentrations. However, different parameters also present a central role in the selectivity, such as the

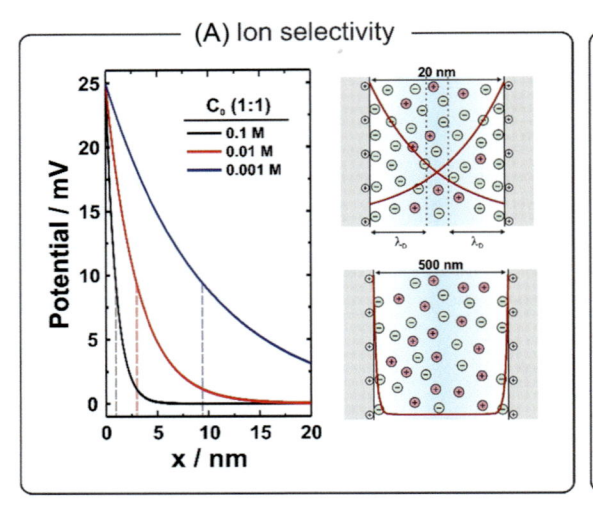 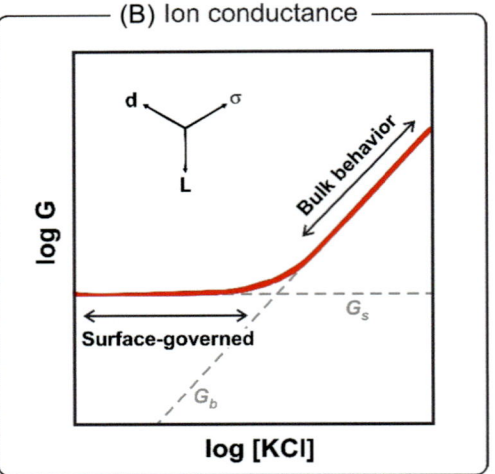

FIGURE 3.3

(A) Electrostatic potential derived from Debye–Hückel equation versus the distance to the surface (x), for different electrolyte concentrations. Dashed lines denotes the Debye length for the different electrolyte concentrations. The 2D scheme illustrates the major relevance of the electrical double-layer for nanochannels compared to microchannels. (B) Typical dependence of the nanochannel conductance (G) as a function of the electrolyte concentration. It is possible to distinguish two different regimes: one of them governed by the surface (low [KCl]) and the other governed by the bulk behavior (high [KCl]).

channel length and pH [32−35]. For instance, the system selectivity increases for longer channel lengths [33]. The pH of the electrolyte is especially important for those membranes with pH-dependent surface charge and, in that case, an increase of the selectivity is achieved for those pH conditions where the surface charge is maximized (e.g., alkaline conditions in carboxylic-terminated surfaces or acidic conditions in amino-terminated surfaces) [32].

Considering the RED systems, the selectivity directly impacts the maximum energy-conversion efficiency, η_{max}, which is defined as the ratio between the extracted power and ΔG_m and can be roughly estimated with the following equation:

$$\eta_{max} = \frac{(t_+ - t_-)^2}{2} \tag{3.10}$$

Given Eq. (3.10), an increase in the channel selectivity implies an increase in the efficiency, and therefore, from this perspective, smaller channels with highly charged surfaces would represent an ideal condition.

In particular, the trend between the selectivity and experimental parameters, such as surface charge density and the channel size, was studied in several previous reports. For instance, Radenovic and coworkers analyzed the impact of the channel size on the E_m in single-layer molybdenum disulfide (MoS_2) nanopores [36]. By comparing the output voltage obtained for MoS_2 single nanopores with different sizes ranging from $\sim$3 to $\sim$18 nm, they found that E_m decreases as the size increases. Similar results were obtained by our group in studies carried out in PET single bullet-shaped nanochannels [32]. For its part, the relationship between the selectivity and the surface charge has been also addressed in different systems. Usually, the surface charge can be easily modulated by changing the pH media. In this line, membranes in which the surface charge is mainly given by the acid−base equilibrium of carboxylic groups (as track-etched PET membranes) have evidenced that the E_m can be maximized by increasing the pH, which is ascribed to the major proportion of negatively charged carboxylate groups. However, when the surface charge density is excessively high, its further increase can generate a counterproductive effect on the E_m [37]. This peculiar phenomenon has been attributed to the local attenuation of the concentration gradient due to the combination of high surface charge density and nanoconfinement.

3.2.3 Membrane resistance

As Eq. (3.7) illustrates, high selectivity is not enough to obtain high energy conversion, instead, decreasing the membrane resistance (R_m) is another requirement of paramount relevance.

Considering the high surface-to-volume ratio, it is expected that there is a significant influence of the surface on the conductance of nanometric channels. In its simplest model, the conductance G (inverse of membrane resistance, $G = 1/R_m$) is typically given by the contribution of two different magnitudes: the bulk -G_b- and surface conductance -G_s- (Fig. 3.3B) [38−41]

$$G = G_b + G_s = \frac{k\pi d^2}{4L} + \frac{k_s \pi d}{L} \tag{3.11}$$

where k and k_s are the bulk and surface conductivities, respectively, d is the channel diameter, and L is the channel length. The bulk conductivity depends on both the ion mobilities and the salt concentration, whereas k_s is proportional to the counterion mobility and the surface charge. Hence, at high salt concentrations, the terms arising from the bulk conductance dominate the behavior and, on the contrary, when the salt concentration is very low, G is mostly explained by the surface conductance and, therefore, it is independent of the bulk salt concentration. Consequently, in the low concentration regime, the conductance is mainly given by the counterions shielding the surface charges. It is worth mentioning that this simplified model does not consider ion hydration effects (important for small channels <2 nm), the dependence of surface charge on salt concentration (very important at low electrolyte concentrations due to charge regulation effects), or access resistances (particularly relevant in 2D membranes). Moreover, this equation is valid in symmetric conditions of electrolyte concentrations, which is not a representative scenario of RED experiments. However, even with all these limitations, this simple model enables some generalities between the conductance (or membrane resistance) and the experimental variables to be determined. Specifically, it is possible to conclude that bigger channels with a short length are favorable for osmotic energy generation from the conductance perspective. Unfortunately, these conditions compete with the experimental requirements of selectivity where small and long channels are needed to increase E_m. This effect was also experimentally evidenced in the seminal work conducted in single-pore MoS$_2$ membranes, where while the osmotic current was raised three times by increasing the channel from ~3 to ~15 nm, E_m displayed a clear drop from values around 90 to ~60 mV [36]. Thus, the competition between osmotic conductance and selectivity involves one of the most challenging trade-offs in the creation of membranes with high performance in osmotic energy generation. On the other hand, both the conductance and the selectivity can be maximized by increasing the surface charge. However, as happens with E_m, an excessive increase in the surface charge density could generate a counterproductive effect on the osmotic conductance due to the local reduction of the concentration gradient [37].

In addition to the previous analysis, rectifying behavior is another desirable requirement for the membrane since it favors the forward current and, at the same time, suppresses the backflow current [14,42]. To create membranes with ion current rectification properties, a disruption in the symmetry of the electrical potential is needed which could be easily obtained by creating systems with either asymmetric geometry or asymmetric charge distribution (asymmetrically modified).

3.3 Fabrication of advanced reverse electrodialysis nanoporous membranes

In order to make this energy-harvesting technology competitive and practically useful, it is imperative to produce highly efficient membranes with, ideally, several features such as low resistance, high selectivity, good mechanical properties, low cost, and high fouling resistance and durability. This challenge has encouraged the scientific community to combine different materials and synthesis routes for constructing advanced nanoporous membranes.

Throughout this chapter section, some of the materials and synthesis protocols more applied for the creation of nanoporous membranes with applications in RED will be addressed. In particular, three different building blocks have been extensively employed: membranes with discrete aligned nanochannels (hereinafter referred to as multichannel membranes); stacking of 2D materials (2D lamellar membranes); and 3D-nanoporous networks. In the following subsections, basic considerations about each group will be described.

3.3.1 Multichannel membranes

In recent years, both inorganic and organic membranes that contain parallel channels have been widely employed as building blocks to create nanofluidic RED (nRED) systems. Among the inorganic alternatives, the most employed membranes have been anodic aluminum oxide (AAO) [43,44]. These systems are typically created by exposing an aluminum clean sheet to two anodic oxidation steps in an acidic electrolyte (Fig. 3.4A) [45]. Then, the aluminum barrier is removed either by chemical etching or applying a voltage pulse, and the final result is a membrane that contains a honeycomb-like array of channels where their sizes can be modulated between 5 and 250 nm, and channel density up to $\sim 10^{11}$ channels/cm^2 can be obtained by setting the experimental conditions of the anodic oxidation steps [44,46].

On the other hand, among the organic materials, the most employed nanofluidic systems are based on track-etched polymeric membranes (Fig. 3.4B). In the ion-track-etching technology, a polymeric membrane (typically polyethylene terephthalate-, polyimide, or polycarbonate) is irradiated with swift heavy ions (e.g., Au or U) with energies around 10 MeV per nucleon, and then it is exposed to a chemical etching process [51–53]. By handling the irradiation flux, the channel density can be modulated from 1 to 10^9 channels per cm^2, whereas the channel size can be easily

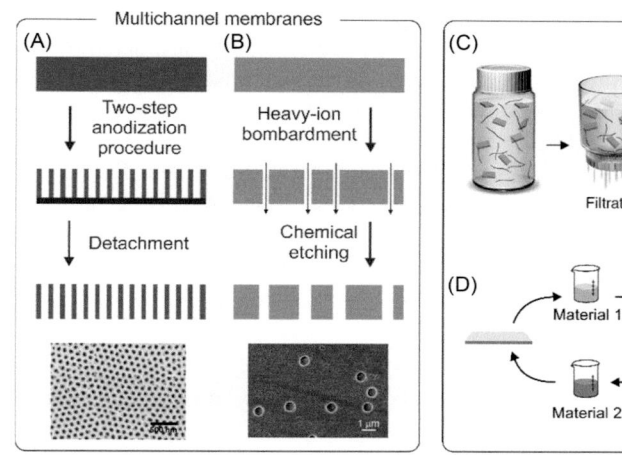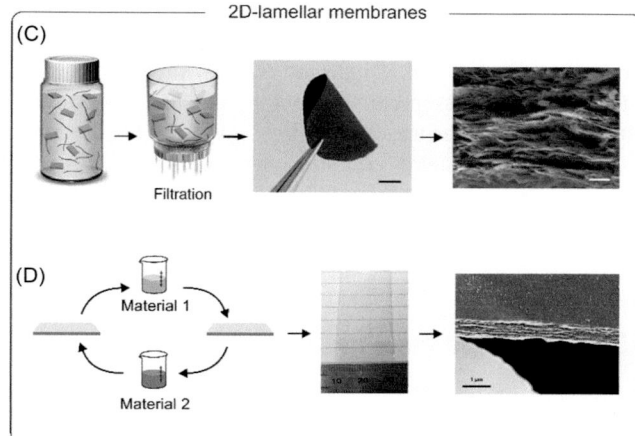

FIGURE 3.4

Scheme of nanofabrication techniques. Synthesis of aligned multichannel membranes by (A) electrochemical anodization and (B) ion-track-etching technology. Synthesis of 2D-lamellar membranes by (C) vacuum filtration and (d) layer-by-layer method.

SEM image in (A) was adapted with permissions from Ref. [47]. Copyright 2019, the Author(s). SEM image in (B) was adapted with permissions from Ref. [48]. Copyright 2019, American Chemical Society. Figure (C) was adapted with permissions from Ref. [49]. Copyright 2019, the Author(s). Figure (D) was adapted with permissions from Ref. [50]. Copyright 2020, American Chemical Society.

fine-tuned from the nanometric (> 10 nm) to the micrometric range by setting the chemical etching conditions [54–57]. In addition, this technology enables the creation of channels with different geometries ranging from cylindrical, conical, cigar, to bullet by simply changing the etching conditions .

In both kinds of membranes, inorganic and organic, the characteristics of the surface provide different chemical and physical routes to integrate a wide variety of building blocks, and thus to create composite membranes. Thus several nRED systems have been reported by modifying these membranes with different charged polymers, and 2D and 3D materials.

3.3.2 **2D-based membranes**

2D-membranes with nanopores or channels are usually created by employing two different approaches: either by exposing ultrathin membranes to an electron beam or ion beam, or by stacking nanosheets to create 2D lamellar membranes. The former approach has been widely used to create systems with precise control of pore size and pore-to-pore distances, which have enabled the basic study of the effects of different experimental variables on energy conversion performance [36,58]. However, their difficulties related to scalability compromise its applicability [25]. Instead, the stacking strategy has shown very promising results and, for this reason, it will be addressed in more detail in this section.

In the case of 2D lamellar membranes, the stacked nanosheets generate interstitial spaces between neighboring sheets that work as nanofluidic channels. These systems are typically constructed by two approaches: vacuum filtration and layer-by-layer. In the former, a stable colloidal dispersion is vacuum filtrated to assemble the membrane and then, detached from the support to form the free-standing 2D lamellar membrane (Fig. 3.4C) [59,60]. In the latter, usually employed to create composite membranes, a stable colloidal dispersion of each component is generated and then the support (e.g., a glass slide) is sequentially immersed into such dispersions (Fig. 3.4D) [61]. Both protocols have been employed to create 2D lamellar membranes based on different materials such as graphene [59], carbonitrides of early transition metals (MXene) [49,62], and black phosphorous [63].

3.3.3 3D-based membranes

In recent years, 3D-based membranes have gained considerable impetus since their interconnected nanoporosity, added to the possibility to modulate the membrane thickness during the synthesis, offers an ideal scenario to generate systems with a good balance between resistance and ion selectivity [64–67]. In particular, for this kind of membrane, different synthesis methods have been proposed depending on the chemical nature of the desired membrane. In addition, different works have shown the possibility of combining these materials with other building blocks, such as multipore membranes, metal–organic frameworks, etc., which position 3D membranes as a versatile scaffold for the construction of nRED systems [66,68].

3.4 Recent advances in upscaled membranes

In the last decade, vast knowledge has been gained by studying the osmotic energy performance of low-porosity membranes (e.g., single-pore membranes) under different experimental and structural conditions. However, to show the technological potential of such advanced membranes, the scaling up to high-porosity systems is highly necessary. In this context, this section will briefly discuss the recent advances in the design, construction, and performance of upscaled membranes based on a wide variety of materials, such as graphene, polymers, MXene, and metal–organic frameworks, among others (tables addressing most of the recent advances related to the use of nanomembranes for RED can be found in Refs. [11,25]). Considering that RED has been mostly thought to exploit the energy available in the salinity gradient arising from the mixing of river and seawater, the experimental laboratory conditions typically involve a 50-fold gradient of NaCl. For this reason, the maximum power density (PD_{max}) reported in the works reviewed in this chapter referred to that gradient magnitude (otherwise other conditions are specified). However, it is worth mentioning that the employment of other gradient sources has been positioned as an interesting alternative to expand the horizons of this technology.

For instance, a seminal work by Ji et al. shows the creation of a full RED cell by the fabrication of an anion and cation-selective membrane based on stacked graphene nanosheets [59]. The native graphene membrane acts as a cation-selective membrane due to the presence of negatively charged sites. However, they demonstrate that it is also possible to create anion-selective graphene-based membranes by performing a previous modification on the nanosheets with 1-aminopropyl-3-methylimidazolium bromide to form positively charged imidazole groups. Thus the system exhibited a maximum power density (PD_{max}) of 0.7 W/m^2 by stacking positively and negatively charged 2D membranes. Moreover, by employing a tandem arrangement of 13 pairs of membranes, the output voltage (E_m) reached values around 2.7 V. Since then, several 2D laminar membranes for osmotic power generation have been created in different materials such as MoS$_2$, MXene, boron nitride, etc. [49,69−74].

More interestingly, the integration of different building blocks into the nanosheets has given rise to novel 2D membranes with superior properties. In particular, the generation of stacked composite membranes by combining 2D materials with aramid or cellulose nanofibers allows improvement of several features of the membrane, such as mechanical stability, ion selectivity, and interlayer spaces [75]. For instance, Zhu et al. reported the construction of a stacked membrane based on metallic phase-MoS$_2$ nanosheets and cellulose nanofibers (Fig. 3.5A) [73]. The presence of cellulose nanofibers in the membrane not only provides the required mechanical strength and stability in the aqueous medium but also increases the number of negatively charged sites which play an important role in the system selectivity. Thus under optimized conditions and applying a

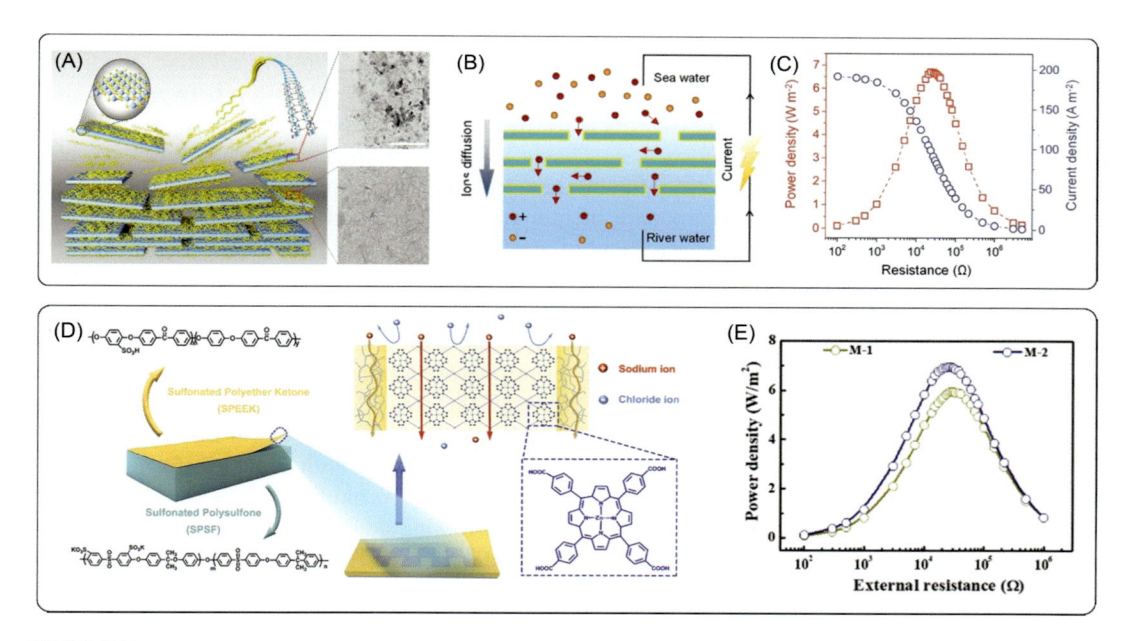

FIGURE 3.5

Scheme of (A) MoS$_2$/cellulose nanofibers lamellar membrane and (B) the experimental setup. (C) P-R and I-R curve for the 2D-composite membrane. The system exhibited a PD_{max} of 6.7 W/m^2. (D) Scheme of the 3D-composite membrane formed by sulfonated polyether ketone, sulfonated polysulfone and metal-organic frameworks. (E) P-R curves for the composite membrane without (M-1) and with (M-2) the metal-organic framework. The inclusion of the metal-organic framework enables an increment of PD_{max}.

Parts (A)−(C) were adapted with permissions from C. Zhu, P. Liu, B. Niu, Y. Liu, W. Xin, W. Chen, et al., J. Am. Chem. Soc. 143 (2021)1932. Copyright 2021 American Chemical Society. Parts (D) and (E) were adapted with permissions from X. Zhao, C. Lu, L. Yang, W. Chen, W. Xin, X.-Y. Kong, et al., Nano Energy 81 (2021) 105657. Copyright 2020 Elsevier Ltd. All rights reserved.

gradient concentration given by the China East Sea and river water, a PD_{max} of 6.7 W/m^2 was achieved (Fig. 3.5B and C). Similar improvements in the performance were obtained by combining MXene and boron nitride nanosheets with aramid nanofibers [70].

The use of etched nanosheets with discrete porosity to create a perforated and lamellar-structured 2D membrane has been reported as another novel option to improve the balance between ion conductance and selectivity. Under this configuration, the energy barrier to the ion diffusion is lowered and, consequently, the ion conduction can be maximized. For instance, Hong et al. reported the construction of a 2D lamellar membrane by stacking perforated $Ti_3C_2T_x$ (MXene) nanosheets [76]. The nanoholes were introduced into the MXene sheets via the exposition of MXene nanosheets suspension to 3 M H_2SO_4 under vacuum at 40°C. Remarkably, this chemical procedure neither affected the functionality nor the crystallinity of the MXene material. Under a gradient of 100-fold KCl, the nanoporous lamellar membrane evidenced a PD_{max} of 17.5 W/m^2, which involved an enhancement of 38% regarding the nonporous lamellar membrane.

Multichannel membranes based on AAO or track-etched foils have been also extensively used for osmotic energy generation. For instance, Xin et al. reported the creation of a heterogeneous membrane based on the asymmetric integration of a silk fibroin film onto an AAO membrane. Under optimized conditions, the system displayed a PD_{max} of 2.86 W/m^2 in alkaline conditions [47]. Besides, the authors emphasize the importance to fine-tune the relationship between the sizes of silk pores and AAO channels. Excessive increase in the AAO channel size reduces the ion storage capability of the membrane, whereas a significant reduction in the channel size limits the efficiency in the contact between the AAO membrane and the silk film. Chen et al. showed the modification of a track-etched polycarbonate membrane that contained cylindrical channels (pore density 5×10^{12} pores per m^{-2}) with a hydrogel of acrylic acid–co-acrylamide–co-methyl methacrylate [77]. The hydrogel hybrid membrane displayed a PD_{max} of 4.08 W/m^2 under an artificial seawater/river water gradient. Also, the membrane exhibited prominent mechanical properties, which is an interesting feature with regard to the creation of membranes with real applications.

The creation of 3D nanoporous networks has been proposed as a viable alternative to increase the ion conductance of the membrane without affecting the ion selectivity. In such systems, the presence of interconnected nanopores allows a more efficient flux of ions due to the diminution of the interfacial resistance. Moreover, these systems can be combined with other platforms such as metal–organic frameworks or multipore membranes, as they can be modified in their structural and chemical features such as length, surface charge, or cross-linking degree by adjusting the synthesis conditions, thus providing them with high potentiality and versatility. For instance, Zhao et al. demonstrated the creation of a hybrid heterogeneous inorganic/organic membrane (Fig. 3.5D) [78]. The system consisted of a sulfonated polysulfone porous support with a top layer composed of a composite of sulfonated poly(ether etherketone) and MOF nanosheets (zinc(II) tetrakis(4-carboxy-phenyl)porphyrin). The formed membrane, with a total thickness of 4.9 μm, displayed a PD_{max} of 6.96 W/m^2 which exceeded the performance exhibited by the separated components (Fig. 3.5E).

Taking advantage of a different approach, Xu et al. showed the combination of a track-etched membrane that contained conical nanochannels with a 3D network constituted by negatively charged oxidized cellulose nanofibers [68]. The asymmetrical modification provided not only an ion current rectification behavior which is a favorable condition for osmotic energy conversion but also allowed weak ion concentration polarization which enabled the maximization of the effective gradient. Thus under optimized conditions, the system demonstrated a PD_{max} of 0.96 W/m^2.

3.5 **Current limitations and challenges**

In the last few years, considerable advances both in the understanding of osmotic power generation at the nanoscale as well as in the creation of upscaled membranes have been consolidated. In addition, the osmotic power generation exhibited by several upscaled membranes of different nature has shown values around or exceeding the proposed commercial benchmark (5 W/m^2). However, most of the reported results are in laboratory conditions which, usually, are far from the operating conditions of this technology. Thus to move ahead in this field, different challenges have to be faced. In principle, it is possible to highlight three main challenges: development of strategies to reduce the effects from ion concentration polarization, testing (and, eventually, improvement) of mechanical and antifouling properties of advanced nanoporous membranes, and creation of full cells.

When the results obtained in single-pore or low-pore density are linearly scaled to hypothetical multipore systems or referred to its cross-section area, PD$_{max}$ takes values that are very promising for the technology. As an example, the PD$_{max}$ reported for a single boron nitride nanotube reaches a value around 4.10^4 W/m^2 [79]. However, the membrane conductance does not follow a linear relationship with the pore density, instead, the relationship is sublinear. Consequently, results arising from such linear estimation overestimate the real performance achievable in multipore membranes (Fig. 3.6A) [26,28,58]. For instance, Ma et al. compared the performance of hydrogel-functionalized nanochannels in membranes with different pore-densities. By incrementing the pore-density from 2×10^5 pores per cm^2 to 3×10^9 channels per cm^2 ($\sim$15,000 times), the current only increased 51 times yielding a power output far below that expected by the linear estimation [80]. This effect is ascribed to the larger flux of ions that produces a local decrease in the ion concentration close to the membrane surface in the side facing to the concentrated solution (previously referred to as ion concentration polarization). As it was mentioned, ion concentration polarization reduces the driving force of the method, which negatively impacts on the extracted power (Fig. 3.6B) [26,27,58]. Then it becomes imperative to reduce the concentration polarization. For this aim, different authors have proposed the creation of heterogeneous membranes with smaller pores [26,81]. Another option relies on decreasing the boundary layers (Fig. 3.2B) by enhancing the hydrodynamic conditions [28].

Furthermore, most of the reported works have been performed with very low working areas (<0.1 mm^2). Under this condition, the amount of mixing and, consequently, the extractable power is very low, which does not represent a real scenario for the application. In addition, as it has been reported, there is a trade-off between the working area and PD$_{max}$: an increment in the former leads to an abrupt drop in PD$_{max}$ [10,11,74]. Moreover, an increase in the mixing process produces a loss in the concentration gradient which makes necessary the renewal of the solutions (and gradient) with a pump system that consumes energy.

While notable advances were consolidated in only a few years, there are still several challenges for the creation of advanced nanomembranes [25]. For instance, most of the new membranes reported are cation-selective [11]. The creation of anion-selective membranes with a good balance between selectivity and ion conductance typically involves a more challenging scenario [82]. Moreover, membranes are commonly susceptible to fouling which hinders the energy conversion properties [83]. This fact, especially important in anion-selective membranes, makes necessary the

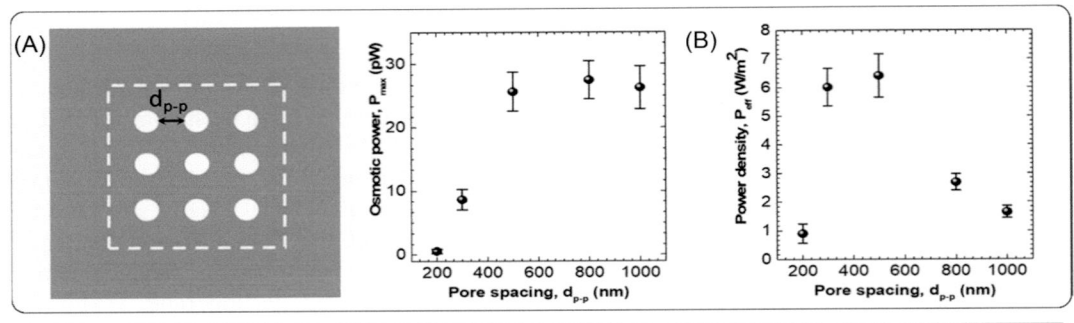

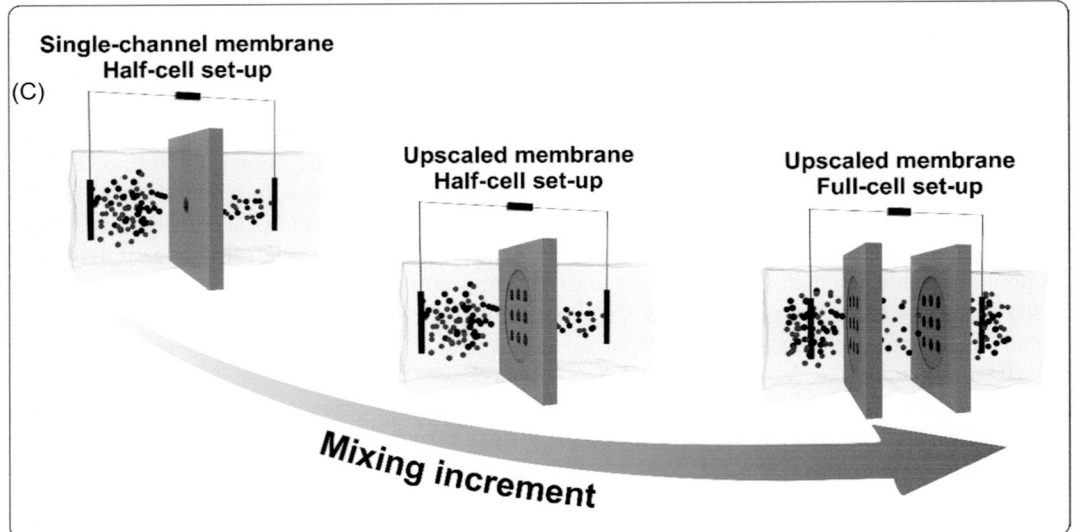

FIGURE 3.6

(A) Comparison of the estimated power density by single-pore linear growth and multipore systems. Linear estimation of single-pore systems produces a strong overestimation of the power capability. (B) Scheme depicting the increase of ion concentration polarization (ICP) as the pore-density is increased. (C) Scheme illustrating the increase in the mixing as the setup is scaled up.

Parts (A) and (B) were adapted with permissions from K. Yazda, K. Bleau, Y. Zhang, X. Capaldi, T. St-Denis, P. Grutter, et al., Nano Lett. 21 (2021) 4152. Copyright 2021 American Chemical Society.

pretreatment of water sources that consume energy and, concomitantly, diminishes the efficiency of the method. In particular, the inclusion of antifouling scaffolds in the membrane has been proposed as a viable alternative to mitigate this effect. Until now, there is not enough information available about the impact of this phenomenon on nanoporous membranes.

Last but not least, most of the systems reported so far are based on half-cell setups. Under this configuration, the transport of counterions across the membrane is favored, and thus the net charge transport is converted to a current by redox reactions on the electrode surfaces. This configuration results in the simplest way to understand the principles that operate in RED and, for this reason, it is preferred when a new nanoporous membrane is synthesized, but it is not the most efficient from an applied perspective because it only enables the harvesting of half of the Gibbs free energy available [84]. To achieve large-scale applications, the development of full-cell devices is imperative based on the intercalation of half-cell units composed of both cation- and anion-selective membranes (Fig. 3.6C). By using this arrangement, it is possible to harvest the net flux charge promoted by both ions, cations (in the cation-selective membrane) and anions (in the anion-selective membrane), which enhances the mixing and therefore, the energy efficiency. Naturally, this aim is not a trivial issue; instead, it makes necessary the development of both cation- and anion-selective membranes and the accurate design of other cell parameters (electrodes, working area, reservoir sizes, electrolytes, etc.).

3.6 Conclusions

Throughout this chapter, the different concepts related to the conversion of SGE by employing advanced nanoporous membranes have been addressed. In the last decade, the interest in this topic has grown very rapidly, positioning it at the forefront of materials science. As has been shown, the use of these systems seems to have changed the paradigm of membrane-based salinity gradient energies due to the promising results in laboratory-scale devices. Moreover, the possibility to combine osmotic power generators with other technologies, such as desalination systems and batteries, expands the frontiers of these devices (for further details see Refs. [11,12,85]). In particular, these alternatives not only would allow for a more efficient energy use but also a more adequate management of brines and the exploitation of other saline sources.

Nowadays, a great number of synthesis methods to create membranes with a wide variety of pore sizes, chemical groups, and mechanical properties are available. In particular, the ion selectivity provided by the surface and space charges added to the high ion conductance of nanochannel and nanopores has provided very interesting results in terms of power density. As an example, single-pore membranes have shown PD_{max} up to 4 kW/m^2, which represents a value 1000 times above the commercial benchmark (5 W/m^2). For its part, upscaled membranes based on different materials such as polymer networks, 2D-layered materials, and AAO multichannel membranes have also shown encouraging PD_{max} that exceeds the 5 W/m^2 value under either simulated or real river/seawater gradient. The last scenario involves a more representative situation of the real application of the technology; however, there are still several crucial challenges to become nanoporous membranes for osmotic power generation in viable platforms. In particular, the search for strategies to alleviate the effect of ion concentration polarization, and thus to boost the output power, is positioned as one of the most important and challenging issues. Moreover, most of the results have been obtained by employing very small testing areas, which is not the ideal condition from an applied perspective. Finally, the development of new advanced membranes with desirable chemical, mechanical, and antifouling properties is also an important requirement. Undoubtedly, to address these challenges, interdisciplinary efforts in different fields of science, such as materials science, environmental science, engineering, and chemistry, will play a key role.

References

[1] S. Chu, Y. Cui, N. Liu, Nat. Mater. 16 (2017) 16.
[2] Z. Jia, B. Wang, S. Song, Y. Fan, Renew. Sustain. Energy Rev. 31 (2014) 91.
[3] B.E. Logan, M. Elimelech, Nature 488 (2012) 313.
[4] G.Z. Ramon, B.J. Feinberg, E.M.V.v Hoek, Energy Env. Sci. 4 (2011) 4423.
[5] A. Cipollina, M. Giorgio, Sustainable Energy from Salinity Gradients, Elsevier/Woodhead Publishing, 2016.
[6] K. Nijmeijer, S. Metz, Sustain. Sci. Eng. 2 (2010) 95−139.
[7] J. Veerman, D.A. Vermaas, Sustainable Energy from Salinity Gradients, Elsevier, 2016, pp. 77−133.
[8] S. Pawlowski, J. Crespo, S. Velizarov, Electrokinetics Across Discip. Cont, Springer International Publishing, Cham, 2016, pp. 57−80.
[9] E. Güler, K. Nijmeijer, J. Membr. Sci. Res. 4 (2018) 108.
[10] X. Tong, S. Liu, J. Crittenden, Y. Chen, ACS Nano 15 (2021) 5838.
[11] G. Laucirica, M.E. Toimil-Molares, C. Trautmann, W. Marmisollé, O. Azzaroni, Chem. Sci. 12 (2021) 12874.
[12] Z. Zhang, L. Wen, L. Jiang, Nat. Rev. Mater. 6 (2021) 622.
[13] Y. Zhou, L. Jiang, Joule 4 (2020) 2244.
[14] A. Siria, M.-L. Bocquet, L. Bocquet, Nat. Rev. Chem. 1 (2017) 0091.
[15] R.E. Pattle, Nature 174 (1954) 660.
[16] R.S. Norman, Sci. (1979) 186 (1974) 350.
[17] S. Loeb, *Method and Apparatus for Generating Power Utilizing Pressure-Retarded-Osmosis*, US3906250A, 1975.
[18] S. Loeb, *Method and Apparatus for Generating Power Utilizing Reverse Electrodialysis*, US4171409A, 1979.
[19] J.W. Post, J. Veerman, H.V.M.M. Hamelers, G.J.W.W. Euverink, S.J. Metz, K. Nymeijer, et al., J. Memb. Sci. 288 (2007) 218.
[20] J.W. Post, H.V.M. Hamelers, C.J.N. Buisman, Env. Sci. Technol. 42 (2008) 5785.
[21] H. Strathmann, Ion-Exchange Membrane Separation Processes, Vol. 9, Elsevier Science, 2004.
[22] A. Campione, L. Gurreri, M. Ciofalo, G. Micale, A. Tamburini, A. Cipollina, Desalination 434 (2018) 121.
[23] N. Lakshminarayanaiah, Equations of Membrane Biophysics, Academic Press Inc, UK, 1984.
[24] Y. Tanaka, Ion Exchange Membranes - Fundamentals and Applications, Elsevier, 2007.
[25] M. Macha, S. Marion, V.V.R. Nandigana, A. Radenovic, Nat. Rev. Mater. 4 (2019) 588.
[26] J. Su, D. Ji, J. Tang, H. Li, Y. Feng, L. Cao, et al., Chin. J. Chem. 36 (2018) 417.
[27] J. Gao, X. Liu, Y. Jiang, L. Ding, L. Jiang, W. Guo, Small 15 (2019) 1804279.
[28] L. Wang, Z. Wang, S.K. Patel, S. Lin, M. Elimelech, ACS Nano 15 (2021) 4093.
[29] A.J. Bard, L.R. Faulkner, Electrochemical Methods. Fundamentals and Applications, Wiley, USA, 2001.
[30] P.C. Hiemenz, R. Rajagopalan, Principles of Colloid and Surface Chemistry, Marcel Dekker, 1997.
[31] J.N. Israelachvili, Intermolecular and Surface Forces, Elsevier, 2011.
[32] G. Laucirica, A.G. Albesa, M.E. Toimil-Molares, C. Trautmann, W.A. Marmisollé, O. Azzaroni, Nano Energy 71 (2020) 104612.
[33] L. Cao, F. Xiao, Y. Feng, W. Zhu, W. Geng, J. Yang, et al., Adv. Funct. Mater. 27 (2017) 1604302.
[34] Y.-S. Su, S.-C. Hsu, P.-H. Peng, J.-Y. Yang, M. Gao, L.-H. Yeh, Nano Energy 84 (2021) 105930.
[35] G. Laucirica, J.A. Allegretto, M.F. Wagner, M.E. Toimil-Molares, C. Trautmann, M. Rafti, et al., Adv. Mater. 34 (2022) 2207339.
[36] J. Feng, M. Graf, K. Liu, D. Ovchinnikov, D. Dumcenco, M. Heirianian, et al., Nature 536 (2016) 197.

[37] J.-P. Hsu, T.-C. Su, P.-H. Peng, S.-C. Hsu, M.-J. Zheng, L.-H. Yeh, ACS Nano 13 (2019) 13374.

[38] G. Laucirica, Y. Toum Terrones, V. Cayón, M.L. Cortez, M.E. Toimil-Molares, C. Trautmann, et al., TrAC. Trends Anal. Chem. 144 (2021) 116425.

[39] R.B. Schoch, H.van Lintel, P. Renaud, Phys. Fluids 17 (2005) 100604.

[40] R.B. Schoch, P. Renaud, Appl. Phys. Lett. 86 (2005) 253111.

[41] R.B. Schoch, J. Han, P. Renaud, Rev. Mod. Phys. 80 (2008) 839.

[42] K. Xiao, L. Jiang, M. Antonietti, Joule 3 (2019) 2364.

[43] H. Masuda, K. Fukuda, Sci. (1979) 268 (1995) 1466.

[44] R.C. Furneaux, W.R. Rigby, A.P. Davidson, Nature 337 (1989) 147.

[45] J.H. Yuan, F.Y. He, D.C. Sun, X.H. Xia, Chem. Mater. 16 (2004) 1841.

[46] L.A. Baker, P. Jin, C.R. Martin, Crit. Rev. Solid. State Mater. Sci. 30 (2005) 183.

[47] W. Xin, Z. Zhang, X. Huang, Y. Hu, T. Zhou, C. Zhu, et al., Nat. Commun. 10 (2019) 3876.

[48] G. Laucirica, W.A. Marmisollé, M.E. Toimil-Molares, C. Trautmann, O. Azzaroni, ACS Appl. Mater. Interfaces 11 (2019) 30001.

[49] Z. Zhang, S. Yang, P. Zhang, J. Zhang, G. Chen, X. Feng, Nat. Commun. 10 (2019) 2920.

[50] C. Chen, D. Liu, X. Qing, G. Yang, X. Wang, W. Lei, ACS Appl. Mater. Interfaces 12 (2020) 52771.

[51] M.E. Toimil-Molares, Beilstein J. Nanotechnol. 3 (2012) 860.

[52] R. Spohr, Ion Tracks and Microtechnology, Vieweg + Teubner Verlag, Wiesbaden, 1990.

[53] P.Y. Apel, D. Fink, in: D. Fink (Ed.), Transport Processes in Ion-Irradiated Polymers, Springer-Verlag, Berlin Heidelberg, 2004, pp. 147–202.

[54] G. Pérez-Mitta, M.E. Toimil-Molares, C. Trautmann, W.A. Marmisollé, O. Azzaroni, Adv. Mater. 31 (2019) 1901483.

[55] G. Pérez-Mitta, W.A. Marmisollé, C. Trautmann, M.E. Toimil-Molares, O. Azzaroni, J. Am. Chem. Soc. 137 (2015) 15382.

[56] G. Laucirica, Y. Toum Terrones, V.M. Cayón, M.L. Cortez, M.E. Toimil-Molares, C. Trautmann, et al., Nanoscale 12 (2020) 18390.

[57] M. Ali, P. Ramirez, H.Q. Nguyen, S. Nasir, J. Cervera, S. Mafe, et al., ACS Nano 6 (2012) 3631.

[58] K. Yazda, K. Bleau, Y. Zhang, X. Capaldi, T. St-Denis, P. Grutter, et al., Nano Lett. 21 (2021) 4152.

[59] J. Ji, Q. Kang, Y. Zhou, Y. Feng, X. Chen, J. Yuan, et al., Adv. Funct. Mater. 27 (2017) 1603623.

[60] G.-R. Xu, J.-M. Xu, H.-C. Su, X.-Y. Liu, Lu-Li, H.-L. Zhao, et al., Desalination 451 (2019) 18.

[61] Z. Jakšić, O. Jakšić, Biomimetics 5 (2020) 24.

[62] J. Li, X. Li, B. van der Bruggen, Env. Sci. Nano 7 (2020) 1289.

[63] Z. Zhang, P. Zhang, S. Yang, T. Zhang, M. Löffler, H. Shi, et al., Proc. Natl Acad. Sci. 117 (2020) 13959.

[64] X. Zhu, J. Zhong, J. Hao, Y. Wang, J. Zhou, J. Liao, et al., Adv. Energy Mater. 10 (2020) 2001552.

[65] W. Chen, Q. Wang, J. Chen, Q. Zhang, X. Zhao, Y. Qian, et al., Nano Lett. 20 (2020) 5705.

[66] S. Hou, Q. Zhang, Z. Zhang, X. Kong, B. Lu, L. Wen, et al., Nano Energy 79 (2021) 105509.

[67] Z. Zhang, L. He, C. Zhu, Y. Qian, L. Wen, L. Jiang, Nat. Commun. 11 (2020) 875.

[68] Y. Xu, Y. Song, F. Xu, Nano Energy 79 (2021) 105468.

[69] H. Gao, W. Chen, C. Xu, S. Liu, X. Tong, Y. Chen, Env. Sci. Technol. 54 (2020) 2931.

[70] G. Yang, D. Liu, C. Chen, Y. Qian, Y. Su, S. Qin, et al., ACS Nano (2021). acsnano.0c09845.

[71] S. Hong, F. Ming, Y. Shi, R. Li, I.S. Kim, C.Y. Tang, et al., ACS Nano 13 (2019) 8917.

[72] L. Ding, D. Xiao, Z. Lu, J. Deng, Y. Wei, J. Caro, et al., Angew. Chem. 132 (2020) 8798.

[73] C. Zhu, P. Liu, B. Niu, Y. Liu, W. Xin, W. Chen, et al., J. Am. Chem. Soc. 143 (2021) 1932.

[74] C. Chen, D. Liu, L. He, S. Qin, J. Wang, J.M. Razal, et al., Joule 4 (2020) 247.

[75] Y. Wu, W. Xin, X.-Y. Kong, J. Chen, Y. Qian, Y. Sun, et al., Mater. Horiz. 7 (2020) 2702.

[76] S. Hong, J.K. El-Demellawi, Y. Lei, Z. Liu, F. al Marzooqi, H.A. Arafat, et al., ACS Nano 16 (2022) 792.
[77] W. Chen, Q. Zhang, Y. Qian, W. Xin, D. Hao, X. Zhao, et al., ACS Cent. Sci. 6 (2020) 2097.
[78] X. Zhao, C. Lu, L. Yang, W. Chen, W. Xin, X.-Y. Kong, et al., Nano Energy 81 (2021) 105657.
[79] A. Siria, P. Poncharal, A. Biance, R. Fulcrand, X. Blasé, S.T. Purcell, et al., Nature 494 (2013) 455.
[80] T. Ma, E. Balanzat, J.-M. Janot, S. Balme, ACS Appl. Mater. Interfaces 11 (2019) 12578.
[81] H. Li, F. Xiao, G. Hong, J. Su, N. Li, L. Cao, et al., Chin. J. Chem. 37 (2019) 469.
[82] Y.-C. Liu, L.-H. Yeh, M.-J. Zheng, K.C.-W. Wu, Sci. Adv. 7 (2021) eabe9924.
[83] D.A. Vermaas, D. Kunteng, J. Veerman, M. Saakes, K. Nijmeijer, Env. Sci. Technol. 48 (2014) 3065.
[84] Z. Wang, L. Wang, M. Elimelech, Engineering 9 (2022) 51.
[85] H. Tian, Y. Wang, Y. Pei, J.C. Crittenden, Appl. Energy 262 (2020) 114482.

Biointerfacial nanoarchitectonics: layer-by-layer assembly as a versatile technique for the fabrication of highly functional nanocoatings of biological interest

4

Miguel Ángel Pasquale and Omar Azzaroni

Instituto de Investigaciones Fisicoquímicas Teóricas y Aplicadas (INIFTA)—Departamento de Química—Facultad de Ciencias Exactas—Universidad Nacional de La Plata—CONICET, La Plata, Argentina

4.1 Introduction

The synthesis of new polyfunctional materials with high versatility is of upmost importance for the scientific community, involving increasing interdisciplinary work. In many applications, as is the case of tissue engineering, the bulk properties of a material as well as its surface properties play a key role in the overall material performance. For instance, prostheses require materials with appropriate strength, but their surface properties play a crucial role as they influence the biological response of the organism through different events related to protein adsorption, cell adhesion and migration, and immune system reactions [1–3]. Thus a big effort is devoted to the development of new functional materials with increasing biocompatibility [4]. The current demand, not only for biological applications but for many others, namely, optics, electronics, sensor designing, and energy, involves the coexistence of properties and functions for achieving high performance.

Interfacial supramolecular assemblies generated in a layer-by-layer (LbL) fashion have sparked great interest owing to their potential applications in a wide range of research fields [5–13]. An important cornerstone for the construction of multicomponent interfacial architectures is the development of methods for integrating molecular building blocks into well-defined organized assemblies [14–19]. Research efforts on this matter are often referred to as "nanoarchitectonics," a term popularized by Ariga and coworkers [20–22]. It is now widely accepted that the LbL of polyelectrolytes represents a valuable technique for the fabrication of thin functional films of a composition controlled at the nanometer scale [23–26].

It is for this reason that the LbL technique plays a key role for fabricating supramolecular interphases with potential applications in the field of biology and medicine due to its versatility and the possibility of integrating different functions, thus increasing the resemblance to real biosystems [27–31]. Different chemical species, with a particular distribution and dynamically interacting at the nanoscale level, are combined and, according to the fabrication conditions, supramolecular systems with advanced properties can be obtained. This nanometric control of the physicochemical properties may be an outstanding possibility for handling biological responses of biosystems. In particular, biosystems can be tackled as dynamic systems with multifunctional properties originated from a concerted occurrence of compatible functions and structures at the nanoscale. These systems can inspire the design of new supramolecular materials for performing certain specific processes under the application of external stimuli or due to the interactions of the different parts of the system [32]. In this vein, living cell culture has played a crucial role in many research areas due to its applications, namely, for testing new drugs, in gene therapy, the production of vaccines and pharmaceutical drugs, and for basic studies of cell biology. The latter include research work related to cell proliferation, adhesion, migration, and differentiation mechanisms, as well as cell−cell interaction and cell−matrix interactions [33]. Cells interact with the extracellular matrix (ECM) through a large number of receptors, such as integrins, syndecanes, and receptors for hormones and growth factors [34,35]. These interactions are continuously modified as cells, individually or collectively, perturb the microenvironment, and trigger a signal cascade guiding cell fate and defining its phenotype. ECM properties play a key role in tissue regeneration as they define the tissue characteristics, rendering either a homeostatic or a pathological state [36,37]. Furthermore, the ECM contains bioactive chemical compounds that regulate cell adhesion, proliferation, migration, differentiation, and survival events [38]. In all these events, cells and their environment execute a number of interrelated actions at different time- and space-scales that coexist along the event. For instance, the stiffness change in the ECM modifies cell properties, such as motility ability, and a more malignant phenotype sets in, as is the case of the propagation of some tumors [39].

Along this chapter, examples of LbL fabricated materials following a nanoarchitectonic fashion, and capable of performing controlled functions to conveniently affect the behavior of biological systems interacting with them, will be given. The time- and space scale of interactions is managed by appropriate synthesis design of these artificial materials, including postfabrication treatments or the application of external stimuli in a three-element arrangement: artificial material, external stimulus, and the system under study. The performance of this type of material seems to follow, in some aspects, similar rules to biological systems, in which concerted or harmonized functions set in. In this vein, the design of materials for a certain application could take advantage of natural systems by either approaching a similar synthetic pathway or triggering a set of functions that allow a proper interaction with the environment.

First, we introduce basic aspects of the LbL assembly technique and depict some relevant construction blocks, as well as the main steps of procedures most commonly employed for performing the LbL assembly. Moreover, the combination of the LbL assembly technique with other fabrication strategies is briefly reviewed. Some general characteristics of the LbL-obtained films are assessed and some strategies for tuning them are indicated. Then, relevant properties affecting cell behavior are described and exemplified. Finally, selected applications for biological and medical sciences, particularly for controlling cell phenotype, and of interest in tissue engineering and regeneration, including comments on the use of LbL-fabricated materials for sensing purposes and for the delivery of proteins, cells, and drugs, are presented. In Fig. 4.1, the basic principles of the LbL strategy, possible building blocks, supporting substrates, assembly strategies, and techniques to increase versatility are depicted.

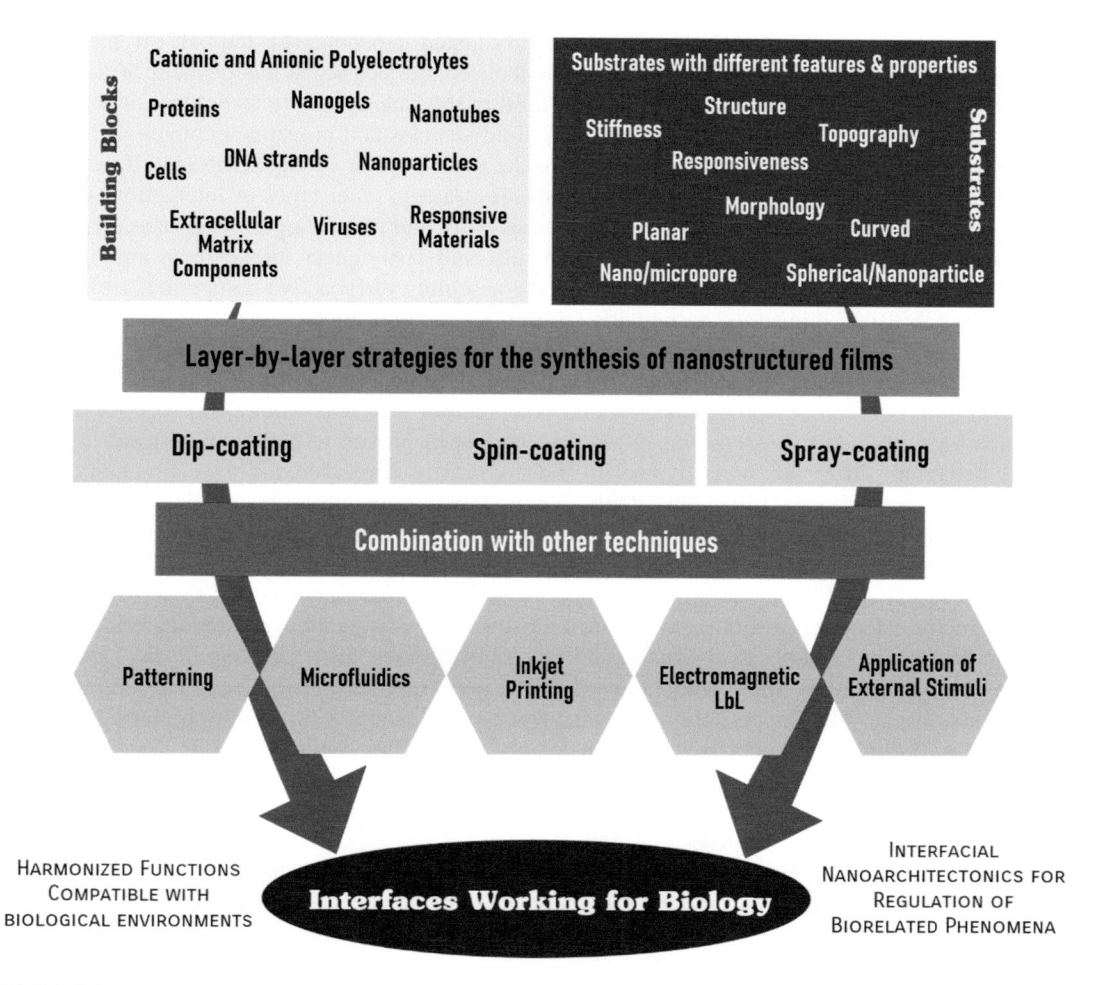

FIGURE 4.1

Scheme showing the characteristics of the layer-by-layer strategy for the fabrication of coatings.

4.2 The layer-by-layer technique

4.2.1 Building blocks and assembly procedures

The LbL assembly technique was introduced in the 1990s and is based on the alternate interaction of positive and negative polyelectrolytes (PEs), the electrostatic interaction between PE being the main driving force [40]. In this technique, polycations and polyanions are sequentially deposited on charged surfaces with either planar or curved geometry, that is, planar substrates or colloidal particles, to generate polyelectrolyte multilayers (PEMs) [41]. Strong and weak PE can be used for the assembly of PEMs. Weak PEs are macromolecules bearing acidic or basic groups with small dissociation constants, which are protonated or deprotonated depending on the pH. Strong PEs are protonated or deprotonated at almost any pH, thus their charge density is independent of the pH. In addition to PEs, other building blocks have been used in the LbL assembly technique: nanoparticles [42], conductive graphene-based compounds [43], lipid vesicles and bioactive compounds [44], and even cells [45,46], which can be grafted on top of the multilayer or by replacing inner layers. With this broad set of building blocks for the LbL synthesis strategy, other types of interactions different than electrostatic appear to play a key role in the assembly and reordering process, namely, coordination chemistry, hydrogen bonding, covalent bonds, and host—guest interactions, among others [40,47]. This variety of interactions is useful for the assembly of bioactive compounds and the stability of the fabricated thin films, particularly for biological applications.

Some synthetic PEs have been widely used for the assembly of multilayers with different applications in biological sciences. Among the most frequently used PE, poly(acrylic acid) (PAA) combined with poly(allylamine) (PAH) [48,49] and poly(sodium 4-styrenesulphonate) (PSS) with PAH [50] have been utilized for assembling films that have been proven to have a significant influence on cell behavior. There are other synthetic PEs commonly employed for different applications, namely, polyethylenimine (PEI), poly(diallyldimethylammonium chloride) (PDDAC), and polymethacrylic acid (PMAA), which will be considered in the following sections.

Building blocks of natural origin are appealing due to their biocompatibility, often easy biodegradability, rather significant solubility in water, the ability to mimic biological systems conferring similar structural features and bioactivity, such as specific cell signaling and recognition, and furthermore they are susceptible to chemical modification. However, these building blocks have disadvantages associated with their limited availability, low monodispersity, and limited range of working conditions. Multilayers fabricated with this type of building block may be unstable in the environment of application, which may differ from the assembly conditions, as is the case of biological applications. Chitosan (CHI), poly-L-lysine (PLL), hyaluronic acid (HA), alginic acid (ALG), chondroitin sulfate (CS), dextran sulfate (DEX), polyglutamic acid (PGA), and heparin (HEP) are examples of natural PEs.

The LbL technology has been supported by several deposition methods, that is, dip coating, spray coating, and spin coating [51,52]. The innovations in the assembly technology and the availability of materials that can be used in LbL technology have been recently reviewed [53]. Some of the main deposition methods are shown in Fig. 4.2.

Briefly, dip-coating methods are the most popular as they may be carried out by employing very simple equipment for dipping the substrate alternately into dilute solutions of the construction blocks separated by rinsing steps. Spray coating LbL assembly consists of aerosolizing polymer solutions and sequentially spraying them onto substrates. It is much faster than immersive assembly and fulfills the industrial scale requirements. In the spin-coating LbL assembly, the deposition of the material is assisted by spinning the substrate. The time required for the multilayer assembly is significantly reduced in comparison with dipping methods, and often more homogeneous films are obtained. Electromagnetic LbL assembly employs an electric or magnetic field to control the layering process. Electrodeposition can be used for coating electrode substrates with redox-active materials, and building blocks with magnetic properties can be assembled in the presence of a magnetic field. Details of the different LbL assembly categories can be found in the literature [51,53] and some examples of the application of microfluidic-assisted LbL will be given in the next section.

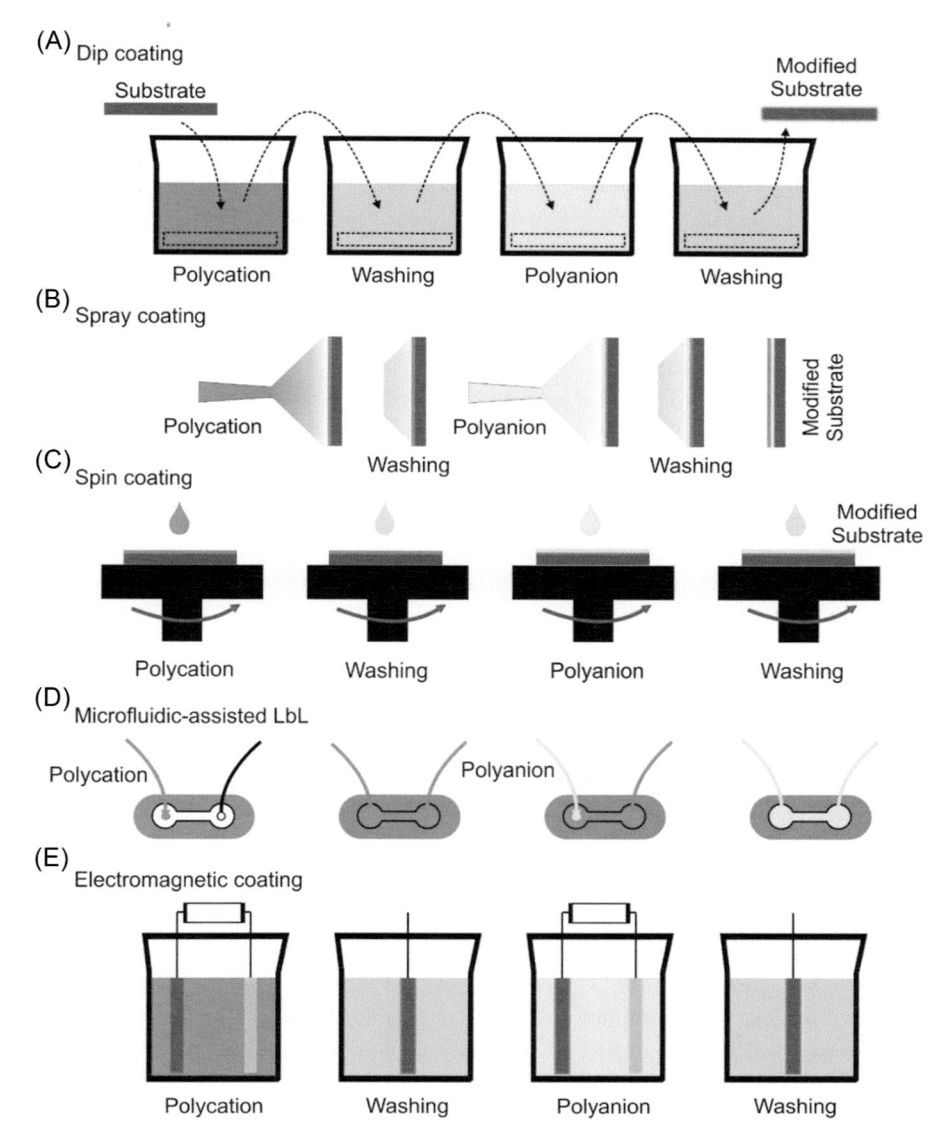

FIGURE 4.2

Different methods employed in the layer-by-layer assembly technique. The main steps of each method are depicted.

4.2.2 Combination of the layer-by-layer assembly technique with other fabrication procedures

The LbL technique can be easily combined with other fabrication procedures increasing its versatility. For instance, microfluidic assembly has been combined with the LbL technique to obtain synthetic platforms with spatially controlled characteristics [27]. Strings made of PEI/PSS and PAH/PSS multi-layers were patterned on poly(dimethylsiloxane) (PDMS) surfaces by flowing the polyelectrolyte through a microchannel network in contact with the PDMS. The obtained platform allowed the adhesion of neural cells only on the LbL-obtained strings [54]. Recently, the combination of LbL and microfluidic assembly to fabricate multilayer capsule-based supramolecular scaffolds to protect and control the release and localization of biomolecules with potential use for drug delivery or tissue engineering has been reviewed [55].

The LbL has been combined with inkjet printing to obtain a spatial control of the assembly of the synthetic blocks of each layer. The first reported investigations described the printing of gold nanoparticles (NPs) and PDDAC [56]. A similar approach has been recently reported to construct nanofilms for controlling the release of bioactive molecules such as ovalbumin and fibroblast growth factor for biomedical applications [57]. In another approach bacteria cells were hosted in an LbL-constructed silk nest fabricated by printing silk chemically modified with either PLL or PGA [58]. Furthermore, an on-chip liver tissue model for drug screening was printed employing GepG2 (from human hepatocarcinoma) and HUVECs (human endothelial cells), and it was proposed for drug screening. The hierarchical assembly of these two cells was accomplished by repeating the process of printing a cell line followed by an LbL assembly of a bilayer of gelatin (G)/fibronectin (FN) on top of the cells, and the printing of the other cell line in coculture. The protein-based nano-layer assists in cell interactions increasing viability and better mimics the liver tissue [59]. More recently, a three-dimensional cardiac tissue using human-induced pluripotent stem cell-derived cardiomyocytes coated with a FN/G nanofilm was fabricated by a bioprinting technique consisting of a microscopic painting device, improving printing resolution in comparison with conventional methods [60]. The fabricated heart model was made of multiple layers, reaching a thickness of 60 μm and a diameter of 1.1 mm, and showed synchronous beating.

Inkjet printing combined with LbL coating has been utilized to fabricate perfusable vascular networks in a photopolymerizable matrix encompassing biocompatible and biodegradable glycidyl methacrylated xanthan gum (XG-GMA). In a first step, ALG was utilized as bioink to produce the sacrificial customized network structure, followed by LbL coating with chitosan (CHI)/argini-ne—glycine—aspartic acid (RGD)-grafted ALG bilayers. Then, the ALG-based structure was embedded in XG-GMA and treated with ethylenediaminetetraacetic acid (EDTA) solution to liquefy the ALG structure, rendering a network that supports the culture of human umbilical vein endothelial cells (HUVECs) [61]. The main steps of the procedure to generate the biocompatible embedded patterned structure are schematized in Fig. 4.3.

Material patterning methodologies are usually combined with the LbL synthesis technique to obtain substrates with spatially controlled topographical properties at different size scales. This is of uppermost importance in the design of devices for biomedical applications such as tissue engineering or cell control behavior. Different tissues of an organism are characterized by certain topographical features. For instance, cardiomyocytes generate myofibrils upon their organization into aligned sheets, rendering the myocardium tissue. The topography of substrates and scaffolds interacting with cells

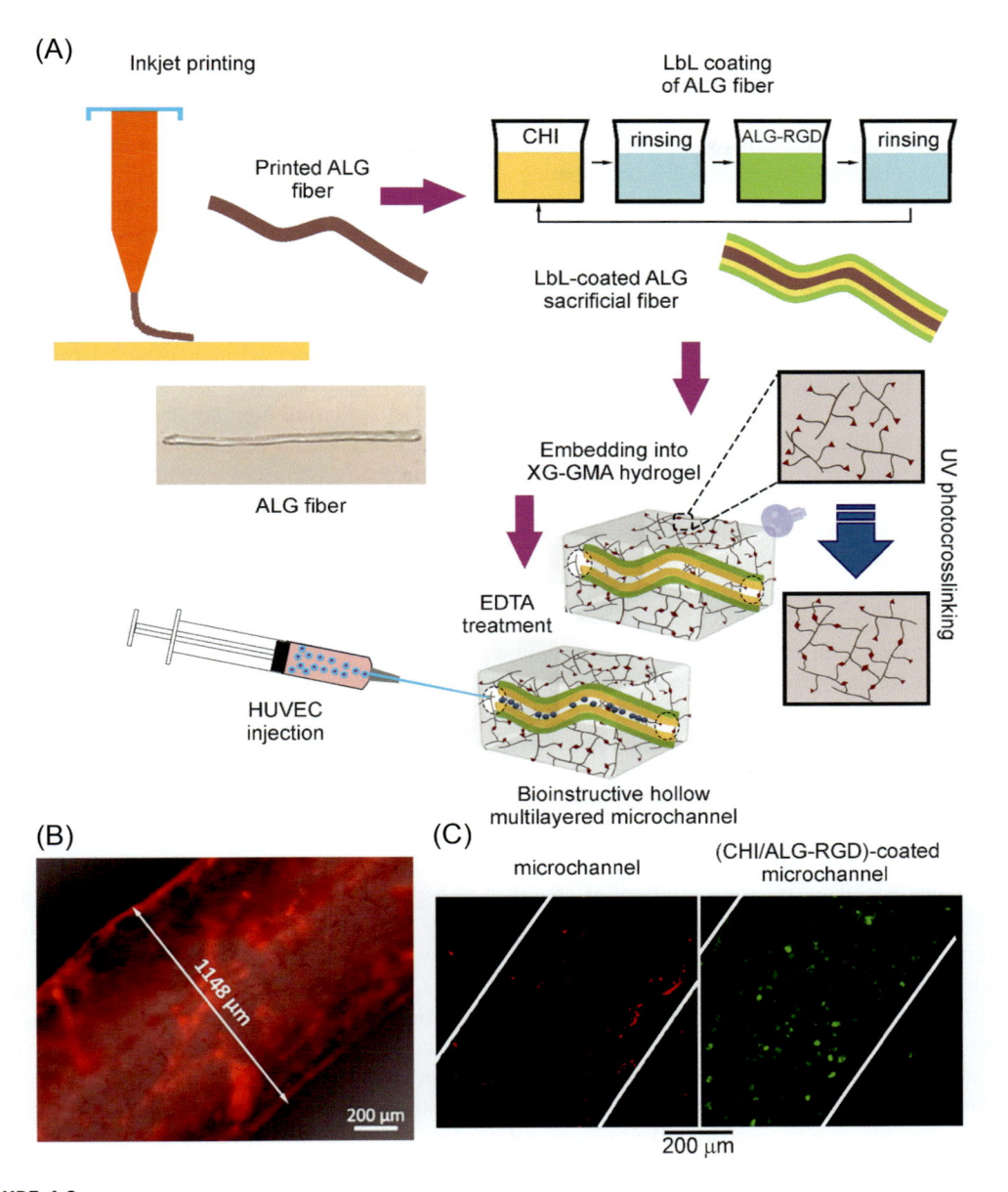

FIGURE 4.3

(A) Scheme of the procedure for fabricating of 3D perfusable constructs with $(CHI/ALG-RGD)_6$ multilayers templated on liquefied ALG microchannels embedded in XG-GMA hydrogel. (B) Fluorescence microimage of a 3D-printed ALG sacrificial microfiber coated with $(RITC-CHI/ALG)_6$ thin film. (C) Live/dead stained HUVECs 3 days postseeding from fetal bovine serum-free medium in layer-by-layer -treated and untreated microchannels, as indicated in the figure. The white lines indicate the channel walls.

Adapted from C.F.V. Sousa, C.A. Saraiva, T.R. Correia, T. Pesqueira, S.G. Patrício, M.I. Rial-Hermida, et al., Bioinstructive layer-by-layer-coated customizable 3d printed perfusable microchannels embedded in photocrosslinkable hydrogels for vascular tissue engineering, Biomolecules 11 (2021). https://doi.org/10.3390/biom11060863.

can stimulate specific pathways to control their cytoskeletal organization [29,62]. An increased functionality is obtained from LbL-obtained coatings when deposited on patterned substrates. The LbL technique is suitable to coat substrates with topographical features in both the nano- and microscale. Thus when combining LbL technique with lithography, different well-defined structures can be generated. The LbL technique, in a properly designed fabrication process, is suitable to graft specific active chemical compounds with spatial control, to further mimic natural biological systems. A recent review describes the combination of LbL assembly with different substrate patterning procedures [55]. Lithographic techniques are top-down approaches that have received increasing interest in the field of nano- and microengineering of materials due to their high resolution and the large variety of shapes that can be generated. Among lithographic techniques, photolithography and soft lithography can be distinguished. Photolithography employs a photoreactive material coated onto a substrate. This material is selectively exposed to ultraviolet radiation by means of precisely mechanized masks, producing its polymerization, degradation or cross-linking reactions. The unwanted remaining material is then dissolved by organic solvents [63]. The obtained substrate can be used as a master for molding polymeric substrates. Soft lithography takes advantage of a flexible elastomer material to create topographic features [64]. PDMS is commonly used as an elastomer in this lithography technique. For instance, microstructured PDMS substrates have been tailored with PLL/HA LbL-assembled multilayers to explore cell orientation and differentiation [65].

To give an example of the combination of LbL and photolithography, an application to neuronal cell studies is depicted. An engineered substrate for primary cortical neurons has been reported [66]. In this work, microscope cover glasses were coated with a precursor layer of PDDAC attached onto a plain microscope slide by means of a photoresist. Then, the set was photopatterned employing a photomask and ultraviolet radiation to render a chip configuration of square regions. LbL assembly was performed in the developed patterned substrate to form a (PSS/PDDA)-based precursor film made of three bilayers, that is, $(PSS/PDDAC)_3$. In a following step, half of the chips were modified by four bilayers of secreted phospholipase A_2 ($sPLA_2$) labeled with fluorescein isothiocyanate (FITC) assembled with PEI, and a $sPLA_2$-FITC top layer. On the remaining chip regions of the substrate, gelatin or bovine serum albumin or PLL labeled with tetramethylrhodamine-5-(and-6)-isothiocyanate (TRITC) was assembled according to a similar fabrication design. Finally, a lift-off process was performed to remove the photoresist, and the cover glass was detached from the slides. Neurons were seeded on the fabricated substrates, and a relatively enhanced adhesion was observed for $sPLA_2$ modified square microregions. In conclusion, this work described a simple method to generate a comparison chip modified substrate applicable to neuron cell fundamental studies.

The above paragraphs show the enhanced versatility of the LbL technique when combined with other techniques, allowing temporal and spatial control of the physicochemical and biological properties of the fabricated materials in complex devices. These materials bear a set of functionalities that are orchestrated and may be triggered upon interaction with the surroundings.

4.2.3 Structural characteristics of layer-by-layer multilayers

The LbL assembly technique to fabricate coatings from the adsorption of opposite charge PEs involves the formation of complexation sites and the presence of a charge excess. Thus upon the alternate assembly of the polycation and the polyanion, the sign of the surface charge alternates between negative and positive, respectively. The overlapping of layers was demonstrated in relatively early works [47,67]. LbL coatings obtained with PEs and even other building blocks, exhibit an internal structure far from a simple and ordered layered one. The transport of the whole building block or the displacement of part of it is likely to occur. Interdiffusion of the film components and exchange of PEs within the film can be very important in determining the physicochemical properties of the coating.

For certain polyelectrolyte pairs, both the polycation and the polyanion can diffuse, as is the case of PLL/PGA film. Diffusion processes may also occur when a film obtained by the assembly of a certain polycation/polyanion pair is in contact with a third type of PE. For instance, the diffusion of PSS chains within a PLL/PGA film and exchange with PGA chains have been reported [68]. Another research work evidenced diffusion of PGA into a (PGA/PAH)-based multilayer containing multivalent ferrocyanide anions when put in contact with a solution containing PGA. The multivalent anions diffuse out of the film as PGA molecules diffuse into it [69]. In this work, the release of ferrocyanide anions from PGA/PAH multilayers with a PAH ending layer when put in contact with a solution of PHA has also been analyzed. In this case, PAH, which strongly interacts with the anions, does not diffuse into the film but acts as a sink for ferricyanide anions. In the same vein, it has been reported that a quantitative exchange of PLL with PAH occurred when a PLL/HA multilayer was put in contact with a solution of PAH, whereas PAH/HA multilayers remained stable in a PLL solution [70].

Neutron and X-ray reflectivity have been utilized to study the internal structure of PEM [71]. For instance, data from PAH/PSS and PDDAC/PSS multilayers with an architecture containing blocks of hydrogenated (h) and deuterated (d) bilayers, maintaining the total number of layers constant and systematically varying the position of the separation between blocks, have been reported [72]. Three regions can be distinguished in the multilayer films: a precursor region located near the substrate, a core region, and an outer region. The bilayers of the precursor region were thinner than the average value, while the outer bilayers were thicker than the average. These values depend on the polyelectrolyte nature and the assembly conditions. Results from this work, also indicated that the roughness increases in going from the film-air interface into the film, a fact that has been interpreted by the interdiffusion of the supporting layer upon each deposition step, generating an increase of the internal roughness. Furthermore, for PDDAC-containing multilayers, with a greater ability to diffuse than PAH, blended layers near the substrate were inferred.

The phenomenon of polyelectrolyte interdiffusion, which is more pronounced for LbL-obtained multilayers following supralinear growth dynamics, has been demonstrated by charged virus patterning [73]. In this work, negatively charged M13 viruses were assembled in a thin (linear polyethylenimine (LPEI)/PAA)-based multilayer (less than 10 nm) with an LPEI top layer, and upon assembly of additional LPEI/PAA multilayers, viruses were exchanged for PAA due to polymer interdiffusion (Fig. 4.4A). The latter provides the required motility for virus particles to assemble into packed arrangement onto the top surface. In another approach, viruses ordered spontaneously upon adsorption onto an (LPEI/PAA)-based multilayer with more than 10 bilayers (about 100 nm) (Fig. 4.4B−D). In this case, the motility of the polycation in the inner layers assists in virus ordering mainly by an increase in the lateral displacement of polymeric chains [74]. In this work, a strategy to obtain packed virus arrangements on top of patterned (LPEI/PAA)-based multilayers has been proposed (Fig. 4.4F). Atomic force microscopy (AFM) microimages showing negatively charged M13 viruses disorderedly adhered to a thin (LPEI/PAA)-based multilayer (Fig. 4.4G),

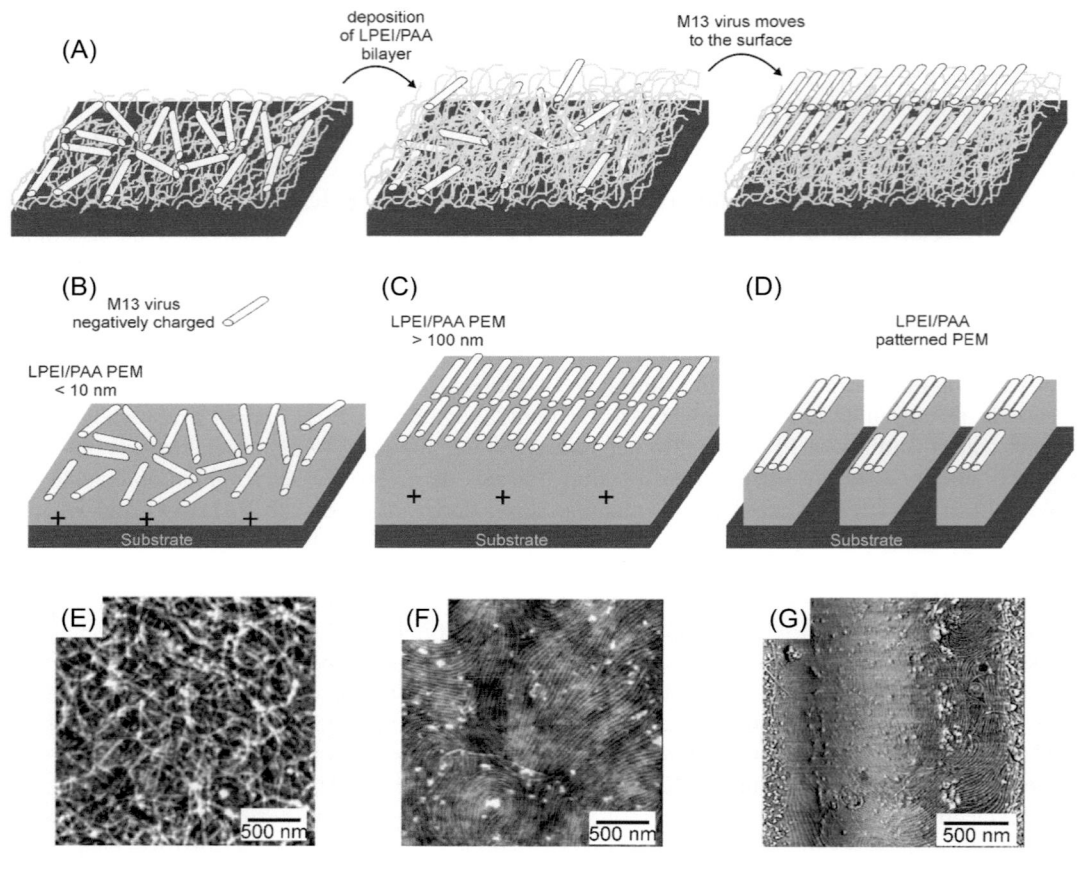

FIGURE 4.4

Schemes of (LPEI/PAA)-based PEM dynamic structure and AFM images of M13 deposited virus. (A) Scheme of M13 virus deposited on LPEI/PAA thin film and subsequent diffusion and ordering after the assembly of a new bilayer. (B) M13 virus deposited on LPEI/PAA nanothin film; (C) Spontaneously ordered M13 virus on thick (LPEI/PAA)-based film; (D) Ordered M13 virus on solvent-assisted capillary molded (LPEI/PAA) ridges. AFM microimages of M13 virus deposited on (E) LPEI/PAA PEM (<10 nm thick), (F) LPEI/PAA PEM (>100 nm thick), and (G) ordered M13 virus on PEMs with ridges. Scale bars are indicated in the figures. *LPEI/PAA*, linear polyethylenimine.

Adapted with permission from P.J. Yoo, K.T. Nam, A.M. Belcher, P.T. Hammond, Solvent-assisted patterning of polyelectrolyte multilayers and selective deposition of virus assemblies, Nano Lett. 8 (2008) 1081–1089. Copyright 2008, American Chemical Society.

spontaneously ordered M13 viruses onto a relatively thick (LPEI/PAA)-based multilayer of about 100 nm (Fig. 4.4H), and M13 viruses patterned on rigged (LPEI/PAA)-based multilayer are depicted.

4.2.4 Effect of assembly conditions and postassembly treatment on the multilayer properties

4.2.4.1 Variables for the assembly process

Polyelectrolyte assembly conditions can hugely affect film properties. Key variables include the type or nature of PEs, the solution pH, and ionic strength, which allow the tuning of film properties, that is, wettability, charge, surface roughness, and rigidity.

Type of polyelectrolyte. The polyelectrolyte multilayer structure can be modified by changing the type of polyelectrolyte in the assembly process. Natural PEs are known to render soft films with poor adhesion properties toward adherent cells that are not able to expand on the surface, as is further described in the following sections. The increase in the film rigidity when PAH was assembled with λ-carrageenan polysaccharide instead of *l*-carrageenan has been reported. This claim highlights the key role played by the polyelectrolyte structure: *l*-carrageenan forms a helical structure, while λ-carrageenan has a random coil structure [75,76].

Acid/base concentrations. The dissociation of weak PEs is controlled by the pH of the medium, thus it affects the magnitude of interaction between opposite charge PEs. This fact increases the film versatility due to the pH-dependent properties. As indicated previously, electrostatic interactions between cationic and anionic PEs are the main driving force for the assembly process. The complexation between anionic and cationic PEs produces an increase in entropy due to counterion release. The rigidity of the film depends on the density number of complexation sites. Thus the polyelectrolyte charge density is an important factor that greatly determines the system rigidity. For weak PEs, the degree of cross-linking sites depends on the pH of the assembly solutions. When cross-linking is low, the generated multilayer will swell and hydrate, resulting in a low stiffness material. The great influence of pH on the properties of (PAH/PAA)-based films has been reported. At acidic pH, these films are relatively thicker and softer than those obtained at neutral pH [77]. The synthesis, characteristics, and applications of multilayers containing weak PEs, that is, systems made of weak polycations and polyanions, as well as those obtained by assembling weak PEs and other components, namely, strong PEs, neutral polymers, and nanomaterials, have been recently reviewed [78].

Ionic strength. The ionic strength affects the PEMs structure by enhancing extrinsic polyelectrolyte charge compensation by counterions. In certain cases, polyelectrolyte complexes may be destroyed. The abovementioned effect can be more pronounced for counterions with a low charge density, such as Cs^+ or Br^-. Thus these ions have a small hydration atmosphere, and can easily interact with the opposite charge of the polyelectrolyte chains. Consequently, the density of complexation sites decreases, leading to thicker films with mobile polyelectrolyte chains [79]. In another work, the effect of ionic strength and type of ions on the structure of PSS/PDDAC multilayers has been reported [80]. In this work, PEMs were prepared employing solutions containing different concentrations of NaF, NaCl, and NaBr, and the resulting PEM structure was analyzed by neutron reflectometry. For the range of salt concentrations employed, the increase in the ionic strength produces thicker multilayers with increased roughness. These facts are explained by the extrinsic charge compensation by counterions and the screening of electrostatic attraction between oppositely charged polyelectrolyte chains, changing from a planar conformation to a coil one. Results from this work also indicate that no exchange of hydration water or replacement of H by D is observed when the multilayers prepared in water are measured in dry or previously incubated in D_2O. Data from neutron reflectometry indicated that the total water content of the multilayers can be divided into "void water" and "swelling water." The former fills the empty spaces between polyelectrolyte chains and decreases with the ionic strength and size of counterions. In contrast, the latter contributes to swelling of the multilayer, and increases with the ionic strength and size of counterions. In a more recent paper, the effect of the ionic strength on PEM formation at different growth regimes has been studied and the results interpreted on the basis of the phase diagram of aqueous solutions of mixtures of oppositely charged PEs [81]. In this work, (PDDAC/PSS)-based multilayers are the model systems, and the ionic strength was adjusted by NaCl concentration from 0 to 1 M. The growth kinetics was followed by optical reflectometry in combination with controlled transport conditions by an impinging-jet stagnation-point flow cell. At the lowest NaCl concentrations (0, 0.001, and 0.01 M), the growth is linear with the number of polyelectrolyte layers, and the slope of the deposited mass versus the number of layers increases with salt concentrations. In these conditions the multilayers can be considered as a glassy solid undergoing the adsorption of a new polyelectrolyte layer at each assembly cycle, with an enhanced adsorption as the ionic strength increases. For the highest NaCl concentrations (0.5 and 1 M) the multilayer grows exponentially, at least for the initial assembly steps. The increase of mass in these conditions is consistent with the assumption of PEs as fluids that undergo diffusion, rather than glassy solids. For this growth regime, the increase in the ionic strength reduces the accumulated mass at each assembly step.

Assembly of rigid material. The incorporation of rigid particles into the polyelectrolyte-based films is utilized to increase their stiffness. It has been reported that composites with structural and mechanical properties similar to those of nacres can be obtained by assembling PDDAC polyelectrolyte and montmorillonite clays in free-standing films [82]. In other works, the incorporation of carbon nanotubes [83] and metallic nanoparticles was reported to increase the stiffness of films by more than two orders of magnitude [84]. In this paper, the optical properties of the obtained hybrid materials were highlighted, and their potential uses for biomedical applications indicated. Recently, the key role of inorganic nanoparticles in the LbL assembling process with the aim to obtain hybrid coating for applicative devices has been reviewed [85].

Another strategy for increasing the film rigidity is based on the combination of soft polyelectrolyte blocks with a capping block of hard PEs. For instance, the Young's modulus of (PLL/HA)-based polyelectrolyte block has been increased by capping the underlying natural block with a (PSS/PAH)-based block [86].

4.2.4.2 Postassembly treatments

Most of the variables used to control polyelectrolyte-based multilayer structure in the assembly process can affect their characteristics in a postassembly treatment. We shall pay particular attention to some aspects of the thermal treatment and chemical modifications.

Thermal annealing. PEMs are kinetically stabilized systems, and it is expected that an increase in temperature can produce polyelectrolyte rearrangements. Polyelectrolyte microcapsules obtained by employing a colloid template consist of a free-standing multilayer film undergoing considerable changes in volume upon thermal annealing, in contrast to multilayers deposited on planar substrates. For instance, PDDAC/PSS microcapsules with a different number of layers were incubated at different temperatures in the $20°C-90°C$ range, and their volume and wall thickness were analyzed [87]. Upon heating $(PDDAC/PSS)_4$ microcapsules at a temperature higher than $30°C$, their spherical shape is retained, although they shrink by a magnitude that depends on the temperature (Fig. 4.5A), reaching a constant value at about $75°C$. This shrinkage appears to be irreversible as

(A)

(B)

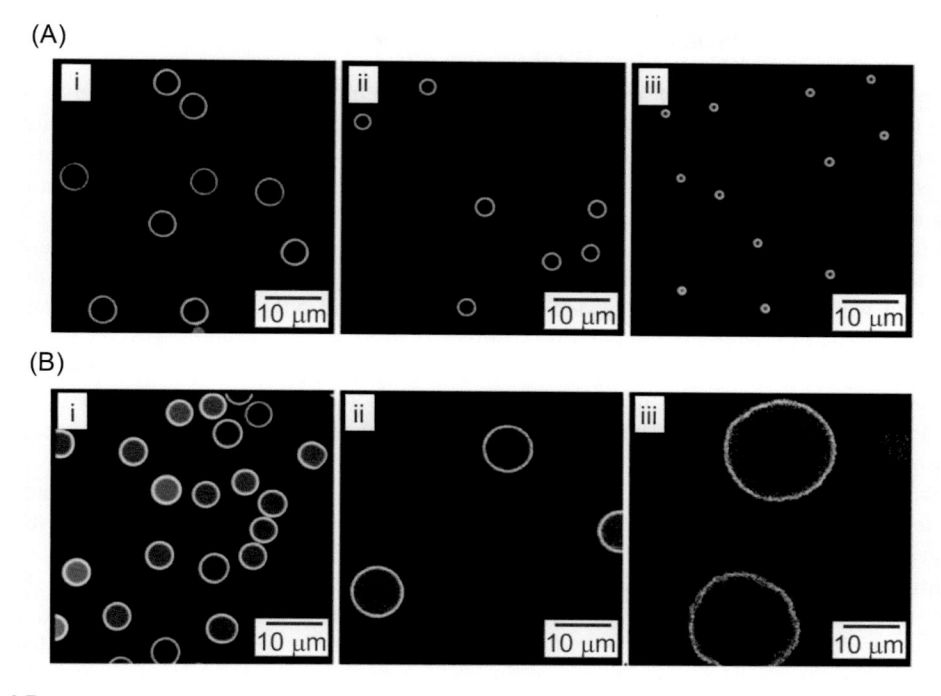

FIGURE 4.5

Effect of temperature on polyelectrolyte microcapsules. (A) (PDADAC/PSS)$_4$ polyelectrolyte microcapsules. Confocal laser images of microcapsules made on 4.55 μm silica core particles (i), after incubation for 20 min at 50°C (ii), and at 70°C (iii). (B) (PDADAC/PSS)$_4$PDADAC polyelectrolyte microcapsules. Confocal laser images of microcapsules made on 4.55 mm silica core particles (i), after incubation for 20 min at 45°C (ii), and at 55°C (iii).

Adapted with permission from K. Köhler, D.G. Shchukin, H. Möhwald, G.B. Sukhorukov, Thermal behavior of polyelectrolyte multilayer microcapsules. 1. The effect of odd and even layer number, J. Phys. Chem. B 109 (2005) 18250–18259. Copyright 2005, American Chemical Society.

no reswelling is observed when they are stored at low temperatures. Scanning electron microscopy (SEM) and AFM results indicate that the stabilization process is accompanied by an increase in the microcapsule wall thickness. In contrast, microcapsules of (PDDAC/PSS)$_4$PDDAC (with the polycation as the uppermost layer) swell during heating. A fivefold increase in their initial volume was observed before microcapsule rupture (Fig. 4.5B). The authors interpreted these observations taking into account surface tension and repulsive interactions, the latter originated by the excess charge. Upon heating, polyelectrolyte chain motility would assist in the multilayer rearrangement according to the prevalence of the abovementioned forces. Thus for microcapsules with an even number of layers (equal number of polycations and polyanions) the tendency to reduce the polymer/water interface due to surface tension forces is the driving force for microcapsule shrinkage and wall thickening. For multilayers with an odd number of layers, the repulsions due to the excess of positive charge overcompensates the surface tension effects, and the increase in the mutual separation distance produces a swelling of the structure. The microcapsule wall becomes so thin that the spherical structure is no longer stable, and its rupture takes place.

More recently, the effect of temperature on the structure of thin (20−21 layers) and thick (36−37 layers) (PDDAC/PSS)-based multilayers with either the polycation or the polyanions as the top layer has been reported [88]. In this work, neutron reflectometry (NR) and ellipsometry were employed to investigate the multilayer structure and thickness upon thermal treatment. For NR measurements, multilayers were prepared with deuterated PSS assembled at the first six bilayers. The number of layers used ensures a block structure, as for a lower number of layers, interdiffusion or interdigitation would render a mixed structure. Measurements were carried out for dry multilayers or in water or in deuterated water for ellipsometry and NR, respectively, and the thermal treatment consists in raising the temperature from 25°C to 65°C and going back to 25°C. Data analysis requires the distinction in the behavior of the thin multilayers compared with the thick ones. Taking all results together, the authors indicated that both loss of material and densification take place during thermal treatment. The former process is ascribed to occur at the outer zone, and densification at the inner part of either the thin or thick multilayers.

Chemical modifications. Chemical reactions involving polyelectrolyte cross-linking are used to modulate multilayer mechanical properties, a subject of uppermost relevance to achieve stability in different conditions. This is particular important for natural PEs such as polysaccharides.

Chemical reactions between polyelectrolyte moieties can be used to create cross-links within the multilayer to change its mechanical properties. For instance, carbodiimide chemistry-based reactions have been employed for cross-linking carboxyl groups with amine groups, yielding covalent amide bonds [89]. This strategy was applied to different polyelectrolyte pairs including (PLL/HA)- and (CHI/HA)-based multilayers, increasing the rigidity of the multilayers, measured by AFM indentation. To characterize the rigidity, the Young's modulus (E) has been determined, and a change from about 5 kPa for pristine multilayers to about 500 kPa for cross-linked materials has been reported.

Polypeptide-based multilayers have been stabilized by cross-links of polypeptides by disulfide bonds in a reversible process [90]. This strategy requires at least one polypeptide with the amino acid cysteine for controlling the cross-linking. In a more recent paper, in order to increase the biocompatibility of materials and perform more nature friendly reactions, cross-linking with genipin (a natural and nontoxic compound) has been carried out in (CHI/HA)- and (CHI/Alg)-based films [91]. In this work, differences in both cross-linked systems have been observed by AFM measurements and quartz crystal microbalance with dissipation (QCM-D), in agreement with the growth dynamics, exhibited by both LbL-assembled systems.

4.3 Interactions of materials with living systems

In this section we will briefly describe some aspects related to the effect of the material properties on the interactions with living cells. Material properties, such as chemical composition, surface energy and wettability, surface charge, material conductivity, morphological characteristics, and stiffness, affect cell functionalities. For many eukaryote cells, adhesion to the substrate is the first step for performing any function and is mediated by the adsorption of extracellular proteins present in the medium. A concise description of cell adhesion modulation by the abovementioned material properties is given below.

The chemical composition of a material, particularly at the surface, plays a key role in determining the surface energy, polarity, charge, and wettability. For a material the surface energy is the excess energy at the surface compared to that of the bulk of the material. The surface energy can be expressed by a polar component and a nonpolar one, and can be calculated by employing contact angle measurements for liquids with different polarity [92]. The large spreading of a polar liquid, that is, small contact angle, is an indication of a hydrophilic surface with a large polar contribution to the surface energy. Cell adhesion characteristics are optimal for surfaces with moderate wettability, indicated by contact angles in the 60 degrees—80 degrees range. It has been suggested that this behavior is due to the proper state of the adhesion-mediating ECM proteins, such as fibronectin, vitronectin, collagen, or laminin, which allow the interaction of cell receptors with specific protein motifs. Highly hydrophobic polymers (contact angle larger than 90 degrees) adsorb protein with a conformation in a very rigid state that may result in being inappropriate for binding membrane cell receptors. Furthermore, in highly hydrophobic surfaces the amount of adsorbed protein is relatively large, but the intraprotein interaction and the protein—substrate interaction may produce denaturing of the proteins or prevent protein structure remodeling from cells. This is a drawback of the many polymeric materials employed in medicine, requiring the application of a process to increase wettability. This is accomplished by physical methods involving irradiation with ions, plasma, or ultraviolet light and eventually followed by chemical functionalization. On increased hydrophilic surfaces, proteins preserve their native state. The conformation of proteins appears to be more important than the amount absorbed, as claimed in the literature [93]. The amount of fibronectin adsorbed on functionalized self-assembled monolayers decreased for the different functionalization groups according to $NH_2 > CH_3 > COOH > OH$. In turn, the adhesion of MC3T3 osteoblastic cells increased according to the sequence $CH_3 < NH_2 \cong COOH < OH$, indicating that protein conformation is a relevant aspect for cell adhesion [93].

On the other hand, in other research works a limited or null cell adhesion on highly hydrophilic surfaces has been observed [94,95]. This fact is explained by the high lability of adsorbed cell adhesion-promoting proteins, resulting in very low cell-surface interactions.

The charge of the coated substrate is another relevant factor for cell colonization of a material surface. There is evidence indicating that positively charged surfaces exhibit a better performance toward cell adhesion than surfaces with negative charge, a fact which is explained by the increased polar interaction between the material surface and adhesion-mediating extracellular proteins, the latter often bearing a positive charge. It is often observed that COOH groups generate negative species and give the appropriate wettability [96] or allow specific grafting of compounds enhancing cell adhesion [97], thus a surface with good adhesion characteristics is commonly produced.

The electrical conductivity of the material also plays a crucial role in modeling cell functions. Conductive polymers without the application of any stimulus, improve cell functions such as colonization. These functions are further enhanced if stimulation with an electrical current and/or electromagnetic field is performed. For instance, electrical stimulations can increase osteoblast proliferation, and also the switching between proliferation and differentiation programs can be modified [98—100].

In relation to the effect of surface roughness and topography on cell function, it is convenient to distinguish different size scales of the features. The macroroughness involves features larger than 1 μm, the micro-/submicroroughness features from 100 nm to 1 μm, and the nanoroughness features below 100 nm. Geometrical features in the macrometer scales do not affect cell adhesion as cells are able to adhere to or between the surface irregularities that are larger than cells. To the contrary, the nanoroughness has a beneficial influence on cell adhesion as the fabricated material with nanoroughness better mimics the natural ECM made of nanofibers, nanocrystals, and nanometric supramolecular complexes. Cell adhesion-promoting molecules acquire the appropriate conformation upon interaction with geometrical features at the nanoscale, resembling ECM components arrangement [101].

The effect of microroughness is well recognized, although reported data are rather contradictory as it is not clear if it improves or hampers cell performance. For instance, it has been reported that rat osteoblasts on the microporous surface of titanium dental implants exhibited an increased average cytoplasm spreading compared to the cell spreading area on flat surfaces [102]. But for the case of normal fibroblast, cell adhesion was decreased on PDMS-based substrates with features of $2-6$ μm produced by PDMS casting on SiC paper in comparison to flat PDMS [103]. This difference has been explained, for the initial cell adhesion process, by the change in surface area the cell encounters when spreading on the substrate. Fibroblast on flat substrate spreads immediately on the surface, becoming flat. On patterned substrates, cells encounter geometrical features out of the plane, and cell spreading is hindered, rendering cells with a relatively larger height. Cell cytoskeleton is influenced by the microroughness and also by the distribution of fibronectin produced by the adhering fibroblast.

The stiffness of the substrate is also a parameter of great importance in the modulation of cell functions. Cells performing different functions require exerting forces that need to be properly balanced by the environment. Moreover, by varying the stiffness of an LbL-coated substrate, the trigger of distinct cell functions has been observed and mechanistically described [104]. In order to exemplify the effect of the stiffness of LbL-coated substrates on cell adhesion, a few cases are depicted. For instance, chondrosarcoma antiadhesive properties have been reported for (PLL/HA)-based multilayers [105]. These nonadhesive properties have been attributed to the high water content and gel-like structure, due to the presence of HA. Contrarily, (CHI/HA)-based multilayer exhibited a relatively enhanced adhesion to chondrosarcoma cells, probably due to the increase in stiffness [106]. In this work the effect of the composition of the upper layer, being CHI or HA, has been indicated. The presence of HA on both the cell surface and the top of the film would inhibit cell adhesion due to repulsive interactions.

4.3.1 Cell adhesion and its relation to proliferation, motility and differentiation

Cell cytoplasm spreading is a measure of cell adhesion to a substrate, and may stimulate cell proliferation. Biochemical signals are triggered in the cell, promoting the change of the cell phase from G0 to G1, and from this to S phase, with the synthesis of new DNA followed by cell division. During cell growth, changes in tension of the actin fibers anchored to the structure of the adhesion sites take place. These fibers are associated with the membrane of the nucleus and other organelles, and cell spreading can stimulate proliferation by nuclear expansion and enlargement of nuclear pores with the increase of DNA accessibility to the replication machinery and the change in the synthesis and transport of various extracellular cell cycle-regulating factors. It is worth noting that there is no linear relation between adhesion and cell proliferation. It has been reported that rat aortic smooth muscle cells seeded on ion-implanted highly adhesive substrates, exhibit a decrease in proliferation but an increase in markers associated with the differentiation toward the contractile phenotype.

Other works suggest that the cell motility ability of different cell lines seeded on various substrates is enhanced for intermediate adhesion, while relatively higher adhesion appears to be related to quiescence and maturation [[101] and reference therein]. The adhesive interaction patterns of

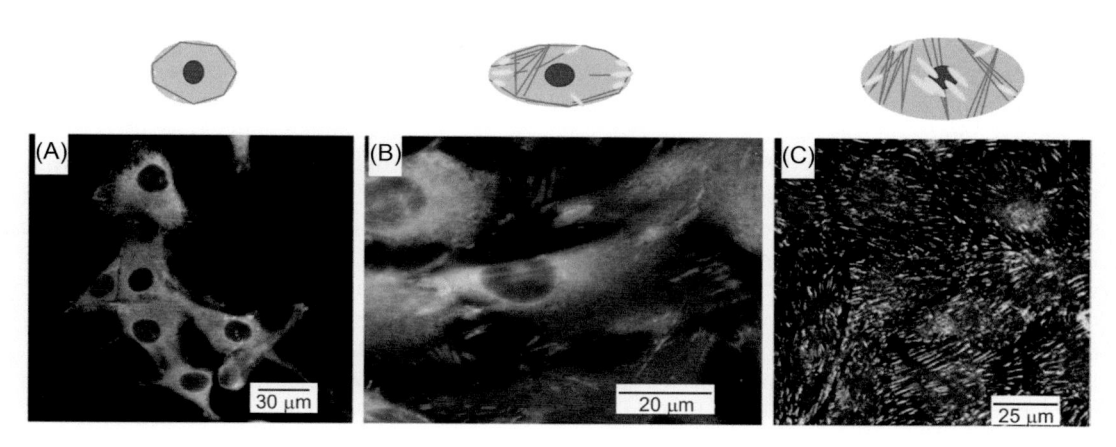

FIGURE 4.6

Different maturation stages of cell-substrate adhesion entities. Immunofluorescence staining against adhesion proteins (vinculin and talin). (A) Focal complexes; (B) focal adhesions; (C) fibrillar adhesions. Scale bars are indicated in the figure.

Adapted with permission from L. Bacakova, E. Filova, M. Parizek, T. Ruml, V. Svorcik, Modulation of cell adhesion, proliferation and differentiation on materials designed for body implants, Biotechnol. Adv. 29 (2011) 739–767. Copyright 2011, Elsevier.

cells in contact with the extracellular environment have been systemized into three types [107]. (1) Adhesion mediated by relatively less mature focal complexes located at the periphery and assisted by $\alpha_V\beta_3$ integrins. (2) Mature focal adhesion sites, allowing a strong cell-ECM interaction. The adhesion sites are larger than focal adhesion complexes and extend into the cell cytoplasm. Beside $\alpha_V\beta_3$ integrins, focal adhesion contains a relatively high amount of talin, zyxin, and vinculin. (3) Fibrillar adhesions, their morphology being long thin streaks localized even in the central regions of the cell. Fig. 4.6 shows the different maturation stages of cell-ECM adhesions. In tissue engineering, a fine control between cell proliferation, migration and differentiation is of uppermost importance, particularly for the construction of devices for vascular replacement. An excessive proliferation and migration of vascular smooth muscle cells can produce stenosis and occlusion of prostheses. Thus it is necessary to properly manage cell behavior through the incorporation of soluble factors, controlling cell-cell interactions and the physicochemical properties of the ECM. All these factors are not fully understood and deserve intensive research.

4.4 Selected examples of the application of the layer-by-layer technique in biosciences

4.4.1 Platforms for cell behavior modulation

4.4.1.1 Tailoring physicochemical properties of the substrates

Among the devices obtained employing the LbL assembly technique, those to modulate cell behavior are relevant not only for basic research on cell mechanisms but also for assisting in the design of devices for biomedical applications. As indicated in previous sections, the composition of the substrate plays a key role in determining its physical and chemical properties as the substrate interacts with biological systems. Many basic research investigations related to cell processes employ fabricated polymeric substrates.

As indicated above, the stiffness of the substrate plays a key role in determining cell phenotype. In this vein, a polyelectrolyte multilayer-based device with controlled stiffness has been employed for studying cell processes at the cell nucleus level, namely, the effect of mechanical properties of the ECM on replication and transcription [104]. For this purpose, diblock $(PLL/HA)_{24}(PSS/PAH)_n$ PEMs with $n = 0$, 2, 5, and 12 were utilized. The first block exhibited an elastic modulus below 1 KPa, which increased to 50, 200, and 500 KPa after $(PSS/PAH)_n$ capping with $n = 2$, 5, and 12, respectively. Accordingly, substrates with different n were named E0, E50, E200, and E500, respectively. In this work, Ptk2 cells (from potorous tridactylus kidney) 4 hours postseeding, showed a decrease in cell adhesion in going from E500 to E0 (Fig. 4.7). For E200 and E500, immunofluorescence-staining microimages indicated peripheral αv-integrin spots at the tip of actin microfilaments and vinculin structures similarly to those observed for cells seeded on glass. In contrast, for E0 and E50 the described fluorescence pattern was not observed, although for E50 only αv-integrin spots were observed. Furthermore, results from cell synchronization by the shake-off method suggested that inhibition of replication for soft substrates was strongly correlated with the inhibition of vinculin assembly, requiring an elastic modulus larger than 200 kPa (Fig. 4.7). This work showed that Ptk2 cells seeded on soft substrates are induced to perform uncouple transcription-replication processes. The authors suggest that the transcriptional machinery alone is not enough to initiate replication but requires the assistance of actin fibers and vinculin structure formation. Thus for E0, transcription was inhibited and active for E50. For this elasticity modulus range, a relation between transcription and αv-integrin engagement and low cell spreading can be inferred.

In another research work, LbL-assembled polyelectrolyte-based thin films with different controlled composition have been utilized to evaluate the adhesion of A549 cells (from human lung cancer) [108]. In this work, PEMs assembled with different natural and synthetic polycation/polyanion pairs were obtained by the LbL technique. Cell adhesion was evaluated by cell shape,

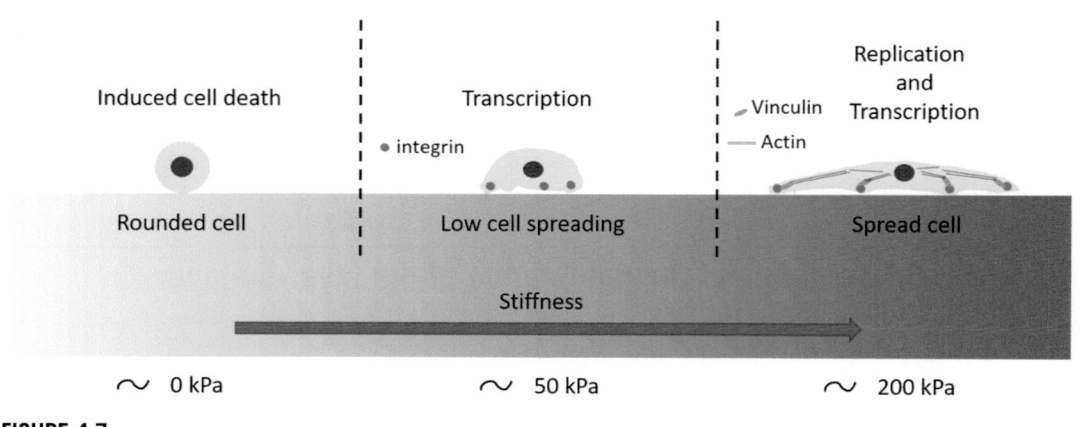

FIGURE 4.7

Scheme of the effect of substrate stiffness on cell functions at the whole cell and nucleus level.

spreading area and characteristics of focal contacts were assessed by immunostaining, and cell proliferation was evaluated by thiazolyl blue tetrazolium bromide (MTT) assay. The PEs PLL, CHI, PEI, PAH, and PDDAC were used as polycations and were combined with the polyanions Alg, DEX, HA, PAA, and PSS to obtain (polycation/polyanion)$_7$polycation multilayers. Results indicate that with the exception of (CHI/HA)$_7$CHI all other tested combinations of natural polyelectrolyte pairs yielded a relatively poor adhesion of A549 cells in comparison with multilayers containing synthetic PEs, particularly PSS. Nevertheless, multilayers containing PSS exhibited a linear increase in cell number, while for some combinations such as PLL/DEX or PLL/Alg, a supralinear increase was observed, indicating an increased biocompatibility for natural polyelectrolyte combinations. In order to increase cell adhesion on multilayers with natural PEs exhibiting exponential proliferation, they were assembled onto an initial block of PSS and PLL for increasing the whole material rigidity. The deposited PLL/DEX or PLL/Alg blocks show a linear growth dynamics when deposited onto a PLL/PSS block, but an exponential one when deposited on glass.

The above-presented strategy to increase the rigidity of the whole film differs from that previously reported by other authors, in which a more rigid block was assembled onto a softer one [86]. The presented diblock strategy for increasing the film stiffness maintains the top layer made of natural PEs and is based on the polyelectrolyte chain reordering and the concomitant changes in the interactions between polyelectrolyte layers assembled by successive steps. The result is a supramolecular interphase offering a cushion of specific interaction to cell membrane components and mediating proteins, enhancing a certain cell phenotype, that is, favoring adhesion and cytoplasm expansion for the appropriate type and number of polyelectrolyte layers. It is worth mentioning that the above observations are cell type dependent, and experiments were performed in fetal bovine serum (FBS) supplemented media, thus proteins with different characteristics deposit onto the surface depending on its composition and other physicochemical properties. In any case, the effect of the chemical composition of the assembled interface can be sensed by cells. Diblock PEMs have different properties than each block. A549 cell adhesion is poorer, similar and larger for (PLL/Alg)$_n$PLL (with $2 \leq n \leq 7$), (PLL/PSS)$_7$PLL, and (PLL/PSS)$_6$(PLL/Alg)$_2$PLL, respectively, than that measured on glass. This behavior is in agreement with the different physicochemical properties of diblock multilayers, due to the polyelectrolyte interdiffusion process. Fig. 4.8 briefly summarizes the above-described results.

As indicated in previous sections, the postassembly treatment affects the physicochemical properties of the polyelectrolyte-based films, and set up new interactions between medium components and cell membrane components. A postassembly thermal treatment consisting in heating the samples at 37°C for 3 days has been utilized for increasing cell adhesion of (PLL/ALG)$_7$PLL multilayers to a variety of cells (Fig. 4.9) [109]. The reordering of polyelectrolyte chains yields changes in the physicochemical properties of the coating assessed by QCM-D, AFM, atomic force spectroscopy (AFS), Z-potential and contact angle measurements [110]. Results from these measurements are depicted in Table 4.1. Upon annealing, surfaces become more hydrophobic, with the contact angle increasing from about 40 degrees to 85 degrees and the Z-potential decreasing from about -1 to -14 mV. The roughness of the surface in between protrusions decreases, and the coating becomes significantly stiffer after thermal annealing at 37°C. Furthermore, in this work, no effect of the thermal annealing when samples are immersed in water is observed. These facts point to a reordering of polyelectrolyte chains yielding a larger number density of complexation sites, diffusion of PLL chains inside the film, and an irreversible decrease in water content. Both A549 and C2C12 (murine myoblasts) improve their adhesion characteristics (Fig. 4.9A), as indicated by the

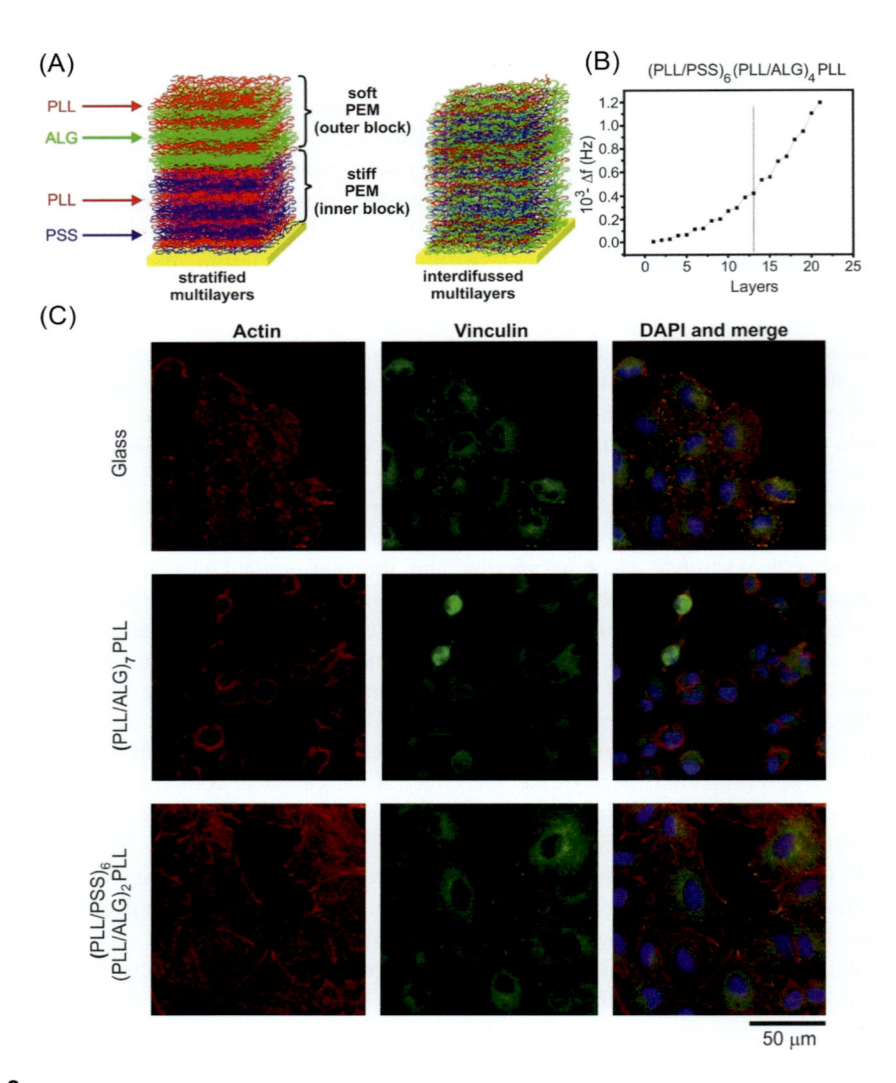

FIGURE 4.8

Diblock PEM design for modulating cell adhesion. (A) Scheme of a diblock multilayer made up of an initial $(PLL/PSS)_6$ block and a second $(PLL/ALG)_n PLL$ block on top, and the possible interdiffusion between the blocks. (B) Data obtained from quartz crystal microbalance measurements. The absolute frequency as a function of the number of layers is plotted. The mass of PLL/ALG increases linearly with the number of layers when deposited on the PLL/PSS block. (C) Fluorescence microimages of stained A549 cells (from human lung carcinoma) seeded on glass, $(PLL/ALG)_7 PLL$ and $(PLL/PSS)_6(PLL/ALG)_2 PLL$, as indicated in the figure. Red stands for actin staining, green for vinculin and blue for DAPI (4′,6-diamidino-2-fenilindol).

Adapted with permission from N.E. Muzzio, M.A. Pasquale, D. Gregurec, E. Diamanti, M. Kosutic, O. Azzaroni, et al., Polyelectrolytes multilayers to modulate cell adhesion: a study of the influence of film composition and polyelectrolyte interdigitation on the adhesion of the A549 cell line, Macromol. Biosci. 16 (2016) 482–495. Copyright 2016, Wiley.

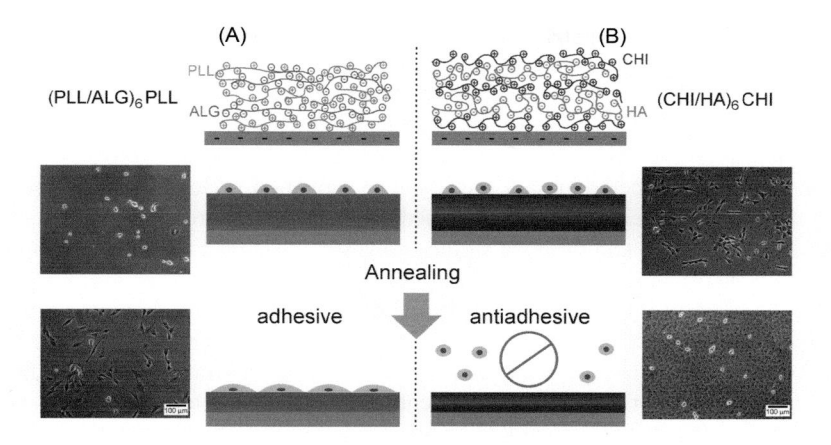

FIGURE 4.9

Effect of thermal annealing of (A) (PLL/ALG)- and (B) (CHI/HA)-based PEMs on cell adhesion. Contrast phase images of C2C12 cells adhered to unannealed and annealed PEMs are included. Thermal annealing enhances and prevents cell adhesion to PLL/ALG PEMs and CHI/HA PEMs, respectively.

Table 4.1 Changes in physicochemical properties of (PLL/Alg)$_7$PLL and (CHI/HA)$_7$CHI upon annealing.

	(PLL/ALG)$_7$PLL		(CHI/HA)$_7$CHI	
	Unannealed	**Annealed**	**Unannealed**	**Annealed**
Roughness (nm)	4.3	2.3	9.3	3.1
Young's modulus (MPa)	50 ± 5	450 ± 15	75 ± 8	73 ± 4
Zeta potential (mV)	0	-16	-24	-31
Contact angle ($^\circ$)	36 ± 5	92 ± 4	30 ± 0.3	20.6 ± 1.8
FN exchangeability	High	Low	High	High

increase in the cell spreading area, and size and distribution of immuno-stained focal contacts. This postassembly treatment has been observed to be useful for improving cell adhesion to coatings obtained by a supramolecular reaction between PAH and phosphate ions [111].

In another research work it has been reported that the same thermal annealing applied to (CHI/HA)$_7$CHI multilayers enhances eukaryote cell and bacteria strain antiadhesive properties (Fig. 4.9B) [112]. Different adherent eukaryote cells, that is, A549, BHK fibroblasts, C2C12 myoblasts, and MC-3T3-E1 osteoblasts, have been tested. A549 cells exhibit almost no adhesion to either unannealed or annealed multilayers, while for the other cell lines a significant decrease in the adhesion properties was observed for annealed multilayers in comparison to unannealed ones. Furthermore, *Escherichia coli* and *Staphylococcus aureus* were cultured on the multilayers, confirming the antifouling properties of CHI-based coatings and enhanced antiadherent properties toward *S. aureus* strain, Gram-positive bacterium. The characterization of both unannealed and annealed multilayers, according to the measured properties depicted in Table 4.1, indicates no significant changes in rigidity, surface charge, and wettability, but the roughness of the multilayers is reduced upon annealing. On the other hand, circular dichroism data show a small adhesion of FN to both annealed and unannealed PEMs, although with a tendency to decrease for annealed multilayers, accompanied by a change in protein conformation. Protein adsorption plays a fundamental role in cell adhesion. Besides the quantity of deposited proteins, their conformational state can influence the cell adhesion process. The formation of focal contacts for cell adhesion and expansion require that adhesive proteins adsorbed on a surface acquire the appropriate conformation [113,114]. It has been reported that different surface roughness changes FN conformation [115,116].

It has been indicated that cell adhesion in culture is mediated by proteins from the medium, and as suggested in the literature, prior to any influence of the substrate's mechanical properties, a good protein adsorption allowing cells to exert tension for spreading is required [114]. In this vein, the importance of tethering ECM proteins to the substrate for guiding stem cell fate has been highlighted [117]. In this work, stem cell differentiation yielded different outcomes for substrates with the same rigidity but different chemical composition, pointing to differences in the characteristics of extracellular component adsorption.

In the above paragraphs LbL multilayer-based devices for either enhancing or impairing cell adhesion were presented. Results can be rationalized considering the adsorption of FN and bovine serum albumin (BSA), an adhesive and an antiadherent protein model, respectively [118]. As indicated, cells are able to expand on an interface due to appropriate forces exerted by the substrate to compensate cell cytoplasm tension. This interaction between cells and the substrate is mediated by proteins from the medium. Results from gamma counting experiments and exchange assays of radiolabeled proteins indicated that thermal annealing of PLL/Alg PEMs increases the stability of preadsorbed FN, regardless of whether the adsorption was performed employing a solution containing either FN alone or FN and BSA. Furthermore, circular dichroism studies suggested that upon annealing, a larger number of arginine-glycine-aspartate (RGD) groups from FN are exposed for cell membrane receptor binding, as is suggested by the improvement of cell adhesion properties [118]. Thus providing FN conformation is suitable, cell transmembrane integrins are able to interact tightly with the PEM surface, and then cells can sense the increase in the Young's modulus of PLL/Alg films upon annealing (Fig. 4.10A). In fact, after annealing, both BSA and FN exhibit an increased interaction with the substrate, and FN molecules are able to adsorb tightly onto either the pristine substrate or BSA-coated substrate [114]. Moreover, dichroism data indicate that FN adsorbed on annealed PLL/Alg adopts an elongated tertiary structure, favoring the exposure of RGD adhesive groups that enhance cell adhesion [118]. For CHI/HA PEMs, results indicate a combined effect of a larger deposition of BSA (nonadhesive protein), and lower deposition of FN, the latter with an increased exchangeability, which renders the PEM surface unfavorable for stable interactions with integrins from the cell membrane (Fig. 4.10B).

4.4.2 Simple tunable multifunctional devices

One of the most important factors defining the success of an engineered scaffold or prosthesis for in vivo applications is the inflammatory and healing response upon implantation. This response can be reduced by controlling biological fouling, that is, the deposition of proteins or other biomolecules and the formation of a biofilm, often a problem in the design of interfaces for biomedical devices in contact with biological fluids. The presence of proteins or bacteria can lead to undesired

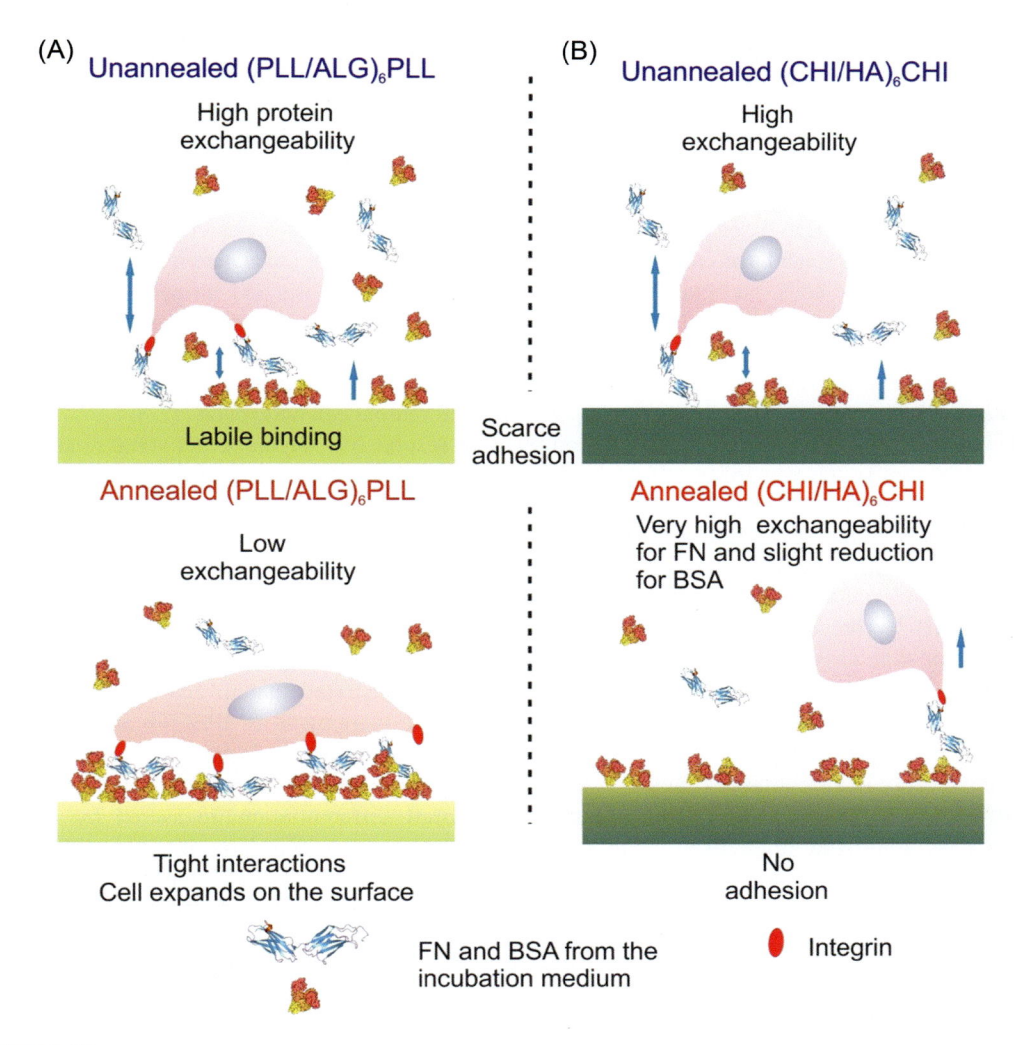

FIGURE 4.10

Scheme of protein and cell interactions with unannealed and annealed (PLL/ALG)- and (CHI/HA)-based PEMs. (A) For (PLL/ALG)-based films the annealing produces a decrease in both BSA and FN exchangeability, reflecting the larger protein-substrate interaction. (B) For (CHI/HA)-based films, the annealing increases FN exchangeability. Cells adhere to and spread onto the substrate provided forces exerted by the cytoplasm are compensated by the substrate via adhesion proteins.

biological responses or infections [119]. In other cases, the adherence of cells to medical device surfaces must be restricted, for example, in surgery devices either during or after intervention [120]. The use of LbL multilayers for controlling the adhesion of proteins, bacteria and mammalian cells has been recently reviewed [28,121]. An emerging strategy for fabricating optimal coatings for tissue engineering consists in combining antithrombotic, antibacterial, and specific cell adhesive properties in the same systems. This has been achieved by grafting bioactive molecules such as matrix components, proteins, growth factors, peptide sequences, etc., to an antifouling substrate.

Several research works have been devoted to substrates that combine nonfouling properties with specific adhesive biomolecules [28]. For example, the Arg-Glu-Asp-Val peptide sequence (REDV) from the CS-5 domain of FN adhesive protein has been grafted on cell antiadhesive glass- and polyethylene terephthalate (PET)-based surfaces [122]. The resulting substrates appeared to be adhesive for endothelial cells that spread and colonized the material surface but were antiadherent for fibroblast, vascular smooth muscle cells, and platelets. As indicated above, thrombosis and restenosis are the main problems of the healing process following cardiovascular surgical interventions. A possible strategy to overcome this problem is the enhancement of the endothelialization of implants. (HEP/CHI)-based multilayers are known to be cell resistant, and after the grafting of REDV sequence, yielded a surface with enhanced adhesion to endothelial cells but impaired smooth muscle cell adherence [123].

It is not an objective of this chapter to comment on responsive polymers, but some related concepts to exemplify the versatility of LbL multilayer-based devices to bear controllable functionalities are given below, in this or in the next sections.

Taking advantage of the LbL assembly technique to conformally coat responsive materials onto substrates with nanotopographic features capable of loading bioactive compounds, tunable multifunctional devices have been proposed. For instance, titania nanotubes (TNTs) loaded with the osteogenic BMP-2 protein were modified with a pH-responsive multilayer film deposited by the LbL assembly of dialdehyde-gentamicin and CHI yielding a composite material with antibacterial and osteogenic properties [124]. In this work, it has been demonstrated that in acidic environmental conditions, both the trigger of the antibiotic release and the increase in BMP-2 protein release are produced. Excellent antibacterial activity was inferred from *E. coli* and *S. aureus* assays, and high biocompatibility of osteoblast extracted from neonatal calvaria rat even in coculture with the bacteria strands was observed. Moreover, enhanced osteoblast differentiation, measured by alkaline phosphatase, improved mineralization and osteogenic-relative gene expression, was observed.

Another work reported the assembly and characterization of an LbL fabricated film that becomes cell-adhesive through exhibition of arginine—glycine—aspartic acid (RGD) adhesion peptides under stretching [125]. The designing concept consists of the use of a stretchable silicon substrate tailored with different PEMs. A precursor film made of two PAH/PSS bilayers with a PAH layer on top was first deposited onto the substrate. Then, a bilayer of polyacrilic acid PAA bearing 5% of RGD adhesion peptide assembled with PAH was deposited onto the precursor and capped with a PAA modified with phosphorylcholine (PC) moieties PAA-PC/PAH multilayer ending with the polyanion. PC moieties are known to render antifouling surfaces when grafted at sufficient densities. In the present work, it is reported that antifouling surfaces toward primary gingival fibroblast were obtained with two or more PAA-PC/PAH layers, and upon stretching (1.5 times), cell adhesion was induced through specific interactions with the RGD peptide sequence.

4.4.3 Gradients in physicochemical properties

Coatings with gradients in their properties appear to be very appealing for applications in biomedical research, and many strategies have been developed for gradually changing their physicochemical properties [27,126]. The presence of gradients increases the fabricated material resemblance to the ECM, inducing a spatial dependence of cell functionalities as well as cooperative effects. A few examples are briefly introduced below.

A high-throughput cellular survival screening platform has been fabricated based on physico-chemical property gradients engineered in PEMs assembled employing PAH and PAA as polyca-tions and polyanion, respectively [127]. Both polyelectrolytes are weak, and their charge depends on the solution pH, which provides control over the polyelectrolyte conformation (i.e., planar or coiled) used for the film assembly. Thus by changing the pH of both polyelectrolyte solutions, the physicochemical properties of the LbL fabricated multilayer as well as their behavior toward cell survival have been tuned. The designed device for the assembly process uses two pumps, one for pumping the solution with the desired pH into the reservoir with the polyelectrolyte solution, and the other to simultaneously pump the polyelectrolyte solution into a container with the substrate vertically placed (Fig. 4.11). Thus the polyelectrolyte solution with continuously changing pH flows

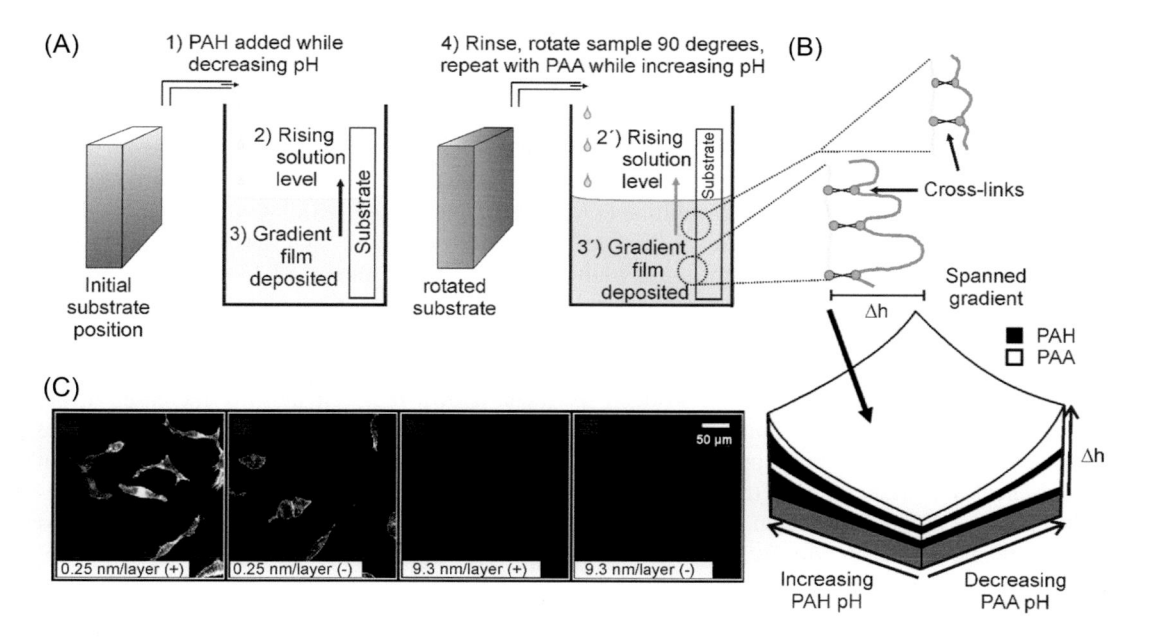

FIGURE 4.11

Gradient in PEM thickness induced by the pH of the assembly solutions. (A) Main steps for the assembly process. PAH and PAA are deposited forming a film with a structure changing with the pH, the latter changes with the level of the solution. The blocks with gradient in red indicate the amount of polymer deposited. The rotation of the sample by 90 degrees can also be appreciated. (B) The thickness of the PEM varies along the substrate. The resulting polymeric multilayer with enhanced differences in thickness is depicted in a 3D scheme. (C) HEC 293 cells on a different region of the LbL-modified substrate, as indicated in the figure. (+) and (−) indicate the surface charge of the film. Scale bar is included.

Adapted with permission from M. Sailer, K. Lai Wing Sun, O. Mermut, T.E. Kennedy, C.J. Barrett, High-throughput cellular screening of engineered ECM based on combinatorial polyelectrolyte multilayer films, Biomaterials 33 (2012) 5841–5847. Copyright 2012, Elsevier.

into the sample container, and the height of the solution level corresponds to the position of adsorption on the substrate. This technique is only suitable for systems where the adsorption is irreversible, and the underlying coating structure is unaffected. A scheme of the assembly protocol is depicted in Fig. 4.11A. At the first stage, PAH is added at the desired velocity while decreasing the pH (from 11 to 7), then once the adsorption takes place, the solution is replaced by rinsing water. In the subsequent stage, the sample is rotated by 90°, and PAA is incorporated at the desired velocity while increasing the pH (from 3 to 6.5). At the highest solution levels, which correspond to the limit of pH, each polyelectrolyte bears the highest charge density and the assembly process yields a relatively compact film. Sample rotation by 90° allows increasing the gradient in the deposited film (Fig. 4.11B). The obtained films were used to analyze the survival of both human embryonic kidney 293 (HEK 293) cells and spinal commissural neurons isolated from Sprague-Dawley rat embryos (Fig. 4.11C).

In another work, the physicochemical properties of PSS/PDDAC were continuously changed by postassembly treatment with solutions of NaCl with varying concentration (c_{NaCl}) in the $1-5$ M range [128]. At relatively higher salt concentrations ($c_{NaCl} > 3$ M) the surface properties are dominated by PSS chains due to the increase in PDDAC loss, and the swelling ratio increases along with the salt concentration. A linear salt gradient was used for the postassembly treatment of PSS/PDDAC multilayers. A solution of increasing salt concentration was pumped at a controlled rate into the sample container, and the sites at increasing levels of the multilayers were exposed to decreasing salt concentrations. Thus multilayer sectors at the bottom exhibited a larger swelling ratio than those sectors at a higher level. Due to the greater density of solutions with larger c_{NaCl}, the gradient remained stable in time, and multilayers remained immersed in the gradients for 2 hours. It has been observed that smooth muscle cells (SMCs) adhered better to and their motility was enhanced in PSS/PDDAC multilayers dominated by negative charges and relatively more hydrated. Directional cell motility was observed to be dependent on cell density, being the highest at an appropriate cell density. In this work, it has been shown that cell motility increases in the presence of gradients. Results pointed to the key role played by cell-cell interaction in sensing the environmental physicochemical properties at a distance higher than the average cell size. Thus in the presence of a gradient, cells (at a certain density) are able to sense relatively larger differences in the substrate properties, and an enhanced directional cell motility takes place. These observations are in agreement with previously reported investigations indicating that cell-cell contact can promote cell migration [129].

Based on the results from cell adhesion experiments on thermally annealed and unannealed (PLL/ALG)- and (CHI/HA)-based PEMs, a temperature gradient was applied to those multilayers after assembly [130]. For this purpose, samples were placed in a temperature gradient with one end maintained at 0°C and the other one at 55°C, using thermostated chambers. A linear temperature gradient was obtained for samples of 2 cm length, with the upper and lower temperature limit being 50°C and 10°C. A continuous change in the physicochemical properties appears to set in along the substrate upon the application of the thermal gradient, as indicated by the measured values of contact angles and the adsorption of FITC-labeled BSA protein, both of which decreased in going from the side at the lowest temperature to the side at the highest temperature. For the substrate region held at the highest temperature, C2C12 cell adhesion was close to that observed on glass. In contrast, at the coldest region, cell adhesion was very poor (Fig. 4.12C).

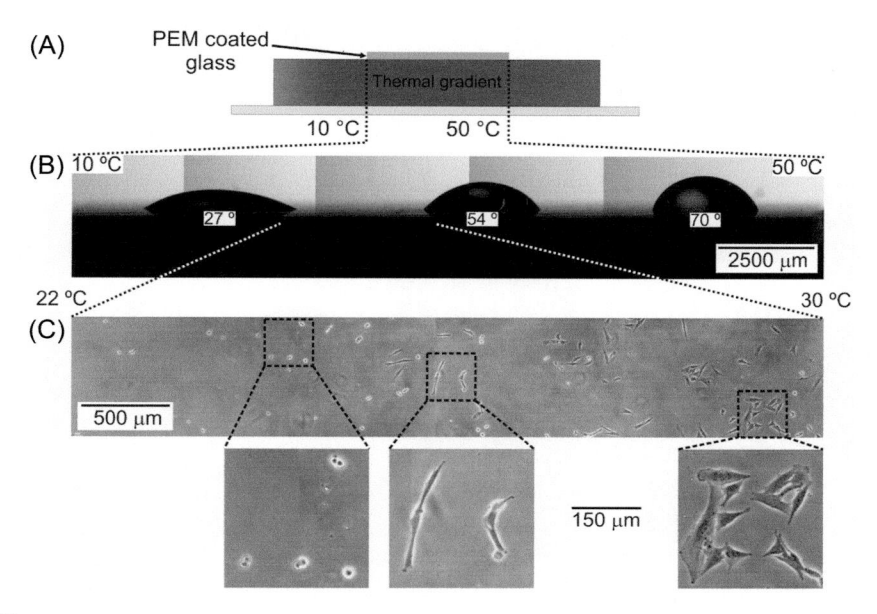

FIGURE 4.12

Thermal gradient effect on physicochemical properties of (PLL/ALG)-based PEMs and modulation of C2C12 cell adhesion. (A) Scheme of the applied thermal gradient. (B) Contact angle variation in the sample. (C) Spatial variation of C2C12 cell adhesion characteristics.

Adapted with permission from N.E. Muzzio, M.A. Pasquale, S.E. Moya, O. Azzaroni, Tailored polyelectrolyte thin film multilayers to modulate cell adhesion, Biointerphases 12 (2017) 04E403. Copyright 2017, American Vacuum Society.

4.5 Cell encapsulation and cell modification by layer-by-layer technique

Cell encapsulation refers to the immobilization of living cells with polymeric semipermeable membranes with appropriate conditions to maintain cell viability, that is, offering sufficient oxygen and nutrient transport, as well as the elimination of cell metabolism by-products. Cell encapsulation by the LbL technique has been recently reviewed, and perspectives on biomedical applications in relation to transplantation, biosensing, tissue engineering, and cell or molecule delivery have been envisaged [131]. In this review, different material used for single cell or cell aggregate encapsulation, that is, synthetic and natural PEs and nanomaterials, as well as the direct and indirect encapsulation strategies of the LbL technique, are described.

In the direct encapsulation by the LbL technique (Fig. 4.13A), PEs or other appropriate assembly blocks interact with the negatively charged cell membrane through electrostatic, hydrogen, or covalent binding, among others. The electrostatic-based LbL strategy is the most widely employed, and the assembly starts with the deposition of a polycation and is followed by the deposition of an oppositely charged polyelectrolyte. This process is repeated until the encapsulation design is achieved. The direct contact of positively charged materials with the membrane of mammalian cells

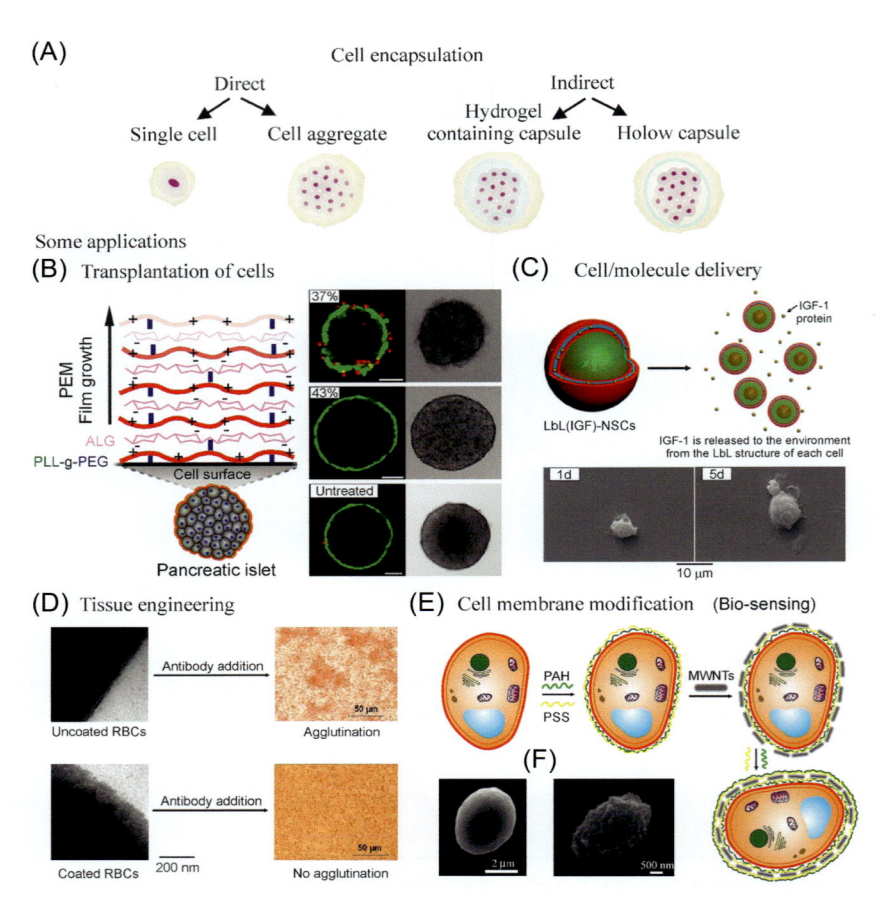

FIGURE 4.13

Layer-by-layer-based cell encapsulation strategies and applications. (A) Scheme of direct single cell and cell aggregate encapsulation and indirect cell aggregate encapsulation. (B) Scheme of cell surface-supported PEM thin films employing PLL-grafted-PEG (PLL-g-PEG), and representative confocal and bright field microimages of pancreatic islets stained with calcein AM and ethidium homodimer after incubation with the copolymer. (C) Schematic representation of neural stem cells coated with gelatin/ALG bilayers containing IGF-1 growth factor and its release to the environment. SEM microimages of encapsulated neuronal progenitor cells (NSCs) after 1 and 5 days of the LbL-assembly procedure are included (1d and 5d in the figure). (D) TEM and optical images of uncoated and coated red blood cells (RBs). (E) Scheme of *Saccharomyces cerevisiae* cell wall tailored with multiwalled carbon nanotubes (MWNTs) and SEM images of untailored and tailored yeast cells (F).

Part (B) Adapted with permission from Ref. [132]. Copyright 2011, American Chemical Society. (C) Adapted with permission from Ref. [133]. Copyright 2015, American Chemical Society. (D) Adapted with permission from Ref. [134]. Copyright 2011, American Chemical Society. (F) Adapted with permission from Ref. [135]. Copyright 2010, American Chemical Society.

usually results in an increased cytotoxicity, being more pronounced for materials with high charge density. Thus natural PEs or synthetic PEs conjugated with neutral poly(ethylene glycol) (PEG) increase cytocompatibility.

In the indirect encapsulation strategy (Fig. 4.13A), the LbL assembly is performed on either bulk hydrogel matrix or hollow vesicles encapsulating cells. The main objective of the LbL deposition is the stabilization of the encapsulated system, assisting in the control of material transport or enhancing the biocompatibility at the implantation site.

A few examples concerning cell encapsulation and cell surface modification for different typical applications will be briefly described below. And in Table 4.2, different encapsulation systems, the employed cell line, and the specific objectives of the encapsulations are listed.

Table 4.2 Typical examples of layer-by-layer single cell and cell aggregate encapsulation for different applications in medicine.

Cell line	Encapsulation material	Function	References	Application
Murine pancreatic islets	PLL-grafted-PEG/ALG	Cytocompatible encapsulation	[132]	Cell transplantation
Murine pancreatic islets	PLL-grafted-PEG (biotin)/streptavidin	Encapsulation with minimized void volume for transplantation into the portal vein of the liver	[136]	
Syrian hamster pancreatic islets	Poly(vinyl alcohol)/PEG-lipids	Cytocompatible encapsulation	[137]	
Red blood cells	CHI-grafted-phosphorylcholine/PLL-PEG/ALG	Immuno-camouflage of blood group antigen	[134]	
Human dermal fibroblast cells and endothelial cells	Fibronectin/gelatin	Fabrication of 3D tissues and human skin patches	[46,138]	Tissue engineering
Insulin-secreting MIN6 cells	Fibronectin/gelatin	Fabrication of pancreatic cell spheroids	[139]	
Human primary hepatocytes, endothelial cells and dermal fibroblast cells	Fibronectin/gelatin	Production of human liver tissue	[140]	
L929 cells and endothelial cells	Fibronectin/gelatin	Fabrication of cellular multilayers	[45]	
Mouse dermal papilla cells	Gelatin/ALG	Fabrication of dermal papilla spheroids for hair follicle regeneration	[141]	
Human myoblast cells	Gelatin/CHI/ALG	Fabrication of tissue by paper-stacking strategy	[142]	
Mouse adipose-derived stem cells	Gelatin/polycaprolactone	Fabrication of paper-stacking tissues for bone regeneration	[143,144]	

(*Continued*)

Table 4.2 Typical examples of layer-by-layer single cell and cell aggregate encapsulation for different applications in medicine. *Continued*

Cell line	Encapsulation material	Function	References	Application
Lactobacillus acidophilus	CHI/ carboxymethyl cellulose	Protection of microbes for probiotic delivery	[145]	Cell/molecule delivery
Rat neural stem cells	Gelatin/ALG	IGF-1 loading for enhancing cell proliferation	[133]	
Mesenchymal stem cells	Gelatin/ALG	VEGF loading for enhancing cardiac function and promoting angiogenesis	[146]	
MELN cells	PDADCMA/PSS	Protection of mammalian cells used in biosensors	[147]	Cell-based biosensing
Yeast cells	PAH-magnetic nanoparticles	Magnetic immobilization of yeast cells in biosensor devices	[148]	

For diabetes treatment, the transplantation of pancreatic cell islets has emerged as a promising therapeutic strategy. Nevertheless, limitations emerge due to the side effects of the immune-depressive therapy required to prevent host rejection of the implanted islets. A large scientific effort has been devoted to the development of microcapsules that protect implantable cells from the immune system and are suitable for nutrient and waste compounds transport. The LbL assembly with cells or cell aggregates as substrates appears as a promising strategy. One important subject is related to the toxicity of many PEs employed as building blocks for capsule fabrication. The toxicity of PLL, as well as that of many other polycations, has been reported [149,150]. Thus to reduce PLL cytotoxicity, N-hydroxysuccinimide (NHS)-poly(ethylene glycol) $(PEG)_{3.4 \text{ kDa}}$ (biotin) was grafted to primary amines on the PLL backbone [136]. The increase in the grafting ratio dramatically reduces cytotoxicity. The adsorption of PLL-grafted-PEG (biotin) copolymer (PBB) to the cell surface was evaluated by measuring the fluorescence of Cy3-labeled streptavidin (SA). Moreover, when encapsulation was performed with the grafted copolymer and labeled SA, a high fluorescence signal was detected at the islet surface region. Contrarily, when unmodified PLL was employed, fluorescence appeared as concentrated domains, suggesting cell membrane permeabilization. Measurements of absorbance by solid-state UV−Vis spectroscopy indicated that the assembly dynamics of the PBB/Cy3-labeled SA followed a linear increase with the number of layers, and fluorescence images obtained with FITC-labeled PBB and Cy3-SA depicted a conformal coating of islets. Function maintenance is another determinant aspect for cell transplantation. In this case, PBB/SA encapsulated pancreatic islet functionality was assessed by measuring insulin secretion in response to a step change in glucose concentration. It was shown that encapsulated islets retain similar functionality to untreated ones. In the same work, encapsulated islets were transplanted into the portal vein of the liver of mice glycemia models, obtaining promising results for the conversion to euglycemia in a slightly larger number of cases in comparison to the treatment with nonencapsulated islets [136]. A trend that, according to the authors' statements, could be increased with the number of layers and the loading of bioactive compounds.

In a more recent research work, a subset of polycation copolymer based on PEG-grafted PLL was investigated [132]. The grafting ratio and the PEG graft length exhibit a synergism in reducing cytotoxicity, but an appropriate formulation is required for maintaining interacting sites for performing direct assembly onto the cell surface. In this work PEG-grafted polycation has been assembled with ALG polyanion to provide a new platform for pancreatic islet encapsulation (Fig. 4.13B) with increased versatility and modularity relative to conventional approaches.

Mammalian cell protection is also required for the design of cell-based sensors, as is the case of devices for sensing the presence of xenoestrogens such as pesticides or estrogenic drugs. This subject is of uppermost importance as the association of some environmental estrogen with hormone-dependent diseases, that is, testicular and breast cancers, among others, has been envisaged by epidemiological investigations. The encapsulation of the cell line MELN with high viability and function maintenance has been reported [147]. MELN cell line is derived from MCF7 breast cancer cells upon transfection with a construct expressing the luciferase gene under the control of an estrogen-regulated promoter. MELN cells were encapsulated by the LbL strategy with the PDDAC/PSS polyelectrolyte pair. The PDDAC/PSS-based encapsulation yielded good cell viability when assembled in Ringer/KCl buffer but not in HEPES/NaCl buffer [147]. Furthermore, cell metabolic functions were shown to be preserved, and the diffusion of small molecules (i.e., hormones) through the multilayer appeared to be efficient. These facts were inferred from luminescence measurements of the luciferase synthetized by estradiol induced encapsulated MELN cells, which exhibit a similar synthesis increase to uncoated cells.

Another example involves the nanoencapsulation of single neural stem cells (NSCs) likely to be implanted for dealing with neurological diseases (Fig. 4.13C) [133]. In this investigation, NSCs were encapsulated with a gelatin/ALG/gelatin multilayer that resulted biocompatible, as inferred from the proliferation and differentiation preservation. Cell function improvement was also achieved by incorporation of insulin-like growth factor 1 (IGF-1) in the ALG-layer, an effect that persisted for a relatively long period. This was in agreement with the time- and pH-dependent release of IGF-1. It could be concluded that the presented encapsulation system appeared to be suitable as a drug carrier, providing a potential treatment strategy for nervous system disorders.

LbL films are appealing as carriers of bioactive molecules, their general transport properties mainly depend on the film porosity and mesh. These properties can be controlled by the number density of polycation-polyanion complexation sites. The greater the number density of complexation sites the larger the rigidity of the film and the lower the mesh size. In comparison with inorganic materials, porous LbL films have the advantage of exhibiting a flexible mesh with certain control. Complexation sites can be incorporated in a smart design and triggered by external stimuli, for example, pH, ionic strength or temperature. Versatile systems can be obtained provided they show an appropriate combination of stability and responsiveness to environmental conditions. These characteristics can be tuned by a number of variables, as described in the literature [79,151]. LbL-based drug delivery systems have proven to be promising in cancer therapy, bacterial infection prevention and treatment, and directing cellular responses [9,152]. In the same vein, cell encapsulation for drug delivery has also been recently reviewed [153].

The LbL assembly technique has also been employed to obtain knowledge related to the production of universal red blood cells (RBCs). Nonimmunogenic PEs have been assembled to decrease the inherent immunogenicity toward foreign live RBCs by suppressing antibody binding (Fig. 4.13D) [134]. Cell immuno-camouflage is achieved by cell membrane manipulation, which is likely to prevent the antibody recognition by concealing the epitope via grafting nonimmunogenic molecules that are not recognized by the immune system. The multilayered shell was composed of a protective block containing five bilayers of CHI-grafted-phosphorylcholine (CHI-PC) and HA, and an outer camouflage block made up of five PLL-grafted-PEG and HA PEs, and the incubation and rinsing steps were performed using a filtration membrane [134]. Coated RBCs in suspension retained their viability and functionality, the latter was assessed by measuring the ability to take up oxygen. The immune-camouflage of RBCs was tested by agglutination assays of anti-A, anti-B, and anti-D sera with uncoated and coated cells. The absence of agglutination for coated RBCs in comparison with untreated ones was observed (Fig. 4.13D).

In the examples described above, different artificial substances were made to interact with living cells, yielding systems with integrated properties. Cells not only retained their intrinsic biological properties but were able to promote specific cooperative functions in a realistic environment due to appropriate modifications performed by the integrated artificial substances. These features are rather common in complex systems of biological origin, appearing hierarchically structured and multifunctional. Processes taking place in these systems, often involve multiple steps and different parts that accomplish certain functions with high efficiency, maintaining their homeostatic state.

Following this conceptual frame, it would be possible to fabricate bio-like highly functional materials for a wide range of applications. The emerging related approaches include living cells as complex entities that can be properly modified for managing their functions. The modification of living cells with conventional polymers, functional polymers, bioactive materials, nanoparticles and inorganic substances has been recently reviewed [154]. One example of this approach is depicted in Fig. 4.13E,F. The assembly of MWNT on living yeast cells that maintain biological functions is schematized [135]. First, individual yeast cells were coated with a PAH/PSS/PAH nanofilm, then the polyelectrolyte-modified yeast cells were immersed in negatively charged MWNT aqueous dispersion. Finally, another PAH/PSS bilayer was deposited onto the MWNT/polyelectrolyte modified yeast cells for increasing stability. Optical microscopy, TEM and SEM were employed to prove yeast cell membrane modification. Results from the intercellular esterase activity assay indicated that cell viability was preserved after coating. Furthermore, electrochemical studies, using voltammetry and electrochemical impedance measurements, were performed with polyelectrolyte/MWNT coated cells. Results indicated that they sufficiently affected the charge transfer process of $[Fe(CN)_6]^{3-}/[Fe(CN)_6]^{4-}$ couple and enhanced the difference between the electrochemical signal from live yeast cells and thermal inactivated ones.

In a more recent investigation, human cells were engineered using PAH-stabilized magnetic nanoparticles (MNPs) [155]. The polycationic-coated MNP were directly deposited onto A549 cells, producing a mesoporous semipermeable layer, and rendering magnetic-responsible cells. In this work, a detailed cell viability study was performed, and results suggested that MNP modification maintained cell membrane integrity, cell enzymatic activity, adhesion, proliferation, and cytoskeleton formation, and no increase in apoptosis in either cancer or primary cell culture was observed. The engineered cells were employed for fabricating 2D monolayers as well as 3D multicellular spheroids with a single cell type or in coculture.

4.6 Conclusions

The LbL assembly technique for fabricating appealing coating materials for biological applications has been briefly reviewed in this chapter. It appears to be a versatile technique for the fabrication of biointerphases and complex systems with multifunctional activity of interest not only for basic studies but also for biomedical applications.

A wide variety of building blocks can be used for the assembly of multilayers, and, consequently, different interactions contribute to the assembly driving force besides the entropic effects. The LbL technique can be used for conformal coating of substrates with different geometry and topographical features, and also combined with other procedures, such as microfluidics, inkjet printing, and lithography, to yield systems with increased versatility. Thus considering these beneficial characteristics, as well as the possibility of changing the assembly conditions, that is, the chemical nature of the building blocks, temperature, pH, ionic strength, among the most important factors, and postassembly treatments, a precise control of the properties and functions of the materials at the nanoscale level can be obtained. By properly designing the assembly procedure it is possible to generate gradients in the physicochemical and biological properties of the material, that is, stiffness, wettability, charge, topography, and protein adsorption to the material. These features allow increasing materials' resemblance to biological environments, thus facilitating the biocompatible interaction between the fabricated materials and natural living systems.

Throughout the chapter, particular attention was paid to the fabrication of platforms for managing cell phenotype and fate. Applications for basic studies of processes at the cell level, as well as in the biomedical field, were presented. Cell encapsulation approaches were briefly described, a subject also related to drug/protein and even cell delivery, as is required in emerging strategies for treating several diseases or for tissue engineering.

The LbL strategy plays a key role in obtaining complex multifunctional materials of interest for biological applications, with distinctive general features resembling the behavior of living systems, characterized by the harmonization of coexisting specific structures and functions.

References

[1] M. Drobota, S. Ursache, M. Aflori, Surface functionalities of polymers for biomaterial applications, Polymers (Basel) 14 (2022) 2307. Available from: https://doi.org/10.3390/polym14122307.

[2] C.-Y. Chen, P.-H. Tsai, Y.-H. Lin, C.-Y. Huang, J.H.Y. Chung, G.-Y. Chen, Controllable graphene oxide-based biocompatible hybrid interface as an anti-fibrotic coating for metallic implants, Mater. Today Bio. 15 (2022) 100326. Available from: https://doi.org/10.1016/j.mtbio.2022.100326.

[3] X. Han, W. Gao, Z. Zhou, S. Yang, J. Wang, R. Shi, et al., Application of biomolecules modification strategies on PEEK and its composites for osteogenesis and antibacterial properties, Colloids. Surf. B: Biointerfaces 215 (2022). Available from: https://doi.org/10.1016/j.colsurfb.2022.112492.

[4] D. Cai, Z. Zhang, Z. Feng, J. Song, X. Zeng, Y. Tu, et al., A lipophilic chitosan-modified self-nanoemulsifying system influencing cellular membrane metabolism enhances antibacterial and anti-biofilm efficacy for multi-drug resistant Pseudomonas aeruginosa wound infection, Biomater. Adv. (2022) 213029. Available from: https://doi.org/10.1016/j.bioadv.2022.213029.

[5] M.L. Cortez, A. Lorenzo, W.A. Marmisollé, C. von Bilderling, E. Maza, L. Pietrasanta, et al., Highly-organized stacked multilayers: via layer-by-layer assembly of lipid-like surfactants and polyelectrolytes. Stratified supramolecular structures for (bio)electrochemical nanoarchitectonics, Soft Matter 14 (2018) 1939–1952. Available from: https://doi.org/10.1039/c8sm00052b.

[6] E. Piccinini, J.S. Tuninetti, J. Irigoyen Otamendi, S.E. Moya, M. Ceolín, F. Battaglini, et al., Surfactants as mesogenic agents in layer-by-layer assembled polyelectrolyte/surfactant multilayers: nanoarchitectured "soft" thin films displaying a tailored mesostructure, Phys. Chem. Chem. Phys. 20 (2018) 9298–9308. Available from: https://doi.org/10.1039/c7cp08203g.

[7] E. Piccinini, G.A. González, O. Azzaroni, F. Battaglini, Mass and charge transport in highly mesostructured polyelectrolyte/electroactive-surfactant multilayer films, J. Colloid Interface Sci. 581 (2021) 595−607. Available from: https://doi.org/10.1016/j.jcis.2020.07.060.

[8] E. Piccinini, M. Ceolín, F. Battaglini, O. Azzaroni, Mesostructured electroactive thin films through layer-by-layer assembly of redox surfactants and polyelectrolytes, ChemPlusChem 85 (2020) 1616−1622. Available from: https://doi.org/10.1002/cplu.202000358.

[9] A. Escobar, N.E. Muzzio, P. Andreozzi, S. Libertone, E. Tasca, O. Azzaroni, et al., Antibacterial layer-by-layer films of poly(acrylic acid)−gentamicin complexes with a combined burst and sustainable release of gentamicin, Adv. Mater. Interfaces 6 (2019). Available from: https://doi.org/10.1002/admi.201901373.

[10] S.E. Herrera, M.L. Agazzi, M.L. Cortez, W.A. Marmisollé, C. von Bilderling, O. Azzaroni, Layer-by-Layer formation of polyamine-salt aggregate/polyelectrolyte multilayers. loading and controlled release of probe molecules from self-assembled supramolecular networks, Macromol. Chem. Phys. 220 (2019). Available from: https://doi.org/10.1002/macp.201900094.

[11] E. Maza, J.S. Tuninetti, N. Politakos, W. Knoll, S. Moya, O. Azzaroni, PH-responsive ion transport in polyelectrolyte multilayers of poly(diallyldimethylammonium chloride) (PDADMAC) and poly(4-styrenesulfonic acid-co-maleic acid) (PSS-MA) bearing strong- and weak anionic groups, Phys. Chem. Chem. Phys. 17 (2015) 29935−29948. Available from: https://doi.org/10.1039/c5cp03965g.

[12] M. Lorena Cortez, N. de Matteis, M. Ceolín, W. Knoll, F. Battaglini, O. Azzaroni, Hydrophobic interactions leading to a complex interplay between bioelectrocatalytic properties and multilayer mesoorganization in layer-by-layer assemblies, Phys. Chem. Chem. Phys. 16 (2014) 20844−20855. Available from: https://doi.org/10.1039/c4cp02334j.

[13] J. Irigoyen, S.E. Moya, J.J. Iturri, I. Llarena, O. Azzaroni, E. Donath, Specific ζ-potential response of layer-by-layer coated colloidal particles triggered by polyelectrolyte ion interactions, Langmuir 25 (2009) 3374−3380. Available from: https://doi.org/10.1021/la803360n.

[14] A.P. Mártire, G.M. Segovia, O. Azzaroni, M. Rafti, W. Marmisollé, Layer-by-layer integration of conducting polymers and metal organic frameworks onto electrode surfaces: enhancement of the oxygen reduction reaction through electrocatalytic nanoarchitectonics, Mol. Syst. Des. Eng. 4 (2019) 893−900. Available from: https://doi.org/10.1039/c9me00007k.

[15] G.E. Fenoy, B. van der Schueren, J. Scotto, F. Boulmedais, M.R. Ceolín, S. Bégin-Colin, et al., Layer-by-layer assembly of iron oxide-decorated few-layer graphene/PANI:PSS composite films for high performance supercapacitors operating in neutral aqueous electrolytes, Electrochim. Acta 283 (2018) 1178−1187. Available from: https://doi.org/10.1016/j.electacta.2018.07.085.

[16] G.E. Fenoy, E. Maza, E. Zelaya, W.A. Marmisollé, O. Azzaroni, Layer-by-layer assemblies of highly connected polyelectrolyte capped-Pt nanoparticles for electrocatalysis of hydrogen evolution reaction, Appl. Surf. Sci. 416 (2017) 24−32. Available from: https://doi.org/10.1016/j.apsusc.2017.04.086.

[17] W.A. Marmisollé, E. Maza, S. Moya, O. Azzaroni, Amine-appended polyaniline as a water dispersible electroactive polyelectrolyte and its integration into functional self-assembled multilayers, Electrochim. Acta 210 (2016) 435−444. Available from: https://doi.org/10.1016/j.electacta.2016.05.182.

[18] W.A. Marmisollé, O. Azzaroni, Recent developments in the layer-by-layer assembly of polyaniline and carbon nanomaterials for energy storage and sensing applications. from synthetic aspects to structural and functional characterization, Nanoscale 8 (2016) 9890−9918. Available from: https://doi.org/10.1039/c5nr08326e.

[19] M. Coustet, J. Irigoyen, T.A. Garcia, R.A. Murray, G. Romero, M. Susana Cortizo, et al., Layer-by-layer assembly of polymersomes and polyelectrolytes on planar surfaces and microsized colloidal particles, J. Colloid. Interface Sci. 421 (2014) 132−140. Available from: https://doi.org/10.1016/j.jcis.2014.01.038.

[20] K. Ariga, J.P. Hill, Q. Ji, Layer-by-layer assembly as a versatile bottom-up nanofabrication technique for exploratory research and realistic application, Phys. Chem. Chem. Phys. 9 (2007) 2319−2340. Available from: https://doi.org/10.1039/b700410a.

[21] M. Aono, K. Ariga, The Way to Nanoarchitectonics and the way of nanoarchitectonics, Adv. Mater. 28 (2016) 989–992. Available from: https://doi.org/10.1002/adma.201502868.

[22] K. Ariga, Y. Yamauchi, Nanoarchitectonics from Atom to Life, Chem. Asian J. 15 (2020) 718–728. Available from: https://doi.org/10.1002/asia.202000106.

[23] T. Berninger, C. Bliem, E. Piccinini, O. Azzaroni, W. Knoll, Cascading reaction of arginase and urease on a graphene-based FET for ultrasensitive, real-time detection of arginine, Biosens. Bioelectron. 115 (2018) 104–110. Available from: https://doi.org/10.1016/j.bios.2018.05.027.

[24] E. Maza, C. von Bilderling, M.L. Cortez, G. Díaz, M. Bianchi, L.I. Pietrasanta, et al., Layer-by-Layer assembled microgels can combine conflicting properties: switchable stiffness and wettability without affecting permeability, Langmuir 34 (2018) 3711–3719. Available from: https://doi.org/10.1021/acs.langmuir.8b00047.

[25] E. Piccinini, C. Bliem, C. Reiner-Rozman, F. Battaglini, O. Azzaroni, W. Knoll, Enzyme-polyelectrolyte multilayer assemblies on reduced graphene oxide field-effect transistors for biosensing applications, Biosens. Bioelectron. 92 (2017) 661–667. Available from: https://doi.org/10.1016/j.bios.2016.10.035.

[26] O. Azzaroni, K.H.A. Lau, Layer-by-layer assemblies in nanoporous templates: nano-organized design and applications of soft nanotechnology, Soft Matter 7 (2011) 8709–8724. Available from: https://doi.org/10.1039/c1sm05561e.

[27] C. Monge, J. Almodóvar, T. Boudou, C. Picart, Spatio-temporal control of LbL films for biomedical applications: from 2D to 3D, Adv. Healthc. Mater. 4 (2015) 811–830. Available from: https://doi.org/10.1002/adhm.201400715.

[28] S. Guo, X. Zhu, X.J. Loh, Controlling cell adhesion using layer-by-layer approaches for biomedical applications, Mat. Sci. Eng. C. 70 (2017) 1163–1175. Available from: https://doi.org/10.1016/j.msec.2016.03.074.

[29] M.P. Sousa, E. Arab-Tehrany, F. Cleymand, J.F. Mano, Surface micro- and nanoengineering: applications of layer-by-layer technology as a versatile tool to control cellular behavior, Small 15 (2019). Available from: https://doi.org/10.1002/smll.201901228.

[30] Y. Zhu, D. Zhou, X. Zan, Q. Ye, S. Sheng, Engineering the surfaces of orthopedic implants with osteogenesis and antioxidants to enhance bone formation in vitro and in vivo, Colloids Surf. B: Biointerfaces 212 (2022). Available from: https://doi.org/10.1016/j.colsurfb.2022.112319.

[31] F. Kazemi-Andalib, M. Mohammadikish, U. Sahebi, A. Divsalar, pH-sensitive and targeted core-shell and yolk-shell microcarriers for in vitro drug delivery, J. Drug. Deliv. Sci. Technol. 75 (2022) 103633. Available from: https://doi.org/10.1016/j.jddst.2022.103633.

[32] Y. Wang, S.E. Naleway, B. Wang, Biological and bioinspired materials: structure leading to functional and mechanical performance, Bioact. Mater. 5 (2020) 745–757. Available from: https://doi.org/10.1016/j.bioactmat.2020.06.003.

[33] A. Verma, M. Verma, A. Singh, Animal tissue culture principles and applications, Animal Biotechnology: Models in Discovery and Translation, Elsevier, 2020, pp. 269–293. Available from: https://doi.org/10.1016/B978-0-12-811710-1.00012-4.

[34] S.H. Kim, J. Turnbull, S. Guimond, Extracellular matrix and cell signalling: the dynamic cooperation of integrin, proteoglycan and growth factor receptor, J. Endocrinolo 209 (2011) 139–151. Available from: https://doi.org/10.1530/JOE-10-0377.

[35] B. Yue, Biology of the extracellular matrix: an overview, Glaucoma 23 (2014) S20–S23. Available from: https://doi.org/10.1097/IJG.0000000000000108.

[36] C. Bonnans, J. Chou, Z. Werb, Remodelling the extracellular matrix in development and disease, Nat. Rev. Mol. Cell Biol. 15 (2014) 786–801. Available from: https://doi.org/10.1038/nrm3904.

[37] M.W. Pickup, J.K. Mouw, V.M. Weaver, The extracellular matrix modulates the hallmarks of cancer, EMBO Rep. 15 (2014) 1243–1253. Available from: https://doi.org/10.15252/embr.201439246.

[38] A.D. Theocharis, S.S. Skandalis, C. Gialeli, N.K. Karamanos, Extracellular matrix structure, Adv. Drug. Deliv. Rev. 97 (2016) 4−27. Available from: https://doi.org/10.1016/j.addr.2015.11.001.

[39] P. Beri, B.F. Matte, L. Fattet, D. Kim, J. Yang, A.J. Engler, Biomaterials to model and measure epithelial cancers, Nat. Rev. Mater. 3 (2018) 418−430. Available from: https://doi.org/10.1038/s41578-018-0051-6.

[40] J. Borges, J.F. Mano, Molecular interactions driving the layer-by-layer assembly of multilayers, Chem. Rev. 114 (2014) 8883−8942. Available from: https://doi.org/10.1021/cr400531v.

[41] J. Zhou, G. Romero, E. Rojas, L. Ma, S. Moya, C. Gao, Layer by layer chitosan/alginate coatings on poly(lactide-co-glycolide) nanoparticles for antifouling protection and Folic acid binding to achieve selective cell targeting, J. Colloid Interface Sci. 345 (2010) 241−247. Available from: https://doi.org/10.1016/j.jcis.2010.02.004.

[42] S. Schmidt, N. Madaboosi, K. Uhlig, D. Köhler, A. Skirtach, C. Duschl, et al., Control of cell adhesion by mechanical reinforcement of soft polyelectrolyte films with nanoparticles, Langmuir 28 (2012) 7249−7257. Available from: https://doi.org/10.1021/la300635z.

[43] J. Lipton, G.M. Weng, J.A. Röhr, H. Wang, A.D. Taylor, Layer-by-layer assembly of two-dimensional materials: meticulous control on the nanoscale, Matter 2 (2020) 1148−1165. Available from: https://doi.org/10.1016/j.matt.2020.03.012.

[44] D. Volodkin, A. Skirtach, H. Möhwald, LbL films as reservoirs for bioactive molecules, Adv. Polym. Sci. 240 (2010) 135−161. Available from: https://doi.org/10.1007/12_2010_79.

[45] M. Matsusaki, K. Kadowaki, Y. Nakahara, M. Akashi, Fabrication of cellular multilayers with nanometer-sized extracellular matrix films, Angew. Chem. Int. Ed. 46 (2007) 4689−4692. Available from: https://doi.org/10.1002/anie.200701089.

[46] A. Nishiguchi, H. Yoshida, M. Matsusaki, M. Akashi, Rapid construction of three-dimensional multilayered tissues with endothelial tube networks by the cell-accumulation technique, Adv. Mater. 23 (2011) 3506−3510. Available from: https://doi.org/10.1002/adma.201101787.

[47] F. Caruso, H. Lichtenfeld, E. Donath, H. Möhwald, Investigation of electrostatic interactions in polyelectrolyte multilayer films: binding of anionic fluorescent probes to layers assembled onto colloids, Macromolecules 32 (1999) 2317−2328. Available from: https://doi.org/10.1021/ma980674i.

[48] A. Jaklenec, A.C. Anselmo, J. Hong, A.J. Vegas, M. Kozminsky, R. Langer, et al., High throughput layer-by-layer films for extracting film forming parameters and modulating film interactions with cells, ACS Appl. Mater. Interfaces 8 (2016) 2255−2261. Available from: https://doi.org/10.1021/acsami.5b11081.

[49] S.R. Pattabhi, A.M. Lehaf, J.B. Schlenoff, T.C.S. Keller, Human mesenchymal stem cell osteoblast differentiation, ECM deposition, and biomineralization on PAH/PAA polyelectrolyte multilayers, J. Biomed. Mater. Res. A 103 (2015) 1818−1827. Available from: https://doi.org/10.1002/jbm.a.35322.

[50] L.M. Petrila, F. Bucatariu, M. Mihai, C. Teodosiu, Polyelectrolyte multilayers: an overview on fabrication, properties, and biomedical and environmental applications, Materials 14 (2021). Available from: https://doi.org/10.3390/ma14154152.

[51] J.J. Richardson, M. Björnmalm, F. Caruso, Technology-driven layer-by-layer assembly of nanofilms, Science 348 (2015). Available from: https://doi.org/10.1126/science.aaa2491.

[52] P. Schaaf, J.C. Voegel, L. Jierry, F. Boulmedais, Spray-assisted polyelectrolyte multilayer buildup: from step-by-step to single-step polyelectrolyte film constructions, Adv. Mater. 24 (2012) 1001−1016. Available from: https://doi.org/10.1002/adma.201104227.

[53] J.J. Richardson, J. Cui, M. Björnmalm, J.A. Braunger, H. Ejima, F. Caruso, Innovation in Layer-by-Layer Assembly, Chem. Rev. 116 (2016) 14828−14867. Available from: https://doi.org/10.1021/acs.chemrev.6b00627.

[54] D.R. Reyes, E.M. Perruccio, S.P. Becerra, L.E. Locascio, M. Gaitan, Micropatterning neuronal cells on polyelectrolyte multilayers, Langmuir 20 (2004) 8805−8811. Available from: https://doi.org/10.1021/la049249a.

[55] A.M. Yola, J. Campbell, D. Volodkin, Microfluidics meets layer-by-layer assembly for the build-up of polymeric scaffolds, Appl. Surf. Sci. Adv. 5 (2021). Available from: https://doi.org/10.1016/j.apsadv.2021.100091.

[56] C.M. Andres, N.A. Kotov, Inkjet deposition of layer-by-layer assembled films, J. Am. Chem. Soc. 132 (2010) 14496−14502. Available from: https://doi.org/10.1021/ja104735a.

[57] M. Choi, J. Heo, M. Yang, J. Hong, Inkjet printing-based patchable multilayered biomolecule-containing nanofilms for biomedical applications, ACS Biomater. Sci. Eng. 3 (2017) 870−874. Available from: https://doi.org/10.1021/acsbiomaterials.7b00138.

[58] R. Suntivich, I. Drachuk, R. Calabrese, D.L. Kaplan, V.v Tsukruk, Inkjet printing of silk nest arrays for cell hosting, Biomacromolecules 15 (2014) 1428−1435. Available from: https://doi.org/10.1021/bm500027c.

[59] M. Matsusaki, K. Sakaue, K. Kadowaki, M. Akashi, Three-dimensional human tissue chips fabricated by rapid and automatic inkjet cell printing, Adv. Healthc. Mater. 2 (2013) 534−539. Available from: https://doi.org/10.1002/adhm.201200299.

[60] S. Chikae, A. Kubota, H. Nakamura, A. Oda, A. Yamanaka, T. Akagi, et al., Three-dimensional bioprinting human cardiac tissue chips of using a painting needle method, Biotechnol. Bioeng. 116 (2019) 3136−3142. Available from: https://doi.org/10.1002/bit.27126.

[61] C.F.V. Sousa, C.A. Saraiva, T.R. Correia, T. Pesqueira, S.G. Patrício, M.I. Rial-Hermida, et al., Bioinstructive layer-by-layer-coated customizable 3d printed perfusable microchannels embedded in photocrosslinkable hydrogels for vascular tissue engineering, Biomolecules 11 (2021). Available from: https://doi.org/10.3390/biom11060863.

[62] H. Miyoshi, T. Adachi, Topography design concept of a tissue engineering scaffold for controlling cell function and fate through actin cytoskeletal modulation, Tissue Eng. Part. B Rev. 20 (2014) 609−627. Available from: https://doi.org/10.1089/ten.teb.2013.0728.

[63] K.T.M. Tran, T.D. Nguyen, Lithography-based methods to manufacture biomaterials at small scales, J. Sci.: Adv. Mater. Dev. 2 (2017) 1−14. Available from: https://doi.org/10.1016/j.jsamd.2016.12.001.

[64] D. Qin, Y. Xia, G.M. Whitesides, Soft lithography for micro- and nanoscale patterning, Nat. Protoc. 5 (2010) 491−502. Available from: https://doi.org/10.1038/nprot.2009.234.

[65] C. Monge, K. Ren, K. Berton, R. Guillot, D. Peyrade, C. Picart, Engineering muscle tissues on microstructured polyelectrolyte multilayer films, Tissue Eng. Part. A 18 (2012) 1664−1676. Available from: https://doi.org/10.1089/ten.tea.2012.0079.

[66] J.S. Mohammed, M.A. DeCoster, M.J. McShane, Micropatterning of nanoengineered surfaces to study neuronal cell attachment in vitro, Biomacromolecules 5 (2004) 1745−1755. Available from: https://doi.org/10.1021/bm0498631.

[67] G. Decher, Fuzzy nanoassembly: toward layered polymeric multicomposites, Science 277 (1997) (1979) 1232−1237.

[68] F. Boulmedais, M. Bozonnet, P. Schwinté, J.C. Voegel, P. Schaaf, Multilayered polypeptide films: secondary structures and effect of various stresses, Langmuir 19 (2003) 9873−9882. Available from: https://doi.org/10.1021/la0348259.

[69] E. Hübsch, G. Fleith, J. Fatisson, P. Labbé, J.C. Voegel, P. Schaaf, et al., Multivalent ion/polyelectrolyte exchange processes in exponentially growing multilayers, Langmuir 21 (2005) 3664−3669. Available from: https://doi.org/10.1021/la047258d.

[70] V. Ball, E. Hübsch, R. Schweiss, J.C. Voegel, P. Schaaf, W. Knoll, Interactions between multivalent ions and exponentially growing multilayers: dissolution and exchange processes, Langmuir 21 (2005) 8526−8531. Available from: https://doi.org/10.1021/la050866o.

[71] J. Schmitt, T. Grünewald, G. Decher, P. Pershan, K. Kjaer, M. Lösche, Internal structure of layer-by-layer adsorbed polyelectrolyte films: a neutron and X-ray reflectivity study, Macromolecules 26 (1993) 7058−7063.

[72] O. Soltwedel, O. Ivanova, P. Nestler, M. Müller, R. Köhler, C.A. Helm, Interdiffusion in polyelectrolyte multilayers, Macromolecules 43 (2010) 7288−7293. Available from: https://doi.org/10.1021/ma101279q.

[73] P.J. Yoo, K.T. Nam, J. Qi, S.K. Lee, J. Park, A.M. Belcher, et al., Spontaneous assembly of viruses on multilayered polymer surfaces, Nat. Mater. 5 (2006) 234−240. Available from: https://doi.org/10.1038/nmat1596.

[74] P.J. Yoo, K.T. Nam, A.M. Belcher, P.T. Hammond, Solvent-assisted patterning of polyelectrolyte multilayers and selective deposition of virus assemblies, Nano Lett. 8 (2008) 1081−1089. Available from: https://doi.org/10.1021/nl073079f.

[75] B. Schoeler, N. Delorme, I. Doench, G.B. Sukhorukov, A. Fery, K. Glinel, Polyelectrolyte films based on polysaccharides of different conformations: effects on multilayer structure and mechanical properties, Biomacromolecules 7 (2006) 2065−2071. Available from: https://doi.org/10.1021/bm060378a.

[76] M. Schönhoff, V. Ball, A.R. Bausch, C. Dejugnat, N. Delorme, K. Glinel, et al., Hydration and internal properties of polyelectrolyte multilayers, Colloids Surf. A: Physicochem. Eng. Asp. 303 (2007) 14−29. Available from: https://doi.org/10.1016/j.colsurfa.2007.02.054.

[77] P.v Pavoor, A. Bellare, A. Strom, D. Yang, R.E. Cohen, Mechanical characterization of polyelectrolyte multilayers using quasi-static nanoindentation, Macromolecules 37 (2004) 4865−4871. Available from: https://doi.org/10.1021/ma049777t.

[78] W. Yuan, G.M. Weng, J. Lipton, C.M. Li, P.R. van Tassel, A.D. Taylor, Weak polyelectrolyte-based multilayers via layer-by-layer assembly: approaches, properties, and applications, Adv. Colloid Interface Sci. 282 (2020). Available from: https://doi.org/10.1016/j.cis.2020.102200.

[79] D. Volodkin, R. von Klitzing, H. Moehwald, Polyelectrolyte multilayers: toward single cell studies, Polym. (Basel) 6 (2014) 1502−1527. Available from: https://doi.org/10.3390/polym6051502.

[80] S. Dodoo, R. Steitz, A. Laschewsky, R. von Klitzing, Effect of ionic strength and type of ions on the structure of water swollen polyelectrolyte multilayers, Phys. Chem. Chem. Phys. 13 (2011) 10318−10325. Available from: https://doi.org/10.1039/c0cp01357a.

[81] K. Tang, N.A.M. Besseling, Formation of polyelectrolyte multilayers: ionic strengths and growth regimes, Soft Matter 12 (2016) 1032−1040. Available from: https://doi.org/10.1039/c5sm02118a.

[82] Z. Tang, N.A. Kotov, S. Magonov, B. Ozturk, Nanostructured artificial nacre, Nat. Mater. 2 (2003) 413−418. Available from: https://doi.org/10.1038/nmat906.

[83] M.K. Gheith, V.A. Sinani, J.P. Wicksted, R.L. Matts, N.A. Kotov, Single-walled carbon nanotube polyelectrolyte multilayers and freestanding films as a biocompatible platform for neuroprosthetic implants, Adv. Mater. 17 (2005) 2663−2670. Available from: https://doi.org/10.1002/adma.200500366.

[84] S. Srivastava, N.A. Kotov, Composite Layer-by-Layer (LBL) assembly with inorganic nanoparticles and nanowires, Acc. Chem. Res. 41 (2008) 1831−1841. Available from: https://doi.org/10.1021/ar8001377.

[85] E.v Lengert, S.I. Koltsov, J. Li, A.v Ermakov, B.v Parakhonskiy, E.v Skorb, et al., Nanoparticles in polyelectrolyte multilayer layer-by-layer (Lbl) films and capsules—key enabling components of hybrid coatings, Coatings 10 (2020) 1−28. Available from: https://doi.org/10.3390/coatings10111131.

[86] G. Francius, J. Hemmerlé, V. Ball, P. Lavalle, C. Picart, J.-C. Voegel, et al., Stiffening of soft polyelectrolyte architectures by multilayer capping evidenced by viscoelastic analysis of AFM indentation measurements, J. Phis. Chem. C. 111 (2007) 8299−8306. Available from: https://doi.org/10.1021/jp070435 + .

[87] K. Köhler, D.G. Shchukin, H. Möhwald, G.B. Sukhorukov, Thermal behavior of polyelectrolyte multilayer microcapsules. 1. The effect of odd and even layer number, J. Phys. Chem. B 109 (2005) 18250−18259. Available from: https://doi.org/10.1021/jp052208i.

[88] M. Zerball, A. Laschewsky, R. Köhler, R. von Klitzing, The effect of temperature treatment on the structure of polyelectrolyte multilayers, Polym. (Basel) 8 (2016) 1−16. Available from: https://doi.org/10.3390/polym8040120.

[89] L. Richert, A.J. Engler, D.E. Discher, C. Picart, L. Pasteur, Elasticity of native and cross-linked poly-electrolyte multilayer films, Biomacromolecules 5 (2004) 1908−1916. Available from: https://doi.org/10.1021/bm0498023.

[90] B. Li, D.T. Haynie, Multilayer biomimetics: reversible covalent stabilization of a nanostructured bio-film, Biomacromolecules 5 (2004) 1667−1670. Available from: https://doi.org/10.1021/bm0496155.

[91] A.L. Hillberg, C.A. Holmes, M. Tabrizian, Effect of genipin cross-linking on the cellular adhesion properties of layer-by-layer assembled polyelectrolyte films, Biomaterials 30 (2009) 4463−4470. Available from: https://doi.org/10.1016/j.biomaterials.2009.05.026.

[92] W. Youssef, R.R. Wickett, S.B. Hoath, Surface free energy characterization of vernix caseosa. Potential role in waterproofing the newborn infant, Skin. Res. Technol. 7 (2001) 10−17.

[93] B.G. Keselowsky, D.M. Collard, A.J. García, Surface chemistry modulates fibronectin conformation and directs integrin binding and specificity to control cell adhesion, J. Biomed. Mater. Res. 66 (2003) 247−259.

[94] L. Grausova, A. Kromka, L. Bacakova, S. Potocky, M. Vanecek, V. Lisa, Bone and vascular endothe-lial cells in cultures on nanocrystalline diamond films, Diam. Relat. Mater. 17 (2008) 1405−1409. Available from: https://doi.org/10.1016/j.diamond.2008.02.008.

[95] L. Grausova, L. Bacakova, A. Kromka, M. Vanecek, B. Rezek, V. Lisa, Molecular markers of adhe-sion, maturation and immune activation of human osteoblast-like MG 63 cells on nanocrystalline dia-mond films, Diam. Relat. Mater. 18 (2009) 258−263. Available from: https://doi.org/10.1016/j.diamond.2008.10.023.

[96] M.R. Bet, G. Goissis, S. Vargas, H.S. Selistre-De-Araujo, Cell adhesion and cytotoxicity studies over polyanio-nic collagen surfaces with variable negative charge and wettability, Biomaterials 24 (2003) 131−134.

[97] H. von Recum, A. Kikuchi, M. Yamato, Y. Sakurai, T. Okano, S.W. Kim, Growth factor and matrix molecules preserve cell function on thermally responsive culture surfaces, Tissue Eng. 5 (1999) 251−265.

[98] L. Fassina, L. Visai, M.G. Cusella De Angelis, F. Benazzo, G. Magenes, Surface modification of a porous polyurethane through a culture of human osteoblasts and an electromagnetic bioreactor, Technol. Health Care 5 (2007) 33−45.

[99] L. Khatib, D.E. Golan, M. Cho, Physiologic electrical stimulation provokes intracellular calcium increase mediated by phospholipase C activation in human osteoblasts, FASEB J. 18 (2004) 1903−1905. Available from: https://doi.org/10.1096/fj.04-1814fje.

[100] S. Sun, Y. Liu, S. Lipsky, M. Cho, Physical manipulation of calcium oscillations facilitates osteodiffer-entiation of human mesenchymal stem cells, FASEB J. 21 (2007) 1472−1480. Available from: https://doi.org/10.1096/fj.06-7153com.

[101] L. Bacakova, E. Filova, M. Parizek, T. Ruml, V. Svorcik, Modulation of cell adhesion, proliferation and differentiation on materials designed for body implants, Biotechnol. Adv. 29 (2011) 739−767. Available from: https://doi.org/10.1016/j.biotechadv.2011.06.004.

[102] R.L. Sammons, N. Lumbikanonda, M. Gross, P. Cantzler, Comparison of osteoblast spreading on microstructured dental implant surfaces and cell behaviour in an explant model of osseointegration: a scanning electron microscopic study, Clin. Oral. Implant. Res. 16 (2005) 657−666. Available from: https://doi.org/10.1111/j.1600-0501.2005.01168.x.

[103] M.M. Stanton, R.E. Ducker, J.C. MacDonald, C.R. Lambert, W. Grant McGimpsey, Super-hydrophobic, highly adhesive, polydimethylsiloxane (PDMS) surfaces, J. Colloid Interface Sci. 367 (2012) 502−508. Available from: https://doi.org/10.1016/j.jcis.2011.07.053.

[104] L. Kocgozlu, P. Lavalle, G. Koenig, B. Senger, Y. Haikel, P. Schaaf, et al., Selective and uncoupled role of substrate elasticity in the regulation of replication and transcription in epithelial cells, J. Cell Sci. 123 (2010) 29−39. Available from: https://doi.org/10.1242/jcs.053520.

[105] L. Richert, F. Boulmedais, P. Lavalle, J. Mutterer, E. Ferreux, G. Decher, et al., Improvement of stability and cell adhesion properties of polyelectrolyte multilayer films by chemical cross-linking, Biomacromolecules 5 (2004) 284−294. Available from: https://doi.org/10.1021/bm0342281.

[106] J. Zhang, B. Senger, D. Vautier, C. Picart, P. Schaaf, J.C. Voegel, et al., Natural polyelectrolyte films based on layer-by layer deposition of collagen and hyaluronic acid, Biomaterials 26 (2005) 3353−3361. Available from: https://doi.org/10.1016/j.biomaterials.2004.08.019.

[107] R. Zaidel-Bar, R. Milo, Z. Kam, B. Geiger, A paxillin tyrosine phosphorylation switch regulates the assembly and form of cell-matrix adhesions, J. Cell Sci. 120 (2007) 137−148. Available from: https://doi.org/10.1242/jcs.03314.

[108] N.E. Muzzio, M.A. Pasquale, D. Gregurec, E. Diamanti, M. Kosutic, O. Azzaroni, et al., Polyelectrolytes multilayers to modulate cell adhesion: a study of the influence of film composition and polyelectrolyte interdigitation on the adhesion of the A549 cell line, Macromol. Biosci. 16 (2016) 482−495.

[109] N.E. Muzzio, D. Gregurec, E. Diamanti, J. Irigoyen, M.A. Pasquale, O. Azzaroni, et al., Thermal annealing of polyelectrolyte multilayers: an effective approach for the enhancement of cell adhesion, Adv. Mater. Interfaces 4 (2017). Available from: https://doi.org/10.1002/admi.201600126.

[110] E. Diamanti, N. Muzzio, D. Gregurec, J. Irigoyen, M. Pasquale, O. Azzaroni, et al., Impact of thermal annealing on wettability and antifouling characteristics of alginate poly-l-lysine polyelectrolyte multilayer films, Colloids Surf. B: Biointerfaces 145 (2016) 328−337. Available from: https://doi.org/10.1016/j.colsurfb.2016.05.013.

[111] N.E. Muzzio, M.A. Pasquale, W.A. Marmisollé, C. von Bilderling, M.L. Cortez, L.I. Pietrasanta, et al., Self-assembled phosphate-polyamine networks as biocompatible supramolecular platforms to modulate cell adhesion, Biomater. Sci. 6 (2018) 2230−2247. Available from: https://doi.org/10.1039/c8bm00265g.

[112] N.E. Muzzio, M.A. Pasquale, E. Diamanti, D. Gregurec, M.M. Moro, O. Azzaroni, et al., Enhanced antiadhesive properties of chitosan/hyaluronic acid polyelectrolyte multilayers driven by thermal annealing: low adherence for mammalian cells and selective decrease in adhesion for Gram-positive bacteria, Mater. Sci. Eng. C. 80 (2017) 677−687. Available from: https://doi.org/10.1016/j.msec.2017.07.016.

[113] D. Yang, X. Lü, Y. Hong, T. Xi, D. Zhang, The molecular mechanism of mediation of adsorbed serum proteins to endothelial cells adhesion and growth on biomaterials, Biomaterials 34 (2013) 5747−5758. Available from: https://doi.org/10.1016/j.biomaterials.2013.04.028.

[114] C.J. Arias, R.L. Surmaitis, J.B. Schlenoff, Cell adhesion and proliferation on the "living" surface of a polyelectrolyte multilayer, Langmuir 32 (2016) 5412−5421. Available from: https://doi.org/10.1021/acs.langmuir.6b00784.

[115] M.B. Hovgaard, K. Rechendorff, J. Chevallier, M. Foss, F. Besenbacher, Fibronectin adsorption on tantalum: the influence of nanoroughness, J. Phys. Chem. B 112 (2008) 8241−8249. Available from: https://doi.org/10.1021/jp801103n.

[116] M.S. Lord, M. Foss, F. Besenbacher, Influence of nanoscale surface topography on protein adsorption and cellular response, Nano Today 5 (2010) 66−78. Available from: https://doi.org/10.1016/j.nantod.2010.01.001.

[117] B. Trappmann, J.E. Gautrot, J.T. Connelly, D.G.T. Strange, Y. Li, M.L. Oyen, et al., Extracellular-matrix tethering regulates stem-cell fate, Nat. Mater. 11 (2012) 642−649. Available from: https://doi.org/10.1038/nmat3339.

[118] N.E. Muzzio, M.A. Pasquale, X. Rios, O. Azzaroni, J. Llop, S.E. Moya, Adsorption and exchangeability of fibronectin and serum albumin protein corona on annealed polyelectrolyte multilayers and their consequences on cell adhesion, Adv. Mater. Interfaces 6 (2019). Available from: https://doi.org/10.1002/admi.201900008.

[119] S. Franz, S. Rammelt, D. Scharnweber, J.C. Simon, Immune responses to implants - a review of the implications for the design of immunomodulatory, biomaterials, Biomater. 32 (2011) 6692–6709. Available from: https://doi.org/10.1016/j.biomaterials.2011.05.078.

[120] M. Posaric-Bauden, K. Isaksson, D. Åkerberg, R. Andersson, B. Tingstedt, Novel anti-adhesive barrier Biobarrier reduces growth of colon cancer cells, J. Surg. Res. 191 (2014) 196–202. Available from: https://doi.org/10.1016/j.jss.2014.04.002.

[121] H. Hartmann, R. Krastev, Biofunctionalization of surfaces using polyelectrolyte multilayers, BioNanoMaterials 18 (2017). Available from: https://doi.org/10.1515/bnm-2016-0015.

[122] J.A. Hubbell, S.P. Massia, N.P. Desai, P.D. Drumheller, Endothelial selective cell adhesion to non fouling surfaces, Biothechnol 9 (1991) 568–572.

[123] Q.K. Lin, Y. Hou, K.F. Ren, J. Ji, Selective endothelial cells adhesion to Arg-Glu-Asp-Val peptide functionalized polysaccharide multilayer, Thin Solid. Films 520 (2012) 4971–4978. Available from: https://doi.org/10.1016/j.tsf.2012.03.041.

[124] B. Tao, Y. Deng, L. Song, W. Ma, Y. Qian, C. Lin, et al., BMP2-loaded titania nanotubes coating with pH-responsive multilayers for bacterial infections inhibition and osteogenic activity improvement, Colloids Surf. B: Biointerfaces 177 (2019) 242–252. Available from: https://doi.org/10.1016/j.colsurfb.2019.02.014.

[125] J. Davila, A. Chassepot, J. Longo, F. Boulmedais, A. Reisch, B. Frisch, et al., Cyto-mechanoresponsive polyelectrolyte multilayer films, J. Am. Chem. Soc. 134 (2012) 83–86. Available from: https://doi.org/10.1021/ja208970b.

[126] J.M. Silva, S.G. Caridade, N.M. Oliveira, R.L. Reis, J.F. Mano, Chitosan-alginate multilayered films with gradients of physicochemical cues, J. Mater. Chem. B 3 (2015) 4555–4568. Available from: https://doi.org/10.1039/c5tb00082c.

[127] M. Sailer, K. Lai Wing Sun, O. Mermut, T.E. Kennedy, C.J. Barrett, High-throughput cellular screening of engineered ECM based on combinatorial polyelectrolyte multilayer films, Biomaterials 33 (2012) 5841–5847. Available from: https://doi.org/10.1016/j.biomaterials.2012.05.001.

[128] L. Han, Z. Mao, J. Wu, Y. Guo, T. Ren, C. Gao, Directional cell migration through cell-cell interaction on polyelectrolyte multilayers with swelling gradients, Biomaterials 34 (2013) 975–984. Available from: https://doi.org/10.1016/j.biomaterials.2012.10.041.

[129] N. Itano, F. Atsumi, T. Sawai, Y. Yamada, O. Miyaishi, T. Senga, et al., Abnormal accumulation of hyaluronan matrix diminishes contact inhibition of cell growth and promotes cell migration, Proc. Natl. Sci. USA 99 (2002) 3609–3614. Available from: https://doi.org/10.1073/pnas.052026799.

[130] N.E. Muzzio, M.A. Pasquale, S.E. Moya, O. Azzaroni, Tailored polyelectrolyte thin film multilayers to modulate cell adhesion, Biointerphases 12 (2017) 04E403. Available from: https://doi.org/10.1116/1.5000588.

[131] T. Liu, Y. Wang, W. Zhong, B. Li, K. Mequanint, G. Luo, et al., Biomedical applications of layer-by-layer self-assembly for cell encapsulation: current status and future perspectives, Adv. Healthc. Mater. 8 (2019). Available from: https://doi.org/10.1002/adhm.201800939.

[132] J.T. Wilson, W. Cui, V. Kozlovskaya, E. Kharlampieva, D. Pan, Z. Qu, et al., Cell surface engineering with polyelectrolyte multilayer thin films, J. Am. Chem. Soc. 133 (2011) 7054–7064. Available from: https://doi.org/10.1021/ja110926s.

[133] W. Li, T. Guan, X. Zhang, Z. Wang, M. Wang, W. Zhong, et al., The effect of layer-by-layer assembly coating on the proliferation and differentiation of neural stem cells, ACS Appl. Mater. Interfaces 7 (2015) 3018–3029. Available from: https://doi.org/10.1021/am504456t.

[134] S. Mansouri, Y. Merhi, F.M. Winnik, M. Tabrizian, Investigation of layer-by-layer assembly of polyelectrolytes on fully functional human red blood cells in suspension for attenuated immune response, Biomacromolecules 12 (2011) 585–592. Available from: https://doi.org/10.1021/bm101200c.

[135] A.I. Zamaleeva, I.R. Sharipova, Av Porfireva, G.A. Evtugyn, R.F. Fakhrullin, Polyelectrolyte-mediated assembly of multiwalled carbon nanotubes on living yeast cells, Langmuir 26 (2010) 2671−2679. Available from: https://doi.org/10.1021/la902937s.

[136] J.T. Wilson, W. Cui, E.L. Chaikof, Layer-by-layer assembly of a conformal nanothin PEG coating for intraportal islet transplantation, Nano Lett. 8 (2008) 1940−1948. Available from: https://doi.org/10.1021/nl080694q.

[137] Y. Teramura, Y. Kaneda, H. Iwata, Islet-encapsulation in ultra-thin layer-by-layer membranes of poly (vinyl alcohol) anchored to poly(ethylene glycol)-lipids in the cell membrane, Biomaterials 28 (2007) 4818−4825. Available from: https://doi.org/10.1016/j.biomaterials.2007.07.050.

[138] M. Matsusaki, K. Fujimoto, Y. Shirakata, S. Hirakawa, K. Hashimoto, M. Akashi, Development of full-thickness human skin equivalents with blood and lymph-like capillary networks by cell coating technology, J. Biomed. Mater. Res. A 103 (2015) 3386−3396. Available from: https://doi.org/10.1002/jbm.a.35473.

[139] Y. Fukuda, T. Akagi, T. Asaoka, H. Eguchi, K. Sasaki, Y. Iwagami, et al., Layer-by-layer cell coating technique using extracellular matrix facilitates rapid fabrication and function of pancreatic β-cell spheroids, Biomaterials 160 (2018) 82−91. Available from: https://doi.org/10.1016/j.biomaterials.2018.01.020.

[140] K. Sasaki, T. Akagi, T. Asaoka, H. Eguchi, Y. Fukuda, Y. Iwagami, et al., Construction of three-dimensional vascularized functional human liver tissue using a layer-by-layer cell coating technique, Biomaterials 133 (2017) 263−274. Available from: https://doi.org/10.1016/j.biomaterials.2017.02.034.

[141] J. Wang, Y. Miao, Y. Huang, B. Lin, X. Liu, S. Xiao, et al., Bottom-up nanoencapsulation from single cells to tunable and scalable cellular spheroids for hair follicle regeneration, Adv. Healthc. Mater. 7 (2018). Available from: https://doi.org/10.1002/adhm.201700447.

[142] J. Chen, X. Qiu, L. Wang, W. Zhong, J. Kong, M.M.Q. Xing, Free-standing cell sheet assembled with ultrathin extracellular matrix as an innovative approach for biomimetic tissues, Adv. Funct. Mater. 24 (2014) 2216−2223. Available from: https://doi.org/10.1002/adfm.201302949.

[143] W. Wan, S. Zhang, L. Ge, Q. Li, X. Fang, Q. Yuan, et al., Layer-by-layer paper-stacking nanofibrous membranes to deliver adipose-derived stem cells for bone regeneration, Int. J. Nanomed. 10 (2015) 1273−1290. Available from: https://doi.org/10.2147/IJN.S77118.

[144] L. Ge, Q. Li, Y. Huang, S. Yang, J. Ouyang, S. Bu, et al., Polydopamine-coated paper-stack nanofibrous membranes enhancing adipose stem cells' adhesion and osteogenic differentiation, J. Mater. Chem. B 2 (2014) 6917−6923. Available from: https://doi.org/10.1039/c4tb00570h.

[145] A.J. Priya, S.P. Vijayalakshmi, A.M. Raichur, Enhanced survival of probiotic Lactobacillus acidophilus by encapsulation with nanostructured polyelectrolyte layers through layer-by-layer approach, J. Agric. Food Chem. 59 (2011) 11838−11845. Available from: https://doi.org/10.1021/jf203378s.

[146] G. Liu, L. Li, D. Huo, Y. Li, Y. Wu, L. Zeng, et al., A VEGF delivery system targeting MI improves angiogenesis and cardiac function based on the tropism of MSCs and layer-by-layer self-assembly, Biomaterials 127 (2017) 117−131. Available from: https://doi.org/10.1016/j.biomaterials.2017.03.001.

[147] M. Germain, P. Balaguer, J.C. Nicolas, F. Lopez, J.P. Esteve, G.B. Sukhorukov, et al., Protection of mammalian cell used in biosensors by coating with a polyelectrolyte shell, Biosens. Bioelectron. 21 (2006) 1566−1573. Available from: https://doi.org/10.1016/j.bios.2005.07.011.

[148] J. García-Alonso, R.F. Fakhrullin, V.N. Paunov, Z. Shen, J.D. Hardege, N. Pamme, et al., Microscreening toxicity system based on living magnetic yeast and gradient chips, Anal. Bioanal. Chem. 400 (2011) 1009−1013. Available from: https://doi.org/10.1007/s00216-010-4241-3.

[149] S. Hong, P.R. Leroueil, E.K. Janus, J.L. Peters, M.M. Kober, M.T. Islam, et al., Interaction of polycationic polymers with supported lipid bilayers and cells: nanoscale hole formation and enhanced membrane permeability, Bioconjug. Chem. 17 (2006) 728−734. Available from: https://doi.org/10.1021/bc060077y.

[150] D. Fischer, Y. Li, B. Ahlemeyer, J. Krieglstein, T. Kissel, In vitro cytotoxicity testing of polycations: influence of polymer structure on cell viability and hemolysis, Biomaterials 24 (2003) 1121−1131.

[151] D. Volodkin, R. von Klitzing, Competing mechanisms in polyelectrolyte multilayer formation and swelling: polycation-polyanion pairing vs. polyelectrolyte-ion pairing, Curr. Opin. Colloid Interface Sci. 19 (2014) 25−31. Available from: https://doi.org/10.1016/j.cocis.2014.01.001.

[152] D. Alkekhia, P.T. Hammond, A. Shukla, Layer-by-Layer biomaterials for drug delivery, Annu. Rev. Biomed. 22 (2020) 1−24. Available from: https://doi.org/10.1146/annurev-bioeng-060418.

[153] W. Li, X. Lei, H. Feng, B. Li, J. Kong, M. Xing, Layer-by-layer cell encapsulation for drug delivery: the history, technique basis, and applications, Pharmaceutics 14 (2022). Available from: https://doi.org/10.3390/pharmaceutics14020297.

[154] K. Ariga, R. Fakhrullin, K. Ariga, Nanoarchitectonics on living cells, RSC Adv. 11 (2021) 18898−18914. Available from: https://doi.org/10.1039/d1ra03424c.

[155] M.R. Dzamukova, E.A. Naumenko, Ev Rozhina, A.A. Trifonov, R.F. Fakhrullin, Cell surface engineering with polyelectrolyte-stabilized magnetic nanoparticles: a facile approach for fabrication of artificial multicellular tissue-mimicking clusters, Nano Res. 8 (2015) 2515−2532. Available from: https://doi.org/10.1007/s12274-015-0759-1.

Charged porphyrins as building blocks of π-electronic ion-pairing assemblies

Kazuhisa Yamasumi and Hiromitsu Maeda

Department of Applied Chemistry, College of Life Sciences, Ritsumeikan University, Kusatsu, Shiga, Japan

5.1 Introduction

Noncovalent interactions such as $\pi-\pi$ and hydrogen-bonding interactions play important roles in determining the arrangements of components in diverse fields, such as nanoarchitectonics [1]. Such interactions are represented by the fundamental intermolecular forces, namely long-range ones such as electrostatic, induction, and dispersion and short-range ones such as exchange-repulsion and charge-transfer forces (Fig. 5.1A) [2]. Most organic compounds assemble using van der Waals interactions, which mainly originate from dispersion forces that are indiscriminately attractive with a low selectivity. In contrast, electrostatic forces are attractive between oppositely charged species and, in most cases, repulsive between identically charged species [3,4]. The introduction of charges is an effective way to increase the contribution of electrostatic forces, which in turn cause drastic changes in intermolecular interactions, and thus in assembling behaviors. Additionally, multiple ion pair combinations can be derived from a small selection of oppositely charged species, e.g., 100 types of ion pairs can be formed from only 10 types of cations and anions.

Electronically neutral π-electronic systems form assemblies based on $\pi-\pi$ and $CH-\pi$ interactions, which are mainly due to dispersion force [5]. The introduction of charge into π-electronic systems increases the contribution of electrostatic forces, resulting in interactions between the charged π-electronic systems that cannot be strictly characterized as ordinary $\pi-\pi$ interaction; hence, we proposed the concept of $^i\pi-^i\pi$ interactions [6]. π-Electronic ion pairs can exhibit various assembling modes depending on their $^i\pi-^i\pi$ interactions. In general, attractive interactions between oppositely charged species result in charge-by-charge assemblies, where oppositely charged species alternately stack. Conversely, increased dispersion and reduced electrostatic forces promote the stacking of identically charged species, resulting in the formation of charge-segregated assemblies (Fig. 5.1B). Using a rational design, noncovalent interactions can be manipulated by modifying the constituent charged π-electronic systems, providing desired assembling modes and properties.

A variety of π-electronic cations have been investigated. Tropylium [7] and triazatriangulenium ($TATA^+$) [8] cations (Fig. 5.2A(i)) can be studied directly, as they form assemblies with neighboring entities, whereas stable π-electronic anions, e.g., pentacyanocyclopentadienide ($PCCp^-$) (Fig. 5.2A(ii) left) [9], form ion-pairing assemblies with countercations. However, electron-rich π-electronic anions are generally less stable due to their intrinsic reactivity toward electrophiles.

Materials Nanoarchitectonics. DOI: https://doi.org/10.1016/B978-0-323-99472-9.00012-2

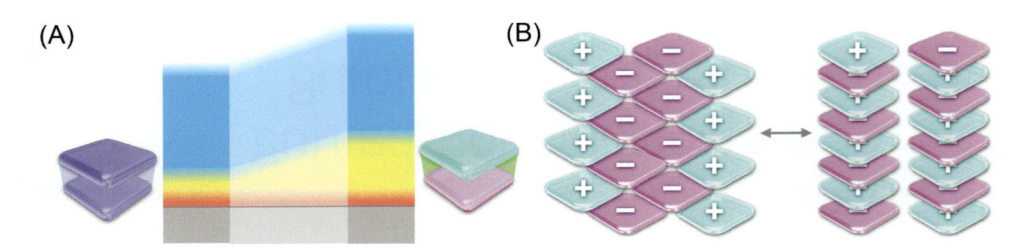

FIGURE 5.1

(A) Fundamental intermolecular forces constituting $\pi-\pi$ and $^i\pi-^i\pi$ interactions and (B) assembling modes of the π-electronic ion pairs. The blue, yellow, red, and gray parts in (A) indicate the contributions of dispersion, electrostatic, charge-transfer, and exchange repulsion forces to the interactions between two electrically neutral species (left) and two charged π-electronic systems (right).

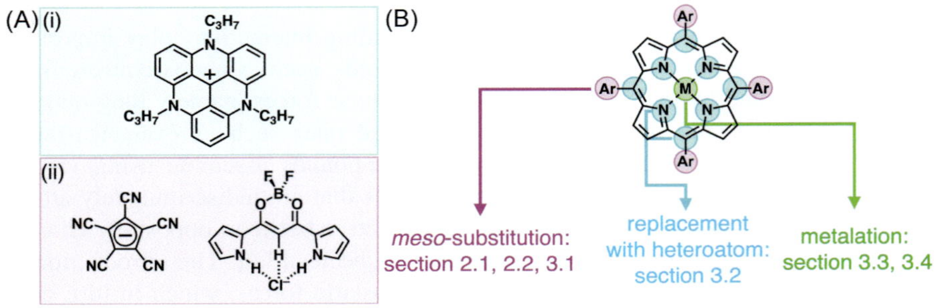

FIGURE 5.2

(A) Selected charged π-electronic systems as (i) triazatriangulenium cation (TATA$^+$) and (ii) pentacyanocyclopentadienyl anion (PCCp$^-$) and 1,3-dipyrrolyl-1,3-propandione BF$_2$ complex in the anion complex form and (B) porphyrin framework that is available for the charged π-systems by various modifications.

Therefore reports on stable π-electronic anions are limited, prompting the development of novel π-electronic anion systems. One approach utilizes anion complexation by anion-responsive π-electronic systems. For example, 1,3-dipyrrolyl-1,3-propandione BF$_2$ complexes form anion complexes via hydrogen bonding at the pyrrole NH and bridging CH units (Fig. 5.2A(ii) right) [4,10,11]. The anion complexes can be recognized as pseudo π-electronic anions, which form assemblies with countercations. Inspired by the geometries of the anion complexes, π-electronic anions whose anionic site is covalently linked to the π-systems have also been designed [12]. The obtained anions are stabilized by the hydrogen bonding of the deprotonated hydroxy group with two adjacent pyrrole NH units and the delocalization of the negative charge over entire π-systems.

Further effective charge delocalization over extended π-systems can result in more stable π-electronic ions. Porphyrin is a representative extended π-electronic system that forms various assemblies, mainly through $\pi-\pi$ interactions [13]. Porphyrin exhibits absorption bands, known as Soret and Q bands, in the UV−Vis region and emissions for some species in the visible region. Such electronic and photophysical properties are used in biological systems that exhibit photosynthesis. In addition, the electronic structures of porphyrins can be tuned by peripheral substituents, metal centers, and heteroatoms in the core macrocycles. Owing to their photophysical properties, porphyrin derivatives have also been adopted for nanoarchitectures [14]. The introduction of charged substituents, appropriate heteroatoms, and metal ions can afford charged porphyrinoids (Fig. 5.2B). The proximal location of the π-electronic systems with proper orientation, especially identical species, produces drastic changes in the photophysical properties due to exciton coupling [15]. Several review articles have summarized the assemblies of porphyrin derivatives with charged *meso*-aryl groups [16], and we have also recently reported the account on the charged porphyrin systems, wherein charges are delocalized over macrocycles [17]. This chapter includes the recent advancements in π-electronic ion-pairing assemblies comprising charged porphyrins.

5.2 **Porphyrins with charged substituents at the *meso* positions**

The introduction of charged substituents is the simplest approach to obtaining charged porphyrins. These species exhibit increased solubilities in aqueous media, enabling their application in biotic systems. Charged porphyrins are associated with various organs and biomolecules based on the amphiphilicity of charged and hydrophobic units [18]. It is worth noting that charged porphyrins also aggregate via electrostatic and π−π interactions along with hydrophobic effects. In this section, the self-aggregates of zwitterionic and charged porphyrins are summarized.

5.2.1 **Ion-pairing assemblies comprising zwitterionic porphyrins**

Sulfonatophenyl-substituted porphyrin derivatives exist as anionic species $1a^{4-}$ and $1b^{3-}$ in neutral aqueous media and undergo protonation at the two core N sites ($1a^{2-}$ and $1b^{-}$) in hydrochloric acid solutions, forming aggregates (Fig. 5.3A) [19,20]. Zwitterionic $1b^{-}$, consisting of a dicationic

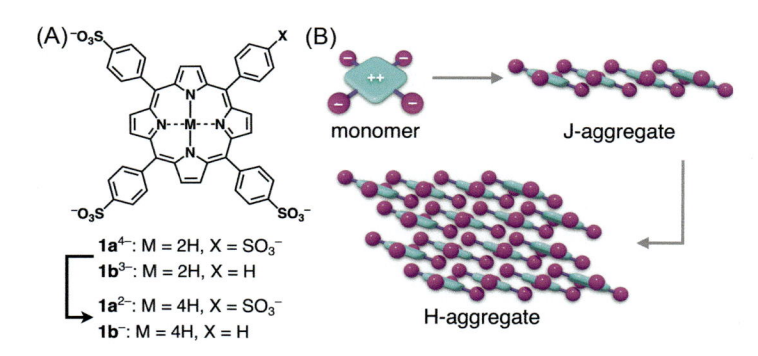

$1a^{4-}$: M = 2H, X = SO_3^-
$1b^{3-}$: M = 2H, X = H
$1a^{2-}$: M = 4H, X = SO_3^-
$1b^{-}$: M = 4H, X = H

FIGURE 5.3

(A) Sulfonatophenyl-substituted porphyrins and (B) schematic structures of the aggregates of $1a^{2-}$.

porphyrin core and three sulfonatophenyl groups, provided a J-aggregate with a red-shifted Soret band (Fig. 5.3B) [19]. $1a^{2-}$ also formed a J-aggregate, which was less thermodynamically stable owing to the larger electrostatic repulsion of the net charges [19]. Interestingly, the absorption band of $1b^-$ aggregate was blue-shifted at higher concentrations owing to the formation of higher-order H-aggregates [19]. In CH_2Cl_2, the tetrabutylammonium (TBA^+) salt of $1a^{4-}$ was protonated at the sulfonate groups and inner N sites, displaying complicated aggregation behaviors [19].

Bulk-state assemblies can be obtained by removing dissolution solvents. Ribó et al. reported film formation when evaporating drops of a $1a^{2-}$ aqueous solution, which was acidified by hydrochloric acid, at r.t [20]. The water contents in the film depended on the water vapor pressure, with higher vapor pressures resulting in a gel-like material. Notably, rotary evaporation of a dilute acidic aqueous solution of $1b^-$ resulted in J-aggregates, whose chirality depended upon the rotation direction [20]. The preparation of $1b^-$ aggregates under Na^+-free conditions decreased the aggregation rates, and helically folded ribbon-like structures and shape evolution were observed by atomic force microscopy (AFM) observation [20]. Knoester et al. reported J-aggregate formation on addition of hydrochloric acid into an aqueous solution of $1a^{4-}$, with a tubular structure being observed by cryogenic transmission electron microscopy (TEM) [20]. de Paula et al. reported the formation of rod-like aggregates upon the addition of hydrochloric acid into a phosphate buffer solution of $1a^{2-}$ [20]. The latter aggregates exhibited photoconductivity derived from the tightly packed arrangement of the porphyrin skeleton [20].

5.2.2 Ion-pairing assemblies comprising oppositely charged porphyrin derivatives

Various ion pairs comprising tetrakis(sulfonatophenyl)porphyrin with positively charged porphyrin derivatives have been reported for a variety of nanostructures [21]. An ion pair of $1a^{4-}$ and tetra(4-pyridiniumyl)porphyrin Sn^{IV} complex $2a_{Sn}^{4+}$ formed nanotubes having diameters in the range of 50−70 nm and wall thicknesses of ∼20 nm (Fig. 5.4A) [21]. The nanotubes catalyzed the reduction of Au^I species to Au^0 on the internal surface, causing gold nanowires to grow along the nanotubes [21]. Compared to the monomeric constituents, the suspensions of ion pairs exhibited a red-shifted absorption band indicating the formation of J-aggregates. The ion pairs of $1a^{4-}$ and tetrakis(4-N-(hydroxyethyl)pyridiniumyl)porphyrin $2b^{4+}$ metal complexes were also prepared, and they formed nanostructures with clover-like morphologies (Fig. 5.4B) [21]. In addition to the *meso*-substituents and preparation condition, metal centers (Zn^{II} and $Sn^{IV}(OH)_2$) of porphyrin skeletons also slightly affected the morphologies of the nanostructures. Suspensions of $2b_{Sn}^{4+}$-$1a_{Zn}^{4-}$ showed J-aggregate-like red-shifted Soret bands and photoconductivity in contrast to no such observation for $2b_{Zn}^{4+}$-$1a_{Sn}^{4-}$. The electron transfer was facilitated by the combinations of electron-deficient porphyrin Sn^{IV} complex with cationic substituents and electron-rich porphyrin Zn^{II} complex with anionic substituents, resulting in high photoconductivity [21]. The ion pair of tetrakis(N-methylpyridiniumyl)porphyrin $2c_{Sn}^{4+}$ and $1a_{Zn}^{4-}$ formed crystalline nanosheets (Fig. 5.5A) [21]. In the crystal structure, coordination of water and hydroxide to the metal centers hindered the stacking of π-systems, resulting in the absence of J-aggregate bands and photoconductivity. Free-base ion pair $2c^{4+}$-$1a^{4-}$ formed a

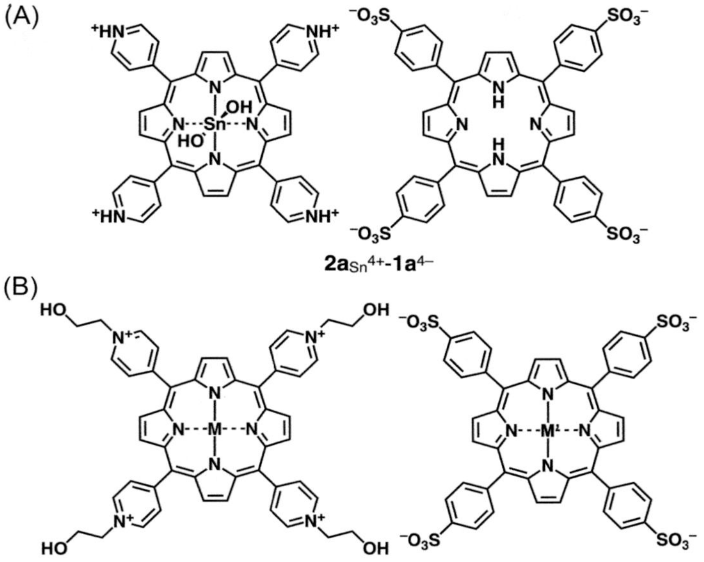

$2b_{Zn}^{4+}$-$1a_{Zn}^{4-}$: M = M' = Zn
$2b_{Sn}^{4+}$-$1a_{Sn}^{4-}$: M = M' = $Sn(OH)_2$
$2b_{Zn}^{4+}$-$1a_{Sn}^{4-}$: M = Zn, M' = $Sn(OH)_2$
$2b_{Sn}^{4+}$-$1a_{Zn}^{4-}$: M = $Sn(OH)_2$ M' = Zn

FIGURE 5.4

Chemical structures of (A) pyridiniumyl- and sulfonatophenyl-substituted porphyrins $\mathbf{2a}_{Sn}^{4+}$ and $\mathbf{1a}^{4-}$ and hydroxyethylpyridiniumyl- and sulfonatophenyl-substituted porphyrins metal complexes $\mathbf{2b}_{M}^{4+}$ and $\mathbf{1a}_{M'}^{4-}$ (M,M′ = Zn or Sn).

charge-by-charge assembly in the crystalline state (Fig. 5.5B) [21]. The absorption spectrum of the ion pair was similar to the sum of the individual spectra. The luminescence spectrum exhibited a significant shift, implying a charge transfer between charged species, causing photoconductivity.

5.3 Porphyrins charged at the porphyrin skeleton

In the abovementioned systems, the twisted geometries of *meso*-aryl groups and porphyrin skeletons hinder the effective delocalization of charges on the *meso*-aryl groups over the entire π-systems; however, the introduction of charges into porphyrin cores can modulate the electronic states. Charges can be introduced in the porphyrin cores by introducing (1) charged peripheral substituents, (2) heteroatom(s) into the macrocycles, and (3) valence-mismatched metal ions into the inner coordination site (Fig. 5.6).

(A) (i)
(ii)
2cSn4+-1aZn4−
1aZn4−
2cSn4+
(B) (i)
(ii)
2c4+-1a4−
3.76 Å
2c4+
1a4−

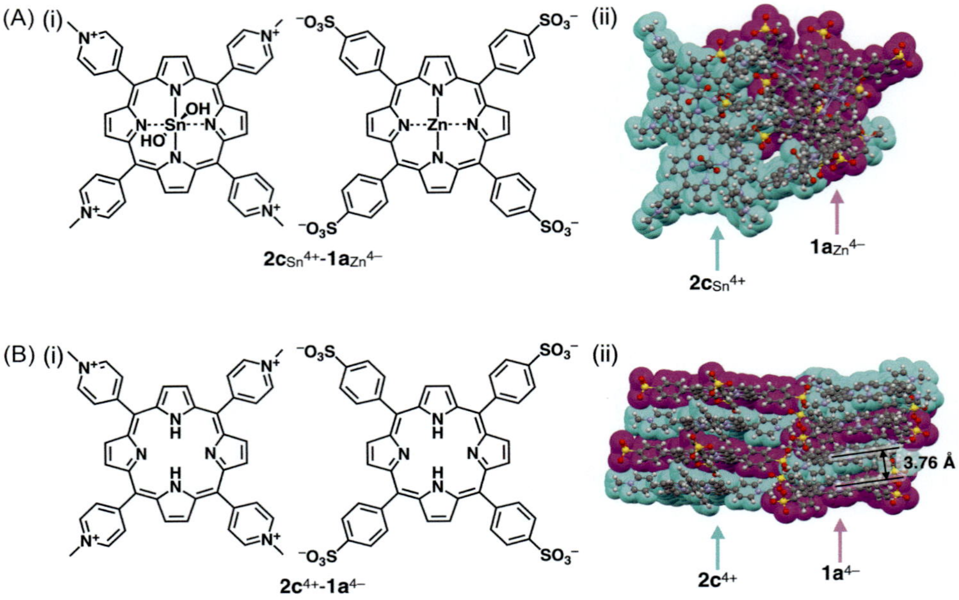

FIGURE 5.5

(A) (i) Chemical structures of methylpyridiniumyl- and sulfonatophenyl-substituted porphyrin metal complexes $\mathbf{2c}_{Sn}{}^{4+}$ and $\mathbf{1a}_{Zn}{}^{4-}$ and (B) free-base form $\mathbf{2c}^{4+}$ and $\mathbf{1a}^{4-}$ and (ii) the crystal structure, wherein cyan and magenta represent cationic and anionic parts, respectively. The color labels are also applied for the following figures.

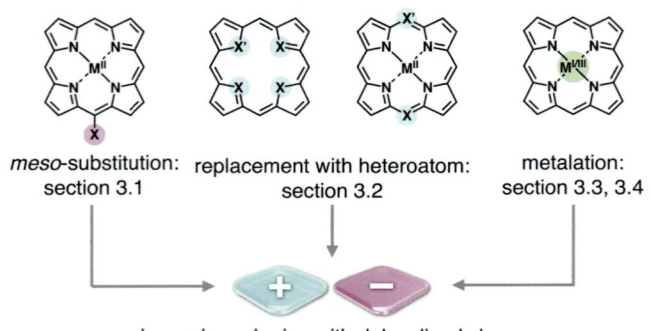

meso-substitution: section 3.1
replacement with heteroatom: section 3.2
metalation: section 3.3, 3.4
charged porphyrins with delocalized charge

FIGURE 5.6

Structural modifications of porphyrins to introduce charges into π-systems.

5.3.1 Negatively charged porphyrins formed by deprotonation

Deprotonation of acid units, including hydroxy groups, attached to π-systems, such as porphyrins, affords π-electronic anions, with negative charge delocalized over the π-systems (Fig. 5.7) [22]. Upon treatment with tetrabutylammonium hydroxide (TBAOH), *meso*-hydroxyporphyrin Ni^{II} complex $3a_{Ni}$ was deprotonated, affording the ion pair $TBA^+\text{-}3a_{Ni}^-$ (Fig. 5.8A) [22]. Changes in

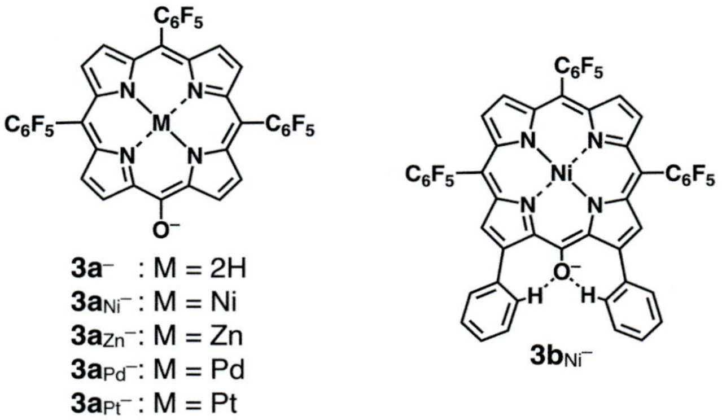

3a⁻ : M = 2H
3a_Ni⁻ : M = Ni
3a_Zn⁻: M = Zn
3a_Pd⁻: M = Pd
3a_Pt⁻ : M = Pt

3b_Ni⁻

FIGURE 5.7

Deprotonated *meso*-hydroxy-substituted porphyrin derivatives.

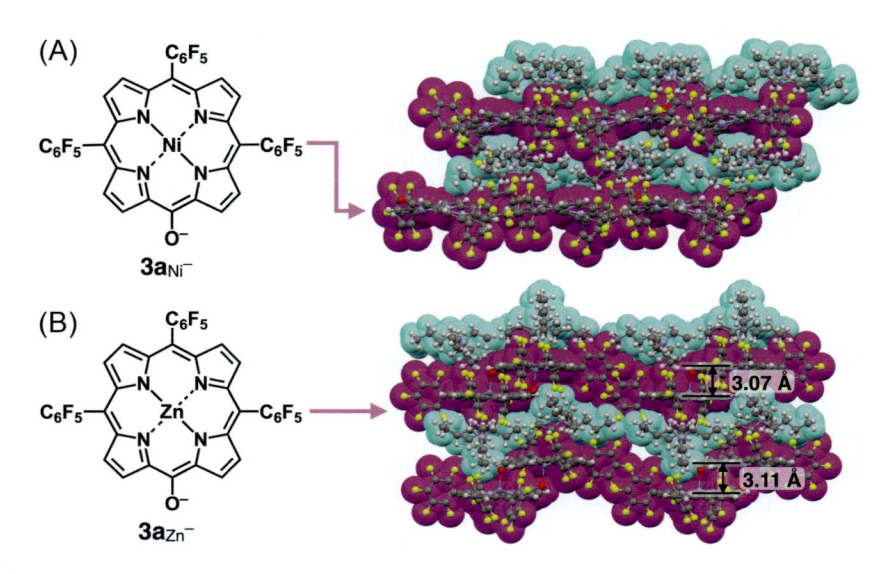

FIGURE 5.8

Single-crystal X-ray structures of ion-pairing assemblies comprising deprotonated *meso*-hydroxyporphyrins: (A) $TBA^+\text{-}3a_{Ni}^-$ and (B) $TBA^+\text{-}3a_{Zn}^-$.

UV−Vis absorption and ^{1}H NMR spectra and the electrostatic potential (ESP) map indicated the delocalization of negative charge and significant effects on the electronic states. Each component was arranged linearly, and the resulting belt-like structures had an alternating pattern. The ion pairs comprising Pd^{II} and Pt^{II} complexes $3a_{Pd}^-$ and $3a_{Pt}^-$ displayed assembling modes similar to that of TBA^+-$3a_{Ni}^-$, whereas the Zn^{II} complex $3a_{Zn}^-$ was stacked onto another $3a_{Zn}^-$ unit by the coordination of the oxide site with the Zn^{II} center (Fig. 5.8B) [6]. The stacked dimer and two TBA^+ entities were alternately arranged, forming a [2 + 2]-type charge-by-charge assembly. The introduction of phenyl groups at the neighboring β-positions as hydrogen-bond donors enabled the stabilization of the anion form $3b_{Ni}^-$, whose ion pair with TBA^+ formed a charge-by-charge assembly [22].

5.3.2 Heteroatom-containing charged porphyrin analogs

Incorporating heteroatoms such as oxygen and sulfur into the macrocyclic core resulted in positively charged porphyrin analogs with 18π aromaticity (Fig. 5.9) [23]. Tetraoxaporphyrin dication ($4a^{2+}$) was prepared as a moderately stable ClO_4^- ion pair. The X-ray result of $4a^{2+}$-$2ClO_4^-$ indicated a highly planar structure and straight columnar structure, showing $4a^{2+}$ slip-stacked with two neighboring $4a^{2+}$ units on both faces. Moreover, the aromatic nature of $4a^{2+}$ was supported by its diamagnetic ring-current effect in the ^{1}H NMR spectrum [23]. The reaction of $4a^{2+}$ with H_2S under acidic conditions provided tetrathiaporphyrin $4b^{2+}$. The crystal structure of $4b^{2+}$-$2ClO_4^-$ deviated more than that of $4a^{2+}$, culminating in lower aromaticity, evidently observed by ^{1}H NMR [23]. Further structural deviation of $4c^{2+}$ resulted in weaker aromaticity. Improved stability and solubility were observed for *meso*-phenyl ion pair $4d^{2+}$-$2ClO_4^-$ [23]. Ion-pair metathesis with $LiB(C_6F_5)_4$ afforded the ion pair $4d^{2+}$-$B(C_6F_5)_4^-$, which showed no stacking of $4d^{2+}$ due to the bulky counteranion. Trioxaporphyrin $4e^{2+}$ existed as a dication with an NH proton forming hydrogen bonds with oxygens [23]. In the ion-pairing assembly with $SbCl_6^-$, two $SbCl_6^-$ were located on both sides of the π-plane of $4e^{2+}$, affording a [2 + 1]-type charge-by-charge assembly.

Replacement of *meso*-carbons with heteroatoms also afforded positively charged 18π systems [24]. 5-Oxa-octaethylporphyrin cation 5^+ was initially obtained as an ion pair with OTf^-, but exchange via ion-pair metathesis with $NaBAr^F_4$ ($BAr^F_4^-$: tetrakis(3,5-bis(trifluoromethyl)phenyl) borate) afforded the ion pair 5^+-$BAr^F_4^-$ [24]. The replacement of the two *meso*-carbons of the tetraarylporphyrin Ni^{II} complex with nitrogen yielded dicationic 6_{Ni}^{2+}, which provided crystal-state charge-by-charge assemblies in the ion pairs with PF_6^- and $SbCl_6^-$ [24].

5.3.3 Positively charged porphyrin metal complexes with valence mismatch

Porphyrins contain two NH protons, thus serving as divalent ligands for square planar metal complexes. This is particularly true for electronically neutral species with divalent metal ions. However, metalation of porphyrins with trivalent metal ions results in the formation of cationic species because of the partial compensation of the positive charge of the metal ions (Fig. 5.10). The remaining positive charge can be compensated by axial ligands that coexist as counteranions. For example, porphyrin Fe^{III} complex 7_{Fe}^+, with a metal in the d^6 electronic state, forms pentacoordinated complexes via an axial ligand [25]. Therefore bulky substituents and counteranions are required to form π-electronic cations by hampering the axial coordination. Subporphyrin is a macrocycle with a divalent tridentate coordination site and is obtained as a boron complex with a

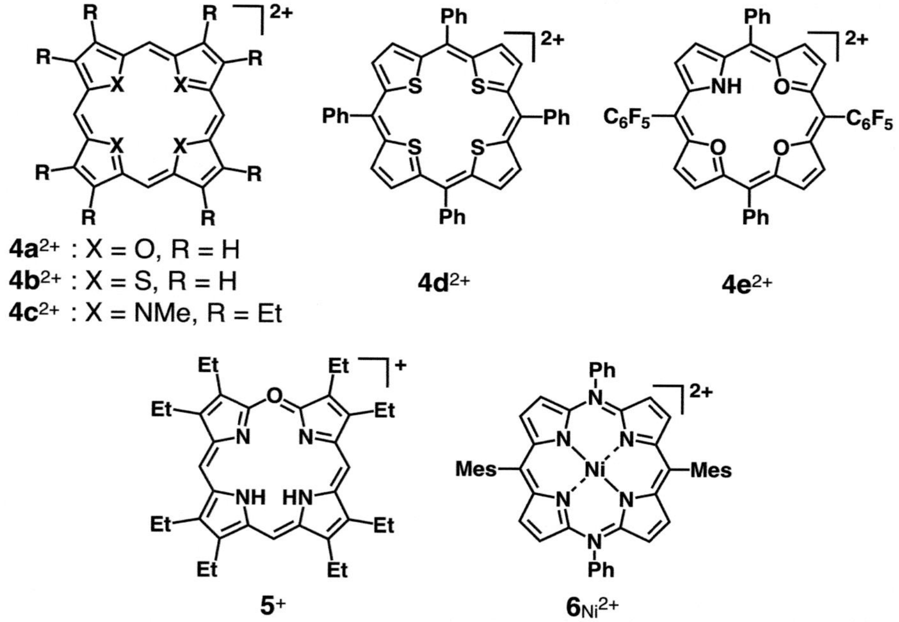

4a²⁺ : X = O, R = H
4b²⁺ : X = S, R = H
4c²⁺ : X = NMe, R = Et

4d²⁺

4e²⁺

5⁺

6ₙᵢ²⁺

FIGURE 5.9

Heteroatom-containing positively charged porphyrin analogs.

negatively charged axial ligand [26]. The methoxy group on the B^{III} center can be removed by treatment with $Et_3Si[CH_6B_{11}Br_6]$, giving a planar subporphyrin cation as the $CH_6B_{11}Br_6^-$ ion pair 8_B^+-$CH_6B_{11}Br_6^-$ [26]. Thiaporphyrin, a porphyrin analog with one internal NH replaced by sulfur, serves as monovalent ligand, and metal complexes with divalent metal ion in a d^8 electronic state can form π-electronic cations. Tetraphenylthiaporphyrin **9** formed a Ni^{II} complex with the thiophene unit slightly distorted from the mean plane of the macrocycle, providing a pseudo-square pyramidal complex with an axial Cl^- ligand [27]. Cl^- was removed by treatment with $LiB(C_6F_5)_4$, and the π-electronic ion pair 9_{Ni}^+-$B(C_6F_5)_4^-$ was obtained.

Trivalent gold ions in d^8 electronic state form tetracoordinated square planar complexes without axial ligands. Therefore porphyrin Au^{III} complexes behave as planar π-electronic cations [28]. Notably, various functions based on the porphyrin core and peripheral modifications of Au^{III} complexes can be added to ion-pairing assemblies [29]. Tetraphenylporphyrin Au^{III} complex $10a_{Au}^+$ was prepared as an ion pair with Cl^- [29]. The ESP map of $10a_{Au}^+$ showed the delocalization of the positive charge over the macrocycle. The counteranion was exchanged via ion-pair metathesis, utilizing salts containing Li^+, Na^+, and Ag^+. The ^1H NMR spectra of the ion pairs with Cl^-, BF_4^-, and PF_6^- were almost identical, whereas that of the ion pair with $PCCp^-$ was slightly shifted downfield. In two crystal pseudo-polymorphs for $10a_{Au}^+ \cdot Cl^-$, the $Au \cdots Cl$ distances (3.00 and 3.12 Å), which were larger than the sum of the ionic radii (2.66 Å), and the deviation of Cl^- from the axial position of the Au^{III} center suggested the formation of a contact ion pair without

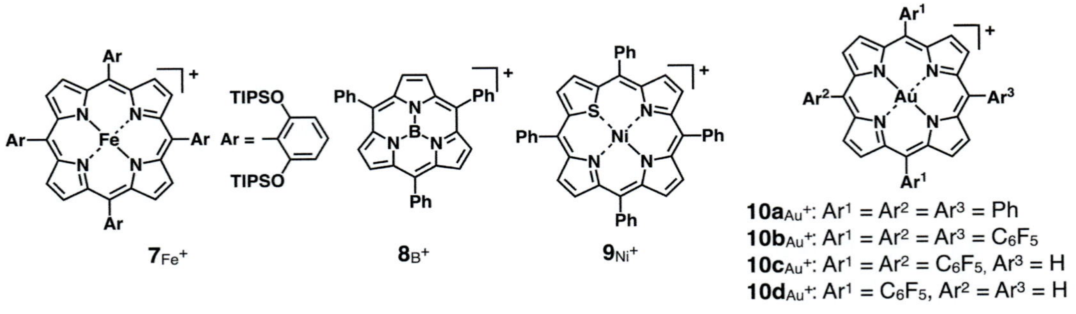

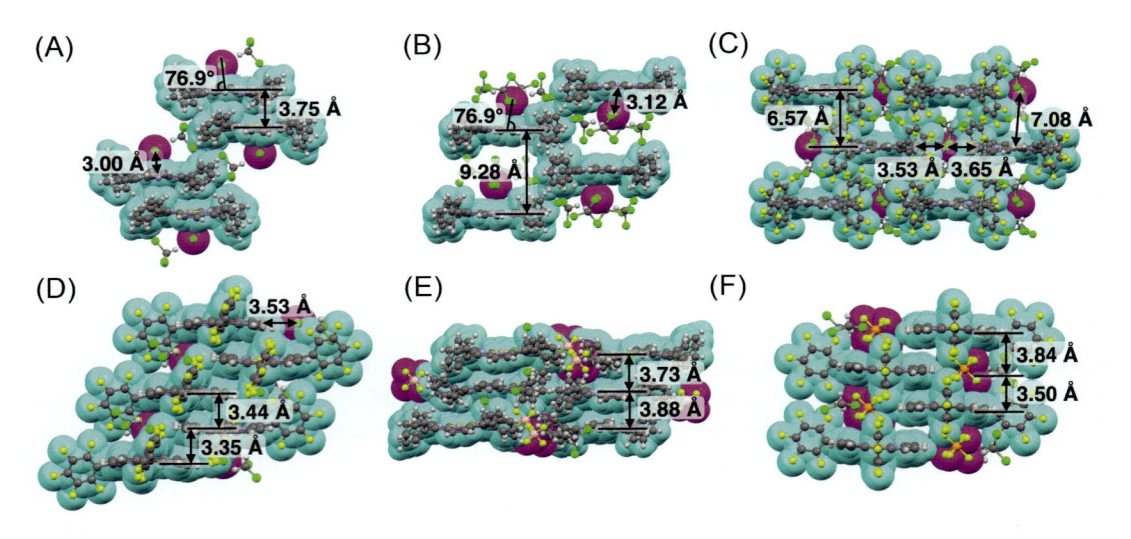

FIGURE 5.10

Positively charged porphyrin formed by valence-mismatch.

FIGURE 5.11

Single-crystal X-ray structures of ion-pairing assemblies comprising porphyrin Au^{III} complexes: (A,B) $10a_{Au}{}^+$-Cl^- (two pseudo-polymorphs), (C) $10b_{Au}{}^+$-Cl^-, (D) $10c_{Au}{}^+$-Cl^-, (E) $10a_{Au}{}^+$-$BF_4{}^-$, and (F) $10c_{Au}{}^+$-$PF_6{}^-$.

Au−Cl coordination (Fig. 5.11A,B). One and four $CHCl_3$ molecules, included in the pseudo-polymorphs, were located around Cl^- via C−H···Cl hydrogen bonding. In the former structure, $10a_{Au}{}^+$ was stacked with another $10a_{Au}{}^+$ with an interplanar distance of 3.75 Å, whereas in the latter structure $10a_{Au}{}^+$ alternated with one Cl^- that was surrounded by four $CHCl_3$ molecules to form a charge-by-charge assembly [29].

Assembling modes can be tuned by *meso*-substituents. Pentafluorophenyl-substituted porphyrin Au^{III} complexes $10b−d_{Au}{}^+$ were prepared as OTf^- ion pairs, which were converted to the Cl^- ion pairs by ion-exchange resin [29]. The crystal structure of the tetrasubstituted $10b_{Au}{}^+$ as a Cl^- ion pair contained four $CHCl_3$ molecules hydrogen bonded to Cl^- (Fig. 5.11C). Cl^- was located at the side of $10b_{Au}^+$ with a $C(-H)\cdots Cl$ distance of 3.648 Å at β-CH and was completely isolated from the Au^{III} center due to $CHCl_3$. Contact of $10b_{Au}^+$ with adjacent $10b_{Au}^+$ at the *meso*-C_6F_5 group occurred without the direct stacking of the π-systems. In the crystal structure of $10c_{Au}^+$-Cl^-, trisubstituted $10c_{Au}^+$ stacked with an adjacent $10c_{Au}^+$ shows relatively short contacts of 3.35 and 3.44 Å in the π-systems due to the absence of a *meso*-C_6F_5 group, resulting in a columnar structure (Fig. 5.11D) [29].

With respect to the hard and soft acids and bases (HSAB) theory, ion-pair metathesis of the Cl^- ion pairs with salts containing soft anions with hard cations resulted in porphyrin Au^{III} complexes as ion pairs with softer counteranions due to the removal of hard inorganic salts [29]. Ion-pairing assemblies with bulky counteranions changed the assembling behaviors: for $10a_{Au}^+$-BF_4^- and $10a_{Au}^+$-PF_6^-, the ion-pairing assemblies formed slip-stacked structures of $10a_{Au}^+$ (Fig. 5.11E) [29]. In contrast, $10c_{Au}^+$-PF_6^- showed the edge-on stacking of $10c_{Au}^+$ with the charge-by-charge arrangement of PF_6^- (Fig. 5.11F) [29].

5.3.4 Negatively charged porphyrin metal complexes with valence mismatch

The positive charges of high-valent metals in porphyrin complexes are only partially compensated by the divalent porphyrin ligands on formation with π-electronic cations. The negative charges of the divalent porphyrin ligands are also only partially compensated by low-valent metal complexes, resulting in π-electronic anions (Fig. 5.12). For example, the Li^I complex of octaethylporphyrin 11_{Li}^- serves as a π-electronic anion that is included in the ion-pairing assembly with $[Li(thf)_4]^+$, which has an alternating arrangement with 11_{Li}^- [30]. Similarly, corrole, a contracted porphyrin analog whose one *meso*-carbon was removed, serves as a trivalent ligand. Thus divalent metal complexes can behave as π-electronic anions. Pd^{II} metalation of tris(pentafluorophenyl)corrole in the presence of pyridine afforded the corresponding Pd^{II} complex 12_{Pd}^- as a pyridinium ion pair, which was unstable and gradually decomposed [31]. Stability was improved by the methylation of pyridine, giving the corresponding ion-pairing assembly of $PyCH_3^+$-12_{Pd}^-. Deprotonation at the *meso*-hydroxy group of **3a**, in addition to the two NH protons, afforded a trivalent ligand. Therefore divalent metal complexes can be recognized as π-electronic anions formed by valence mismatch of the metal centers and the ligands.

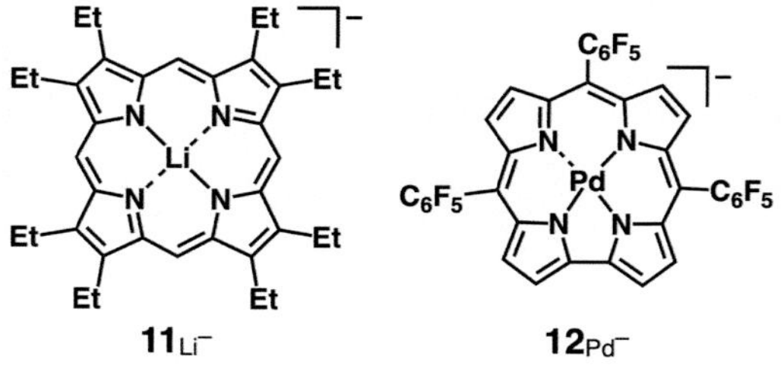

FIGURE 5.12

Negatively charged porphyrin formed by valence-mismatch.

5.4 π-Electronic ion-pairing assemblies of charged porphyrins

Assembling behaviors of ion pairs containing charged π-electronic systems depend on the charged constituents. Positively and negatively charged π-electronic systems are significantly electron-deficient and electron-rich, respectively. The characteristic electronic states can increase the contributions from electrostatic forces into the interactions between charged π-electronic systems, inducing unique assembling behaviors. In this section, recent advancements on π-electronic ion pairs comprising deprotonated *meso*-hydroxyporphyrin and porphyrin Au[III] complexes are shown.

5.4.1 Negatively charged porphyrins with π-electronic cations

TATA[+] ion pairs with **3a**[−] and **3a**$_{Pd}$[−] formed charge-by-charge assemblies in crystalline states, whereas the ion-pairing assembly with **3a**$_{Ni}$[−] showed an intermediate structure of charge-segregated and charge-by-charge assemblies (Fig. 5.13) [6]. Energy decomposition analysis (EDA) of the crystal structure of TATA[+]-**3a**$_{Ni}$[−] suggested the presence of significant attractive interactions between oppositely charged species (E_{tot} = −192.4 kcal/mol) consisting of electrostatic (E_{es} = −69.7 kcal/mol) and dispersion forces (E_{disp} = −141.9 kcal/mol). The total interaction in TBA[+]-**3a**$_{Ni}$[−] (E_{tot} = −169.0 kcal/mol) was less than that in TATA[+]-**3a**$_{Ni}$[−] due to the decreased contribution of dispersion forces (E_{disp} = −111.1 kcal/mol, E_{es} = −68.3 kcal/mol), suggesting the advantage of using the interaction between oppositely charged π-electronic systems ($^i\pi-^i\pi$ interaction). The absorption spectrum of TATA[+]-**3a**$_{Ni}$[−] in solution state was similar to the additive spectra of TATA[+]-Cl[−] and TBA[+]-**3a**$_{Ni}$[−]. In contrast, in the crystalline state, the absorption bands derived from TATA[+] and **3a**$_{Ni}$[−] were red- and blue-shifted, respectively, due to the exciton coupling between identically charged species (arrangement of transition dipole moments) (Fig. 5.14).

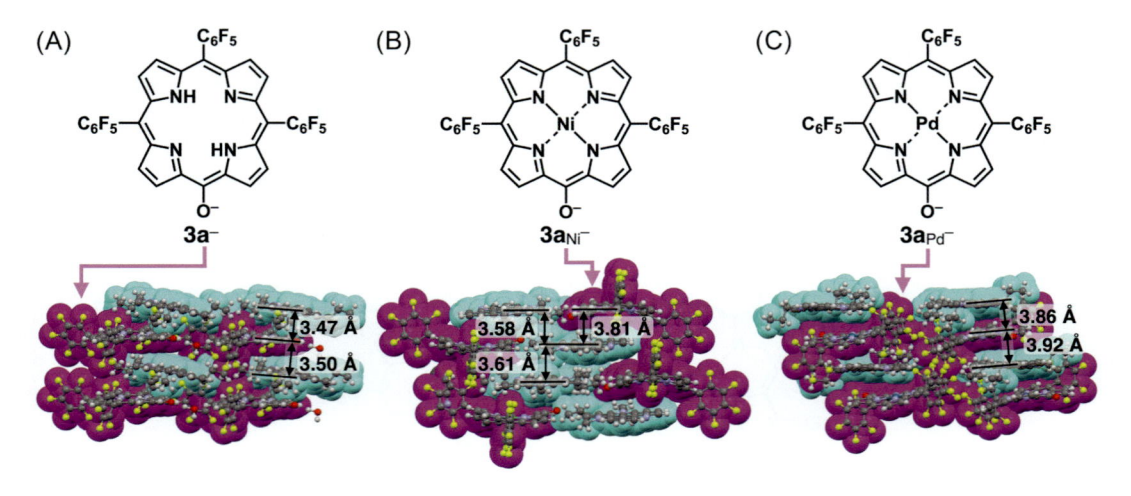

FIGURE 5.13

Single-crystal X-ray structures of ion-pairing assemblies comprising deprotonated hydroxyporphyrins and TATA[+]: (A) TATA[+]-**3a**[−], (B) TATA[+]-**3a**$_{Ni}$[−], and (C) TATA[+]-**3a**$_{Pd}$[−].

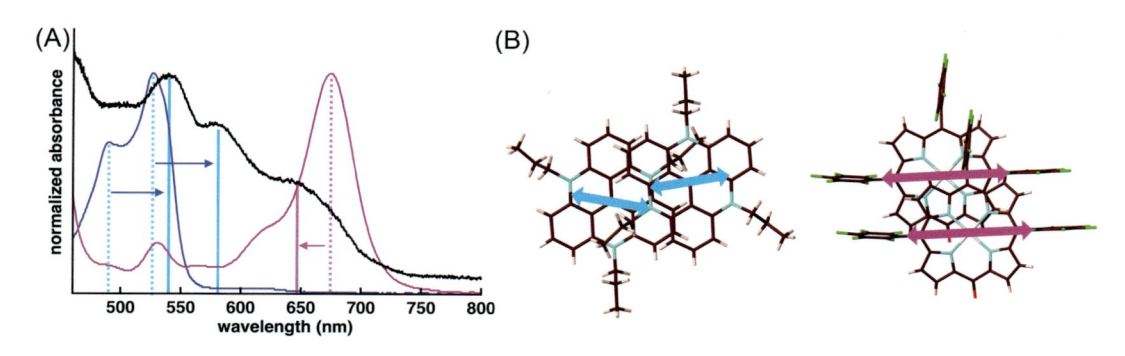

FIGURE 5.14

(A) UV–Vis absorption spectrum of TATA$^+$-**3a**$_{Ni}^-$ in the solid state (black) and those of TATA$^+$-Cl$^-$ (blue) and TBA$^+$-**3a**$_{Ni}^-$ (magenta) in CH$_2$Cl$_2$ and (B) the stacking dimers of **3a**$_{Ni}^-$ and TATA$^+$ in the crystal and the orientations of the transition dipole moments of each ion.

In addition, upon photoexcitation, transient absorption spectrum indicated the production of **3a**$_{Ni}^-$, suggesting photoinduced electron transfer between π-electronic ions [6].

5.4.2 Porphyrin Au^{III} complexes with π-electronic receptor–Cl$^-$ complexes

Ion-pair metathesis is an effective approach to forming π-electronic ion pairs. However, it requires the preparation of inorganic ion salts of desired π-electronic ions in advance. On the other hand, combination of anion receptors, such as dipyrrolyldiketone boron complexes, and π-electronic cations as a Cl$^-$ ion pair causes the formation of π-electronic ion pairs with anion complexes, comprising π-electronic receptors and Cl$^-$, and π-electronic cations (Fig. 5.15). The ion-pairing assemblies of TATA$^+$ and Cl$^-$ complexes were obtained by the treatment of TATA$^+$-Cl$^-$ with **13a,c,d,e**, exhibiting diverse assembling modes (Fig. 5.16) [11]. The synthesis strategy of the TATA$^+$ ion pairs can be applied to porphyrin Au^{III} complexes [32]. Ion-pairing assembly of **10a**$_{Au}^+$-**13a** · Cl$^-$ was prepared by mixing **10a**$_{Au}^+$-Cl$^-$ and **13a** (Fig. 5.16A). This mixing formed a [2 + 2]-type tetrad due to the two **10a**$_{Au}^+$ and two **13a** · Cl$^-$ species stacked in an almost parallel arrangement. In the case of β-methyl **13b**, **10a**$_{Au}^+$ was stacked with **13b** · Cl$^-$, and the stacked ion pair was further slip-stacked, resulting in a brickwork-like structure (Fig. 5.16B). β-Fluorinated **13c** · Cl$^-$ was alternately stacked with **10a**$_{Au}^+$, forming a charge-by-charge assembly (Fig. 5.16C). In contrast to **13a–c** · Cl$^-$ positioned just above **10a**$_{Au}^+$ in the crystal structures, phenylethynyl and phenyltriazolyl **13d,e** · Cl$^-$ were slip-stacked with **10a**$_{Au}^+$ on both sides, yielding tilted columnar structures (Fig. 5.16D,E) [14,32]. EDA for **10a**$_{Au}^+$-**13e** · Cl$^-$ revealed the significant interaction between the oppositely charged π-electronic systems (E_{tot} = −232.65 kcal/mol), mainly derived from large dispersion and small electrostatic forces (E_{disp} = −187.63 kcal/mol, E_{es} = −66.66 kcal/mol) [14]. Interestingly, **13c** formed a [2 + 2]-type anion complex in the ion-pairing assembly with **10c**$_{Au}^+$-Cl$^-$, where **10c**$_{Au}^+$ was stacked with another **10c**$_{Au}^+$ on the unsubstituted side (Fig. 5.16F). The stacked **10c**$_{Au}^+$ dimer and the dianionic anion complex were alternately stacked to form a [2 + 2]-type charge-by-charge assembly [32].

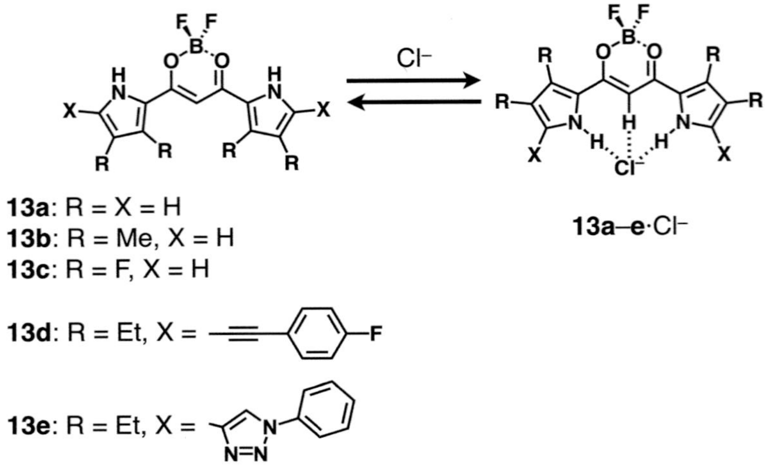

13a: R = X = H
13b: R = Me, X = H
13c: R = F, X = H

13a–e·Cl⁻

13d: R = Et, X = ═══⟨⟩─F

13e: R = Et, X =

FIGURE 5.15

Anion-binding behavior of dipyrrolyldiketone BF$_2$ complexes.

The diketone unit of 1,3-dipyrrolyl-1,3-propandiones formed various metal complexes as seen in Pt^{II} complexes, which also served as anion receptors providing Cl⁻ complexes that were adopted as the counterparts of the porphyrin Au^{III} complexes in ion-pairing assemblies (Fig. 5.17) [33,34]. An ion pair of **14a · Cl⁻** and **10b$_{Au}$⁺** formed a charge-by-charge columnar structure containing two independent structures (Fig. 5.17A). Interestingly, the phenylpyridine unit of **14a · Cl⁻** participated in stacking, and the anion-binding site was located at the side of the column. Short contact between Cl⁻ and the β-CH sites of **14a** and **10b$_{Au}$⁺** in adjacent columns indicated the interactions between the columns, with the C(−H)···Cl⁻ distances of 3.35 and 3.60/3.67 Å, respectively. Contrastingly, the ion-pairing assembly of **10a$_{Au}$⁺-14b · Cl⁻** contained two independent structures of **10a$_{Au}$⁺**, one of which was stacked with the phenylpyridine units of **14b · Cl⁻** on both faces and the other was stacked with Cl⁻-complex units on both faces (Fig. 5.17B) [33].

5.4.3 Porphyrin Au^{III} complexes with π-electronic anions

Porphyrin Au^{III} complex ion pairs having π-electronic anions were formed by ion-pair metathesis with Na⁺ salts. Dipyrrolylnitrophenoxide **15a⁻** was prepared as a Na⁺ ion pair by treating the corresponding phenol **15a** with an excess amount of NaOH [12]. The hard countercation (Na⁺) was excluded from the system by pairing with the hard anion in the **10a$_{Au}$⁺** ion pair verified by HSAB theory. In the crystal structure, **15a⁻** was stacked just above **10a$_{Au}$⁺** in relatively short interplanar distances (3.21 and 3.27 Å) in a charge-by-charge assembly (Fig. 5.18A). All ion pairs of **15a–c⁻** with **10a$_{Au}$⁺** formed charge-by-charge assemblies, and the interplanar distances were elongated by the introduction of substituents on the pyrrole units (Fig. 5.18A−C). The differences in the interplanar distances were small (0.01−0.06 Å) compared to those in the ion-pairing assemblies of

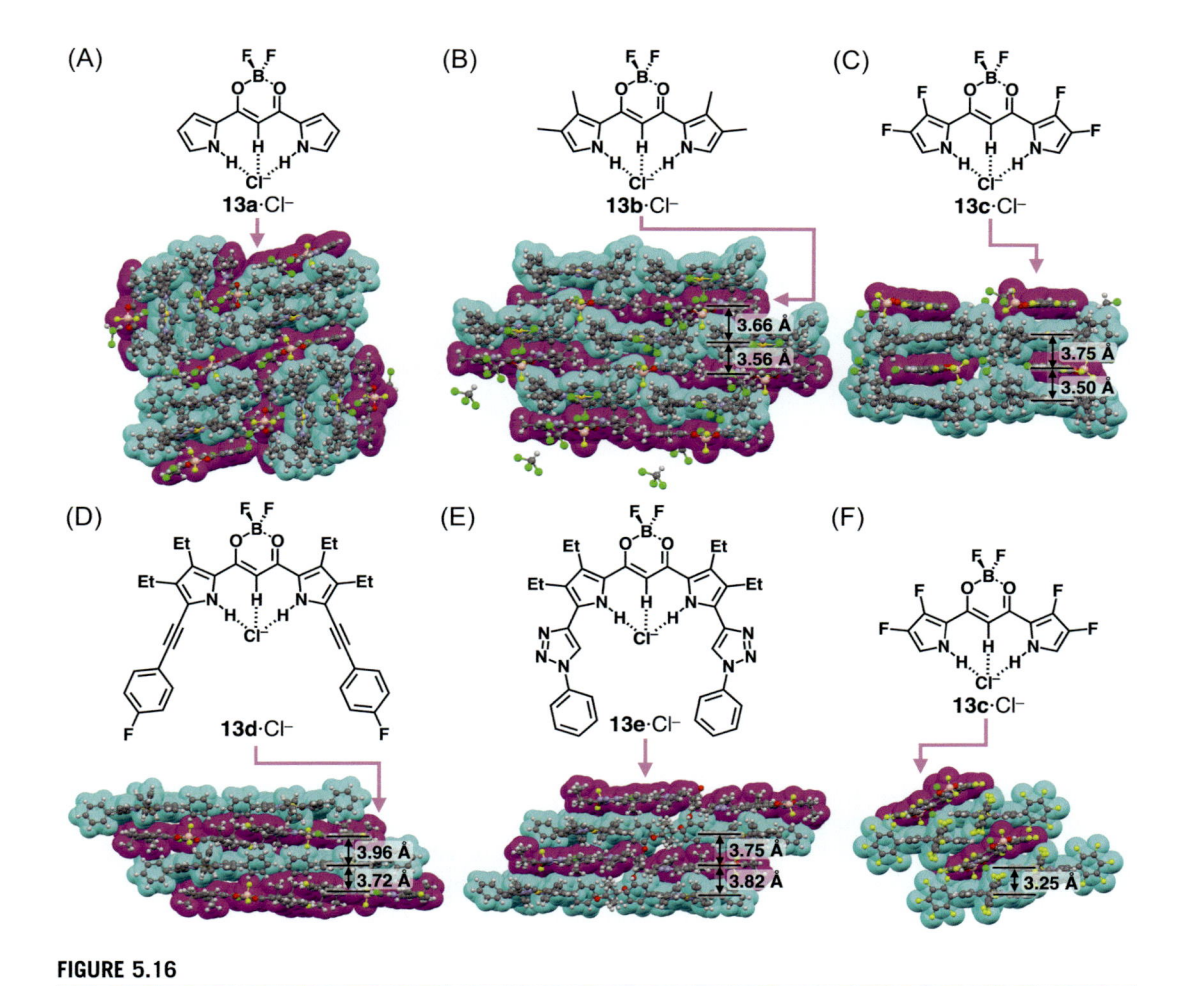

FIGURE 5.16

Single-crystal X-ray structures of ion-pairing assemblies comprising the porphyrin Au^{III} complexes and anion complexes of dipyrrolyldiketone BF_2 complexes: (A) $10a_{Au}^+$-$13a \cdot Cl^-$, (B) $10a_{Au}^+$-$13b \cdot Cl^-$, (C) $10a_{Au}^+$-$13c \cdot Cl^-$, (D) $10a_{Au}^+$-$13d \cdot Cl^-$, (E) $10a_{Au}^+$-$13e \cdot Cl^-$, and (F) $10c_{Au}^+$-$13c \cdot Cl^-$.

$15a-e \cdot Cl^-$ (0.10–0.25 Å) [12]. A similar procedure for $10a_{Au}^+$-Cl^- with Na^+-$3a_{Ni}^-$ yielded the ion pair $10a_{Au}^+$-$3a_{Ni}^-$, forming a crystal structure containing alternately arranged $10a_{Au}^+$ and $3a_{Ni}^-$ (Fig. 5.18D) [29]. A *meso*-phenyl group of $10a_{Au}^+$, located on the π-system of $3a_{Ni}^-$, hindered the stacking of $10a_{Au}^+$ and $3a_{Ni}^-$, creating a parallel arrangement. Consequently, $10a_{Au}^+$ was in contact with the nearest $10a_{Au}^+$, giving a partially charge-segregated assembly. In addition, the ^1H NMR spectrum of $10a_{Au}^+$-$3a_{Ni}^-$ differed significantly from those of the ion pairs containing bulky counterions, such as $10a_{Au}^+$-PF_6^- and TBA^+-$3a_{Ni}^-$, due to the ring-current effect derived from the stacked charged π-electronic systems in close proximity, even in solution state [29].

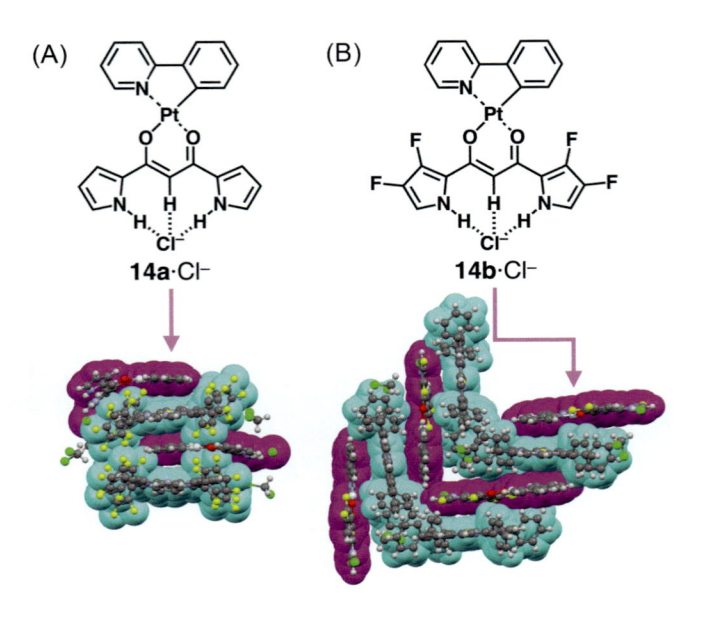

FIGURE 5.17

Single-crystal X-ray structures of ion-pairing assemblies consisting of porphyrin Au^{III} complexes and anion complexes of Pt^{II}-coordinated receptors: (A) $10b_{Au}^{+}$-**14a** · Cl^{-} and (B) $10a_{Au}^{+}$-**14b** · Cl^{-}.

PCCp^{-} is a substantially stabilized π-electronic anion, with respect to aromatic nature, and electron-withdrawing cyano groups that cause assembling behavior to differ significantly from ion pairs containing bulky anions (Fig. 5.19). In the assemblies of $10a_{Au}^{+}$-PCCp^{-} and $10d_{Au}^{+}$-PCCp^{-}, porphyrin Au^{III} complexes and PCCp^{-} were alternately arranged in charge-by-charge assemblies (Fig. 5.19A,C) [29]. Interestingly, PCCp^{-} was stacked with another PCCp^{-}, and the stacked PCCp^{-} dimer and $10b_{Au}^{+}$ were alternately stacked, resulting in a [2 + 1]-type charge-by-charge columnar structure (Fig. 5.19B) [29]. Another $10b_{Au}^{+}$ unit was located between the columns with the interactions of PCCp^{-} cyano groups with β-CH and C$_6$F$_5$ groups.

Fig. 5.20 shows that a perpendicularly directed ordered columnar structure of $10a_{Au}^{+}$-PCCp^{-} can be applied to dimension-controlled assemblies [29]. Tetraarylporphyrins with aliphatic chains were prepared as Cl^{-} ion pairs. $10e_{Au}^{+}$-Cl^{-}, as an example, was converted to the PCCp^{-} ion pair by ion-pair metathesis with NaPCCp (Fig. 5.20A). $10e_{Au}^{+}$-PCCp^{-} provided a supramolecular gel in octane (10 mg/mL), containing the fibers with the diameters of 1−3 μm and lengths of >100 μm. Differential scanning calorimetry (DSC) and polarized optical microscopy (POM) revealed a stable mesophase with dendritic textures over a significantly wide temperature range of 43−293 °C (Fig. 5.20B). Synchrotron XRD of $10e_{Au}^{+}$-PCCp^{-} showed the formation of a hexagonal columnar structure (a = 3.46, c = 0.71 nm, and Z = 1 ($ρ$ = 0.88) at 280 °C (cooling)). The c value with double the π−π stacking distance was consistent with the formation of a charge-by-charge assembly, which was effective in forming thermally stable liquid crystals (Fig. 5.20C) [29].

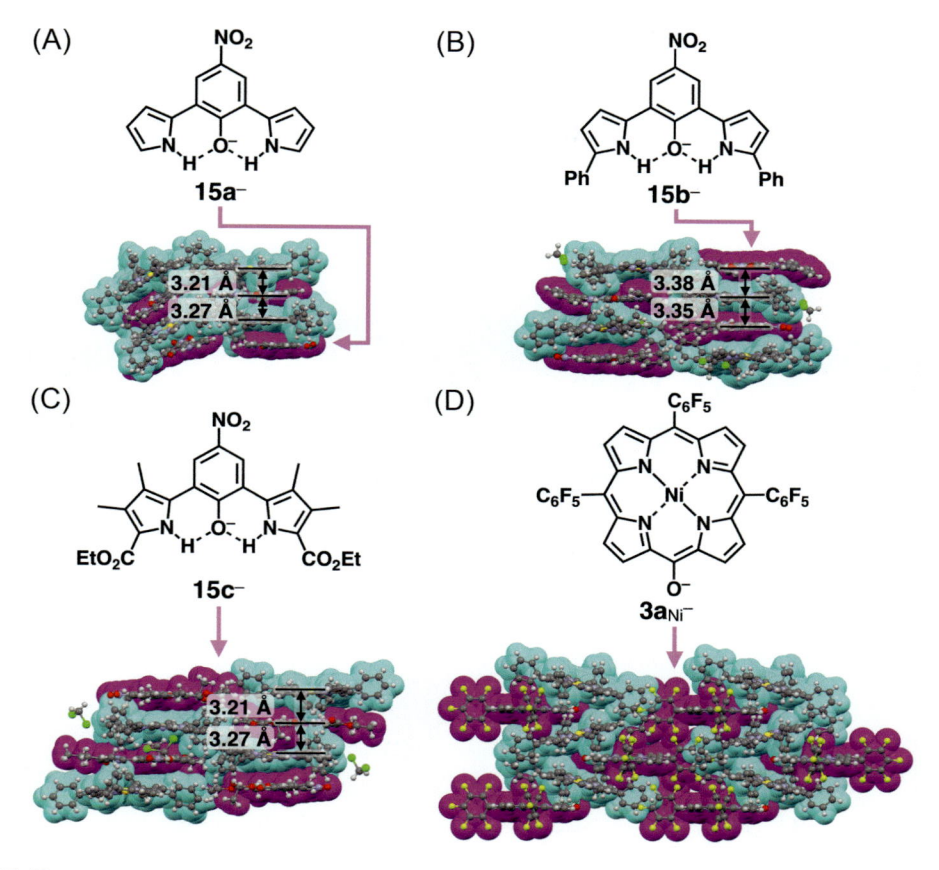

FIGURE 5.18

Single-crystal X-ray structures of ion-pairing assemblies comprising porphyrin Au^{III} complexes and π-electronic anions formed by deprotonation: (A) $10a_{Au}^+$-$15a^-$, (B) $10a_{Au}^+$-$15b^-$, (C) $10a_{Au}^+$-$15c^-$, and (D) $10a_{Au}^+$-$3a_{Ni}^-$.

In addition to porphyrin Au^{III} complexes, cationic porphyrin analogs as ion pairs with $PCCp^-$ have been reported. Ion-pair metathesis of 5^+-OTf^-, 5_{Ni}^+-OTf^-, and 5_{Zn}^+-Cl^- with NaPCCp yielded the corresponding $PCCp^-$ ion pairs, 5^+-$PCCp^-$, 5_{Ni}^+-$PCCp^-$, and 5_{Zn}^+-$PCCp^-$. Crystal structures of 5^+-$PCCp^-$ showed $PCCp^-$ stacking with 5^+ on both faces, with short contacts (3.34 and 3.38 Å) (Fig. 5.21A) [24]. The stacked triad in contact with another triad on the edge of 5^+ exhibited the interplanar distances of 3.44 and 3.54 Å. 5_{Ni}^+ and $PCCp^-$ were stacked with identically charged species on one face with interplanar distances of 3.31 and 3.28 Å, respectively, while β-ethyl groups hindered the direct stacking of other π-systems on another face (Fig. 5.21B). However, the cyano groups of $PCCp^-$ are in close proximity with Zn^{II}, suggesting the coordination rather than the stacking of the π-systems (Fig. 5.21C). EDA calculation revealed the significant contribution of electrostatic and dispersion forces between the oppositely charged species of 5^+-$PCCp^-$ and 5_{Zn}^+-$PCCp^-$ and strong

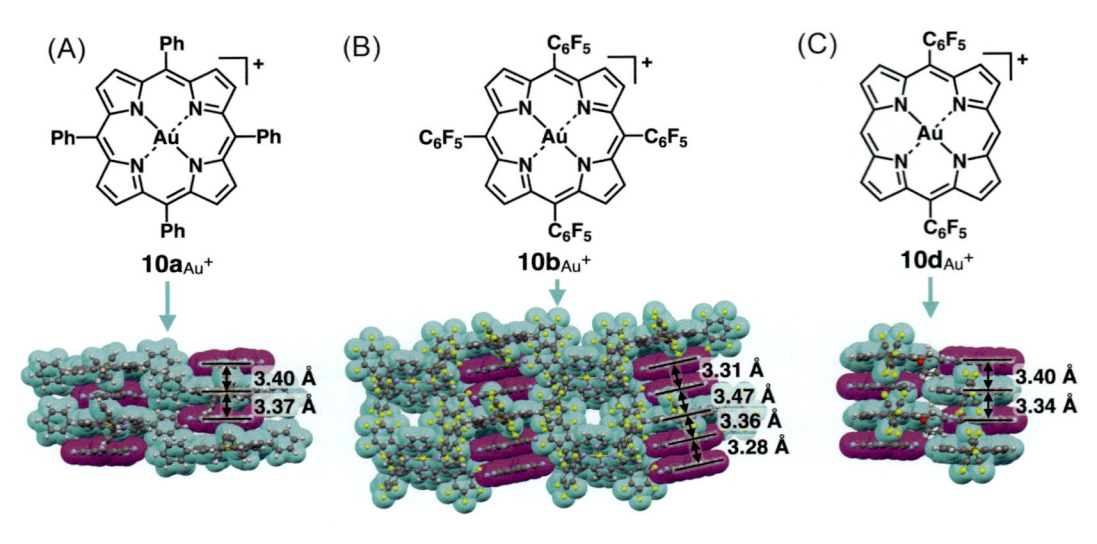

FIGURE 5.19

Single-crystal X-ray structures of ion-pairing assemblies comprising porphyrin Au^{III} complexes and $PCCp^-$: (A) $10a_{Au}^+$-$PCCp^-$, (B) $10b_{Au}^+$-$PCCp^-$, and (C) $10d_{Au}^+$-$PCCp^-$.

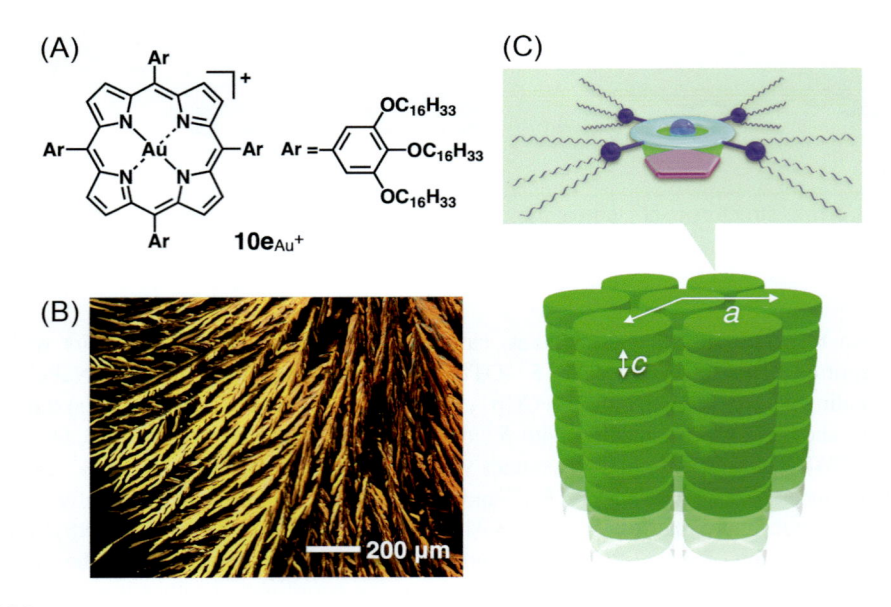

FIGURE 5.20

(A) $10e_{Au}^+$-$PCCp^-$, (B) POM texture, and (C) schematic representation of the Col_h mesophase.

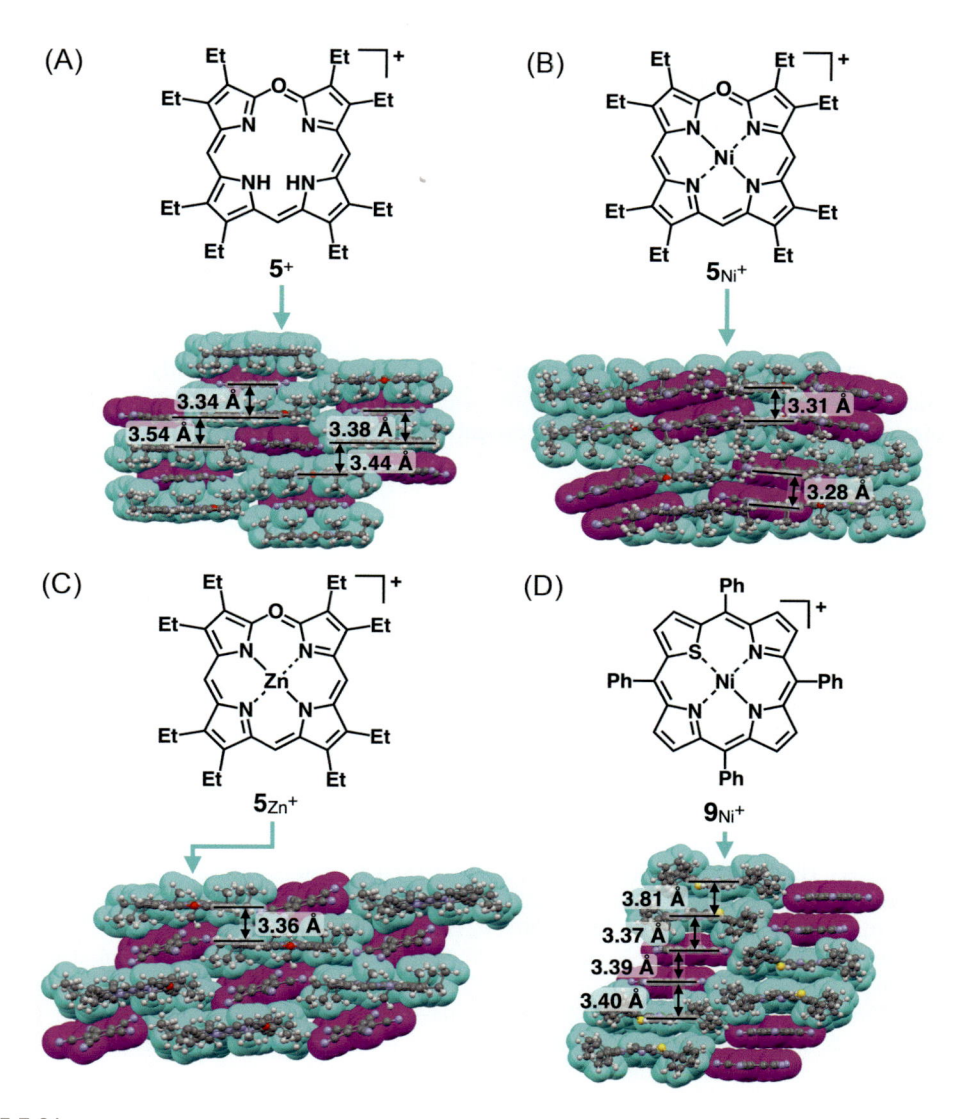

FIGURE 5.21

Single-crystal X-ray structures of ion-pairing assemblies comprising cationic porphyrin analogs and π-electronic anions formed by deprotonation: (A) 5^+-PCCp$^-$, (B) 5_{Ni}^+-PCCp$^-$, (C) 5_{Zn}^+-PCCp$^-$, and (D) 9_{Ni}^+-PCCp$^-$.

dispersion forces between identically charged species for 5_{Ni}^+-PCCp$^-$ [24]. The crystal structure of 9_{Ni}^+-PCCp$^-$ shows two independent structures were observed for both 9_{Ni}^+ and PCCp$^-$ (Fig. 5.21D) [27]. Each independent structure was stacked with another identically charged but different independent structure, providing a [2 + 2]-type charge-by-charge assembly.

5.5 **Summary and future perspective**

As summarized in this chapter, charged π-electronic systems form ordered structures in the crystalline states and mesophases with two-component systems, as well as stacked dimers in solution. Thus the interactions between the charged π-electronic systems are promising building blocks for nanoarchitectonics. Various porphyrin derivatives have been adopted as the frameworks of π-electronic ions that form ion-pairing assemblies. Moreover, the direct introduction of charges into π-electronic systems changes the electronic states more dramatically than the introduction of peripherally charged substituents, resulting in more redox-active systems. High redox activity can induce fascinating phenomena, such as photoinduced electron transfer and radical-pair formation, which can be affected by the arrangement of the charged components. Currently, the effects of the electron-deficient and electron-rich natures of the π-electronic cations and anions on the properties of ion pairs and their assemblies are under investigation.

References

[1] (a) K. Ariga, Q. Ji, J.P. Hill, Y. Bando, M. Aono, NPG Asia Mater. 4 (2012) e17.
(b) K. Ariga, X. Jia, L.K. Shrestha, Mol. Syst. Des. Eng. 4 (2019) 49−64.
(c) K. Ariga, Nanoscale Horiz. 6 (2021) 364−378.
(d) K. Ariga, Y. Lvov, G. Decher, Phys. Chem. Chem. Phys. 24 (2022) 4097−4115.
(e) K. Ariga, Nanoscale 14 (2022) 10610−10629. Available from: https://doi.org/10.1039/d2nr02513b.
(f) V. Karthick, L.K. Shrestha, V.G. Kumar, P. Pranjali, D. Kumar, A. Pal, et al., Nanoscale 14 (2022) 10630−10647. Available from: https://doi.org/10.1039/d2nr02293a.
[2] A.J. Stone, The Theory of Intermolecular Forces, Oxford University Press, 2013.
[3] (a) C.F.J. Faul, M. Antonietti, Adv. Mater. 15 (2003) 673−683.
(b) C.F.J. Faul, Acc. Chem. Res. 47 (2014) 3428−3438.
[4] (a) Y. Haketa, H. Maeda, Bull. Chem. Soc. Jpn. 91 (2018) 420−436.
(b) Y. Haketa, K. Urakawa, H. Maeda, Mol. Syst. Des. Eng. 5 (2020) 757−771.
[5] C.R. Martinez, B.L. Iverson, Chem. Sci. 3 (2012) 2191−2201.
[6] Y. Sasano, H. Tanaka, Y. Haketa, Y. Kobayashi, Y. Ishibashi, T. Morimoto, et al., Chem. Sci. 12 (2021) 9645−9657.
[7] (a) G. Merling, Ber. Deut. Chem. Ges. 24 (1891) 3108−3126.
(b) W. von, E. Doering, L.H. Knox, J. Am. Chem. Soc. 76 (1954) 3203−3206.
[8] (a) B.W. Laursen, F.C. Krebs, Angew. Chem. Int. Ed. 39 (2000) 3432−3434.
(b) B.W. Laursen, F.C. Krebs, Chem. Eur. J. 7 (2001) 1773−1783.
[9] (a) O.W. Webster, J. Am. Chem. Soc. 87 (1965) 1820−1821.
(b) T. Sakai, S. Seo, J. Matsuoka, Y. Mori, J. Org. Chem. 78 (2013) 10978−10985.
(c) Y. Bando, Y. Haketa, T. Sakurai, W. Matsuda, S. Seki, H. Takaya, et al., Chem. Eur. J. 22 (2016) 7843−7850.
[10] (a) H. Maeda, Y. Kusunose, Chem. Eur. J. 11 (2005) 5661−5666.
(b) H. Maeda, Y. Haketa, T. Nakanishi, J. Am. Chem. Soc. 129 (2007) 13661−13674.
(c) H. Maeda, K. Naritani, Y. Honsho, S. Seki, J. Am. Chem. Soc. 133 (2011) 8896−8899.
(d) Y. Haketa, Y. Bando, K. Takaishi, M. Uchiyama, A. Muranaka, M. Naito, et al., Angew. Chem. Int. Ed. 51 (2012) 7967−7971.
(e) B. Dong, T. Sakurai, Y. Honsho, S. Seki, H. Maeda, J. Am. Chem. Soc. 135 (2013) 1284−1287.

(f) A. Kuno, N. Tohnai, N. Yasuda, H. Maeda, Chem. Eur. J. 23 (2017) 11357−11365.

(g) S. Sugiura, Y. Kobayashi, N. Yasuda, H. Maeda, Chem. Commun. 55 (2019) 8242−8245.

[11] (a) Y. Haketa, S. Sasaki, N. Ohta, H. Masunaga, H. Ogawa, N. Mizuno, et al., Angew. Chem. Int. Ed. 49 (2010) 10079−10083.

(b) Y. Haketa, Y. Honsho, S. Seki, H. Maeda, Chem. Eur. J. 18 (2012) 7016−7020.

(c) Y. Haketa, M. Takayama, H. Maeda, Org. Biomol. Chem. 10 (2012) 2603−2606.

(d) R. Yamakado, T. Sakurai, W. Matsuda, S. Seki, N. Yasuda, S. Akine, et al., Chem. Eur. J. 22 (2016) 626−638.

(e) H. Maeda, T. Nishimura, Y. Haketa, H. Tanaka, M. Fujita, N. Yasuda, J. Org. Chem. 87 (2022) 7818−7825.

[12] (a) H. Maeda, A. Fukui, R. Yamakado, N. Yasuda, Chem. Commun. 51 (2015) 17572−17575.

(b) H. Maeda, Y. Takeda, Y. Haketa, Y. Morimoto, N. Yasuda, Chem. Eur. J. 24 (2018) 8910−8916.

(c) N. Fumoto, Y. Haketa, H. Tanaka, N. Yasuda, H. Maeda, Org. Lett. 23 (2021) 3897−3901.

[13] (a) K.M. Kadish, K.M. Smith, R. Guilard (Eds.), The Porphyrin Handbook, Academic Press, 1999.

(b) K.M. Kadish, K.M. Smith, R. Guilard (Eds.), Handbook of Porphyrin Science, World Scientific, 2010.

[14] (a) S. Ishihara, J. Labuta, W.V. Rossom, D. Ishikawa, K. Minami, J.P. Hill, et al., Phys. Chem. Chem. Phys. 16 (2014) 9713−9746.

(b) K. Ariga, J. Porphyrins Phthalocyanines 25 (2021) 897−916.

[15] M. Kasha, H.R. Rawls, M.A. El-Bayoumi, Pure Appl. Chem. 11 (1965) 371−392.

[16] (a) C.J. Medforth, Z. Wang, K.E. Martin, Y. Song, J.L. Jacobsen, J.A. Shelnutt, Chem. Commun. (2009) 7261−7277.

(b) A. D'Urso, M.E. Fragalà, R. Purrello, Chem. Commun. 48 (2012) 8165−8176.

[17] K. Yamasumi, H. Maeda, Bull. Chem. Soc. Jpn. 94 (2021) 2252−2262.

[18] (a) J. Winkelman, Cancer Res. 22 (1962) 589−596.

(b) R.J. Fiel, J.C. Howard, E.H. Mark, N.D. Gupta, Nucleic Acids Res. 6 (1979) 3093−3118.

(c) S. Ikeda, T. Nezu, G. Ebert, Biopolymers 31 (1991) 1257−1263.

(d) D.W. Celander, J.M. Nussbaum, Biochemistry 35 (1996) 12061−12069.

(e) R.T. Wheelhouse, D. Sun, H. Han, F.X. Han, S.H. Hurley, J. Am. Chem. Soc. 120 (1998) 3261−3262.

(f) V. Bogoeva, L. Petrova, P. Kubát, J. Photochem. Photobiol. B 153 (2015) 276−280.

[19] (a) R.F. Pasternack, P.R. Huber, P. Boyd, G. Engasser, L. Francesconi, E. Gibbs, et al., J. Am. Chem. Soc. 94 (1972) 4511−4517.

(b) J.M. Ribó, J. Crusats, J.-A. Farrera, M.L. Valero, J. Chem. Soc., Chem. Commun. (1994) 681−682.

(c) R.F. Pasternack, K.F. Schaefer, P. Hambright, Inorg. Chem. 33 (1994) 2062−2065.

(d) G. de Luca, A. Romeo, L.M. Scolaro, J. Phys. Chem. B 110 (2006) 7309−7315.

[20] (a) J.M. Ribó, R. Rubires, Z. El-Hachemi, J.-A. Farrera, L. Campos, G.L. Pakhomov, et al., Mater. Sci. Eng. C 11 (2000) 107−115.

(b) J.M. Ribó, J. Crusats, F. Sagués, J. Claret, R. Rubires, Science 292 (2001) 2063−2066.

(c) A.D. Schwab, D.E. Smith, C.S. Rich, E.R. Young, W.F. Smith, J.C. de Paula, J. Phys. Chem. B 107 (2003) 11339−11345.

(d) C. Escudero, J. Crusats, I. Díez-Pérez, Z. El-Hachemi, J.M. Ribó, Angew. Chem. Int. Ed. 45 (2006) 8032−8035.

(e) A.L. Yeats, A.D. Schwab, B. Massare, D.E. Johnston, A.T. Johnson, J.C. de Paula, et al., J. Phys. Chem. C 112 (2008) 2170−2176.

(f) S.M. Vlaming, R. Augulis, M.C.A. Stuart, J. Knoester, P.H.M. van Loosdrecht, J. Phys. Chem. B 113 (2009) 2273−2283.

[21] (a) Z. Wang, C.J. Medforth, J.A. Shelnutt, J. Am. Chem. Soc. 126 (2004) 15954−15955.

(b) Z. Wang, C.J. Medforth, J.A. Shelnutt, J. Am. Chem. Soc. 126 (2004) 16720−16721.

(c) K.E. Martin, Z. Wang, T. Busani, R.M. Garcia, Z. Chen, Y. Jiang, et al., J. Am. Chem. Soc. 132 (2010) 8194−8201.

(d) Y. Tian, C.M. Beavers, T. Busani, K.E. Martin, J.L. Jacobsen, B.Q. Mercado, et al., Nanoscale 4 (2012) 1695−1700.

(e) M. Adinehnia, B. Borders, M. Ruf, B. Chilukuri, K.W. Hipps, U. Mazur, J. Mater. Chem. C 4 (2016) 10223−10239.

[22] (a) Y. Sasano, N. Yasuda, H. Maeda, Dalton Trans 46 (2017) 8924−8928.

(b) Y. Sasano, Y. Haketa, H. Tanaka, N. Yasuda, I. Hisaki, H. Maeda, Chem. Eur. J. 25 (2019) 6712−6717.

[23] (a) E. Vogel, W. Haas, B. Knipp, J. Lex, H. Schmickler, Angew. Chem. Int. Ed. Engl. 27 (1988) 406−409.

(b) E. Vogel, P. Röhrig, M. Sicken, B. Knipp, A. Herrmann, M. Pohl, et al., Angew. Chem. Int. Ed. Engl. 28 (1989) 1651−1655.

(c) M. Kon-No, J. Mack, N. Kobayashi, M. Suenaga, K. Yoza, T. Shinmyozu, Chem. Eur. J. 18 (2012) 13361−13371.

(d) S.P. Panchal, S.C. Gadekar, V.G. Anand, Angew. Chem. Int. Ed. 55 (2016) 7797−7800.

[24] (a) T. Satoh, M. Minoura, H. Nakano, K. Furukawa, Y. Matano, Angew. Chem. Int. Ed. 55 (2016) 2235−2238.

(b) A. Takiguchi, S. Kang, N. Fukui, D. Kim, H. Shinokubo, Angew. Chem. Int. Ed. 60 (2021) 2915−2919.

(c) A. Takiguchi, H. Tanaka, H. Maeda, H. Shinokubo, Bull. Chem. Soc. Jpn. 95 (2022) 796−801.

[25] M. Fang, S.R. Wilson, K.S. Suslick, J. Am. Chem. Soc. 130 (2008) 1134−1135.

[26] (a) Y. Inokuma, J.H. Kwon, T.K. Ahn, M.-C. Yoo, D. Kim, A. Osuka, Angew. Chem. Int. Ed. 45 (2006) 961−964.

(b) E. Tsurumaki, S.Y. Hayashi, F.S. Tham, C.A. Reed, A. Osuka, J. Am. Chem. Soc. 133 (2011) 11956−11959.

[27] M. Fujita, Y. Haketa, H. Tanaka, N. Yasuda, H. Maeda, Chem. Commun. 58 (2022) 9870−9873.

[28] (a) A. MacCragh, W.S. Koski, J. Am. Chem. Soc. 87 (1965) 2496−2497.

(b) E.B. Fleischer, A. Laszlo, Inorg. Nucl. Chem. Lett. 5 (1969) 373−376.

(c) R. Timkovich, A. Tulinsky, Inorg. Chem. 16 (1977) 962−963.

(d) M.E. Jamin, R.T. Iwamoto, Inorg. Chim. Acta 27 (1978) 135−143.

[29] (a) Y. Haketa, Y. Bando, Y. Sasano, H. Tanaka, N. Yasuda, I. Hisaki, et al., iScience 14 (2019) 241−256.

(b) H. Tanaka, Y. Haketa, N. Yasuda, H. Maeda, Chem. Asian J. 14 (2019) 2129−2137.

[30] J. Arnold, J. Chem. Soc., Chem. Commun. (1990) 976−978.

[31] Q.-C. Chen, N. Fridman, Y. Diskin-Posner, Z. Gross, Chem. Eur. J. 26 (2020) 9481−9485.

[32] H. Tanaka, Y. Haketa, Y. Bando, R. Yamakado, N. Yasuda, H. Maeda, Chem. Asian J. 15 (2020) 494−498.

[33] A. Kuno, G. Hirata, H. Tanaka, Y. Kobayashi, N. Yasuda, H. Maeda, Chem. Eur. J. 27 (2021) 10068−10076.

[34] (a) A. Schmidt, B. Heinrich, G. Kirscher, A. Chaumont, M. Henry, N. Kyritsakas, et al., Inorg. Chem. 59 (2020) 12802−12816.

(b) Y. Haketa, H. Maeda, Molecules 26 (2021) 861.

Layered structures in soft nanoarchitectonics: towards functional photonic materials

Youfeng Yue[1,2]

[1]*Research Institute for Advanced Electronics and Photonics, National Institute of Advanced Industrial Science and Technology (AIST), Tsukuba, Japan* [2]*PRESTO, Japan Science and Technology Agency (JST), Kawaguchi, Japan*

6.1 Introduction

Soft materials are widely found in nature and industrial production, including surfactants, gels, polymers, liquid crystals, colloids, biomolecules, membranes, and live tissues. One of the differences between soft and hard materials is their response to the external stimuli. Soft materials generally exhibit high response to external stimuli. For example, they deform significantly to the mechanical stress compared to hard materials.

Soft materials with ordered nanoscale structures widely exist in nature and industry. Among them, nanoscale layered materials, consisting of alternating layers of different reflective indices, have attracted increasing attention owing to their excellent chemical, physical, optical, and mechanical properties. For example, in nature, the iridescent colors in the feathers of peacocks can be attributed to light diffracting from nanoscale layered structures. Furthermore, many biological functions of soft tissues originate from the unique arrangement of their hierarchical structures, which has inspired researchers to design and fabricate soft materials with layered structures. However, fabricating ordered nanoscale layered structures on a macroscopic scale in soft materials remains a challenge.

Specific strategies (either bottom-up or top-down) have been developed to fabricate layered structures in soft materials. Bottom-up strategies involving the self-assembly of various building blocks, such as amphiphilic molecules, have been used to build nanoscale layered structures. Top-down approaches, such as multilayer coextrusion and holographic photopolymerization, have been applied to fabricate nanoscale layered materials.

The ordered layered structures of soft materials impart excellent chemical, physical, and optical functions, which contribute to the extension of their applications across various fields. Additionally, soft materials with layered structures exhibit unique mechanical properties. For example, nacre naturally possesses significantly greater fracture resistance than monolithic aragonite owing to its alternating hard inorganic and soft polymeric layered structures. Soft materials with ordered layered structures, similar to nacre, also contribute to the excellent mechanical strength.

In this chapter, we summarize the current progress made in developing soft materials with alternating nanoscale layered structures, mainly focusing on bio-inspired layered structures, bottom-up and top-down fabrication strategies, and their applications in chemical/biological sensors, anisotropic molecular release, and optical devices. Moreover, the unique mechanical properties of these soft materials (e.g., high fracture and crack resistance and strength anisotropic moduli) are highlighted. Finally, future development and prospects of soft materials with layered structures are discussed.

6.2 Naturally existing layered structures

Biological systems have used nanoscale architectures to produce iridescent optical features for millions of years. Nature is filled with iridescent structural colors (e.g., in the feathers of peacocks), which result from the interaction of light with ordered nanoscale structures. Structural colors in nature are mainly for signaling, mimicry, and mate-choice. In many cases, structural colors originate from the layered structures of materials with different refractive indices. The stacking of nanoscale layers with high and low refractive indices leads to photonic bandgap structures, in which the incident light is selectively enhanced because it is prevented from propagating within certain parts of the structure, thereby producing the corresponding colors. Light with frequencies in the bandgaps is reflected. Nanoscale structures with alternating layers of refractive indices, also known as photonic crystals or Bragg stacks, have attracted considerable attention. When compared with other two- or three-dimensional (2D or 3D) photonic nanostructures, the structural colors from the alternating stacking of multiple layers can enhance the reflections from the multiple interfaces. Therefore, the structural colors arising from a large number of stacked layers are bright, stable, colorful, and of high quality. The embedding of these layered structures in soft and responsive materials further enables the materials to have stimuli-sensitive properties. Thus the colors can be tuned by changing the interlayer distances or refracting angles and exhibit angle-dependent iridescent features, which can be widely found in natural organisms such as beetles, birds, tropical fish, mollusks, and fruits.

For example, longhorn beetles exhibit structural color arising from nanoscale layered structures (Fig. 6.1A and B) [1] Under dry weather, they display a gold color; when the ambient humidity increases, the gold color rapidly changes to red; when the weather becomes dry again, the color reverts to gold. This means that the color changes reversibly in response to changes in humidity. The layered structure of the beetle was characterized using scanning electron microscopy, with cross-sectional images showing a periodic multilayer structure (Fig. 6.1C) [1]. The change in color from gold to red in the wet state is due to the swelling of the layers as a result of water infiltration.

Such structural color changes from layered structures are often observed in animals, but rarely seen in plants. However, there is an interesting example of strong iridescent coloration based on multiple layers in the fruit of *Pollia condensata* (Fig. 6.1D and E) [2]. Although the fruit of *P. condensata* has no blue pigment, it exhibits a bright blue color. This color is caused by the Bragg diffraction of helicoidally stacked cellulose microfibrils that form multilayers in the cell walls of the epicarp (Fig. 6.1F). They have a layered structure similar to that of beetles but exhibit more intense reflectivity. Interestingly, they also retain a highly intense blue color even after falling from the plant. In low-light forests, this blue color is visible and attracts birds to help them disperse seeds widely.

FIGURE 6.1

Iridescent structural colors in (A–C) beetles and (D–F) *Pollia condensata* and their corresponding nanoscale layered structures.

Panels A–C were reprinted with permission from F. Liu, B. Dong, X. Liu, Y. Zheng, J. Zi, Structural color change in longhorn beetles Tmesisternus isabellae, *Opt. Express. 17 (2009) 16183–16191, Copyright (2009) Optical Society of America. Panels D–F were reprinted with permission from S. Vignolini, P.J. Rudall, A.V. Rowland, et al., Pointillist structural color in Pollia fruit, Proc. Natl Acad. Sci. 109 (2012) 15712–15715, Copyright (2012) National Academy of Sciences.*

Thus natural systems have used nanoscale architectures to produce striking optical features. These natural photonic structures provide inspiration for technological fabrication and applications in our daily lives and production.

6.3 Fabrication of nanoarchitectonics with layered structures in soft materials

Inspired by this, researchers have aimed to develop various artificial strategies for synthesizing layered materials using inorganic, polymeric, and hybrid components. Both bottom-up and top-down fabrication strategies have been used to construct layered structures in soft materials. Different preparation routes can affect the periodicity of the structures and the physicochemical characteristics of the layers [3]. Bottom-up approaches typically involve the self-assembly of nanoscale building blocks into periodic layered structures. The building blocks can be small amphiphilic molecules, block copolymers, or 2D nanoparticles. Self-assembly is an efficient way to fabricate nanoscale layered structures. It is an energetically driven or energy-dissipation process that arrives at the lowest energy equilibrium structure after self-assembly. However, a highly ordered layered structure cannot be obtained by self-assembly alone. Therefore, external forces, such as shear and magnetic forces, are applied to direct the orientation of the layered structures [4].

Top-down nanofabrication approaches also have been developed for layered nanomaterials with high structural and optical properties. They are widely used for light management purposes, including optical filters, light-emitting diodes, optical components, and lasers. Layered structures in these applications are commercially fabricated from inorganic dielectric materials that are based on vacuum evaporation and sputtering. However, these materials are neither soft nor stretchable; therefore, there is a growing demand for the development of flexible optoelectronic devices based on organic semiconductors. In this case, mild vacuum chemical vapor deposition has been employed for the fabrication of polymeric layered structures [5]. Additionally, layered structures in soft materials can also be fabricated by other top-down approaches, such as cyclic deposition of two materials, multilayer coextrusion, and holographic photopolymerization.

Cyclic deposition of two materials can be achieved by alternating the deposition of nanoparticles, sol—gel precursors, or their polymer solutions with different reflective indices (n). These types of depositions can be performed via spin-coating, dip-coating, casting, etc. These strategies offer precise manufacturing but require complex apparatus and multiple processing steps [6].

The refractive index contrast of pure polymer materials is usually smaller than that of other layered structures that are based on inorganic materials. However, this limitation can be compensated for by building up several hundred stacked layers. Furthermore, the refractive index can be adjusted through the chemical structure and composition of the materials. For example, increasing the porosity of a polymeric material can effectively reduce the refractive index, while doping layers with high-index inorganic nanoparticles into polymeric materials can significantly increase the refractive index.

6.3.1 Self-assembly of amphiphilic molecules

In this section, the self-assembly processes used to fabricate nanoscale layered structures are discussed. First, the self-assembly of amphiphilic molecules in water was used as a strategy to fabricate nanoscale layered structures.

Amphiphilic molecules have hydrophilic heads and hydrophobic tails. Hydrophilic heads can be either ionic or nonionic, whereas hydrophobic tails may consist of a long saturated or unsaturated hydrocarbon chain. They can be synthetic or natural molecules with the ability to self-assemble into various packing structures, such as micelles, vesicles, nanotubes, and lamellae. Thus the self-assembly of amphiphilic molecules into different structures in water offers a unique opportunity to design soft materials with specific nanoarchitectures.

For example, dodecyl glyceryl itaconate (DGI: $n\text{-}C_{12}H_{25}\text{-}OCOCH_2C(=CH_2)COOCH_2CH(OH)CH_2OH$) is a type of synthetic amphiphilic molecule (Fig. 6.2) [7]. Unlike other amphiphilic molecules, DGI has a double bond that can be polymerizable. DGI is insoluble in water at 25°C; however, when its concentration exceeds the critical micelle concentration, and the temperature rises to the Krafft point (defined as the temperature below which amphiphilic molecules are insoluble in

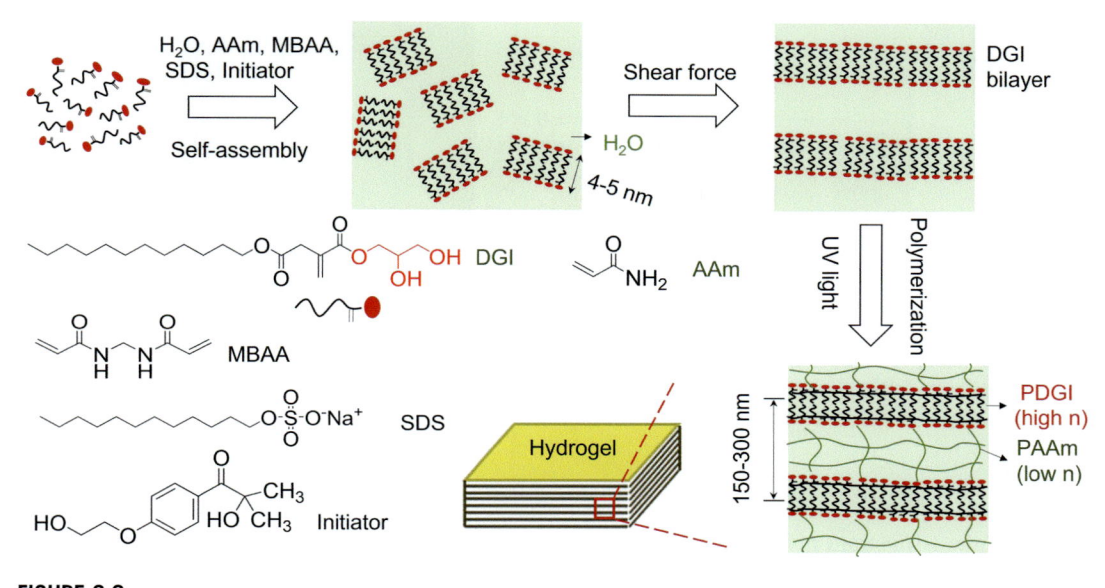

FIGURE 6.2

Self-assembly of amphiphilic molecules into nanoscale layered structures in water and further polymerization to obtain ordered layered structures in a hydrogel.

Reprinted with permission from Y. Yue, J.P. Gong, Structure and unique functions of anisotropic hydrogels comprising uniaxially aligned lamellar bilayers. Bull. Chem. Soc. Jpn. 94 (2021) 2221–2234, Copyright (2021) The Chemical Society of Japan.

water) of DGI ($T_k = 43$ °C), DGI molecules self-assemble into an aggregated packing bilayer structure in water [8].

DGI molecules in water self-assemble into lamellar bilayer structures with an interlayer spacing of a few nanometers. To obtain iridescent color effects through the light diffraction from the layered structures, the layer spacing needs to be increased with a periodicity equal to half the wavelength of visible light. For this purpose, a small amount of an ionic surfactant, such as sodium dodecyl sulfate (SDS), is added to the DGI/water system. The interlayer spacing dramatically increases to several hundred nanometers owing to electrostatic repulsion [9]. When the layer spacing is in the range of approximately 170–350 nm, the aqueous DGI solution exhibits structural colorations that change from red to blue with a slight increase in the mole ratio of SDS-to-DGI.

The layered structures in water, with an interlayer spacing of several hundred nanometers, can be further polymerized to build polymeric layered hydrogels (Fig. 6.2). The layered structures are retained when other nonionic monomers, crosslinkers, and initiators are added to the iridescent DGI/water system. Tsujii et al. polymerized a DGI solution in the presence of acrylamide (AAm) monomers, crosslinkers, and initiators to form a hydrogel [10,11]. This process successfully embedded nanoscale layered structures into soft materials. However, the structural colorations exhibited by this hydrogel were not uniform, because the domain structures of the bilayers were not unidirectionally aligned. To produce ordered layered structures in the hydrogel, Haque et al. significantly improved the system by using a strong shear force when injecting the precursor solution before

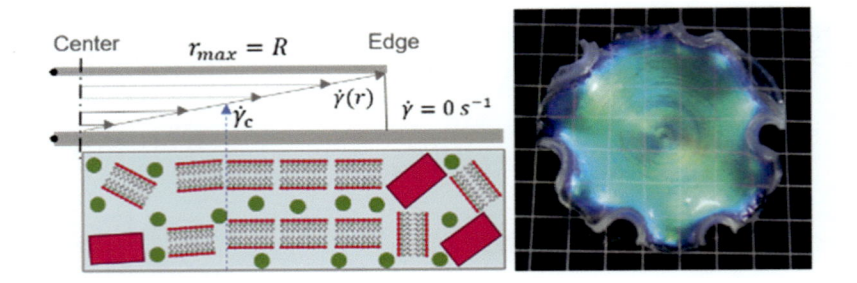

FIGURE 6.3

The effect of shear force on the layered structures in a hydrogel.

Reprinted (adapted) with permission from Y.N. Ye, M.A. Haque, A. Inoue, Y. Katsuyama, T. Kurokawa, J.P. Gong, Flower-like photonic hydrogel with superstructure induced via modulated shear field, ACS Macro Lett. 10 (2021) 708–713, Copyright (2021) American Chemical Society.

polymerization [12]. The shear force caused the bilayer microdomains of DGI to align uniaxially along the glass substrate at the macroscopic scale. Finally, these alternating layered structures of poly(dodecyl glyceryl itaconate) (PDGI) and polyacrylamide (PAAm) remain highly ordered and stable after photopolymerization, as the AAm monomers polymerized to form chemically cross-linked PAAm gel layers, and the DGI molecules polymerized to form PDGI layers.

External forces, such as shear, are important for achieving high-degree layered structures. Ye et al. introduced a gradual shear force to layered structures using a customized rheometer. The distribution of the shear field gave rise to different orientations of the structures, which caused swelling mismatch of the gel and, consequently, the formation of flower-like photonic hydrogels (Fig. 6.3) [13].

6.3.2 Self-assembly of block copolymers

In addition to the self-assembly of amphiphilic molecules, the self-assembly of block copolymers can also be used as an efficient strategy for building soft nanoarchitectures. Block copolymers are macromolecules consisting of chemically distinct sequences of repeating units (referred to as "blocks").

Advanced polymerization techniques that allow control of chain architectures make block copolymers attractive for the fabrication of various nanostructures. Block copolymers are composed of two or more covalently linked, chemically distinct polymers. For example, the simplest molecular architecture is a linear AB diblock copolymer consisting of two polymer chains (i.e., A and B). When the molecular weight of the component blocks is sufficiently large to drive microphase separation, they have the inherent ability to self-assemble into various nanoscale architectures.

Linear and brush block copolymers are capable of self-assembling into ordered layered structures (or lamellar structures) with minimum free energy configurations (Fig. 6.4) [6,14]. In particular, when the interlayer distance matches the wavelength of visible light, the layered structures of block copolymers also exhibit iridescent colors, similar to the self-assembly of amphiphilic molecules [15,16]. These nanoscale layered structures of block copolymers are favored because it is

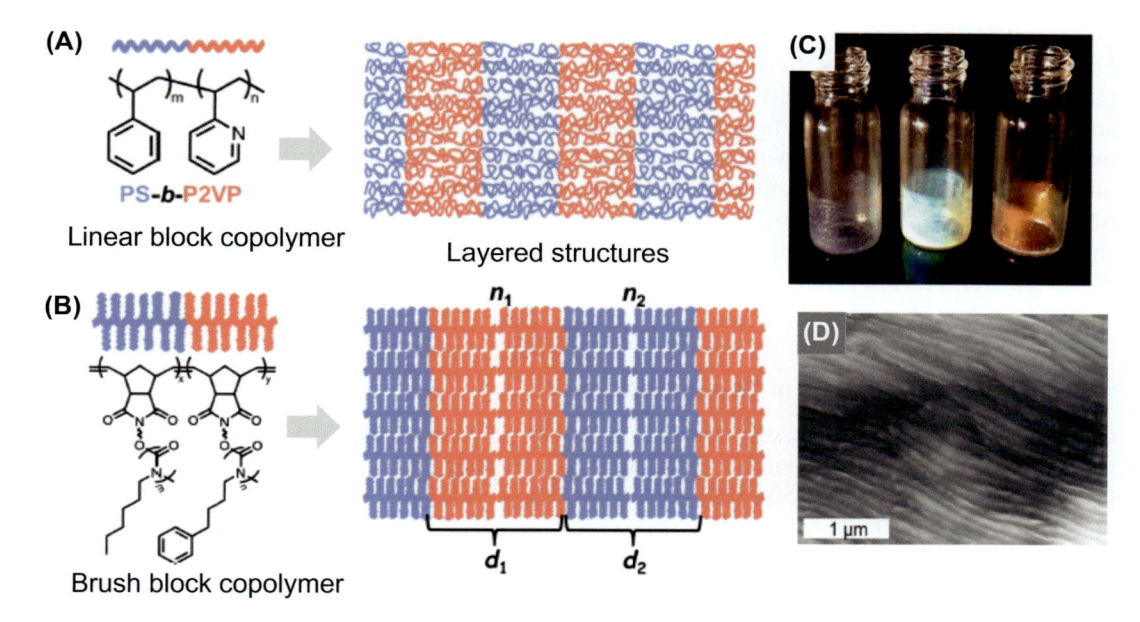

FIGURE 6.4

Chemical structures of common monomers used to produce (A) linear and (B) brush block copolymers and schematics showing their self-assembly into nanoscale layered structures. (C and D) The iridescent structural colors from the layered structures and a scanning electron microscopy image of the cross-sections of brush block copolymers with Mw = 1512.

Panels A and B were reprinted with permission from Z. Wang, C.L.C. Chan, R.M. Parker, S. Vignolini, The limited palette for photonic block-copolymer materials: a historical problem or a practical limitation? Angew. Chem. Int. Ed. 61 (2022) e202117275, Copyright 2022 by the authors and licensed under CC BY 4.0. Panels C and D were reprinted (adapted) with permission from G.M. Miyake, R. A. Weitekamp, V.A. Piunova, R.H. Grubbs, Synthesis of isocyanate-based brush block copolymers and their rapid self-assembly to infrared-reflecting photonic crystals, J. Am. Chem. Soc. 134 (2012) 14249–14254, Copyright (2012) American Chemical Society.

simple to obtain 1D structures that provide optimal optical properties, such as maximum reflectivity from the smallest size [14].

For block copolymers, the self-assembly process is driven by the block chain length (N) and the Flory–Huggins parameter (χ) because $\chi \cdot N$ describes the strength of the repulsive interaction between blocks [17]. The morphology of the nanostructures is related to the volume ratio fraction of the two polymer blocks. For instance, a layered structure typically requires a volume fraction of $f = 0.5$ [18], and if either χ or N is decreased substantially, the entropic factors will lead to a compositionally disordered phase [19]. To improve the quality of the ordered layered structure at a macroscale, some external forces, such as electric and magnetic fields, can be used during the self-assembly process.

To date, it has remained a challenge to achieve a well-ordered layered structure that can reflect near infrared (NIR) light ($\lambda = 750-1500$ nm) [6] because the slow self-assembly of block copolymers inhibits their ability to assemble into domain sizes large enough to reflect long wavelengths.

One of the primary antagonists of self-assembly is chain entanglement, which hinders high-molecular-weight polymers from achieving large domains of up to several hundred nanometers [20]. In the case of layered structures derived from the self-assembly of amphiphilic molecules, layered structures reflecting NIR light can be achieved through postmodifications (swelling) of polymer networks. For example, with some chemical treatment, the polymer network swells and the interlayer spacing is increased to more than 300 nm, which enables reflection close to the red and NIR wavelengths [21]. However, the bandwidth of these materials is large and broadens with a red-shift in the maximum wavelength (λ_{max}), indicating poor ordering of the layered structures. Thus producing well-ordered layered materials for NIR-reflection is an ongoing challenge.

6.3.3 Cyclic deposition of two materials

Ordered layered structures in soft materials can also be fabricated by the repeated deposition of polymeric or sol−gel precursors to form thin films. Among the commonly used polymers, poly (methyl methacrylate) ($n = 1.50$), [22] polyacrylic acid (PAAc, $n = 1.51$), [23] and polyvinyl alcohol ($n = 1.52$) [23] are widely used as low-reflective index media, while polystyrene ($n = 1.58$) [24] and poly(N-vinylcarbazole) ($n = 1.68$) [25] are used as media with relatively high reflective indices. [5] In terms of manufacture, they can mainly be fabricated via spin-coating, dip-coating, casting, or sol−gel chemistry depositions. Spin-coating is one of the simplest methods for obtaining alternating layered nanostructures based on two different materials with different reflective indices.

In addition to polymers, cyclic spin-coating of inorganic nanoparticles and polymeric materials has been widely used for the synthesis of layered soft materials. However, using only inorganic materials, such as silica (SiO_2) and titania (TiO_2), yields materials with a high reflective index that lack the softness to deform under chemical or physical stimuli. Pure polymeric materials show great variation in structural color through swelling or deswelling but suffer from low refractive index contrasts, with values ranging from 1.4 to 1.6. Therefore, the incorporation of inorganic particles into responsive polymeric materials provides a useful route for obtaining layered nanostructures with tunable structural colors. For example, Jeon et al. prepared thermochromic layered structures using benzophenone-containing photocrosslinkable polymers with zirconia nanoparticles via spin-coating and ultraviolet (UV) crosslinking [26]. The inorganic nanoparticles greatly increased the refractive index of the polymer, enhancing its reflectivity by a factor of 2.5 when compared with the pure polymer material used.

Howell et al. fabricated organic/inorganic layered composites via spin-coating [27]. The refractive index of the polymeric material was altered by incorporating high-refractive-index nanoparticles, thereby reducing the number of layers required to generate sufficient reflection. These composite materials were used to construct a strain-tunable layered structure with a reflectivity above 40% and maximum mechanochromic sensitivity of over 6 nm/% strain [27].

The cyclic spin-coating of polymers provides a means of constructing layered materials with different compositions and interesting properties. In this case, it is often necessary to increase the porosity of one type of alternating layer to increase the reflective index contrast. Ito et al. fabricated multilayer structures via the spin-coating of crosslinked and noncrosslinked layers (porous) [28]. This layered material exhibited structural color across the full visible spectrum, and the colors could be tuned by varying the temperature and solvent conditions. By using lithographic and masking tools, they produced a color printing process that generated images with a resolution of up to 14,000 dots per inch [28].

6.3.4 Multilayer coextrusion

Multilayer coextrusion is a top-down method for fabricating high-performance layered nanostructures in a continuous melt process that yields large areas of layered polymeric films in a single roll-to-roll process [29]. It has become an attractive and industrially scalable technology that is used to form multilayered films for a wide range of products from food packaging to dielectric materials [30].

For example, two polymer films (e.g., SAN25 (RI = 1.57) and THV 220 G (RI = 1.37)) from two or more extruders were combined [29]. The films were then cut and the two halves were overlapped and pressed together. This overlapping process, which was automated, increased the number of layers.

This method allows for the production of multilayer films with hundreds of alternating nanometer-thin layers. Additionally, this method enables rapid production of large areas of high-quality films in a one-step roll-to-roll melting process. The ease of tunability of the polymer used and the possibility of tuning coupled structures in different regions of the sunlight spectrum make layered film coatings attractive for their specific properties and functions, such as thermal shielding [5].

6.3.5 Holographic photopolymerization

Holographic photopolymerization based on soft lithography is a simple method for creating layered polymeric nanostructures. A typical holographic lithography process uses a photopolymer as the starting material. This photopolymer produces periodic index modulations determined by the interference pattern of the coherent light beams. Thus relatively large areas of layered structures are formed.

Hsiao et al. reported holographic photopatterning of a formamide polar phase and an acrylate monomer-containing nonpolar phase to produce layered polymer structures [31]. Formamide is a highly polar solvent that forms well-dispersed, nonaqueous emulsion droplets within the monomer-containing nonpolar phase before holographic patterning. Photopolymerization of the nonpolar phase produces an ordered structure. Formamide evaporates from the multilayered nanostructures with high optical reflectivity and wide reflection bandwidths.

6.4 Potential applications

In this section, we discuss the applications of layered soft materials in various fields such as sensors, actuators, anisotropic molecular diffusion, optical devices, and soft templates for controllable syntheses. All of these potential applications are closely related to their iridescent structural colors, nanoscale layered structures, and soft features.

6.4.1 Layered structures for chemical/physical sensing

To meet the growing needs in the health and environmental fields, significant research efforts have been made to fabricate low-cost, high-performance sensors capable of responding to different chemical (e.g., target molecules) or physical stimuli (e.g., stress/strain) from the environment. Soft materials with tunable structural colors can provide high-level optically detectable readouts in response to external stimuli. When compared with other dye-based or plasmonic nanoparticle-based sensors, soft materials with layered structures are very cost-effective and offer advantages such as nonfading, environment-friendly, reversible, and direct readouts.

Another advantage of using soft materials to sense analytes is that they can be transferred to any substrate, even onto curved surfaces for instant measurements [32,33]. Layered materials undergo a change in their interlayer distances in response to external stimuli and exhibit a color change, which allows qualitative or quantitative detection of the analytes.

6.4.1.1 Strain sensitivity

Color tuning of layered materials in response to environmental stimuli is based on two mechanisms. One is based on the change in volume caused by the swelling or deswelling of the soft material, such as a hydrogel in response to chemical stimuli. The other is based on shape changes triggered by external stimuli such as mechanical stress. They change the structural color in response to mechanical stress/strain owing to modulation of the interlayer distance. Under mechanical stress, the color blue-shifts and returns to its original color after the stress/strain is removed. The change in the color or reflection spectrum indicates how much stress/strain they have been subjected to. Thus they are called mechanochromic materials and have potential applications as stress/strain sensors.

The three main important characteristics required for mechanochromic soft materials are high color saturation, wide color tunability, and fast color switching. For example, a layered PDGI/PAAm hydrogel exhibited a color change in response to mechanical strain [12]. As the interlayer distance decreased from 260 to 180 nm, the red-colored gel underwent a blue shift when an increasing strain (ε) was loaded onto the material. However, because of the hydrophobic PDGI bilayers, such layered materials were highly viscoelastic; therefore, they exhibited a slow color recovery (of several minutes) upon removal of the mechanical stress.

However, in nature, many tropical fishes exhibit iridescent structural colorations that can undergo extremely rapid transformations in response to environmental stimuli. To improve the response speed and expand the color shift in hydrogels, Yue et al. split the rigid and continuous bilayers into smaller domains. As a result, the bilayer domains did not deform during compression but only facilitated light reflection, resulting in an ultrafast response (of approximately 0.1 Ms) and an extremely wide wavelength band (340−640 nm) in response to external stress/strain [34].

6.4.1.2 Temperature sensitivity

In addition to their response to mechanical stress, layered hydrogels can be designed to demonstrate swelling or deswelling responses to changes in environmental conditions by embedding responsive materials into photonic crystal structures.

For example, a PDGI/PAAm-PAAc layered hydrogel can respond to changes in the temperature or pH of water, resulting in color tuning [35]. In the soft PAAm-PAAc layers, the polymer networks form hydrogen bonds that strengthen the gel. These hydrogen bonds are weakened at elevated temperatures, giving the soft layers a positive thermosensitivity, that is, they swell in response to increasing temperature and vice versa. Furthermore, the dissociation of the carboxyl groups of PAAc at high pH causes significant swelling of the soft layers, resulting in a responsive layered gel that can detect changes in the pH in water [35].

6.4.1.3 Ionic-strength sensitivity

Soft materials that are sensitive to ionic strength are usually synthesized from ionizable polymers, such as sodium polyacrylate, which exhibit different swelling behaviors in solutions with different ionic strengths. Most of the prepared layered hydrogels with neutral polymer networks do not respond to ionic strength. To overcome this, hydrolysis is employed to modulate the degree of ionization of the hydrogel by changing some of the amide groups in the neutral polymer network to carboxyl groups, thereby producing sodium polyacrylate within the gel.

The degree of ionization and charge density in the hydrogels have important effects on the swelling and deswelling behaviors. The developed layered hydrogels are sensitive to ionic strength and exhibit different structural colors in different sodium chloride solutions [36]. The reflection spectrum of the hydrogel is blue-shifted when the ionic strength increases. For example, when the ionic strength of the salt solution was increased from 0.001 to 0.05 mol/L, the maximum wavelength was blue-shifted from 680 to 500 nm (Fig. 6.5A, B).

Furthermore, based on the integration of laser power and chemical modifications, patterned photonic hydrogels were fabricated using laser engraving. They were embedded with different polymer composites (PAAcNa polyelectrolyte and PAAm neutral polymers) along a prescribed laser-printed path (Fig. 6.5C) [36]. The laser defects on the layered gels facilitated the rapid diffusion of molecules into or out of the gel, along with a dynamic color change on the patterns.

6.4.1.4 Electric-field sensitivity

As mentioned above, layered materials can display structural color changes by reflecting specific wavelength from the environment without any light-emitting elements. Therefore, the development of layered photonic materials for paper-like full-color displays has attracted considerable attention. They can be integrated into scalable, flexible substrates, and the displayed colors are visible under ambient conditions in bright rooms or outdoor sunlight, which meets the basic requirements for practical mobile display applications.

An electrically tunable soft material was developed based on a polyelectrolyte-layered hydrogel with high water content (Fig. 6.6) [21]. This photonic hydrogel exhibited versatile color tuning in response to an electric field applied either parallel or perpendicular to the direction of the gel layers. When an electric field was applied parallel to the layers, this layered gel exhibited rainbow-like structural colors.

Some photonic patterns in layered materials can appear or disappear in response to external stimuli, which means that hidden information can be revealed in specific environments such as mechanical stress or changes in pH. For example, layered hydrogels exhibit a green butterfly pattern under ambient conditions, which disappears under mechanical stress [36]. Similarly, some hydrogels possess a uniform red color in solutions with a pH above 5, where the pattern changes to green or blue as the pH decreases [37]. This feature makes them potentially useful in security applications.

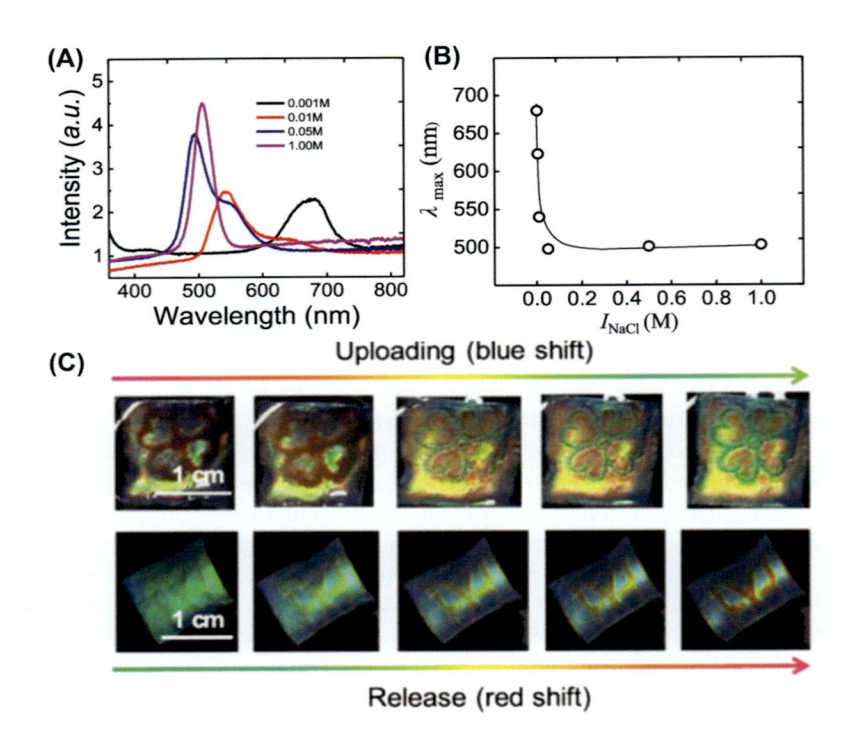

FIGURE 6.5

Ionic strength sensitive layered hydrogels. (A) Reflection spectra of the layered hydrogel in NaCl solutions of different concentrations and (B) their maximum reflection wavelength plotted as a function of the ionic strength. (C) Dynamic molecular uploading and release in the patterned photonic hydrogels.

Reprinted (adapted) with permission from Y. Yue, T. Kurokawa, Designing responsive photonic crystal patterns by using laser engraving, ACS Appl. Mater. Interfaces. 11 (2019) 10841–10847, Copyright (2019) American Chemical Society.

6.4.2 Layered structures for anisotropic molecular diffusion

Conventional soft materials, such as hydrogels, absorb solvents at an identical rate through their interfaces, that is, the solvent/molecules diffuse isotropically into the polymer networks. However, in layered soft materials, the diffusion of water molecules in and out of the polymer networks is influenced by the layered structure. For example, in a platelet gel with alternating hydrophobic and hydrophilic layers, anisotropic diffusion is observed because of the presence of thousands of hydrophobic layers (Fig. 6.7). More specifically, a PAAm hydrogel allows the same diffusion at every interface, whereas a layered hydrogel exhibits anisotropic diffusion, where the diffusion coefficient in the direction parallel to the layers ($D_\parallel$) is much larger than that in the direction perpendicular to the layers ($D_\perp$) [7]. As a result, external molecules can only cross the gel in the direction parallel to the layers, whereas in the direction perpendicular to the layers, molecular diffusion is restricted. Li et al. reported physically layered hydrogels that exhibited tunable isotropic/anisotropic

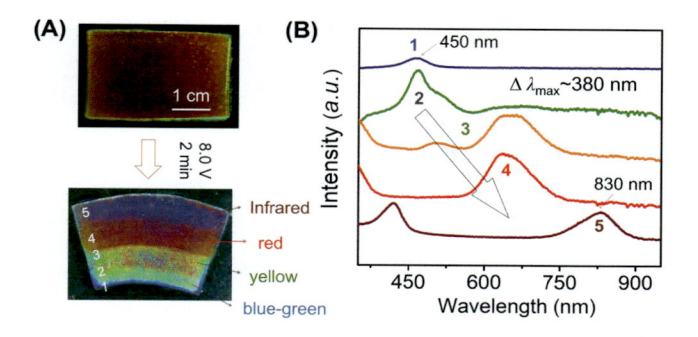

FIGURE 6.6

Electric field induced color tuning in layered hydrogels. (A) Color change and (B) the reflection spectra of the layered gels in response to an electric field applied parallel to the layers.

Reprinted with permission from Y. Yue, Y. Norikane, J.P. Gong, Ultrahigh-water-content photonic hydrogels with large electro-optic responses in visible to near-infrared region, Adv. Opt. Mater. 9 (2021) 2002198. Copyright 2021 Wiley-VCH Verlag GmbH & Co. KGaA, Weinheim.

structures. Accompanied by structural transition, the layered hydrogels exhibited isotropic and anisotropic swelling [38].

This alternating hydrophobic/hydrophilic layered structure resulted in unique molecular uptake and release behavior. In contrast to isotropic materials, the hydrophobic layers can be considered as protective membranes that separate the interior of the materials from the external environment. The hydrophilic gel layers tend to retain their shapes during swelling, whereas hydrophobic bilayers tend to retain their areas upon swelling-induced deformation to maintain a low surface energy [39]. When the water in the hydrophilic layers is evaporated, the layered microstructure is also well preserved owing to the existence of thousands of hydrophobic layers [40].

Based on anisotropic structures and diffusion, these layered gels can be used to control molecular uptake/release and thus have potential applications in controllable drug release systems. In plate-like hydrogels, molecules or drugs can be released from defects on the surface of the gel via specific predesigned routes. For example, laser defects on the layers facilitate the diffusion of molecules or water-soluble drugs in or out of the gel (Fig. 6.7A, B) [36].

In gels with multicylindrical structures, the molecules or water-soluble drugs can undergo rapid diffusion along the axial direction of the gel; however, the diffusion that occurs along the radial direction of the gels is hindered by their alternating hydrophobic layers (Fig. 6.7C) [41]. Thus these tubelike gels allow unique quasi-one-dimensional diffusion when compared with common gels without layered structures (Fig. 6.7D).

From the above examples, a targeted release window for wound healing can be designed with an extended duration of drug release as it is released from a single direction, which helps to maintain a constant, desired drug release concentration to the wound site. Moreover, the release of target molecules or drugs can be evaluated based on the color change of the periodic layered materials used.

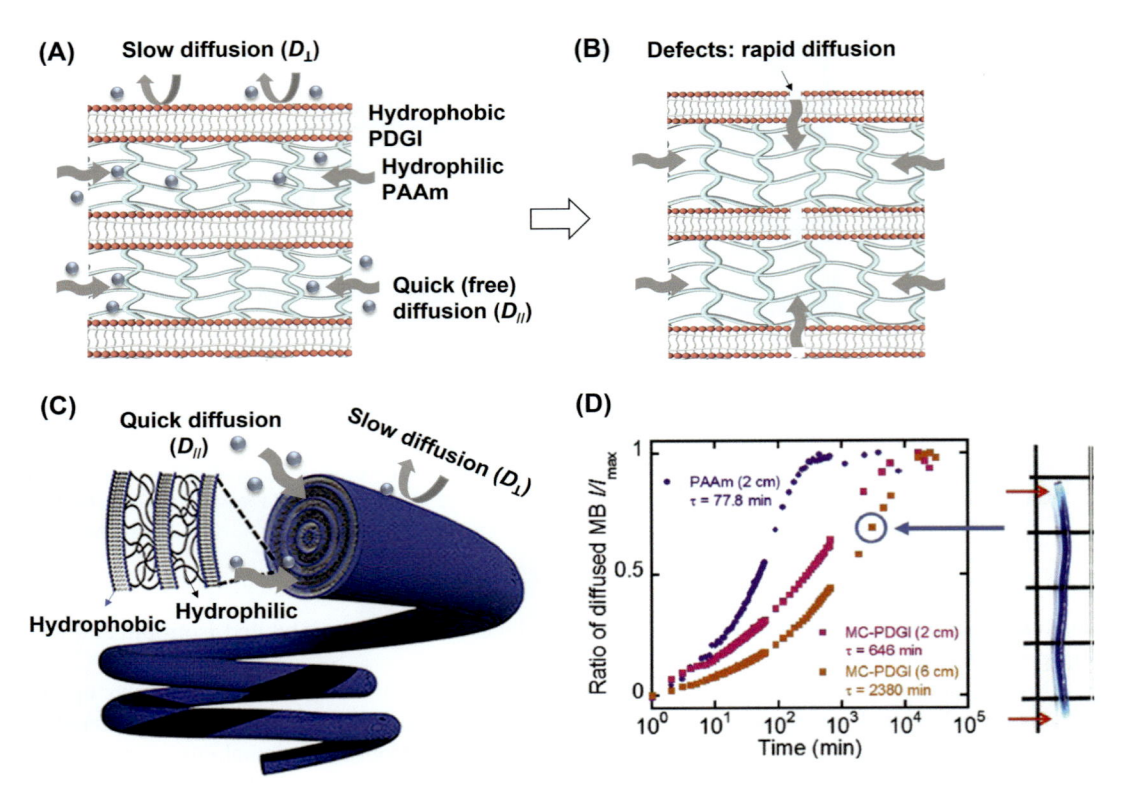

FIGURE 6.7

Anisotropic diffusion in plate-like and multicylindrical layered gels. (A and B) The external molecules diffuse into the plate-like gel in the direction parallel to the layers. The diffusion is also facilitated by the defects on the gel surface. (C) The external molecules diffuse into multicylindrical gel along the axial direction of the gel. (D) Quasi-one-dimensional diffusion in multicylindrical gels.

Panels A and B were reprinted (adapted) with permission from Y. Yue, T. Kurokawa, Designing responsive photonic crystal patterns by using laser engraving, ACS Appl. Mater. Interfaces. 11 (2019) 10841–10847, Copyright (2019) American Chemical Society. Panels C and D were reprinted with permission from K. Mito, M.A. Haque, T. Nakajima, et al., Supramolecular hydrogels with multi-cylindrical lamellar bilayers: swelling-induced contraction and anisotropic molecular diffusion, Polym. 2017;128:373-378. Copyright 2017 Elsevier Ltd.

6.4.3 Layered structures used for light management purposes in devices

There are many examples of nanoscale layered materials used in optical filters and devices for controlling light propagation. In this section, we summarize the technological applications of layered soft materials in devices such as optical filters and photovoltaics.

6.4.3.1 Optical filters

Optical filters based on periodic multilayer systems are well known and highly developed in various fields. Such layered structures acquire specific optical properties through the periodic modulation of the interlayer spacing or refractive index. Many applications of such materials involve selective filters or antireflective coatings [42].

The generally employed multilayer systems fabricated by alternating the deposition of inorganic-only materials, such as TiO_2 and SiO_2, result in highly rigid coatings that are not suitable for adaptation on surfaces of arbitrary curvature or different physicochemical properties. Calvo et al. presented a series of hybrid layered materials fabricated by integrating TiO_2 and SiO_2 nanoparticle multilayers in a soft polydimethylsiloxane film [42]. These layered materials could be employed as flexible and freestanding optical filters that shield against undesirable radiation, including the harmful UV range.

6.4.3.2 Photovoltaics

Owing to the current energy-related problems, there has been considerable research interest in thin-film photovoltaic devices based on solution-processable technologies. These systems use photoactive materials such as semiconducting polymers, organic molecules, and inorganic colloids. However, photovoltaic devices suffer some drawbacks such as high charge recombination and limited spectral absorption.

Thus a number of methods have been employed to improve light harvesting. Among them, a periodic layered nanostructure known as a dielectric mirror layer with photonic crystal properties has been used to improve the optical properties. For example, a multilayered film was inserted on the backside of the cell to reflect unabsorbed light back into the device, thereby allowing the transmitted photons to pass through the semiconductor device for a second time. For such applications, inorganic materials with vastly different dielectric constants are combined with photovoltaic devices. For example, Míguez et al. used porous SiO_2/TiO_2 ($nSiO_2 = 1.45$; $nTiO_2 = 2.44$) together with a dye-sensitized solar cell to achieve amplified light absorption for enhanced power-conversion efficiency [43].

As mentioned above, pure inorganic nanomaterials are not adaptable on different types of surfaces, which limits their applications in flexible devices. Thus Míguez et al. developed TiO_2 and SiO_2 nanoparticle multilayers in a polydimethylsiloxane film, which was soft, environment-friendly, and adhesive. This flexible inorganic—organic nanolayered hybrid film was further used as a back reflector in dye-sensitized solar cells to improve the energy-conversion efficiency of the device [42].

6.4.4 Layered structures used as soft templates for controllable synthesis

Iridescent layered structures in water can be used as soft organic templates for the controllable synthesis of inorganic nanomaterials. In some cases, multilayered structures can function as reducing, shape-controlling, and stabilizing agents. Layered structures can also trigger the lateral growth of metallic nanoparticles on the surface of bilayer membranes, while the growth along the thickness direction is hindered due to the lack of reducing agents. Thus it is possible to synthesize large-size 2D nanoparticles.

For example, thin gold nanoplates can be synthesized in the presence of multilayered structures in water (Fig. 6.8) [44]. The original layered structures composed of thousands of bilayer membranes with water arise from the self-assembly of amphiphilic molecules, and they show a greenish structural color change due to Bragg diffraction of light on the periodic bilayers (with layer distances of up to approximately 200 nm). When gold ions were added, the metallic ions were absorbed on the surface of the bilayers and reduced to gold nanoparticles. The nanoparticles in solution changed to a dark-wine color owing to surface plasmon resonances as a result of their interaction with light. Here, the hydrophilic heads of the amphiphilic molecules played a role in reducing the effects. The layered structures triggered the growth of the gold nanoparticles to form 2D shapes; therefore, many thin and semitransparent gold nanoplates can be synthesized using this approach. In particular, for nanoparticles exhibiting a large shape anisotropy, the diameters of the nanoplates can be more than 6 µm with a thickness of only 10 nm. The assembly of such large-sized gold nanosheets can be further deformed into freestanding and highly conductive 3D metallic architectures.

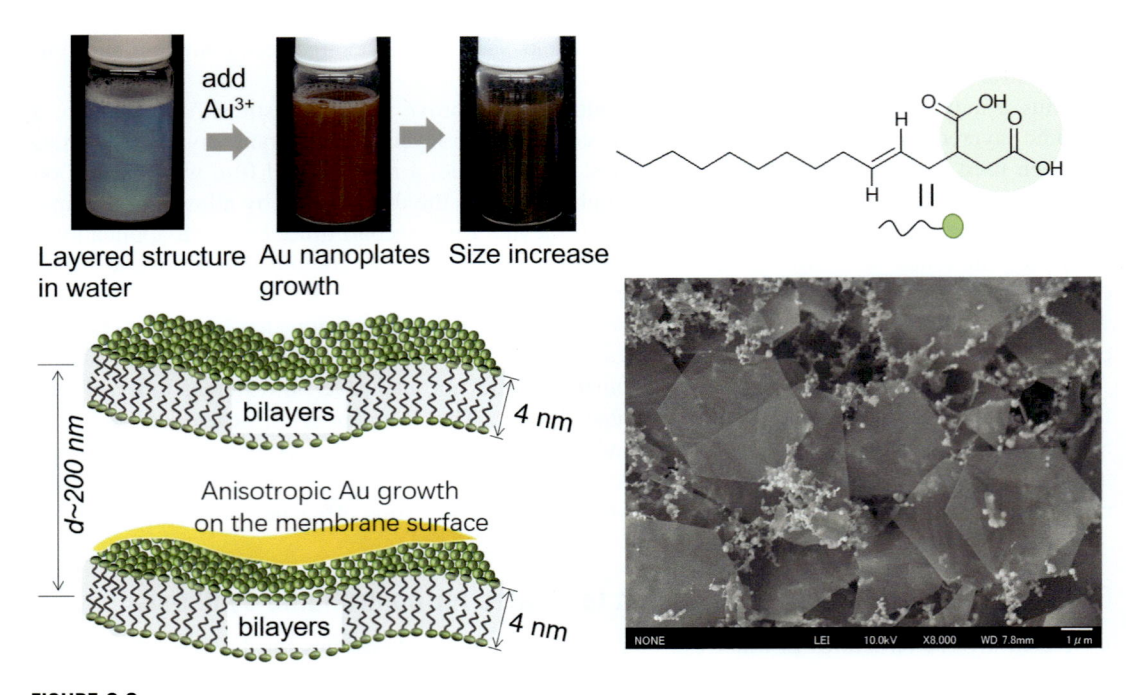

FIGURE 6.8

Flexible layered membranes in water that work as soft templates for the controlled synthesis of large-scale, thin gold nanoplates.

6.5 Excellent mechanical properties of nanoscale layered structures in soft materials

Nanoscale layered structures in soft materials also provide excellent mechanical properties. This principle has been demonstrated in various natural materials. For example, the cross-section of an abalone shell shows a layered accumulation of calcium carbonate ($CaCO_3$) platelets separated by thin layers of conchiolin (an elastic biopolymer). The abalone shells comprise 95 wt.% crystalline ceramic and 5 wt.% polymeric components [45]. Due to these alternating hard and soft layers, they exhibit ultrahigh fracture resistance that greatly exceeds that of pure $CaCO_3$. Thus abalone shells have adapted to their living conditions to protect their soft bodies from various external aggressions. In addition to abalone shells, bones are also known for their high strength, which is attributed to the layered platelet structures in a ductile organic matrix.

Both polymeric and ceramic materials have mechanical properties within their own specific applications but cover only a fraction of the required properties. Nanoscale layered materials are designed to overcome the weaknesses of one material and to obtain desirable mechanical properties based on the nature of the composite.

6.5.1 High fracture strength

Similar to the abovementioned hard materials, the same principle applies to soft materials. Polymers with alternating soft- and hard-layered structures exhibit high fracture strengths. For example, a PDGI/PAAm (hard/soft) layered gel experienced fracture stress and strain of approximately 600 kPa and 22, respectively, under tensile elongation, whereas a pure PAAm gel experienced lower fracture stress and strain of 38 kPa and 11, respectively (Fig. 6.9A) [46]. Therefore, the fracture stress and strain experienced by a layered gel were 15.8- and twofold higher than those of the gel with simple constituents, respectively.

High fracture strengths have also been observed for other layered materials. For example, Zhao et al. demonstrated a strategy to produce nanocomposite films with highly ordered layered structures using shear flow to induce the alignment of 2D graphene oxide or clay nanosheets (hard constituent) at an immiscible hydrogel/oil interface (soft constituent) [4]. As a result, the nanolayered hybrid films exhibited a tensile strength of up to 1215 ± 80 MPa and a Young's modulus of 198.8 ± 6.5 GPa, which are 9.0 and 2.8 times higher than that of natural nacre, respectively.

6.5.2 Anisotropic mechanical properties

The layered structure in soft materials also provides anisotropic mechanical properties (e.g., anisotropic modulus). For example, the elastic modulus of the gel was significantly higher in the parallel to the layer direction ($E_\parallel$) than that in the perpendicular direction ($E_\perp$). Common PAAm gels, similar to most hydrogels that lack an ordered structure, always exhibit an isotropic elastic modulus ($E_\parallel = E_\perp$). However, for layered hydrogels, $E_\parallel$ can be one order of magnitude larger than $E_\perp$, resulting in a large anisotropy in the elastic modulus (Fig. 6.9B) [12]. This is because the elastic modulus of the hard layers ($\sim$ MPa) is much higher than that of the soft layers ($\sim$ kPa).

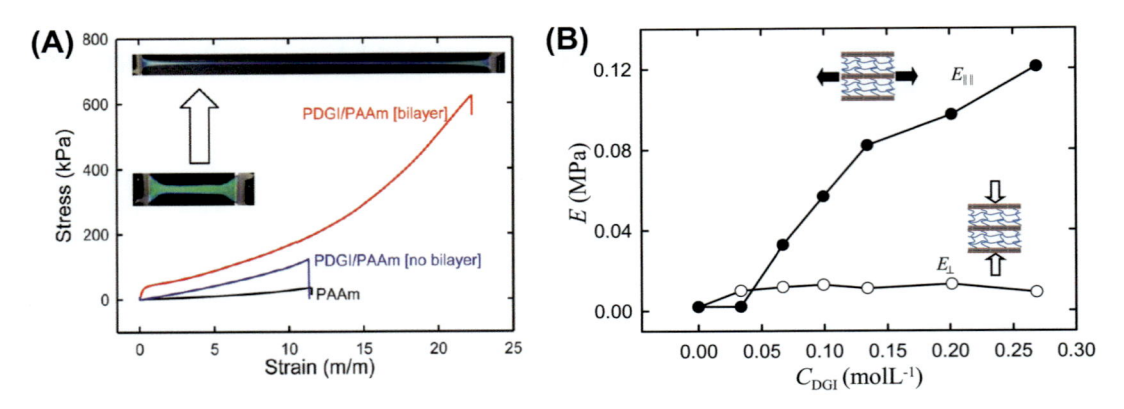

FIGURE 6.9

Mechanical properties: (A) high fracture stress and (B) anisotropic elastic moduli.

Panel A was reprinted (adapted) with permission from M.A. Haque, T. Kurokawa, G. Kamita, J.P. Gong, Lamellar bilayers as reversible sacrificial bonds to toughen hydrogel: hysteresis, self-recovery, fatigue resistance, and crack blunting, Macromolecules 44 (2011) 8916–8924, Copyright (2011) American Chemical Society. Panel B was reprinted with permission from M.A. Haque, G. Kamita, T. Kurokawa, K. Tsujii, J.P. Gong, Unidirectional alignment of lamellar bilayer in hydrogel: one-dimensional swelling, anisotropic modulus, and stress/strain tunable structural color, Adv. Mater. 22 (2010) 5110–5114, Copyright 2010 Wiley-VCH Verlag GmbH & Co. KGaA, Weinheim.

Liu et al. prepared a composite hydrogel with a layered structure of titanate nanosheets embedded in a gel that aligned cofacially in a magnetic flux [47]. This layered material also exhibited anisotropic mechanical properties. For example, compressing the hydrogel orthogonal ($\perp$) to the plane of the titanate nanosheets yielded an $E_\perp$ value of 66 kPa at 40% strain, which is 2.6 times greater than that of $E_\parallel = 25$ kPa obtained when compressing the hydrogel parallel ($\parallel$) to the titanate nanosheets [47].

6.5.3 High crack resistance

Tensile strength represents the deformation energy of a material, and fracture strength describes the ability of a material containing a crack to resist fracturing in units of J/m^2. The fracture strength of the natural nacre can reach 1.5 kJ/m^2, which is approximately 3000 times higher than that of pure $CaCO_3$ platelets (0.5 J/m^2) [48]. As mentioned above, this high fracture strength is attributed to the alternating nanoscale inorganic/organic layered structure.

Common gels, such as PAAm, exhibit extremely rapid crack propagation because of their single isotropic polymer networks with a low critical tearing energy of 2.8 J/m^2. However, for a hydrogel with a layered structure, it is difficult to propagate a crack through a preexisting crack. The hard layers with lipid-like character dramatically reduce the maximum stress at the crack tip and further increase the crack resistance of the layered material. Thus when the hydrogel is elongated perpendicular to the crack direction, significant blunting occurs at the crack tip (Fig. 6.10A) [46].

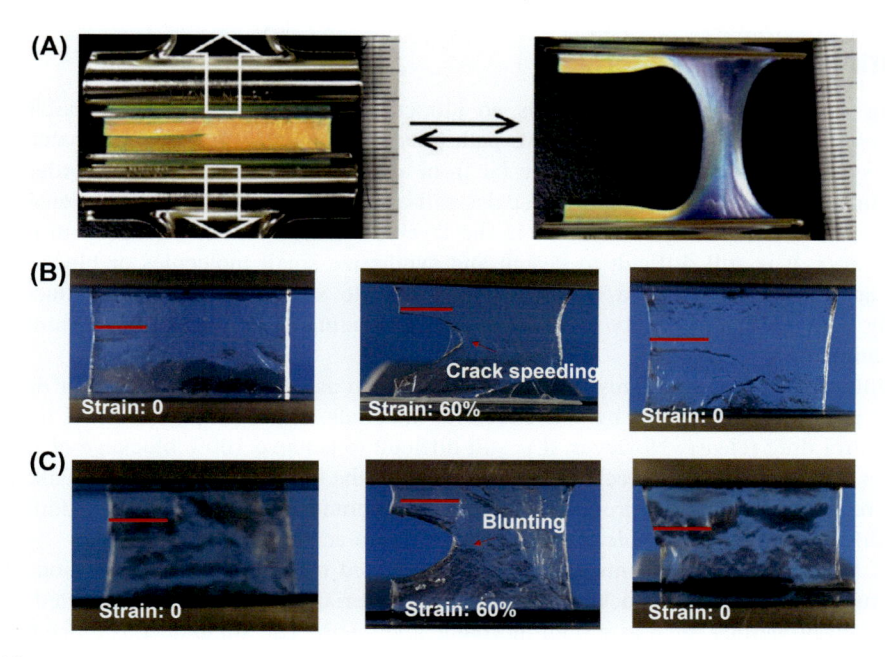

FIGURE 6.10

High crack resistance arising from the alternating layered structures and strong interfaces between the layers. (A) Crack resistance of a layered hydrogel by forming a large blunting of the crack tip. (B) Crack speeding in a polyelectrolyte-layered gel. (C) Improved crack resistance of the polyelectrolyte-layered gel after loading an electric field in the direction perpendicular to the layers.

Panel A was reprinted with permission from M.A. Haque, T. Kurokawa, G. Kamita, J.P. Gong, Lamellar bilayers as reversible sacrificial bonds to toughen hydrogel: hysteresis, self-recovery, fatigue resistance, and crack blunting, Macromolecules 44 (2011) 8916–8924, Copyright (2011) American Chemical Society. Panel B and C were reprinted with permission from Y. Yue, Y. Yokota, G. Matsuba, Polyelectrolyte-layered hydrogels with electrically tunable toughness, viscoelasticity, hysteresis, and crack resistance, Macromolecules 55 (2022) 1230–1238, Copyright (2022) American Chemical Society.

In addition to the layered structures, the interface strength between the layers is another important parameter that must be considered in layered materials. The interfacial strength of layered materials arises from three phenomena: hydrogen bonding, ionic bonding, and covalent bonding. Introducing such interfacial forces into layered materials can significantly improve fracture resistance [49]. For example, in high-water-content polyelectrolyte-layered hydrogels, weak interfaces exist between the layers, resulting in a low critical tearing energy (12.94 J/m^2). Owing to the weak interfaces, delamination could easily occur in these layered gels during crack propagation (Fig. 6.10B) [50]. However, when a voltage was applied to the gel in the direction perpendicular to the layers, the crack resistance was largely increased to 221.8 J/m^2, owing to the increase in the interfacial strength that arose from the physical bonding formed under the electric field (Fig. 6.10C).

6.6 Outlook and perspectives

Mimicking nanostructures from nature is an attractive research topic. The hierarchical design approach observed in nature is an effective path toward designing soft structural materials [51,52] However, synthetic structural materials are far from achieving the same degree of order in the layered structures. In some cases, small molecules self-assemble into thousands of layered structures, thereby achieving a high degree of order in the layered structures from molecular to macroscopic scale. However, it is still difficult to design and synthesize small molecules or block copolymers that can achieve a high degree of order in large-scale self-assemblies. Sometimes, the self-assembling behavior can be highly sensitive to the surrounding environment, which limits its practical applications.

In addition to layered structures, some properties, such as anisotropic swelling or diffusion, are interesting phenomena exhibited by soft structural materials. These properties could potentially be used to control drug release. However, it is still difficult to achieve 100% blockage of the diffusion of water or molecules in the direction perpendicular to the hydrophobic bilayers. Full anisotropic diffusion requires perfectly hydrophobic layered membranes without defects. Additionally, the selective diffusion of specific molecules is still difficult to achieve in soft materials.

One of the future developments in bio-inspired layered materials is the integration of multiple types of functional building blocks into the same material, which can be expected to achieve unprecedented properties, such as a high mechanical strength. The simultaneous embedding of two or more building blocks into layered materials with a robust interfacial design will provide new opportunities for obtaining high-performance artificial layered materials.

Acknowledgments

This work was supported by the Japan Society for the Promotion of Science (JSPS) KAKENHI (grant nos. JP20H05237 and JP20H04683) and by Japan Science and Technology Agency (JST), Precursory Research for Embryonic Science and Technology (PRESTO), grant no. JPMJPR209A.

References

[1] F. Liu, B. Dong, X. Liu, Y. Zheng, J. Zi, Structural color change in longhorn beetles *Tmesisternus isabellae*, Opt. Express 17 (2009) 16183−16191.
[2] S. Vignolini, P.J. Rudall, A.V. Rowland, et al., Pointillist structural color in Pollia fruit, Proc. Natl Acad. Sci. 109 (2012) 15712−15715.
[3] Y. Yue, J.P. Gong, Tunable one-dimensional photonic crystals from soft materials, J. Photochem. Photobiol. C. Photochem Rev. 23 (2015) 45−67.
[4] C. Zhao, P. Zhang, J. Zhou, et al., Layered nanocomposites by shear-flow-induced alignment of nanosheets, Nature. 580 (2020) 210−215.
[5] P. Lova, G. Manfredi, D. Comoretto, Advances in functional solution processed planar 1D photonic crystals, Adv. Opt. Mater. 6 (2018) 1800730.

[6] A.L. Liberman-Martin, C.K. Chu, R.H. Grubbs, Application of bottlebrush block copolymers as photonic crystals, Macromol. Rapid Commun. 38 (2017) 1700058.

[7] Y. Yue, J.P. Gong, Structure and unique functions of anisotropic hydrogels comprising uniaxially aligned lamellar bilayers, Bull. Chem. Soc. Jpn. 94 (2021) 2221−2234.

[8] K. Naitoh, Y. Ishii, K. Tsujii, Iridescent phenomena and polymerization behaviors of amphiphilic monomers in lamellar liquid crystalline phase, J. Phys. Chem. 95 (1991) 7915−7918.

[9] X. Chen, H. Mayama, G. Matsuo, T. Torimoto, B. Ohtani, K. Tsujii, Effect of ionic surfactants on the iridescent color in lamellar liquid crystalline phase of a nonionic surfactant, J. Colloid Interface Sci. 305 (2007) 308−314.

[10] M. Hayakawa, T. Onda, T. Tanaka, K. Tsujii, Hydrogels containing immobilized bilayer membranes, Langmuir. 13 (1997) 3595−3597.

[11] K. Tsujii, M. Hayakawa, T. Onda, T. Tanaka, A novel hybrid material of polymer gels and bilayer membranes, Macromolecules. 30 (1997) 7397−7402.

[12] M.A. Haque, G. Kamita, T. Kurokawa, K. Tsujii, J.P. Gong, Unidirectional alignment of lamellar bilayer in hydrogel: one-dimensional swelling, anisotropic modulus, and stress/strain tunable structural color, Adv. Mater. 22 (2010) 5110−5114.

[13] Y.N. Ye, M.A. Haque, A. Inoue, Y. Katsuyama, T. Kurokawa, J.P. Gong, Flower-like photonic hydrogel with superstructure induced via modulated shear field, ACS Macro Lett. 10 (2021) 708−713.

[14] Z. Wang, C.L.C. Chan, R.M. Parker, S. Vignolini, The limited palette for photonic block-copolymer materials: a historical problem or a practical limitation? Angew. Chem. Int. Ed. 61 (2022) e202117275.

[15] G.M. Miyake, R.A. Weitekamp, V.A. Piunova, R.H. Grubbs, Synthesis of isocyanate-based brush block copolymers and their rapid self-assembly to infrared-reflecting photonic crystals, J. Am. Chem. Soc. 134 (2012) 14249−14254.

[16] G.M. Miyake, V.A. Piunova, R.A. Weitekamp, R.H. Grubbs, Precisely tunable photonic crystals from rapidly self-assembling brush block copolymer blends, Angew. Chem. Int. Ed. 51 (2012) 11246−11248.

[17] J.A. Dolan, B.D. Wilts, S. Vignolini, J.J. Baumberg, U. Steiner, T.D. Wilkinson, Optical properties of gyroid structured materials: from photonic crystals to metamaterials, Adv. Opt. Mater. 3 (2015) 12−32.

[18] G.M. Whitesides, J.P. Mathias, C.T. Seto, Molecular self-assembly and nanochemistry: a chemical strategy for the synthesis of nanostructures, Science 254 (1991) 1312−1319.

[19] K. Almdal, J.H. Rosedale, F.S. Bates, G.D. Wignall, G.H. Fredrickson, Gaussian-to stretched-coil transition in block copolymer melts, Phys. Rev. Lett. 56 (1990) 1112.

[20] B.R. Sveinbjörnsson, R.A. Weitekamp, G.M. Miyake, Y. Xia, H.A. Atwater, R.H. Grubbs, Rapid self-assembly of brush block copolymers to photonic crystals, Proc. Natl Acad. Sci. 109 (2012) 14332−14336.

[21] Y. Yue, Y. Norikane, J.P. Gong, Ultrahigh-water-content photonic hydrogels with large electro-optic responses in visible to near-infrared region, Adv. Opt. Mater. 9 (2021) 2002198.

[22] G. Beadie, M. Brindza, R.A. Flynn, A. Rosenberg, J.S. Shirk, Refractive index measurements of poly (methyl methacrylate)(PMMA) from 0.4−1.6 μm, Appl. Opt. 54 (2015) 139−143. 4946.

[23] G. Manfredi, C. Mayrhofer, G. Kothleitner, R. Schennach, D. Comoretto, Cellulose ternary photonic crystal created by solution processing, Cellulose. 23 (2016) 2853−2862.

[24] G. Manfredi, P. Lova, F. Di Stasio, P. Rastogi, R. Krahne, D. Comoretto, Lasing from dot-in-rod nanocrystals in planar polymer microcavities, RSC Adv. 8 (2018) 13026−13033.

[25] G. Canazza, F. Scotognella, G. Lanzani, S. De Silvestri, M. Zavelani-Rossi, D. Comoretto, Lasing from all-polymer microcavities, Laser Phys. Lett. 11 (2014) 35804.

[26] S.-J. Jeon, M.C. Chiappelli, R.C. Hayward, Photocrosslinkable nanocomposite multilayers for responsive 1D photonic crystals, Adv. Funct. Mater. 26 (2016) 722−728.

[27] I.R. Howell, C. Li, N.S. Colella, K. Ito, J.J. Watkins, Strain-tunable one dimensional photonic crystals based on zirconium dioxide/slide-ring elastomer nanocomposites for mechanochromic sensing, ACS Appl. Mater. Interfaces 7 (2015) 3641−3646.

[28] M.M. Ito, A.H. Gibbons, D. Qin, et al., Structural colour using organized microfibrillation in glassy polymer films, Nature. 570 (2019) 363−367.

[29] H. Song, K. Singer, J. Lott, et al., Continuous melt processing of all-polymer distributed feedback lasers, J. Mater. Chem. 19 (2009) 7520−7524.

[30] D. Langhe, M. Ponting, Manufacturing and Novel Applications of Multilayer Polymer Films, Elsevier, Oxford, UK, 2016.

[31] V.K. Hsiao, K.-T. Yong, A.N. Cartwright, et al., Nanoporous polymeric photonic crystals by emulsion holography, J. Mater. Chem. 19 (2009) 3998−4003.

[32] M. Däntl, S. Guderley, K. Szendrei-Temesi, et al., Transfer of 1D photonic crystals via spatially resolved hydrophobization, Small. 17 (2021) 2007864.

[33] W. Ma, S. Li, D. Kou, J.L. Lutkenhaus, S. Zhang, B. Tang, Flexible, self-standing and patternable P (MMA-BA)/TiO2 photonic crystals with tunable and bright structural colors, Dye. Pigment. 160 (2019) 740−746.

[34] Y. Yue, T. Kurokawa, M.A. Haque, et al., Mechano-actuated ultrafast full-colour switching in layered photonic hydrogels, Nat. Commun. 5 (2014) 4659. Available from: https://doi.org/10.1038/ncomms5659.

[35] Y.F. Yue, M.A. Haque, T. Kurokawa, T. Nakajima, J.P. Gong, Lamellar hydrogels with high toughness and ternary tunable photonic stop-band, Adv. Mater. 25 (22) (2013) 3106−3110. Available from: https://doi.org/10.1002/adma.201300775.

[36] Y. Yue, T. Kurokawa, Designing responsive photonic crystal patterns by using laser engraving, ACS Appl. Mater. Interfaces 11 (2019) 10841−10847.

[37] Y. Yue, X. Li, T. Kurokawa, M.A. Haque, J.P. Gong, Decoupling dual-stimuli responses in patterned lamellar hydrogels as photonic sensors, J. Mater. Chem. B 4 (2016) 4104−4109.

[38] X. Li, T. Kurokawa, R. Takahashi, et al., Polymer adsorbed bilayer membranes form self-healing hydrogels with tunable superstructure, Macromolecules. 48 (2015) 2277−2282.

[39] T. Nakajima, C. Durand, X.F. Li, M.A. Haque, T. Kurokawa, J.P. Gong, Quasi-unidirectional shrinkage of gels with well-oriented lipid bilayers upon uniaxial stretching, Soft Matter 11 (2015) 237−240.

[40] M. Ilyas, M.A. Haque, Y. Yue, et al., Water-triggered ductile−brittle transition of anisotropic lamellar hydrogels and effect of confinement on polymer dynamics, Macromolecules. 50 (2017) 8169−8177.

[41] K. Mito, M.A. Haque, T. Nakajima, et al., Supramolecular hydrogels with multi-cylindrical lamellar bilayers: swelling-induced contraction and anisotropic molecular diffusion, Polym. (Guildf.) 128 (2017) 373−378.

[42] M. Calvo, H. Míguez, Flexible, adhesive, and biocompatible Bragg mirrors based on polydimethylsiloxane infiltrated nanoparticle multilayers, Chem. Mater. 22 (2010) 3909−3915.

[43] S. Colodrero, A. Mihi, L. Häggman, et al., Porous one-dimensional photonic crystals improve the power-conversion efficiency of dye-sensitized solar cells, Adv. Mater. 21 (2009) 764−770.

[44] Y. Yue, Y. Norikane, Gold clay from self-assembly of 2D microscale nanosheets, Nat. Commun. 11 (2020) 568−576.

[45] M. Sarikaya, K. Gunnison, M. Yasrebi, I. Aksay, Mechanical property-microstructural relationships in abalone shell, MRS Online Proc. Libr. 174 (1989) 109−116.

[46] M.A. Haque, T. Kurokawa, G. Kamita, J.P. Gong, Lamellar bilayers as reversible sacrificial bonds to toughen hydrogel: hysteresis, self-recovery, fatigue resistance, and crack blunting, Macromolecules. 44 (2011) 8916−8924.

[47] M. Liu, Y. Ishida, Y. Ebina, et al., An anisotropic hydrogel with electrostatic repulsion between cofacially aligned nanosheets, Nature. 517 (2015) 68−72.

[48] Q. Cheng, L. Jiang, Z. Tang, Bioinspired layered materials with superior mechanical performance, Acc. Chem. Res. 47 (2014) 1256−1266.

[49] Y. Yue, R. Hayashi, Y. Yokota, Co-self-assembly of amphiphiles into nanocomposite hydrogels with tailored morphological and mechanical properties, ACS Appl. Mater. Interfaces 15 (2023) 21507−21516. Available from: https://doi.org/10.1021/acsami.3c01862.

[50] Y. Yue, Y. Yokota, G. Matsuba, Polyelectrolyte-layered hydrogels with electrically tunable toughness, viscoelasticity, hysteresis, and crack resistance, Macromolecules. 55 (2022) 1230−1238.

[51] U.G. Wegst, H. Bai, E. Saiz, A.P. Tomsia, R.O. Ritchie, Bioinspired structural materials, Nat. Mater. 14 (2015) 23−36.

[52] K. Ariga, X. Jia, J. Song, et al., Nanoarchitectonics beyond self-assembly: challenges to create bio-like hierarchic organization, Angew. Chem. Int. Ed. 59 (2020) 15424−15446.

Metal Nanoarchitectonics: Fabrication of Sophisticated Gold Nanostructures for Functional Plasmonic Devices

7

Hideyuki Mitomo and Kuniharu Ijiro

Research Institute for Electronic Science, Hokkaido University, Sapporo, Japan

7.1 Introduction

The questions of how we can create useful materials and fabricate highly functional devices are important issues at the moment. Of course, we can learn from nature, particularly from living creatures. Living cells possess many amazing functional nano-systems composed of biomolecules, such as lipids, nucleic acids, and proteins. Although they are made of limited elements, such as C, H, O, Pi, N, and S, they build unimaginably excellent but complex systems involving sophisticated nanostructures formed via self-organization. For example, the cell membrane consists mostly of lipids that form a lipid bilayer via self-assembly. Proteins are composed of sequential amino acids as a primary structure and spontaneously construct hierarchical nanostructures, known as secondary, tertiary, and quaternary structures, through self-organization, including multistep folding processes and the assembly of various components. Further, both dynamic changes in conformation based on flexible structures and structural rearrangements based on reversible assembly/disassembly systems have led to emergent functions, such as vesicular transportation by the budding and fusion of lipid vesicles, molecular recognition and catalysis with enzymes, and cytoskeletal formation with actins and microtubules. In other words, not only fine structures but also spatiotemporally changeable flexible structures are thought to be critical factors in the tremendous functionalities observed in living systems, and these sophisticated nanostructures are realized by self-organization and the soft properties of the constituent materials.

Recently, "nanoarchitectonics" has come under the spotlight in some research fields as a new paradigm that is based on the arrangement of nanoscale structural components into a predefined configuration to create novel functionality through the mutual interactions of the constituent components [1,2]. That is, there is now an emphasis on the construction of hierarchical structures via the self-organization of various artificial nanoscale building blocks through the multiple interactions to afford novel functions. According to this concept, many functional materials or systems with elegant nanostructures were produced, based mostly on supra-molecular chemistry and polymer science [3,4]. From the viewpoint of material scientists, this can be regarded as an artificial expansion of bio-molecular systems composed of organic molecules. As mentioned above, living things only use limited elements for their survival. On the other hand, we can handle various kinds of elements including metals and semiconductors, and may further expand this concept to inorganic materials.

Materials Nanoarchitectonics. DOI: https://doi.org/10.1016/B978-0-323-99472-9.00004-3

There are various inorganic materials, some of which show unique physical properties, such as electrical, magnetic, optical, and photonic properties, on a nanoscale but not in their bulk state [5,6]. Typical examples can be found in semiconductors and metals. Some semiconductors of a few nanometers in size emit strong fluorescence, known as quantum dots (QDs), which have useful applications in quantum information science [7] as well as bio-related applications [8]. Some noble metals; e.g., gold and silver, show significant nanoscale size- and structure-dependent properties, such as localized surface plasmonic resonance (LSPR), via strong light-matter interactions [9,10]. As plasmonic properties in metal nanostructures, in particular, are sensitive to the surrounding environment and also have the ability to enhance fluorescence and Raman signals, metal nanostructures have received considerable attention across a variety of research fields due to their great potential for applications in sensing, biomedicals, energy harvesting, and so on [6,11−18]. Thus, the development of nanofabrication technologies with a focus on inorganic materials also has received a good deal of attention in the research fields of nanoscience and nanotechnologies. Nanofabrication techniques can be roughly divided into two categories, known as top-down and bottom-up approaches. Generally speaking, top-down approaches can afford freely designed two-dimensional (2D) structures of a little as several tens of nanometers in size. However, the fabrication of three-dimensional (3D) or conformation changeable nanostructures exhibiting four-dimensional (4D) behavior remains a challenge. On the other hand, self-organization as a bottom-up type technology can produce sophisticated nanostructures with a minimal energy, once the preparation design through self-organization is identified, although this in itself is a significant issue in terms of the fabrication and integration of complex structures. As with molecular self-assembly [4], the fabrication of self-assembled metal nanoparticles is of great interest to researchers due to fact that the plasmon coupling effect derived from the closely packed structures obtained can provide greatly increased electric fields and further enhanced signals at their narrow gaps [19−21]. Plasmonic nanoparticles also demonstrate significant properties, even in a single dispersed state or a simple 2D-assembled static structure. Thus, the main interest now lies in how much greater are the properties found in 3D- or 4D-controlled plasmonic nanostructures.

It is clear that the fabrication of sophisticated inorganic nanostructures affords one answer to the creation of useful materials and the production of highly functional devices. So, the question arises as to how they can be successfully fabricated. We have studied this topic for a decade [22−24] and many well-written reviews on the fabrication of static 3D nanostructures have already been published [25−27]. Thus, in this chapter, we introduce the topic of the fabrication of sophisticated 4D-metal nanostructures for functional plasmonic devices as a part of "metal nanoarchitectonics" with a particular focus on our research.

7.2 Stimuli-responsive metal nanoparticles for configurable structures via assembly/disassembly

Both conformational changes based on flexible structures and structural rearrangement based on reversible assembly/disassembly provide elegant functions in living creatures. Thus, changes in metal nanostructures have become a particular focus of attention [28−30]. One of the widely studied approaches is the stimuli-responsive assembly/disassembly of metal nanoparticles dispersed in solution driven by various stimuli, such as pH [31,32], temperature [33−35], and light [36−38]. Surface modification with stimuli-responsive molecules is the key to this approach. For pH- and light-responsiveness, small molecules are usually used for surface modification. On the other hand, for temperature-responsiveness, polymers, such as poly(N-isopropyl acrylamide) (pNIPAM), poly(propylene glycol), or oligo(ethylene glycol), hereinafter abbreviated as OEG, -attached polymers as a pendant group, are mostly used, although only a limited number of reports on thermo-responsive small molecules have been published to date [39]. Surface modification is fundamental to the stable dispersion of nanoparticles [40]. As polymer coatings can not only reduce inter-particle interactions by providing distance between particles but also confer specific surface properties, they have been widely applied to nanoparticles [41]. On the other hand, for plasmonic nanoparticles, the distance between the nanoparticles on their assembly is critical to the provision of strongly enhanced electric fields, known as hot spots, from plasmon coupling effects [42,43]. To put it more simply, the shorter the distance, the stronger is the electric field. From this viewpoint, surface modification with small molecules appears to have greater potential. Charged molecules, which have carboxyl or amino groups, can provide good dispersity in a polar solvent due to electrostatic repulsion and also demonstrate switchable surface properties due to the solvent conditions. While temperature change is a useful stimulus based on its ease of delivery and/or removal in a non-invasive manner, few reports on thermo-responsive nanoparticles coated with small molecules exist, despite its great potential. In this section, we introduce an example of thermo-responsive gold nanoparticles coated with small molecules and present its unique characteristics.

We focused on the OEG-attached surfactants. While these surfactants are widely used as nonionic detergents or surface modifiers providing, via immobilization, bioinert surfaces that prevent non-specific adsorption of biomacromolecules in biosensors [44], they have another use due to the thermo-responsive molecules in their micelle structures. To link these bio-friendly properties with thermo-responsiveness, we developed thermo-responsive AuNPs coated with OEG-derivatives (Fig. 7.1A) [45−47]. Interestingly, although hexa(ethylene glycol) undecane thiol (**OH**-terminated ligand) molecules, which are widely used for the modification of gold surfaces for application in biosensing devices, show a clouding at 49°C, **OH**-ligand-modified AuNPs do not show any thermo-responsiveness in the temperature range from 5 to 70°C. The lower critical solution temperature

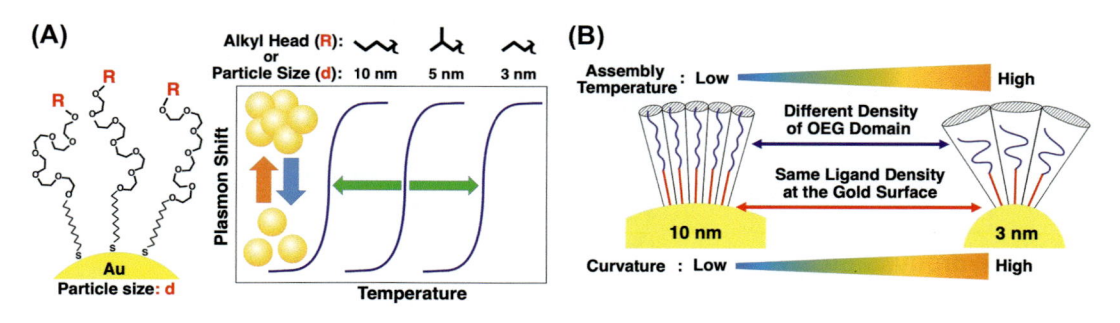

FIGURE 7.1

Thermo-responsive AuNPs coated with OEG-derivatives. (A) A schematic illustration of the AuNP surface and thermo-responsive assemblies and (B) a schematic representation of the relationship between assembly temperature and surface curvature [45].

(LCST) or clouding point of thermo-responsive molecules is tunable via the attachment of hydrophobic substrates; thus, the terminus of the OEG-ligands was modified with an ethyl (**C2**), isopropyl (**iC3**), or propyl (**C3**) group. The three resultant OEG-derivative-modified AuNPs, all having a diameter of 5 nm, demonstrated thermo-responsive assembly/disassembly at 19, 33, and 56°C, respectively. Significant changes in assembly temperature (T_A) resulted from small changes in chemical structure. Importantly, the size of the core particle also had a significant effect on T_A, with **C2**-ligand-coated AuNPs showing a T_A of 39, 56, and 67°C at diameters of 3, 5, and 10 nm, respectively. From this, it can be seen that surface curvature also represents a key factor in the tuning of their T_A. It is expected that surface curvature contributes to differences in the local density of the OEG moieties, despite the fact that there is no difference in ligand density at the gold surface (ca. 5 molecules/nm^2) (Fig. 7.1B). Dynamic light scattering (DLS) measurements show that their thermal-responses were sudden, with size changes from the assembly/disassembly state completed within a temperature change of 1°C (Fig. 7.2A) [47]. This could be the result of the uniformity of the surface molecules in terms of molecular weight and homogenous ligand density derived from the well-packed self-assembled monolayer. Molecular dynamics (MD) simulation also provided molecular information regarding the AuNP surfaces, such as their configuration (Fig. 7.2B). Although the details remain unclear, these thermo-responsive gold nanoparticles have unique characteristics, such as curvature dependence, which appear to be difficult to realize by polymer-coating.

OEG-derivatives confer unique thermo-responsive properties to the nanoparticles. How does this lead to sophisticated assemblies? As is well known, metal nanoparticles of various shapes, such as rods, plates, and cubes, have been prepared by wet synthesis [48,49], and these nanoparticles show additional unique plasmonic properties [50]. Thus, the fabrication of ordered structures via the self-assembly of anisotropic nanoparticles is also an important research topic. The formation of various ordered assemblies, such as static structures with anisotropic nanoparticles, was reported. However, the dynamic control of these structures is limited. Here, we introduce particle shape (surface curvature)-mediated ordered assembly using OEG-derivative-modified thermo-responsive gold nanoparticles.

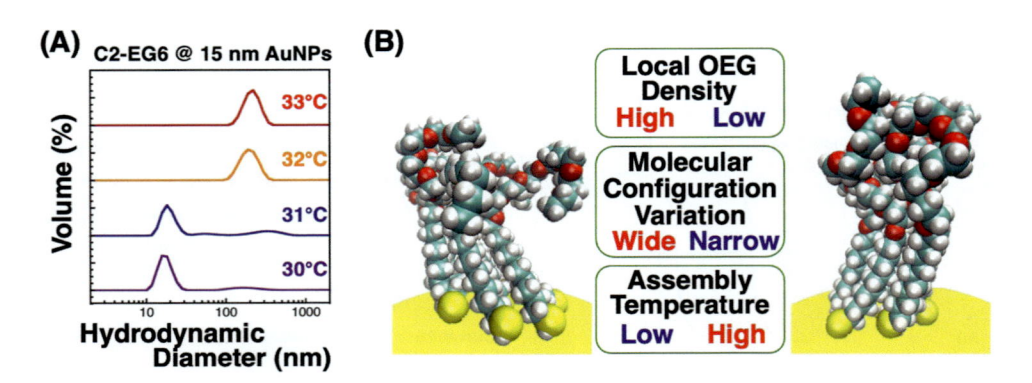

FIGURE 7.2

The sudden responsiveness to temperature change conferred by surface ligands. (A) Size changes (assembly formation) of **C2-EG6** ligand-modified AuNPs upon heating and (B) molecular configuration of the ligands on the curved surfaces calculated by MD simulation [47].

First, we applied the above to rod-shaped gold nanoparticles (gold nanorods: AuNRs) [46]. It should be noted that AuNRs have two different curvatures on each particle; i.e., the side and the edge portion. This means that OEG-derivative-modified AuNRs should show two different thermo-responsive properties after simple surface modification with a single kind of ligand. Thus, we prepared **C2**-ligand-modified AuNRs and investigated their thermo-responsive properties (Fig. 7.3). As a result of the lower degree of curvature at the side portion, dehydration of the OEG element first occurred at T_{A1}, leading to side-by-side assembly via hydrophobic interactions. Further heating resulted in a completely dehydrated state at T_{A2}, which led to assembled-assemblies. Based on this result, an anisotropic shape was expected to provide hierarchical self-assembly, even after simple surface modification with a uniform surface ligand. Our first report using spherical nanoparticles was unable to exclude volume effects related to inter-particle forces, as the volume of the AuNPs changes with changes in curvature. On the other hand, the surface curvature of AuNRs can be changed while maintaining their volume or their volume can be changed while maintaining their surface curvature through the tuning of their diameter and length. To confirm the curvature effect on T_A, three AuNRs of different sizes, 33×8, 33×14, and 54×15 nm (length x diameter), were prepared (Fig. 7.4). The first pair of AuNRs were of equivalent length but different diameter, while the latter pair of AuNRs were of different lengths but similar diameter. The results clearly support our hypothesis that curvature (in this case diameter) rather than volume (in this case length) is the dominant factor in tuning T_A [46]. This report demonstrates that not only the molecules' chemical structure but also the nanoparticle shape contributes to the tunability of their thermo-responsive properties, with anisotropic nanoparticles providing hierarchical assemblies due to their stepwise assembly, even in the case of uniform modification with a single kind of surface ligand.

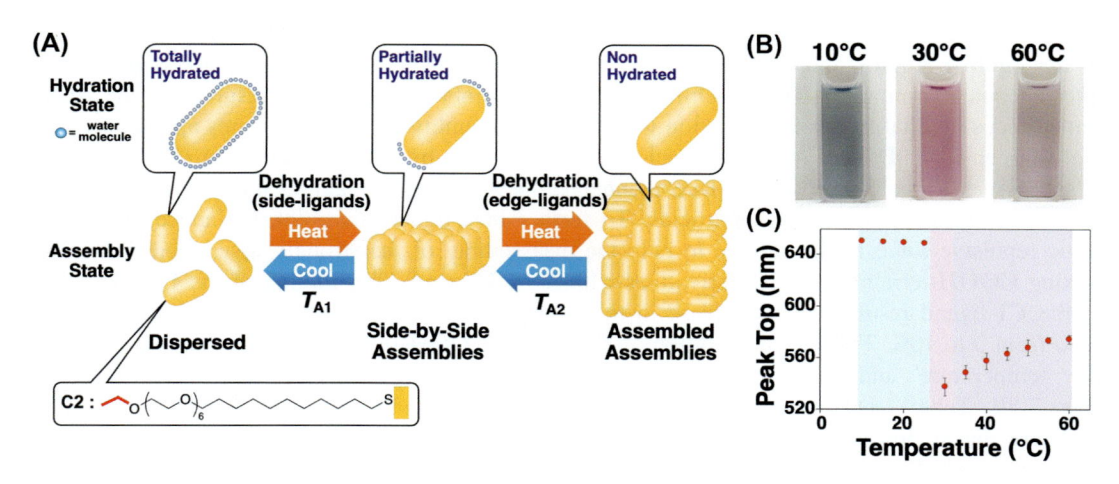

FIGURE 7.3

Two-step assembly of thermo-responsive AuNRs. (A) A schematic illustration of the temperature-dependent changes in AuNR hydration states and assembled states. (B) Photographs of dispersed **C2**-modified AuNRs at 10, 30, and 60 °C. (C) Temperature-dependent changes in the longitudinal LSPR peak, showing a sudden blue-shift at around 30°C and a red-shift over 35°C.

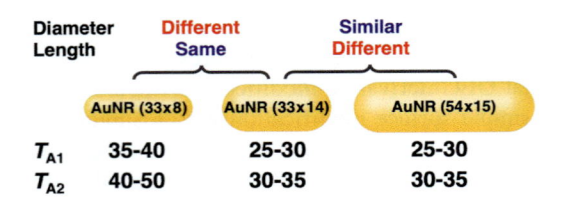

FIGURE 7.4

Effects of size on the T_{A1} and T_{A2} of the **C2**-modified AuNRs [46].

Next, we applied the same approach to disk-shaped gold nanoparticles (gold nanodisks: AuNDs) to investigate their thermo-responsive assembly [51]. Similar to the AuNRs, AuNDs also have two different curvatures on each particle, the flat faces and the circular edge. Thus, according to our previous report [46], it is easy to imagine the face-to-face assembly of the primary form upon heating (Fig. 7.5A). Experimental results support this expectation. However, things do not work so simply in reality and we found significant differences between AuNRs and AuNDs in terms of their thermo-responsive assembly/disassembly. The thermo-responsive experiments using **C1**-ligand-modified AuNDs, which had a diameter of 105 nm and a thickness of ca. 6 nm, showed a plasmonic peak shift between 30 and 40°C upon heating due to assembly formation. This temperature range is much lower than that of spherical gold nanoparticles (AuNSs) (73°C for 10 nm, 63°C for 15 nm, and 53°C for 20 nm in diameter) [47] or AuNRs (ca. 65°C for 33 × 14 nm) [46]. These findings are in line with the curvature dependence of the T_A, in which the lower curvature of their flat surfaces results in a lower T_A. A significant difference was observed in the disassembly process in that, unlike AuNPs and AuNRs, **C1**-ligand-modified AuNDs did not show disassembly even after adequate cooling. The irreversibility of AuND assembly is thought to result from the AuND flat surfaces facilitating strong attraction to counter disassembly. Although spheres attach at a point (zero dimension; 0D) and rods can attach linearly in a side-by-side configuration (1D), discs can attach across a plane in a face-to-face manner (2D), which is thought to induce strong van der Waals interactions between nanoparticles. To further investigate this point, we added an electrostatic repulsive force to the surface to help the disassembly of the strongly assembled AuNDs by mixing **COOH**-terminated OEG derivatives with the **C1**-ligand. This mixing of the **COOH**-ligand to the **C1**-ligand resulted in a decrease in the zeta-potential of the AuNDs in accordance with a ratio from 0 to 10%. Further, the addition of 1 to 5% of **COOH**-ligand led to an increase in assembly temperature and aided disassembly, supporting their reversible assembly/disassembly (Fig. 7.5B, Table 7.1). On the other hand, 10% of **COOH**-ligand inhibited assembly formation, probably due to too strong repulsion or increased responsive temperature over the experimental range. It should be noted that the introduction of only 1% of **COOH**-ligand totally diminished the thermo-responsive assembly of AuNSs with a diameter of 40 nm. Further, **COOH**-ligand-introduced AuNDs also showed large hysteresis on assembly/disassembly (T_A-T_D) depending on the **COOH**-ligand content (Fig. 7.5C, Table 7.1). This effect could be due to the suppression of the increase in T_D against T_A, probably as a result of the stabilization of their assembly via hydrogen bonding between carboxylic groups and/or the carboxylic acid and oxygen atom in the ethylene glycol units under a hydrophobic environment between well-attached flat surfaces [52,53]. This

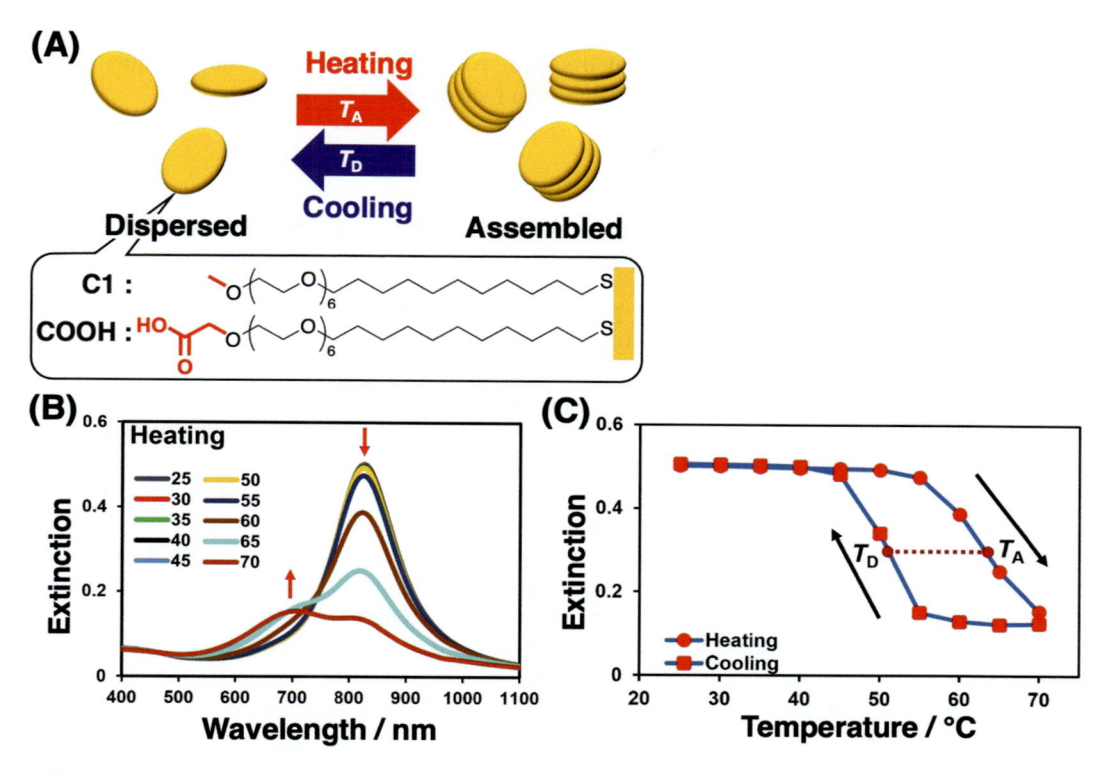

FIGURE 7.5

Thermo-responsive assembly/disassembly of AuNDs. (A) Schematic illustration and (B)(C) temperature-mediated spectral changes of 105 nm-AuNDs modified with 97% **C1**- and 3% **COOH**-ligands [51].

Table 7.1 The assembly (T_A) and disassembly (T_D) temperatures of AuNDs (105 nm and 60 nm in diameter) and of AuNSs (40 nm in diameter) modified with a mixture of C1 and COOH-ligands [51].

	105-AuND			60-AuND			40-AuNS		
C1: COOH	T_A	T_D	T_A -T_D	T_A	T_D	T_A-T_D	T_A	T_D	T_A-T_D
100:0	36	N.D.	N.D.	38	N.D.	N.D.	45	39	6
99:1	57	39	18	54	44	10	N.D.	N.D.	N.D.
97:3	67	52	15	66	53	13	-	-	-
95:5	74	49	25	76	52	24	-	-	-

N.D.; not determined.

phenomenon appears similar to the hydrogen bonding in the double-stranded DNA. As shown above, the flat surfaces of the AuNDs, not their size or volume, are critical to their unique thermo-responsive behaviors. These also support the significant shape effect on their assembly.

7.3 Nanoparticle assembly control with the aid of polymers as an additive

Due to the sensitivity of plasmon coupling phenomena to the assembled structure, it is essential to develop a system that allows on-demand structural changes through AuNP assembly/disassembly in a controlled configuration that includes AuNP-assembly size and gap distances among the AuNPs. Conventional approaches involve the direct attachment of stimuli-responsive molecules to the AuNP surfaces to produce stimuli-responsive assembly/disassembly systems. As their assembly formation originates from direct colloidal surface-surface interactions, the assembly behavior is difficult to control, with precipitation observed in many cases due to the formation of larger assemblies. Lequeux and coworkers reported that the uncontrolled aggregation of thermo-responsive AuNPs is prevented in the presence of Pluronic polymer in solution, thereby providing a totally reversible thermo-responsive assembly/disassembly system with well-defined assembly size [54]. Further, polymer-mediated self-assembly afforded both 2D and 3D nanoparticle networks [55]. These findings indicate that AuNP co-assembly in the presence of certain additives provides a means of solving the abovementioned issue. On the other hand, Klajn and coworkers reported that photo-switchable spiropyran derivatives can mediate the light-responsive assembly/disassembly of non-photo-responsive AuNPs [37,56]. Taken together, the above findings support the hypothesis that nanoparticle assembly can be controlled with the aid of additives, particularly stimuli-responsive polymers.

To obtain better controlled nanostructures with enhanced properties, more and more complex systems are required. For future expansion of these technologies, a simplified system based on clear mechanisms that allows greater refinement is preferable. From this perspective, we have pursued the self-assembly of AuNPs modified with OEG-derivatives as a small, simple molecule with a definite chemical structure. Based on the above, we fabricated a system for the stimuli-responsive assembly/disassembly of AuNPs possessing inert OEG-surfaces via co-assembly with anionic polymers (Fig. 7.6) [52]. In this, we mixed hexa(ethylene glycol) undecane thiol-modified AuNPs with poly(acrylic acid) (PAA) as an anionic polymer, of which the protonation upon changes in pH triggering hydrogen bond formation between the OEG-AuNPs and polymers to afford sensitive pH-responsive assemblies. The plasmonic properties and assembly size were regulated by a combination of the mixing ratio of PAA to AuNPs and the molecular weight of the PAA. Further, the attachment of the hydrophobic moiety to the OEG-ligand (**C1**-ligand) or anionic polymer terminus (poly(methyl acrylate)) changed the responsive pH range, likely through the enhanced hydrogen bonding between the ethylene glycol moieties and carboxylic acid under the local hydrophobic environment. In short, although these approaches show only limited controllability and need further improvement, the findings presented herein demonstrate that this co-assembly system with an external mediator allows the stimuli-responsive assembly of OEG-coated AuNPs with tunable plasmonic properties, assembly size, and stimuli-responsivity.

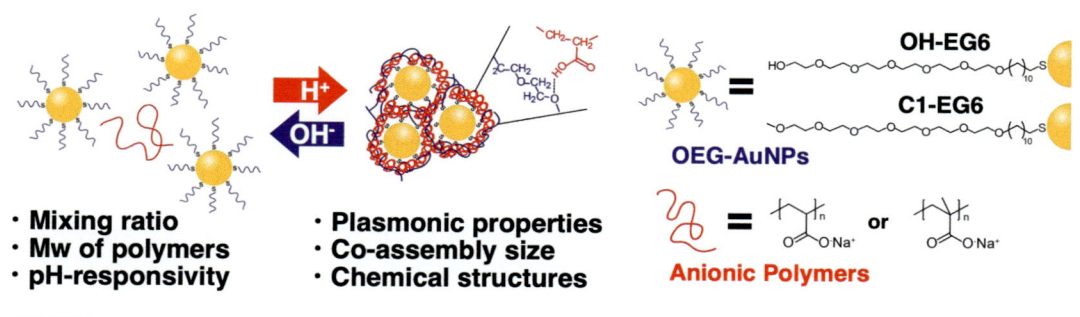

FIGURE 7.6

A schematic illustration of the pH-responsive assembly/disassembly of OEG-coated AuNPs with external anionic polymers via hydrogen bonding.

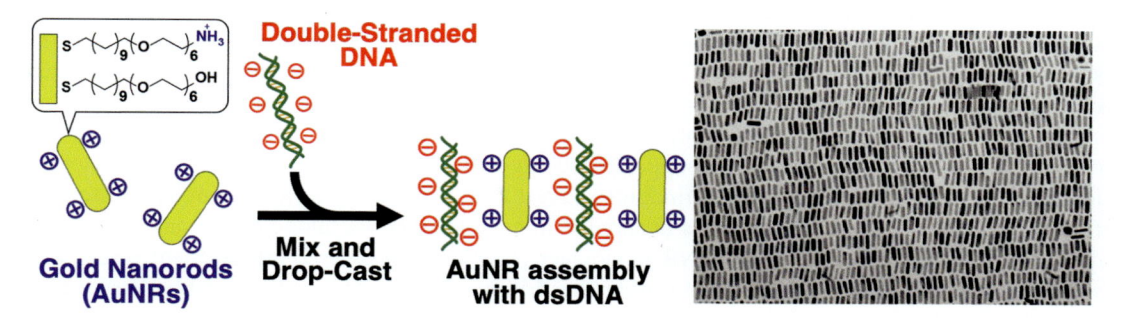

FIGURE 7.7

A schematic illustration and TEM image of dsDNA-assisted AuNR assembly formation by drop-casting [57].

As mentioned somewhere else, anisotropic plasmonic nanoparticles, such as AuNRs, have gained considerable attention as a building block owing to the important properties originating from their shape. Thus, the controlled assembly of anisotropic particles with the aid of additives is expected to progress as a research topic. The anisotropic-shaped nanoparticles can afford shape-specific interactions and, here, we introduce AuNR assembly with the aid of double-stranded DNA as a rigid anionic polymer of definite length [57]. The idea is that site-specific adsorption of rigid dsDNA along the longer axis of the AuNR leads to the formation of ordered side-by-side assemblies (Fig. 7.7). Electrostatic interactions were conferred by the modification of the AuNRs, 40 nm x 9 nm in size, with a mixture of cationic and nonionic alkanethiol ligands at a ratio of X:100-X, allowing control of their surface positive charge densities. After removal of excess free CTAB and alkanethiol ligands by repeat centrifugal purification, the AuNRs were mixed with the 148 bp dsDNA solution. The length of this DNA is ca. 50 nm, which is equivalent to the Kuhn length and slightly longer than the length of the AuNRs. Dynamic light scattering (DLS) experiments showed a size increase upon mixing, indicating assembly formation in solution. Interestingly, the weak cationic AuNRs, which were modified with 5% cationic ligand, provided larger assemblies than did strongly cationic (100% cationic ligand-modified) AuNRs. This is thought to be due to the strong electrostatic interactions between the 100% cationic ligand-modified AuNRs and dsDNAs forming tight complexes and inhibiting interconnection among complexes, resulting in the suppression of larger sized assemblies. Further investigation of the extinction spectra of the cationic AuNR solutions focused on changes in the L-LSPR peak, which affords information on AuNR assembly formation and morphology. Theoretical studies showed that side-by-side assembly in which the sides of the AuNRs are brought into close proximity results in a peak shift of the L-LSPR toward a shorter wavelength. Conversely, edge-to-edge assembly with the AuNR tips in close proximity results in a peak shift toward a longer wavelength [58,59]. The L-LSPR of the 5% cationic AuNRs was observed to spread toward both a shorter and longer wavelength (although the peak was shifted by only 2 nm toward a shorter wavelength), indicating that their assembly involves both edge-to-edge as well as the side-by-side interactions. These edge-to-edge interactions occur due to the presence of multiple contact points on the edge-portion of the AuNRs in side-by-side assemblies. Next, the 5% cationic AuNR solution mixed with dsDNA was drop-cast onto a grid for transmission electron microscopic (TEM) observation and then dried. TEM images revealed AuNR monolayer sheets with the AuNRs in parallel orientation and arranged in a well-ordered shape that resembled a 2D smectic structure. In contrast, the 100% cationic AuNRs showed a random orientation in small side-by-side assemblies consisting of just 3−5 AuNRs. This result indicates that a moderately charged AuNR surface is necessary for the formation of an ordered structure. Meanwhile, using a single-stranded (ss)DNA (148-base poly(dT)) in place of 148 bp dsDNA, resulted in the formation of random AuNR aggregates, which supported our hypothesis that the dsDNA rigidity is essential for this self-assembly. Overall, these findings suggest that dsDNA mediated side-by-side assembly formation as well as further assemblies in an edge-to-edge manner via tuning of the surface cationic density of the AuNRs to afford a AuNR monolayer sheet with an ordered arrangement resembling a 2D smectic structure on a substrate during the drying process. Although assembly/disassembly control in a 4D-structure was not examined in this study, the insights gained afforded a new approach to the assembly of anisotropic colloidal NPs with rigid ionic polymers into well-ordered NP monolayers via tuning of electrostatic interactions.

7.4 Precise active control of plasmonic nanostructures on polymer gels as a substrate

Dynamic change in plasmonic nanostructures represents one of the most important topics in the field of nanotechnologies, and tremendous efforts have been focused on the fabrication of 3D structures, such as hollow capsules, as well as their reversible assembly/disassembly in solution. Nevertheless, the reversible formation of 3D nanostructures with active and precise tunability remains a significant challenge. To this end, actively and precisely tunable 2D systems were first targeted. In this context, the placement of nanostructures "on" or "in" soft materials composed of polymers, such as elastomers and gels, provided an important approach. These materials can be stretched and expanded to a large degree, so that the gap distances in nanopatterns can be adjusted. To date, the results obtained using these polymer-supported approaches have been superior, in terms of reproducibility and adjustability, to those obtained using the assembly/disassembly control of dispersions in solutions. Due to the beneficial properties displayed by poly(dimethyl siloxane)

(PDMS), such as transparency, extensibility, and ease of handling, PDMS-based active plasmonic substrates have been developed for fundamental research on SPR [60−63] and surface-enhanced Raman scattering (SERS) [11,64,65], as well as for subsequent applications, such as in tunable coloring materials [66−68].

Notwithstanding the expansion of the applications of PDMS-based active plasmonic substrates, there have been few reports on practical SERS measurements for the label-free detection of biomolecules such as proteins, despite the potential benefits of applications with high sensitivity through the extreme enhancement of Raman scattering signals. One reason for this is related to the unsuitable affinity between PDMS and biomolecules. Hydrogel represents another soft material with hydrophilic properties that provides a number of bio-friendly features. Some hydrogels show large volume changes in response to stimuli or changes in external conditions and are, therefore, thought to provide a suitable platform for SERS biosensing on active plasmonic substrates. As part of our work on hydrogel-based active plasmonic devices [69−71] we found the precise control of ordered nanostructures to be important to achieving improved performance in terms of active plasmonics. However, we experienced a number of technical issues in relation to the preparation and observation of nanostructures on hydrogels. Therefore, we first developed a method for the preparation of metal nanostructures on hydrogels [69] involving gold patterns fabricated on a silicon substrate by conventional lithographic techniques prior to their transfer onto the gel surface by tuning of the gold-gel interactions (Fig. 7.8A). Active changes in the gap distances in the gold patterns on the PAA through volume changes were confirmed by optical microscope imaging on a micrometer scale and structural color changes on a sub-micrometer scale (Fig. 7.8B) [69]. Although diffraction experiments showed uniform changes in the gap distances, analyses using light generally afford only sub-micrometer accuracy. Further, several reports on the inhomogeneity of the polymer networks in the gels, prepared using a radical polymerization technique, provided data on a scale of tens to hundred nanometers [72,73].

As structural configurations on a nanoscale are valuable for plasmonic applications, we next examined their more precise tunability [70] using sub-100 nm gold nanodot arrays prepared by electron beam lithography. Atomic force microscopy (AFM) showed uniform gel-volume change-induced interval changes on the hydrogel. Nevertheless, the resolution was insufficient due to technical limitations associated with AFM analysis on the polymer gel surfaces. To obtain clear

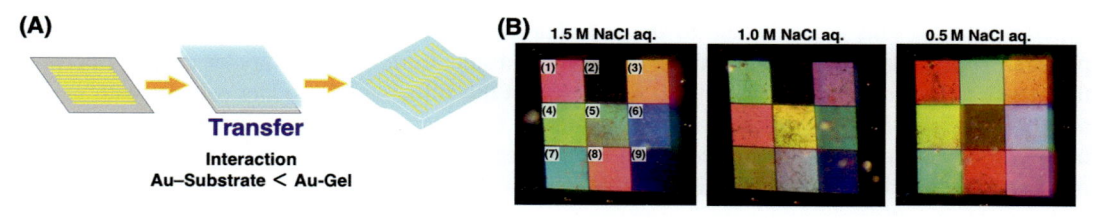

FIGURE 7.8

Active distance tuning of Au-patterns on PAA hydrogels (A) A schematic illustration of the fabrication of Au-patterns on a PAA hydrogel. (B) Structural color changes for Au-patterns on a PAA gel immersed in different concentrations of NaCl. Numbers denote the spacing between the Au dot: (1) 500 nm, (2) 150 nm, (3) 400 nm, (4) 350 nm, (5) 550 nm, (6) 250 nm, (7) 300 nm, (8) 450 nm and (9) 200 nm.

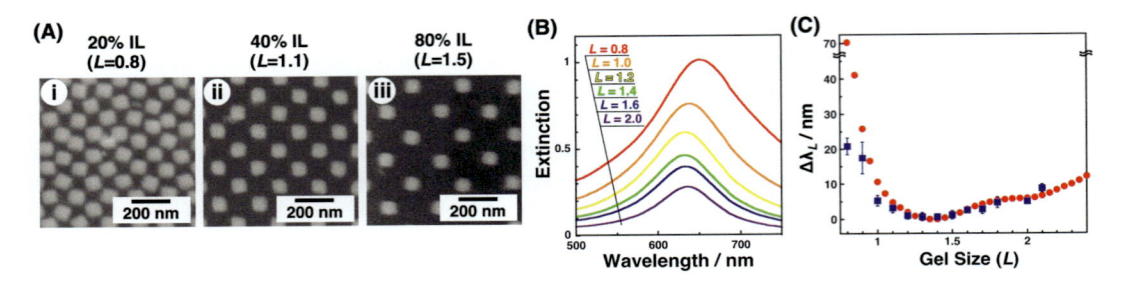

FIGURE 7.9

Nanoscale distance tuning of Au-nanopatterns on gels (A) SEM images of IL gels at various degrees of swelling, (i) 20% IL gels ($L = 0.8$), (ii) 40% IL gels ($L = 1.1$), and (iii) 80% IL gels ($L = 1.5$). Here, L refers to the relative gel size defined by the diagonal length of the total patterned area. (B) Extinction spectra of nanodot arrays on a hydrogel at various degrees of swelling. (C) Plots of spectral peak shifts ($\Delta\lambda_L$) against the relative gel size (L). $\Delta\lambda_L$ values determined from experimental spectra are shown in blue, with $\Delta\lambda_L$ values calculated from FDTD simulations shown in red.

Reproduced from [70] Copyright (2019) Royal Society of Chemistry.

nanoscale images, we obtained SEM images after solvent exchange in the gel from water to an ionic liquid (IL). As a hydrogel swollen with water easily shrinks under a vacuum, direct SEM observation is difficult. On the other hand, no changes in the size of gels swollen with an ionic liquid are observed under high vacuum conditions due to the non-volatile property. In addition, as there is no increase in charge as a result of electron beam exposure without gold or carbon deposition during SEM observation due to their conductive properties, clear images can be obtained at a nanoscale resolution (Fig. 7.9A). Good agreement was found between the plasmonic spectra derived from finite difference time domain (FDTD) simulations based on the SEM images and the experimentally determined spectra, which supports our expectations that the active changes in plasmonic patterns via gel volume changes are quite homogenous (Fig. 7.9B,C). Interesting plasmonic shifts were revealed by these spectral analyses: a reduction in the gap distances between gold nanodots caused a slight initial blue shift followed by a larger red shift. Although the red shift of the plasmon peak is a commonly observed effect of plasmon coupling, the observed blue shift is not generally observed by experimentalists. This is likely due to the fact that structural variations, mainly resulting from the preparation procedure, obscure this slight shift, which results from radiative dipolar coupling between the nanodots and retardation effects [74]. As a result of this effect, gold nanodot patterns with moderately sized gaps show a weaker surface plasmon in comparison to that of independently existing nanodots spaced at adequate distances from each other. The uniformly tunable plasmonic system in our experiments clearly showed this illegible effect. Our research, thus, showed the potential of hydrogel-based active plasmonic devices possessing high tunability.

We next introduce a novel approach to biosensing applications by SERS utilizing the above hydrogel-based actively tunable plasmonic device. In relation to the plasmonic properties, advanced SERS substrates have been fabricated with a particular focus on nanogap structures as a hot spot [75,76]. In general, narrower gaps provide stronger electromagnetic fields, which in turn are expected to afford greater enhancement. On the other hand, narrow gaps restrict analyte entry into the hot spots. Thus, a trade-off exists between the two conditions. When small molecules such as crystal violet, a well-known Raman-active dye, are used as analytes, this trade-off may not be critical. In the case of macromolecules, however, gap distance becomes critical as steric hindrance can limit the approach of analytes to the narrow gaps acting as hot spots. From this perspective, while wider gaps are desirable for target insertion, narrower gaps are desirable for the greater enhancement of detection. This trade-off can be resolved using actively gap tunable plasmonic systems. We, therefore, examined the effects of active gap control using a hydrogel-based tunable plasmonic system [71]. We prepared highly ordered plasmonic structures by the self-assembly of AuNPs via drying on the substrate according to our previous reports [77]. These thin films were then transferred onto PAA hydrogels and active plasmonic tuning was confirmed by spectral analysis upon changes in NaCl concentration. The effects of the active gap control during SERS measurements were subsequently evaluated through the detection of crystal violet (CV; molecular weight (MW) of 408 Da when containing a chloride ion) as a small molecule and the hemoprotein cytochrome c (Cyto c; MW of 12 kDa) as a relatively large molecule. SERS intensities were increased with increases in the salt concentration in both cases, demonstrating that gap distance is a significant factor. The effect of active gap control on the insertion of target molecules into hot spots was next evaluated through a comparison of the three approaches shown in Fig. 7.10A. When the gaps were maintained in an "open form," no signals were detected for CV or Cyto c (Fig. 7.10A-i open form), with weak signals detected due to the enhancement by plasmonic fields at the hot spots when the gaps were maintained in a "closed form" (Fig. 7.10A-ii closed form). On the other hand, when target molecules were introduced in an open state and SERS measurement was performed in a closed state, significantly increased signals were observed in comparison to those in the closed form

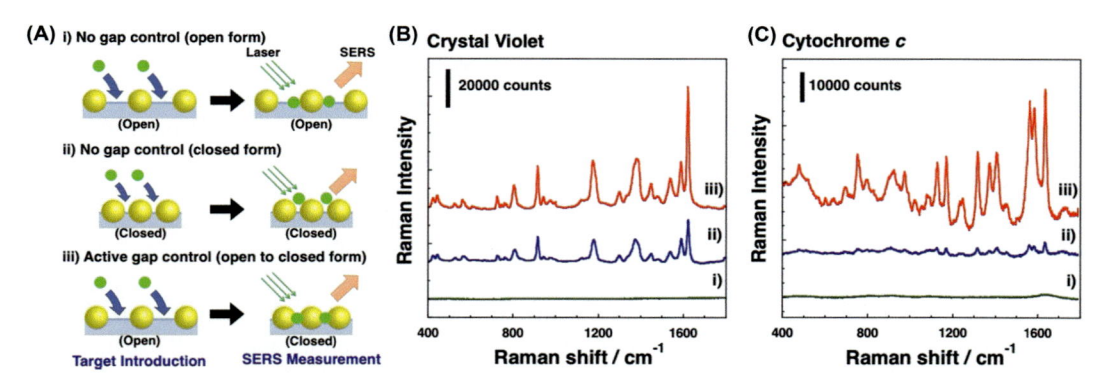

FIGURE 7.10

SERS measurements of cytochrome c and crystal violet using tunable plasmonic substrates. (A) Schematic representations of the three approaches: i) no gap control in an "open form" - in this approach, analytes were injected onto a gel in an expanded state (in MilliQ water) and analyzed as is; ii) no gap control in a "closed form" - here, the analytes were injected onto a gel in a contracted state (in 1 M NaCl solution) and analyzed as is; and iii) active gap control in an "open-to-closed form"- in this approach, the target molecules were injected onto a gel in an expanded state (in MilliQ water) and analyzed after contraction (in 1 M NaCl solution). (B)(C) SERS spectra of and crystal violet (B) and cytochrome c (C) for each of the approaches.

without gap control, despite the fact that the conditions for SERS measurement were the same (Fig. 7.10A-iii open-to-closed form). These findings indicate that active gap control promoted the introduction of target molecules into the gaps, which then became hot spots. The differences in signal intensities, ca. two-fold for CV and ca. 10-fold for Cyto c, also support this notion, with the larger molecules showing the effect of greater steric hindrance. Thus, this active gap system can be observed to provide an improved SERS platform, particularly for large analytes such as biomacromolecules, affording a win-win situation through the combination of wider gaps for target insertion and narrower gaps for enhanced detection. The effect, here, is like that of a conformation change in enzymes on their reactions.

7.5 Active alignment control of gold nanorods with the aid of polymers brushes

As mentioned above, actively tunable plasmonic structures demonstrate enhanced functionalities; however, the changes in macroscopic shape on stretching and stimuli-mediated changes in volume are not suitable for specific applications, such as sensing systems for integration into microfluidic devices. In this respect, polymer brushes consisting of a collection of polymer chains end-attached to a substrate have shown great potential. One unique feature of polymer brushes is their polymer chain configuration in that they assume a vertically stretched chain conformation under dense conditions through osmotic pressure and excluded volume effects [78]. As these polymer brushes possess soft material properties and the osmotic pressure is controllable through changes in solvent properties, such as ionic strength, pH, and solvent species, microscopic scale active changes in configuration can be achieved on solid substrates. To date, various studies have examined the assembly of AuNPs with polymer brushes. Further, the active control of their plasmonic properties has been pursued using mainly spherical AuNPs [79−82]. The plasmonic properties of noble metal nanoparticles show a significant dependence on their size, shape, and surrounding conditions [6,48,83,84]. Rod-shaped nanoparticles (AuNRs), in particular, exhibit unique plasmon modes of transverse and longitudinal localized SPR (T- and L-LSPR). Due to the strong, sensitive plasmonic absorption shown by L-LSPR in the near-IR region, the utilization of such L-LSPR in various applications has received a good deal of attention [85−87]. A small number of papers have been published on nanorod assembly using polymer brushes as a scaffold. Uhlmann et al, for example, reported the fabrication of assemblies consisting of Au or Au/Ag core-shell nanorods and P2VP brushes [88], while Mangeney et al achieved the nanoassembly of AuNRs and PNIPAM brushes on nano-sized triangle arrays [89]. In both reports, surfactant-stabilized nanorods were just adsorbed onto the brushes and were found to act as SERS substrates for Raman sensing applications. Importantly, the excitation of this L-LSPR is strongly dependent on its orientation to the incident light [90,91]. Thus, the control of nanorod orientation is crucial to harnessing their properties. A vertically aligned polymer configuration of polymer brushes is considered to be well-suited for rod-shaped nanoparticle assembly; however, the spontaneous alignment of nanorods in the desired direction has not yet been achieved [23].

Our group first achieved the fabrication of vertically aligned AuNR arrays on a polymer brush template [92,93]. To this end, DNA was utilized in the fabrication of negatively charged polymer

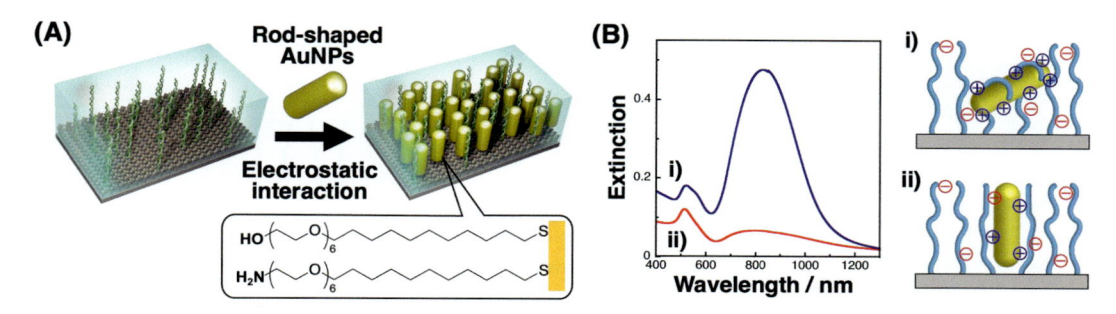

FIGURE 7.11

Vertical alignment of AuNRs adsorbed on DNA brushes (A) Schematic illustration of AuNR adsorption on the DNA brushes. (B) Extinction spectra and schematic illustrations of AuNRs (34 nm × 10 nm) adsorbed on 148 bp double-stranded DNA brushes; (i) AuNRs coated with 100% cationic ligands and (ii) 20% cationic ligands.

Reprinted from ref [92,93]. Copyright (2017 and 2020) American Chemical Society.

brushes based on benefits associated with complete tunability and the mono-dispersibility of polymer length which reduces the complexity of the nanoparticle/polymer brush composite systems derived from polydispersity and makes it easier to grasp the nanoparticle adsorption mechanism. In brief, we grafted DNA strands onto a streptavidin-immobilized substrate through the introduction of biotin at one end. Cationic AuNRs modified with amine-terminated alkane thiol ligands were then adsorbed onto the DNA brushes through electrostatic interactions (Fig. 7.11A). AuNRs modified with only a cationic ligand showed a strong L-LSPR peak at around 800 nm, as shown in previous papers (Fig. 7.11B-i) [88,94]. As L-LSPR is excited by electric fields oscillating along the longer axis, this result indicated that the AuNRs were adsorbed in a tilted but random orientation. This tilted orientation is thought to be result from a too-strong interaction between the AuNRs and DNA polymers. We, therefore, adjusted the cationic density on the AuNR surface through surface modification with two kinds of surface ligands; a cationic ligand and a nonionic ligand. AuNRs with a reduced cationic charge demonstrated a weak L-LSPR peak in comparison to the T-LSPR peak at around 530 nm, indicating their vertical orientation (Fig. 7.11B-ii). The fact of the vertical orientation of the AuNRs was strongly supported by the angle-dependent optical properties observed with polarized light (p- and s-polarized light). A significant advantage of this approach, based on self-assembly, is the ability to control inter-particle distances; i.e., the vertically aligned nanorod array density, through control of the adsorption conditions.

The advantages of this AuNR/polymer brush composite system are related to its lack of complexity, the ability to control inter-particle distances; i.e., vertically aligned nanorod array density, and polymer brush flexibility as a template. As shown in Fig. 7.11B, AuNR orientation varies in accordance with the cation density of the nanorod surface due to the changes in polymer brush conformation resulting from the strong electrostatic interactions between the AuNRs and DNA polymers [93]. This suggests that active tuning of AuNR surface charges can provide active plasmonic tuning through changes in AuNR orientation. Active changes in the orientation of the DNA brush-attached AuNRs were successfully accomplished by simple pH tuning (Fig. 7.12A) [95]. To

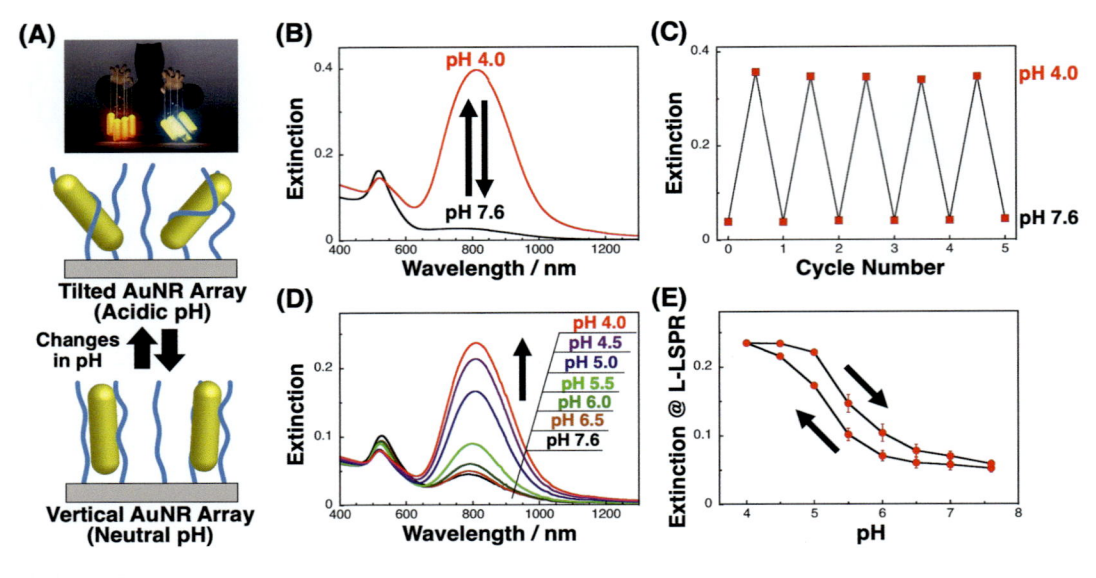

FIGURE 7.12

Gold nanorods animated with polymer brushes in response to changes in pH as a stimulus. (A) Schematic illustrations of the pH-induced changes in the orientation of nanorods adsorbed on polymer brushes. (B) Spectral changes in AuNRs on ds-DNA brushes between pH 7.6 (black) and 4.0 (red). (C) Extinction values for the L-LSPR peak on ds-DNA for repeated changes in pH between 7.6 and 4.0. (D) pH-Responsive spectral changes in AuNRs on ssDNA[poly(dT)] brushes. (E) Extinction values for the L-LSPR peak on ss-DNA with changes in pH.

achieve this, the mixing ratio of the primary amine-terminated surface ligand was adjusted to allow the vertically oriented immobilization of AuNRs at a neutral pH range ($\sim$7.6). When the solution pH changed to acidic ($\sim$4.0), an unprecedented increase in L-LSPR intensity was observed without any peak shift, indicating a significant change in AuNR orientation (Fig. 7.12B). The positive charge on the nanorod surface was also increased due to the protonation of the amine group and the double-stranded structure of the DNA was affected. Significantly, these spectral changes; that is, the changes in AuNR orientation, were highly repeatable (Fig. 7.12C). The flexibility of the polymer chain is also considered an important factor. Fortunately, single- and double-stranded DNAs can be employed as flexible and rigid polymers, respectively. When single-stranded (ss-) DNA is used as a template, L-LSPR on AuNRs gradually increases with decreases in pH (Fig. 7.12D), which is in good agreement with the gradual changes in the surface potential of the AuNRs observed in this pH range. These spectral changes were reversible, although some hysteresis was noted (Fig. 7.12E). On the other hand, for double-stranded (ds-)DNA brushes, no spectral change was observed between pH 7.6 and 5.0, but a sudden was observed below pH 4.5. This can be explained by the rigidity of the ds-DNA inhibiting structural changes. These results suggest that the various polymer properties allow the construction of diverse active plasmonic systems. It should, however, be noted that, unlike the systems based on plasmon coupling, this system was able to induce repeated extremely large plasmon changes.

The use of salt solutions at various concentrations also provides a simple means of changing electrostatic interactions. This technique can be applied to produce changes in AuNR assembly/dispersion state due to a reduction in electrostatic repulsion and an increase in hydrophobic interaction. We adjusted salt concentration to trigger the aggregation of highly cationic AuNRs adsorbed onto DNA brushes [96]. First, the adsorbed AuNRs (as adsorbed) showed strong L-LSPR and relatively weak T-LSPR, indicating a random though generally horizontal attachment (Fig. 7.13A-black and C-ii). An increase in the NaCl concentration resulted in a large decrease and blue-shift (ca. 130 nm) in the L-LSPR peak (Fig. 7.13A-blue), indicating the assembly of AuNRs in a side-by-side manner (Fig. 7.13C-iii) [58,59]. However, when the NaCl concentration was subsequently decreased, a red-shift in L-LSPR peak was observed, which did not return to the original position (Fig. 7.13A-red). The exposure of this AuNR/DNA brush composite to a solution with a high salt concentration resulted in the spectral changes caused by changes in the salt concentration becoming repeatable (Fig. 7.13B). This phenomenon is thought to derive from the nanorod position moving

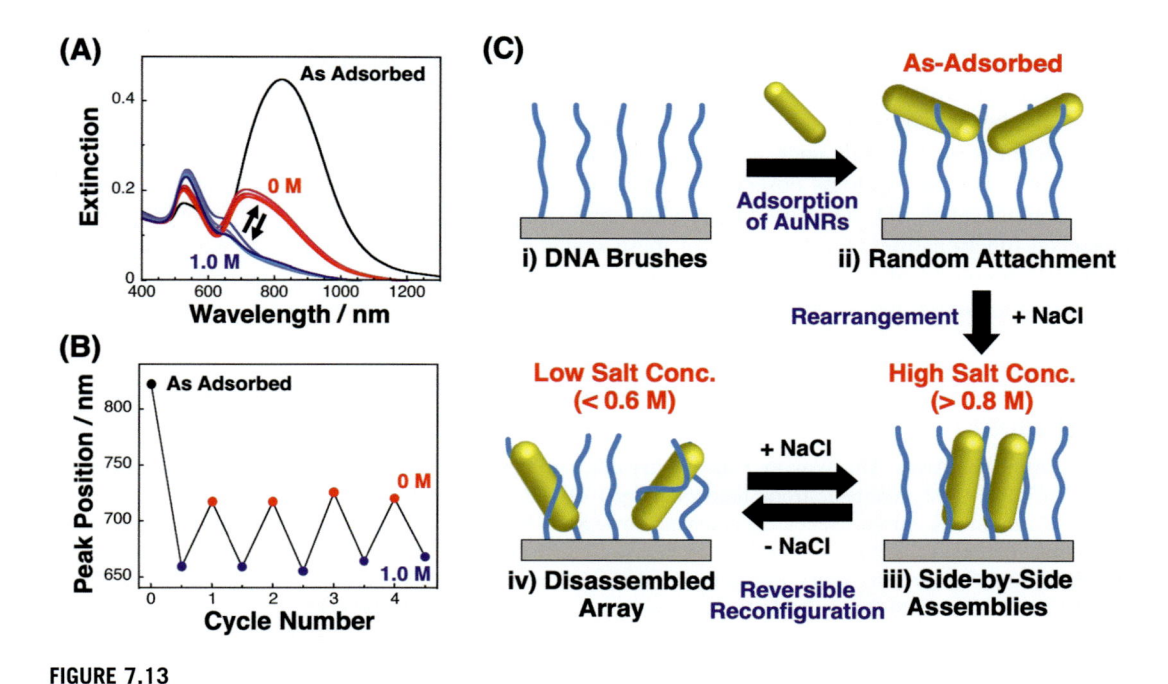

FIGURE 7.13

Gold nanorods animated with polymer brushes in response to changes in NaCl concentration as a stimulus. (A) Extinction spectra of AuNRs adsorbed on a DNA brush layer in response to changes in NaCl concentration between 0 M and 1.0 M for 4 cycles. (B) A plot of the L-LSPR peak position as a function of salt concentration cycle number. (C) Schematic representation the of changes in the AuNR configuration on DNA brushes.

from the outer surface to the inside of the brush. In short, in this system, AuNR position and assembly/dispersion state changed with changes in the salt concentration of the solutions.

Further, the application of thermo-responsive AuNRs coated with OEG-derivatives, as shown above, to this polymer brush system realized the thermo-responsive assembly/disassembly of AuNRs with a uniform (vertical) orientation in the polymer brush layers over an extensive surface (manuscript under preparation). As these plasmonic nanoparticles have good photo-thermal conversion ability and light stimulus has high controllability in terms of spatiotemporal irradiation without direct contact, they could activate on-target on demand. Moreover, it is expected that polymer brushes composed of synthetic polymers could greatly expand the tunability and applications of this system. That is, this combination of AuNRs with polymer brush substrates has great potential for application to plasmonic devices.

7.6 Conclusion

In this chapter, we introduced the recent advances in the sophisticated fabrication of metal nanostructures via the self-assembly of gold nanoparticles with a particular focus on our work on topics including the stimuli-responsive reversible assembly/disassembly and precise tuning of nanostructures using polymer materials, as a part of "metal nanoarchitechtonics". Over the last three decades, there have been remarkable advances in nanotechnology, in particular nanofabrication techniques. The potentials of materials such as metals and semiconductors have also been expanded. On the other hand, the nano-systems observed in living cells continue to indicate that the development of truly sophisticated 4D nanostructures with spatiotemporal controllability will provide further enhanced functionalities. Further study of these systems will allow the creation of excellent nanostructures composed of a wide range of materials, providing meta-material systems not previously existing in nature, a goal which we are currently well on the way to achieving.

References

[1] M. Aono, K. Ariga, The Way to Nanoarchitectonics and the Way of Nanoarchitectonics, Adv. Mater. 28 (2016) 989—992. Available from: https://doi.org/10.1002/adma.201502868.

[2] O. Azzaroni, K. Ariga, Nanoarchitectonics, now, Mol. Syst. Des. Eng. 4 (2019) 9—10. Available from: https://doi.org/10.1039/c9me90001b.

[3] K. Ariga, Nanoarchitectonics Revolution and Evolution: From Small Science to Big Technology, Small Sci. 1 (2021) 2000032. Available from: https://doi.org/10.1002/smsc.202000032.

[4] K. Ariga, M. Shionoya, Nanoarchitectonics for Coordination Asymmetry and Related Chemistry, Bull. Chem. Soc. Jpn. 94 (2021) 839—859. Available from: https://doi.org/10.1246/bcsj.20200362.

[5] E. Roduner, Size matters: why nanomaterials are different, Chem. Soc. Rev. 35 (2006) 583—592. Available from: https://doi.org/10.1039/b502142c.

[6] K.A. Willets, R.P. Van Duyne, Localized Surface Plasmon Resonance Spectroscopy and Sensing, Annu. Rev. Phys. Chem. 58 (2007) 267—297. Available from: https://doi.org/10.1146/annurev.physchem.58.032806.104607.

[7] C.R. Kagan, L.C. Bassett, C.B. Murray, S.M. Thompson, Colloidal Quantum Dots as Platforms for Quantum Information Science, Chem. Rev. 121 (2021) 3186—3233. Available from: https://doi.org/10.1021/acs.chemrev.0c00831.

[8] S. Kargozar, S.J. Hoseini, P.B. Milan, S. Hooshmand, H.W. Kim, M. Mozafari, Quantum Dots: A Review from Concept to Clinic, Biotechnol. J. 15 (2020) 2000117. Available from: https://doi.org/10.1002/biot.202000117.

[9] M. Grzelczak, L.M. Liz-Marzán, Colloidal nanoplasmonics: From building blocks to sensing devices, Langmuir 29 (2013) 4652−4663. Available from: https://doi.org/10.1021/la4001544.

[10] H. Wei, X. Yan, Y. Niu, Q. Li, Z. Jia, H. Xu, Plasmon−Exciton Interactions: Spontaneous Emission and Strong Coupling, Adv. Funct. Mater. 31 (2021) 1−47. Available from: https://doi.org/10.1002/adfm.202100889.

[11] K. Xu, R. Zhou, K. Takei, M. Hong, Toward Flexible Surface-Enhanced Raman Scattering (SERS) Sensors for Point-of-Care Diagnostics, Adv. Sci. (2019) 6. Available from: https://doi.org/10.1002/advs.201900925.

[12] E.S. Cho, J. Kim, B. Tejerina, T.M. Hermans, H. Jiang, H. Nakanishi, et al., Ultrasensitive detection of toxic cations through changes in the tunnelling current across films of striped nanoparticles, Nat. Mater. 11 (2012) 978−985. Available from: https://doi.org/10.1038/nmat3406.

[13] W. Ma, H. Kuang, L. Xu, L. Ding, C. Xu, L. Wang, et al., Attomolar DNA detection with chiral nanorod assemblies, Nat. Commun. 4 (2013) 2689. Available from: https://doi.org/10.1038/ncomms3689.

[14] L. Dykman, N. Khlebtsov, Gold nanoparticles in biomedical applications: Recent advances and perspectives, Chem. Soc. Rev. 41 (2012) 2256−2282. Available from: https://doi.org/10.1039/c1cs15166e.

[15] P. Huang, J. Lin, W. Li, P. Rong, Z. Wang, S. Wang, et al., Biodegradable Gold Nanovesicles with an Ultrastrong Plasmonic Coupling Effect for Photoacoustic Imaging and Photothermal Therapy, Angew. Chem. Int. Ed. 52 (2013) 13958−13964. Available from: https://doi.org/10.1002/anie.201308986.

[16] Y. Niidome, A.T. Haine, T. Niidome, Anisotropic gold-based nanoparticles: Preparation, properties, and applications, Chem. Lett. 45 (2016) 488−498. Available from: https://doi.org/10.1246/cl.160124.

[17] M.E. Stewart, C.R. Anderton, L.B. Thompson, J. Maria, S.K. Gray, J.A. Rogers, et al., Nanostructured Plasmonic Sensors, Chem. Rev. 108 (2008) 494−521.

[18] H. Abramczyk, B. Brozek-Pluska, Raman imaging in biochemical and biomedical applications, Diagnosis Treat. breast cancer. Chem. Rev. 113 (2013) 5766−5781. Available from: https://doi.org/10.1021/cr300147r.

[19] M.B. Ross, C.A. Mirkin, G.C. Schatz, Optical Properties of One-, Two-, and Three-Dimensional Arrays of Plasmonic Nanostructures, J. Phys. Chem. C. 120 (2016) 816−830. Available from: https://doi.org/10.1021/acs.jpcc.5b10800.

[20] K.J. Si, Y. Chen, Q. Shi, W. Cheng, Nanoparticle Superlattices: The Roles of Soft Ligands, Adv. Sci. 5 (2018) 1700179. Available from: https://doi.org/10.1002/advs.201700179.

[21] A. Klinkova, R.M. Choueiri, E. Kumacheva, Self-assembled plasmonic nanostructures, Chem. Soc. Rev. 43 (2014) 3976−3991. Available from: https://doi.org/10.1039/c3cs60341e.

[22] H. Mitomo, Fabrication of sophisticated metal nanostructures for actively tunable plasmonic devices, Impact 2020 (2020) 54−56. Available from: https://doi.org/10.21820/23987073.2020.1.54.

[23] S. Nakamura, H. Mitomo, K. Ijiro, Assembly and Active Control of Nanoparticles using Polymer Brushes as a Scaffold, Chem. Lett. 50 (2021) 361−370. Available from: https://doi.org/10.1246/cl.200767.

[24] H. Mitomo, K. Ijiro, Controlled nanostructures fabricated by the self-assembly of gold nanoparticles via simple surface modifications, Bull. Chem. Soc. Jpn. 94 (2021) 1300−1310. Available from: https://doi.org/10.1246/bcsj.20210031.

[25] M.A. Boles, M. Engel, D.V. Talapin, Self-assembly of colloidal nanocrystals: From intricate structures to functional materials, Chem. Rev. 116 (2016) 11220−11289. Available from: https://doi.org/10.1021/acs.chemrev.6b00196.

[26] M.S. Lee, D.W. Yee, M. Ye, R.J. MacFarlane, Nanoparticle Assembly as a Materials Development Tool, J. Am. Chem. Soc. 144 (2022) 3330−3346. Available from: https://doi.org/10.1021/jacs.1c12335.

[27] V. Linko, H. Zhang, Nonappa, M.A. Kostiainen, O. Ikkala, From Precision Colloidal Hybrid Materials to Advanced Functional Assemblies, Acc. Chem. Res. 55 (2022) 1785−1795. Available from: https://doi.org/10.1021/acs.accounts.2c00093.

[28] N. Jiang, X. Zhuo, J. Wang, Active Plasmonics: Principles, Structures, and Applications, Chem. Rev. 118 (2018) 3054−3099. Available from: https://doi.org/10.1021/acs.chemrev.7b00252.

[29] M. Grzelczak, L.M. Liz-Marzán, R. Klajn, Stimuli-responsive self-assembly of nanoparticles, Chem. Soc. Rev. 48 (2019) 1342−1361. Available from: https://doi.org/10.1039/c8cs00787j.

[30] Z. Qian, D.S. Ginger, Reversibly Reconfigurable Colloidal Plasmonic Nanomaterials, J. Am. Chem. Soc. 139 (2017) 5266−5276. Available from: https://doi.org/10.1021/jacs.7b00711.

[31] C.J. Orendorff, P.L. Hankins, C.J. Murphy, pH-triggered assembly of gold nanorods, Langmuir 21 (2005) 2022−2026. Available from: https://doi.org/10.1021/la047595m.

[32] Z. Tian, C. Yang, W. Wang, Z. Yuan, Shieldable Tumor Targeting Based on pH Responsive Self-Assembly/Disassembly of Gold Nanoparticles, ACS Appl. Mater. Interfaces 6 (2014) 17865−17876. Available from: https://doi.org/10.1021/am5045339.

[33] L. Wang, G.J. Exarhos, A.D.Q. Li, Thermosensitive Gold Nanoparticles, J. Am. Chem. Soc. 126 (2004) 2656−2657. Available from: https://doi.org/10.1021/ja038544z.

[34] Y. Liu, X. Han, L. He, Y. Yin, Thermoresponsive Assembly of Charged Gold Nanoparticles and Their Reversible Tuning of Plasmon Coupling, Angew. Chem. Int. Ed. 51 (2012) 6373−6377. Available from: https://doi.org/10.1002/anie.201201816.

[35] Y. Lin, X. Xia, M. Wang, Q. Wang, B. An, H. Tao, et al., Genetically programmable thermoresponsive plasmonic gold/silk-elastin protein core/shell nanoparticles, Langmuir 30 (2014) 4406−4414. Available from: https://doi.org/10.1021/la403559t.

[36] R. Klajn, K.J.M. Bishop, B.A. Grzybowski, Light-controlled self-assembly of reversible and irreversible nanoparticle suprastructures, Proc. Natl. Acad. Sci. 104 (2007) 10305−10309. Available from: https://doi.org/10.1073/pnas.0611371104.

[37] P.K. Kundu, D. Samanta, R. Leizrowice, B. Margulis, H. Zhao, M. Börner, et al., Light-controlled self-assembly of non-photoresponsive nanoparticles, Nat. Chem. 7 (2015) 646−652. Available from: https://doi.org/10.1038/nchem.2303.

[38] L. Kong, L. Wang, Y. Shi, L. Peng, X. Liang, G. Wang, et al., DNA-Functionalized Silver Nanoparticles in an Alcoholic Solvent for Environment-Dictated Multimodal Actuation, ACS Appl. Nano Mater. (2022). Available from: https://doi.org/10.1021/acsanm.2c01493.

[39] D. Roy, W.L.A. Brooks, B.S. Sumerlin, New directions in thermoresponsive polymers, Chem. Soc. Rev. 42 (2013) 7214−7243. Available from: https://doi.org/10.1039/c3cs35499g.

[40] K.J.M. Bishop, C.E. Wilmer, S. Soh, B.A. Grzybowski, Nanoscale forces and their uses in self-assembly, Small 5 (2009) 1600−1630. Available from: https://doi.org/10.1002/smll.200900358.

[41] N.N. Zhang, X. Shen, K. Liu, Z. Nie, E. Kumacheva, Polymer-Tethered Nanoparticles: From Surface Engineering to Directional Self-Assembly, Acc. Chem. Res. 55 (2022) 1503−1513. Available from: https://doi.org/10.1021/acs.accounts.2c00066.

[42] P.K. Jain, W. Huang, M. El-Sayed, On the Universal Scaling Behavior of the Distance Decay of Plasmon Coupling in Metal Nanoparticle Pairs: A Plasmon Ruler Equation, Nano Lett. 7 (2007) 2080−2088. Available from: https://doi.org/10.1021/nl071008a.

[43] X. Ben, H.S. Park, Size dependence of the plasmon ruler equation for two-dimensional metal nanosphere arrays, J. Phys. Chem. C. 115 (2011) 15915−15926. Available from: https://doi.org/10.1021/jp2055415.

[44] W. Senaratne, L. Andruzzi, C.K. Ober, Self-assembled monolayers and polymer brushes in biotechnology: Current applications and future perspectives, Biomacromolecules 6 (2005) 2427−2448. Available from: https://doi.org/10.1021/bm050180a.

[45] R. Iida, H. Mitomo, Y. Matsuo, K. Niikura, K. Ijiro, Thermoresponsive Assembly of Gold Nanoparticles Coated with Oligo(Ethylene Glycol) Ligands with an Alkyl Head, J. Phys. Chem. C. 120 (2016) 15846—15854. Available from: https://doi.org/10.1021/acs.jpcc.5b11687.

[46] R. Iida, H. Mitomo, K. Niikura, Y. Matsuo, K. Ijiro, Two-Step Assembly of Thermoresponsive Gold Nanorods Coated with a Single Kind of Ligand, Small 14 (2018) 1704230. Available from: https://doi.org/10.1002/smll.201704230.

[47] K. Xiong, H. Mitomo, X. Su, Y. Shi, Y. Yonamine, S. Sato, et al., Molecular configuration-mediated thermo-responsiveness in oligo(ethylene glycol) derivatives attached on gold nanoparticles, Nanoscale Adv. 3 (2021) 3762—3769. Available from: https://doi.org/10.1039/d1na00187f.

[48] M. Grzelczak, J. Pérez-Juste, P. Mulvaney, L.M. Liz-Marzán, Shape control in gold nanoparticle synthesis, Chem. Soc. Rev. 37 (2008) 1783—1791. Available from: https://doi.org/10.1039/b711490g.

[49] N.D. Burrows, A.M. Vartanian, N.S. Abadeer, E.M. Grzincic, L.M. Jacob, W. Lin, et al., Anisotropic Nanoparticles and Anisotropic Surface Chemistry, J. Phys. Chem. Lett. 7 (2016) 632—641. Available from: https://doi.org/10.1021/acs.jpclett.5b02205.

[50] K.L. Kelly, E. Coronado, L.L. Zhao, G.C. Schatz, The Optical Properties of Metal Nanoparticles: The Influence of Size, Shape, and Dielectric Environment, J. Phys. Chem. B 107 (2003) 668—677. Available from: https://doi.org/10.1021/jp026731y.

[51] J.C. Mba, H. Mitomo, Y. Yonamine, G. Wang, Y. Matsuo, K. Ijiro, Hysteresis in the Thermo-Responsive Assembly of Hexa(ethylene glycol) Derivative-Modified Gold Nanodiscs as an Effect of Shape, Nanomaterials 12 (2022) 1421. Available from: https://doi.org/10.3390/nano12091421.

[52] Y. Torii, N. Sugimura, H. Mitomo, K. Niikura, K. Ijiro, PH-Responsive Coassembly of Oligo(ethylene glycol)-Coated Gold Nanoparticles with External Anionic Polymers via Hydrogen Bonding, Langmuir 33 (2017) 5537—5544. Available from: https://doi.org/10.1021/acs.langmuir.7b01084.

[53] H.J. Kim, W. Wang, A. Travesset, S.K. Mallapragada, D. Vaknin, Temperature-Induced Tunable Assembly of Columnar Phases of Nanorods, ACS Nano 14 (2020) 6007—6012. Available from: https://doi.org/10.1021/acsnano.0c01540.

[54] C. Durand-Gasselin, N. Sanson, N. Lequeux, Reversible controlled assembly of thermosensitive polymer-coated gold nanoparticles, Langmuir 27 (2011) 12329—12335. Available from: https://doi.org/10.1021/la2023852.

[55] M.M. Abul Kashem, D. Patra, J. Perlich, A. Rothkirch, A. Buffet, S.V. Roth, et al., Two- and Three-Dimensional Network of Nanoparticles via Polymer-Mediated Self-Assembly, ACS Macro Lett. 1 (2012) 396—399. Available from: https://doi.org/10.1021/mz200141q.

[56] D. Samanta, R. Klajn, Aqueous Light-Controlled Self-Assembly of Nanoparticles, Adv. Opt. Mater. 4 (2016) 1373—1377. Available from: https://doi.org/10.1002/adom.201600364.

[57] S. Nakamura, H. Mitomo, S. Suzuki, Y. Torii, Y. Sekizawa, Y. Yonamine, et al., Self-assembly of Gold Nanorods into a Highly Ordered Sheet via Electrostatic Interactions with Double-stranded DNA, Chem. Lett. 51 (2022) 529—532. Available from: https://doi.org/10.1246/cl.220069.

[58] M. Gluodenis, C.A. Foss, The effect of mutual orientation on the spectra of metal nanoparticle rod-rod and rod-sphere pairs, J. Phys. Chem. B 106 (2002) 9484—9489. Available from: https://doi.org/10.1021/jp014245p.

[59] P.K. Jain, S. Eustis, M.A. El-Sayed, Plasmon coupling in nanorod assemblies: Optical absorption, discrete dipole approximation simulation, and exciton-coupling model, J. Phys. Chem. B 110 (2006) 18243—18253. Available from: https://doi.org/10.1021/jp063879z.

[60] Y. Cui, J. Zhou, V.A. Tamma, W. Park, Dynamic tuning and symmetry lowering of fano resonance in plasmonic nanostructure, ACS Nano 6 (2012) 2385—2393. Available from: https://doi.org/10.1021/nn204647b.

[61] P. Guo, D. Sikdar, X. Huang, K.J. Si, B. Su, Y. Chen, et al., Large-scale self-assembly and stretch-induced plasmonic properties of core-shell metal nanoparticle superlattice sheets, J. Phys. Chem. C. 118 (2014) 26816—26824. Available from: https://doi.org/10.1021/jp508108a.

[62] H.S. Ee, R. Agarwal, Tunable Metasurface and Flat Optical Zoom Lens on a Stretchable Substrate, Nano Lett. 16 (2016) 2818–2823. Available from: https://doi.org/10.1021/acs.nanolett.6b00618.

[63] F. Huang, J.J. Baumberg, Actively tuned plasmons on elastomerically driven Au nanoparticle dimers, Nano Lett. 10 (2010) 1787–1792. Available from: https://doi.org/10.1021/nl1004114.

[64] K.D. Alexander, K. Skinner, S. Zhang, H. Wei, R. Lopez, Tunable SERS in Gold Nanorod Dimers through Strain Control on an Elastomeric Substrate, Nano Lett. 10 (2010) 4488–4493. Available from: https://doi.org/10.1021/nl1023172.

[65] M. Kang, J.-J. Kim, Y.-J. Oh, S.-G. Park, K.-H. Jeong, A Deformable Nanoplasmonic Membrane Reveals Universal Correlations Between Plasmon Resonance and Surface Enhanced Raman Scattering, Adv. Mater. 26 (2014) 4510–4514. Available from: https://doi.org/10.1002/adma.201305950.

[66] L. Gao, Y. Zhang, H. Zhang, S. Doshay, X. Xie, H. Luo, et al., Optics and Nonlinear Buckling Mechanics in Large-Area, Highly Stretchable Arrays of Plasmonic Nanostructures, ACS Nano 9 (2015) 5968–5975. Available from: https://doi.org/10.1021/acsnano.5b00716.

[67] S. Song, X. Ma, M. Pu, X. Li, K. Liu, P. Gao, et al., Actively Tunable Structural Color Rendering with Tensile Substrate, Adv. Opt. Mater. 5 (2017) 1600829. Available from: https://doi.org/10.1002/adom.201600829.

[68] M.L. Tseng, J. Yang, M. Semmlinger, C. Zhang, P. Nordlander, N.J. Halas, Two-Dimensional Active Tuning of an Aluminum Plasmonic Array for Full-Spectrum Response, Nano Lett. 17 (2017) 6034–6039. Available from: https://doi.org/10.1021/acs.nanolett.7b02350.

[69] N. Shimamoto, Y. Tanaka, H. Mitomo, R. Kawamura, K. Ijiro, K. Sasaki, et al., Nanopattern Fabrication of Gold on Hydrogels and Application to Tunable Photonic Crystal, Adv. Mater. 24 (2012) 5243–5248. Available from: https://doi.org/10.1002/adma.201201522.

[70] S. Hamajima, H. Mitomo, T. Tani, Y. Matsuo, K. Niikura, M. Nayad, et al., Nanoscale uniformity in the active tuning of a plasmonic array using polymer gel volume change, Nanoscale Adv. 1 (2019) 1731–1739. Available from: https://doi.org/10.1039/c8na00404h.

[71] H. Mitomo, K. Horie, Y. Matsuo, K. Niikura, T. Tani, M. Naya, et al., Active Gap SERS for the Sensitive Detection of Biomacromolecules with Plasmonic Nanostructures on Hydrogels, Adv. Opt. Mater. 4 (2016) 259–263. Available from: https://doi.org/10.1002/adom.201500509.

[72] M. Shibayama, Small-angle neutron scattering on polymer gels: Phase behavior, inhomogeneities and deformation mechanisms, Polym. J. 43 (2011) 18–34. Available from: https://doi.org/10.1038/pj.2010.110.

[73] S. Seiffert, Origin of nanostructural inhomogeneity in polymer-network gels, Polym. Chem. 8 (2017) 4472–4487. Available from: https://doi.org/10.1039/c7py01035d.

[74] C.L. Haynes, A.D. McFarland, L. Zhao, R.P. Van Duyne, G.C. Schatz, L. Gunnarsson, et al., Nanoparticle Optics: The Importance of Radiative Dipole Coupling in Two-Dimensional Nanoparticle Arrays, J. Phys. Chem. B 107 (2003) 7337–7342. Available from: https://doi.org/10.1021/jp034234r.

[75] Y.-J. Oh, K.-H. Jeong, Glass Nanopillar Arrays with Nanogap-Rich Silver Nanoislands for Highly Intense Surface Enhanced Raman Scattering, Adv. Mater. 24 (2012) 2234–2237. Available from: https://doi.org/10.1002/adma.201104696.

[76] K. Jung, J. Hahn, S. In, Y. Bae, H. Lee, P.V. Pikhitsa, et al., Hotspot-Engineered 3D Multipetal Flower Assemblies for Surface-Enhanced Raman Spectroscopy, Adv. Mater. 26 (2014) 5924–5929. Available from: https://doi.org/10.1002/adma.201401004.

[77] T. Nishio, K. Niikura, Y. Matsuo, K. Ijiro, Self-lubricating nanoparticles: self-organization into 3D-superlattices during a fast drying process, Chem. Commun. (Camb). 46 (2010) 8977–8979. Available from: https://doi.org/10.1039/c0cc03538f.

[78] W.J. Brittain, S. Minko, A structural definition of polymer brushes, J. Polym. Sci. Part. A Polym. Chem. 45 (2007) 3505–3512. Available from: https://doi.org/10.1002/pola.22180.

[79] Y. Roiter, I. Minko, D. Nykypanchuk, I. Tokarev, S. Minko, Mechanism of nanoparticle actuation by responsive polymer brushes: from reconfigurable composite surfaces to plasmonic effects, Nanoscale 4 (2012) 284−292. Available from: https://doi.org/10.1039/c1nr10932d.

[80] S. Gupta, M. Agrawal, P. Uhlmann, F. Simon, U. Oertel, M. Stamm, Gold Nanoparticles Immobilized on Stimuli Responsive Polymer Brushes as Nanosensors, Macromolecules 41 (2008) 8152−8158.

[81] D. Boyaciyan, P. Krauseb, R.von Klitzing, Making strong polyelectrolyte brushes pH-sensitive by incorporation of gold nanoparticles, Soft Matter 14 (2018) 4029−4039. Available from: https://doi.org/10.1039/c8sm00411k.

[82] I. Tokareva, I. Tokarev, S. Minko, E. Hutter, J.H. Fendler, Ultrathin molecularly imprinted polymer sensors employing enhanced transmission surface plasmon resonance spectroscopy, Chem. Commun. (2006) 3343. Available from: https://doi.org/10.1039/b604841b.

[83] P.C. Ray, Size and shape dependent second order nonlinear optical properties of nanomaterials and their application in biological and chemical sensing, Chem. Rev. 110 (2010) 5332−5365. Available from: https://doi.org/10.1021/cr900335q.

[84] H. Chen, T. Ming, L. Zhao, F. Wang, L.-D. Sun, J. Wang, et al., Plasmon−molecule interactions, Nano Today 5 (2010) 494−505. Available from: https://doi.org/10.1016/j.nantod.2010.08.009.

[85] H. Chen, L. Shao, Q. Li, J. Wang, Gold nanorods and their plasmonic properties, Chem. Soc. Rev. 42 (2013) 2679−2724. Available from: https://doi.org/10.1039/c2cs35367a.

[86] L. Vigderman, B.P. Khanal, E.R. Zubarev, Functional gold nanorods: Synthesis, self-assembly, and sensing applications, Adv. Mater. 24 (2012) 4811−4841. Available from: https://doi.org/10.1002/adma.201201690.

[87] R. Takahata, T. Tsukuda, Ultrathin gold nanowires and nanorods, Chem. Lett. 48 (2019) 906−915. Available from: https://doi.org/10.1246/cl.190313.

[88] R. Contreras-Caceres, C. Dawson, P. Formanek, D. Fischer, F. Simon, A. Janke, et al., Polymers as Templates for Au and Au@Ag Bimetallic Nanorods: UV−Vis and Surface Enhanced Raman Spectroscopy, Chem. Mater. 25 (2013) 158−169. Available from: https://doi.org/10.1021/cm3031329.

[89] M. Nguyen, A. Kanaev, X. Sun, E. Lacaze, S. Lau-Truong, A. Lamouri, et al., Tunable Electromagnetic Coupling in Plasmonic Nanostructures Mediated by Thermoresponsive Polymer Brushes, Langmuir 31 (2015) 12830−12837. Available from: https://doi.org/10.1021/acs.langmuir.5b03339.

[90] L. Fu, Y. Liu, W. Wang, M. Wang, Y. Bai, E.L. Chronister, et al., A pressure sensor based on the orientational dependence of plasmonic properties of gold nanorods, Nanoscale 7 (2015) 14483−14488. Available from: https://doi.org/10.1039/C5NR03450G.

[91] B.M.I. Van Der Zande, G.J.M. Koper, H.N.W. Lekkerkerker, Alignment of rod-shaped gold particles by electric fields, J. Phys. Chem. B 103 (1999) 5754−5760. Available from: https://doi.org/10.1021/jp984737a.

[92] S. Nakamura, H. Mitomo, M. Aizawa, T. Tani, Y. Matsuo, K. Niikura, et al., DNA Brush-Directed Vertical Alignment of Extensive Gold Nanorod Arrays with Controlled Density, ACS Omega 2 (2017) 2208−2213. Available from: https://doi.org/10.1021/acsomega.7b00303.

[93] S. Nakamura, H. Mitomo, Y. Sekizawa, T. Higuchi, Y. Matsuo, H. Jinnai, et al., Strategy for Finely Aligned Gold Nanorod Arrays Using Polymer Brushes as a Template, Langmuir 36 (2020) 3590−3599. Available from: https://doi.org/10.1021/acs.langmuir.9b03835.

[94] A.R. Ferhan, Y. Huang, A. Dandapat, D. Kim, Surface-floating gold nanorod super-aggregates with macroscopic uniformity, Nano Res. 11 (2018) 2379−2391.

[95] Y. Sekizawa, H. Mitomo, M. Nihei, S. Nakamura, Y. Yonamine, A. Kuzuya, et al., Reversible changes in the orientation of gold nanorod arrays on polymer brushes, Nanoscale Adv. 2 (2020) 3798−3803. Available from: https://doi.org/10.1039/d0na00315h.

[96] S. Nakamura, H. Mitomo, Y. Yonamine, K. Ijiro, Salt-Triggered Active Plasmonic Systems Based on the Assembly/Disassembly of Gold Nanorods in a DNA Brush Layer on a Solid Substrate, Chem. Lett. 49 (2020) 749−752. Available from: https://doi.org/10.1246/cl.200185.

Molecular Imprinting as Key Technology for Smart Nanoarchitectonics

Makoto Komiyama

Research Center for Advanced Science and Technology (RCAST), The University of Tokyo, Komaba, Meguro, Tokyo, Japan.

8.1 Introduction

"Molecular imprinting" is a very powerful strategy to construct artificial receptors for target guest on demand [1−14]. Eminent receptors which strongly bind a target molecule with high selectivity can be straightforwardly obtained in a tailor-made fashion. The fundamental concept of "molecular imprinting" is to take a snap-shot of host-guest system by polymerizing the solution containing both functional monomer(s) and target guest molecule, and freezing its dynamic motion in polymeric structure. The resultant "molecularly imprinted polymer (MIP)" memorizes the template in terms of its size, shape, and other physicochemical properties. Accordingly, upon subsequent encounter with the template molecule, the MIP binds it selectively and strongly. There are no specific limitations in the kind of target guest molecule employed (size, electrostatic charges, flexibility, and other properties). When necessary, even large molecular assemblies can be used as templates to construct still larger ordered assemblies on demand.

It is noteworthy that molecular imprinting strategy usually employs only the commercially available reagents (of course, still more complicated nanostructures for sophisticated functions can be constructed by synthesizing the reagents according to detailed molecular design). No specific experimental techniques and apparatus are necessary. The solvent employed widely ranges from water to various organic solvents. Reaction conditions are also almost freely chosen. Thus, even in non-chemical laboratories, this useful and powerful method is successfully accomplishable without difficulty. Apparently, the scope of applications of molecular imprinting strategy to nanoarchitectonics as interdisciplinary concept is very wide and deep, and almost unlimited.

This chapter overviews the fundamental aspects of molecular imprinting approaches to nanoarchitectonics. The primary emphasis is that this method is very convenient and useful to provide a variety of smart materials. In Section 8.2, some experimental details of this technique are first presented mainly for the readers who are not very familiar with this method. Then, recent developments of molecular imprinting methodology for more strict "imprinting" are surveyed in Section 8.3. Guest-binding activity of the first-round MIP can be further promoted by post-imprinting strategy. Other elegant tactics have been also developed for higher selectivity and binding affinity. In Section 8.4, typical examples of practical applications of the molecular imprinting methods are presented (e.g., eminent sensors, highly selective catalysts, and biological applications). Lastly, the potential of "molecular imprinting" for further developments of nanoarchitectonics is discussed.

8.2 Some technical details of molecular imprinting

8.2.1 Experimental procedures and chemicals employed for imprinting

As depicted in Fig. 8.1, the reaction mixture for molecular imprinting should involve both (1) a template (target molecule to be recognized or its analog) and (2) functional monomer(s) which interacts with the template either covalently or non-covalently (top in the right). Either water or organic solvents are used as the media. In order to control the rigidity of the resultant MIP (and its porosity) and strengthen its molecular memory, a crosslinking agent is usually added. The crosslinker can also interact with the template to promote the binding activity of MIP. The concept of molecular imprinting is applicable to both organic polymers and inorganic polymers.

This mixture is polymerized (and crosslinked) so that the molecular movements in the solution are frozen in the polymeric structures (bottom in the right). Finally, the template is removed from the polymer by appropriate washing procedures. As the result. vacant cavities are behind in the MIP as guest-binding sites which remember the template molecule (bottom in the left). Because of limited flexibility of the crosslinked polymer, the molecular memory of MIP lasts for a long time.

For molecular imprinting, radical polymerization is widely employed because of its experimental convenience. In "non-covalent imprinting", functional monomers interact with template molecule through hydrogen-bonding, electrostatic interaction, apolar binding, and other non-covalent interactions. Commercially obtainable monomers having polar functional groups (e.g., acrylamide, acrylic acid, and 4-vinylpyridine) are often used. Crosslinkers are, in many cases, also commercially

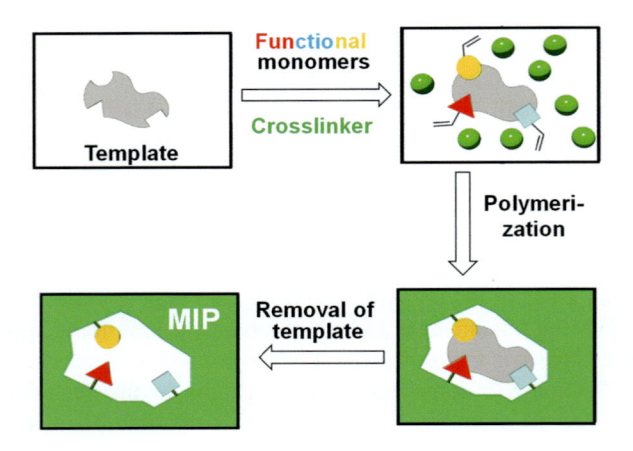

FIGURE 8.1

Scheme of molecular imprinting procedure. Functional monomers (ocher, red, and light blue) and crosslinkers (green) are mixed with template molecule (gray), and polymerized. Afterwards, the template is removed from the MIP polymer, leaving cavities which memorize the template for its selective binding.

available (ethylene glycol dimethacrylate, divinylbenzene, and others). These two components are mixed with template molecule (or its derivative), and the polymerization is started with the use of radical initiator such as 2,2'-azobis(isobutyronitrile) (AIBN) and benzoyl peroxide. After the polymerization, the products are thoroughly washed with appropriate solvent(s) to remove the template from the polymer, together with unreacted functional monomer(s) and crosslinker. As the result, the space which is originally occupied by the template in the polymerization mixture is left behind as cavities in the MIP. These cavities remember the template in terms of its spatial structure and the positions of functional groups, and thus efficiently and preferentially bind this molecule over other molecules. In "covalent imprinting", the functional monomer is in advance synthesized by covalently connecting target template to a polymerizable unit (acrylate group and others). After the polymerization, the template is removed through the scission of the covalent linkage to the polymer.

In order to synthesize MIP of uniform structures and properties, controlled living polymerizations are very attractive, since their living nature autonomously regulates the molecular weights of polymeric products and their compositions [15−17]. Either "atom transfer radical polymerization (ATRP)" or "reversible addition−fragmentation chain transfer polymerization (RAFT)" is very useful. The experimental procedures for these advanced polymerizations are very simple and easy to carry out. For example, ATRP is accomplishable simply by using a unique initiating system (a mixture of commercially available alkyl halide, copper salt, and ligand), in place of conventional free-radical initiator. For RAFT, commercially available dithioester compounds (-C(=S)-S-) are employed as additives. From wide repertoire available, we can choose the most appropriate system for our purpose (more details of these two methods should be referred to the reviews 15−17).

8.2.2 Coverage of the surface of nanoarchitectures by MIP

In many applications to nanoarchitectonics, MIP is prepared in situ on the surface of nanostructures (nanoparticles, nanoplates, nanofibers, and others) to provide molecular recognition activity to target compound. The most common preparation method is surface polymerization. One of the components for molecular imprinting (functional monomer, crosslinker, or template) is covalently or noncovalently connected to their surface. For example, acrylate monomers are bound to the surface of polystyrene beads. Then, functional monomer(s) for molecular imprinting is added, together with template and crosslinker. Even when the polymerization starts in solutions, the layer of MIP is covalently fixed to the surface of substrate. In order to cover gold nanoparticles with MIP layer, various monomers are attached to its surface by using -S-Au linkage.

In order to prepare MIP layer on the surface of nanoparticles in water, polydopamine technique is very convenient and attractive [18,19]. Simply by soaking a substrate in slightly alkaline solution containing both dopamine (2-(3,4-dihydroxyphenyl)ethylamine) and template, MIP layers can be fabricated on the surface of various kinds of substrates. Polymerization of dopamine (and also crosslinking) spontaneously proceeds under mild alkaline conditions (e.g., pH 8), and the MIP layer is directly bound to the surface of substrates without any additional glue. Neither initiator nor crosslinker is necessary for the preparation of MIP layer. In Fig. 8.2, the surface of Fe_3O_4 particles were imprinted with a protein template. Guest-binding cavities are formed in the polydopamine layer of crosslinked polymeric structure. In addition to this guest-binding activity, the polydopamine functions as glue to stick the MIP layer to the nanoparticles. The thickness of polydopamine layer is controllable in terms of dopamine concentration and reaction time. Note that this method is

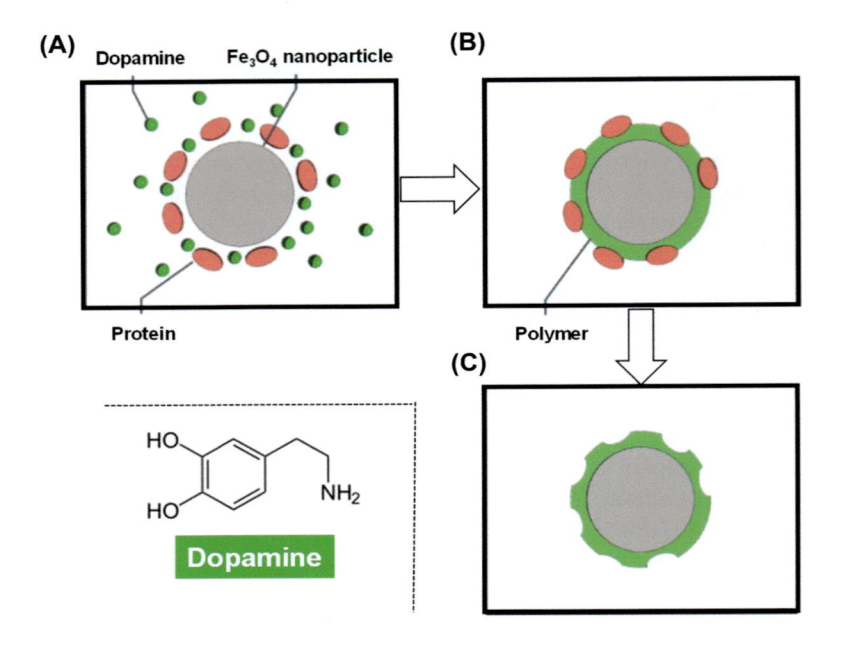

FIGURE 8.2

Molecular imprinting using dopamine (below left) as both functional polymer and glue. In slightly alkaline solutions (pH 8), dopamine and template protein are simply incubated with Fe_3O_4 nanoparticles. Thin layer of imprinted polydopamine is straightforwardly obtained on the surface of nanoparticles.

Adapted with permission from ref. [6].

especially powerful to imprint biomolecules (e.g., proteins and viruses), since the polymerization is achieved in water under mild conditions.

8.3 Tactics to improve the guest-binding activity and selectivity of MIP

8.3.1 Post-imprinting modification to provide still more advanced functions

In all the molecular imprinting methods described above, the functional groups for guest-recognition are derived from the reagents used for the imprinting (functional monomers and crosslinkers). With the use of only these functional groups, however, the strictness of guest-recognition (and its strength) is not necessarily sufficient. In order to bind the guest with larger selectivity and strength, alternative or additional functional groups should be incorporated to the MIPs. The most convenient solution to these challenges is post-imprinting modifications of MIPs [20−22]. In this methodology, MIPs obtained by the first-round molecular imprinting are further treated by some chemical means to replace predetermined functional groups with other ones. As the result, supplemental functional groups are additionally provided to desired sites in the MIP to improve guest-binding activity. The resultant nanodevices show far better performances than the first-round MIP. Entirely new and unique functions (e.g., catalysis, photo-responsibility, dynamic motions, and others) can be also envisioned [21,23]. Some of successful examples are presented in the following sections.

8.3.2 Cyclodextrins as functional monomers to memorize large-sized template by molecular imprinting

Although molecular imprinting is applicable to versatile templates, very large-sized templates are still difficult to memorize precisely. In these imprintings, many functional groups (from functional monomer and crosslinker) must be precisely placed around the large template during the imprinting process, which accompanies a lot of entropy loss. In these cases, cyclodextrins (CyDs) are very useful as the functional monomers for molecular imprinting [24]. These cyclic oligosaccharides have cylindrical structures of nanometer dimensions, which are stabilized by the rings of intramolecular hydrogen bonding between adjacent D-glucose units. The internal diameters of cavities of α-, β- and γ-CyDs (composed of 6, 7, and 8 D-glucose units) are about 4.5−6, 6−8, and 8−9.5 Å, respectively. At both sides of the cavity, two kinds of hydroxyl groups (either the primary or the secondary ones) are precisely placed. One of the most important characteristics of CyDs is the formation of inclusion complexes with a variety of guests in water [25,26]. Bulky and apolar guest compounds, which well fit the cavity, are preferentially accommodated in the apolar cavities. In molecular imprinting, CyDs behave as preorganized assemblies of multiple functional monomers showing unique properties (inclusion complex formation, hydrogen bondings, and others). Accordingly, only several CyD molecules (in place of many functional monomers in conventional molecular imprinting) are necessary to memorize large-sized templates. The entropy loss is much reduced, resulting in far more efficient molecular imprinting.

By using CyDs for molecular imprinting, eminent receptors for large-sized template (e.g., cholesterol derivatives and oligopeptides) can be obtained. CyDs are directly crosslinked by diisocyanate compounds or other crosslinkers, or acryl groups are covalently incorporated to CyDs for radical polymerization. For example, an octapeptide hormone angiotensin II for blood pressure regulation in our body (Asp-Arg-Val-Tyr-Ile-His-Pro-Phe) was precisely recognized by molecularly imprinted β-CyD polymer (Fig. 8.3) [1,27]. In the molecular imprinting, several CyD molecules bind to apolar side chains of the template (the aromatic ring of Phe, the phenol of Tyr, and others), and the whole solution structure is frozen in the MIP. Although each of these CyD molecules in the MIP binds only small part of the large template, they are arranged precisely in a predetermined position (and orientation) through the molecular imprinting. Thus, their assembly as a whole binds the large-guest strongly and selectivity. In other words, the MIP recognizes the solution conformations of the template oligopeptides (the secondary- and the tertiary-structures), rather than their primary structures. Consistently, another oligopeptide of the same size, which is different from the template oligopeptide by only one amino acid, is clearly differentiated from the template and hardly bound by the MIP prepared for angiotensin II. With the use of similar methodology, many MIPs to bind important drugs were prepared and used to deliver these drugs to target site [28−30]. Alternatively, CyD-based hydrogel MIP showing pH-responsive guest-binding was prepared by using poly(L-lysine) as connecting residues [31]. When the pH was changed, the poly(L-lysine) portions show conformational changes between a random coil and α-helix structures, and the mutual conformations of β-CyDs in the gel were changed.

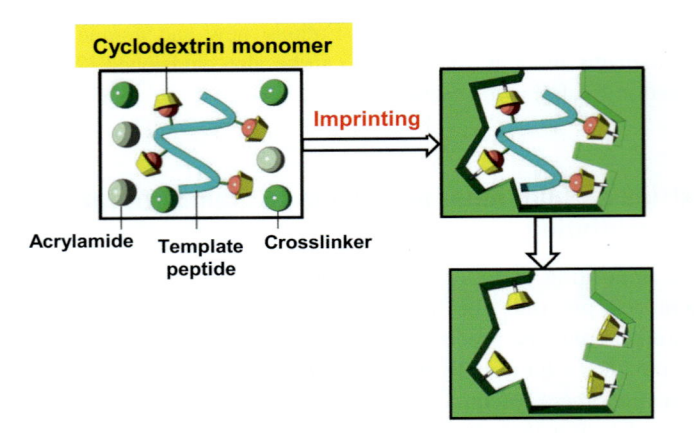

FIGURE 8.3

Molecular imprinting of large-sized template using CyD as functional monomer. The pink circles are apolar and bulky groups in the side-chain of oligopeptide template (light blue ribbon), which efficiently form inclusion complexes with each of the CyDs (yellow cones) for molecular imprinting.

Adapted with permission from ref. [6].

8.4 Examples of practical applications of MIP to nanoarchitectonics

8.4.1 Highly selective sensors

By combining superior guest-binding selectivity of MIP with appropriate analytical means, a variety of sensors were straightforwardly prepared [32–38]. As discussed in Section 8.2, MIP can be prepared for any analyte in desired format (layers, particles, and others) in tailor-made fashion, and thus the applications of this strategy are so versatile. The coverages by thin MIP layers on field-effect capacitors [39] and on Mn-doped ZnS quantum dots [40] are classical examples. Furthermore, MIP nanoparticles were prepared on glass beads as substructure, and chemically bound to the sensor-chip for surface plasmon resonance detector [41]. Uniformity of MIP nanoparticles (and the resultant homogeneous guest-binding activity) facilitates detailed analysis through the output of clear-cut signals. Furthermore, still more complicated and advanced sensors have been recently fabricated to show higher selectivity and sensitivity for more versatile analytes.

8.4.1.1 Sophisticated sensors prepared by post-imprinting strategy

Through the introduction of desired functional groups to desired sites by post-modification (post-treatment of the first-round MIP; see 3.1), the performance of MIP-based sensors can be greatly improved and the scope of applications are widened [42]. In conventional molecular imprinting, the chromophores to emit the signals are provided by the functional monomers, and thus widely distributed in the MIP to induce notable background signal. In Fig. 8.4, however, fluorophores were selectively introduced by post-modification only to the guest-binding cavities, and thus highly sensitive sensor of minimal background signal was manipulated [43]. A newly designed functional monomer

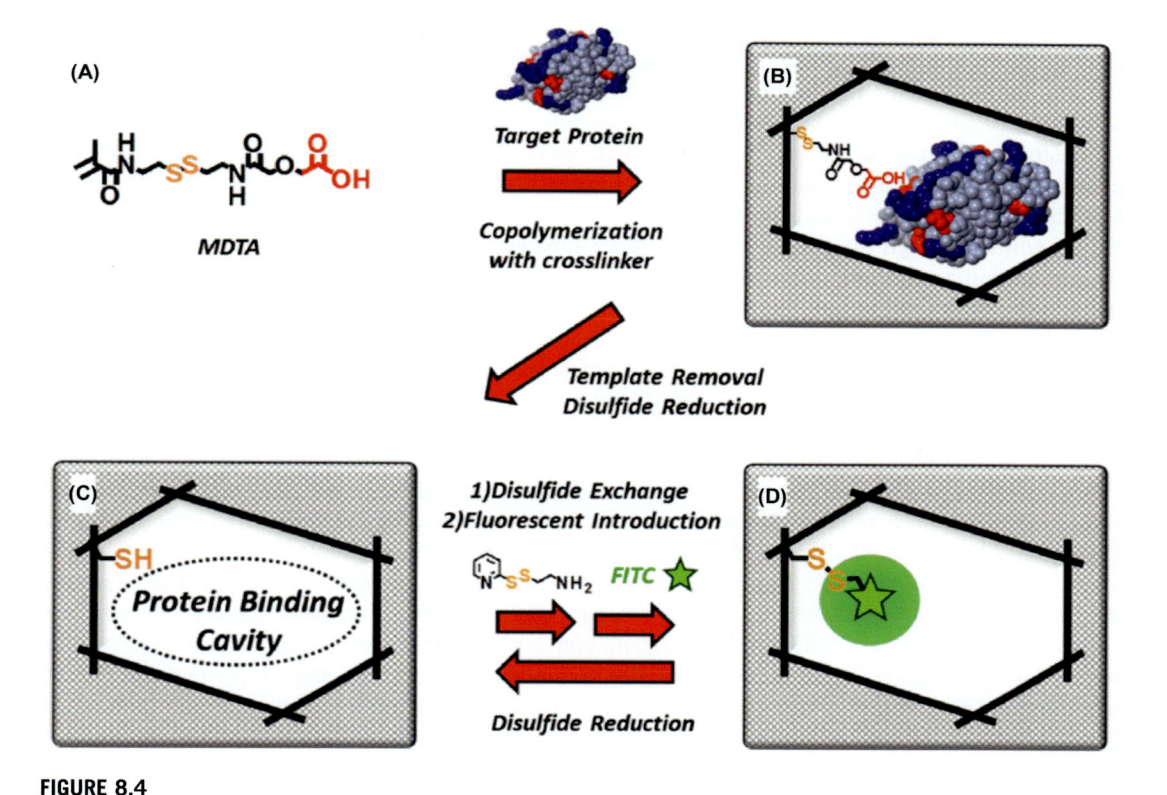

FIGURE 8.4

Preparation of highly sensitive sensor of target protein through selective introduction of fluorophore (FITC) to the guest-binding cavity by post-imprinting modification.

Reproduced with permission from ref. [43].

MDTA (Fig. 8.4a) involved both an S-S residue for group-exchange and a carboxylate to interact with template protein (e.g., lysozyme). In the presence of a template protein, this monomer was copolymerized with acrylamide and N,N'-methylenebis(acrylamide). After the template was removed, the S-S linkage in the MDTA unit was reduced to leave SH residue (orange in Fig. 8.4c) which was used to introduce a fluorophore (fluorescein isothiocyanate: FITC) by disulfide-exchange reaction (green star in Fig. 8.4d). Note that this reaction occurred only in the cavity where the protein was originally bound in the imprinting process, and thus the background fluorescence signals of this sensor from non-specifically bound fluorophores were intrinsically minimized. When necessary, the FITC can be replaced by another chromophore to construct another sensor of different emission property. There exist so many combinations of functional monomers, templates, and crosslinkers for the first-round imprinting. The choice of post-imprinting reaction is also almost completely free. Thus, the scope of applications of this post-imprinting strategy should be enormously wide.

Furthermore, post-imprinted sensors are applicable to selective detection of exosomes (small biological vesicles) which are released from human cells and reliable biomarkers for cancer [44,45]. By introducing the antibody towards target exosome to the binding cavity of MIP, this exosome was precisely recognized. The sensor was prepared as follows. First, a lipid monolayer was fabricated on gold plate by using a lipid which involves 2-bromoisobutyryl group. The template exosome was immobilized onto the monolayer through the binding of surface protein CD9 to anti-CD9 antibody which was also bound to the gold plate. Then, acrylate groups were introduced to the exosomes through the formation of disulfide linkages, and molecular imprinting was achieved by surface-initiated ATRP from the 2-bromoisobutyryl group (see Section 8.2.1). After the polymerization, the exosome and the antibody were removed from the nanostructure, and the antibody was rebound to the cavity. Furthermore, the chromophores for signal-output were introduced into the cavity also by post-imprinting. The binding of target exosome to the MIP-based sensing system was detected in terms of quenching of the fluorescence from the reporter fluorophore.

8.4.1.2 Combination of MIP with DNA aptamer

DNA aptamers are single-stranded DNA molecules that bind their target molecules strongly and selectively. They are selected from a large pool of so many oligonucleotides by using SELEX (Systematic Evolution of Ligands by EXponential enrichment). With the use of cooperation of this artificial antibody with MIP, highly selective and sensitive sensors were developed. For example, acrylate residues are connected to a DNA aptamer, and incubated with silica particles bearing the corresponding template (cocaine) [46]. Then, conventional functional monomers for molecular imprinting (acrylamide derivatives, acrylic acid, and others) were added and copolymerized by radical initiator. The MIP was obtained by removing the products from the composite. The rebinding of cocaine was efficient, although a control polymer (prepared without the aptamer) shows minimal binding under the same conditions. Combination of molecular imprinting and DNA aptamer is essential for the selective guest binding.

In Fig. 8.5, a target protein (PSA; green) was bound to DNA aptamer (black line) which was immobilized onto the surface of a gold electrode [47]. Then, polydopamine strategy (2.2) was employed for molecular imprinting, and the DNA aptamer/PSA complex was trapped in the polydopamine MIP layer (b). After the removal of the DNA aptamer/PSA complex from the resultant polymer, the aptamer for PSA was accommodated in the cavity which preferentially binds this protein (c). Exactly as designed, this sensor showed a high sensitivity towards PSA (d). Interestingly and importantly, human Kallikrein 2, which is 80% homologous to PSA, was hardly detected by this sensor. Through the cooperation of two recognition moieties towards PSA (the MIP and the DNA aptamer), these two analogous proteins were strictly differentiated from each other.

8.4.2 Biological Applications of MIP

MIPs, synthesized from biocompatible monomers, are successfully applied to a variety of biological purposes [48,49]. For example, MIP-based nano-vehicles are prepared for drug delivery systems (DDS) in which medicines are selectively delivered to target site in our body and released at desired timing. In order to target tumor cells, for example, MIP was constructed to bind a membrane protein p32, which is overexpressed on the surface of these malignant cells [50]. When acrylamide and N,N'-methylenebis(acrylamide) were copolymerized in the presence of this template, the

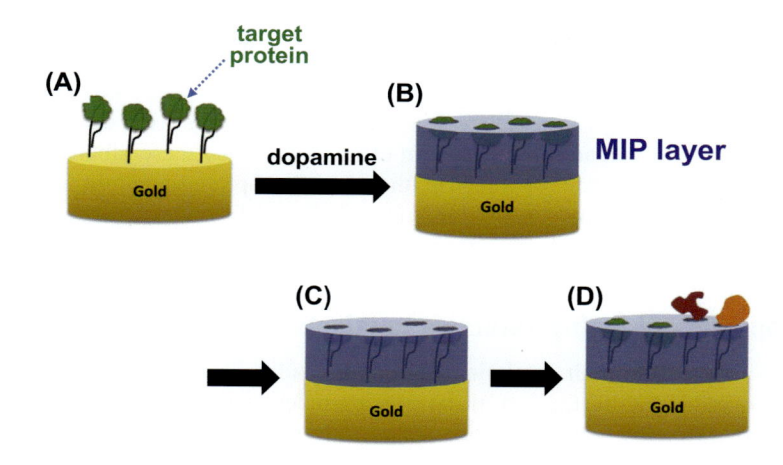

FIGURE 8.5

Combination of MIP (blue layer) with a DNA aptamer (black line) on a gold electrode to construct a highly specific sensor of PSA (green). The polydopamine-based MIP layer and the PSA-binding aptamer cooperate for precise recognition of target protein.

Adapted with permission from Ref. [47].

resultant polymeric nanoparticles satisfactorily recognized p32-positive tumour cells, and can deliver anticancer drugs thereto. There are so many other examples of MIP-based DDS, and comprehensive coverage of them is beyond the scope of this chapter. Thus, the readers who are interested in this topic are advised to refer to recent reviews [6,51,52].

MIP has a strong potential as a drug to combat against various viruses (e.g., SARS-CoV-2 which causes COVID-19 pandemic) [53,54]. This virus infects human beings through the binding of its spike protein to ACE2 protein on the surface of human cells [55,56]. The receptor-binding domain (RBD) is primarily responsible for the contact with ACE2. Thus, artificial antibodies, which bind the spike protein of the virus, should competitively suppress the infection. To synthesize ACE2-binding MIP, a peptide of 17 residues in the RBD of the spike of SARS-CoV-2 was used as the template [57]. First, this template was bound to silica particles which were in advance functionalized with glutaraldehyde. Then, the composite was treated at pH 7.4 and 80°C with a mixture of tetraethylorthosilicate, phenyltriethoxysilane, 3-aminopropyltriethoxysilane, and ureidopropyltrimethoxysilane to form oligopeptide-imprinted organosilica layers on the surface of silica particles. The template was removed by washing with 0.1 M HCl solution. The imprinted nanoparticles efficiently bound to the spike protein of SARS-CoV-2 virus. Assumedly, the tertiary structure of the template (and thus of the RBD of spike protein) is memorized by the MIP on the particle surface [27].

Molecular imprinting technique is also very useful to fabricate polymeric nanogels which acquire stealth nature from immune systems in our body by forming a corona of human serum albumin [58]. In drug delivery system, stealth capabilities against non-specific binding of opsonic proteins and their complement compounds are critically important to avoid undesired immunological response by macrophages, since otherwise the drugs are rapidly eliminated from the reticuloendothelial system. To solve this critical issue, pyrrolidyl acrylate, which strongly interacts with human serum albumin (the most abundant protein in blood), was used as the functional monomer for molecular imprinting. This monomer was polymerized in the presence of albumin, together with *N*-isopropylacrylamide, 2-methacryloyloxyethylphosphorylcholine, and *N,N'*-methylenebis (acrylamide) in pH 7.4 buffer. The template was removed to provide MIP nanogels of 35 nm size having the cavities to bind human serum albumin efficiently. When these albumin-imprinted nanogels were injected into the tail vein of a mouse, they rapidly captured albumin in the body to form a corona of this protein. As the result, the binding of IgG and fibrinogen was suppressed, and the immunological response by macrophages was minimized.

8.4.3 Developments of highly selective catalysts through molecular imprinting

Nanoparticles of Fe_3O_4 have high catalytic activity for oxidation, but their substrate specificity is poor [59]. By covering the surface of these nanoparticles with MIP and providing molecule-recognizing activity, a highly active peroxidase mimic was obtained (Fig. 8.6a) [60]. First, a template 3,3',5,5'-tetramethylbenzidine (TMB) was adsorbed onto Fe_3O_4 nanoparticles through electrostatic interactions. On these nanoparticles, a mixture of acrylamide, *N*-isopropylacrylamide, and *N, N'*-methylenebis(acrylamide) was polymerized to form layers of MIP. As expected, the resultant imprinted nanoparticles showed high catalytic activity on the oxidation of TMB. For the purpose of comparison, another imprinted Fe_3O_4 catalyst was prepared with the use of 2,2'-azinobis(3-ethylbenzothiazoline-6-sulfonic acid) (ABTS) as template, in place of TMB. Importantly, for the oxidation of TMB, TMB-imprinted catalyst (T-MIP) was much more active than ABTS-imprinted

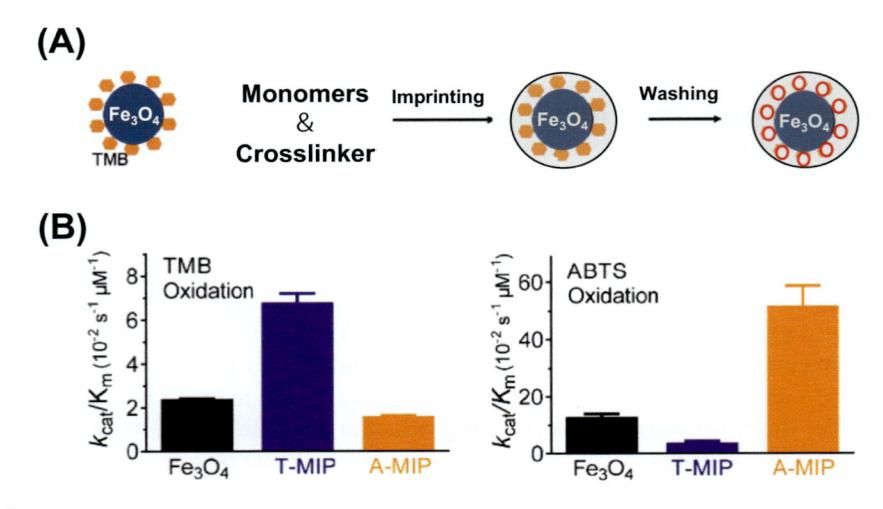

FIGURE 8.6

(a) Molecular imprinting to modify the surface of Fe_3O_4 nanoparticles to provide substrate-specificity to these highly active catalysts for oxidation (the template used for molecular imprinting is colored in orange). (b) Catalytic activity of TMB- and ABTS-imprinted catalysts (T-MIP and A-MIP, respectively) for the oxidation of each of TMB (left) and ABTS (right).

Adapted with permission from ref. [60].

catalyst (A-MIP) (Fig. 8.6b, left). For the oxidation of ABTS, however, the order in the activity was completely reversed (Fig. 8.6b, right). The effect of molecular imprinting was evident. Similarly, substrate specificity was provided to gold nanoparticles (peroxidase mimics) and nanoceria (oxidase mimics) by molecular imprinting.

With the use of surface molecular imprinting strategy, substrate specificity for the hydrogenation of various aromatic molecules was provided to a supported palladium catalyst [61]. First, Pd/SiO_2 catalyst was reduced with H_2 at 200°C. Onto this activated catalyst, a template (e.g., benzene, toluene, or mesitylene) was added to cover the catalyst surface through the adsorption. Then, the template-adsorbed palladium catalyst was poisoned by bubbling dimethylaminopropylamine liquid at 30°C. Finally, the template was removed from the catalyst by heating the catalyst up to 150°C in H_2 flow. As the result of these treatments, non-poisoned catalytically active islands of predetermined shape and size are formed on the surface of the Pd layer. Because of steric constraints, these active islands exhibit high selectivity in the hydrogenation of aromatic molecules. With the use of benzene as template, for example, benzene was hydrogenated predominantly over toluene and mesitylene.

8.4.4 Adsorbents to recover uranium from seawater

Uranium is an essential fuel for nuclear reactors, and exists in natural seawater in the form of $[UO_2(CO_3)_3]^{4-}$. Although the total amount in seawater is sufficiently large as energy resource, its concentration is extraordinarily small (3-4 ppb). Accordingly, enormously effective methods to isolate uranium ion are required for practical applications. In addition, the selectivity for uranium adsorption must be very high, since many interfering metal ions (e.g., Co^{2+}, Ni^{2+}, Fe^{3+}, and VO_3^-) coexist in seawater [62−64]. Selective adsorbents of uranium were developed by employing the molecular imprinting concept to porous aromatic frameworks in which phenyl fragments are periodically connected and characterized by large surface area [63]. Onto the porous aromatic framework, uranyl-specific salicylaldoxime ligands were bound to increase the binding activity [64]. First, acetyl groups were grafted onto porous aromatic frameworks and condensed with dibromosuccinic acid, and then 2:1 salicylaldoxime/UO_2^{2+} complex was tethered through hydrogen-bonding interactions. The UO_2^{2+} ion was removed by washing with aqueous solution of $NaHCO_3$. Upon rebinding, the adsorbent efficiently bound UO_2^{2+} with minimal binding of other metal ions. The stability during recycle experiments was satisfactorily high. As control experiments, the porous aromatic framework was directly treated with salicylaldoxime ligand (without the formation of 2:1 salicylaldoxime/UO_2^{2+} complex). However, the selectivity for the binding of UO_2^{2+} was far poorer, confirming the predominant role of molecular imprinting.

Alternatively, more practically useful adsorbents of uranium ion were constructed from metal organic framework (MOF) [65,66] by using molecular imprinting technique [67]. In this MOF, the ligands for uranium coordination are precisely placed in desired sites during the imprinting process, so that only this metal ion forms stable coordination complex (other metal ions require different coordination modes). The molecularly imprinted MOF was prepared as follows. As the template for the imprinting, $UO_2(CH_3COO^-)_2$ was used. In the presence of this template, $ZrCl_4$, 2,5-dihydroxyterephthalic acid, and 1,2,4,5-benzenetetracarboxylic acid were heated at 95°C in N,N-dimethylformamide. After the reaction, the template was removed by rinsing with 0.1 M nitric acid solution. The molecularly imprinted MOF obtained adsorbed uranyl ion very strongly and selectively from seawater, since the functional groups from the monomers were placed in the MOF appropriate for the coordination of uranyl ion. Two carboxylate groups of the MOF were coordinated to the uranyl ion. Furthermore, two hydroxyl groups of the phenols formed hydrogen bonds with the axial oxygens from the uranyl ion. The coordinations of others metal ions were minimized. The present MOF adsorbents (and thus the positions of all the ligands used for the coordination) were sufficiently stable, and the adsorption capacity and selectivity on uranium ion remained high even after repeated adsorption/release cycles.

8.5 Conclusions

Molecular imprinting is a very convenient and powerful technology to fabricate eminent receptors towards desired compounds. Simply by polymerizing the solutions of functional monomer(s) and target compound (template), sophisticated host-guest systems are produced in tailor-made fashion. The products are obtained in situ in the formats which are directly compatible with nanotechnology (thin layers, nanoparticles, and many others). As the results, various useful nano-parts can be straightforwardly manipulated to show unique functions which are otherwise difficult to accomplish. It is especially noteworthy that all the experimental procedures are simple and easy, and successfully achievable in most cases with the use of commercially available reagents. Furthermore, post-imprinting treatments of the first-round MIP allows further improvements of guest-recognizing activity and selectivity. By incorporating new entities in place of original functional groups, entirely new functions are envisioned. Undoubtedly, molecular imprinting should be one of the key technologies for further developments of nanoarchitectonics.

One of the most promising applications of this method is to facilitate spontaneous construction of complicated and otherwise hardly obtainable nanostructures from smaller parts. With the use of molecular imprinting, complementary features are provided to the desired positions in smaller parts (molecules and their assemblies), so that they are exclusively bound each other in a predetermined conformation and provide desired nanoassemblies. The molecular recognition by molecular imprinting is so strict that even multiple components in solutions are precisely and spontaneously assembled in one-pot in designed hierarchical order. Thus, even complicated nanostructures can be straightforwardly constructed, opening a new way to nanoarchitectonics.

References

[1] H. Asanuma, T. Hishiya, M. Komiyama, Adv. Mater. 12 (2000) 1019–1030.
[2] B. Sellergren, Molecularly Imprinted Polymers: Man-Made Mimics of Antibodies and Their Applications in Analytical Chemistry. Techniques and instrumentation in analytical chemistry, Elsevier Science, Amsterdam, 2001.
[3] M. Komiyama, T. Takeuchi, T. Mukawa, H. Asanuma, Molecular Imprinting: From Fundamentals to Applications, Wiley-VCH, Weinheim, 2003.
[4] M.J. Whitcombe, I. Chianella, L. Larcombe, S.A. Piletsky, J. Noble, R. Porter, et al., Chem. Soc. Rev. 40 (2011) 1547–1571.
[5] G. Aragay, F. Pino, A. Merkoci, Chem. Rev. 112 (2012) 5317–5338.

[6]   M. Komiyama, T. Mori, K. Ariga, Bull. Chem. Soc. Jpn. 91 (2018) 1075−1111.

[7]   O.I. Parisi, F. Francomano, M. Dattilo, F. Patitucci, S. Prete, F. Amone, et al., J. Funct. Biomater. 13 (2022) 12.

[8]   J.J. BelBruno, Chem. Rev. 119 (2019) 94−119.

[9]   L. Chen, X. Wang, W. Lu, X. Wu, J. Li, Chem. Soc. Rev. 45 (2016) 2137−2211.

[10] N.W. Turner, C.W. Jeans, K.R. Brain, C.J. Allender, V. Hlady, D.W. Britt, Polymers 13 (2021) 2657.

[11] K.A. Sarpong, W. Xu, W. Huang, W. Yang, Am. J. Anal. Chem. 10 (2019) 202−226.

[12] M. Guć, G. Schroeder, World J. Res. Rev. 5 (2017) 36−47.

[13] G.Z. Kyzas, D.N. Bikiaris, Adv. Mater. Sci. Eng. 2014 (2014) 932637.

[14] M. Fizir, A. Richa, H. He, S. Touil, M. Brada, L. Fizir, Rev. Environ. Sci. BioTechnol. 19 (2020) 241−258.

[15] H. Zhang, Eur. Polym. J. 49 (2013) 579−600.

[16] H. Zhang, Polymer 55 (2014) 699−714.

[17] S. Beyazit, B.T.S. Bui, K. Haupt, C. Gonzato, Prog. Polym. Sci. 62 (2016) 1−21.

[18] W.H. Zhou, C.H. Lu, X.C. Guo, F.R. Chen, H.H. Yang, X.R. Wang, J. Mater. Chem. 20 (2010) 880−883.

[19] T. Chen, M.W. Shao, H.Y. Xu, S.J. Zhou, S.S. Liu, S.T. Lee, J. Mater. Chem. 22 (2012) 3990−3996.

[20] H. Sunayama, T. Takeuchi, Chromatography 42 (2021) 73−81.

[21] Y. Suga, H. Sunayama, T. Ooya, T. Takeuchi, Chem. Commun. 49 (2013) 8450−8452.

[22] T. Takeuchi, T. Mori, A. Kuwahara, T. Ohta, A. Oshita, H. Sunayama, et al., Angew. Chem. Int. Ed. 53 (2014) 12765−12770.

[23] L. Nováková, H. Vlcková, Anal. Chim. Acta 656 (2009) 8−35.

[24] H. Asanuma, M. Kakazu, M. Shibata, T. Hishiya, M. Komiyama, J. Chem. Soc., Chem. Commun. (1997) 1971−1972.

[25] M.L. Bender, M. Komiyama, Cyclodextrin Chemistry, Springer-Verlag, Berlin, 1978.

[26] G. Fang, X. Yang, S. Chen, Q. Wang, A. Zhang, B. Tang, Coord. Chem. Rev. 454 (2022) 214352.

[27] T. Hishiya, H. Asanuma, M. Komiyama, J. Am. Chem. Soc. 124 (2002) 570−575.

[28] X. Zhao, Y. Wang, P. Zhang, Z. Lu, Y. Xiao, Macromol. Rapid Commun. 42 (2021) 2100004.

[29] T. Higashi, Chem. Pharm. Bull. 67 (2019) 289−298.

[30] F. Trotta, F. Caldera, R. Cavalli, M. Soster, C. Riedo, M. Biasizzo, et al., Expert. Opin. Drug. Delivery 13 (2016) 1671.

[31] K. Matsumoto, A. Kawamura, T. Miyata, Macromolecules 50 (2017) 2136−2144.

[32] O.S. Ahmad, T.S. Bedwell, C. Esen, A. Garcia-Cruz, S.A. Piletsky, Trends Biotechnol. 37 (2019) 294−309.

[33] R. D'Aurelio, I. Chianella, J.A. Goode, I.E. Tothill, Biosensors 10 (2020) 22.

[34] J. Xu, E. Prost, K. Haupt, B.T. Sum Bui, Sens. Actuators B 258 (2018) 10−17.

[35] A. Adumitrăchioaie, M. Tertiş, A. Cernat, R. Săndulescu, C. Cristea, Int. J. Electrochem. Sci. 13 (2018) 2556−2576.

[36] D. Xiao, Y. Jiang, Y. Bi, Microchimica Acta 185 (2018) 247.

[37] M. Gao, Y. Gao, G. Chen, X. Huang, X. Xu, J. Lv, et al., Front. Chem. 8 (2020) 616326.

[38] F. Canfarotta, J. Czulak, K. Betlem, A. Sachdeva, K. Eersels, B. van Grinsven, et al., Nanoscale 10 (2018) 2081−2089.

[39] E. Hedborg, F. Winquist, I. Lundström, L.I. Andersson, K. Mosbach, Sens. Actuators A 37-38 (1993) 796−799.

[40] H. Wang, Y. He, T. Ji, X. Yan, Anal. Chem. 81 (2009) 1615−1621.

[41] J. Ashley, Y. Shukor, R. D'Aurelio, L. Trinh, T.L. Rodgers, J. Temblay, et al., ACS Sens. 3 (2018) 418−424.

[42] T. Takeuchi, H. Sunayama, Chem. Commun. 54 (2018) 6243−6251.

[43] H. Sunayama, T. Takeuchi, ACS Appl. Mater. Interfaces 6 (2014) 20003. 2009.

[44] K. Mori, M. Hirase, T. Morishige, E. Takano, H. Sunayama, Y. Kitayama, et al., Angew. Chem. Int. Ed. 58 (2019) 1612−1615.

[45] T. Takeuchi, K. Mori, H. Sunayama, E. Takano, Y. Kitayama, T. Shimizu, et al., J. Am. Chem. Soc. 142 (2020) 6617–6624.
[46] A. Poma, H. Brahmbhatt, H.M. Pendergraff, J.K. Watts, N.W. Turner, Adv. Mater. 27 (2015) 750–758.
[47] P. Jolly, V. Tamboli, R.L. Harniman, P. Estrela, C.J. Allender, J.L. Bowen, Biosens. Bioelectron. 75 (2016) 188–195.
[48] Z. El-Schich, Y. Zhang, M. Feith, S. Beyer, L. Sternbæk, L. Ohlsson, et al., BioTechniques 69 (2020) 407–419.
[49] M.I. Neves, M.E. Wechsler, M.E. Gomes, L. Rui, R.L. Reis, P.L. Granja, et al., Tissue Eng. Part. B 23 (2017) 27–43.
[50] Y. Zhang, C.Y. Deng, S. Liu, J. Wu, Z.B. Chen, C. Li, et al., Angew. Chem. Int. Ed. 54 (2015) 5157–5160.
[51] M. Komiyama, N. Shigi, K. Ariga, Adv. Funct. Mater. 32 (2022) 2200924.
[52] X. Liang, L. Li, J. Tang, M. Komiyama, Bull. Chem. Soc. Jpn. 93 (2020) 581–603.
[53] S.P.B. Teixeira, R.L. Reis, N.A. Peppas, M.E. Gomes, R.M.E. Domingues, Sci. Adv. 7 (2021) eabi9884.
[54] A.F. Nahhas, T.J. Webster, J. Nanobiotechnol 19 (2021) 305.
[55] M. Komiyama, Bull. Chem. Soc. Jpn. 94 (2021) 1478–1490.
[56] M. Komiyama, Bull. Chem. Soc. Jpn. 95 (2022) 1308–1317.
[57] A.D. Batista, S. Rajpal, B. Keitel, S. Dietl, B. Fresco-Cala, M. Dinc, et al., Adv. Mater. Interfaces 9 (2022) 2101925.
[58] T. Takeuchi, Y. Kitayama, R. Sasao, T. Yamada, K. Toh, Y. Matsumoto, et al., Angew. Chem. Int. Ed. 56 (2017) 7088–7092.
[59] M. Komiyama, K. Ariga, Mol. Catal. 475 (2019) 110492.
[60] Z. Zhang, X. Zhang, B. Liu, J. Liu, J. Am. Chem. Soc. 139 (2017) 5412–5419.
[61] D. Wu, W. Baaziz, B. Gu, M. Marinova, W.Y. Hernandez, W. Zhou, et al., Nat. Cat. 4 (2021) 595–606.
[62] A.S. Ivanov, C.J. Leggett, B.F. Parker, Z. Zhang, A. Arnold, S. Sheng Dai, et al., Nat. Commun. 8 (2017) 1560.
[63] Y. Yuan, G. Zhu, ACS Cent. Sci. 5 (2019) 409–418.
[64] Y. Yuan, Q. Meng, M. Faheem, Y. Yang, Z. Li, Z. Wang, et al., ACS Cent. Sci. 5 (2019) 1432–1439.
[65] L. Feng, K. Wang, J. Willman, H. Zhou, ACS Cent. Sci. 6 (2020) 359–367.
[66] Y. Yuan, Y. Yang, G. Zhu, ACS Cent. Sci. 6 (2020) 1082–1094.
[67] L. Feng, H. Wang, T. Feng, B. Yan, Q. Yu, J. Zhang, et al., Angew. Chem. Int. Ed. 61 (2022) 82–86.

Further Reading

Bai et al., 2013   W. Bai, N.A. Gariano, D.A. Spivak, J. Am. Chem. Soc. 135 (2013) 6977–6984.
Bai and Spivak, 2014   W. Bai, D.A. Spivak, Angew. Chem. Int. Ed. 53 (2014) 2095–2098.
Hu et al., 2007   X. Hu, G. Li, J. Huang, D. Zhang, Y. Qiu, Adv. Mater. 19 (2007) 4327–4332.

Self-assembled structures as emerging cellular scaffolds

Divya Gaur[1], Nidhi C. Dubey[2] and Bijay P. Tripathi[1]
[1]Department of Materials Science & Engineering, Indian Institute of Technology Delhi, New Delhi, India
[2]Department of Molecular Medicine, Jamia Hamdard, New Delhi, India

9.1 Introduction

Self-assembly is a process whereby building blocks/individual units of a material spontaneously assemble into large, well-organized, thermodynamically stable structures. Inspired by the self-assembled nanostructures found in nature, for example, cells and the subcellular organelles, a range of synthetic scaffolds have been constructed to mimic simple living systems. Self-assembly is the most practical method for creating nanosized structures because it is highly flexible and allows control over the generated structures by combining different, comparable self-assembling building components. Due to the noncovalent nature of interaction, a careful balancing of the intra- and intermolecular bonds within the assemblies can produce smart materials that can swiftly respond to environmental stimuli. Moreover, such bottom-up assembled *in vitro* systems aid in understanding isolated, compartmentalized reactions with a lower level of biological complexity (Fig. 9.1). The vesicular structures are characterized by the boundary material, which in certain cases have also provided insights into the dynamics of the cellular membrane in a defined system. The foremost vesicular architectures were lipid vesicles, followed by vesicles constituted of boundary material like polymers, proteins, and dendrimers that have allowed the extension of vesicular properties, and their fascinating applications in materials science.

Recently, various interesting materials have been developed, some of which include hydrogels and polymers. With significant interest focused on the encapsulation of small organic molecule, genetic, or enzymatic systems, inside the vesicles and membraneless structures, the self-assembly of complex soft materials into compartmentalized systems has opened ways to exciting new applications, including drug delivery systems (DDS), nanoreactors, and synthetic cell.

9.2 Self-assembled structures

Emulating the simple biological processes involved in the formation of cells and their various organelles, various membrane-enclosed (vesicles) and membraneless structures (coacervates) have been constructed by employing a variety of building blocks, such as lipids, polymers, organic/inorganic

Materials Nanoarchitectonics. DOI: https://doi.org/10.1016/B978-0-323-99472-9.00023-7

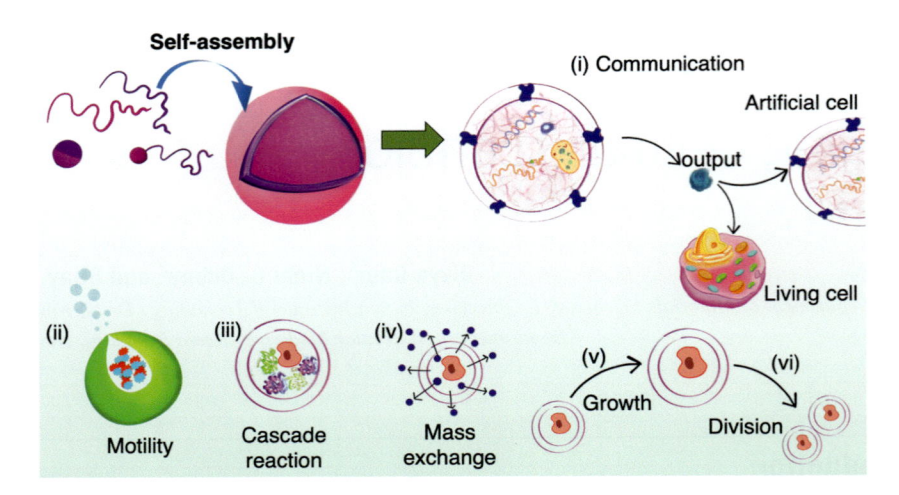

FIGURE 9.1

Schematic summarizing the cellular phenomena that have been studied using cell mimics prepared via self-assembly of nanostructures.

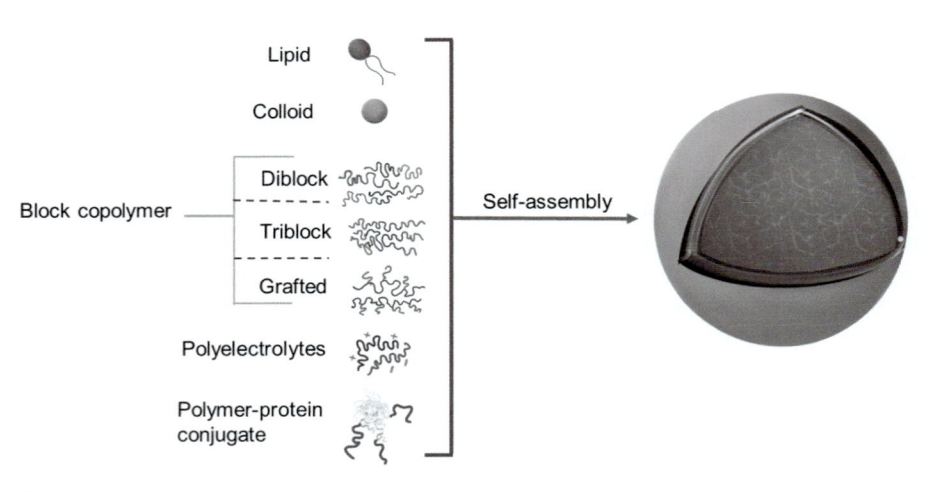

FIGURE 9.2

Different molecular components used for the construction of self-assembled vesicles.

nanoparticles, dendrimers, and conjugates of protein—polymer, and lipid—polymer (Fig. 9.2). These structures exhibit distinct properties, depending on their constituents and the preparation method adopted for their construction, and accordingly have been used for applications like DDS and cellular archetypes.

9.2.1 Self-assembled vesicular structures

Vesicles are supramolecular assemblies containing an aqueous interior that is separated from the bulk solution by one or more bilayers of amphiphiles. This amphiphilic membrane forms a semipermeable barrier that regulates the transfer of small molecules across the cell and encapsulates a range of biomolecules while providing spatiotemporal segregation from the outer environment. Based on the building block, size, and membrane lamellarity, vesicles cover a rich and diverse landscape. Classified in terms of the building block, vesicles can be derived from lipids (liposomes), amphiphilic block copolymers (polymersomes), amphiphilic Janus dendrimers (dendrimersomes), inorganic/organic nanoparticles (colloidosomes), and protein—polymer conjugates (proteinosomes).

Ranging from nano- to micrometer diameters, vesicles can exist as unilamellar vesicles (UVs, single bilayer) or multilamellar vesicles (MLVs, several bilayers) (Fig. 9.3). Based on the size, unilamellar vesicles have been further classified as small unilamellar vesicles (SUVs, < 100 nm), large unilamellar vesicles (LUVs, 100—1000 nm), and giant unilamellar vesicles (GUVs, > 1 μm) [4,5]. In the context of cell mimics, GUVs have been significantly investigated for their striking resemblance to eukaryotic cells, that is, their macroscopic size (10—100 μm) and cell volume ($\sim 10^{-4}$ μL).

Vesicle systems have characteristic properties, such as (1) membrane lamellarity, (2) membrane thickness and its dynamics, (3) dynamics of the constituting amphiphiles, (4) membrane permeability, (5) chemical stability of the amphiphiles, (6) membrane surface charge, and (7) vesicle stability under different conditions. All these properties are controlled by the chemical structure of the amphiphiles and the method of vesicle formation [6]. The latter is significant as most vesicles are kinetically trapped systems, thereby dictating the vesicle stability [7]. In addition, vesicles are required to possess high mechanical properties in terms of bending and stretching elasticity, which provides resistance to the *in vivo* shear forces experienced by the vesicles [8].

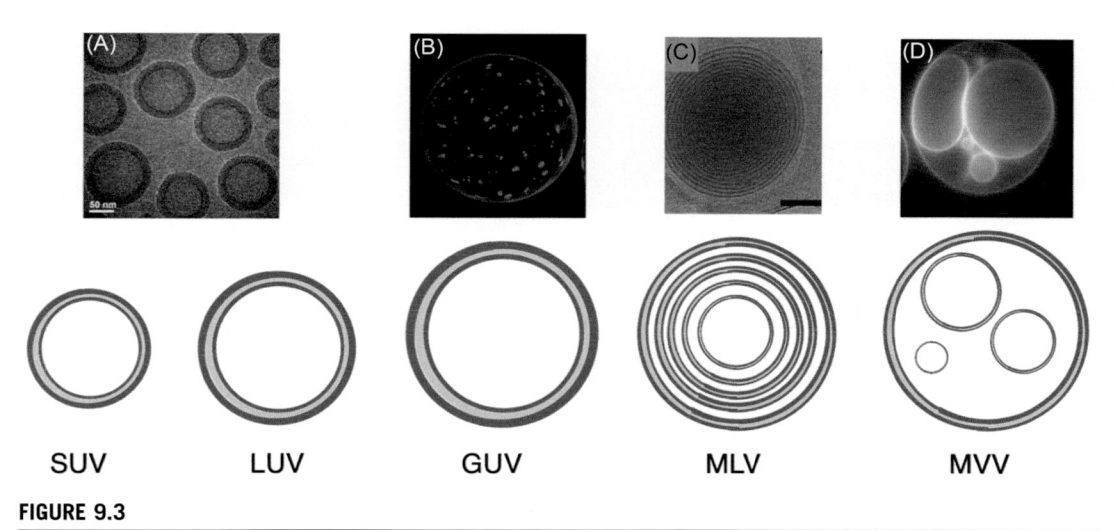

FIGURE 9.3

Classification of self-assembled vesicles. SUV, LUV, GUV: Small, large, giant unilamellar vesicles; MLV, multilamellar vesicles; MVV, multivesicular vesicles. Microscopy images of (A) SUV and LUV [1]; (B) GUV [2]; (C) MLV [3]; (D) MVV [2].

9.2.1.1 Principles of vesicle formation

The formation of nanostructures can be achieved by two methods: (1) breakdown of macrophasic phase of matter and (2) molecular aggregation into clusters [9]. However, the former method requires an external force for breaking interparticle bonds, while the latter occurs in a sufficiently high solution concentration, resulting in stable structures like micelles and vesicles, which are difficult to achieve using the first method. Similar to lipids, amphiphilic block copolymers aggregate in solution to produce vesicular structures, though the stability of lipids and polymer vesicles inevitably varies due to the difference in their chemical composition. However, the principle of their formation remains essentially the same: both are held together solely by noncovalent interactions. While natural lipids can self-assemble only into vesicles, amphiphilic polymers can self-assemble into a range of morphologies like polymersomes (bilayer), spherical micelles (monolayer), cylindrical micelles, rods, lamellae (bilayer), hexagonal cylinders, etc., depending on the ratios of the blocks [10]. However, unlike lipids, amphiphilic block copolymers show microphase separation due to the chemical incompatibility of the hydrophilic and hydrophobic blocks. This rearrangement is entropically favored, as it reduces unfavorable interactions in aqueous media by aggregating the hydrophobic groups away from the polar solvent.

Vesicle formation is a two-step self-assembly process, in which the amphiphile first forms the bilayer, which, in the second step, closes to form a vesicle. The ratio of the size of hydrophobic to hydrophilic moiety drives the self-assembly of amphiphiles into closed structures (Fig. 9.4A) [11]. This ratio is significant to determine the curvature of the hydrophobic–hydrophilic interface, described by its mean curvature H and its Gaussian curvature K, which are given by the two radii of curvature R_1 and R_2, as shown in (Fig. 9.4B). The two curvatures are related to the critical packing parameter (P) by Eq. (9.1) [12]:

$$P = \frac{v}{al} = 1 + Hl + \frac{Kl^2}{3} \tag{9.1}$$

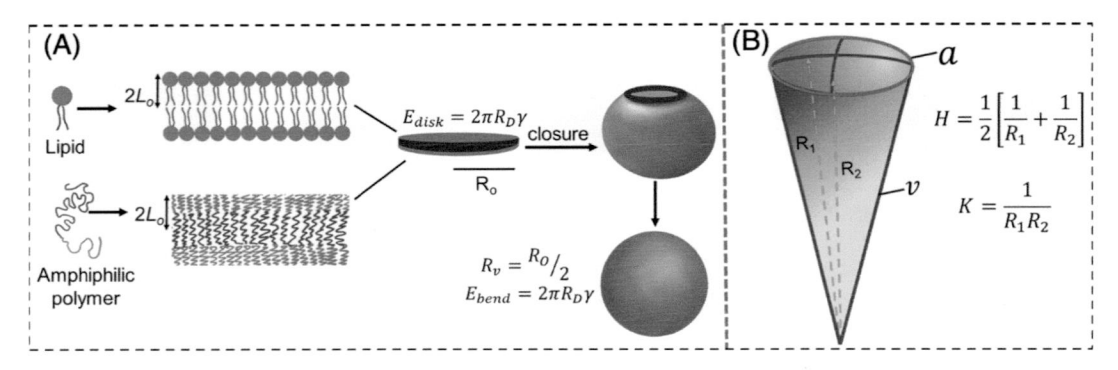

FIGURE 9.4

(A) Illustration of amphiphile shape with reference to surfactant parameter and its relationship with interfacial mean curvature (H) and Gaussian curvature (K). (B) Stepwise formation of bilayers and subsequent closure into vesicles [11].

Redrawn from M. Antonietti, S. Förster, Vesicles and liposomes: a self-assembly principle beyond lipids, Adv. Mater. 15 (2003) 1323–1333.

Table 9.1 Critical packing parameter (P), and curvatures for different morphologies of self-assembled structures [11].

Shape	v/al	H	K
Sphere	1/3	$1/R$	$1/R^2$
Cylinder	½	½ R	0
Bilayer	1	0	0

where v is the hydrophobic volume of the amphiphile, a the interfacial area, and l the chain length normal to the interface. The simplest shapes observed during the self-assembly process are spheres, cylinders, and bilayers, which are characterized by certain values of packing parameter and curvature, as shown in Table 9.1. Accordingly, to obtain bilayers for a tecton of length l and volume v, the interfacial area a needs to be optimized until the surfactant parameter approaches unity. For example, a decrease in hydrophilic/hydrophobic block ratio results in morphological changes from sphere to cylindrical micelle and then to vesicle. A similar morphological series is observed for tectons (surfactants, polymers, and dendrons) which are flexible enough to reorganize according to the molecular demands of the membrane curvature [13,14].

Eq. (9.1) holds good for surfactants and lipids, and is generally accepted for amphiphilic block copolymers. However, it is difficult to determine these values, especially for the block copolymer head group, which is comparably indefinite. Lately, hydrophilic weight fraction f_w (Eq. 9.2) has been used to predict the microphase separation in block copolymers. Though the morphology of the aggregates highly depends on the M_n, hydrophilicity/hydrophobicity of the block, steric, degree of freedom, polymer polydispersity, rigidity, etc., as a general rule, $0.25 < f_w < 0.40$ results in polymersome formation [15].

$$f_w = \frac{M_n(\text{hydrophilic block })}{M_n(\text{hydrophobic block })} \tag{9.2}$$

9.2.1.2 Liposomes

Liposomes, the synthetic analogs of living cells have been extensively investigated as artificial cells since the 1960s [16]. Typically composed of natural phospholipids like phosphatidylcholine (PC), phosphatidylethanolamine (PE), and phosphatidylglycerol (PG) [7,17], liposomes consist of an aqueous core enclosed by a phospholipid bilayer that readily mimics the cell membrane. Phospholipids with a low molecular weight of 100−1000 g/mol, form a bilayer membrane of thickness, 3−5 nm [15,18−20]. This results in high lateral fluidity of the liposomal membrane, contributing to their leaky nature, and hence showing poor retention capacity [20]. The encapsulated volume for liposomal SUVs and LUVs is independent of membrane thickness and roughly falls between 10^{-15} and 10^{-9} µL [8]. The problematic stability of liposomes owes largely to the chemical instability of the lipids [7,21,22]. The unsaturated moieties and ester linkages present in the lipids are prone to oxidative attack, which alters overall liposomal properties. Due to their metastable nature, liposomes have to be stored in dark and inert conditions, at low temperatures ($-20°C$), in a highly pure state, with antioxidants as stabilizers [7,22].

The relatively low chemical diversity of liposomes can be owed to their lipid constituents, which are commonly natural phospholipids or their mixtures (soy-PC or hydro soy-PC). As these are fairly similar in structure and size (MW $\sim$ 800 g/mol) and only vary in aliphatic chain lengths (13−17Cs long; 0−2 degrees of saturation, and always cis) with end groups (primary and tertiary amine, diol, or carboxylic acid) that dictate the overall charge of the lipids [8]. The low chemical versatility of lipids restricts their physical and chemical properties as these are intrinsically highly specialized, offering little scope for further chemical modifications. Even small chemical changes can significantly alter the lipid properties affecting their self-assembly. Though the modification of the lipid head has been achieved, it is strictly restricted to hydrophilic moieties to retain the self-assembling properties. Despite their structural and functional similarity, lipids exhibit a wide range of gel−liquid crystal transition temperatures (T_m) (−22 to 60°C), significant for liposome formation. However, liposome formation occurs at temperatures above the T_m of the lipid used [7,23], which limits the encapsulation of temperature-sensitive, biomolecules (proteins, DNA, etc.) in liposomes.

9.2.1.3 Polymersomes

With the lipid-like amphiphilic structure (a hydrophilic head covalently attached to a hydrophobic tail), block copolymers have been employed for the construction of polymeric vesicles, called polymersomes [11,24]. Block copolymers (BCPs) consist of distinct polymer chains covalently linked in a series of segments, the mean molecular weight of which ranges from 2 to 20 kDa, considerably higher than those of lipids. The high molecular weight of BCPs results in thicker polymersome membranes from 5 to 50 nm, which affords polymersomes with better structural and physical properties, for example, stability, fluidity, intermembrane dynamics, permeability, viscosity, and elasticity [25]. In comparison to liposomes, well-designed polymersomes much better resemble natural cells, which possess 8−10 nm thick membranes. Moreover, by modifying the chemical properties of either segment of BCPs, it is possible to construct polymersomes with the desired characteristics [25].

Despite the thickness, the polymeric membrane formed with triblock copolymer or grafted polymer chains exhibits a high degree of adaptability to form thinner membrane rafts that can easily integrate transmembrane proteins and are thus biocompatible [19,26]. Moreover, the membrane thickness also dominates the cell volume for polymeric SUVs and LUVs, which is small ($> 10^{-9}$ µL) and highly variable [8]. In addition, the thicker polymersome membranes exhibit low lateral diffusivity making them more thermodynamically inert than liposomes and thus they show low membrane permeability [15,27]. This property varies with the polymer chain entanglement, overall M_n, nature, and glass-transition temperature (T_g) of the hydrophobic block [28−30]. However, the low permeability of polymersome membranes can be problematic for applications (e.g., nanoreactors) as this also restricts the passive diffusion of small molecules [20,31]. Interestingly, the chemical versatility of polymer chemistry allows one to design polymersomes for tunable permeability, for example, the grafted polymer PDMS-g-PEO generates vesicles of thickness (5 nm) and high fluidity comparable to liposomes and cells [19].

Depending on the membrane thickness, polymersome membranes can exhibit significant stiffness as evident from the values of the bending rigidity modulus. Polymersomes have high lysis tension (> 20 mN/m) [32], which is directly related to the critical transmembrane potential leading to the poration of vesicles. Thus the pores in polymersomes reseal at a much slower rate and are generally arrested in gel-phase membranes [33,34]. The low molecular mobility in the polymersome membranes can be deemed a consequence of high shear viscosity, which is reported to be ~ 500 times more than the lipid membranes.

9.2.1.4 Hybrid vesicles

These vesicles possess a membrane with a mixture of lipids, polymers, peptides, surfactants, and nanoparticles. Other lipid−polymer hybrid systems that are not bilayer vesicles are not discussed here. Most hybrid vesicles of block/graft copolymers and lipids are motivated by obtaining a combination of the biocompatibility and fluidity of lipid vesicles, and the mechanical/thermal stability and chemical variability of polymer vesicles [35−45]. Hydrophobic nanoparticles with appropriate sizes are embedded into the bilayer membrane for imaging and stimuli-responsive purposes [46−54]. Peptides are integrated with a lipid layer to reconstitute the biological machinery, including endocytosis, fusion, secretion, and recognition [55−58].

Lipid−copolymer hybrid vesicles have unique properties. For instance, hybrid vesicles with a small fraction of lipids show a significant reduction in mechanical rigidity as opposed to pure polymersomes, but the change in lateral diffusion is indistinct [43]. Hybrid vesicles with intermediate lipid/copolymer ratios can be more permeable to ion transport than pure lipid and polymer vesicles, arising from the dynamic equilibrium of vesicles and tubular micelle formations [45,59]. Hybrid vesicles allow the accommodation of hydrophobic proteins and nanoparticles more easily than pure lipid vesicles [51,60], however, hydrophilic nanoparticles cannot be accommodated since they appear either in the interior or the exterior of the vesicular bilayer. The size and surface charge of the hydrophobic nanoparticle affects the topology of the bilayer vesicle differently [48,49,51−54,61]. Peptides can also be introduced into lipid vesicle membranes. In addition, charged stimuli-responsive surfactants, polymers, and nanoparticles are often introduced to interact with lipid bilayers, either to disrupt and release their contents [36,46,60−63], or to localize biomacromolecules [38,64]. The surfactant loses its surface activeness upon exposure to UV light, leading to a higher permeability for the controlled release of interior cargo [65,66]. The fatty acid−lipid hybrid vesicles and surfactant−lipid vesicles have also been investigated to test the hypothesis that the protocell of fatty acid vesicle transits into phospholipid vesicles through intermediate hybrid vesicles [67]. The result shows that the hybrid vesicle is stable in salty water, and permeable to ions and nucleotides. In principle, the complex morphology of the hybrid bilayer, constituted of lipid, copolymer, surfactant, nanoparticle, and peptide, is influenced by several intermolecular forces between these components, including electrostatic, hydrogen bonding, and hydrophobic interaction forces. The morphology of the hybrid bilayer dictates the functions of the vesicle, changing their porosity, permeability, curvature and how they interact with other cellular membranes [68].

9.2.1.5 Colloidosomes

Colloidosomes, also known as Pickering emulsions, are microcapsules that are produced by the spontaneous self-assembly of inorganic or organic colloids in a water/oil emulsion [69−71]. First reported by Weitz et al. in 2002 [69], a variety of colloidosomes have been constructed using inorganic particles, such as silica, clay, $CaCO_3$, gold nanoparticles, and metal−organic frameworks [72−75]. The formation of colloidosomes is a two-step process: first, the colloidal particles are dispersed in the continuous phase of a w/o emulsion. The particles self-assemble at the oil−water interface and form a continuous colloidal membrane. Subsequently, the colloidal assembly is then transferred to an aqueous phase [55,69]. These water-filled colloidosomes have been employed for the construction of compartmentalized systems to house various biomimetic functions [60,76]. Tens of micrometers in diameter, colloidosomes consist of a monolayer of closely packed nanoparticles that exhibit selective permeability based on the solute size-exclusion and can be used to encapsulate macromolecules [35]. Biomolecules like enzymes, nucleic acid, ribosomes, and plasmids have been successfully encapsulated and employed to develop a cell-free gene expression system for spatially confined in vitro protein synthesis [36,62]. Interestingly, the colloidosome membrane can be tailored to tune the membrane permeability by using either smart polymeric microgel particles, crosslinking the membrane with a stimuli-responsive crosslinker, or varying the particles' size and shape.

However, these approaches are effective at restricting molecule diffusion up to a critical value, and thus one of the main challenges is to regulate the membrane permeability that severely impacts their potential as artificial cells. To overcome this limitation, colloidosomes were developed from inorganic colloids grafted to a stimuli-responsive polymer to integrate electrostatically-gated/thermal-gated inorganic protocells which can inhibit the transport of small molecules across the semipermeable colloidosome membrane [62]. Further, colloidosomes with molecularly crowded interiors were constructed that could mimic the structural and functional properties of the cytoskeletal matrix [60]. In addition, growth and division in colloidosomes were obtained by using organosilane-mediated methanol formation inside the microcompartments [77].

9.2.1.6 Proteinosomes

Proteinosomes are a class of self-assembled vesicles characterized by a semipermeable, smart, biocatalytic, and elastic membrane composed of a closely packed monolayer of protein—polymer nanoconjugates. Compared to lipid and polymeric vesicles, proteinosomes exhibit enhanced intrinsic permeability due to the irregular packing of the constituting protein—polymer nanoconjugates [46]. Similar to colloidosomes, these constructs are permeable to macromolecules with molecular masses of less than ca. 40 kDa [78]. Their membrane permeability can be controlled to some extent by the use of stimuli-responsive blocks that reversibly alter size in response to pH and temperature [46,62]. Further, these are highly stable, can withstand temperatures of 70°C, and can be partially dried and rehydrated without loss of structural integrity [46]. Hence, they have been used to exhibit protocellular properties like encapsulation, gene-directed transcription, preferential permeability [46], molecularly-crowded cytosol-like matrix, and membrane-gated internalized enzyme-catalyzed reactions or membrane-mediated cascade reactions [79]. With an internal volume, ca. 1 pL, proteinosomes provides an excellent scaffold to encapsulate thermophilic enzymes and to construct artificial cells that can efficiently undergo thermal cyclic processes like PCR-mediated DNA amplification. Different approaches have been employed to construct proteinosomes like interfacial self-assembly, layer-by-layer, and swelling/extrusion technique [80]. A significant advantage of proteinosomes as synthetic protocells is that the properties of the proteinosome membrane can be altered to construct different protocells, by controlling the chemistry of the protein—polymer nanoconjugates.

9.2.1.7 Dendrimersomes

Dendrimersomes are monodispersed vesicles delineated by a selectively permeable bilayer composed of amphiphilic Janus dendrimers (AJDs) as building blocks [81]. Composed of distinct hydrophilic and hydrophobic domains, AJDs possess a highly branched structure that renders a globular shape to the dendrimers [47]. Based on the branch topology, dendrimers can be synthesized via divergent or convergent techniques into varying morphologies like micelles, tubular, disk-shaped, cubosomes, spherical (dendrimersomes), and onion-like multilamellar vesicles [81].

Additionally, these can be coassembled phospholipids, block copolymers, as well as can integrate membrane proteins, offering a potential alternative to liposomes and polymersomes [82]. Apart from liposomes, the dendrimersome is the only other scaffold that could be reconstituted with cell division machinery [83]. Dendrimersomes consist of three distinct regions, namely, a central core, branches, and surface decorated with functional groups [84]. While the core is used for the encapsulation of biomolecules like enzymes, etc., the branches and their terminals have been employed for the guest molecule immobilization. However, dendrimersomes exhibit very low encapsulation efficiency—one enzyme molecule per cavity [85]—and require specific conditions for preparations [86]. Interestingly, the hydrophobic and hydrophilic segments of dendrimers can be judiciously tailored to encapsulate biomolecules like enzymes and genes to construct biocatalytic nanoreactors and delivery systems, and to exhibit the dynamic behavior of cellular membranes to engineer emergent biological behavior from the bottom-up [83].

9.2.2 Coacervate droplets

Coacervation is a spontaneous process in which a homogeneous colloidal solution phase separates into two immiscible liquid phases consisting of a colloid-rich phase, called coacervate, existing in equilibrium with a colloid-poor phase. In general, coacervate microdroplet is a highly unstable and polydisperse system and exhibit large variation in size and volume. Microfluidic technology is commonly employed to produce uniform and monodispersed coacervate microdroplets enabling facile and precise control over their size, composition, and structure [87,88]. Based on the formation process, coacervation can be classified as either simple or complex coacervation. Simple coacervation requires only one macromolecular species and can be initiated by the addition of dehydrating agents like salts, alcohols, etc. or a temperature change that can increase the intramacromolecular interactions over macromolecule—solvent interactions (Fig. 9.5A(I)). On the other hand, complex coacervation is driven by the electrostatic interactions between oppositely charged polyelectrolytes, including a wide range of nucleotides, peptides, and saccharides (Fig. 9.5A(II)).

With a sponge-like structure, often referred as the "anomalous phase" or "L_3 phase" [90], the coacervate phase exists as metastable amorphous droplets in a turbid solution, which upon resting coalesce with time into a bulk coacervate phase. Due to its resemblance to the molecularly rich cytosol, Oparin suggested the use of coacervate microdroplets as compartmentalization models to construct protocells to explain the evolution of cells [91]. Five aspects of membraneless organelle (MLOs), namely, physicochemical properties, hierarchical organization, selective sequestration, regulation, and liquid—solid transition, are increasingly being studied using coacervates as membraneless models (Fig. 9.5B(I—IV)) [92—96]. The significant high internal volume of the charged electrolytes expediates the active and selective sequestration of a wide range of components, like enzymes and polymers. The sequestration ability allows encapsulation of molecules varying from macro- to nanometer in size [68,97—100], and thus enables study of the biological phenomena in a cell-like environment. This encapsulation approach provides significant protein stabilization and protection inside a molecularly enriched environment with controlled release of cargo by inducing disintegration of the coacervate phase [68,101].

Coacervation has enabled preferential compartmentalization of a variety of guest molecules such as cell-free gene expression [102], enzyme-mediated predation [103], photosynthetic machinery [104,105], and ribozyme activity [106,107]. However, a major limitation of these constructs arises from the intrinsic instability of the coacervate microdroplets to changes in pH, ionic strength,

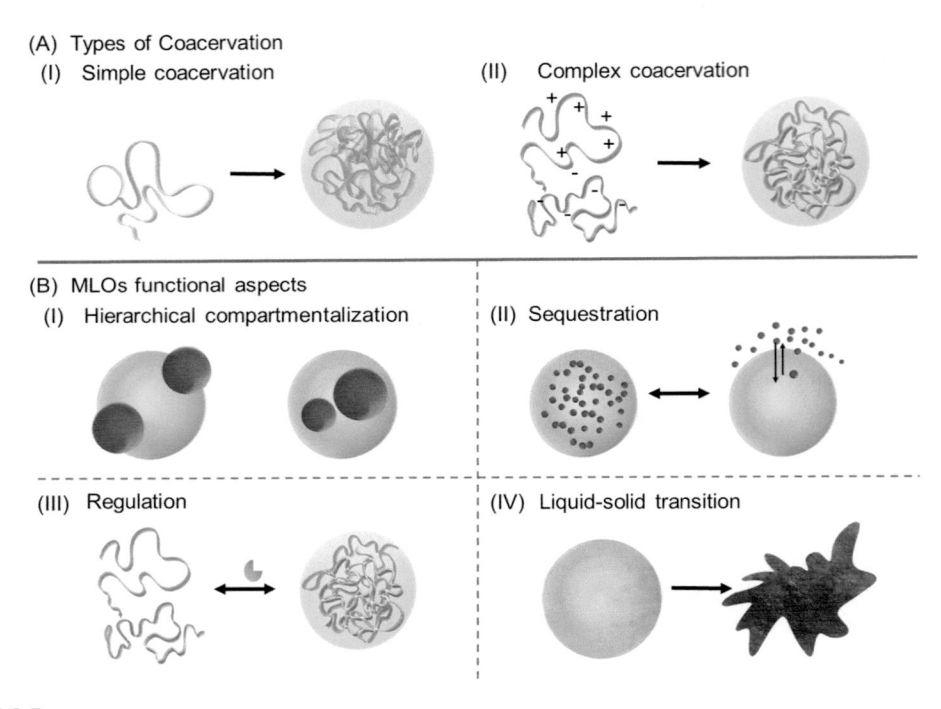

FIGURE 9.5

(A) Types of coacervation: simple and complex coacervation. (B) Various aspects of membraneless organelles investigated using coacervate microdroplets. (I) Hierarchical compartmentalization. (II) Selective sequestration. (III) Regulation. (IV) Liquid—solid transition [89].

Redrawn from N.A. Yewdall, A.A.M. André, T. Lu, E. Spruijt, Coacervates as models of membraneless organelles, Curr. Opin. Colloid Interface Sci. 52 (2021) 101416.

and temperature. Additionally, these systems exhibit a high tendency to coalesce because of low interfacial energies [108]. Alternatively, polymer-stabilized complex coacervate microdroplets have emerged as a more controllable and robust system in synthetic biology. Their peculiar core—shell morphology not only prevents coalescence of microdroplets, but allows sequestration of a high volume of biomacromolecules with their subsequent release via a selectively permeable polymeric membrane [101]. This modular system advances the development of synthetic cells for applications in cell biology, artificial cell engineering, and biomedical disciplines like drug delivery systems, sensors, and biomimetic adhesives.

9.2.3 Multicompartment self-assembled structures

A living cell mainly comprises organelles that spatially separate the biomolecules while maintaining an intercompartment communication network that runs the cellular machinery [108]. Assembling synthetic cell models into distinct multilevel hybrid microstructures with nested organization provides advantages including (1) programmed release of encapsulated molecules, (2) encapsulation of different/incompatible molecules in a single vesicle, and (3) subcompartments of different compositions can be loaded into a single vesicle, allowing facile control over the physico-chemical properties of the vesicle. Based on the vesicle constituents, these multicompartments are designated as capsosomes (LbL assembly of liposomes), vesosomes (liposomes-in-liposome), polymersomes-in-polymersome, coacervates-in-coacervate, and mixed multicompartment systems.

Capsosomes are subcompartmentalized systems, formed by the alternate deposition of oppositely charged polyelectrolytes on the liposomes-loaded core [109,110]. The assembly of multi-layers is dictated by several interacting forces like electrostatic interactions, DNA hybridization, hydrogen bonding, or covalent bonding, which offers a range of building blocks like charged and neutral polymers, polysaccharides, nanoparticles, and biomolecules, including nucleic acids, proteins, and viruses [111−118]. The low stability of liposomes is thus circumvented by enclosing them within a more stable and less permeable polymeric membrane [109,110]. By tuning the number of polyelectrolyte layers, the thickness, and hence the permeability of the polymeric membrane can be controlled. The membrane thickness also depends on the conditions of the LbL method, like solvent polarity, ionic strength and pH of the solution, and temperature conditions [119−125]. The method offers easy tailoring of vesicle properties like size, composition, porosity, stability, surface functionality, and colloidal stability, besides the integration of stimuli-responsiveness in capsosomes [69,126−128].

Vesosomes (vesicles-in-vesicle) are nested lipid vesicles in which multiple liposomes are encapsulated within a larger liposome. These can be used for applications in drug delivery systems, bioreactors, and synthetic protocells [129−131]. Other multicompartment vesicles, including polymersomes-in-polymersome [106,132,133], coacervates-in coacervate [134], and mixed multicompartment vesicles like liposomes-in-polymersome [135], coacervates-in-polymersome [136,137], coacervates-in-liposome [138,139], and coacervates-in-proteinosome [140], have been constructed to compensate the limitations of the individual constituting vesicle, summarized in Table 9.2, thereby engineering vesicles with cell-like structural and functional properties. Among these, hybrid systems containing coacervates as subcompartments have gained significant interest as these mimic the MLOs found in cells, showing the *in situ* phase separation mechanism of living cells [136].

9.3 Methods for the construction of self-assembled structures

Strategically mimicking nature's way to compartmentalize biomolecules within cells, various techniques have been designed to construct prototypes of cells and their organelles. By using naturally found lipids and chemically synthesized amphiphilic polymers, cellular scaffolds like membrane-bound liposomes, polymersomes, proteinosomes, and membraneless coacervates have been prepared.

9.3.1 Film rehydration method

In the film rehydration method, a thin film of amphiphiles is formed by drying a solution of amphiphiles in the organic solvent on a solid surface [156,157]. This is followed by the addition of an aqueous medium that results in swelling of thin film, which finally buds off from the surface in the

Table 9.2 Different self-assembled structures and their characteristic properties.

Self-assembled structures	Liposomes	Polymersomes	Colloidosomes	Proteinosomes	Dendrimersomes	Coacervates droplet
Building blocks	Phospholipids	Amphiphilic block copolymers	Inorganic/organic nanoparticles	Protein–polymer conjugates	Amphiphilic janus dendrimers	Proteins, nucleic acids, neutral/charged polymers
Structure	Spherical lipid bilayer with hollow interior	Spherical mono/bilayer of polymer with hollow interior	Hollow sphere stabilized by nanoparticles	Supramolecular assembly of protein–polymer conjugate	Unilamellar vesicles of amphiphilic janus dendrimers in aqueous media	Concentrated polymer phase-separated within polymer depleted phase
Advantages	• Simple preparation method • Free-standing membrane • Compatibility with multiple preparation methods • Incorporation of cell division machinery • Incorporation of integral proteins	• High stability • Good mechanical and chemical properties • High chemical versatility	• Good control over physical and chemical properties of the vesicle • High chemical versatility • High stability • High temperature resistant • Low mass transfer resistance	• High loading capacity • Multicompartmentalization • Size-selective permeability • biocompatible • functional membranes • mechanically robust	• High stability • Multicompartmentalization • Reconstitution of cell division machinery • High chemical versatility	• Mimic biological condensed phase • High loading capacity • Offers high stability and activity of encapsulated biomolecules.
Disadvantages	• High polydispersity in terms of size and lamellarity. • Low stability • High membrane permeability • Low chemical versatility • Highly sensitive	• Low membrane permeability • Low lateral mobility	• Limited encapsulation efficiency • Crosslinking process affects the activity of encapsulated biomolecules.	• Denaturation of building block	• Limited encapsulation efficiency • Stringent conditions required for vesicle formation	• Low Structural stability • Time dependent stability.
References	[141–143]	[18,20,30–32]	[62,122,144–148]	[78,148–151]	[81–85,152]	[36,137,153–155]

Partially adapted from Advances in Colloid and Interface Science 299 (2022) 102566, Copyright © 2022, Elsevier Inc.

form of vesicles (Fig. 9.6A). Most commonly used for vesicle preparation, the method is straightforward—no special equipment is required—and it produces a high yield of vesicles free from an organic solvent. However, the method generates MLVs and SUVs with high polydispersity, which can be solved by using sequential extrusion by repeatedly passing the mixture of vesicles through a polycarbonate membrane with small pores (typically 100 nm diameter but can be any desired size) to obtain homogenously distributed SUVs. Compared to film rehydration, its other variants like electroformation or gel-assisted hydration generate GUVs with low polydispersity [22,156,159].

9.3.2 Electroformation method

A more controlled method to produce low dispersity GUVs, with desired microstructures is aided film hydration methods, which include electroformation and gel-assisted film rehydration (Fig. 9.6B). Electroformation results in vesicle formation by swelling the amphiphilic thin layer deposited on electrodes [glass, Pt, or indium titanium oxide (ITO)] on the application of electric current. The concept behind the process is that when current is applied, it generates fluctuation in the film along with interlayer repulsions, and thereby results in the release of the film from the surface in the form of vesicles. Due to the use of electric current, charged amphiphiles cannot be employed in the process. Also, it is challenging to establish the nature of BCPs and the conditions to obtain vesicles through this methodology [160].

9.3.3 Solvent displacement techniques

In solvent displacement techniques, before removing the organic solvent, the BCPs are treated simultaneously in both organic and aqueous phases to facilitate the self-assembly process in amphiphiles. This technique provides efficient encapsulation of water-soluble molecules but poses problems in the complete removal of the organic solvent from the vesicular system. Interaction of organic solvent with the hydrophilic molecules can deactivate them and limit the applications of the vesicles. Besides, surfactants are also used in the process, which gets embedded in the membrane, altering their properties [59,161].

In the case of solvent injection, amphiphiles dissolved in an organic solvent are added to an aqueous medium and finally, the organic solvent is removed from the system using dialysis or centrifugation methods. Whereas this technique produces SUVs, often MLVs of broad size distribution are also formed. The double emulsion method using a microfluidic setup has proved advantageous in the fabrication of monodispersed SUVs to GUVs, especially highly organized multicompartmental vesicular structures [156,162,163]. Initially, a water-in-oil emulsion of copolymers is prepared to produce water droplets coated with copolymers. This is then added to the oil-in-water emulsion, where the amphiphile-coated water droplets migrate to the o/w system generating a w/o/w double emulsion system, which results in the formation of a second amphiphilic layer on the mono-coated droplet (Fig. 9.6C). Dewetting is performed as a workup using ethanol in the aqueous phase and surfactants, and octanol in the oil phase. As the layers deposit in a sequential and controlled manner, this method provides remarkable control over the size, dispersity, and membrane properties. In addition, high encapsulation of cargo is possible, irrespective of the nature of the encapsulating moieties. However, complete solvent removal to date is not reported, and trace amounts of solvents and other used components (surfactants, additives) are found in the vesicles' membranes and

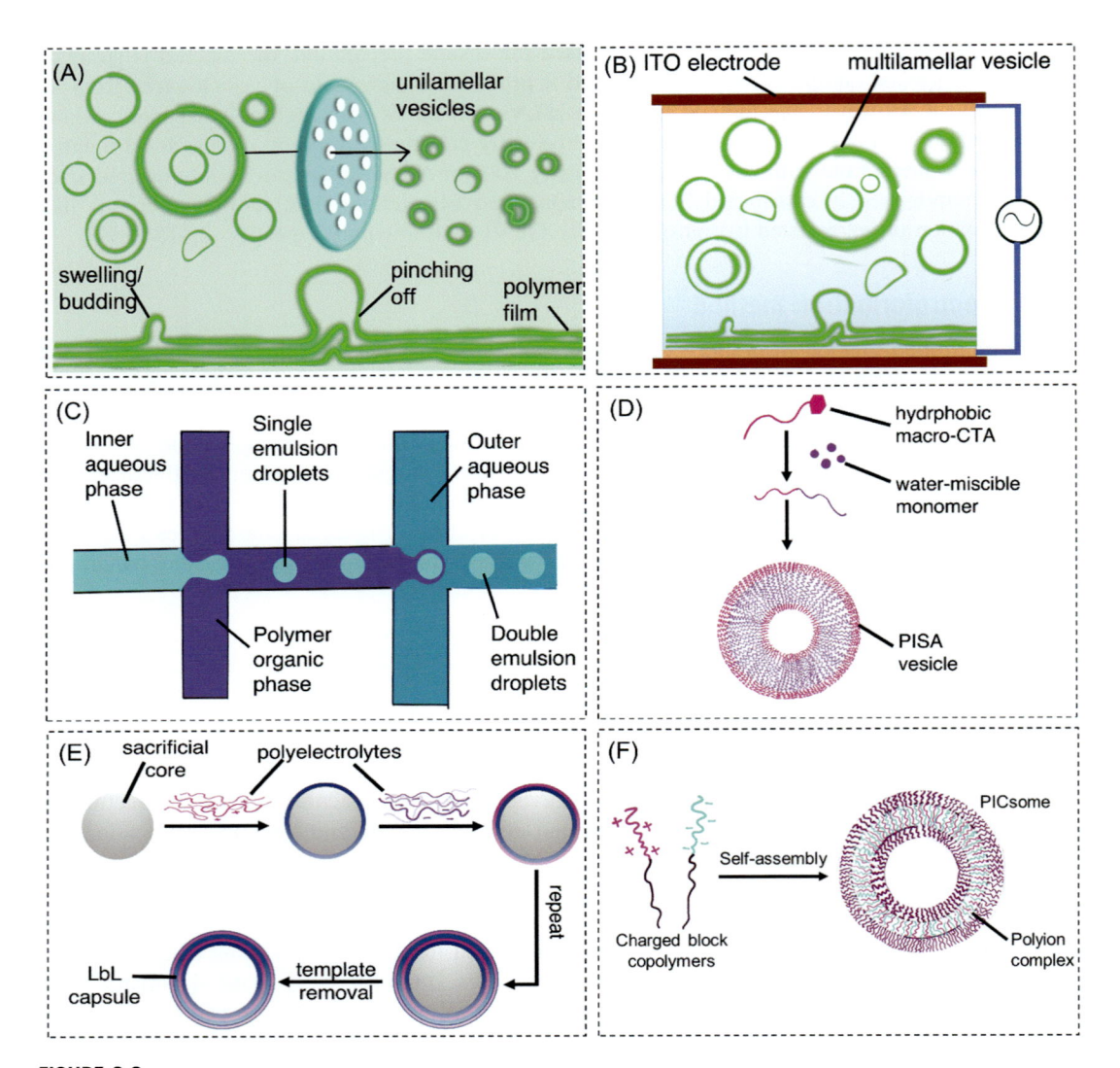

FIGURE 9.6

Methods for vesicle formation: (A) film rehydration; (B) electroformation; (C) microfluidics; (D) polymerization-induced self-assembly; (E) layer-by-layer assembly; and (F) self-assembly of charged block copolymers into PIC (polyion complexes)-somes [158].

Redrawn from L. Heuberger, M. Korpidou, O.M. Eggenberger, M. Kyropoulou, C.G. Palivan, Current perspectives on synthetic compartments for biomedical applications, Int. J. Mol. Sci. 23 (2022).

lumen, which affects the properties of the vesicle and the encapsulated molecules [162,164]. Besides, considerable rearrangement is required as the inner layer is much smaller than the outer layer, leading to aggregation when there is an excess of copolymers [10].

9.3.4 **Polymerization-induced self-assembly**

Another popular approach for the preparation of self-assembled structures is polymerization-induced self-assembly (PISA). First introduced in 2002 [165], PISA is a type of emulsion polymerization, in which an insoluble polymer is formed from a soluble monomer. In contrast to emulsion polymerization, which involves the use of polymer stabilizer/surfactant for the particle formation [166], PISA involves chain extension of an initial soluble precursor block, which acts as a block stabilizer, chemically bonded to an insoluble polymer block that forms the nanoparticle core *in situ* (Fig. 9.6D). On reaching an initial critical polymer concentration, self-assembly of the diblock copolymer is triggered, resulting in the formation of spheres, then worms, and finally vesicles [160]. In addition, PISA can occur at a range of final polymer concentration (5%−50% w/w) [167]. Targeted morphologies can be obtained by optimizing the amphiphile concentration or the extent of polymerization of the stabilizer block. Though diverse polymerization methods like atom transfer radical polymerization, nitroxide-mediated polymerization, and ring-opening metathesis polymerization have been employed for PISA, RAFT (reversible addition-fragmentation chain transfer) is most commonly used to form diblock copolymer self-assemblies. In general, nanometer-range (SUVs ∼ 10−100 nm) assemblies are produced using PISA, however, recent works have successfully reported the construction of GUVs (∼ 10 μm) [168]. PISA is efficient as both polymer synthesis and self-assembly occur simultaneously in one pot, thus explicitly exemplifying the versatility of block copolymers over lipids, and allows bypassing of the multiple stages of assembly. However, the dependency of the method on hydrophilicity modulation limits PISA to limited block copolymers (especially 2-hydroxypropyl methacrylate).

Another useful approach to obtain molecular self-assemblies is via polyion complexes (PIC) fabrication technology that can be performed in aqueous media without the organic solvent. PIC formation is a simple method based on the electrostatic interactions between charged polymers, to produce nanosized aggregates like PIC micelles and PIC hollow capsules and has been employed for applications such as gene delivery, charged drug molecules, proteins, and nucleic acid [169−171]. Two approaches can be employed to prepare hollow PIC architecture via a self-assembly process. The most common approach is a layer-by-layer (LbL) method, which involves the layer formation of oppositely charged polyelectrolytes onto a charged colloid surface/template particle (Fig. 9.6E) [109]. Although the LbL method allows facile control over membrane properties like permeability, thickness, etc., it can be a tedious method as it requires multiple layer formation, and sometimes template removal can become a serious concern. A second approach is the direct mixing of charged block copolymers or block-homo ionomers to prepare hollow capsules [37]. Typically, diblock copolymers are mixed to neutralize their charges, resulting in the formation of nano−mesoscopic hollow assemblies (Fig. 9.6F). Further, crosslinking of PICsomes membranes provides enhanced stability for *in vivo* applications. However, due to having less control over the PICsomes architectures, this method has not been well developed.

9.4 **Applications**

The self-assembly of molecules into biomimetic structures with the ability to show rapid adaptation in response to environmental stimuli, with simultaneous response to energy inputs, provides an elegant strategy in the field of bottom-up synthetic biology. By encapsulation/immobilization of functional molecules into membrane-bound single-compartment to multicompartment systems, and membraneless compartments, scientists have been able to emulate the structural and functional complexities of the cell and the cellular matrix, generating a broader purview for the development of artificial organelles (AOs), eventually leading to artificial cells. Apart from their application in drug delivery and biosensors, these engineered biomimetic structures are interestingly being used to develop AOs and protocell systems that can simplify the complexities of a living cell. Owing to their resemblance to cell organelles that compartmentalize reaction spaces for complex metabolic reactions, self-assembled structures exhibit a hollow interior enclosed within a membrane with intrinsic/engineered permeability (via the insertion of membrane proteins) similar to the cell membrane. In this context, synthetic compartments have been designed to construct synthetic minimal cells encapsulating genetic circuits, and mimics of cell organelles like mitochondria, endoplasmic reticulum, peroxisomes, and functional aspects of MLOs Fig. 9.7(A-E) [104,172−175].

9.4.1 Cell models and cell mimics

Compartmentalized soft matter systems are simplified and controllable tools to quantitatively and analytically explore the phenomenology of living cells. These synthetic compartments have been designed to exhibit cellular processes like growth and division, polymerase chain reactions, protein synthesis, *in vitro* transcription/translation, and inter-/intracellular communication, and thus can be used to envisage the cellular machinery experimentally by constructing cell replicas. The construction of synthetic compartments in a bottom-up approach provides autonomous control of physicochemical and biological factors and thus assists in understanding their roles in sculpting the cellular machinery. Cellular mechanisms, such as cytokinesis mechanism, the role of membrane curvature in the insertion of transmembrane proteins, and cell motility, have also been investigated using these synthetic compartments Fig. 9.8A−F [38−41,176,177].

The development of cell mimics has enabled a deeper understanding of many cellular processes. Mimics of cellular membranes and composited lipid membranes, have been used to investigate the thermodynamic forces involved in the folding and insertion of membrane proteins [42]. Liposomes were used to establish the role of membrane curvatures in promoting the redistribution of membrane-anchored proteins [38]. Lipid-coated microdroplets reconstituted with Min proteins and FtsZ filament structures successfully reproduced the pole-to-pole oscillations of Min proteins and localized the FtsZ rings in the center, and can be used as promising models to study the complex protein interactions existing in a cell [177]. Further, the direct contribution of myosin to cytokinesis was elucidated via actin polymerization inside water-in-oil droplets [176]. Bridging the gap between natural and synthetic biology, the construction of cell-free systems highlights the significance of macromolecular crowding in gene expression [68]. On-chip two-dimensional DNA compartments as artificial cells offer a controlled environment to study the synchrony and pattern formation of gene expression in space and time [43,44]. Vesicles reconstituted with protein-complexes are used to design genetic circuits and cascades, which can further fuse together, combining the transcriptional and translational machinery in one compartment [172].

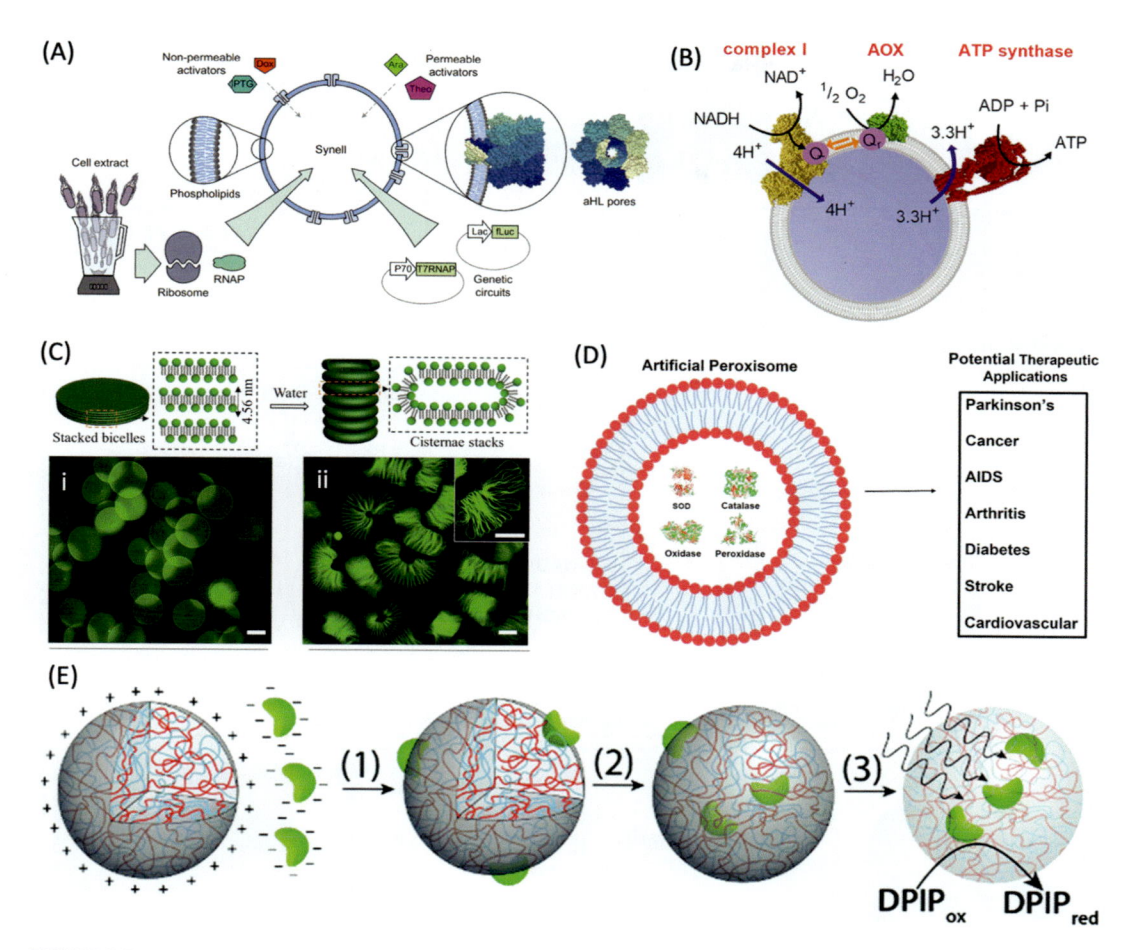

FIGURE 9.7

Artificial organelles: (A) rational design for the construction of synthetic minimal cells (synells) via encapsulation of genetic circuits and cascades inside semipermeable liposomes [172]; (B) reconstitution of mitochondrial respiratory system in liposomes to construct energy regenerating vesicles [173]; (C) schematics for the preparation cisternae from bicelles; fluorescence image of (i) stacked bicelles (ii) cisternae stacks [174]; (D) schematics showing the formation of artificial peroxisomes by encapsulation of different enzymes [175]; (E) schematics for the preparation of photosynthetically-active coacervate microdroplets: (1) electrostatic attraction of negatively-charged chloroplasts towards positively-charged coacervate droplets; (2) subsequent internalization of chloroplasts into coacervates; (3) photoinduced conversion of DPIP by chloroplast containing coacervate microdroplet [104].

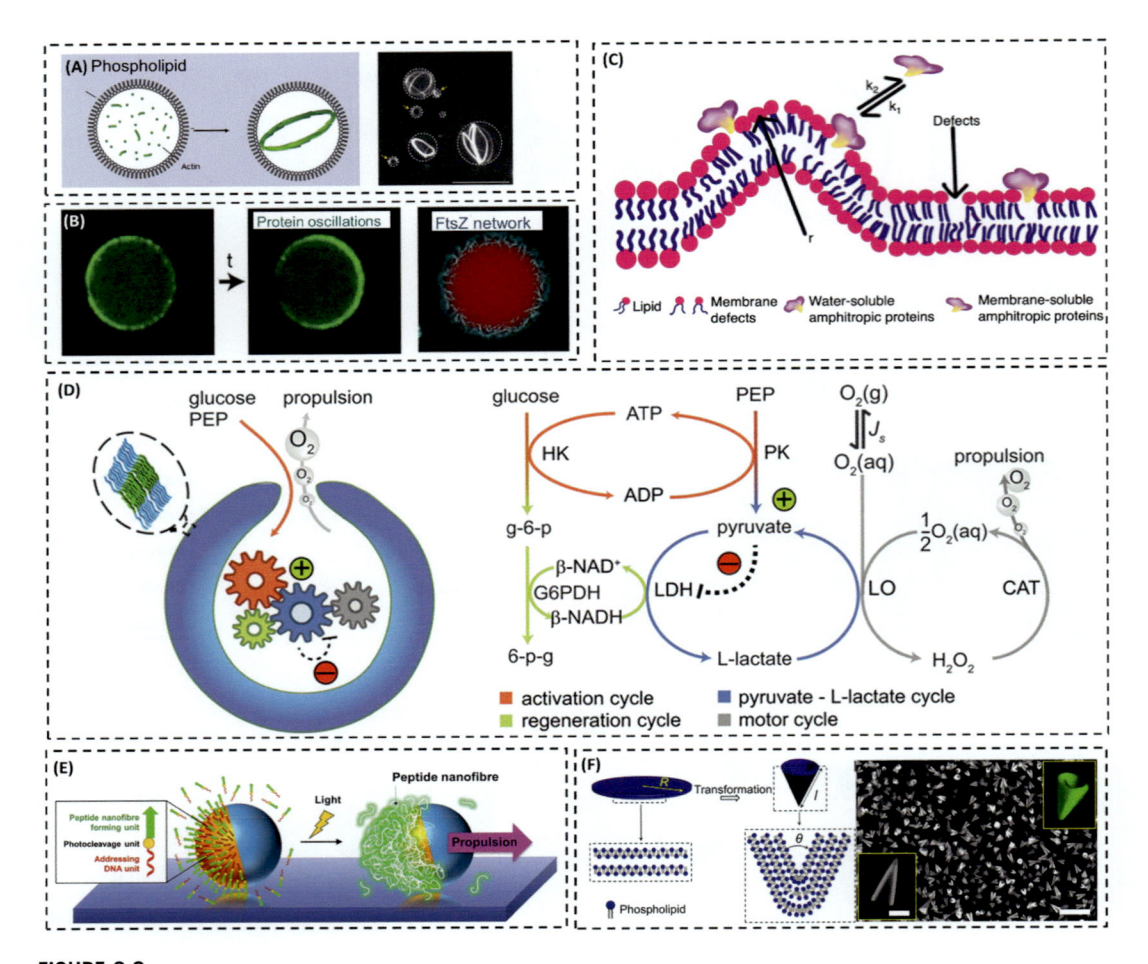

FIGURE 9.8

Cellular aspects studied using cell models: (A) schematics showing the self-organization of actin bundles inside water-in-oil droplets, confocal image of actin polymerization in vitro models [176]; (B) reconstitution of Min proteins and FtsZ filaments in lipid droplets to investigate interaction of Min/ FtsZ during cell division [177]; (C) role of membrane curvature in insertion of transmembrane proteins: bending a bilayer generates packing defects that becomes recruitment site for amphiphilic molecules [38]. Micromotors: (D) illustration of catalytic nanomotors powered by metabolic pathway of glucose to phosphoenolpyruvate by four enzymes [39]; (E) light-driven propulsion of phase-separated giant liposomes using light-induced cleavage and subsequent formation of peptide—DNA conjugates [40]; (F) formation of microcones from bicelles. Fluorescence image of microcones formed (right) [41].

Redrawn part (C) from N.S. Hatzakis, V.K. Bhatia, J. Larsen, K.L. Madsen, P.-Y. Bolinger, A.H. Kunding, et al., How curved membranes recruit amphipathic helices and protein anchoring motifs, Nat. Chem. Biol. 5 (2009) 835–841.

9.4.2 **Intracellular delivery vehicles**

Intracellular delivery is generally required for very large or hydrophilic therapeutics that cannot cross the cell membrane by diffusion. These therapeutics include nucleic acid, proteins, peptides, impermeable drugs, probes, and nanoparticles [45,48,49]. Significant progress has been achieved in the context of gene delivery of plasmid DNA and messenger RNA (mRNA), and small interfering RNA (siRNA) and microRNA for gene silencing, while intracellular delivery of proteins like antibodies and transcription factors, sensors, and probes are advancing rapidly [45,48−51]. Intercellular delivery of these active materials is carried out by nanosized carriers, which perform timely release of the cargo at the targeted site and prevent its early degradation. Thus cellular mimics, especially bilayer vesicles, show fusogenic potential, allowing fusion with the cell membrane and releasing the cargo inside the cytoplasm. Several carriers have been designed by combining natural and synthetic lipids, polymer and inorganic nanoparticles, and decorated by ligands and proteins to accelerate specific take-ups and endocytosis [50,52−54,56−58,65,178−180].

Among these carriers, liposomes and lipid−nucleic acid complexes have been extensively investigated, as the cationic lipid/polymers readily conjugate with anionic nucleic acids, facilitating cellular uptake and endosomal escape [58,61,63,178,181−184]. The nanoconjugates with/without enfolding vesicles have successfully transfected multiple cell lines and primary cells. Polyethyleneimine (PEI) with low molecular weight (<25 Da, highly branched) is an extensively used polymer for nucleic acid delivery [185]. Research studies have increased on synthesizing biocompatible polymers with improved endosomal escape [64,66,67,185−188]. For instance, cationic oligo(carbonate-b-α-aminoester)s, termed as charge-altering releasable transporters (CART) complexes with mRNA and the cationic segment, undergo rapid degradation at physiological pH. Thus upon entry into a cell, CART/mRNA complexes disassemble, and release mRNA into the cytoplasm for protein translation [189].

The therapeutic efficiency of the nanocarriers is by far limited by the low endosomal escape ($\sim 1\%$) [188], which has created the need to improve the delivery vesicles. High-throughput screens are being used to design cationic ionizable lipids [57] and cell-penetrating peptides [185,190], and stimuli-responsive molecules [191] were incorporated to facilitate cellular uptake and endosomal escape. Though some of the lipid-based nanocarriers and membraneless complexes are in human clinical trials for the delivery of siRNAs for gene silencing [48,192], the intracellular delivery of mRNA and DNA is challenging [193]. Gene delivery aims to elucidate the structure−function relationship of protocell-like compartments with different cargoes, and design nanocarriers with high site-specificity and tunability. With the advancement in gene-editing, there is an ever-increasing demand for the cytoplasmic delivery of endonuclease and recombinase, such as clustered regularly interspaced short palindromic repeats, zinc finger nucleases, and Cre recombinase [194]. Unlike DNA and RNA, the charge and conformation of proteins are not constant, and therefore the protein−vehicles can disassemble before reaching the intracellular cytosol [195]. To prevent this, proteins like Cas9 and Cre are genetically modified with negatively supercharged proteins, like GFP, and anionic peptides to enhance their binding with the vehicle [195−199]. Alternatively, dendritic polymers can be employed that show effective complexes of both anionic and cationic proteins, enabling the codelivery of multiple proteins into cytosol [200].

9.4.3 Micro-/nanoreactors for catalytic cascades

The cellular subcompartments offer spatiotemporal control over the biocatalytic reactions, thus increasing their efficiency and selectivity. To reproduce similar catalytic cascades inside compartments [201,202], various synthetic scaffolds encapsulating biocatalytic components have been designed, such as emulsion droplets, vesicles, colloidosomes, and polymeric hollow capsules, that can localize catalytic steps, while multicompartment systems can carry out a series of orchestrated catalytic cascades [203–207].

These self-assembled structures have been engineered to generate AOs like peroxisome mimics [208,209], self-propelling vesicles, and autonomous vesicles as cytomimics of mitochondria and chloroplast. Peroxisome mimics have been demonstrated using the self-assembly of amphiphilic PMOXA-PDMS-PMOXA triblock copolymer, poly(ethylene glycol)-*b*-poly(caprolactone-*g*-trimethylene)carbonate (PEG-PCLgTMC) copolymers, and liposomes assembled in core–shell microreactors, loaded with enzymes like superoxide dismutase (SOD)-lactoperoxidase (LPO), SOD-LPO-CAT (catalase), or Cu, Zn-SOD that catalyze the reactive oxygen species (ROS) [208–210]. Moreover, by integrating the synthetic scaffolds with cellular components like the photosynthetic reaction center [173,211,212], ATP synthase [19,136,213–215], bacteriorhodopsin [209,213,216–218], bo3 oxidase and F1F0 ATP synthase [219], or mammalian respiratory protein mito-Cl and Ec-F1F2, a variety of autonomous AOs have been generated [173,211–213,219]. Further, the construction of lipid and polymer-based self-propelling vesicles has been achieved at nano- and microscales, such as bowl-shaped stomatocytes or tubular motors that exhibit dynamic motion in response to enzyme/nanoparticle-catalyzed reactions, actin formation, or to stimuli [40,41,220,221]. These systems have shown potential applications in programmed drug delivery under different conditions [39,222–224]. Utilizing the characteristic sequestration and molecular-crowding properties of coacervate microdroplets, models for minimal cells have been constructed that can sequester enzymes and biological machinery, and exhibit inter- and intracellular communication, predation, and cell division [104,136].

9.5 Conclusion

Advances in synthetic biology and materials science over the past 50 years have accelerated progress in the field of artificial life through the design and construction of cell-like microarchitectures via bottom-up approaches. Cytomimetic self-assembled structures have been constructed in the form of GUVs, giant polymer vesicles, proteinosomes, colloidosomes, dendrimersomes, and coacervate microdroplets, as well as hybrid combinations of these compartmentalized systems. Typical biomimetic functions such as enzyme cascades, growth and division, motility, predation, phagocytosis, cell-free gene expression, membrane-gating, and intracellular communication have been demonstrated. However, the sophistication of these synthetic protocellular systems is rudimentary compared with the functional integration of living cells, and we are milestones away from realizing the efficiency and self-sufficiency of these systems. Also, the integration of higher-level cell processes in these constructs, like self-replication, autonomy, and self-production, remain elusive. Important criteria need to be addressed including cellular communication, signal flow, removal of synthetic and biological barriers, preservation of in situ activity, biocompatibility, degradation profiles, and scalability. Additionally, the majority of synthetic constructions have only been used *in vivo* in organelles like lysosomes and endosomes; a significant advancement is needed to enable their use in the cytoplasm of living cells.

References

[1] S.A. Meeuwissen, S.M.C. Bruekers, Y. Chen, D.J. Pochan, J.C.M. van Hest, Spontaneous shape changes in polymersomes via polymer/polymer segregation, Polym. Chem. 5 (2014) 489–501.

[2] K. Oglecka, J. Sanborn, A. Parikh, R. Kraut, Osmotic gradients induce bio-reminiscent morphological transformations in giant unilamellar vesicles, Front. Physiol. (2012) 3.

[3] S. Allen, O. Osorio, Y.-G. Liu, E. Scott, Facile assembly and loading of theranostic polymersomes via multi-impingement flash nanoprecipitation, J. Control. Release 262 (2017) 91–103.

[4] P. Walde, S. Ichikawa, Enzymes inside lipid vesicles: preparation, reactivity and applications, Biomol. Eng 18 (2001) 143–177.

[5] M.L. Immordino, F. Dosio, L. Cattel, Stealth liposomes: review of the basic science, rationale, and clinical applications, existing and potential, Int. J. Nanomed 1 (2006) 297–315.

[6] I.A. Chen, P. Walde, From self-assembled vesicles to protocells, Cold Spring Harbor Perspect. Biol. (2010) 2.

[7] F. Szoka, D. Papahadjopoulos, Comparative properties and methods of preparation of lipid vesicles (Liposomes), Ann. Rev. Biophys. Bioeng 9 (1980) 467–508.

[8] E. Rideau, R. Dimova, P. Schwille, F.R. Wurm, K. Landfester, Liposomes and polymersomes: a comparative review towards cell mimicking, Chem Soc Rev 47 (2018) 8572–8610.

[9] A.I. Rusanov, Striking world of nanostructures, Russ. J. Gen. Chem 72 (2002) 495–511.

[10] H. Che, J.C.M. van Hest, Stimuli-responsive polymersomes and nanoreactors, J. Mater. Chem. B. 4 (2016) 4632–4647.

[11] M. Antonietti, S. Förster, Vesicles and liposomes: a self-assembly principle beyond lipids, Adv. Mater. 15 (2003) 1323–1333.

[12] S.T. Hyde, Curvature and the global structure of interfaces in surfactant-water systems, J Phys Colloques 51 (1990) C7–209-C7–28.

[13] J.C.M. van Hest, D.A.P. Delnoye, M.W.P.L. Baars, M.H.P. van Genderen, E.W. Meijer, Polystyrene-dendrimer amphiphilic block copolymers with a generation-dependent aggregation, Science. 268 (1995) 1592–1595.

[14] S. Förster, M. Zisenis, E. Wenz, M. Antonietti, Micellization of strongly segregated block copolymers, J. Chem. Phys 104 (1996) 9956–9970.

[15] D.E. Discher, F. Ahmed, Polymersomes, Ann. Rev. Biomed. Eng 8 (2006) 323–341.

[16] A.D. Bangham, R.W. Horne, Negative staining of phospholipids and their structural modification by surface-active agents as observed in the electron microscope, J. Mol. Biol 8 (1964) 660–IN10.

[17] J. Li, X. Wang, T. Zhang, C. Wang, Z. Huang, X. Luo, et al., A review on phospholipids and their main applications in drug delivery systems, Asian J. Pharm. Sci 10 (2015) 81–98.

[18] R. Chandrawati, F. Caruso, Biomimetic liposome- and polymersome-based multicompartmentalized assemblies, Langmuir. 28 (2012) 13798–13807.

[19] L. Otrin, N. Marušič, C. Bednarz, T. Vidaković-Koch, I. Lieberwirth, K. Landfester, et al., Toward artificial mitochondrion: mimicking oxidative phosphorylation in polymer and hybrid membranes, Nano Lett 17 (2017) 6816–6821.

[20] J.F. Le Meins, C. Schatz, S. Lecommandoux, O. Sandre, Hybrid polymer/lipid vesicles: state of the art and future perspectives, Mater. Today 16 (10) (2013) 397–402.

[21] A. Jesorka, O. Orwar, Liposomes: technologies and analytical applications, Ann. Rev. Analyt. Chem 1 (2008) 801−832.

[22] S. Vemuri, C.T. Rhodes, Preparation and characterization of liposomes as therapeutic delivery systems: a review, Pharm. Acta Helvetiae 70 (1995) 95−111.

[23] M.I. Angelova, D.S. Dimitrov, Liposome electroformation, Faraday Discuss. Chem. Soc 81 (1986) 303−311.

[24] E. Discher Dennis, A. Eisenberg, Polymer vesicles, Science. 297 (2002) 967−973.

[25] R. Dimova, C. Marques, The Giant Vesicle Book, CRC Press, 2019.

[26] X. Zhang, P. Tanner, A. Graff, C.G. Palivan, W. Meier, Mimicking the cell membrane with block copolymer membranes, J. Polym. Sci. Part A: Polym. Chem 50 (2012) 2293−2318.

[27] S. Winzen, M. Bernhardt, D. Schaeffel, A. Koch, M. Kappl, K. Koynov, et al., Submicron hybrid vesicles consisting of polymer−lipid and polymer−cholesterol blends, Soft. Matter 9 (2013) 5883−5890.

[28] V. Malinova, S. Belegrinou, D. de Bruyn Ouboter, W.P. Meier, Biomimetic block copolymer membranes. Berlin, Heidelberg in: W.P. Meier, W. Knoll (Eds.), Polymer Membranes/Biomembranes, Springer, Berlin Heidelberg, 2010, pp. 87−111.

[29] A. Napoli, M. Valentini, N. Tirelli, M. Müller, J.A. Hubbell, Oxidation-responsive polymeric vesicles, Nat. Mater 3 (2004) 183−189.

[30] L. Messager, J. Gaitzsch, L. Chierico, G. Battaglia, Novel aspects of encapsulation and delivery using polymersomes, Curr. Opin. Pharmacol 18 (2014) 104−111.

[31] L. Klermund, S.T. Poschenrieder, K. Castiglione, Biocatalysis in polymersomes: improving multienzyme cascades with incompatible reaction steps by compartmentalization, ACS Catal 7 (2017) 3900−3904.

[32] H. Aranda-Espinoza, H. Bermudez, F.S. Bates, D.E. Discher, Electromechanical limits of polymersomes, Phys. Rev. Lett 87 (2001) 208301.

[33] R. Dimova, U. Seifert, B. Pouligny, S. Förster, H.G. Döbereiner, Hyperviscous diblock copolymer vesicles, Eur. Phys. J. E 7 (2002) 241−250.

[34] R.L. Knorr, M. Staykova, R.S. Gracià, R. Dimova, Wrinkling and electroporation of giant vesicles in the gel phase, Soft. Matter 6 (2010) 1990−1996.

[35] K.L. Thompson, M. Williams, S.P. Armes, Colloidosomes: synthesis, properties and applications, J. Colloid Interface Sci 447 (2015) 217−228.

[36] S. Sun, M. Li, F. Dong, S. Wang, L. Tian, S. Mann, Chemical signaling and functional activation in colloidosome-based protocells, Small. 12 (2016) 1920−1927.

[37] A. Koide, A. Kishimura, K. Osada, W.-D. Jang, Y. Yamasaki, K. Kataoka, Semipermeable polymer vesicle (PICsome) self-assembled in aqueous medium from a pair of oppositely charged block copolymers: physiologically stable micro-/nanocontainers of water-soluble macromolecules, J. Am. Chem. Soc 128 (2006) 5988−5989.

[38] N.S. Hatzakis, V.K. Bhatia, J. Larsen, K.L. Madsen, P.-Y. Bolinger, A.H. Kunding, et al., How curved membranes recruit amphipathic helices and protein anchoring motifs, Nat. Chem. Biol 5 (2009) 835−841.

[39] M. Nijemeisland, L.K.E.A. Abdelmohsen, W.T.S. Huck, D.A. Wilson, J.C.M. van Hest, A Compartmentalized out-of-equilibrium enzymatic reaction network for sustained autonomous movement, ACS Central Sci. 2 (2016) 843−849.

[40] H. Inaba, A. Uemura, K. Morishita, T. Kohiki, A. Shigenaga, A. Otaka, et al., Light-induced propulsion of a giant liposome driven by peptide nanofibre growth, Sci. Rep 8 (2018) 6243.

[41] Q. Li, C. Li, W. Mu, X. Han, Topological defect-driven buckling of phospholipid bicelles to cones for micromotors with modulated heading pathways, ACS Nano 13 (2019) 3573−3579.

[42] F. Cymer, G. von Heijne, S.H. White, Mechanisms of integral membrane protein insertion and folding, J. Mol. Biol 427 (2015) 999−1022.

[43] A.M. Tayar, E. Karzbrun, V. Noireaux, R.H. Bar-Ziv, Synchrony and pattern formation of coupled genetic oscillators on a chip of artificial cells, Proc. Natl. Acad. Sci. 114 (2017) 11609–11614.

[44] E. Karzbrun, A.M. Tayar, V. Noireaux, R.H. Bar-Ziv, Programmable on-chip DNA compartments as artificial cells, Science. 345 (2014) 829–832.

[45] M.P. Stewart, A. Sharei, X. Ding, G. Sahay, R. Langer, K.F. Jensen, In vitro and ex vivo strategies for intracellular delivery, Nature. 538 (2016) 183–192.

[46] X. Huang, M. Li, D.C. Green, D.S. Williams, A.J. Patil, S. Mann, Interfacial assembly of protein–polymer nano-conjugates into stimulus-responsive biomimetic protocells, Nat. Commun 4 (2013) 2239.

[47] V. Postupalenko, T. Einfalt, M. Lomora, I.A. Dinu, C.G. Palivan, Chapter 11 – Bionanoreactors: from confined reaction spaces to artificial organelles, in: S. Sadjadi (Ed.), Organic Nanoreactors, Academic Press, Boston, 2016, pp. 341–371.

[48] R. Kanasty, J.R. Dorkin, A. Vegas, D. Anderson, Delivery materials for siRNA therapeutics, Nat. Mater 12 (2013) 967–977.

[49] H. Yin, R.L. Kanasty, A.A. Eltoukhy, A.J. Vegas, J.R. Dorkin, D.G. Anderson, Non-viral vectors for gene-based therapy, Nat. Rev. Genet 15 (2014) 541–555.

[50] K.A. Whitehead, R. Langer, D.G. Anderson, Knocking down barriers: advances in siRNA delivery, Nat. Rev. Drug Discov. 8 (2009) 129–138.

[51] D.T. Gonzales, N. Yandrapalli, T. Robinson, C. Zechner, T.Y.D. Tang, Cell-free gene expression dynamics in synthetic cell populations, ACS Synth. Biol. 11 (2022) 205–215.

[52] M. Morille, C. Passirani, A. Vonarbourg, A. Clavreul, J.-P. Benoit, Progress in developing cationic vectors for non-viral systemic gene therapy against cancer, Biomaterials. 29 (2008) 3477–3496.

[53] V. Noireaux, A. Libchaber, A vesicle bioreactor as a step toward an artificial cell assembly, Proc. Natl. Acad. Sci. 101 (2004) 17669–17674.

[54] K.T. Love, K.P. Mahon, C.G. Levins, K.A. Whitehead, W. Querbes, J.R. Dorkin, et al., Lipid-like materials for low-dose, in vivo gene silencing, Proc. Natl. Acad. Sci. 107 (2010) 1864–1869.

[55] Y. Lin, H. Skaff, T. Emrick, A.D. Dinsmore, T.P. Russell, Nanoparticle assembly and transport at liquid-liquid interfaces, Science. 299 (2003) 226–229.

[56] S.C. Semple, A. Akinc, J. Chen, A.P. Sandhu, B.L. Mui, C.K. Cho, et al., Rational design of cationic lipids for siRNA delivery, Nat. Biotechnol 28 (2010) 172–176.

[57] L. Miao, L. Li, Y. Huang, D. Delcassian, J. Chahal, J. Han, et al., Delivery of mRNA vaccines with heterocyclic lipids increases anti-tumor efficacy by STING-mediated immune cell activation, Nat. Biotechnol 37 (2019) 1174–1185.

[58] K.A. Whitehead, J.R. Dorkin, A.J. Vegas, P.H. Chang, O. Veiseh, J. Matthews, et al., Degradable lipid nanoparticles with predictable in vivo siRNA delivery activity, Nat. Commun 5 (2014) 4277.

[59] N.-N. Deng, M. Yelleswarapu, W.T.S. Huck, Monodisperse uni- and multicompartment liposomes, J. Am. Chem. Soc 138 (2016) 7584–7591.

[60] J.-P. Douliez, N. Martin, T. Beneyton, J.-C. Eloi, J.-P. Chapel, L. Navailles, et al., Preparation of swellable hydrogel-containing colloidosomes from aqueous two-phase pickering emulsion droplets, Angew. Chem. Int. Ed 57 (2018) 7780–7784.

[61] R. Fraley, S. Subramani, P. Berg, D. Papahadjopoulos, Introduction of liposome-encapsulated SV40 DNA into cells, J. Biol. Chem 255 (1980) 10431–10435.

[62] M. Li, R.L. Harbron, J.V.M. Weaver, B.P. Binks, S. Mann, Electrostatically gated membrane permeability in inorganic protocells, Nat. Chem 5 (2013) 529–536.

[63] P.L. Felgner, T.R. Gadek, M. Holm, R. Roman, H.W. Chan, M. Wenz, et al., Lipofection: a highly efficient, lipid-mediated DNA-transfection procedure, Proc. Natl. Acad. Sci. 84 (1987) 7413–7417.

[64] I. Lostalé-Seijo, J. Montenegro, Synthetic materials at the forefront of gene delivery, Nat. Rev. Chem 2 (2018) 258–277.

[65] S.-I. Yun, S.-K. Lee, E.-A. Goh, O.S. Kwon, W. Choi, J. Kim, et al., Weekly treatment with SAMiRNA targeting the androgen receptor ameliorates androgenetic alopecia, Sci. Rep 12 (2022) 1607.

[66] M. Breunig, U. Lungwitz, R. Liebl, A. Goepferich, Breaking up the correlation between efficacy and toxicity for nonviral gene delivery, Proc. Natl. Acad. Sci. 104 (2007) 14454−14459.

[67] M. Thomas, A.M. Klibanov, Enhancing polyethylenimine's delivery of plasmid DNA into mammalian cells, Proc. Natl. Acad. Sci. 99 (2002) 14640−14645.

[68] C. Tan, S. Saurabh, M.P. Bruchez, R. Schwartz, P. LeDuc, Molecular crowding shapes gene expression in synthetic cellular nanosystems, Nat. Nanotechnol 8 (2013) 602−608.

[69] A.D. Dinsmore, M.F. Hsu, M.G. Nikolaides, M. Marquez, A.R. Bausch, D.A. Weitz, Colloidosomes: selectively permeable capsules composed of colloidal particles, Science. 298 (2002) 1006−1009.

[70] A. Böker, J. He, T. Emrick, T.P. Russell, Self-assembly of nanoparticles at interfaces, Soft Matter 3 (2007) 1231−1248.

[71] D. Lee, D.A. Weitz, Double emulsion-templated nanoparticle colloidosomes with selective permeability, Adv. Mater 20 (2008) 3498−3503.

[72] Z. Pan, X. Zhu, H. Jiang, Y. Liu, R. Chen, Flexible hierarchical Pd/SiO$_2$-TiO$_2$ nanofibrous catalytic membrane for complete and continuous reduction of p-nitrophenol, J. Exp. Nanosci 16 (2021) 62−80.

[73] G. Zhu, M. Zhang, Y. Bu, L. Lu, X. Lou, L. Zhu, Enzyme-embedded metal−organic framework colloidosomes via an emulsion-based approach, Chem. −Asian J 13 (2018) 2891−2896.

[74] M. Williams, S.P. Armes, D.W. York, Clay-based colloidosomes, Langmuir. 28 (2012) 1142−1148.

[75] J.M. López-de-Luzuriaga, M. Monge, J. Quintana, M. Rodríguez-Castillo, Single-step assembly of gold nanoparticles into plasmonic colloidosomes at the interface of oleic acid nanodroplets, Nanoscale Adv 3 (2021) 198−205.

[76] Y. Gong, A.M. Zhu, Q.G. Zhang, M.L. Ye, H.T. Wang, Q.L. Liu, Preparation of cell-embedded colloidosomes in an oil-in-water emulsion, ACS Appl. Mater. Interfaces 5 (2013) 10682−10689.

[77] M. Li, X. Huang, S. Mann, Spontaneous growth and division in self-reproducing inorganic colloidosomes, Small. 10 (2014) 3291−3298.

[78] X. Huang, A.J. Patil, M. Li, S. Mann, Design and construction of higher-order structure and function in proteinosome-based protocells, J. Am. Chem. Soc 136 (2014) 9225−9234.

[79] G. Wu, L. Wang, P. Zhou, P. Wen, C. Ma, X. Huang, et al., Design and construction of hybrid microcapsules with higher-order structure and multiple functions. Adv, Sci. 5 (2018) 1700460.

[80] Y. Li, L. Liu, H. Zhao, Enzyme-catalyzed cascade reactions on multienzyme proteinosomes, J. Colloid Interface Sci 608 (2022) 2593−2601.

[81] V. Percec, D.A. Wilson, P. Leowanawat, C.J. Wilson, A.D. Hughes, M.S. Kaucher, et al., Self-assembly of janus dendrimers into uniform dendrimersomes and other complex architectures, Science. 328 (2010) 1009.

[82] I. Buzzacchera, Q. Xiao, H. Han, K. Rahimi, S. Li, N.Y. Kostina, et al., Screening libraries of amphiphilic janus dendrimers based on natural phenolic acids to discover monodisperse unilamellar dendrimersomes, Biomacromolecules. 20 (2019) 712−727.

[83] A.M. Wagner, H. Eto, A. Joseph, S. Kohyama, T. Haraszti, R.A. Zamora, et al., Dendrimersome synthetic cells harbor cell division machinery of bacteria, Adv. Mater 34 (2022) 2202364.

[84] V. Gajbhiye, V.K. Palanirajan, R.K. Tekade, N.K. Jain, Dendrimers as therapeutic agents: a systematic review, J. Pharm. Pharmacol 61 (2010) 989−1003.

[85] I. Gitsov, J. Hamzik, J. Ryan, A. Simonyan, J.P. Nakas, S. Omori, et al., Enzymatic nanoreactors for environmentally benign biotransformations. 1. Formation and catalytic activity of supramolecular complexes of laccase and linear − dendritic block copolymers, Biomacromolecules. 9 (2008) 804−811.

[86] D. Gaur, N.C. Dubey, B.P. Tripathi, Biocatalytic self-assembled synthetic vesicles and coacervates: from single compartment to artificial cells, Adv. Colloid Interface Sci (2021) 102566.

[87] D. van Swaay, T.Y.D. Tang, S. Mann, A. de Mello, Microfluidic formation of membrane-free aqueous coacervate droplets in water, Angew. Chem. Int. Ed 54 (2015) 8398−8401.

[88] C. Martino, A.J. deMello, Droplet-based microfluidics for artificial cell generation: a brief review, Interface Focus 6 (2016) 20160011.

[89] N.A. Yewdall, A.A.M. André, T. Lu, E. Spruijt, Coacervates as models of membraneless organelles, Curr. Opin. Colloid Interface Sci. 52 (2021) 101416.

[90] F.M. Menger, A.V. Peresypkin, K.L. Caran, R.P. Apkarian, A sponge morphology in an elementary coacervate, Langmuir. 16 (2000) 9113−9116.

[91] A.I. Oparin, S. Morgulis, The origin of life, Macmillan, New York, 1938, p. 270.

[92] T. Lu, E. Spruijt, Multiphase complex coacervate droplets, J. Am. Chem. Soc 142 (2020) 2905−2914.

[93] G.A. Mountain, C.D. Keating, Formation of multiphase complex coacervates and partitioning of biomolecules within them, Biomacromolecules. 21 (2020) 630−640.

[94] J.R. Simon, S.A. Eghtesadi, M. Dzuricky, L. You, A. Chilkoti, Engineered ribonucleoprotein granules inhibit translation in protocells, Mol. Cell 75 (2019) 66−75. e5.

[95] Y. Huang, X. Wang, J. Li, Y. Lin, H. Chen, X. Liu, et al., Reversible light-responsive coacervate microdroplets with rapid regulation of enzymatic reaction rate, ChemSystemsChem 3 (2021) e2100006.

[96] J.R. Vieregg, M. Lueckheide, A.B. Marciel, L. Leon, A.J. Bologna, J.R. Rivera, et al., Oligonucleotide−peptide complexes: phase control by hybridization, J. Am. Chem. Soc 140 (2018) 1632−1638.

[97] W.J. Altenburg, N.A. Yewdall, D.F.M. Vervoort, M.H.M.E. van Stevendaal, A.F. Mason, J.C.M. van Hest, Programmed spatial organization of biomacromolecules into discrete, coacervate-based protocells, Nat. Commun. 11 (2020) 6282.

[98] D.S. Williams, S. Koga, C.R.C. Hak, A. Majrekar, A.J. Patil, A.W. Perriman, et al., Polymer/nucleotide droplets as bio-inspired functional micro-compartments, Soft. Matter 8 (2012) 6004−6014.

[99] T.Y.D. Tang, M. Antognozzi, J.A. Vicary, A.W. Perriman, S. Mann, Small-molecule uptake in membrane-free peptide/nucleotide protocells, Soft. Matter 9 (2013) 7647−7656.

[100] K.K. Nakashima, M.A. Vibhute, E. Spruijt, Biomolecular chemistry in liquid phase separated compartments, Front. Mol. Biosci (2019) 6.

[101] K.A. Black, D. Priftis, S.L. Perry, J. Yip, W.Y. Byun, M. Tirrell, Protein encapsulation via polypeptide complex coacervation, ACS Macro Lett 3 (2014) 1088−1091.

[102] T.Y.D. Tang, D. van Swaay, A. deMello, J.L. Ross Anderson, S. Mann, In vitro gene expression within membrane-free coacervate protocells, Chem. Commun 51 (2015) 11429−11432.

[103] Y. Qiao, M. Li, R. Booth, S. Mann, Predatory behaviour in synthetic protocell communities, Nat. Chem 9 (2017) 110−119.

[104] B.V.V.S. Pavan Kumar, J. Fothergill, J. Bretherton, L. Tian, A.J. Patil, S.A. Davis, et al., Chloroplast-containing coacervate micro-droplets as a step towards photosynthetically active membrane-free protocells, Chem. Commun 54 (2018) 3594−3597.

[105] Z. Xu, S. Wang, C. Zhao, S. Li, X. Liu, L. Wang, et al., Photosynthetic hydrogen production by droplet-based microbial micro-reactors under aerobic conditions, Nat. Commun 11 (2020) 5985.

[106] S. Thamboo, A. Najer, A. Belluati, C. von Planta, D. Wu, I. Craciun, et al., Mimicking cellular signaling pathways within synthetic multicompartment vesicles with triggered enzyme activity and induced ion channel recruitment, Adv. Funct. Mater 29 (2019) 1904267.

[107] K.P. Adamala, A.E. Engelhart, J.W. Szostak, Collaboration between primitive cell membranes and soluble catalysts, Nat. Commun 7 (2016) 11041.

[108] V. Mukwaya, S. Mann, H. Dou, Chemical communication at the synthetic cell/living cell interface, Commun. Chem 4 (2021) 161.

[109] E. Donath, G.B. Sukhorukov, F. Caruso, S.A. Davis, H. Möhwald, Novel hollow polymer shells by colloid-templated assembly of polyelectrolytes, Angew. Chem. Int. Ed 37 (1998) 2201−2205.

[110] A.P.R. Johnston, C. Cortez, A.S. Angelatos, F. Caruso, Layer-by-layer engineered capsules and their applications, Curr. Opin. Colloid Interface Sci. 11 (2006) 203−209.

[111] G.K. Such, A.P.R. Johnston, F. Caruso, Engineered hydrogen-bonded polymer multilayers: from assembly to biomedical applications, Chem. Soc. Rev 40 (2011) 19−29.

[112] A.N. Zelikin, J.F. Quinn, F. Caruso, Disulfide cross-linked polymer capsules: en route to biodeconstructible systems, Biomacromolecules. 7 (2006) 27−30.

[113] C.-J. Huang, F.-C. Chang, Using click chemistry to fabricate ultrathin thermoresponsive microcapsules through direct covalent layer-by-layer assembly, Macromolecules. 42 (2009) 5155−5166.

[114] A.P.R. Johnston, H. Mitomo, E.S. Read, F. Caruso, Compositional and structural engineering of DNA multilayer films, Langmuir. 22 (2006) 3251−3258.

[115] J.F. Quinn, A.P.R. Johnston, G.K. Such, A.N. Zelikin, F. Caruso, Next generation, sequentially assembled ultrathin films: beyond electrostatics, Chem. Soc. Rev 36 (2007) 707−718.

[116] A.P. Johnston, E.S. Read, F. Caruso, DNA multilayer films on planar and colloidal supports: sequential assembly of like-charged polyelectrolytes, Nano Lett. 5 (2005) 953−956.

[117] B.G. De Geest, W. Van Camp, F.E. Du Prez, S.C. De Smedt, J. Demeester, W.E. Hennink, Biodegradable microcapsules designed via 'click' chemistry, Chem Commun (2008) 190−192.

[118] C.J. Ochs, G.K. Such, B. Städler, F. Caruso, Low-fouling, biofunctionalized, and biodegradable click capsules, Biomacromolecules. 9 (2008) 3389−3396.

[119] C. Xu, S. Hu, X. Chen, Artificial cells: from basic science to applications, Mater. Today 19 (2016) 516−532.

[120] S. Li, B.A. Moosa, J.G. Croissant, N.M. Khashab, Electrostatic assembly/disassembly of nanoscaled colloidosomes for light-triggered cargo release, Angew. Chem. Int. Ed 54 (2015) 6804−6808.

[121] R.K. Shah, J.-W. Kim, D.A. Weitz, Monodisperse stimuli-responsive colloidosomes by self-assembly of microgels in droplets, Langmuir. 26 (2010) 1561−1565.

[122] Z. Wang, M.C.M. van Oers, F.P.J.T. Rutjes, J.C.M. van Hest, Polymersome colloidosomes for enzyme catalysis in a biphasic system, Angew. Chem. (Int. Ed Engl.) 51 (2012) 10746−10750.

[123] W. Wang, A.H. Milani, L. Carney, J. Yan, Z. Cui, S. Thaiboonrod, et al., Doubly crosslinked microgel-colloidosomes: a versatile method for pH-responsive capsule assembly using microgels as macro-crosslinkers, Chem. Commun 51 (2015) 3854−3857.

[124] R. Tamate, T. Ueki, R. Yoshida, Evolved colloidosomes undergoing cell-like autonomous shape oscillations with buckling, Angew. Chem. Int. Ed 55 (2016) 5179−5183.

[125] Y.S. Kim, R. Tamate, A.M. Akimoto, R. Yoshida, Recent developments in self-oscillating polymeric systems as smart materials: from polymers to bulk hydrogels, Mater. Horiz 4 (2017) 38−54.

[126] M.F. Hsu, M.G. Nikolaides, A.D. Dinsmore, A.R. Bausch, V.D. Gordon, X. Chen, et al., Self-assembled shells composed of colloidal particles: fabrication and characterization, Langmuir. 21 (2005) 2963−2970.

[127] G. Sobczak, T. Wojciechowski, V. Sashuk, Submicron colloidosomes of tunable size and wall thickness, Langmuir. 33 (2017) 1725−1731.

[128] F.J. Rossier-Miranda, C.G.P.H. Schroën, R.M. Boom, Colloidosomes: versatile microcapsules in perspective, Colloids Surf. A: Physicochem. Eng. Asp 343 (2009) 43−49.

[129] E.T. Kisak, B. Coldren, C.A. Evans, C. Boyer, J.A. Zasadzinski, The vesosome − a multicompartment drug delivery vehicle, Curr. Med. Chem 11 (2004) 199−219.

[130] N.-N. Deng, M. Yelleswarapu, L. Zheng, W.T.S. Huck, Microfluidic assembly of monodisperse vesosomes as artificial cell models, J. Am. Chem. Soc 139 (2017) 587−590.

[131] C.M. Paleos, D. Tsiourvas, Z. Sideratou, A. Pantos, Formation of artificial multicompartment vesosome and dendrosome as prospected drug and gene delivery carriers, J. Contr. Rel 170 (2013) 141−152.

[132] A. Belluati, S. Thamboo, A. Najer, V. Maffeis, C. von Planta, I. Craciun, et al., Multicompartment polymer vesicles with artificial organelles for signal-triggered cascade reactions including cytoskeleton formation, Adv. Funct. Mater 30 (2020) 2002949.

[133] R.J.R.W. Peters, M. Marguet, S. Marais, M.W. Fraaije, J.C.M. van Hest, S. Lecommandoux, Cascade reactions in multicompartmentalized polymersomes, Angew. Chem. Int. Ed. 53 (2014) 146—150.

[134] Y. Chen, M. Yuan, Y. Zhang, S. Liu, X. Yang, K. Wang, et al., Construction of coacervate-in-coacervate multi-compartment protocells for spatial organization of enzymatic reactions, Chem. Sci 11 (2020) 8617—8625.

[135] M. Schulz, W.H. Binder, Mixed hybrid lipid/polymer vesicles as a novel membrane platform, Macromol. Rapid Commun 36 (2015) 2031—2041.

[136] A.F. Mason, N.A. Yewdall, P.L.W. Welzen, J. Shao, M. van Stevendaal, J.C.M. van Hest, et al., Mimicking cellular compartmentalization in a hierarchical protocell through spontaneous spatial organization, ACS Central Sci. 5 (2019) 1360—1365.

[137] S. Koga, D.S. Williams, A.W. Perriman, S. Mann, Peptide—nucleotide microdroplets as a step towards a membrane-free protocell model, Nat. Chem 3 (2011) 720—724.

[138] M.G. Last, S. Deshpande, C. Dekker, pH-Controlled coacervate—membrane interactions within liposomes, ACS Nano 14 (2020) 4487—4498.

[139] S. Deshpande, F. Brandenburg, A. Lau, M.G.F. Last, W.K. Spoelstra, L. Reese, et al., Spatiotemporal control of coacervate formation within liposomes, Nat. Commun 10 (2019) 1800.

[140] J. Li, M. Zhu, S. Wang, Z. Tao, X. Liu, X. Huang, Construction of coacervates in proteinosome hybrid microcompartments with enhanced cascade enzymatic reactions, Chem Commun 57 (2021) 11713—11716.

[141] A. Samad, Y. Sultana, M. Aqil, Liposomal drug delivery systems: an update review, Curr. Drug Deliv 4 (2007) 297—305.

[142] A.H. Kunding, M.W. Mortensen, S.M. Christensen, D. Stamou, A fluorescence-based technique to construct size distributions from single-object measurements: application to the extrusion of lipid vesicles, Biophys. J 95 (2008) 1176—1188.

[143] P. Szwedziak, Q. Wang, T.A.M. Bharat, M. Tsim, J. Löwe, Architecture of the ring formed by the tubulin homologue FtsZ in bacterial cell division, eLife. 3 (2014) e04601.

[144] C. Wu, S. Bai, M.B. Ansorge-Schumacher, D. Wang, Nanoparticle cages for enzyme catalysis in organic media, Adv. Mater 23 (2011) 5694—5699.

[145] R. Varshney, S. Kumar, K. Ghosh, D. Patra, Fabrication of dual catalytic microcapsules by mesoporous graphitic carbon nitride (mpg-C3N4) nanoparticle—enzyme conjugate stabilized emulsions, New J. Chem 44 (2020) 3097—3102.

[146] R. Lv, S. Lin, S. Sun, H. He, F. Zheng, D. Tan, et al., Cascade cycling of nicotinamide cofactor in a dual enzyme microsystem, Chem. Commun 56 (2020) 2723—2726.

[147] B. Li, H.C. Zeng, Architecture and preparation of hollow catalytic devices, Adv. Mater 31 (2019) 1801104.

[148] Z. Liu, W. Zhou, C. Qi, T. Kong, Interface engineering in multiphase systems toward synthetic cells and organelles: from soft matter fundamentals to biomedical applications, Adv. Mater. 32 (2020) 2002932.

[149] M. Li, X. Huang, T.Y.D. Tang, S. Mann, Synthetic cellularity based on non-lipid micro-compartments and protocell models, Curr. Opin. Chem. Biol 22 (2014) 1—11.

[150] X. Liu, P. Zhou, Y. Huang, M. Li, X. Huang, S. Mann, Hierarchical proteinosomes for programmed release of multiple components, Angew. Chem. Int. Ed 55 (2016) 7095—7100.

[151] F. Liu, Y. Cai, H. Wang, X. Yang, H. Zhao, Polymerization-induced proteinosome formation, J. Mater. Chem. B. 9 (2021) 1406—1413.

[152] M. Peterca, V. Percec, P. Leowanawat, A. Bertin, Predicting the size and properties of dendrimersomes from the lamellar structure of their amphiphilic janus dendrimers, J. Am. Chem. Soc 133 (2011) 20507–20520.

[153] N.A. Yewdall, A.F. Mason, J.C.M. van Hest, The hallmarks of living systems: towards creating artificial cells, Interface Focus 8 (2018) 20180023.

[154] J. Fu, J.B. Schlenoff, Driving forces for oppositely charged polyion association in aqueous solutions: enthalpic, entropic, but not electrostatic, J. Am. Chem. Soc 138 (2016) 980–990.

[155] W.C. Blocher McTigue, S.L. Perry, Protein encapsulation using complex coacervates: what nature has to teach us, Small. (2020) 1907671.

[156] P. Walde, K. Cosentino, H. Engel, P. Stano, Giant vesicles: preparations and applications, ChemBioChem. 11 (2010) 848–865.

[157] K.S. Horger, D.J. Estes, R. Capone, M. Mayer, Films of agarose enable rapid formation of giant liposomes in solutions of physiologic ionic strength, J. Am. Chem. Soc 131 (2009) 1810–1819.

[158] L. Heuberger, M. Korpidou, O.M. Eggenberger, M. Kyropoulou, C.G. Palivan, Current perspectives on synthetic compartments for biomedical applications, Int. J. Mol. Sci. 23 (2022).

[159] L.D. Mayer, M.J. Hope, P.R. Cullis, Vesicles of variable sizes produced by a rapid extrusion procedure, Biochim. Biophys. Acta (BBA) – Biomembr. 858 (1986) 161–168.

[160] N.J. Warren, S.P. Armes, Polymerization-induced self-assembly of block copolymer nano-objects via RAFT aqueous dispersion polymerization, J. Am. Chem. Soc 136 (2014) 10174–10185.

[161] J.C.M. Lee, H. Bermudez, B.M. Discher, M.A. Sheehan, Y.-Y. Won, F.S. Bates, et al., Preparation, stability, and in vitro performance of vesicles made with diblock copolymers, Biotechnol. Bioeng 73 (2001) 135–145.

[162] K. Göpfrich, I. Platzman, J.P. Spatz, Mastering complexity: towards bottom-up construction of multifunctional eukaryotic synthetic cells, Trends Biotechnol 36 (2018) 938–951.

[163] H.C. Shum, Y.-j Zhao, S.-H. Kim, D.A. Weitz, Multicompartment polymersomes from double emulsions, Angew. Chem. Int. Ed 50 (2011) 1648–1651.

[164] M. Weiss, J.P. Frohnmayer, L.T. Benk, B. Haller, J.-W. Janiesch, T. Heitkamp, et al., Sequential bottom-up assembly of mechanically stabilized synthetic cells by microfluidics, Nat. Mater 17 (2018) 89–96.

[165] C.J. Ferguson, R.J. Hughes, B.T.T. Pham, B.S. Hawkett, R.G. Gilbert, A.K. Serelis, et al., Effective ab initio emulsion polymerization under RAFT control, Macromolecules. 35 (2002) 9243–9245.

[166] G. Riess, C. Labbe, Block copolymers in emulsion and dispersion polymerization, Macromol. Rapid Commun 25 (2004) 401–435.

[167] M.J. Derry, L.A. Fielding, S.P. Armes, Industrially-relevant polymerization-induced self-assembly formulations in non-polar solvents: RAFT dispersion polymerization of benzyl methacrylate, Polym. Chem. 6 (2015) 3054–3062.

[168] A.N. Albertsen, J.K. Szymański, J. Pérez-Mercader, Emergent properties of giant vesicles formed by a polymerization-induced self-assembly (PISA) reaction, Sci. Rep 7 (2017) 41534.

[169] S.M. Hartig, R.R. Greene, M.M. Dikov, A. Prokop, J.M. Davidson, Multifunctional nanoparticulate polyelectrolyte complexes, Pharm. Res 24 (2007) 2353–2369.

[170] K. Miyata, N. Nishiyama, K. Kataoka, Rational design of smart supramolecular assemblies for gene delivery: chemical challenges in the creation of artificial viruses, Chem. Soc. Rev 41 (2012) 2562–2574.

[171] Y. Lee, K. Kataoka, Biosignal-sensitive polyion complex micelles for the delivery of biopharmaceuticals, Soft. Matter 5 (2009) 3810–3817.

[172] K.P. Adamala, D.A. Martin-Alarcon, K.R. Guthrie-Honea, E.S. Boyden, Engineering genetic circuit interactions within and between synthetic minimal cells, Nat. Chem 9 (2017) 431–439.

[173] O. Biner, J.G. Fedor, Z. Yin, J. Hirst, Bottom-up construction of a minimal system for cellular respiration and energy regeneration, ACS Synth. Biol. 9 (2020) 1450−1459.

[174] Q. Li, X. Han, Self-assembled rough endoplasmic reticulum-like proto-organelles, iScience 8 (2018) 138−147.

[175] A. Bachhuka, T. Chand Yadav, A. Santos, L.F. Marsal, S. Ergün, S. Karnati, Emerging nanomaterials for targeting peroxisomes, Mater. Today Adv 15 (2022) 100265.

[176] M. Miyazaki, M. Chiba, H. Eguchi, T. Ohki, Si Ishiwata, Cell-sized spherical confinement induces the spontaneous formation of contractile actomyosin rings in vitro, Nat. Cell Biol 17 (2015) 480−489.

[177] K. Zieske, G. Chwastek, P. Schwille, Protein patterns and oscillations on lipid monolayers and in microdroplets, Angew. Chem. Int. Ed 55 (2016) 13455−13459.

[178] K.J. Kauffman, J.R. Dorkin, J.H. Yang, M.W. Heartlein, F. DeRosa, F.F. Mir, et al., Optimization of lipid nanoparticle formulations for mRNA delivery in vivo with fractional factorial and definitive screening designs, Nano Lett 15 (2015) 7300−7306.

[179] M.S.D. Kormann, G. Hasenpusch, M.K. Aneja, G. Nica, A.W. Flemmer, S. Herber-Jonat, et al., Expression of therapeutic proteins after delivery of chemically modified mRNA in mice, Nat. Biotechnol 29 (2011) 154−157.

[180] A. Akinc, W. Querbes, S. De, J. Qin, M. Frank-Kamenetsky, K.N. Jayaprakash, et al., Targeted delivery of RNAi therapeutics with endogenous and exogenous ligand-based mechanisms, Mol. Ther 18 (2010) 1357−1364.

[181] A.-T. Dinh, C. Pangarkar, T. Theofanous, S. Mitragotri, Understanding intracellular transport processes pertinent to synthetic gene delivery via stochastic simulations and sensitivity analyses, Biophys. J 92 (2007) 831−846.

[182] T. Bus, A. Traeger, U.S. Schubert, The great escape: how cationic polyplexes overcome the endosomal barrier, J. Mater. Chem. B. 6 (2018) 6904−6918.

[183] U.K. Laemmli, Characterization of DNA condensates induced by poly(ethylene oxide) and polylysine, Proc. Natl. Acad. Sci. 72 (1975) 4288−4292.

[184] G.Y. Wu, C.H. Wu, Receptor-mediated in vitro gene transformation by a soluble DNA carrier system, J. Biol. Chem 262 (1987) 4429−4432.

[185] A. Bolhassani, Potential efficacy of cell-penetrating peptides for nucleic acid and drug delivery in cancer, Biochim. Biophys. Acta (BBA) − Rev. Cancer 1816 (2011) 232−246.

[186] Y. Dong, J.R. Dorkin, W. Wang, P.H. Chang, M.J. Webber, B.C. Tang, et al., Poly(glycoamidoamine) brushes formulated nanomaterials for systemic siRNA and mRNA delivery in vivo, Nano Lett 16 (2016) 842−848.

[187] J.C. Kaczmarek, A.K. Patel, K.J. Kauffman, O.S. Fenton, M.J. Webber, M.W. Heartlein, et al., Polymer−lipid nanoparticles for systemic delivery of mRNA to the lungs, Angew. Chem. Int. Ed 55 (2016) 13808−13812.

[188] A. Wittrup, A. Ai, X. Liu, P. Hamar, R. Trifonova, K. Charisse, et al., Visualizing lipid-formulated siRNA release from endosomes and target gene knockdown, Nat. Biotechnol 33 (2015) 870−876.

[189] C.J. McKinlay, J.R. Vargas, T.R. Blake, J.W. Hardy, M. Kanada, C.H. Contag, et al., Charge-altering releasable transporters (CARTs) for the delivery and release of mRNA in living animals, Proc. Natl. Acad. Sci. 114 (2017) E448−E456.

[190] S.H. Medina, S.E. Miller, A.I. Keim, A.P. Gorka, M.J. Schnermann, J.P. Schneider, An intrinsically disordered peptide facilitates non-endosomal cell entry, Angew. Chem. Int. Ed 55 (2016) 3369−3372.

[191] M.S. Shim, Y.J. Kwon, Stimuli-responsive polymers and nanomaterials for gene delivery and imaging applications, Adv. Drug Deliv. Rev 64 (2012) 1046−1059.

[192] K. Fitzgerald, M. Frank-Kamenetsky, S. Shulga-Morskaya, A. Liebow, B.R. Bettencourt, J.E. Sutherland, et al., Effect of an RNA interference drug on the synthesis of proprotein convertase subtilisin/kexin type 9 (PCSK9) and the concentration of serum LDL cholesterol in healthy volunteers: a randomised, single-blind, placebo-controlled, phase 1 trial, The Lancet 383 (2014) 60−68.

[193] K.A. Hajj, K.A. Whitehead, Tools for translation: non-viral materials for therapeutic mRNA delivery, Nat. Rev. Mater. 2 (2017) 17056.

[194] S. Mitragotri, P.A. Burke, R. Langer, Overcoming the challenges in administering biopharmaceuticals: formulation and delivery strategies, Nat. Rev. Drug Discov. 13 (2014) 655−672.

[195] M. Akishiba, T. Takeuchi, Y. Kawaguchi, K. Sakamoto, H.-H. Yu, I. Nakase, et al., Cytosolic antibody delivery by lipid-sensitive endosomolytic peptide, Nat. Chem 9 (2017) 751−761.

[196] J.A. Zuris, D.B. Thompson, Y. Shu, J.P. Guilinger, J.L. Bessen, J.H. Hu, et al., Cationic lipid-mediated delivery of proteins enables efficient protein-based genome editing in vitro and in vivo, Nat. Biotechnol 33 (2015) 73−80.

[197] Y. Li, T. Yang, Y. Yu, N. Shi, L. Yang, Z. Glass, et al., Combinatorial library of chalcogen-containing lipidoids for intracellular delivery of genome-editing proteins, Biomaterials. 178 (2018) 652−662.

[198] M. Wang, J.A. Zuris, F. Meng, H. Rees, S. Sun, P. Deng, et al., Efficient delivery of genome-editing proteins using bioreducible lipid nanoparticles, Proc. Natl. Acad. Sci. 113 (2016) 2868−2873.

[199] X. Gao, Y. Tao, V. Lamas, M. Huang, W.-H. Yeh, B. Pan, et al., Treatment of autosomal dominant hearing loss by in vivo delivery of genome editing agents, Nature. 553 (2018) 217−221.

[200] C. Liu, T. Wan, H., Wang, S. Zhang, Y. Ping, Y. Cheng, A boronic acid−rich dendrimer with robust and unprecedented efficiency for cytosolic protein delivery and CRISPR-Cas9 gene editing. Sci. Adv. 5, eaaw8922.

[201] A.B. Grommet, M. Feller, R. Klajn, Chemical reactivity under nanoconfinement, Nat. Nanotechnol 15 (2020) 256−271.

[202] L. Schoonen, J.C.M. van Hest, Compartmentalization approaches in soft matter science: from nanoreactor development to organelle mimics, Adv. Mater. 28 (2016) 1109−1128.

[203] A. Küchler, M. Yoshimoto, S. Luginbühl, F. Mavelli, P. Walde, Enzymatic reactions in confined environments, Nat. Nanotechnol 11 (2016) 409−420.

[204] D.C. Dewey, C.A. Strulson, D.N. Cacace, P.C. Bevilacqua, C.D. Keating, Bioreactor droplets from liposome-stabilized all-aqueous emulsions, Nat. Commun 5 (2014) 4670.

[205] M. Vázquez-González, C. Wang, I. Willner, Biocatalytic cascades operating on macromolecular scaffolds and in confined environments, Nat. Catal 3 (2020) 256−273.

[206] K.S. Rabe, J. Müller, M. Skoupi, C.M. Niemeyer, Cascades in compartments: en route to machine-assisted biotechnology, Angew. Chem. Int. Ed 56 (2017) 13574−13589.

[207] F. Fernandez-Trillo, L.M. Grover, A. Stephenson-Brown, P. Harrison, P.M. Mendes, Vesicles in nature and the laboratory: elucidation of their biological properties and synthesis of increasingly complex synthetic vesicles, Angew. Chem. Int. Ed 56 (2017) 3142−3160.

[208] P. Tanner, O. Onaca, V. Balasubramanian, W. Meier, C.G. Palivan, Enzymatic cascade reactions inside polymeric nanocontainers: a means to combat oxidative stress, Chem. − A Eur. J 17 (2011) 4552−4560.

[209] P. Tanner, V. Balasubramanian, C.G. Palivan, Aiding nature's organelles: artificial peroxisomes play their role, Nano Lett 13 (2013) 2875−2883.

[210] F. Axthelm, O. Casse, W.H. Koppenol, T. Nauser, W. Meier, C.G. Palivan, Antioxidant nanoreactor based on superoxide dismutase encapsulated in superoxide-permeable vesicles, J. Phys. Chem. B 112 (2008) 8211−8217.

[211] E. Altamura, F. Milano, R.R. Tangorra, M. Trotta, O.H. Omar, P. Stano, et al., Highly oriented photosynthetic reaction centers generate a proton gradient in synthetic protocells, Proc. Natl. Acad. Sci. 114 (2017) 3837−3842.

[212] E. Altamura, P. Albanese, R. Marotta, F. Milano, M. Fiore, M. Trotta, et al., Chromatophores efficiently promote light-driven ATP synthesis and DNA transcription inside hybrid multicompartment artificial cells, Proc. Natl. Acad. Sci. U S A 118 (2021). e2012170118.

[213] H.-J. Choi, C.D. Montemagno, Artificial organelle: ATP synthesis from cellular mimetic polymer-somes, Nano Lett 5 (2005) 2538–2542.

[214] R. Seneviratne, S. Khan, E. Moscrop, M. Rappolt, S.P. Muench, L.J.C. Jeuken, et al., A reconstitution method for integral membrane proteins in hybrid lipid-polymer vesicles for enhanced functional durability, Methods. 147 (2018) 142–149.

[215] K.Y. Lee, S.-J. Park, K.A. Lee, S.-H. Kim, H. Kim, Y. Meroz, et al., Photosynthetic artificial organelles sustain and control ATP-dependent reactions in a protocellular system, Nat. Biotechnol 36 (2018) 530–535.

[216] M. Kumar, M. Grzelakowski, J. Zilles, M. Clark, W. Meier, Highly permeable polymeric membranes based on the incorporation of the functional water channel protein Aquaporin Z, Proc. Natl. Acad. Sci. 104 (2007) 20719–20724.

[217] P. Broz, S. Driamov, J. Ziegler, N. Ben-Haim, S. Marsch, W. Meier, et al., Toward intelligent nanosize bioreactors: a pH-switchable, channel-equipped, functional polymer nanocontainer, Nano Lett 6 (2006) 2349–2353.

[218] M. Kumar, J.E.O. Habel, Y.-x Shen, W.P. Meier, T. Walz, High-density reconstitution of functional water channels into vesicular and planar block copolymer membranes, J. Am. Chem. Soc 134 (2012) 18631–18637.

[219] S. Khan, M. Li, S.P. Muench, L.J.C. Jeuken, P.A. Beales, Durable proteo-hybrid vesicles for the extended functional lifetime of membrane proteins in bionanotechnology, Chem. Commun 52 (2016) 11020–11023.

[220] L.K.E.A. Abdelmohsen, F. Peng, Y. Tu, D.A. Wilson, Micro- and nano-motors for biomedical applications, J. Mater. Chem. B. 2 (2014) 2395–2408.

[221] J. Solon, P. Streicher, R. Richter, F. Brochard-Wyart, P. Bassereau, Vesicles surfing on a lipid bilayer: self-induced haptotactic motion, Proc. Natl. Acad. Sci. 103 (2006) 12382–12387.

[222] R. Chandrawati, M.T.J. Olesen, T.C.C. Marini, G. Bisra, A.G. Guex, M.G. de Oliveira, et al., Enzyme prodrug therapy engineered into electrospun fibers with embedded liposomes for controlled, localized synthesis of therapeutics, Adv. Healthcare Mater 6 (2017) 1700385.

[223] Y. Tu, F. Peng, A.A.M. André, Y. Men, M. Srinivas, D.A. Wilson, Biodegradable hybrid stomatocyte nanomotors for drug delivery, ACS Nano 11 (2017) 1957–1963.

[224] D.A. Wilson, R.J.M. Nolte, J.C.M. Van Hest, Autonomous movement of platinum-loaded stomatocytes, Nat. Chem 4 (2012) 268–274.

2D materials-based nanoarchitectonics for metal-ion batteries

10

Maria K. Ramos and Aldo J.G. Zarbin

Department of Chemistry, Federal University of Paraná (UFPR), Curitiba, Paraná, Brazil

10.1 Introduction: novel materials for battery electrodes

The development of sustainable energy technologies, such as solar cells and energy storage devices, is imperative to attend to the growing global demand for energy observed in recent years. One of the significant challenges is preparing and processing new, safe, and cost-effective materials to improve the efficiency of those devices without compromising the environment. Regarding energy storage devices, rechargeable metal-ion batteries are among the most attractive technologies for large-scale applications [1].

In a straightforward way, a rechargeable battery is composed of two electrodes connected through a porous membrane (separator) immersed in the appropriate electrolyte, which is responsible for supplying the ions for the battery operation (Fig. 10.1A). Materials to be used as electrodes can have a specific structure to store the metal ions. During the discharging process (which is spontaneous), the active material at the negative electrode (anode) oxidizes, releasing electrons to the external circuit (providing the electrical current) and metal ions to migrate to the positive electrode (cathode). Conversely, the charging is carried out using external electricity to promote the nonspontaneous reverse process, in which the ions flow in the opposite direction [2–5]. The specific energy of a battery is determined by the specific capacities (SC) of both the cathode and anode materials, determined by the product between the applied current (ampere) and the discharge time (hours), normalized by the weight of the material (Ah/g).

Lithium-ion batteries (LIBs) were responsible for the growth of portable electronic devices, due to their high relative energy density [10]. Several reasons explain the excellent performance of LIBs, including the relatively small ionic size of Li^+ (0.76 Å); the ease of Li^+ intercalation in the electrode materials; the high energy and power densities; and the low reduction potential (-3.04 V vs standard hydrogen electrode—SHE) [11,12]. However, several safety issues related to LIBs have received attention in recent years, such as: (1) the traditional electrolytes for LIBs are composed of volatile and flammable organic solvents, many of them containing the toxic lithium hexafluorophosphate; (2) the well-known growth of lithium dendrites during the charge/discharge process, which can contact the electrodes and generate a short circuit in the cell, which can lead to an ignition of organic solvents with a low flash point; and (3) lithium is a scarce resource and concentrated in only few countries around the world. Sodium- and potassium-ion batteries (SIBs and KIBs,

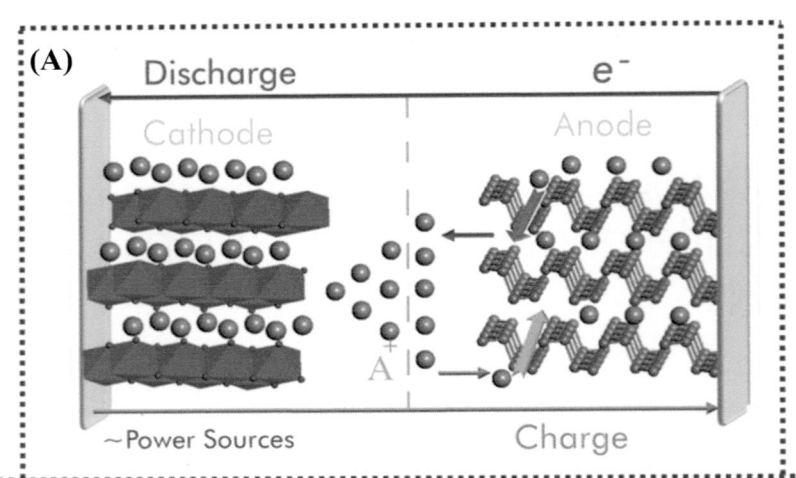

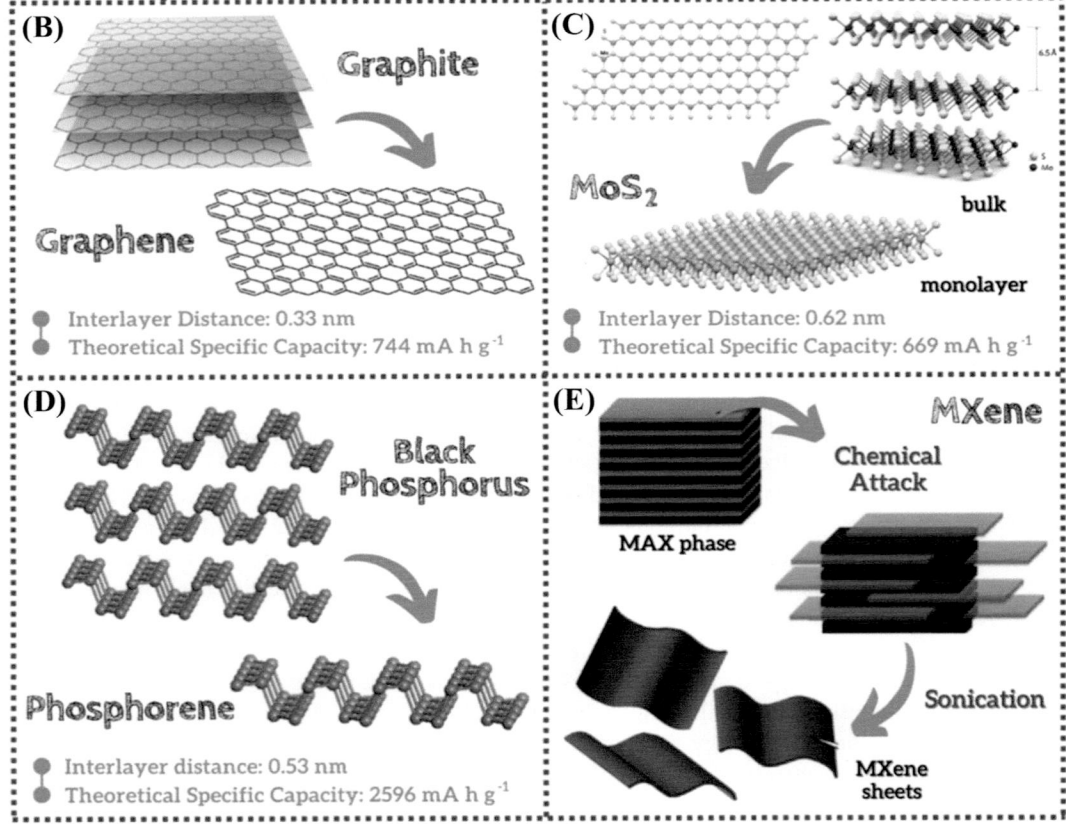

FIGURE 10.1

(A) Schematic representation of a metal-ion battery representation with a phosphorene as an anode and a $LiCoO_2$ as a cathode; (B) schematic representation of graphite and graphene structure; (C) schematic representation of bulk and monolayer MoS_2; (D) schematic representation of black phosphorus and phosphorene; (E) schematic of MXene preparation process.

respectively) are emerging as alternatives to LIBs and aiming to get around these drawbacks. Sodium and potassium are cheaper, abundant, and homogeneously distributed around the planet. Lithium, sodium, and potassium have several electrochemical similarities, for example, their low reduction potential: Li^+/Li (-3.040 V vs SHE), Na^+/Na (-2.714 V vs SHE), and K^+/K (-2.936 V vs SHE), which indicates that all the knowledge acquired by LIBs can be extended to SIBs and KIBs [3,5]. Otherwise, the size of the ionic radius of these ions is different in organic solvents: $Li^+ = 0.76$ Å; $Na^+ = 1.02$ Å; and $K^+ = 1.38$ Å, which means that the intercalation process of Na^+ and K^+ in the electrode structure is more difficult than that for Li^+. However, novel materials processing higher and facilitated intercalation capacity can be developed for these applications [4,13].

Among the numerous candidates for electrodes in all those metal-ion batteries, some two-dimensional (2D) materials have emerged as promising alternatives. The adequate interlayer spaces and the possibility of expansion of the interlayer distance characteristic of 2D structures facilitate the ion intercalation. In addition, the exfoliation of those 2D materials allows the accommodation of larger amounts of ions due to the increase in the active surface area and the production of surface defects in the basal planes, providing additional channels and facilitating intercalation [11,12]. The performance of the 2D materials can be strongly improved if they are nanoarchitected with different compounds.

Nanoarchitected materials can be prepared by several different approaches, aiming the controlled assembly of two or more materials at a nanometric scale. Through an appropriate combination of selected materials, it is possible to minimize the individual limitations and maximize the intrinsic properties of each compound. Also, novel and unique properties can emerge from the synergism coming from the intimate contact at the nanometric level [14].

The purpose of this chapter is present the recent advances (last 5 years) in the preparation of 2D compounds-based nanoarchitected materials specifically prepared for application as electrodes in metal-ion batteries. It will be shown that many works were focused on the preparation of those materials aiming for direct application in LIBs, trying to minimize the safety and environmental impacts of this well-established technology. Otherwise, several other works were interested in novel materials for the replacement of LIBs batteries, that is, SIBs and KIBs. The chapter starts with a brief summary of the four 2D materials that won the preference of researchers in the field of batteries: graphene (Section 10.1.1); molybdenum disulfide (Section 10.1.2); black phosphorous (Section 10.1.3); and metallic nitrides and carbides (Section 10.1.4), followed by examples of architected nanostructures based on them: 2D/2D (Section 10.2.1); 2D/transition metal oxides (Section 10.2.2); 2D/carbon nanotubes (Section 10.2.3); and 2D/conjugated polymers (Section 10.2.4). Each section is organized to show selected examples, a short overview of the reasons for the preparation of that specific kind of nanoarchitected material, and examples of real devices prepared based on them. Finally, a general conclusion, with some perspectives, is presented and discussed.

10.1.1 Graphene

Graphene is a one-atom-thick material constituted of sp^2-hybridized carbon atoms bonded together in a honeycomb-like hexagonal crystal lattice structure (Fig. 10.1B). Graphene is the building block of graphite, and it was isolated for the first time by Andre Geim and Konstantin Novoselov, who received the Nobel Prize in Physics in 2010. Graphene has superlative and unique properties:

defect-free graphene has a theoretical specific surface area of 2630 m^2/g; its electrical conductivity at room temperature is 1.5×10^4 cm^2/Vs, presenting an intrinsic charge carriers mobility of 20×10^4 cm^2/Vs; graphene has one of the highest thermal conductivity ever reported, of 3000 W/mK and can reach up to 5000 W/mK; graphene has one of the highest values of Young's modulus among the known materials (around 1 TPa), with a breaking strength 200 times higher than steel; the opacity of graphene is $\sim 2.3\%$ over a wide wavelength, and experimental values have already reached $\sim 97\%$ transparency in the visible range [1,15,16]. These incredible properties are strongly dependent on the structural characteristics and quality of the graphene, such as the lateral size, the presence of defects and the number of stacked layers, which can be tunable according to the synthesis method [1,15,16]. Perfectly structured graphene usually interacts with other components via physical adsorption (e.g., $\pi-\pi$ interaction). However, real graphene samples have structural defects (vacancies, edges, and deformations), the presence of heteroatoms (H, F, N, B, etc.), and surface functional groups (carboxyl, carbonyl or amine groups) [1], which increase the reactivity and the possibilities of chemical/physical interactions with different compounds. Among the different ways to produce graphene, the most versatile and mass-production route is based on the chemical oxidation of graphite, followed by exfoliation to yield the graphene oxide (GO, an oxidized graphene sheet containing several oxygenated functional groups as epoxides, carboxylic acids, phenols), which is afterwards chemically reduced to eliminate these oxygenated groups, yielding the so-called reduced graphene oxide (rGO) material.

Similarly, to graphite, graphene (and all the graphene-like materials, such as GO and rGO) is a high-potential candidate to be applied as battery anodes due to the capacity of the host cations. Single-layer graphene has a theoretical specific capacity of 744 mAh/g, twice that of graphite (372 mAh/g). Additionally, inherent defects in graphene can also store ions, like a "house of cards" model [1]. Finally, the graphene edges have a specific capacity, electron mobility rate, and catalytic power higher than the ones observed in the basal plane [15], being significant regions for ion intercalation. Those theoretical values have not been experimentally reached until now, due to the spontaneous restacking of the graphene layers during the charge/discharge cycles, leading to a significant reduction of the active sites [1].

10.1.2 Molybdenum disulfide

One of the most studied classes of materials belonging to the so-called 2D materials beyond graphene is the transition metal dichalcogenides (TMDs), which have the general chemical formula MX_2, where M is a transition metal (Ti, V, Nb, Mo, W, etc.) and X is a chalcogenide (S, Se, Te) [11−13]. MoS_2 is the most representative member of the TMD family (Fig. 10.1C), and its structure consists of stacked layers connected by van der Waals forces with an interlayer distance of 0.62 nm (higher than the 0.335 nm observed for graphite). Each MoS_2 layer is three atoms thick (S-Mo-S covalently bonded) with the central Mo atoms coordinated to six sulfurs. There are two main structural phases for MoS_2, the most stable trigonal prismatic (2H) or the less stable octahedral (1T) [17]. MoS_2 is a pseudocapacitive material with surface redox reaction and it is known to allow rapid and reversible intercalation/deintercalation reactions of different ions. Furthermore, it is demonstrated that MoS_2 has high electrochemical activity due to the presence of sulfur vacancies, presenting highly exposed sites of catalytic activity. Due to these characteristics, MoS_2 has a theoretical specific capacity of 669 mAh/g, but it has the potential to reach an even higher SC through changes in morphology, alteration of the spacing between layers, and formation of nanoarchitected structures with other materials [11−13].

As observed also for graphene, achieving the high theoretical value of SC for MoS_2 is not trivial, due to three main reasons: (1) the volume changes during the charge/discharge processes that cause mechanical stresses and lead to a degradation of the material structure and loss of contact between the active material and the current collector; (2) the restacking tendency aiming to minimize the surface energy, causing a slow transport of electrons/ions in the material (especially at high currents) and reducing the stability after several cycles; and (3) its low electronic conductivity, which limits its electrochemical performance [11,18,19].

10.1.3 Black phosphorus

Black phosphorus (BP) is a phosphorous-based material that is a structural analog of graphite (Fig. 10.1D). Despite being a material that has been known for a long time (it was first synthesized in 1914), the scientific and technological interest related to BP started in 2014 when it was successfully exfoliated to generate phosphorene, the phosphorous-based 2D material analog to graphene [2]. BP has a thickness-dependent bandgap ranging from 0.3 to 2.0 eV, electrical conductivity around $1000\ cm^2/Vs$, and its electronic properties have a direct dependence on the number of stacked layers [11]. Phosphorene is essentially one or a few layers of BP, exhibits a characteristic hexagonal ("wrinkled honeycomb") structure, and the few layers are held together by weak van der Waals attractive forces [20]. The crystal structure of phosphorene is orthorhombic, and the phosphorus atom forms covalent bonds with the three nearest neighbors. The spacing between two phosphorene layers is 0.53 nm, due to its wrinkled and stacking structure, which is greater than the spacing between graphene layers (0.335 nm) and slightly smaller than those of MoS_2 (0.62 nm) [2]. The hybridization of phosphorous atoms in BP is sp^3, formed by the three σ bonds and a pair of electrons, giving rise to its nonplanar wrinkled structure, and these electrons can be considered the basis for the chemical reactivity of BP [2].

BP has several properties to be considered one of the most promising materials to be used as electrodes in rechargeable batteries, mainly related to its extremely high theoretical specific capacity (2596 mAh/g) and high electron and ion mobility [21]. However, as described before for MoS_2, phosphorene has a very large volume expansion (values up to 300%) during the charge/discharge process, which weakens the interactions between the layers [2,20]. Nonetheless, a major obstacle in the implementation of this material in practical applications is its high instability under environmental conditions: BP degrades in the presence of oxygen, water, and visible light [2,20].

10.1.4 Metallic nitrides and carbides (MXenes)

Transition metal nitrides, carbides, or carbonitrides are 2D compounds generally known as MXenes (Fig. 10.1F). MAX is a general term for ternary layered compounds with a unified chemical formula $M_{n+1}AX_n$, in which M is a transition metal (Ti, Sr, V, Cr, Ta, Nb, Zr, Mo, Hf), A is an element of main groups 13 or 14 (Al, Ga, In, Ti, Si, Ge, Sn, Pb), X is C or N, and n varies from 1, 2, or 3. Chemical treatment of MAX phase yields a 2D structure that can be exfoliated similarly to the exfoliation of graphite to graphene, and this structure is known MXene. MAX has a hexagonal structure composed of alternating sheets of $M_{n+1}X_n$ and packed layers of atoms. The M-A bond has the weakest binding strength when compared to the M-X bond, so the A layer has the highest atomic reactivity. Layer A is usually removed from the MAX structure by chemical etching, to obtain the new MXene layered structure, with the chemical structural formula of $M_{n+1}X_nT_x$, in which T represents an active functional group (oxygen, fluorine, or hydroxyl) attached to the surface [11,22,23].

MXenes have been considered promising materials for anodes in metal-ion batteries due to their layered structure, remarkable chemical durability, variety of potential windows, and green aspects. The electrochemical properties and conductivity of MXenes can be improved by delamination of this material in a few layers. However, aggregation and restacking of MXene due to van der Waals interactions and hydrogen bonds are some problems for many applications. Also, the oxidation and degradation of MXene when excessively exposed to air decreases the material performance and prevents ion access to active materials [11,22,23].

As can be easily seen, these four 2D materials have incredible properties for their application in battery electrodes, but they have also some limitations that need to be minimized for real application. Preparing novel nanoarchitected structures is a viable and exciting approach to improving the performance of all those materials. These architected nanostructures often include nanosized building blocks, which have synergistic effects on the performance of electrode materials, due to some specific characteristics: first, the nanoarchitected structures have a high surface area, which can ensure sufficient contact between active materials and electrolytes, obtaining many regions for redox reactions; second, the structure resulting from combined materials can shorten the ion diffusion path and improve the kinetic performance of the battery; third, proper architecture can serve as a buffer against large volume changes of materials during intercalation/deintercalation processes and still avoid restacking problems of many materials; and finally, the specific properties of each material can introduce unique characteristics to the final architecture, such as high conductivity, high SC values, and chemical and physical stability[11,13,21−26]. Some of those mechanisms will be detailed in several examples in the next sections, highlighting different nanoarchitected combinations between the 2D nanomaterials above mentioned and their applications as electrodes for LIBs, SIBs, and KIBs.

10.2 Nanoarchitected structures applied to metal-ion batteries

10.2.1 2D/2D nanoarchitected structures

Combining different 2D materials is one of the main ways to produce smart architected nanostructures. Meng et al. [27], Li et al. [28], and Zhao et al. [29] have produced different BP/MXene nanoarchitected materials. In the first case, black phosphorus quantum dots (BPQDs) were obtained by BP milling and ultrasonic treatment, and Ti_3C_2 (TNSs) nanosheets were synthesized starting from Ti_3AlC_2 as a precursor, followed by the selective etching of Al layers. BPQD/TNS compound was obtained by a physical mixture of both components under sonication and mechanical stirring, as schematically represented in Fig. 10.2A. and it is characterized by thin and flat TNS plates (average size of 5 μm) decorated by spheres of BPQDs with sizes between 1.6 and 6.5 nm (inset in Fig. 10.2A) [27]. The performance of the BPQD/TNS as anode for LIBs presented an SC of 1730 mAh/g in the first cycle, which was decreased to 1124 mAh/g after 15 cycles, and a 90% retention rate after 100 cycles [27]. When applied as an anode in SIBs the SC was 723 mAh/g at 50 mA/g with a retention rate of 100%. Coulomb efficiency (CE) evaluated in LIBs increases rapidly from 42% to 90% in the initial five cycles and then remains stable at 99%, due to the activation of anodic materials in the structure [27]. In the work of Li et al. [28] the Ti_3C_2 was also obtained from Ti_3AlC_2 by selective Al etching followed by exfoliation, mixed with BP nanoparticles (NPs), sonicated and filtered to get the BP/Ti_3C_2 composite [28]. When applied as an anode for SIBs, the BP/Ti_3C_2 combination showed higher SC (332 mAh/g to 100 mA/g) and better stability than neat BP (146 mAh/g) or neat Ti_3C_2 (99 mAh/g) [28]. The BP/MXene nanostructured material reported by Zhao et al. [29] was obtained from a material previously prepared between BP and the cationic polymer poly(diallyl dimethyl ammonium chloride) (PDDA). The PDDA aims to improve the electronegativity of BP and passivate it against oxidation, increasing stability and improving its dispersion. Due to the surface functional groups, Ti_3C_2 is negatively charged, which allows the preparation of a thin film by electrostatic interaction with the positive PDDA-BP material [29]. Studied as an anode for SIBs, this PDDA-BP/MXene material exhibits an SC of 1112 mAh/g at 100 mA/g and only 0.05% of degradation after 2000 cycles at 1 A/g [29].

The BP/MXene nanostructured materials described before presented the following advantages for application as anodes in metal-ion batteries: (1) both BP and MXene have layered structures, which allow the insertion of ions and provide a large specific surface area for energy storage; (2) BP has abundant active sites for Li^+/Na^+ diffusion; (3) MXene is primarily responsible for increasing the electrical conductivity of the nanostructure; (4) the nanostructures contribute to shortening the ion/electron diffusion pathways; and (5) BP is stabilized and confined within the MXene layers, which damps its volume expansion and prevents its aggregation and degradation [27−29].

Nanoarchitected materials obtained from BP and graphene are also very relevant in this field. For example, a vacuum filtration method was used to build sandwich structure anodes with BP, graphene, and graphene oxide (GO) [32]. A BP/GO nanostructure was initially prepared by solvothermal reaction and the components were stabilized by P−C and P−O−C chemical bonding [32]. The graphene/BP/GO nanostructure was obtained by vacuum filtration from dispersions of the components. As an anode in SIBs, the first SC of 2587 mAh/g is very close to the theoretical SC of BP (2596 mAh/g). On the other hand, a high SC of 1401 mAh/g at 100 mA/g is obtained after 200 cycles, higher than the BP/GO nanostructure not coated with graphene (750 mAh/g) [32]. Graphene and GO have also a protective function against BP degradation. It was observed after several charge/discharge cycles that the small portion of BP directly exposed to the electrolyte has been degraded, but most of the BP remains intact [32].

A phosphorene/graphene sandwiched nanomaterial was also prepared by Shuai et al. for application as an anode in SIBs [30]. The schematic for the preparation of this material is summarized in Fig. 10.2B. Briefly, phosphorene and graphene have been previously electrochemically exfoliated and chemically activated to originate P-C and P−O−C bonds. The phosphorene−graphene hybrid sandwiched structure shows a high SC of 2311 mAh/g at 100 mA/g, reduced to 83.9% after 100 cycles. The high performance was attributed to the flexible space of the graphene layers that relieve the volumetric expansion of the phosphorene [30].

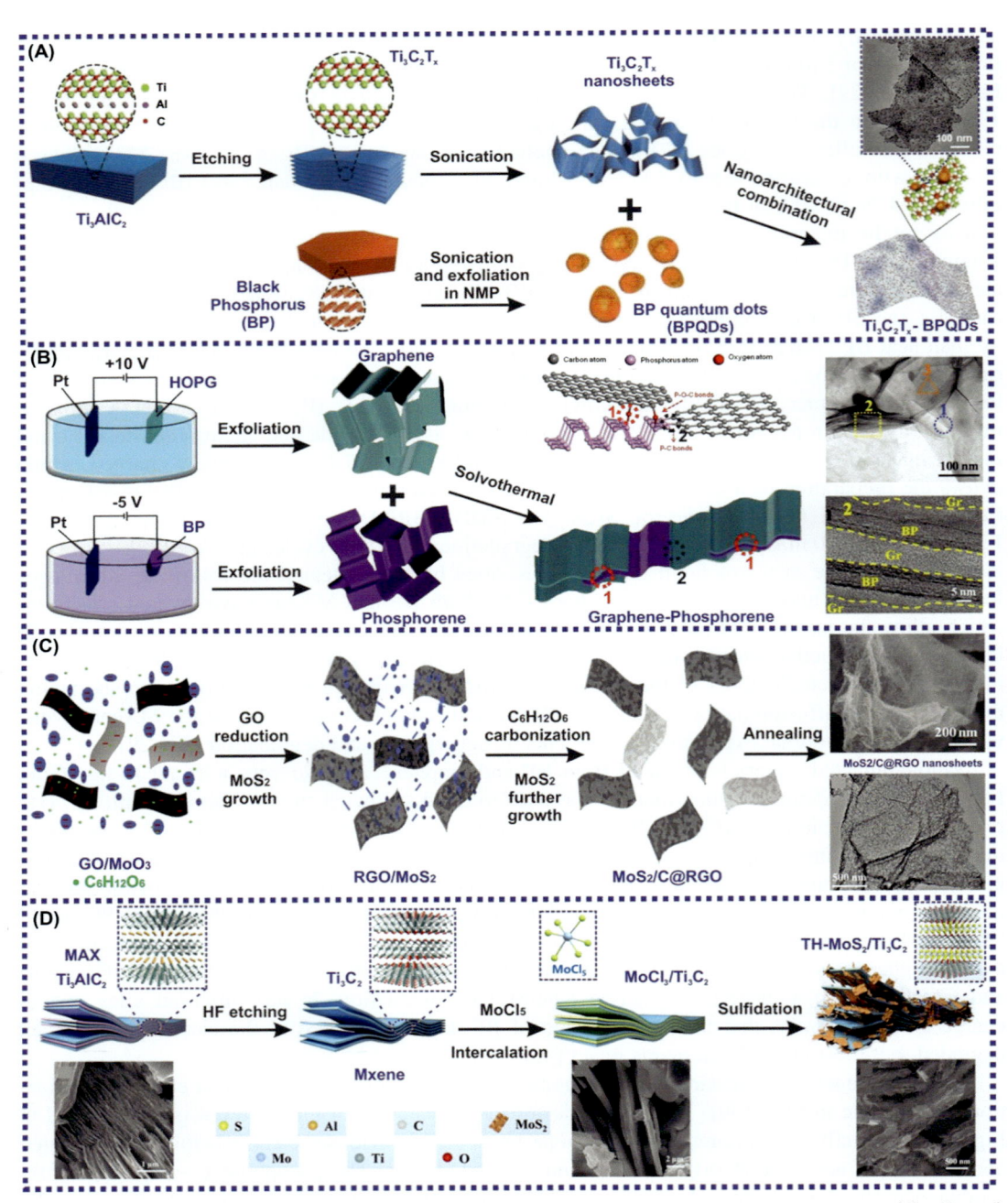

(*Continued*)

Zhang et al. recently described the preparation of phosphorene—graphene heterostructures by a top-down method [33], and their electrochemical performances as anodes for LIBs were systematically compared with a three-dimensional phosphorus—graphite material [33]. The three-dimensional material presented SC of 1980 mAh/g at 2 A/g, and the capacity dropped to 942 mAh/g after 10 cycles and to 271 mAh/g after 100 cycles. Otherwise, the 2D/2D nanoarchitected structure showed excellent electrochemical performance even at high current rates, with initial reversible SC of 2030 mAh/g at 2 A/g [33]. Li et al. produced a BP/graphene composite, through the coexfoliation of graphene and BP in N-methyl-pyrrolidone (NMP) [34], presenting an SC of 2365 mAh/g at 100 mA/g, and maintaining a reversible SC of 1297 mAh/g after 100 cycles and CE of 75.6% for SIBs. The authors attribute these high values to the resulting BP/graphene network which is effective in accommodating volume expansion [34]. An innovative BP/graphene hydrogel was produced by Mei et al. [35], presenting SC values in LIBs of 1864 and 1770 mAh/g at 10 and 20 mA/g, which are higher than the anode containing only graphene (1235 and 1083 mAh/g at 10 and 20 mA/g, respectively) [35]. BP/graphene composite was also easily produced by a high-energy ball-milling process, giving rise to a BP@irGO nanostructure for application in LIBs [36]. BP@irGO exhibited excellent storage performance with an initial SC of 1962 mAh/g, and the reversible SC of 1515 mAh/g at 0.1C ($\sim$260 mA/g) was maintained over 100 cycles [36]. Finally, a layered BP/rGO composite was synthesized by Yihang Liu et al. by pressurization at room temperature [37]. After 500 cycles applied as anode for SIBs, the BP/rGO electrode presents an SC of 1250 mAh/g at 1 A/g, and CE of 89.5%. Surprisingly, it still reaches values of 640 mAh/g and CE of 86.6% after 500 cycles at 40 A/g, and only at 60 A/g does the charging rate exceed the kinetic capacity of the anode. This is one of the best results reported in this chapter for SIBs [37].

The occurrence of P-C and P—O—C chemical bonds in BP/graphene architectural structures highlighted here is responsible for the improved performance of those materials because they immobilize the phosphorene/BP and protect it from breakage and degradation during the charge/discharge processes. Also, the encapsulation of phosphorene or BP into the graphene/graphite network shortens the diffusion distance for ions and the graphene can relieve the volumetric expansion of phosphorene/BP, in addition, to improving the electronic conductivity [30,32−37].

Among the four 2D materials considered in this review, the most common combination found in the literature is between MoS_2 and graphene, for example, Brown et al. [38] presented a MoS_2/rGO aerogel obtained by a 3D freeze printing method. The precursor of MoS_2 (ammonium

◀ **FIGURE 10.2**

Schematic representation and electron microscopy images of some architected nanostructures between 2D/2D materials: (A) MXene/BP, scheme of synthesis of black phosphorus quantum dots (BPQDs) and Ti_3C_2 (TNSs) nanosheets to obtain of $Ti_3C_2T_x$/BPQDs; (B) Graphene/BP, scheme of combination between phosphorene and graphene previously exfoliated by electrochemical methods, and deposited as a sandwich electrode; (C) MoS_2/Graphene, scheme of synthesis of aerogels with ultrathin nanosheets of MoS_2/graphene on a carbon fiber paper substrate; (D) MXene/MoS_2, scheme of synthesis of TH-MoS_2/Ti_3C_2 composite through the intercalation of the molybdenum chloride precursor and subsequent sulfidation reaction.

From (A) Ref. [27], copyright Wiley-VCH Verlag GmbH & Co. KGaA 2018; (B) Ref. [30], Copyright Wiley-VCH Verlag GmbH & Co. KGaA 2019; (C) Ref. [31], copyright Elsevier 2020; (D) Ref. [17], Copyright American Chemical Society 2022.

thiomolybdate) was mixed with graphene oxide (GO) and thermally converted into a highly porous hybrid structure, composed of MoS_2 nanoparticles anchored on the surface of rGO. MoS_2/graphene aerogels applied to SIBs presented an SC of 800 mAh/g during the first 10 charge/discharge cycles, which drops to 429 mAh/g at 100 A/g (after the 10th cycle) and with a CE of 91% [38]. Another MoS_2/graphene combination was demonstrated by Yuan et al. [39], who prepared aerogels based on ultrathin nanosheets of MoS_2/graphene on a carbon fiber paper substrate, based on a freeze-drying and drop-coating thermal reduction method. This nanostructured material exhibited an SC of 1709 mAh/g at 200 mA/g and a CE of 82.7% for LIBs. When applied as an electrode for SIBs, the SC of 984 mAh/g at 100 mA/g and a CE of 96.5% were obtained [39]. Yafeng et al. [31] prepared a very interesting nanoarchitected material between MoS_2, carbon, and rGO. They intercalated carbon into a MoS_2 structure and employed those extended MoS_2 layers to stabilize the rGO, as schematically represented in Fig. 10.2C [31]. The sample presented a large interlayer distance and reversible SC of 417 mAh/g at 100 mA/g after 150 cycles and CE of 97% for SIBs [31]. MoS_2/graphene hybrids were also produced through a cheap, high-yield, single-step, and solvent-free bead-milling process, starting from bulk MoS_2 and graphite [40]. As an anode in LIBs, the highest SC observed was 553 mAh/g at 250 mA/g after 100 cycles, and CE of 85% [40].

The thermal decomposition of a mixture of GO and $(NH_4)_2MoS_4$ yielded a MoS_2/graphene composite that reached an SC of 1730 mAh/g at 100 mA/g as an anode in LIB [41]. The high SC value was attributed to the presence of both Mo and S species anchored and dispersed on and between the defective graphene layers [41]. Xiao et al. [42] anchored a few-layer MoS_2 in nitrogen-doped graphene nanoribbons through multifunctional groups ($-SH$, $-NH_2$, and $-COO-$). This material presented 1151 mAh/g at 100 mA/g and a retention rate of 92.6% after 600 cycles in LIBs, reaching a maximum specific energy of 360 Wh/kg to 81.7 W/kg [42].

The excellent performance of the MoS_2/graphene nanoarchitectonics as electrodes in metal-ion batteries can be attributed to the synergistic effect of these materials on the following points: (1) the high surface area of MoS_2/graphene has a favorable effect on electrolyte penetration and allows optimal contact between the electrolyte and the active material; (2) the large interlayer distance in MoS_2 and the presence of several active sites allow a fast transport in the insertion/extraction of ions; (3) graphene increases electron mobility and dampens MoS_2 volume change during charge/discharge cycles; and (4) the combination prevents the restacking of both materials and stabilizes the electrode structure, resulting in an extended life cycle [31,38,39].

Finally, we present some examples of MoS_2/MXene nanostructures. One of the most interesting works on this combination was published in 2022 by Wen et al. [17]. A MoS_2/Ti_3C_2 composite was prepared through the intercalation of molybdenum chloride (employed as MoS_2 precursor), followed by a sulfidation reaction (Fig. 10.2D). Briefly, in this architecture, mixed-phase (1T and 2H) MoS_2 nanosheets grew steadily on the surface and at the Ti_3C_2 interlayer space, which caused the Ti_3C_2 layer distance to be expanded as a result of MoS_2 intercalation. The so-called TH-MoS_2/Ti_3C_2 anode showed SC for SIBs of 744 mAh/g at 200 mA/g and CE of 83.4%, higher than neat MoS_2 (80.1%) and neat Ti_3C_2 (41.6%) [17].

Luan et al. [43] reported the nanoarchitecture of MoS_2/Ti_3C_2 synthesized through a hydrothermal method, presenting an SC of 355 mAh/g at 1 A/g at 1300 cycles and CE of 91.8%, notably higher than both Ti_3C_2 and MoS_2 (120 mAh/g and 52 mAh/g, respectively). Chen et al. [44] synthesized MoS_2/MXene heterostructures through an in situ sulfidation. $Mo_2TiC_2T_x$ containing sulfur particles was incorporated between the MXene flakes followed by heat treatments at 500°C or

700°C for 4 hours under argon atmosphere, producing two different nanoarchitected materials. The amount of MoS_2 differed in each sample, and the one containing the lower amount presented the best specific capacitance and CE. The researchers attributed this behavior to the decrease of the effective contact between the nanostructure components.

Li et al. [45] designed different $Ti_3C_2T_x$ MXene/MoS_2 heterostructures, prepared by hydrothermal methods, to be applied as electrodes for KIBs. The effect of different amounts of MXene on the morphology of the final material and on the properties of electrodes in KIB was investigated. All the composites exhibited better SC than pure MXene (237 mAh/g and CE of 44.7%), and optimal proportions of the components were demonstrated to get better performance [45]. Zhang et al. [46] proposed a nanoarchitected material constituted of MoS_2 nanoflowers, MXene, and hollow carbonized kapok fiber (CKF) [46]. The SC of MoS_2/MXene/CKF was 584 mAh/g and the CE was 80% in SIBs. The authors showed that MoS_2 plays a leading role in the entire capability of these materials, and the advantage of the MoS_2/MXene/CKF nanostructure lies in its stability and performance [46].

As discussed before for the BP/MXene nanostructures, the MXene serves also as a mechanical support in the MoS_2/MXene nanoarchitectures, providing damping for volume expansion and conductivity. As both materials have good layer spacing, high specific surface area, and active sites, the resistance to ion diffusion is lowered and favored [17,44−46].

Some real devices based on 2D/2D combination were prepared, such as the one described by Yuan et al. [47], based on films of the MXene Nb_2CT_x combined with MoS_2, using carbon spheres (CS) as supports. The material was obtained through hydrothermal and electrostatic methods and then applied as electrodes for SIBs. Different amounts of MXene in the nanostructures were studied and the best sample presented SC of 210, 196, and 173 mAh/g at high rates of 10, 20, and 25 A/g, respectively. The possibility to apply this film as a flexible electrode was demonstrated, showing that the film could be bent by 90 degrees, 180 degrees, and 210 degrees without breaking (Fig. 10.3A), maintaining the uniform distribution of MoS_2@CS between the layers of Nb_2CT_x. Finally, the authors assembled a fully flexible battery using NVP as the cathode and Nb_2CT_x/MoS_2@CS as an anode, showing good stability. The efficiency of the device was demonstrated by shining a red light-emitting diode (LED), including when it was bent by 90 degrees (Fig. 10.3A) [47].

Pan et al. [48] prepared aqueous SIBs using a BP/MXene nanostructure as electrodes. The BP/MXene provided 91.2% of retention capacity, which is 1.7 times higher than neat MXene film (60.9% of retention) [48], a high energy density of 72.6 Wh/L, close to lead-acid batteries' energy (50 − 90 Wh/L), and long-term stability (90.6% retention after 50,000 cycles and 99% CE) [48]. This amazing electrochemical performance is attributed to a strong chemical bond (Ti−O−P) between BP and MXene that maintains stability and by the distribution of BP on the surface of the MXene, which expands the MXene layers, allowing quick charge/ion transport [48]. The device was manufactured with compact BP/MXene films (10 mg/cm^2) in two different sizes (80 × 110 mm^2 and 50 × 80 mm^2) (Fig. 10.3B). A LED "Christmas Tree" (2.0−3.5 V) was successfully illuminated for 1 hour using two fully charged and series-integrated devices (Fig. 10.3B), demonstrating the capability of this nanostructure in energy storage applications [48].

Yang et al. [49] combined positively charged BP with negatively charged MXene, resulting in different BP/MXene nanocomposites stabilized by electrostatic interaction. The best sample was employed to prepare a flexible and symmetrical device (Fig. 10.3C) that showed a 91.0% retention

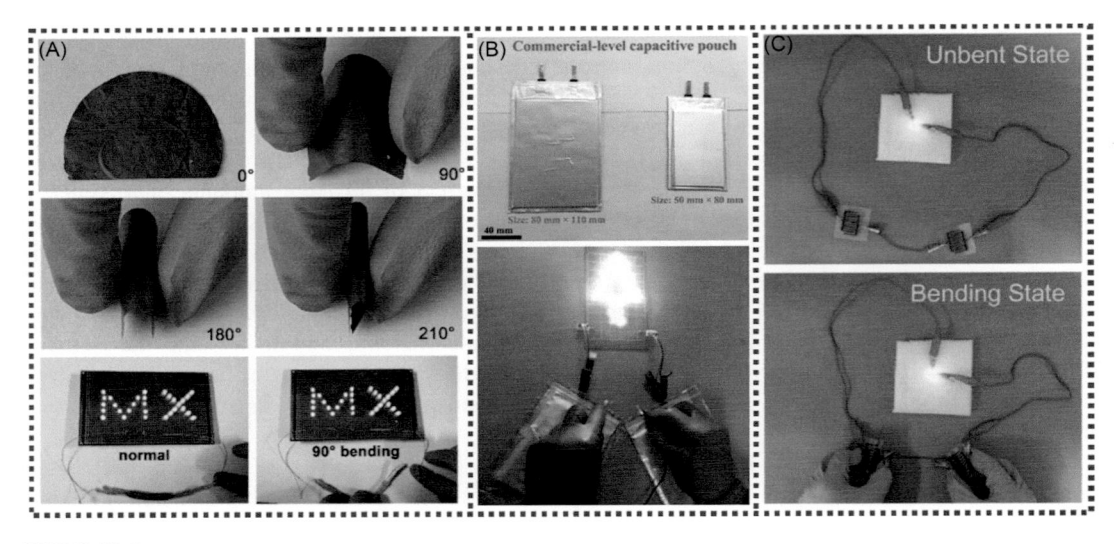

FIGURE 10.3

Devices based in nanostructures between 2D/2D materials: (A) Nb_2CT_x/MoS_2@CS//NVP flexible film bend at 90 degrees, 180 degrees, 210 degrees and the flexible battery lighting a LED with 0 degree or 90 degrees bending; (B) two different sizes of devices produced with BP/MXene film, and two devices connected in series to power a "Christmas Tree" logo with colored LEDs; (C) devices of two charged BP/MXene connected in series at different bending angles to light up a red LED.

From A Ref. [47], copyright Elsevier 2022; (B) Ref. [48], copyright American Chemical Society, 2021; (C) Ref. [49], copyright Elsevier, 2021.

rate, an energy density of 28 mW/ h cm^3 at a power density of 0.043 W/cm^3, and also a high energy of 25.46 mW/h cm^3 at a power density of 4.14 W/cm^3. After 40,000 charge/discharge cycles, the retention was 90.9%, and after 1000 cycles of 90 degrees bending, 89.2% retention was achieved. Fig. 10.3C shows the devices connected in parallel and in series used to trigger a red LED in both unbent and bent states (Fig. 10.3C) [49].

As final considerations regarding this subsection, it is interesting to note that some architected nanostructures presented and discussed here show some instability during the first charge/discharge cycles, which has been attributed to the inevitable generation of the solid electrolytic interface layer (SEI). SEI is a passivation layer formed on the surface of the anodic materials of metal-ion batteries, produced by electrolyte decomposition, and plays a critical role in the cyclability, rate capability, irreversible loss of capacity, and safety of the batteries. In all papers referring to this topic, the formation of the SEI layer is mentioned mainly for the SIBs, and this demonstrates that this is one of the difficulties in the application of these batteries in relation to the LIBs (although many works also report the formation of SEI in LIBs).

Few data are presented in the literature regarding the comparison between the performance of the same material used in different metal-ion batteries (LIBs, SIBs, or KIBs). With regard to 2D/2D architected materials, only two works reported here present the comparison between LIBs and

SIBs [27,39] and the data indicate that both SC and CE have lower values for sodium storage, due to the larger size of Na^+ ions compared with the Li^+ ions [27,39]. In contrast, it is worth mentioning that most of the SIBs have their SC increased little by little, which can be attributed to the process of gradual activation of active sites in the materials that make up these batteries and demonstrate great progress in nanoarchitected materials for this application [44]. Finally, the synergistic effects demonstrated with the combinations described here are mainly responsible for the impressive electrochemical performance of these materials as battery electrodes.

Table 10.1 compiles the main data related to the 2D/2D nanoarchitected materials applied as anodes in metal-ion batteries. It is important to notice that combinations between BP/MoS_2 and graphene/MXene were not found for this specific application.

Table 10.1 Summary of architected nanostructures based on combination of 2D/2D materials (graphene, MoS_2, MXene, and BP) applied as electrodes in metal-ion batteries.

Nanoarchitectonics	Material 1	Material 2	Battery	Specific capacity (mAh/g)	Current (mA/g)	Coulombic efficiency (%)	References
BPQD/TNS	BP	MXene	LIBs SIBs	1124 723	50	90 100	[27]
BP/Ti$_3$C$_2$	BP	MXene	SIBs	332	100	–	[28]
PDDA-BP/Ti$_3$C$_2$	BP	MXene	SIBs	1112	100	95	[29]
G-BPGO	BP	Graphene	LIBs	1401	100	71	[32]
Phosphorene-Graphene	BP	Graphene	SIBs	2311	100	83.9	[30]
BPG_soni	BP	Graphene	LIBs	2030	2000	82.1	[33]
BP/rGO	BP	Graphene	SIBs	1250	1000	89.5	[37]
BP/G	BP	Graphene	SIBs	2365	100	75.6	[34]
PGH	BP	Graphene	LIBs	1864	10	–	[35]
BP@irGO	BP	Graphene	LIBs	1962	260	–	[36]
MoS_2/graphene	MoS_2	Graphene	SIBs	429	100	91	[38]
MoS_2/GS-A	MoS_2	Graphene	LIBs SIBs	1709 984	200 100	82.7 96.5	[39]
MoS_2/C@RGO	MoS_2	Graphene	SIBs	417	100	97	[31]
MoS_2-G (40 h)	MoS_2	Graphene	LIBs	553	250	85	[40]
MoS_2/G	MoS_2	Graphene	LIBs	1730	100	–	[41]
N-GRs/MoS_2	MoS_2	Graphene	LIBs	1151	100	92.6	[42]
MoS_2@Ti$_3$C$_2$	MoS_2	MXene	LIBs	355	1000	91.8	[43]
TH MoS_2/Ti$_3$C$_2$	MoS_2	MXene	SIBs	744	200	83.4	[17]
MoS_2/Mo$_2$TiC$_2$T$_x$-500	MoS_2	MXene	LIBs	509	100	92	[44]
MXene/MoS_2–0.1	MoS_2	MXene	KIBs	454	100	66.5	[45]
MoS_2/MXene/CKF	MoS_2	MXene	SIBs	584	100	80	[46]

10.2.2 **2D/Oxides nanoarchitectonics**

Transition metal oxides (TMOs) are another important class of materials that have been extensively investigated for application as electrodes in metal-ion batteries due to their multiple chemical valence states, diverse morphological characteristics, and high theoretical storage capacity [24]. Some examples include iron, cobalt, copper, and molybdenum oxides. Iron oxides, mainly Fe_2O_3 and Fe_3O_4, are great materials to be used as anodes due to their high theoretical SC (1007 mAh/g), redox reactions, abundance, nontoxicity, corrosion resistance, and low cost. However, these oxides have low electrical conductivity (10^{-14} S/cm), leading to low energy and low power density [24,26]. Cobalt oxides including CoO and Co_3O_4 have many advantages like low toxicity and high capacitance. CoO has a theoretical SC of 716 mAh/g and Co_3O_4 has a theoretical SC of 890 mAh/g. However, the main disadvantage of cobalt oxides is the cost of this material [24,26]. Copper oxide is a promising candidate for commercial applications because copper is one of the most abundant metallic elements on Earth and can be applied with high stability in various classes of energy storage devices. CuO, for example, has attracted a lot of attention due to its high theoretical SC (674 mAh/g) and low toxicity [26]. Molybdenum oxides (MoO_2 and MoO_3) have pseudocapacitive properties that are mainly related to reversible redox reactions at the surface and ion diffusion. MoO_3 is one of the most attractive pseudocapacitor materials due to its theoretical SC of 1117 mAh/g, however, its low conductivity greatly limits its electrochemical performance [26].

In addition to some of the aforementioned disadvantages, the majority of the TMOs suffer from large volume changes during conversion reactions, resulting in electrodes with low stability [11,24]. One way to solve these problems is through the formation of nanoarchitected structures with 2D materials.

It is well known that TMOs can be prepared in different morphologies and dimensionalities, which directly affect their intrinsic properties. Both zero- (nanoparticles) and unidimensional structures are the most commonly found in literature combined with 2D materials for application in metal-ion batteries.

The MXene $Ti_3C_2T_x$ combined with Fe_2O_3 nanoparticles were synthesized by Ali et al. [50] through three different experimental methodologies: hydrothermal, wet sonication, and ball milling, varying the amount of Fe_2O_3 between 25% and 50% by weight. The results demonstrate again the importance of the synthetic method on the characteristics and properties of the nanoarchitected materials: samples prepared both by hydrothermal and sonication methods resulted in severe oxidation of the MXene surface, decreasing the specific surface area. On the other hand, the ball milling-based method provided a simple, environment-friendly, and economical way to produce high-performance $Ti_3C_2T_x/Fe_2O_3$ composite. SCs of 460 mAh/g at 300 mA/g were achieved for this sample in LIBs [50]. Kong et al. [51] successfully fabricated nanoarchitected structures of $Fe_3O_4@Ti_3C_2$ using an in situ growth method (Fig. 10.4A). The initial SC of Ti_3C_2 for electrodes in LIBs was 486 mAh/g, which corresponds to a CE of 45%. Otherwise, SCs for $Fe_3O_4@Ti_3C_2$ were 569, 779, and 857 mAh/g as the Fe_3O_4 content increases, which corresponds to CEs of 64%, 67%, and 66%, respectively. The architected materials were afterwards coated with carbon, which caused improved reversibility and cyclic stability. For example, the carbon-coated anode maintains a reversible SC of 304 mAh/g and CE of 88.1% after 100 charge/discharge cycles, which is better than the 260 mAh/g and CE of 77% of the uncoated material [51].

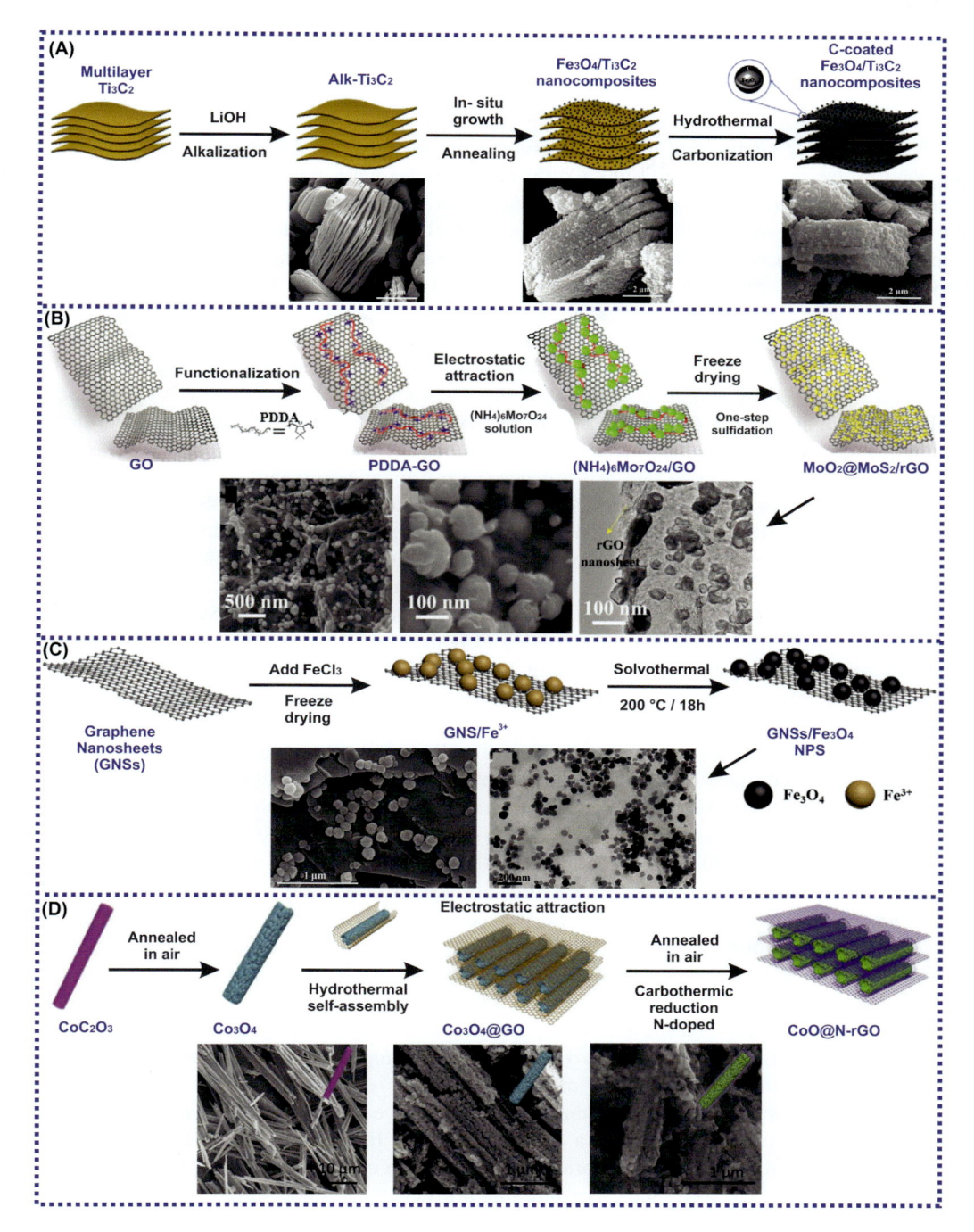

FIGURE 10.4

Schematic representation and electron microscopy images of some architected nanostructures between 2D and TMOs: (A) method to produce Fe_3O_4/Ti_3C_2; (B) synthesis of $MoO_2@MoS_2/rGO$ nanostructures; (C) synthesis of graphene nanosheets with Fe_3O_4 nanoparticles; (D) synthesis of architected graphene nanostructures with unidimensional CoO.

From (A) Ref. [51], copyright Elsevier 2018; (B) Ref. [52], copyright Elsevier 2021; (C) Ref. [53], copyright Elsevier 2020; (D) Ref. [54], copyright Elsevier 2019.

$Ti_3C_2T_x$/Co_3O_4 sandwiched nanoarchitectures were prepared by Zhang et al. [55]. $Ti_3C_2T_x$ was prepared by the chemical method, the Co_3O_4 nanoparticles were synthesized following the hydrothermal method, and the composite was obtained after mixing the components, sonicating and filtering. The synergistic effect of Co_3O_4 and $Ti_3C_2T_x$ and the effective architecture allow a remarkable performance for the Co_3O_4/$Ti_3C_2T_x$ composite with a component content of 50%. The capacity remains at 999 mAh/g after 900 cycles at 500 mA/g in LIBs, while for $Ti_3C_2T_x$ or Co_3O_4 the SC are lower at the same experimental conditions (311 and 145 mAh/g, respectively) [55]. Guo et al. [56] reported a combination of MXene and MoO_3 nanorods prepared using a simple and green method. Prefabricated α-MoO_3 nanorods and MXene nanosheets were dispersed in a bath ultrasound, followed by simple vacuum filtration. The final product presented a high SC for LIBs of 1008 mAh/g at 100 mA/g and an excellent rate capability of 172 mAh/g at 10 A/g. Furthermore, a complete cell was manufactured using commercial $LiFePO_4$ as the cathode, which showed a high SC of 160 mAh g^{-1} with good rate performance (48 mAh/g at 1 A/g) [56].

Interesting nanoarchitected materials arise from the combination of MoS_2 and some oxides. Wang et al. [57] prepared Co_3O_4 dodecahedrons nanoparticles through direct pyrolysis of the metal−organic framework ZIF-67, which were used as substrates for the growth of MoS_2 nanosheets. The final product is characterized by Co_3O_4 uniformly covered by MoS_2 nanosheets. Co_3O_4/MoS_2 was applied as an anodic material for LIBs and showed an SC of 1136 mAh/g at 100 mA/g and a CE of 80.3%, greater than that of both neat Co_3O_4 and MoS_2 which exhibit rapid decay in capacity during cycling [57]. Cheng-Yu et al. [58] synthesized MoS_2@Fe_2O_3 by an ultrasonic vibration process. The nanoarchitected material showed high stability and a reversible SC of 600 mAh/g after 70 cycles, which is higher than that obtained for two different neat MoS_2 samples. The results indicate that an appropriate amount of α-Fe_2O_3 added to MoS_2 can improve the capacity at various current rates due to the reduction of the Li$^+$ diffusion pathway [58]. Luo et al. [52] described an easy in situ localized phase transformation strategy to transform MoO_3 nanoparticles anchored on rGO nanosheets into MoO_2@MoS_2 heterostructures, and the synthesis scheme is shown in Fig. 10.4B. MoO_2@MoS_2/rGO exhibits an SC of 604 mAh/g at 100 A/g when evaluated as an anode for SIBs, and a CE of 90.3% after 300 cycles at 1.0 A/g [52]. In this heterostructure, the phase transformation that occurs on the surface of the MoO_2 nanoparticles results in an unbalanced charge distribution at the interface, which brings greater electronic conductivity and kinetics of facilitated ion transfer within the electrode. Meanwhile, MoS_2 nanosheets on the MoO_2 surface provide extra active sites for Na$^+$ storage, and the graphene layer increases electronic conductivity [52]. Zhao et al. [59] synthesized a MoO_3 exhibiting a uniquely shaped nanostructure that was completely different from conventional α-MoO_3. A nanocomposite between those MoO_3 and MoS_2 was produced by the in situ growth of few-layer MoS_2 nanoflakes on the surface of MoO_3. MoO_3@MoS_2 has a high SC of 1545 mAh/g and CE above 98% after 150 cycles. In addition, it features stable rate performance under a current density of 100−1000 mA/g [59]. The much better electrochemical performance of the hybrid nanostructures in comparison to neat MoS_2 and α-MoO_3 was attributed due to the core−shell structure that prevents the volume variation during the electrochemical cycles, as well as due to the presence of increased active sites on the MoS_2 uniformly covering the surface of MoO_3.

Several graphene/oxide nanoparticles nanoarchitected materials have also been prepared for specific applications in metal-ion batteries. Actually, graphene/TMOs consist of one of the most studied classes of composite materials involving 2D nanostructures, for application in batteries and in many other fields. Graphene sheets loaded with Fe_3O_4 nanoparticles were synthesized by Gu et al. [53] through a thermal synthesis method (Fig. 10.4C). The size of Fe_3O_4 from 20 to 100 nm was regulated by controlling the initial ratio between the precursors. The LIB anode based on this material without any other conductive agent exhibits a high SC of 1145 mAh/g after 120 cycles at 100 mA/g and CE of 83.2%. The authors demonstrated that both the Fe_2O_3 nanoparticles size and amount have a strong influence on the electrode performance. A microwave-assisted synthesis of rGO/Fe_3O_3 material was described by Kumar et al. [60]. Benefiting from the synergistic effect, the hybrid reached initial values of SC of 1625 mAh/g, keeping an SC of 540 mAh/g after 120 cycles at 100 mA/g and better stability in LIBs [60].

Li et al. [61] presented different porous CuO/rGO nanostructures, in which CuO was synthesized by hydrothermal method starting from different copper precursors (nitrate, chloride, sulfate, or acetate). SC of 639, 597, 433, and 618 mAh/g at 200 mA/g was obtained and it was demonstrated that the addition of graphene improved the charge/discharge stability and the final results are dependent on the type of interactions between the components in each nanostructure [61]. Similarly, a nanoarchitectural porous and conductive structure based on CuO/Cu_2O/graphene was built using a solvothermal self-assembly method [62] presenting a reversible SC of 767 mAh/g at 100 mA/g and 586 mAh/g at 1 A/g with impressive stability for LIBs.

Composites of Co_3O_4/graphene with a three-dimensional porous structure were synthesized by Sheng et al. [63] showing SC of 1367 mAh/g after 80 cycles at 100 mA/g, SC of 1910 mAh/g after 300 cycles at 500 mA/g and CE of 78.8% in LIBs [63]. The reason for the increase in capacity during cycling is that the electrolyte penetrates the electrode pores providing more reaction sites to store more Li^+ as the number of cycles gradually increases [63]. An elegant nanoarchitected structure combining N-doped graphene with CoO nanowires was reported by Wu et al. [54]. Fig. 10.4D presents the summary of the electrostatic process of carbothermic adsorption-reduction employed to prepare CoO/N-rGO, architected as CoO microrods encapsulated by the graphene layers through electrostatic adsorption. A high reversible SC of 1588 mAh/g, and excellent long-term cycling stability under high current density (948 mAh/g after 500 cycles at 2 A/g) were observed for LIBs. Furthermore, the pseudocapacitive contribution induced by the hierarchical structure, oxygen vacancies, and modified graphene is dominated in the lithium storage process. This is essential to avoid undesirable side reactions, especially at elevated operating temperatures, in which the CoO/N-rGO maintains a reversible SC of 1164 mAh/g after 1000 cycles under 1 A/g at 55°C [54].

Nanoarchitected structures between graphene and unidimensional MoO_3 have been demonstrated by Naresh et al. [64] and Almodóvar et al. [65], respectively. The MoO_3/rGO nanocomposite was prepared through an easy high-energy ball-milling process followed by an ultrasound method. The milling process produces MoO_3 in a one-dimensional nanostructure mixed with the rGO in different proportions. The best electrochemical performance was observed for the sample containing 10 wt.% of MoO_3, with an SC of 568 mAh/g after 100 cycles and CE greater than 90% at a current density of 500 mA/g. Moreover, the MoO_3/rGO nanocomposite exhibits excellent rating capability with a specific capacity of 502 mAh/g even at a higher current density of 1000 mA/g [64]. The MoO_3 nanotubes/rGO composites prepared by Almodóvar et al. [60] showed excellent performance as anodes for LIBs with SC of 789 mAh/g after 100 cycles at 1000 mA/g and initial CE of 69%, with remarkable stability at very high current densities, with SC of 665 mAh/g and 490 mAh/g at 2000 and 3000 mA/g [65].

Some devices have also been prepared using 2D/TMOs nanostructures as electrodes. Li et al. [66] prepared MoS_2 nanosheets wrapped with Fe_2O_3 nanospheres through a hydrothermal reaction, which was used as an anode for a flexible SIB device, obtaining an energy density of 2 mW/h cm^3 and a power density of 16 mW/cm^3. This asymmetric device was bent 200 times at 90 degrees, and the results remain unchanged, and after 3000 cycles the device maintains 78% of capacitance, which indicates that the supercapacitor has good stability. Li et al. [67] obtained an MXene/Fe_3O_4/MXene sandwich film on a flexible Ni tape. This material was tested directly on the device and has a retention of 96.3%. To test the flexibility, the device was bent at different angles (0 degree−180 degrees) and was able to maintain 98.4% of capacitance retention even at a bending angle of 180 degrees. The device exhibits an energy density of 0.970 μWh/cm^2 at a power density of 0.176 mW/cm^2, and even at a higher power density of 1.399 mW/cm^2, the energy density still maintains 0.548 μWh/cm^2. The device in a sandwich design and with a thin and very porous nanostructure is crucial to facilitate the fast transport of ions and electrons. In this unique architecture, ions/electrons are quickly delivered from MXene to the Fe_3O_4 film (most responsible for the capacitance), the Fe_3O_4 film to the MXene, and then to the electrolyte and/or current collector. In other words, the excellent conductivity of the MXene and the strong chemical interaction between the components allow a faster response. The performance was demonstrated by connecting five devices in series, resulting in a voltage of 3.5 V able to trigger a red LED [67].

As a general behavior, architected nanostructures between 2D materials and TMOs nanoparticles present improved performance as electrodes in metal-ion batteries due to the synergism and the combination of the intrinsic properties of each component. It improves conductivity, provides abundant active sites, reduces volume expansion, has a high specific surface area and they are easy and cheap to be produced. In which concerns the specific combinations discussed here: (1) for MXene and TMOs, the oxide NPs distributed over the MXene sheets prevent the restacking of the layers. Otherwise, the MXene prevents large volume changes and structure breakage of the TMOs nanoparticles, increasing the stability of the material. MXene usually acts as a substrate for the TMOs nanoparticles growth, it improves the overall electronic conductivity and provides a high specific surface area, while nanoparticles from different oxides are mainly responsible for the electrochemical response; (2) in which regard the MoS_2/TMOs nanomaterials, the active sites in MoS_2, provide a large surface area for nucleation and attachment of TMOs nanoparticles without agglomeration. Also, MoS_2 is able to contain volumetric changes during charge/discharge cycles, while the presence of TMOs nanoparticles prevents MoS_2 restacking. The porous structures resulting from these nanoarchitectonics increase the contact area with the electrolytes and provide more electrochemically active surfaces to accommodate ions; (3) graphene is the best material to dampen the expansion of oxides during the redox cycles and to improve the conductivity of the composite, as well as being an amazing substrate for TMOs nanoparticle growth. Other important characteristics that affect the nanocomposite performance are the ratio between the components, the nature of their interaction, the homogeneity of the resulting material, the morphology of the TMOs nanoparticles and the degree of exfoliation of the 2D material.

The SEI layer is also very present in these materials, but TMOs often exhibit a different phenomenon. Some TMOs show an upward trend after decreasing in the first few cycles (or after hundreds of cycles) and a tendency to be stable in the following cycles. The rapid fading of capacity in the early stage is generally attributed to unstable SEI film formation and intercalation-induced mechanical degradation. However, in the subsequent reactivation process, morphological and microstructural evolutions can occur, and refining architectures generally contribute to excellent electrochemical performance. In addition to the SEI layer, at the beginning of the process, many nanoparticles are transferred to the space between the layers of 2D materials during the charge/discharge process where the expansion/contraction of the layers occurs, decreasing the access to these materials and causing the capacity decline. As the cycles increase, the nanostructures form an integrated architecture with several pores, increasing again the diffusion of ions/electrons through these sites, returning access to these materials. These phenomena generally contribute to excellent electrochemical performance [55,57]. Finally, the nanostructures between 2D and TMOs have been in their large majority applied only for LIBs, with few examples for SIBs or KIBs, which means that obtaining new 2D/TMOs structures for applications in SIBs and KIBs is a hole yet to be filled in the literature.

A summary of the different combinations of the 2D/TMOs nanoarchitected materials discussed here are presented in Table 10.2.

10.2.3 2D/Carbon nanotubes nanoarchitectonics

Carbon nanotubes (CNTs) are a family of materials among the most reported in combinations with 2D materials for electrodes in ion-metal batteries, in spite of these materials don't have great storage capacities by themselves. The main explanation for this apparent paradox is that the integration

Table 10.2 Summary of architected nanostructures between 2D and some TMOs applied as electrodes in metal-ion batteries.

Nanoarchitectonics	Material 1	Material 2	Battery	Specific capacity (mAh/g)	Current (mA/g)	Coulombic efficiency (%)	References
BM25	MXene	Fe_2O_3	LIBs	460	300	–	[50]
$Fe_3O_4@Ti_3C_2$	MXene	Fe_3O_4	LIBs	857	0.277	66	[51]
$Co_3O_4/Ti_3C_2T_x$-3	MXene	Co_3O_4	LIBs	999	500	–	[55]
α-MoO_3/MXene	MXene	MoO_3	LIBs	1008	100	–	[56]
Co_3O_4/MoS_2	MoS_2	Co_3O_4	LIBs	1137	100	80	[57]
S-$MoS_2@\alpha$-Fe_2O_3	MoS_2	Fe_2O_3	LIBs	600	0.055	–	[58]
$MoO_2@MoS_2$/rGO	MoS_2	MoO_2	SIBs	604	100	90	[52]
$MoO_3@MoS_2$	MoS_2	MoO_3	LIBs	1545	100	98	[59]
GNSs/Fe_3O_4	Graphene	Fe_3O_4	LIBs	1145	100	83	[53]
Fe_3O_4-ONCs@rGO	Graphene	Fe_3O_4	LIBs	1625	100	99	[60]
CuO/rGO (D)	Graphene	CuO	LIBs	313	200	–	[61]
CCO/CRGO	Graphene	CuO	LIBs	767	100	65	[62]
Co_3O_4/G	Graphene	Co_3O_4	LIBs	1910	500	79	[63]
MoO_3/rGO	Graphene	MoO_3	LIBs	568	500	90	[64]
h-MoO_3/GO	Graphene	MoO_3	LIBs	789	1000	69	[65]
h-MoO_3/Graphene	Graphene	MoO_3	LIBs	665	2000	–	[65]
CoO@N-rGO	Graphene	CoO	LIBs	948	2000	–	[54]

of CNTs with 2D materials improves the electrochemical property of the 2D material, due to the inhibition of the volume change during the charge/discharge cycles, the improvement of the overall conductivity and the avoidance of the restacking of the exfoliated 2D materials [68]. Also, CNTs are cheap and easy to be obtained and they have exceptional mechanical resistance and flexibility, which allows the preparation of nanoarchitectures with multifunctionalities [34,35,69].

The first work describing a multiwalled carbon nanotubes/black phosphorous (MWCNT/BP) hybrid structure was reported by Zhao et al. in 2018 [70]. MWCNTs were dispersed and lyophilized to be conductive backbones for phosphorus nucleation. Black phosphorus was grown in situ on MWCNT structure by vaporizing amorphous and crystalline red phosphorous, originating P-C bound. A fraction of the red phosphorus was afterwards converted into BP. The electrochemical performances of the materials were significantly improved by the BP wrap and extremely high cycle stability was achieved by hybrids as anodes for LIBs [70]. An SC of 560 mAh/g was observed, maintained after 250 cycles at a current rate of 500 A/g. Another BP/CNT combination was performed by Haghighat-Shishavan et al. [71]. The composite was prepared by a surface oxidation-assisted chemical bonding procedure, using a sodium carboxymethyl cellulose-poly (acrylic acid) binary polymeric binder (NaCMC-PAA), resulting in cross-linked structures. Expressive SC of 1681 mAh/g after 400 cycles at 0.2 C (1 C = 2596 mA/g) and CE of 87.5% were observed for LIBs. This material also successfully provided an SC initial of 2073 and 850 mAh/g at 0.2 C and 2 C for SIBs, with excellent capacity retentions after 200 cycles [71]. A composite between BP, graphite and N-doped CNTs was prepared for application in SIBs, showing improved electrochemical performance for SIBs, with a high SC of 1791 mAh/g at 0.2 C and providing a high stable SC of 1665 mAh/g at 0.5 C after 100 cycles. In addition, the composite features good rate performance (744 mA/g at 4.5C), high CE of 89.6% and excellent rate capability due to the small particle size of BP and the intimate connection between BP and the double carbon conductive network [72].

Yupu et al. [68] combined BP and CNTs, and the material showed good stability and high conductivity for LIBs. The BP/CNTs (380 mAh/g) offer SC 4.2 times and 16.7 times greater than CNTs (90 mAh/g) and BP (22 mAh/g) at 2.5 A/g, respectively. Finally, the nanoarchitected material demonstrated excellent cyclic stability with an SC of 522 mAh/g after 650 cycles and a CE of 100%. The BP/CNTs anodes were used to produce a battery device according to the diagram in the left panel of Fig. 10.5A, and these devices were able to lighten a LED display as illustrated in the right side of Fig. 10.5A [68].

The combinations between BP and carbon nanotubes (or also other carbon structures with different dimensionalities) yield nanostructured materials with incredible stability due to the formation of bonds between the carbon materials and the BP, showing the best electrochemical results for application as battery electrodes.

Several reports on the preparation of graphene/carbon nanotube nanostructures have been described in recent years, but few of them aiming application in metal-ion batteries. Feng et al. [73] designed some rGO/CNTs composites through a simple liquid exfoliation technology and the materials were directly evaluated in a coin cell for LIBs (Fig. 10.5B). The resulting material presented an initial SC of 926 mAh/g at 50 mA/g and a CE of 27.5%, and this low CE is attributed to the SEI layer. The SC decreased and remained stable at 251 mAh/g and CE of 100% in the following cycles. A vacuum treatment to decrease the number of oxygenated species was further carried out aiming to improve the electrochemical response, producing a material with an SC of 668 mAh/

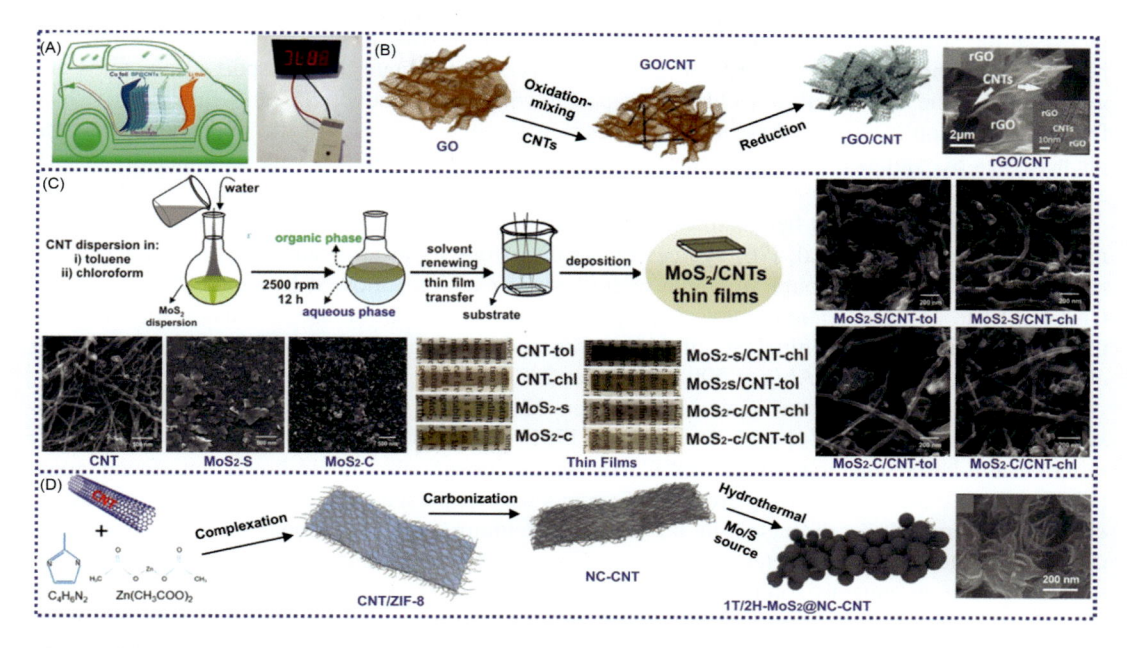

FIGURE 10.5

Scheme of obtaining some architected nanostructures between 2D materials and carbon nanotubes: (A) schematic diagram of the BP/CNTs hybrids-based LIBs and the application for power devices (left) and the photograph of a LED lightened by two BP/CNTs hybrids-based LIBs (right); (B) schematic illustration of the fabrication of rGO/CNTs and SEM image of the GO/CNTs; (C) synthesis of thin films of different molybdenum-based materials with carbon nanotubes (CNTs) prepared by the liquid-liquid interfacial route, photography and SEM images of the films; (D) schematic illustration of the preparation process of MoS_2/N-CNT.

From (A) Ref. [68], copyright Wiley-VCH Verlag GmbH & Co. KGaA 2020; (B) Ref. [73], copyright Elsevier 2019; (C) Ref. [74], copyright Elsevier 2021; (D) Ref. [75], copyright Elsevier 2022.

g at 50 mA/g and a CE of 42.7%. A porous foam obtained by hydrothermal reduction reaction between GO and CNTs aqueous suspensions was also described, presenting an SC of 780 mAh/g at 50 mA/g when applied as anodes for SIBs [73]. Self-assembled films of 3D sandwich-like structures based on rGO and carbon nanotubes were prepared by vacuum filtration and thermal reduction to be used as LIBs, KIBs and SIBs electrodes [69]. The electrical conductivity of the films is high and the electrochemical performance depended on the weight ratio of CNT to rGO. The best results were obtained for rGO/CNT-30%, with reversible SCs of 351 and 223 mAh/g at 10 and 50 mA/g, respectively for KIBs; 260 mAh/g after 60 cycles at 100 mA/g for LIBs; and inexpressive SC of 40 mAh/g to 50 mA/g for SIBs, demonstrating that the same material presents different responses for each kind of metal-ion battery, and good material for one kind should not be for another one [69].

Graphene/CNTs nanostructures are usually simple and easy to be obtained. Regarding their application in metal-ion batteries, the incorporation of CNTs avoids graphene restacking and increases the spacing between graphene/graphite layers to provide more storage locations for ions, besides providing a bridge between the graphene layers to facilitate the transport of electrons and ions. Finally, the combination yields a composite with high conductivity and flexibility, which facilitates the application of these nanostructures in flexible devices. On the other hand, the graphene/CNTs SC values are not as significant as those of CNTs combined with the other 2D materials [69,73].

MoS_2/carbon nanotubes heterostructures correspond also to a very interesting combination for metal-ion batteries. Ma et al. [76] synthesized a 3D porous foam integrated with MoS_2, CNT and rGO (MoS_2/CNT/rGO) through a simple hydrothermal reaction followed by lyophilization. MoS_2/CNT/rGO had an optimization study of the proportions before can be used as anode for LIBs and the best sample exhibited initial SC of 913 mAh/g and reversible SCs of 745 mAh/g at 100 mA/g after 170 cycles and 420 mAh/g at 500 mA/g after 200 cycles.

Interesting nanoarchitected materials between MoS_2/CNT have been more recently demonstrated by Zarbin et al. [74] and Wang et al. [75]. In the first one, thin films of different 2D molybdenum-based materials and MWCNTs were through the liquid-liquid interfacial route (LLIR), Fig. 10.5C [74]. Starting from CNTs previously dispersed in an organic phase (toluene or chloroform) and MoS_2 dispersed in an aqueous/acetonitrile phase, nanoarchitected films were directly obtained at the L/L interface, deposited over transparent electrodes and studied as anodes for KIBs using an aqueous solution as electrolyte [74]. Low SC was obtained for the control samples (neat CNTs or neat MoS_2) at 7 A/g (64 and 278 mAh/g, respectively). However, the well-defined network structures of the MoS_2 interconnected with the CNTs nanoarchitected sample presented a recovery range from 70% to 100% and high SC values (517 mAh/g) [74]. Wang et al. [75] used a carbon/CNTs structure as a backbone to deposit a mixed-phase of MoS_2 (1T/2H), Fig. 10.5D. The obtained nanostructure has a reversible SC of 765 mAh/g after 100 cycles at 200 mA/g, and a CE of 68.6% as an anode for LIBs [75].

MoS_2 and CNTs combination yields nanoarchitected materials which present excellent transport channels for electrolyte penetration and rapid transfer of ions/cations. Also, the carbon structures are mostly supporting pillars of the nanostructures, responsible for conductivity, and flexibility, and are responsible for avoiding aggregation. These nanostructures are obtained with high homogeneity, which is responsible for the best electrochemical responses, mainly attributed to the active sites of MoS_2 nanosheets [74−81].

Architectural nanostructures between 2D materials and carbon nanotubes (and other forms of carbon, such as carbon fibers, carbon foams, and amorphous carbon) allow the design of materials such as foams, hydrogels, thin films and complex structures. All these structures present high contact between the components and a surprising homogeneity, which is responsible for their improved electrochemical responses. On the other hand, the combinations of graphene with carbon nanotubes or other allotropes of carbons are not so common for metal-ion batteries, with few reports describing poor electrochemical performance. Combinations of MXene and CNTs (or either other nanocarbon structures) were not found in recent years for application in metal-ion batteries.

A summary of the main results found in this topic is presented in Table 10.3.

10.2.4 **2D/conjugated polymer nanoarchitectonics**

Another interesting class of architected nanostructure that is often found in the literature is the combination of 2D materials with conjugated polymers, such as polyaniline (PANI) and polypyrrole

Table 10.3 Summary of architected nanostructures based on the combination of 2D materials and carbon nanotubes applied as electrodes in metal-ion batteries.

Nanoarchitectonics	Material 1	Material 2	Battery	Specific capacity (mAh/g)	Current (mA/g)	Coulombic efficiency (%)	References
MWCNT@f-RP@BP	BP	MWCNT	LIBs	560	500	—	[70]
BPC4−c-NaCMC−PAA	BP	MWCNT	LIBs	1681	0.055	87.5	[71]
BPC4−c-NaCMC−PAA	BP	MWCNT	SIBs	2073	0.055	—	[71]
BP/G/CNT	BP	CNT	SIBs	1791	0.055	89.6	[72]
T-rGO/CNT	BP	Graphene	SIBs	926	50	27.5	[73]
V-T-rGO/CNT	BP	Graphene	SIBs	668	50	42.7	[73]
H-rGO/CNT	BP	Graphene	SIBs	780	50	—	[73]
rGO/CNT-30%	BP	Graphene	KIBs	351	10	—	[69]
rGO/CNT-30%	BP	Graphene	LIBs	260	100	—	[69]
MoS_2@CNFIG	MoS_2	Carbon nanofiber/graphene	SIBs	598	100	97.8	[77]
MoS_2/C	MoS_2	Porous carbon	SIBs	1121	200	43.9	[78]
MoS_2/CNT/rGO	MoS_2	CNT/graphene	LIBs	913	100	—	[76]
OMSCF-400	MoS_2	Carbon nanofiber	SIBs	673	100	54	[79]
NCNFs/MoS_2	MoS_2	Carbon nanofiber	LIBs	1592	5000	80.5	[80]
MoS_2-s/CNT-tol	MoS_2	CNT	KIBs	517	7000	100	[74]
MoS_2-s/CNT-chl	MoS_2	CNT	KIBs	492	7000	99	[74]
MoX/CNT-tol	MoS_2	CNT	KIBs	517	7000	100	[74]
1T/2H-MoS_2@NC-CNT	MoS_2	CNT	LIBs	765	200	68.6	[75]

(PPy). PPy is obtained from the oxidative (chemical or electrochemical) polymerization of pyrrole and has high conductivity, intrinsic flexibility, reversible redox electrochemical properties, and a well-known doping/dedoping mechanism. In addition, PPy is widely used to modify electrode materials for batteries and supercapacitors due to its low cost, simple preparation, and good ductility to adapt to the volume changes of electrode materials [82−85]. PANI is one of the most studied and applied conductive polymers, obtained from the oxidative polymerization of aniline, either by chemical or electrochemical routes. PANI is a potential candidate for electrode material in metal-ion batteries due to its high conductivity, high energy storage capacity, easy synthesis, low cost, adjustable physical and chemical properties, and optimal environmental stability. Finally, PANI can effectively reduce the possibility of aggregation of active materials and alleviate the side reactions between the electrolyte and the electrode, bringing better reversibility and cyclic stability for the electrodes [86,87]. However, these pure polymers tend to agglomerate, forming a dense and/or disordered structure, which prevents the penetration of electrolytes and the diffusion of ions, and also aggravates the structural spraying of the polymer during the charge−discharge cycles, resulting in poor stability. The preparation of nanoarchitected structures with other materials (either by in situ polymerization or by a mixture between the components) should be a way to overcome these issues. This concerns specifically the nanoarchitected structures between 2D materials, they offer a large and uniform surface area that can immobilize conductive polymers resulting in incredible nanostructures for charge storage [82−87].

One important point to be highlighted here is that conjugated polymers present an electrochemical profile close to supercapacitors (which means that the energy storage capacity is usually expressed as specific capacitance ($F\ g^{-1}$) instead of the specific capacity characteristic of batteries). This way, several 2D/conjugated polymer nanostructures reported in the literature as electrodes for energy storage devices are also characterized like that, and they will be not considered in this section.

The combinations of conjugated polymers with BP are a very recent topic in literature. Zarbin et al. [88,89] reported the synthesis of a nanoarchitected thin, transparent, and flexible film between BP and PANI using the LLIR [90,91]. The material was obtained by the in situ polymerization of polyaniline over a BP dispersion, in a biphasic liquid/liquid system (Fig. 10.6A). The structure of the nanoarchitected film consisted of PANI-coated BP flakes, which act as an efficient degradation inhibitor, resulting in an impressive increase in the BP lifetime under moisture and oxygen atmosphere, which allows the use of this material as an electrode for SIBs under ambient conditions [88,89]. BP/PANI presented an SC of almost 200 mAh/g at 500 mA/g even after 50 cycles in oxygen/water and a CE of 100%, which remains impressive in a high rate of 3 A/g with an SC of 120 mAh/g, while pure BP was so unstable that no measurements could be recorded. The integrity of the BP was monitored and attested in the composite, certifying its performance in the processes. This was the first report on the possibility of utilization of a BP-based nanoarchitected structure in a water/oxygen-rich environment, and open the possibility to prepare transparent batteries [88,89].

The synergistic effect was remarkable in this nanoarchitected structure: (1) the BP sheets act as a nucleation seed for the polymerization of PANI and/or PPy, resulting in polymer recovering the BP with a more ordered structure, as well as an intimate contact between components; (2) the polymers avoid volume changes in the BP structure, and they help in the conductivity and transport of ions/electrons in the films; (3) both materials have storage capacity, resulting in a high capacity/capacitance; (4) the presented syntheses are easy to obtain, in practically one step and allow the

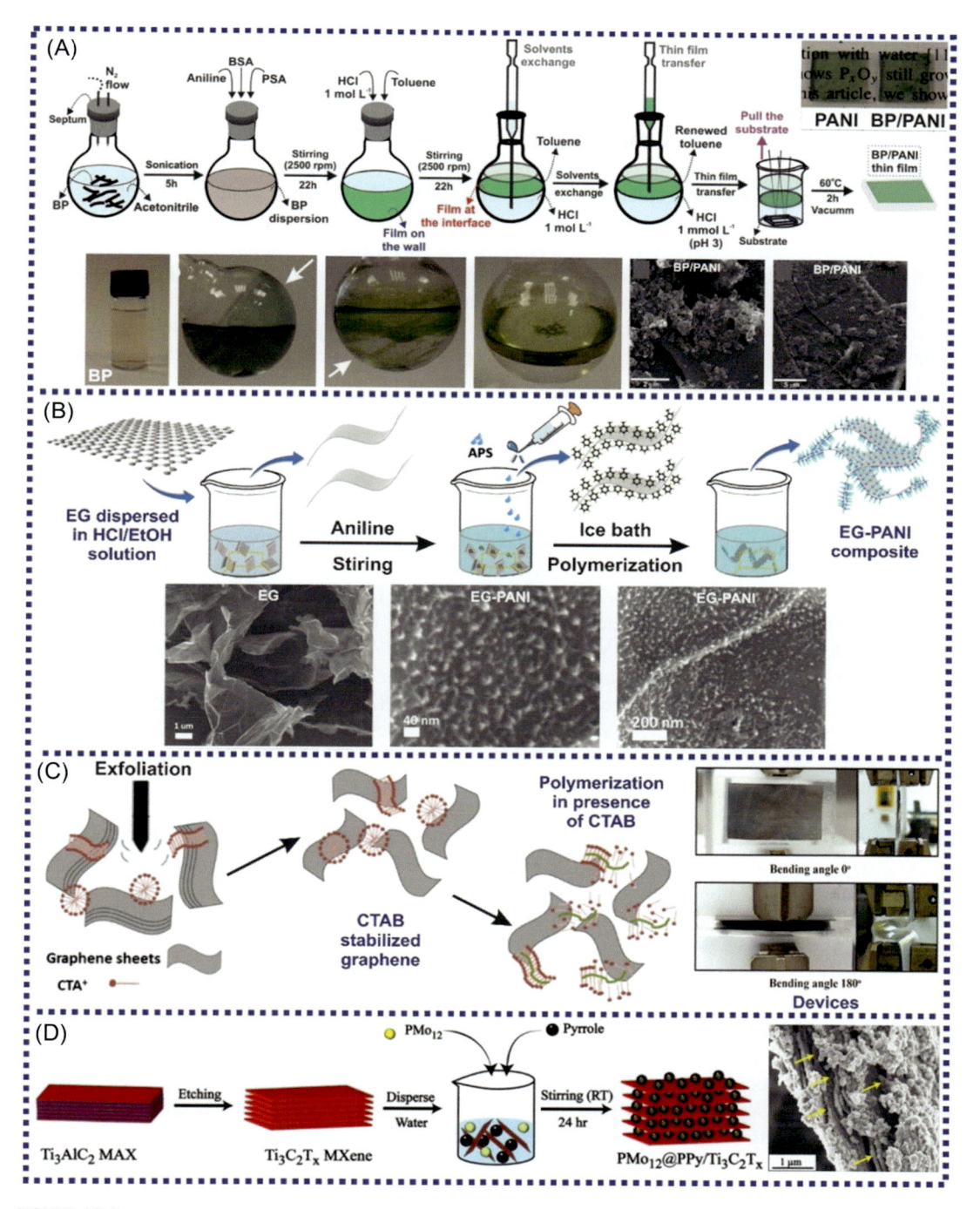

FIGURE 10.6

Scheme of obtaining some architected nanostructures between 2D materials and conjugated polymers, and images of electron microscopy: (A) thin films of BP/PANI prepared by the liquid-liquid interfacial route; (B) electrochemical exfoliated graphene (EG) and PANI; (C) graphene/polypyrrole film preparation (left) and flexible device (right); (D) PMo$_{12}$@PPy/Ti$_3$C$_2$T$_x$ nanocomposite.

From (A) Refs. [88,89], copyright Springer 2017; (B) Ref. [92], copyright Elsevier 2020; (C) Ref. [93], copyright American Chemical Society 2018; (D) Ref. [94], copyright American Chemical Society 2021.

formation of self-supporting thin films; and (5) the coating with polymers blocks the degradation of BP even after many CD cycles or an aggressive environment [88,89].

Nanostructures between needle-type PANI and electrochemically exfoliated graphene were described by Sun et al. (Fig. 10.6B) [92]. The polyaniline was directly polymerized over a graphene dispersion, and reached an SC close to 64 mAh/g at 100 mA/g in full sodium, and noted that different PANI contents have significant effects on electrochemical performance. Devices based on a low amount of PANI presented unsatisfactory results. Otherwise, higher amounts of PANI presented satisfactory results, demonstrating that there is an optimal proportion between the components to obtain a synergistic effect [92]. The electrochemical performance of this sample was superior to that observed for neat graphene or neat polyaniline, and it was attributed to the high conductivity of the graphene, the synergistic effect between the components, and the needle-like morphology of PANI that is beneficial for electron/ion transport in SIBs [92].

Shu et al. [93] presented a new approach for the fabrication of flexible graphene/PPy film using a surfactant (CTAB) as a template for the growth of PPy nanofibers (Fig. 10.6C). Two samples prepared with different graphene/PPy ratios maintained 91% or 98% of the initial capacitance when the current increased from 1 to 4 A/g, better than the observed for neat PPy nanofibers (80%) [88], keeping a retention rate of 85% and 83% after 5000 cycles. To investigate the flexibility, one of the films was bent consecutively at an angle of 180 degrees for 5000 cycles (Fig. 10.6C), then this film was used as an electrode in a device maintaining 92% of retention even after many bending cycles. The device was charged to 1 V using a current of 0.5 A/g, and it was able to maintain 0.8 V after 1.2 hour, 0.6 V after 5.7 hours, and 0.5 V after 15.5 hours [93].

Interesting combinations were also obtained between MoS_2 and PPy. Xie et al. [95] developed a two-step strategy to synthesize MoS_2@PPy core—shell microspheres, in which MoS_2 is homogeneously enveloped by PPy. As a preliminary test, the obtained MoS_2@PPY microspheres exhibit greatly improved electrochemical performance for LIBs, showing SC of 1012 mAh/g after 200 cycles at 200 mA/g, high-rate performance (600 mAh/g at 4 A/g) and initial CE of 79%. On the other hand, worse retention and lower SC are observed for neat MoS_2 (305 mAh/g after 200 cycles at 200 mA/g, 103 mAh/g at 4 A/g, and CE of 56.4%). The superior electrochemical performances of the nanocomposite were attributed to the advantageous combination of hierarchical microsphere structure and to the conductivity of the PPy coating [95]. Yu et al. [96] reported hydrothermally synthesized MoS_2 nanosheets coated with PPy by an in situ vaporization technique. The MoS_2/PPy offers reversible SC of 600, 464, 383, 323, and 200 mAh/g, respectively, at rates of 0.2, 0.5, 1, 2, and 5 A/g for SIBs. This capacity is slightly higher than that of MoS_2, and after two rounds of rate cycling, it still offers a capacity of 300 mAh/g at 2 A/g, indicating electrode stability [96]. The core—shell morphology exposes a large interface between the active material and the electrolyte, thus enhancing the electrochemical reaction and ion transport. PPy shell is conductive and resilient and helps to delay the dissolution of polysulfides, maintaining good contact between MoS_2 and electrolyte. Also, PPy effectively dampens volume expansion and prevents MoS_2 aggregation. The electrochemical performance of the combination of MoS_2 with PPy demonstrates a storage capacity that comes from MoS_2, but the synergism obtained with the presence of the polymer makes the difference in the final result [95,96].

Combinations between MXene and PPy were recently reported by Wu et al. [84]. The synthesis of Ti_3C_2/PPy was performed via in situ polymerization of pyrrole resulting in PPy nanoparticles homogeneously dispersed into the Ti_3C_2 nanosheets, presenting a good performance for SIBs [84].

Mahajan et al. [94] synthesized a tricomponent nanoarchitected structure characterized by a polyoxomolybdate encapsulated into PPy decorated with $Ti_3C_2T_x$ ($PMo_{12}@PPy/Ti_3C_2T_x$), by an electrostatic intercalation method (Fig. 10.6D). $PMo_{12}@PPy/Ti_3C_2T_x$ exhibits SC initial of 800 mAh/g at 100 mA/gfor LIBs, with an initial CE of 50%, due to SEI layer formation. The reversible SC of $PMo_{12}@PPy/Ti_3C_2T_x$ shows an upward trend from 327 mAh/g after 50 cycles to 764 mAh/g after 300 cycles, with a high CE of 100%. The continuous increase of the capacity indicates an electrode reactivation induced by the intercalation processes. In contrast, low SC values were obtained for the bicomponent $PPy/Ti_3C_2T_x$ or to the neat $Ti_3C_2T_x$ (200 and 138 mAh/g at 100 mA/g). The higher performance of the $PMo_{12}@PPy/Ti_3C_2T_x$ was attributed to the combined effects of the capacitive properties of PPy and $Ti_3C_2T_x$ and the reversible redox reactions of molybdenum ions during the charge/discharge cycles [94]. The combinations of PPy with MXene use the idea of intercalating the polymer in the layers of the 2D material. The electrochemical performance of these nanostructures is favored due to the presence of PPy expanding the spaces between the MXene layers, increasing the sites for intercalation and avoiding aggregation and large volume variations. In addition, the polymeric chains that are aligned and the MXene layers prevent the leaching of PPy nanoparticles, which can provide more paths for ion/electron diffusion, reducing the resistance to charge transfer. Finally, the nanostructures obtained in these works allow alteration of the structure during several cycles for better accommodation, increasing the storage capacity [84,94].

The ternary combination of materials, as the one described before, is an emerging approach to efficient electrodes for metal-ion batteries, and a deeper discussion about this interesting and modern class of nanoarchitected materials is out of the scope of this chapter. Table 10.4 summarizes some information regarding the 2D/conjugated polymer materials presented in this section.

Table 10.4 Summary of architected nanostructures between 2D materials and conjugated polymers applied as electrodes in metal-ion batteries.

Nanoarchitectonics	Material 1	Material 2	Battery	Specific capacity (mAh/g)	Current (mA/g)	Coulombic efficiency (%)	References
BP-PANI	BP	PAni	SIBs	200	500	100	[88]
BP/PPy	BP	PPy	–	497 F/g	1000	–	[82]
EG-PANI	Graphene	PAni	SIBs	736 F/g	200	82	[92]
MoS_2@PPY	MoS_2	PPy	LIBs	1012	200	79	[95]
MoS_2-PPy	MoS_2	PPy	SIBs	600	200	–	[96]
Ti_3C_2/PPy	MXene	PPy	–	139 F/g	10	83.3	[84]
$PMo_{12}@PPy/$ $Ti_3C_2T_x$	MXene	PPy	LIBs	800	100	100	[94]
ICN/MoS_2/PANI	MoS_2/ carbon	PAni	LIBs	1293	50	53	[87]
C@MoS_2@PPy	MoS_2/ carbon	PPy	SIBs	1100	100	60,7	[85]
CNT/MoS_2@PPy	MoS_2/ CNT	PPy	LIBs	1683	200	62	[83]

10.3 Conclusions

We have provided here a summary of some examples on the utilization of different 2D materials-based nanoarchitected structures for specific applications as electrodes in metal-ion batteries. It is evident that the properties and performance of the multicomponent electrodes can be strongly modulated by a judicious selection of the materials to be combined, followed by fine control of the nanoarchitecture. Novel and superlative properties which lead to improved performance arise from the synergism between the architected components. All those characteristics are strongly dependent on the way in which these materials are synthesized and processed, which means that variations in the experimental conditions can produce different materials, even when they have the same composition. This is an emerging field with many opportunities for chemists, physicists, engineers, and material scientists.

Although we have focused here on the recent developments (last 5 years) of nanoarchitected structures based on four families of 2D materials and TMOs, two conjugated polymers, and carbon nanotubes, it is important to note that several other combinations can be found with other materials, such as Prussian blue and analogs, other carbonaceous compounds, metal−organic frameworks, covalent organic frameworks, other classes of polymers, metal nanoparticles, and hydroxide nanoparticles, among others. The possibilities for altering their combinations as well as the way in which they can be nanoarchitected is enormous.

In which concerns the specific application as electrodes, almost 60% of the papers discussed in this chapter have focused on LIBs. The 2D/2D and the 2D/carbon nanotubes combinations are those in which SIBs and KIBs are most evident. In addition, a huge potential for aqueous batteries arises from these materials, as well as for new technologies requiring transparency and flexibility.

Finally, the majority of the energy storage devices prepared with these nanoarchitected materials were engineered in a symmetrical sandwich structure, or in other words, the same electrodes were used as cathodes and anodes. Certainly, the combination of different electrodes, nanoarchitected to improve the specific characteristics of the cathode and the anode, will afford modern devices with high energy and power density.

Acknowledgments

The authors thank CNPq, CAPES, INCT-Nanocarbon, and INCT NanoVida for the financial support. MKR thanks CAPES for the fellowship.

References

[1] B. Wang, T. Ruan, Y. Chen, F. Jin, L. Peng, Y. Zhou, et al., Graphene-based composites for electrochemical energy storage, Energy Storage Mater. 24 (2020) 22−51. Available from: https://doi.org/10.1016/j.ensm.2019.08.004.

[2] J. Pang, A. Bachmatiuk, Y. Yin, B. Trzebicka, L. Zhao, L. Fu, et al., Applications of phosphorene and black phosphorus in energy conversion and storage devices, Adv. Energy Mater. 8 (2018) 1−43. Available from: https://doi.org/10.1002/aenm.201702093.

[3] R. Rajagopalan, Y. Tang, X. Ji, C. Jia, H. Wang, Advancements and challenges in potassium ion batteries: a comprehensive review, Adv. Funct. Mater. 30 (2020) 1−35. Available from: https://doi.org/10.1002/adfm.201909486.

[4] K. Khan, A.K. Tareen, M. Aslam, A. Mahmood, Q. khan, Y. Zhang, et al., Going green with batteries and supercapacitor: two dimensional materials and their nanocomposites based energy storage applications, Prog. Solid. State Chem. 58 (2020) 100254. Available from: https://doi.org/10.1016/j.progsolidstchem.2019.100254.

[5] T. Perveen, M. Siddiq, N. Shahzad, R. Ihsan, A. Ahmad, M.I. Shahzad, Prospects in anode materials for sodium ion batteries − a review, Renew. Sustain. Energy Rev. 119 (2020) 109549. Available from: https://doi.org/10.1016/j.rser.2019.109549.

[6] V. Georgakilas, J.N. Tiwari, K.C. Kemp, J.A. Perman, A.B. Bourlinos, K.S. Kim, et al., Noncovalent functionalization of graphene and graphene oxide for energy materials, biosensing, catalytic, and biomedical applications, Chem. Rev. 116 (2016) 5464−5519. Available from: https://doi.org/10.1021/acs.chemrev.5b00620.

[7] F. Wypych, Dissulfeto de molibdênio, um material multifuncional e surpreendente: doze anos depois, Quim. Nova. 37 (2014) 1220−1226. Available from: https://doi.org/10.5935/0100-4042.20140150.

[8] S. Deng, A.V. Sumant, V. Berry, Strain engineering in two-dimensional nanomaterials beyond graphene, Nano Today 22 (2018) 14−35. Available from: https://doi.org/10.1016/j.nantod.2018.07.001.

[9] M. Naguib, O. Mashtalir, J. Carle, V. Presser, J. Lu, L. Hultman, et al., Two-dimensional transition metal carbides, ACS Nano 6 (2012) 1322−1331.

[10] X. Jiang, Y. Chen, X. Meng, W. Cao, C. Liu, Q. Huang, et al., The impact of electrode with carbon materials on safety performance of lithium-ion batteries: a review, Carbon N. Y. 191 (2022) 448−470. Available from: https://doi.org/10.1016/j.carbon.2022.02.011.

[11] S. Mukherjee, Z. Ren, G. Singh, Beyond graphene anode materials for emerging metal ion batteries and supercapacitors, Nano-Micro Lett. 10 (2018) 1−27. Available from: https://doi.org/10.1007/s40820-018-0224-2.

[12] B. Chen, D. Chao, E. Liu, M. Jaroniec, N. Zhao, S.-Z. Qiao, Transition metal dichalcogenides for alkali metal ion batteries: engineering strategies at atomic level, Energy Environ. Sci. 13 (2020) 1096−1131. Available from: https://doi.org/10.1039/C9EE03549D.

[13] Q. Zhang, L. Mei, X. Cao, Y. Tang, Z. Zeng, Intercalation and exfoliation chemistries of transition metal dichalcogenides, J. Mater. Chem. A 8 (2020) 15417−15444. Available from: https://doi.org/10.1039/d0ta03727c.

[14] A.T. Lawal, Graphene-based nano composites and their applications. A review, Biosens. Bioelectron. 141 (2019) 111384. Available from: https://doi.org/10.1016/j.bios.2019.111384.

[15] X.J. Lee, B.Y.Z. Hiew, K.C. Lai, L.Y. Lee, S. Gan, S. Thangalazhy-Gopakumar, et al., Review on graphene and its derivatives: synthesis methods and potential industrial implementation, J. Taiwan. Inst. Chem. Eng. 98 (2019) 163−180. Available from: https://doi.org/10.1016/j.jtice.2018.10.028.

[16] N. Kumar, R. Salehiyan, V. Chauke, O. Joseph Botlhoko, K. Setshedi, M. Scriba, et al., Top-down synthesis of graphene: a comprehensive review, FlatChem 27 (2021) 100224. Available from: https://doi.org/10.1016/j.flatc.2021.100224.

[17] B. Wen, N. Kong, M. Huang, L. Fu, D. Yan, Y. Tian, et al., 1T − 2H MoS_2/Ti_3C_2 MXene heterostructure with high-rate and high- capacity performance for sodium-ion batteries, Energy Fuels 36 (2022) 11234−11244.

[18] D. Gupta, V. Chauhan, R. Kumar, A comprehensive review on synthesis and applications of molybdenum disulfide (MoS_2) material: past and recent developments, Inorg. Chem. Commun. 121 (2020) 108200. Available from: https://doi.org/10.1016/j.inoche.2020.108200.

[19] R. Kumar, S. Sahoo, E. Joanni, R.K. Singh, R.M. Yadav, R.K. Verma, et al., A review on synthesis of graphene, h-BN and MoS_2 for energy storage applications: recent progress and perspectives, Nano Res. 12 (2019) 2655−2694. Available from: https://doi.org/10.1007/s12274-019-2467-8.

[20] S. Kuriakose, T. Ahmed, S. Balendhran, V. Bansal, S. Sriram, M. Bhaskaran, et al., Black phosphorus: ambient degradation and strategies for protection, 2D Mater. 5 (2018) 032001. Available from: https://doi.org/10.1088/2053-1583/aab810.

[21] Y. Sui, J. Zhou, X. Wang, L. Wu, S. Zhong, Y. Li, Recent advances in black-phosphorus-based materials for electrochemical energy storage, Mater. Today. 42 (2021) 117−136. Available from: https://doi.org/10.1016/j.mattod.2020.09.005.

[22] A. Zhang, R. Liu, J. Tian, W. Huang, J. Liu, MXene-based nanocomposites for energy conversion and storage applications, Chem. − A Eur. J. 26 (2020) 6342−6359. Available from: https://doi.org/10.1002/chem.202000191.

[23] T. Kshetri, D.T. Tran, H.T. Le, D.C. Nguyen, H. Van Hoa, N.H. Kim, et al., Recent advances in MXene-based nanocomposites for electrochemical energy storage applications, Prog. Mater. Sci. 117 (2021) 100733. Available from: https://doi.org/10.1016/j.pmatsci.2020.100733.

[24] M. Zheng, H. Tang, L. Li, Q. Hu, L. Zhang, H. Xue, et al., Hierarchically nanostructured transition metal oxides for lithium-ion batteries, Adv. Sci. 5 (2018) 1700592. Available from: https://doi.org/10.1002/advs.201700592.

[25] Y. Liang, W.H. Lai, Z. Miao, S.L. Chou, Nanocomposite materials for the sodium−ion battery: a review, Small 14 (2018) 1−20. Available from: https://doi.org/10.1002/smll.201702514.

[26] G. Zhang, X. Xiao, B. Li, P. Gu, H. Xue, H. Pang, Transition metal oxides with one-dimensional/one-dimensional-analogue nanostructures for advanced supercapacitors, J. Mater. Chem. A 5 (2017) 8155−8186. Available from: https://doi.org/10.1039/c7ta02454a.

[27] R. Meng, J. Huang, Y. Feng, L. Zu, C. Peng, L. Zheng, et al., Black phosphorus quantum dot/Ti_3C_2 MXene nanosheet composites for efficient electrochemical lithium/sodium-ion storage, Adv. Energy Mater. 8 (2018) 1−10. Available from: https://doi.org/10.1002/aenm.201801514.

[28] H. Li, A. Liu, X. Ren, Y. Yang, L. Gao, M. Fan, et al., A black phosphorus/Ti_3C_2 MXene nanocomposite for sodium-ion batteries: a combined experimental and theoretical study, Nanoscale 11 (2019) 19862−19869. Available from: https://doi.org/10.1039/c9nr04790e.

[29] R. Zhao, Z. Qian, Z. Liu, D. Zhao, X. Hui, G. Jiang, et al., Molecular-level heterostructures assembled from layered black phosphorene and Ti_3C_2 MXene as superior anodes for high-performance sodium ion batteries, Nano Energy 65 (2019) 104037. Available from: https://doi.org/10.1016/j.nanoen.2019.104037.

[30] H. Shuai, P. Ge, W. Hong, S. Li, J. Hu, H. Hou, et al., Electrochemically exfoliated phosphorene−graphene hybrid for sodium-ion batteries, Small Methods 3 (2019) 1−12. Available from: https://doi.org/10.1002/smtd.201800328.

[31] Y. Li, H. Mao, C. Zheng, J. Wang, Z. Che, M. Wei, Compositing reduced graphene oxide with interlayer spacing enlarged MoS_2 for performance enhanced sodium-ion batteries, J. Phys. Chem. Solids 136 (2020) 109163. Available from: https://doi.org/10.1016/j.jpcs.2019.109163.

[32] H. Liu, Y. Zou, L. Tao, Z. Ma, D. Liu, P. Zhou, et al., Sandwiched thin-film anode of chemically bonded black phosphorus/graphene hybrid for lithium-ion battery, Small 13 (2017) 23−25. Available from: https://doi.org/10.1002/smll.201700758.

[33] J. Zhang, H. Shin, W. Lu, Top-down ultrasonication-assisted exfoliation for prebonded phosphorene-graphene heterostructures enabling fast lithiation/delithiation, ACS Appl. Mater. Interfaces. 13 (2021) 25946−25959. Available from: https://doi.org/10.1021/acsami.1c03583.

[34] M. Li, N. Muralidharan, K. Moyer, C.L. Pint, Solvent mediated hybrid 2D materials: black phosphorus-graphene heterostructured building blocks assembled for sodium ion batteries, Nanoscale 10 (2018) 10443−10449. Available from: https://doi.org/10.1039/c8nr01400k.

[35] J. Mei, T. He, Q. Zhang, T. Liao, A. Du, G.A. Ayoko, et al., Carbon − phosphorus bonds-enriched 3D graphene by self- sacrificing black phosphorus nanosheets for elevating capacitive lithium storage, ACS Appl. Mater. Interfaces. 12 (2020) 21720−21729.

[36] Y. Shi, Z. Yi, Y. Kuang, H. Guo, Y. Li, C. Liu, et al., Constructing stable covalent bonding in black phosphorus/reduced graphene oxide for lithium ion battery anodes, Chem. Commun. 56 (2020) 11613−11616. Available from: https://doi.org/10.1039/d0cc03698f.

[37] Y. Liu, Q. Liu, A. Zhang, J. Cai, X. Cao, Z. Li, et al., Room-temperature pressure synthesis of layered black phosphorus-graphene composite for sodium-ion battery anodes, ACS Nano 12 (2018) 8323−8329. Available from: https://doi.org/10.1021/acsnano.8b03615.

[38] E. Brown, P. Yan, H. Tekik, A. Elangovan, J. Wang, D. Lin, et al., 3D printing of hybrid MoS_2-graphene aerogels as highly porous electrode materials for sodium ion battery anodes, Mater. Des. 170 (2019) 107689. Available from: https://doi.org/10.1016/j.matdes.2019.107689.

[39] J. Yuan, J. Zhu, R. Wang, Y. Deng, S. Zhang, C. Yao, et al., 3D few-layered MoS_2/graphene hybrid aerogels on carbon fiber papers: a free-standing electrode for high-performance lithium/sodium-ion batteries, Chem. Eng. J. 398 (2020) 125592. Available from: https://doi.org/10.1016/j.cej.2020.125592.

[40] S. Mateti, M.M. Rahman, P. Cizek, Y. Chen, In situ production of a two-dimensional molybdenum disulfide/graphene hybrid nanosheet anode for lithium-ion batteries, RSC Adv. 10 (2020) 12754−12758. Available from: https://doi.org/10.1039/d0ra01503b.

[41] V.O. Koroteev, S.G. Stolyarova, A.A. Kotsun, E. Modin, A.A. Makarova, Y.V. Shubin, et al., Nanoscale coupling of MoS_2 and graphene via rapid thermal decomposition of ammonium tetrathiomolybdate and graphite oxide for boosting capacity of Li-ion batteries, Carbon N. Y. 173 (2021) 194−204. Available from: https://doi.org/10.1016/j.carbon.2020.10.097.

[42] Z. Xiao, L. Sheng, L. Jiang, Y. Zhao, M. Jiang, X. Zhang, et al., Nitrogen-doped graphene ribbons/MoS_2 with ultrafast electron and ion transport for high-rate Li-ion batteries, Chem. Eng. J. 408 (2021) 127269. Available from: https://doi.org/10.1016/j.cej.2020.127269.

[43] C. Shen, L. Wang, A. Zhou, H. Zhang, Z. Chen, Q. Hu, et al., MoS_2-decorated 2D Ti3C2 (MXene): a high-performance anode material for lithium-ion batteries, J. Electrochem. Soc. 164 (2017) A2654−A2659. Available from: https://doi.org/10.1149/2.1421712jes.

[44] C. Chen, X. Xie, B. Anasori, A. Sarycheva, T. Makaryan, M. Zhao, et al., MoS_2-on-MXene heterostructures as highly reversible anode materials for lithium-ion batteries, Angew. Chem. − Int. Ed. 57 (2018) 1846−1850. Available from: https://doi.org/10.1002/anie.201710616.

[45] J. Li, B. Rui, W. Wei, P. Nie, L. Chang, Z. Le, et al., Nanosheets assembled layered MoS_2/MXene as high performance anode materials for potassium ion batteries, J. Power Sources 449 (2020) 227481. Available from: https://doi.org/10.1016/j.jpowsour.2019.227481.

[46] X. Zhang, H. Shi, L. Liu, C. Min, S. Liang, Z. Xu, et al., Construction of MoS_2/Mxene heterostructure on stress-modulated kapok fiber for high-rate sodium-ion batteries, J. Colloid Interface Sci. 605 (2022) 472−482. Available from: https://doi.org/10.1016/j.jcis.2021.07.097.

[47] Z. Yuan, J. Cao, S. Valerii, H. Xu, L. Wang, W. Han, MXene-bonded hollow MoS_2/carbon sphere strategy for high-performance flexible sodium ion storage, Chem. Eng. J. 430 (2022) 132755. Available from: https://doi.org/10.1016/j.cej.2021.132755.

[48] Z. Pan, L. Kang, T. Li, M. Waqar, J. Yang, Q. Gu, et al., Black phosphorus@$Ti_3C_2T_x$MXene composites with engineered chemical bonds for commercial-level capacitive energy storage, ACS Nano 15 (2021) 12975−12987. Available from: https://doi.org/10.1021/acsnano.1c01817.

[49] J. Yang, Z. Pan, J. Zhong, S. Li, J. Wang, P.Y. Chen, Electrostatic self-assembly of heterostructured black phosphorus−MXene nanocomposites for flexible microsupercapacitors with high rate performance, Energy Storage Mater. 36 (2021) 257−264. Available from: https://doi.org/10.1016/j.ensm.2020.12.025.

[50] A. Ali, K. Hantanasirisakul, A. Abdala, P. Urbankowski, M.Q. Zhao, B. Anasori, et al., Effect of synthesis on performance of MXene/iron oxide anode material for lithium-ion batteries, Langmuir 34 (2018) 11325−11334. Available from: https://doi.org/10.1021/acs.langmuir.8b01953.

[51] F. Kong, X. He, Q. Liu, X. Qi, D. Sun, Y. Zheng, et al., Further surface modification by carbon coating for in-situ growth of Fe_3O_4 nanoparticles on MXene Ti_3C_2 multilayers for advanced Li-ion storage, Electrochim. Acta. 289 (2018) 228−237. Available from: https://doi.org/10.1016/j.electacta.2018.09.007.

[52] Y. Luo, X. Ding, X. Ma, D. Liu, H. Fu, X. Xiong, Constructing $MoO_2@MoS_2$ heterostructures anchored on graphene nanosheets as a high-performance anode for sodium ion batteries, Electrochim. Acta. 388 (2021) 138612. Available from: https://doi.org/10.1016/j.electacta.2021.138612.

[53] S. Gu, A. Zhu, Graphene nanosheets loaded Fe_3O_4 nanoparticles as a promising anode material for lithium ion batteries, J. Alloy. Compd. 813 (2020) 152160. Available from: https://doi.org/10.1016/j.jallcom.2019.152160.

[54] N. Wu, J. Shen, L. Sun, M. Yuan, Y. Shao, J. Ma, et al., Hierarchical N-doped graphene coated 1D cobalt oxide microrods for robust and fast lithium storage at elevated temperature, Electrochim. Acta. 310 (2019) 70−77. Available from: https://doi.org/10.1016/j.electacta.2019.04.115.

[55] Z. Zhang, H. Guo, W. Li, G. Liu, Y. Zhang, Y. Wang, Sandwich-like Co_3O_4/MXene composites as high capacity electrodes for lithium-ion batteries, N. J. Chem. 44 (2020) 5913−5920. Available from: https://doi.org/10.1039/c9nj06072c.

[56] Z. Guo, D. Wang, Z. Wang, Y. Gao, J. Liu, A free-standing α-MoO_3/MXene composite anode for high-performance lithium storage, Nanomaterials 12 (2022) 1422. Available from: https://doi.org/10.3390/nano12091422.

[57] J. Wang, H. Zhou, M. Zhu, A. Yuan, X. Shen, Metal-organic framework-derived Co_3O_4 covered by MoS_2 nanosheets for high-performance lithium-ion batteries, J. Alloy. Compd. 744 (2018) 220−227. Available from: https://doi.org/10.1016/j.jallcom.2018.02.086.

[58] C.-Y. Wu, W.-E. Chang, Y.-G. Sun, J.-M. Wu, J.-G. Duh, Three-dimensional S-$MoS_2@$α-Fe_2O_3 nanoparticles composites as lithium- ion battery anodes for enhanced electrochemical performance, Mater. Chem. Phys. (2018).

[59] S. Zhao, Z. Zha, X. Liu, H. Tian, Z. Wu, W. Li, et al., Core − sheath structured $MoO_3@MoS_2$ composite for high- performance lithium-ion battery anodes, Energy Fuels (2020).

[60] R. Kumar, R.K. Singh, A.V. Alaferdov, S.A. Moshkalev, Rapid and controllable synthesis of Fe_3O_4 octahedral nanocrystals embedded-reduced graphene oxide using microwave irradiation for high performance lithium-ion batteries, Electrochim. Acta. 281 (2018) 78−87. Available from: https://doi.org/10.1016/j.electacta.2018.05.157.

[61] Y. Li, C.N. Duan, Z. Jiang, X. Bin Zhou, Y. Wang, CuO/rGO nanocomposite as an anode material for high-performance lithium-ion batteries, Mater. Res. Exp. 8 (2021) 055505. Available from: https://doi.org/10.1088/2053-1591/abfe2f.

[62] Y. Zhang, W. Liu, Y. Zhu, Y. Zhang, R. Zhang, K. Li, et al., Facile self-assembly solvothermal preparation of CuO/Cu_2O/coal-based reduced graphene oxide nanosheet composites as an anode for high-performance lithium-ion batteries, Energy Fuels 35 (2021) 8961−8969. Available from: https://doi.org/10.1021/acs.energyfuels.1c00473.

[63] Y. Sheng, Y. Wang, B. Lan, X. Zhang, C. Wei, G. Wen, A high capacity porous $Co_3O_4@$graphene composite as lithium battery anode, Vacuum 203 (2022) 111266. Available from: https://doi.org/10.1016/j.vacuum.2022.111266.

[64] N. Naresh, P. Jena, N. Satyanarayana, Facile synthesis of MoO_3/rGO nanocomposite as anode materials for high performance lithium-ion battery applications, J. Alloy. Compd. 810 (2019) 151920. Available from: https://doi.org/10.1016/j.jallcom.2019.151920.

[65] P. Almodóvar, M.L. López, J. Ramírez-Castellanos, S. Nappini, E. Magnano, J.M. González-Calbet, et al., Synthesis, characterization and electrochemical assessment of hexagonal molybdenum trioxide (h-MoO_3) micro-composites with graphite, graphene and graphene oxide for lithium ion batteries, Electrochim. Acta. 365 (2021) 137355. Available from: https://doi.org/10.1016/j.electacta.2020.137355.

[66] M. Li, Z. Li, X. Wang, N. Fu, Z. Yang, MoS_2 nanosheet loaded Fe_2O_3@ carbon cloth flexible composite electrode material for quasi-solid asymmetric supercapacitors, J. Electroanal. Chem. 919 (2022) 116556. Available from: https://doi.org/10.1016/j.jelechem.2022.116556.

[67] H. Li, Y. Liu, S. Lin, H. Li, Z. Wu, L. Zhu, et al., Laser crystallized sandwich-like MXene/Fe_3O_4/MXene thin film electrodes for flexible supercapacitors, J. Power Sources 497 (2021) 229882. Available from: https://doi.org/10.1016/j.jpowsour.2021.229882.

[68] Y. Zhang, L. Wang, H. Xu, J. Cao, D. Chen, W. Han, 3D chemical cross-linking structure of black phosphorus@CNTs hybrid as a promising anode material for lithium ion batteries, Adv. Funct. Mater. 30 (2020) 1−12. Available from: https://doi.org/10.1002/adfm.201909372.

[69] S. Peng, L. Wang, Z. Zhu, K. Han, Electrochemical performance of reduced graphene oxide/carbon nanotube hybrid papers as binder-free anodes for potassium-ion batteries, J. Phys. Chem. Solids 138 (2020) 109296. Available from: https://doi.org/10.1016/j.jpcs.2019.109296.

[70] D. Zhao, L. Zhang, C. Fu, J. Huang, H. Huang, Z. Li, et al., Hierarchical phosphorus hybrids with carbon nanotube veins and black phosphorus skins: structure and lithium storage properties, Carbon N. Y. 139 (2018) 1057−1062. Available from: https://doi.org/10.1016/j.carbon.2018.08.005.

[71] S. Haghighat-Shishavan, M. Nazarian-Samani, M. Nazarian-Samani, H.K. Roh, K.Y. Chung, B.W. Cho, et al., Strong, persistent superficial oxidation-assisted chemical bonding of black phosphorus with multi-wall carbon nanotubes for high-capacity ultradurable storage of lithium and sodium, J. Mater. Chem. A 6 (2018) 10121−10134. Available from: https://doi.org/10.1039/c8ta02590h.

[72] L. Dang, J. He, H. Wei, Black phosphorus/nanocarbons constructing a dual-carbon conductive network for high-performance sodium-ion batteries, Trans. Tianjin Univ. (2021).

[73] J. Feng, L. Dong, X. Li, D. Li, P. Lu, F. Hou, et al., Hierarchically stacked reduced graphene oxide/carbon nanotubes for as high performance anode for sodium-ion batteries, Electrochim. Acta. 302 (2019) 65−70. Available from: https://doi.org/10.1016/j.electacta.2019.02.008.

[74] A. Schmidt, M.K. Ramos, C.M. Ferreira, B.A. Braz, A.J.G. Zarbin, Molybdenum-based materials/carbon nanotubes nanocomposites prepared as thin and transparent films for aqueous K-ion batteries, Electrochim. Acta. 387 (2021) 138500. Available from: https://doi.org/10.1016/j.electacta.2021.138500.

[75] J. Wang, L. Sun, H. Tan, F. Xie, Y. Qu, J. Hu, et al., Double-phase 1T/2H−MoS_2 heterostructure loaded in N-doped carbon/CNT complex carbon for efficient and rapid lithium storage, Mater. Today Energy. 29 (2022) 101103. Available from: https://doi.org/10.1016/j.mtener.2022.101103.

[76] H. Ma, S. Du, H. Tao, T. Li, Y. Zhang, Three-dimensionally integrated carbon tubes/MoS_2 with reduced graphene oxide foam as a binder-free anode for lithium ion battery, J. Electroanal. Chem. 823 (2018) 307−314. Available from: https://doi.org/10.1016/j.jelechem.2018.06.033.

[77] M. Liu, P. Zhang, Z. Qu, Y. Yan, C. Lai, T. Liu, et al., Conductive carbon nanofiber interpenetrated graphene architecture for ultra-stable sodium ion battery, Nat. Commun. 10 (2019). Available from: https://doi.org/10.1038/s41467-019-11925-z.

[78] Y.Y. Hu, Y.L. Bai, X.Y. Wu, X. Wei, K.X. Wang, J.S. Chen, MoS_2 nanoflakes integrated in a 3D carbon framework for high-performance sodium-ion batteries, J. Alloy. Compd. 797 (2019) 1126−1132. Available from: https://doi.org/10.1016/j.jallcom.2019.05.142.

[79] Y. Zhang, H. Tao, T. Li, S. Du, J. Li, Y. Zhang, et al., Vertically oxygen-incorporated MoS_2 nanosheets coated on carbon fibers for sodium-ion batteries, ACS Appl. Mater. Interfaces. 10 (2018) 35206−35215. Available from: https://doi.org/10.1021/acsami.8b12079.

[80] X. Ni, H. Chen, C. Liu, F. Zeng, H. Yu, A. Ju, A freestanding nitrogen-doped carbon nanofiber/MoS_2 nanoflowers with expanded interlayer for long cycle-life lithium-ion batteries, J. Alloy. Compd. 818 (2020) 152835. Available from: https://doi.org/10.1016/j.jallcom.2019.152835.

[81] A.J.G. Zarbin, Liquid-liquid interfaces: a unique and advantageous environment to prepare and process thin films of complex materials, Mater. Horizons. 8 (2021) 1409−1432. Available from: https://doi.org/10.1039/d0mh01676d.

[82] S. Luo, J. Zhao, J. Zou, Z. He, C. Xu, F. Liu, et al., Self-standing polypyrrole/black phosphorus laminated film: promising electrode for flexible supercapacitor with enhanced capacitance and cycling stability, ACS Appl. Mater. Interfaces. 10 (2018) 3538−3548. Available from: https://doi.org/10.1021/acsami.7b15458.

[83] J. Wang, L. Sun, Y. Gong, L. Wu, C. Sun, X. Zhao, et al., A CNT/ MoS_2@PPy composite with double electron channels and boosting charge transport for high-rate lithium storage, Appl. Surf. Sci. 566 (2021) 150693. Available from: https://doi.org/10.1016/j.apsusc.2021.150693.

[84] W. Wu, D. Wei, J. Zhu, D. Niu, F. Wang, L. Wang, et al., Enhanced electrochemical performances of organ-like Ti_3C_2 MXenes/polypyrrole composites as supercapacitors electrode materials, Ceram. Int. 45 (2019) 7328−7337. Available from: https://doi.org/10.1016/j.ceramint.2019.01.016.

[85] G. Wang, X. Bi, H. Yue, R. Jin, Q. Wang, S. Gao, et al., Sacrificial template synthesis of hollow C@MoS_2@PPy nanocomposites as anodes for enhanced sodium storage performance, Nano Energy 60 (2019) 362−370. Available from: https://doi.org/10.1016/j.nanoen.2019.03.065.

[86] C. Li, S. Wang, Y. Cui, X. Wang, Z. Yong, D. Liang, et al., Sandwich-like high-load MXene/polyaniline film electrodes with ultrahigh volumetric capacitance for flexible supercapacitors, J. Colloid Interface Sci. 620 (2022) 35−46. Available from: https://doi.org/10.1016/j.jcis.2022.03.147.

[87] W. Li, Y. Liu, S. Zheng, G. Hu, K. Zhang, Y. Luo, et al., Hybrid structures of sisal fiber derived interconnected carbon nanosheets/ MoS_2/polyaniline as advanced electrode materials in lithium-ion batteries, Molecules 26 (2021) 3710. Available from: https://doi.org/10.3390/molecules26123710.

[88] J.E.S. Fonsaca, S.H. Domingues, E.S. Orth, A.J.G. Zarbin, A black phosphorus-based cathode for aqueous Na-ion batteries operating under ambient conditions, Chem. Commun. 56 (2020) 802−805. Available from: https://doi.org/10.1039/c9cc09279j.

[89] J.E.S. Fonsaca, S.H. Domingues, E.S. Orth, A.J.G. Zarbin, Air stable black phosphorous in polyaniline-based nanocomposite, Sci. Rep. 7 (2017) 1−9. Available from: https://doi.org/10.1038/s41598-017-10533-5.

[90] S. Husmann, M.K. Ramos, A.J.G. Zarbin, Transparent aqueous rechargeable sodium-ion battery, Electrochim. Acta. 422 (2022) 140548. Available from: https://doi.org/10.1016/j.electacta.2022.140548.

[91] A. Schmidt, M.K. Ramos, C.S. Pinto, A.F. Pereira, V.H.R. Souza, A.J.G. Zarbin, Electrode fabrication at liquid interfaces: towards transparency and flexibility, Electrochem. Commun. 134 (2022) 107183. Available from: https://doi.org/10.1016/j.elecom.2021.107183.

[92] Y. Sun, L. Jiao, D. Han, F. Wang, P. Zhang, H. Li, et al., Hierarchical architecture of polyaniline nanoneedle arrays on electrochemically exfoliated graphene for supercapacitors and sodium batteries cathode, Mater. Des. 188 (2020) 108440. Available from: https://doi.org/10.1016/j.matdes.2019.108440.

[93] K. Shu, Y. Chao, S. Chou, C. Wang, T. Zheng, S. Gambhir, et al., A "tandem" strategy to fabricate flexible graphene/polypyrrole nanofiber film using the surfactant-exfoliated graphene for supercapacitors, ACS Appl. Mater. Interfaces. 10 (2018) 22031−22041. Available from: https://doi.org/10.1021/acsami.8b03901.

[94] M. Mahajan, G. Singla, S. Ogale, Polypyrrole-encapsulated polyoxomolybdate decorated MXene as a functional 2D/3D nanohybrid for a robust and high performance Li-ion battery, ACS Appl. Energy Mater. 4 (2021) 4541−4550. Available from: https://doi.org/10.1021/acsaem.1c00175.

[95] D. Xie, M. Zhang, F. Cheng, H. Fan, S. Xie, P. Liu, et al., Hierarchical MoS_2@Polypyrrole core-shell microspheres with enhanced electrochemical performances for lithium storage, Electrochim. Acta. 269 (2018) 632−639. Available from: https://doi.org/10.1016/j.electacta.2018.03.068.

[96] H. Yu, G. Jiang, J. Ni, L. Li, Architecting core-shell nanosheets of MoS_2-polypyrrole on carbon cloth as a robust sodium anode, Sustain. Mater. Technol. 28 (2021) e00255. Available from: https://doi.org/10.1016/j.susmat.2021.e00255.

Thin film nanoarchitectonics via Langmuir—Blodgett and layer-by-layer methods

11

Katsuhiko Ariga[1,2]

[1]*Research Center for Materials Nanoarchitectonics (MANA), National Institute for Materials Science (NIMS), Namiki, Tsukuba, Japan* [2]*Graduate School of Frontier Sciences, The University of Tokyo, Kashiwanoha, Kashiwa, Japan*

11.1 Introduction

Successful development of functional materials depends on contributions from many fields of chemistry, physics, materials science, and so on. Interestingly, not only the intrinsic natures and properties of the materials themselves, but also their internal nanoscale structures, including component configurations, are important for the development of new materials with high functionality. Nanotechnology, as a science and technology that examines structures and phenomena at the nanoscopic level, has developed particularly since the late 20th century. It is a technology (actually science) that enables the observation of structures and the analysis of phenomena at the so-called nanoscale. Based primarily on the development of high-resolution microscopes and related high-level instruments, nanotechnology has stimulated nanoscale research [1,2]. This historical trend in materials science requires further advanced innovation through the fusion of nanotechnology with other materials science disciplines. As a postnanotechnology [3], nanoarchitectonics is a concept that seeks to integrate nanotechnology with other materials creation fields [4,5] (Fig. 11.1).

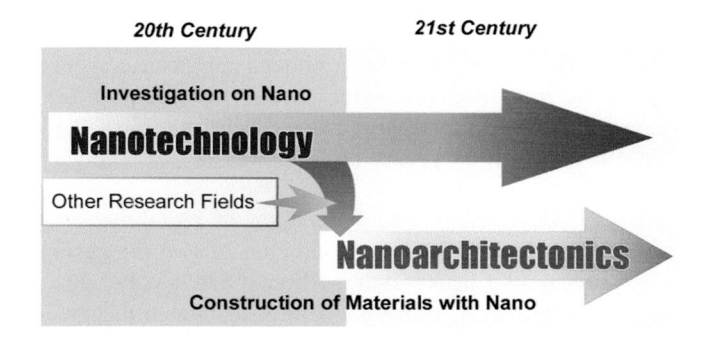

FIGURE 11.1

Outline and history of nanoarchitectonics concept.

Materials Nanoarchitectonics. DOI: https://doi.org/10.1016/B978-0-323-99472-9.00005-5

Nanoarchitectonics combines processes involving human manipulation as well as self-assembly of components, such as the Langmuir—Blodgett (LB) method [6—8] and layer-by-layer (LbL) assembly [9—11]. This makes it particularly suitable for hierarchical structure fabrication [12]. The ability to fabricate hierarchical functional structures through nanoarchitectonics shares the same characteristics of biological systems, where a vast number of functional components are hierarchically organized and their functions are coordinated. Realistically, it is difficult to organize such hierarchical functional systems in a freely deployable three-dimensional medium. From this perspective, nanoarchitectonics offers an attractive methodology to develop functional structures in a defined and dimensionally limited environment. One such attractive environment is the fabrication of structures as thin films [13,14]. According to this background, this chapter introduces the development of the LB and LbL methods and presents an example of building up a hierarchical structure. The use of these thin-film fabrication methods opens the way to form hierarchical structures through nanoarchitectonics.

11.2 Langmuir—Blodgett nanoarchitectonics

The LB method is an old technique with a history going back over 100 years [15,16]. Its principle is that a thin film of water-insoluble molecules spreads on the surface of water as mainly monolayers, just like an everyday event such as oil falling on the surface of water. The first discussion of the relationship between surface tension and monolayer compression is said to have been conducted by a self-taught German woman scientist, Agnes Pockels, using a handmade tub in her kitchen. This was followed by systematic studies of monolayers at the air—water interface by Irving Langmuir. Katharine Blodgett then demonstrated multilayer formation by the simple transfer of monolayers from the water surface to the solid substrate. Thus began the LB technique.

The LB method is considered to be a well-developed traditional method, but several new developments have been recently reported. The LB process is considered to require a vibration-free environment because it deals with delicate monolayers spread over a liquid subphase. Therefore, the LB trough is usually placed on a vibration-proof stabilizer, and the compression and expansion of the monolayer is performed by carefully controlled movement of the compression barrier. Traditionally, a clean and quiet environment has been considered essential for the LB process. However, intense motion at the water—air interface may promote the formation of more organized structures with appropriate shear flow. Therefore, the LB method, in which the aqueous phase is subjected to vigorous vortex motion, was devised and named the vortex LB method [17].

The vortex LB method is applicable to the fabrication of two-dimensional nanocarbon films from molecular precursors [18] (Fig. 11.2). For preparation of two-dimensional carbon materials, the vortex LB method requires only a beaker and a stirrer. Carbon nanorings as molecular precursors were vortex-diffused over water. A uniform two-dimensional film was then transferred onto a solid silicon substrate. The transferred film was further carbonized at 850°C for 3 hours in an N_2 gas atmosphere. The thickness of the film was about 10 nm, and the uniform film morphology with partially porous structure was maintained after carbonization. High-resolution transmission

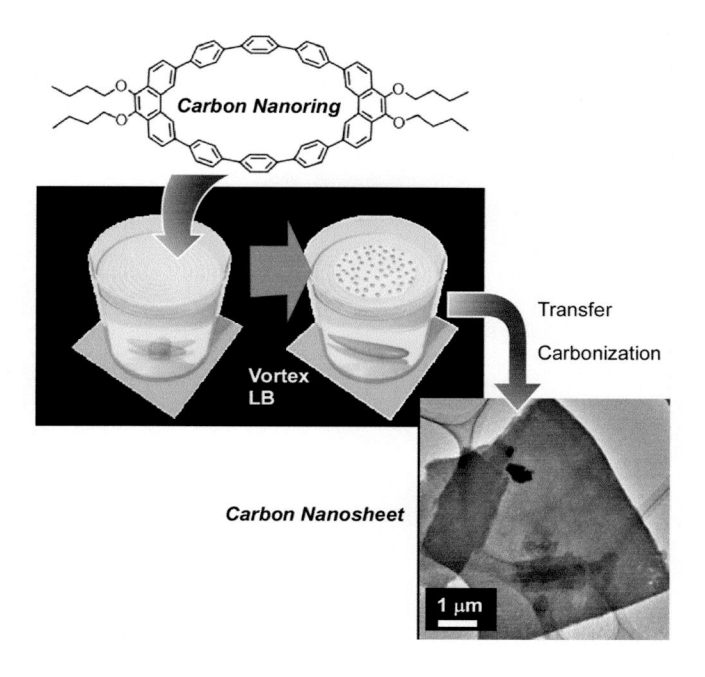

FIGURE 11.2

Vortex Langmuir–Blodgett method for fabrication of two-dimensional nanocarbon films from molecular precursors, carbon nanoring.

electron microscopy observation revealed a turbostratic carbon structure in two dimensions after carbonization.

Water is commonly used as a subphase of the conventional LB method because of its high surface tension, nontoxicity, and harmlessness. The LB method, with water as its main component, is a human-friendly process. At the same time, however, the operating temperature window is only about room temperature to avoid disturbances due to water evaporation and condensation. This temperature limitation is a disadvantage when preparing uniform membranes containing highly cohesive components. In the latter case, it is essential to apply thermal energy to increase the operating temperature and promote the kinetic degrees of freedom of the membrane components. Therefore, a high-temperature LB process (high-temp LB) using nonaqueous subphases is necessary for the preparation of high-quality membranes with aggregation-prone components. As a recent example of a high-temp LB process, the fabrication of highly oriented polymer semiconductor molecules by LB process was demonstrated at around 100°C using ethylene glycol as a subphase [19] (Fig. 11.3).

Highly uniform oriented semiconductor films of about 10 nm thickness were prepared by the high-temperature LB method, and edge-on orientation with high anisotropic conductivity was achieved. Ultrathin films with edge-on orientation are desirable for field effect transistor (FET) applications of polymer semiconductor molecules, but only face-on structures can be

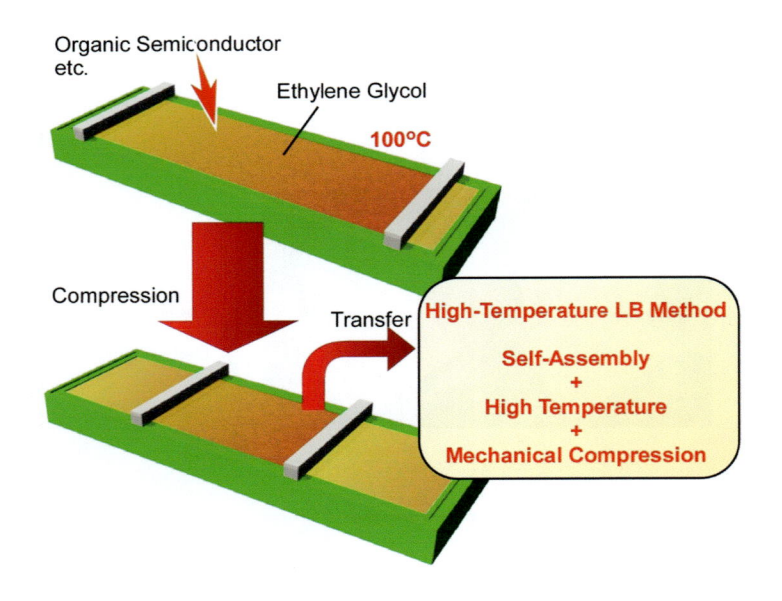

FIGURE 11.3

High-temperature Langmuir—Blodgett process conducted at around 100ₒC using ethylene glycol as a subphase.

obtained by the conventional hot rubbing method, and the uniformity of edge-on orientation is low for films thicker than 100 nm when deposited on ionic liquids at high temperatures. Polymer semiconductor films with both 10 nm level thickness and edge-on orientation can be realized only by the high-temperature LB method. Thus, by expanding the applicable temperature range of the high-temperature LB method, the LB method can be applied to more materials and targets. Recently, automated equipment that can handle temperatures up to 200°C has been manufactured [20].

Analogous to the high-temperature LB method, LB methods at low temperatures (low-temperature LB methods) and at ultrahigh vacuum (high-vacuum LB methods) may also be possible with the appropriate selection of the subphase liquid. Current state-of-the-art science is investigating precise molecular manipulation on solid substrates, at low temperatures, and under ultrahigh vacuum, with minimal disturbance. If low-temperature LB and high-vacuum LB methods are realized, it may be possible to conduct such molecular manipulation studies on a liquid surface. Nanocar racing, the manipulation of molecular-sized cars, is an advanced example of the current science on solid substrates. If low-temperature and high-vacuum LB methods enable molecular manipulation on liquid surfaces under extreme conditions, nanoboat racing and nanosailing racing could become cutting-edge advanced technologies. In addition, the development of the LB concept in biology would have to be considered. While attempts at LB methods using living cells have been reported, more advanced in vivo LB methods and other methods of LB-like manipulation within living organisms may be devised.

11.3 Layer-by-layer nanoarchitectonics

11.3.1 Layer-by-layer basics

We see research papers on LbL assembly in a wide range of fields, from basic science in chemistry and physics to quite practical applications. This confirms that this method is used by researchers with various research backgrounds, but its history is less than 30 years old. In other words, this method is a young but rapidly expanding material organization method. The first person to experimentally demonstrate this method was Gero Decher [21−25], who was at the University of Mainz. It was in the early 1990s that he reported the "self-assembly process" for the preparation of multilayers of charged surfactants and polyelectrolytes. His proposal was subsequently taken up by researchers in Japan, Europe, the United States, China, and other countries, and attracted the attention of many researchers. However, according to Decher, who is credited as the originator, the original concept dates back to Iler's idea of stepwise layering of colloidal particles in 1966 [24].

The reason why the alternating adsorption method has been used by many researchers is due to the simplicity of the method and the diversity of target materials used. Basically, this method is an artificial organization method that skillfully utilizes a simple self-assembly process. Fig. 11.4 shows

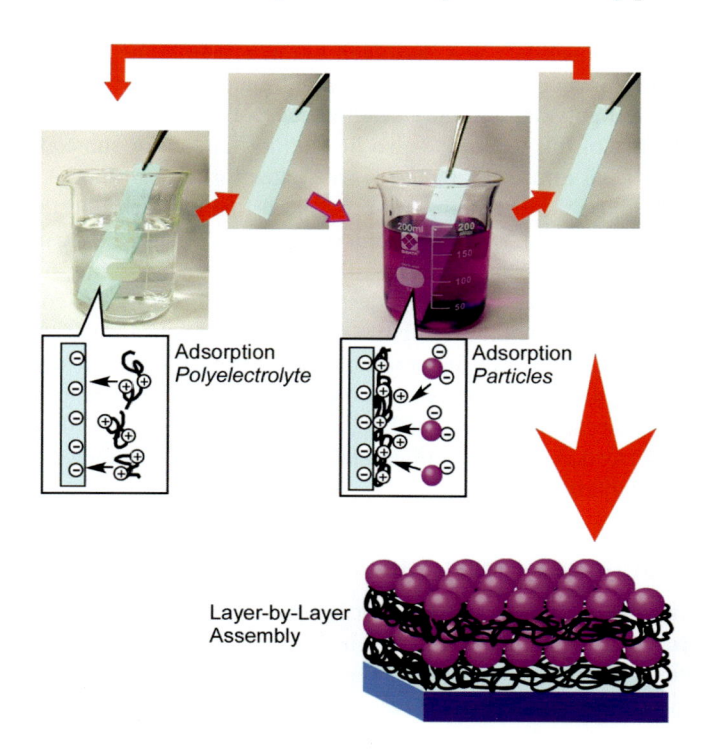

FIGURE 11.4

A simple principle of the alternating layer-by-layer adsorption method for layering organization between a polyelectrolyte and particles.

a simple principle of the alternating adsorption method. The example here is the layering and organization between a polyelectrolyte and nanoparticles having opposite charges to it. A positively charged polyelectrolyte is self-assembled on a negatively charged substrate surface via electrostatic interaction, resulting in charge inversion and resaturation due to charge neutralization and overadsorption, respectively. In this process, the surface charge turns positive, followed by the adsorption of negatively charged particles in a separate solution. The adsorption process reaches equilibrium in a few minutes to several tens of minutes, and the film thickness is spontaneous and relatively uniform. The operation is extremely simple, requiring only a beaker and tweezers. The number of layers is determined by the number of times the layers are dipped into the solution, and the sequence of layers can be freely designed by changing the type of solution. The most commonly used adsorption is based on electrostatic interactions and is applied to thin films of many charged substances, for example, colloidal particles, proteins, DNA, dyes, viruses, and a wide variety of other materials [25].

The self-assembly process of substance adsorption, which is the basis of this LbL method, is not limited to electrostatic interactions, but has been expanded to use biorecognition, hydrogen bonding, charge transfer interactions, inclusion-type molecular recognition, and stereocomplex formation [26]. For example, an alternately adsorbed membrane consisting of isotactic polymethacrylic acid and syndiotactic methyl polymethacrylate was stereocomplexed, and only the syndiotactic methyl polymethacrylate was dissolved out as a membrane reactor [27]. When methyl methacrylate is added to this membrane and polymerized, the membrane remembers the steric form of the previously dissolved polymer and only syndiotactic polymethyl methacrylate is synthesized.

The broad applicability of this technique also lies in the freedom of substrate selection. As indicated above, there is no need to use flat plates as substrates for laminations. A major technological advance in this regard will be the alternating adsorption on fine particles and the creation of hollow capsules [28]. As shown in Fig. 11.5, a hollow microcapsule is obtained by using microparticles as the solid carrier for laminating the membrane as the core material, creating a polyelectrolyte film around it by alternating adsorption, and then destroying or removing the core material in some manner. Since the material used as the shell is not limited to polyelectrolytes, the creation of reactors with a large specific surface area incorporating enzymes or reactors incorporating magnetic nanoparticles that can be collected by an external magnet have also been proposed.

The simplicity and affordability of the method, as well as the diversity of applicable substances and the high degree of freedom in interactions and structures, have attracted many researchers to alternating adsorption methods.

11.3.2 Hierarchical layer-by-layer nanoarchitectonics with microfabrication

The LbL assembly is an organization method that is strongly influenced by artificial manipulation. For example, the process of adsorption of a single layer is due to self-assembling forces such as electrostatic interaction, but which objects are stacked and organized and in what order is left to the intention of the producer. In other words, the order in which the substrates are immersed in the solutions determines the organizational structure of the film to be fabricated. The adsorption of substances in each elemental process is based on the principle of self-assembly, but the process of creating a meaningful organization with functionality is not based on natural principles at all, but on human intention. In this sense, the alternating adsorption method is a nanostructure fabrication

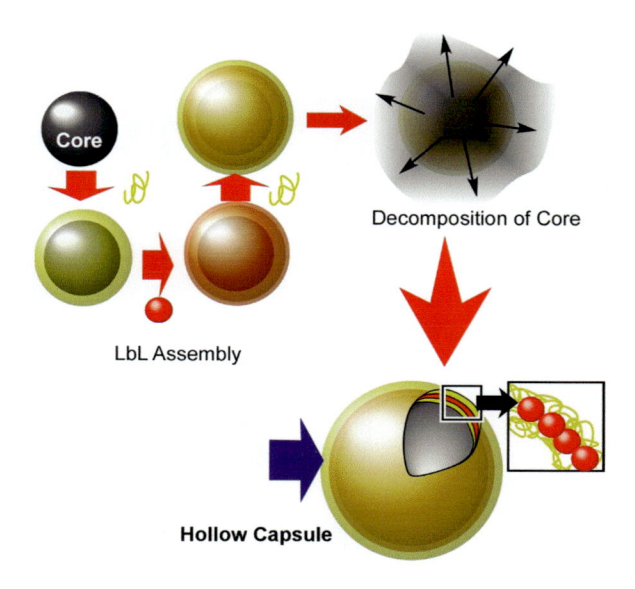

FIGURE 11.5

Alternating layer-by-layer adsorption on fine particles and the creation of hollow capsules.

method in which self-assembly and organization are clearly differentiated. Self-assembly, which is a completely equilibrium process, is not, in a sense, actively alive. To develop functional structures, it is important to create structures by self-assembly through nonequilibrium processes or by artificial perturbation. Alternately adsorbed membranes have a particularly good organizing ability in the latter.

Because of such characteristics, the simple thin-film fabrication technique by alternating adsorption is often combined with an artificial structure fabrication method called microfabrication. For example, fabrication of alternately adsorbed film structures patterned with stamped polydimethylsiloxane using microcontact printing techniques, and patterned with alternately adsorbed films using inkjet and photolithography methods have been reported. Fig. 11.6 shows a reported example that straightforwardly represents an aspect where alternately adsorbed films are suitable for microfabrication [29]. Here, inorganic clay or magnetite is alternately adsorbed with a polycation on a photoresist film patterned on a silicon wafer using the photolithography method. By sequentially dissolving the underlying photoresist film, microstructures (in this case, cantilevers) consisting only of alternately adsorbed inorganic films are fabricated.

11.3.3 Hierarchical layer-by-layer nanoarchitectonics with artificial cell

The LbL assembly is suitable for creating hierarchical structures, which are difficult to achieve by simple self-assembly processes. The most typical hierarchical structure is biological tissue, which is formed by the hierarchical organization of cellular units. Therefore, if liposomes that mimic living cells are alternately adsorbed, they will contribute to the creation of artificial biological tissues.

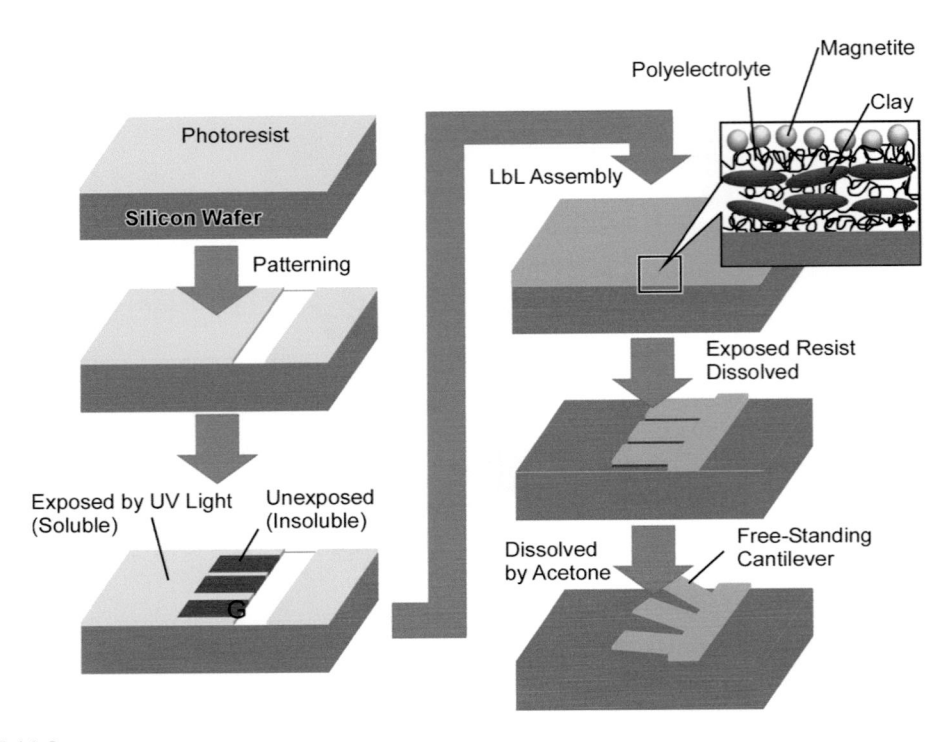

FIGURE 11.6

Layer-by-layer method with inorganic clay or magnetite alternately adsorbed with polycation on a photoresist film patterned on a silicon wafer using the photolithography method.

However, liposomes and vesicles composed of ordinary lipid molecules are not suitable for LbL assembly because their vesicular structure often breaks down and becomes a flat membrane when they interact with polyelectrolytes having opposite charges. An interesting approach in this regard is the stacking of cerasomes (Fig. 11.7). Cerasome is an organic—inorganic hybrid vesicle whose surface is structurally reinforced by a silica backbone [30]. It was found that artificial cells supported by this inorganic backbone can be hierarchically stacked without destroying their structure by LbL assembly with polyelectrolytes having opposite charges [31]. It was also confirmed that cerasomes with positive and negative surface charges can be stacked alternately [32]. Such a hierarchical structure is a model of multicellular organization and is expected to be applicable to the study of intercellular communication and the development of complex drug delivery systems.

11.3.4 Hierarchical layer-by-layer nanoarchitectonics with graphene and ionic liquid

Fig. 11.8 shows an example of a sensing system consisting of an LbL assembly membrane with graphene and an ionic liquid [33]. First, graphite is oxidized to obtain a water dispersion of

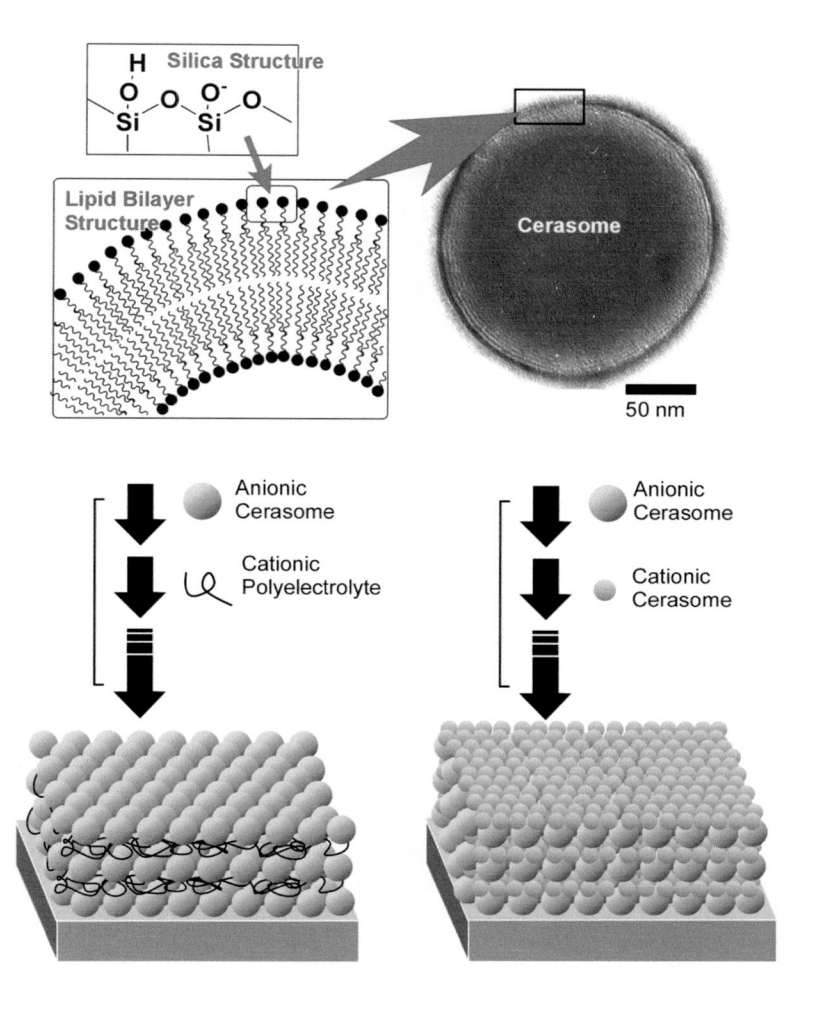

FIGURE 11.7

Layer-by-layer (LbL) assembly of cerasomes as organic—inorganic hybrid vesicles with polyelectrolytes having opposite charges and LbL assembly of cerasomes with positive and negative surface charges.

graphene oxide nanosheets. Then, imidazolium salt, which forms an ionic liquid, is added and reduced to convert graphene oxide nanosheets into graphene nanosheets, resulting in a complex in which an ionic liquid is enclosed between the layers of graphene nanosheets. Since this complex is charged, it can be multilayered by the LbL assembly with an oppositely charged polyelectrolyte. The multilayered structures were immobilized on a quartz crystal microbalance (QCM) sensor and used for gas detection of various organic solvents, showing very high selectivity for aromatic compounds. The thin layer of aromatic ionic liquid sandwiched between graphene nanosheets is

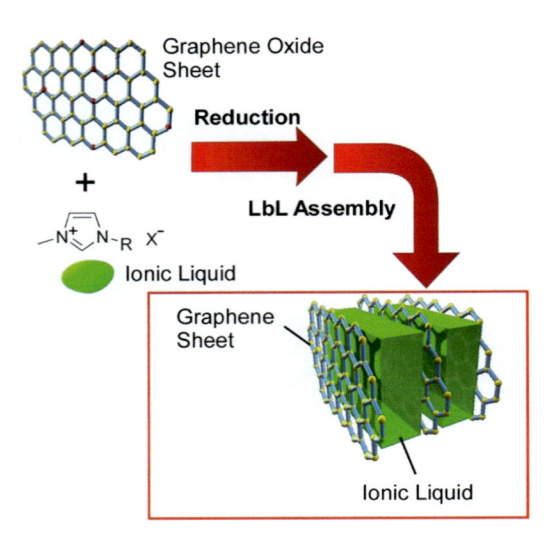

FIGURE 11.8

Layer-by-layer assembly film with graphene and an ionic liquid.

considered to be an electron-rich nanospace, which may be responsible for the selective adsorption of aromatic molecules.

11.3.5 Hierarchical layer-by-layer nanoarchitectonics with mesoporous carbon

As a material with internal nanospaces, a hierarchical structure can also be obtained by LbL assembly of mesoporous carbon (CMK-3) [34]. Weak oxidation of mesoporous carbon can introduce negatively charged carboxyl groups onto its surface without destroying its nanostructure. By alternately adsorbing this modified mesoporous carbon with a positively charged polyelectrolyte, a thin film retaining the nanopore structure of mesoporous carbon was immobilized on the QCM sensor. This sensor was placed in water and various tea components (caffeine, catechins, and tannins) were injected into the aqueous phase, and their adsorption behavior was observed in real time. The results showed that the detection sensitivity for tannins was high, being 13.6 and 3.9 times higher than that for caffeine and catechins, respectively. Compared to the size of the molecules by molecular modeling, the high sensitivity was suggested to be due to the stacking and accumulation of tannin molecules in the pores of the mesoporous carbon.

As a further multilayered structure with a hierarchical structure, the sensing function of thin films with alternately adsorbed mesoporous carbon capsules was also investigated (Fig. 11.9) [35]. These mesoporous carbon capsules are synthesized in several steps using zeolite crystals as a template, and have a hollow capsule-like structure with a mesoporous structure developed on the capsule wall. The mesoporous carbon capsule surface is charged with a surfactant and then alternately charged with a polyelectrolyte having an opposite charge. When this hierarchical carbon structure was immobilized on a QCM sensor and exposed to various solvent vapors, it was found to be

FIGURE 11.9

Layer-by-layer assembly film with mesoporous carbon capsules as a hierarchical structure.

generally more sensitive to nonpolar solvents than to highly polar solvents, especially to aromatic solvents. Comparing cyclohexane and benzene, it was found that benzene was about five times more sensitive than cyclohexane, even though there was no significant difference between molecular size, molecular weight, and vapor pressure. Thus the mesoporous carbon capsule sensor shows higher sensitivity to aromatic gases, but its selectivity can be changed by a second component sealed inside the capsule. When lauric acid is introduced into the inner capsule space, sensitivity to aliphatic amines such as ammonia and butylamine is higher. On the other hand, higher sensitivity to acetic acid was also found when dodecylamine was encapsulated. These sensitivities are determined by a combination of interactions, electrostatic interactions, and hydrogen bonds. A method that can freely change the selectivity of the sensor by inserting a second component into the hierarchical structure would be beneficial in the development of sensors.

11.3.6 Hierarchical layer-by-layer nanoarchitectonics, future

The LbL assembly is recognized as one of the few self-assembly methods suitable for substantial applications due to its wide range of applications. For example, nanoreactors consisting of multiple enzymes can be obtained very easily [36,37]. In principle, such reactors can be fabricated by changing the enzymes, thus enabling the creation of higher value-added reactors and biological sensors. The simplicity of the fabrication technique also makes it suitable for automating the self-assembly process, which is important for industrialization. In fact, the development of an automatic alternating adsorption device coupled with a QCM sensor has been reported.

The greatest feature of the alternating adsorption method is that it is an organization technology based on an extremely simple and flexible principle, and it is a friendly technology onto which anyone can easily jump on board. Therefore, there is a high possibility of technological development based on free ideas, and (although the traditional original form may be lost) it is highly likely to survive as a useful microfabrication technology in the future. For example, the day may come when anyone will be able to make nano thin films of their desired thickness and sequence simply by purchasing a package containing various functional adsorbents and attaching it to an LbL machine.

11.4 Short perspective

This chapter presents several examples of methodologies to control nanostructures and their hierarchical structures to produce a variety of functions in terms of the Langmuir—Blodgett and layer-by-layer assembly methods. Hierarchical structures can be intentionally fabricated by synthesizing and layering materials containing precise nanostructures through template synthesis, and, at the same time, hierarchical structures can be spontaneously formed by the strange self-assembling process of molecules. As seen in the various functions of living organisms, sophisticated functions are based on the coordination of a number of actions. Because of this necessity, many structures in living organisms are hierarchical. To develop artificial material systems with higher functions, it is important to develop methodologies to construct such hierarchical structures. For example, in optical functions, the concept of hierarchical structuring will be important for functions such as light energy harvesting and transfer through rational structures, as well as for the fabrication of advanced metamaterials.

References

[1] R.P. Feynman, There's plenty of room at the bottom, Calif. Inst. Technol. J. Eng. Sci. 4 (1960) 23—36.

[2] M. Roukes, Plenty of room, indeed, Sci. Am. 285 (2001) 48—57.

[3] K. Ariga, Nanoarchitectonics: what's coming next after nanotechnology? Nanoscale Horiz. 6 (2021) 364—378.

[4] W. Chaikittisilp, Y. Yamauchi, K. Ariga, Material evolution with nanotechnology, nanoarchitectonics, and materials informatics: what will be the next paradigm shift in nanoporous materials? Adv. Mater. 34 (2022) 2107212.

[5] K. Ariga, R. Fakhrullin, Materials nanoarchitectonics from atom to living cell: a method for everything, Bull. Chem. Soc. Jpn. 95 (2022) 774–795.

[6] K. Ariga, Y. Yamauchi, T. Mori, J.P. Hill, 25th Anniversary article: what can be done with the Langmuir-Blodgett method? Recent developments and its critical role in materials science, Adv. Mater. 25 (2013) 6477–6512.

[7] K. Ariga, T. Mori, J. Li, Langmuir nanoarchitectonics from basic to frontier, Langmuir 35 (2019) 3585–3599.

[8] K. Ariga, Don't forget Langmuir–Blodgett films 2020: interfacial nanoarchitectonics with molecules, materials, and living objects, Langmuir 36 (2020) 7158–7180.

[9] G. Rydzek, Q. Ji, M. Li, P. Schaaf, J.P. Hill, F. Boulmedais, et al., Electrochemical nanoarchitectonics and layer-by-layer assembly: from basics to future, Nano Today 10 (2015) 138–167.

[10] M. Akashi, T. Akagi, Composite materials by building block chemistry using weak interaction, Bull. Chem. Soc. Jpn. 94 (2021) 1903–1921.

[11] K. Ariga, Y. Lvov, G. Decher, There is still plenty of room for layer-by-layer assembly for constructing nanoarchitectonics-based materials and devices, Phys. Chem. Chem. Phys. 24 (2022) 4097–4115.

[12] K. Ariga, X. Jia, J. Song, J.P. Hill, D.T. Leong, Y. Jia, et al., Nanoarchitectonics beyond self-assembly: challenges to create bio-like hierarchic organization, Angew. Chem. Int. Ed. 59 (2020) 15424–15446.

[13] K. Ariga, Liquid interfacial nanoarchitectonics: molecular machines, organic semiconductors, nanocarbons, stem cells, and others, Curr. Opin. Colloid Interface Sci 63 (2023) 101656.

[14] K. Ariga, Materials nanoarchitectonics in a two-dimensional world within a nanoscale distance from the liquid phase, Nanoscale 14 (2022) 10610–10629.

[15] O.N. Oliveira Jr., L. Caseli, K. Ariga, The past and the future of Langmuir and Langmuir–Blodgett films, Chem. Rev. 122 (2022) 6459–6513.

[16] K. Ariga, Langmuir–Blodgett nanoarchitectonics, out of the box, Acc. Mater. Res. 3 (2022) 404–410.

[17] V. Krishnan, Y. Kasuya, Q. Ji, M. Sathish, L.K. Shrestha, S. Ishihara, et al., Vortex-aligned fullerene nanowhiskers as a scaffold for orienting cell growth, ACS Appl. Mater. Interfaces 7 (2015) 15667–15673.

[18] T. Mori, H. Tanaka, A. Dalui, N. Mitoma, K. Suzuki, M. Matsumoto, et al., Carbon nanosheets by morphology-retained carbonization of two-dimensional assembled anisotropic carbon nanorings, Angew. Chem. Int. Ed. 57 (2018) 9679–9683.

[19] M. Ito, Y. Yamashita, Y. Tsuneda, T. Mori, J. Takeya, S. Watanabe, et al., 100 oC-Langmuir − Blodgett method for fabricating highly oriented, ultrathin films of polymeric semiconductors, ACS Appl. Mater. Interfaces 12 (2020) 56522–56529.

[20] M. Ito, Y. Yamashita, T. Mori, M. Chiba, T. Futae, J. Takeya, et al., Hyper 100 oC Langmuir − Blodgett (Langmuir − Schaefer) technique for organized ultrathin film of polymeric semiconductors, Langmuir 38 (2022) 5237–5247.

[21] G. Decher, Fuzzy nanoassemblies: toward layered polymeric multicomposites, Science 277 (1997) 1232–1237.

[22] G. Decher, J.D. Hong, Buildup of ultrathin multilayer films by a self-assembly process, 1 consecutive adsorption of anionic and cationic bipolar amphiphiles on charged surfaces, Makromol. Chem. Makromol. Symp. 46 (1991) 321–327.

[23] G. Decher, J.D. Hong, Buildup of ultrathin multilayer films by a self-assembly process: II. Consecutive adsorption of anionic and cationic bipolar amphiphiles and polyelectrolytes on charged surfaces, Ber. Bunsenges. Phys. Chem. 95 (1991) 1430–1434.

[24] R.K. Iler, Multilayers of colloidal particles, J. Colloid Interface Sci. 21 (1966) 569–594.

[25] K. Ariga, J.P. Hill, Q. Ji, Layer-by-layer assembly as a versatile bottom-up nanofabrication technique for exploratory research and realistic application, Phys. Chem. Chem. Phys. 9 (2007) 2319–2340.

[26] T. Serizawa, K. Hamada, T. Kitayama, N. Fujimoto, K. Hatada, M. Akashi, Stepwise stereocomplex assembly of stereoregular poly(methyl methacrylate)s on a substrate, J. Am. Chem. Soc. 122 (2000) 1891—1899.

[27] T. Serizawa, K. Hamada, M. Akashi, Polymerization within a molecular-scale stereoregular template, Nature 429 (2004) 52—55.

[28] F. Caruso, R.A. Caruso, H. Möhwald, Nanoengineering of inorganic and hybrid hollow spheres by colloidal templating, Science 282 (1998) 1111—1114.

[29] F. Hua, T. Cui, Y.M. Lvov, Ultrathin cantilevers based on polymer-ceramic nanocomposite assembled through layer-by-layer adsorption, Nano Lett. 4 (2004) 823—825.

[30] K. Katagiri, K. Ariga, J. Kikuchi, Preparation of organic-inorganic hybrid vesicle "Cerasome" derived from artificial lipid with alkoxysilyl head, Chem. Lett. (1999) 661—662.

[31] K. Katagiri, R. Hamasaki, K. Ariga, J. Kikuchi, Layer-by-layer self-assembling of liposomal nanohybrid "cerasome" on substrates, Langmuir 18 (2002) 6709—6711.

[32] K. Katagiri, R. Hamasaki, K. Ariga, J. Kikuchi, Layered paving of vesicular nanoparticles Formed with cerasome as a bioinspired organic — inorganic hybrid, J. Am. Chem. Soc. 124 (2002) 7892—7893.

[33] Q. Ji, I. Honma, S.-M. Paek, M. Akada, J.P. Hill, A. Vinu, et al., Layer-by-layer films of graphene and ionic liquids for highly selective gas sensing, Angew. Chem. Int. Ed. 49 (2010) 9737—9739.

[34] K. Ariga, A. Vinu, Q. Ji, O. Ohmori, J.P. Hill, S. Acharya, et al., A layered mesoporous carbon sensor based on nanopore-filling cooperative adsorption in the liquid phase, Angew. Chem. Int. Ed. 47 (2008) 7254—7257.

[35] Q. Ji, S.B. Yoon, J.P. Hill, A. Vinu, J.-S. Yu, K. Ariga, Layer-by-layer films of dual-pore carbon capsules with designable selectivity of gas adsorption, J. Am. Chem. Soc. 131 (2009) 4220—4221.

[36] M. Onda, Y. Lvov, K. Ariga, T. Kunitake, Sequential actions of glucose oxidase and peroxidase in molecular films assembled by layer-by-layer alternate adsorption, Biotechnol. Bioeng. 51 (1996) 163—167.

[37] M. Onda, Y. Lvov, K. Ariga, T. Kunitake, Sequential reaction and product separation on molecular films of glucoamylase and glucose oxidase assembled on an ultrafilter, J. Ferment. Bioeng. 82 (1996) 502—506.

Langmuir films—a universal method for fabricating organized monolayers from nanomaterials

12

Michal Bodik[1,2] and Peter Siffalovic[1,2]

[1]*Institute of Physics, Slovak Academy of Sciences, Bratislava, Slovakia* [2]*Centre for Advanced Materials Application, Bratislava, Slovakia*

Outline

This chapter will discuss the formation and applications of Langmuir films. First, Langmuir films and other deposition techniques will be compared with an emphasis on the advantages of Langmuir films. Then, we will discuss the origin of Langmuir films and the experimental details required for a successful self-assembled, Langmuir—Blodgett, and Langmuir—Schaefer deposition processes. In the final section, we will present a series of devices prepared from Langmuir films of nanomaterials in different fields to demonstrate the practical usefulness of the technique.

12.1 Introduction

For many applications, devices require a continuous film consisting of a nanomaterial. In some cases, nanomaterials can be either directly grown on the desired substrate [1—4] or grown on a different substrate and subsequently transferred [5—7]. However, growth of nanomaterials directly on a substrate is experimentally limited. A major issue with direct growth is that the growth substrate typically cannot be any arbitrary substrate. Therefore, a subsequent transfer from a growth substrate to a desired arbitrary substrate is required. However, the transfer can be experimentally complicated or too expensive for large-scale processes. The transfer process is even more complicated if the nanomaterial has already been synthesized as an aligned or patterned film. The risk of breaking the required symmetry/alignment complicates such an approach.

A large variety of nanomaterials can (or must) be synthesized in solution. For example, one of the easiest ways to produce 0D nanoparticles is by nucleation of precursors in solutions [8]. Even in the case of 1D materials, solution-based methods are used [9], and in the case of 2D materials, liquid-phase exfoliation is a widely used cost-effective alternative to bottom-up growth methods [10]. The nanomaterial is dispersed in a solvent, and film formation/patterning is performed separately from the synthesis process. Separating the synthesis and deposition processes allows for easier optimization of both processes. On top of that, the patterning/alignment can be made directly on the desired substrate.

Fabricating a continuous monolayer film from nanomaterial dispersed in solution requires a stable, reproducible, and cost-effective processing technique. The straightforward methods rely on

Materials Nanoarchitectonics. DOI: https://doi.org/10.1016/B978-0-323-99472-9.00003-1

applying the nanomaterial-containing solution directly to the desired sample. In this process, the nanomaterial is assembled on the sample surface by evaporation of the solvent. Such methods include dip coating [11], spin coating [12], convective assembly [13], or blade/slot-die coating [14]. These methods are generally inexpensive, require minimal equipment, and can be modified to yield patterned films of nanomaterials [15,16]. However, even though these direct assembly methods do not require very specialized equipment, the assembly process is experimentally complicated. The mechanism behind such assembly consists of a precise balance between the solvent's evaporation rate, particle mobility in the solvent, and the concentration of nanoparticles. Even though all of these parameters could be monitored and controlled, the process of tuning the experimental conditions must be reset for every new batch of a nanomaterial.

In contrast to a solid substrate, a liquid one enables improved mobility of nanoparticles/nanotubes/nanosheets, and the process of film formation can be better controlled than assemblies resulting from solvent evaporation on a solid substrate.

Instead of directly assembling nanomaterials on the desired substrate, Langmuir films are assembled on a liquid surface that provides mobility for the nanomaterial. If the mobility is maintained, the assembly can be controlled more precisely, reducing the number of structural defects.

Upon application of a nanomaterial onto a surface of a liquid in which the nanomaterial is not soluble, the surface tension traps it at the interface. The nanomaterial remains in a potential well, and its removal/immersion from/into the subphase does not occur spontaneously. In contrast, the lateral movement along the interface is not limited. The depth of the energy well can be estimated by a simple model, as shown by Pieranski [17]. His model consists of a hard-sphere with radius R and a defined finite surface energy σ. The sphere is placed at the air/water interface, and the water level is at a distance z from the center of the nanoparticle (Fig. 12.1).

The total surface energy of a nanoparticle on water is composed of three main contributions. First, the energy of the interface between nanoparticle and air:

$$E_{PA} = \sigma_{PA} \times 2\pi R^2(1 + \tilde{z}) \tag{12.1}$$

The energy of the interface between nanoparticle and water:

$$E_{PW} = \sigma_{PW} \times 2\pi R^2(1 - \tilde{z}) \tag{12.2}$$

And the separation energy of the water and air interface:

$$E_{WA} = -\sigma_{WA} \times 2\pi R^2(1 - \tilde{z}^2) \tag{12.3}$$

where σ is the corresponding surface energy of the interface and $\tilde{z} = \dfrac{z}{R}$ is a dimensionless number representing the nanoparticle's immersion depth. If we define a dimensionless energy

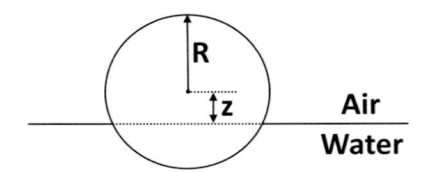

FIGURE 12.1

A schematic illustration of the nanoparticle-on-water model.

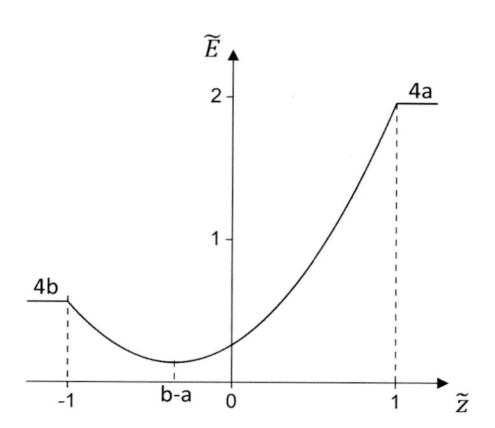

FIGURE 12.2

The estimation of an energy well of a polystyrene nanoparticle [17].

$\tilde{E} = E/\pi R^2 \sigma_{WA}$ and dimensionless surface energies $a = \frac{\sigma_{PA}}{\sigma_{WA}}$ and $b = \frac{\sigma_{PW}}{\sigma_{WA}}$, we obtain the total dimensionless energy of the system as:

$$\tilde{E} = \tilde{z}^2 + 2(a - b)\tilde{z} + 2a + 2b - 1 \tag{12.4}$$

In his work, Pieranski [17] used the known values of σ_{PA} and σ_{PW} of 122.5 nm large polystyrene nanoparticles and calculated the shape of the nanoparticle's energy well (Fig. 12.2). The calculations revealed, that the polystyrene nanoparticles are trapped in a potential well, minimizing the total energy of the system when immersed in water with 1.35 times the radius.

In practice, it is difficult to measure the σ_{PA} and σ_{PW} of each nanomaterial. For that reason, only some conclusions can be drawn for a general formula. The natural energy unit ($\tilde{E} = 1$) depends on the radius of the nanoparticle. For example, a nanoparticle with a radius of 10 nm on water has a natural energy unit of roughly 2×10^{-17} J, which is four orders of magnitude larger than the thermal energy $k_B T$ at room temperature. Therefore, it can be concluded that the energy unit would still be 100 times the thermal energy even for a nanoparticle with a diameter of only 1 nm. In summary, the thermal fluctuations cannot cause any perturbations to the nanoparticle trapped at the air−water interface.

Minimizing the energy results in the nanoparticle partially immersing into the water to the depth $\tilde{z} = b - a$. Thus if the surface energy σ_{PA} is greater than σ_{PW}, the nanoparticle will be immersed deeper, and less energy is required to sink the nanoparticle than lift it from the interface. In the opposite case ($\sigma_{PA} < \sigma_{PW}$), a larger portion of the nanoparticle will float above the water level, and less energy is required to lift it than sink it.

We can calculate the stability limit from the heights of the energy required for sinking/lifting the nanoparticle from the interface. The energy minimum must be within these two values. Otherwise, the floating film will either sink ($\tilde{z}_{min} < -1$) or aggregate ($\tilde{z}_{min} > 1$). Hence, the condition of stability is:

$$(b - a)\epsilon(-1; 1) \tag{12.5}$$

From Eq. (12.5), it is evident that the surface energy of the nanoparticle/air or nanoparticle/subphase cannot exceed the surface energy of the subphase/air.

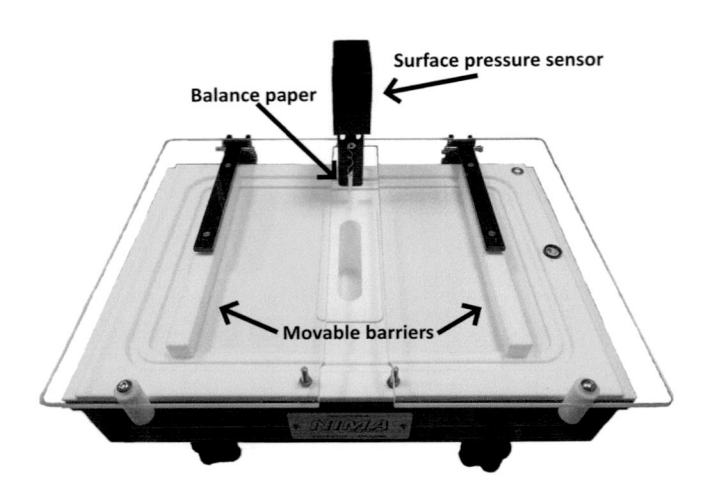

FIGURE 12.3

A photograph of a modern Langmuir—Blodgett trough.

12.2 History and present of Langmuir films

The origin of Langmuir films dates back to the end of the 19th century. Agnes Pockels studied the behavior of oil molecules on the water surface. She constructed her first balance to measure surface tension, and her work attracted the attention of Dr. Rayleigh [18]. In the early 20th century, Dr. Langmuir constructed the first "Langmuir trough" inspired by Pockel's early experiments [19]. Joined later by Dr. Katherine Blodgett, they developed the first "Langmuir—Blodgett deposition trough" (a modern Langmuir—Blodgett trough is shown in Fig. 12.3) and showed that using this technique, monolayers of amphiphilic molecules can be quickly deposited on an arbitrary substrate [20—22]. The technique involves dropping an amphiphilic or hydrophobic material onto a water surface. The nonwater-soluble material will not sink but spreads over the water surface, forming a monolayer thin film. The total water surface area is subsequently reduced by moving macroscopic barriers on the water surface, limiting the area available for the floating material. At some point, the area of the water surface equals the total area occupied by the floating material, and a monolayer is formed. Afterward, the monolayer can be transferred from the water surface to any substrate*

In the first experiments, Dr. Blodgett and Dr. Langmuir worked with amphiphilic molecules [20]. Such molecules provided "excellent results" [20] in terms of monolayer homogeneity. They were able to prepare monolayers on both acidic and basic water and transfer them onto a glass slide. They took advantage of the different adherence of fatty acids monolayers on acidic and basic water. At the time, the transfer method from the water subphase was limited to one of the two

*An archive video of Dr. Langmuir and Dr. Blodgett explaining the basic principle of Langmuir films can be found on YouTube: Surface Chemistry—Thin Film Experiments with Dr Irving Langmuir https://www.youtube.com/watch? v = yn4fuWM007c.

options. First, the monolayer was prepared on an acidic water subphase. Then the glass slide was rinsed with basic water, leaving a thin layer of it on its surface. If such a glass slide gently touched the monolayer surface, the film transferred itself from acidic water onto the glass slide. The second transfer possibility presented is having the entire substrate immersed vertically in the subphase during the film formation. After film formation, the substrate is continuously withdrawn from the subphase and the film adheres to the substrate.

The work of Dr. Vincent Schaefer later extended the experimental work of Dr. Langmuir and Dr. Blodgett. His work focused on preparing multilayer films of proteins [21]. He showed that by slightly altering the original procedure of Langmuir, each deposition process could add new layers. In this way, "30 layers were built up without difficulty." [21] In his work, Dr. Shaefer developed another transfer method. This "Method B" [21] consisted of immersing the substrate horizontally into the subphase or pulling it out through the formed film. Multilayer films can be prepared by repeatedly immersing and withdrawing the substrate through the already-formed monolayer film. Multilayers can be formed by changing the order of immersion and withdrawal, whether ABAB or BABA (where A is the hydrophilic and B is the hydrophobic part of the protein).

At that point, mono- or multilayer films were prepared from stearic acid, egg-albumin, pepsin, and insulin. No dramatic difference in behavior was observed for any of the hydrocarbons mentioned when the pH of the aqueous subphase was changed.

The same mechanism that worked for amphiphilic molecules in the first half of the 20th century also applies to today's nanomaterials. The nanomaterial prepared in a colloid solution can be dropped onto a water surface. The solvent evaporates, leaving individual nanoparticles floating on the water surface. Then, the total surface area is reduced using movable barriers, that is, the nanomaterial is compressed until the point where the surface area of the Langmuir trough equals the total area covered by the nanomaterial (Fig. 12.4). This method is now known as Langmuir—Blodgett (LB) or Langmuir—Schaefer (LS) deposition, depending on which method is used for the subsequent transfer of the film to a substrate. The advantage of such a process is that it does not depend on the concentration of nanomaterial in the solution nor on the solvent itself. Furthermore, the process is highly reproducible and stable since the experimental conditions such as temperature or air humidity play virtually no role.

The second considerable advantage is the controllability of the entire process. Modern Langmuir troughs are equipped with a Wilhelmy plate that records the surface pressure during the entire experiment. By analyzing the surface pressure, the deposition process can be fine-tuned to produce a desired density of nanomaterial. Finally, and most importantly, the method is applicable to 0D, 1D, and 2D materials without any modifications to the experimental process [23].

Nowadays, Langmuir films, that is, a floating monolayer of material on a liquid subphase, are the basis for two techniques used for colloidal nanomaterial deposition. LB (or LS) deposition takes advantage of the Langmuir trough and a so-called "self-assembly" deposition [24].

The self-assembly method follows essentially the same principle. Suppose we have a nanomaterial dispersed in a volatile solvent. Dropping the nanomaterial-containing solvent onto a clean, nonvolatile solvent results in the formation of a free-floating film of nanomaterial. The evaporation of the volatile solvent leaves the nanomaterial floating on the nonvolatile solvent, and a monolayer film can be afterward transferred onto a substrate. This process differs from LB and LS deposition

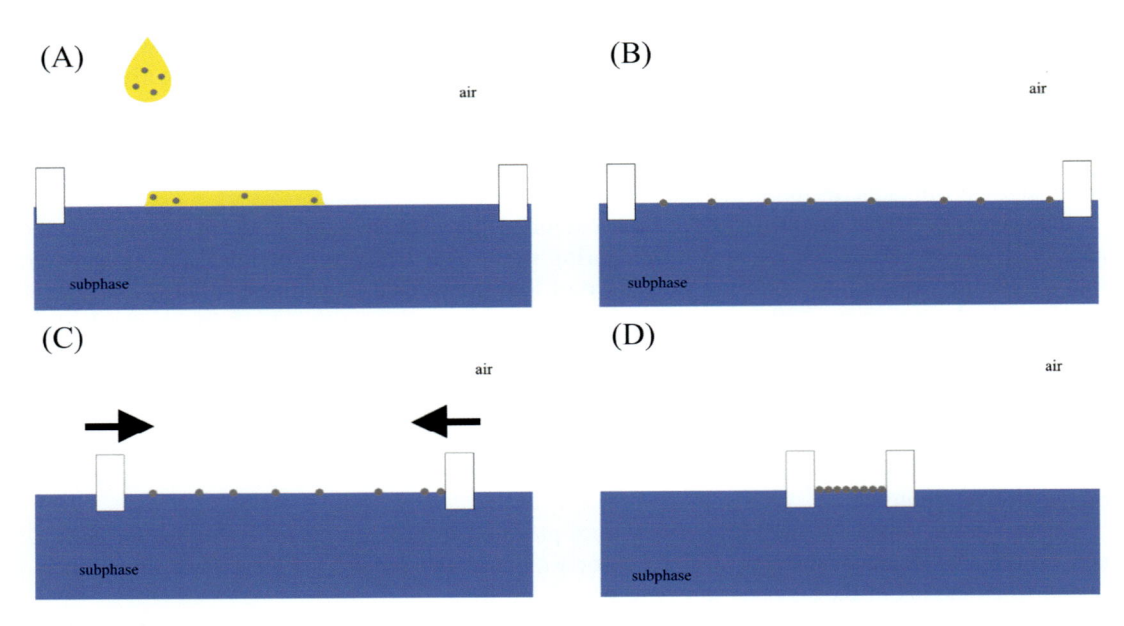

FIGURE 12.4

A schematic illustration of the formation of Langmuir films on a Langmuir–Blodgett trough: (A) dropping the colloid solution containing the nanomaterial; (B) evaporation of the solvent leaving free-floating nanomaterial on the subphase; (C) compression of the nanomaterial; and (D) formation of a monolayer.

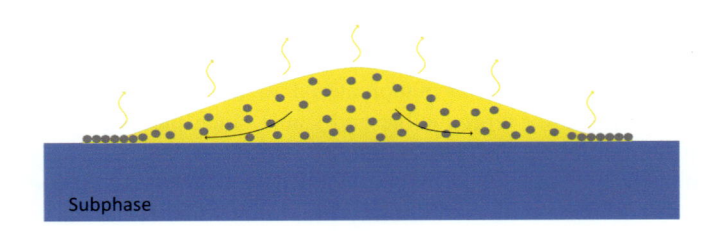

FIGURE 12.5

A schematic illustration of the convective flow-induced nanoparticle self-assembly.

only in the mediator of the "compression" of the nanomaterial. In the case of LB and LS deposition, the volatile solvent evaporates quickly, leaving the nanomaterial freely floating on a large surface area of the Langmuir trough. Compression is performed by moving barriers from the edges of the trough toward the center, eventually forming a monolayer film. On the other hand, the self-assembly method does not require moving barriers. The colloid solution is usually dropped onto a much smaller surface area, and compression is caused by the convective flow and capillary forces of the evaporating solvent (Fig. 12.5).

12.3 **Experimental details**

The processes of self-assembled Langmuir films and Langmuir films on an LB trough share common prerequisites. First, the nanomaterial used in the experiment must not be soluble in the subphase. Second, the nanomaterial must be in the form of a stable colloid solution. Third, the solvent in which the nanomaterial is dispersed must either evaporate quickly or mix with the subphase.

This section will discuss the usual procedure of Langmuir film formation and address some details required for a successful film formation.

12.3.1 **Suitable subphases**

The two ways to prepare a stable colloidal solution of a nanomaterial are by stabilizing the nanomaterial in solution by either steric or Coulomb repulsion. By sterically stabilizing the nanomaterial, a ligand layer is attached to its surface. Typically, the ligand used for stabilizing nanomaterials orchestrates insolubility in water, as ligands with alkane terminating groups are usually used. In the case of Coulomb stabilization, the nanomaterial intrinsically contains a surface charge or is coated with short polar ligands. In this case, water is no longer suitable due to its high polarity.

A typical example of a Coulomb repulsion stabilized nanomaterial is graphene oxide (GO). GO flakes contain carboxyl groups at their edges that are deprotonated (i.e., negatively charged) in neutral or basic water, and protonated (i.e., neutral) in acidic water [25]. Therefore, GO is soluble in neutral or basic water, where the surface charge on two GO flakes causes the repulsive force. However, in acidic water, the surface charge is neutralized by the excess hydrogen ions, and the solubility of GO in water is decreased. If we use acidic water as a subphase, GO flakes can be trapped at the air−water interface, even though GO is soluble in neutral or basic water. Therefore, altering the total charge of the water subphase is commonly used. Typically, the pH is altered [26], and salt [27], or less polar solvents, are added [28,29] to improve the Langmuir film formation.

Of course, Langmuir films are not limited to water as a subphase. High-boiling point organic solvents are also frequently used. The negligible evaporation rate of the subphase allows longer experiments, and therefore such organic solvents are often used for self-assembled Langmuir films. Typically, ethylene glycol (EG) [30] or diethylene glycol (DEG) [31] are used. Apart from the high boiling point, these organic solvents provide another considerable advantage over water. Since most of the commonly used capping ligands (oleylamine, thiols, etc.) are soluble in EG and DEG, there is no need to purify the colloid solution from excess ligands after nanomaterial synthesis [30].

12.3.2 **Suitable solvents**

For the successful formation of Langmuir films, it is essential that the colloidal solution of nanomaterial is stable and that no aggregation is present. If the solution is stable, the second condition is that the solvent must spread on the subphase.

Large et al. [32] performed an extensive experimental study investigating the spreading behavior of commonly used solvents on water. In their work, they used deionized water as a subphase and 21 different solvents as spreading solvents and developed a model to predict the spreading

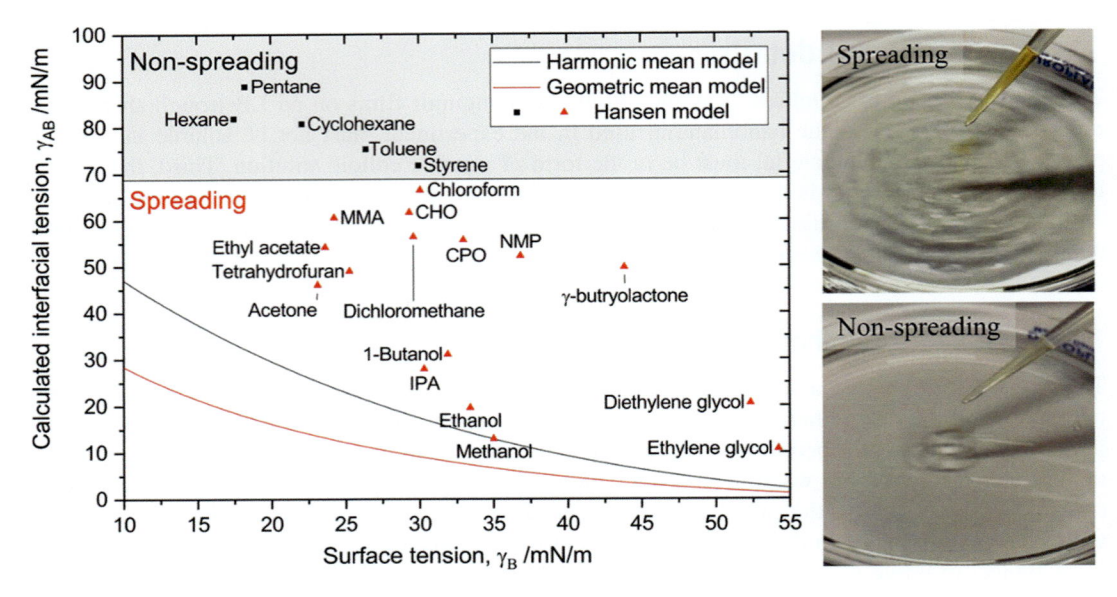

FIGURE 12.6

Interfacial tensions calculated by Large et al. for the most common solvents used in Langmuir films. The photographs on the right illustrate the difference between spreading (top) and nonspreading (bottom) solvent.

Reprinted with permission from M.J. Large, S.P. Ogilvie, A.A.K. King, A.B. Dalton, Understanding solvent spreading for Langmuir deposition of nanomaterial films: a Hansen solubility parameter approach, Langmuir 33 (2017) 14766–14771. Copyright 2017 American Chemical Society.

behavior (Fig. 12.6). Their model was based on Hansen solubility parameters rather than surface tensions, unlike a classical approach to calculate the spreading coefficient as the difference in surface tensions of the system [32].

If a stable colloid solution is capable of spreading the nanomaterial over the subphase surface, the final condition is that the solvent must not occupy the surface. Therefore, the solvent has to either evaporate or mix with the subphase. Examples of suitable spreading solvents on a water subphase are chloroform and N-methyl-2-pyrrolidone (NMP). Both chloroform and NMP spread well on water, but unlike chloroform, NMP has a high boiling point and does not evaporate quickly at room temperature. However, the NMP is a polar organic solvent, and after spreading on a water surface, the two solvents mix. Mixing NMP with water changes the surface tension of the subphase, which must be considered. However, in a standard LB trough, the total volume of the subphase is several orders of magnitude larger than the volume of the spreading solvent. For this reason, the change is negligible.

The same requirements apply in the case of nonaqueous subphases, such as ethylene glycol, diethylene glycol, and acetonitrile, commonly used for self-assembled Langmuir films for an aqueous subphase. Usually, the same solvents are used for both self-assembled and standard Langmuir films since the spreading behavior does not change significantly.

12.3.3 **Experimental procedure for preparation of Langmuir films** [†]

One of the main issues in preparing Langmuir films is cleanliness. Ideally, the trough should be covered or enclosed to minimize environmental contamination.

Prior to film preparation, the trough has to be thoroughly cleaned. Typical Langmuir troughs are made of Teflon or similar polymers that can be cleaned with most solvents. Typically, the trough is first cleaned with chloroform, toluene, or a similar nonpolar solvent. Then, the trough is cleaned with a polar solvent such as ethanol. Finally, before the nanomaterial is introduced, the LB trough is rinsed several times by filling the trough with deionized water and consecutively removing the water. Such a rinsing process will remove the polar solvents' residues and also eliminate dust particles. The extent of cleaning depends on the frequency of use. If the trough is used daily, the cleaning process can be shortened, and rinsing with water can be omitted. If a substrate needs to be immersed into the subphase first, the rinsing process is performed with the substrate already present in the trough.

A Brewster-angle microscope [33,34] can monitor the cleanliness of the subphase surface. A second option is to perform a "clean water test" before the nanomaterial is dropped onto the subphase. The "clean water test" is performed as follows. First, the barriers of the LB trough are opened to the maximum position, and the surface pressure is set to zero. Afterward, the barriers are closed until the end position. If the surface pressure does not change, the surface of the subphase is sufficiently clean. However, if the surface pressure changes by more than 0.2 mN/m, the surface is contaminated.

In such a case, additional cleaning is recommended. If the surface pressure change is small (0.2–1 mN/m), a suction pump can be used to carefully remove the contaminants from the subphase surface between the barriers. If the change is larger than 1 mN/m, it is recommended to change the entire subphase volume.

If the LB trough has not been used for a longer period, a longer cleaning may be necessary. However, if cleaning is insufficient even after several cleanings, the contamination can also originate from the bottle used for DI water. In such cases, a new, clean bottle should be reserved exclusively for the DI water.

After the subphase surface is clean and the surface pressure is set to zero, the nanomaterial can be applied to the subphase between the opened barriers. It may require some skill to place a solvent droplet onto a subphase. Special care must be taken when the solvent is denser than the subphase. A typical example is chloroform and water. Chloroform spreads well on the surface of water, but dropping a droplet that is too large or that falls from a greater height can easily break the surface tension of water and cause the droplet to sink. The droplet will remain on the bottom of the trough, and the nanomaterial in the droplet will be lost. This can be avoided by carefully placing small droplets on the water's surface [35]. Nie et al. showed that the radius of droplets could be significantly reduced by electrospraying. Together with charges generated on the droplets, a nanomaterial is deposited from water (as a solvent) on water (as a subphase) [35].

[†]An instructive video tutorial can be found on YouTube made by Adaptive & Responsive Nanomaterials Group from University College London: Langmuir–Blodgett Trough Tutorial by Alaric Taylor https://www.youtube.com/watch?v = 9Sm8MvlYINg&list = PLNeZqanqUQRSh4mfHKkAb2q_z8mE2jInV.

After applying the nanomaterial, a relaxation period is required. During this time (generally 15 minutes), the solvent should completely evaporate and the flow of the subphase stops. After the stabilization period, the movement of barriers is initiated. In general, the slowest possible barrier speed is preferred. The slower the barriers move, the less perturbation is induced to a floating film, and fewer defects are created. In the theoretical limit of an infinitely slow barrier moving speed, the nanomaterial has infinite time to accommodate for a smaller surface area, resulting in an infinitesimal stress in the film. However, this does not always have to be the case. According to simulations performed by Bodik et al., faster barrier speed can improve the alignment of 1D nanomaterials during the compression thanks to larger velocity gradients [36].

The essential part of a modern Langmuir–Blodgett trough is a surface pressure sensor. Usually, a Wilhelmy plate is installed directly on the trough, and the control software records changes in surface pressure [37,38]. A standard chromatography paper with known dimensions is typically used as a Wilhelmy plate in the Langmuir–Blodgett trough. During compression, the surface pressure and surface area are monitored (Fig. 12.7). Such a graph is called a pressure-area isotherm or Π-A isotherm. Understanding the Π-A isotherm is a basic prerequisite for optimizing Langmuir films. Π-A isotherm of amphiphilic molecules is well understood and can serve as a good initial understanding of a Π-A isotherm of nanomaterials. However, there are several very excellent reviews [39,40] of Langmuir films, therefore, it will not be discussed here.

It is essential to know, in the case of nanomaterials, that the Π-A isotherm consists of several parts. The first part (Fig. 12.7, section I) is at the beginning of the compression. The surface pressure changes very slowly or not at all. Here, the surface area of the trough is too large, and the nanomaterial has a lot of empty surface area to float freely on. Compression only densifies the

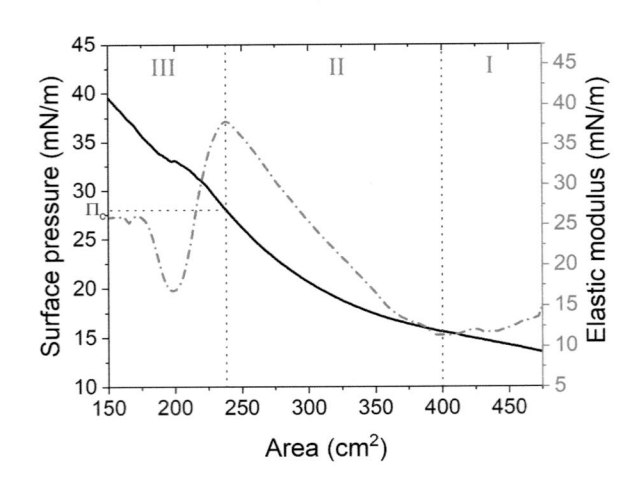

FIGURE 12.7

An example of a Π-A isotherm with (solid black line), with a calculated elastic modulus (*dash-dot red line*) and graphically shown.

unconnected islands of nanomaterial. In the second part (Fig. 12.7, section II), the surface pressure rises rapidly together with the elastic modulus. Here, the film already consists of an interconnected network of nanomaterial, and the compression is forcing the film to densify and fill in the empty water surface. The third part (Fig. 12.7, section III) begins immediately after the critical pressure Π_C, where the maximum nanomaterial density is reached, and the maximum elastic modulus is observed. Any further compression after Π_C results in forming a second layer [41−43].

As a general rule, an experiment is performed in which the film is compressed to a large surface pressure. From this experiment, the Π_C is calculated, and the basic knowledge of the film density and target surface pressure is obtained.

12.3.4 Film deposition techniques

After the barriers are fully closed or the required surface pressure is reached, the film can be transferred onto a substrate [23]. A typical transfer method is LB deposition, where the substrate is attached to a motorized holder perpendicular to the subphase surface. The sample is immersed in the subphase before the film formation and then slowly lifted through the film (Fig. 12.8A). Alternatively, the sample is immersed through the already formed film and then lifted again.

A second option is the so-called LS deposition. In this process, the sample is attached to a motorized holder, but parallel to the subphase. After the film formation, the sample is lowered to the point it touches the film and then slowly retracted up again.

Both processes rely on a good film-to-substrate adhesion. The Langmuir−Schaefer method can be modified by first immersing the substrate into the subphase and lowering the formed film (Fig. 12.8B). Typically, the substrate is placed on the bottom of the trough, one side slightly tilted, and after the film is formed, the water is slowly removed from the LB trough. The level of the subphase will move down with the nanomaterial film moving with it. After the level of the subphase reaches the substrate, the film has to land on it, and strong adhesion is not required.

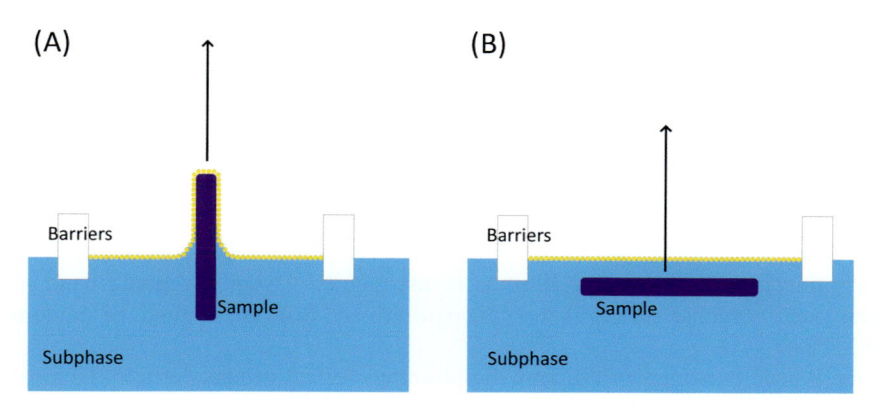

FIGURE 12.8

A sketch of (A) Langmuir−Blodgett and (B) modified Langmuir−Schaefer deposition method.

12.3.5 Experimental procedure for preparation of self-assembled Langmuir films

Self-assembled Langmuir films have one significant advantage over the films prepared on an LB trough, namely the size of the experiment. Unlike the LB trough, typically several hundred square centimeters in size, the self-assembled films can be prepared as an arbitrary small area. Typically, a few square centimeter-large troughs are used [31,44,45], fabricated from PTFE, which is chemically inert, soft, and therefore easy to process. Alternatively, a glass Petri dish can be used since it is standard equipment in a laboratory, and glass is also chemically stable for most commonly solvents.

Such a small size of the trough permits the use of more dangerous and expensive solvents such as diethylene glycol or acetonitrile. Also, the experiment preparation is easier, as the trough can simply be rinsed several times before the experiment.

The self-assembly process usually proceeds follows (Fig. 12.9A). First, the trough is cleaned and rinsed several times with the subphase. Then, the nanomaterial is carefully dropped onto the subphase, similar to the standard Langmuir films described above. The main difference is that in the case of self-assembled Langmuir films, the solvent should not be left to evaporate quickly. Typically, a glass slide or a similar lid is placed onto a trough immediately after the solvent was dropped [31,46,47]. Under the lid, the atmosphere quickly saturates with the solvent's vapors, and further evaporation is slowed. The self-assembly process is then triggered by the convective flow of the evaporating solvent [44]. Typically, the self-assembly process is left untouched for a couple of hours or even overnight. After the solvent evaporates and the film is formed (Fig. 12.9B), the film can be transferred to a substrate by the same methods as the standard Langmuir films. However, due to the small dimensions of the trough, the most widely used method is inserting a substrate into the subphase and then slowly lifting it through the film or touching the film with a substrate (similar to the modified or standard Langmuir–Schaefer method).

12.4 Applications

For applications that require a monolayer film of a nanomaterial on an arbitrary substrate, using Langmuir films can have several advantages over alternative preparation techniques. The deposition process can be scaled up from several millimeters squared to a large-scale roll-to-roll deposition process [48]. Another potential advantage is that using the LB technique, an alignment of 1D nanomaterials can be achieved, and devices can take advantage of this alignment [49].

12.4.1 Plasmonics

A common application for Langmuir films is in plasmonics. Using plasmonic nanoparticles, the resulting plasmonic response differs if the nanoparticles are isolated, aligned, or randomly distributed [50,51].

One of the typical examples of utilizing the plasmonic effect of nanoparticle (NPs) films is as a substrate for Surface-Enhanced Raman Scattering (SERS). Tahghighi et al. have demonstrated two different approaches to optimize the SERS response. First, they deposited a monolayer of gold nanoparticles with a 10-nm diameter via LB deposition. In their work [52], the authors investigated

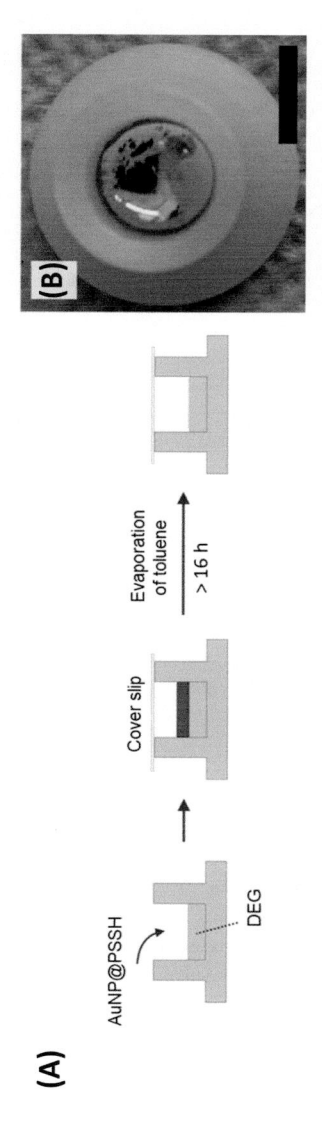

FIGURE 12.9

(A) A schematic illustration of a self-assembly process. (B) A photograph of the resulting self-assembled film of gold nanoparticles. The scale bar is 1 cm.

Reprinted from F. Schulz et al., Structural order in plasmonic superlattices, Nat. Commun. 111(11) (2020) 1–9.

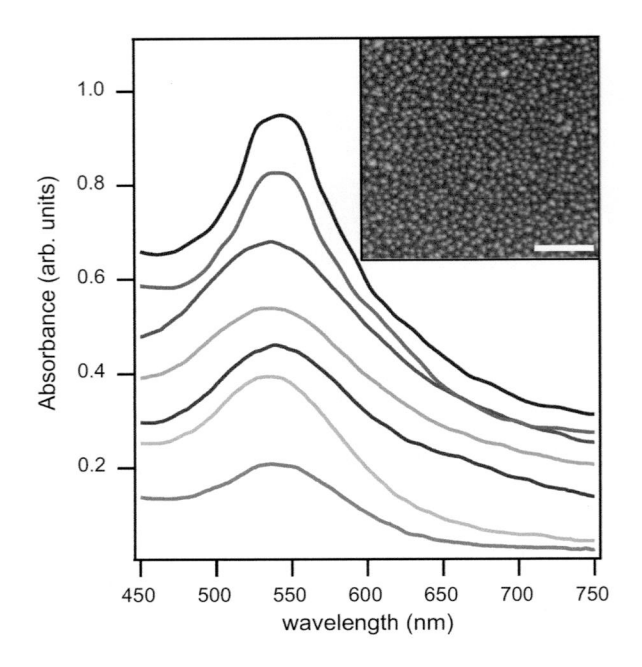

FIGURE 12.10

Vis—NIR absorption spectra for Langmuir—Blodgett (LB) films of 10 nm gold nanoparticles, prepared under different surface pressures. From bottom to top, $\Pi = 3$, 6, 8, 10, 12, 15, and 17 mN/m. In the inset, an scanning electron microscopy image for an LB film obtained at 17 mN/m. The scale is 200 nm.

the possibility of fine-tuning the plasmonic response of Langmuir films using a postdeposition gold electroplating treatment. The monolayer collapse of the 10 nm gold nanoparticles was observed at about 15 mN/m, and the deposition on a glass/fused silica substrates was performed at different surface pressures. The different deposition surface pressure resulted in different NPs packing densities, thus changing the absorption spectrum (Fig. 12.10). As a result, the plasmonic peak (measured by absorption) is broader at low packing densities, and the Raman signal consists of weaker and less symmetric peaks.

On the other hand, at higher packing densities, the absorption and Raman peaks become more pronounced, symmetric, and narrower. In the second step, the size of individual NPs was increased by electroplating with gold. As shown, the diameter of the NPs changed from the initial 10 nm to approx. 80 nm after 12 minutes of electroplating. Such a change in the NP's diameter resulted in a redshift of the plasmonic peak by about 50 nm. As an analyte, they used 4-mercaptobenzoic acid and recorded the SERS signal at different NPs films. The strongest SERS signal was observed in samples prepared at a surface pressure of 13 mN/m followed by 9 minutes of electroplating.

Tahghighi et al. investigated the possibility of using nonspherical NPs as SERS substrates [53]. Their work compared the SERS response of a water contaminant Thiram on spherical and urchin-shaped gold NPs. The authors focused on optimizing the deposition surface pressure (i.e., NPs packing density) and showed that the optimized packing density could lead to a 20% increase in SERS sensitivity. Furthermore, they showed that the nonspherical shape of gold NPs leads to an improved SERS sensitivity compared to spherical ones and demonstrated detection of 240 ppb of Thiram in water.

Saha et al. investigated the SERS potential of larger gold NPs (55 nm in diameter) [54]. However, their work used LB deposition to preprepare substrates for subsequent self-assembly of NPs. LB deposition was used as a precise tool to prepare monolayer poly(methyl methacrylate) substrates. The substrates were used afterward as a template for self-assembly gold nanoparticles. The substrate was immersed in a colloid solution of gold NPs for 24 hours, resulting in the formation of a self-assembled film of NPs. The as-prepared substrates were used to detect the 4-mercaptopyridine molecules.

Similarly, Das et al. [55] prepared substrates with stearic acid bilayers that were subsequently immersed into a colloid solution of gold NPs (50 nm diameter) for 24 hours. Their work focused on the possibility of distinguishing between cancerous and normal cell lines associated with prostate and breast cancer.

The SERS active substrates are not limited to 0D nanoparticles. Tim et al. [56] investigated the SERS response of gold nanorods (53 nm length, 26 nm width) films prepared by LB deposition. As in the case of nanoparticles, the packing density was a critical parameter for optimizing the SERS enhancement factor. At the surface pressure of 12 mN/m (approx. surface coverage of 25%), the Raman enhancement factor was 2×10^5 for the detection of 4-mercaptobenzoic acid and rhodamine 6G.

12.4.2 Gas sensing

A monolayer of nanomaterial can act as a high-performance gas sensor. A large surface-to-area ratio and a small thickness allow fabrication of cost-effective and sensitive gas sensors. Nanomaterials with different dimensionality from 0D (nanoparticles) to 2D (nanosheets) are used to prepare gas sensors; each type brings its advantages. Nanoparticles (0D) can create a sensing surface with a larger roughness and a larger active sensing surface area. Nanotubes and nanowires (1D), on the other hand, can benefit from the alignment achieved by the LB method, thus improving the charge transfer between electrodes. Nanosheets (2D) have the ultimate surface-to-volume ratio since all the atoms in a monolayer are on the surface. It is not uncommon to take advantage of multiple nanomaterials and create a composite sensor.

A typical example of 0D nanomaterial is iron-based nanoparticles, which have excellent sensing properties [57] (Fig. 12.11). Langmuir films of Fe_2O_3 and $CoFe_2O_4$ have been used for sensing oxidizing (nitric dioxide) and reducing (carbon monoxide, acetone) gases [58−60]. Monolayers and multilayers were investigated in terms of their sensitivity. Multilayers were found to be more sensitive to all gases studied [58]. Interestingly, a change in sensing type was observed if multilayers were used. For example, a monolayer of Fe_3O_4/γ-Fe_2O_3 nanoparticles exhibited a p-type response. However, all multilayers (up to 10 layered films) had an n-type sensing response.

One of the most typical 1D nanomaterials is carbon nanotubes [61]. Aligned carbon nanotubes prepared by LB deposition were used for ammonia gas sensing [62]. An excellent sensitivity (21% at 2 ppm) and response time in seconds at room temperature were demonstrated. In order to

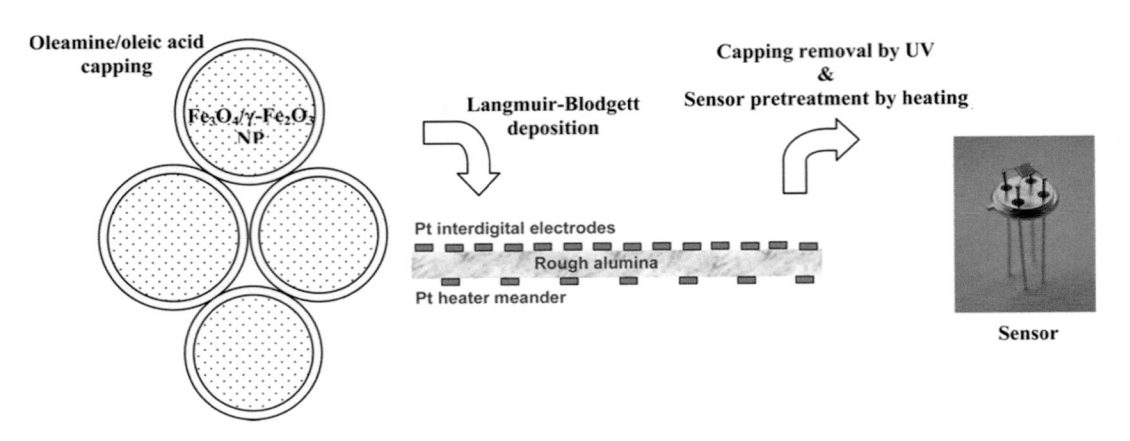

FIGURE 12.11

An schematic illustration of the composition of an Langmuir–Blodgett-based gas sensor.

Reprinted with permission from S. Capone et al., Fe₃O₄/γ-Fe₂O₃ nanoparticle multilayers deposited by the Langmuir-Blodgett technique for gas sensors application, Langmuir 30 (2014) 1190–1197. Copyright 2014 American Chemical Society.

compare the results obtained on the aligned film of nanotubes, the authors prepared a control sample of spin-coated nanotubes that had no alignment in the resulting film. The sensing response was observed to be approximately twice as good in the case of an aligned film as in the spin-coated film of carbon nanotubes.

Graphene is probably the best known 2D nanomaterial and is widely used in gas sensing [63–65]. Although preparation of graphene Langmuir films and subsequent use of such films in gas sensing is possible [66,67], recent developments in chemical vapor deposition (CVD) synthesis [68,69] seem to be a more promising route for application in gas sensors [70]. Another example of such a development is MoS_2 [71–73]. Langmuir films of MoS_2 are possible [74,75], but CVD synthesis leads to better quality films in terms of charge carrier mobility [76]. Therefore, thin films of 2D nanomaterials that are not easily produced by other methods are more typically used for gas sensing. Such examples are mainly graphene oxide (GO), a chemically more reactive derivative of graphene. GO consists of a graphene base plane functionalized with various oxygen functional groups that play an essential role in its gas-sensing mechanism [77]. Büyükkabasakal et al. showed a good response of GO sensors to hazardous organic vapor (benzene, carbon tetrachloride, dichloromethane). Qian et al. used a less traditional 2D nanomaterial, black phosphorus (BP), to prepare gas sensors [78]. In their work, they used 4-azidobenzoic acid to stabilize BP nanosheets and later prepared Langmuir films of such functionalized nanomaterial. However, as a subphase, they used water mixed with Safranine T or Congo Red dyes to prepare composite BP/dye films. Such films showed a photometric response to acidic (HCl) exposure or alkali (NH_3) water vapors.

Nevertheless, the Langmuir films are a helpful technique for preparing composite nanomaterials. The high repeatability and the possibility of a facile way to produce multilayers enable the production of more complicated, multilayer gas sensors. Ly et al. [79] used LB and LS deposition techniques to prepare hierarchically self-assembled films of GO and gold NPs. Their sensor consisted of a monolayer of gold NPs covered with GO nanosheets. The authors investigated the influence of gold NPs and GO nanosheets' surfactant layer size to optimize the sensor's sensing properties. Their sensor showed a response of almost 10% to 70 ppm of NH_3 at room temperature.

Miao and Lin [80] demonstrated another possibility of using Langmuir films to prepare nano-composite sensors. In their work, aligned ZnO nanowires decorated with Au nanoparticles served as an acetylene sensor. LB deposition was used to fabricate a dense, aligned film of ZnO nano-wires, and Au NPs were deposited onto such film by DC sputtering. The detection limit of acetylene was measured at 3 ppb with high selectivity towards CO, H_2, and CH_4.

Decorating relatively chemically inert graphene with more reactive NPs can lead to a substantial increase in gas sensitivity. Kostiuk et al. [81] prepared H_2 and NO_2 sensors by decorating few-layer graphene films (prepared by the LS deposition) with Pd NPs. Spin-coating with Pd NPs improved the NO_2 sensitivity of graphene films twice and allowed sensing the H_2 gas, which was impossible with pure graphene.

12.4.3 Electronics

Langmuir films are of great use in electronics since preparing closed-packed nanomaterial films is often desired in that field. Also, the ability to align 1D nanomaterial is highly beneficial as it provides a very good uniaxial conductivity [82].

Aligning carbon nanotubes was used by Cao et al. [83] to produce a 40-nm footprint transistor. The authors used a careful multiple-isotherm cycle variation of LB deposition in their work. After applying the nanotubes onto a water surface and evaporating the solvent (dichloroethane), they performed 30 isotherm cycles before transferring the nanotubes to a substrate. Each cycle consisted of first compressing to an integer value of the surface pressure (i.e., 1 mN/m in the first cycle, 2 mN/m in the second cycle, etc.), followed by opening the barriers up until the surface pressure dropped back to 0 mN/m. Continuously decreasing surface pressure minimizes stresses in the Langmuir film and ultimately results in a well-aligned film of carbon nanotubes that serves as a template to prepare a high-density array of transistors.

A memristor is another electronic device that can be fabricated from Langmuir films. Assembled films of gold NPs are commonly used to prepare memristors [84,85]. A multilayer LS film of polyaniline-coated gold NPs was used as an active channel. The resulting device showed lower hysteresis but high reproducibility [85]. Xu et al. [84] used the self-assembled Langmuir films method to prepare a triple-layer SiO_2 shell over gold NPs memristors. The memristors exhibited a negative differential resistance behavior, which can be modulated by illumination.

Another application that takes advantage of the 1D nanomaterial alignment is preparing high-performance light-emitting diodes (LED) [86]. Rhee et al. fabricated CdSe/CdS quantum rod (30 nm long and 5 nm thick) LS film devices in their work. Such devices were able to produce linearly polarized electroluminescence. Such LEDs were comparable to state-of-the-art quantum rods LEDs with the addition of linearly polarized light.

Not only can LEDs be prepared from Langmuir films, but light detectors can be as well. Norton et al. [87] produced fully solution-processed photodetectors of SnS thin films by LB deposition (Fig. 12.12). The detectors they prepared could detect AM1.5 (daylight) and monochromatic illumination. The fully solution-processed photodetector production method has a good potential for producing flexible photodetectors.

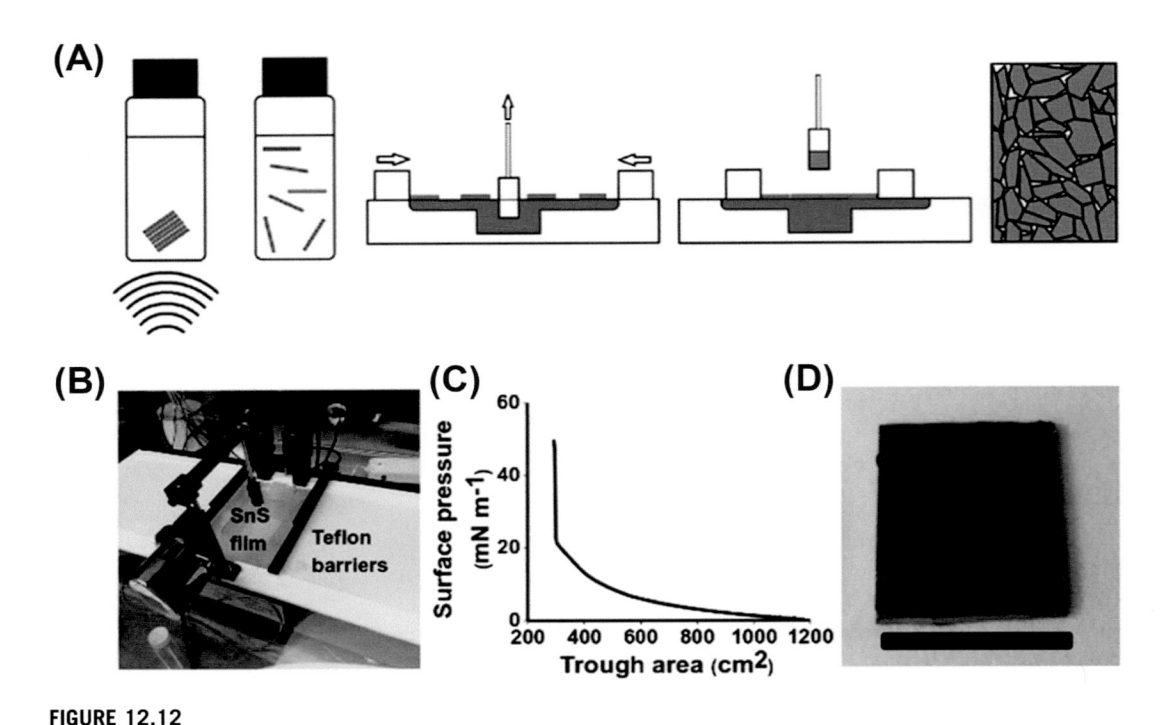

FIGURE 12.12

(A) A schematic of Langmuir–Blodgett film preparation; (B) a photo of Langmuir–Blodgett trough with compressed SnS film; (C) Π-A isotherm during film compression; (D) a photo of a sample prepared on Si/ SiO$_2$ substrate with edges masked (scale bar 1.5 cm).

Reprinted from K. Norton et al., Preparation of solution processed photodetectors comprised of two-dimensional tin(ii) sulfide nanosheet thin films assembledviathe Langmuir-Blodgett method, RSC Adv. 11 (2021) 26813–26819 [87].

12.4.4 **Substrate patterning**

Langmuir films can also be used indirectly as patterns for lithography. One of the possible techniques is nanosphere lithography [88]. In this technique, the desired material is deposited over a film of densely packed spheres [89]. After a desired thickness of the material is deposited, the spheres are removed. Thus only the material deposited into the interparticle areas will remain on the sample. The simplicity of their preparation and the possibility of large-scale sample production favor Langmuir films as a template for the nanosphere lithography [90]. Alternatively, Langmuir films can be used as templates for the precise growth of nanopillars or nanocones (Fig. 12.13) [91]. An LB film of nanoparticles was first treated by reactive ion etching (RIE) to optimize their diameter. Afterward, the sample was treated by RIE, which etched away the silicon substrate. In the final step, additional HF treatment can remove the nanoparticles on top of the pillars. The spacing and height of the nanopillars depend on the size and spacing of deposited NPs, and the etching duration.

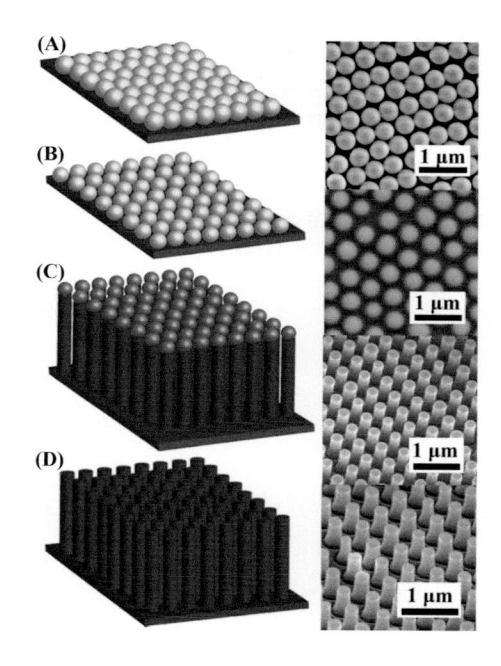

FIGURE 12.13

Fabrication process of Si nanopillars: (A) Deposition of the silica nanoparticles by Langmuir–Blodgett; (B) shrinking of the mask by isotropic RIE of SiO_2; (C) anisotropic etching of Si into pillars by RIE; (D) removal of the residual mask by HF etching. *RIE*, Reactive ion etching.

Reprinted from C.M. Hsu, S.T. Connor, M.X. Tang, Y. Cui, Wafer-scale silicon nanopillars and nanocones by Langmuir-Blodgett assembly and etching, Appl. Phys. Lett. 93 (2008), with the permission of AIP Publishing.

12.4.5 Batteries

A promising application of Langmuir films is in battery research [92]. Langmuir films can be used as templates for later processing and patterning or to prepare layered structures. Chen et al. prepared Li-ion batteries with a cathode coated with LB films of graphene [92]. The graphene coating improved Li^+ ion diffusion and reduced the oxidation of the cathode and the accumulation of Li^+ ions on the cathode's surface.

Ramasamy et al. used the LB deposition method to fabricate Li-ion battery anodes [92]. First, a GO-coated copper was prepared by LB deposition with subsequent thermal treatment to give reduced graphene oxide (rGO). Next, a layer of SnO_2 nanoparticles was deposited on the rGO film, followed by a second LB deposition. Repeating such a process, the authors created nanocomposites with excellent adaptability to volume changes while maintaining a good conductivity.

Kim et al. used an alternative approach to produce CNT/graphene nanocomposites (Fig. 12.14) [93]. In their work, the authors first coated CNTs with a positively charged cetyltrimethylammonium 4-vinylbenzoate and prepared a standard Langmuir film from such a solution. After film formation, they injected negatively charged graphene nanosheets into the subphase. The opposite charges on CNTs and graphene nanoflakes caused a Coulomb attraction between them. As a result,

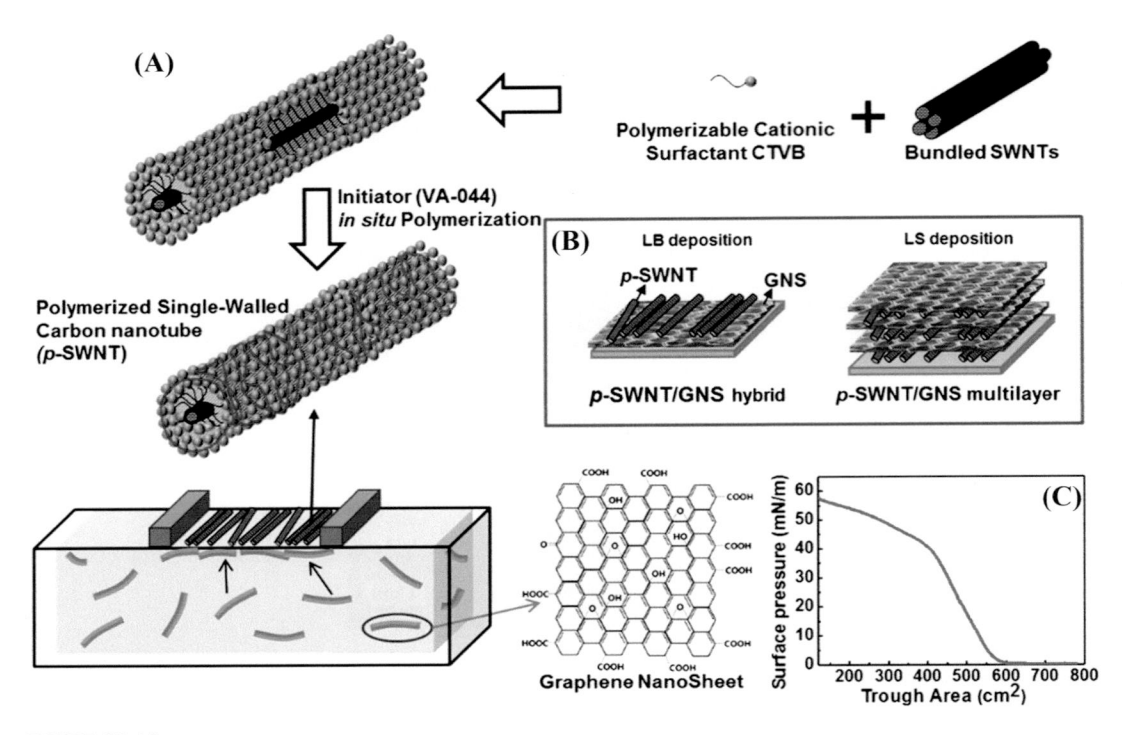

FIGURE 12.14

(A) A schematic illustration of the nanocomposite anode preparation; (B) preparation of multilayered anodes by a repeating Langmuir–Schaefer deposition; (C) the Π-A isotherm of the CNTs compression.

Reproduced from H. Kim et al. Spontaneous hybrids of graphene and carbon nanotube arrays at the liquid-gas interface for Li-ion battery anodes, Chem. Commun. 54 (2018) 5229–5232 with permission from the Royal Society of Chemistry.

a hybrid film of CNT-decorated graphene nanoflakes was obtained. A second approach by the same group consisted of preparing a wrinkled CNT film to increase the surface area [94] of Li-ion anodes. A larger deposition compression was used, resulting in an above-monolayer Π. The above-monolayer Π forces the nanosheets to wrinkle as a smaller surface area is available.

Langmuir films can be useful for battery characterization as well. Liu et al. have developed an on-chip diagnostic device for Zn-ion batteries [95]. In their work, the authors used MnO_2 nanowires self-assembled films transferred onto a current collector as a working electrode during electrochemical processes.

12.5 Summary

Langmuir and Langmuir–Blodgett films, as developed in the late 19th century for amphiphilic molecules, are also beneficial in today's research. A deeper understanding of the underlying physical and chemical principles allows for optimizing the process for a wide range of nanomaterials. Invented more than 130 years ago, the technique can be used to produce highly ordered monolayers of nanoparticles, aligned nanotubes, or interconnected nanosheets. Today, Langmuir films provide a platform for fundamental research, but also for a range of applications, from more straightforward and cost-effective self-assembled Langmuir films to advanced large-area roll-to-roll Langmuir–Blodgett deposition.

References

[1] M. Ahmad, S.R.P. Silva, Low temperature growth of carbon nanotubes – a review, Carbon N. Y. 158 (2020) 24–44.

[2] Y. Shi, et al., Hierarchical growth of Au nanograss with intense built-in hotspots for plasmonic applications, J. Mater. Chem. C. 8 (2020) 16073–16082.

[3] C.H. Lee, M. Xie, V. Kayastha, J. Wang, Y.K. Yap, Patterned growth of boron nitride nanotubes by catalytic chemical vapor deposition, Chem. Mater. 22 (2010) 1782–1787.

[4] P. Novikov, et al., Space arrangement of Ge nanoislands formed by growth of Ge on pit-patterned Si substrates, J. Cryst. Growth 323 (2011) 198–200.

[5] Y. Song, et al., Graphene transfer: paving the road for applications of chemical vapor deposition graphene, Small 17 (2021) 2007600.

[6] J. Yang, et al., MoS_2 liquid cell electron microscopy through clean and fast polymer-free MoS_2 transfer, Nano Lett. 19 (2019) 1788–1795.

[7] M. Sharma, A. Singh, R. Singh, Monolayer MoS_2 transferred on arbitrary substrates for potential use in flexible electronics, ACS Appl. Nano Mater. 3 (2020) 4445–4453.

[8] J. Quinson, K.M.Ø. Jensen, From platinum atoms in molecules to colloidal nanoparticles: a review on reduction, nucleation and growth mechanisms, Adv. Colloid Interface Sci. 286 (2020) 102300.

[9] S. Qiu, et al., Solution-processing of high-purity semiconducting single-walled carbon nanotubes for electronics devices, Adv. Mater. 31 (2019) 1800750.

[10] J.N. Coleman, et al., Two-dimensional nanosheets produced by liquid exfoliation of layered materials, Sci. 331 (2011) 568–571.

[11] C.J. Brinker, Dip Coating, in: T. Schneller, R. Waser, M. Kosec, D. Payne (Eds.), Chemical Solution Deposition of Functional Oxide Thin Films, Springer, Vienna, 2013, pp. 233–261. Available from: https://doi.org/10.1007/978-3-211-99311-8_10.

[12] D.P. Birnie, Spin coating: art and science. in, in: T. Schneller, R. Waser, M. Kosec, D. Payne (Eds.), Chemical Solution Deposition of Functional Oxide Thin Films, Springer, Vienna, 2013, pp. 263–274. Available from: https://doi.org/10.1007/978-3-211-99311-8.

[13] Y. Wang, W. Zhou, A review on inorganic nanostructure self-assembly, J. Nanosci. Nanotechnol. 10 (2010) 1563–1583.

[14] J. Cheng, F. Liu, Z. Tang, Y. Li, Scalable blade coating: a technique accelerating the commercialization of perovskite-based photovoltaics, Energy Technol. 9 (2021) 2100204.

[15] C. Farcau, et al., High-sensitivity strain gauge based on a single wire of gold nanoparticles fabricated by stop-and-go convective self-assembly, ACS Nano 5 (2011) 7137–7143.

[16] C. Wang, et al., Precise patterning of single crystal arrays of organic semiconductors by a patterned microchannel dip-coating method for organic field-effect transistors, J. Mater. Chem. C. 9 (2021) 5174–5181.

[17] P. Pieranski, Two-dimensional interfacial colloidal crystals, Phys. Rev. Lett. 45 (1980) 569.

[18] A. Pockels, Surface tension, Nature 43 (1891) 437–439.

[19] I. Langmuir, Monomolecular layers of fatty acids at air–water interfaces, J. Am. Chem. Soc. 39 (1917) 1848.

[20] K.B. Blodgett, Monomolecular films of fatty acids on glass, J. Am. Chem. Soc. 56 (1934) 495.

[21] I. Langmuir, V.J. Schaefer, Built-up films of proteins and their properties, Sci. 85 (1937) 76–80.

[22] K.B. Blodgett, I. Langmuir, Built-up films of barium stearate and their optical properties, Phys. Rev. 51 (1937) 964–982.

[23] M. Bodik, M. Jergel, E. Majkova, P. Siffalovic, Langmuir films of low-dimensional nanomaterials, Adv. Colloid Interface Sci. 283 (2020) 102239.

[24] S. Kinge, M. Crego-Calama, D.N. Reinhoudt, Self-assembling nanoparticles at surfaces and interfaces, ChemPhysChem 9 (2008) 20–42.

[25] J. Kim, et al., Graphene oxide sheets at interfaces, J. Am. Chem. Soc. 132 (2010) 8180–8186.

[26] M. Bodik, et al., The collapse mechanism in few-layer MoS_2 Langmuir films, J. Phys. Chem. C. 124 (2020) 15856–15861.

[27] R. Wang, et al., Facile preparation of self-assembled black phosphorus-dye composite films for chemical gas sensors and surface-enhanced Raman scattering performances, ACS Sustain. Chem. Eng. 8 (2020) 4521–4536.

[28] D. Wang, Y.L. Chang, Z. Liu, H. Dai, Oxidation resistant germanium nanowires: bulk synthesis, long chain alkanethiol functionalization, and Langmuir-Blodgett assembly, J. Am. Chem. Soc. 127 (2005) 11871–11875.

[29] H. Kaur, et al., Large area fabrication of semiconducting phosphorene by Langmuir-Blodgett assembly, Sci. Rep. 6 (2016) 1–8.

[30] J. Lv, et al., Gold nanowire chiral ultrathin films with ultrastrong and broadband optical activity, Angew. Chem. – Int. Ed. 56 (2017) 5055–5060.

[31] F. Schulz, et al., Structural order in plasmonic superlattices, Nat. Commun. 11 (2020) 1–9.

[32] M.J. Large, S.P. Ogilvie, A.A.K. King, A.B. Dalton, Understanding solvent spreading for Langmuir deposition of nanomaterial films: a Hansen solubility parameter approach, Langmuir 33 (2017) 14766–14771.

[33] R. Reiter, H. Motschmann, H. Orendi, A. Nemetz, W. Knoll, Ellipsometric microscopy. imaging monomolecular surfactant layers at the air–water interface, Langmuir 8 (1992) 1784–1788.

[34] D. Vollhardt, Brewster angle microscopy: a preferential method for mesoscopic characterization of monolayers at the air/water interface, Curr. Opin. Colloid Interface Sci. 19 (2014) 183–197.

[35] H.L. Nie, X. Dou, Z. Tang, H.D. Jang, J. Huang, High-yield spreading of water-miscible solvents on water for langmuir-blodgett assembly, J. Am. Chem. Soc. 137 (2015) 10683–10688.

[36] M. Bodik, et al., Langmuir–Scheaffer technique as a method for controlled alignment of 1D materials, Langmuir 36 (2020) 4540–4547.

[37] L. Wilhelmy, Ueber die Abhängigkeit der Capillaritäts-Constanten des Alkohols von Substanz und Gestalt des benetzten festen Körpers, Ann. Phys. 195 (1863) 177–217.

[38] J.F. Padday, D.R. Russell, The measurement of the surface tension of pure liquids and solutions, J. Colloid Sci. 15 (1960) 503–511.

[39] M.C. Petty, Langmuir-Blodgett Films: An Introduction, Cambridge University Press, 1996.

[40] G. Roberts, Langmuir-Blodgett Films, Springer Science & Business Media, 2013.

[41] K. Vegso, et al., Nonequilibrium phases of nanoparticle Langmuir films, Langmuir 28 (2012) 10409–10414.

[42] G. Barnes, I. Gentle, Interfacial Science: An Introduction, Oxford University Press, 2011.

[43] M. Bodik, et al., Collapse Mechanism in few-layer MoS_2 Langmuir films, J. Phys. Chem. C. 15 (2020). Available from: https://doi.org/10.1021/acs.jpcc.0c02365.

[44] D.M. Balazs, T.A. Dunbar, D.M. Smilgies, T. Hanrath, Coupled dynamics of colloidal nanoparticle spreading and self-assembly at a fluid-fluid interface, Langmuir 36 (2020) 6106−6115.

[45] A. Dong, Y. Jiao, D.J. Milliron, Electronically coupled nanocrystal superlattice films by in situ ligand exchange at the liquid-air interface, ACS Nano 7 (2013) 10978−10984.

[46] A. Dong, J. Chen, P.M. Vora, J.M. Kikkawa, C.B. Murray, Binary nanocrystal superlattice membranes self-assembled at the liquid−air interface, Nat. 2010 466 (2010) 474−477.

[47] T. Wen, S.A. Majetich, Ultra-large-area self-assembled monolayers of nanoparticles, ACS Nano 5 (2011) 8868−8876.

[48] M. Parchine, J. McGrath, M. Bardosova, M.E. Pemble, Large area 2D and 3D colloidal photonic crystals fabricated by a roll-to-roll, Langmuir-Blodgett Method. Langmuir 32 (2016) 5862−5869.

[49] A. Tao, et al., Langmuir-Blodgett silver nanowire monolayers for molecular sensing using surface-enhanced Raman spectroscopy, Nano Lett. 3 (2003) 1229−1233.

[50] T. Maurer, P.M. Adam, G. Lévêque, Coupling between plasmonic films and nanostructures: from basics to applications, Nanophotonics 4 (2015) 363−382.

[51] V. Amendola, R. Pilot, M. Frasconi, O.M. Maragò, M.A. Iatì, Surface plasmon resonance in gold nanoparticles: a review, J. Phys. Condens. Matter 29 (2017) 203002.

[52] M. Tahghighi, I. Mannelli, D. Janner, J. Ignés-Mullol, Tailoring plasmonic response by Langmuir−Blodgett gold nanoparticle templating for the fabrication of SERS substrates, Appl. Surf. Sci. 447 (2018) 416−422.

[53] M. Tahghighi, D. Janner, J. Ignés-Mullol, Optimizing gold nanoparticle size and shape for the fabrication of SERS substrates by means of the Langmuir−Blodgett technique, Nanomater 10 (2020) 2264.

[54] S. Saha, M. Ghosh, J. Chowdhury, Infused self-assembly on Langmuir−Blodgett film: fabrication of highly efficient SERS active substrates with controlled plasmonic aggregates, J. Raman Spectrosc. 50 (2019) 330−344.

[55] S.K. Das, K. Pal, T.S. Bhattacharya, P. Karmakar, J. Chowdhury, Fabrication of SERS active Langmuir−Blodgett Film substrate for screening human cancer cell lines: experimental observations supported by multivariate data analyses, Sens. Actuators B Chem. 299 (2019) 126962.

[56] B. Tim, P. Błaszkiewicz, A.B. Nowicka, M. Kotkowiak, Optimizing SERS performance through aggregation of gold nanorods in Langmuir-Blodgett films, Appl. Surf. Sci. 573 (2022) 151518.

[57] N.V. Long, et al., Iron oxide nanoparticles for next generation gas sensors, Int. J. Metall. Mater. Eng. 1 (2015).

[58] S. Capone, et al., Fe_3O_4/γ-Fe_2O_3 nanoparticle multilayers deposited by the langmuir-blodgett technique for gas sensors application, Langmuir 30 (2014) 1190−1197.

[59] J. Ivančco, et al., Nitric dioxide and acetone sensors based on iron oxide nanoparticles, Sens. Lett. 11 (2013) 2322−2326.

[60] S. Luby, et al., Sensitivity and long-term stability of γ-Fe_2O_3 and $CoFe_2O_4$ nanoparticle gas sensors for NO_2, CO and acetone sensing − a comparative study, Conf. Proc. − 10th Int. Conf. Adv. Semicond. Devices Microsystems, ASDAM 2014 (2014) 245−246. Available from: https://doi.org/10.1109/ASDAM.2014.6998691.

[61] I.V. Zaporotskova, N.P. Boroznina, Y.N. Parkhomenko, L.V. Kozhitov, Carbon nanotubes: sensor properties. A review, Mod. Electron. Mater. 2 (2016) 95−105.

[62] B. Pullithadathil, S. Abdulla, Unidirectional Langmuir-Blodgett-mediated alignment of polyaniline-functionalized multiwalled carbon nanotubes for NH3gas sensor applications, Langmuir 36 (2020) 11618−11628.

[63] C. Wang, Y. Wang, Z. Yang, N. Hu, Review of recent progress on graphene-based composite gas sensors, Ceram. Int. 47 (2021) 16367−16384.

[64] D.J. Buckley, et al., Frontiers of graphene and 2D material-based gas sensors for environmental monitoring, 2D Mater. 7 (2020) 032002.

[65] S. Basu, P. Bhattacharyya, Recent developments on graphene and graphene oxide based solid state gas sensors, Sens. Actuators B Chem. 173 (2012) 1−21.

[66] D. Kostiuk, et al., Few-layer graphene langmuir-schaefer nanofilms for H_2 gas sensing, Procedia Eng. 168 (2016) 243−246.

[67] S. Nufer, et al., Edge-selective gas detection using Langmuir films of graphene platelets, ACS Appl. Mater. Interfaces 10 (2018) 21740−21745.

[68] B. Deng, et al., Toward mass production of CVD graphene films, Adv. Mater. 31 (2019) 1800996.

[69] G. Deokar, et al., Towards high quality CVD graphene growth and transfer, Carbon N. Y. 89 (2015) 82−92.

[70] J.H. Choi, et al., Graphene-based gas sensors with high sensitivity and minimal sensor-to-sensor variation, ACS Appl. Nano Mater. 3 (2020) 2257−2265.

[71] H. Tabata, H. Matsuyama, T. Goto, O. Kubo, M. Katayama, Visible-light-activated response originating from carrier-mobility modulation of NO_2 gas sensors based on MoS_2 monolayers, ACS Nano 15 (2021) 2542−2553.

[72] Y. Zhao, et al., Low-temperature synthesis of 2D MoS_2 on a plastic substrate for a flexible gas sensor, Nanoscale 10 (2018) 9338−9345.

[73] B. Cho, et al., Charge-transfer-based gas sensing using atomic-layer MoS_2, Sci. Rep. 5 (51) (2015) 1−6.

[74] A. Kalosi, et al., Tailored Langmuir−Schaefer deposition of few-layer MoS_2 nanosheet films for electronic applications, Langmuir 35 (2019) 9802−9808.

[75] M. Bodik, et al., Collapse mechanism in few-layer MoS_2 Langmuir films, J. Phys. Chem. C. (2020). Available from: https://doi.org/10.1021/acs.jpcc.0c02365.

[76] N. Huo, et al., High carrier mobility in monolayer CVD-grown MoS_2 through phonon suppression, Nanoscale 10 (2018) 15071−15077.

[77] S.E. Zaki, et al., Role of oxygen vacancies in vanadium oxide and oxygen functional groups in graphene oxide for room temperature CO_2 gas sensors, Sens. Actuators, A Phys. 294 (2019) 17−24.

[78] C. Qian, et al., Facile preparation of self-assembled black phosphorus-based composite LB films as new chemical gas sensors, Colloids Surf. A Physicochem. Eng. Asp. 608 (2021) 125616.

[79] T.N. Ly, S. Park, Highly sensitive gas sensor using hierarchically self-assembled thin films of graphene oxide and gold nanoparticles, J. Ind. Eng. Chem. 67 (2018) 417−428.

[80] J. Miao, J.Y.S. Lin, Nanometer-thick films of aligned ZnO nanowires sensitized with Au nanoparticles for few-ppb-level acetylene detection, ACS Appl. Nano Mater. 3 (2020) 9174−9184.

[81] D. Kostiuk, et al., Graphene Langmuir-Schaefer films decorated by Pd nanoparticles for NO_2 and H_2 gas sensors, Meas. Sci. Rev. 19 (2019) 64−69.

[82] Q. Cao, et al., Arrays of single-walled carbon nanotubes with full surface coverage for high-performance electronics, Nat. Nanotechnol. 8 (2013) 180−186.

[83] Q. Cao, J. Tersoff, D.B. Farmer, Y. Zhu, S.J. Han, Carbon nanotube transistors scaled to a 40-nanometer footprint, Sci. 356 (2017) 1369−1372.

[84] C. Xu, C. Li, Y. Jin, Programmable organic-free negative differential resistance memristor based on plasmonic tunnel junction, Small 16 (2020) 2002727.

[85] T. Berzina, K. Gorshkov, A. Pucci, G. Ruggeri, V. Erokhin, Langmuir−Schaefer films of a polyaniline−gold nanoparticle composite material for applications in organic memristive devices, RSC Adv. 1 (2011) 1537−1541.

[86] S. Rhee, et al., Polarized electroluminescence emission in high-performance quantum rod light-emitting diodes via the Langmuir-Blodgett technique, Small 17 (2021) 2101204.

[87] K. Norton, et al., Preparation of solution processed photodetectors comprised of two-dimensional tin(ii) sulfide nanosheet thin films assembled via the Langmuir-Blodgett method, RSC Adv. 11 (2021) 26813−26819.

[88] J.C. Hulteen, R.P. Van Duyne, Nanosphere lithography: a materials general fabrication process for periodic particle array surfaces, J. Vac. Sci. Technol. A Vacuum, Surfaces, Film. 13 (1998) 1553.

[89] X. Wang, C.J. Summers, Z.L. Wang, Large-scale hexagonal-patterned growth of aligned ZnO nanorods for nano-optoelectronics and nanosensor arrays, Nano Lett. 4 (2004) 423–426.

[90] M. Thangamuthu, C. Santschi, O.J.F. Martin, Reliable Langmuir Blodgett colloidal masks for large area nanostructure realization, Thin Solid. Films 709 (2020).

[91] C.M. Hsu, S.T. Connor, M.X. Tang, Y. Cui, Wafer-scale silicon nanopillars and nanocones by Langmuir-Blodgett assembly and etching, Appl. Phys. Lett. 93 (2008).

[92] C. Fang, et al., Recent applications of Langmuir−Blodgett technique in battery research, ACS Appl. Mater. Interfaces 14 (2022) 2431–2439.

[93] H. Kim, et al., Spontaneous hybrids of graphene and carbon nanotube arrays at the liquid-gas interface for Li-ion battery anodes, Chem. Commun. 54 (2018) 5229–5232.

[94] H.S. Jeong, et al., Oriented wrinkle textures of free-standing graphene nanosheets: application as a high-performance lithium-ion battery anode, Carbon Lett. 31 (2021) 277–285.

[95] Q. Liu, et al., Langmuir−Blodgett nanowire devices for in situ probing of zinc-ion batteries, Small 15 (2019) 1902141.

MXenes and their applications in sensors

13

Jun-Ge Liang[1,2] and Lijia Pan[2]

[1]*Engineering Research Center of IoT Technology Applications (Ministry of Education), Department of Electronic Engineering, Jiangnan University, Wuxi, P.R. China* [2]*School of Electronic Science and Engineering, Collaborative Innovation Center of Advanced Microstructures, Nanjing University, Nanjing, P.R. China*

The development of nanotechnology and new materials, especially the two-dimensional (2D) nano-materials, is bringing increased vitality to the sensors field. 2D materials generally have single or multiatomic layer structures, including graphene, transition metal disulfides (TMDs), and covalent organic frameworks (COFs). In 2011, Naguib et al. prepared $Ti_3C_2T_x$ 2D material (T_x is the group on the surface of Ti_3C_2) for the first time by etching the aluminum atom layer in Ti_3AlC_2 with HF [1,2]. Its atomic structure is similar to that of carbon atoms in graphene, so it is named MXene, derived from $M_{n+1}X_n$ and graphene, respectively. MXenes, referring to the early transition metal carbides, nitrides, and carbonitrides, gained attention for their superb features such as high conductivity, hydrophilicity, and flexibility. The morphology can be adjusted by the control of etching, by which the nanoparticle, single-, or multiple layers nanosheets can be prepared. Besides, various functional groups can be easily modified by chemical reactions.

The high specific surface areas, abundant functional groups, specific structures, and mechanical properties endow MXenes with great potential as sensor applications, such as gas sensors, biosensors, strain sensors, and pressure sensors. After mixing with other substances, the MXenes composites could enhance their sensing performances. As gas-sensitive materials, MXenes feature a large specific surface area and abundant surface functional groups, and hence can easily absorb gas molecules. The absorption reduces the system enthalpy and induces the electron or hole depletion, which shows a linear conductivity decrease according to the gas concentration increase. As biosensors, MXenes show good biocompatibility and provide superior sites for biomolecule fixation. As flexible sensors, including in tensile and pressure applications, due to the unique accordion-like shape and stable mechanical properties, MXenes are applied as supersensitive piezoresistive sensors and express the conductivity change regarding external forces.

Materials Nanoarchitectonics. DOI: https://doi.org/10.1016/B978-0-323-99472-9.00008-0

13.1 MXenes' properties

13.1.1 Chemical properties

MXenes are a class of two-dimensional inorganic compound materials with a layered structure similar to graphene and the properties of metal compounds. The general formula of the structure of MXene is $M_{n+1}X_n T_x$ ($n = 1-3$), in which T represents the surface functional groups, generally referring to the fluorine ($-F$), oxygen ($-O$), and hydroxyl ($-OH$), and x is their number; M represents an early transition metal, such as Ti, V, Cr, Zr, Nb, Ta, Mo, etc.; X is the C or N element. The differences in the chemical properties of MXenes are mainly due to the different functional groups inserted on their surfaces.

In 2000, Barsoum et al. collectively named the ternary compound with the general chemical formula of $M_{n+1}AX_n$ ($n = 1-3$) as MAX phase, in which M is the transition metal element, A is mainly the third and fourth main group element, and X is the carbon or nitrogen element. Up to now, the MAX phase family has been expanded to more than 100 species, the M position has been developed to lanthanide rare earth metal, the A position has added subgroup elements containing less than D electrons, and the X position has added boron element, as shown in Fig. 13.1. In the crystal structure of MAX phase, $M_{n+1}X_n$ unit layer has a strong covalent bond. In contrast, the electron cloud overlap between the atom of the A-layer and the atom of the adjacent M layer is low, resulting in weak binding. Based on this structure, $M_{n+1}X_n$ can be separated from MAX, and a large class of two-dimensional transition metal-carbon/nitride materials can be prepared, as shown in Fig. 13.2.

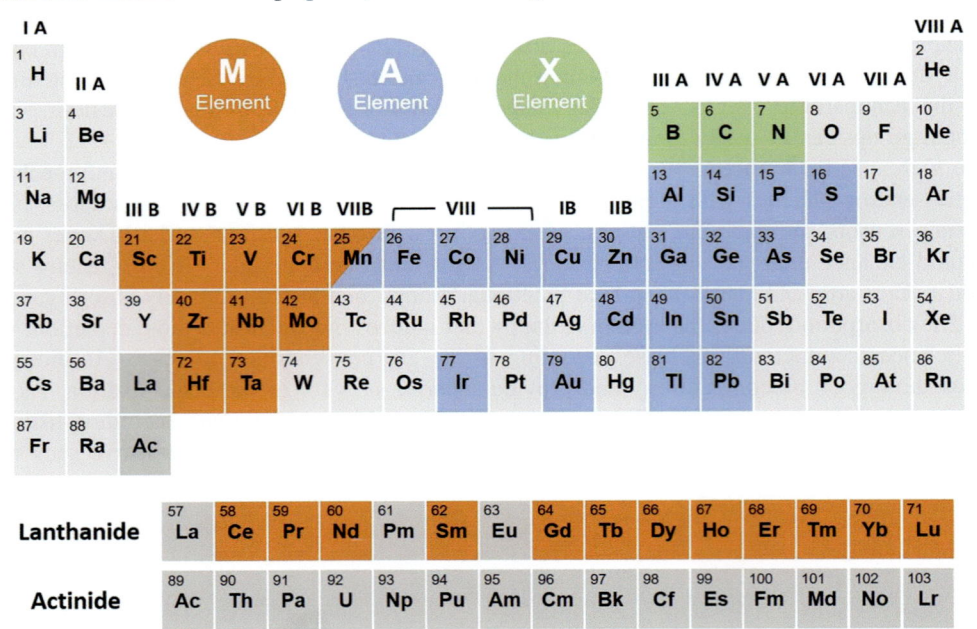

FIGURE 13.1

MAX phases $M_{n+1}AX_n$ forming elements [3].

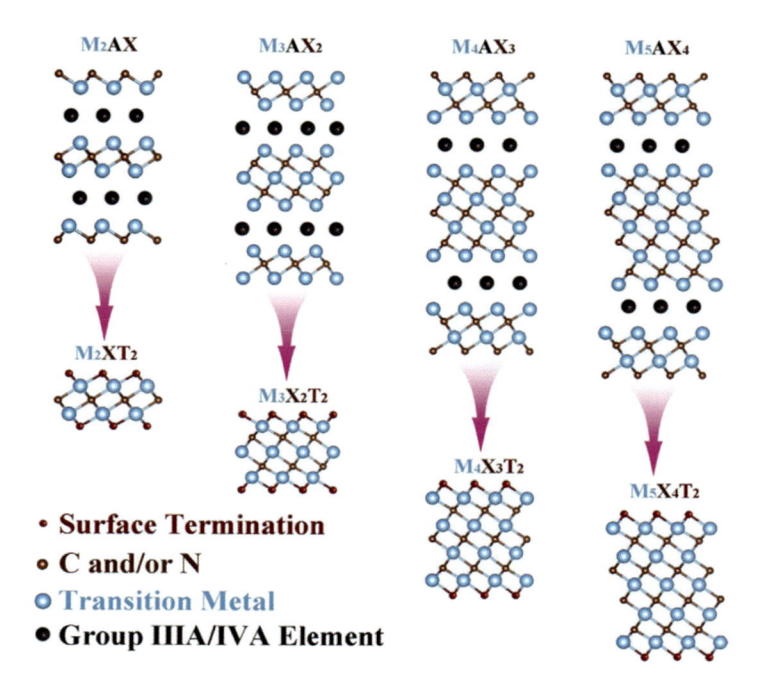

FIGURE 13.2

Atomic structure of MXenes [1].

13.1.2 Electric properties

Electronic properties are the core properties of MXene, and similar to other 2D materials, MXenes have high conductivity. The difference is that the electrical properties of MXenes are related to the ordered arrangement of metal atoms within them, which makes them conductors or semiconductors. The required surface groups of MXenes can be prepared by chemical reactions, which leads to differences in electron attraction ability. The increase of surface defects formed during the preparation of MXenes restricts the free movement of electrons, resulting in the unbalanced distribution of electron density in the metal layer, which will significantly affect its electrical properties [4]. Fig. 13.3 presents experimental values of electrical conductivities of different MXenes. Owing to variations in the (1) surface functionalities, (2) defect concentration, (3) d-spacing betwixt MXenes particulates, (4) delamination yield, and (5) lateral dimensions caused due to the etching mechanism, the electrical conductivities of $Ti_3C_2T_x$ vary from 850 to 9870 S/cm [5].

MXenes exhibit a metallic character with high electron density near the Fermi level [7]. However, the researchers found that the electrical properties of MXenes could be weakened by changing the species and orientation of functional groups to achieve the semiconductor properties. The outermost transition metal atoms can be replaced to form the double transition metal M′M″Xene, such as $Mo_2TiC_2T_x$ and $Mo_2Ti_2C_3T_x$. In addition, surface groups also greatly influence the electrical properties of MXene. According to first-principle calculation, Mo-MXene modified

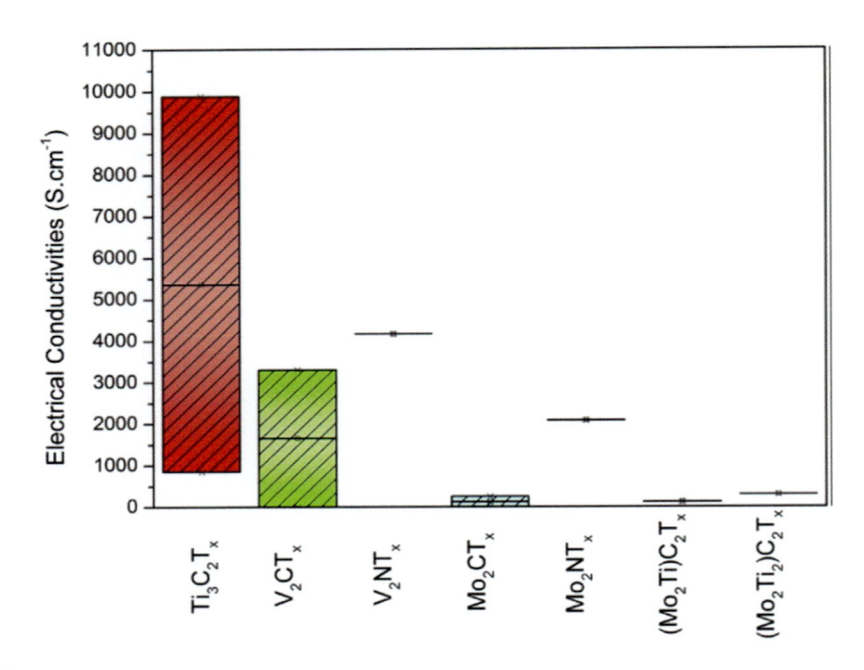

FIGURE 13.3

Experimental electrical conductivities of different MXenes [6].

by −OH and −F functional groups presents metallic properties, while Mo-MXene modified by −O functional groups expresses semiconductor properties. In addition to adjusting the surface groups (−OH, −F, −O), the electrical properties of MXene can also be tuned by a covalent graft of functional groups. For example, monolayer $Ti_3C_2T_x$ MXene was chemically modified by diazo salt of 4-nitrobenzene tetrafluoroborate. After the 4-nitrophenyl group is grafted to the surface of MXene, the conductivity and mobility decrease with the increase of the concentration of the group grafted to the surface [8]. Similarly, installing and removing surface groups by performing substitution and elimination reactions in molten inorganic salts will also change the electrical properties of MXene, such as the conductivity, work function, and electron mobility [9].

13.1.3 Optical properties

The linear optical properties (such as absorption, photoluminescence, etc.) and nonlinear optical properties (such as saturation absorption, nonlinear refractive index, etc.) of MXenes are highly dependent on their band structure (such as bandgap, direct/indirect bandgap, topological insulation properties, etc.). When MXenes interact with the light quanta, if the optical bandgap of MXenes is equal to the light quanta's power, the light quanta can be absorbed. As mentioned, the surface functional groups highly influence the MXenes' properties, which can serve in tuning the optical properties.

(A) **(B)**

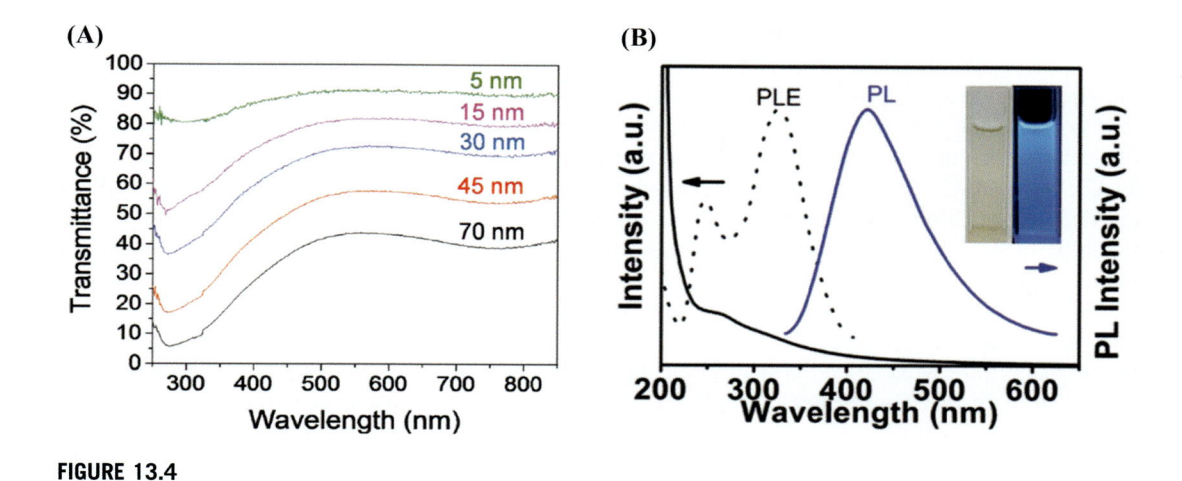

FIGURE 13.4

(A) UV–Vis spectra of $Ti_3C_2T_x$ films with different thicknesses [10]; (B) UV–Vis spectra (solid line), PLE (dashed line), and PL spectra (solid line, Ex = 320 nm) of MQD-100 in aqueous solutions [12].

In the visible spectral range, the −O terminal samples show stronger light absorption than the pure MXenes, while the fluorinated and hydroxylated samples show weaker absorption ability. $M_{n+1}X_n$ with a lower n is generally considered with higher optical transparency due to its lower state density. Pure phase MXenes mostly have a bandgap around 2.7 eV, accordingly expressing poor light absorption. However, pure MXene has a high carrier recombination rate. Therefore, it can be utilized as an efficient cocatalyst to form a Schottky junction or heterojunction by coupling with other semiconductor materials. The compounds can effectively inhibit the recombination of photogenerated electron−hole pairs in the photocatalytic reaction process and improve the utilization efficiency of photogenerated charge. For example, pure Zr_2CO_2 has an absorption in the wavelength range of 300−500 nm, and the coupling effect between Zr_2CO_2 and blue phosphorus is induced by van der Waals force. Under biaxial compression strain, the bandgap of Zr_2CO_2 will change so that its light absorption range will shift to the visible region.

In addition to compounding with other semiconductors, the optical properties of MXenes can also be changed by the thickness control. As shown in Fig. 13.4A, the transmittance of $Ti_3C_2T_x$ rises with the film thickness reduction [10]. In general, the photoluminescence intensity of MXenes in an aqueous solution is extremely low, which significantly limits their application in biology and optics. The preparation of quantum dots (QDs) effectively reveals new physical properties caused by quantum confinement and edge effects [11]. Unlike nanosheets, QDs have an ultrasmall size, making them more suitable for biological imaging. As shown in Fig. 13.4B, due to the strong quantum confinement effect, MXenes exhibit luminescence properties with peaks at 250 and 320 nm [12].

13.1.4 Mechanical properties

MXenes, as a typical two-dimensional nanomaterial, has intense M-X binding energy, which endows MXenes with excellent mechanical properties. They are a more formidable structure than multilayer graphene of the same thickness. Generally speaking, an elastic body changes shape (called deformation) when an external force is applied. The general definition of elastic modulus is the stress in a unidirectional state of stress divided by the strain in that direction; "Young's modulus" is one of the manifestations. Kurtoglu and his colleagues used density functional theory to simulate the elastic modulus of several MXene materials, including Ta_2C, Ta_3C_2, Ta_4C_3, Ti_2C, Ti_3C_2, Ti_4C_3, V_2C, Cr_2C, Zr_2C, and Hf_2C. The results show that all the above MXenes exhibit excellent elastic modulus along the datum plane, ranging from 523 to 788 GPa [13]. Of course, the elastic modulus of MXenes is still lower than graphene [14], but the flexural stiffness is several times higher [15].

The experiment research also proved that MXene has high mechanical strength and can maintain the stability of the structure under significant stress. Ling et al. demonstrated that MXene and the mixed film of MXene and polyvinyl alcohol (PVA) possess excellent mechanical strength. The tensile strength of 3.3-μm-thick $Ti_3C_2T_x$ MXene film is about 22 MPa, and Young's modulus is about 3.5 GPa. Such properties are comparable to those of GO paper (GO) and Barker paper based on carbon nanotubes (CNT). Besides, Lipatov et al. reported the most robust 2D sheets with Young's modulus of 0.33 ± 0.03 TPa [16].

13.1.5 Magnetic properties

Similar to electrical properties, the functional groups also influence the magnetic properties of MXene. Theoretical calculations show that only pure MXenes, such as $Ti_{n+1}C_n$ and $Ti_{n+1}N_n$, ($n = 1-9$), are magnetic, primarily due to surface Ti atoms [17]. However, the magnetic properties of most MXenes will disappear after the functionalization, such as $Ti_3C_2F_2$ and $Ti_3C_2(OH)_2$ [18]. There are rigid covalent linkages among M and X, which render MXene's nonmagnetic properties. Theoretical calculations show that of all terminating MXenes, only Cr_2CT_x and Cr_2NT_x MXene possess magnetic domains, meaning that only Cr_2CT_x and Cr_2NT_x MXene are magnetic. In addition, through functional group modification of Cr_2CT_x, MXene can realize the ferromagnetic and antiferromagnetic transition, accompanied by the transition from metallic to insulating properties [19].

The magnetic properties of MXenes are also significantly different with different calculation methods. For example, LDA (local-density approximation) is used to predict the magnetic properties of Ta_2C and Ta_4C_3, and the results are nonmagnetic and antiferromagnetic materials. If they are predicted by PBEsol (Perdew–Burke–Ernzerhof) method, they are antiferromagnetic and nonmagnetic, respectively. Besides, the stress enhancement will also change the magnetic properties of MXenes, gradually increasing the tensile stress, and the nonmagnetic MXenes will also undergo the magnetic transformation.

13.1.6 Stability

The stability of materials in the environment limits their application field and practical value, so it is of great significance to study the stability of materials. Single MXene materials do not exist stably in an environment containing oxygen and water, and light also promotes the oxidation of MXenes. For example, the crystal structure of Ti_3C_2 MXenes after oxidation is that Ti atoms are oxidized into TiO_2, forming a composite structure of the TiO_2 layer and C layer [20]. The thermal stability of MXenes is closely related to their environment. In an Ar gas environment, Ti_3C_2 MXene shows different properties when the heating temperature is 25°C−1200°C (Fig. 13.5): At 500°C, the overall structure of Ti_3C_2 MXene is relatively stable, but a small number of TiO_2 crystals appear, which may be formed by the reaction of Ti_3C_2 MXene with the adsorbed −OH and =O functional groups at high temperature. When the temperature is 900°C, the crystal structure of Ti_3C_2 changes and TiC crystal appears. When the temperature is 1200°C, Ti_3C_2 completely transforms into the TiC phase, and it is speculated that the −F functional group has been eliminated from MXene at this time [21].

It can be seen from the above that a single MXene will undergo decomposition in the presence of oxygen and water. However, compared with the aqueous phase, the oxidation rate of MXenes in organic solvents, air and solid media is slower. Therefore, to inhibit the oxidation rate of MXenes, many antioxidants, such as L-ascorbate, have appeared in recent years. As shown in Fig. 13.6, $Ti_3C_2T_x$ retained its appearance after 6 months with the addition of sodium L-ascorbate. Sodium L-ascorbate as an antioxidant can be added to the colloidal $Ti_3C_2T_x$ MXene to avoid oxidation. However, as mentioned above, $Ti_3C_2T_x$ was oxidized and degraded to form TiO_2 and carbon without antioxidants. Under high temperature or ultraviolet radiation, the oxidation decomposition rate of MXenes flakes will be accelerated. Hence, the oxygen-less or dry-air storage will improve the MXenes' stability. In addition, setting MXenes in the colloidal solution in a dark, cool, vacuum environment will also effectively reduce the oxidation rate.

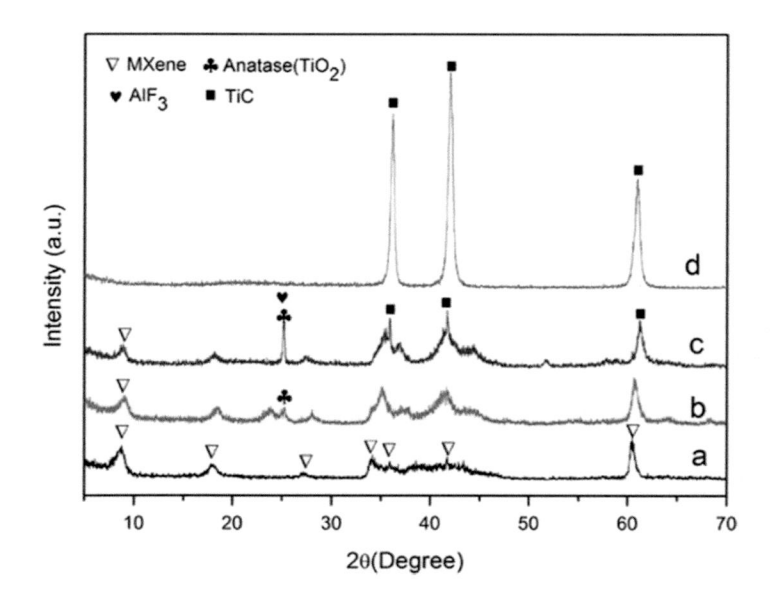

FIGURE 13.5

X-ray diffraction (XRD) patterns of as-prepared Ti_3C_2 nanosheets [21] (A) before and after heat treatment for (B) 500°C, (C) 900°C, (D) 1200°C in Ar.

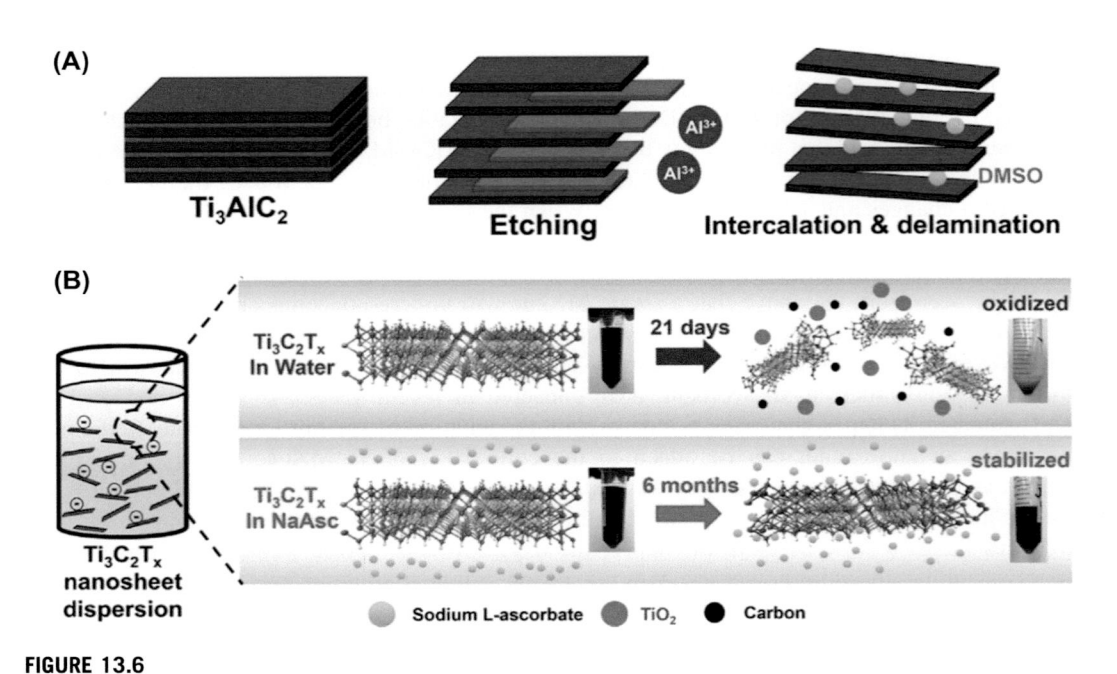

FIGURE 13.6

Schematic of shelf-stable Ti$_3$C$_2$T$_x$ nanosheet dispersion enabled by the antioxidant sodium L-ascorbate. (A) Schematic representation of Ti3C2Tx nanosheet synthesis, (B) Shelf-stable Ti3C2Tx nanosheets stabilized by sodium L-ascorbate (NaAsc) [22].

13.2 MXenes synthesis

13.2.1 Hydrofluoric etching

The most used method for MXenes preparation is the selective etching of the A atom layer in the MAX phase by hydrofluoric (HF) aqueous acid solution. In 2011, Barsoum et al. [2] used HF acid etching to prepare MXenes for the first time. Compared with the chemical bond of M-X, M-A is weaker and easier to be etched with HF acid. (Fig. 13.7). After etching A in $M_{n+1}AX_n$ with HF acid, a multilayer of $M_{n+1}X_n$ T_x containing terminal functional groups is obtained, among which the terminal functional groups T are usually $-O$, $-F$, and OH groups [23]. The chemical process of MXenes' preparation by HF acid etching is speculated to be as follows:

$$M_{n+1}AX_n + 3HF \rightarrow AF_3 + M_{n+1}X_n + 1.5H_2$$

$$M_{n+1}X_n + 2H_2O \rightarrow M_{n+1}X_n(OH)_2 + H_2$$

$$M_{n+1}X_n + 2HF \rightarrow M_{n+1}X_nF_2 + H_2$$

This method is simple and easy to implement. MXenes with different defects states and element sizes can be obtained by controlling the acid concentration, etching time, and operating temperature. In addition to HF etching, LiF/HCl, NH$_4$HF$_2$ and other fluorides can also work as the etching

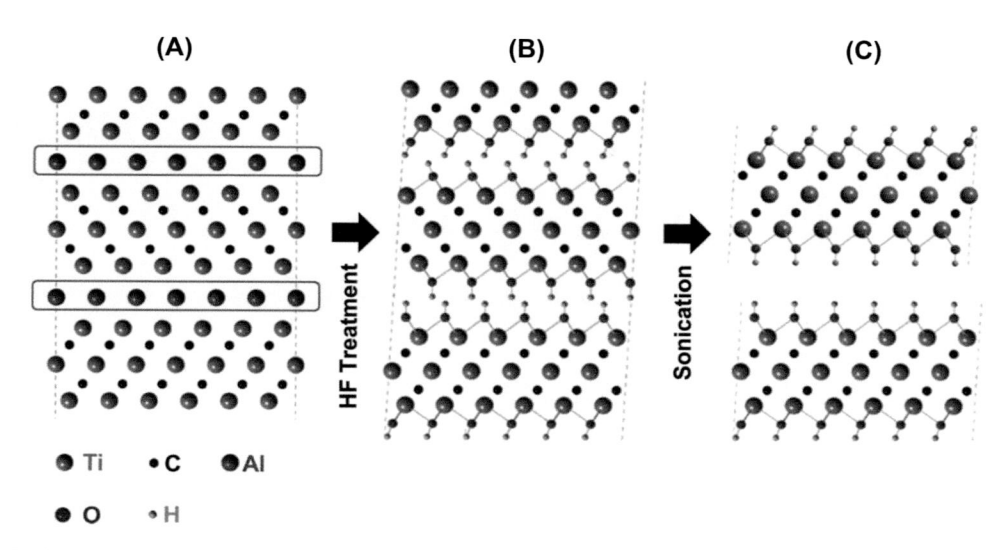

FIGURE 13.7

Schematic of the exfoliation process for Ti_3AlC_2. (A) Ti_3AlC_2 structure. (B) OH replaces Al atoms after reaction with hydrofluoric. (C) Breakage of the hydrogen bonds and separation of nanosheets after sonication in methanol [2].

agent. However, these methods also have disadvantages such as a low etching efficiency, uneven etching state, a mass of contaminants including unreacted HF acid solution, and precipitation of heavy metals, etc.

13.2.2 Fluoride etching

It has been found that ammonium hydrogen fluoride (NH_4HF_2), which shows a milder chemical property, can work as the substitute for an etching agent for MXenes preparation, and the MXenes are of high purity and offer higher thermal stability. Using HF acid to etch the Ti_3C_2 will generate TiO_2 at 200°C, while with NH_4HF_2 as the etching agent, the TiO_2 won't generate until the temperature reaches 500°C. Even if the temperature rises to 650°C and 900°C, only a small amount of TiO_2 generates on the surface, and the structure of Ti_3C_2 remains intact and shows higher thermal stability [1]. Halim [24] synthesized $Ti_3C_2T_x$ by etching Ti_3AlC_2 film with NH_4HF_2. By comparison, the final product features a more uniform layer spacing (Fig. 13.8). However, the etching process is slow due to the mild chemical properties of NH_4HF_2; and it might be incomplete, which affects the homogeneity of MXenes.

Chidiu [25] used lithium fluoride (LiF) and hydrochloric acid (HCl) as etching agents to achieve etching and intercalation simultaneously. This method shows high efficiency and productivity, besides it also proves that only the coexistence of protons and fluoride ions can successfully prepare the laminar MXenes. During the etching process, the presence of metal halides will insert cations (such as Li^+) and water into the MXenes, increasing the layer spaces and weakening the

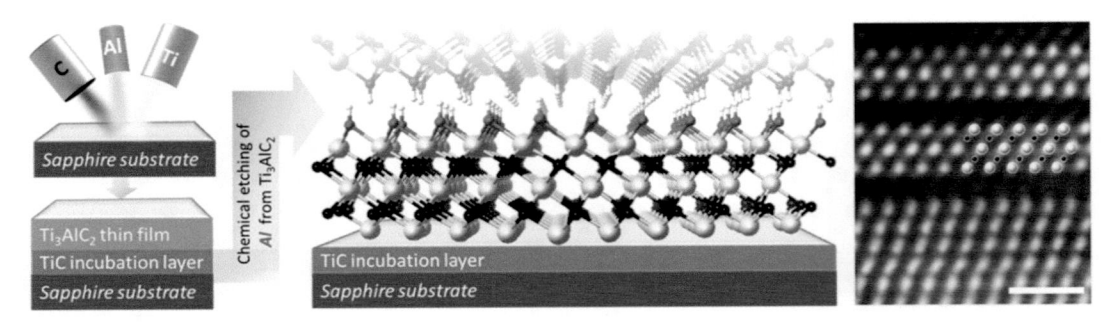

FIGURE 13.8

Steps used to produce epitaxial MXene films [24].

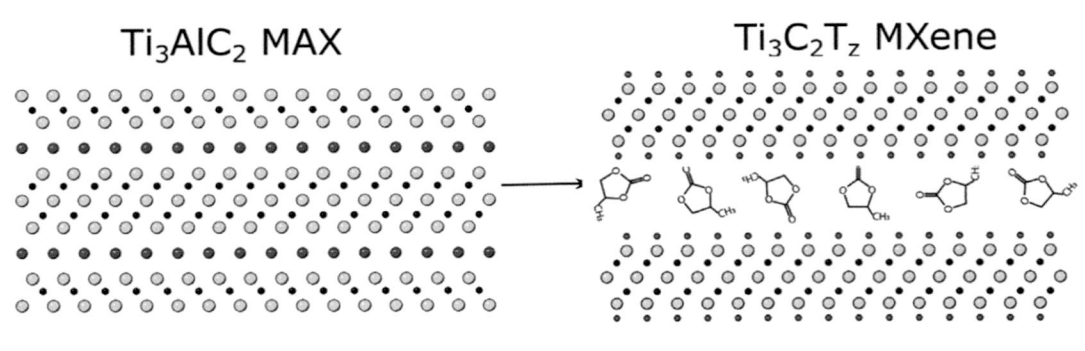

FIGURE 13.9

Water-free etching using polar organic solvents and ammonium bifluoride [26].

mutual interaction between the MXenes layers to achieve a relative more complete etching. Due to the intense surface activity of MXene, it has a noticeable adsorption effect, while solution etching will introduce trace water. Barsoum's team used polar organic solvents to replace water fluorinated solutions to achieve etching results (Fig. 13.9). NH_4HF_2 dissolved in organic solvents decomposed into HF and NH_4F can etch the A-layer to form fluorine-containing MXene [26].

13.2.3 Molten mixed fluoride

In 2013, Naguib et al. [13] prepared TiC_x by placing Ti_3AlC_2 in LF and heating it at 900°C, then processed Ti_3AlN_3 in molten mixed fluoride under argon at 550°C to prepare Ti_4N_3 which showed partial structural damage. $Ti_{n+1}N_n$ cannot be obtained by HF etching, since the stability of nitride MXene is poor in HF acid solution. The internal bond energy of nitride MAX is relative higher, hence the etching conditions are challenging to deploy. In 2016, Gogotsi's team obtained the first nitride MXene ($Ti_4N_3T_x$) through molten salt etching. As shown in Fig. 13.10, the reaction principle is similar to the solution method. F reacts with A-layer atoms in a molten salt containing

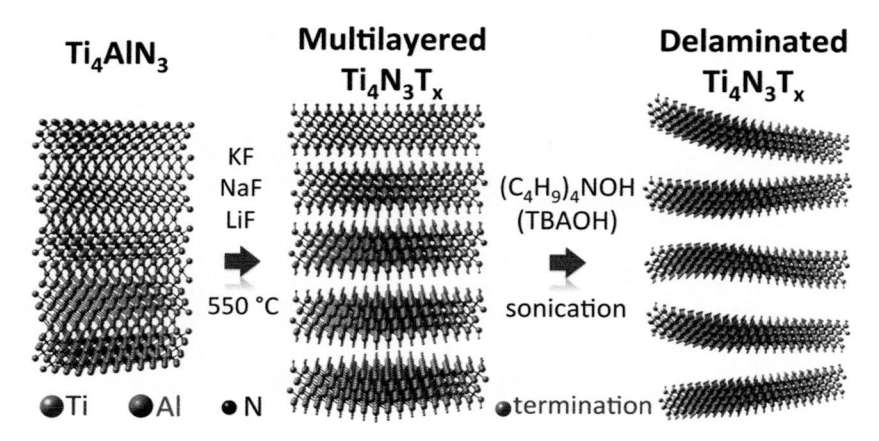

FIGURE 13.10

Schematic of the synthesis of $Ti_4N_3T_x$ by molten salt treatment of Ti_4AlN_3 at 550°C under Ar, followed by delamination of the multilayered MXene by TBAOH [27].

fluorine [27]. Due to the high temperature and high concentration, $Ti_4N_3T_x$ (T = F, O, OH) nanosheets with few or single layers can be obtained. The shortcoming of the molten salt method is apparent: it cannot guarantee good purity and crystallization. The heat treatment process also produces by-products, such as K_2NaAlF_6, K_3AlF_6, and AlF_3, among others. Therefore, additional work is required to remove these fluorine-containing impurities.

13.2.4 In situ electrochemical method

Li et al. [28] prepared V_2CT_x by the in situ electrochemical method using a typical MAX (V_2AlC) as cathode and 21 mol/L LiTFSI and 1 mol/L $Zn(OTf)_2$ mixed solution as electrolyte. V_2AlC is confined to a closed cell during the spalling process. No leakage happens, demonstrating it to be a green process excluding acid/base. The prepared zinc-ion battery is superior to HF acid-etched V_2CT_x in terms of rate performance and capacity. As shown in Fig. 13.11, the MAX V_2AlC was completely etched to V_2CT_x MXene with tiny Al residual and F- and O-functionalized surface terminations by an F-contained electrolyte. Then, V_2CT_x MXene is oxidized to be V_2O_5.

13.2.5 Acidic and alkaline etching

In addition to acidic etching agents, alkaline solutions can also be used, in which Al can be dissolved by the alkaline. MXenes with oxygen-containing functional groups are obtained by the NaOH-assisted hydrothermal method [22]. The experimental conditions are an ultrahigh concentration of NaOH solution and hydrothermal temperature of 270°C. High temperature can promote the rapid flow of OH$^-$ and obtain high-purity $Ti_3C_2T_x$ and fluoride-free MXene with better capacitance performance. Li et al. [29] prepared $Ti_3C_2T_x$ by NaOH-assisted hydrothermal etching of Ti_3AlC_2, as shown in Fig. 13.12. It was found that the ideal sample could not be obtained at low temperature (or high temperature) and

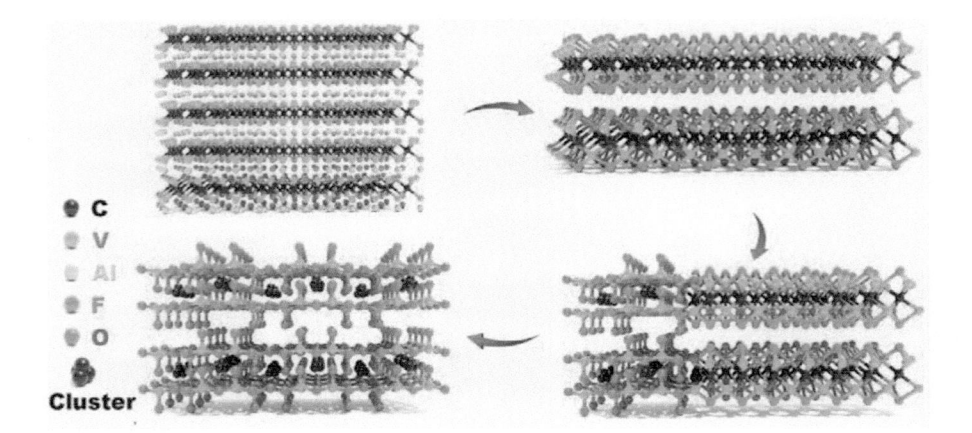

FIGURE 13.11

Illustration of phase and structure transition of V_2AlC cathode during the cycle at 5 A/g [28].

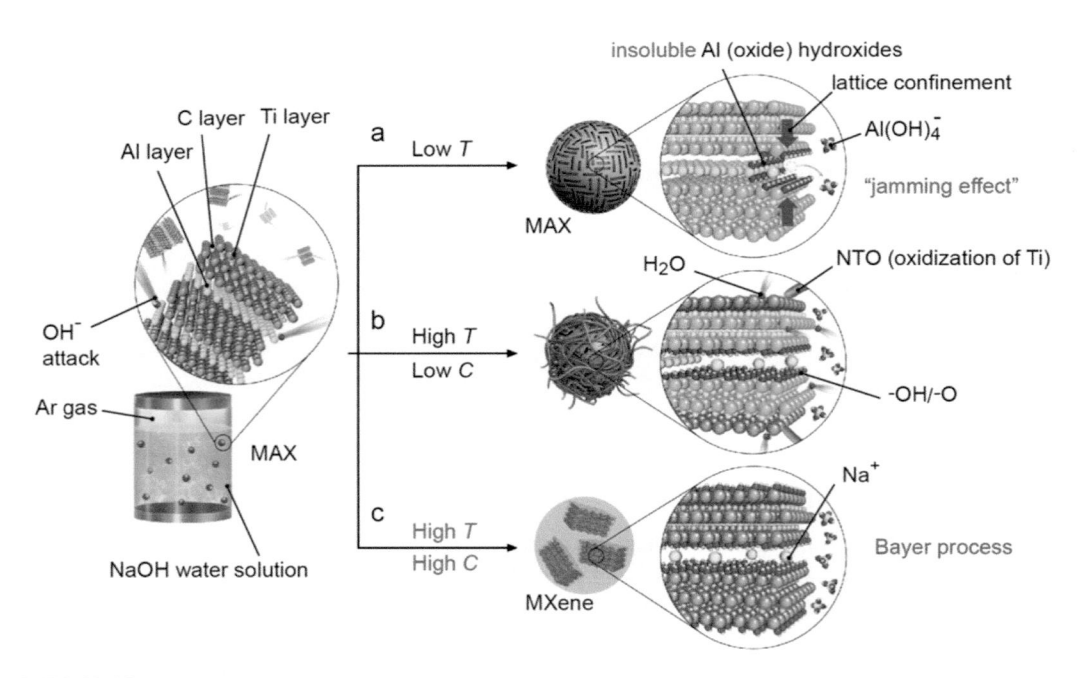

FIGURE 13.12

Schematic of the reaction between Ti_3AlC_2 and NaOH water solution under different conditions [29].

low concentration of NaOH solution etching. Only at high temperature and high concentration of NaOH solution (270°C, 27.5 mol/L) could the Al layer in Ti_3AlC_2 be removed to prepare $Ti_3C_2T_x$. The purity of the obtained material was about 92%, and the electrochemical performance of the sample is better than that of the sample etched by HF acid.

13.2.6 **Electrochemical etching**

HF acid solution greatly harms the environment and human body, and most of the MXene etched by HF acid is a fluorine-containing functional group [30]. In 2017, the CREEN team first reported fluoride-free electrochemical etching using a low concentration of HCl solution and low voltage for etching aluminum titanium carbide (Ti_2AlC) into Ti_2CT_x. However, since the reaction is incomplete, aluminum carbide (AlC) by-products are generated. Subsequent researchers used 1.0 mol/L ammonium chloride (NH_4Cl) and 0.2 mol/L tetramethylammonium hydroxide (TMAOH) as the electrolyte. During the reaction, Cl^- first bonded with layer A, and NH_4OH generated by the reaction intercalated the surface Ti_3C_2, gradually deepening the reaction [31]. The final MXene ($Ti_3C_2T_x$) shows oxygen-containing functional groups on its surface. The ion intercalation method can enlarge the distance between layers to obtain a large size and a few layers of nanosheets. To avoid the toxicity caused by the etching agent, HCl can also be used as the electrolyte, and the temperature can be appropriately raised to achieve the same effect.

As shown in Fig. 13.13, Ti_2CT_x can be prepared by electrochemical etching Ti_2AlC in an aqueous HCl electrolyte. The Ti_2CT_x obtained was terminated by $-Cl$, $-O$, and $-OH$ groups instead of

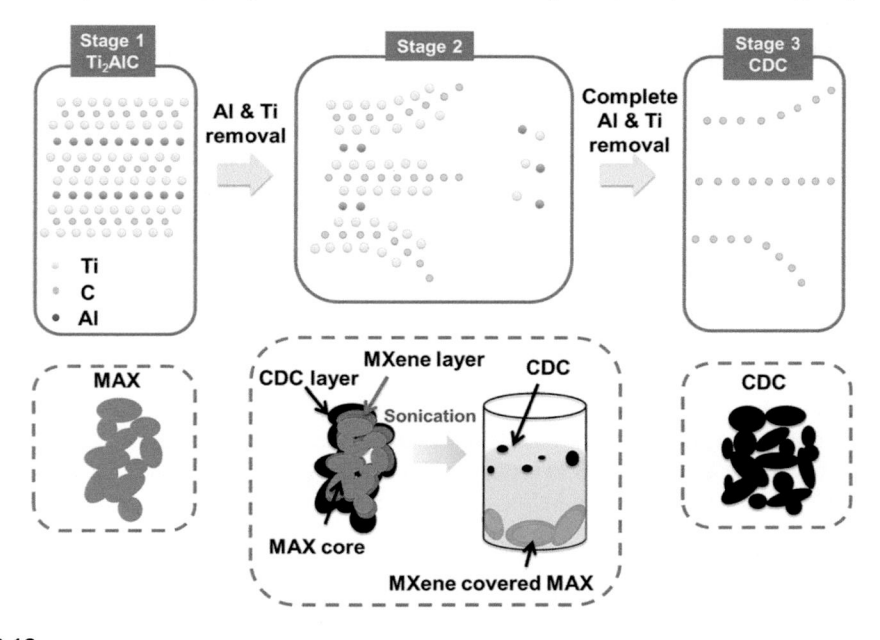

FIGURE 13.13

Proposed mechanism for electrochemical etching of Ti_2AlC in HCl aqueous electrolyte [32].

—F. In addition, carbide-derived carbon could be obtained by electrochemical transition etching from MAX [32]. Lukatskaya et al. [11] in a three-electrode system, with Ag/AgCl as the reference electrode, Pt as the counter electrode, MAX as the working electrode, and 3 mol/L sodium chloride solution as the electrolyte, Ti_3SiC_2, $TiAlC_2$, and Ti_2AlC were etched into CDC with theoretical yields of 12.3%, 12.3%, and 8.9%, respectively.

13.2.7 Chemical vapor deposition

The MXenes obtained by Chemical vapor deposition show a larger transverse size and few lattice defects, which is beneficial to study their intrinsic properties. However, there are few studies on the synthesis of monolayer MXenes by this method. 2D MXenes with monolayer structure were prepared by chemical vapor deposition. The bimetallic laminates of upper copper foil and bottom molybdenum foil are utilized as the substrate. Under methane atmosphere and high temperature, carbon atoms are generated via a catalytic cracking reaction and react with molybdenum atoms diffused to the copper surface. Finally, a high-quality, ultrathin 2D Mo_2C crystal with a thickness of only 1.07 nm is grown. Due to its low thickness, high toughness, and high reliability, it has good mechanical properties [33]. Recently, the team introduced elemental silicon (Si) into unlayered molybdenum nitride (MoN_2) using CVD technology to passivate the surface of unlayered 2D MoN_2, allowing $MoSi_2N_4$ to grow in centimeter-level monolayers (Fig. 13.14). Materials prepared by this method show excellent semiconductor properties and stability [34].

13.2.8 MXenes compounds preparation

Compared with other 2D nanomaterials, the electrochemical performance of MXene nanosheets is relatively limited since the structure tends to aggregate or reaccumulate, which impedes the infiltration and diffusion of ions in the electrolyte [36]. To fully use MXene nanosheets for energy storage, a series of strategies to solve restacking are proposed. The most common approach is introducing the interlayer spacers, such as surfactants, CNTs, polymers, metal ions, transition metal oxides, and reduced GO [37], which can increase interlayer spacers to prevent stacking, thus enhancing the electrochemical performance of MXene nanosheets. Crumpled MXene nanosheets can expand their interlayer space and provide more active sites [38].

Mashtalir et al. [35] used anhydrous salt solution to spontaneously insert cations (Na^+, K^+, NH^{4+}, Mg^{2+} and Al^{3+}) into the 2D Ti_3C_2 MXene layer, and $Ti_3C_2T_x$ material had insertion capacitance performance in an anhydrous electrolyte solution (Fig. 13.14). The $Ti_3C_2T_x$ paper electrode in an alkaline solution such as KOH or NaOH solution has the best performance, with flexible characteristics and a volume capacity of 350 F/cm^3, which is much higher than the porous carbon electrode. The electrical conductivity of MXene can be improved by using carbon nanofibers (for transporting electrolyte ions).

Lin et al. [39] prepared multilayer $Ti_3C_2T_x$/carbon nanofiber (CNFs) composites. As CNFs grow in the interval of $Ti_3C_2T_x$ accordion structure, it provides ion and electron transport channels perpendicular to the layer, improving the materials' conductivity and electrochemical performance. Rakhi et al. [40] annealed the etched MXene at 500K and argon for 2 hours to prepare ε-MnO_2 nanocrystals on the surface of $Ti_3C_2T_x$ by direct chemical synthesis and obtained ε-MnO_2/$Ti_3C_2T_x$

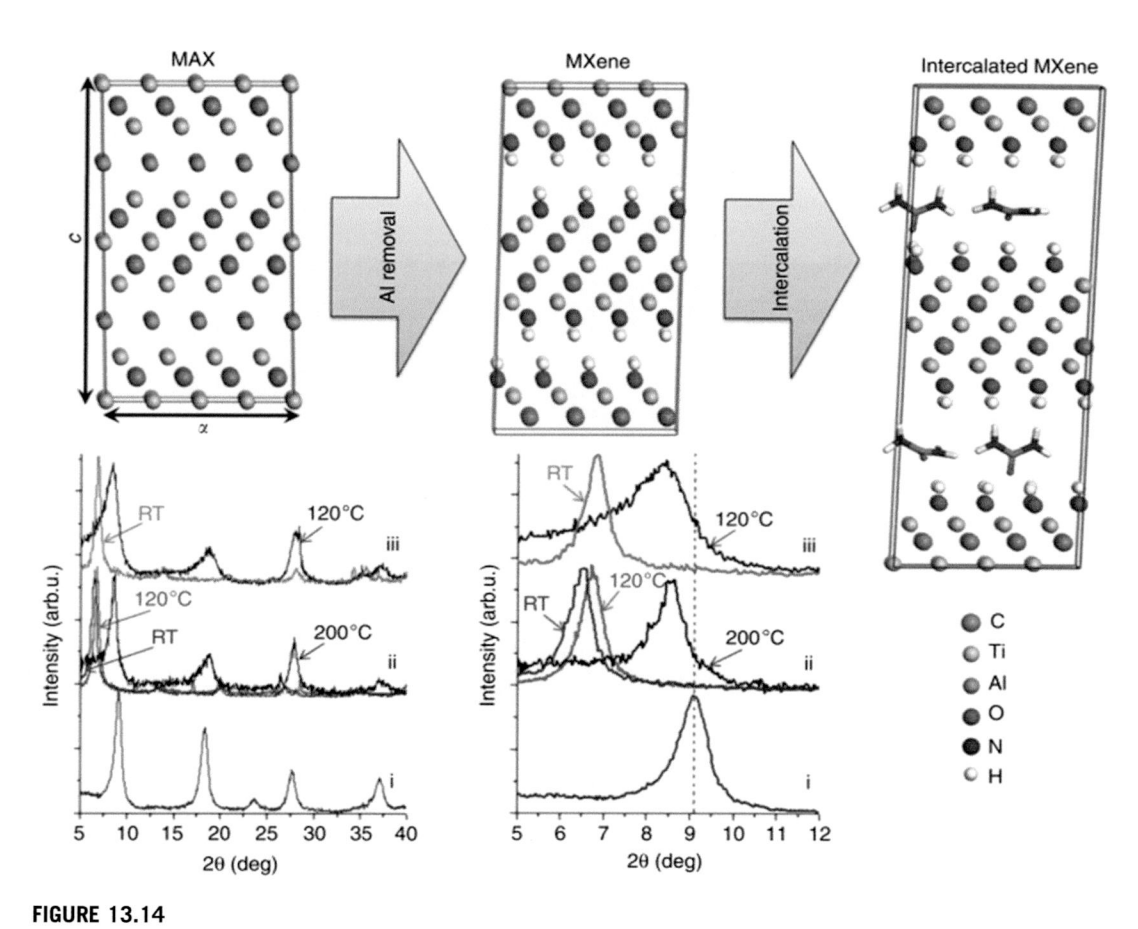

FIGURE 13.14

Evidence of MXene intercalation [35].

composite. Wang et al. [41] prepared three-dimensional multi-LDH/$Ti_3C_2T_x$ composites by growing nickel–aluminum layered bimetallic hydroxide (LDH) on the surface of $Ti_3C_2T_x$.

Polymers can spontaneously insert between layers of MXene to prevent stacking and significantly improve the mechanical properties of MXene material, but rarely improve its electrochemical properties [42]. When preparing polymer/MXene composites, the solubility of polymer in electrolyte solution and the interaction between polymer and negatively charged MXene should be considered. Boota et al. [42] arranged polypyrrole (PPy) chains between $Ti_3C_2T_x$ layers and obtained composite materials with good electrical conductivity. Zhu et al. [43] deposited HF-etched $Ti_3C_2T_x$ onto the substrate by electrophoresis, immersed it in organic electrolyte PPy for 600s, and prepared unsupported PPy/$Ti_3C_2T_x$ film with excellent electrochemical performance (Fig. 13.15).

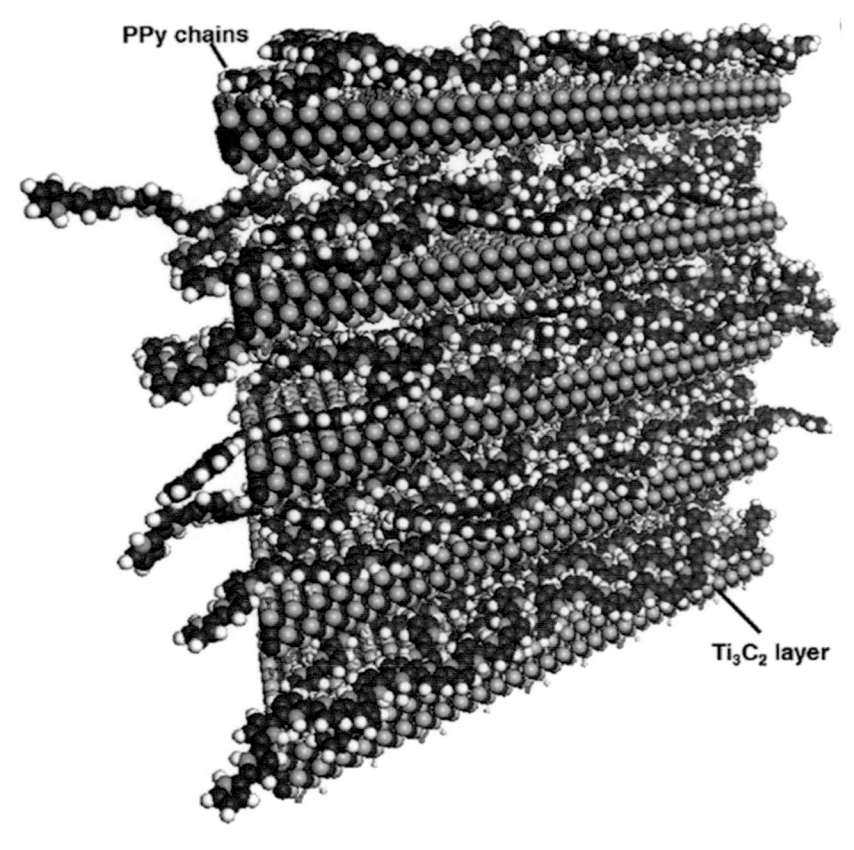

FIGURE 13.15

Schematic of the intercalated polypyrrole in the interlayers of l-Ti_3C_2 [43].

13.2.9 MXenes 3D porous structure

Another method of preventing MXene restack is integrating 2D nanosheets into 3D macrostructures, which is effective for many 2D materials, such as graphene and molybdenum disulfide (MoS_2) [44]. The 3D porous structure can provide efficient ion and electron transport channels in electrode materials, resulting in high-performance electrodes. Among the different methods of constructing 2D materials into 3D porous structures, template sacrifice is one of the most promising because it allows easy control of pore size [45].

Gogotsi reported hollow MXene spheres and 3D microporous MXene frames made from sacrificial polymethyl methacrylate (PMMA) spherical templates [46]. After removing the PMMA template at 450°C, 3D microporous MXene membranes were obtained (Fig. 13.16). When used as an anode for Na ion batteries, they show significantly enhanced capacity, long-term cycling stability, and rate performance compared with MXene and MXene CNT composites, indicating the

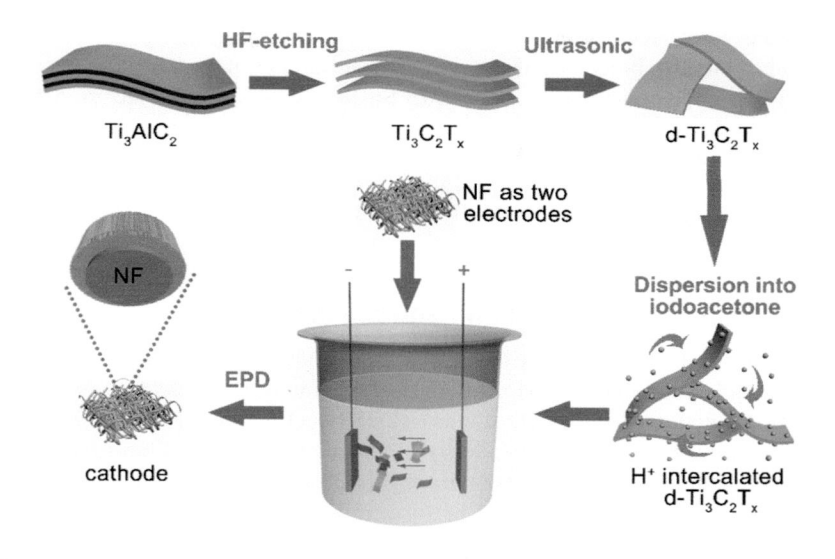

FIGURE 13.16

Schematic of the preparation route for fabricating $Ti_3C_2T_x$ -based electrode film on the nanofiber substrate by the MEPD method [46].

superiority of 3D microporous MXene architecture. The 3D porous shape of $Ti_3C_2T_x$ MXene foam can effectively prevent the restack of $Ti_3C_2T_x$ flakes and shorten the diffusion distance of ions. Therefore, $Ti_3C_2T_x$ MXene foam electrode shows improved electrochemical performance, with high surface capacitance, good rate performance, and stable cycle life.

13.3 MXenes and their applications in biosensors

The global virus pandemic has profoundly affected people's daily lives, urging and witnessing the transition for fast, sensitive, and reliable pathogens detection approaches from need-to-do to must-do. Here, biosensors have been put in an unprecedented position to detect pathogens owing to their convenient use and immediate readout, and they are considered practical alternatives to accurate yet time-consuming laboratory analysis. Biosensors not only contribute to solving the problems of the virus pandemic but also have the ability to detect human diseases, which are unexpectedly increasing due to environmental pollution such as metal ion contaminants, overusage of pesticides, and high dosages of drugs. Therefore, the application prospect of biosensors is wide based on the above needs, covering all aspects of life.

MXene has the advantages of high specific surface area, hydrophilicity, biocompatibility, and low cytotoxicity. Two-dimensional inorganic compounds, MXenes have a thickness of several atomic layers, and are composed of transition metal carbides, nitrides, or carbonitrides. MXene is a novel material with special properties, including high electrical conductivity and superior fluorescence, optical, and plasma properties [47]. It has been widely used in catalyst generation, energy storage, and biosensors in recent years owing to its excellent conductivity and semiconductor properties [48]. In addition, the biocompatibility of MXene enables its application in the biomedical field with superiority over other materials [49,50]. Since it was first reported in 2011, MXene has been widely used to develop various biosensors, including electrochemical and optical biosensors, by utilizing their different characteristics. It could also be combined with other nanomaterials and provides enhanced performance [51−53]. Recent studies have verified that this novel nanomaterial is ideal for biosensor application.

13.3.1 The application prospect of MXene in biosensors

The research and development of biosensors is important and meaningful as they detect the biological and chemical molecules that are harmful [54]. The aim of biosensors research is to develop accurate and rapid detection of target molecules, which benefits disease prevention and early treatment. Herein, the high sensitivity and selectivity as instructive indicators possess a dominant position in the development of biosensors. Nanomaterials provide a favorable microenvironment for biosensors due to their biocompatibility and perform improved transfer ability, benefiting from the large specific surface area and catalytic properties. Therefore, many nanomaterials, from metal nanoparticles and graphene to more recently transition metal dichalcogenides (TMD) nanomaterials, have been studied and applied to biosensors. Graphene, MoS_2, etc. are two-dimensional (2D) materials and have been widely used. They possess dimensional advantages but are limited by a lack of functional groups, easy aggregation, or unsatisfactory electrical transmission characteristics on the established two-dimensional nanoplatform. Among them, the lack of functional groups and easy aggregation constraints of 2D materials restrict the biorecognition elements loading and bioactivities maintenance, reducing the sensitivity and stability of biosensors. At the same time, the unsatisfactory electrical transmission characteristics will limit the application of the biosensor. The novel material MXene, which is also a two-dimensional material, has overcome these constraints with its excellent biocompatibility, good dispersibility, and the abundance of anchoring sites, thus attracting a lot of attention in the field of biosensing (Table 13.1).

13.3.1.1 Synthesis of 2D MXene for biosensing

In the general preparation process of MXene, liquid phase stripping and CVD approaches inevitably introduce toxic fluorine groups, which limits the use of MXene in biosensing applications. Therefore, the alkali-assisted hydraulic method [71] is used to prepare the MXene for biosensing, adopting sodium hydroxide as an etching agent. The process did not contain fluorine, and the purity of fabricated multilayer $Ti_3C_2T_x$ MXene was 92%. The results of high-angle annular dark field scanning transmission electron microscope (HAADF-STEM) verified that the interlayer spacing of etched $Ti_3C_2T_x$ is larger than the original Ti_3AlC_2, and performs consistently with that of HF-etched Ti_3AlC_2. In addition, the Zn-based MAX phase provides another synthesis method for the preparation of fluorine-free MXene. This top-down method first replaces the A element in MAX phase with Zn in $ZnCl_2$ through a replacement reaction between MAX and molten $ZnCl_2$, producing Cl-terminated MXene. Owing to the strong Lewis acidity of molten $ZnCl_2$, $Ti_3C_2Cl_2$ MXene is exfoliated and synthesized. It exhibits good ordering along the basal planes and is predicted to be more stable with enhanced electrochemical behavior, showing great potential in biosensors.

Table 13.1 MXenes and their applications in biosensors.

Probing method	MXene species materials	Target	Limit of detection	Sensitivity	Linear range	Preparation scheme	Chemical material properties	References
Electrochemical	$Ti_3C_2T_x$ -MXene	ACOP	0.048×10^{-6} M	$0.095\ \mu AuM^{-1}$	$0.25-2000 \times 10^{-6}$ M	Hydrofluoric acid etching	Abundant active sites F, OH, and O functional group	[55]
Electrochemical	$Ti_3C_2T_x$ -MXene	INZ	65×10^{-6} M	10.55 μA/mM	$0.1-4.6 \times 10^{-3}$ M	Hydrofluoric acid etching	Abundant active sites F, OH, and O functional group	[55]
Electrochemical	AuPt NPs-MXene	Ascorbic Acid (AA) and monosaccharides	0.2×10^{-6} M	8 μA/mM	$0.4-9.5 \times 10^{-6}$ M	Hydrofluoric acid etching	Electrode surface catalyzes the reaction	[56]
Electrochemical	$Ti_3C_2T_x$ -MXene	Urea	0.02×10^{-3} M	15.22 μA (1 of UA) mM^{-1}	$0-3 \times 10^{-3}$ M	Hydrofluoric acid etching	Electrocatalytic REDOX reaction	[56]
Electrochemical	$Ti_3C_2T_x$ -MXene	Creatinine	1.2×10^{-6} M	0.015 (1 of Cu^+) μAuM^{-1}	$10-100 \times 10^{-6}$ M	Hydrofluoric acid etching	Electrocatalytic REDOX reaction	[57]
Fluorescence	Ti_3C_2/ssDNA	HPV-18	1×10^{-10}M	–	$5 \times 10^{-10}-5 \times 10^{-8}$ M	–	MXene nanosheets	[58]
Fluorescence	Aptamer/Ti_3C_2	Exosome	1.4 particles μL^{-1}	–	$10-10^6$ particles μL^{-1}	Hydrofluoric acid etching	MXene nanosheets	[1]
Fluorescence	Chimeric DNA/Ti_3C_2	MUC1miR-21	6×10^{-9} M 8×10^{-10} M	–	$0-6 \times 10^{-8}$ M $0-2.5 \times 10^{-8}$ M	Fluoride preparation	dcDNA-Ti_3C_2 probe	[3]
ECL	$Ru(dcbpy)_3^{2+}$/ AuNPs@Ti_3C_2	Siglec-5	2.936×10^{-11} M	20.22 fM	$2 \times 10^{-11^{-1}} \times 10^{-9}$ M	–	$Ru(dcbpy)32 + /$ AuNPs@Ti_3C_2 nanocomposites	[4]
ECL	$g-C_3N_4-$MXenes	PKA	1.0 U/mL	–	$0.015-40$ U/mL	Fluoride preparation	MXenes nanosheets	[4]
ECL	Ti_3C_2	MCF-7 exosomes	125 particles/μL	–	$5 \times 10^2-5 \times 10^6$ particles/μL	Hydrofluoric acid etching	Ti_3C_2-MXenes nanosheets	[59]
SPR	Ti_3C_2-APTES/ AuNPs	CEA	1.5×10^{-16} M	0.15 fM	$1 \times 10^{-15}-1 \times 10^{-9}$ M	Hydrofluoric acid etching	N-Ti_3C_2-MXene nanosheets	[60]
SPR	Ti_3C_2/AuNPs	CEA	7×10^{-16} M	0.07 fM	$2 \times 10^{-16^{-2}} \times 10^{-5}$ M	Hydrofluoric acid etching	Ti_3C_2-MXene/ AuNPs nanocomposites	[61]
Photothermal	Ti_3C_2 QDs/ liposomes	PSA	4×10^{-10} g/mL	–	$1 \times 10^{-9}-5 \times 10^{-8}$ g/mL	–	Ti_3C_2 QDs-encapsulated liposome	[62]
SERS	Ti_3C_2/Au$-$Ag nanoshuttles	Carbendazim	1×10^{-8} M	–	$3.3 \times 10^{-8}-1 \times 10^{-5}$M	Hydrofluoric acid etching	Ti_2C-MXene/Au-Ag NSs	[63]
SERS	MXene/Au$-$Ag NPs	Ochratoxin A	1.28×10^{-12} M	–	$1 \times 10^{-11^{-5}} \times 10^{-8}$ M	Hydrofluoric acid etching	MXene/Au$-$Ag NPs	[64]

(Continued)

Table 13.1 MXenes and their applications in biosensors. *Continued*

Probing method	MXene species materials	Target	Limit of detection	Sensitivity	Linear range	Preparation scheme	Chemical material properties	References
SERS	MXene/MoS$_2$/AuNPs	miRNA-182	6.61×10^{-10} M	–	$10^{-9}-10^{-17}$ M	–	With controllable topography MXene/MoS$_2$) @AuNPs	[65]
SERS	Ti$_3$C$_2$/Au-Ag nanoshuttles	Carbendazim	10.3×10^{-9} M	0.024 μA/M	$50-100 \times 10^{-6}$ M	Hydrofluoric acid etching	Abundant active sites F, OH and O functional group	[66]
Colorimetric	Ti$_3$C$_2$/CuS	Cholesterol	1.9×10^{-6} M	–	$1 \times 10^{-5}-1 \times 10^{-4}$ μM	–	–	[67]
Colorimetric	Ti$_3$C$_2$T$_x$ @NiFe-LDH	Glutathione	8.4×10^{-8} M	–	$1 \times 10^{-7}-3 \times 10^{-5}$M	HCl-LiF selective etching in etching solution	MXenes nanosheets	[68]
Colorimetric	TiO$_2$@MXene/C-QDs	Glutathione	2×10^{-7} M	–	$5 \times 10^{-7}-2.5 \times 10^{-5}$ M	Hydrofluoric acid etching	Ti$_3$C$_2$T$_x$ -MXene nanosheets	[69]
Colorimetric	Prussian blue-MXene	Glucose	0.33×10^{-6} M	35.3 μAmM^{-1} cm^{-2}	$10 \times 10^{-6}-1.5 \times 10^{-3}$ M	–	Glucoses catalyzes the reaction	[70]
Colorimetric	Prussian blue-MXene	lactate	0.67×10^{-6} M	11.4 μA/mMcm^2	$0-22 \times 10^{-3}$ M	–	Lactate oxidase	[70]

13.3.1.2 Functionalization and intercalation: manipulation of MXenes properties

Generally, surface modification (functionalization) and intercalation are adopted to manipulate MXenes properties. Surface functionalization is used to modify the affinity of MXene with recognition elements or target molecules. And intercalation expands the interlayer spacing of MXene through ions and small molecules' inclusion or insertion, effectively controlling the electrical and optical properties of MXene. Both approaches contribute to overcoming the instability of MXene and promote biosensing performance.

The surface modification approaches of MXene are mainly classified into organic modification, inorganic modification, and organic−inorganic hybrid modification. Organic modification usually refers to utilizing coupling agents, polymers, and other organic substances to interact with MXene and combine by chemical bonding. The process introduces new chemical functional groups to MXene, thus improving its stability and reactivity.

Common organic modifications include coupling agent modification and polymer modification. Coupling agent modification has been widely used to adjust the surface properties of various materials. The coupling agent modification of MXene prevents its structural degradation owing to spontaneous oxidation and also effectively improves its dispersion and interface bonding with the polymer matrix. Polymer modification generates a covalent bond, hydrogen bond, or van der Waals force on the surface of MXene, followed by a combination to form polymer-modified MXene. It improves the dispersion and thermal stability of MXene in solution without destroying the original structure and also endows MXene with some new properties.

Inorganic modification usually refers to the surface functionalization of MXene utilizing inorganic nanomaterials. This modification approach can effectively prevent MXene from laminated stacking and endow MXene with some optical, electrical, and magnetic properties, which facilitate synthesis of high-performance and multifunctional composites. Inorganic oxide and inorganic sulfide modifications are commonly used. They possess similar structures but various properties due to different bond types, thus playing distinct functions in modifying MXene. The organic−inorganic hybrid modification method combines the advantages of organic modification and inorganic modification, contributing to greater flexibility and designability of modified materials. It functionally modifies the surface of MXene with improved performance and expands the application range of MXene in different fields.

Like other 2D nanomaterials, oxidized MXene will aggregate and precipitate in physiological media, which adversely affects its sensing performance in complex biological samples. This problem can be solved by polymer-based functionalization. For example, polymer polyethylene glycol and polyvinylpyrrolidone (PVP) were used to modify Ti_3C_2 and Nb_2C, respectively, displaying successfully promoted dispersion stability of MXene under physiological conditions [72]. Also, utilizing polymer soybean phospholipid (SP) generates an organic chain, bringing the steric hindrance to enhance the robustness of MXene material, thus resulting in good stability in the biosensing process. Moreover, the functionalization of surface groups or nanohybrids of MXene material also contributes to biosensing performance. For example, amino groups were introduced to the MXene surface by (3-aminopropyl)triethoxysilane (APTES), generating $-NH_2$-terminated Ti_3C_2 MXene. The $-NH_2$ group covalently supports bioreceptors immobilization, facilitating ultrasensitive biomarkers detection. In situ growth of metal nanoparticles (NPs) on MXene has also been reported and shows extraordinary sensitivity in monitoring superoxide secreted by living cells. In addition,

$Ti_3C_2T_x$/graphene and MoS_2/Ti_3C_2 hybrid membranes generate abundant biosensing active sites by providing a hydrophilic microenvironment [56].

The intercalation approaches of MXene manipulation highly benefit from its significantly larger interlayer distances than other graphene-like materials (15 Å of $Ti_3C_2T_x$ MXene, 3.35 Å of graphene, and 8.27 Å of reduced graphene oxide), facilitating manipulation of the structural and macrooptoelectrical properties [73]. In 2013, Michael Naguib et al. [74] first reported the work of expanding the MXene interlayer distance by intercalation technology. It was found that the interlayer distance increases when hydrofluoric acid-etched MXene is mixed with a certain proportion of small molecules or ions, including DMF, Na^+, K^+, Mg^{2+}, urea, and other small molecules and metal cations. The principle of interpolation is that the mixing induces transitions between metals and semiconductor-like transport via the interflake effect [75], increasing the interlayer distance of MXene. And the interaction of interlayer spaces will promote the transport of thermally excited electrons between sheets, which greatly improves the electrochemical performance of MXene. The test results show that, after intercalation, the unique nanochannel of MXene generates a convenient path for ion migration, providing a specific high capacitance of 300 F/cm^3 [76]. When doped with different cations, the transmittance of $Ti_3C_2T_x$ MXene films varies from negative 2% (dimethyl sulfoxide) to positive 17% (tetramethylammonium hydroxide) [35]. The reversible transmissivity of MXene thin film indicates its good potential and wide application prospects in transparent conductive electrodes, electrochromic electrodes, and other sensors.

The discovery that the interlayer distance of MXene can be expanded through small molecule or ion intercalation not only promotes the research of MXene intercalation but also inspires high-quality single-layer MXene nanosheets fabrication. Michael Naguib's team added dimethyl sulfoxide (DMSO) as an intercalation molecule into the etched multilayer $Ti_3C_2T_x$ MXene, then treated it by mechanical vibration stripping method, and finally obtained MXene nanosheets with single-layer thickness [77].

13.3.1.3 Advantages of MXene for biosensing

Biorecognition and signal transducing elements are two crucial roles in biosensors configurations. Biorecognition elements provide selectivity and improve the sensitivity of biosensors, and involve the anchoring of biomaterials, such as proteins, enzymes, and nucleic acids. Thus, interfaces with good biocompatibility and high loading capacities are important for loading recognition elements. Transduction elements convert biological signals to electrical signals or optical signals and require interfaces with good conductivity to obtain high signal output and low noise. The appearance of MXene makes up for the limitations of graphene and TMD in the biosensing field. Graphene possesses good electrical conductivity but few functional groups for biorecognition, while TMD has the problem of low aggregation and conductivity, which affects the electrical signal conduction and output stability. MXene provides abundant hydrophilic terminals for electrode modification [10], stable dispersion characteristics due to the polarity termination surface, and good electrical conductivity. It has been verified that $Ti_3C_2T_x$ colloid can be uniformly dispersed in ethanol for several weeks. Moreover, MXene has a highly tunable bandgap and can be used in ultrasensitive photoelectrochemical biosensors [78].

Many MXene-based biosensor types of research have been investigated with good performance. Zheng et al. reported a Ti_3C_2−GO sensor for H_2O_2 detection and exhibited a low detection limit of 1.95 μM [79]. Wang et al. prepared TiO_2−Ti_3C_2 nanocomposites by in situ pyrolysis and hydrothermal method for H_2O_2 detection. The suspension with Hb and Nafion was prepared and dropped on the glassy carbon electrode (GCE) to obtain Nafion/Hb/TiO_2–Ti_3C_2/GCE. The MXene-based H_2O_2 sensor exhibited a linear detection range was $0.1 \sim 380$ mM, the detection limit was as low as 14 nM, and the sensitivity was 447.3 mA/(mM•cm^2) [80]. Rakhi et al. dispersed the prepared Au–Ti_3C_2 complex in Nafion solution to form a uniform suspension and dripped it onto the GCE surface, followed by glucose oxidase (GOx) immobilization, forming GOx/Au-MXene/Nafion/GCE for glucose detection. The detection limit of glucose was 5.9 μM, the sensitivity was 4.2 μA/(mM•cm^2), and the detection range was $0-18$ mm [81]. Li et al. designed a MXene/Nico-LDH nanocomposite as a glucose sensor, with a linear detection range of $0.002-4.096$ mM and a detection limit as low as 0.53 mM [82]. Zheng et al. prepared MXene/DNA/Pd/Pt nanocomposites for sensitive dopamine detection, showing a wide response range of $0.2-1000$ μm to dopamine and high selectivity [83]. As a promising electrode material, MXene has high biological binding ability and high signal transduction ability, which is beneficial to improving the sensitivity and detection limit of electrochemical biosensors. In addition, MXene has broad-spectrum adjustable photoelectric characteristics, including electronic band structure, work function, transmittance, and light absorption. The extensive photoelectric characteristics of MXenes contribute to the generation and transportation of carriers sensitive to biological reactions or biological binding in optoelectronic devices. Therefore, MXenes are being extensively studied in electrochemical and optical biosensors.

Also, MXene possesses some excellent properties that are beneficial to different biosensing techniques separately. Many MXene-based biosensor types of research have been investigated with good performance, especially electrochemical biosensors and optical biosensors. The former technique benefits from MXene's high biological binding ability and good signal transduction ability, exhibiting improved sensitivity and limit of detection (LOD), while the latter approach possesses more sensitive generation and transportation based on the broad-spectrum adjustable photoelectric characteristics of MXene. Therefore, the MXene studies in electrochemical and optical biosensors are well studied and illustrated below.

13.3.2 Application of MXene in electrochemical biosensors

Electrochemical biosensors have been widely used as a novel detection approach benefiting from their low cost, simple operation, fast response speed, and high sensitivity. They can quickly detect a small volume of samples with high selectivity and sensitivity owing to the loaded and modified electrodes. MXene is an ideal modification material because it contains complete metal atomic layers and abundant oxygen or hydroxyl functional groups, which can load and interact with biomolecules. The combination of MXene and electrochemical biosensor using MXene as nano–biological interface materials for biosensing has become a research hotspot in recent years.

13.3.2.1 Electrochemical biosensing introduction

Electrochemical detection methods have many advantages, including low detection limit, high sensitivity, fast analysis speed, etc., and have been widely used in the biosensing field [84]. They are classified into the potential and amperometric methods, providing rapid, economical, and high-sensitivity platforms for the detection of many diseases with high selectivity. Among the different electrochemical techniques, amperometry and time-amperometry assisted with enzymes are the most common biosensing methods because enzymes can selectively perform REDOX reactions with simple potentiostatic structure, contributing to the rapid electrochemical analysis. However, the amperometric method is limited by the inability to provide enough biological detection information for complex REDOX reactions. Therefore, cyclic voltammetry (CV) is proposed to broaden the sweep potential information and distinguish different targets by using different REDOX peaks. More approaches, including differential pulse voltammetry (DPV) and square wave voltammetry (SWV), are proposed to remove background current signals, exhibiting a large improved LOD. Therefore, DPV and SWV can be applied to detect nucleic acid and immunosensing.

13.3.2.2 Electrochemical biosensing applications based on MXene

Benefiting from electrochemical biosensors' wide applications and MXenes superiority properties, MXene-based electrochemical biosensors have been widely investigated. MXene provides abundant functional groups, contributing to loading and interacting with biomolecules. Many researchers have verified the good performance of electrochemical biosensors with MXene. A $Ti_3C_2T_x$ MXene-based electrochemical biosensor fabricated by screen printing can detect acetaminophen (ACOP) and isoniazid (INZ). Figs. 13.17 and 13.18 illustrate the results, where two oxidation potentials correspond to 0.53 V for ACOP and 0.95 V for INZ. The current peaks are linearly related to the concentration of the respective substances, exhibiting low LODs of 0.048×10^{-6} M for INZ and 65×10^{-6} M for ACOP. A nitrite biosensor is based on hemoglobin (Hb)-immobilized substrates using Ti_3C_2 MXene [85]. The MXene contributes to the Hb carrier and facilitates substrate encapsulation and enzyme acquisition, promoting electrochemical signal and enzyme stability. Wu et al. [86]. immobilized enzymes on MXene nanosheets for phenol detection. Kumara et al. [87] immobilized a carcinoembryonic antigen (CEA) directly on biofunctionalized $Ti_3C_2T_x$ MXene nanosheets for the detection of cancer biomarkers.

Moreover, combining nanomaterials, including noble metal nanoparticles (AgNPs, PtNPs), CNT etc., further facilitates sensing performance with extended properties. Noble metal nanoparticles are commonly used to expand the electroactive surface area and increase electron mobility [88]. MXene combined with the above nanoparticles could further improve its electrical conductivity,

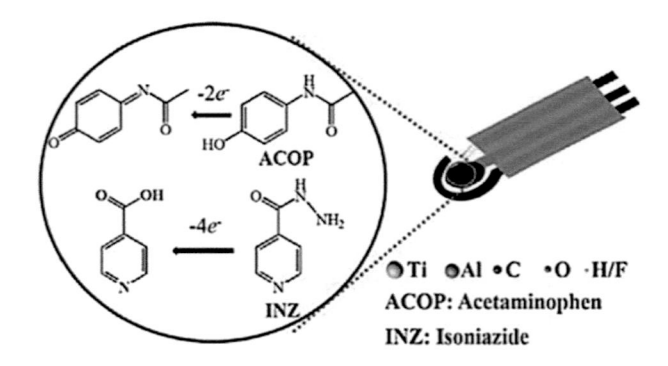

FIGURE 13.17

Schematic diagram of electrocatalytic oxidation mechanism [55].

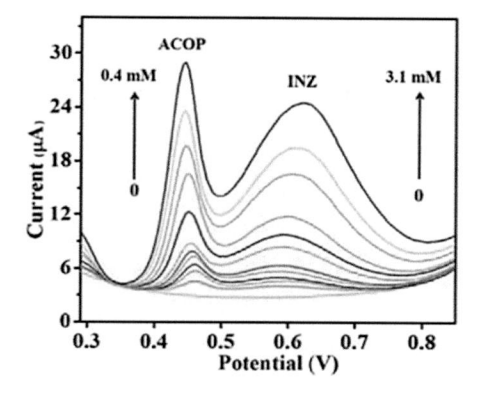

FIGURE 13.18

Relationship between peak current and acetaminophen and isoniazid concentration [55].

promoting electrochemical biosensor performance. The nanocomposite (MXene-Au-MB) is loaded with AuNPs and MB with good electrical conductivity, where the AuNPs are used to trap biomolecules containing sulfhydryl terminations, and the MB molecules are used to generate electrochemical signals. The nanocomposite-modified electrochemical biosensor operates over a wide detection range from 5 pg/mL to 10 ng/mL, and a low LOD of 0.83 pg/mL for thrombin detection. Similarly, the AgNPs modified a Ti_3C_2 electrochemical biosensor to detect malathion, where the nanoplatform enabled electron transfer between acetylcholinesterase (AChE) and the electrode [89].

Also, many electrochemical studies have investigated and presented composite materials combining MXene and nanomaterials such as CNT and graphene. An electrochemical biosensor based on a composite of lactate oxidase and glucose oxidase-modified CNT/MXene/Prussian blue modified electrodes [70] was integrated into a wearable sweat sensor, as shown in Fig. 13.19. Glucose reacts catalytically with glucose oxidase, producing hydrogen peroxide, which then ionizes Prussian Blue. After ionization, the ions undergo a redox reaction with the MXene electrode, thereby increasing the electrochemical response. Fig. 13.20 illustrates the sensing results, showing the different lactic acid and glucose detection ranges. When the sensor ranges from 10×10^{-6} M to 1.5×10^{-3} M, it shows a linear correlation with glucose concentration with a LOD of 0.33×10^{-6} M; while when it ranges from 0 M to 22×10^{-3} M, the results are linearly correlated with lactic acid concentrations with a LOD of 0.67×10^{-6}M.

Graphene assisted with MXene has been also investigated and constructed on the FET channel for biosensing. The gating effect of FET devices can selectively capture biomolecules or analytes and can convert them into electrical signals through biofunctionalization of the dielectric layer. Ti_3C_2 MXene−graphene composite is used to construct an FET channel with ample binding sites for antibodies immobilization by MXene and promoted charge transportation by graphene. The bio-FET responds to H1N1 virus and SARS-CoV-2 spike protein with distinctive changes in source-drain current and gate voltage, thus facilitating the sensitive point-of-care testing for the environmental virus of low level.

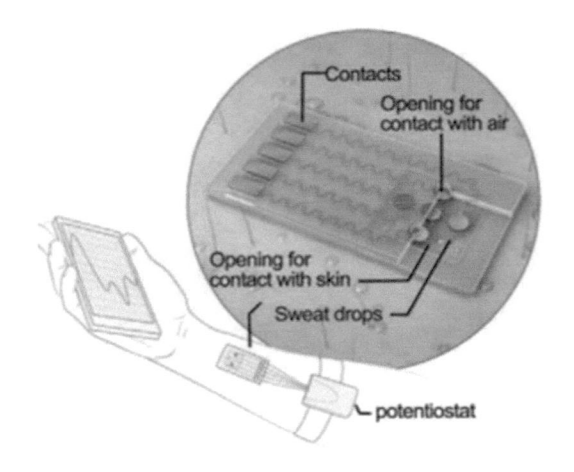

FIGURE 13.19

Schematic diagram of external perspiration analysis system [70].

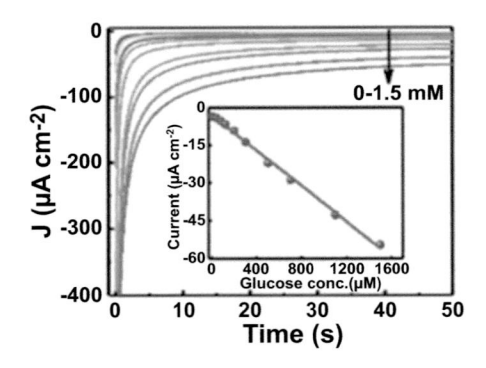

FIGURE 13.20

The amperometric response of glucose sensor varies with glucose concentration [70].

In addition, some nanoparticles are utilized as secondary analytes to further enable better performance of electrochemical detection. An electrochemical biosensor based on a composite of MXene and gold-platinum NPs (Au Pt NPs) for superoxide detection, where Au Pt NPs are introduced as secondary analytes. Superoxide is produced when antigen reacts with antibody, which causes damage to the organism when in excess. Fig. 13.21 depicts the superoxide concentration sensing results in the simultaneous presence of the secondary analytes monosaccharide and ascorbic acid (AA), exhibiting good selectivity, low LOD of 0.2×10^{-6}M and linear correlation for concentration ranging from 0.4×10^{-6} to 9.5×10^{-6} M. Excellent performance is obtained from the dense Au Pt NPs forming on the electrode surface, which thus promotes the catalytic reaction. Also, in vitro measurements of the electrochemical biosensor were applied practically by injecting zymosan into

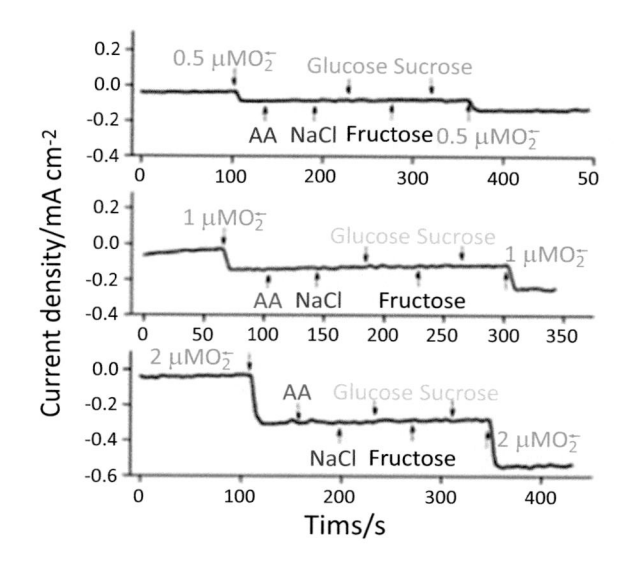

FIGURE 13.21

Selective response of Au Pt NP/MXene to AA, NaCl, glucose, fructose, sucrose, and O_2 [56].

Hep-G2 cells. The results show that the electrochemical biosensor can detect small amounts of superoxide produced by Hep-G2 cells in response to 5 μL of zymosan. Using a variety of different NPs as secondary analytes, various biomolecules excluding superoxide, such as fexofetamine, piroxicam, and dopamine, can be effectively detected.

13.3.3 Application of MXene in optical biosensors

Optical biosensors mainly contain biorecognition elements for specific binding as well as transduction layers that convert biosignals into optical signals. The biorecognition elements are used for specific binding with measured molecules, which require high biocompatibility and abundant active sites. The transduction layers convert the biological signal to optical signals, requiring good conductivity to obtain a high signal output with low noise. Optical biosensors are classified into fluorescence sensors, electrochemiluminescence sensors—exhibiting high sensitivity and low LOD—and plasma biosensors, comprising surface plasmon resonance (SPR) and surface-enhanced Raman scattering (SERS). Many 2D nanomaterials have been widely applied in optical biosensing, such as graphene and TMD. Here, graphene has excellent electrical conductivity, but its gapless band structure limits its application in optical biosensing. The TMD has an engineered bandgap and better photoresponsivity than grapheme [90,91]. However, the signal cannot be transmitted stably due to the low conductivity and aggregation of TMD [92], also limiting its optical biosensing applications. The rise of MXene has brought a new revolution in optical biosensors, benefiting from solving the above problems and its excellent light adsorption capacity, as ideal photon—electron mediators and light-to-heat conversion interfaces. It provides abundant functional groups for anchoring biomolecules and possesses excellent conductivity and stable dispersion. The tunable bandgap of MXene is widely used in ultrasensitive photoelectrochemical biosensors.

13.3.3.1 Electrochemical luminescence biosensor based on MXene

Electrochemiluminescence (ECL) assay measures light emission from electrogenerated luminous species redox reactions and establishes the quantification relationship between ECL intensity and target concentration. Therefore, the utilization of MXene can greatly improve the REDOX efficiency of luminescent body and accelerates electron transfer owing to excellent electrical conductivity and catalytic performance. For example, the optical sensor's nanoprobe modified by MXene performs significant acceleration in the redox of luminol and light emitting, realizing selective detection of McF-7 exosomes and exhibiting a linear relationship between ECL intensity and McF-7 exosomes concentration. Also, the ECL biosensor modified by a combination of MXene and Ru $(Bpy)_3^{2+}$ modification on GCE performs significantly enhanced ECL strength due to the MXenes abundant binding sites for luminescent and oligonucleotide as shown in Fig. 13.22. It also achieved label-free single nucleotide detection.

The combination of MXene and specific nanoparticles contributes to stronger luminescence intensity and brings additional benefits. For example, the utilization of G-C$_3$N$_4$ nanoparticles combined with MXene is used for single nucleotide mismatch detection in human urine matrix, where it shows 3.5 times higher luminescence intensity of ECL at 100 nM than that of the complete complementary DNA sequence. This group contains a large number of laminators and inhibits the passivation of G-C$_3$N$_4$. The involvement of kinase and adenosine triphosphate causes the phosphorylated kemptide chelate hence to enrich Ti defects in MXene/g-C$_3$N$_4$ hybrid, which indicates the concentration of phosphorylated kemptide via the ECL signal. MXene could mediate injected electrons to g-C$_3$N$_4$, prolonging system stability. Furthermore, the specific structure

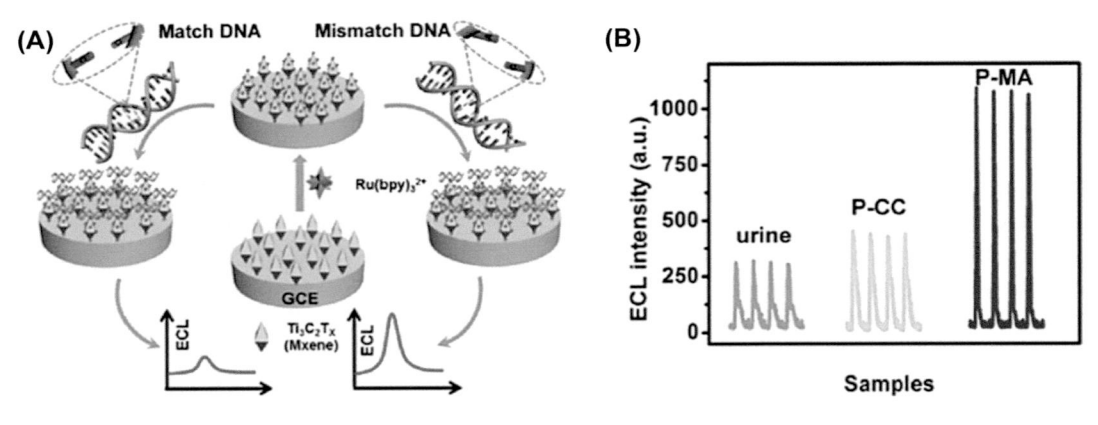

FIGURE 13.22

MXene-based electrochemiluminescence (ECL) sensors. The solid-state Nafion/MXene/Ru ECL sensor for single-nucleotide mismatch discrimination: (A) samples detection in human urine matrix; (B) P-CC and P-MA refer to p53 gene sequence with completely complementary sequence (P-CC) or with single-nucleotide mismatch (P-MA) [93].

modification of MXene could extend its properties. MXene fabricated into QDs performed extended ECL properties originating from its aligned structure, ample edge sites, and quantum confinement effect. Song et al. prepared a novel fluorine-free Nb_2CTx nanosheet by a simple electrochemical etching stripping strategy and constructed an AChE biosensor for fluorescence detection, exhibiting an extremely amplified low detection limit of 0.046 ng/mL [94]. Zhang et al. constructed a 2D ultrathin $Ti_3C_2T_x$ MXene-based ECL biosensor for Siglec-5 detection. The biosensor realized a sensitive amplification with a sensitivity of 20.22 fM through cleavage of CRISPR-Cas12a, which was combined with CHA-mediated isothermal amplification [95].

13.3.3.2 Fluorescence biosensor based on MXene

MXene contributes to fluorescence biosensor sensing in solution, benefiting from its hydrophilic nature, good dispersion, and adjustable structure. Take $Ti_3C_2T_x$ -MXene exosome fluorescence detection as an example: $Ti_3C_2T_x$ QDs of several nanometers emit light under light excitation, while $Ti_3C_2T_x$ nanosheets have light absorption characteristics in a wide wavelength range. Therefore, $Ti_3C_2T_x$ -MXene is used to emit fluorescence as a luminescent emitter, where fluorescence labeling is eliminated by the fluorescence resonance energy transfer effect brought by MXene nanosheets. The sensor realizes fluorescent-labeled DNA sensing through the fluorescence resonance energy transfer (FRET) effect. As shown in Fig. 13.23, Cy3-labeled CD63 aptamers selectively adsorb on Ti_3C_2 MXene by hydrogen bonding, followed by fluorescent nanoprobes being quenched due to the FERT effect. Then, the fluorescence is restored after the combination between the apterer and the exon, presenting a stable self-fluorescence signal at 650 nm.

Moreover, additional groups are added to perform extended properties. Amino groups were added to the Ti_3C_2-MXene surface, acting as pronated or depronated in response to pH variations. This produces pH-sensitive tunable fluorescent peaks at 460 nm responding to different pHs, owing to Ti_3C_2 QDs, and pH-insensitive fluorescent intensity of $[Ru(dpp)_3]Cl_2$ at 615 nm, which facilitates the intracellular pH detection. Also, nitrogen-doping enables H_2O_2 detection. The fluorescence

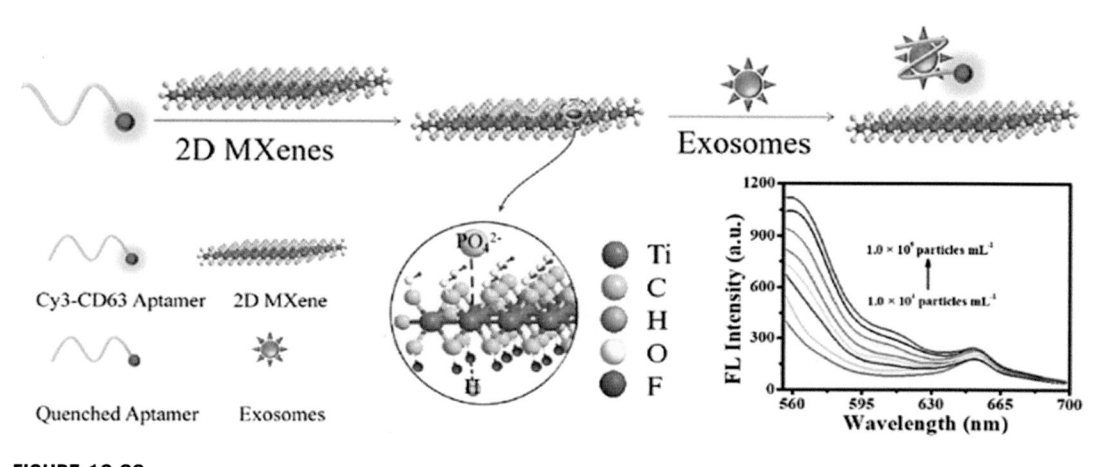

FIGURE 13.23

Concept diagram of exosomes capture and fluorescent intensities sensing results [96].

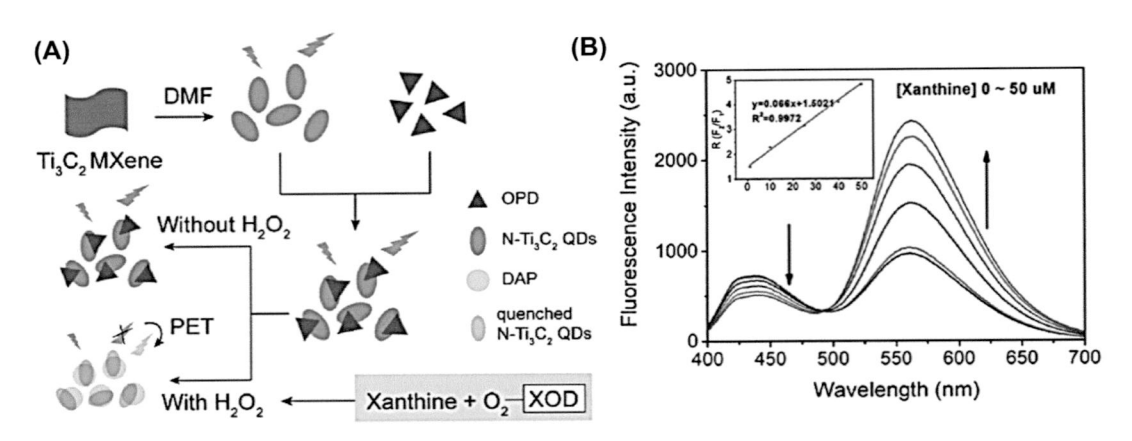

FIGURE 13.24

MXene-based fluorescent biosensors for H_2O_2 monitoring. (A) The configuration of the detection process. (B) The fluorescent intensities sensing results for different H_2O_2 concentrations [97].

peak at 448 nm of nitrogen-doped Ti_3C_2 QDs is quenched under the excitation at 369 nm and H_2O_2's presence, generating a new emission peak at 560 nm at the same time, as shown in Fig. 13.24. The ratio of the two fluorescence peaks is determined by H_2O_2 concentration, providing a determination of H_2O_2 produced by metabolic activity

13.3.3.3 Plasma biosensor based on MXene

Coating MXene on metals significantly enhances electromagnetic waves coupling to collective electron oscillations, thus improving the SPR effect, enabling sensing by optical signals when biomolecules bind to them. The sensing mechanism of SERS technology is also an SPR effect, and the introduction of MXene improves the plasma coupling, thereby enhancing the Raman scattering intensity. When biomolecules were attached to the detection area, the optical signal of the sensor changed. Sensing with SPR and SERS techniques relied on changes in resonance angle and Raman scattering intensity, respectively [60].

The Ti_3C_2 MXene modified with AuNPs possesses a rough oxygen-enriched surface, providing abundant binding sites and oriented immobilization of antibodies. The electromagnetic coupling between the AgNPs and Au layer enhances the SPR signal [61], which offers a feasible scheme for the ultrasensitive detection of CEA. Due to its large specific surface area and its own negative charge, MXene can be combined with metal to improve SERS intensity. MXene modified by Nb_2C and Ta_2C had higher sensitivity for biosensor applications [98]. There were more excited electrons outside the nuclei of Nb^{5+} and Ta^{5+}, which was conducive to cation exchange in SERS, enabling it to detect SARS-CoV-2 spike protein with high sensitivity as shown in Fig. 13.25.

13.3.3.4 Photothermal immunoassay biosensor based on MXene

MXene possesses excellent light absorption capacity and internal light-to-heat conversion efficiency, showing potential in photothermal ablation of tumors [53]. Its energy bandgap is tunable by

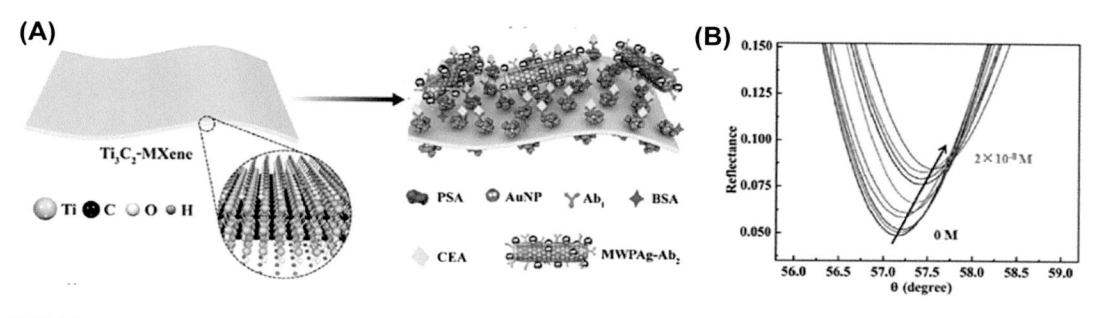

FIGURE 13.25

MXene-based plasmonic sensors: (A) Ti_3C_2 MXene-based surface plasmon resonance (SPR) sensing platform combined with MWCNTs-AgNPs signal enhancer for ultrasensitive carcinoembryonic antigen (CEA) detection; (B) SPR spectra shift in the presence of different CEA concentrations [61].

changing surface terminations and transition metal categories. Therefore, MXene could be widely applied in photothermal therapy and photothermal immunoassay by enhanced light absorption and heat conversion process through surface modification.

Photothermal immunoassay converts photoexcitation into heat generation by utilizing the thermal effect of photothermal materials, and the amount is closely related to the content of photothermal materials absorbed at the interface. Biometric identification units modified by photothermal materials convert biological combinations or reactions into intuitive and quantitative readouts. Photothermal immunoassay requires significantly lower instrument sensitivity than fluorescence sensing. Two important parameters in promoting the sensing performance of photothermal immunoassay are irradiation-absorption match and proper energy bandgap. The adsorption peak of Ti_3C_2 MXene is located in the near-infrared region of 808 nm, while semiconductor-like MXene with a wider bandgap has better near-infrared absorption capacity. A photothermal biosensor based on MXene for PSA detection integrated an 808 nm laser for absorbing matched excitation and a near-infrared imaging camera for temperature output recording, and possessed a large number of QDs with good sensitivity at the ng level. When PSA concentration is lower than 80 ng/mL, the temperature output has a good linear relationship with PSA concentration, and the theoretical LOD is 0.4 ng/mL.

13.4 MXenes and their applications in strain sensor

13.4.1 Detection principle of strain sensor based on MXene

The MXene strain sensor is generally composed of conductive nanosheets of MXene and an elastic substrate. The MXene conductive network transforms the sensed small deformations into changes in electrical signals (capacitance or resistance), and the elastic substrate provides the mechanical properties of a sensor that can be stretched, telescoped, etc. [99]. Fig. 13.26A and B illustrate the structural change and scanning electron microscopy (SEM) images of an MXene/elastic material

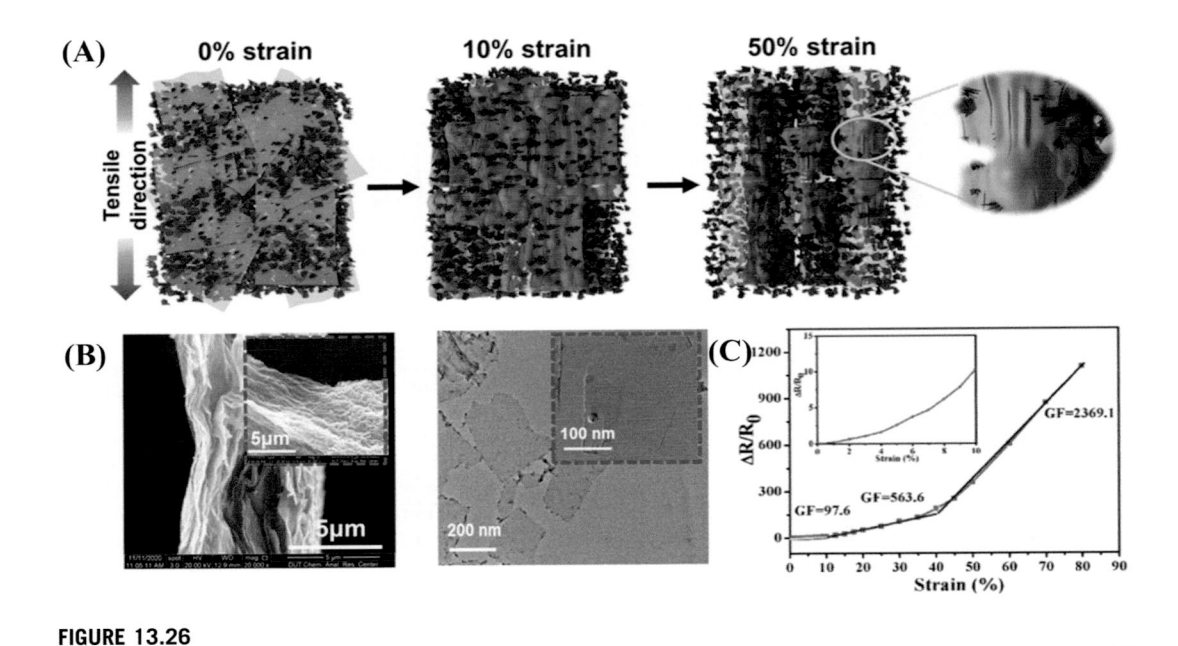

FIGURE 13.26

(A) Schematic diagram of MXene material under stress [100]; (B) SEM and transmission electron microscope of MXene material [101]; (C) sensitivity results versus stress [99].

subjected to tension, respectively. There are two main sensing mechanisms of strain sensors using MXene as a sensitive material: crack propagation mechanism and breaking mechanism.

The mechanism of crack propagation refers to the sensitive materials dispersing the stress by generating cracks during the stress stretching process of the sensor, resulting in changes in the resistance. For MXene, cracks first occur in the area where stress is concentrated, and gradually expand with stress increases. The increase of cracks decreases the MXene internal connectivity path and increases its resistance. Conversely, the cracks close gradually when the stress decreases, leading to an increase in the internal connectivity path of MXene, resulting in the resistance to restoration to the appropriate size.

The mechanism of disconnection refers to small relative slips that occur between MXene layers when the external stress is applied. The overlap area between the MXene layers decreases as the stress increases, leading to a decrease in the path of connectivity and resulting in resistance increasing. Both sensing mechanisms are effective within the critical strain range. When the applied stress exceeds the critical level or the sensitive material is completely torn apart due to cracks or is completely disconnected due to interlaminar sliding, the resistance of the flexible strain sensor reaches infinity, that is, to the limit of its operating range. The sensing range of the MXene flexible sensor can be expanded by reducing the interaction between MXene layers to construct a new conductive network. The common approach is to add other suitable conductive materials to the MXene to form composites. Especially, one-dimensional materials can be used as bridges to connect the layers of MXene and contribute to a wider sensing range as well as higher sensitivity, thus they are often used.

An important indicator of sensor performance is the strain sensitivity factor (f_g). The strain sensitivity factor is the ratio of the physical quantity (stress/strain) measured by the sensor to the output electrical signal (resistance/voltage, etc.). For example, for resistive sensors that measure strain, the formula for calculating the strain sensitivity factor is $f_g = (\Delta R/R_0)/\varepsilon$; that is, the strain sensitivity factor is the ratio of the resistance change ($\Delta R/R_0$) to the strain (ε). Sensors with different strain sensitivity factors are suitable for different application scenarios. Fig. 13.26C shows a diagram of the relationship between strain sensitivity and strain.

13.4.2 The critical process and preparation method of MXene strain sensors

13.4.2.1 Strain sensors using electrostatic spinning method

There are usually two methods for preparing MXene flexible sensors by electrostatic spinning, as shown in Fig. 13.27. One is to use electrostatic spinning to spin the mixture of MXene and elastic material directly into a two-dimensional (2D) film for flexible sensing. While the other is to use electrostatic spinning to form a stretchable elastic substrate and then decorate MXene on the surface of this substrate by hydrogen bonding or electrostatic interactions to make the sensor. Yang et al. [102] used the second method to first spin the polyurethane dispersion into stretchable polyurethane (PU) felt under ambient temperature of 30°C and relative humidity of 45%. Then, MXene solution was dropped onto PU to form an MXene−PU interlocked conductive network, from which wires are drawn out from both ends with carbon double-sided adhesive to obtain MXene/PU (Network-M/P) strain sensors. When subjected to tensile stress, the adjacent MXene layers covering the PU slip between each other, and the internal conductive paths are disconnected or connected,

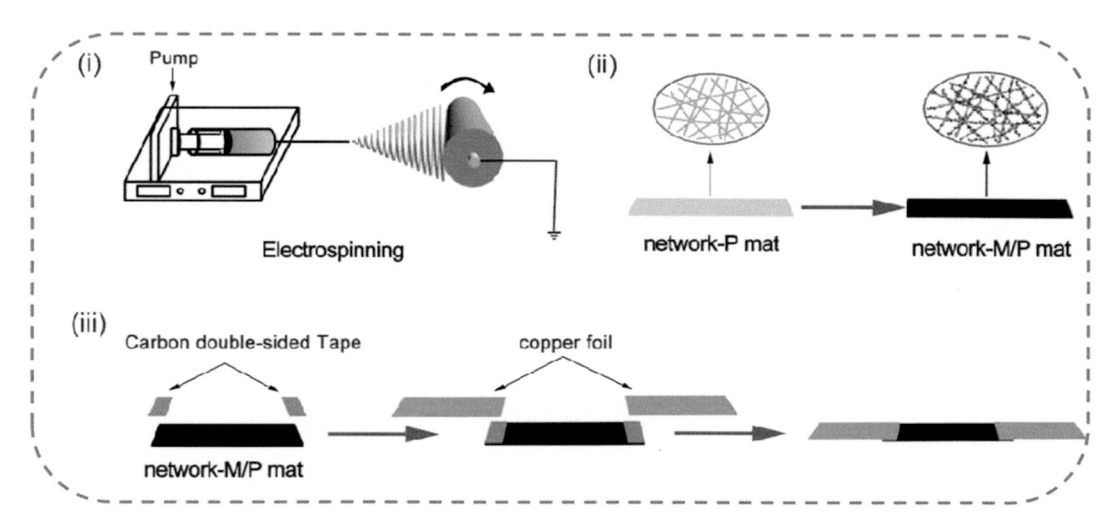

FIGURE 13.27

Schematic diagram of the manufacturing process of a typical network-MXene/PU strain sensor [102].

resulting in changes in the output resistance. The strain detection range of the sensor is 1%−100%, which can be used to detect signals such as pulse, heart rate, finger bending, and arm bending.

13.4.2.2 Strain sensors using the filter coating method

The filter coating method refers to precipitating a mixture of MXene colloidal solution or MXene composite system onto an elastic substrate through suction, filtration, layer-by-layer spraying, or natural evaporation for flexible strain sensing. Ma et al. [103] prepared a sandwich-structured MXene/PI strain sensor by dispersing MXene powder in 1 mL ethanol, followed by dropping it on PI film printed with cross finger electrodes. The sensor exhibits a fast response time of less than 30 ms and can detect joint bending and facial expression changing. The preparation process is simple and low cost. However, pure MXene nanosheets are difficult to assemble into continuous macroscopic structures with ideal mechanical and sensing properties because of their low aspect ratio. Therefore, Yang et al. [100]. obtained a mixture of MXene nanoparticles−nanosheets colloidal solution containing tetramethylammonium hydroxide by treating MAX phase (Ti_3AlC_2) with hydrofluoric acid (HF) and tetramethylammonium hydroxide (TMAOH), respectively. Then, the colloidal solution was filtered onto the microporous membrane to form MXene conductive film, which was then dried in a vacuum drying chamber at 25°C for 12 hours. After drying, the film was cut into a 0.6×2.0 cm rectangle. After removing the filter, the rectangular sheets of MXene were encapsulated with Polydimethylsiloxane (PDMS), and silver electrodes were coated on both ends of the MXene to obtain a 2D planar MXene strain sensor. The sensor can detect joint movements, clench and release of the fist, walking, and breathing, and distinguish between signal changes such as word pronunciation within its deformation range. Fig. 13.28 illustrates the preparation of the MXene sensor based on the filter method and its sensing signal output.

Chao et al. used layer-by-layer spraying technology to add polyaniline fibers (PANIF) into $Ti_3C_2T_x$ MXene thin flakes to form a composite conductive network, which was further assembled into a flexible strain sensor for real-time monitoring of human motion. This flexible sensor can accurately detect voice vibration, pulse pulsation, facial expression changes, fingers bending, and elbows in the human throat [99].

13.4.2.3 Strain sensors using the impregnation method

The impregnation approach refers to immersing elastomer substrates (e.g., porous materials, foams, sponges, etc.) into conductive solutions, so that the elastic base and sensor materials can be fully combined to prepare sensors. Elastomer substrate such as sponge has a rich porous structure, which can be compressed to 95% of the original volume under external forces and then can be restored to the initial state with the gradual removal of external forces. MXene nanosheets possess a large contact area and strong van der Waals forces, and can be well attached to the sponge. Therefore, Yue et al. [105] immersed the melamine sponge with a porosity of more than 97% in different concentrations of MXene colloidal solution until the sponge completely blackened, thus building a three-dimensional (3D) MXene−sponge mesh structure. The MXene−sponge was deposited on crossfinger electrode deposited with PVA, sequentially along with a polyimide substrate and a PVA membrane to assemble an MXene−sponge−PVA pressure sensor. The sensor has a response time of 138 ms and a minimum detection limit of 9 Pa. It can detect physiological signals such as finger bending, breathing, and pulse. This method is simple and cost-effective for large-scale production but limited by thermal and chemical instability between MXene−sponges for some applications. Li

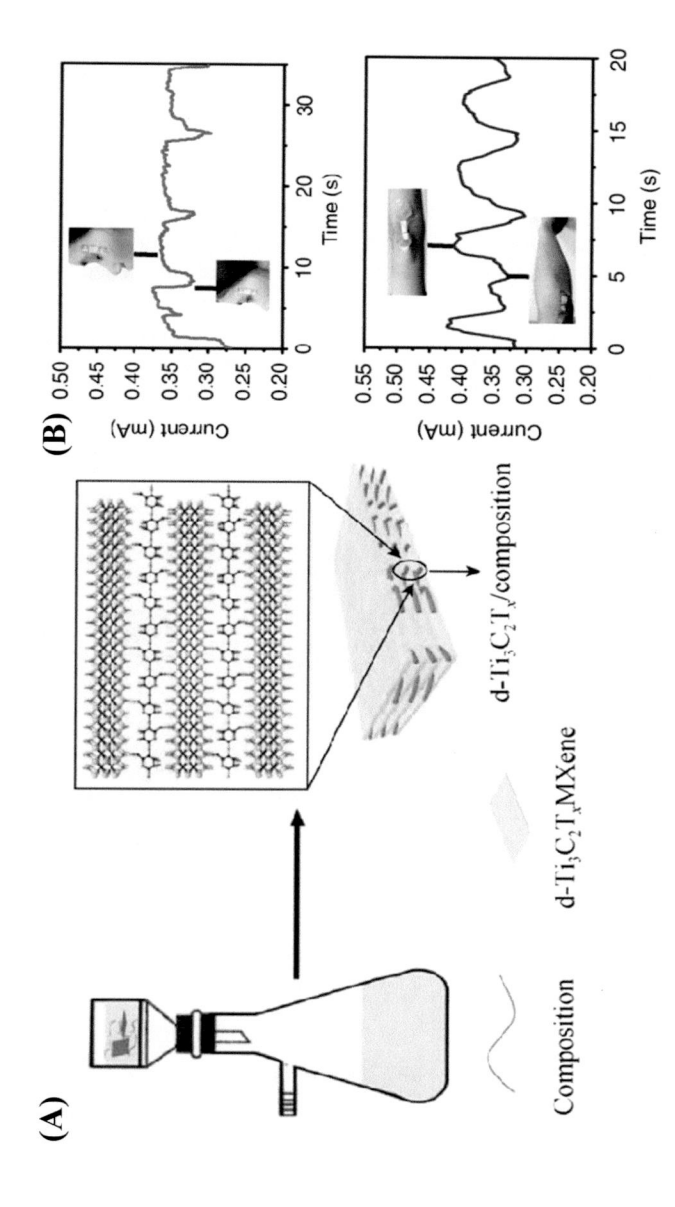

FIGURE 13.28

(A) Process method for the preparation of MXene strain sensors by the filter coating method [104]. (B) Sensing results of monitoring facial expression and joint flexion [102].

et al. [106] immersed a PU sponge in a positively charged chitosan (CS) solution and obtained CS@PU sponge. The sponge was then immersed in a negatively charged MXene colloid solution, and after being fully immersed, it was taken out and transferred to a vacuum dryer at 80°C. The above steps were repeated to prepare the MXene@CS@PU pressure sensor. Within a certain pressure range, the sensor has good fatigue resistance and can be used for strain (breathing, pulse, etc.) detection when compressed to less than 95% of the original volume. Fig. 13.29 illustrates the preparation of MXene sensors by two different impregnation methods.

13.4.2.4 Strain sensors using the screen-printing method

Screen printing refers to transferring the MXene ink made by MXene composite to the elastic substrate through the holes in the substrate to prepare sensors, as shown in Fig. 13.30. Shi et al. [107] first mixed MXene/silver nanowires (AgNWs)/polydopamine (PDA) to prepare gel ink. Then the gel ink was printed on a polyurethane substrate using screen printing technology to obtain the MXene/AgNWs/PDA strain sensor. The composite sensing materials of MXene and AgNWs enable the sensor to have a higher strain sensitivity factor ($f_g = 8700$), and PDA enables the sensor to have better flexibility. The sensor exhibits different strain sensitivity factors under different strains. At 0%−55% strain, the strain sensitivity factor is greater than 100, while at 55%−83% strain, the strain sensitivity factor is greater than 200. These sensors can track human activity in different

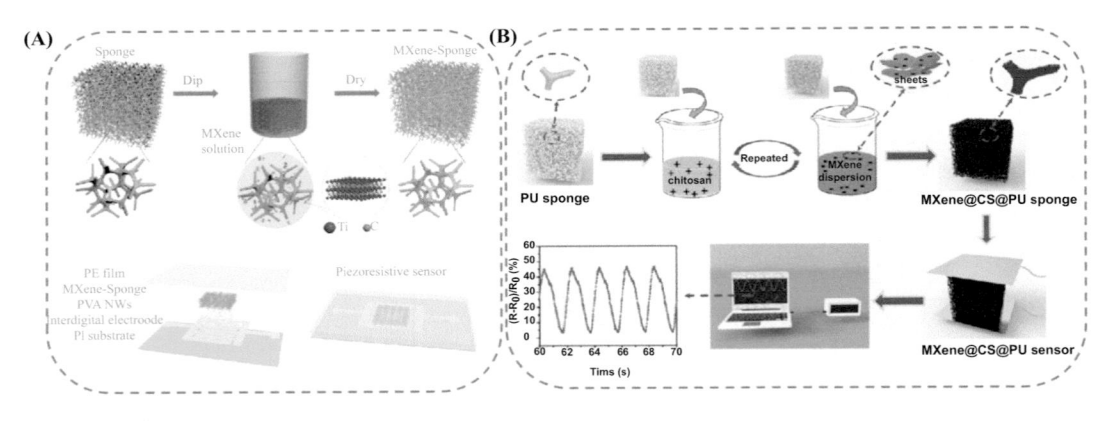

FIGURE 13.29

(A) Fabrication process of MXene-sponge material [105]; (B) fabrication process of MXene@CS@PU sponge sensor [106].

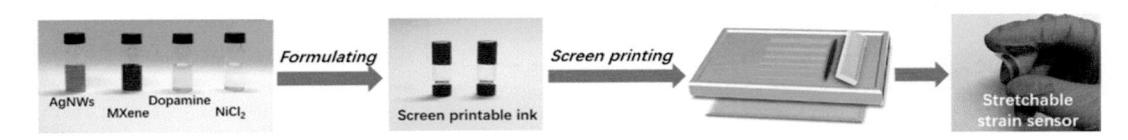

FIGURE 13.30

Schematic diagram of the screen-printing preparation process for MXene strain sensors [107].

motion states. This approach successfully achieves a high strain sensitivity factor under large deformation but is restricted by easy shedding between layers from its multilayer structure.

13.4.2.5 Strain sensors using the freeze-drying method

The freeze-drying method refers to the freeze-drying of MXenes complex into an aerogel for assembling into sensors. Fig. 13.31 illustrates the preparation process of an aerogel sensor by freeze-drying. Hu et al. [108] prepared aerogels by mixing MXene with CS in the freeze-dryer. Since MXene is negatively charged and CS is positively charged, the positive and negative charges of MXene and CS attract each other and grow into ordered layers during freezing. After freeze-drying, CS is converted to CS and its derived carbon; these carbonized CS connected the MXene layers as a continuous conductive path. During compression, the distance between holes in the micromorphology decreases and the conductive contact area increases, increasing the number of conductive paths per unit area and a larger output current. Therefore, the prepared MXene/CS aerogels exhibit certain mechanical properties and linear strain sensitivity factors and can be assembled into strain sensors to detect signal changes in the behavior of expression, breathing, pulse, swallowing, etc. This combination of MXene and CS achieves a high linear strain sensitivity factor over a wide pressure range, but the CS tends to agglomerate during the synthesis process, resulting in reduced performance. Another combination of MXene and aramid nanofibers (ANFs) is implemented by Wang et al. through freeze-drying to form MXene/ANFs aerogels [109]. MXene/ANFs is a mortar brick structure with a deformation range of 2%−80% and a minimum detection limit of 100 Pa, which can detect small and large movements of human body at 200°C. This method achieves a highly compressible and thermally stable sensor that can be used at very high or very low temperatures but has poor conductivity and strain sensitivity factor at room temperature.

Liu et al. [111] froze the mixture of MXene and polyamic acid (PAA) for 24 hours, then moved it to a freeze-dryer for 72 hours, and annealed it in argon at 300°C for 2 hours. PAA was converted to polyimide (PI), resulting in an MXene/PI aerogel strain sensor, which can be used for flexible

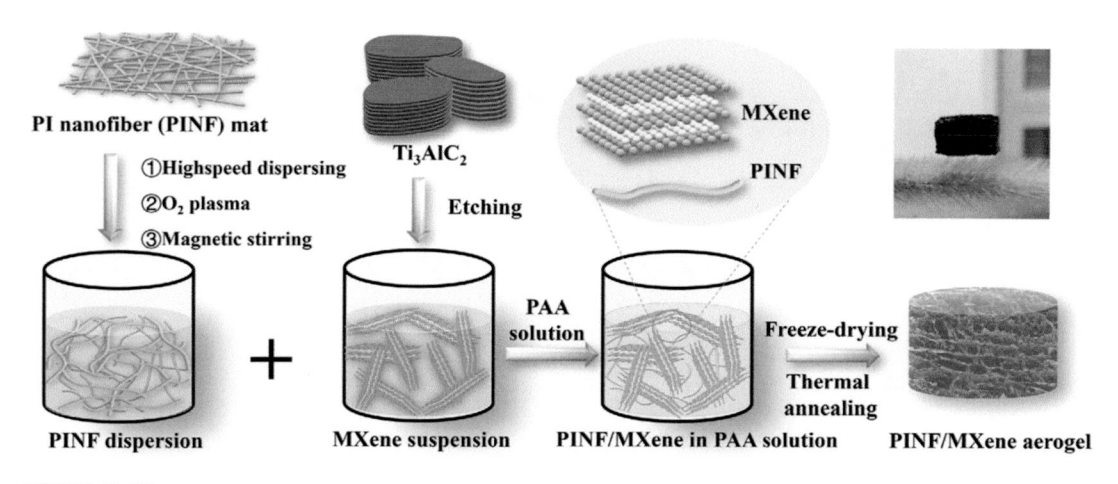

FIGURE 13.31

Schematic diagram of the preparation of aerogel sensors by freeze-drying method [110].

strain sensing. This method integrates 2D MXenes into a 3D macroframe to form an independent, highly conductive 3D MXenes monolith, but the sensor cannot detect small pressures. Liu et al. [13] proposed a MXene/PINF suspension prepared by adding MXene colloidal solution, polyimide nanofibers (PINF), and adhesive polyamide-triethylamine salt solution (PAATEA) to deionized water. The suspension was put in the refrigerator at $-20°C$ for 12 hours, then moved to the freeze-dryer for 72 hours, and finally subjected to thermal imidization to assemble the MXene/PINF aerogel piezoresistive sensor. MXene/PINF is a layer-column porous spider web structure with 0.5%–90.0% compressive strain and retains stable sensing properties in liquid nitrogen and high temperature environment. It can detect human movements such as pulse, inspiration and expiration, pronunciation, shaking head, knee bending, and finger bending.

13.4.2.6 Strain sensors using the freeze–thaw method

Freeze–thaw is also known as the physical crosslinking method. After repeatedly freezes–thaws at low and room temperatures, the MXene composite materials form microcrystalline regions inside the materials as physical crosslinking points, resulting in conductive hydrogels with a 3D network structure. Lu et al. [112] added MXene to the solution of PVA and polyvinylpyrrolidone (PVP), then moved it to a refrigerator at $-20°C$ for 12 hours and thawed at room temperature for 3 hours. This was repeated three times to obtain the MXene/PVA/PVP composite dual-network hydrogel sensor. In this hydrogel, the "hard segment" of PVP with larger side chain groups is incorporated into the "soft segment" of PVA. The network structure of "soft" and "hard" segments makes the hydrogel exhibit good mechanical properties. The sensor has a response time of 33.5 ms and a minimum detection limit of 0.87 Pa, and can detect a variety of physiological signals in the human body. Fig. 13.32 is a diagram of a hydrogel sensor prepared by the freeze–thaw method.

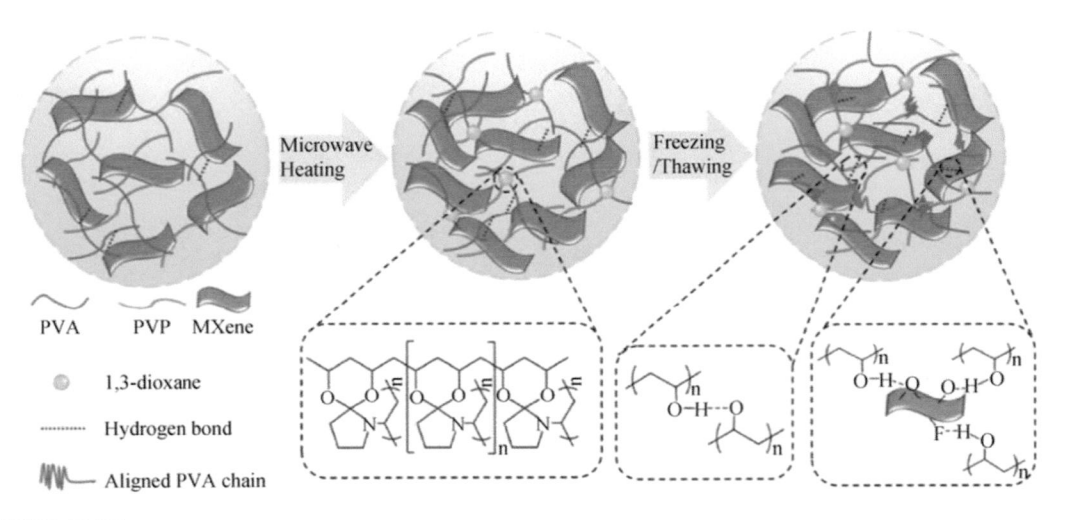

FIGURE 13.32

Schematic diagram of the preparation of hydrogel sensors by freeze–thawing [112].

13.5 **MXenes and their applications in pressure sensor**

With the rapid development of consumer electronics, artificial intelligence, and clinical medicine, the demand for pressure sensors is increasing. At present, pressure sensors have gradually developed toward having good flexibility, being lightweight, having high sensitivity, and a fast response speed, and the performance of pressure sensors largely depends on both the high flexibility and good electrical conductivity of sensitive materials. The functional groups on the surface of the two-dimensional MXene material make it have strong electrical conductivity, excellent hydrophilicity, stability, and mechanical properties. Because of its unique structure and outstanding performance, MXene has a great prospects for application in preparing pressure sensors. Also, the microstructure of MXenes is adjustable to combine with other functional compounds to form composite materials, exhibiting additional excellent properties. Therefore, MXene is an ideal material for preparing pressure sensors.

13.5.1 **Pressure sensor based on Pure MXene**

Direct use of pure MXene with an adjustable structure as the sensitive material is a simple and effective method to prepare flexible pressure sensors, such as encapsulating multilayer $Ti_3C_2T_x$ into piezoresistive sensors, as shown in Fig. 13.33A [103]. However, when the pressure reaches high levels, the MXene's interlayer spacing approaches saturation, making further compression difficult and resulting in a loss of sensitivity. Therefore, a simple multilayer flat MXene cannot fully satisfy the needs of the sensor. To solve this problem, a wrinkled 3D structure of MXene is modified and used for sensor construction. The built sensor exhibited a wider response range, higher sensitivity, and good stability at 3000 cycles [113].

As can be seen from Fig. 13.33, the performance of the pressure sensor depends on the microstructure of the pure MXene material, which could be a precise adjustment to get better sensing performance. In addition, laser technology can be used to precisely design the shape and size of pure MXene layer microstructures. Gao [114] et al. prepared an interdigital sensor based on pure MXene material through laser printing technology. Benefiting from the microchannel confined effect of the sensor's fingerprint structure, the sensor provides multiple parameters detection, including pressure, sounds, and even small movement accelerations, with high sensitivity and reproducibility, as shown in Fig. 13.33B. Subsequently, a method to optimize the microstructure of MXene was proposed, and the MXene piezoresistive sensor shown in Fig. 13.33C was prepared by jet printing technology [115]. In addition, the obtained pressure sensor can realize effective recognition of braille and possesses broad application prospects in medical monitoring and human−computer interaction, as shown in Fig. 13.33D.

13.5.2 **MXene pressure sensor based on flexible substrate**

The MXene material can be deposited on the flexible substrate by many approaches, including spin coating and dip coating, to prepare the flexible pressure sensor. The obtained flexible pressure sensor possesses good ductility and flexibility. In addition, the MXene deposition on the flexible substrate also promotes the pressure sensor's relative resistance value and dielectric constant,

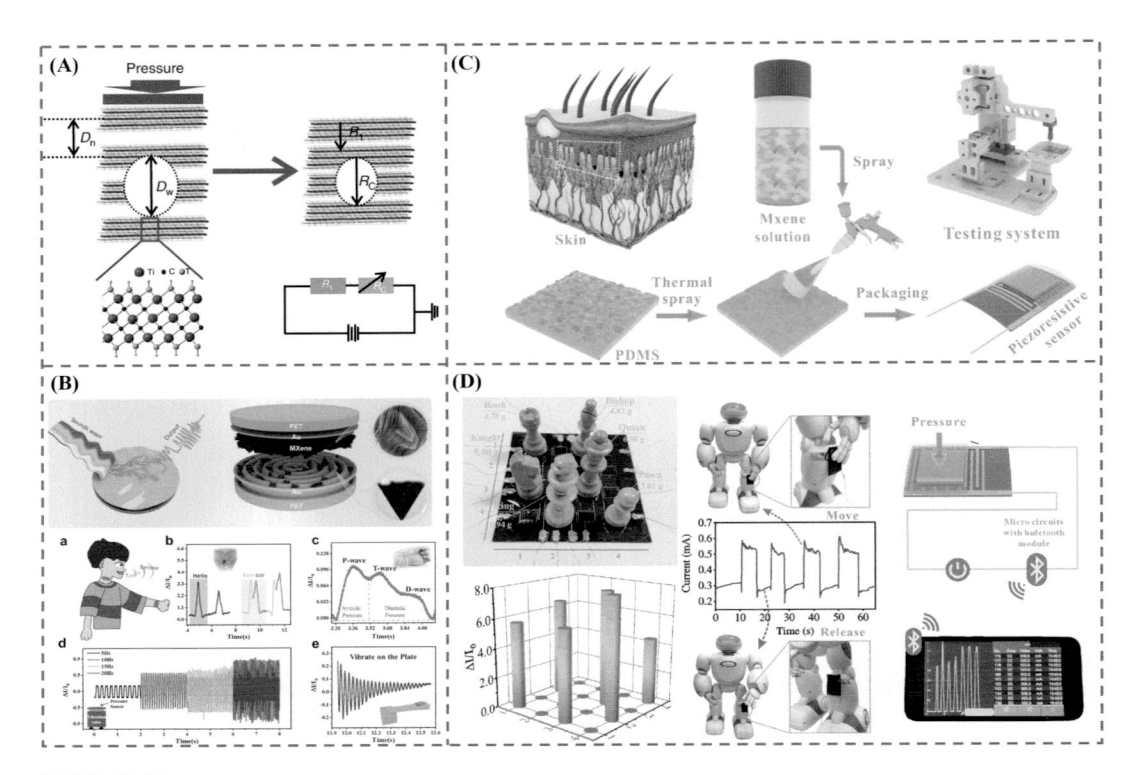

FIGURE 13.33

Pressure sensor based on pure MXene. (A) Multilayer MXene sensor mechanism diagram [103]. (B) MXene pressure sensor based on laser printing [114]. (C) Principle diagram of MXene pressure sensor fabricated by jet printing technology. (D) Application of MXene pressure sensor based on jet printing technology [115].

facilitating sensing performance, as shown in Fig. 13.34A [116,117]. Therefore, the MXene pressure sensor based on the flexible substrate has the advantages of fast response time, high sensitivity, and wide detection range, and exhibits excellent application potential in artificial skin and next-generation healthcare monitoring, as shown in Fig. 13.34B [118]. Moreover, MXene nanosheets are connected to form a well-conducting pathway; thus the MXene nanosheet-based flexible pressure sensor has excellent mechanical properties and normally works in the range of 0 degree−180 degrees bending angle, as shown in Fig. 13.34C [119].

Wearable cotton fabric has good human compatibility and is ideal for wearable pressure sensors. MXene nanosheets were sprayed, dried, and deposited on cotton fabric to obtain composite materials, as shown in Fig. 13.34D. The formed composite material maintains the fabric's original network and flexibility and effectively prevents the buildup of MXene sheets, exhibiting a continuous conductive path. Moreover, the obtained smart textiles are used for pressure sensing and show Joule effect and electromagnetic shielding functions, which are capable of monitoring human movement (Fig. 13.34E). Guo et al. [118] constructed a transient pressure sensor based on

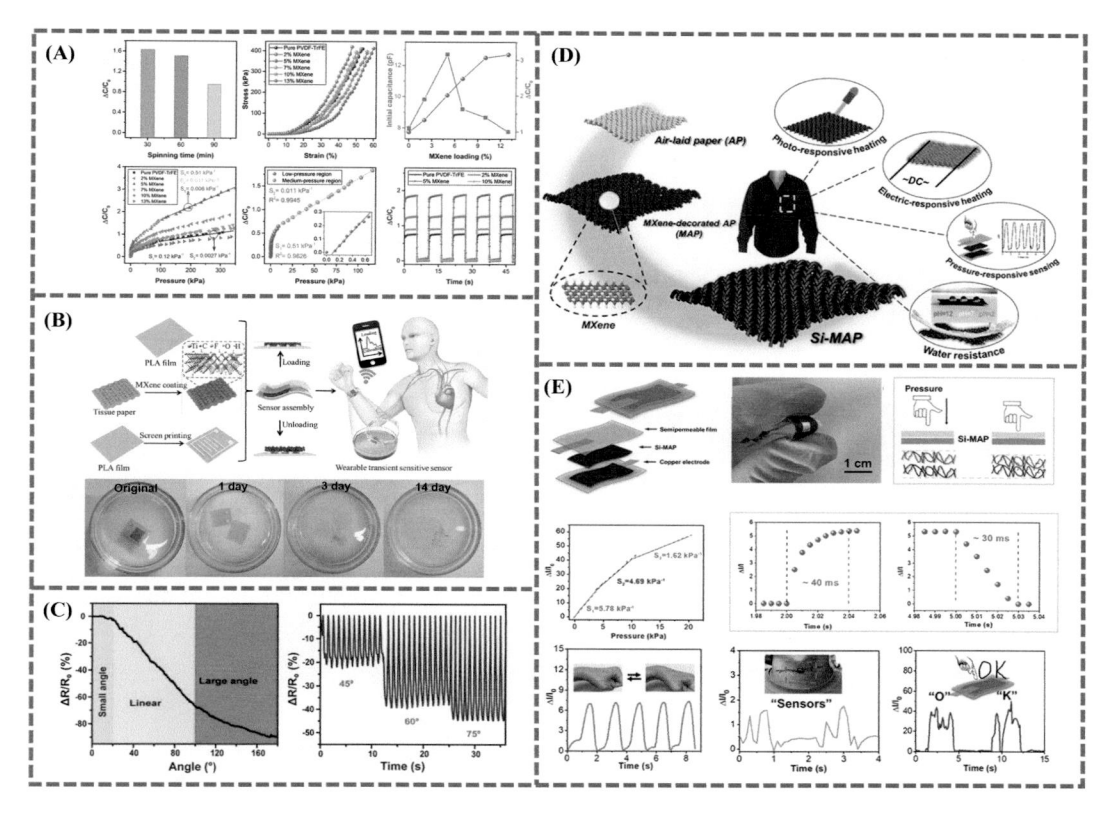

FIGURE 13.34

(A) Detection performance of MXene pressure sensor based on flexible substrate [116]. (B) Application of MXene pressure sensor based on flexible substrate in medical monitoring [118]. (C) Mechanical performance detection of flexible pressure sensor [119]. (D) Preparation of MXene/cotton fabric composites and (E) Sensing performance testing of flexible pressure sensors [120].

sandwich structure. The top and bottom are degradable polylactic acid sheets, and the middle layer is an MXene sheet impregnated in porous fabric. More than 14 days' soaking in sodium hydroxide can completely degrade the sensor.

Polyurethane (PU), due to its good stability and mechanical properties, is often used as a stretchable substrate. MXene/PU composites with high sensitivity and good stability can be prepared by the deposition method, as shown in Fig. 13.35A [121]. Using spraying technology with a PU substrate can obtain a $Ti_3C_2T_x$ /PU pressure sensor with a microwrinkle structure, which is capable of significant resistance response quickly under a tiny force (Fig. 13.35B [122]). Additional positive charged substances such as chitosan can be added to MXene/PU composites to enhance their binding force, contributing to the sensor's stability, as shown in Fig. 13.35C [106]. The same method was used by Li et al. [106] to evenly and tightly load MXenes lamellae onto the

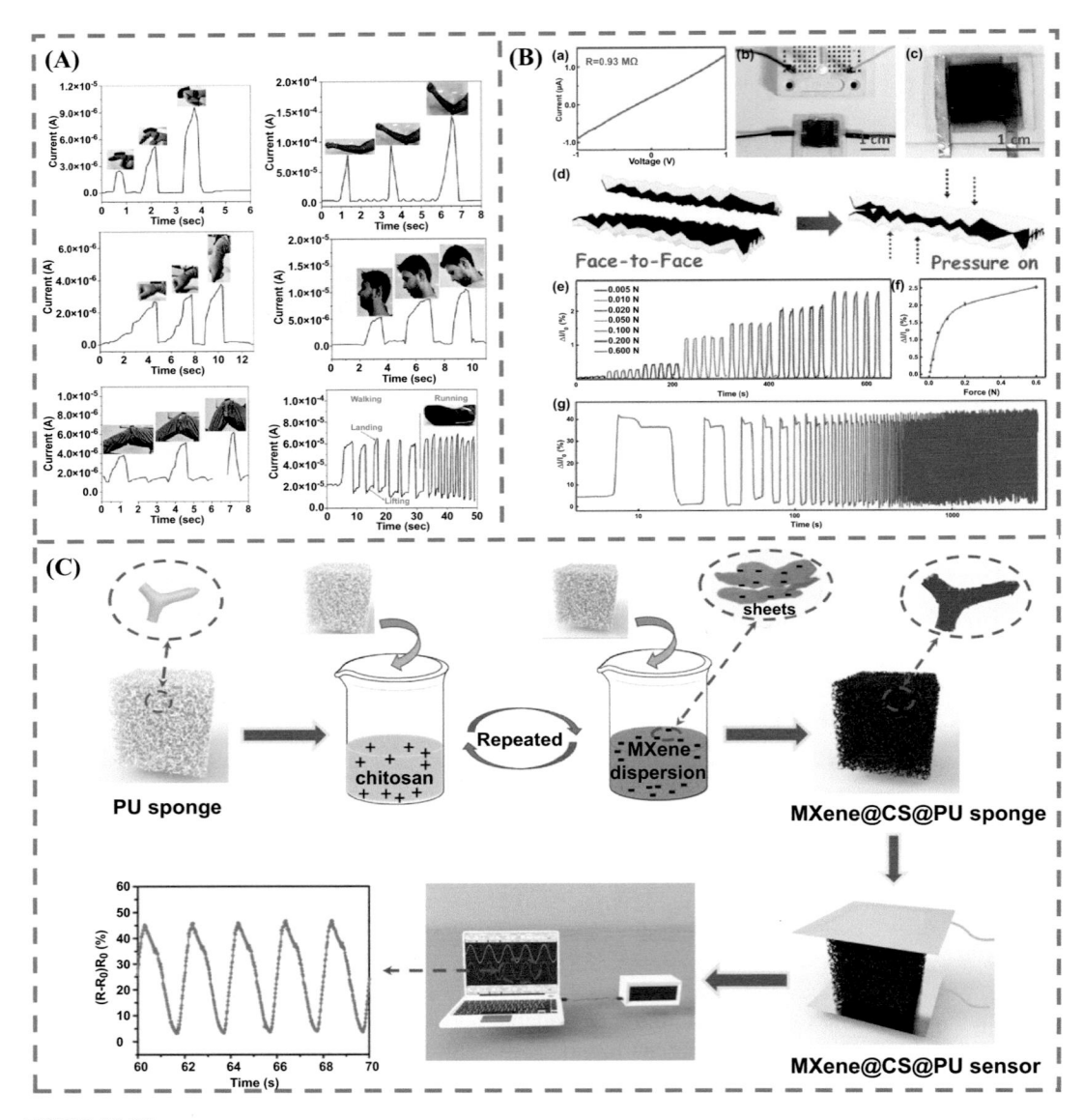

FIGURE 13.35

(A) MXene/polyurethane composites [121]. (B) MXene/polyurethane composites [122]. (C) Preparation of MXene/polyurethane composite sponge [106].

skeleton of a chitosan-treated PU sponge, owing to MXenes lamellae's negative surface and chitosan's positive property.

Highly conductive silver nanowires (AgNWs) can still maintain good sensing ability under immense strain/pressure. Combining them with MXene to form a flexible sensing material has wide applicability. Pu [123] et al. adapted layer-by-layer self-assembly to construct an AgNWs/ MXene/waterborne polyurethane (WPU) sensing layer, as shown in Fig. 13.36A. It provides slippage propagation mechanisms owing to the effective network slippage of AgNWs, and cracks originated from MXene flakes generation in the whole operating range. This combination promotes the sensor's high sensitivity (GF > 100) (Fig. 13.36B). Moreover, the rebound stability of the sensor under repeated stress is also essential, which can be achieved by using the high elastic rubber material., For example, the $Ti_3C_2T_x$/rubber sensors manufactured by the mechanical mixing process keep good signal cycle stability during high tensile/compression cycles [124]. This is because the interaction between epoxy rubber and MXene can form a supramolecular hydrogen bond interface, which make gives the pressure sensor a good self-healing capability and mechanical flexibility [125]. As shown in Fig. 13.36C, the sensor can accurately detect the human physiological motion signal (including speech, facial expression, pulse, and heartbeat), exhibiting excellent application potential in intelligent robots and wearable electronic devices.

Although MXene pressure sensors based on flexible substrates have a series of advantages, such as good flexibility and mechanical properties, the sensor sensitivity is still low due to the poor compatibility between substrates and MXene. Therefore, preparing a flexible pressure sensor with high sensitivity is still challenging.

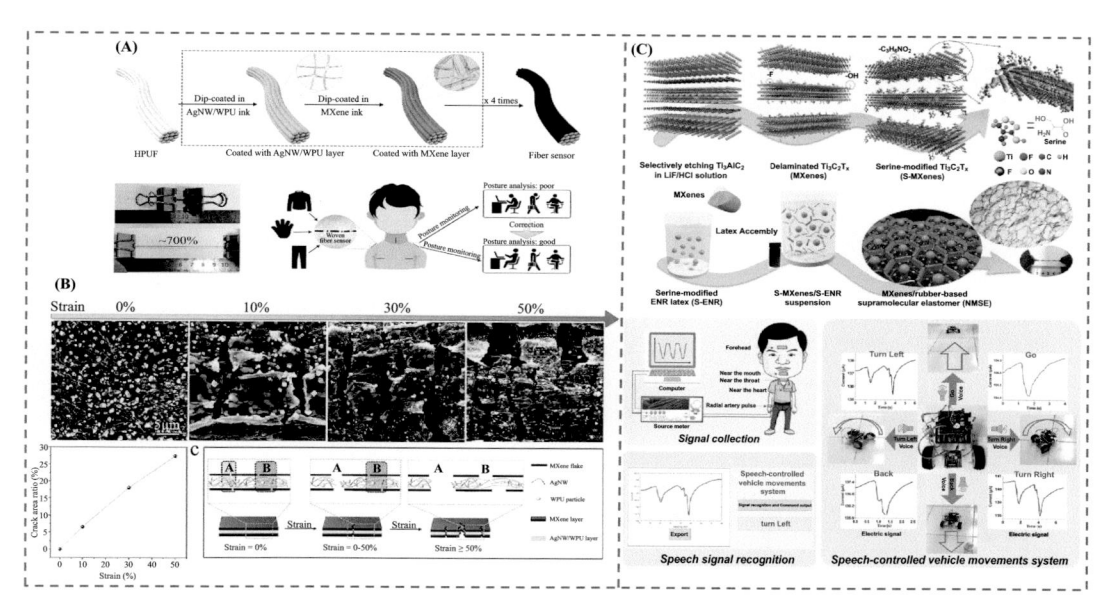

FIGURE 13.36

(A) Preparation of AgNWs/WPU/MXene composite fiber; and (B) sensing mechanism diagram [123]. (C) Preparation and application of MXene composite elastomer [125].

13.5.3 Flexible pressure sensor based on MXene composite aerogel

Aerogel has the characteristics of high porosity, good elastic properties, and lightweight features. It is widely used as a highly sensitive active material for preparing piezoresistive sensors because the internal conductivity will change when it is compressed. The gel network formed by an MXene itself is easy to collapse during compression but can hardly form aerogel alone; it is necessary to combine MXene with other materials to form aerogel to obtain a flexible pressure sensor.

MXenes aerogels can use nanocellulose with a good aspect ratio and flexibility as a gel skeleton via interentanglement between lines and surfaces. Nanocellulose has an excellent aspect ratio and flexibility and provides hydroxyl groups for hydrogen bonding with MXene, effectively preventing MXene stacking. Liquid nitrogen directional freezing and high-temperature carbonization are used to construct an oriented wave-shaped lamellar architecture for final composite aerogels, exhibiting high flexibility, fatigue resistance, and excellent cycle stability (Fig. 13.37A) [126]. These aerogels

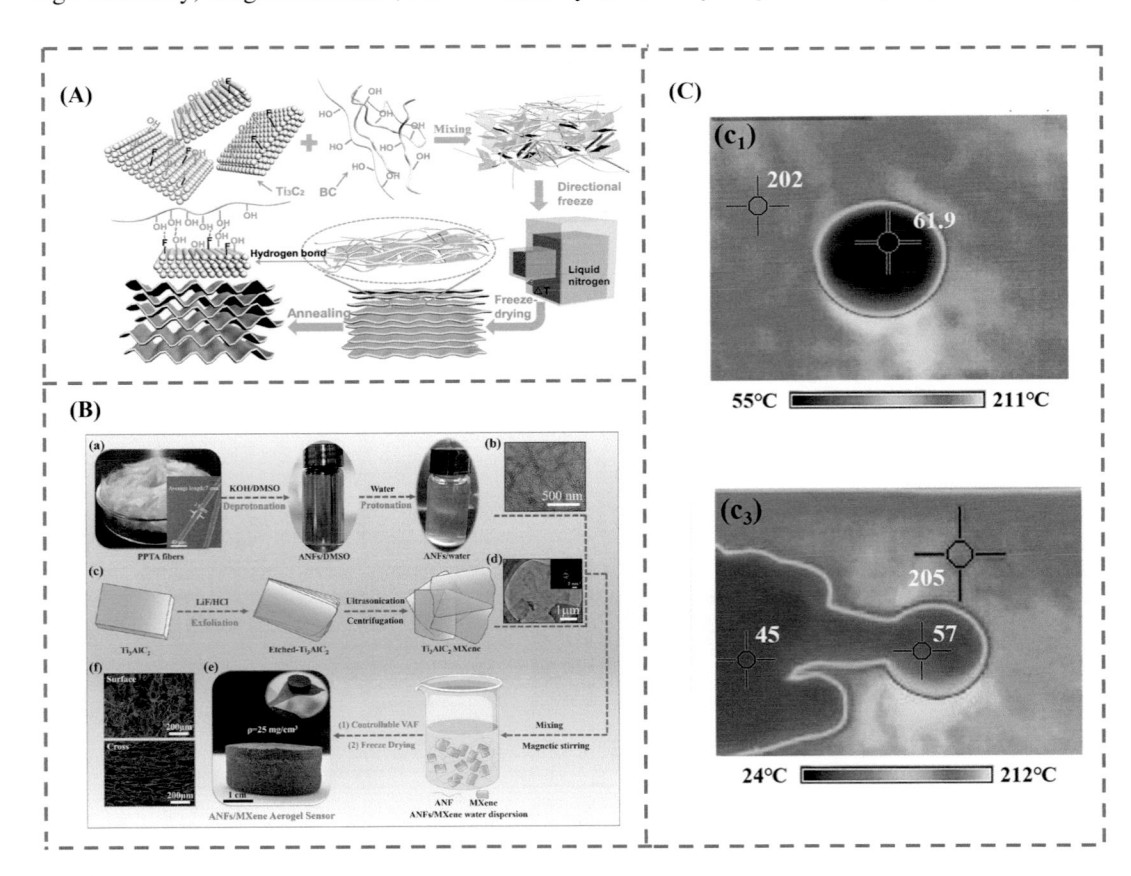

FIGURE 13.37

(A) Preparation of MXene/nanocellulosic composite aerogel by directional freeze drying [126]. (B) Preparation of MXene composite aerogel [109]. (C) Flame retardant ability of MXene composite aerogel [109].

can be used to prepare highly sensitive pressure sensors that detect human voice and movement. In addition, polyamide (PI) fiber and Arlene nanofiber (ANF) serve as the skeleton and antioxidant protective layer of MXene-based aerogel, which form a uniform porous structure with MXene, making the pores of the aerogel more compact, as shown in Fig. 13.37B, thus widening the conductive path [109]. Chen [127] et al. constructed a compressible elastic carbon aerogel, which adopted bacterial cellulose fibers to join MXenes (Ti_3C_2) nanosheets into a continuous wavy sheet. This kind of aerogel possesses high elasticity and conductivity owing to the combination between MXene and mechanically strong material, enabling its application in high-performance wearable MXenes piezoresistive sensors. Moreover, Fig. 13.37C shows aerogel application in microwave absorbents and flame retardants.

As a two-dimensional support surface, GO can form large nanosheets of aerogel with MXene. The synergistic effect of graphene and MXene makes aerogel possess excellent sensing properties. Utilizing PS microspheres as templates, Zhu et al. [128] prepared hollow spherical MXenes and introduced reduced graphene oxide (rGO) to prepare MXene/rGO composite aerogel. The contact between spherical MXene and rGO increased the conductive path and significantly improved the sensor's sensitivity. It shows a bright prospect for application in real-time monitoring of human activities and measuring pressure distribution. Ma et al. [129] prepared an MXene/reduced GO (MX/rGO) hybrid structure by combining GO with $Ti_3C_2T_x$. The synergistic interaction includes the rGO layer providing a high mechanical strength skeleton for aerogels from its larger surface area and conductive $Ti_3C_2T_x$ promoting the sensor's resistance effect, which endows better sensing performance.

13.5.4 Flexible pressure sensor based on MXene composite film

Compared with the MXene composite aerogel sensor, the sensor based on MXene flexible film material has a wider range of applications, higher sensitivity, and thinner shape [130]. Aqueous solutions of MXene possess good dispersion and can maintain their original properties without any surfactants. In addition, MXene can be easily processed to form a flexible conductive film through its aqueous solution.

Many studies have been inspired by natural materials and biological forms to construct MXene-based composite for more suitable flexible applications. Shen et al. [131] prepared a biocomposite film with a bionic interlocked structure like human skin (Fig. 13.38A) by mixing MXene with *Helianthus annuus* L. spores, similar to a sea urchin-like microstructure. The interlocking structure in the film can effectively adjust the stress distribution under the action of external forces and avoid the cracking caused by pressure concentration. The sensitivity of the pressure sensor prepared by the film is 9.4 times higher than that of the flat MXene sensor. Zhang et al. [132] developed a composite film based on MXene, which presents a wrinkled ridge shape on its surface. The inhomogeneous strain in the layered film allows microcracks to develop in the valleys between the ridges under stress; at the same time, the propagation of these cracks is effectively limited due to the deflection and passivation effects of the layered structure (Fig. 13.38B). The film can be used for electromagnetic interference (EMI) shielding and pressure sensing, showing the potential for the synthesis of fully flexible sensor electrodes and circuits with high cyclic stability/pressure sensitivity (Fig. 13.38C). Natural silk cellulose (SF) can be used as a bridging agent to assemble MXene nanosheets and construct continuous wavy layered structures [133]; the obtained sensors exhibit biocompatibility and reduce risk in the human body, which offers the possibility of developing safe and high-performance flexible electronics (Fig. 13.38D).

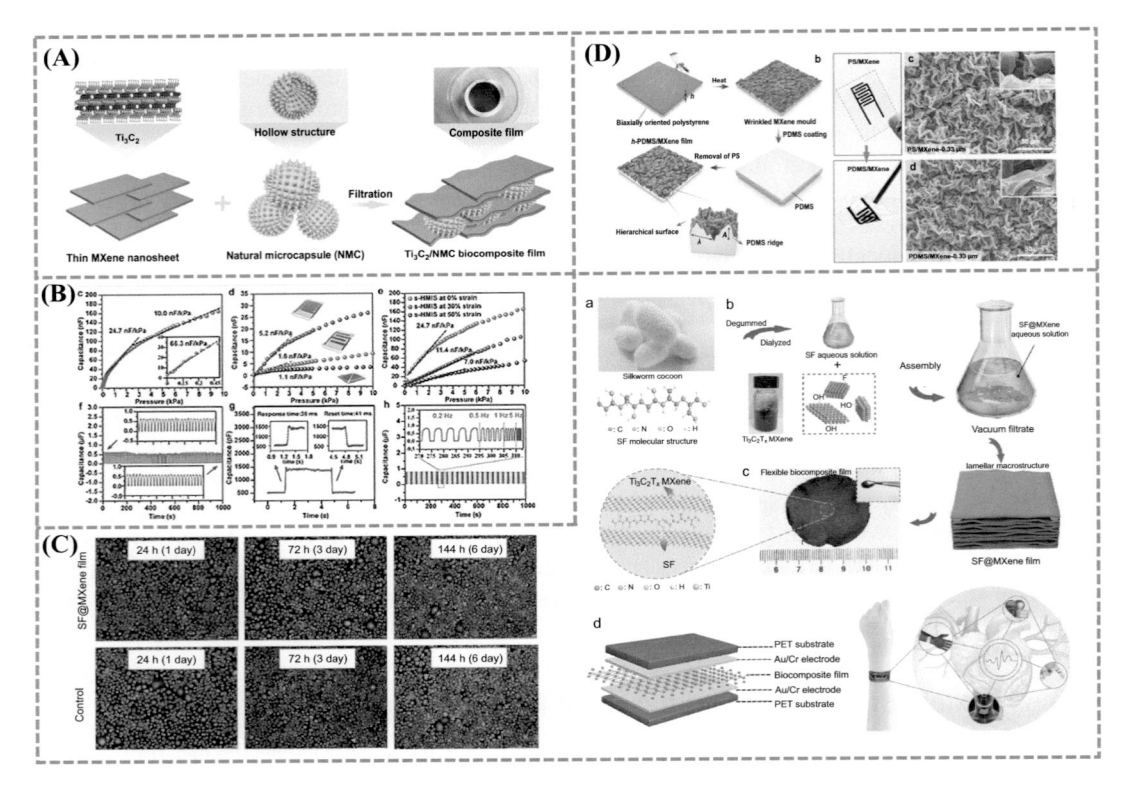

FIGURE 13.38

(A) Bionic interconnected biocomposite membrane [131]. (B) Preparation and properties of MXene composite membrane [134]. (C) Performance of the PDMS/MXene-based touch sensors [134]. (D) Schematic diagram of biocomposite membrane [133].

13.5.5 Flexible pressure sensor based on MXene composite hydrogel

As a polymer, the hydrogel is often used as the carrier of active materials for pressure sensors due to its good flexibility and tensile properties. However, active substances tend to slide in hydrogels, which severely destabilizes the assay. Benefiting from the MXene hydrophilic surface groups, more hydrogen bonds are generated in the system for a stable crosslinked structure with polymer. Therefore, this composite hydrogel possesses enhanced mechanical and self-healing properties.

Polyvinyl alcohol is a water-soluble, nontoxic long-chain polymer. PVA hydrogel is a colloidal dispersion with a three-dimensional network structure utilizing polymer chain crosslinking and swelling. It has attracted much attention due to its low toxicity, high water absorption, good mechanical properties (i.e., high elastic modulus and high mechanical strength), and biocompatibility. The hydroxyl group of PVA forms a hydrogen bond with MXene to enhance the self-healing property of hydrogel. Zhang [135] et al. introduced MXene nanosheets into PVA-based hydrogels to prepare MXene composite hydrogels (M-hydrogels). The M-hydrogel has different MXene interconnection modes under external

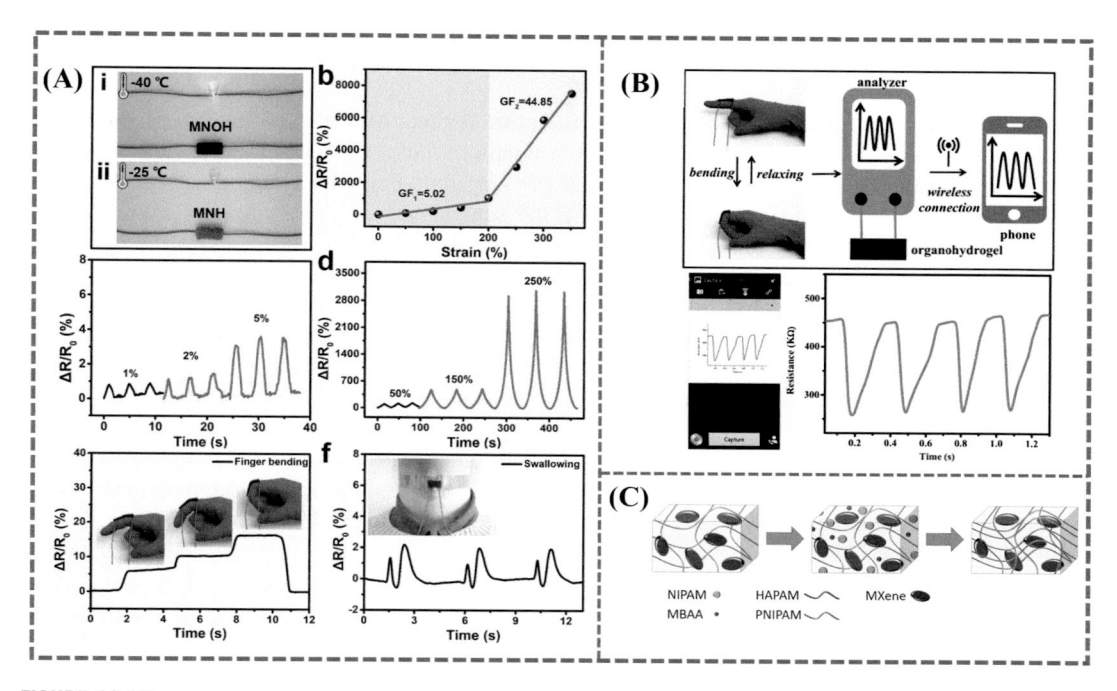

FIGURE 13.39

(A) Flexibility test of antifreeze hydrogel [136]. (B) Sensing performance of antifreeze hydrogel [137]. (C) Preparation of thermosensitive hydrogel [138].

forces in different directions, so it has good anisotropy and can detect the direction, velocity, and trajectory of external forces acting on the surface. It also performs extremely excellent stretchability, instantaneous self-repair ability, and can tightly adhere to various surfaces such as human skin.

Adding substances with environmental response functions can yield smart hydrogel sensors. Antifreeze hydrogel was prepared by adding MXene into a polymer composed of polyacrylamide (PAM) and PVA and using ethylene glycol (EG) as solvent. The hydrogel pressure sensor can be repeatedly bent, even at $-40°C$, as shown in Fig. 13.39A [136]. A composite hydrogel constructed by alginate MXene with PAM-g-sodium possesses self-healing ability and long-lasting water retention (10 days) [137]. It could be assembled in a wearable pressure sensor with a wireless transmitter to monitor human activity, as shown in Fig. 13.39B. MXene hydrogels constructed based on poly(N-isopropylacrylamide), a derivative polymer of PAM, impart temperature sensitivity, facilitating the sensor's good thermal conductivity (Fig. 13.39C) [138].

13.6 MXenes in gas sensors

As a two-dimensional material, MXenes have inherent lamellar structure, a large specific surface area, easily decorated surface functional groups, and abundant active sites, therefore absorbing the gas molecules and indicating the gas concertation via the conductivity change.

13.6.1 MXene gas-sensitive mechanism

13.6.1.1 The principle of MXene adsorption

Gas adsorption or desorption may be affected by active defects and surface functional groups on the MXenes surface [139]. Using first-principles simulation, Yu et al. [140] theoretically predicted MXenes composite's potential in multiple types gas sensing. For film-forming gas sensors, the film's conductivity is the critical factor affecting the sensing performance, and the adsorption of gas is directly related to the conductivity. Due to physical or chemical adsorption, the charge is transferred hence the conductivity is changed. When the electrostatic force dominates gas adsorption, adsorption capacity is not strong, which will cause small changes in the sensor resistance value. However, when the surface functional groups are replaced by gas adsorption, large changes in resistance will be affected [141]. Lee et al. [142] utilized the solution casting method to integrate $Ti_3C_2T_x$ on flexible polyimide platforms, and the gas sensor showed a high sensitivity to NH_3. Xiao et al. [143] believed that NH_3 could interact with O-atom semiconductor MXenes with different charge states, and NH_3 could be adsorbed on M_2CO_2 through charge transfer. Taking $ZrCO_2$ as an example, the adsorption energy increased from -0.81 to -0.20 eV during NH_3 release. The above studies indicate that MXenes can be used as a gas-sensitive material for NH_3 detection.

In 2017, Lee et al. [142] synthesized MXene by removing aluminum atoms in MAX phase and conducted gas sensitivity tests for ethanol, methanol, acetone, and ammonia. They found that $Ti_3C_2T_x$ MXene showed p-type semiconductor behavior to these four reducing gases and had the highest response to ammonia. However, a study in 2018 also demonstrated that the $Ti_3C_2T_x$ MXene sensor had a positive response to both reducing and oxidizing gases [144], in which the mechanism of a P-type semiconductor could not be used to interpret this result. MXenes have high metal conductivity and abundant functional groups on the surface, which is more conductive than traditional semiconductor materials. Therefore, Kim believes that the MXene gas response is due to the high metal conductivity and strong adsorption energy, rather than the semiconductor mechanism [145]. The gas adsorbed on the surface of the sensor replaces the functional groups on the surface of MXene, reducing the number of carriers and increasing the resistance value of the sensor surface.

Koh et al. suggest a hypothesis that the gas sensing is due to the interlayer expansion of MXene [145]. Through Na^+ intercalation of $Ti_3C_2T_x$, the dynamic swelling behavior was measured by in situ XRD, which proves that ethanol has the swelling effect on $Ti_3C_2T_x$ MXene. The thorough cleaning by ethanol for 70 minutes causes the 002 peak of $Ti_3C_2T_x$ film to shift to a smaller Angle, in which the interlayer space increases by 0.82 Å. Blowing N_2 to desorb and decrease the ethanol, the 002 peak recovers to a larger angle. The layer space decreases around 0.51 Å after the ethanol absorption and swelling. The results show that the concentration of sodium ions determines the swelling degree of ethanol after adsorption. In addition, the degree of swelling is proportional to the response strength of the gas in $Ti_3C_2T_x$ MXene film. Therefore, it is essential to control the interlayer distance of $Ti_3C_2T_x$ MXene to improve the selectivity of gas sensing.

Maleski et al. [144] used density functional theory (DFT) to simulate the binding energies of acetone and ammonia on $Ti_3C_2T_x$, MoS_2, RGO, and BP to study the sensing mechanism of acetone and NH_3 gas by $Ti_3C_2T_x$, as shown in Fig. 13.40B. The signal-to-noise ratio of $Ti_3C_2T_x$ sensor is two orders of magnitude higher than that of other two-dimensional materials. Through the comparison of binding energy, with the intense adsorption energy of functional groups on the surface of the $Ti_3C_2T_x$ sensor, acetone and ammonia gases replace hydroxyl groups in

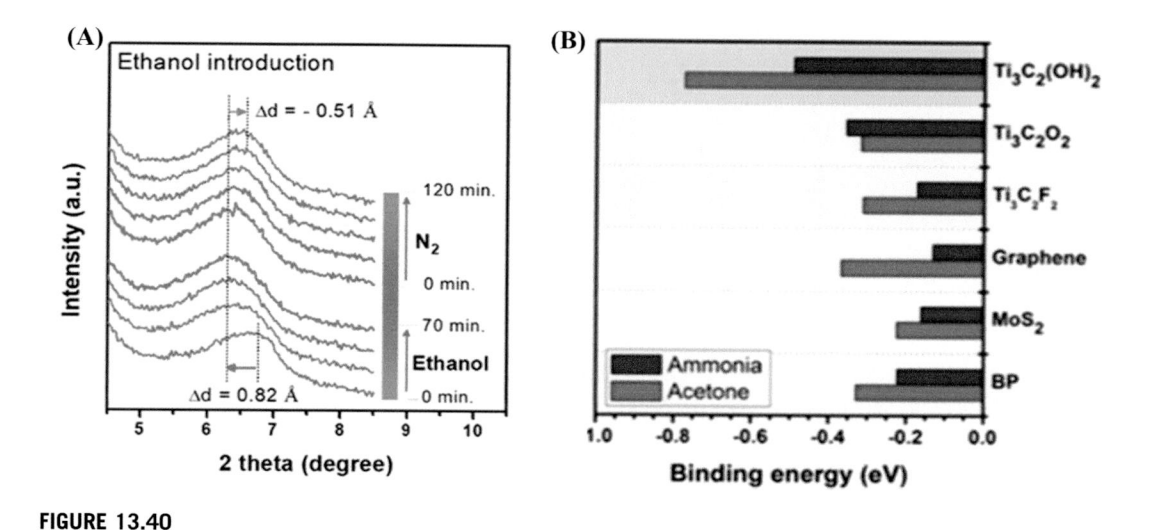

FIGURE 13.40

(A) The peak shift of (002) after $Ti_3C_2T_x$ was purged with 0.1% ethanol for 70 min and N_2 purged for 120 min after the adsorbed ethanol was stripped away [145]. (B) Acetone and ammonia in $Ti_3C_2(OH)_2$, $Ti_3C_2O_2$, $Ti_3C_2F_2$, graphene, MoS_2, the lowest binding energy on BP [144].

MXene by adsorption, change the conductivity, and improve the sensitivity of the sensor. This work demonstrates the charge transfer induced by gas adsorption in the gas sensing mechanism of MXenes.

13.6.1.2 MXene surface adsorption calculation

The gas sensing mechanism in MXene is through the physical or chemical adsorption of gas molecules, resulting in changes in resistance or conductivity. The selectivity of MXene to gas is also determined by the energy and charge transfer characteristics of gas molecules adsorbed on the surface. Therefore, it is crucial to calculate the gas adsorption on the MXene surface. DFT is usually used to calculate the adsorption of two-dimensional materials and gas molecules. The molecular binding energy calculated by DFT is related to the adsorption energy, and the binding energy is proportional to the gas response. Therefore, the first principle of density functions is generally used to calculate the absorption energy and the charge transfer under different types of gases.

In 2015, Yu et al. [140] used first principles to simulate the adsorption capacity of monolayer Ti_3CO_2 on NH_3, H_2, CH_4, CO, CO_2, N_2, NO_2, O_2, and other gases, and for the first time proved MXenes' potential for gas sensing. Among all gases, only NH_3 can be chemically adsorbed in which the N atom directly connects with the Ti atom on the Ti_3CO_2 nanosheet (Fig. 13.41A). By calculating the adsorption energy and charge transfer on the surface of MXene, the removal and sensing functions of polar and ionic pollutants, organic dyes, radionuclides, and gaseous pollutants can occur, instead of detergents, indicating that MXene material has great potential in the adsorption of environmental pollutants [146].

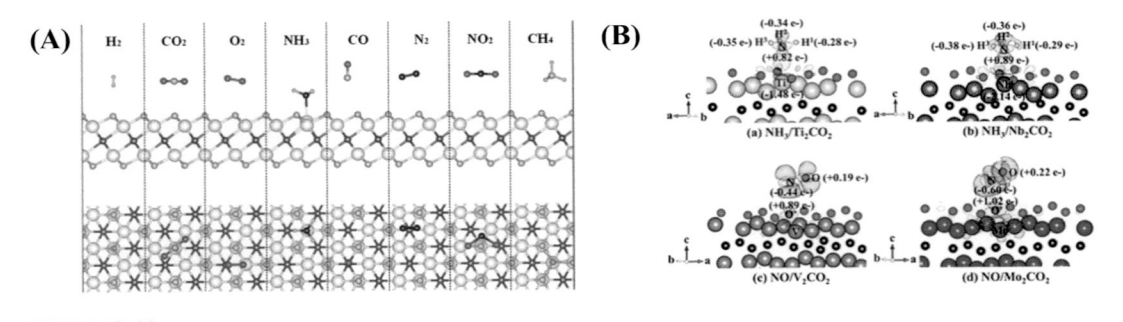

FIGURE 13.41

(A) Side view and top view of adsorption of gas molecules on monolayer Ti_2CO_2 [140]. (B) Charge difference and Bader charge change of MXene corresponding gas [147].

In 2018, Junkaew et al. [147] studied the response and selectivity of four kinds of MXene (M_2C (M = Ti, V, Nb, Mo)). The predominant chemisorption process with low adsorption energy indicates the high reactivity of MXenes but low selectivity towards different gases. Only NH_3 can be molecularly absorbed on four types of MXenes. The oxygen functionalization of MXene improves its selectivity to gases, such as Mo_2CO_2 and V_2CO_2, which have good selectivity to NO molecules. Fig. 13.41B shows the charge difference and Bader charge change of atoms selected by NH_3/ Ti_2CO_2, NH_3/Nb_2CO_2, NO/V_2CO_2, and NO/Mo_2CO_2. The accumulation and depletion of electrons are represented by green and red areas, respectively. Nb has a more positive charge than Ti, and therefore has a stronger adsorption strength (adsorption energies of -0.5 eV of Nb, and -0.37 eV of Ti).

Pourfath et al. [148] used the first principle to calculate the gas sensing performance of $Ti_3C_2T_x$ MXene, and analyzed the influence of different functional group ratios on the adsorption of $Ti_3C_2T_x$ by NH_3, NO, NO_2, N_2O, CO, CO_2, CH_4, and H_2S gas molecules through charge difference calculation. The effect of surface functional groups on charge transfer is different, and the charge transfer amount of fluorine and oxygen atoms is more significant than that of fluorine atoms. It can be deduced that a strong electrostatic attraction between the lone pair electrons of O atoms in functional groups plays a vital role in charge transfer. Controlling the amount of O-functional groups can adjust the NH_3 selectivity. Recently, Naqvi et al. [149] explored charge transfer and adsorption of a variety of gases (such as CH_4, CO, CO_2, NH_3, NO, NO_2, H_2S and SO_2) on nanoscale gas sensors based on M_2NS_2 MXenes (M = Ti, V) through density generalized function theory calculation. The analysis of charge transfer shows that a considerable amount of charge is transferred from NO and NO_2 gas molecules to Ti_2NS_2 and V_2NS_2, indicating that the S-based functional groups in 2D M_2NS_2 MXenes have a high sensitivity to NO and NO_2 gases.

13.6.2 Gas-sensitive properties

13.6.2.1 Gas-sensitive properties of pure MXene

MXene is a novel two-dimensional transition metal carbide or carbon-nitride, and the response to gas comes from the change in resistance at the material surface, which can be described as $S = 100|$

R_g-R_a/R_a [150], where R_g and R_a denote the resistance of the material in the gas and air under test, respectively. Since the gas-sensitive response of the material is based on the surface-catalyzed reaction between gas molecules and MXene. MXene materials with high specific surface area usually have more reactive sites, which contribute to the adsorption and surface reaction of the gas on the material surface. Thus monolayer or layer-less MXene exhibits excellent gas-sensitive performance.

Ti_3C_2 MXene was first applied to the experimental study of gas sensors because it is easier to prepare and more stable than other MXenes. In 2017 Lee et al. [142] first used MXene for gas sensors. As shown in Fig. 13.42A, the $Ti_3C_2T_x$ material was integrated into a flexible polyimide platform and it was successfully measured at room temperature ethanol, methanol, acetone, and ammonia gas. As shown in Fig. 13.42C, the two-dimensional $Ti_3C_2T_x$ presents good selectivity for NH_3 at room temperature. Moreover, for acetone gas, which has the lowest response, the detection limit is still calculated to be about 9.27 ppm. The sensing mechanism (Fig. 13.42B) was proposed regarding the interactions between the majority charge carriers of $Ti_3C_2T_x$ and gas species.

Kim et al. [144] used $Ti_3C_2T_x$ MXene film as a metal channel for a chemiresistive gas sensor, as shown in Fig. 13.43A. VOCs such as acetone and ethanol were detected at 0.05−0.1 ppm at room temperature. By comparing typical two-dimensional materials such as black phosphorus (BP), transition metal disulfide (MoS_2), and reduced rGO, it was found that the $Ti_3C_2T_x$ gas sensor has the lowest lower LOD for VOCs at room temperature, as shown in Fig. 13.43C. The SNR values are displayed in Fig. 13.43B, which shows that the SNR of $Ti_3C_2T_x$ sensors is up to two orders of magnitude higher than those of other 2D material-based sensors.

Yang et al. [151] successfully discovered new ways to improve the chemical sensing property of $Ti_3{}^+C_2T_x$ MXene: Ti_3AlC_2 was used as a raw material to obtain two-dimensional accordion-like $Ti_3C_2T_x$ using the classical hydrofluoric acid (HF) selective etching method $Ti_3C_2T_x$, and further alkali treatment with sodium hydroxide solution obtained alkalized $Ti_3C_2T_x$. Compared with

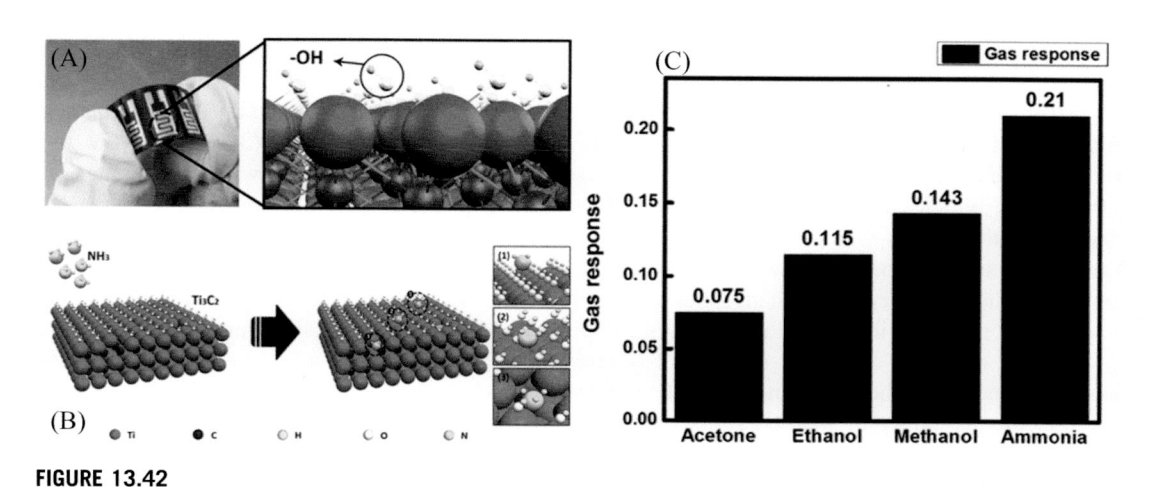

FIGURE 13.42

(A) Surface structure of $Ti_3C_2T_x$ MXene sensor; (B) schematic diagram of the gas-sensitive mechanism of $Ti_3C_2T_x$ MXene to NH_3; (C) average gas response of $Ti_3C_2T_x$ MXene for each gas [142].

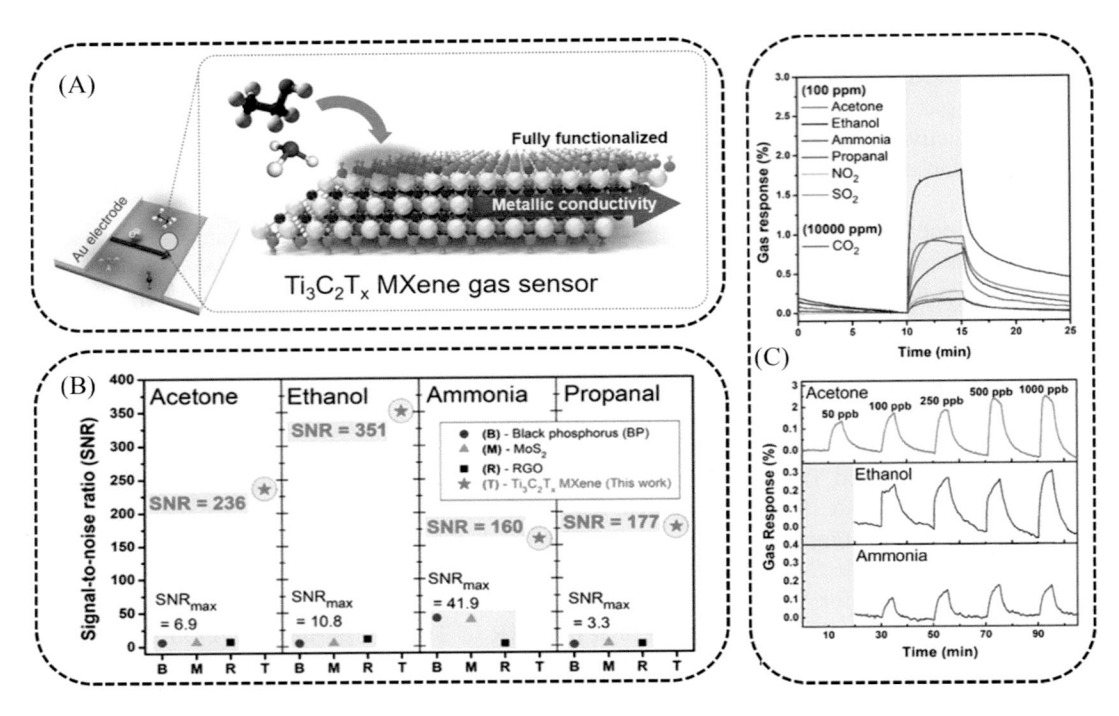

FIGURE 13.43

(A) Thin film and structure of $Ti_3C_2T_x$ MXene sensor. (B) Maximum SNR of sensors exposing to 100 ppm of acetone, ethanol, ammonia, and propanal. (C) Gas response results of $Ti_3C_2T_x$ gas sensor [144].

$Ti_3C_2T_x$, the device based on alkalized $Ti_3C_2T_x$ had opposite response signals with enhanced humidity and NH_3 sensing properties. Such an enhancement is attributed to the intercalation of Na and the increase of $-O$ terminals.

Yuan et al. [152] combined electrostatic spinning and self-assembly techniques to prepare a high-performance flexible VOC sensor based on the 3D MXene framework (3D-M). Due to the intrinsic metallic conductivity of MXene, the resistance of the 3D-M sensor was in the range of 10^3-10^4 Ω, which can be facilely monitored by common practical sensing elements. Test results show that the sensor is highly sensitive to traces of acetone, methanol, and ethanol in the ppb range. The sensor shows high sensitivity (0.10−0.17 ppm^{-1}), low experimental LOD (50 ppb), fast response and recovery (less than 2 minutes), wide sensing range (from ppb level to saturated vapor), and good flexibility and reliability for a wide range of VOC gases simultaneously.

Kim et al. [153] prepared ultrathin ($\sim$10 nm) MXene thin films using an interfacial self-assembly method. The addition of ethyl acetate to dilute aqueous MXene solution allowed the spontaneous assembly of MXene films on aqueous surfaces, which were easily transferred to the target substrates. The self-assembling behavior of MXene flakes resulted in films with a high stacking order and strong plane-to-plane adherence. A signal-to-noise ratio of up to 320 was observed when the film was used as a gas sensor, and the gas response of a sample assembled from a small piece was 10 times higher than that of a large piece (Fig. 13.44B).

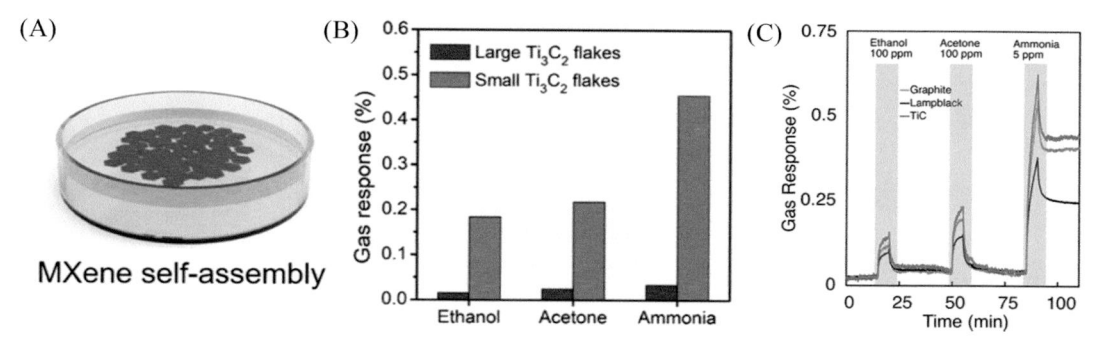

FIGURE 13.44

(A) Ultrathin (nanoscale) MXene films fabricated by interfacial self-assembly method [153]. (B) Gas responses as a function of detected gases [153]. (C) Gas sensing properties of the $Ti_3C_2T_x$ produced from different MAX precursors [154].

Shuck et al. [154] suggested that the difference in gas-sensitive properties of two-dimensional $Ti_3C_2T_x$ is related to the components of the MAX phase precursors. Ti_3AlC_2, as the precursor for $Ti_3C_2T_x$ MXene, was synthesized by three carbon sources (graphite, carbon lampblack, and titanium carbide (TiC)) at 1650°C for two hours. The results show that the $Ti_3C_2T_x$ products obtained by selective etching exhibit significant differences in sheet thickness, chemical composition, chemical stability, and electrical conductivity, which in turn affect the gas-sensitive performance of the 2D $Ti_3C_2T_x$ samples. As shown in Fig. 13.44C, the best gas-sensitive performance was obtained for the sample with TiC as the carbon source, followed by the graphite carbon source sample.

Wu et al. [150] selectively etched Al atoms between Ti_3AlC_2 layers by an in situ etching method to obtain multilayer Ti_3C_2, then used dimethyl sulfoxide (DMSO) intercalation to obtain monolayer Ti_3C_2 after ultrasonic dispersion. Gas sensor devices fabricated by coating a colloidal suspension of monolayer Ti_3C_2 on the surface of ceramic tubes were detected at room temperature at concentrations of 500 ppm of CH_4, H_2S, H_2O, NH_3, NO, ethanol, methanol, and acetone at room temperature, and the results are shown in Fig. 13.45A. The monolayer Ti_3C_2 sensor showed excellent gas response and selectivity to NH_3 at room temperature (the response to NH_3 was 6.13%). To understand the humidity sensing performance of Ti_3C_2, the sensor was tested in different relative humidity, as shown in Fig. 13.45B.

Koh et al. [145] used in situ XRD testing to study the gas-induced interlayer swelling of $Ti_3C_2T_x$ MXene thin films and its influence on gas-sensing performance. It was found that the degree of swelling caused by the sodium ions embedded in the sample can form a good correspondence with the intensity of its response to the gas, and the results showed that the treatment of $Ti_3C_2T_x$ with 0.3 mmol/L NaOH was the maximum swelling observed between the layers (swelling close to 35), and the sample responded sensitively to ethanol and did not respond to CO_2 (Fig. 13.45C).

V_2CT_x MXene [155] and Mo_2CT_x MXene [156] are able to recognize certain nonpolar gases and have promising applications in environmental monitoring. Lee et al. [155] coated V_2CT_x colloidal solution onto a polyimide film to make a sensor with ultrahigh sensitivity for nonpolar gases. As shown in Fig. 13.46, it was able to detect trace amounts of many gases, especially nonpolar

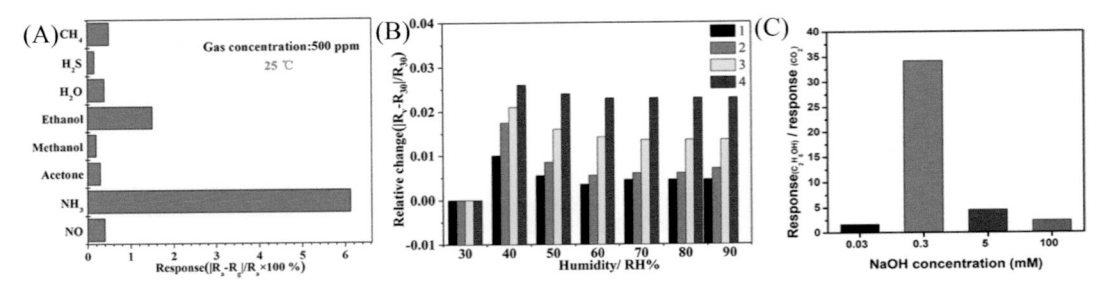

FIGURE 13.45

(A) Gas sensing performance of Ti$_3$C$_2$ MXene sensors toward 500 ppm CH$_4$, H$_2$S, H$_2$O, ethanol, methanol, acetone, NH$_3$, and NO at room temperature [150]. (B) Effects of environmental humidity on the relative resistance with different Ti$_3$C$_2$ slayers of sensitive films [150]. (C) Selectivity, calculated by dividing the response toward ethanol over the response toward CO$_2$ [145].

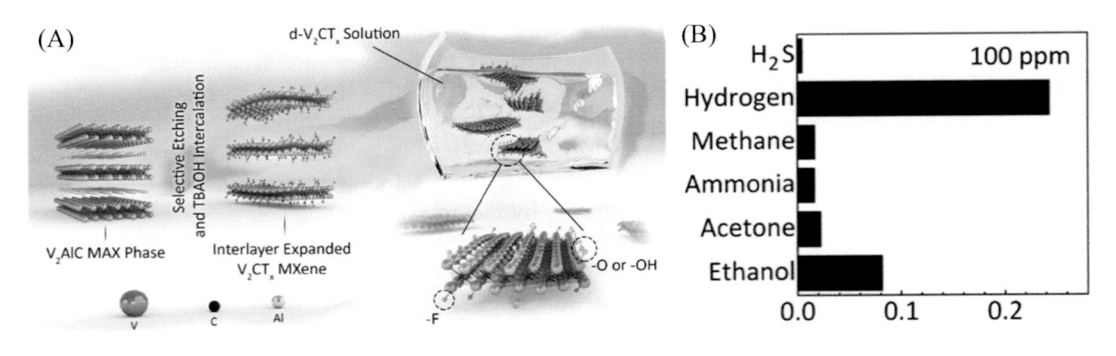

FIGURE 13.46

(A) Schematic of the synthesis and delamination of V$_2$CT$_x$ MXene [155]. (B) Gas response toward 100 ppm of hydrogen, ethanol, acetone, methane, ammonia, and hydrogen sulfide [155].

gases (hydrogen and methane) (Fig. 13.46B). The prepared V$_2$CT$_x$ gas sensor outperforms previously reported gas sensors based on other 2D materials in the detection of nonpolar gases. The theoretical detection limits at room temperature are 1 ppm for hydrogen and 9 ppm for methane.

Mo$_2$CT$_x$ MXene is another MXene that has been applied as a gas-sensitive material. Guo et al. [156] prepared a chemiresistive device, Mo$_2$CT$_x$ MXene sensor, on Si/SiO$_2$ substrate using photolithography. The sensor has sensitive performance for VOCs (toluene, benzene, ethanol, methanol and acetone) and has the highest response sensitivity for toluene at the same concentration. The sensor has a linear response to toluene in the concentration range of 35−175 ppm with a theoretical detection limit of 0.22 ppm.

13.6.2.2 Gas-sensitive properties of MXene composites

It is generally believed that the energy level interleaving generated by heterogeneous composites can effectively strengthen the free load separation, and the defects formed through the heterogeneous interfaces can increase the active sites of gas—solid surface reactions, so multicomponent heterogeneous composites are an effective means to improve the gas-sensitive properties of materials. As far as MXene composites are concerned, the more studied $Ti_3C_2T_x$ compounded with metal oxides and TMDs inorganic materials have been reported in the field of gas sensing, and good results have been achieved in improving the gas-sensitive performance of pure MXene materials (Table 13.2).

Table 13.2 Summary of gas-sensitive properties of MXene composites.

Sensitive material	Synthesis	Gas species	Response % ($\Delta R/R_i$)	LoD	Test condition	References
$TiO_2/$ $Ti_3C_2T_x$	Spraying	NH_3	3.1	0.50 ppm	25°C	[157]
$CuO/$ $Ti_3C_2T_x$	Electrostatic self-assembly	Toluene	11.4	–	250°C	[158]
$W_{18}O_{49}/$ $Ti_3C_2T_x$	Solvothermal process	Acetone	1.67	0.17 ppm	25°C	[159]
$SnO_2/$ $Ti_3C_2T_x$	Hydrothermal synthesis	NH_3	40	4.29 ppb	25°C	[160]
$Fe_2(MoO_4)_3/$ $Ti_3C_2T_x$	Hydrothermal synthesis	n-butyl alcohol/ xylene	43.1/39.5	–	120°C	[172]
$SnO-SnO_2/$ $Ti_3C_2T_x$	Hydrothermal synthesis	acetone	12.1	–	25°C	[161]
$WO_3/$ $Ti_3C_2T_x$	Ultrasonic technology	NH_3	22.3	–	25°C	[162]
$WSe_2/$ $Ti_3C_2T_x$	Liquid phase stripping and Inkjet printing	Ethyl alcohol	-9.2	–	25°C	[163]
$ZnO/$ $Ti_3C_2T_x$	ultrasonic spray pyrolysis method	NO_2	41.9	–	25°C	[164]
PEDOT: PSS/$Ti_3C_2T_x$	In situ polymerization	NH_3	36.6	–	25°C	[165]
(PANI)/ $Ti_3C_2T_x$	In situ self-assembly	NH_3	–	–	10−40°C 60% RH	[167]
PANI/ $Ti_3C_2T_x$	Cryogenic in situ polymerization	Ethyl alcohol	41.1	300 ppm	25°C	[168]
rGO/N-MXene/ (PEI)	Reduction	CO_2	–	–	25°C 62% RH	[169]
Co_3O_4@PEI/ $Ti_3C_2T_x$	Hydrothermal synthesis	NO_x	–	300 ppm	25°C 26% RH	[170]
PANI/ Nb_2CT_x	In situ polymerization	NH_3	205.4	200 ppm	25°C 62% RH	[171]

Tai et al. [157] prepared $TiO_2/Ti_3C_2T_x$ composites using a spray method and designed a gas sensor with a response value of 3.1% for 10 ppm NH_3 at room temperature (25°C). Compared with pure $Ti_3C_2T_x$, the response of $TiO_2/Ti_3C_2T_x$ composite to NH_3 was 1.63 times higher than that of pure $Ti_3C_2T_x$, and the response/recovery time was 0.65/0.52 times that of pure $Ti_3C_2T_x$. Hermawan et al. [158] prepared composite films of CuO nanoparticles/$Ti_3C_2T_x$ by electrostatic self-assembly. The 7-nm-thick CuO nanoparticles were uniformly dispersed on the surface and the interlayers of the $Ti_3C_2T_x$ MXene forming hybrid heterostructures (Fig. 13.47). As MXene acts as a support layer for p-type semiconductor CuO nanoparticles, it plays a key role in improving the gas response and recovery time of CuO nanoparticles by increasing the mobility of charge carriers, as shown in Fig. 13.4. The gas sensor fabricated by this composite achieved a gas response (R_g/R_a) of up to 11.4 at 250°C for 50 ppm toluene, which is approximately five times high comparing to the starting material of CuO nanoparticles.

Sun et al. [159] presented the preparation of the $W_{18}O_{49}/Ti_3C_2T_x$ composites based on the in situ growth of 1D $W_{18}O_{49}$/nanorods (NRs) on the surfaces of the 2D $Ti_3C_2T_x$ MXene sheets via a facile solvothermal process. The $W_{18}O_{49}/Ti_3C_2T_x$ composite containing 2% (mass fraction) $Ti_3C_2T_x$ exhibited the best acetone sensing performance, as shown in Fig. 13.48B, C. As shown in Fig. 13.48A, due to the uniform distribution of $W_{18}O_{49}$ nanorods on the surface of $Ti_3C_2T_x$, the fluorine-containing groups on the surface of the two-dimensional material can be effectively removed by solvent heat treatment, and there are synergistic interfacial interactions between the $W_{18}O_{49}$ NRs and the $Ti_3C_2T_x$ sheets. The composites have gas-sensitive response values up to 1.67 for 11.6−20 ppm acetone and detection limits as low as 0.17 ppm for acetone, with fast response/recovery times (5.6/6 seconds). The gas-sensitive performance of the composites is superior to that of pure $Ti_3C_2T_x$ and $W_{18}O_{49}$.

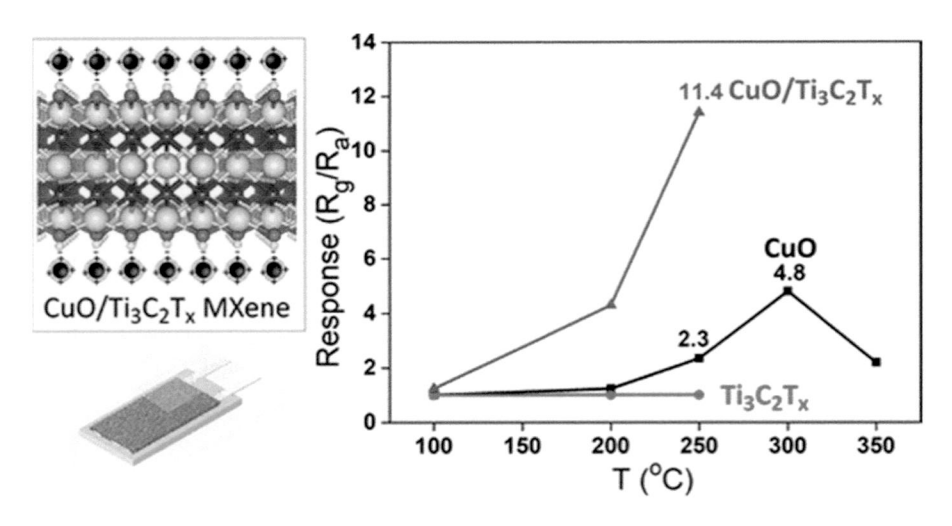

FIGURE 13.47

Gas sensor of CuO nanoparticles/$Ti_3C_2T_x$ and gas-sensitive response to toluene [158].

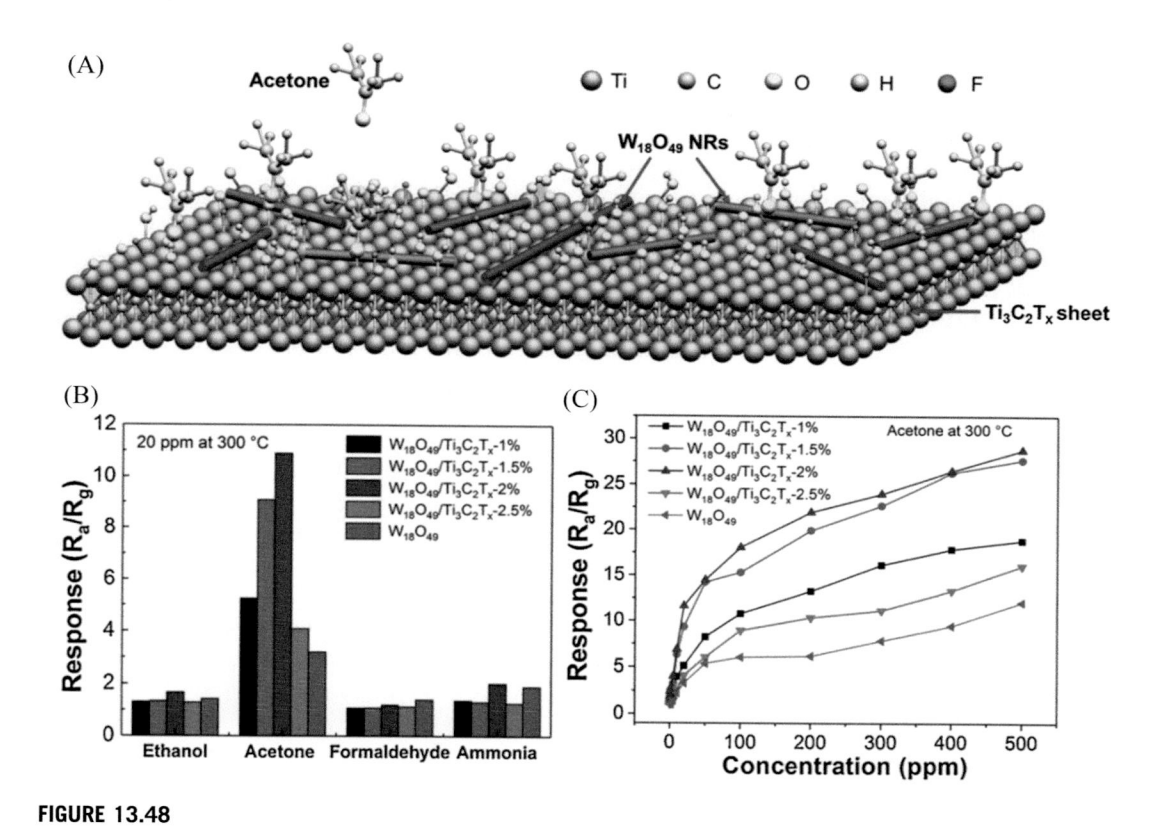

FIGURE 13.48

(A) Schematic diagram of the complex reaction of acetone with $W_{18}O_{49}/Ti_3C_2T_x$ [159]. (B) Comparison of responses of different sensors to gases [159]. (C) Response plots of different sensors as a function of acetone concentration [159].

He et al. [160] synthesized two-dimensional $Ti_3C_2T_x$ modified by SnO_2 nanoparticles by an hydrothermal method. Gas sensors with $Ti_3C_2T_x$ SnO_2 heterostructures showed a good gas-sensitive response and selectivity for NH_3 from 50–100 ppm, and the enhanced gas-sensitive performance was attributed to the fact that two-dimensional $Ti_3C_2T_x$ is a good conductivity and adsorption substrate. Besides, there is intense charge transfer happening at the heterojunctions due to the difference of the Fermi level of the 2D MXene and SnO_2. Wang et al. [161] synthesized $SnO–SnO_2/Ti_3C_2T_x$ nanocomposites in situ by a one-step hydrothermal method. Due to the existence of a small amount of oxygen during the hydrothermal conditions, part of the p-type SnO was oxidized to n-type SnO_2, forming in situ p-n junctions on the surface. $SnO–SnO_2$ (p-n junction) were evenly distributed on the surface of $Ti_3C_2T_x$ MXene nanosheets to build an effective sensor pathway. Compared with $SnO–SnO_2$ and/$Ti_3C_2T_x$ alone, the $SnO–SnO_2/Ti_3C_2T_x$ sensor has a response value of up to 12.1 for 100 ppm acetone gas at room temperature and has a faster recovery time (9 seconds).

Guo et al. [162] used an ultrasonic technique to composite WO_3 nanoparticles with $Ti_3C_2T_x$ nanosheets to obtain a $Ti_3C_2T_x/WO_3$ composite gas sensor with a gas response of 22.3% to 1 ppm NH_3 at room temperature, which is 15.4 times higher than that of the pure $Ti_3C_2T_x$ sensor. Chen et al. [163] fabricated flexible nanosensors consisting of MXene ($Ti_3C_2T_x$) and TMD (WSe_2) by liquid-phase exfoliation and inkjet printing, which were sensitive to 40 ppm ethanol due to the formation of heterojunctions. By contrast, the $Ti_3C_2T_x/WSe_2$ composite sensor demonstrated a 12 times higher sensitivity.

Yang et al. [164] prepared 3D crumpled MXene $Ti_3C_2T_x$ spheres and spherical ZnO composites by ultrasonic spray pyrolysis technology. Because of the defects, the p−n heterojunction shows between the MXene sphere and ZnO nanoparticles (Fig. 13.49A, B). 3D-folded MXene spheres/ZnO achieved reversible NO_2 sensing while significantly improving the selectivity and response to NO_2 (Fig. 13.49C). The response to 100 ppm NO_2 was enhanced from 27.27%−41.93% accompanied by the substantial increase in recovery rate from 30% to 100%.

Studies have shown that organic conductive polymers such as polyaniline (PANI), PPy, poly (3,4-ethylenedioxythiophene), and their derivatives have received wide attention in gas sensor applications because of their easy synthesis, low working temperature, and good electrical conductivity. However, gas sensors made from organic conducting polymers alone often suffer from low sensitivity, long response/recovery time, and poor long-term device stability. In contrast, MXene materials with high surface area, low electronic noise, and flexible surface chemistry can exhibit fast response to low concentration analytes at room temperature, and the heterogeneous interfacial synergistic effect generated by the compounding of MXene with organic conducting polymers is a feasible strategy to improve the sensitivity and selectivity of gas sensors.

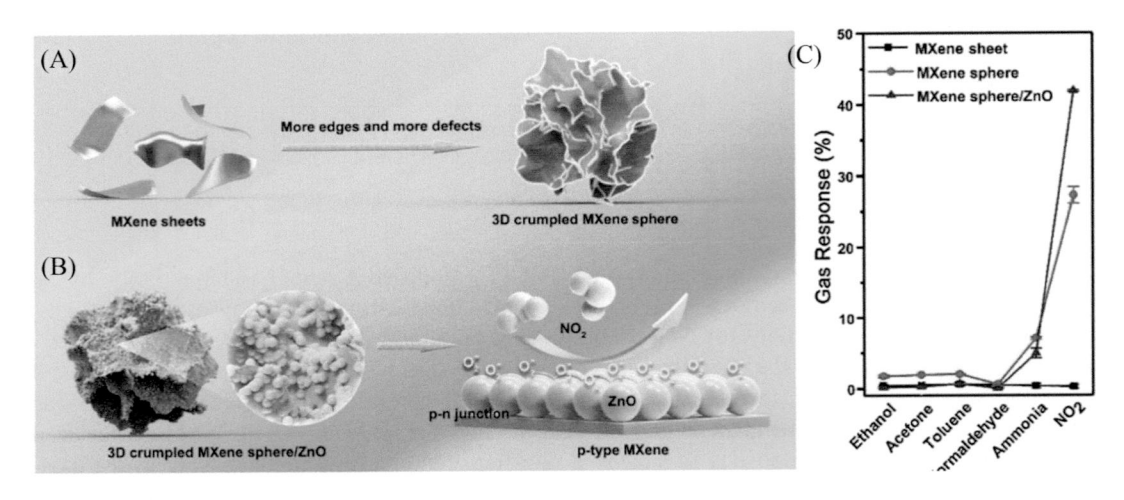

FIGURE 13.49

(A) The increase of the edges and defects of 3D crumpled MXene sphere [164]. (B) The effect of ZnO nanoparticles on NO_2 sensing of 3D crumpled MXene sphere/ZnO [164]. (C) The gas response of sensors based on MXene sheet, MXene sphere, and MXene sphere/ZnO to various target gases at the concentration of 100 ppm [164].

Jin et al. [165] prepared PEDOT:PSS/Ti$_3$C$_2$T$_x$ nanocomposites using in situ polymerization of 3,4-ethylenedioxythiophene (EDOT) and poly(4-styrenesulfonate) (PSS). The PEDOT:PSS/MXene composites were developed into a gas sensor using a dip-coating technique. The sensor of the PEDOT: PSS/Ti$_3$C$_2$T$_x$ composite has a gas-sensitive response of 36.6% to 100 ppm NH$_3$ at room temperature due to the presence of a large number of reactive sites in the Ti$_3$C$_2$T$_x$ layer and the direct charge transfer between PEDOT:PSS and Ti$_3$C$_2$T$_x$ MXene. In addition, the composite sensor exhibited enhanced sensing performance compared to both pure PEDOT:PSS- and Ti$_3$C$_2$T$_x$ MXene-based sensors, demonstrating a synergistic enhancement effect.

Hou et al. [166] prepared MXene-derived TC-CN heterojunctions by combining MXene with g-C$_3$N$_4$ using an in situ growth method, and investigated the gas-sensitive properties of the synthesized TC-CN heterojunctions under UV irradiation. The results show that due to the layered structure of the material, TC-CN has a good response to 10 ppm ethanol at room temperature and UV light irradiation, and exhibits superior gas-sensitive properties than the conventional ethanol sensor. Li et al. [167] prepared PANI/Ti$_3$C$_2$T$_x$ hybrid sensitive film for flexible chemical resistive NH$_3$ sensor using an in situ self-assembly method on flexible Au digital interelectrode (IDEs), and developed a flexible chemiresistive gas sensor based on this film. Due to the gas-sensitive enhancement effect of the polyaniline/Ti$_3$C$_2$T$_x$ Schottky junction and the increased protonation of polyaniline, the sensor showed excellent performance for NH$_3$ at 10°C−40°C and 60% relative humidity (Fig. 13.50A).

Zhao et al. [168] synthesized PANI/Ti$_3$C$_2$T$_x$ nanocomposites by modifying polyaniline (PANI) nanoparticles on the surface of Ti$_3$C$_2$T$_x$ nanosheets using a low-temperature in situ polymerization method, which provided an open structure between MXene layers. Because of the synergistic properties of composites and highly active Ti$_3$C$_2$T$_x$ MXene, this composite sensing material displayed both high ethanol sensitivity (41.1%, 200 ppm) (Fig. 13.50B) and rapid response/recovery time (0.4/0.5 second) at room temperature. Zhou et al. [169] employed a nitrogen-doped MXene Ti$_3$C$_2$T$_x$ (N-MXene)/polyethyleneimine (PEI) composite film decorated with reduced rGO nanosheets to detect ppm-level CO$_2$ gas at room temperature (20°C) in the presence of humidity.

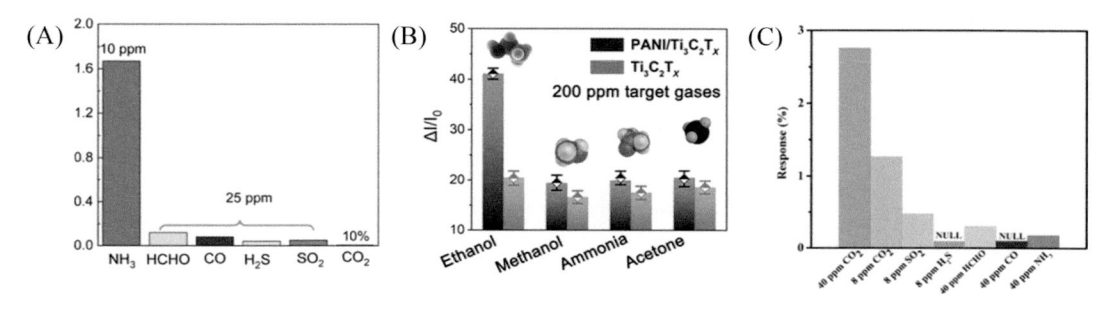

FIGURE 13.50

(A) The selectivity of the hybrid sensor to NH$_3$ and other interference gases in agricultural fields at room temperature [167]. (B) Selectivity of Ti$_3$C$_2$T$_x$ and PANI/Ti$_3$C$_2$T$_x$-based flexible sensors exposed to 200 ppm of ethanol, methanol, ammonia, and acetone at room temperature (n = 3 measurements) [168]. (C) Selectivity of the ternary sensor [169].

The results show that the ternary sensor at room temperature, 62% humidity, and PEI concentration of 0.01 mg/mL has good selectivity for CO_2 gas at 40 ppm (Fig. 13.50C) and has long-term operational stability.

Sun et al. [170] used a simple noncovalent chemical approach and hydrothermal method for effectively riveting Co_3O_4 nanocrystals to branched polyethylenimine (PEI)-functionalized Co_3O_4 MXene sheets to fabricate Co_3O_4@PEI/Co_3O_4 MXene composites (Fig. 13.51A). In the gas-sensitive process, Co_3O_4 nanoparticles were homogeneously dispersed as the active central

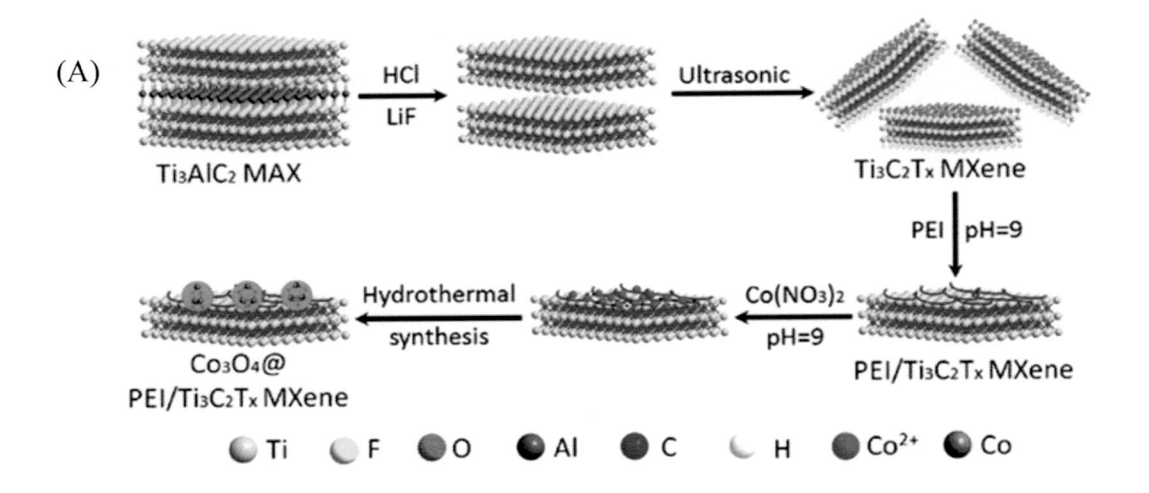

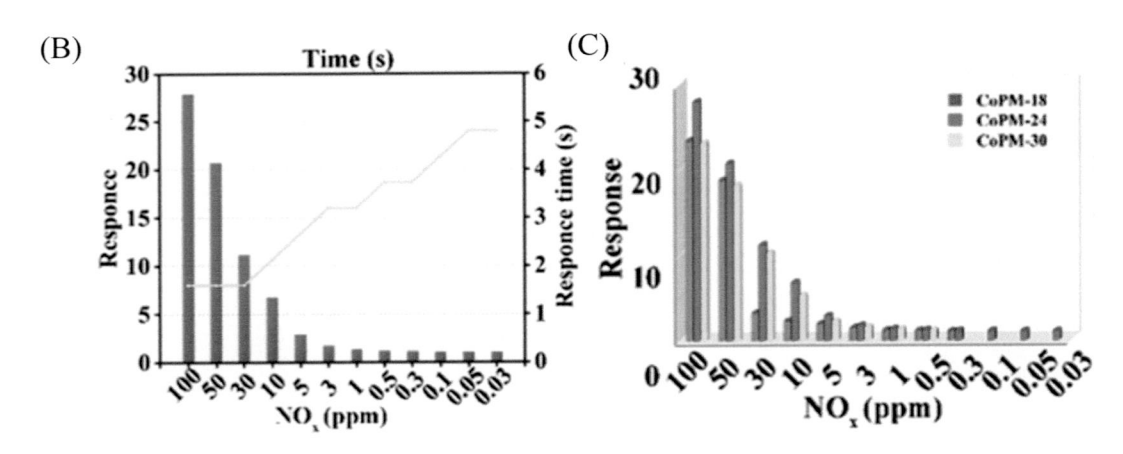

FIGURE 13.51

(A) Schematic illustration of the Co_3O_4@PEI/$Ti_3C_2T_x$ MXene composites [170]; (B) response and response time of CoPM-24 to 100–0.03 ppm NO_x [170],; (C) responses of all the sensors (CoPM-18, CoPM-24 and CoPM-30) to NO_x at RT [170].

component in $Ti_3C_2T_x$ MXene composites with PEI acting as a channel for electron transfer, which effectively improves the interaction between $Ti_3C_2T_x$ and Co_3O_4. The composite sensor exhibits fast recovery (27.9 seconds) and low response time (<2 seconds) for NO_x gas at room temperature (Fig. 13.51B) and 26% relative humidity, and the detection limit for NO_x can be as low as 0.03 ppm (Fig. 13.51C).

Wang et al. [171] synthesized Nb_2CT_x nanosheet solutions by HF aqueous etching, obtained a few layers of Nb_2CT_x by TPAOH intercalation, and prepared Nb_2CT_x films by spraying Nb_2CT_x solutions (1 mg/mL) on hydrophilic pretreated PI substrates. PANI was synthesized on the Nb_2CT_x layer by in situ polymerization of HCl-doped aniline. The PANI/Nb_2CT_x sensor presents a high sensing response of 205.39% toward 50 ppm NH_3 at 62.0% RH, indicating a potential to achieve environment NH_3 detection. Meanwhile, the PANI/Nb_2CT_x sensor exhibits excellent selectivity, high sensitivity (74.68%/10 ppm), low detection limit (20 ppb), and good long-term reliability at 87.1% RH, demonstrating a superior detection capacity for NH_3.

13.6.3 Humidity sensitivity of MXene

Sensitivity to atmospheric water molecules is an essential factor in the MXene-based sensor. Since the surface of MXenes is hydrophilic, and the interaction between layers is relatively weak, water molecules can be inserted spontaneously under environmental humidity, which has excellent potential as a humidity sensor. Experiments show that the resistivity of Ti_3C_2 film increases linearly by 15%−80% with the increase of relative humidity [173]. Metal ion intercalation has great influence on the structure and inner surface hydrodynamics of MXenes [75].

Muckley et al. [174] first studied water vapor's gravimetric and electrical responses when K^+ and Mg^{2+} were added to Ti_3C_2 at 20%−80% RH. On this basis, the gravity response of MXenes to water was further studied by adding MXenes, and it was concluded that the gravity response of MXenes to water was 10 times faster than the electrical response. This is explained as the expansion/contraction of the channel between MXenes sheets induced by water molecules results in the capture of water molecules as charge-consuming dopants. Yang et al. [151] synthesized $Ti_3C_2T_x$ with sodium hydroxide solution and Ti_3AlC_2 and treated it with alkali. The increase of oxygen/fluorine ratio and alkali metal ions on the surface significantly improved the humidity and gas sensitivity at room temperature.

An et al. [175] used LbL technology to assemble and prepare a multimembrane humidity sensor with MXene/polyelectrolyte with ultrafast response and recovery time. When the humidity changes, water molecules disperse into the MXene/polyelectrolyte multilayers, resulting in an increase in the thickness and distance between the membranes, thus changing the tunnel resistance between the MXene multilayers, and the change of the resistance can be used to characterize the change of humidity. Li et al. [176] prepared sea urchin-like Ti_3C_2/TiO_2 composite by in situ growth of TiO_2 nanowire on Ti_3C_2 by alkaline oxidation method. The staggered dendritic nanowire structure has excellent adsorption performance at low RH, which is conducive to the formation of a continuous water layer. Liu et al. [177] developed vacuum-assisted LbL assembly technology using AgNW and MXenes flakes, and prepared leaf-like composites with high conductivity on silk substrate to achieve efficient humidity sensing. The MXenes layer protects AgNW from oxidation and endows textiles with high sensitivity to humidity. This research has great potential for smart clothing and sensor applications. The result shows in the Fig. 13.52.

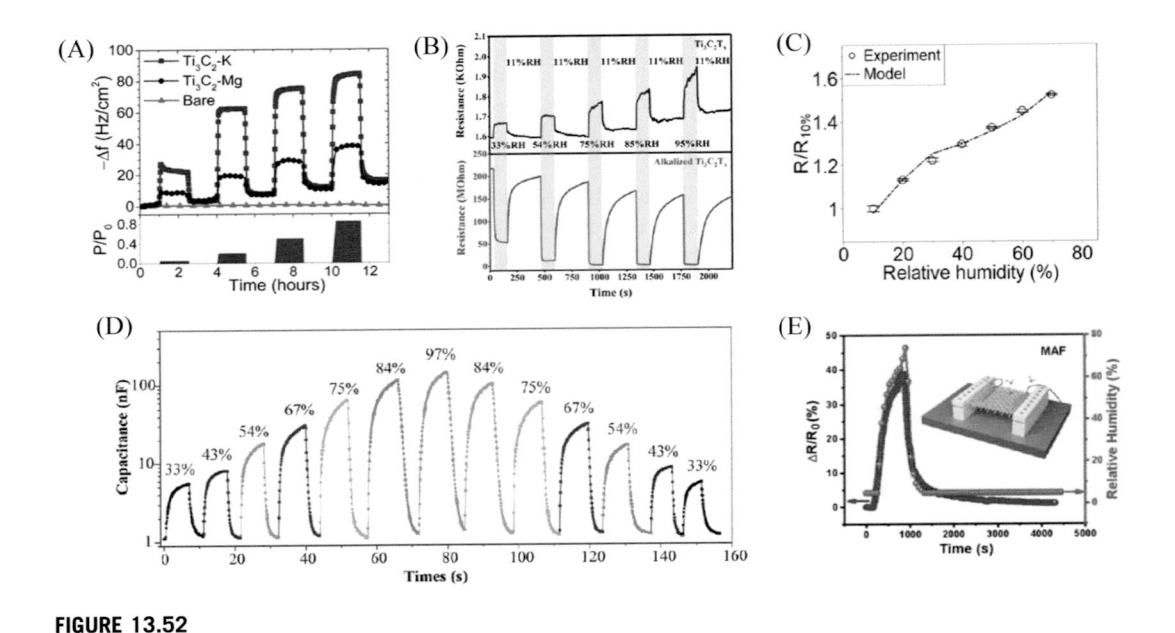

FIGURE 13.52

(A) Normalized frequency shift under the same RH conditions. For reference, frequency shift of a bare Au-coated crystal is shown in red [174]. (B) Dynamic response–recovery curve of the devices based on $Ti_3C_2T_x$ and alkalized $Ti_3C_2T_x$ to different relative humidity [151]. (C) Response of the multilayer to the varying relative humidity by experimental measurements and numerical modeling [175]. (D) Response and recovery characteristics of sensor exposed to a series of RH conditions [176]. (E) Similar humidity response of MAF silk with a commercial sensor. The illustration presents a homemade humidity sensor. The humidity response is measured at 20°C [177].

Bi Mingpan et al. [178] integrated Nb_2CT_X nanosheets with microfiber through an optical deposition method, and realized a humidity sensor based on fiber. The water molecules around the detection cavity can be absorbed and inserted into Nb_2CT_X film to adjust the transmission spectrum of the microfiber interferometer. The experimental results show that when the humidity increases from 18.5% RH to 72.4% RH, the transmission spectrum starts to blueshift due to the change of effective refractive index of Nb_2CT_X, and the sensitivity of transmission spectrum is 86 PM/% RH. When the humidity increases from 72.4% RH to 95.4% RH, the transmission spectrum redshifts in the range due to the structural change of Nb_2CT_X, and the sensitivity of the transmission spectrum is as high as 585 PM/% RH. The combination of Nb_2CT_X MXene and optical fiber shows broad prospects in RH sensing field and brings innovative ideas for gas sensing applications.

Wang et al. [179] introduced the preparation method and structural characteristics of MXenes, as well as the working mechanism of the humidity response driver. Recent important advances of MXene materials in actuators are objectively reviewed and evaluated, and existing issues are discussed. The development of these systems is summarized from the aspects of MXene preparation, structure control, design and assembly, and application, which provides a new idea and guidance for the development of the next generation of high-performance humidity response actuators based on MXene.

References

[1] J. Peng, X. Chen, W.J. Ong, X. Zhao, N. Li, Surface and heterointerface engineering of 2D MXenes and their nanocomposites: insights into electro- and photocatalysis, Chem 5 (1) (2019) 18−50. Available from: https://doi.org/10.1016/j.chempr.2018.08.037.

[2] M. Naguib, et al., Two-dimensional nanocrystals produced by exfoliation of Ti 3AlC 2, Adv. Mater. 23 (37) (2011) 4248−4253. Available from: https://doi.org/10.1002/adma.201102306.

[3] S.H.W. Kok, J. Lee, L.L. Tan, W.J. Ong, S.P. Chai, MXene - a new paradigm toward artificial nitrogen fixation for sustainable ammonia generation: synthesis, properties, and future outlook, ACS Mater. Lett. 4 (2) (2022) 212−245. Available from: https://doi.org/10.1021/acsmaterialslett.1c00673.

[4] Z. Huang, et al., Two-dimensional MXene-based materials for photothermal therapy, Nanophotonics 9 (8) (2020) 2233−2249. Available from: https://doi.org/10.1515/nanoph-2019-0571.

[5] L. Wang, et al., Loading actinides in multilayered structures for nuclear waste treatment: the first case study of uranium capture with vanadium carbide MXene, ACS Appl. Mater. Interfaces 8 (25) (2016) 16396−16403. Available from: https://doi.org/10.1021/acsami.6b02989.

[6] R.M. Ronchi, J.T. Arantes, S.F. Santos, Synthesis, structure, properties and applications of MXenes: current status and perspectives, Ceram. Int. 45 (15) (2019) 18167−18188. Available from: https://doi.org/10.1016/j.ceramint.2019.06.114.

[7] H. Wang, et al., Clay-inspired MXene-based electrochemical devices and photo-electrocatalyst: state-of-the-art progresses and challenges, Adv. Mater. 30 (12) (2018) 1−28. Available from: https://doi.org/10.1002/adma.201704561.

[8] H. Jing, et al., Modulation of the electronic properties of mxene (Ti3C2Tx) via surface-covalent functionalization with diazonium, ACS Nano 15 (1) (2021) 1388−1396. Available from: https://doi.org/10.1021/acsnano.0c08664.

[9] D.V.T. Vladislav Kamysbayev, Alexander S. Filatov, Huicheng Hu, Xue Rui, Francisco Lagunas, Di Wang, et al., Covalent surface modifications and superconductivity of two-dimensional metal carbide MXenes, Sci. (80-.) 369 (6506) (2020) 979−984.

[10] K. Hantanasirisakul, et al., Fabrication of Ti3C2Tx MXene transparent thin films with tunable optoelectronic properties, Adv. Electron. Mater. 2 (6) (2016) 1−7. Available from: https://doi.org/10.1002/aelm.201600050.

[11] I.L. Medintz, H.T. Uyeda, E.R. Goldman, H. Mattoussi, Quantum dot bioconjugates for imaging, labelling and sensing, Nat. Mater. 4 (6) (2005) 435−446. Available from: https://doi.org/10.1038/nmat1390.

[12] Q. Xue, et al., Photoluminescent Ti_3C_2 MXene quantum dots for multicolor cellular imaging, Adv. Mater. 29 (15) (2017). Available from: https://doi.org/10.1002/adma.201604847.

[13] M. Kurtoglu, M. Naguib, Y. Gogotsi, M.W. Barsoum, First principles study of two-dimensional early transition metal carbides, MRS Commun. 2 (4) (2012) 133−137. Available from: https://doi.org/10.1557/mrc.2012.25.

[14] K.H. Michel, B. Verberck, Theory of the elastic constants of graphite and graphene, Phys. Status Solidi Basic. Res. 245 (10) (2008) 2177−2180. Available from: https://doi.org/10.1002/pssb.200879604.

[15] V.N. Borysiuk, V.N. Mochalin, Y. Gogotsi, Bending rigidity of two-dimensional titanium carbide (MXene) nanoribbons: a molecular dynamics study, Comput. Mater. Sci. 143 (2018) 418−424. Available from: https://doi.org/10.1016/j.commatsci.2017.11.028.

[16] A. Lipatov, et al., Elastic properties of 2D Ti3C2Tx MXene monolayers and bilayers, Sci. Adv. 4 (6) (2018) eaat0491.

[17] Y. Xie, P.R.C. Kent, Hybrid density functional study of structural and electronic properties of functionalized $Ti_{n+1}Xn$ (X = C, N) monolayers, Phys. Rev. B − Condens. Matter Mater. Phys. 87 (23) (2013). Available from: https://doi.org/10.1103/PhysRevB.87.235441, pp. 235441(1−10).

[18] Q. Tang, Z. Zhou, P. Shen, Are MXenes promising anode materials for Li ion batteries? Computational studies on electronic properties and Li storage capability of Ti_3C_2 and $Ti_3C_2X_2$ (X = F, OH) monolayer, J. Am. Chem. Soc. 134 (40) (2012) 16909–16916. Available from: https://doi.org/10.1021/ja308463r.

[19] C. Si, J. Zhou, Z. Sun, Half-metallic ferromagnetism and surface functionalization-induced metal-insulator transition in graphene-like two-dimensional Cr_2C crystals, ACS Appl. Mater. Interfaces 7 (31) (2015) 17510–17515. Available from: https://doi.org/10.1021/acsami.5b05401.

[20] M. Naguib, et al., One-step synthesis of nanocrystalline transition metal oxides on thin sheets of disordered graphitic carbon by oxidation of MXenes, Chem. Commun. 50 (56) (2014) 7420–7423. Available from: https://doi.org/10.1039/c4cc01646g.

[21] K. Wang, Y. Zhou, W. Xu, D. Huang, Z. Wang, M. Hong, Fabrication and thermal stability of two-dimensional carbide Ti_3C_2 nanosheets, Ceram. Int. 42 (7) (2016) 8419–8424. Available from: https://doi.org/10.1016/j.ceramint.2016.02.059.

[22] X. Zhao, et al., Antioxidants unlock shelf-stable Ti3C2Tx (MXene) nanosheet dispersions, Matter 1 (2) (2019) 513–526. Available from: https://doi.org/10.1016/j.matt.2019.05.020.

[23] W. Wang, Z.D. Hood, X. Zhang, I.N. Ivanov, Z. Bao, Construction of 2D BiVO 4 À CdS À $Ti_3C_2T_x$ heterostructures for enhanced photo-redox activities, ChemCatChem 26506 (2020) 3496–3503. Available from: https://doi.org/10.1002/cctc.202000988.

[24] J. Halim, et al., Transparent conductive two-dimensional titanium carbide epitaxial thin films, Chem. Mater. 26 (7) (2014) 2374–2381. Available from: https://doi.org/10.1021/cm500641a.

[25] M. Ghidiu, M.R. Lukatskaya, M.Q. Zhao, Y. Gogotsi, M.W. Barsoum, Conductive two-dimensional titanium carbide 'clay' with high volumetric capacitance, Nature 516 (7529) (2015) 78–81. Available from: https://doi.org/10.1038/nature13970.

[26] V. Natu, R. Pai, M. Sokol, M. Carey, V. Kalra, M.W. Barsoum, 2D $Ti_3C_2T_z$ MXene synthesized by water-free etching of Ti_3AlC_2 in Polar Organic Solvents, Chem 6 (3) (2020) 616–630. Available from: https://doi.org/10.1016/j.chempr.2020.01.019.

[27] P. Urbankowski, et al., Synthesis of two-dimensional titanium nitride Ti_4N_3 (MXene), Nanoscale 8 (22) (2016) 11385–11391. Available from: https://doi.org/10.1039/c6nr02253g.

[28] X. Li, et al., In situ electrochemical synthesis of MXenes without acid/alkali usage in/for an aqueous zinc ion battery, Adv. Energy Mater. 10 (36) (2020) 1–8. Available from: https://doi.org/10.1002/aenm.202001791.

[29] T. Li, et al., Fluorine-free synthesis of high-purity Ti3C2Tx (T = OH, O) via alkali treatment, Angew. Chem. − Int. Ed. 57 (21) (2018) 6115–6119. Available from: https://doi.org/10.1002/anie.201800887.

[30] T. Yin, et al., Synthesis of $Ti_3C_2F_x$ MXene with controllable fluorination by electrochemical etching for lithium-ion batteries applications, Ceram. Int. 47 (20) (2021) 28642–28649. Available from: https://doi.org/10.1016/j.ceramint.2021.07.023.

[31] L.Y. Xiu, Z.Y. Wang, J.S. Qiu, General synthesis of MXene by green etching chemistry of fluoride-free Lewis acidic melts, Rare Met. 39 (11) (2020) 1237–1238. Available from: https://doi.org/10.1007/s12598-020-01488-0.

[32] W. Sun, et al., Electrochemical etching of Ti_2AlC to $Ti_2CT:X$ (MXene) in low-concentration hydrochloric acid solution, J. Mater. Chem. A 5 (41) (2017) 21663–21668. Available from: https://doi.org/10.1039/c7ta05574a.

[33] C. Xu, et al., Large-area high-quality 2D ultrathin Mo_2C superconducting crystals, Nat. Mater. 14 (11) (2015) 1135–1141. Available from: https://doi.org/10.1038/nmat4374.

[34] Y.L. Hong, et al., Chemical vapor deposition of layered two-dimensional $MoSi_2N_4$ materials, Science (80-.) 369 (6504) (2020) 670–674. Available from: https://doi.org/10.1126/science.abb7023.

[35] O. Mashtalir, et al., Intercalation and delamination of layered carbides and carbonitrides, Nat. Commun. 4 (2013) 1–7. Available from: https://doi.org/10.1038/ncomms2664.

[36] L. Qin, et al., High-performance ultrathin flexible solid-state supercapacitors based on solution process-able Mo1.33C MXene and PEDOT:PSS, Adv. Funct. Mater. 28 (2) (2018) 1−8. Available from: https://doi.org/10.1002/adfm.201703808.

[37] Z. Lin, et al., Capacitance of Ti3C2Tx MXene in ionic liquid electrolyte, J. Power Sources 326 (2016) 575−579. Available from: https://doi.org/10.1016/j.jpowsour.2016.04.035.

[38] Z. Lin, et al., Electrochemical and in situ X-ray diffraction studies of Ti3C2Tx MXene in ionic liquid electrolyte, Electrochem. commun. 72 (2016) 50−53. Available from: https://doi.org/10.1016/j.elecom.2016.08.023.

[39] Z. Lin, D. Sun, Q. Huang, J. Yang, M.W. Barsoum, X. Yan, Carbon nanofiber bridged two-dimensional titanium carbide as a superior anode for lithium-ion batteries, J. Mater. Chem. A 3 (27) (2015) 14096−14100. Available from: https://doi.org/10.1039/c5ta01855b.

[40] R.B. Rakhi, B. Ahmed, D. Anjum, H.N. Alshareef, Direct chemical synthesis of MnO_2 nanowhiskers on transition-metal carbide surfaces for supercapacitor applications, ACS Appl. Mater. Interfaces 8 (29) (2016) 18806−18814. Available from: https://doi.org/10.1021/acsami.6b04481.

[41] Y. Wang, et al., Three-dimensional porous MXene/layered double hydroxide composite for high performance supercapacitors, J. Power Sources 327 (2016) 221−228. Available from: https://doi.org/10.1016/j.jpowsour.2016.07.062.

[42] M. Boota, B. Anasori, C. Voigt, M.Q. Zhao, M.W. Barsoum, Y. Gogotsi, Pseudocapacitive electrodes produced by oxidant-free polymerization of pyrrole between the layers of 2D titanium carbide (MXene), Adv. Mater. 28 (7) (2016) 1517−1522. Available from: https://doi.org/10.1002/adma.201504705.

[43] M. Zhu, et al., Highly flexible, freestanding supercapacitor electrode with enhanced performance obtained by hybridizing polypyrrole chains with MXene, Adv. Energy Mater. 6 (21) (2016). Available from: https://doi.org/10.1002/aenm.201600969.

[44] R. Li, L. Zhang, L. Shi, P. Wang, MXene Ti3C2: an Effective 2D Light-to-Heat Conversion Material, ACS Nano 11 (4) (2017) 3752−3759. Available from: https://doi.org/10.1021/acsnano.6b08415.

[45] M. Alhabeb, et al., Selective Etching of Silicon from Ti 3 SiC 2 (MAX) To Obtain 2D Titanium Carbide (MXene), Angew. Chem. 130 (19) (2018) 5542−5546. Available from: https://doi.org/10.1002/ange.201802232.

[46] S. Xu, et al., Binder-free Ti3C2Tx MXene electrode film for supercapacitor produced by electrophoretic deposition method, Chem. Eng. J. 317 (2017) 1026−1036. Available from: https://doi.org/10.1016/j.cej.2017.02.144.

[47] D.P. Chen, J.L. Tong, P. Huang, L.T. Yu, Z.H. Tang, MXene nanosheet-based capacitance immunoassay with tyramine-enzyme repeats to detect prostate-specific antigen on interdigitated micro-comb electrode, Electrochim 7 (1) (2015) 37−72 [Online]. Available from: https://www.researchgate.net/publication/269107473_What_is_governance/link/548173090cf22525dcb61443/download.

[48] Y. Sarycheva, A. Makaryan, T. Maleski, K. Satheeshkumar, E. Melikyan, A. Minassian, et al., Two-dimensional titanium carbide (MXene) as surface-enhanced Raman scattering substrate, J. Phys. Chem, J. Phys. Chem. C. 1999 (December) (2006) 1−6.

[49] C.-Z.L. Subbiah Alwarappan, Noel Nesakumar, Dali Sun, Tony Y. Hu, 2D metal carbides and nitrides (MXenes) for sensors and biosensors, Biosens. Bioelectron. 1999 (December) (2022) 1−6.

[50] J. Yu, R. Xue, J. Wang, Y. Qiu, J. Huang, X. Chen, et al., Novel Ti3C2Tx MXene nanozyme with manage-able catalytic activity and application to electrochemical biosensor, J. Nanobiotechnol. 20 (1) (2022) 119.

[51] X. Wang, Y. Sun, W. Li, Y. Zhuang, X. Tian, C. Luan, et al., Imidazole metal-organic frameworks embedded in layered Ti3C2Tx MXene as a high-performance electrochemiluminescence biosensor for sensitive detection of HIV-1 protein, Microchem. J. 167 (2021) 106332.

[52] J. Liu, Y. Fan, G. Chen, Y. Liu, Highly sensitive glutamate biosensor based on platinum nanoparticles decorated MXene-Ti3C2Tx for L-glutamate determination in foodstuffs, Lwt 148 (May) (2021) 111748. Available from: https://doi.org/10.1016/j.lwt.2021.111748.

[53] K. Huang, Z. Li, J. Lin, G. Han, P. Huang, Two-dimensional transition metal carbides and nitrides (MXenes) for biomedical applications, Chem. Soc. Rev. 47 (14) (2018) 5109−5124. Available from: https://doi.org/10.1039/c7cs00838d.

[54] Y. Guan, Q. Ma, J. Yang, W. Zhang, R. Zhang, X. Dong, et al., Highly fluorescent Ti3C2 MXene quantum dots for macrophage labeling and Cu^{2+} ion sensing, Nanoscale 11 (2019) 14123−14133 [Online]. Available from: https://www.researchgate.net/publication/269107473_What_is_governance/link/548173090cf22525dcb61443/download.

[55] Y. Zhang, X. Jiang, J. Zhang, H. Zhang, Y. Li, Simultaneous voltammetric determination of acetaminophen and isoniazid using MXene modified screen-printed electrode, Biosens. Bioelectron. 130 (December 2018) (2019) 315−321. Available from: https://doi.org/10.1016/j.bios.2019.01.043.

[56] J.P.Y. Yao, L. Lan, X. Liu, Y. Ying, Spontaneous growth and regulation of noble metal nanoparticles on flexible biomimetic MXene paper for bioelectronics, Biosens. Bioelectron. 148 (2020) 111799. Available from: https://doi.org/10.1016/j.bios.2019.111799.

[57] X. Jiang, et al., Broadband nonlinear photonics in few-layer MXene Ti3C2Tx (T = F, O, or OH), Laser Photonics Rev. 12 (2) (2018) 1−10. Available from: https://doi.org/10.1002/lpor.201700229.

[58] X. Peng, Y. Zhang, D. Lu, Y. Guo, S. Guo, Ultrathin Ti3C2 nanosheets based 'off-on' fluorescent nanoprobe for rapid and sensitive detection of HPV infection, Sens. Actuators B Chem. 286 (2019) 222−229. Available from: https://doi.org/10.1016/j.snb.2019.01.158.

[59] H. Zhang, Z. Wang, Q. Zhang, F. Wang, Y. Liu, Ti_3C_2 MXenes nanosheets catalyzed highly efficient electrogenerated chemiluminescence biosensor for the detection of exosomes, Biosens. Bioelectron. 124−125 (2019) 184−190. Available from: https://doi.org/10.1016/j.bios.2018.10.016.

[60] Q. Wu, et al., Ultrasensitive and selective determination of carcinoembryonic antigen using multifunctional ultrathin amino-functionalized Ti3C2-MXene nanosheets, Anal. Chem. 92 (4) (2020) 3354−3360. Available from: https://doi.org/10.1021/acs.analchem.9b05372.

[61] Q. Wu, et al., A 2D transition metal carbide MXene-based SPR biosensor for ultrasensitive carcinoembryonic antigen detection, Biosens. Bioelectron. 144 (July) (2019) 111697. Available from: https://doi.org/10.1016/j.bios.2019.111697.

[62] G. Cai, Z. Yu, P. Tong, D. Tang, Ti3C2 MXene quantum dot-encapsulated liposomes for photothermal immunoassays using a portable near-infrared imaging camera on a smartphone, Nanoscale 11 (33) (2019) 15659−15667. Available from: https://doi.org/10.1039/c9nr05797h.

[63] X. Zhu, et al., A novel graphene-like titanium carbide MXene/Au−Ag nanoshuttles bifunctional nanosensor for electrochemical and SERS intelligent analysis of ultra-trace carbendazim coupled with machine learning, Ceram. Int. 47 (1) (2021) 173−184. Available from: https://doi.org/10.1016/j.ceramint.2020.08.121.

[64] F. Zheng, W. Ke, L. Shi, H. Liu, Y. Zhao, Plasmonic Au-Ag Janus nanoparticle engineered ratiometric surface-enhanced Raman scattering aptasensor for ochratoxin a detection, Anal. Chem. 91 (18) (2019) 11812−11820. Available from: https://doi.org/10.1021/acs.analchem.9b02469.

[65] L. Liu, et al., Ultrasensitive SERS detection of cancer-related miRNA-182 by MXene/MoS₂@AuNPs with controllable morphology and optimized self-internal standards, Adv. Opt. Mater. 8 (23) (2020) 1−10. Available from: https://doi.org/10.1002/adom.202001214.

[66] D. Wu, et al., Delaminated Ti3C2Tx (MXene) for electrochemical carbendazim sensing, Mater. Lett. 236 (2019) 412−415. Available from: https://doi.org/10.1016/j.matlet.2018.10.150.

[67] A. Depeursinge, et al., Fusing visual and clinical information for lung tissue classification in HRCT data, Artif. Intell. Med. 229 (2010) ARTMED1118. Available from: https://doi.org/10.1016/j.

[68] M. Mojtabavi, A. Vahidmohammadi, W. Liang, M. Beidaghi, M. Wanunu, Single-molecule sensing using nanopores in two-dimensional transition metal carbide (MXene) membranes, ACS Nano 13 (3) (2019) 3042−3053. Available from: https://doi.org/10.1021/acsnano.8b08017.

[69] Z. Jin, et al., Ti3C2Tx MXene-derived TiO_2/C-QDs as oxidase mimics for the efficient diagnosis of glutathione in human serum, J. Mater. Chem. B 8 (16) (2020) 3513−3518. Available from: https://doi.org/10.1039/c9tb02478f.

[70] Y. Lei, et al., A MXene-based wearable biosensor system for high-performance in vitro perspiration analysis, Small 15 (19) (2019) 1−10. Available from: https://doi.org/10.1002/smll.201901190.

[71] H. Xu, A. Ren, J. Wu, Z. Wang, Recent advances in 2D MXenes for photodetection, Adv. Funct. Mater. 30 (24) (2020) 1−16. Available from: https://doi.org/10.1002/adfm.202000907.

[72] D. Chimene, D.L. Alge, A.K. Gaharwar, Two-dimensional nanomaterials for biomedical applications: emerging trends and future prospects, Adv. Mater. 27 (45) (2015) 7261−7284. Available from: https://doi.org/10.1002/adma.201502422.

[73] G.S.H. Gu, Y. Xing, P. Xiong, H. Tang, C. Li, S. Chen, et al., Three dimensional porous Ti3C2Tx MXene−graphene hybrid films for glucose biosensing, ACS Appl. Nano Mater. 2 (10) (2019) 6537−6545. Available from: https://doi.org/10.1021/acsanm.9b01465.

[74] H.J. Shin, et al., Efficient reduction of graphite oxide by sodium borohydride and its effect on electrical conductance, Adv. Funct. Mater. 19 (12) (2009) 1987−1992. Available from: https://doi.org/10.1002/adfm.200900167.

[75] M. Ghidiu, J. Halim, S. Kota, D. Bish, Y. Gogotsi, M.W. Barsoum, Barsoum, Ion-exchange and cation solvation reactions in Ti3C2MXene, Chem. Mater. 28 (10) (2016) 3507−3514. Available from: https://doi.org/10.1021/acs.chemmater.6b01275.

[76] J.L. Hart, et al., Control of MXenes' electronic properties through termination and intercalation, Nat. Commun. 10 (1) (2019). Available from: https://doi.org/10.1038/s41467-018-08169-8.

[77] M.R. Lukatskaya, et al., Two-dimensional titanium carbide cation intercalation and high volumetric capacitance of, Science (80-.) 341 (6153) (2013) 1502−1505. Available from: https://doi.org/10.1126/science.1241488.

[78] G.H. Jeong, S.P. Sasikala, T. Yun, G.Y. Lee, W.J. Lee, S.O. Kim, Nanoscale assembly of 2D materials for energy and environmental applications, Adv. Mater. 32 (35) (2020) 1−23. Available from: https://doi.org/10.1002/adma.201907006.

[79] J. Zheng, et al., An inkjet printed Ti_3C_2 -GO electrode for the electrochemical sensing of hydrogen peroxide, J. Electrochem. Soc. 165 (5) (2018) B227−B231. Available from: https://doi.org/10.1149/2.0051807jes.

[80] F. Wang, C.H. Yang, M. Duan, Y. Tang, J.F. Zhu, TiO_2 nanoparticle modified organ-like Ti3C2 MXene nanocomposite encapsulating hemoglobin for a mediator-free biosensor with excellent performances, Biosens. Bioelectron. 74 (2015) 1022−1028. Available from: https://doi.org/10.1016/j.bios.2015.08.004.

[81] R.B. Rakhi, P. Nayak, C. Xia, H.N. Alshareef, Erratum: novel amperometric glucose biosensor based on MXene nanocomposite (Scientific Reports (2016) 6(36422) DOI: 10.1038/srep36422), Sci. Rep. 6 (2016) 38465. Available from: https://doi.org/10.1038/srep38465.

[82] M. Li, et al., Three-dimensional porous MXene/NiCo-LDH composite for high performance non-enzymatic glucose sensor, Appl. Surf. Sci. 495 (2019). Available from: https://doi.org/10.1016/j.apsusc.2019.143554.

[83] J. Zheng, B. Wang, A. Ding, B. Weng, J. Chen, Synthesis of MXene/DNA/Pd/Pt nanocomposite for sensitive detection of dopamine, J. Electroanal. Chem. 816 (January) (2018) 189−194. Available from: https://doi.org/10.1016/j.jelechem.2018.03.056.

[84] Y.H. Wang, K.J. Huang, X. Wu, Recent advances in transition-metal dichalcogenides based electrochemical biosensors: a review, Biosens. Bioelectron. 97 (2017) 305−316. Available from: https://doi.org/10.1016/j.bios.2017.06.011.

[85] H. Liu, C. Duan, C. Yang, W. Shen, F. Wang, Z. Zhu, A novel nitrite biosensor based on the direct electrochemistry of hemoglobin immobilized on MXene-Ti3C2, Sens. Actuators B Chem. 218 (2015) 60−66. Available from: https://doi.org/10.1016/j.snb.2015.04.090.

[86] L. Wu, et al., 2D transition metal carbide MXene as a robust biosensing platform for enzyme immobilization and ultrasensitive detection of phenol, Biosens. Bioelectron. 107 (February) (2018) 69−75. Available from: https://doi.org/10.1016/j.bios.2018.02.021.

[87] S. Kumar, Y. Lei, N.H. Alshareef, M.A. Quevedo-Lopez, K.N. Salama, Biofunctionalized two-dimensional Ti3C2 MXenes for ultrasensitive detection of cancer biomarker, Biosens. Bioelectron. 121 (August) (2018) 243−249. Available from: https://doi.org/10.1016/j.bios.2018.08.076.

[88] Y. Shao, J. Wang, H. Wu, J. Liu, I.A. Aksay, Y. Lin, Graphene based electrochemical sensors and biosensors: a review, Electroanalysis 22 (10) (2010) 1027−1036. Available from: https://doi.org/10.1002/elan.200900571.

[89] Y. Jiang, et al., Silver nanoparticles modified two-dimensional transition metal carbides as nanocarriers to fabricate acetycholinesterase-based electrochemical biosensor, Chem. Eng. J. 339 (November 2017) (2018) 547−556. Available from: https://doi.org/10.1016/j.cej.2018.01.111.

[90] K. Kalantar-Zadeh, J.Z. Ou, Biosensors based on two dimensional MoS, ACS Sens. 1 (1) (2016) 5−16. Available from: https://doi.org/10.1021/acssensors.5b00142.

[91] Z. Yin, et al., Single-layer MoS_2 phototransistors, ACS Nano 6 (1) (2012) 74−80.

[92] H. Hu, et al., Recent advances in two-dimensional transition metal dichalcogenides for biological sensing, Biosens. Bioelectron, Biosens. Bioelectron. 142 (2019) 111573. Available from: https://doi.org/10.1016/j.bios.2019.111573.

[93] Y. Fang, et al., Two-dimensional titanium carbide (MXene)-based solid-state electrochemiluminescent sensor for label-free single-nucleotide mismatch discrimination in human urine, Sens. Actuators B Chem. 263 (2018) 400−407. Available from: https://doi.org/10.1016/j.snb.2018.02.102.

[94] M. Song, S.Y. Pang, F. Guo, M.C. Wong, J. Hao, Fluoride-free 2D niobium carbide MXenes as stable and biocompatible nanoplatforms for electrochemical biosensors with ultrahigh sensitivity, Adv. Sci. 7 (24) (2020) 1−8. Available from: https://doi.org/10.1002/advs.202001546.

[95] K. Zhang, et al., Exploring the trans-cleavage activity of CRISPR-Cas12a for the development of a MXene based electrochemiluminescence biosensor for the detection of Siglec-5, Biosens. Bioelectron. 178 (December 2020) (2021) 113019. Available from: https://doi.org/10.1016/j.bios.2021.113019.

[96] Q. Zhang, F. Wang, H. Zhang, Y. Zhang, M. Liu, Y. Liu, Universal Ti_3C_2 MXenes based self-standard ratiometric fluorescence resonance energy transfer platform for highly sensitive detection of exosomes, Anal. Chem. 90 (21) (2018) 12737−12744. Available from: https://doi.org/10.1021/acs.analchem.8b03083.

[97] Q. Lu, et al., Dual-emission reverse change ratio photoluminescence sensor based on a probe of nitrogen-doped Ti3C2 quantum dots@DAP to detect H_2O_2 and xanthine, Anal. Chem. 92 (11) (2020) 7770−7777. Available from: https://doi.org/10.1021/acs.analchem.0c00895.

[98] Y. Peng, et al., Charge-transfer resonance and electromagnetic enhancement synergistically enabling MXenes with excellent SERS sensitivity for SARS-CoV-2 S protein detection, Nano-Micro Lett. 13 (1) (2021). Available from: https://doi.org/10.1007/s40820-020-00565-4.

[99] M. Chao, et al., Wearable MXene nanocomposites-based strain sensor with tile-like stacked hierarchical microstructure for broad-range ultrasensitive sensing, Nano Energy 78 (2020). Available from: https://doi.org/10.1016/j.nanoen.2020.105187.

[100] Y. Yang, L. Shi, Z. Cao, R. Wang, J. Sun, Strain sensors with a high sensitivity and a wide sensing range based on a $Ti_3C_2T_x$ (MXene) nanoparticle−nanosheet hybrid network, Adv. Funct. Mater. 29 (14) (2019) 1−10. Available from: https://doi.org/10.1002/adfm.201807882.

[101] L. Wu, et al., Lotus root structure-inspired Ti3C2-MXene-Based flexible and wearable strain sensor with ultra-high sensitivity and wide sensing range, Compos. Part. A Appl. Sci. Manuf. 152 (August 2021) (2022). Available from: https://doi.org/10.1016/j.compositesa.2021.106702.

[102] K. Yang, F. Yin, D. Xia, H. Peng, J. Yang, W. Yuan, A highly flexible and multifunctional strain sensor based on a network-structured MXene/polyurethane mat with ultra-high sensitivity and a broad sensing range, Nanoscale 11 (20) (2019) 9949–9957. Available from: https://doi.org/10.1039/c9nr00488b.

[103] Y. Ma, et al., A highly flexible and sensitive piezoresistive sensor based on MXene with greatly changed interlayer distances, Nat. Commun. 8 (1) (2017) 1–7. Available from: https://doi.org/10.1038/s41467-017-01136-9.

[104] W.T. Cao, et al., Binary strengthening and toughening of MXene/cellulose nanofiber composite paper with nacre-inspired structure and superior electromagnetic interference shielding properties, ACS Nano 12 (5) (2018) 4583–4593. Available from: https://doi.org/10.1021/acsnano.8b00997.

[105] Y. Yue, et al., 3D hybrid porous MXene-sponge network and its application in piezoresistive sensor, Nano Energy 50 (May) (2018) 79–87. Available from: https://doi.org/10.1016/j.nanoen.2018.05.020.

[106] X.P. Li, et al., Highly sensitive, reliable and flexible piezoresistive pressure sensors featuring polyurethane sponge coated with MXene sheets, J. Colloid Interface Sci. 542 (2019) 54–62. Available from: https://doi.org/10.1016/j.jcis.2019.01.123.

[107] X. Shi, et al., Bioinspired ultrasensitive and stretchable MXene-based strain sensor via nacre-mimetic microscale 'brick-and-Mortar' architecture, ACS Nano 13 (1) (2019) 649–659. Available from: https://doi.org/10.1021/acsnano.8b07805.

[108] Y. Hu, et al., Biomass polymer-assisted fabrication of aerogels from MXenes with ultrahigh compression elasticity and pressure sensitivity, J. Mater. Chem. A 7 (17) (2019) 10273–10281. Available from: https://doi.org/10.1039/c9ta01448a.

[109] L. Wang, M. Zhang, B. Yang, J. Tan, X. Ding, Highly compressible, thermally stable, light-weight, and robust aramid nanofibers/Ti$_3$AlC$_2$MXene composite aerogel for sensitive pressure sensor, ACS Nano 14 (8) (2020) 10633–10647. Available from: https://doi.org/10.1021/acsnano.0c04888.

[110] H. Liu, et al., Lightweight, superelastic, and hydrophobic polyimide nanofiber/MXene composite aerogel for wearable piezoresistive sensor and oil/water separation applications, Adv. Funct. Mater. 31 (13) (2021) 1–12. Available from: https://doi.org/10.1002/adfm.202008006.

[111] J. Liu, et al., Multifunctional, superelastic, and lightweight MXene/polyimide aerogels, Small 14 (45) (2018) 1–10. Available from: https://doi.org/10.1002/smll.201802479.

[112] Y. Lu, et al., Highly stretchable, elastic, and sensitive MXene-based hydrogel for flexible strain and pressure sensors, Research 2020 (2020) 1–13. Available from: https://doi.org/10.34133/2020/2038560.

[113] Y. Xiang, et al., 3D crinkled alk-Ti3C2 MXene based flexible piezoresistive sensors with ultra-high sensitivity and ultra-wide pressure range, Adv. Mater. Technol. 6 (6) (2021) 1–11. Available from: https://doi.org/10.1002/admt.202001157.

[114] Y. Gao, et al., Microchannel-confined MXene based flexible piezoresistive multifunctional micro-force sensor, Adv. Funct. Mater. 30 (11) (2020) 1–8. Available from: https://doi.org/10.1002/adfm.201909603.

[115] Y. Cheng, et al., Bioinspired microspines for a high-performance spray Ti3C2Tx MXene-based piezoresistive sensor, ACS Nano 14 (2) (2020) 2145–2155. Available from: https://doi.org/10.1021/acsnano.9b08952.

[116] S. Sharma, A. Chhetry, M. Sharifuzzaman, H. Yoon, J.Y. Park, Wearable capacitive pressure sensor based on MXene composite nanofibrous scaffolds for reliable human physiological signal acquisition, ACS Appl. Mater. Interfaces 12 (19) (2020) 22212–22224. Available from: https://doi.org/10.1021/acsami.0c05819.

[117] P. Sobolčiak, A. Tanvir, K.K. Sadasivuni, I. Krupa, Piezoresistive sensors based on electrospun mats modified by 2D Ti3C2Tx MXene, Sens. (Switz.) 19 (20) (2019) 1–8. Available from: https://doi.org/10.3390/s19204589.

[118] Y. Guo, M. Zhong, Z. Fang, P. Wan, G. Yu, A wearable transient pressure sensor made with MXene nanosheets for sensitive broad-range human-machine interfacing, Nano Lett. 19 (2) (2019) 1143–1150. Available from: https://doi.org/10.1021/acs.nanolett.8b04514.

[119] D. Song, X. Li, X.P. Li, X. Jia, P. Min, Z.Z. Yu, Hollow-structured MXene-PDMS composites as flexible, wearable and highly bendable sensors with wide working range, J. Colloid Interface Sci. 555 (2019) 751–758. Available from: https://doi.org/10.1016/j.jcis.2019.08.020.

[120] C. Ma, Q. Yuan, H. Du, M.G. Ma, C. Si, P. Wan, Multiresponsive MXene ($Ti_3C_2T_x$)-decorated textiles for wearable thermal management and human motion monitoring, ACS Appl. Mater. Interfaces 12 (30) (2020) 34226–34234. Available from: https://doi.org/10.1021/acsami.0c10750.

[121] V. Adepu, V. Mattela, P. Sahatiya, A remarkably ultra-sensitive large area matrix of MXene based multifunctional physical sensors (pressure, strain, and temperature) for mimicking human skin, J. Mater. Chem. B 9 (22) (2021) 4523–4534. Available from: https://doi.org/10.1039/d1tb00947h.

[122] S. Han, et al., Tunable fabrication of conductive Ti3C2Tx MXenes via inflating a polyurethane balloon for acute force sensing, Langmuir 36 (5) (2020) 1298–1304. Available from: https://doi.org/10.1021/acs.langmuir.9b03281.

[123] J.H. Pu, et al., Multilayer structured AgNW/WPU-MXene fiber strain sensors with ultrahigh sensitivity and a wide operating range for wearable monitoring and healthcare, J. Mater. Chem. A 7 (26) (2019) 15913–15923. Available from: https://doi.org/10.1039/c9ta04352g.

[124] Q. Li, et al., Ti3C2 MXene as a new nanofiller for robust and conductive elastomer composites, Nanoscale 11 (31) (2019) 14712–14719. Available from: https://doi.org/10.1039/c9nr03661j.

[125] Q. Guo, et al., Protein-inspired self-healable Ti3C2 MXenes/rubber-based supramolecular elastomer for intelligent sensing, ACS Nano 14 (3) (2020) 2788–2797. Available from: https://doi.org/10.1021/acsnano.9b09802.

[126] H. Zhuo, et al., A carbon aerogel with super mechanical and sensing performances for wearable piezoresistive sensors, J. Mater. Chem. A 7 (14) (2019) 8092–8100. Available from: https://doi.org/10.1039/c9ta00596j.

[127] Z. Chen, et al., Compressible, elastic, and pressure-sensitive carbon aerogels derived from 2D titanium carbide nanosheets and bacterial cellulose for wearable sensors, Chem. Mater. 31 (9) (2019) 3301–3312. Available from: https://doi.org/10.1021/acs.chemmater.9b00259.

[128] M. Zhu, et al., Hollow MXene sphere/reduced graphene aerogel composites for piezoresistive sensor with ultra-high sensitivity, Adv. Electron. Mater. 6 (2) (2020) 1–9. Available from: https://doi.org/10.1002/aelm.201901064.

[129] Y. Ma, et al., 3D Synergistical MXene/reduced graphene oxide aerogel for a piezoresistive sensor, ACS Nano 12 (4) (2018) 3209–3216. Available from: https://doi.org/10.1021/acsnano.7b06909.

[130] G.Y. Gou, et al., Flexible two-dimensional Ti3C2 MXene films as thermoacoustic devices, ACS Nano 13 (11) (2019) 12613–12620. Available from: https://doi.org/10.1021/acsnano.9b03889.

[131] K. Wang, et al., Bioinspired interlocked structure-induced high deformability for two-dimensional titanium carbide (MXene)/natural microcapsule-based flexible pressure sensors, ACS Nano 13 (8) (2019) 9139–9147. Available from: https://doi.org/10.1021/acsnano.9b03454.

[132] C. Antoine, Tensions des vapeurs: nouvelle relation entre les tensions et les temperatures, C. R. Hebd. Seances Acad. Sci. 107 (681) (1888).

[133] D. Wang, et al., Biomimetic, biocompatible and robust silk Fibroin-MXene film with stable 3D cross-link structure for flexible pressure sensors, Nano Energy 78 (July) (2020) 105252. Available from: https://doi.org/10.1016/j.nanoen.2020.105252.

[134] W. Chen, L.X. Liu, H. Bin Zhang, Z.Z. Yu, Kirigami-inspired highly stretchable, conductive, and hierarchical Ti3C2Tx MXene films for efficient electromagnetic interference shielding and pressure sensing, ACS Nano 15 (4) (2021) 7668–7681. Available from: https://doi.org/10.1021/acsnano.1c01277.

[135] J. Zhang, et al., Highly stretchable and self-healable MXene/polyvinyl alcohol hydrogel electrode for wearable capacitive electronic skin, Adv. Electron. Mater. 5 (7) (2019) 1−10. Available from: https://doi.org/10.1002/aelm.201900285.

[136] H. Liao, X. Guo, P. Wan, G. Yu, Conductive MXene nanocomposite organohydrogel for flexible, healable, low-temperature tolerant strain sensors, Adv. Funct. Mater. 29 (39) (2019) 1−9. Available from: https://doi.org/10.1002/adfm.201904507.

[137] X. Wu, et al., A wearable, self-adhesive, long-lastingly moist and healable epidermal sensor assembled from conductive MXene nanocomposites, J. Mater. Chem. C. 8 (5) (2020) 1788−1795. Available from: https://doi.org/10.1039/c9tc05575d.

[138] Y. Zhang, et al., High-strength, self-healable, temperature-sensitive, MXene-containing composite hydrogel as a smart compression sensor, ACS Appl. Mater. Interfaces 11 (50) (2019) 47350−47357. Available from: https://doi.org/10.1021/acsami.9b16078.

[139] R. Ghosh, A. Singh, S. Santra, S.K. Ray, A. Chandra, P.K. Guha, Highly sensitive large-area multi-layered graphene-based flexible ammonia sensor, Sens. Actuators B Chem. 205 (2014) 67−73. Available from: https://doi.org/10.1016/j.snb.2014.08.044.

[140] X.F. Yu, et al., Monolayer Ti_2CO_2: a promising candidate for NH_3 sensor or capturer with high sensitivity and selectivity, ACS Appl. Mater. Interfaces 7 (24) (2015) 13707−13713. Available from: https://doi.org/10.1021/acsami.5b03737.

[141] H. Geistlinger, Electron theory of thin-film gas sensors, Sens. Actuators B. Chem. 17 (1) (1993) 47−60. Available from: https://doi.org/10.1016/0925-4005(93)85183-B.

[142] E. Lee, A. Vahidmohammadi, B.C. Prorok, Y.S. Yoon, M. Beidaghi, D.J. Kim, Room temperature gas sensing of two-dimensional titanium carbide (MXene), ACS Appl. Mater. Interfaces 9 (42) (2017) 37184−37190. Available from: https://doi.org/10.1021/acsami.7b11055.

[143] B. Xiao, Y.C. Li, X.F. Yu, J.B. Cheng, MXenes: reusable materials for NH3 sensor or capturer by controlling the charge injection, Sens. Actuators B Chem. 235 (2016) 103−109. Available from: https://doi.org/10.1016/j.snb.2016.05.062.

[144] S.J. Kim, et al., Metallic Ti3C2Tx MXene gas sensors with ultrahigh signal-to-noise ratio, ACS Nano 12 (2) (2018) 986−993. Available from: https://doi.org/10.1021/acsnano.7b07460.

[145] H.J. Koh, et al., Enhanced selectivity of MXene gas sensors through metal ion intercalation: in situ X-ray diffraction study, ACS Sens. 4 (5) (2019) 1365−1372. Available from: https://doi.org/10.1021/acssensors.9b00310.

[146] Y. Zhang, L. Wang, N. Zhang, Z. Zhou, Adsorptive environmental applications of MXene nanomaterials: a review, RSC Adv. 8 (36) (2018) 19895−19905. Available from: https://doi.org/10.1039/c8ra03077d.

[147] A. Junkaew, R. Arróyave, Enhancement of the selectivity of MXenes (M2C, M = Ti, V, Nb, Mo) via oxygen-functionalization: promising materials for gas-sensing and -separation, Phys. Chem. Chem. Phys. 20 (9) (2018) 6073−6082. Available from: https://doi.org/10.1039/c7cp08622a.

[148] P. Khakbaz, et al., Titanium carbide MXene as NH_3 sensor: realistic first-principles study, J. Phys. Chem. C. 123 (49) (2019) 29794−29803. Available from: https://doi.org/10.1021/acs.jpcc.9b09823.

[149] S.R. Naqvi, V. Shukla, N.K. Jena, W. Luo, R. Ahuja, Exploring two-dimensional M2NS2 (M = Ti, V) MXenes based gas sensors for air pollutants, Appl. Mater. Today 19 (2020) 100574. Available from: https://doi.org/10.1016/j.apmt.2020.100574.

[150] M. Wu, et al., Ti3C2 MXene-based sensors with high selectivity for NH_3 detection at room temperature, ACS Sens. 4 (10) (2019) 2763−2770. Available from: https://doi.org/10.1021/acssensors.9b01308.

[151] Z. Yang, et al., Improvement of gas and humidity sensing properties of organ-like mxene by alkaline treatment, ACS Sens. 4 (5) (2019) 1261−1269. Available from: https://doi.org/10.1021/acssensors.9b00127.

[152] W. Yuan, K. Yang, H. Peng, F. Li, F. Yin, A flexible VOCs sensor based on a 3D MXene framework with a high sensing performance, J. Mater. Chem. A 6 (37) (2018) 18116−18124. Available from: https://doi.org/10.1039/c8ta06928j.

[153] S.J. Kim, et al., Interfacial assembly of ultrathin, functional mxene films, ACS Appl. Mater. Interfaces 11 (35) (2019) 32320−32327. Available from: https://doi.org/10.1021/acsami.9b12539.

[154] C.E. Shuck, et al., Effect of Ti_3AlC_2 MAX phase on structure and properties of resultant $Ti_3C_2T_x$ MXene, ACS Appl. Nano Mater. 2 (6) (2019) 3368−3376. Available from: https://doi.org/10.1021/acsanm.9b00286.

[155] E. Lee, A. Vahidmohammadi, Y.S. Yoon, M. Beidaghi, D.J. Kim, Two-dimensional vanadium carbide MXene for gas sensors with ultrahigh sensitivity toward nonpolar gases, ACS Sens. 4 (6) (2019) 1603−1611. Available from: https://doi.org/10.1021/acssensors.9b00303.

[156] W. Guo, et al., Selective toluene detection with Mo_2CT_x MXene at room temperature, ACS Appl. Mater. Interfaces 12 (51) (2020) 57218−57227. Available from: https://doi.org/10.1021/acsami.0c16302.

[157] H. Tai, et al., Enhanced ammonia response of Ti3C2Tx nanosheets supported by TiO_2 nanoparticles at room temperature, Sensors Actuators, B Chem. 298 (May) (2019) 126874. Available from: https://doi.org/10.1016/j.snb.2019.126874.

[158] A. Hermawan, et al., CuO nanoparticles/Ti3C2Tx MXene hybrid nanocomposites for detection of toluene gas, ACS Appl. Nano Mater. 3 (5) (2020) 4755−4766. Available from: https://doi.org/10.1021/acsanm.0c00749.

[159] S. Sun, et al., W18O49/Ti3C2Tx MXene nanocomposites for highly sensitive acetone gas sensor with low detection limit, Sens. Actuators B Chem. 304 (August 2019) (2020) 127274. Available from: https://doi.org/10.1016/j.snb.2019.127274.

[160] T. He, et al., MXene/SnO_2 heterojunction based chemical gas sensors, Sens. Actuators B Chem. 329 (December 2020) (2021) 129275. Available from: https://doi.org/10.1016/j.snb.2020.129275.

[161] Z. Wang, et al., SnO-SnO_2 modified two-dimensional MXene Ti3C2Tx for acetone gas sensor working at room temperature, J. Mater. Sci. Technol. 73 (2021) 128−138. Available from: https://doi.org/10.1016/j.jmst.2020.07.040.

[162] X. Guo, et al., Enhanced ammonia sensing performance based on MXene-Ti3C2Tx multilayer nanoflakes functionalized by tungsten trioxide nanoparticles, J. Colloid Interface Sci. 595 (2021) 6−14. Available from: https://doi.org/10.1016/j.jcis.2021.03.115.

[163] W.Y. Chen, X. Jiang, S.N. Lai, D. Peroulis, L. Stanciu, Nanohybrids of a MXene and transition metal dichalcogenide for selective detection of volatile organic compounds, Nat. Commun. 11 (1) (2020) 1−10. Available from: https://doi.org/10.1038/s41467-020-15092-4.

[164] Z. Yang, et al., Flexible resistive NO_2 gas sensor of three-dimensional crumpled MXene Ti3C2Tx /ZnO spheres for room temperature application, Sens. Actuators B Chem. 326 (2) (2021) 128828. Available from: https://doi.org/10.1016/j.snb.2020.128828.

[165] L. Jin, et al., Polymeric Ti3C2TxMXene composites for room temperature ammonia sensing, ACS Appl. Nano Mater. 3 (12) (2020) 12071−12079. Available from: https://doi.org/10.1021/acsanm.0c02577.

[166] M. Hou, J. Gao, L. Yang, S. Guo, T. Hu, Y. Li, Room temperature gas sensing under UV light irradiation for Ti3C2Tx MXene derived lamellar TiO_2-C/g-C_3N_4 composites, Appl. Surf. Sci. 535 (July 2020) (2021). Available from: https://doi.org/10.1016/j.apsusc.2020.147666.

[167] X. Li, et al., Toward agricultural ammonia volatilization monitoring: a flexible polyaniline/Ti3C2Tx hybrid sensitive films based gas sensor, Sens. Actuators B Chem. 316 (December 2019) (2020). Available from: https://doi.org/10.1016/j.snb.2020.128144.

[168] L. Zhao, K. Wang, W. Wei, L. Wang, W. Han, High-performance flexible sensing devices based on polyaniline/MXene nanocomposites, InfoMat 1 (3) (2019) 407−416. Available from: https://doi.org/10.1002/inf2.12032.

[169] Y. Zhou, Y. Wang, Y. Wang, X. Li, Humidity-enabled ionic conductive trace carbon dioxide sensing of nitrogen-doped Ti3C2Tx MXene/polyethyleneimine composite films decorated with reduced graphene oxide nanosheets, Anal. Chem. 92 (24) (2020) 16033−16042. Available from: https://doi.org/10.1021/acs.analchem.0c03664.

[170] B. Sun, et al., Co_3O_4@PEI/Ti3C2Tx MXene nanocomposites for a highly sensitive NOxgas sensor with a low detection limit, J. Mater. Chem. A 9 (10) (2021) 6335−6344. Available from: https://doi.org/10.1039/d0ta11392a.

[171] S. Wang, et al., Ultrathin Nb_2CT_x nanosheets-supported polyaniline nanocomposite: enabling ultrasensitive NH3 detection, Sens. Actuators B Chem. 343 (May) (2021) 130069. Available from: https://doi.org/10.1016/j.snb.2021.130069.

[172] S. Gasso, M.K. Sohal, A. Mahajan, MXene modulated SnO_2 gas sensor for ultra-responsive room-temperature detection of NO_2, Sens. Actuators B Chem. 357 (2) (2022) 131427. Available from: https://doi.org/10.1016/j.snb.2022.131427.

[173] F.M. Römer, et al., Controlling the conductivity of Ti3C2 MXenes by inductively coupled oxygen and hydrogen plasma treatment and humidity, RSC Adv. 7 (22) (2017) 13097−13103. Available from: https://doi.org/10.1039/c6ra27505b.

[174] E.S. Muckley, et al., Multimodality of structural, electrical, and gravimetric responses of intercalated MXenes to water, ACS Nano 11 (11) (2017) 11118−11126. Available from: https://doi.org/10.1021/acsnano.7b05264.

[175] H. An, et al., Water sorption in MXene/polyelectrolyte multilayers for ultrafast humidity sensing, ACS Appl. Nano Mater. 2 (2) (2019) 948−955. Available from: https://doi.org/10.1021/acsanm.8b02265.

[176] N. Li, et al., High-performance humidity sensor based on urchin-like composite of Ti3C2 MXene-derived TiO_2 nanowires, ACS Appl. Mater. Interfaces 11 (41) (2019) 38116−38125. Available from: https://doi.org/10.1021/acsami.9b12168.

[177] L.X. Liu, W. Chen, H. Bin Zhang, Q.W. Wang, F. Guan, Z.Z. Yu, Flexible and multifunctional silk textiles with biomimetic leaf-like MXene/silver nanowire nanostructures for electromagnetic interference shielding, humidity monitoring, and self-derived hydrophobicity, Adv. Funct. Mater. 29 (44) (2019) 1−10. Available from: https://doi.org/10.1002/adfm.201905197.

[178] M. Bi, Y. Miao, W. Li, J. Yao, Niobium carbide MXene-optics fiber-sensor for high sensitivity humidity detection, Appl. Phys. Lett. 120 (2) (2022). Available from: https://doi.org/10.1063/5.0064005.

[179] J. Wang, H. Ma, Y. Liu, Z. Xie, Z. Fan, MXene-based humidity-responsive actuators: preparation and properties, ChemPlusChem 86 (3) (2021) 406−417, doi:10.1002/cplu.202000828.

Composite materials based on mesoporous oxides and noble metal nanoparticles

14

Ianina L. Violi[1], M. Cecilia Fuertes[2] and Paula C. Angelomé[2]

[1]*Nanosystems Institute, University of San Martín, Buenos Aires, Argentina* [2]*Chemistry Department & Nanoscience and Nanotechnology Institute, CAC, CNEA, CONICET, Buenos Aires, Argentina*

14.1 Introduction

14.1.1 Mesoporous oxides

Mesoporous oxides are custom-designed materials that have been investigated, since their first synthesis 30 years ago, due to their versatility, very high specific surface, and pore sizes in the mesoscale [1]. These materials are synthesized using the sol−gel method in conjunction with the self-assembly of surfactants [2−4]. The sol−gel method allows to obtain oxides with highly controlled compositions: pure oxides (mainly silica and titania, but also zirconia, alumina, etc.), mixed oxides, and hybrid materials, with a diverse organic content, directly from the synthesis [5]. The addition of structure-directing agents is used to generate mesoporous materials, with monodisperse pores with sizes between 2 and 50 nm, accessible porosities between 10% and 50%, and diverse pore arrangements: cubic, hexagonal, channels, etc., in a very controlled manner [4,6,7].

Mesoporous materials are commonly produced in the form of powders, nanoparticles, and supported films (see some examples in Fig. 14.1). The synthesis is carried out from solutions under ambient conditions, employing simple equipment, without the need to use vacuum or high temperatures.

Mesoporous oxide powders (Fig. 14.1A−C) can be synthesized using aerosol [9,10] or precipitation methods [2,3,11,12]. Among the wide variety of mesoporous materials, powdered mesoporous silica is one of the most studied. The first ordered mesoporous silica was synthesized in 1992 by Beck et al. and it was called MCM-41. It presents a highly ordered hexagonal array of monodisperse pores, obtained using hexadecyltrimethylammonium bromide (CTAB) as a template [2,3]. A few years later, the synthesis of SBA-15, another famous mesoporous silica, was reported [11,12]. SBA-15 is particularly interesting due to the thick walls of oxide between pores, the accessible hexagonal array of mesopores, the simplicity of its synthesis and functionalization, the high specific surface, and its high thermal and mechanical stability. Moreover, the structure-directing agent employed for the synthesis of SBA-15, the triblock copolymer Pluronic P123, is nontoxic, low-cost, and biodegradable [13]. The synthetic approach employed in these seminal works was later expanded to obtain other ordered mesoporous oxides [14] mainly TiO_2 and ZrO_2 [15]. In the last three decades, there have also been great advances in the understanding of the synthetic strategies to produce mesoporous oxides as powders [16], nanoparticles [17,18], and thin films [19].

Materials Nanoarchitectonics. DOI: https://doi.org/10.1016/B978-0-323-99472-9.00007-9

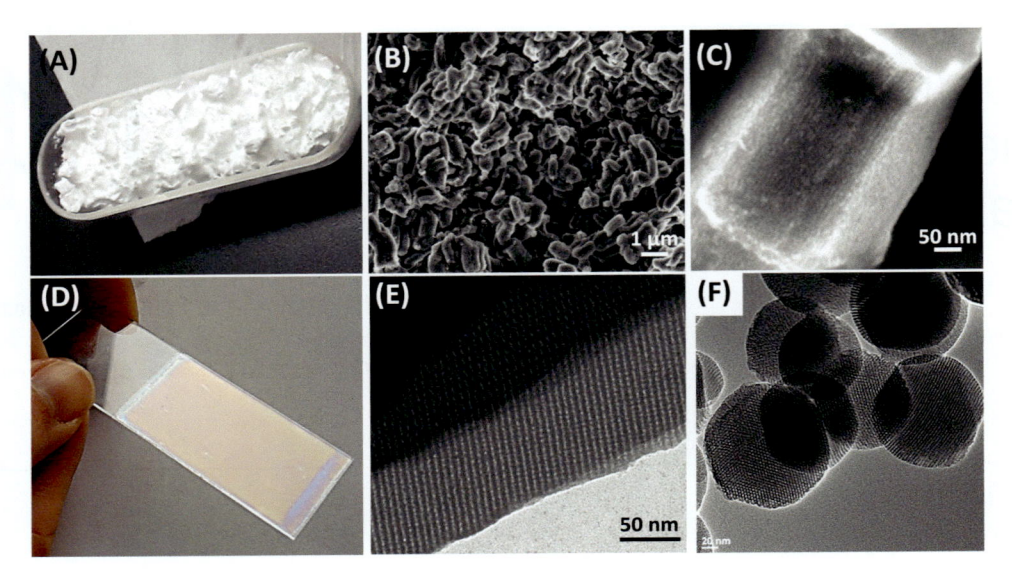

FIGURE 14.1

(A) Macroscopic optical image of SBA-15; (B) scanning electron microscopy (SEM) of SBA-15 (micron-scale); (C) SEM image of SBA-15 (nanoscale); (D) macroscopic optical image of a mesoporous TiO_2 thin film templated with the triblock copolymer Pluronic F127; (E) transmission electron microscopy (TEM) image of a mesoporous SiO_2 thin film templated with the diblock copolymer Brij 58; (F) TEM image of mesoporous MCM-41 silica nanoparticles.

Adapted with permission from J.L. Vivero-Escoto, I.I. Slowing, C.-W. Wu, V.S.Y. Lin, Photoinduced intracellular controlled release drug delivery in human cells by gold-capped mesoporous silica nanosphere, J. Am. Chem. Soc.131 (10) (2009) 3462–3463. Copyright 2009 American Chemical Society. Online panel (F) was taken from this reference. The rest of the panels were prepared by the authors.

In the case of nanoparticles (Fig. 14.1F), the focus has been the production of the so-called Mesoporous Silica Nanoparticles or MSNs, whose main applications are the controlled release of drugs. The synthesis of these particles is based on the MCM-41 and SBA-15 production procedures, modified to diminish the particle size. The modifications include the manipulation of pH during the synthesis and the addition of cosolvents and other additives [18,20].

Thin films with ordered arrays of mesopores, on the other hand, are generally obtained by spin or dip coating, with thicknesses between 20 and 300 nm depending on the spin or withdrawal speed and the precursor solution viscosity (Fig. 14.1D−E) [21]. They can be deposited on a wide variety of planar substrates, including temperature-sensitive substrates such as plastics or paper, thanks to the low processing temperatures that the method involves [22].

In the last years, significant progress has also been made in their direct use and in the integration of these materials in various platforms and devices [23−26]. The applications of these materials cover a number of areas of technological interest, including photocatalysis [15,27,28], adsorption of contaminants [29], drug-delivery [30], and sensing or detection of analytes, both in liquid and gas phases [31−33].

The high reactivity generated due to the presence of accessible mesoporosity allows these oxides to be used as platforms to synthesize other nanocomposite materials with novel properties [31]. These composites include the combination of the oxide matrix with metal nanoparticles, semiconductors, and polymers and the addition of numerous organic functions to the surface of the pores.

14.1.2 Metallic nanoparticles

Metallic nanoparticles (NPs) are arrangements of elementary metal atoms, having at least one of their three dimensions in the range of 1–100 nm [34]. Nanoparticles composed by noble metals like Au, Ag, Pt, and Pd have awakened the interest of the scientific community over the last three decades due to the properties that differentiate them from the bulk material, and the applications that derive from these distinctive features. For example, the widespread use of metallic nanoparticles as catalysts is a consequence of their increased specific surface and reactivity, promoting a wide variety of chemical reactions [35,36]. A particular set of applications such as nanomedicine, sensing and optics, rely on a unique characteristic of metallic nanoparticles: their special interaction with the light. When a metallic nanoparticle is illuminated, the light excites the free electrons of the conduction band, and localized oscillations take place. They are the so-called localized surface plasmon resonances (LSPRs), which are confined to the nanoparticle dimensions [37]. A strong optical response can be observed in some nanoparticles, mostly if they have large electron densities, as in the case of Au. These resonances are highly tunable, and the frequency of the LSPRs can cover the UV, visible, and near infrared region of the electromagnetic spectrum. The position of the LSPRs in the spectrum depends on structural parameters of the nanoparticles, for example, their size, shape, and composition, and it changes with the environment in which the nanoparticles are immersed. To depict the influence of shape, size, and composition, it can be seen in Fig. 14.2A, B that there is only one LSPR for spherical Ag and Au nanoparticles, two LSPRs when the particle is a nanorod [38], and several for Ag with a tetrahedral shape. Also, every LSPR is unique for each kind of shape and size [39,40]. The most widely used plasmonic nanoparticles are made of Au and Ag because their resonances are tunable in a very useful part of the spectrum, that is, the visible and near infrared, and they are also very stable in most solvents, especially in water. For these reasons, there is a plethora of colloidal synthesis reported to obtain different shaped and sized Au and Ag nanoparticles [41]. Colloidal synthesis consists of the controlled reduction of metallic precursors like $HAuCl_4$ or $AgNO_3$ with different reagents, such as sodium citrate, sodium borohydride, and ascorbic acid. In some cases, light is employed to promote the NPs formation [42]. When the excess of reducing agent don't stabilize the obtained NPs through electrostatic repulsion, like in the case of citrate ions, other molecules such as CTAB or polyvinylpyrrolidone (PVP) must be employed to prevent their aggregation though steric stabilization [43,44]. NPs functionalization with organic ligands that have strong affinity for their surface, such as thiols for Au, can also suppress NPs aggregation [45]. It is often necessary to use shape-directing agents for obtaining specific morphologies through anisotropic growth, like the use of CTAB and silver ions for gold nanorods preparation [46,47], or CTAC and bromide ions when synthetizing gold nanocubes [48]. Some examples of transmission electron microscopy images of the different shapes that can be obtained through colloidal synthesis are shown in Fig. 14.2C.

The strong optical response is not the only feature that makes plasmonic nanoparticles very attractive. There are at least three more interesting phenomena taking place in the vicinity of a

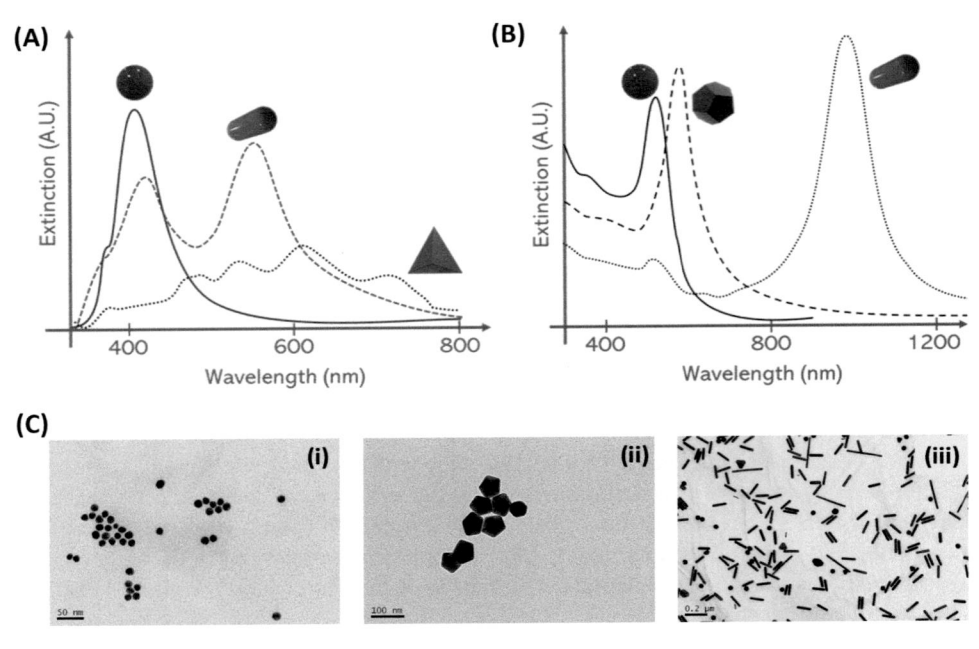

FIGURE 14.2

(A) Extinction efficiency of an Ag sphere (40 nm, solid line, simulated), Ag nanorod (50 × 108 nm, long dash-dot line, experimental) and an Ag tetrahedra (effective diameter of 40 nm, short dash-dot line, simulated) in water. Data were taken from Refs [38,39]. (B) Experimental extinction spectra of Au spheres (15 nm diameter, solid line) and Au nanorods (26 × 140 nm, short dash-dot line) in water and Au decahedrons (45 nm side, long dash-dot line) in ethanol. (C) TEM images of the nanoparticles presented in (B): (i) Au nanospheres; (ii) Au decahedrons; (iii) Au nanorods.

nanoparticle when it is illuminated near the LSPRs: a dramatic intensification of the electromagnetic field, the generation of energetic carriers (electrons and holes), and the local increase of the nanoparticle temperature [49,50]. The presence of these unique features gave birth to a whole new area called nanoplasmonics [51,52]. Many biological processes and chemical reactions can benefit from having energetic carriers, photons, and heat confined at the nanoscale that are also available in short response times. This is why metallic nanoparticles are envisioned to be applied in areas as diverse as ultrasensitive detection of molecules [53], nanoscopic heat sources [49,54,55], to promote solar to chemical energy conversion [56], to perform noninvasive biological studies at the cellular and subcellular level [57] and as light antennas [58,59], among others. Nanomedicine is the area that has exploited most the heat produced by the illumination of nanoparticles, finding applications in drug delivery [60], photothermal cancer therapy [61], and nanosurgery [62]. Plasmon-assisted chemistry is a fairly new area [63] that is gaining considerable attention due to the potential applications in controlled and greener approaches for synthesis and degradation of organic molecules [51,64−66].

14.1.3 Combination of metallic nanoparticles and mesoporous oxides

14.1.3.1 Why combine them?

As described in the previous sections, both noble metal nanoparticles and mesoporous oxides present distinctive and interesting properties, derived from their sizes, shapes, and compositions. Interestingly, the size of the nanoparticles and the size of the pores in the oxides are in the same range. Moreover, the synthetic paths most frequently used to obtain both families of materials are compatible.

Thus the idea of combining these nanomaterials in a single composite is straightforward. The obtained composite, if prepared correctly, will present the properties of the components and the properties derived from their vast contact area and consequent interaction. Because of this synergy, new or enhanced applications can be envisioned. In particular, incorporating metallic nanoparticles inside the pores of mesoporous oxides can generate or improve the oxides catalytic and sensing properties [33,67]. On the other hand, the wide variety of shapes in which mesoporous oxides can be prepared, thanks to the versatility of the sol–gel method, allows the simple obtention of supported and, more importantly, immobilized metallic nanoparticles, making their use simpler [68,69]. In addition, the nanoparticles resistance to harsh physicochemical conditions can be dramatically increased if a proper mesoporous oxide is used to support or encapsulate them, preventing their dissolution, decomposition or aggregation and incrementing their durability and reusability [67,69–73].

14.1.3.2 How to combine them?

Three main architectures can be found in the literature for noble metal nanoparticles–mesoporous oxides composites, which are depicted in Scheme 14.1: powders, nanoparticles, and thin films.

In the case of powders, most of the examples are based on the well-known mesoporous silica MCM-41 [2,3] and SBA-15 [11,12], although other oxides have been used [33]. In these composites, the metallic nanoparticles are generally included within the pores of the oxide, but can also be located

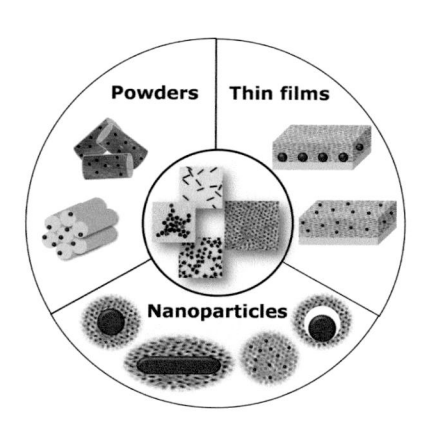

SCHEME 14.1

Main reported architectures of composites based on metallic nanoparticles and mesoporous oxides.

in the oxide walls or used as pores caps [67]. In either case, the size and the shape of the nanoparticles are limited by the oxide features: sub-10 nm spherical particles are the most usual [69,73].

In the case of nanoparticles, several architectures have been presented, including: core—shell particles with single or multiple cores, nanorattles or yolk—shell particles, and mesoporous particles with metallic nanoparticles within the pores or the walls [67,69,73—76]. The latter are very similar to the powders described before, with only a smaller particle size. In the other three cases, the variety of nanoparticles sizes and shapes is wider; in fact, anisotropic nanoparticles such as nanorods and nanostars can be adequately covered with mesoporous shells.

Finally, mesoporous thin films have also been used as supports or protective layers for metallic nanoparticles. Depending on the chosen architecture, the nanoparticles are deposited inside the pores or the walls of the oxide or are attached to the substrate and covered by the oxide. As expected, depending on the chosen architecture, the available nanoparticles compositions, sizes, and shapes vary [77—79].

In all cases, the shape, size, and architecture of the composites define their final properties and, consequently, their possible applications. This includes the spatial position of each component but also the physicochemical interactions between them and with the medium.

In summary, having a large surface area, accessibility, the possibility of giving them a defined function by modifying their surface with organic molecules, and in some cases intrinsic catalytic functionalities, mesoporous oxides not only can act as stabilizers for NPs but also can produce a synergistic effect for many applications [80]. Also, the mesoporous framework imposes a very ordered and well-known distance for the nanoparticles to be arranged, adding an extra degree of control in the final properties of the composite materials. Finally, the mesoporous oxide can act as a filter for larger molecules when the architecture is cleverly designed, which makes them great platforms for sensing applications [81]. Next, several examples of noble metal nanoparticles—mesoporous oxides composites will be presented with the focus on the role the porous oxide plays in the composite synthesis and properties, as introduced before: support, active host for NPs or molecules, filter, and stabilizer. The reported and projected applications of these advanced materials will be also described.

14.2 Mesoporous oxides as support for metallic nanoparticles

Metallic nanoparticles have found many applications in catalysis, sensing, nanomedicine, and renewable energies, among others. To be able to employ them as a long-term stable, affordable, and robust material, it is important to disperse these nanoparticles in a matrix, so they can be recovered and reused, and to stabilize them and hinder any potential damage that can occur in *operando* conditions. An interesting approach to achieve these goals is to support them on high-surface-area materials with increased thermal stability like mesoporous oxides. Their pore size makes them excellent templates for creating predefined sized NPs as well as to direct their final shape by modulating the nanoarchitecture of the mesoporous array. The mesopores are also large enough to trap presynthesized smaller NPs, but both materials can also be in contact in different architectures. For example, with the NPs grafted or impregnated in the walls or by deposition of the porous oxide on top of the NPs (as a shell that covers the NPs or a thin film in a solid substrate), among others [67,70].

Next, selected examples of different mesoporous oxides acting as the nanoparticles support and/ or template and some of their applications will be addressed.

14.2.1 Mesoporous oxides as simple supports

The simplest and more widespread method to obtain metallic NPs hosted in mesoporous oxides is to adsorb the metal precursor in the oxide walls, which is afterwards reduced. This reduction is usually performed by a thermal treatment in vacuum or in the presence of H_2, or in solution by employing reducing agents, such as hydrazine or sodium borohydride. The only prerequisite to success when employing this approach is to have electrostatically compatible surfaces, meaning the oxide support must have the opposite charge than the metallic precursor at the impregnation pH. This method is commonly known as impregnation—reduction or deposition—precipitation. Many catalytic materials have been prepared by employing this approach, and they have been applied for water gas-shift reactions [82], oxidation of carbon monoxide [83], total oxidation of toluene [84], and glycerol oxidation [85], among many others.

Another strategy to produce this kind of composite material is to prepare the mesoporous oxide in the presence of a dispersion of presynthesized NPs [70]. When employing this method, obtaining well dispersed NPs within the pores in high loadings is not straightforward [86]. It is also challenging to match the stability and reactivity of both materials, because the presence of the NPs can interfere with the synthesis of the mesoporous oxide. Besides, this synthetic path can modify the particles morphology. Finally, it is also possible that the mesoporous framework gets clogged by the NPs as they are obtained within the oxide walls, modifying its overall accessibility.

Decoupling the nanoparticles synthesis from the mesoporous material preparation is an interesting approach [87]. The main advantage of this synthetic procedure is that a large number of well-known synthetic methods may be employed for the design of both types of materials. After synthesis and purification, the NPs can be infused into the pores of a presynthesized mesoporous material by a physical technique without modifying the structure of either material. Various methods such as supercritical CO_2 infiltration, ultrasonication, electrophoretic, and direct deposition methods have been used to infuse the metal NPs into the mesoporous framework [70].

It has to be noted that for efficient infusion of NPs into the mesopores the dispersion should flow through the pores by capillary wetting and the nanoparticles must not block the pores, which can be hard to achieve. In addition, the nanoparticles need to be properly adsorbed to the substrate, so they do not leach when the composite material is employed.

14.2.2 Mesopores as templates for nanoparticles formation

Among the different ordered mesoporous oxides, silica is one of the most studied for hosting metallic NPs. It is usually an inert support, only acting as a stabilizer of the nanoparticles. By changing the synthetic pathway and the surfactant employed to obtain the mesopores array, different types of porous silica can be prepared. As mentioned before, among the wide variety of silica-based structures that have been reported, SBA-15 and MCM-41 are the most explored. Depending on the composites application, different kinds of metallic NPs can be included inside the mesoporous channels, which direct their final shape. In the case of SBA-15, the obtained NPs are nanospheres, nanorods, or nanowires, depending on the synthetic conditions used [88−91]. For example, by combining a microwave heating process and taking advantage of the confined pore structure of SBA-15, Mori et al. have developed a synthetic pathway to obtain Ag nanoparticles with different plasmonic properties, as shown in Fig. 14.3A [92]. The different types of NPs were obtained by varying the time of the microwave thermal treatment and by using surface-modifying organic ligands. The authors demonstrated that the presence of Ag nanoparticles unambiguously enhances the catalytic activity in the LSPR-assisted H_2 production from ammonia borane under visible−light irradiation. The observed enhancement of the catalytic performance was in close agreement with the increased plasmonic absorption at each specific wavelength.

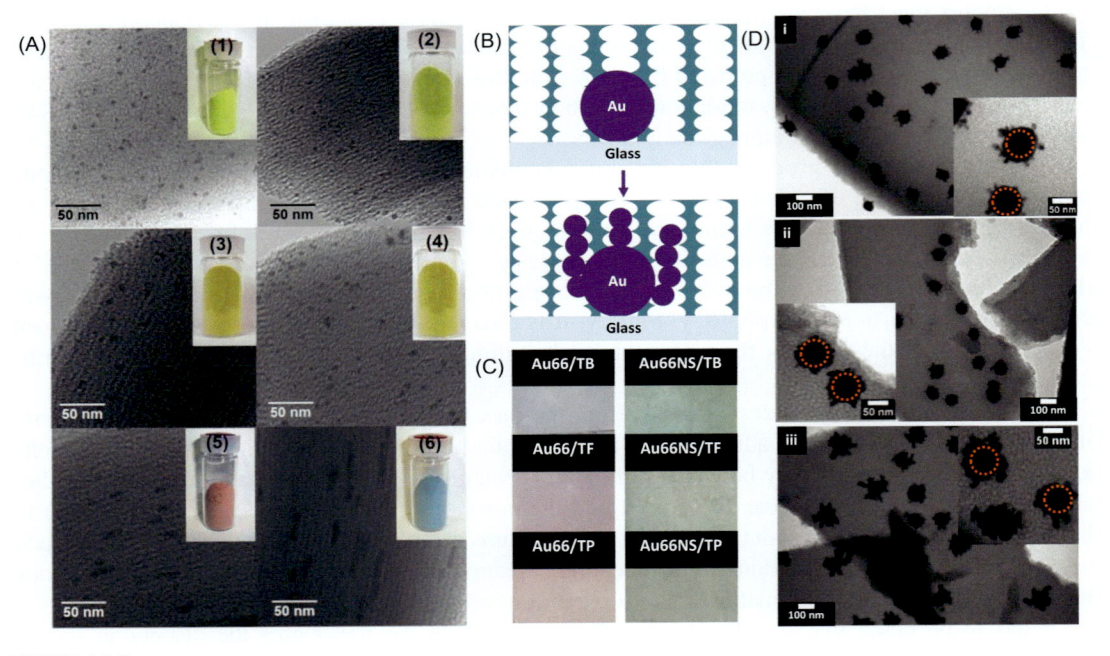

FIGURE 14.3

(A) TEM images and samples photographs (inset) of the different Ag catalysts inside SBA-15 channels prepared by Mori et al. (B) Scheme of the Au nanoparticle growth through mesoporous titania thin films developed by Steinberg et al. (C) Optical images of the mesoporous titania thin films on top of Au 66 nm nanoparticles before (left) and after (right) growth through different mesopores obtained by employing three surfactants: *TB*, titania Brij 58; *TF*, titania Pluronic F127; *TP*, titania Pluronic P123. (D) TEM images of the growth nanoparticles through the pores: (i) TB; (ii) TF; (iii) TP.

(A) Adapted with permission from K. Mori, P. Verma, R. Hayashi, K. Fuku, H. Yamashita, Color-controlled Ag nanoparticles and nanorods within confined mesopores: microwave-assisted rapid synthesis and application in plasmonic catalysis under visible-light irradiation, Chem. Euro. J. 21 (33) (2015) 11885−11893. Copyright 2015 John Wiley and Sons. (B−D) Reproduced from P.Y. Steinberg, M.M. Zalduendo, G. Giménez, G.J.A.A. Soler-Illia, P.C. Angelomé, TiO₂ mesoporous thin films architecture as a tool to control Au nanoparticles growth and sensing capabilities, Phys. Chem. Chem. Phys. 21 (2019) 10347−10356 with permission from the PCCP Owner Societies.

In some cases, a previous functionalization of the mesoporous walls can improve the fabrication of metallic NPs inside the pores. A nice example was presented by Rafti and coworkers [94] that employed a positively charged polymer brush to confine and trap the platinum precursor $PtCl_6^{2-}$ inside the mesopores of a silica thin film. The precursor was then reduced to Pt by using $NaBH_4$. The reason for employing such an approach lies in the low surface charge present in pristine silica at the slightly acidic conditions in which the impregnation is performed. The silica isoelectric point is ~ 2, and it therefore has almost a neutral charge at $pH = 2$. When functionalized by a cationic polymer brush, the surface presents a highly positive charge at lower pH, leading to an increased loading of the Pt precursor. The obtained nanocomposite presents a well-controlled dispersion of ~ 3 nm Pt nanoparticles, and was employed in the gas-phase oxidation of ammonia at low temperatures.

Another way to employ the mesopore structure as a template for NPs is by overgrowing colloidal nanoparticles in contact with the mesoporous oxide. Angelomé et al. showed this concept in different works [79,81,93,95−98]. They attached spherical Au NPs to a substrate, then deposited a mesoporous thin film on top, and finally overgrew the nanoparticles, which led to branched structures. A scheme of this process is shown in Fig. 14.3B. This kind of materials was employed in surface-enhanced Raman spectroscopy (SERS)-based detection, showing better activity when compared with similar sized nanospheres. In one of the works [93], the group compared the obtained branching through the pores by performing the growth on different mesopore structures, that is, diverse pore sizes and ordering obtained by changing the templating agent. Optical images of the mesoporous thin films before and after the growth are shown in Fig. 14.3C, where the differences in the optical properties are evident to the naked eye. In the TEM images of the Fig. 14.3D, a clearer picture of the results is shown, evidencing that the smaller pores (the ones obtained by employing Brij 58 as template) lead to shorter branches, and the bigger and more connected pores (obtained by employing Pluronic P123) lead to the largest branches. In the case of Pluronic F127, an intermediate pore size is obtained, and the branching is also demonstrated to be in-between the other two. This example shows the power of employing different mesoporous nanoarchitectures to template metallic NPs, resulting in differential optical properties.

Employing the same concepts explained before, but a different starting material, Sanz-Ortiz and coworkers [99] have developed a method to overgrow Au nanoparticles through the channels of a mesoporous silica shell. In this case, the NPs are trapped inside the shell, but they remain accessible through the radial mesoporous channels. They have shown the success of this approach by growing spherical NPs as well as different shaped NPs, like single-crystal or pentatwinned nanorods and triangles, following the process through 2D and 3D TEM. The radial silica channels acted as a template, and therefore Au branches appeared from the surface of the Au cores, regardless of the initial core shape, as can be observed in Fig. 14.4. Also, the pores remained stable and accessible after the branch growth. The authors have also shown that these hybrid nanosystems can be employed in SERS detection of different analytes, demonstrating the superior performance of branched NPs.

14.3 **Mesoporous oxides as stabilizers for nanoparticles**

In the previous section it was described how the mesoporous oxides are used as support of metallic nanoparticles. In this section, several examples of the stabilizing role of the mesoporous matrix will be presented and discussed.

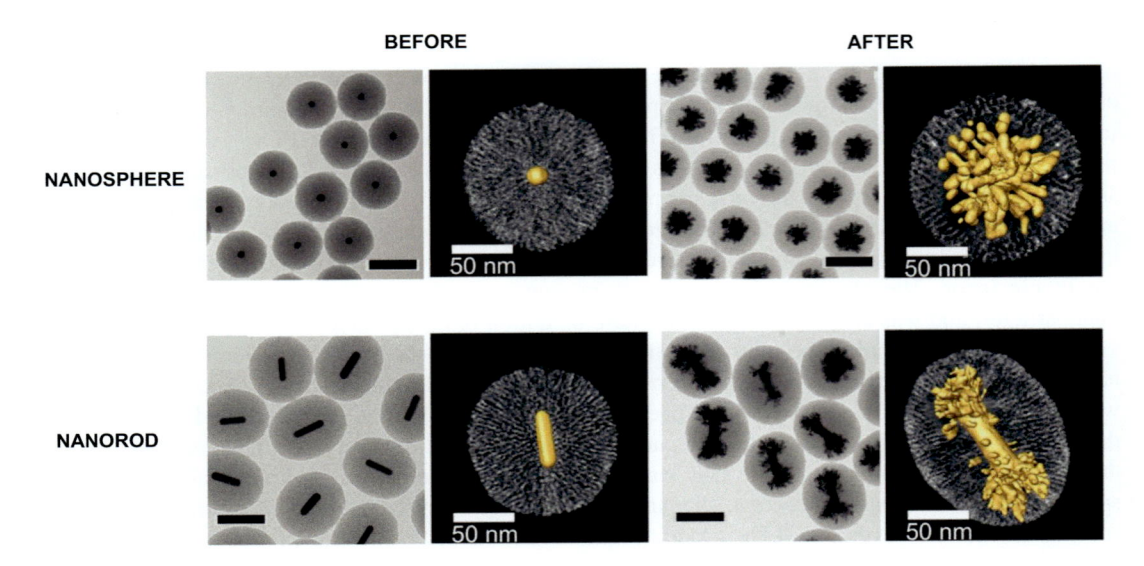

FIGURE 14.4

2D and 3D TEM characterization of hybrid nanostructures containing gold nanoparticles with various shapes: nanospheres and single-crystal nanorods. The radial mesopores in the silica shells can be observed in the visualizations of the 3D reconstructions in the right panel (the Au nanoparticle and the silica shell are displayed in yellow and white, respectively). Scale bar in TEM images: 100 nm.

Adapted with permission from M.N. Sanz-Ortiz, K. Sentosun, S. Bals, L.M. Liz-Marzán, Templated growth of surface enhanced Raman scattering-active branched gold nanoparticles within radial mesoporous silica shells, ACS Nano 9 (10) (2015) 10489–10497 (https://pubs.acs.org/doi/full/10.1021/acsnano.5b04744, further permission should be directed to the ACS).

Numerous applications including metallic NPs involve demanding reaction conditions, such as elevated temperatures or pressures, harsh solvents, radiation, etc. In some of these conditions, highly reactive metallic NPs can change their size and/or shape, with undesirable consequences. For example, nanoscale particles submitted to moderate temperatures can suffer deactivation by mass migration; the aggregation process generates larger and less reactive particles [69]. Besides, in some environments, NPs are easily dissolved also due to their high reactivity arising from their high specific area. Finally, in the case of nanocomposite materials that contain metallic NPs, the NPs may leave the porous matrix that supports them; the leaching process that can occur in some environments generates a big loss of the nanocomposite functionality [100]. Therefore, to have a stable and still reactive system, metallic NPs must be protected for its use in certain applications.

Among the applications in harsh conditions, the use of metallic NPs as catalysts is the most studied. Catalytic reactions generally need high operating temperatures (usually above 300°C) [101−103], so the metallic NPs need to be stabilized to avoid aggregation effects caused by these extreme conditions [104] (see an example in Fig. 14.5). The use of nanocomposite materials that include protected metal NPs in heterogeneous catalysis is wide and varied, and several examples of reactions catalyzed using metallic NPs—mesoporous oxides composites can be found in the literature, in both gas and liquid phase. Some of the most studied reactions are gas-phase CO oxidation

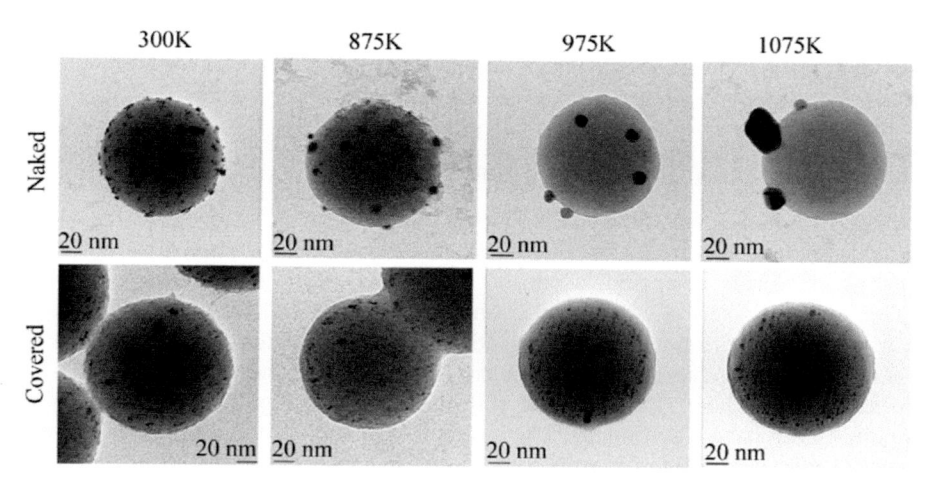

FIGURE 14.5

TEM images of Pt NPs/SiO$_2$ uncovered (top row) and covered with a layer of mesoporous SiO$_2$ (bottom row), after calcination at different temperatures between 300 and 1075K. Sintering of the Pt NPs occurs in the naked samples, but no appreciable sintering is observed in the samples covered with the mesoporous shell.

Reprinted by permission from Springer Nature from I. Lee, Q. Zhang, J. Ge, Y. Yin, F. Zaera, Encapsulation of supported Pt nanoparticles with mesoporous silica for increased catalyst stability, Nano Res. 4 (1) (2011) 15–123.

[100,103,105,106], methanol [80] and *n*-hexane reforming [101], ethylene hydrogenation [103], reduction of nitrates [107] or 4-nitrophenol [108], and hydrodeoxygenation of phenolic compounds [102], among others.

Another application that requires stable NPs is for their use as SERS-based detection platforms. Aggregation of metal NPs reduces the reproducibility and the reliability of these measurements. This is particularly remarkable for Ag NPs which are easily oxidized and present a limited stability at high pH values and temperatures [72]. Moreover, after the SERS detection, it is desirable to detach the adsorbed molecules from the metallic surfaces, leading to clean SERS substrates. This process is usually done using thermal annealing (up to 300°C), so one basic prerequisite to perform this cleaning is to have a thermally robust SERS platform [109].

The NPs resistance to thermal degradation or aggregation is highly increased if materials with high thermal stability such as mesoporous silica [110], titania [101], or alumina [107,111] are used to support or encapsulate the NPs [67,69–73]. For applications that involve the interaction of the analyte with the metal cores, direct access of target molecules to the particles is required, so a fully dense oxide layer surrounding the metallic NPs is not of interest. Therefore, only the architectures in which the metallic NPs are protected but still accessible will be discussed in this section. It is worth noting that beyond the stabilizing role, the presence of mesopores near the metallic cores provides a medium in which the preconcentration of analytes can occur, improving the efficiency of conversion in the catalytic reaction [112,113] or the SERS detection [72,114].

Depending on the composite architecture, different performances can be achieved. Several designs have been presented in the literature considering the stabilizing role of the mesoporous oxide, but all of them include one of the following concepts: to have the NPs surrounded by a mesoporous shell or to include the NPs inside the mesoporous structure of the oxide. Next, the key features and some examples of both kinds of nanocomposites will be described.

14.3.1 Metallic nanoparticles inside the mesopores

One way to stabilize highly reactive metallic NPs is to include them inside the pores of a mesoporous oxide. This strategy has been widely applied for the use of these nanocomposites in several applications, as previously discussed, using the mesoporous oxide in the form of thin films and powders. This method is very useful for designing catalytic materials due to the precise control of NPs sizes as the mesopores act as directing agents for the NPs growth. However, it must be considered that in composites produced by including the NPs within the mesopores of the oxide support, the surface area and the pore sizes decrease with the metal loading. This reduction of the mesopores accessibility can have a detrimental effect on the efficiency of the catalytic reactions [100].

The effect of pore structures of mesoporous silica was evaluated in the stabilization of Au NPs synthesized inside the pores by Bore and coworkers [105]. Silica matrices with 2D-hexagonal, 3D-hexagonal, and cubic arrays of mesopores were studied, with pore sizes varying between 2.2 and 6.5 nm. The Au NPs were prepared inside the mesopores after the functionalization of the silica surface with amino groups and then the nanocomposite material was tested for CO oxidation up to 400°C. By comparing the leaching behavior of the Au NPs in these systems, the authors observed for samples with 2D-hexagonal structure that the Au NPs got larger after the applied thermal treatment but remained within the pores. In contrast, the Au NPs migrated outside the pores in the silica samples having 3D interconnected pores when treated equivalently. Besides, among the 2D structures studied, the SBA-15 sample, which had the thickest pore walls, was effective at controlling the growth of the Au NPs with the temperature.

The nature of the oxidic support was also investigated to design thermally stable and deactivation resistant nanocatalysts, while keeping their selectivity and catalytic activity. A hierarchical TiO_2 support was prepared by An et al. using Pluronic P123 to generate mesopores and polystyrene beads to include microporosity [101]. For comparison purposes, SBA-15 was used as a mesoporous silica support. Pt NPs with an average diameter of 2.7 nm were then loaded into both supports by sonication, and the obtained materials were evaluated for the catalytic reforming of h-hexane up to 500°C. After testing both nanocomposites, it was observed that the Pt/SiO_2 catalyst sinters above 400°C while the Pt/TiO_2 catalyst did not exhibit TiO_2 overgrowth or deactivation by Pt sintering up to 600°C. The authors proposed that the strong metal support interaction presented for the Pt/TiO_2 catalyst not only promoted the extremely high thermal stability but also the higher reaction rates and isomer selectivity found for this system, in contrast with the behavior found for the silica-based composite.

In a different approach, the intercalation of metallic NPs inside the walls between the pores of a thermally stable oxide can also enhance the NPs thermal stability. In the interesting work of Gage et al., Pd NPs were encapsulated in the walls or inside the pores of SBA-15, and the performance of both architectures to phenol hydrodeoxygenation reactions was compared [102]. The Pd NPs intercalated into mesoporous silica walls were synthesized by sol—gel chemistry in the presence of Pd and a metal-directing agent. The authors showed that these NPs are stable up to 900°C, and no particle migration is observed within the support. Besides, the encapsulated Pd active sites remain catalytically active, and this catalyst is regenerated even after several reaction cycles. Pd NPs inside the pores were produced via direct precipitation within a previously prepared SBA-15 matrix. The authors show that the NPs prepared in this way sinter at the explored temperatures and do not recover prior activity after the same regeneration procedure applied to the former catalyst.

14.3.2 Metallic nanoparticles covered with a mesoporous shell

In this architecture, a mesoporous oxide shell surrounds individual metallic cores or a group of (supported or unsupported) metallic NPs. The inorganic shell stabilizes the NPs but also provides a direct access to the metal core, keeping the protected NPs as catalytically active as bare metal.

This strategy was used by Joo et al. to stabilize Pt NPs up to 750°C in air, for ethylene hydrogenation and CO oxidation [103]. In this case, the final nanocomposite consisted of 14 nm Pt cores covered with 17-nm thick mesoporous silica shells. In this architecture, the mesopores of the silica shell allow the reacting molecules to access the Pt cores and the product molecules to exit. Using an analogous approach, 25−40 nm Au NPs were surrounded by 50-nm-thick silica shells with accessible radial mesopores [108]. This nanocomposite exhibits good catalytic activity in the reduction of 4-nitrophenol even after calcination at 950°C.

An alternative architecture with similar performance is the formation of mesoporous hollow spheres with several metallic NPs inside each sphere. A highly active and stable nanoreactor with Pd NPs was built with this design by Vhen and coworkers [115]. 5 nm Pd NPs were distributed on the surface of 200 nm diameter carbon nanospheres and then they were coated with a 40 nm layer of CTAB templated SiO_2. A final calcination step eliminated the carbon spheres and the surfactant, generating the mesopores in the silica shell and leaving the Pd NPs inside the hollow shell. These nanoreactors show a high activity for Pd-catalyzed C−C coupling reactions. The authors proposed that the 2 nm mesopores the silica shell possess produce a preconcentration of reagents around the Pd nanoparticles, leading to the high reaction rates found for this material. Besides, Pd leaching is not observed for this nanocomposite, which assures a good catalytic performance.

A related synthetic strategy was developed by Lee et al. to encapsulate Pt NPs, with the aim of prevent their sintering during heterogeneous catalysis at high temperatures. Three-nanometer Pt NPs supported on 120-nm silica beads were thermally treated at different temperatures before and after being covered with a thin mesoporous silica layer [104]. The authors demonstrated that the unprotected catalysts already sinter at 600°C whereas there is no appreciable sintering upon calcination of the protected NPs up to 800°C (see Fig. 14.5).

Finally, it is remarkable that among the different encapsulated metallic NPs that can be used as catalysts, those based in magnetic NPs are of high interest due to the possibility of simple separation and reuse [116,117].

14.4 Mesoporous oxides as hosts for functional molecules

One of the interesting outcomes of the mesoporous materials' high specific surface is the fact that they can be used to host drugs or other small molecules. Such species can be subsequently released to generate a therapeutic effect. However, if the release is only based on concentration differences, poor specificity can be achieved. In fact, constructing drug delivery carriers with a high control over the location and timing of drug release under physiological conditions is a big challenge [8]. Therefore, different alternatives have been proposed to ensure site-specific release of the drugs, increasing the effectivity of the delivery [30].

One of the promising alternatives to improve mesoporous oxides-based drug delivery platforms is the use of nanoparticles as gatekeepers. Thanks to the size match between the pores diameter and the particles diameter, pores can be effectively blocked by the particles. Furthermore, the union between the particles and the oxide, in general SiO_2, can be controlled to allow site-specific detachment of the gatekeepers and subsequent cargo release, taking advantage of specific chemical reactions. Among all the NPs that can be used for such purpose, noble metal-based nanoparticles and particularly Au ones stand out [67]. These NPs present a high chemical stability, are not susceptible to acidic or enzymatic degradation and can be easily functionalized with a wide variety of functional and cleavable organic molecules. Moreover, they induce contrast enhancement in X-ray-based computed tomography [118] and present their already discussed optical properties, two features that can be exploited to obtain theranostic platforms [119,120]. The most usual size and structure of the obtained composites is exemplified in the TEM image shown in Fig. 14.6A.

Vivero-Escoto and coworkers presented an interesting example of the development of these composites. They prepared Au nanoparticles capped mesoporous silica nanoparticles for the photoinduced intracellular controlled release of an anticancer drug (paclitaxel) inside of human fibroblast

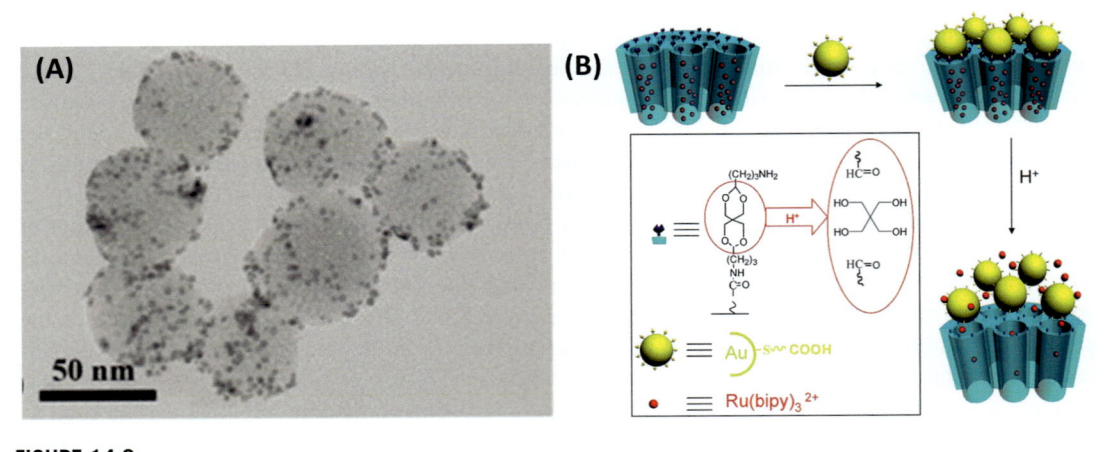

FIGURE 14.6

(A) TEM image of mesoporous SiO_2 nanoparticles covered with Au nanoparticles, which act as gatekeepers. (B) Schematic Illustration of pH-responsive Au nanoparticles gated mesoporous SiO_2 through acid-labile acetal linker.

Reprinted (A) from S. Xu, Y. Li, Z. Chen, C. Hou, T. Chen, Z. Xu, et al., Mesoporous silica nanoparticles combining Au particles as glutathione and pH dual-sensitive nanocarriers for doxorubicin, Mater. Sci. Eng. C. 59 (2016) 258–264. Copyright 2016, with permission from Elsevier. (B) Reproduced with permission from R. Liu, Y. Zhang, X. Zhao, A. Agarwal, L.J. Mueller, P. Feng, pH-responsive nanogated ensemble based on gold-capped mesoporous silica through an acid-labile acetal linker, J. Am. Chem. Soc. 132 (5) (2010) 1500–1501. Copyright 2010 American Chemical Society.

and liver cells [8]. For this purpose, the Au nanoparticles were functionalized with a photolabile linker and were subsequently attached to the negatively charged silica particles by electrostatic interactions. Afterwards, they demonstrated the feasibility of the release in both water (using a fluorescent dye as probe molecule) and in live human cells under low-power UV irradiation. In this case, the plasmonic properties of the Au nanoparticles are not involved in the release mechanism, a feature that would be interesting to exploit in future developments.

Another alternative to induce the cleavage of the nanoparticle−oxide union involves taking advantage of enzymatic species that specifically cut the union between the composite's components. For example, the use of peptide-cleavable linkers sensitive to matrix metalloproteinases has been described [123]. Such proteolytic enzymes are overexpressed in cancerous tissue and thus the Au-S union between the Au NPs and the peptide functionalized MP silica nanoparticles can be preferentially cut in such an environment, allowing the cargo release. Similarly, Au-S cleavage-based drug delivery due to the presence of glutathione has been tested [121], taking advantage of the high concentration of such molecules within the cytosol. An equivalent approach was used to release Ag NPs gatekeepers from a more complex system based on Au nanorods covered with mesoporous SiO_2 [124].

Alternatively, the Au NPs detachment and the consequent drug delivery could be triggered by pH changes [122,125]. The main goal in this case is to have controlled release in low pH tissues, such as tumors and inflammatory sites. Liu et al. followed this approach and developed a pH-responsive nanogated ensemble by capping mesoporous silica with Au nanoparticles through an acid-labile acetal linker [122]. At neutral pH, the linker remains intact and the pores are blocked; at acidic pH, the hydrolysis of the acetal group induces the Au nanoparticles cap removal and the consequent escape of the entrapped molecules (see schematic representation of the process in Fig. 14.6B).

14.5 **Mesoporous oxides as active hosts**

In this section, the focus will be on examples of composite materials whose characteristics and applications are defined by the mesoporous oxide's physical and chemical properties. In other words, systems in which the mesoporous oxide has an effect over the final characteristics or properties of the composites and is not merely a support or an inert host, such as the systems described in the previous sections.

14.5.1 **Mesoporous oxides as synthetic hosts**

The first area in which the characteristics of the oxide could determine the properties of the composites is in their synthesis itself. In this sense, the chemical identity of both hybrid and transition metal oxides has been used to reduce metal salts and allow metallic particles to be obtained within the pore array.

For example, Martínez and coworkers demonstrated that the photocatalytic activity of TiO_2 is key to obtain Ag NPs within its pores, when the sample is irradiated with UV light [126]. The authors used mesoporous thin films as base of the composites, that were immersed in an Ag (I) solution and subsequently exposed to UV light for different periods of time. A material with particles distributed in all the film thickness was obtained. These composites were successfully applied as SERS-based sensors, for the detection of organic molecules [24,127]. Using the same systems as a base, Coneo-Rodríguez et al. demonstrated the possibility of obtaining Ag-Au bimetallic particles taking advantage of galvanic replacement; the new composites presented a superior catalytic activity compared with the Ag based ones [128].

Moreover, since the particles formation is based on irradiation, patterns can be obtained using appropriate masks. In that way, very well-defined empty or full with nanoparticles regions in the micrometer range can be obtained, with interesting conductive properties and a high potentiality for devices design [129].

Similarly, the different reactivity of silica and titania-based systems towards the Ag(I) reduction when using formaldehyde as a reductant can be exploited. This approach has been used to preferentially incorporate Ag NPs inside TiO_2 mesoporous films in bilayered [130] or multilayered [131] systems that combine them with SiO_2 mesoporous layers. An example of such behavior, for the case of multilayered devices, can be seen in Fig. 14.7: Ag nanoparticles incorporation is only observed inside TiO_2 mesoporous thin films, while SiO_2 layers remain empty. Also, the amount of Ag nanoparticles incorporated within mixed TiO_2-SiO_2 oxides can be controlled by changing the proportion of each oxide in the mixture [132].

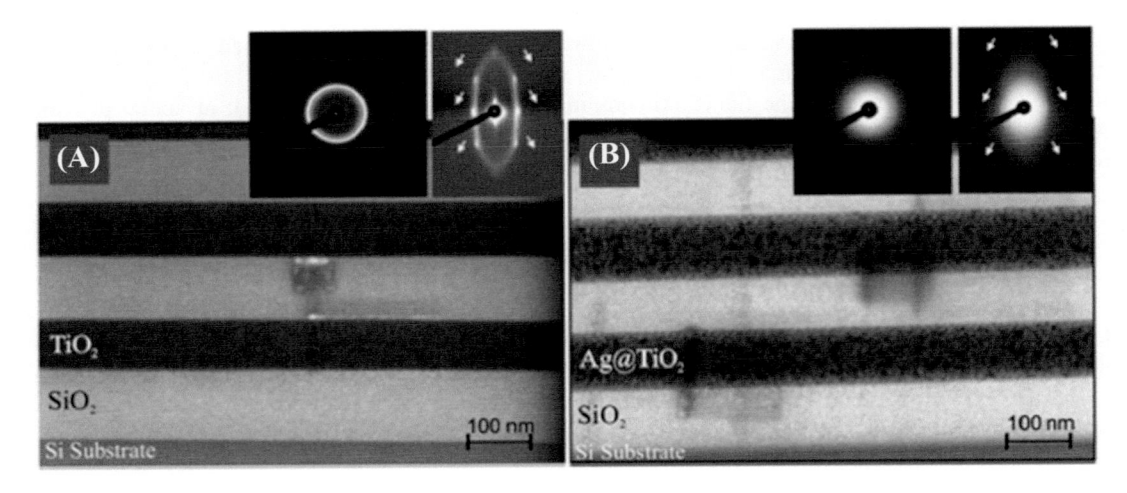

FIGURE 14.7

Scanning TEM images of a six layers system built by alternate deposition of mesoporous TiO_2 and SiO_2 thin films. (A) System prior to infiltration; inset: small angle X-Ray scattering (SAXS) patterns obtained in transmission mode at 90 degrees and 3 degrees incidence. Arrows indicate the mesoporous SiO_2 structure. (B) Equivalent system after 45 min infiltration with Ag (I) followed by 20 min reduction with formaldehyde; inset: SAXS patterns obtained at 90 degrees and 3 degrees incidence. Due to the random formation of Ag NPs inside mesoporous TiO_2, the signal of its structure is no longer present, while the SiO_2 signal remains unaltered.

Reproduced from R. Martínez Gazoni, M.G. Bellino, M. Cecilia Fuertes, G. Giménez, G.J.A.A. Soler-Illia, M.L.M. Ricci, Designed nanoparticle—mesoporous multilayer nanocomposites as tunable plasmonic—photonic architectures for electromagnetic field enhancement, J. Mater. Chem. C. 5 (14) (2017) 3445—3455 with permission from the Royal Society of Chemistry.

The specific reactivity of Ag with TiO_2 has another useful aspect: photochromism, that is, the reversible change of color upon exposure to light [77−79]. Under UV irradiation, electrons at TiO_2 valence band are excited into the conduction band and then reduce Ag (I) to metallic Ag nanoparticles. Since such particles absorb visible light through plasmonic resonances, total or partial oxidation of these nanoparticles can occur by illumination, since the transference of the photoexcited electrons to oxygen molecules is facilitated by the titania. Both processes are accompanied by color changes, and therefore this property can be exploited to obtain optically active devices. For example, it is possible to form Ag nanoparticles inside mesoporous TiO_2 thin films with specific patterns using laser irradiation, which can also be erased depending on the "writing" conditions [133−135]. Interestingly, the final color and the photochromic behavior of the composites depend on the reduction procedure [136] and the incident laser intensity and polarization [137].

It is also possible to take advantage of the difference in reactivity between Al and Si-based oxides to obtain Ag NPs [33]. In this sense, Zhang et al. demonstrated that the identity of the Ag species incorporated inside mesoporous powders depends on the Al:Si proportion, with a consequent effect over the catalytic properties of the composites [138]. In particular, they showed that a low proportion of Al leads to the formation of Ag nanoparticles while higher proportions of Al favor the formation of Ag_2O. Similarly, the obtaining of Cu or Cu oxide nanoparticles inside mesoporous SiO_2 thin films can be controlled by including carboxylic acid or amino groups on the oxide framework, respectively. For these composites, again, the chemical identity of the encapsulated particles leads to different catalytic activity towards a model reduction reaction [139].

14.5.2 Synergy between the mesoporous oxide and the nanoparticles

The second aspect in which the characteristics of the mesoporous oxide define the final material properties involves the use of the composites in applications in which both components are essential to the ultimate goal.

The main area in which these characteristics has been extensively exploited is the preparation of catalysts. As described before, the use of metallic nanoparticles for catalysis is well-known, and supporting them onto mesoporous oxides is an excellent strategy to have stable NPs and a great activity concentrated in a small space [69,75,140]. A step beyond is choosing a mesoporous support that can make the composite more effective that their separated components [70,141].

In this sense, Somorjai group has demonstrated that the identity of the mesoporous oxide that forms composites with Pt NPs defines the rate of the oxidation of CO [142]. In particular, they prepared 2.5-nm Pt nanoparticles supported onto mesoporous Co_3O_4, NiO, MnO_2, Fe_2O_3, and CeO_2 powders, obtained using a hard templating approach based on KIT-6 mesoporous silica (see Fig. 14.8A,B). Afterwards, they demonstrated that the turnover frequencies (TOFs) of the Pt/metal oxide systems are orders of magnitude greater than those of the pure oxides or the silica-supported nanoparticles (see Fig. 14.8C). In fact, the TOFs are significantly higher than those expected by simply adding the contributions from the oxide and the pure Pt nanoparticles. The authors attributed the enhancement to activity at the oxide − metal interface, in which the redox behavior of the oxide plays a key role. The same group studied several factors that influence the catalytic activity of supported Pt nanoparticles towards methanol oxidation reactions [143]. By performing a careful study comparing several composites in different experimental conditions, they demonstrated that

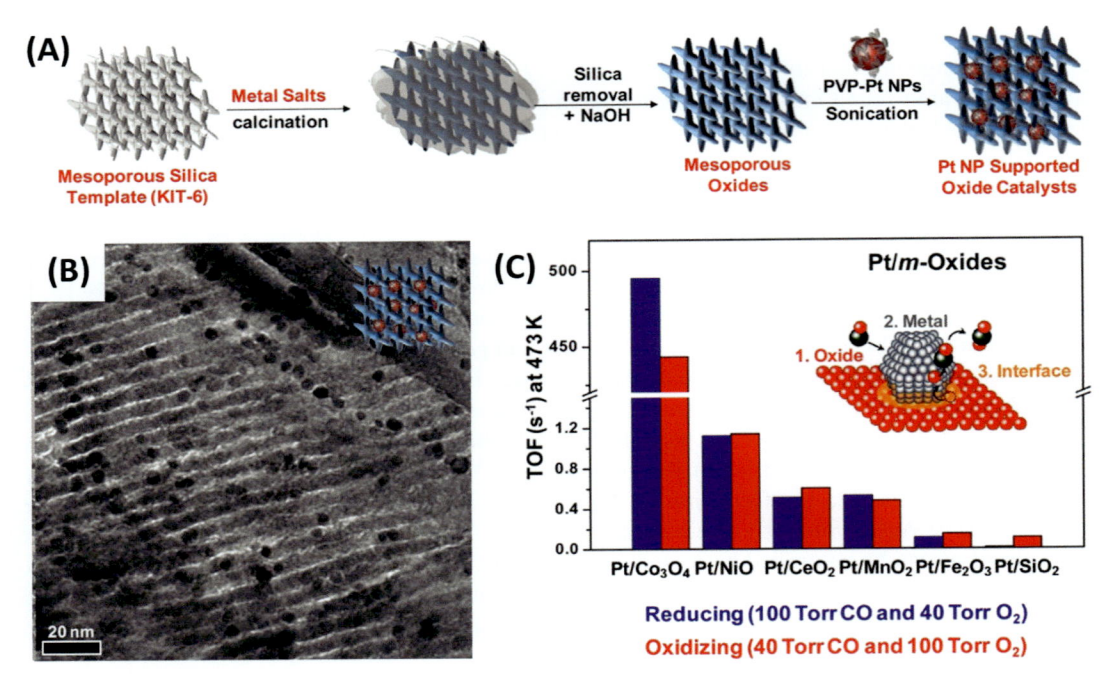

FIGURE 14.8

(A) Illustration of the hard template approach used for the preparation of Pt nanoparticles supported onto different mesoporous oxides. (B) TEM image of Pt/Co_3O_4 catalyst and (C) TOFs measured at 473K for different catalysts under either net reducing or net oxidizing conditions. The inset in (C) is an illustration showing the potential reaction sites during CO oxidation.

Reproduced with permission from K. An, S. Alayoglu, N. Musselwhite, S. Plamthottam, G. Melaet, A.E. Lindeman, et al., Enhanced CO oxidation rates at the interface of mesoporous oxides and Pt nanoparticles, J. Am. Chem. Soc. 135 (44) (2013) 16689–16696. Copyright 2013 American Chemical Society.

interactions of Pt NPs with transition metal oxide supports drastically alter the reaction rates in the gas phase, but less influence was observed for reactions that occur in the liquid phase.

This synergy between components to enhance the composite catalytic activity has also been exploited in the case of core–shell particles [75,76]. For example, Xu and coworkers [144] tested the catalytic activity of Au spheres coated with mesoporous SnO_2 shells and compared it with the activity of the same particles surrounded by SiO_2 shells. They observed that the former present lower induction times and higher catalytic activity, indicating that the intrinsic properties of SnO_2 have a clear influence over the composite activity.

Although the use of metallic oxides seems to be the most promising approach to increase catalytic activity, the effect of hybrid silica as support has also been tested with interesting results [145]. For example, Canhaci et al. studied the conversion of xylose to furfuryl alcohol using Pt nanoparticles supported onto SBA-15 silica and the same oxide functionalized with sulfonic acid groups. They demonstrated that the presence of SO_3H surface acid sites increases both the percentage of xylose conversion and the selectivity towards the desired product [146]. Similarly, Zhang and coworkers studied the bromate catalytic reduction in aqueous medium using Pt nanoparticles supported on SBA-15. The results indicated significantly higher catalytic reaction efficiencies when amino groups were incorporated within the silica framework. The authors attributed the difference to the improved electrostatic attractive interaction between NH_2-SBA-15 and the negative bromate ions [147].

Alternatively, the metallic NPs could help to increase the catalytic properties of the oxide support; thus the already active oxides became more efficient [69,77,78,148]. In particular, this approach has been used for enhancing the photocatalytic activity of semiconductor oxides such as TiO_2 [149−156] and WO_3 [157]. The presence of metallic species could improve either the absorption in the visible region of the spectrum or the electrons flow in the composite [158]. Consequently, the catalytic activity results higher under visible irradiation and/or the undesired electron-hole recombination is minimized. A clear example of the effect of plasmonic nanoparticles was presented by Sun et al. using Ag NPs doped mesoporous WO_3, prepared by hard templating method [157]. They studied the photodegradation of acetaldehyde, a common air pollutant, under visible-light irradiation, determined by measuring the production of CO_2. The authors demonstrated that the photocatalytic activity of Ag/WO_3 composite is about three times higher than that of the mesoporous WO_3 alone. These results were attributed to the improved charge separation of the photogenerated electrons and holes in the Ag−WO_3 heterojunction, allowing both electrons and holes participating in the overall photocatalytic reaction. On the other hand, Ismail and coworkers added nonplasmonic Pt nanoparticles onto mesoporous TiO_2 powders for the photooxidation of dichloroacetic acid [154]. They demonstrated that the addition of Pt increases the photonic efficiency by a factor of two, compared with pure TiO_2. The authors propose that a transfer of excitation energy or of photogenerated charge carriers through the network could explain the observed results.

Finally, another area that can benefit from this synergy strategy is the development of highly selective sensors [78]. For such use, it is interesting to have composites that concentrate or specifically bind the target analyte, to detect it properly. Since the chemical characteristics of oxides define their interactions with molecular or ionic species, changing the oxide may have an impact over the species that can be adsorbed onto it. Although there are not many examples of such approach, a recent work demonstrated that using Au nanoparticles embedded in mesoporous ZrO_2 films as electrochemical As(III) sensors is more efficient, in terms of analytical sensitivity, than using porous SiO_2 films as hosts [159]. The difference can be ascribed to the specific interaction between As(III) and ZrO_2, that leads to the analyte adsorption and accumulation onto the oxide surface. The effect of the analyte chemisorption has also been observed in the case of ammonia sensors based on WO_3 mesoporous thin films filled with Pt NPs [160]. However, in this case, the effect of the metal identity seems to be more crucial over the sensing properties than the support effect.

14.6 **Mesoporous oxides as filters**

Composite materials in which the mesoporous oxide acts as a molecular sieve will be presented in this section, and the main features that the porous structure can offer to improve the filtering function will be discussed.

Mesoporous materials can filter molecules by size exclusion, by charge, or using selective adsorption through the modification of the pores. One of the areas that have mostly explored this filtering action is the detection of analytes using SERS. In particular, the detection of small molecules in biological samples is challenging due to the signal contamination by ubiquitous macromolecules present in biological fluids: proteins, lipids, etc. The adsorption of these molecules can hinder the selectivity and sensitivity of SERS-based detectors, and therefore having the possibility of filter such molecules is of great interest for obtaining clean signals [161,162].

Four different architectures in which the mesoporous material acts as a molecular sieve are found in the literature: (1) supported NPs covered with a mesoporous thin film [81,95,161,163]; (2) NPs coated with a mesoporous shell [164,165]; (3) NPs embedded into the pores of mesoporous particles [109]; and (4) NPs inside hollow mesoporous shells [162]. Some examples in SERS detection considering the different architectures will be presented next, focused on the filtering mechanism.

The already discussed composite based on supported Au NPs covered with mesoporous films [81] were tested as SERS platforms for the detection of 4-nitrobenzenethiol (4-NBT) in a simulated biological sample containing bovine serum albumin (BSA). In this case, the mesoporous layer allowed only the diffusion of the small 4-NBT molecules toward the plasmonic Au NPs while preventing the contamination of the optical sensor with the macromolecules present in the biological solution. Besides, the stability of films based on different oxides was tested in the biological medium, and the titania films showed an improved chemical stability compared with the silica mesoporous films. In a similar approach, Bodelon et al. designed a mesoporous silica-coated Au nanorod supercrystals SERS platform [161]. The authors focused on the detection of pyocyanin released from bacterial biofilms and small clusters of cells. They found that the porous substrates restrict the contact of the plasmonic component with (bio)molecules that contaminate the SERS signal and thereby hinder the selectivity and sensitivity of the platform.

The filtering process can be tuned by the functionalization of the mesoporous films. The addition of specific functional groups is an interesting tool to control the surface chemistry of the pores and, as a consequence, to adjust the selectivity of the pore array for specific molecules beyond their sizes. Lopez Puente and coworkers investigated the selective diffusion properties of amino-functionalized ordered mesoporous films to discern the presence of two organic dyes [95]. The authors synthesized a hybrid amino-modified TiO_2 mesoporous film by cocondensation, to ensure chemical stability during the SERS measurements and the presence of well-distributed functional groups within the entire film. The films were deposited on top of glass supported 80-nm Au NPs. The selective SERS detection of methylene blue in mixtures with acid blue was achieved with this material, based on the different affinities of molecules toward the pore surface chemistry, containing both titanol (Ti − OH) sites and amino functional surface groups.

The selectivity in the detection can be also tuned using the preconcentration and condensation control that the mesopores provide to the nanocomposite. Stanzel et al. proposed a functionalization and localized sensing strategy using precisely placed plasmonic metal NPs inside mesoporous films with pore accessibility control [163]. Amine-functionalized mesostructured silica films were incubated in an Au NPs dispersion, the Au NPs functionalized films were then covered with a second mesostructured silica film. The bilayered mesoporous composite material was consolidated and calcined to obtain the final porous architecture. Afterwards, the Au NPs were used as nanolocal light sources to induce a photopolymerization of preconcentrated monomers between the two layers of mesoporous silica thin films. The authors proposed that the placement of, for example, responsive polymers into nanopores using this synthetic strategy could offer a simultaneous filtering and read-out function.

Fathima et al. designed core—shell nanocomposites for SERS sensing that employ both the exclusion for size and charge to detect analytes that are only placed in the vicinity of the plasmonic NPs [164]. Fifty-five-nanometer Ag NPs were coated with a 40-nm-thick shell of mesoporous silica. The mesoporous channels in the SiO_2 shell, with an average pore size of 2.4 nm, enable the passage of small molecules and sieve away large molecules from the vicinity of the plasmonic field. Besides, the large negative surface charge of this nanocomposite eliminates negatively charged molecules from SERS sensing due to strong electrostatic repulsion, providing an additional level of selectivity. The authors demonstrated the applicability of these nanostructures for the selective SERS sensing of neutral and positively charged small analytes in the presence of large molecules.

Gao and coworkers designed a related filtering architecture, coating Au nanorods with a mesoporous silica shell. With this platform, the authors studied the SERS sensing dependence on the pore size and the analyte diffusion to the Au cores [165]. The SERS signals of seven thiolated analytes were detected when the molecules transported through the shell and reached the Au core; it was observed that shells with pores of around 4 nm only allow the transport of analytes smaller than 1.5 nm. The authors proposed that by using different surfactants it would be possible to develop SERS-active core—shell composites with particular size cutoffs to evaluate mixtures of analytes.

Finally, Liu et al. presented a novel architecture with several NPs inside a hollow mesoporous silica shell [162]. This shell improves the metallic nanoparticles stability and diminishes the contamination from undesirable chemical species of large sizes, as was previously presented for other architectures (Fig. 14.9A). The production of this material includes several steps. First, 400-nm polystyrene (PS) particles were synthesized. Then, a layer of metal (Au, Ag or Pt, 10 − 20 nm thick) was deposited onto the PS particles. As a final step, the surface of the covered particles was encapsulated by a mesostructured silica shell 20 nm thick and the templates were removed with a thermal treatment. The metallic layers evolve from shells to nanodots after calcination (Fig. 14.9B). The irregular-shaped Au nanodots and aggregates offer hotspots for SERS sensing applications. The nanocomposites were tested as smart probes that can screen between small analytes [crystal violet (CV) and rhodamine 6 G (R6G)] and contaminating macromolecules (BSA) (Fig. 14.9C,D). The results the authors obtained demonstrated that the designed nanocomposites are able to prevent macromolecules from entering into the cavity; they envision that this isolation effect can be effective to avoid the protein corona effect in the study of biological samples.

14.7 **Conclusions and perspectives**

In this chapter, noble metal nanoparticles—mesoporous oxides composites shaped as powders, discrete nanoparticles, and thin films were presented. Due to the diversity of these composites, it was chosen to group them focusing on the role the porous oxide plays in their final properties. Examples in which the porous oxides act as simple supports for the metal nanoparticles, stabilize the nanoparticles towards harsh conditions, host molecules that became entrapped thanks to the

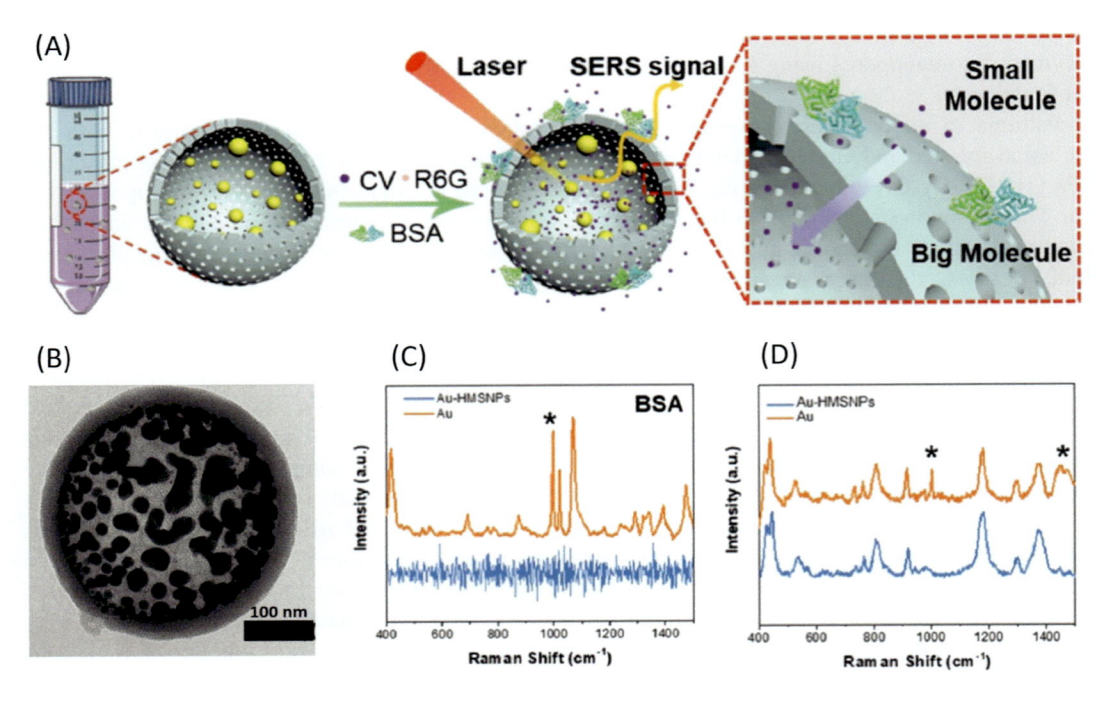

FIGURE 14.9

(A) Scheme presenting the surface-enhanced Raman spectroscopy (SERS) detection by Au NPs inside hollow mesoporous silica in the presence of two small analytes (CV and R6G) and a biomacromolecule (BSA). (B) TEM image of the nanocomposite. (C) BSA detection by SERS: the signals are observed when bare Au NPs are used as sensors and nondetected when the composites are used. (D) SERS spectra of the bare Au and the composite suspension in the presence of both CV and BSA. By using the composites, CV signals are detected, while peaks of BSA (marked with stars) disappear.

Adapted with permission from X. Liu, S. Yang, Y. Li, B. Wang, J. Guo, X. Ma, Mesoporous nanostructures encapsulated with metallic nanodots for smart SERS sensing, ACS Appl. Mater. Interfaces 13 (1) (2021) 186–195. Copyright 2021 American Chemical Society.

nanoparticles, act as active hosts that define the metal nanoparticles properties, or have a filtering action that controls what can reach the metal surface were presented and analyzed. In all cases, examples of applications in several fields were discussed, highlighting the contribution of each composite component to make the application successful.

Interestingly, despite being shaped in different forms and applied for different uses, most of the systems reviewed are based on spherical Au nanoparticles and SiO_2 porous oxides. The reasons behind this prevalence are varied and include the availability of simple and reliable methods to obtain both components, the high chemical stability of Au nanoparticles, and the low toxicity of both. However, a very important thing to consider when employing mesoporous silica is its low stability in neutral and alkaline aqueous solutions, showing a pronounced dissolution in relative short times (between hours and days depending on the conditions). Moreover other metals, other particles shapes, and other porous oxides present interesting features that have not been adequately exploited. Therefore, it is crucial to develop more robust and stable composite materials based on these less explored components. Such synthesis definitely represents a challenge, but could lead to an improvement of the already tested applications or new uses that have not been explored so far.

Another interesting point is the fact that plenty of work has been done in probing the potential of the composites as catalyst, drug delivery platforms, and sensors, but some basic features about them remain unclear or require further investigation. For example, the accessibility of the mesoporous oxides after being filled with metallic particles has not been studied in depth, particularly for their use as catalysts. This is a key point since the particles formation could lead to pore clogging and consequently reduce the activity of the whole composite. Also, the electronic interaction between the oxides and the metals has not been adequately covered, another point that can have an impact over the composites properties and uses.

It has been highlighted that the reviewed composites present the intrinsic properties of both components and also acquire new features derived from the synergy between them. In this sense, it is observed that the use of the metallic nanoparticles' plasmonic properties is still in its early development. Novel systems that take advantage of the light concentration, temperature, and reactive electron formation that occurs upon plasmonic particles excitation with light are expected to be developed in the near future.

Finally, it can be envisioned that the next step for some of the presented composites is to move forward to commercial applications. However, as mentioned before, the need for basic knowledge about the composites' physicochemical and structural properties is a key point to ensure their success.

Acknowledgments

We thank Paula Borovik, María Jose Arenas, Gonzalo Rumi, and M. Verónica Lombardo for their help with the elaboration of Fig. 14.1.

This work was supported by Agencia I + D + I (PICT 2019−01615, PICT 2017−1133).

References

[1]   M. Antonietti, G.A. Ozin, Promises and problems of mesoscale materials chemistry or why meso? Chem. Euro. J. 10 (1) (2004) 28−41.

[2]   J.S. Beck, J.C. Vartuli, W.J. Roth, M.E. Leonowicz, C.T. Kresge, K.D. Schmitt, et al., A new family of mesoporous molecular sieves prepared with liquid crystal templates, J. Am. Chem. Soc. 114 (27) (1992) 10834−10843.

[3]   C.T. Kresge, M.E. Leonowicz, W.J. Roth, J.C. Vartuli, J.S. Beck, Ordered mesoporous molecular sieves synthesized by a liquid-crystal template mechanism, Nature 359 (1992) 710−712.

[4]   C.J. Brinker, Y. Lu, A. Sellinger, H. Fan, Evaporation-induced self-assembly: nanostructures made easy, Adv. Mater. 11 (7) (1999) 579−585.

[5]   C.J. Brinker, G.W. Scherer, Sol−gel Science: The Physics and Chemistry of Sol−gel Processing, Elsevier, 1990.

[6] C. Sánchez, C. Boissière, D. Grosso, C. Laberty, L. Nicole, Design, synthesis, and properties of inorganic and hybrid thin films having periodically organized nanoporosity, Chem. Mater. 20 (3) (2008) 682–737.

[7] G.J.A.A. Soler-Illia, C. Sanchez, B. Lebeau, J. Patarin, Chemical strategies to design textured materials: from microporous and mesoporous oxides to nanonetworks and hierarchical structures, Chem. Rev. 102 (11) (2002) 4093–4138.

[8] J.L. Vivero-Escoto, I.I. Slowing, C.-W. Wu, V.S.Y. Lin, Photoinduced intracellular controlled release drug delivery in human cells by gold-capped mesoporous silica nanosphere, J. Am. Chem. Soc. 131 (10) (2009) 3462–3463.

[9] N. Baccile, D. Grosso, C. Sanchez, Aerosol generated mesoporous silica particles, J. Mater. Chem. 13 (12) (2003) 3011–3016.

[10] A. Zelcer, E.A. Franceschini, M.V. Lombardo, A.E. Lanterna, G.J.A.A. Soler-Illia, A general method to produce mesoporous oxide spherical particles through an aerosol method from aqueous solutions, J. Sol–gel Sci. Technol. 94 (1) (2020) 195–204.

[11] P. Yang, D. Zhao, D.I. Margolese, B.F. Chmelka, G.D. Stucky, Block copolymer templating syntheses of mesoporous metal oxides with large ordering lengths and semicrystalline framework, Chem. Mater. 11 (10) (1999) 2813–2826.

[12] D. Zhao, J. Feng, Q. Huo, N. Melosh, G.H. Fredrickson, B.F. Chmelka, et al., Triblock copolymer syntheses of mesoporous silica with periodic 50 to 300 angstrom pores, Science 279 (5350) (1998) 548–552.

[13] S. Singh, R. Kumar, H.D. Setiabudi, S. Nanda, D.-V.N. Vo, Advanced synthesis strategies of mesoporous SBA-15 supported catalysts for catalytic reforming applications: a state-of-the-art review, Appl. Catal., A 559 (2018) 57–74.

[14] P. Yang, D. Zhao, D.I. Margolese, B.F. Chmelka, G.D. Stucky, Generalized syntheses of large-pore mesoporous metal oxides with semicrystalline frameworks, Nature 396 (1998) 152–155.

[15] A.A. Ismail, D.W. Bahnemann, Mesoporous titania photocatalysts: preparation, characterization and reaction mechanisms, J. Mater. Chem. 21 (32) (2011) 11686–11707.

[16] C.T. Kresge, W.J. Roth, The discovery of mesoporous molecular sieves from the twenty year perspective, Chem. Soc. Rev. 42 (9) (2013) 3663–3670.

[17] R.K. Kankala, Y.-H. Han, J. Na, C.-H. Lee, Z. Sun, S.-B. Wang, et al., Nanoarchitectured structure and surface biofunctionality of mesoporous silica nanoparticles, Adv. Mater. 32 (23) (2020) 1907035.

[18] B.G. Trewyn, I.I. Slowing, S. Giri, H.-T. Chen, V.S.Y. Lin, Synthesis and functionalization of a mesoporous silica nanoparticle based on the sol–gel process and applications in controlled release, Acc. Chem. Res. 40 (9) (2007) 846–853.

[19] P. Innocenzi, L. Malfatti, Mesoporous thin films: properties and applications, Chem. Soc. Rev. 42 (9) (2013) 4198–4216.

[20] S. Zhou, Q. Zhong, Y. Wang, P. Hu, W. Zhong, C.-B. Huang, et al., Chemically engineered mesoporous silica nanoparticles-based intelligent delivery systems for theranostic applications in multiple cancerous/non-cancerous diseases, Coord. Chem. Rev. 452 (2022) 214309.

[21] K.J. Edler, Formation of ordered mesoporous thin films through templating, in: L. Klein, M. Aparicio, A. Jitianu (Eds.), Handbook of Sol–gel Science and Technology: Processing, Characterization and Applications, Springer International Publishing, Cham, 2018, pp. 917–983.

[22] J.J. Mikolei, L. Neuenfeld, S. Paech, M. Langhans, M. Biesalski, T. Meckel, et al., Mechanistic understanding and three-dimensional tuning of fluid imbibition in silica-coated cotton linter paper sheets, Adv. Mater. Interfaces 9 (19) (2022) 2200064.

[23] C. Sanchez, P. Belleville, M. Popall, L. Nicole, Applications of advanced hybrid organic-inorganic nanomaterials: from laboratory to market, Chem. Soc. Rev. 40 (2) (2011) 696–753.

[24] A. Wolosiuk, N.G. Tognalli, E.D. Martínez, M. Granada, M.C. Fuertes, H. Troiani, et al., Silver nanoparticle-mesoporous oxide nanocomposite thin films: a platform for spatially homogeneous SERS-active substrates with enhanced stability, ACS Appl. Mater. Interfaces 6 (7) (2014) 5263–5272.

[25] B. Auguié, M.C. Fuertes, P.C. Angelomé, N.L. Abdala, G.J.A.A. Soler Illia, A. Fainstein, Tamm plasmon resonance in mesoporous multilayers: toward a sensing application, ACS Photonics 1 (9) (2014) 775–780.

[26] C.D.S. Brites, M.C. Fuertes, P.C. Angelomé, E.D. Martínez, P.P. Lima, G.J.A.A. Soler-Illia, et al., Tethering luminescent thermometry and plasmonics: light manipulation to assess real-time thermal flow in nanoarchitectures, Nano Lett. 17 (8) (2017) 4746–4752.

[27] S.S. Soni, M.J. Henderson, J.-F. Bardeau, A. Gibaud, Visible-light photocatalysis in titania-based mesoporous thin films, Adv. Mater. 20 (8) (2008) 1493–1498.

[28] P.C. Angelomé, L. Andrini, M.E. Calvo, F.G. Requejo, S.A. Bilmes, G.J.A.A. Soler-Illia, Mesoporous anatase TiO_2 films: use of Ti K XANES for the quantification of the nanocrystalline character and substrate effects in the photocatalysis behavior, J. Phys. Chem. C. 111 (29) (2007) 10886–10893.

[29] I. Sierra, D. Perez-Quintanilla, Heavy metal complexation on hybrid mesoporous silicas: an approach to analytical applications, Chem. Soc. Rev. 42 (9) (2013) 3792–3807.

[30] M. Vallet-Regí, F. Schüth, D. Lozano, M. Colilla, M. Manzano, Engineering mesoporous silica nanoparticles for drug delivery: where are we after two decades? Chem. Soc. Rev. 51 (13) (2022) 5365–5451.

[31] K. Lan, D. Zhao, Functional ordered mesoporous materials: present and future, Nano Lett. 22 (8) (2022) 3177–3179.

[32] J. Gangadhar, B. Tirumuruhan, R. Sujith, Applications and future trends in mesoporous materials, in: A. Uthaman, S. Thomas, T. Li, H. Maria (Eds.), Advanced Functional Porous Materials: From Macro to Nano Scale Length, Springer International Publishing, Cham, 2022, pp. 235–258.

[33] P. Verma, Y. Kuwahara, K. Mori, R. Raja, H. Yamashita, Functionalized mesoporous SBA-15 silica: recent trends and catalytic applications, Nanoscale 12 (21) (2020) 11333–11363.

[34] M. Vert, Y. Doi, K.-H. Hellwich, M. Hess, P. Hodge, P. Kubisa, et al., Terminology for biorelated polymers and applications (IUPAC recommendations 2012), Pure Appl. Chem. 84 (2) (2012) 377–410.

[35] M. Haruta, When gold is not noble: catalysis by nanoparticles, Chem. Rec. 3 (2) (2003) 75–87.

[36] K. Fuku, R. Hayashi, S. Takakura, T. Kamegawa, K. Mori, H. Yamashita, The synthesis of size- and color-controlled silver nanoparticles by using microwave heating and their enhanced catalytic activity by localized surface plasmon resonance, Angew. Chem. Int. Ed. 52 (29) (2013) 7446–7450.

[37] S.A. Maier, Plasmonics: Fundamentals and Applications, Springer, US, 2007.

[38] B. Pietrobon, M. McEachran, V. Kitaev, Synthesis of size-controlled faceted pentagonal silver nanorods with tunable plasmonic properties and self-assembly of these nanorods, ACS Nano 3 (1) (2009) 21–26.

[39] A.-Q. Zhang, D.-J. Qian, M. Chen, Simulated optical properties of noble metallic nanopolyhedra with different shapes and structures, Eur. Phys. J. D. 67 (11) (2013) 231.

[40] C. Noguez, Surface plasmons on metal nanoparticles: the influence of shape and physical environment, J. Phys. Chem. C. 111 (10) (2007) 3806–3819.

[41] Y. Xia, Y. Xiong, B. Lim, S.E. Skrabalak, Shape-controlled synthesis of metal nanocrystals: simple chemistry meets complex physics? Angew. Chem. Int. Ed. 48 (1) (2009) 60–103.

[42] M. Grzelczak, L.M. Liz-Marzán, The relevance of light in the formation of colloidal metal nanoparticles, Chem. Soc. Rev. 43 (7) (2014) 2089–2097.

[43] J. Polte, Fundamental growth principles of colloidal metal nanoparticles – a new perspective, CrystEngComm 17 (36) (2015) 6809–6830.

[44] M. Wuithschick, A. Birnbaum, S. Witte, M. Sztucki, U. Vainio, N. Pinna, et al., Turkevich in new robes: key questions answered for the most common gold nanoparticle synthesis, ACS Nano 9 (7) (2015) 7052–7071.

[45] J. Park, J. Joo, S.G. Kwon, Y. Jang, T. Hyeon, Synthesis of monodisperse spherical nanocrystals, Angew. Chem. Int. Ed. 46 (25) (2007) 4630–4660.

[46] N.R. Jana, L. Gearheart, C.J. Murphy, Seed-mediated growth approach for shape-controlled synthesis of spheroidal and rod-like gold nanoparticles using a surfactant template, Adv. Mater. 13 (18) (2001) 1389–1393.

[47] X. Huang, S. Neretina, M.A. El-Sayed, Gold nanorods: from synthesis and properties to biological and biomedical applications, Adv. Mater. 21 (48) (2009) 4880–4910.

[48] J.-E. Park, Y. Lee, J.-M. Nam, Precisely shaped, uniformly formed gold nanocubes with ultrahigh reproducibility in single-particle scattering and surface-enhanced Raman scattering, Nano Lett. 18 (10) (2018) 6475–6482.

[49] G. Baffou, F. Cichos, R. Quidant, Applications and challenges of thermoplasmonics, Nat. Mater. 19 (9) (2020) 946–958.

[50] G. Baffou, I. Bordacchini, A. Baldi, R. Quidant, Simple experimental procedures to distinguish photothermal from hot-carrier processes in plasmonics, Light. Sci. Appl. 9 (2020) 108.

[51] G. Baffou, R. Quidant, Nanoplasmonics for chemistry, Chem. Soc. Rev. 43 (11) (2014) 3898–3907.

[52] M. Pelton, J. Aizpurua, G. Bryant, Metal-nanoparticle plasmonics, Laser Photonics Rev. 2 (3) (2008) 136–159.

[53] C. Rosman, J. Prasad, A. Neiser, A. Henkel, J. Edgar, C. Sönnichsen, Multiplexed plasmon sensor for rapid label-free analyte detection, Nano Lett. 13 (7) (2013) 3243–3247.

[54] R. Lachaine, É. Boulais, M. Meunier, From thermo- to plasma-mediated ultrafast laser-induced plasmonic nanobubbles, ACS Photonics 1 (4) (2014) 331–336.

[55] G. Baffou, P. Berto, E. Bermudez Urena, R. Quidant, S. Monneret, J. Polleux, et al., Photoinduced heating of nanoparticle arrays, ACS Nano 7 (8) (2013) 6478–6488.

[56] S. Linic, P. Christopher, D.B. Ingram, Plasmonic-metal nanostructures for efficient conversion of solar to chemical energy, Nat. Mater. 10 (12) (2011) 911–921.

[57] J.-F. Masson, Surface plasmon resonance clinical biosensors for medical diagnostics, ACS Sens. 2 (1) (2017) 16–30.

[58] Z. Li, S. Butun, K. Aydin, Touching gold nanoparticle chain based plasmonic antenna arrays and optical metamaterials, ACS Photonics 1 (3) (2014) 228–234.

[59] B. Kenens, M. Rybachuk, J. Hofkens, H. Uji-i, Silver nanowires terminated by metallic nanoparticles as effective plasmonic antennas, J. Phys. Chem. C. 117 (6) (2013) 2547–2553.

[60] J. Liu, C. Detrembleur, M.-C. De Pauw-Gillet, S. Mornet, C. Jérôme, E. Duguet, Gold nanorods coated with mesoporous silica shell as drug delivery system for remote near infrared light-activated release and potential phototherapy, Small 11 (19) (2015) 2323–2332.

[61] S.J. Chadwick, D. Salah, P.M. Livesey, M. Brust, M. Volk, Singlet oxygen generation by laser irradiation of gold nanoparticles, J. Phys. Chem. C. 120 (19) (2016) 10647–10657.

[62] E.Y. Lukianova-Hleb, D.S. Wagner, M.K. Brenner, D.O. Lapotko, Cell-specific transmembrane injection of molecular cargo with gold nanoparticle-generated transient plasmonic nanobubbles, Biomaterials 33 (21) (2012) 5441–5450.

[63] J.R. Adleman, D.A. Boyd, D.G. Goodwin, D. Psaltis, Heterogenous catalysis mediated by plasmon heating, Nano Lett. 9 (12) (2009) 4417–4423.

[64] H. Ren, J.-L. Yang, W.-M. Yang, H.-L. Zhong, J.-S. Lin, P.M. Radjenovic, et al., Core–shell–satellite plasmonic photocatalyst for broad-spectrum photocatalytic water splitting, ACS Mater. Lett. 3 (1) (2021).

[65] A. Gellé, T. Jin, L. de la Garza, G.D. Price, L.V. Besteiro, A. Moores, Applications of plasmon-enhanced nanocatalysis to organic transformations, Chem. Rev. 120 (2) (2020).

[66] S. Vásquez-Céspedes, R.C. Betori, M.A. Cismesia, J.K. Kirsch, Q. Yang, Heterogeneous catalysis for cross-coupling reactions: an underutilized powerful and sustainable tool in the fine chemical industry? Org. Process. Res. & Dev. (2021).

[67] R.K. Kankala, H. Zhang, C.-G. Liu, K.R. Kanubaddi, C.-H. Lee, S.-B. Wang, et al., Metal species—encapsulated mesoporous silica nanoparticles: current advancements and latest breakthroughs, Adv. Funct. Mater. 29 (43) (2019) 1902652.

[68] R.J. White, R. Luque, V.L. Budarin, J.H. Clark, D.J. Macquarrie, Supported metal nanoparticles on porous materials. Methods and applications, Chem. Soc. Rev. 38 (2009) 481−494.

[69] C. Gao, F. Lyu, Y. Yin, Encapsulated metal nanoparticles for catalysis, Chem. Rev. 121 (2) (2021) 834−881.

[70] B.P. Bastakoti, D. Kuila, C. Salomon, M. Konarova, M. Eguchi, J. Na, et al., Metal-incorporated mesoporous oxides: synthesis and applications, J. Hazard. Mater. 401 (2021) 123348.

[71] M. Lukosi, H. Zhu, S. Dai, Recent advances in gold-metal oxide core-shell nanoparticles: synthesis, characterization, and their application for heterogeneous catalysis, Front. Chem. Sci. Eng. 10 (1) (2016) 39−56.

[72] P. Innocenzi, L. Malfatti, Mesoporous materials as platforms for surface-enhanced Raman scattering, TrAC, Trends Anal. Chem. 114 (2019) 233−241.

[73] M. Davidson, Y. Ji, G.J. Leong, N.C. Kovach, B.G. Trewyn, R.M. Richards, Hybrid mesoporous silica/noble-metal nanoparticle materials—synthesis and catalytic applications, ACS Appl. Nano Mat. 1 (9) (2018) 4386−4400.

[74] C. Hanske, M.N. Sanz-Ortiz, L.M. Liz-Marzán, Silica-coated plasmonic metal nanoparticles in action, Adv. Mater. 30 (27) (2018) 1707003.

[75] M.B. Gawande, A. Goswami, T. Asefa, H. Guo, A.V. Biradar, D.-L. Peng, et al., Core−shell nanoparticles: synthesis and applications in catalysis and electrocatalysis, Chem. Soc. Rev. 44 (21) (2015) 7540−7590.

[76] M.J. Penelas, S. Poklepovich-Caride, P.C. Angelomé, Metallic nanoparticles with mesoporous shells: synthesis, characterization and applications, in: E.A. Franceschini (Ed.), Nanostructured Multifunctional Materials. Synthesis, Characterization, Applications and Computational Simulation, 1st ed., CRC Press, USA, 2021, pp. 137−160.

[77] P.C. Angelomé, M.C. Fuertes, Metal nanoparticles-mesoporous oxide nanocomposite thin films, in: L. Klein, M. Aparicio, A. Jitianu (Eds.), Handbook of Sol−gel Science and Technology, Springer International Publishing, Cham, 2018, pp. 2507−2533.

[78] P. Innocenzi, L. Malfatti, Nanoparticles in mesoporous films, a happy marriage for materials science, J. Nanopart. Res. 20 (6) (2018) 167.

[79] P.C. Angelomé, L.M. Liz-Marzán, Synthesis and applications of mesoporous nanocomposites containing metal nanoparticles, J. Sol−gel Sci. Technol. 70 (2) (2014) 180−190.

[80] V.G. Deshmane, S.L. Owen, R.Y. Abrokwah, D. Kuila, Mesoporous nanocrystalline TiO$_2$ supported metal (Cu, Co, Ni, Pd, Zn, and Sn) catalysts: effect of metal-support interactions on steam reforming of methanol, J. Mol. Catal. A: Chem. 408 (2015) 202−213.

[81] V. López-Puente, S. Abalde-Cela, P.C. Angelomé, R.A. Alvarez-Puebla, L.M. Liz-Marzán, Plasmonic mesoporous composites as molecular sieves for SERS detection, J. Phys. Chem. Lett. 4 (16) (2013) 2715−2720.

[82] V. Idakiev, T. Tabakova, A. Naydenov, Z.Y. Yuan, B.L. Su, Gold catalysts supported on mesoporous zirconia for low-temperature water−gas shift reaction, Appl. Catal. B: Environ. 63 (3) (2006) 178−186.

[83] S. Huang, K. Hara, Y. Okubo, M. Yanagi, H. Nambu, A. Fukuoka, Preferential oxidation of carbon monoxide in excess hydrogen over platinum catalysts supported on different-pore-sized mesoporous silica, Appl. Catal., A 365 (2) (2009) 268−273.

[84] H.L. Tidahy, M. Hosseni, S. Siffert, R. Cousin, J.F. Lamonier, A. Aboukaïs, et al., Nanostructured macro-mesoporous zirconia impregnated by noble metal for catalytic total oxidation of toluene, Catal. Today 137 (2) (2008) 335−339.

[85] S. Schünemann, G. Dodekatos, H. Tüysüz, Mesoporous silica supported Au and AuCu nanoparticles for surface plasmon driven glycerol oxidation, Chem. Mater. 27 (22) (2015) 7743−7750.

[86] Z. Kónya, V.F. Puntes, I. Kiricsi, J. Zhu, J.W. Ager, M.K. Ko, et al., Synthetic insertion of gold nanoparticles into mesoporous silica, Chem. Mater. 15 (6) (2003) 1242−1248.

[87] G. Gupta, C.A. Stowell, M.N. Patel, X. Gao, M.J. Yacaman, B.A. Korgel, et al., Infusion of presynthesized iridium nanocrystals into mesoporous silica for high catalyst activity, Chem. Mater. 18 (26) (2006) 6239−6249.

[88] Y.-J. Han, J.M. Kim, G.D. Stucky, Preparation of noble metal nanowires using hexagonal mesoporous silica SBA-15, Chem. Mater. 12 (8) (2000) 2068−2069.

[89] K.J. Lin, L.J. Chen, M.R. Prasad, C.Y. Cheng, Core−shell synthesis of a novel, spherical, mesoporous silica/platinum nanocomposite: Pt/PVP@MCM-41, Adv. Mater. 16 (20) (2004) 1845−1849.

[90] J.-J. Li, X.-Y. Xu, Z. Jiang, Z.-P. Hao, C. Hu, Nanoporous silica-supported nanometric palladium: synthesis, characterization, and catalytic deep oxidation of benzene, Environ. Sci. Technol. 39 (5) (2005) 1319−1323.

[91] H. Song, R.M. Rioux, J.D. Hoefelmeyer, R. Komor, K. Niesz, M. Grass, et al., Hydrothermal growth of mesoporous SBA-15 silica in the presence of PVP-stabilized Pt nanoparticles: synthesis, characterization, and catalytic properties, J. Am. Chem. Soc. 128 (9) (2006) 3027−3037.

[92] K. Mori, P. Verma, R. Hayashi, K. Fuku, H. Yamashita, Color-controlled Ag nanoparticles and nanorods within confined mesopores: microwave-assisted rapid synthesis and application in plasmonic catalysis under visible-light irradiation, Chem. Euro. J. 21 (33) (2015) 11885−11893.

[93] P.Y. Steinberg, M.M. Zalduendo, G. Giménez, G.J.A.A. Soler-Illia, P.C. Angelomé, TiO$_2$ mesoporous thin films architecture as a tool to control Au nanoparticles growth and sensing capabilities, Phys. Chem. Chem. Phys. 21 (2019) 10347−10356.

[94] M. Rafti, A. Brunsen, M.C. Fuertes, O. Azzaroni, G.J.A.A. Soler-Illia, Heterogeneous catalytic activity of platinum nanoparticles hosted in mesoporous silica thin films modified with polyelectrolyte brushes, ACS Appl. Mater. Interfaces 5 (18) (2013) 8833−8840.

[95] V. López-Puente, P.C. Angelomé, G.J.A.A. Soler-Illia, L.M. Liz-Marzán, Selective SERS sensing modulated by functionalized mesoporous films, ACS Appl. Mater. Interfaces 7 (46) (2015) 25633−25640.

[96] M.M. Zalduendo, J. Langer, J.J. Giner-Casares, E.B. Halac, G.J.A.A. Soler-Illia, L.M. Liz-Marzán, et al., Au nanoparticles−mesoporous TiO$_2$ thin films composites as SERS sensors: a systematic performance analysis, J. Phys. Chem. C. 122 (24) (2018) 13095−13105.

[97] P.C. Angelomé, I. Pastoriza-Santos, J. Pérez Juste, B. Rodríguez-González, A. Zelcer, G.J.A.A. Soler-Illia, et al., Growth and branching of gold nanoparticles through mesoporous silica thin films, Nanoscale 4 (2012) 931−939.

[98] M.M. Zalduendo, P.Y. Steinberg, T. Prudente, E.J. Di Liscia, J. Morrone, P.C. Angelomé, et al., Gold semicontinuous thin-film-coated mesoporous TiO$_2$ for SERS substrates, SN Appl. Sci. 2 (11) (2020) 1809.

[99] M.N. Sanz-Ortiz, K. Sentosun, S. Bals, L.M. Liz-Marzán, Templated growth of surface enhanced Raman scattering-active branched gold nanoparticles within radial mesoporous silica shells, ACS Nano 9 (10) (2015) 10489−10497.

[100] J. Zhu, X. Xie, S.A.C. Carabineiro, P.B. Tavares, J.L. Figueiredo, R. Schomäcker, et al., Facile one-pot synthesis of Pt nanoparticles /SBA-15: an active and stable material for catalytic applications, Energy Environ. Sci. 4 (6) (2011) 2020−2024.

[101] K. An, Q. Zhang, S. Alayoglu, N. Musselwhite, J.-Y. Shin, G.A. Somorjai, High-temperature catalytic reforming of n-hexane over supported and core−shell Pt nanoparticle catalysts: role of oxide−metal interface and thermal stability, Nano Lett. 14 (8) (2014) 4907−4912.

[102] S.H. Gage, J. Engelhardt, M.J. Menart, C. Ngo, G.J. Leong, Y. Ji, et al., Palladium intercalated into the walls of mesoporous silica as robust and regenerable catalysts for hydrodeoxygenation of phenolic compounds, ACS Omega 3 (7) (2018) 7681−7691.

[103] S.H. Joo, J.Y. Park, C.-K. Tsung, Y. Yamada, P. Yang, G.A. Somorjai, Thermally stable Pt/mesoporous silica core−shell nanocatalysts for high-temperature reactions, Nat. Mater. 8 (2) (2009) 126−131.

[104] I. Lee, Q. Zhang, J. Ge, Y. Yin, F. Zaera, Encapsulation of supported Pt nanoparticles with mesoporous silica for increased catalyst stability, Nano Res. 4 (1) (2011) 115−123.

[105] M.T. Bore, H.N. Pham, E.E. Switzer, T.L. Ward, A. Fukuoka, A.K. Datye, The role of pore size and structure on the thermal stability of gold nanoparticles within mesoporous silica, J. Phys. Chem. B 109 (7) (2005) 2873−2880.

[106] T. Zhang, H. Zhao, S. He, K. Liu, H. Liu, Y. Yin, et al., Unconventional route to encapsulated ultra-small gold nanoparticles for high-temperature catalysis, ACS Nano 8 (7) (2014) 7297−7304.

[107] Z. Gao, Y. Zhang, D. Li, C.J. Werth, Y. Zhang, X. Zhou, Highly active Pd−In/mesoporous alumina catalyst for nitrate reduction, J. Hazard. Mater. 286 (2015) 425−431.

[108] J. Chen, R. Zhang, L. Han, B. Tu, D. Zhao, One-pot synthesis of thermally stable gold@mesoporous silica core-shell nanospheres with catalytic activity, Nano Res. 6 (12) (2013) 871−879.

[109] M. Chen, W. Luo, Z. Zhang, R. Wang, Y. Zhu, H. Yang, et al., Synthesis of multi-Au-nanoparticle-embedded mesoporous silica microspheres as self-filtering and reusable substrates for SERS detection, ACS Appl. Mater. Interfaces 9 (48) (2017) 42156−42166.

[110] A. Cao, G. Veser, Exceptional high-temperature stability through distillation-like self-stabilization in bimetallic nanoparticles, Nat. Mater. 9 (1) (2010) 75−81.

[111] X. Xu, S.K. Megarajan, Y. Zhang, H. Jiang, Ordered mesoporous alumina and their composites based on evaporation induced self-assembly for adsorption and catalysis, Chem. Mater. 32 (1) (2020) 3−26.

[112] C. Wu, Z.-Y. Lim, C. Zhou, W. Guo Wang, S. Zhou, H. Yin, et al., A soft-templated method to synthesize sintering-resistant Au−mesoporous-silica core−shell nanocatalysts with sub-5 nm single-cores, Chem. Comm. 49 (31) (2013) 3215−3217.

[113] C.-H. Lin, X. Liu, S.-H. Wu, K.-H. Liu, C.-Y. Mou, Corking and uncorking a catalytic yolk-shell nanoreactor: stable gold catalyst in hollow silica nanosphere, J. Phys. Chem. Lett. 2 (23) (2011) 2984−2988.

[114] M.M. Zalduendo, V. Oestreicher, J. Langer, L.M. Liz-Marzán, P.C. Angelomé, Monitoring chemical reactions with SERS-active Ag-loaded mesoporous TiO$_2$ films, Anal. Chem. 92 (20) (2020) 13656−13660.

[115] Z. Chen, Z.-M. Cui, F. Niu, L. Jiang, W.-G. Song, Pd nanoparticles in silica hollow spheres with mesoporous walls: a nanoreactor with extremely high activity, Chem. Comm. 46 (35) (2010) 6524−6526.

[116] J.C. Park, J.U. Bang, J. Lee, C.H. Ko, H. Song, Ni@SiO$_2$ yolk-shell nanoreactor catalysts: high temperature stability and recyclability, J. Mater. Chem. 20 (7) (2010) 1239−1246.

[117] J. Ge, Q. Zhang, T. Zhang, Y. Yin, Core−satellite nanocomposite catalysts protected by a porous silica shell: controllable reactivity, high stability, and magnetic recyclability, Angew. Chem. Int. Ed. 47 (46) (2008) 8924−8928.

[118] M. Babaei, K. Abnous, S.M. Taghdisi, S.A. Farzad, M.T. Peivandi, M. Ramezani, et al., Synthesis of theranostic epithelial cell adhesion molecule targeted mesoporous silica nanoparticle with gold gate-keeper for hepatocellular carcinoma, Nanomedicine 12 (11) (2017) 1261−1279.

[119] Q. Gao, J. Zhang, J. Gao, Z. Zhang, H. Zhu, D. Wang, Gold nanoparticles in cancer theranostics, Front. Bioeng. Biotechnol. 9 (2021) 647905.

[120] P. Kaur, M.L. Aliru, A.S. Chadha, A. Asea, S. Krishnan, Hyperthermia using nanoparticles − promises and pitfalls, Int. J. Hyperth. 32 (1) (2016) 76−88.

[121] S. Xu, Y. Li, Z. Chen, C. Hou, T. Chen, Z. Xu, et al., Mesoporous silica nanoparticles combining Au particles as glutathione and pH dual-sensitive nanocarriers for doxorubicin, Mater. Sci. Eng. C. 59 (2016) 258−264.

[122] R. Liu, Y. Zhang, X. Zhao, A. Agarwal, L.J. Mueller, P. Feng, pH-responsive nanogated ensemble based on gold-capped mesoporous silica through an acid-labile acetal linker, J. Am. Chem. Soc. 132 (5) (2010) 1500−1501.

[123] P. Eskandari, B. Bigdeli, M. Porgham Daryasari, H. Baharifar, B. Bazri, M. Shourian, et al., Gold-capped mesoporous silica nanoparticles as an excellent enzyme-responsive nanocarrier for controlled doxorubicin delivery, J. Drug. Target. 27 (10) (2019) 1084−1093.

[124] Z. Zhang, C. Liu, J. Bai, C. Wu, Y. Xiao, Y. Li, et al., Silver nanoparticle gated, mesoporous silica coated gold nanorods (AuNR@MS@AgNPs): low premature release and multifunctional cancer theranostic platform, ACS Appl. Mater. Interfaces 7 (11) (2015) 6211−6219.

[125] E. Aznar, M.D. Marcos, R. Martínez-Máñez, F. Sancenón, J. Soto, P. Amorós, et al., pH- and photo-switched release of guest molecules from mesoporous silica supports, J. Am. Chem. Soc. 131 (19) (2009) 6833−6843.

[126] E.D. Martínez, M.G. Bellino, G.J.A.A. Soler-Illia, Patterned production of silver − mesoporous Titania nanocomposite thin films using lithography-assisted metal reduction, ACS Appl. Mater. Interfaces 1 (4) (2009) 746−749.

[127] S. Mura, G. Greppi, P. Innocenzi, M. Piccinini, C. Figus, M.L. Marongiu, et al., Nanostructured thin films as surface-enhanced Raman scattering substrates, J. Raman Spectrosc. 44 (1) (2013) 35−40.

[128] R. Coneo Rodríguez, H. Troiani, S.E. Moya, M.M. Bruno, P.C. Angelomé, Bimetallic Ag-Au nanoparticles inside mesoporous Titania thin films: synthesis by photoreduction and galvanic replacement, and catalytic activity, Eur. J. Inorg. Chem. 2020 (6) (2020) 568−574.

[129] E.D. Martinez, L. Granja, M.G. Bellino, G.J.A.A. Soler-Illia, Electrical conductivity in patterned silver-mesoporous titania nanocomposite thin films: towards robust 3D nano-electrodes, Phys. Chem. Chem. Phys. 12 (43) (2010) 14445−14448.

[130] M.C. Fuertes, M. Marchena, M.C. Marchi, A. Wolosiuk, G.J.A.A. Soler-Illia, Controlled deposition of silver nanoparticles in mesoporous single- or multilayer thin films: from tuned pore filling to selective spatial location of nanometric objects, Small 5 (2) (2009) 272−280.

[131] R. Martínez Gazoni, M.G. Bellino, M. Cecilia Fuertes, G. Giménez, G.J.A.A. Soler-Illia, M.L.M. Ricci, Designed nanoparticle−mesoporous multilayer nanocomposites as tunable plasmonic−photonic architectures for electromagnetic field enhancement, J. Mater. Chem. C. 5 (14) (2017) 3445−3455.

[132] P.C. Angelomé, L. Andrini, M.C. Fuertes, F.G. Requejo, G.J.A.A. Soler-Illia, Large-pore mesoporous titania-silica thin films ($Ti1-xSixO_2$, $0.1 \leq x \leq 0.9$) with highly interdispersed mixed oxide frameworks, Comptes Rendus Chimie 13 (2010) 256−269.

[133] N. Crespo-Monteiro, N. Destouches, L. Bois, F. Chassagneux, S. Reynaud, T. Fournel, Reversible and irreversible laser microinscription on silver-containing mesoporous titania films, Adv. Mater. 22 (29) (2010) 3166−3170.

[134] N. Destouches, N. Crespo-Monteiro, G. Vitrant, Y. Lefkir, S. Reynaud, T. Epicier, et al., Self-organized growth of metallic nanoparticles in a thin film under homogeneous and continuous-wave light excitation, J. Mater. Chem. C. 2 (31) (2014) 6256−6263.

[135] N. Crespo-Monteiro, N. Destouches, L. Nadar, S. Reynaud, F. Vocanson, J.Y. Michalon, Irradiance influence on the multicolor photochromism of mesoporous TiO_2 films loaded with silver nanoparticles, Appl. Phys. Lett. 99 (17) (2011) 173106.

[136] L. Nadar, R. Sayah, F. Vocanson, N. Crespo-Monteiro, A. Boukenter, S. Sao Joao, N. Destouches, Influence of reduction processes on the colour and photochromism of amorphous mesoporous TiO_2 thin films loaded with a silver salt, Photochem. Photobiol. Sci. 10 (11) (2011) 1810−1816.

[137] L. Nadar, N. Destouches, N. Crespo-Monteiro, R. Sayah, F. Vocanson, S. Reynaud, et al., Multicolor photochromism of silver-containing mesoporous films of amorphous or anatase TiO_2, J. Nanopart. Res. 15 (11) (2013) 1−10.

[138] X. Zhang, H. Dong, Y. Wang, N. Liu, Y. Zuo, L. Cui, Study of catalytic activity at the Ag/Al-SBA-15 catalysts for CO oxidation and selective CO oxidation, Chem. Eng. J. 283 (2016) 1097−1107.

[139] R. Coneo Rodríguez, L. Yate, E. Coy, Á.M. Martínez-Villacorta, A.V. Bordoni, S. Moya, et al., Copper nanoparticles synthesis in hybrid mesoporous thin films: controlling oxidation state and catalytic performance through pore chemistry, Appl. Surf. Sci. 471 (2019) 862–868.

[140] S. Liu, M.D. Regulacio, S.Y. Tee, Y.W. Khin, C.P. Teng, L.D. Koh, et al., Preparation, functionality, and application of metal oxide-coated noble metal nanoparticles, Chem. Rec. 16 (4) (2016) 1965–1990.

[141] T.W. van Deelen, C. Hernández Mejía, K.P. de Jong, Control of metal-support interactions in heterogeneous catalysts to enhance activity and selectivity, Nat. Catal. 2 (11) (2019) 955–970.

[142] K. An, S. Alayoglu, N. Musselwhite, S. Plamthottam, G. Melaet, A.E. Lindeman, et al., Enhanced CO oxidation rates at the interface of mesoporous oxides and Pt nanoparticles, J. Am. Chem. Soc. 135 (44) (2013) 16689–16696.

[143] H. Wang, K. An, A. Sapi, F. Liu, G.A. Somorjai, Effects of nanoparticle size and metal/support interactions in Pt-catalyzed methanol oxidation reactions in gas and liquid phases, Catal. Lett. 144 (11) (2014) 1930–1938.

[144] N. Zhou, L. Polavarapu, Q. Wang, Q.-H. Xu, Mesoporous SnO_2-coated metal nanoparticles with enhanced catalytic efficiency, ACS Appl. Mater. Interfaces 7 (8) (2015) 4844–4850.

[145] G. Mohammadi Ziarani, S. Rohani, A. Ziarati, A. Badiei, Applications of SBA-15 supported Pd metal catalysts as nanoreactors in C–C coupling reactions, RSC Adv. 8 (71) (2018) 41048–41100.

[146] S.J. Canhaci, R.F. Perez, L.E.P. Borges, M.A. Fraga, Direct conversion of xylose to furfuryl alcohol on single organic–inorganic hybrid mesoporous silica-supported catalysts, Appl. Catal. B: Environ. 207 (2017) 279–285.

[147] Z. Zhang, J. Cheng, Y. Luo, W. Shi, W. Wang, B. Zhang, et al., Pt nanoparticles supported on amino-functionalized SBA-15 for enhanced aqueous bromate catalytic reduction, Catal. Commun. 105 (2018) 11–15.

[148] M.S.A. Che Mansor, M.N.I. Amir, N. Muhd Julkapli, A. Ma'amor, Gold hybrid nanomaterials: prospective on photocatalytic activities for wastewater treatment application, Mater. Chem. Phys. 241 (2020) 122415.

[149] M. Andersson, H. Birkedal, N.R. Franklin, T. Ostomel, S. Boettcher, A.E.C. Palmqvist, et al., Ag/AgCl-loaded ordered mesoporous anatase for photocatalysis, Chem. Mater. 17 (6) (2005) 1409–1415.

[150] I. Bannat, K. Wessels, T. Oekermann, J. Rathousky, D. Bahnemann, M. Wark, Improving the photocatalytic performance of mesoporous Titania films by modification with gold nanostructures, Chem. Mater. 21 (8) (2009) 1645–1653.

[151] M.V. Roldán, Y. Castro, N. Pellegri, A. Durán, Enhanced photocatalytic activity of mesoporous SiO_2/TiO_2 sol–gel coatings doped with Ag nanoparticles, J. Sol–gel Sci. Technol. 76 (1) (2015) 180–194.

[152] M.V. Roldán, P. de Oña, Y. Castro, A. Durán, P. Faccendini, C. Lagier, et al., Photocatalytic and biocidal activities of novel coating systems of mesoporous and dense TiO_2-anatase containing silver nanoparticles, Mater. Sci. Eng. C. 43 (2014) 630–640.

[153] J. Zhao, S. Sallard, B.M. Smarsly, S. Gross, M. Bertino, C. Boissiere, et al., Photocatalytic performances of mesoporous TiO_2 films doped with gold clusters, J. Mater. Chem. 20 (14) (2010) 2831–2839.

[154] A.A. Ismail, D.W. Bahnemann, Mesostructured Pt/TiO_2 nanocomposites as highly active photocatalysts for the photooxidation of dichloroacetic acid, J. Phys. Chem. C. 115 (13) (2011) 5784–5791.

[155] M.A. Ghanem, P. Arunachalam, M.S. Amer, A.M. Al-Mayouf, Mesoporous titanium dioxide photoanodes decorated with gold nanoparticles for boosting the photoelectrochemical alkali water oxidation, Mater. Chem. Phys. 213 (2018) 56–66.

[156] N. Crespo-Monteiro, A. Cazier, F. Vocanson, Y. Lefkir, S. Reynaud, J.-Y. Michalon, et al., Microstructuring of mesoporous Titania films loaded with silver salts to enhance the photocatalytic degradation of methyl blue under visible light, Nanomaterials 7 (10) (2017) 334.

[157] S. Sun, W. Wang, S. Zeng, M. Shang, L. Zhang, Preparation of ordered mesoporous Ag/WO$_3$ and its highly efficient degradation of acetaldehyde under visible-light irradiation, J. Hazard. Mater. 178 (1) (2010) 427−433.

[158] G. Pacchioni, H.-J. Freund, Controlling the charge state of supported nanoparticles in catalysis: lessons from model systems, Chem. Soc. Rev. 47 (22) (2018) 8474−8502.

[159] R. Coneo Rodríguez, M.M. Bruno, P.C. Angelomé, Au nanoparticles embedded in mesoporous ZrO$_2$ films: multifunctional materials for electrochemical detection, Sens. Actuators B Chem. 254 (2018) 603−612.

[160] M. D'Arienzo, M. Crippa, P. Gentile, C.M. Mari, S. Polizzi, R. Ruffo, et al., Sol−gel derived mesoporous Pt and Cr-doped WO$_3$ thin films: the role played by mesoporosity and metal doping in enhancing the gas sensing properties, J. Sol−gel Sci. Technol. 60 (3) (2011) 378−387.

[161] G. Bodelón, V. Montes-García, V. López-Puente, E.H. Hill, C. Hamon, M.N. Sanz-Ortiz, et al., Detection and imaging of quorum sensing in Pseudomonas aeruginosa biofilm communities by surface-enhanced resonance Raman scattering, Nat. Mater. 15 (11) (2016) 1203−1211.

[162] X. Liu, S. Yang, Y. Li, B. Wang, J. Guo, X. Ma, Mesoporous nanostructures encapsulated with metallic nanodots for smart SERS sensing, ACS Appl. Mater. Interfaces 13 (1) (2021) 186−195.

[163] M. Stanzel, L. Zhao, R. Mohammadi, R. Pardehkhorram, U. Kunz, N. Vogel, et al., Simultaneous nano-local polymer and in situ readout unit placement in mesoporous separation layers, Anal. Chem. 93 (13) (2021) 5394−5402.

[164] H. Fathima, L. Paul, S. Thirunavukkuarasu, K.G. Thomas, Mesoporous silica-capped silver nanoparticles for sieving and surface-enhanced Raman scattering-based sensing, ACS Appl. Nano Mat. 3 (7) (2020) 6376−6384.

[165] Z. Gao, N.D. Burrows, N.A. Valley, G.C. Schatz, C.J. Murphy, C.L. Haynes, In solution SERS sensing using mesoporous silica-coated gold nanorods, Analyst 141 (17) (2016) 5088−5095.

Nanoarchitectonics of Metal–Organic Frameworks (MOFs) for energy and sensing applications

15

Melina Arcidiácono, Ana Paula Mártire, Juan A. Allegretto, Matías Rafti, Waldemar A. Marmisollé and Omar Azzaroni

Instituto de Investigaciones Fisicoquímicas Teóricas y Aplicadas (INIFTA)—Departamento de Química—Facultad de Ciencias Exactas—Universidad Nacional de La Plata—CONICET, La Plata, Argentina

15.1 Introduction

Metal–Organic Frameworks (MOFs) constitute a whole family of emergent materials [1], which were first reported a few decades ago and have been intensively employed since then for numerous and shockingly diverse applications [2–10]. MOFs can be defined as coordination networks assembled by noncovalent interactions between metallic ions and multidentate organic linkers featuring permanent porosity [11]. By combining appropriate building blocks, either resorting to liquid phase, or via solvent-free (gas or solid phase) synthesis procedures [12–14], MOFs nano- or microparticles consisting of ordered networks capable of X-ray diffraction can be straightforwardly obtained [15–17]. A careful selection of linkers and metallic ions define important features of the particular MOF obtained, such as pore size and pore wall surface chemistry [18–21]. Among the unique characteristics responsible for the multiplication of MOF-related research projects worldwide, it must be stressed that their relatively high thermal and chemical stability, record-high surface areas and pore volumes, and the possibility of employing them as heterogeneous catalysts are due to the presence of coordinatively unsaturated metal sites or active sites in linkers [22–25]. During an incipient development stage, the main drive for the observed bloom was the consideration of MOFs as prospects for next-generation tailored adsorbent materials, both polar and nonpolar (i.e., as replacement for traditionally employed zeolites and carbon-derived materials). MOFs allow an interesting alternative: a "smart" ad hoc predesigned porous material with tunable affinity toward selected adsorbates [26–32]. Moreover, slight modification of the common synthesis procedures allows fabrication of MOF nanoparticles with porosity (i.e., both macro-, meso-, and micropores), which in turn can be assembled into films with tunable permeability, a crucial parameter in many technological applications [33–36].

Soon enough it became clear that, although MOFs excel in some key applications and there is room for further improvement in many other interesting examples, unavoidable issues related to economic feasibility must be addressed for enabling large-scale use for industrial applications; specifically, production scale-up and susceptibility upon exposure to harsh operating conditions that compromise cyclability and life span. Possible solutions for the above mentioned issues are being intensively

Materials Nanoarchitectonics. DOI: https://doi.org/10.1016/B978-0-323-99472-9.00017-1

explored; it is worth mentioning the various reported high-throughput "green" synthesis methods as cost-effective and environment-friendly alternatives, and several attempts aiming to increase durability by including MOFs in composites with other materials acting as stable scaffolds [37–44].

Recently, a new approach was intensively explored in the mainstream literature, which deals with the opportunities arising from the potential synergy between MOFs and other materials, given that suitable integration using nanodesign is achieved. Such integration can be attained through nanoarchitectonics, which is a discipline that determines strategies allowing an impressive level of control, that is, rendering nanometer-sized objects into functional materials, a concept introduced by Masakazu Aono back in 2000 [45,46]. The tools needed for this task are diverse and require a strongly multidisciplinary approach due to the kind of interactions which come into play, for example, chemical coordination, self-assembly of molecules and/or nanoobjects through soft interactions, and even DNA complementary base-pairing or enzymatic molecular recognition [47,48]. To produce such complex architectures, often involving nonsymmetric and hierarchical motifs, it is mandatory a strong interplay between expertise related to nanotechnology, supramolecular assembly, biotechnology, and chemistry of materials [49,50].

Remarkable examples are the combination of MOFs with, for example, metallic nanoparticles, carbon-derived materials, mesoporous scaffolds, biomolecules, and polymers to form nanoarchitectonic objects with interesting features, such as catalytic and optical activity, tailored conductivity, enhanced biocompatibility, or stimuli-responsiveness [47,51–57].

In this chapter, we will describe common strategies for the integration of MOFs into nanoarchitectonic objects. Then, some selected recent examples of advanced devices based on such composites, which are suitable for energy storage/conversion and sensing applications, will be discussed.

15.2 Common strategies for the integration of metal–organic frameworks into nanoarchitectonic objects

Synthesis methods for MOFs were extensively reviewed in the past, and the interested reader is referred to the abundant specific literature [16,58–64]. Briefly, the choice of a given metal-linker combination restricts the required synthetic conditions; while some MOFs can be synthesized using green or biologically compatible conditions (i.e., near-neutral pH aqueous solutions and room temperatures), other materials can only be obtained using solvothermal methods (high pressure and temperature or the addition of strong acids as nucleation promoters). It must be stressed that even when a given material could be obtained using a wide range of different conditions and/or chemical modulators, the actual choice has a great impact on the final morphology and size distributions, and therefore restricts the possible uses. Despite recent efforts which resorted to the use of computer-assisted methods like machine learning and ad hoc curated databases, the task of rationalizing the overwhelmingly large number of reported synthetic conditions (even for fairly similar MOFs) is far from complete [65].

MOFs integration into nanoarchitectonic objects can be discussed in terms of the synthesis stage in which such integration actually occurs, that is, either during MOF formation ("one-pot" procedure) or in a postsynthetic step. Each strategy has advantages and shortcomings; a one-pot procedure typically ensures robust and efficient integration of the different materials, but the control is somehow limited, while postsynthetic modification usually requires additional chemicals for ensuring robust binding of the MOF particle with the other(s) component(s) of the intended nanoarchitectonic object. One-pot procedures impose a challenging requisite, namely, the compatibility of the whole set of components with the synthesis conditions; this becomes an issue if biomolecules are present. For such scenario, postsynthetic modification would be the smart choice, because allows the integration of MOFs with labile components under mild conditions. For one-pot strategies, it is important to keep in mind that the presence of all the constitutive parts of the final nanoarchitecture might alter the expected MOF structure, and thus thorough characterization must be carried after each modification step.

Having established general strategies, that is, one-pot or postsynthetic integration, all the different well-established techniques such as drop-casting, spin coating, layer-by-layer (LbL), liquid phase epitaxy, seeded growth, and even stamping, constitute the toolbox from which the selection must be made depending on the intended nanoarchitectonic design. Later on, we will discuss several examples where one or several of the abovementioned *tools* were employed, to cope with specific issues related to the units composing the assembled nanoarchitectonic object [66–71].

15.3 Nanoarchitectonics for the integration of metal—organic frameworks into electrochemical devices

A particularly interesting example of MOFs' integration through nanoarchitectonics into devices is their use in electrochemical applications. Early studies focused on exploiting their large surface areas and ordered pore networks for adsorption and separation applications, however, due to their large bandgap semiconductor or even insulating nature, applications requiring electrical conductivity were not envisaged. However, several relatively recent reports in which successful strategies for conferring MOFs and related analogs with electrical conductivity opened the path for developments in this new subfield [7]. Approaches proposed for this end are quite diverse, for example, to include redox-active guests or organic linkers [72,73], to integrate conductive guests in the framework [74,75], or to employ nanofabrication techniques to induce conductivity to thin MOF films [76,77], see Fig. 15.1.

Integration requires usually multiple steps to obtain a material with the desired features. According to the intended function, MOF-containing and MOF-derived materials can play two main roles: increasing capacitance and/or providing (electro)catalytic activity, which is key for its use in supercapacitors, batteries, or biosensing. High specificity is an unavoidable requirement for biosensing to avoid interference with chemicals present in real-life samples. In this section, some selected examples of MOF integration strategies will be presented and analyzed in detail.

15.3.1 Metal—organic frameworks in supercapacitors

Supercapacitors (SCs) are devices employed for energy conversion and storage and consist generally of two electrodes separated by an electrolyte. SCs can be classified as symmetric or asymmetric depending on whether the electrodes are equivalent or not. The improvement of SCs performance is at the heart of efficient strategies for sustainable energy storage and conversion [78].

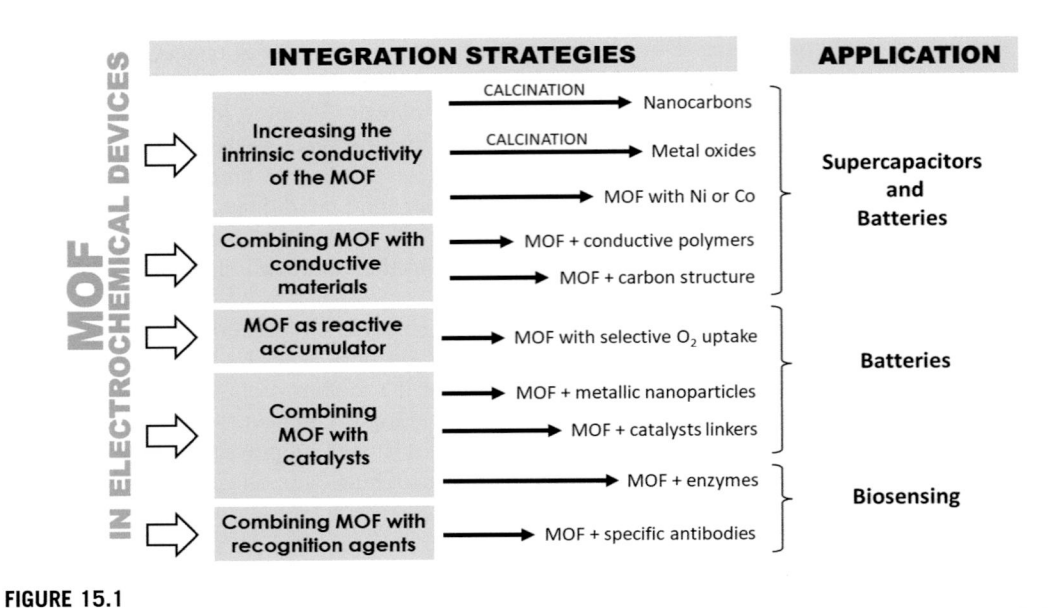

FIGURE 15.1

Scheme of the strategies applied for integrating metal—organic framework into electrochemical devices.

There are some examples of unprocessed MOFs employed for SCs applications, which require the use of MOFs that already contain redox-active species in their structure (e.g., Ni, Fe, and Co) [79—84]. However, MOFs are usually modified in order to increase their conductivity or electroactivity. One of the most employed strategies is to perform controlled pyrolysis, which yields conductive nanoporous and heteroatom-doped carbons (NPCs) [85—88]. By doing so, the obtained material yields improved performance but also gains stability upon exposure to harsh conditions. It is important to bear in mind that during pyrolysis the porous structure might collapse (either partially or completely) which makes the task of obtaining well-defined structure NPCs particularly challenging. Another advantage of pyrolysis is that conditions can be manipulated to yield metal oxide doping, which was shown to boost electroactivity, improving thus the attainable SCs performances [89].

Directly related to the above discussion, Bu Yuan Guan et al. [90]. recently showed interesting performances from onion-like multishelled Ni-Co oxide particles acting as a cathode in a hybrid supercapacitor with graphene anode. This particular structure rendered a material with high specific capacitance and stability, both relevant and necessary features for SCs. Along the same line of thought, MOF-derived NPCs prepared from MOFs assembled in a core—shell morphology were reported by Jing Tang et al. [91]. The authors combined therein two prominent members of the ZIF-MOFs family: Zn-imidazolate-based ZIF-8, and the homologous Co-based ZIF-67. A ZIF-8@ZIF-67 core—shell structure was generated, where each phase brought along their characteristic properties upon pyrolysis. A nitrogen-doped core carbon phase arising from ZIF-8 (NC) surrounded by a highly conductive graphitic carbon (GC) shell phase obtained from ZIF-67 MOF. This work is a nice example of how rational design of nanoarchitectures allows not only to retain key features of individual components but to synergically enhance performances.

As discussed above, synergy can arise from a rational combination of different materials, but it is also possible to apply differential treatment to a single component in order to achieve the same effect. In this regard, Salunkhe et al. [92] reported on the use of ZIF-67 MOF for this purpose; it was shown that NPCs can be obtained by 5-hour pyrolysis at 800$_{\circ}$C under inert N_2 atmosphere, as schematized in Fig. 15.2A. Additionally, by resorting to lower temperatures (500$_{\circ}$C) and shorter pyrolysis time (30 minutes), the authors showed that ZIF-67 can be converted into a porous structure merely composed of Co_3O_4 particles. The combination of these two products of ZIF-67 pyrolysis with KOH as electrolyte was employed to assemble an asymmetric supercapacitor (ASCs, see Fig. 15.2B). The electrochemical performance of the ASCs showed higher specific energy and

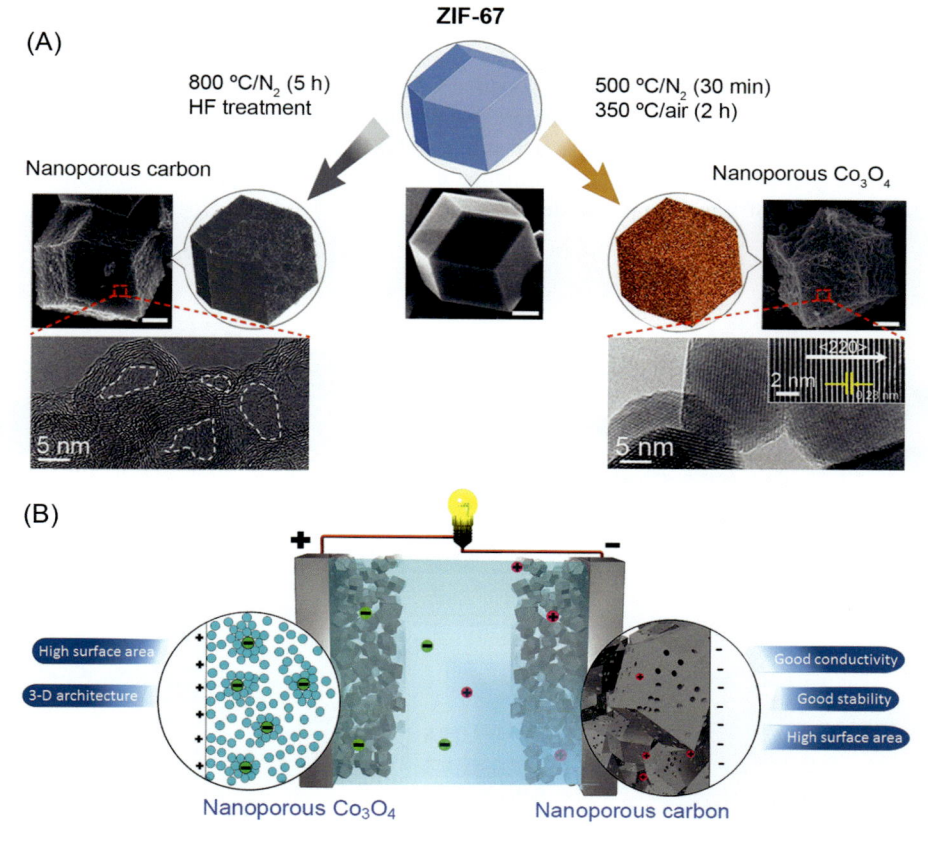

FIGURE 15.2

(A) Preparation procedure of electrode materials with ZIF-67 as a precursor. (B) Asymmetric supercapacitor constructed with ZIF-67 nanoporous carbons and ZIF-67 nanoporous Co_3O_4 metal oxide.

Adapted from R.R. Salunkhe, J. Tang, Y. Kamachi, T. Nakato, J.H. Kim, Y. Yamauchi, Asymmetric supercapacitors using 3D nanoporous carbon and cobalt oxide electrodes synthesized from a single metal-organic framework, ACS Nano. 9 (2015) 6288–6296.

power than the symmetric device (prepared either with NPCs or metal oxide nanoparticles), thus stressing the importance of materials chosen for the nanoarchitectonic design, but also regarding the treatments applied for enhancing a given feature.

As discussed above, to include carbon-derived structures represents an interesting possibility for enhancing properties of a given MOF [93–95]. In this regard, it is unavoidable to consider combinations employing graphene and few-layer graphene due to their unique features such as semiconductive behavior and high chemical stability and versatility [96–98]. For example, Xiao et al. [99] reported on the preparation of an heterolayered Ni-based 2D-MOF by a simple solvothermal process using graphene as nucleation point for MOF growth. As depicted in Fig. 15.3, growth of nearly-2D Ni-based MOF was carried following a three-step assembly process. The first step consisted of a liquid-phase exfoliation of graphene (LEG) with N-methyl-2-pyrrolidinone (NMP) as solvent. In the second step, LEG was modified by adding glucose in a solvothermal process which yields a glucose-modified-LEG (GM-LEG). And finally, the 2D Ni-MOF layer was solvothermally synthesized in the presence of GM-LEG to produce GM-LEG@Ni-MOF composites. The electrode material thus assembled was tested as a supercapacitor in an asymmetric device with commercial active carbon as a negative electrode. As a result, the GM-LEG@Ni-MOF lamellar heterolayered nanoarchitecture yielded an enhanced performance compared to the separated components, arising from an improvement of transport properties.

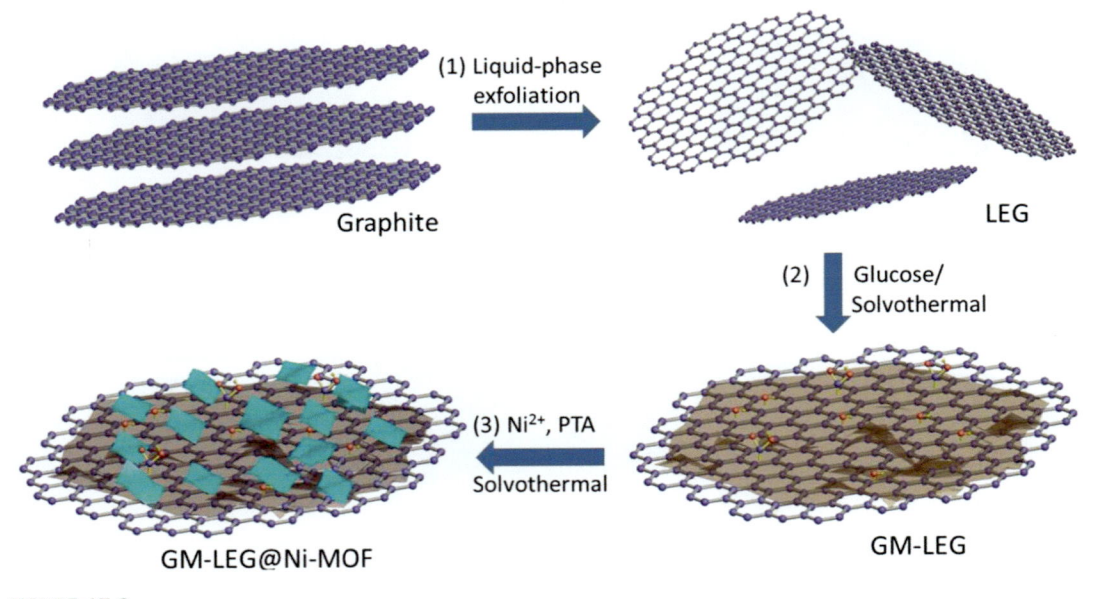

FIGURE 15.3

Preparation procedure of the GM-LEG@Ni-metal—organic framework composites.

From Y. Xiao, W. Wei, M. Zhang, S. Jiao, Y. Shi, S. Ding, Facile surface properties engineering of high-quality graphene: toward advanced Ni-MOF heterostructures for high-performance supercapacitor electrode, ACS Appl. Energy Mater. 2 (2019) 2169–2177.

A different approach for increasing conductivity is to combine MOF with conductive polymers [100]. Hou et al. explored this idea and prepared a symmetric supercapacitor by combining a Cu_3(2,3,6,7,10,11-hexahydroxytriphenylene)$_2$ catecholate MOF (Cu-CAT MOF) with a polypyrrole membrane (PPy) [101]. The choice of these materials was based on the flexibility of the PPy membrane and the improvement in surface area and conductivity provided by the Cu-CAT MOF. Fig. 15.4A shows the procedure followed for the nanointegration of the different components, Cu-CAT MOF nanowire array in the PPy membrane. The PPy membrane was first obtained by electrodeposition on the surface of a super [13]Cr stainless steel plate and peeled off. Then, the Cu-CAT MOF was synthesized in the presence of the PPy membrane yielding nanowire arrays onto the membrane (Cu-CAT-NWAs/PPy). To test this material in supercapacitor applications, a symmetric two-electrode cell was assembled using a gel electrolyte. Fig. 15.4B shows an illustration of the all-solid-state supercapacitor prepared by placing the Cu-CAT-NWAs/PPy membranes as working electrodes. From the electrochemical results, this flexible device showed a synergic effect leading to an exceptional capacitive performance and great stability, even in a wide range of temperatures. This is another example of a nanoarchitectonic design leading to a device with improved features compared to their individual components.

Salunkhe et al. presented an interesting approach for the preparation of three-dimensional core—shell electrode material from Zn-based ZIF-8 MOF NPCs covered with polyaniline (PANI) nanorods [102]. The procedure employed for obtaining the nanoporous carbon consisted of a controlled pyrolysis of ZIF-8 at 800°C in a N_2 atmosphere. In order to grow PANI nanorods onto NPCs, the annealed ZIF-8 was added to the synthesis solution of the conducting polymer. Fig. 15.5A shows the steps leading to the three-dimensional nanoarchitectures. The electrochemical experiments were conducted in a two-electrode cell with a symmetric configuration. To build the electrode, the NPCs-PANI core—shell composites were mixed with poly(vinylidene-difluoride) (PVDF, 20%) in N-methyl 2-pyrrolidinone (NMP) solvent and then coated on graphite substrates. Fig. 15.5B illustrates the electrode constructed. The electrochemical response of the carbon-PANI//carbon-PANI supercapacitor was tested in H_2SO_4 medium, revealing that the device outperforms carbon//carbon or PANI//PANI arrays. The unique 3D nanoarchitecture allowed avoiding problems arising from stacking that are known to occur in 2D and 1D carbon electrode materials. The successful integration strategy for MOFs in supercapacitors with enhanced performances suggests a promising new avenue for design that should be further explored.

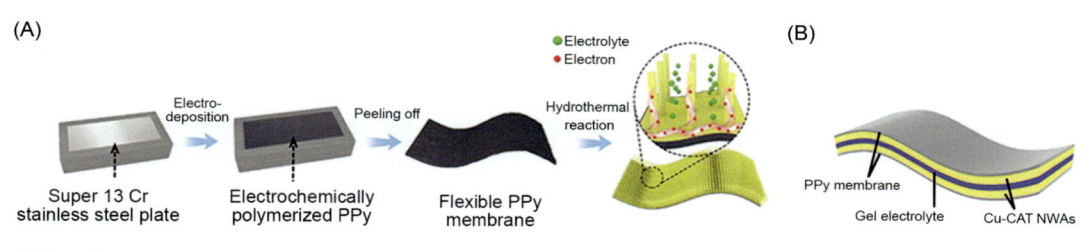

FIGURE 15.4

(A) Preparation procedure of Cu-CAT-NWAs/PPy. (B) Scheme of symmetric Cu-CAT-NWAs/PPy supercapacitor.

Adapted from R. Hou, M. Miao, Q. Wang, T. Yue, H. Liu, H.S. Park, et al., Integrated conductive hybrid architecture of metal—organic framework nanowire array on polypyrrole membrane for all-solid-state flexible supercapacitors, Adv. Energy Mater. 10 (2020) 1901892.

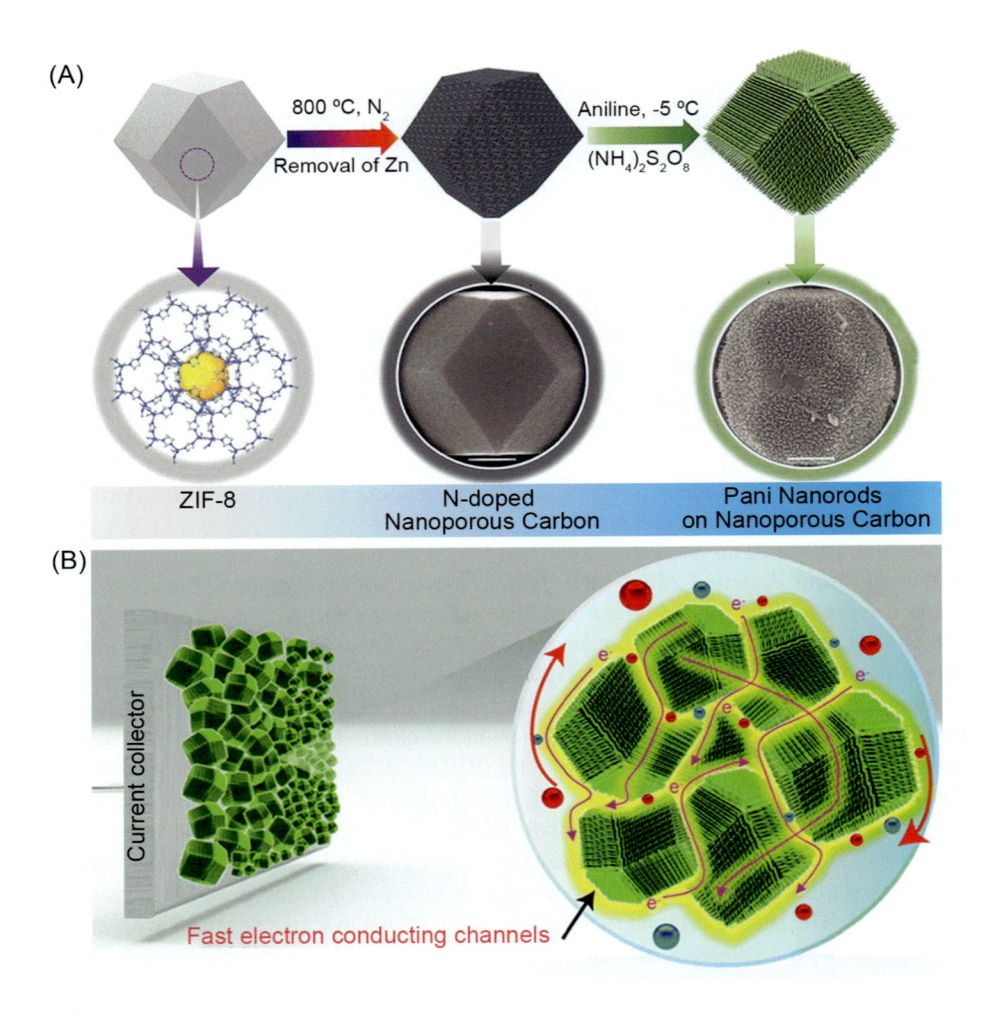

FIGURE 15.5

(A) Scheme of the preparation procedure of 3D core–shell carbon-PANI structure. (B) Illustration of the electrode prepared with the core–shell composites.

Adapted from R.R. Salunkhe, J. Tang, N. Kobayashi, J. Kim, Y. Ide, S. Tominaka, et al., Ultrahigh performance supercapacitors utilizing core–shell nanoarchitectures from a metal–organic framework-derived nanoporous carbon and a conducting polymer, Chem. Sci. 7 (2016) 5704–5713.

15.3.2 **Metal–organic frameworks in electrocatalysis**

Heterogeneous catalysis is another important application for MOFs derived from their high surface areas and the presence of coordinatively unsaturated metallic centers capable of acting as active sites. In this section, the focus will be placed on electrocatalytic reactions, and the new possibilities enabled by nanoarchitectonic assembly for devices such as batteries and biosensors. A first approach would be to make use of the above-discussed strategies for increasing conductivity, for example, MOF pyrolysis or direct combination with conducting materials such as graphite, graphene or conducting polymers.

On a different note, there are recent reports on the assembly of nanoarchitectures based on MOFs that do not need further modification. For example, conductive MOFs (either inherently conductive or endowed with conductivity after ad hoc modifications) [7,103,104], or catalytically active MOFs such as those including porphyrin-based linkers, the so-called PCN-MOFs [105–108], constitute a promising option for the nanoarchitectonic assembly of devices. A particular member of PCN-MOFs, PCN-226 combines metalloporphyrin linkers with Zr metallic centers, the material can be prepared using $ZrCl_4$ and features zig-zag Zr chains as a starting structural motif which later on leads to a structure with high density of active sites. A related material, PCN-226(Co) can be synthesized by controlled mixing of $ZrCl_4$ precursors and [5,10,15,20-Tetrakis(4-metthoxycarbonyl-phenyl)porphyrinato]-Co(II) (CoTCPPCl), being Co^{2+} the porphyrin metalation agent. Although other metal centers were tested, the linker prepared with Co^{2+} (see Fig. 15.6B) showed the best electrochemical activity in both the OER and ORR reaction. This PCN-MOF was then employed as cathode material in Zn-air rechargeable batteries capable of powering a minicar prototype (Fig. 15.6B). The cathode was prepared by depositing a MOF layer on a carbon cloth electrode. It was proposed that the zigzag initial chain allows the proximity of the cations and enhance the reaction kinetics. As a result, this material showed a better performance than other 3D MOFs due to its particular structure and a shockingly similar response compared to high-performance $Pt/C + RuO_2$ catalysts [107].

The examples of MOF-based electrochemical applications discussed above employ a controlled calcination step followed by the combination with electroactive components for conferring an increased conductivity, with the main challenge being the preservation of high porosity on the nanoarchitecture thus assembled. There is, however, an alternative approach which benefits from

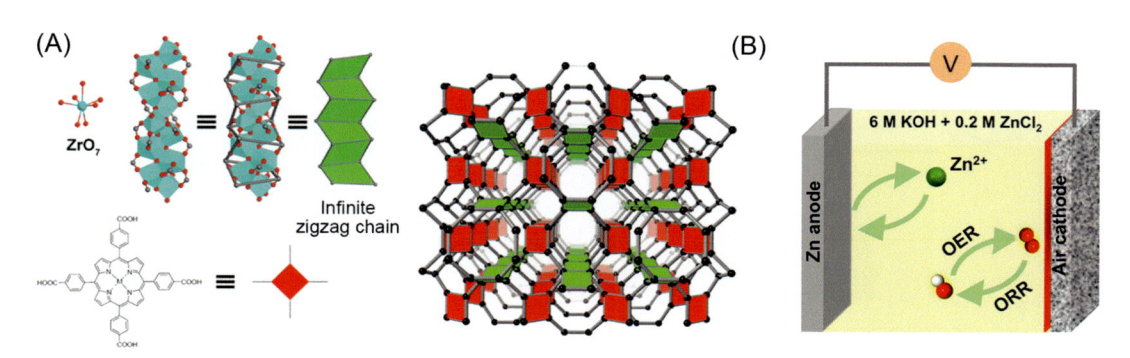

FIGURE 15.6

(A) Schematic of PCN-226(Co) structure. (B) Device prepared including the metal–organic framework in a Zn-air battery.

Adapted from M.O. Cichocka, Z. Liang, D. Feng, S. Back, S. Siahrostami, X. Wang, et al., A Porphyrinic zirconium metal–organic framework for oxygen reduction reaction: tailoring the spacing between active-sites through chain-based inorganic building units, J. Am. Chem. Soc. 142 (2020) 15386–15395.

the nonpyrolyzed MOF structure and the adsorption selectivity featured by these materials. For example, considering ZIF-8 MOF, which is known to adsorb preferentially oxygen, it has been proposed to assemble complex nanoarchitectures which combine such porous material with a conducting polymer to yield an enhancement to the oxygen reduction reaction (ORR) performance even in neutral pH, with obvious implications for biofuel cells-related applications [109]. Although ZIF-8 is electrochemically inert, it can be thought as an in situ O_2 reservoir which increases the current density by relaxing the diffusional barrier existing for the electrode without MOF. There are further examples of this approach in which ZIF-8 was included in electrode materials to improve the ORR performance [47,110,111], specifically, the addition of ZIF-8 layers to a gold electrode previously modified with PABA (poly(allyl benzylamine)) conductive polymer through electropolymerization is schematized in Fig. 15.7A [112]. Furthermore, an electroactive material can be also incorporated in such nanoarchitecture to enable a synergic enhancement of the ORR, as demonstrated recently by Fenoy et al., with the inclusion of Pt nanoparticles (see Fig. 15.7B) [47]. Yet another example on the same line of thought is the use of a top-down LbL strategy for film growth; ZIF-8 nanoparticles were integrated with PANI:PSS polymer to yield a porous ORR-active electrode (Fig. 15.7C) [110]. Altogether, these examples illustrate how the same MOF can be integrated using substantially different procedures to build distinctive interface nanoarchitectonics, all of them with catalytic activity toward the ORR.

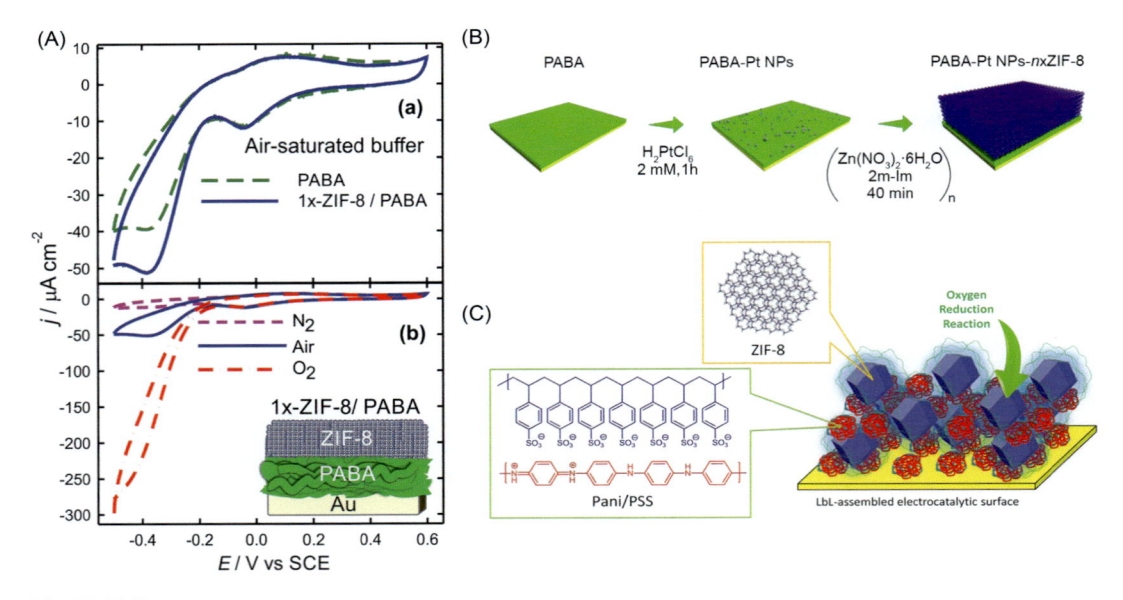

FIGURE 15.7

(A) Representation of substrate modified with PABA and ZIF-8 layers and the electrochemical activity toward the ORR. (B) Scheme of a similar construction including Pt nanoparticles. (C) Scheme of the layer-by-layer assembly made with ZIF-8 microparticles and PANI.

(A) From Ref. [112], (B) From Ref. [47], (C) From Ref. [110].

Inclusion of biologically active moieties (e.g., enzymes and antibodies) as building blocks in the nanoarchitectonic objects bring a new dimension of complexity but allow a dramatic enhancement of selectivity and specific catalytic activity. Such enhanced selectivity is highly desirable when aiming to assemble biosensing devices, and it can be achieved by different strategies. The first option that comes to mind is to employ MOF microporosity as protective scaffolds for enzymes [113—117]. There are many recent reports on the enhanced activity and stability of enzymes when immobilized inside pores; however, it remains a challenge to identify the immobilization strategies that would minimize disruption of both tertiary and quaternary protein structure, and thus ultimately decrease catalytic activity [118—120]. A nice example of this approach is the biosensor for H_2O_2 proposed by Liu et al., which makes use of direct mixing combination of ZIF-67 MOF and multiwalled carbon nanotubes (MWCNT) and horseradish peroxidase (HRP) [114]. The nanoarchitectonic assembly (ZIF-67/MWCNT/HRP) was then deposited on a glassy carbon electrode for electrochemical measurements (Fig. 15.8A) and combined catalytic properties of the ZIF-67 and HRP for the H_2O_2 reaction, with the presence of MWCNT increasing electron conductivity and MOF/enzyme affinity. Hydrogen peroxide detection was achieved by using hydroquinone (HQ) as an electron mediator. The idea behind the design is that, when H_2O_2 is present, the HRP catalyzes reduction to H_2O, and switching from HRP (reduced) to HRP (oxidized). In a subsequent step, HRP (ox) reverts back to HRP (red) oxidizing HQ to benzoquinone (BQ), which then exchanges electrons with the electrode yielding HQ. Considering such a mechanism, a direct correlation between the measured anodic current and H_2O_2 concentration can be derived, as shown in Fig. 15.8B.

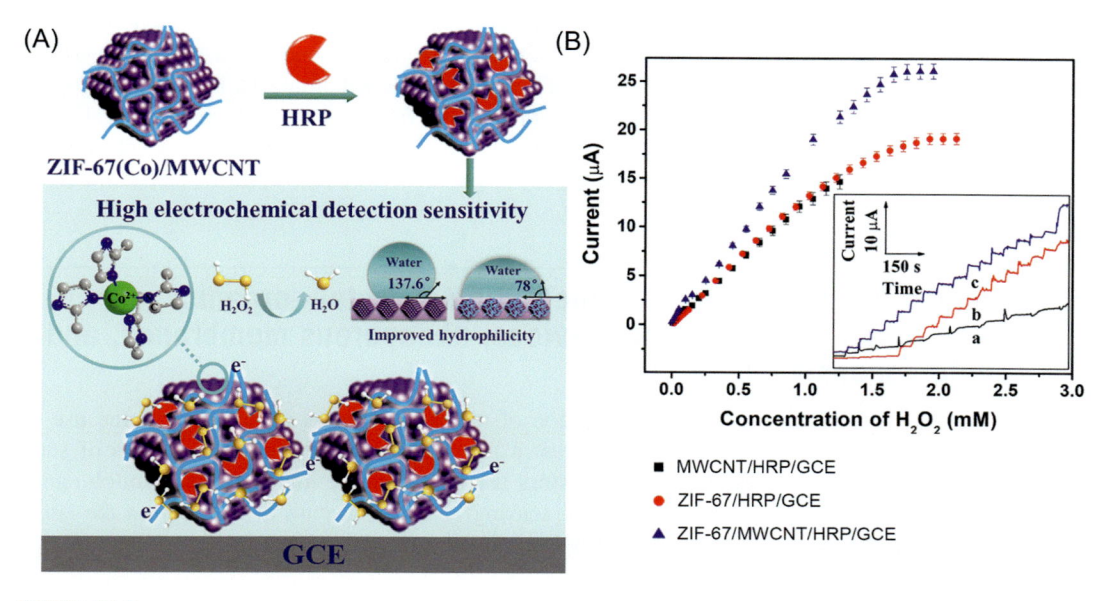

FIGURE 15.8

(A) Scheme of a hydrogen peroxide biosensor by integration of ZIF-67 and horseradish peroxidase. (B) Current response for increasing H_2O_2 concentration.

Adapted from X. Liu, W. Chen, M. Lian, X. Chen, Y. Lu, W. Yang, Enzyme immobilization on ZIF-67/MWCNT composite engenders high sensitivity electrochemical sensing, J. Electroanal. Chem. 833 (2019) 505—511.

Although highly specific, enzymes present some limitations on the operational conditions possible, mainly related to the avoidance of denaturation, as discussed above. For this reason, the combination of different nanomaterials endowing enzyme-like activity, which are known as nanozymes, has gained interest over recent years [121]. Usually, nanozyme preparation is simple, and low cost, and the obtained product is more resistant to extreme environmental conditions than biological enzymes. Considering the versatility of MOFs in terms of chemical and structural flexibility, they have become prospect candidates for the assembly of nanozymes for biosensing applications [122−128]. There are recent examples of nanozymes and MOFs employed for sandwich-type sensing devices, similar to ELISA tests [129]. These devices consists of two separate recognition elements which can only come into contact if the sensed molecule is present in the analyzed sample. The linking is selective because both parts have specific recognition agents for the target analyte. After the assembly of both recognition elements, a signal can be detected and quantified. In electrochemical tests, this signal is usually a current. As an example, a *Staphylococcus aureus* (SA) biosensor was prepared by applying a MOF nanozyme [128]. SA is a pathogen bacteria that causes several diseases and quickly develops antibiotic resistance. A 2D MOF was chosen for this device, to take advantage of the fact that a planar structure increases the exposure of the active sites to the sample medium. This MOF was prepared by mixing $Cu(NO_3)_2$ with [Tetrakis(4-carboxyphenyl)porphyrinato]-Fe(III) (TCPP(Fe)) in the presence of CF_3COOH, PVP, DMF, and ethanol. Then, Au nanoparticles (NPs) were synthesized covering the MOF, by adding $HAuCl_4$ to the obtained product. Finally, an antibody (Ab2) was added leading to the Ab2/AuNPs/MOF nanozyme. The second recognition unit was prepared by modifying a carbon glass electrode with BSA and then with vancomycin (a bactericide), as schematized in Fig. 15.9. Only in the presence of SA, can the MOF-based nanozyme be connected to the electrode. Then, o-phenylenediamine (o-PD) and H_2O_2 are added. The nanozyme decomposes H_2O_2 and generates OH^- free radicals, which then oxidize o-PD to produce o-PD (ox), which finally reduces again at the electrode, generating an electrochemical signal that can be directly related to the presence of SA.

15.4 Nanoarchitectonics for the integration of metal−organic frameworks into electronic devices, nanoporous membranes, and photoresponsive materials

Recent advances in techniques for the manipulation of nanoarchitectures made possible the use of MOFs in applications that seemed not feasible just a few years ago. The main examples of such uses are applications related to field effect transistors (FETs), selective ion transport through nanoporous membranes, and luminescent materials for sensing. We will discuss in this section the latest developments through a selection of model cases in each subcategory.

15.4.1 Metal−organic frameworks with tailored electronic properties and field effect transistors: sensing applications

MOFs with modified electronic properties are appealing for several applications, either direct sensing when assembled in appropriate nanoarchitectures, or when integrated into devices with

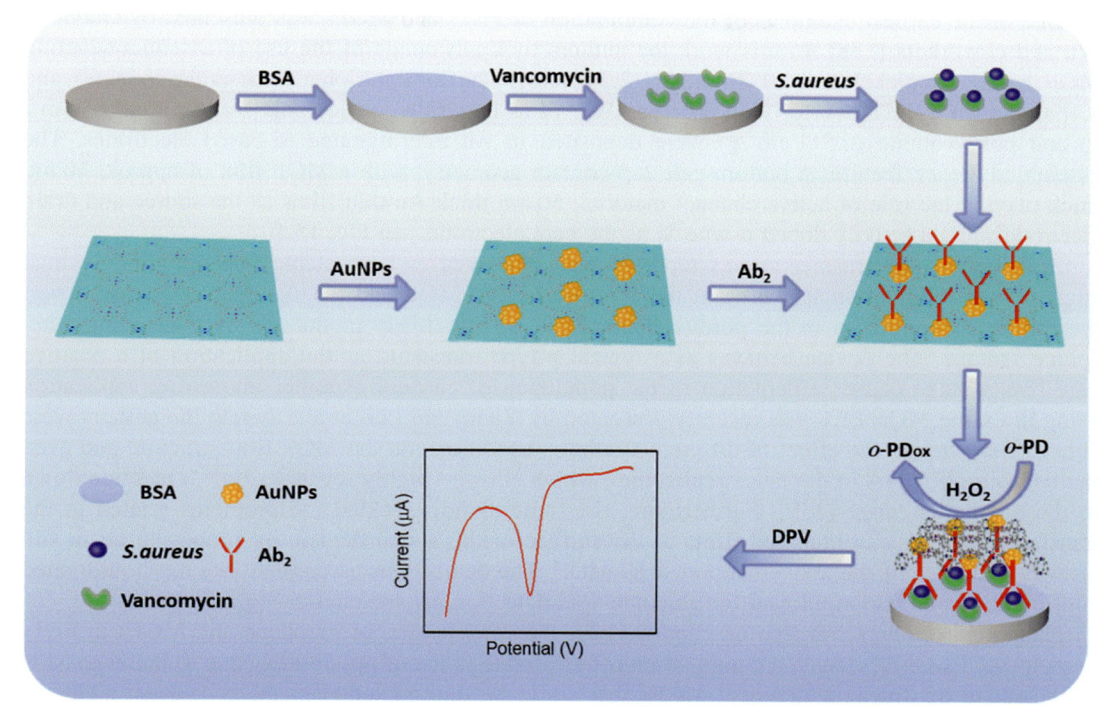

FIGURE 15.9

Representation of a sandwich-type sensor based on a metal—organic frameworks-nanozyme.

From W.-C. Hu, J. Pang, S. Biswas, K. Wang, C. Wang, X.-H. Xia, Ultrasensitive detection of bacteria using a 2D MOF nanozyme-amplified electrochemical detector, Anal. Chem. 93 (2021) 8544—8552.

a higher degree of complexity such as field effect transistors or FETs, and even photochromic MOF-FET devices [130]. FETs are devices with three electrodes, namely, gate, drain, and source; equivalent to base, collector, and emitter present in bipolar junction transistors (BJT) widely employed in modern electronic devices [131]. The design principle of sensors based on FETs is the following: the gate terminal is endowed somehow with responsivity to a certain stimulus (e.g., the concentration of a given analyte, or the variation of an environmental variable), thus controlling the current flow between source and drain terminals. Interesting examples are the use of carbon nanotubes or graphene for direct or indirect modification of gate electrodes (to the so-called CNT-FETs and G-FETs) with extraordinary electrical properties and even flexibility, ideal for wearable electronics which made possible sensing and biosensing in a wide range of challenging conditions [132—137]. In this context and bearing in mind the above-discussed strategies for synthesizing conductive MOFs, a nanoarchitectonic approach toward the combination of MOFs and FETs creates a completely new playground in which porous and tailorable gate electrodes can be employed for enhancing responsivity and selectivity in an unprecedented fashion.

One of the earliest examples of the combination of FETs and MOFs was presented by Guodong Wu and coworkers [138]. In this work the authors took advantage of the use of a semiconducting MOF based on nickel and HITP linker (2,3,6,7,10,11-hexaiminotriphenylenesemiquinonate) and prepared a MOF-FET device. Thin ordered MOF films featuring considerable surface area (650 m^2/g) and pore volume (0.511 cm^3/g) were deposited in Au interdigitated Si-based electrodes. The assembled device features a bottom-gate top-contact geometry, with a MOF film of approx. 10 nm thick playing the role of active channel material, 50 nm thick Au thin films as the source and drain electrodes, and a heavily doped p-type Si as the gate electrode, see Fig. 15.10.

Both output and transfer curves reported were observed to have stable linear regimes, which suggest good ohmic contact between the semiconductive MOF and the Au electrodes. Moreover, the carrier concentration in the porous channel can be effectively modulated by controlling gate-source voltage. The device behaves as a typical p-type transistor, as the application of a positive gate bias voltage causes a depletion in the population of carriers. Another interesting application using the same MOF-FET was recently presented by Yuanyuan Luo et al.; therein the authors were able to characterize the effect of different synthesis conditions on the MOF film structure and even the inclusion of DNA in the nanoarchitecture, which allowed highly sensitive Hg^{2+} detection down to the picomolar range [139]. Furthermore, the same authors tackled the problems related to the transference of ex situ prepared films to the surface of the actual device, by proposing an in situ growth technique of the abovementioned Ni-MOF. The device was assembled in a top liquid-gated configuration and was employed for gluconic acid detection, as depicted in Fig. 15.11.

Nanoarchitectonics can also be employed for the combination of graphene and MOFs in FETs to yield MOFs-GFETs. Such a combination offers a wide range of possibilities due to the expansion of the range of (low-conductivity) MOFs that could be thus included in these devices, adding a dimension of selectivity to the otherwise nonselective graphene-based GFETs. Wöll, Krupke, and coworkers introduced a detailed multistep electron-beam lithography approach to the synthesis of

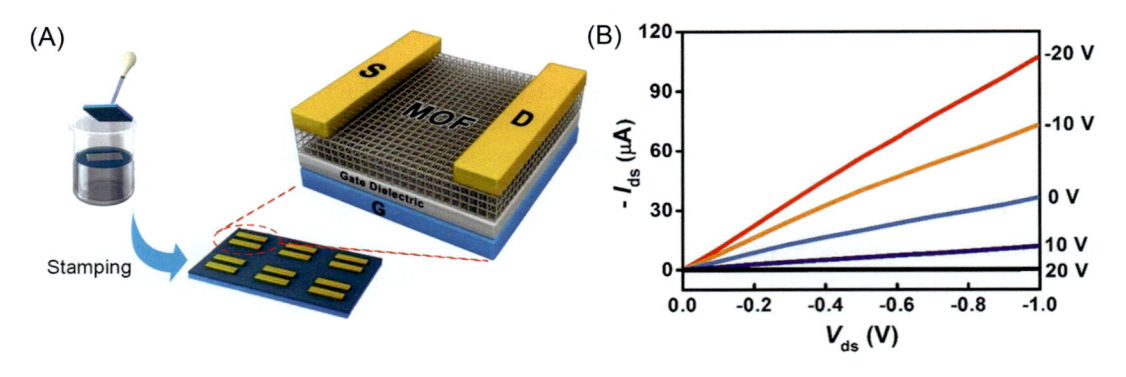

FIGURE 15.10

(A) Procedure for fabrication and final configuration of metal–organic frameworks-field effect transistors based on semiconducting Ni-HITIP framework. (B) Output and transfer curves obtained for the configuration.

Adapted from G. Wu, J. Huang, Y. Zang, J. He, G. Xu, Porous field-effect transistors based on a semiconductive metal-organic framework, J. Am. Chem. Soc. 139 (2017) 1360–1363.

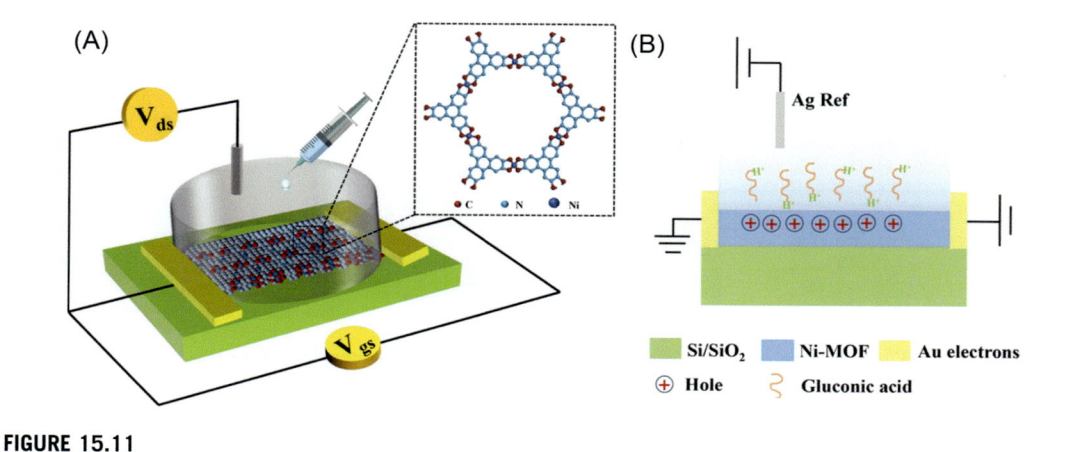

FIGURE 15.11

(A) Top liquid-gated configuration of metal—organic frameworks-field effect transistors based on semiconducting Ni-HITIP framework. (B) Schematic representation of electrodes employed for gluconic acid detection.

Reproduced from S. Shen, P. Tan, Y. Tang, G. Duan, Y. Luo, Adjustable synthesis of Ni-based metal-organic framework membranes and their field-effect transistor sensors for mercury detection, ACS Appl. Electron. Mater. 4 (2022) 622—630.

solid-state MOF-GFETs aiming to create sensors for gas phase [140]. For this end they employed a MOF based on Cu^{2+} and benzene-dicarboxylic acid belonging to an isoreticular series for which liquid phase epitaxy synthesis procedures were proved successful [141]. The detailed nanofabrication procedure is depicted in Fig. 15.12. In the final device obtained, sensing areas are located between the metal source—drain electrodes where graphene can be recognized behind the MOF layer due to the optical transparency of the thin layer.

Regarding applications exploiting changes on electrical conductivity arising from the responsiveness of carefully designed devices including MOFs, there are several recent interesting examples that must be discussed. Leme et al. prepared an impedance-based detection device for highly specific protein-protein interactions using gold interdigitated electrodes [142]. Specifically, the nanoarchitecture proposed exploits the use of a highly versatile and biocompatible Zn-based MOF (ZIF-8) film, which was decorated with an affinity-conferring agent known as Trx-1, a highly specific binding partner of ADAM17cyto protein (see Fig. 15.13).

Another approach to the fabrication of chemiresistive sensing devices was presented by Aggarwal et al. In this contribution the researchers used a Cu-based HKUST-1 MOF and performed a careful integration with paper and MoS_2 substrates. Such flexible nanoarchitecture allows online monitoring of breath dynamics and is thus very useful for the diagnosis of sleep apnea (Fig. 15.14) [143].

Closely related to the above-discussed approach, recently proposed was a method for impedimetric detection of a common post-harvest pesticide, imazalil [144]. Based on a nanoarchitecture that integrates MOFs and conductive polymers, the mechanism for sensing relies on the affinity of the analyte for the adsorption sites in the framework which induce an impedimetric response of the sensor proportional to imazalil concentration (see Fig. 15.15).

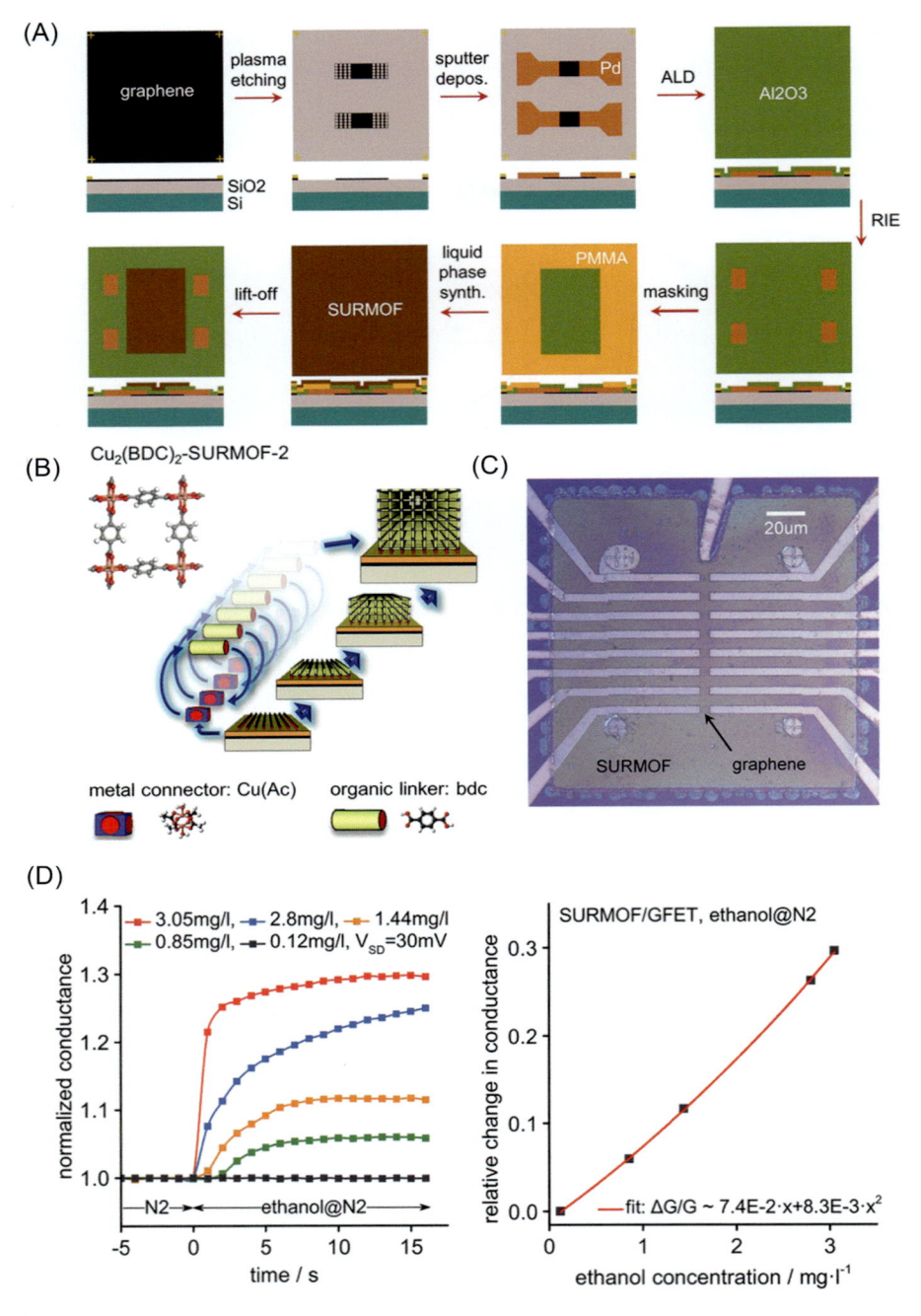

FIGURE 15.12

(A) Procedure for the fabrication of metal−organic frameworks(MOF)-graphene-based field effect transistor (GFETs) devices. Multiple electron beam lithography employed for patterning, etching, and deposition. (B) Liquid-phase epitaxy MOF synthesis on top of the GFET. (C) Optical microscopy image of the device obtained. (D) Time response and sensitivity obtained upon exposure to ethanol vapor in a stream of N_2.

Adapted from J. Liu, B. Lukose, O. Shekhah, H.K. Arslan, P. Weidler, H. Gliemann, et al., A novel series of isoreticular metal organic frameworks: realizing metastable structures by liquid phase epitaxy, Sci. Rep. 2 (2012) 1−5.

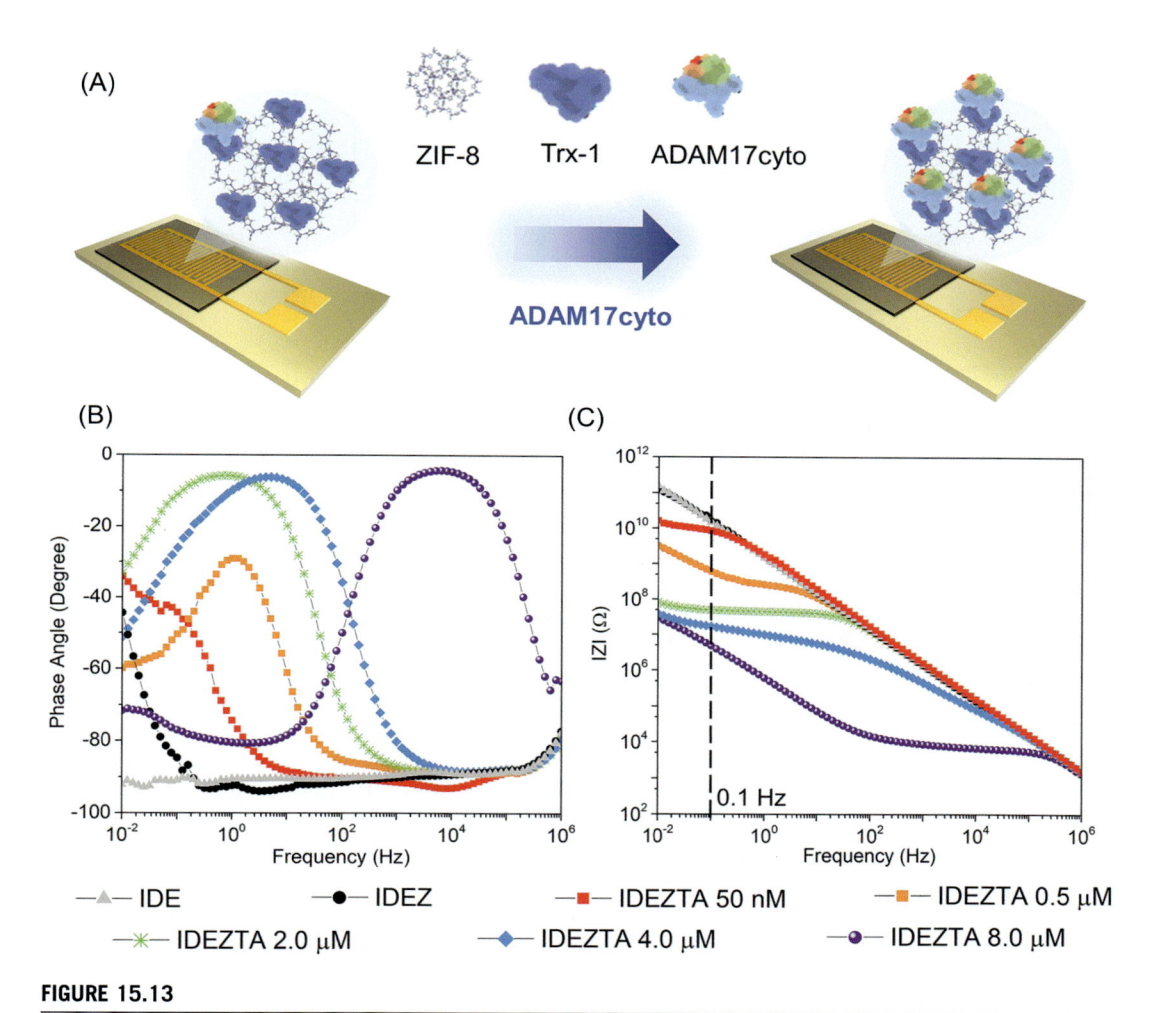

FIGURE 15.13

(A) Illustration indicating the incorporation of ADAM17cyto for the EIS analysis. (B) Changes in Phase angle with concentration of ADAM17cyto. (C) Bode plot for the same conditions.

Reproduced from L.D. Trino, L.G.S. Albano, D.C. Granato, A.G. Santana, D.H.S. De Camargo, C.C. Correa, et al., Framework electrochemical biosensor for the detection of protein-protein interaction, Chem. Mater. 33 (2021) 1293−1306.

15.4.2 Metal−organic frameworks and nanoporous membranes: separation and selective ionic transport

All the examples discussed in this chapter with regard to the integration of MOFs into devices following a nanoarchitectonic approach [145], rely on the separation and/or selective transport of ions or molecules.

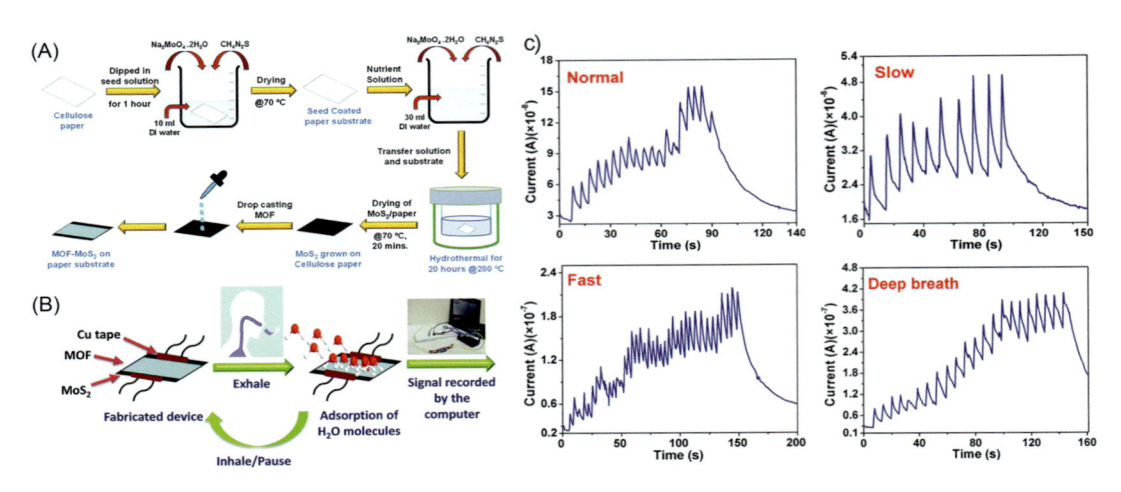

FIGURE 15.14

(A) Schematic illustration of MoS_2 growth on a cellulose paper substrate and drop-casting of metal−organic frameworks powder on the MoS_2 support for device fabrication. (B) HKUST-1−MoS_2 fabricated sensor for breath sensing. (C) Breath sensing experiments showing variation in the current in response to different types of breaths: normal breath, slow breath, fast breath, and deep breath.

Adapted from T. Leelasree, V. Selamneni, T. Akshaya, P. Sahatiya, H. Aggarwal, MOF based flexible, low-cost chemiresistive device as a respiration sensor for sleep apnea diagnosis, J. Mater. Chem. B. 8 (2020) 10182−10189.

Fig. 15.16A illustrates the different porous dimensions required for some typical applications. Pore sizes of MOFs are typically below 5 nm diameter, which makes them suitable for ultra- and nanofiltration, reverse osmosis, and gas/liquid phase separations, especially appealing for applications involving relatively small organic molecules, dissolved ions, and gases [147]. When it comes to separation processes, biological ion channels can be regarded as the gold standard, such systems are highly specific, selective, and efficient toward ion separation, even for ions with very similar charge densities and sizes. To produce synthetic systems capable of mimicking the performances of biological ion channels is something that has driven the interest of the research community around the world in the last few decades [148−151]. One of the most impressive examples of this much-desired biomimetic systems was obtained through the use of ion-track etched membranes. In these membranes it is possible to position a predesigned density of nanochannels with controlled shapes, even reaching single-pore limit; such single-channel membranes have proven to be excellent candidates for testing applications in different fields, ranging from sensing to energy harvesting [152−156]. The discussed examples highlight the fact that track-etched membranes are capable of achieving efficient ion separation and selective transport of monatomic ions, small polyatomic ions, and even larger entities [157,158]. A possible strategy for the creation of further selectivity would be to reduce channel dimensions, alter the chemistry, or both, and this is where the integration of MOFs in such nanoarchitectures becomes interesting. The recently proposed synergic combination of MOFs with ion-track etched nanochannel membranes represents a new step toward biomimetic channels. Following this approach, Li et al. reported the synthesis of UiO-66 derivates (a Zr-based

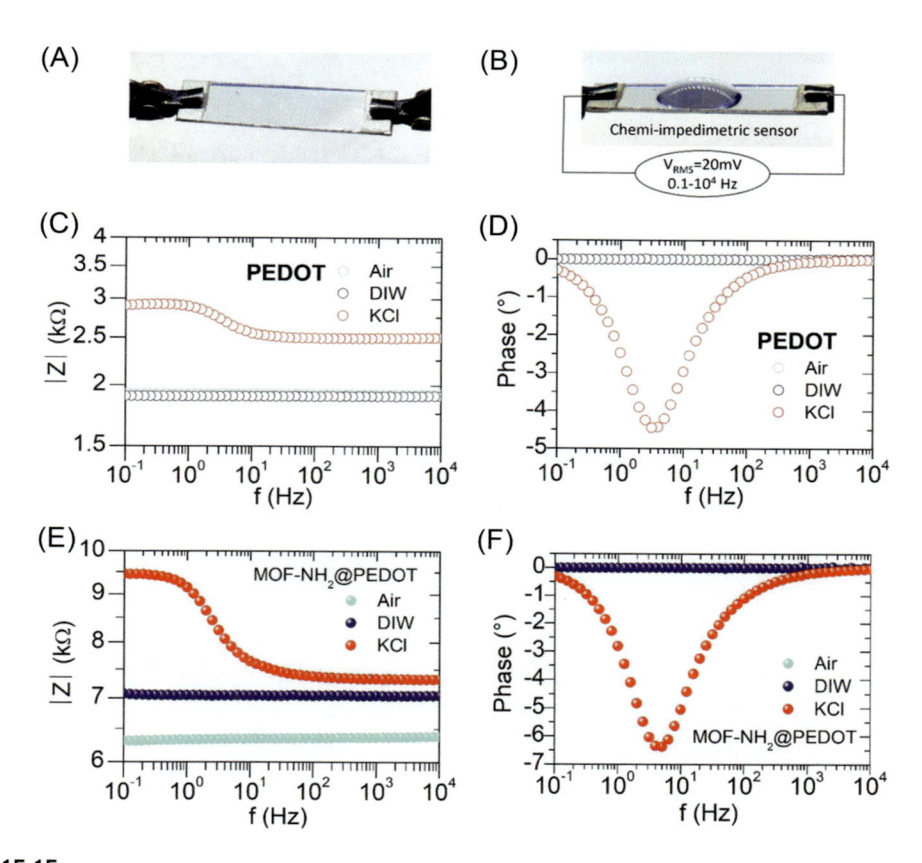

FIGURE 15.15

(A,B) Experimental setup for the impedance experiments applying a 20 mV AC sinusoidal signal from 0.1 Hz to 10 kHz of a metal—organic frameworks(MOF)-NH$_2$@PEDOT composite deposited on a PET substrate. Module and phase of the electrical impedance in air, DIW, and 0.1 m KCl for (C,D) the PEDOT and (E,F) MOF-NH$_2$@PEDOT composite films.

Adapted from L.D. Sappia, J.S. Tuninetti, M. Ceolín, W. Knoll, M. Rafti, O. Azzaroni, MOF@PEDOT Composite films for impedimetric pesticide sensors, Glob. Chall. 4 (2020) 1900076.

MOF) in track-etched nanochannel membranes, yielding selective and fast transport of fluoride ions [146]. As depicted in Fig. 15.16B, UiO-66-based MOFs provide a positively charged porous environment, which excludes cations while allowing the transport of anions. In that work, the nanochannel was completely filled with the UiO-66 phase, as shown in Fig. 15.16C. In addition to the charge state and size-exclusion, the MOF chemistry can be used as an advantage; the strong interaction with the Zr centers and positively charged groups from the MOF structure provided nanochannels with a F$^-$/anion selectivity ranging from ∼50 to ∼250. Other derivates of UiO-66 MOF were successfully integrated into nanochannels, allowing the selective transport of alkali metal ions [159,160].

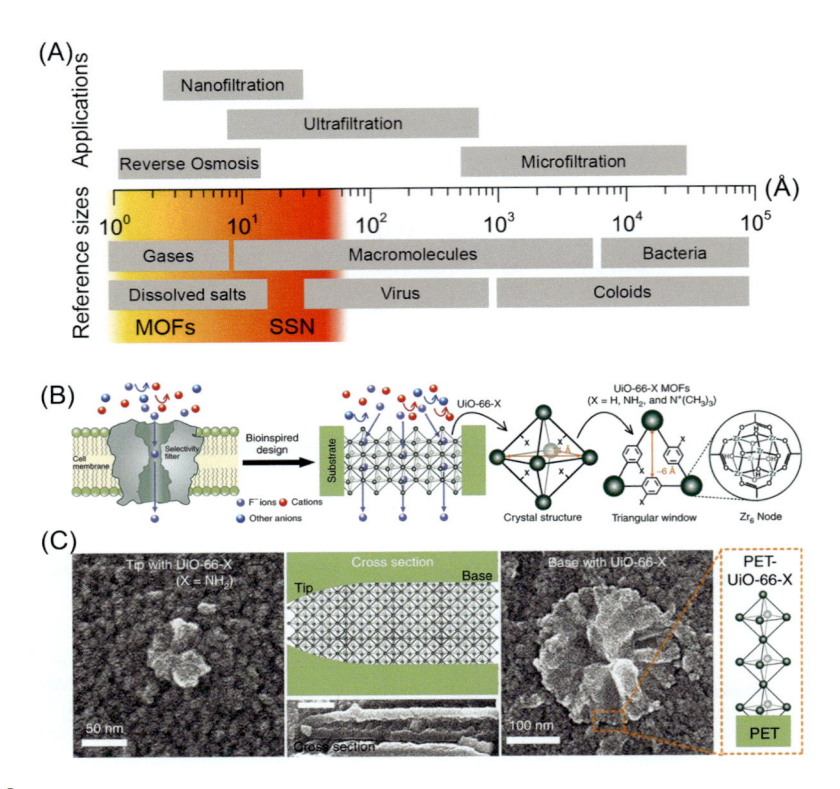

FIGURE 15.16

(A) Pore size range of the most extended metal—organic frameworks across literature and the separation process they can be applied on. (B) Comparison of biological fluoride ion channel and UiO-66-modified track-etched nanochannel membranes. (C) SEM images for tip- and base-sides and cross-section of UiO-66 modified membranes [146].

Adapted from X. Li, H. Zhang, P. Wang, J. Hou, J. Lu, C.D. Easton, et al., Fast and selective fluoride ion conduction in sub-1-nanometer metal-organic framework channels, Nat. Commun. 10 (2019) 1—12.

As for fluoride and alkali metal ions, nanochannels with unidirectional and selective transport of protons were also generated by integration with MOFs [161]. For this work, Li et al. employed benzene-carboxylic acid-based MOFs, namely, MIL-53(Al), MIL-53(Al)-NH_2, and MIL-121. The MOF microporous environment provided a way to rearrange water molecules across the nanochannel, from disordered to ordered states, as depicted in Fig. 15.17A. This rearrangement allows for the selective and unidirectional transport of protons, where the MIL-121 MOF, with carboxylic pending moieties, the most favorable candidate. In a subsequent work, Li et al. also explored the use of UiO-66 derivates for proton transport, see Fig. 15.17B [162]. Particularly, by modifying UiO-66-$(NH_2)_2$ with sulfonic acid groups by postsynthesis modification with sulfoacetic acid, a high selectivity toward H^+ was achieved, that increases with the extent of sulfonic-acid moieties present after chemical modification.

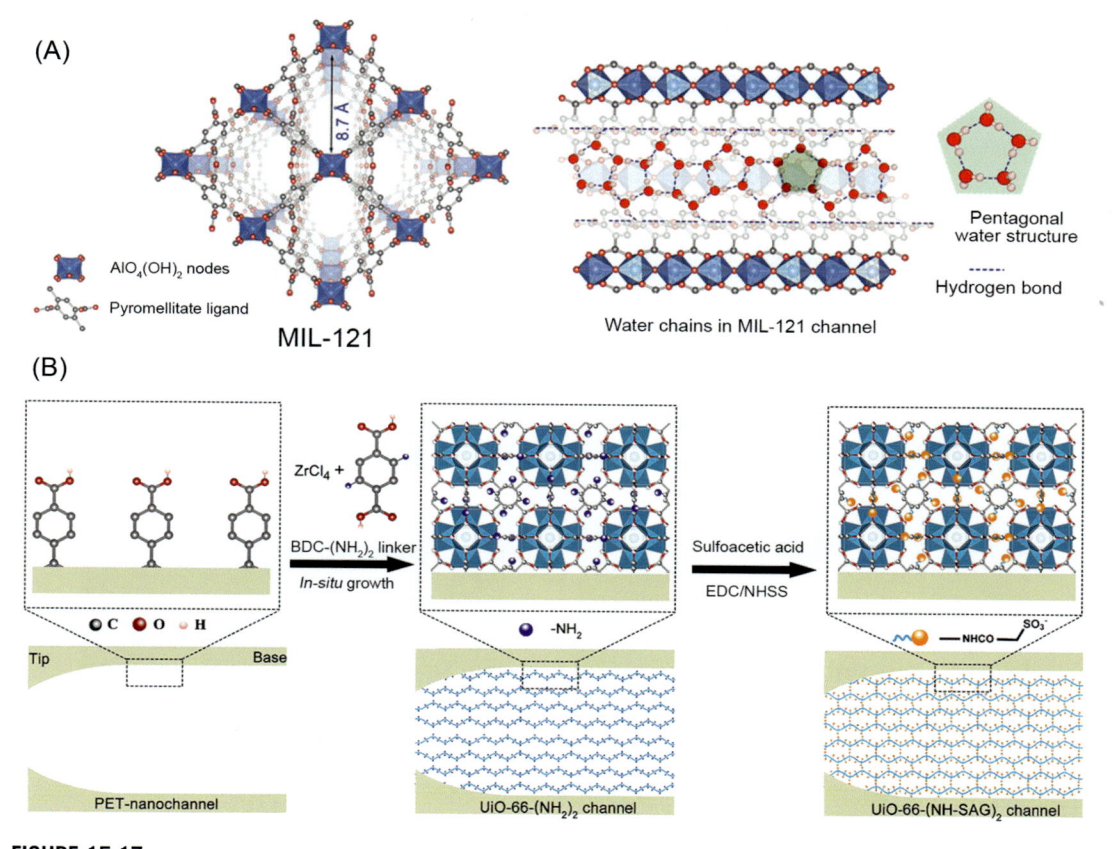

FIGURE 15.17

(A) MIL-121 crystalline structure and ordered water structure that allows for selective and unidirectional transport of H^+. (B) Scheme of PET-nanochannel modification with UiO-66-$(NH_2)_2$ and postmodification with sulfoacetic acid.

(A) Adapted from 10.1002/adma.202001777, (B) Adapted from X. Li, H. Zhang, J. Hou, R. Ou, Y. Zhu, C. Zhao, et al., Sulfonated Sub-1-nm metal-organic framework channels with ultrahigh proton selectivity, J. Am. Chem. Soc. 142 (2020) 9827–9833.

In the previously discussed examples, MOFs@nanochannels were synthesized by interfacial growth, that is, using the nanochannel walls as nucleation points for growth. Those synthesis methods were conducted either in a symmetrical or asymmetrical fashion. For the so-called symmetrical synthesis, both MOFs precursors were mixed in a Teflon autoclave with the membrane vertically placed inside. For asymmetrical interfacial synthesis, the nanochannel membrane is not used only as a template for MOF growth, but also as a barrier between each of the MOF precursors. Typically, the linker solution is placed on the tip-side of the membrane, while the base-side is exposed to the metal ion solution, and the nucleation and growth of the MOF phase takes place upon mixing of the precursors inside the nanochannels. The previously discussed reports presented either ohmic or diode-like behaviors as iontronic output. *Iontronics* is the field dealing with devices where current is governed by ionic transport, rather than electrons, and solid-state nanochannels have proven to be perfect candidates to produce readable signals from biological, chemical, or physical inputs based on ion transport [148,163]. Fairly recently, by following this asymmetric interfacial synthesis approach (see Fig. 15.18A), Laucirica et al. reported a UiO-66-modified single-channel PET membrane, with an ion current saturation regime, not previously observed for

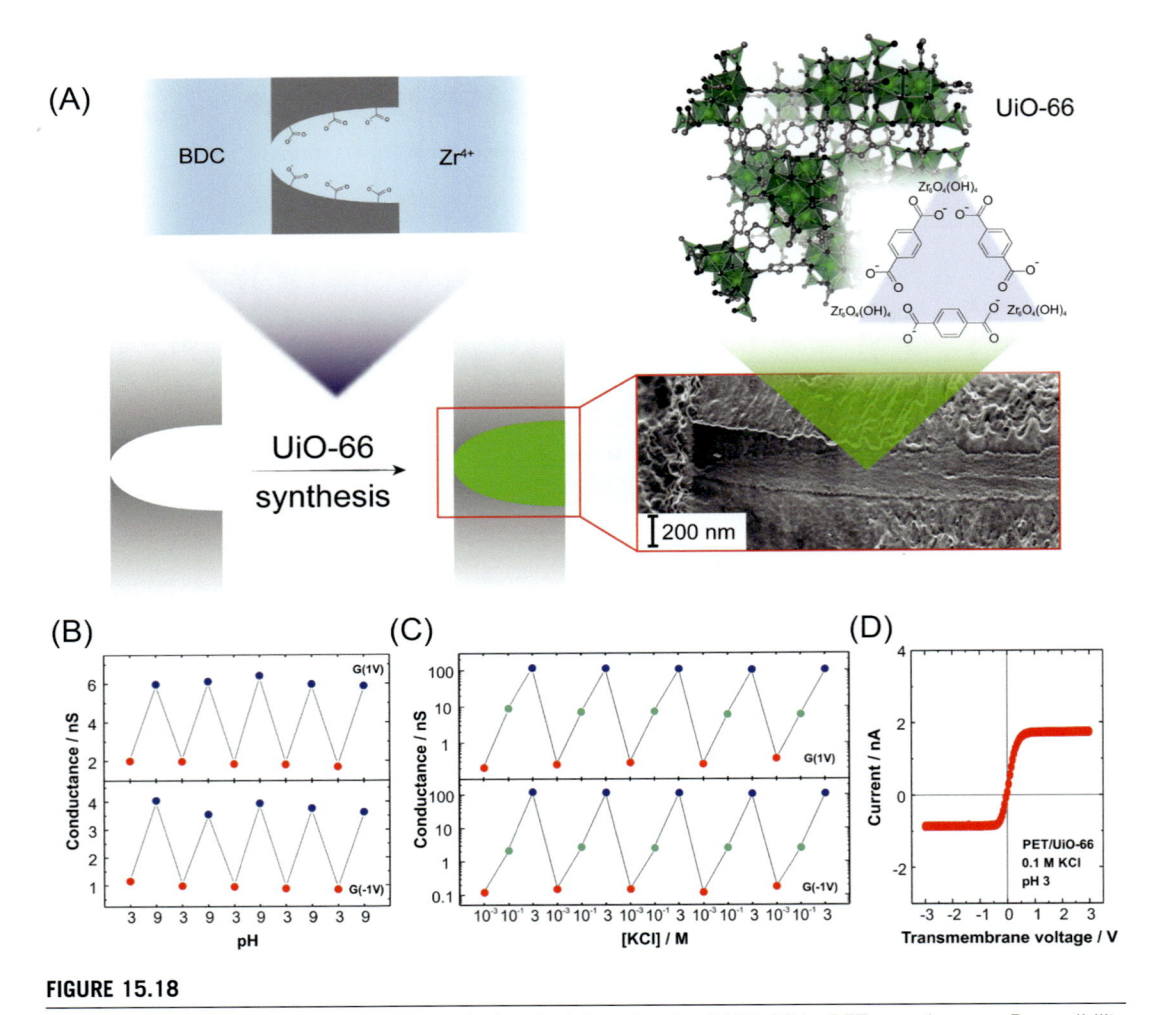

FIGURE 15.18

(A) Schematic representation of asymmetric interfacial synthesis of UiO-66 in PET membranes. Reversibility of iontronic output upon (B) pH and (C) ionic strength changes. (D) Ion current saturation response of the system at pH = 3 in a ± 3 V voltage window.

Adapted from G. Laucirica, J.A. Allegretto, M.F. Wagner, M.E. Toimil-Molares, C. Trautmann, M. Rafti, et al., Switchable ion current saturation regimes enabled via heterostructured nanofluidic devices based on metal−organic frameworks, Adv. Mater. 2207339 (2022) 2207339.

MOF-based solid-state nanochannels [164]. In this work, it is proposed that intrinsic microporosity of the MOF phase is not playing a major role on the ion transport, but rather it is the *constructional porosity* (with wider dimensions), which arises from the coalescence of growth fronts across the crystalline phase. Notoriously, such a transport regime could be reversibly switched on/off by modifying pH values, which indicates a charge–structure interplay dictating the ion transport regime. Even further, the device showcased an excellent reversibility upon changes in pH (Fig. 15.18B) and ionic strength (Fig. 15.18C). Remarkably, and aside from the chemical stability of the device, the UiO-66-modified PET membrane retained the ion current saturation regime in a ± 3 V voltage window (Fig. 15.18C).

Other synthesis strategies can be followed in order to integrate MOFs in nanochannels. In the work of Lu et al., however, a seeded growth approach was proposed: presynthesized UiO-66-$(COOH)_2$ particles were employed to fill the tip-region of PET nanochannels, followed by an asymmetrical interfacial growth process [160].

As can be seen in Fig. 15.19A, bullet-shaped nanochannels were modified by introducing preformed UiO-66 particles (Fig. 15.19B), thus facilitating and directing the subsequent growth of the crystalline phase inside the nanochannel (Fig. 15.19C–D). The resulting membrane presented a rectifying iontronic output and allowed for pH-tunable K^+/Mg^{2+} selectivity and permeability, highlighting again the versatility and potential of these systems as biomimetic devices.

15.4.3 Photoresponsive metal–organic frameworks and applications in photonic crystals

The versatility of both native (i.e., as-synthesized) MOFs, and the possibility of rational design regarding surface chemistry and porosity has led to a great number of developments in terms of tailored interaction with analytes in confined spaces. Recent developments have enabled manipulation of interesting MOF features such as bandgap and semiconducting properties, which has opened a whole new subfield devoted to their possible applications in photovoltaics [165]. Although the potential of MOFs for their integration in photonic devices is evident, the number of reports is considerably lower compared to the other mainstream uses, such as selective adsorption or heterogeneous catalysis. One of the earliest examples of sensors taking advantage of measurable changes in refractive index upon adsorption of an analyte in the porous environment provided by a thin-film MOF was given by Hupp's group [166]. This seminal work exploited the size-selectivity of the MOF employed (ZIF-8); the assembly was capable of sensing n-hexane, while the bulkier isomer cyclohexane was not detected.

The next logic step in the fabrication of an actual porous photonic crystal would be to create alternating layers of porous materials with different refractive index. Along this line of thought, but following rather a bottom-up approach, Takashi et al. demonstrated the feasibility of creating multiple MOF tailored architectures by resorting to the epitaxial growth [167]. For example, by alternating layers of different linkers, the structure of an epitaxial MOF composed of the same metal center can be switched, thus providing the refractive index variation needed in a photonic crystal. The feasibility of such epitaxial growth and the effect on the observed film properties can be quantified by resorting to the crystallographic mismatch ratio between the considered phases.

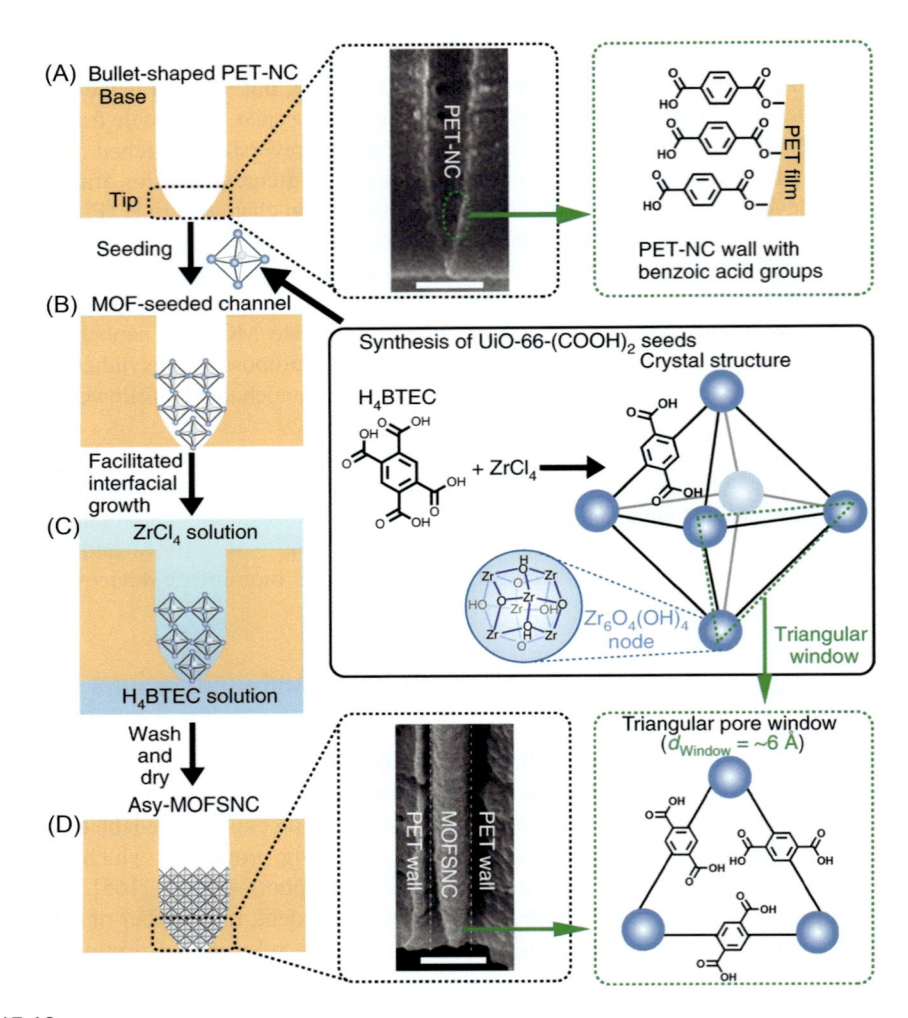

FIGURE 15.19

(A) Pristine PET nanochannel membrane employed for (B) seeded growth of UiO-66-(COOH)2 metal—organic frameworks (MOF) by incorporation of presynthesized nanoparticles and later (C) interfacial growth, to generate (D) an asymmetrically filled MOF@Nanochannel membrane.

From J. Lu, H. Zhang, J. Hou, X. Li, X. Hu, Y. Hu, et al., Efficient metal ion sieving in rectifying subnanochannels enabled by metal—organic frameworks, Nat. Mater. 19 (2020) 767—774.

Top-down approaches also have proven to be useful for the assembly of photonic crystals, Lotsch et al. reported an alternating refractive index structure employing ZIF-8 MOF and TiO_2 nanoparticles giving rise to an uniform nanoarchitecture of Bragg stacks, and used it as label-free optical sensors, one-dimensional photonic crystals (1D-MOF-PC) [168]. A nice example of the above discussed strategy was given also by Lotsch et al., as depicted in Fig. 15.20 [169]. The

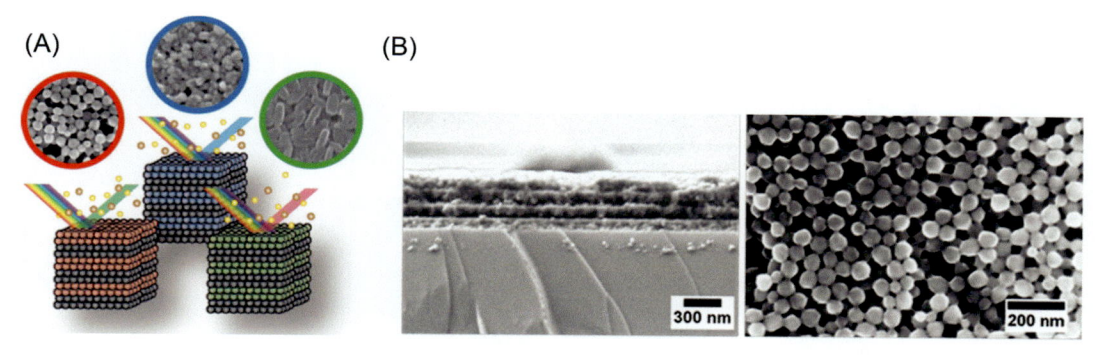

FIGURE 15.20

(A) Schematic of (B) Bragg stacks created using alternate metal–organic frameworks for thin films and detection of vapor.

Adapted from A. Ranft, F. Niekiel, I. Pavlichenko, N. Stock, B. V. Lotsch, Tandem MOF-based photonic crystals for enhanced analyte-specific optical detection, Chem. Mater. 27 (2015) 1961–1970.

authors resorted to a structuration strategy for creating photonic multilayers using different MOFs, which were deposited into films from colloidal suspensions of nanoparticles. Specifically, they employed HKUST-1, ZIF-8, and CAU-1-NH$_2$ MOFs for the label free detection of vapors, these MOFs comply with the following requisites needed in such a proof-of-concept: well-determined structures and synthesis methods, availability of colloidal nanoparticles, and different affinities provided by pore chemistry.

Another approach toward a top-down synthesis of photonic crystals using MOFs was presented recently by Chen [170]. They employed a combination between a dendrimer (polyamidoamine, PAMAM) and ZIF-8 nanoparticles, which upon integration using reduced graphene oxide (rGO) through a postsynthetic method yielded different colors depending on the size of the nanoparticles used for the assembly. And moreover, such a hierarchical nanoarchitecture of PAMAM@ZIF-8 featured the possibility of selective dye adsorption, see Fig. 15.21.

An interesting combination of photonics and porous environments provided by MOFs is the development of *parts-per-billion-sensitive* volatile organic compounds (VOCs) detectors. Similar to the sensing approaches based on whispering gallery mode microresonators, they are even capable of reaching single-molecule limit [171,172]. Zhao et al. employed a microring resonator, in which the light propagation is confined in circulating waveguide modes arising from total internal reflection in a curved interface between two media of different refractive index [173]. The waveguide modes generate evanescent fields that can be employed to sense alterations in the dielectric properties of the surroundings caused by, for example, adsorption of an analyte. The authors employed a nanoarchitecture based on a thin film of ZIF-8 MOF, and through careful integration with the microresonator obtained increased affinity for the analyte to create a highly sensitive sensor, as shown in Fig. 15.22.

In a recent report Ameloot et al. presented a parts-per-million-sensitive detection method for volatile organic compounds using surface plasmon polaritons (SPP) wavelengths shifts caused by adsorption on thin MOF films. It employs a total internal reflection ellipsometry (TIRE) setup and by careful integration of a ZIF-8 MOF capable of adsorption achieves such a low limit resolution. The working principle is presented in Fig. 15.23 below [174].

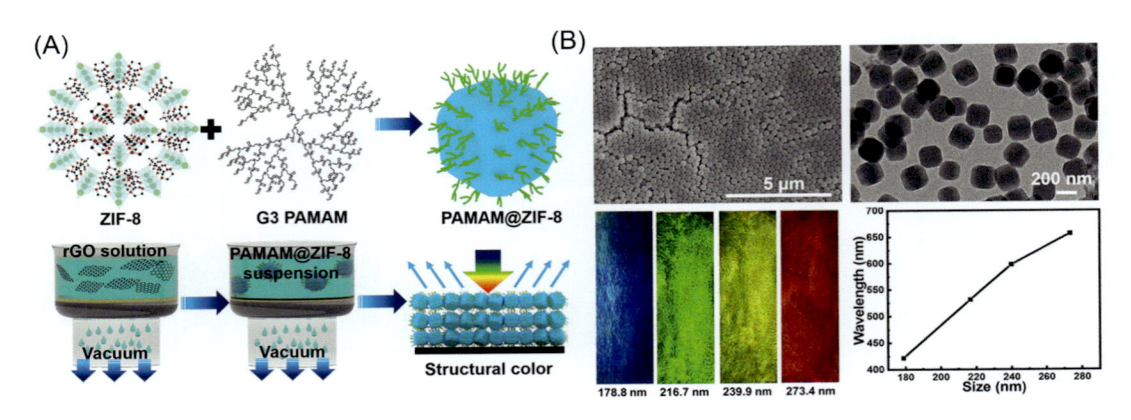

FIGURE 15.21

(A) Schematic illustration of the preparation of PAMAM@ZIF-8 particles (top) and the assembly process of PAMAM@ZIF-8 particles (bottom). (B) SEM and TEM images of ZIF-8 nanoparticles employed (top) to generate PC films with blue, green, yellow, and red structural colors (bottom).

Adapted from C. Liu, Y.L. Tong, X.Q. Yu, H. Shen, Z. Zhu, Q. Li, et al., MOF-based photonic crystal film toward separation of organic dyes, ACS Appl. Mater. Interfaces. 12 (2020) 2816–2825.

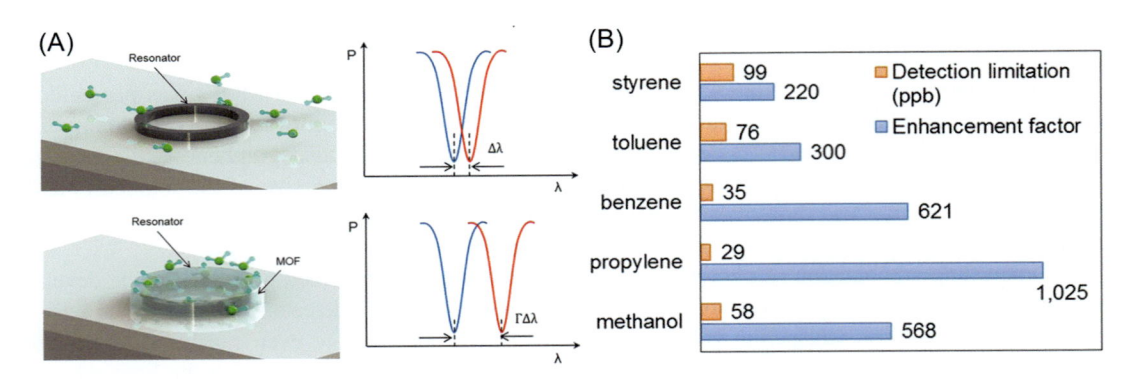

FIGURE 15.22

(A) Schematic representation the gas sensing by use of an optical microring resonator. (B) Enhancement factor and detection limitation of five VOCs at equilibrium.

Adapted from J. Tao, X. Wang, T. Sun, H. Cai, Y. Wang, T. Lin, et al., Hybrid photonic cavity with metal-organic framework coatings for the ultra-sensitive detection of volatile organic compounds with high immunity to humidity, Sci. Rep. 7 (2017) 41640.

Finally, it is worth highlighting a recent development reported by Huang et al. [175]. in which the nanoarchitecture consisted of a MOF microcrystal mounted on a single optical fiber (OF), and through light input−output readout the uptake of an analyte can be sensed. The sensor consists of an OF and an MOF single crystal (OF@MOF) attached to the end-face of the OF, as illustrated in

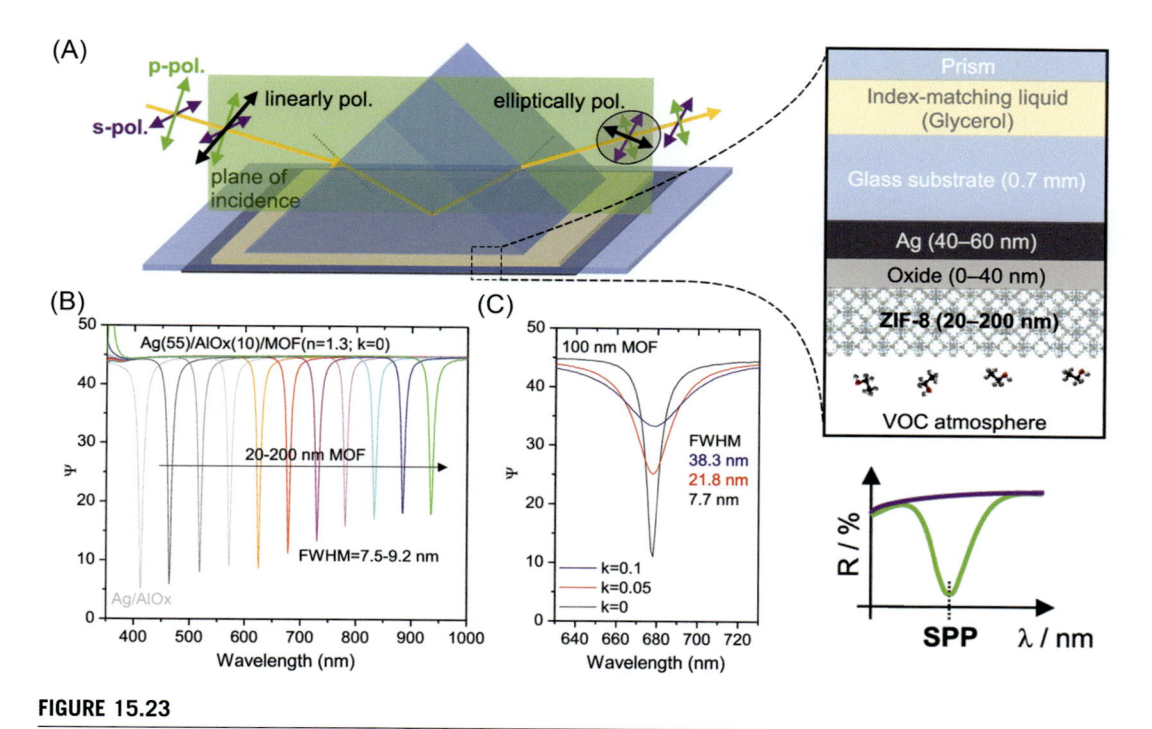

FIGURE 15.23

(A) TIRE measurement geometry and sample layer stack. (B) Modeled SPP spectra for Ag(55 nm)/AlOx (10 nm)/metal–organic frameworks (MOF) stacks with MOF layers in the 20 – 200 nm thickness range. (C) Modeled SPP broadening due to an increase in the MOF k-value from 0 to 0.1 for a 100 nm MOF film.

Adapted from M.L. Tietze, M. Obst, G. Arnauts, N. Wauteraerts, S. Rodríguez-Hermida, R. Ameloot, Parts-per-million detection of volatile organic compounds via surface plasmon polaritons and nanometer-thick metal–organic framework films, ACS Appl. Nano Mater. 5 (2022) 5006–5016.

the schematic diagram in Fig. 15.24A and the optical microscope image of an OF@HKUST-1 in Fig. 15.24B. An FPI is formed by the OF − MOF interface and the MOF − surrounding interface, with the MOF crystal acting as the medium of the Fabry − Pérot cavity. The input light transmitted via the core of the OF is partially reflected (I1) by the OF − MOF interface, and the remaining input light that passes through the crystal is again partially reflected (I2) by the MOF − surrounding interface. These two reflected light beams interfere to generate the output interference signal, see Fig. 15.24C.

In a similar note to what has been discussed in previous sections, also MOFs can be used as a protection matrix for biomolecules, such as enzymes, when the goal is to assemble optical devices. In this way, Vicki Chen et al. designed a microfluidic optical biosensor using a deposit of ZIF-8 MOF biomineralized with glucose oxidase enzyme (GOx) for glucose detection [176]. A chromogen changes its color depending on a cascade reaction directly related to glucose concentration, as schematized in Fig. 15.25.

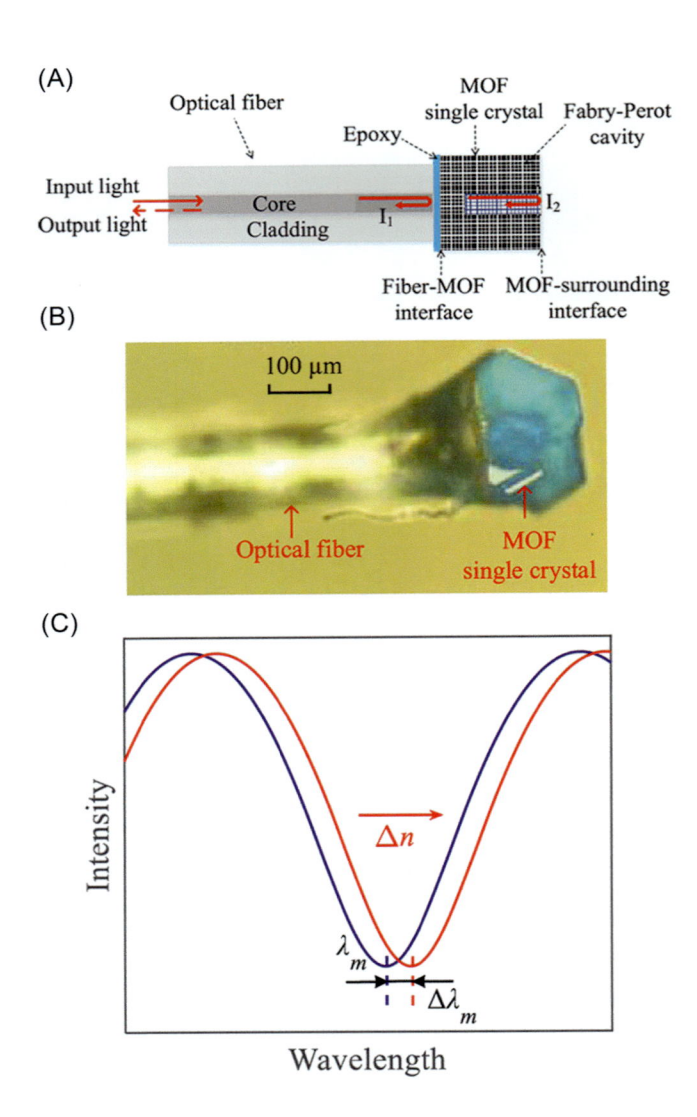

FIGURE 15.24

(A) Schematic diagram of the OF@MOF (metal–organic frameworks) sensor. (B) An optical microscope image of a prototype OF@MOF sensor. (C) Illustration of a section of the original (blue) and shifted (red) interferograms.

Adapted from C. Zhu, J.A. Perman, R.E. Gerald, S. Ma, J. Huang, Chemical detection using a metal-organic framework single crystal coupled to an optical fiber, ACS Appl. Mater. Interfaces. 11 (2019) 4393–4398.

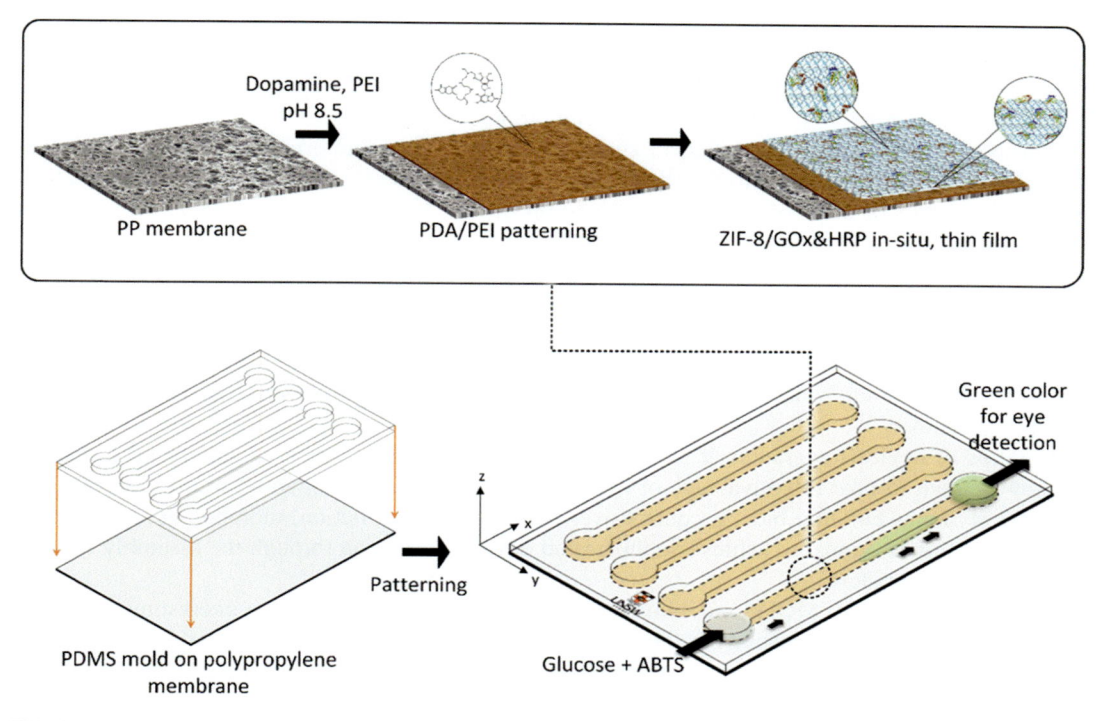

FIGURE 15.25

Assembly of metal—organic frameworks-containing optical biosensor. A polypropylene membrane coated with polydopamine/polyethyleneimine (PDA/PEI) patterning using PDMS mold with microchannels. Subsequently, a composite of ZIF-8/Gox/Horseradish-peroxidase (HRP) is integrated over the PDA/PEI layer.

Adapted from M. Mohammad, A. Razmjou, K. Liang, M. Asadnia, V. Chen, Metal-organic-framework-based enzymatic microfluidic biosensor via surface patterning and biomineralization, ACS Appl. Mater. Interfaces. 11 (2019) 1807–1820.

15.5 **Conclusions and outlook**

Metal—organic frameworks have attracted much interest because of the variety of nanoarchitectures with remarkable porosity that can be controlled on the molecular scale via self-assembly. In comparison to other highly porous materials such as, for example, zeolites or carbon-based materials, the active surface area of MOFs is extremely high. One appealing aspect of MOFs relies on the fact that the size, shape, and composition of the pores can be easily tunable upon adsorption of host organic molecules, whereas the modulation of the pore size can vary between 50% and 235% of the volume of the unit cell, depending on the structure and the length of the linker.

Since its inception in the late 1990s these materials have been reported to provide an ample variety of functional features such as gas sorption, catalysis, and tunable electrical conductivity. While the focus in the past decades has been mostly on their synthetic aspects together with an in-depth study of their outstanding properties, over several years now the core area of research has already started to shift toward its integration into specific devices. For instance, the relative ease of preparation is a special strength of MOFs, which has lent important impetus to the exploration of these nanoarchitectures in different device configurations.

The controlled growth of MOFs on surfaces constitutes an important step toward the successful integration of MOFs into functional devices. As MOFs are essentially solid-state materials and insoluble in any solvent without decomposition, the deposition of MOF materials as homogeneous films or ultrathin crystalline layers is not a trivial task. In this sense, the ability to fully control MOF crystallization processes is crucial for subsequent integration into any functional devices, such as chemical sensors or smart membranes. Within this framework, the production of MOFs in the form of colloids has shed new light on the design and preparation of these microporous nanoarchitectures. Along with their intrinsic interest as porous solids, colloidal MOF particles can additionally be dispersed, shaped, functionalized, and assembled in a controlled manner.

In terms of device integration, it is clear that control over the location of the crystal formation is a key element to take full advantage of the MOF properties. Concomitantly, in the case of using colloidal MOFs as building blocks, control over particle size distribution, morphology, and surface functionalization is critical to achieve sophisticated device fabrication through the assembly of individual nanoscale objects.

The combination of MOFs and biomolecules, such as enzymes, offers interesting routes to porous nanobiocomposites displaying biorecognition properties. In this regard, considerable interest is presently being generated on the potential application of these hybrid nanomaterials in diverse fields such as biosensing and catalysis. The development of field-effect transistors based on MOFs is also a true demonstration of the potential feasibility of integration between electronics and inorganic materials to devise nanosystems with potential benefits for different industrial applications. In a similar way, the development of (bio)electrochemical devices using MOFs as functional units illustrates the capability of these materials to give access to nanostructured electrodes designed for specific applications.

It is clear that MOFs have provided an entirely new way of designing porous materials with the aid of inorganic and coordination chemistry. Our understanding of MOFs at the molecular and nanoscale level together with our ability to control their function and structure has granted access to a range of materials with novel characteristics, functions and potential applications that could lead, in a not-too-distant future, to interesting technologies. At this point, for the sake of fairness, it must be said that device nanomanufacturing with MOFs is still in the conceptual stage of development and will probably have to face many challenging technical barriers in the near future. In this respect, devising highly reproducible processes to grow or deposit MOFs on different surfaces—such as polymers, graphene, self-assembled monolayers, ceramics and so on—would be essential to assure the reproducibility and repeatability of the proposed nanomanufacturing protocols.

In summary, this chapter has provided a broad description of relevant examples of MOF nanoarchitectonics as well as a discussion of applications and potential uses of MOFs in the fields of energy and biosensing. Hopefully, this work will trigger a cascade of new, refreshing ideas in MOF nanoarchitectonics as well as engender an interest in resorting to MOFs as building blocks to create advanced devices.

References

[1]   M. Faustini, L. Nicole, E. Ruiz-hitzky, C. Sanchez, History of organic − inorganic hybrid materials: prehistory, Art, Science, Adv. Appl. 1704158 (2018) 1−30. Available from: https://doi.org/10.1002/adfm.201704158.

[2]   V. Aggarwal, S. Solanki, B.D. Malhotra, Applications of metal−organic framework-based bioelectrodes, Chem. Sci. 13 (2022) 8727−8743. Available from: https://doi.org/10.1039/D2SC03441G.

[3]   I. Stassen, N. Burtch, A. Talin, P. Falcaro, M. Allendorf, R. Ameloot, An updated roadmap for the integration of metal−organic frameworks with electronic devices and chemical sensors, Chem. Soc. Rev. 46 (2017) 3185−3241. Available from: https://doi.org/10.1039/C7CS00122C.

[4]   H. Furukawa, K.E. Cordova, M. O'Keeffe, O.M. Yaghi, The chemistry and applications of metal-organic frameworks, Sci. (80-.) 341 (2013). Available from: https://doi.org/10.1126/science.1230444.

[5]   L. Zhu, X.Q. Liu, H.L. Jiang, L.B. Sun, Metal-organic frameworks for heterogeneous basic catalysis, Chem. Rev. 117 (2017) 8129−8176. Available from: https://doi.org/10.1021/acs.chemrev.7b00091.

[6]   D.T. Bui, K.J. Chua, J.M. Gordon, Comment on "Water harvesting from air with metal-organic frameworks powered by natural sunlight, Sci. (80-.) 358 (2017) 1−10. Available from: https://doi.org/10.1126/science.aao0791.

[7]   L. Sun, M.G. Campbell, M. Dincə, Electrically conductive porous metal-organic frameworks, Angew. Chem. - Int. Ed. 55 (2016) 3566−3579. Available from: https://doi.org/10.1002/anie.201506219.

[8]   C.H. Hendon, A.J. Rieth, M.D. Korzyński, M. Dincǎ, Grand challenges and future opportunities for metal-organic frameworks, ACS Cent. Sci. 3 (2017) 554−563. Available from: https://doi.org/10.1021/acscentsci.7b00197.

[9]   P.-L. Wang, L.-H. Xie, E.A. Joseph, J.-R. Li, X.-O. Su, H.-C. Zhou, Metal−organic frameworks for food safety, Chem. Rev. 119 (2019) 10638−10690. Available from: https://doi.org/10.1021/acs.chemrev.9b00257.

[10] P. Li, Z. Zhou, Y.S. Zhao, Y. Yan, Recent advances in luminescent metal−organic frameworks and their photonic applications, Chem. Commun. 57 (2021) 13678−13691. Available from: https://doi.org/10.1039/D1CC05541K.

[11] S.R. Batten, N.R. Champness, X.M. Chen, J. Garcia-Martinez, S. Kitagawa, L. Öhrström, et al., Terminology of metal-organic frameworks and coordination polymers (IUPAC recommendations 2013), Pure Appl. Chem. 85 (2013) 1715−1724. Available from: https://doi.org/10.1351/PAC-REC-12-11-20.

[12] P.J. Beldon, L. Fábián, R.S. Stein, A. Thirumurugan, A.K. Cheetham, T. Friščić, Rapid room-temperature synthesis of zeolitic imidazolate frameworks by using mechanochemistry, Angew. Chem. - Int. Ed. 49 (2010) 9640−9643. Available from: https://doi.org/10.1002/anie.201005547.

[13] I. Stassen, M. Styles, G. Grenci, H. Van Gorp, W. Vanderlinden, S. De Feyter, et al., Chemical vapour deposition of zeolitic imidazolate framework thin films, Nat. Mater. 15 (2016) 304−310. Available from: https://doi.org/10.1038/nmat4509.

[14] X.-G. Wang, Q. Cheng, Y. Yu, X.-Z. Zhang, Controlled nucleation and controlled growth for size predicable synthesis of nanoscale metal-organic frameworks (MOFs): a general and scalable approach, Angew. Chem. Int. Ed. 57 (2018) 7836−7840. Available from: https://doi.org/10.1002/anie.201803766.

[15] K.A.S. Usman, J.W. Maina, S. Seyedin, M.T. Conato, L.M. Payawan, L.F. Dumée, et al., Downsizing metal−organic frameworks by bottom-up and top-down methods, NPG Asia Mater. 12 (2020). Available from: https://doi.org/10.1038/s41427-020-00240-5.

[16] N. Stock, S. Biswas, Synthesis of metal-organic frameworks (MOFs): routes to various MOF topologies, morphologies, and composites, Chem. Rev. 112 (2012) 933−969. Available from: https://doi.org/10.1021/cr200304e.

[17] W.M. Bloch, N.R. Champness, C.J. Doonan, X-ray crystallography in open-framework materials, Angew. Chem. - Int. Ed. 54 (2015) 12860−12867. Available from: https://doi.org/10.1002/anie.201501545.

[18] M. Sánchez-Sánchez, N. Getachew, K. Díaz, M. Díaz-García, Y. Chebude, I. Díaz, Synthesis of metal-organic frameworks in water at room temperature: salts as linker sources, Green. Chem. 17 (2015) 1500–1509. Available from: https://doi.org/10.1039/c4gc01861c.

[19] C.V. McGuire, R.S. Forgan, The surface chemistry of metal-organic frameworks, Chem. Commun. 51 (2015) 5199–5217. Available from: https://doi.org/10.1039/c4cc04458d.

[20] S. Wang, W. Morris, Y. Liu, C.M. McGuirk, Y. Zhou, J.T. Hupp, et al., Surface-specific functionalization of nanoscale metal-organic frameworks, Angew. Chem. - Int. Ed. 54 (2015) 14738–14742. Available from: https://doi.org/10.1002/anie.201506888.

[21] A.G. Zavyalova, D.V. Kladko, I.Y. Chernyshov, V.V. Vinogradov, Large MOFs: synthesis strategies and applications where size matters, J. Mater. Chem. A. 9 (2021) 25258–25271. Available from: https://doi.org/10.1039/D1TA05283G.

[22] O.K. Farha, I. Eryazici, N.C. Jeong, B.G. Hauser, C.E. Wilmer, A.A. Sarjeant, et al., Metal-organic framework materials with ultrahigh surface areas: is the sky the limit? J. Am. Chem. Soc. 134 (2012) 15016–15021. Available from: https://doi.org/10.1021/ja3055639.

[23] B.M. Connolly, D.G. Madden, A.E.H. Wheatley, D. Fairen-Jimenez, Shaping the future of fuel: monolithic metal-organic frameworks for high-density gas storage, J. Am. Chem. Soc. 142 (2020) 8541–8549. Available from: https://doi.org/10.1021/jacs.0c00270.

[24] Ü. Kökçam-Demir, A. Goldman, L. Esrafili, M. Gharib, A. Morsali, O. Weingart, et al., Coordinatively unsaturated metal sites (open metal sites) in metal-organic frameworks: design and applications, Chem. Soc. Rev. 49 (2020) 2751–2798. Available from: https://doi.org/10.1039/c9cs00609e.

[25] Z. Jiang, L. Ge, L. Zhuang, M. Li, Z. Wang, Z. Zhu, Fine-tuning the coordinatively unsaturated metal sites of metal-organic frameworks by plasma engraving for enhanced electrocatalytic activity, ACS Appl. Mater. Interfaces. 11 (2019) 44300–44307. Available from: https://doi.org/10.1021/acsami.9b15794.

[26] S. Xiang, Y. He, Z. Zhang, H. Wu, W. Zhou, R. Krishna, et al., Microporous metal-organic framework with potential for carbon dioxide capture at ambient conditions, Nat. Commun. 3 (2012) 954–959. Available from: https://doi.org/10.1038/ncomms1956.

[27] Q. Qian, P.A. Asinger, M.J. Lee, G. Han, K. Mizrahi Rodriguez, S. Lin, et al., MOF-based membranes for gas separations, Chem. Rev. 120 (2020) 8161–8266. Available from: https://doi.org/10.1021/acs.chemrev.0c00119.

[28] M. Ding, R.W. Flaig, H.L. Jiang, O.M. Yaghi, Carbon capture and conversion using metal-organic frameworks and MOF-based materials, Chem. Soc. Rev. 48 (2019) 2783–2828. Available from: https://doi.org/10.1039/c8cs00829a.

[29] K.A. Cychosz, A.G. Wong-Foy, A.J. Matzger, Liquid phase adsorption by microporous coordination polymers: removal of organosulfur compounds, J. Am. Chem. Soc. 130 (2008) 6938–6939. Available from: https://doi.org/10.1021/ja802121u.

[30] A.M. Plonka, Q. Wang, W.O. Gordon, A. Balboa, D. Troya, W. Guo, et al., In situ probes of capture and decomposition of chemical warfare agent simulants by Zr-based metal organic frameworks, J. Am. Chem. Soc. 139 (2017) 599–602. Available from: https://doi.org/10.1021/jacs.6b11373.

[31] A.R. Millward, O.M. Yaghi, Metal-organic frameworks with exceptionally high capacity for storage of carbon dioxide at room temperature, J. Am. Chem. Soc. 127 (2005) 17998–17999. Available from: https://doi.org/10.1021/ja0570032.

[32] J.S. Tuninetti, M. Rafti, A. Andrieu-Brunsen, O. Azzaroni, Molecular transport properties of ZIF-8 thin films in aqueous environments: the critical role of intergrain mesoporosity as diffusional pathway, Microporous Mesoporous Mater. 220 (2016) 253–257. Available from: https://doi.org/10.1016/j.micromeso.2015.08.035.

[33] G.M. Segovia, J.S. Tuninetti, O. Azzaroni, M. Rafti, Self-assembled mesoporous zeolitic imidazolate framework-8 (ZIF-8) nanocrystals bearing thiol groups for separations technologies, ACS Appl. Nano Mater. 3 (2020) 11266–11273. Available from: https://doi.org/10.1021/acsanm.0c02376.

[34] S. Yuan, L. Zou, J.S. Qin, J. Li, L. Huang, L. Feng, et al., Construction of hierarchically porous metal-organic frameworks through linker labilization, Nat. Commun. 8 (2017). Available from: https://doi.org/10.1038/ncomms15356.

[35] K. Xie, Q. Fu, P.A. Webley, G.G. Qiao, MOF scaffold for a high-performance mixed-matrix membrane, Angew. Chem. - Int. Ed. 57 (2018) 8597−8602. Available from: https://doi.org/10.1002/anie.201804162.

[36] G.M. Segovia, J.S. Tuninetti, S. Moya, A.S. Picco, M.R. Ceolín, O. Azzaroni, et al., Cysteamine-modified ZIF-8 colloidal building blocks: direct assembly of nanoparticulate MOF films on gold surfaces via thiol chemistry, Mater. Today Chem. 8 (2018) 29−35. Available from: https://doi.org/10.1016/j.mtchem.2018.02.002.

[37] M. Rubio-Martinez, C. Avci-Camur, A.W. Thornton, I. Imaz, D. Maspoch, M.R. Hill, New synthetic routes toward MOF production at scale, Chem. Soc. Rev. 46 (2017) 3453−3480. Available from: https://doi.org/10.1039/c7cs00109f.

[38] Y. Khabzina, J. Dhainaut, M. Ahlhelm, H.J. Richter, H. Reinsch, N. Stock, et al., Synthesis and shaping scale-up study of functionalized UiO-66 MOF for ammonia air purification filters, Ind. Eng. Chem. Res. 57 (2018) 8200−8208. Available from: https://doi.org/10.1021/acs.iecr.8b00808.

[39] H.M. Titi, J.L. Do, A.J. Howarth, K. Nagapudi, T. Friščić, Simple, scalable mechanosynthesis of metal-organic frameworks using liquid-assisted resonant acoustic mixing (LA-RAM), Chem. Sci. 11 (2020) 7578−7584. Available from: https://doi.org/10.1039/d0sc00333f.

[40] N.D.H. Gamage, K.A. McDonald, A.J. Matzger, MOF-5-polystyrene: direct production from monomer, improved hydrolytic stability, and unique guest adsorption, Angew. Chem. - Int. Ed. 55 (2016) 12099−12103. Available from: https://doi.org/10.1002/anie.201606926.

[41] M. Rubio-Martinez, T.D. Hadley, M.P. Batten, K. Constanti-Carey, T. Barton, D. Marley, et al., Scalability of continuous flow production of metal-organic frameworks, ChemSusChem 9 (2016) 938−941. Available from: https://doi.org/10.1002/cssc.201501684.

[42] M.X. Wu, Y. Wang, G. Zhou, X. Liu, Core-shell MOFs@MOFs: diverse designability and enhanced selectivity, ACS Appl. Mater. Interfaces. 12 (2020) 54285−54305. Available from: https://doi.org/10.1021/acsami.0c16428.

[43] S. Dai, A. Tissot, C. Serre, Recent progresses in metal−organic frameworks based core−shell composites, Adv. Energy Mater. 12 (2022) 1−26. Available from: https://doi.org/10.1002/aenm.202100061.

[44] D. Yu, Q. Song, J. Cui, H. Zheng, Y. Zhang, J. Liu, et al., Designing core−shell metal−organic framework hybrids: toward high-efficiency electrochemical potassium storage, J. Mater. Chem. A. 9 (2021) 26181−26188. Available from: https://doi.org/10.1039/D1TA08215A.

[45] M. Aono, Focus on materials nanoarchitectonics, Sci. Technol. Adv. Mater. 12 (2011) 9−11. Available from: https://doi.org/10.1088/1468-6996/12/4/040301.

[46] K. Ariga, Nanoarchitectonics: what's coming next after nanotechnology, Nanoscale Horiz. 6 (2021) 364−378. Available from: https://doi.org/10.1039/d0nh00680g.

[47] G.E. Fenoy, J. Scotto, J. Azcárate, M. Rafti, W.A. Marmisollé, O. Azzaroni, Powering up the oxygen reduction reaction through the integration of O_2-adsorbing metal−organic frameworks on nanocomposite electrodes, ACS Appl. Energy Mater. (2018)acsaem.8b01021. Available from: https://doi.org/10.1021/acsaem.8b01021.

[48] O. Azzaroni, K. Ariga, Concepts and Design of Materials Nanoarchitectonics, The Royal Society of Chemistry, 2022.

[49] K. Ariga, Nanoarchitectonics approach for sensing, Mater. Nanoarchitectonics. (2018) 255−263. Available from: https://doi.org/10.1002/9783527808311.ch15.

[50] O. Azzaroni, K. Ariga, Nanoarchitectonics, now, Mol. Syst. Des. Eng. 4 (2019) 9−10. Available from: https://doi.org/10.1039/c9me90001b.

[51] Q. Yang, Q. Xu, H.-L. Jiang, Metal−organic frameworks meet metal nanoparticles: synergistic effect for enhanced catalysis, Chem. Soc. Rev. 46 (2017) 4774−4808. Available from: https://doi.org/10.1039/C6CS00724D.

[52] E.T. Hardy, Y.J. Wang, S. Iyer, R.G. Mannino, Y. Sakurai, T.H. Barker, et al., Interdigitated microelectronic bandage augments hemostasis and clot formation at low applied voltage: in vitro and in vivo, Lab. Chip 18 (2018) 2985−2993. Available from: https://doi.org/10.1039/c8lc00573g.

[53] S. Dutta, J. Kim, P.H. Hsieh, Y.S. Hsu, Y.V. Kaneti, F.K. Shieh, et al., Nanoarchitectonics of biofunctionalized metal−organic frameworks with biological macromolecules and living cells, Small Methods 3 (2019) 1−15. Available from: https://doi.org/10.1002/smtd.201900213.

[54] P. Cheng, C. Wang, Y.V. Kaneti, M. Eguchi, J. Lin, Y. Yamauchi, et al., Practical MOF nanoarchitectonics: new strategies for enhancing the processability of MOFs for practical applications, Langmuir 36 (2020) 4231−4249. Available from: https://doi.org/10.1021/acs.langmuir.0c00236.

[55] M. Rafti, J.S. Tuninetti, M.P. Serrano, A.H. Thomas, O. Azzaroni, Shelter for biologically relevant molecules: photoprotection and enhanced thermal stability of folic acid loaded in a ZIF-8 MOF porous host, Ind. Eng. Chem. Res. 59 (2020) 22155−22162. Available from: https://doi.org/10.1021/acs.iecr.0c04905.

[56] G.E. Fenoy, M. Rafti, W.A. Marmisollé, O. Azzaroni, Nanoarchitectonics of metal organic frameworks and PEDOT layer-by-layer electrodes for boosting oxygen reduction reaction, Mater. Adv. (2021). Available from: https://doi.org/10.1039/d1ma00747e.

[57] M. Li, S. Yin, M. Lin, X. Chen, Y. Pan, Y. Peng, et al., Current status and prospects of metal−organic frameworks for bone therapy and bone repair, J. Mater. Chem. B. 10 (2022) 5105−5128. Available from: https://doi.org/10.1039/D2TB00742H.

[58] D. Zacher, O. Shekhah, C. Wöll, R.A. Fischer, Thin films of metal−organic frameworks, Chem. Soc. Rev. 38 (2009) 1418−1429. Available from: https://doi.org/10.1039/b805038b.

[59] S. Furukawa, J. Reboul, S. Diring, K. Sumida, S. Kitagawa, Structuring of metal-organic frameworks at the mesoscopic/macroscopic scale, Chem. Soc. Rev. 43 (2014) 5700−5734. Available from: https://doi.org/10.1039/c4cs00106k.

[60] J.D. Rimer, M. Tsapatsis, Nucleation of open framework materials: navigating the voids, MRS Bull. 41 (2016) 393−398. Available from: https://doi.org/10.1557/mrs.2016.89.

[61] M.J. Van Vleet, T. Weng, X. Li, J.R. Schmidt, In Situ, time-resolved, and mechanistic studies of metal-organic framework nucleation and growth, Chem. Rev. 118 (2018) 3681−3721. Available from: https://doi.org/10.1021/acs.chemrev.7b00582.

[62] C.R. Marshall, S.A. Staudhammer, C.K. Brozek, Size control over metal-organic framework porous nanocrystals, Chem. Sci. 10 (2019) 9396−9408. Available from: https://doi.org/10.1039/c9sc03802g.

[63] M. Kalaj, K.C. Bentz, S. Ayala, J.M. Palomba, K.S. Barcus, Y. Katayama, et al., MOF-polymer hybrid materials: from simple composites to tailored architectures, Chem. Rev. 120 (2020) 8267−8302. Available from: https://doi.org/10.1021/acs.chemrev.9b00575.

[64] J. Fonseca, T. Gong, Fabrication of metal-organic framework architectures with macroscopic size: a review, Coord. Chem. Rev. 462 (2022) 214520. Available from: https://doi.org/10.1016/j.ccr.2022.214520.

[65] Y. Luo, S. Bag, O. Zaremba, A. Cierpka, J. Andreo, S. Wuttke, et al., MOF synthesis prediction enabled by automatic data mining and machine learning**, Angew. Chem. - Int. Ed. 61 (2022). Available from: https://doi.org/10.1002/anie.202200242.

[66] J.S. Tuninetti, M. Rafti, O. Azzaroni, Early stages of ZIF-8 film growth: the enhancement effect of primers exposing sulfonate groups as surface-confined nucleation agents, RSC Adv. 5 (2015) 73958−73962. Available from: https://doi.org/10.1039/C5RA12789K.

[67] J.A. Allegretto, J. Dostalek, M. Rafti, B. Menges, O. Azzaroni, W. Knoll, Shedding light on the dark corners of metal−organic framework thin films: growth and structural stability of ZIF-8 layers probed by optical waveguide spectroscopy, J. Phys. Chem. A. 123 (2019) 1100−1109. Available from: https://doi.org/10.1021/acs.jpca.8b09610.

[68] M. Rafti, J.A. Allegretto, G.M. Segovia, J.S. Tuninetti, J.M. Giussi, E. Bindini, et al., Metal−organic frameworks meet polymer brushes: enhanced crystalline film growth induced by macromolecular primers, Mater. Chem. Front. 1 (2017) 2256−2260. Available from: https://doi.org/10.1039/C7QM00235A.

[69] J.A. Allegretto, A. Iborra, J.M. Giussi, C. von Biderling, M. Ceolín, S. Moya, et al., Growth of ZIF-8 MOF films with tuneable porosity by using Poly(1-vinylimidazole) brushes as 3D primers, Chem. − A Eur. J. (2020)chem.202002493. Available from: https://doi.org/10.1002/chem.202002493.

[70] J.A. Allegretto, J.S. Tuninetti, A. Lorenzo, M. Ceolín, O. Azzaroni, M. Rafti, Polyelectrolyte capping as straightforward approach toward manipulation of diffusive transport in MOF films, Langmuir 34 (2018) 425−431. Available from: https://doi.org/10.1021/acs.langmuir.7b03083.

[71] J.A. Allegretto, M. Arcidiácono, P.Y. Steinberg, P.C. Angelomé, O. Azzaroni, M. Rafti, Impact of chemical primers on the growth, structure, and functional properties of ZIF-8 films, J. Phys. Chem. C. 126 (2022) 6724−6735. Available from: https://doi.org/10.1021/acs.jpcc.1c10425.

[72] I. Stassen, J.H. Dou, C. Hendon, M. Dincǎ, Chemiresistive sensing of ambient CO_2 by an autogenously hydrated Cu_3(hexaiminobenzene)2 framework, ACS Cent. Sci. 5 (2019) 1425−1431. Available from: https://doi.org/10.1021/acscentsci.9b00482.

[73] A. Dragässer, O. Shekhah, O. Zybaylo, C. Shen, M. Buck, C. Wöll, et al., Redox mediation enabled by immobilised centres in the pores of a metal−organic framework grown by liquid phase epitaxy, Chem. Commun. 48 (2012) 663−665. Available from: https://doi.org/10.1039/c1cc16580a.

[74] A.A. Talin, A. Centrone, A.C. Ford, M.E. Foster, V. Stavila, P. Haney, et al., Tunable electrical conductivity in metal-organic framework thin-film devices, Sci. (80-.) 343 (2014) 66−69. Available from: https://doi.org/10.1126/science.1246738.

[75] S. Han, S.C. Warren, S.M. Yoon, C.D. Malliakas, X. Hou, Y. Wei, et al., Tunneling electrical connection to the interior of metal-organic, frameworks, J. Am. Chem. Soc. 137 (2015) 8169−8175. Available from: https://doi.org/10.1021/jacs.5b03263.

[76] V. Rubio-Giménez, M. Galbiati, J. Castells-Gil, N. Almora-Barrios, J. Navarro-Sánchez, G. Escorcia-Ariza, et al., Bottom-up fabrication of semiconductive metal−organic framework ultrathin films, Adv. Mater. 30 (2018) 1−8. Available from: https://doi.org/10.1002/adma.201704291.

[77] W.J. Li, J. Liu, Z.H. Sun, T.F. Liu, J. Lü, S.Y. Gao, et al., Integration of metal-organic frameworks into an electrochemical dielectric thin film for electronic applications, Nat. Commun. 7 (2016). Available from: https://doi.org/10.1038/ncomms11830.

[78] B.E. Conway, Theory and Principles of Electrode Processes, AMA 10th e, Ronald Press Co, New York, 1965.

[79] Y. Yan, P. Gu, S. Zheng, M. Zheng, H. Pang, H. Xue, Facile synthesis of an accordion-like Ni-MOF superstructure for high-performance flexible supercapacitors, J. Mater. Chem. A. 4 (2016) 19078−19085. Available from: https://doi.org/10.1039/C6TA08331E.

[80] Y. Liu, Y. Wang, H. Wang, P. Zhao, H. Hou, L. Guo, Acetylene black enhancing the electrochemical performance of NiCo-MOF nanosheets for supercapacitor electrodes, Appl. Surf. Sci. 492 (2019) 455−463. Available from: https://doi.org/10.1016/j.apsusc.2019.06.238.

[81] T. Deng, Y. Lu, W. Zhang, M. Sui, X. Shi, D. Wang, et al., Inverted design for high-performance supercapacitor via Co(OH) 2 -derived highly oriented MOF electrodes, Adv. Energy Mater. 8 (2018) 1702294. Available from: https://doi.org/10.1002/aenm.201702294.

[82] Y. Wang, Y. Liu, H. Wang, W. Liu, Y. Li, J. Zhang, et al., Ultrathin NiCo-MOF nanosheets for high-performance supercapacitor electrodes, ACS Appl. Energy Mater. 2 (2019) 2063−2071. Available from: https://doi.org/10.1021/acsaem.8b02128.

[83] H. Xia, J. Zhang, Z. Yang, S. Guo, S. Guo, Q. Xu, 2D MOF nanoflake-assembled spherical microstructures for enhanced supercapacitor and electrocatalysis performances, Nano-Micro Lett. 9 (2017). Available from: https://doi.org/10.1007/s40820-017-0144-6.

[84] K.M. Choi, H.M. Jeong, J.H. Park, Y.B. Zhang, J.K. Kang, O.M. Yaghi, Supercapacitors of nanocrystalline metal-organic frameworks, ACS Nano 8 (2014) 7451–7457. Available from: https://doi.org/10.1021/nn5027092.

[85] W. Chaikittisilp, M. Hu, H. Wang, H.S. Huang, T. Fujita, K.C.W. Wu, et al., Nanoporous carbons through direct carbonization of a zeolitic imidazolate framework for supercapacitor electrodes, Chem. Commun. 48 (2012) 7259–7261. Available from: https://doi.org/10.1039/c2cc33433j.

[86] R.R. Salunkhe, Y. Kamachi, N.L. Torad, S.M. Hwang, Z. Sun, S.X. Dou, et al., Fabrication of symmetric supercapacitors based on MOF-derived nanoporous carbons, J. Mater. Chem. A. 2 (2014) 19848–19854. Available from: https://doi.org/10.1039/c4ta04277h.

[87] P. Pachfule, D. Shinde, M. Majumder, Q. Xu, Fabrication of carbon nanorods and graphene nanoribbons from a metal–organic framework, Nat. Chem. 8 (2016) 718–724. Available from: https://doi.org/10.1038/nchem.2515.

[88] Z. Zhao, S. Liu, J. Zhu, J. Xu, L. Li, Z. Huang, et al., Hierarchical nanostructures of nitrogen-doped porous carbon polyhedrons confined in carbon nanosheets for high-performance supercapacitors, ACS Appl. Mater. Interfaces. 10 (2018) 19871–19880. Available from: https://doi.org/10.1021/acsami.8b03431.

[89] S. Xiong, S. Jiang, J. Wang, H. Lin, M. Lin, S. Weng, et al., A high-performance hybrid supercapacitor with NiO derived NiO@Ni-MOF composite electrodes, Electrochim. Acta. 340 (2020) 135956. Available from: https://doi.org/10.1016/j.electacta.2020.135956.

[90] B.Y. Guan, A. Kushima, L. Yu, S. Li, J. Li, X.W.D. Lou, Coordination polymers derived general synthesis of multishelled mixed metal-oxide particles for hybrid supercapacitors, Adv. Mater. 29 (2017) 1605902. Available from: https://doi.org/10.1002/adma.201605902.

[91] J. Tang, R.R. Salunkhe, J. Liu, N.L. Torad, M. Imura, S. Furukawa, et al., Thermal conversion of core–shell metal–organic frameworks: a new method for selectively functionalized nanoporous hybrid carbon, J. Am. Chem. Soc. 137 (2015) 1572–1580. Available from: https://doi.org/10.1021/ja511539a.

[92] R.R. Salunkhe, J. Tang, Y. Kamachi, T. Nakato, J.H. Kim, Y. Yamauchi, Asymmetric supercapacitors using 3D nanoporous carbon and cobalt oxide electrodes synthesized from a single metal-organic framework, ACS Nano 9 (2015) 6288–6296. Available from: https://doi.org/10.1021/acsnano.5b01790.

[93] X. Wang, N. Yang, Q. Li, F. He, Y. Yang, B. Wu, et al., Solvothermal synthesis of flower-string-like NiCo-MOF/MWCNT composites as a high-performance supercapacitor electrode material, J. Solid. State Chem. 277 (2019) 575–586. Available from: https://doi.org/10.1016/j.jssc.2019.07.019.

[94] S. Li, K. Yang, P. Ye, K. Ma, Z. Zhang, Q. Huang, Three-dimensional porous carbon/Co_3O_4 composites derived from graphene/Co-MOF for high performance supercapacitor electrodes, Appl. Surf. Sci. 503 (2020) 144090. Available from: https://doi.org/10.1016/j.apsusc.2019.144090.

[95] P. Wen, P. Gong, J. Sun, J. Wang, S. Yang, Design and synthesis of Ni-MOF/CNT composites and rGO/carbon nitride composites for an asymmetric supercapacitor with high energy and power density, J. Mater. Chem. A. 3 (2015) 13874–13883. AvailableG from: https://doi.org/10.1039/C5TA02461G.

[96] L. Chen, Y. Zhang, W. Liu, Z. Liu, Graphene nanoarchitectonics: a new material horizon for reinforcement of sustainable polymers, Front. Mater. 7 (2020) 1–6. Available from: https://doi.org/10.3389/fmats.2020.00276.

[97] J. Zhang, X. Zhang, J. Shen, H. Pan, Z. Chen, Y. Li, et al., Nanoarchitectonics of graphene oxide with functionalized cellulose nanocrystals achieving simultaneous dual connections and defect repair through catalytic graphitization for high thermal conductivity, Carbon N. Y. 201 (2023) 295–306. Available from: https://doi.org/10.1016/j.carbon.2022.09.027.

[98] C. Moreno, M. Vilas-Varela, B. Kretz, A. Garcia-Lekue, M.V. Costache, M. Paradinas, et al., Bottom-up synthesis of multifunctional nanoporous graphene, Sci. (80-.) 360 (2018) 199–203. Available from: https://doi.org/10.1126/science.aar2009.

[99] Y. Xiao, W. Wei, M. Zhang, S. Jiao, Y. Shi, S. Ding, Facile surface properties engineering of high-quality graphene: toward advanced Ni-MOF heterostructures for high-performance supercapacitor electrode, ACS Appl. Energy Mater. 2 (2019) 2169−2177. Available from: https://doi.org/10.1021/acsaem.8b02201.

[100] R. Srinivasan, E. Elaiyappillai, E.J. Nixon, I. Sharmila Lydia, P.M. Johnson, Enhanced electrochemical behaviour of Co-MOF/PANI composite electrode for supercapacitors, Inorganica Chim. Acta. 502 (2020) 119393. Available from: https://doi.org/10.1016/j.ica.2019.119393.

[101] R. Hou, M. Miao, Q. Wang, T. Yue, H. Liu, H.S. Park, et al., Integrated conductive hybrid architecture of metal−organic framework nanowire array on polypyrrole membrane for all-solid-state flexible supercapacitors, Adv. Energy Mater. 10 (2020) 1901892. Available from: https://doi.org/10.1002/aenm.201901892.

[102] R.R. Salunkhe, J. Tang, N. Kobayashi, J. Kim, Y. Ide, S. Tominaka, et al., Ultrahigh performance supercapacitors utilizing core−shell nanoarchitectures from a metal−organic framework-derived nanoporous carbon and a conducting polymer, Chem. Sci. 7 (2016) 5704−5713. Available from: https://doi.org/10.1039/C6SC01429A.

[103] L.S. Xie, G. Skorupskii, M. Dincǎ, Electrically conductive metal-organic frameworks, Chem. Rev. 120 (2020) 8536−8580. Available from: https://doi.org/10.1021/acs.chemrev.9b00766.

[104] M.G. Campbell, M. Dincǎ, Metal−organic frameworks as active materials in electronic sensor devices, Sens. (Switz.) 17 (2017) 1−11. Available from: https://doi.org/10.3390/s17051108.

[105] P.M. Usov, B. Huffman, C.C. Epley, M.C. Kessinger, J. Zhu, W.A. Maza, et al., Study of electrocatalytic properties of metal−organic framework PCN-223 for the oxygen reduction reaction, ACS Appl. Mater. Interfaces. 9 (2017) 33539−33543. Available from: https://doi.org/10.1021/acsami.7b01547.

[106] M.O. Cichocka, Z. Liang, D. Feng, S. Back, S. Siahrostami, X. Wang, et al., A porphyrinic zirconium metal−organic framework for oxygen reduction reaction: tailoring the spacing between active-sites through chain-based inorganic building units, J. Am. Chem. Soc. 142 (2020) 15386−15395. Available from: https://doi.org/10.1021/jacs.0c06329.

[107] S.S. Rajasree, X. Li, P. Deria, Physical properties of porphyrin-based crystalline metal-organic frameworks, Commun. Chem. 4 (2021) 1−14. Available from: https://doi.org/10.1038/s42004-021-00484-4.

[108] A. Schlachter, P. Asselin, P.D. Harvey, Porphyrin-containing MOFs and COFs as heterogeneous photosensitizers for singlet oxygen-based antimicrobial nanodevices, ACS Appl. Mater. Interfaces. 13 (2021) 26651−26672. Available from: https://doi.org/10.1021/acsami.1c05234.

[109] C.H. Kwon, Y. Ko, D. Shin, M. Kwon, J. Park, W.K. Bae, et al., High-power hybrid biofuel cells using layer-by-layer assembled glucose oxidase-coated metallic cotton fibers, Nat. Commun. 9 (2018). Available from: https://doi.org/10.1038/s41467-018-06994-5.

[110] A.P. Mártire, G.M. Segovia, O. Azzaroni, M. Rafti, W. Marmisollé, Layer-by-layer integration of conducting polymers and metal organic frameworks onto electrode surfaces: enhancement of the oxygen reduction reaction through electrocatalytic nanoarchitectonics, Mol. Syst. Des. Eng. 4 (2019) 893−900. Available from: https://doi.org/10.1039/c9me00007k.

[111] F.C.D. López, A.I. Bertoni, J.A. Allegretto, W.A. Marmisollé, O. Azzaroni, M. Rafti, et al., Oxygen on-demand: understanding the oxygen reduction reaction (ORR) performance enhancement of conductive polymer films upon modification with ZIF-8 MOF, ChemCatChem 14 (2022). Available from: https://doi.org/10.1002/cctc.202201015.

[112] M. Rafti, W.A. Marmisollé, O. Azzaroni, Metal-organic frameworks help conducting polymers optimize the efficiency of the oxygen reduction reaction in neutral solutions, Adv. Mater. Interfaces. 3 (2016) 1600047. Available from: https://doi.org/10.1002/admi.201600047.

[113] E. Mahmoudi, H. Fakhri, A. Hajian, A. Afkhami, H. Bagheri, High-performance electrochemical enzyme sensor for organophosphate pesticide detection using modified metal-organic framework sensing platforms, Bioelectrochemistry 130 (2019) 107348. Available from: https://doi.org/10.1016/j.bioelechem.2019.107348.

[114] X. Liu, W. Chen, M. Lian, X. Chen, Y. Lu, W. Yang, Enzyme immobilization on ZIF-67/MWCNT composite engenders high sensitivity electrochemical sensing, J. Electroanal. Chem. 833 (2019) 505−511. Available from: https://doi.org/10.1016/j.jelechem.2018.12.027.

[115] W. Chen, W. Yang, Y. Lu, W. Zhu, X. Chen, Encapsulation of enzyme into mesoporous cages of metal-organic frameworks for the development of highly stable electrochemical biosensors, Anal. Methods. 9 (2017) 3213−3220. Available from: https://doi.org/10.1039/c7ay00710h.

[116] C. Zhang, X. Wang, M. Hou, X. Li, X. Wu, J. Ge, Immobilization on metal−organic framework engenders high sensitivity for enzymatic electrochemical detection, ACS Appl. Mater. Interfaces. 9 (2017) 13831−13836. Available from: https://doi.org/10.1021/acsami.7b02803.

[117] S. Dong, L. Peng, W. Wei, T. Huang, Three MOF-templated carbon nanocomposites for potential platforms of enzyme immobilization with improved electrochemical performance, ACS Appl. Mater. Interfaces. 10 (2018) 14665−14672. Available from: https://doi.org/10.1021/acsami.8b00702.

[118] Z. Zhou, A. Inayat, W. Schwieger, M. Hartmann, Improved activity and stability of lipase immobilized in cage-like large pore mesoporous organosilicas, Microporous Mesoporous Mater. 154 (2012) 133−141. Available from: https://doi.org/10.1016/j.micromeso.2012.01.003.

[119] R.J. Drout, L. Robison, O.K. Farha, Catalytic applications of enzymes encapsulated in metal−organic frameworks, Coord. Chem. Rev. 381 (2019) 151−160. Available from: https://doi.org/10.1016/j.ccr.2018.11.009.

[120] C. Doonan, R. Riccò, K. Liang, D. Bradshaw, P. Falcaro, Metal−organic frameworks at the biointerface: synthetic strategies and applications, Acc. Chem. Res. 50 (2017) 1423−1432. Available from: https://doi.org/10.1021/acs.accounts.7b00090.

[121] J. Wu, X. Wang, Q. Wang, Z. Lou, S. Li, Y. Zhu, et al., Nanomaterials with enzyme-like characteristics (nanozymes): next-generation artificial enzymes (II), Chem. Soc. Rev. 48 (2019) 1004−1076. Available from: https://doi.org/10.1039/c8cs00457a.

[122] Z. Tang, J. He, J. Chen, Y. Niu, Y. Zhao, Y. Zhang, et al., A sensitive sandwich-type immunosensor for the detection of galectin-3 based on N-GNRs-Fe-MOFs@AuNPs nanocomposites and a novel AuPt-methylene blue nanorod, Biosens. Bioelectron. 101 (2018) 253−259. Available from: https://doi.org/10.1016/j.bios.2017.10.026.

[123] J. Feng, H. Wang, Z. Ma, Ultrasensitive amperometric immunosensor for the prostate specific antigen by exploiting a Fenton reaction induced by a metal-organic framework nanocomposite of type Au/Fe-MOF with peroxidase mimicking activity, Microchim. Acta. 187 (2020). Available from: https://doi.org/10.1007/s00604-019-4075-4.

[124] X. Li, X. Li, D. Li, M. Zhao, H. Wu, B. Shen, et al., Electrochemical biosensor for ultrasensitive exosomal miRNA analysis by cascade primer exchange reaction and MOF@Pt@MOF nanozyme, Biosens. Bioelectron. 168 (2020) 112554. Available from: https://doi.org/10.1016/j.bios.2020.112554.

[125] L. Cui, J. Wu, J. Li, H. Ju, Electrochemical sensor for lead cation sensitized with a DNA functionalized porphyrinic metal-organic framework, Anal. Chem. 87 (2015) 10635−10641. Available from: https://doi.org/10.1021/acs.analchem.5b03287.

[126] P. Ling, J. Lei, L. Zhang, H. Ju, Porphyrin-encapsulated metal-organic frameworks as mimetic catalysts for electrochemical DNA sensing via allosteric switch of hairpin DNA, Anal. Chem. 87 (2015) 3957−3963. Available from: https://doi.org/10.1021/acs.analchem.5b00001.

[127] P. Ling, J. Lei, H. Ju, Porphyrinic metal-organic framework as electrochemical probe for DNA sensing via triple-helix molecular switch, Biosens. Bioelectron. 71 (2015) 373−379. Available from: https://doi.org/10.1016/j.bios.2015.04.046.

[128] W.-C. Hu, J. Pang, S. Biswas, K. Wang, C. Wang, X.-H. Xia, Ultrasensitive detection of bacteria using a 2D MOF nanozyme-amplified electrochemical detector, Anal. Chem. 93 (2021) 8544−8552. Available from: https://doi.org/10.1021/acs.analchem.1c01261.

[129] S. Zhand, A. Razmjou, S. Azadi, S.R. Bazaz, J. Shrestha, M.A.F. Jahromi, et al., Metal-Organic Framework-Enhanced ELISA Platform for Ultrasensitive Detection of PD-L1, 2020. https://doi.org/10.1021/acsabm.0c00227.

[130] C.R. Martin, G.A. Leith, P. Kittikhunnatham, K.C. Park, O.A. Ejegbavwo, A. Mathur, et al., Heterometallic actinide-containing photoresponsive metal-organic frameworks: dynamic and static tuning of electronic properties, Angew. Chem. - Int. Ed. 60 (2021) 8072–8080. Available from: https://doi.org/10.1002/anie.202016826.

[131] U. Tietze, C. Schenk, E. Gamm, Field Effect Transistor, Springer Berlin Heidelberg, Berlin, Heidelberg, 2008. Available from: https://doi.org/10.1007/978-3-540-78655-9_3.

[132] B. Zhan, C. Li, J. Yang, G. Jenkins, W. Huang, X. Dong, Graphene field-effect transistor and its application for electronic sensing, Small 10 (2014) 4042–4065. Available from: https://doi.org/10.1002/smll.201400463.

[133] E. Piccinini, C. Bliem, C. Reiner-Rozman, F. Battaglini, O. Azzaroni, W. Knoll, Enzyme-polyelectrolyte multilayer assemblies on reduced graphene oxide field-effect transistors for biosensing applications, Biosens. Bioelectron. 92 (2017) 661–667. Available from: https://doi.org/10.1016/j.bios.2016.10.035.

[134] M.-Z. Li, S.-T. Han, Y. Zhou, Recent advances in flexible field-effect transistors toward wearable sensors, Adv. Intell. Syst. 2 (2020) 2000113. Available from: https://doi.org/10.1002/aisy.202000113.

[135] S. Chong, S.M.J. Rogge, J. Kim, Tunable electrical conductivity of flexible metal-organic frameworks, Chem. Mater. 34 (2022) 254–265. Available from: https://doi.org/10.1021/acs.chemmater.1c03236.

[136] X. Zhang, Q. Jing, S. Ao, G.F. Schneider, D. Kireev, Z. Zhang, et al., Ultrasensitive field-effect biosensors enabled by the unique electronic properties of graphene, Small 16 (2020) 1–24. Available from: https://doi.org/10.1002/smll.201902820.

[137] G.E. Fenoy, W.A. Marmisollé, W. Knoll, O. Azzaroni, Highly sensitive urine glucose detection with graphene field-effect transistors functionalized with electropolymerized nanofilms, Sens. & Diagnostics 1 (2022) 139–148. Available from: https://doi.org/10.1039/d1sd00007a.

[138] G. Wu, J. Huang, Y. Zang, J. He, G. Xu, Porous field-effect transistors based on a semiconductive metal-organic framework, J. Am. Chem. Soc. 139 (2017) 1360–1363. Available from: https://doi.org/10.1021/jacs.6b08511.

[139] S. Shen, P. Tan, Y. Tang, G. Duan, Y. Luo, Adjustable synthesis of Ni-based metal-organic framework membranes and their field-effect transistor sensors for mercury detection, ACS Appl. Electron. Mater. 4 (2022) 622–630. Available from: https://doi.org/10.1021/acsaelm.1c01009.

[140] S. Kumar, Y. Pramudya, K. Müller, A. Chandresh, S. Dehm, S. Heidrich, et al., Sensing molecules with metal−organic framework functionalized graphene transistors, Adv. Mater. (2021). Available from: https://doi.org/10.1002/adma.202103316.

[141] J. Liu, B. Lukose, O. Shekhah, H.K. Arslan, P. Weidler, H. Gliemann, et al., A novel series of isoreticular metal organic frameworks: realizing metastable structures by liquid phase epitaxy, Sci. Rep. 2 (2012) 1–5. Available from: https://doi.org/10.1038/srep00921.

[142] L.D. Trino, L.G.S. Albano, D.C. Granato, A.G. Santana, D.H.S. De Camargo, C.C. Correa, et al., Framework electrochemical biosensor for the detection of protein-protein interaction, Chem. Mater. 33 (2021) 1293–1306. Available from: https://doi.org/10.1021/acs.chemmater.0c04201.

[143] T. Leelasree, V. Selamneni, T. Akshaya, P. Sahatiya, H. Aggarwal, MOF based flexible, low-cost chemiresistive device as a respiration sensor for sleep apnea diagnosis, J. Mater. Chem. B. 8 (2020) 10182–10189. Available from: https://doi.org/10.1039/d0tb01748e.

[144] L.D. Sappia, J.S. Tuninetti, M. Ceolín, W. Knoll, M. Rafti, O. Azzaroni, MOF@PEDOT composite films for impedimetric pesticide sensors, Glob. Chall. 4 (2020) 1900076. Available from: https://doi.org/10.1002/gch2.201900076.

[145] W. Fan, X. Zhang, Z. Kang, X. Liu, D. Sun, Isoreticular chemistry within metal−organic frameworks for gas storage and separation, Coord. Chem. Rev. 443 (2021) 213968. Available from: https://doi.org/10.1016/j.ccr.2021.213968.

[146] X. Li, H. Zhang, P. Wang, J. Hou, J. Lu, C.D. Easton, et al., Fast and selective fluoride ion conduction in sub-1-nanometer metal-organic framework channels, Nat. Commun. 10 (2019) 1−12. Available from: https://doi.org/10.1038/s41467-019-10420-9.

[147] D. Farrusseng, Metal-Organic Frameworks, Wiley, 2011. Available from: https://doi.org/10.1002/9783527635856.

[148] G. Pérez-Mitta, A.G. Albesa, C. Trautmann, M.E. Toimil-Molares, O. Azzaroni, Bioinspired integrated nanosystems based on solid-state nanopores: "iontronic" transduction of biological, chemical and physical stimuli, Chem. Sci. 8 (2017) 890−913. Available from: https://doi.org/10.1039/c6sc04255d.

[149] Z.S. Siwy, Ion-current rectification in nanopores and nanotubes with broken symmetry, Adv. Funct. Mater. 16 (2006) 735−746. Available from: https://doi.org/10.1002/adfm.200500471.

[150] H. Zhang, X. Hou, J. Hou, L. Zeng, Y. Tian, L. Li, et al., Synthetic asymmetric-shaped nanodevices with symmetric pH-gating characteristics, Adv. Funct. Mater. 25 (2015) 1102−1110. Available from: https://doi.org/10.1002/adfm.201403693.

[151] B. Yameen, M. Ali, R. Neumann, W. Ensinger, W. Knoll, O. Azzaroni, Single conical nanopores displaying pH-tunable rectifying characteristics. manipulating ionic transport with zwitterionic polymer brushes, J. Am. Chem. Soc. 131 (2009) 2070−2071. Available from: https://doi.org/10.1021/ja8086104.

[152] G. Laucirica, M.E. Toimil-Molares, C. Trautmann, W. Marmisollé, O. Azzaroni, Nanofluidic osmotic power generators - advanced nanoporous membranes and nanochannels for blue energy harvesting, Chem. Sci. 12 (2021) 12874−12910. Available from: https://doi.org/10.1039/d1sc03581a.

[153] G. Laucirica, Y. Toum Terrones, V. Cayón, M.L. Cortez, M.E. Toimil-Molares, C. Trautmann, et al., Biomimetic solid-state nanochannels for chemical and biological sensing applications, TrAC. - Trends Anal. Chem. 144 (2021). Available from: https://doi.org/10.1016/j.trac.2021.116425.

[154] S. Balme, T. Ma, E. Balanzat, J.M. Janot, Large osmotic energy harvesting from functionalized conical nanopore suitable for membrane applications, J. Memb. Sci. 544 (2017) 18−24. Available from: https://doi.org/10.1016/j.memsci.2017.09.008.

[155] M. Lepoitevin, M. Bechelany, E. Balanzat, J.M. Janot, S. Balme, Non-fluorescence label protein sensing with track-etched nanopore decorated by avidin/biotin system, Electrochim. Acta. 211 (2016) 611−618. Available from: https://doi.org/10.1016/j.electacta.2016.06.079.

[156] A.S. Peinetti, R.J. Lake, W. Cong, L. Cooper, Y. Wu, Y. Ma, et al., Direct detection of human adenovirus or SARS-CoV-2 with ability to inform infectivity using DNA aptamer-nanopore sensors, Sci. Adv. 7 (2021) 1−13. Available from: https://doi.org/10.1126/sciadv.abh2848.

[157] K.B. Jirage, J.C. Hulteen, C.R. Martin, Nanotubule-based molecular-filtration membranes, Science. (80-.) 278 (1997) 655−658. Available from: https://doi.org/10.1126/science.278.5338.655.

[158] R.K. Joshi, P. Carbone, F.C. Wang, V.G. Kravets, Y. Su, I.V. Grigorieva, et al., Precise and ultrafast molecular sieving through graphene oxide membranes, Science (80-.) 343 (2014) 752−754. Available from: https://doi.org/10.1126/science.1245711.

[159] J. Lu, H. Zhang, X. Hu, B. Qian, J. Hou, L. Han, et al., Ultraselective monovalent metal ion conduction in a three-dimensional sub-1 nm nanofluidic device constructed by metal-organic frameworks, ACS Nano 15 (2021) 1240−1249. Available from: https://doi.org/10.1021/acsnano.0c08328.

[160] J. Lu, H. Zhang, J. Hou, X. Li, X. Hu, Y. Hu, et al., Efficient metal ion sieving in rectifying subnano-channels enabled by metal—organic frameworks, Nat. Mater. 19 (2020) 767—774. Available from: https://doi.org/10.1038/s41563-020-0634-7.

[161] X. Li, H. Zhang, H. Yu, J. Xia, Y.B. Zhu, H.A. Wu, et al., Unidirectional and selective proton transport in artificial heterostructured nanochannels with nano-to-subnano confined water clusters, Adv. Mater. 32 (2020) 1—7. Available from: https://doi.org/10.1002/adma.202001777.

[162] X. Li, H. Zhang, J. Hou, R. Ou, Y. Zhu, C. Zhao, et al., Sulfonated sub-1-nm metal-organic framework channels with ultrahigh proton selectivity, J. Am. Chem. Soc. 142 (2020) 9827—9833. Available from: https://doi.org/10.1021/jacs.0c03554.

[163] J.H. Han, K.B. Kim, H.C. Kim, T.D. Chung, Ionic circuits based on poly electrolyte diodes on a micro-chip, Angew. Chem. - Int. Ed. 48 (2009) 3830—3833. Available from: https://doi.org/10.1002/anie.200900045.

[164] G. Laucirica, J.A. Allegretto, M.F. Wagner, M.E. Toimil-Molares, C. Trautmann, M. Rafti, et al., Switchable ion current saturation regimes enabled via heterostructured nanofluidic devices based on metal—organic frameworks, Adv. Mater. 2207339 (2022) 2207339. Available from: https://doi.org/10.1002/adma.202207339.

[165] C.C. Chueh, C.I. Chen, Y.A. Su, H. Konnerth, Y.J. Gu, C.W. Kung, et al., Harnessing MOF materials in photovoltaic devices: recent advances, challenges, and perspectives, J. Mater. Chem. A. 7 (2019) 17079—17095. Available from: https://doi.org/10.1039/c9ta03595h.

[166] G. Lu, J.T. Hupp, Metal-organic frameworks as sensors: a ZIF-8 based fabry-pérot device as a selective sensor for chemical vapors and gases, J. Am. Chem. Soc. 132 (2010) 7832—7833. Available from: https://doi.org/10.1021/ja101415b.

[167] K. Ikigaki, K. Okada, M. Takahashi, Epitaxial growth of multilayered metal-organic framework thin films for electronic and photonic applications, ACS Appl. Nano Mater. 4 (2021) 3467—3475. Available from: https://doi.org/10.1021/acsanm.0c03462.

[168] F.M. Hinterholzinger, A. Ranft, J.M. Feckl, B. Rühle, T. Bein, B.V. Lotsch, One-dimensional metal-organic framework photonic crystals used as platforms for vapor sorption, J. Mater. Chem. 22 (2012) 10356—10362. Available from: https://doi.org/10.1039/c2jm15685g.

[169] A. Ranft, F. Niekiel, I. Pavlichenko, N. Stock, B.V. Lotsch, Tandem MOF-based photonic crystals for enhanced analyte-specific optical detection, Chem. Mater. 27 (2015) 1961—1970. Available from: https://doi.org/10.1021/cm503640c.

[170] C. Liu, Y.L. Tong, X.Q. Yu, H. Shen, Z. Zhu, Q. Li, et al., MOF-based photonic crystal film toward separation of organic dyes, ACS Appl. Mater. Interfaces. 12 (2020) 2816—2825. Available from: https://doi.org/10.1021/acsami.9b18012.

[171] S. Subramanian, H.B.L. Jones, S. Frustaci, S. Winter, M.W. Van Der Kamp, V.L. Arcus, et al., Sensing enzyme activation heat capacity at the single-molecule level using gold-nanorod-based optical whispering gallery modes, ACS Appl. Nano Mater. 4 (2021). Available from: https://doi.org/10.1021/acsanm.1c00176.

[172] J. Ward, O. Benson, WGM microresonators: sensing, lasing and fundamental optics with microspheres, Laser Photonics Rev. 5 (2011) 553—570. Available from: https://doi.org/10.1002/lpor.201000025.

[173] J. Tao, X. Wang, T. Sun, H. Cai, Y. Wang, T. Lin, et al., Hybrid photonic cavity with metal-organic framework coatings for the ultra-sensitive detection of volatile organic compounds with high immunity to humidity, Sci. Rep. 7 (2017) 41640. Available from: https://doi.org/10.1038/srep41640.

[174] M.L. Tietze, M. Obst, G. Arnauts, N. Wauteraerts, S. Rodríguez-Hermida, R. Ameloot, Parts-per-million detection of volatile organic compounds via surface plasmon polaritons and nanometer-thick metal—organic framework films, ACS Appl. Nano Mater. 5 (2022) 5006—5016. Available from: https://doi.org/10.1021/acsanm.2c00012.

[175] C. Zhu, J.A. Perman, R.E. Gerald, S. Ma, J. Huang, Chemical detection using a metal-organic framework single crystal coupled to an optical fiber, ACS Appl. Mater. Interfaces. 11 (2019) 4393–4398. Available from: https://doi.org/10.1021/acsami.8b19775.

[176] M. Mohammad, A. Razmjou, K. Liang, M. Asadnia, V. Chen, Metal-organic-framework-based enzymatic microfluidic biosensor via surface patterning and biomineralization, ACS Appl. Mater. Interfaces. 11 (2019) 1807–1820. Available from: https://doi.org/10.1021/acsami.8b16837.

Ionic nanoarchitectonics for nanochannel-based biosensing devices

Yamili Toum Terrones[1], Gregorio Laucirica[1], Vanina M. Cayón[1,2], M. Lorena Cortez[1], María Eugenia Toimil-Molares[2], Christina Trautmann[2,3], Waldemar A. Marmisollé[1] and Omar Azzaroni[1]

[1]*Instituto de Investigaciones Fisicoquímicas Teóricas y Aplicadas (INIFTA)—Departamento de Química—Facultad de Ciencias Exactas—Universidad Nacional de La Plata—CONICET, La Plata, Argentina* [2]*GSI Helmholtzzentrum für Schwerionenforschung, Darmstadt, Germany* [3]*Technische Universität Darmstadt, Materialwissenschaft, Darmstadt, Germany*

16.1 Introduction

Biological ion channels can be defined as ion-transport proteins located in cell membranes. Through them, the cell is capable of transporting ions across the membrane which is crucial for many vital life processes such as pumping nutrients into cells, generating electrical signals, and regulating cell volume [1]. Although a vast variety of ion channels with different structures and functions is known, they all share one fundamental property: a notorious ability to transport ions in a selective mode [1]. The potassium ion channel is a paradigmatic example: its potassium transport rate is near the diffusion limit without leaving selectivity aside [2]. In the last decades, inspired by the outstanding properties of biological ion channels, many research groups decided to focus their attention on the development of fully abiotic solid-state nanopores and nanochannels for (bio)sensing, energy conversion, filtration, nanoelectronics, and other applications [3,4]. Synthetic solid-state nanochannels (SSNs) (length much larger than diameter) and nanopores (SSNPs) (length and diameter with similar dimensions) can be designed to meet not only key properties that characterize biological ion channels such as selectivity, stimulus responsiveness, and ion rectification but also to be chemically versatile, highly robust, and mechanically resistant [5,6]. Over recent decades, material science and nanofabrication techniques are moving ahead at a staggering speed enabling the construction of nanopores and nanochannels with dimensions comparable to the size of biological molecules that not only can be used as stimuli-responsive nanofluidic elements [7–16], but also as remarkably sensitive molecule biosensors [17]. As the biosensing performance of these nanofluidic devices is strongly linked to the surface characteristics of their inner walls, much effort has been made to establish surface-modification strategies at the nanoscale level [4].

In this chapter, we will first introduce basic notions of biosensing with nanochannel-based nanofluidic devices. Then, we will center the attention on the steady-state approach mainly used for sensing with SSNs and discuss its differences with the resistive-pulse sensing method generally

Materials Nanoarchitectonics. DOI: https://doi.org/10.1016/B978-0-323-99472-9.00010-9

used for sensing with SSNPs. Also, we will explain the principles that govern SSNs' transduction of ionic signals mechanisms. Then, the most common nanofabrication and surface functionalization techniques of abiotic SSNs in order to construct biosensors, including the integration of biological recognition elements, are described. Finally, we will address some remarkable examples of SSNs-based biosensors of relevant analytes.

16.2 Fundamental concepts in ion transport across nanopores and nanochannels for biosensing applications

16.2.1 What is a biosensor?

As stated by Lee et al., a biosensor can be defined as an object formed by two essential components: (1) a biological molecule that specifically recognizes the analyte, that is, the *bioreceptor* or *recognition element*; and (2) a *transducer*, which is an element that transforms one form of energy into another, meaning it is capable of converting the biorecognition event into a measurable signal, usually proportional to the amount of analyte−bioreceptor interactions (Fig. 16.1A) [18,19].

An accurate choice of the bioreceptor is the key to develop a biosensor with the desired selectivity, sensibility, and robustness. Plenty of biological molecules and macromolecules such as proteins, enzymes, DNA, RNA, aptamers, antibodies, among others, have been used as bioreceptors in SSNs-based biosensors [20]. They enable the specific and selective recognition of the target analyte, and at the same time, they bring the possibility of promoting signal amplification by means of its integration into nanotechnological platforms [21].

Besides, SSNs (and SSNPs) act as *transducers* because they reveal the ion transport change triggered by a modification in the physicochemical properties of the channels (pores) as a consequence of the biorecognition event. More specifically, these modifications determine a change in the ion flux across the channels (pores), which means a change in the transmembrane ion current (ions are considered the signal carriers). In 2009, Han et al. introduced for the first time the term "iontronics" to describe ion-based information processes and devices [22]. Analogously to electronic

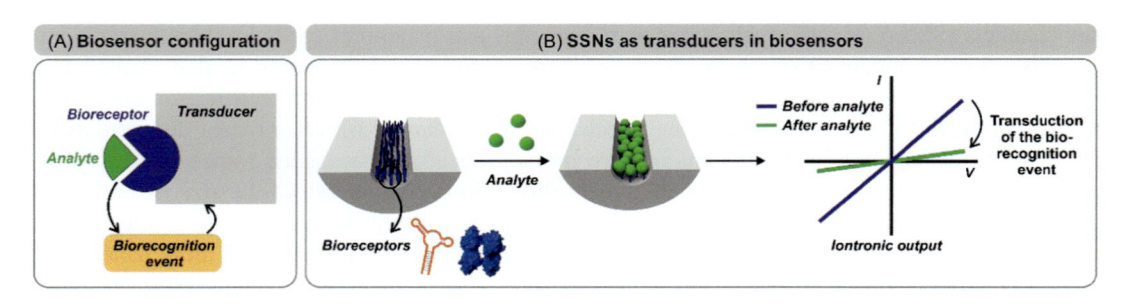

FIGURE 16.1

(A) Scheme summarizing the key components of a biosensor. (B) In this cartoon, a symmetric solid-state nanochannel conveniently functionalized acts as the transducer of the recognition event to yield a change in the ionic transport response ($I-V$ curve, iontronic output).

devices that control the transport of electrons, iontronic devices respond to changes in ionic transport [5]. SSNs and SSNPs are important examples of iontronic devices, and consequently, its response in terms of the current—voltage (I—V) dependence is often called *iontronic output* (Fig. 16.1B).

As systematically addressed by Laucirica et al., a plethora of (bio)sensors based on SSNs for many different analytes including ions, nucleic acids, proteins, cells, drugs, amino acids, sugars, neurotransmitters, pollutants, and gases has been created [4]. In this section, we will focus the attention on the different transduction mechanisms that govern the biosensing processes in SSNs. The appropriate choice of the transduction method is crucial for the construction of biosensors based on these abiotic nanochannels.

16.2.2 Ionic transport measurements and transduction mechanisms

16.2.2.1 Resistive pulse-sensing versus steady state measurements

At this point, it is important to distinguish between the two main methods for sensing via ionic transport measurements with SSNPs and SSNs: the resistive-pulse sensing method and the steady-state approach.

Resistive-pulse sensing is based on the Coulter-counting method. Here, the membrane containing the SSNP is placed between two conductivity cells filled with an electrolyte solution and by using a two-electrode setup, a transmembrane constant voltage is applied [23,24]. Ions from the supporting electrolyte generate a steady-state current that, for a given channel and in high ionic strength conditions (where surface charge effects can be neglected), is mainly determined by the transmembrane voltage magnitude, the conductivity of the bulk solution, and the pore dimensions (diameter and length) [25]. If the nanopore is partially occluded due to the translocation of a bulky analyte (e.g., proteins), the current shuts off (Fig. 16.2A). This perturbation (or event) in the current is usually characterized by an amplitude, duration, and frequency rate [23,24]. The amplitude can be usually related, among other variables, to the shape, volume, and charge of the analyte. The duration of the event also provides information about the analyte such as its diffusion coefficient, the conformation, and the characteristics of its interaction with the pore surface. Notably, by analyzing the frequency of the event, valuable information about the analyte concentration can be obtained. Also, the integration of recognition elements onto the pore surface is a widely used strategy to improve the selectivity [26].

Historically, resistive pulse sensing has been applied to nanopores. In particular, biological and artificial (solid-state) nanopores have been widely used not only for sensing but also for sequencing of nucleic acid and proteins, studying dynamics of aggregation, performing discrimination of viruses, among others [27−32]. Regarding the use of SSNs in combination with resistive-pulse sensing strategies, until now there are only a few examples for the detection of amyloid-β fibrils, which could open up new possibilities for this type of experiments [33,34]. Otherwise, steady-state measurements are one of the most employed methods for the quantification of analytes with track-etched membranes [4]. In this method, a membrane that contains a single or an array of nanochannels is placed between two conductivity cells filled with an electrolyte solution. An arrangement of two or four electrodes is connected to record I—V curves. The ions are the signal carrier, and therefore, their concentration inside the channel determines the final response (iontronic output). In this context, the analyte is quantified by comparing the response before and after the exposition to the analyte

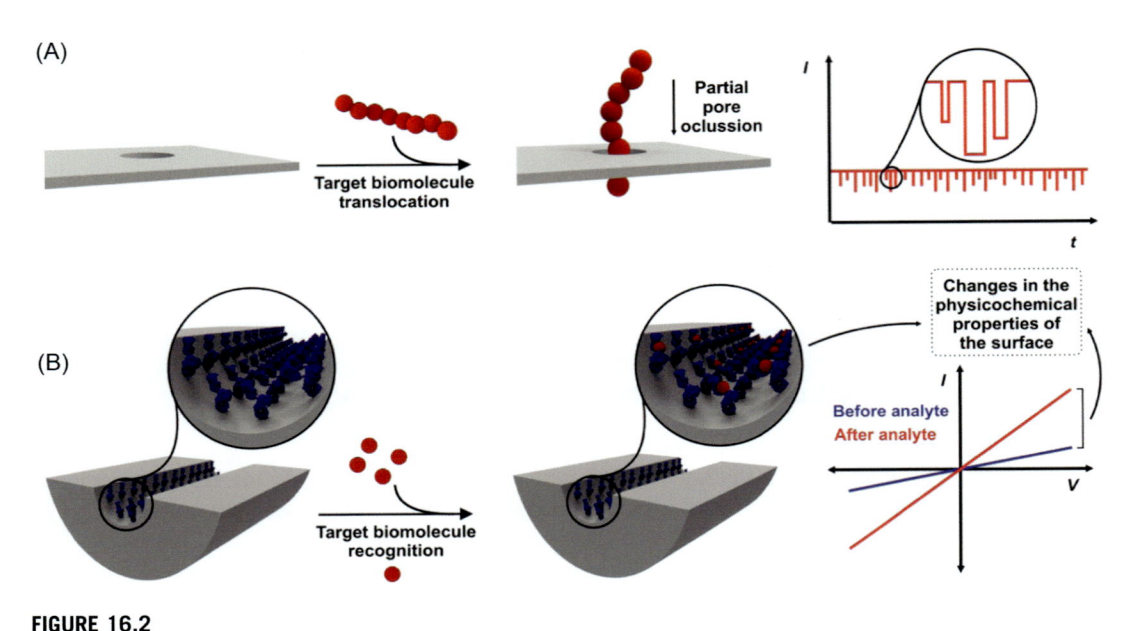

FIGURE 16.2

Scheme illustrating the two main approaches for sensing by ion transport measurements with synthetic solid-state nanochannels (SSNPs) and solid-state nanochannel (SSNs). (A) Sensing by resistive-pulse sensing with SSNPs: the translocation of a bulky analyte produces current blockades with a certain amplitude, duration, and frequency. (B) Sensing by steady-state measurements in SSNs: if the presence of the target analyte interacting with the nanochannel walls produces changes in the physicochemical properties of the channel (surface charge, wettability, or effective size), it can be evidenced by the changes in the $I-V$ curves. In this case, these changes were arbitrary schematized as an increase in the ionic current, but it is important to remark that the ionic current change (increment or decrease) will be determined by the specific changes in the physicochemical properties.

solutions (Fig. 16.2B). In this chapter, we will center our attention on biosensing with SSNs by steady-state measurements. Until now, some review articles regarding sensing with biological and abiotic nanopores via the resistive-pulse technique have been published and could help the reader to acquire a general vision of the subject [24,27,29,35−37].

Regarding the transduction mechanisms, as we will address in the following section, biosensing by steady-state measurements requires the analyte to generate any change in the surface charge density or in the effective size of the channel (steric effects) that could be transduced into changes in the $I-V$ curves. Typically, to tackle this aim, the channel surface is decorated with a recognition element that, in the presence of the analyte, selectively interacts with it and generates the required variation in the mentioned physicochemical properties (Fig. 16.2B). Also, the wettability of the pore/channel can be modified in the presence of the target analyte affecting the iontronic output. Instead, as this mechanism is not common in sensing applications [5], in the next section we will only review the surface charge density and the effective size changes-based signal transduction mechanisms.

16.2.2.2 Most common transduction mechanisms when using the steady-state approach
16.2.2.2.1 Response modulated by changes in the surface charge density

To understand how analyte-promoted surface charge changes can modulate the ion transport across the channel, the concept of electrical double layer (EDL) is key [38]. The electrostatic potential originated by the immersion of a charged surface into an electrolyte solution spontaneously triggers ion redistribution around it, giving rise to the so-called EDL. In other words, ions with the same charge polarity of the surface (coions) are repelled whereas ions with the opposite charge (counter-ions) are enriched at the proximity of the surface. Typically, the characteristic length of EDL is called "Debye length" and it takes values of a few nanometers in electrolyte solutions of low and moderated concentrations (e.g., ~ 3 nm in aqueous 0.1 M KCl at 298K). Then, in the case of charged SSNs with a size comparable to the Debye length exposed to an electrolyte solution, the inner volume of the pore/channel will be enriched with counter-ions and this effect will play a central role in the features of the iontronic output [3,39]. In this context, the ion transport across the pore/channel is surface charge-governed, and it is possible to obtain information about its magnitude by analyzing the $I-V$ curves. Consequently, if the analyte promotes changes in the surface charge density due to its interaction with the recognition elements attached to the surface, it can be transduced into a difference in the iontronic output [4].

As it is possible to anticipate, in the case of SSNs, both the appropriate design of the channel architecture as well as the appropriate selection and immobilization of the recognition element onto its surface are of paramount relevance to determine the features of the biosensor. In the case of the SSN architecture, the iontronic output is highly dependent on the channel geometry (Fig. 16.3) [4,40]. Cylindrical channels present an ohmic response (linear dependence between the current and the transmembrane voltage) where the conductance magnitude (the slope of the $I-V$ curve) depends on the surface charge density (Fig. 16.3A). Concomitantly, if the recognition element-analyte interaction promotes surface charge density changes, it can be evidenced by variations in the curve slope before and after the analyte exposition. On the other hand, asymmetric channel geometries such as conical or bullet-shaped and/or asymmetric distributions of surface charges cause the disruption of the electrical potential symmetry inside of the channel which gives rise to a rectifying behavior (asymmetric iontronic output). Under this regime, the iontronic output is characterized by an enhancement of the ion current at a given polarity (diode-like behavior). Analogously to the ohmic behavior, the ion current rectification (ICR) regime (or rectifying regime) is extremely sensitive to the changes in the surface charge (Fig. 16.3B) [5,41]. In particular, the variations in the efficiency of rectification, usually quantified as the ratio between the currents at ± 1 V (rectification factor, f_{rec}), can be used as a parameter to monitor the differences in the charged state of the surface and therefore, to determine the analyte concentration (Fig. 16.3B). In addition, rectifying behavior is also sensitive to the polarity of the surface charge. The direction of the rectification, given by the voltage polarity where the high conductance branch is placed, is indicative of the sign of the net surface charge (depending on the electrode arrangement with respect to the channel) (Fig. 16.3B).

16.2.2.2.2 Response modulated by steric effects

Beyond the surface charge, the current magnitude also depends on the inner volume of the channel. Thus, if the interaction between the analyte and the recognition element promotes changes in the effective size of the channel by a partial occlusion, it can be typically registered as a diminution in

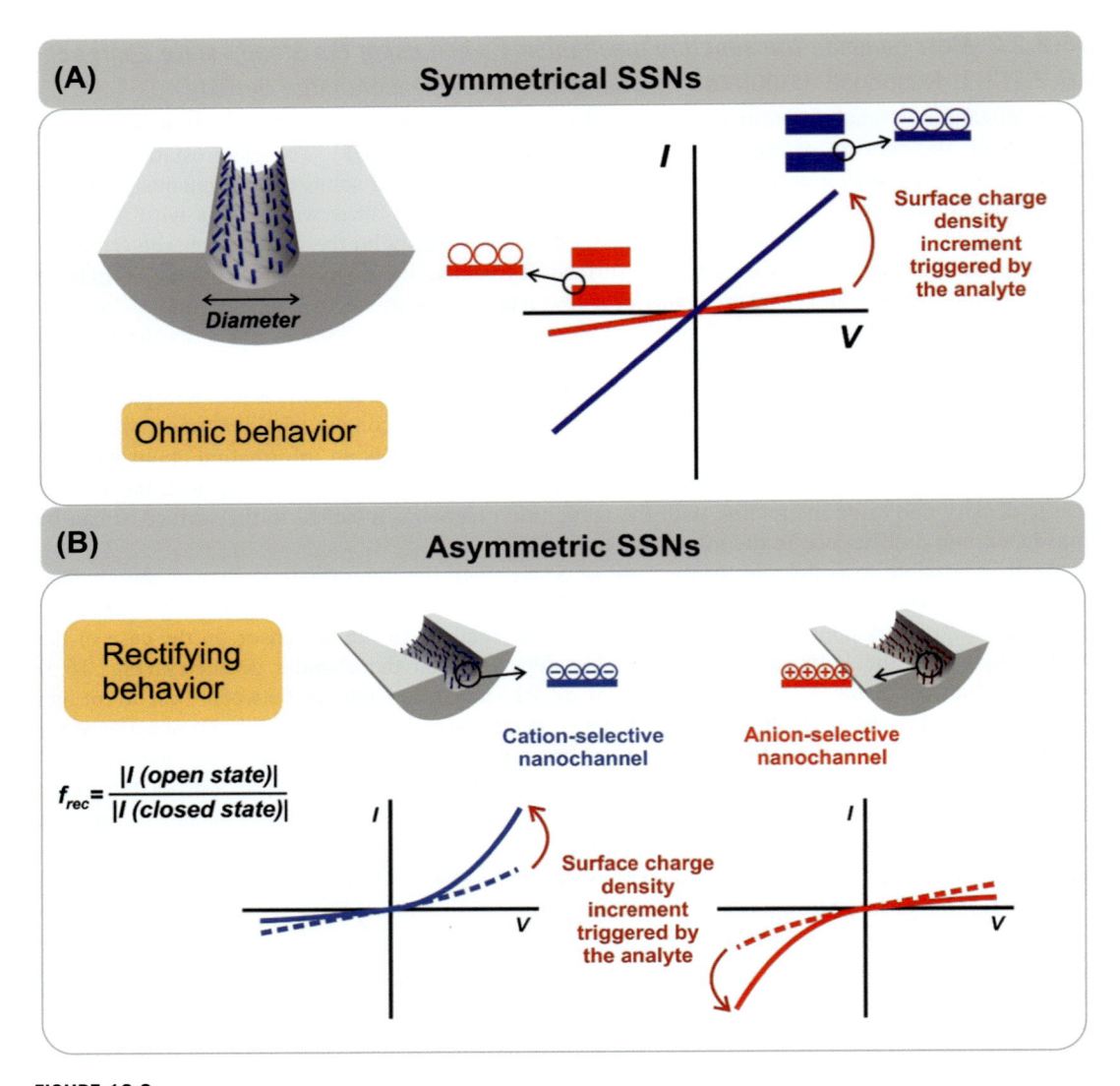

FIGURE 16.3

Surface charge effect transduction mechanism in solid-state nanochannels (SSNs). Correlation between the iontronic response and the surface charge in (A) cylindrical (symmetrical) nanochannels and (B) asymmetrical nanochannels. In asymmetrical SSNs, the ion current rectification phenomenon is dependent on the surface charge density and its sign, which are, in turn, modulated by the presence of an analyte.

the current (Fig. 16.4A). This sensing strategy has been demonstrated to be very advantageous for quantifications of bulky analytes such as proteins and viruses [42,43]. On the contrary, when the analyte promotes the release of the recognition element into the medium, the channel increases its volume which usually implies an increment in the transmembrane ion current (Fig. 16.4B).

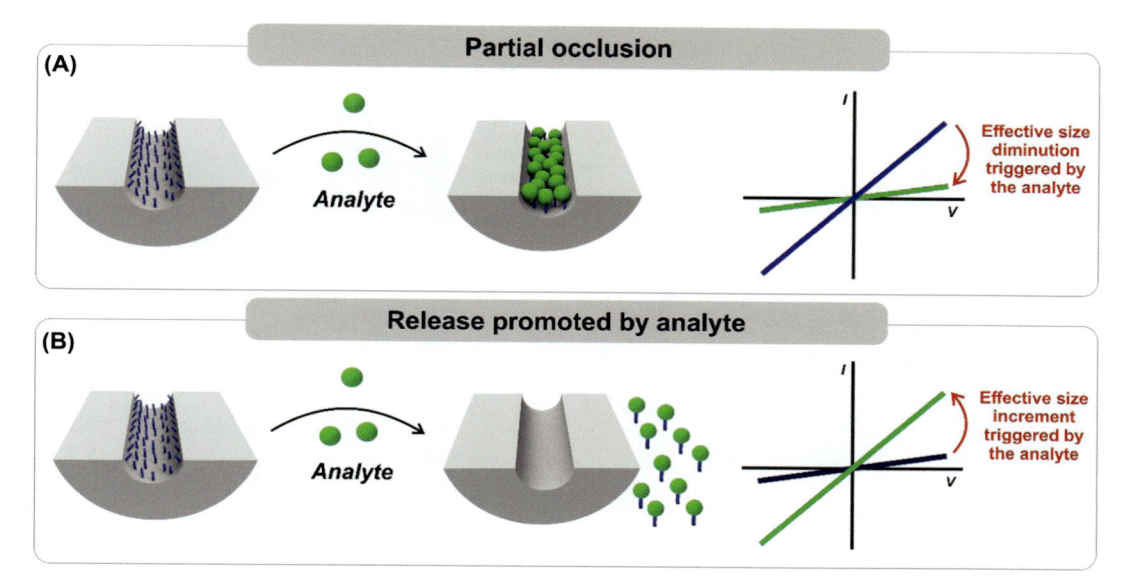

FIGURE 16.4

Scheme depicting the most common situations in which the analyte triggers changes in the effective channel size that determine the iontronic output. (A) The analyte interacts with the recognition element generating partial channel blockage, and therefore, decreasing the effective channel size. (B) The analyte interacts with the recognition element and releases it from the channel, thus increasing the effective channel size.

16.3 Nanofabrication and functionalization techniques

16.3.1 Nanofabrication methods

The word "nanofabrication" evolved from "microfabrication," a term used in the field of microelectronics and integrated circuits [44]. In that context, a nanofabrication procedure implies the design and manufacture of devices with dimensions in the nanometer scale. In the last decades, many research groups have been devoted to develop and refine nanofabrication technologies to make them more reliable, repeatable and with the lowest defect level [44]. Up to date, there are three main nanofabrication techniques to obtain SSNPs and/or SSNs: ion-track etching technology, nanopipettes and glass nanopores fabrication, and electrochemical anodization.

Ion-track-etching technology has been employed extensively for the construction of SSNs not only in polymeric materials, such as polyethylene terephthalate —PET-, polyimide —PI-, and polycarbonate —PC, but also in inorganic materials, such as silica and mica [45,46]. As shown in Fig. 16.5Ai, the ion fluence in ion-track technology—number of ions per area unity—that impacts on the sample can be fine-tuned between a single ion and $\sim 10^{11}$ ions/cm^2 obtaining the so-called single or multichannel membranes [45]. It comprises two main steps: first, a polymeric foil is *irradiated* with swift heavy ions of MeV—GeV energy (^{197}Au, ^{206}Pb, and ^{238}U with energies of up to 11.4 MeV per nucleon) generating a damaged region (called "ion-track") along its trajectory in the material (Fig. 16.5Aii). Then, the ion-track is selectively dissolved during a chemical treatment

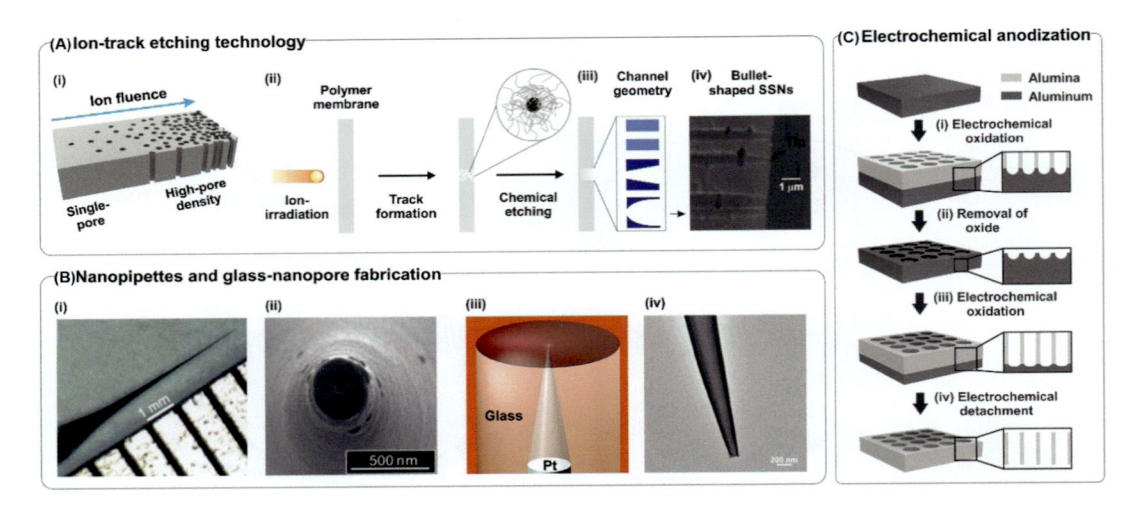

FIGURE 16.5

(A) Nanofabrication by ion-track technology. (i) Different ion fluence regimes are correlated with different membrane pore-densities: from single-pore to multipore membranes. (ii) Scheme of the ion irradiation process and the ion-track formation. (iii) The chemical etching conditions determines some of the channel geometries obtained by this nanofabrication method. (iv) Scanning electron microscopy (SEM) images of membrane cross-sections containing bullet-shaped channels (B) Nanopipettes and glass nanopore fabrication. (i) Optical micrograph of a nanopipette. (ii) Scanning electron micrograph of a nanopipette in end-on view showing the nanopore opening at the tip. (iii) Schematic representation of a glass nanopore electrode (15-nm-orifice radius, 0.9-μm-base, 2.8-μm depth, and 0.75-mm-radius glass shroud, drawn to scale except for the glass shroud). (iv) TEM image of a glass nanopore tip. (C) Steps involved in the electrochemical anodization for the fabrication of AAO nanochannel arrays.

known as the *etching procedure*, which relies on the different properties of the damaged region compared with the bulk material. The appropriate selection of the etching conditions (chemicals, additives, temperature, exposure time, and symmetric or asymmetric treatment) determines the channel geometry (cylindrical, conical, or bullet-shaped) which is crucial for establishing the ion-transport characteristics of the SSN (Fig. 16.5Aiii), as mentioned in the previous section. For example, bullet-shaped nanochannels can be created by exposing one side of the membrane to the etchant solution (a highly concentrated NaOH solution ∼6 M) and the opposite side to the same etchant solution modified with a small concentration of a surfactant (Fig. 16.5Aiv) [47,48].

The fabrication of nanopipettes and glass nanopores does not imply the drilling with ion or electron sources, these nanofabrication methods are considered great options for the construction of biosensors. In particular, nanopipettes have been extensively employed for the creation of electrochemical biosensors, as they offer the combined advantage of nanoscale dimensions and the selectivity and sensitivity of conventional solid-state biosensors [53]. More specifically, "nanopipette" is a term typically used to describe glass/quartz pipettes with pore diameters lower than 200 nm and having a needle-like geometry (Fig. 16.5Bi and ii) [53]. For the fabrication, the starting materials quartz or borosilicate glass capillaries are converted into nanopipettes using a CO_2-laser pipette puller that alternates between cycles of heating and pulling to create two identical nanopipettes [50,53]. On the other hand, the fabrication and electrochemical characterization of glass nanopores were first reported by White and coworkers in 2004 [54]. In their work, they described the fabrication of the truncated cone-shaped nanopore electrodes (pore orifice radii lower than 100 nm) as including the sealing of an atomically sharp Pt wire in a glass capillary, the polishing of the capillary until a nm-dimensioned Pt disk was exposed, and the electrochemical etching of the exposed Pt to obtain the final nanopore in glass (Fig. 16.5Biii and iv) [51,55,56]. As shown in Fig. 16.5Biii, these glass nanopore electrodes are opened to solution through a single orifice and incorporate the transduction element (Pt wire) inside the nanopore allowing detailed investigation of the transport of molecules in nanoconfined structures [54].

Finally, the electrochemical anodization method is a versatile procedure for the creation of highly ordered channels with nanometric dimensions in alumina and titania [4]. As shown in Fig. 16.5C, for the construction of anodic aluminum oxide (AAO) membranes, firstly a clean aluminum foil is used as precursor and electrochemically oxidized in acidic conditions to give a porous structure of aluminum oxide attached to the aluminum foil. Then, by treating the structure with phosphochromic acid, the oxide layer is removed, which produces an aluminum textured pattern that can be used as a substrate in a second electrochemical oxidation process, giving rise to the porous aluminum oxide structure (Fig. 16.5C) [4]. Pore dimensions, pore density, and membrane thickness can be adjusted by selecting the appropriate experimental conditions (temperature, electrolyte concentration, applied voltage, etc.) [57]. A similar process can be used to create TiO_2 nanochannels by electrochemical anodization [58].

16.3.2 Functionalization of solid-state nanochannels

After fabrication, the surface of the SSNs needs to be adequately prepared to anchor the desired recognition element that, in the case of biosensors, consists of a bio(macro)molecule that acts as the bioreceptor. Even though the details of the functionalization method will depend on the nature of the surface and the bioreceptor, we can introduce some general strategies as a starting point.

Firstly, gold-deposition (Fig. 16.6A) and atomic layer deposition (ALD, Fig. 16.6B) are both strategies that imply the coating of the nanochannel surface with a layer of gold or a layer of a certain inorganic oxide, respectively. There are two classical methods to add a gold layer to a SSN: the *electroless method*, which generates a gold film deposition as a result of the spontaneous reduction of a gold precursor (typically $Au(SO_3)_2^{3-}$) in the presence of a catalyst [59], and the *sputtering procedure* that, in contrast to the electroless approach, produces asymmetrically covered SSNs (on the side of the small aperture for example) (Fig. 16.6A) [4]. After the covering with Au, and given the well-known interactions between thiol groups (R-SH) and noble metals, if the bioreceptor is dotted with a thiol moiety, it can be immobilized in the Au-modified SSN-surface to give self-assembled monolayers (SAMs) (Fig. 16.6A) [60,61]. On the other hand, metallization allows further functionalization by electrochemical methods using the Au-coated membrane as an electrode [47].

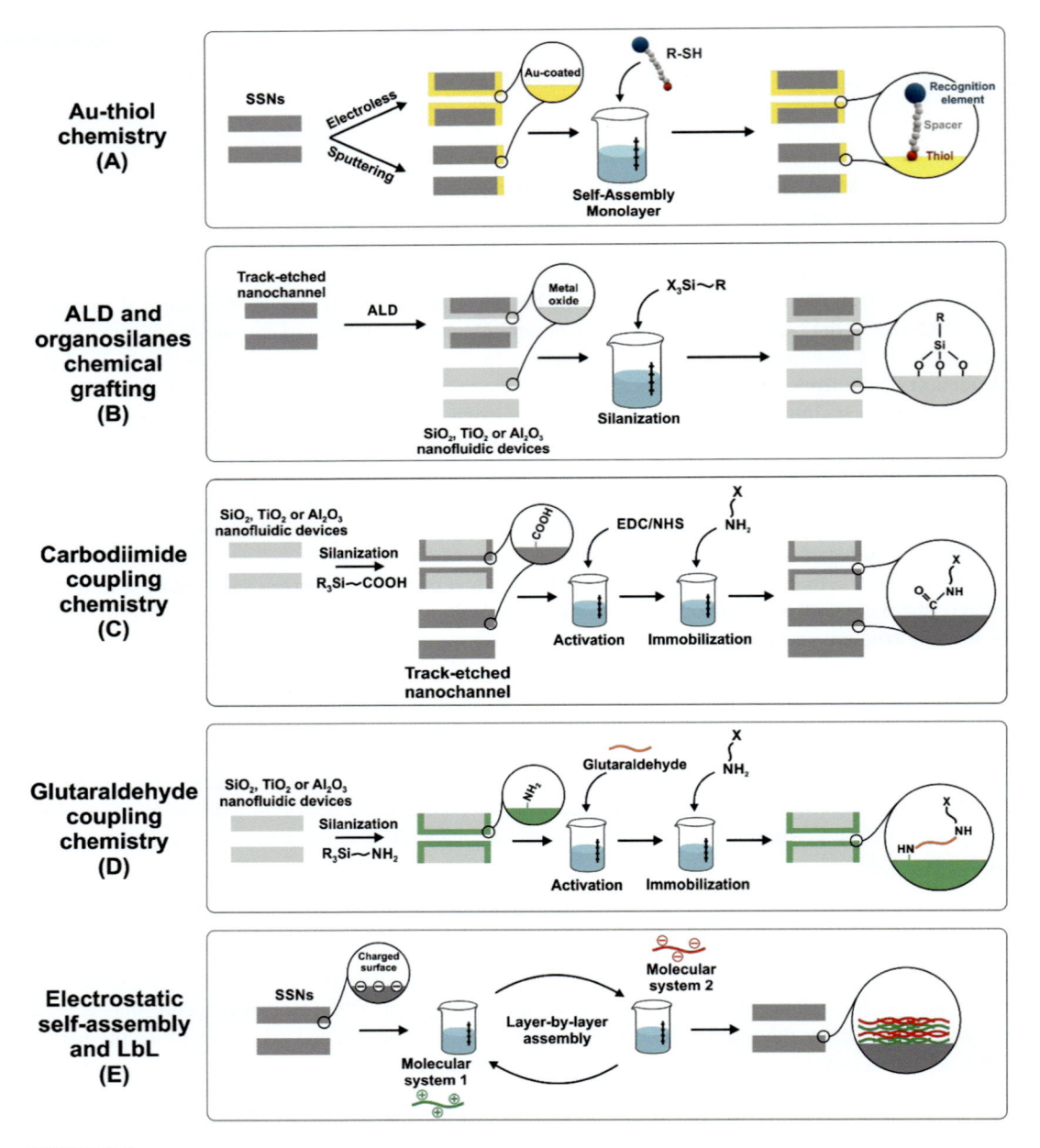

FIGURE 16.6

Illustrative scheme summarizing different functionalization strategies of solid-state nanochannel: (A) self-assembled monolayers via thiol-Au chemistry; (B) atomic layer deposition and grafting with organosilanes; (C) ethylcarbodiimide/N-hydroxysuccinimide coupling reaction; (D) glutaraldehyde coupling chemistry, and (E) electrostatic self-assembly and layer-by-layer.

Also, ALD is a plausible method for introducing inorganic oxide surfaces (SiO_2, TiO_2, Al_2O_3) to noninorganic SSNs as polymeric track-etched SSNs (Fig. 16.6B) [62]. It is based on the sequential and self-limiting gas—solid surface reactions of two reactants in the gas phase, which guarantees the ultraprecise control of the inorganic oxide layer thickness and its composition at the atomic level [63]. Further, inorganic oxide SSNs as AAO-based or glass nanopore membranes as well as ALD-coated track-etched SSNs can be functionalized via silanization reactions to introduce the desired moiety needed to anchor the bioreceptor in a subsequent step (Fig. 16.6B) [4].

On the other hand, *carbodiimide* and *glutaraldehyde couplings* are common methods for the covalent functionalization of chemical surface groups in SSNs. If the surface contains carboxylic acid groups (−COOH) either as a result of the ion track-etching method in polymeric membranes or as a consequence of the silanization process in an inorganic oxide-based nanochannel, using N-(3-dimethylaminopropyl)-N'-ethylcarbodiimide (EDC) together with N-hydroxysuccinimide (NHS) or analogs activates the carboxylic surface groups for their reaction with an amine-containing molecule that might be part of the bioreceptor—as biomolecules, such as proteins, enzymes, and antibodies, usually contain several amino groups in their structures (Fig. 16.6C) [4]. Besides, if the nanochannel surface exposes amine groups (for example, after SiO_2, TiO_2 or Al_2O_3-based nanochannels silanization with an amino-terminal silane, Fig. 16.6D), glutaraldehyde (OHC−$(CH_2)_3$−CHO) can be used to activate them in order facilitate its reaction with an amine-containing bioreceptor [4,64]. While the reaction mechanism remains unclear, it is known that final product contains glutaraldehyde as an spacer between the surface and the bioreceptor, which is an advantage in avoiding steric hindrance issues [64].

Finally, electrostatic self-assembly of polyelectrolytes and biomolecules is one of the most versatile and simple strategies for SSNs functionalization that is not based on covalent modifications [65]. If the native surface is electrically charged, such as, for example, the negative surface charge of ion-track etched SSNs, it can be treated with a solution containing a polyelectrolyte with the opposite charge, and it will be adsorbed due to electrostatic attraction (Fig. 16.6E) [66,67]. This adsorption inverts the surface charge making possible the subsequent modification with another polyelectrolyte or biomolecule with an opposite charge to that of the original polyelectrolyte [68]. Since first reported by Decher on flat surfaces, this protocol called layer-by-layer (LbL) self-assembly has become a very affordable strategy to functionalize SSNs [4,69].

16.4 Biosensors based on solid-state nanochannels

As the chemical nature of the bioreceptor defines not only the analyte to be detected but also the recognition mechanism and the performance of the biosensor, in this section we will address some paradigmatic examples, and recent advances in the development of steady-state biosensors based on SSNs gathered by families of recognition elements: proteins (4.1), enzymes (4.2), and nucleic acids-related molecules (4.3). Further details about these types of biosensors and more examples based on other recognition elements can be found in recently published review articles [3,4,23].

16.4.1 Proteins

A pioneering work that showed the incorporation of multivalent ligands and proteins as recognition elements into SSNs via electrostatic self-assembly was reported by Azzaroni and coworkers [17].

The authors demonstrated that attractive electrostatic interactions allow the incorporation of binding sites into the nanochannels that can be used as recognition elements to create nanobiosensors. In that work, a single-pore PET track-etched nanochannel was functionalized with a bifunctional macromolecular ligand (biotinylated poly(allylamine)) by electrostatic self-assembly (Fig. 16.7i). Then, when the positively charged biotinylated-surface was exposed to the protein streptavidin, even at very low concentrations, a drastic decrease on the rectified current was observed due to steric effects (Fig. 16.7iii). At this point, the nanochannel was functionalized with the protein streptavidine, which served as a new bioreceptor for the recognition of biotin-containing molecules and the creation of multilayered supramolecular assemblies inside SSNs (Fig. 16.7ii-iii).

Besides, an illustrative example of the detection of proteins via protein—protein affinity was presented by Ali et al., who reported the immobilization of the mannose-rich glycoprotein and enzyme horseradish peroxidase (HRP) in track-etched PET SSNs (Fig. 16.7iv) [70]. For that aim, the carboxylic acid groups of the nanochannel PET surface were covalently coupled with the amine groups of HRP by using carbodiimide coupling chemistry. Then, the HRP-PET-SSN was exposed to solutions containing Concanavalin A (Con A), a lectin (carbohydrate-binding protein) that has a high affinity for the mannose residues present in HRP enzyme [70]. In terms of ion transport, the biorecognition event between HRP and Con A produced a reduction of the available area for ionic transport caused by steric effects, which could be exploited to tune the conductance and selectivity of the nanochannel in aqueous solution (Fig. 16.7v) [70]. Regarding this work, it is important to remark that even though the bioreceptor is an enzyme, it was not chosen for its catalytic properties but for its higher content of carbohydrates in its surface.

Another work reported the immobilization of the protein bovine serum albumin (BSA) on the surface of a single asymmetric nanochannel to create a tryptophan (L-Trp) biosensor [71]. In this case, BSA is an appropriate bioreceptor as it enantioselectively recognizes L-Trp over D-Trp through specific BSA—L-tryptophan interactions. A polydopamine (PDA) thin film was deposited onto the channel surface and then by means of a Michael addition reaction BSA amine groups were linked to PDA-functionalized SSNs. Regarding the ion transport, the specific BSA—L-tryptophan interactions caused changes in the surface charge density that tuned the ionic transport through the nanochannel [71].

16.4.2 Enzymes

Enzymes are biopolymers formed in all living cells from the 20 natural amino acids. They regulate the rate at which chemical reactions occur without itself being altered in the process; thus, they act as biological catalysts [72]. What makes enzymes unique for its integration in SSNs is that they bind with remarkable specificity to certain analytes, in accordance to their protein structure. As a result, enzymes are ideal bioreceptors that coupled with SSNs can be used to create specific and ultrasensitive biosensors taking advantage of the SSN surface changes triggered by the confined enzymatic reaction [68].

One of the first reports on the integration of an enzyme to SSNs and the use of the iontronic response to monitor de enzyme activity was provided in 2018 by Pérez-Mitta et al. [68]. The authors studied the mechanism for urea sensing by means of a urease-modified single PET nanochannel. Using the electrostatic self-assembly approach, a positively charged polyelectrolyte (poly (allylamine)) was adsorbed on the surface of a bullet-shaped PET SSN. Then, the negatively

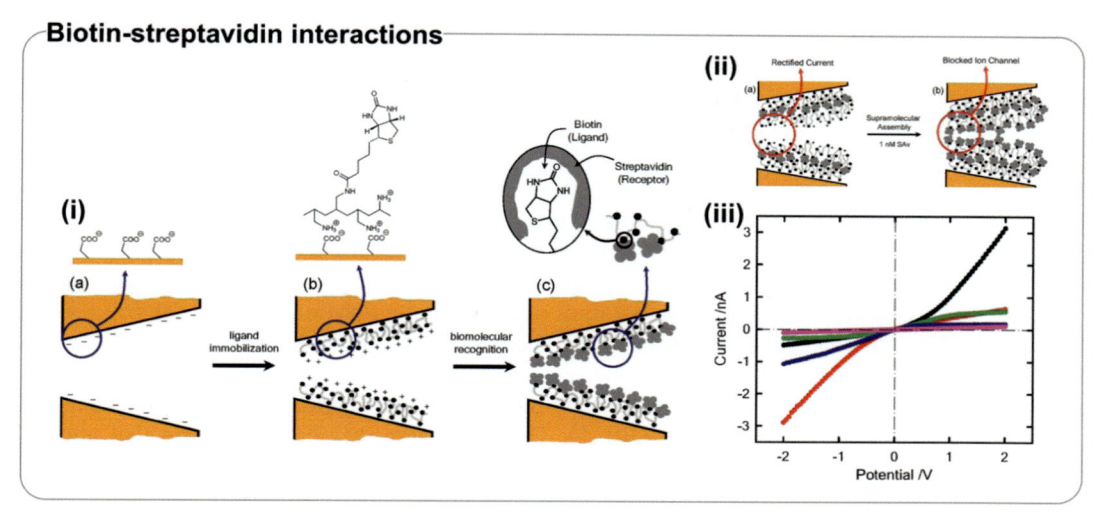

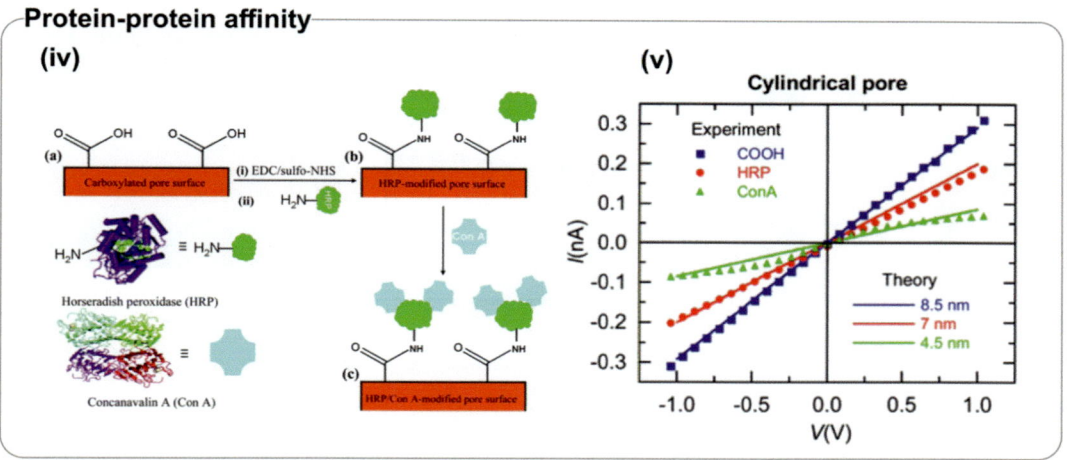

FIGURE 16.7

(i) Scheme of the incorporation of the biorecognition elements into the single conical nanochannel: from the carboxylate-terminated nanochannel to the recognition of streptavidin by the biotin electrostatically immobilized ligand. (ii) Formation of a multilayered supramolecular bioconjugate inside the nanochannel: here, the recognition element is the protein streptavidin. (iii) Current − voltage characteristics of a surface-modified single conical nanochannel in 0.1 M KCl corresponding to (black) carboxylate-terminated pore; (red) b-PAH-modified pore; (green) (SAv)(b-PAH)-modified pore; (dark blue) (SAv)(b-PAH)$_2$-modified pore; (pink) (SAv)$_2$(b-PAH)$_2$-modified pore. (iv) Schematic representation of the synthetic strategy for the incorporation of HRP into the nanochannel surface and the biorecognition event between HRP and ConA. (v) I−V curves for the different functionalization steps. Comparison between theory (lines) and experiment (symbols).

charged urease enzyme was immobilized by electrostatic attractions [68]. In the presence of the analyte urea, the enzymatic hydrolysis occurs, producing ammonia and carbon dioxide, which leads to a local pH increase [68]. As the consequence of the confined local pH change, there is a decrease in the protonation degree of the poly(allylamine), which reduces the positive surface charge density in the channel. With this device, an ultralow limit of detection of 1 nM of urea was achieved [68].

Recently, based on the same approach, Toum Terrones et al. reported for the first time the construction of acetylcholine (Ach) enzymatic biosensors with remarkable sensitivity by immobilizing acetylcholinesterase (AchE) on polyethylenimine-modified bullet-shaped single nanochannels via electrostatic self-assembly (Fig. 16.8A−C) [73]. The signal transduction mechanism that allowed the detection of the neurotransmitter in the nanomolar range (LoD 16 nM) is the change in the

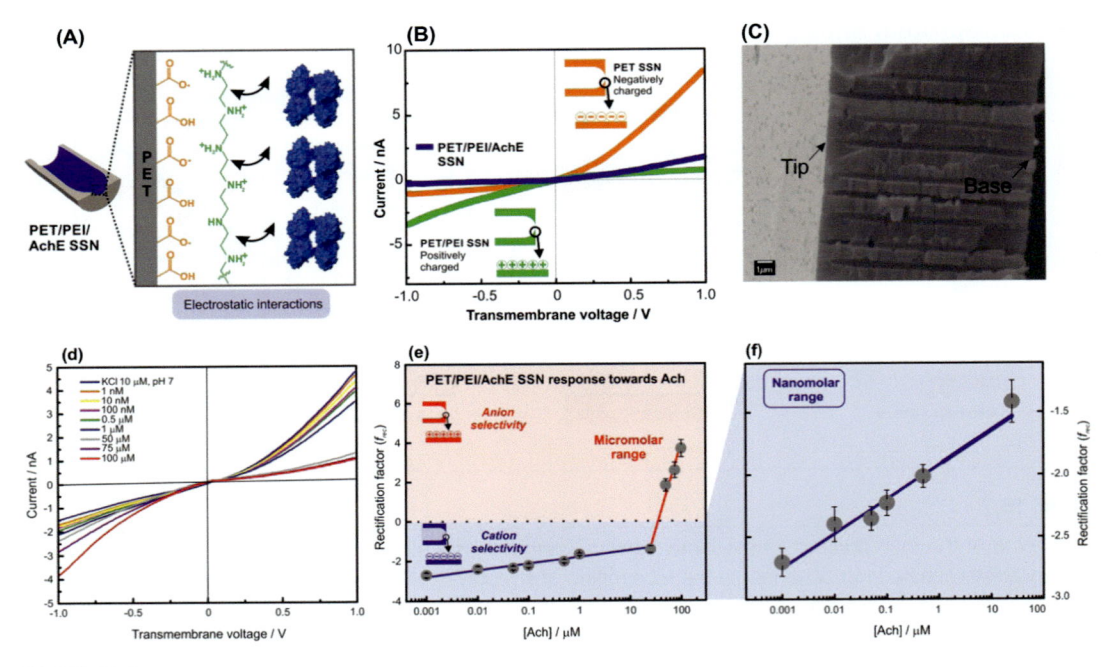

FIGURE 16.8

(A) Illustrative scheme depicting the electrostatic interactions between polyethylene terephthalate (PET) carboxylate groups, PEI amino groups, and charged groups of AchE in the PET/PET/AchE solid-state nanochannels (SSN). (B) $I-V$ curves for the different modification steps measured in 10 mM KCl at pH 7; bare PET nanochannel (orange), after the modification with PEI (green), and finally, after AchE assembly (blue). (C) Scanning electron microscopy (SEM) micrograph of the PET SSNs. (D) $I-V$ curves recorded at different Ach concentrations, from 0 to 100 mM, in 10 mM KCl at pH 7. Every measurement was carried out in situ after a 30-min exposure time. (E) Changes in f_{rec} for different Ach concentrations. (F) The PET/PEI/AchE calibration curve in the nanomolar region.

surface charge density of the functionalized SSN provoked by local pH changes caused by the confined enzyme-catalyzed hydrolysis of Ach (Fig. 16.8D—E) [73]. Furthermore, the biosensor showed a highly reversible iontronic response towards Ach and selectivity in the presence of other analytes such as dopamine, serotonin, L-cysteine, ascorbic acid, and glucose [73].

16.4.3 Nucleic acids-related molecules

In 2004, Harrel et al. reported for the first time the integration of DNA into single conically-shaped polycarbonate-based nanochannels coated with a thin layer of gold to create artificial ion channels [74]. With their experiments, they could demonstrate that the magnitude of rectification could be modulated not only varying the DNA chain length by means of a chemical method but also by altering the nanochannel mouth diameter via the nanofabrication method [74].

After that seminal contribution, many research groups have been devoted to the integration of DNA architectures into SSNs for the goal of meeting new biosensors. In this section, we will review some of those efforts with concrete examples regarding the recognition mechanism between the analyte and the DNA-related molecule that acts as the bioreceptor: in Section 16.4.3.1 we address the DNA hybridization mechanism when using DNA oligonucleotides, DNA superstructures, aptamers, and DNA hydrogels as bioreceptors; then, in 16.4.3.2, we describe the formation of DNA complex structures. For further reading on the topic we recommend the complete review by Pérez-Mitta et al. [3].

16.4.3.1 DNA hybridization reaction

A founding work that used the DNA hybridization process to detect target DNA oligomers with an AAO membrane-based platform was reported by Vlassiouk et al. in 2005 [75]. Since then, many attempts have been made to build nucleic acid sensors based on the hybridization process in different types of nanochannels. The hybridization mechanism consists of the binding of a primer or synthetic DNA probe to a biological DNA target sequence via Watson—Crick base pairing [76].

16.4.3.1.1 DNA oligonucleotides

In 2016, Sun et al. reported the creation of a label-free ultrasensitive highly sequence-specific DNA biosensor based on the immobilization of a probe DNA into track-etched SSNs by electrostatic self-assembly [77]. The native negatively charged surface of the SSN was treated sequentially with polyethyleneimine to invert the surface charge, the negatively charged DNA probe (bioreceptor), and finally, bovine serum albumin to block the surface and avoid further nonspecific interactions with other negatively charged analytes (Fig. 16.9A). This functionalization gave the device an almost zero surface charge density which, in terms of ion transport, was seen as an ohmic behavior. When treating the nanochannel with the target complementary DNA, the hybridization reaction occurred and a strong increment in the net negative surface charge was observed. Furthermore, the magnitude of that increment was related with the concentration of the target DNA (Fig. 16.9B). This biosensor was able of distinguishing between complementary DNA (c-DNA), noncomplementary DNA (nc-DNA) and one-base mismatched DNA (1bm-DNA) samples with high specificity, even in real serum samples.

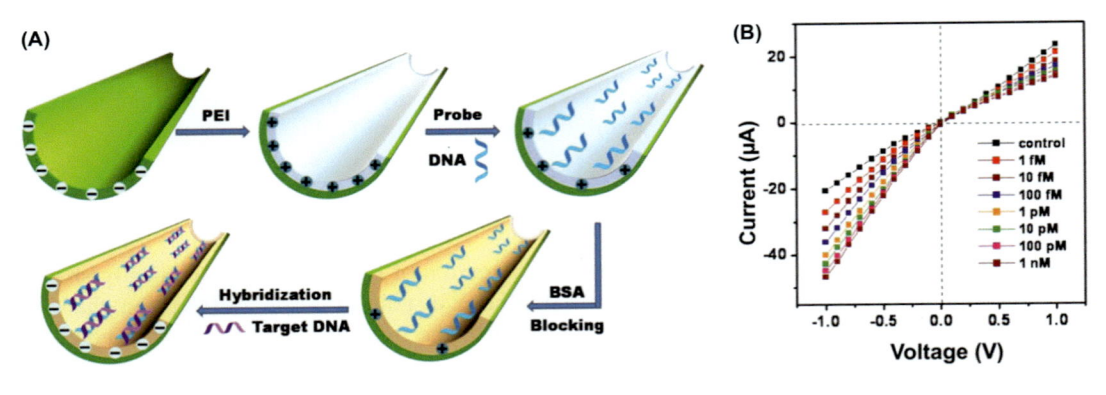

FIGURE 16.9

(A) Schematic illustration of the functionalized polyethylene terephthalate nanochannel biosensor construction for DNA detection. (B) *I−V* curves for different c-DNA concentrations.

Reproduced with permission Sun Z., Liao T., Zhang Y., et al., 2016. Biomimetic nanochannels based biosensor for ultrasensitive and label-free detection of nucleic acids. Biosens. Bioelectron. 86, 194–201. Copyright © 2016, Elsevier.

16.4.3.1.2 DNA superstructures

Lei Jiang and coworkers reported an illustrative example of the development of more efficient gating mechanisms in nanochannels using DNA superstructures [78]. The authors created a subnanomolar sequence-specific DNA (oligonucleotides) sensor using track-etched PET nanochannel membranes decorated with predesigned capture DNA probes [78]. When the device was exposed to the target DNA, alternating units of signal probes (S1 and S2) were consecutively hybridized bringing DNA superstructures in the form of long concatamers that efficiently block ionic transport across the membrane, resulting in a diminution in the magnitude of the transmembrane ionic current.

16.4.3.1.3 Aptamers

Aptamers are nucleic acid derived molecules rationally designed for a specific target [79]. They can be incorporated into sensors and actuators with by simple strategies, which is central to emerging technologies [79]. In this context, Li et al. reported a bioinspired adenosine artificial regulatory receptor based on the self-assembly of predesigned adenosine (AD) aptamers onto gold-modified PET multinanochannels [80]. The selected bioreceptor was a sequence-specific aptamer (SSA) that binds to adenosine with a high affinity. In terms of ionic transport, SSA self-assembly generated very low ionic currents (OFF state) as a consequence of the randomly stretched single-stranded structure of SSA that partly blocked the SSN tip. In the presence of the target analyte, the recognition event between SSA and AD produced a conformational change (beanpod structure) in the aptamer that opened the pore (ON state) and permitted an increase in the transmembrane ionic current.

Recently, Peinetti et al. presented the direct detection of human adenovirus or SARS-CoV-2 with the ability to inform infectivity using DNA aptamers-modified PET track-etched SSNs (Fig. 16.10) [43]. For that aim, the aptamers (bioreceptors) were selected from a DNA library to bind intact infectious but not noninfectious viruses and synthesized using the SELEX technique. Then, they were incorporated into SSNs using the carbodiimide coupling chemistry strategy (EDC/NHS)

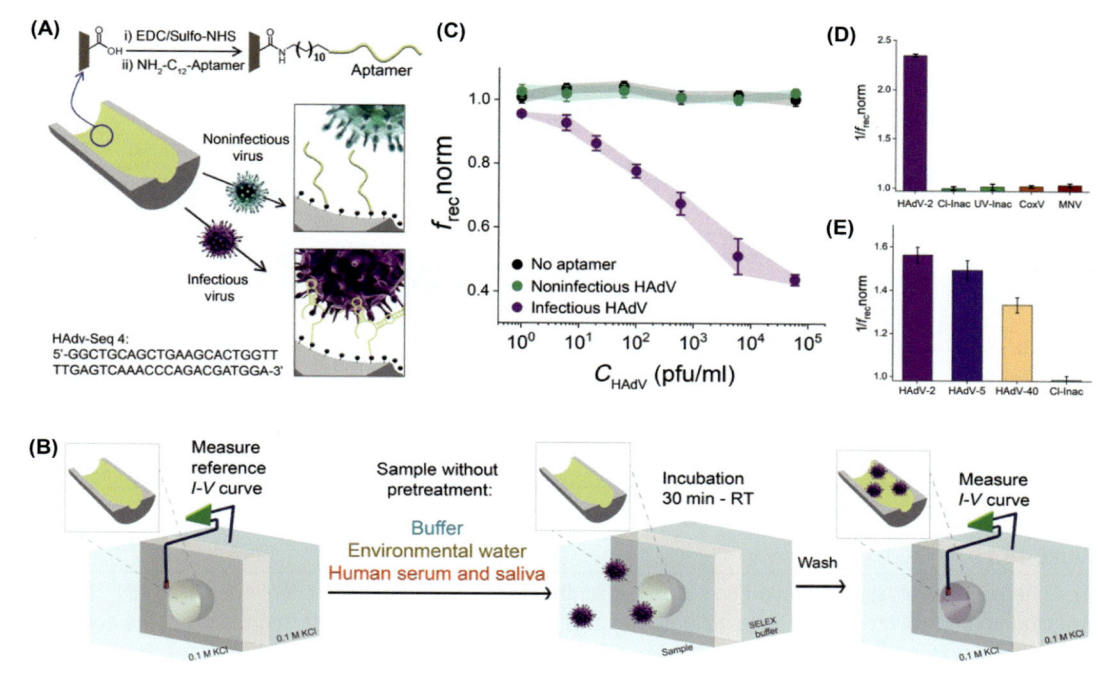

FIGURE 16.10

(A) Scheme depicting the modification of the nanochannel using the carbodiimide coupling strategy and the interaction of the aptamer with infectious HAdV samples. (B) Scheme of infectious HAdV detection by the aptamer-nanochannel system. (C) Normalized rectification efficiencies versus virus concentration. (D) Selectivity assay. Inverse of frecnorm obtained for infectious HAdV (HAdV-2), two noninfectious HAdVs (Cl-inact and UV-inact), and two other viruses (CoxV and MNV). (E) Inverse of *frecnorm* obtained for different serotypes of infectious adenovirus (HAdV-2, HAdV-5, and HAdV-40) and comparison with an inactivated HAdV sample at the same concentration.

(Fig. 16.10A). The grafting of the aptamer on the SSN wall provoked a decrease in the rectification efficiency of the system with increasing amount of the infectious HAdV (Fig. 16.10B−C). Furthermore, this technology was tested in different types of water samples, saliva and serum with undoubtedly promising results.

16.4.3.1.4 DNA hydrogels

As DNA is constituted by a polar sugar−phosphate backbone and has a negatively charged helix structure, it can operate as a strong hydrophilic polyelectrolyte [81]. This hydrophilic nature due to the capacity to interact with water molecules allows DNA the formation of gel-like materials [81]. Wu et al. developed smart DNA hydrogels-functionalized nanochannels for the biosensing of K^+

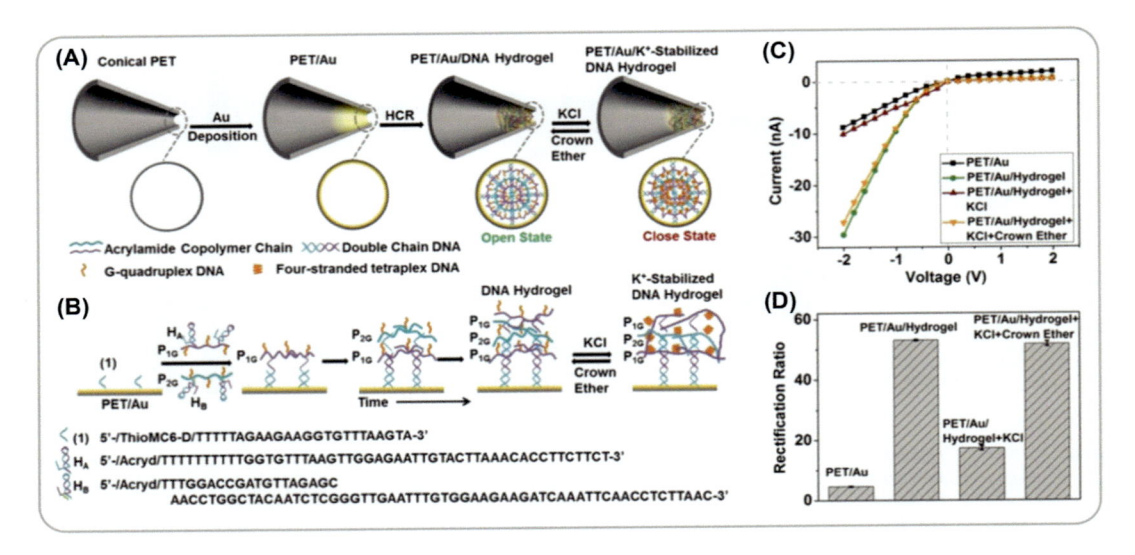

FIGURE 16.11

(A) Scheme illustrating the preparation process of the DNA hydrogels-based ion channels. (B) Detailed formation and transition of DNA hydrogel networks. (C) $I-V$ curves and (D) rectification ratios of the conical nanochannel under different conditions (0.1 M LiCl, 10 mM Tris-HCl buffer, pH 7).

Reproduced with permission Wu Y., Wang D., Willner I., et al., 2018. Smart DNA hydrogel integrated nanochannels with high ion flux and adjustable selective ionic transport. Angew. Chem. Int. Ed. 57, 7790–7794. Copyright © 2018 John Wiley & Sons, Inc.

ions using the concept of the Hybridization Chain Reaction (HCR) [82]. HCR is a mechanism in which stable DNA monomers (called hairpins H1 and H2) act as building blocks and assemble only by being treated with an initiator that is a target single-stranded DNA fragment [83]. To reach K^+ detection, the surface of a PET/Au SSN was first prepared with a thiolated nucleic acid as initiator and then, using a series of hairpins, DNA hydrogels were self-assembled via HCR (Fig. 16.11A−B). The DNA hydrogel-modified SSN showed a high rectification factor owing to the hydrophilicity and negative charge provided by the hydrogel, which constitutes the "open state" (Fig. 16.11C−D). When treating the system with different concentrations of K^+ ions (0.05−1 M), a decrease in the rectification factor was observed due to the formation of a K^+-stabilized hydrogel in the SSN ("close state"). Furthermore, the exposure of the as prepared SSNs to 18-crown-6-ether solutions caused the release of K^+ ions from the DNA-hydrogel structure which caused the softening of the DNA hydrogel structure (Fig. 16.11C−D).

16.4.3.2 Formation of DNA complex structures

There are some examples of the formation of DNA complex structures in nanochannel-based devices that could lead to iontronic changes due to steric effects in the presence of certain stimuli/target analyte. Buchsbaum et al. reported the construction of synthetic DNA-modified nanochannels responsive both to pH and to transmembrane potential [84]. Here, DNA oligomers containing protonable adenine (A) and cytosine (C) were attached to the nanochannel surface (tip side) and the

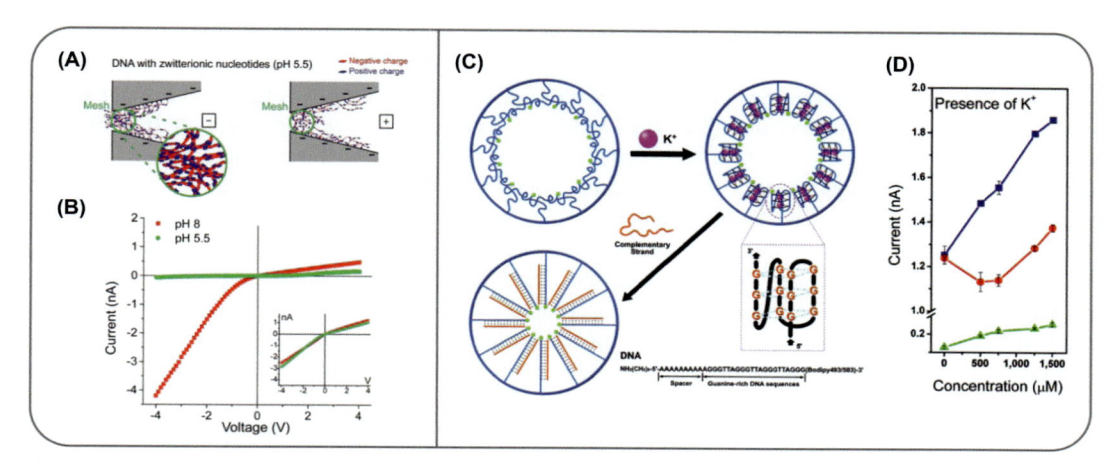

FIGURE 16.12

(A) Scheme depicting the influence of pH and applied voltage in the DNA oligomers-modified solid-state nanochannels. At pH 5.5 an electrostatic mesh is formed and the pore closes (left). Voltage-induced deflection of oligomers toward the wide opening leads to partial pore opening (right). (B) $I-V$ curves of a nanochannel modified with AC-rich oligomers at pH 8 (red) and pH 5.5 (green) in KCl 100 mM. (C) K^+-responsive G4 DNA. (D) Current versus K^+ concentration curve before G4 DNA modification (blue); after G4 DNA modification (red); after the exposure to the complementary DNA strands (green).

(B) Reproduced with permission Buchsbaum SF, Nguyen G, Howorka S, Siwy ZS (2014) DNA-modified polymer pores allow pH- and voltage-gated control of channel flux. J. Am. Chem. Soc. 136:9902–9905. Copyright © 2014 American Chemical Society. (D) Reproduced with permission Wang Y.G., Hou X., Guo W., et al., 2009. A biomimetic potassium responsive nanochannel: G-quadruplex DNA conformational switching in a synthetic nanopore. J. Am. Chem. Soc. 131, 7800–7805. Copyright © 2009 American Chemical Society.

iontronic response could be linked to a pH-dependent reversible closing mechanism [84]. Concretely, at mild pH conditions (pH 5.5), the local positive charges given by A and C are electrostatically attracted to the negatively charged backbone of neighboring DNA strands, which forms an electrostatic mesh that blocks the channel (Fig. 16.12A−B) [84].

Otherwise, a potassium-responsive bionanosensor was created by grafting the walls of an asymmetric single track-etched SSN with a G-rich human telomere strand (bioreceptor), a particular example of a DNA secondary structure called G-quadruplex (G4) [85]. In the absence of K^+ the bioreceptor shows a random single-stranded structure, while after the treatment with K^+ it changes to a four-stranded G4 structure (Fig. 16.12C). Thus, there is a variation in the effective pore size revealed by changes in the ion transport properties of the device (Fig. 16.12D) [85].

16.5 **Conclusions**

The construction of sensing platforms for the selective, sensitive, accurate, and reliable detection of biomarkers is of vital relevance in today's world for contributing to solve many challenges the human population is facing. In this context, sensors based on SSNs have been extensively investigated due to their high sensitivity, reversibility, and simple usage procedure. In close resemblance to its biological counterparts, SSNs present unique ion transport properties that make them excellent candidates for the detection of a wide range of analytes. As we addressed in this chapter, there is a variety of nanofabrication techniques to produce SSNs in different materials, from polymer materials to glass and nanoporous alumina. These nanofabrication techniques in combination with specific surface functionalization strategies lead to the integration of the desired recognition elements into the SSNs in accordance with the analyte of interest. When the recognition elements are biological entities such as proteins, enzymes, and nucleic-acid related molecules, the SSNs-biosensors are tremendously specific for the target analyte and in many cases ultrasensitive biosensors can be obtained. Furthermore, as a result of the influence of the surface charge and the effective channel size on the ionic current through the nanochannels, the anchoring of biorecognition entities on the nanochannel surface and their specific interaction with target analytes affect the transmembrane ionic current, while enabling specific responsiveness and biosensing. In this chapter, we presented the fundamental concepts on biosensing with SSNs, including the nanofabrication and functionalization techniques and the most common transduction mechanisms of iontronic signals. We reviewed some illustrative examples of how the suitable integration of bioreceptors within the inner surface of the channels allows the biosensing, in some cases quantitative, of different target molecules in solution by means of the changes they trigger in the ion transport across the nanochannels.

Acknowledgments

Y.T.T and G.L. acknowledge the scholarship from CONICET. G.L. also acknowledges a scholarship for a research stay from German Academic Exchange Service (DAAD) and the support of the GET_INvolved program of the GSI, Darmstadt (Germany). V. M. C. acknowledges the postdoctoral financial support from Universidad Nacional de La Plata. M.L.C., W.A.M., and O.A. acknowledge the financial support from Universidad Nacional de La Plata (PPID-X867), CONICET (PIP-0370), and ANPCyT (PICT-2020−2468, PICT-2017−1523, PICT2016−1680).

References

[1] E. Gouaux, R. MacKinnon, Principles of selective ion transport in channels and pumps, Science (80-) 310 (2005) 1461−1465.

[2] D.A. Doyle, M. Cabral, R.A. Pfuetzner, et al., The structure of the potassium channel: molecular basis of K^+ conduction and selectivity, Science (80-) 280 (1998) 1−9.

[3] G. Pérez-Mitta, M.E. Toimil-Molares, C. Trautmann, et al., Molecular design of solid-state nanopores: fundamental concepts and applications, Adv. Mater. 31 (2019) 1901483.

[4] G. Laucirica, Y.T. Terrones, V. Cayón, et al., Biomimetic solid-state nanochannels for chemical and biological sensing applications, TrAC. Trends Anal. Chem. 144 (2021) 116425.

[5] G. Pérez-Mitta, A.G. Albesa, C. Trautmann, et al., Bioinspired integrated nanosystems based on solid-state nanopores: "iontronic" transduction of biological, chemical and physical stimuli, Chem. Sci. 8 (2017) 890−913.

[6] L. Jiang, X. Hou, Learning from nature: building bio-inspired smart nanochannels, ACS Nano 3 (2009) 3339−3342.

[7] G. Pérez-Mitta, W.A. Marmisollé, C. Trautmann, et al., Nanofluidic diodes with dynamic rectification properties stemming from reversible electrochemical conversions in conducting polymers, J. Am. Chem. Soc. 137 (2015) 15382−15385.

[8] G. Pérez-Mitta, W.A. Marmisollé, A.G. Albesa, et al., Phosphate-responsive biomimetic nanofluidic diodes regulated by polyamine−phosphate interactions: insights into their functional behavior from theory and experiment, Small 14 (2018) 1702131.

[9] G. Pérez-Mitta, W.A. Marmisolle, L. Burr, et al., Proton-gated rectification regimes in nanofluidic diodes switched by chemical effectors, Small 14 (2018) 1703144.

[10] G. Laucirica, J.A. Allegretto, M.F. Wagner, et al., Switchable ion current saturation regimes enabled via heterostructured nanofluidic devices based on metal−organic frameworks, Adv. Mater. 34 (2022) 2207339.

[11] G. Laucirica, Y. Toum Terrones, M.F.P. Wagner, et al., Electrochemically addressed FET-like nanofluidic channels with dynamic ion-transport regimes, Nanoscale 15 (2023) 1782−1793.

[12] V.M. Cayón, G. Laucirica, Y. Toum Terrones, et al., Borate-driven ionic rectifiers based on sugar-bearing single nanochannels, Nanoscale 13 (2021) 11232−11241.

[13] G. Pérez-Mitta, L. Burr, J.S. Tuninetti, et al., Noncovalent functionalization of solid-state nanopores via self-assembly of amphipols, Nanoscale 8 (2016) 1470−1478.

[14] G. Pérez-Mitta, A. Albesa, F.M. Gilles, et al., Noncovalent approach toward the construction of nanofluidic diodes with pH-reversible rectifying properties: insights from theory and experiment, J. Phys. Chem. C. 121 (2017) 9070−9076.

[15] G. Pérez-Mitta, W.A. Marmisollé, C. Trautmann, et al., An all-plastic field-effect nanofluidic diode gated by a conducting polymer layer, Adv. Mater. 29 (2017) 1700972.

[16] G. Pérez-Mitta, A.G. Albesa, W. Knoll, et al., Host-guest supramolecular chemistry in solid-state nanopores: potassium-driven modulation of ionic transport in nanofluidic diodes, Nanoscale 7 (2015) 15594−15598.

[17] M. Ali, B. Yameen, R. Neumann, et al., Biosensing and supramolecular bioconjugation in single conical polymer nanochannels. Facile incorporation of biorecognition elements into nanoconfined geometries, J. Am. Chem. Soc. 130 (2008) 16351−16357.

[18] Y.H. Lee, R. Mutharasan, Biosensors, in: J.S. Wilson (Ed.), Sensor Technology Handbook, Elsevier Inc, 2005, pp. 161−180.

[19] N. Bhalla, P. Jolly, N. Formisano, P. Estrela, Introduction to biosensors, Essays Biochem. 60 (2016) 1−8.

[20] Y. Toum Terrones, V.M. Cayón, G. Laucirica, et al., Ion track-based nanofluidic biosensors, in: P. Chandra, K. Mahato (Eds.), Miniaturized Biosensing Devices: Fabrication and Applications, Springer Nature Singapore, Singapore, 2022, pp. 57−81.

[21] D. Ding, P. Gao, Q. Ma, et al., Biomolecule-functionalized solid-state ion nanochannels/nanopores: features and techniques, Small 15 (2019) 1804878.

[22] J.-H. Han, K.B. Kim, H.C. Kim, T.D. Chung, Ionic circuits based on polyelectrolyte diodes on a microchip, Angew. Chem. Int. Ed. 48 (2009) 3830−3833.

[23] N. Meyer, I. Abrao-Nemeir, J.-M. Janot, et al., Solid-state and polymer nanopores for protein sensing: a review, Adv. Colloid Interface Sci. 298 (2021) 102561.

[24] L. Xue, H. Yamazaki, R. Ren, et al., Solid-state nanopore sensors, Nat. Rev. Mater. 5 (2020) 931−951.

[25] R.B. Schoch, J. Han, P. Renaud, Transport phenomena in nanofluidics, Rev. Mod. Phys. 80 (2008) 839−883.

[26] R. Hu, X. Tong, Q. Zhao, Four aspects about solid-state nanopores for protein sensing: fabrication, sensitivity, selectivity, and durability, Adv. Healthc. Mater. 9 (2020) 2000933.

[27] B.N. Miles, A.P. Ivanov, K.A. Wilson, et al., Single molecule sensing with solid-state nanopores: novel materials, methods, and applications, Chem. Soc. Rev. 42 (2013) 15−28.

[28] I. Nir, D. Huttner, A. Meller, Direct sensing and discrimination among ubiquitin and ubiquitin chains using solid-state nanopores, Biophys. J. 108 (2015) 2340−2349.

[29] L. Restrepo-Pérez, C. Joo, C. Dekker, Paving the way to single-molecule protein sequencing, Nat. Nanotechnol. 13 (2018) 786−796.

[30] R.J. Yu, S.M. Lu, S.W. Xu, et al., Single molecule sensing of amyloid-β aggregation by confined glass nanopores, Chem. Sci. 10 (2019) 10728−10732.

[31] S. Schmid, C. Dekker, Nanopores: a versatile tool to study protein dynamics, Essays Biochem. 65 (2021) 93−107.

[32] M. Taniguchi, S. Minami, C. Ono, et al., Combining machine learning and nanopore construction creates an artificial intelligence nanopore for coronavirus detection, Nat. Commun. 12 (2021) 3726.

[33] N. Giamblanco, D. Coglitore, A. Gubbiotti, et al., Amyloid growth, inhibition, and real-time enzymatic degradation revealed with single conical nanopore, Anal. Chem. 90 (2018) 12900−12908.

[34] N. Meyer, N. Arroyo, J.M. Janot, et al., Detection of amyloid-β fibrils using track-etched nanopores: effect of geometry and crowding, ACS Sens. 6 (2021) 3733−3743.

[35] Y. Goto, R. Akahori, I. Yanagi, K. Takeda, Solid-state nanopores towards single-molecule DNA sequencing, J. Hum. Genet. 65 (2020) 69−77.

[36] Y. He, M. Tsutsui, Y. Zhou, X.-S. Miao, Solid-state nanopore systems: from materials to applications, NPG Asia Mater. 13 (2021) 48.

[37] M. Rahman, M.J.N. Sampad, A. Hawkins, H. Schmidt, Recent advances in integrated solid-state nanopore sensors, Lab. Chip (2021).

[38] P.C. Hiemenz, R. Rajagopalan, Principles of Colloid and Surface Chemistry, 3rd ed., Marcel Dekker, New York, 1997.

[39] G. Laucirica, M.E. Toimil-Molares, C. Trautmann, et al., Nanofluidic osmotic power generators - advanced nanoporous membranes and nanochannels for blue energy harvesting, Chem. Sci. 12 (2021) 12874−12910.

[40] Z.S. Siwy, Ion-current rectification in nanopores and nanotubes with broken symmetry, Adv. Funct. Mater. 16 (2006) 735−746.

[41] S. Zhang, W. Chen, L. Song, et al., Recent advances in ionic current rectification based nanopore sensing: a mini-review, Sens. Actuators Rep. 3 (2021) 100042.

[42] M. Ali, B. Schiedt, R. Neumann, W. Ensinger, Biosensing with functionalized single asymmetric polymer nanochannels, Macromol. Biosci. 10 (2010) 28−32.

[43] A.S. Peinetti, R.J. Lake, W. Cong, et al., Direct detection of human adenovirus and SARS-CoV-2 with ability to inform infectivity using a DNA aptamer-nanopore sensor, Sci. Adv. (2021) 1−12.

[44] Z. Cui, Nanofabrication: Principles, Capabilities and Limits, second ed, Springer International Publishing, 2017.

[45] M.E. Toimil-Molares, Characterization and properties of micro- and nanowires of controlled size, composition, and geometry fabricated by electrodeposition and ion-track technology, Beilstein J. Nanotechnol. 3 (2012) 860−883.

[46] C. Trautmann, Micro- and nanoengineering with ion tracks, in: R. Hellborg, H.J. Whitlow, Y. Zhang (Eds.), Ion Beams in Nanoscience and Technology, Springer-Verlag, Berlin Heidelberg, 2009, pp. 369−387.

[47] G. Laucirica, V.M. Cayón, Y. Toum Terrones, et al., Electrochemically addressable nanofluidic devices based on PET nanochannels modified with electropolymerized poly-o-aminophenol films, Nanoscale 12 (2020) 6002−6011.

[48] G. Laucirica, W.A. Marmisollé, M.E. Toimil-Molares, et al., Redox-driven reversible gating of solid-state nanochannels, ACS Appl. Mater. Interfaces 11 (2019) 30001−30009.

[49] G. Laucirica, Y. Toum Terrones, V. Cayón, et al., High-sensitivity detection of dopamine by biomimetic nanofluidic diodes derivatized with poly(3-aminobenzylamine), Nanoscale 12 (2020) 18390−18399.

[50] C.A. Morris, A.K. Friedman, L.A. Baker, Applications of nanopipettes in the analytical sciences, Analyst 135 (2010) 2190–2202.

[51] G. Wang, A.K. Bohaty, I. Zharov, H.S. White, Photon gated transport at the glass nanopore electrode, J. Am. Chem. Soc. 128 (2006) 13553–13558.

[52] S. Cao, S. Ding, Y. Liu, et al., Biomimetic mineralization of gold nanoclusters as multifunctional thin films for glass nanopore modification, characterization, and sensing, Anal. Chem. 89 (2017) 7886–7892.

[53] J. Stanley, N. Pourmand, Nanopipettes - the past and the present, APL. Mater. (2020) 8.

[54] B. Zhang, Y. Zhang, H.S. White, The nanopore electrode, Anal. Chem. 76 (2004) 6229–6238.

[55] B. Zhang, J. Galusha, P.G. Shiozawa, et al., Bench-top method for fabricating glass-sealed nanodisk electrodes, glass nanopore electrodes, and glass nanopore membranes of controlled size, Anal. Chem. 79 (2007) 4778–4787.

[56] J.H. Shim, J. Kim, G.S. Cha, et al., Glas. Nanopore-Based Ion-Selective Electrodes 79 (2007) 3568–3574.

[57] A. Ruiz-Clavijo, O. Caballero-Calero, M. Martín-González, Revisiting anodic alumina templates: from fabrication to applications, Nanoscale 13 (2021) 2227–2265.

[58] J. Wang, Z. Lin, Freestanding TiO_2 nanotube arrays with ultrahigh aspect ratio via electrochemical anodization, Chem. Mater. 20 (2008) 1257–1261.

[59] C.R. Martin, Nanomaterials: a membrane-based synthetic approach, Science (80-) 266 (1994) 1961–1966.

[60] M. Tagliazucchi, I. Szleifer, Chemically Modified Nanopores and Nanochannels, 1st ed., Elsevier Inc, 2016.

[61] O. Azzaroni, M. Mir, W. Knoll, Supramolecular architectures of streptavidin on biotinylated self-assembled monolayers. Tracking biomolecular reorganization after bioconjugation, J. Phys. Chem. B 111 (2007) 13499–13503.

[62] N. Ulrich, A. Spende, L. Burr, et al., Conical nanotubes synthesized by atomic layer deposition of Al_2O_3, TiO_2, and SiO_2 in etched ion-track nanochannels, Nanomaterials 11 (2021) 1874.

[63] M. Ritala, M. Leskelä, Atomic layer deposition, Handbook of Thin Films, Elsevier, 2002, pp. 103–159.

[64] P. Zucca, E. Sanjust, Inorganic materials as supports for covalent enzyme immobilization: methods and mechanisms, Molecules 19 (2014) 14139–14194.

[65] M. Coustet, J. Irigoyen, T.A. Garcia, et al., Layer-by-layer assembly of polymersomes and polyelectrolytes on planar surfaces and microsized colloidal particles, J. Colloid Interface Sci. 421 (2014) 132–140.

[66] E. Maza, J.S. Tuninetti, N. Politakos, et al., pH-responsive ion transport in polyelectrolyte multilayers of poly(diallyldimethylammonium chloride) (PDADMAC) and poly(4-styrenesulfonic acid-co-maleic acid) (PSS-MA) bearing strong- and weak anionic groups, Phys. Chem. Chem Phys 17 (2015) 29935–29948.

[67] G.E. Fenoy, E. Maza, E. Zelaya, et al., Layer-by-layer assemblies of highly connected polyelectrolyte capped-Pt nanoparticles for electrocatalysis of hydrogen evolution reaction, Appl. Surf. Sci. 416 (2017) 24–32.

[68] G. Pérez-Mitta, A.S. Peinetti, M.L. Cortez, et al., Highly sensitive biosensing with solid-state nanopores displaying enzymatically reconfigurable rectification properties, Nano Lett. 18 (2018) 3303–3310.

[69] G. Decher, J.D. Hong, J. Schmitt, Buildup of ultrathin multilayer films by a self-assembly process: III. Consecutively alternating adsorption of anionic and cationic polyelectrolytes on charged surfaces, Thin Solid. Films 210–211 (1992) 831–835.

[70] M. Ali, P. Ramirez, M.N. Tahir, et al., Biomolecular conjugation inside synthetic polymer nanopores via glycoprotein-lectin interactions, Nanoscale 3 (2011) 1894–1903.

[71] M. Ali, S. Nasir, W. Ensinger, Stereoselective detection of amino acids with protein-modified single asymmetric nanopores, Electrochim. Acta 215 (2016) 231–237.

[72] D. Abedi, L. Zhang, M. Pyne, C. Perry Chou, in: M.B.T.-C.B. Moo-Young (Ed.), 1.03 - Enzyme Biocatalysis, Second ed., Academic Press, Burlington, 2011, pp. 15–24.

[73] Y. Toum Terrones, G. Laucirica, V.M. Cayón, et al., Highly sensitive acetylcholine biosensing via chemical amplification of enzymatic processes in nanochannels, Chem. Commun. 58 (2022) 10166–10169.

[74] C.C. Harrell, P. Kohli, Z. Siwy, C.R. Martin, DNA - nanotube artificial ion channels, J. Am. Chem. Soc. 126 (2004) 15646−15647.

[75] I. Vlassiouk, P. Takmakov, S. Smirnov, Sensing DNA hybridization via ionic conductance through a nanoporous electrode, Langmuir 21 (2005) 4776−4778.

[76] D. Khodakov, C. Wang, D.Y. Zhang, Diagnostics based on nucleic acid sequence variant profiling: PCR, hybridization, and NGS approaches, Adv. Drug. Deliv. Rev. 105 (2016) 3−19.

[77] Z. Sun, T. Liao, Y. Zhang, et al., Biomimetic nanochannels based biosensor for ultrasensitive and label-free detection of nucleic acids, Biosens. Bioelectron. 86 (2016) 194−201.

[78] N. Liu, Y. Jiang, Y. Zhou, et al., Two-way nanopore sensing of sequence-specific oligonucleotides and small-molecule targets in complex matrices using integrated DNA supersandwich structures, Angew Chemie - Int Ed, 52, 2013, pp. 2007−2011.

[79] M.R. Dunn, R.M. Jimenez, J.C. Chaput, Analysis of aptamer discovery and technology, Nat. Rev. Chem. 1 (2017) 1−16.

[80] P. Li, X.-Y. Kong, G. Xie, et al., Adenosine-activated nanochannels inspired by G-protein-coupled receptors, Small 12 (2016) 1854−1858.

[81] V. Morya, S. Walia, B.B. Mandal, et al., Functional DNA based hydrogels: development, properties and biological applications, ACS Biomater. Sci. Eng. 6 (2020) 6021−6035.

[82] Y. Wu, D. Wang, I. Willner, et al., Smart DNA hydrogel integrated nanochannels with high ion flux and adjustable selective ionic transport, Angew. Chem. Int. Ed. 57 (2018) 7790−7794.

[83] R.M. Dirks, N.A. Pierce, Triggered amplification by hybridization chain reaction, PNAS 101 (2004) 15275−15278.

[84] S.F. Buchsbaum, G. Nguyen, S. Howorka, Z.S. Siwy, DNA-modified polymer pores allow pH- and voltage-gated control of channel flux, J. Am. Chem. Soc. 136 (2014) 9902−9905.

[85] X. Hou, W. Guo, F. Xia, et al., A biomimetic potassium responsive nanochannel: G-quadruplex DNA conformational switching in a synthetic nanopore, J. Am. Chem. Soc. 131 (2009) 7800−7805.

Molecular, supramolecular, and macromolecular engineering at hybrid mesoporous interfaces: choose your own nanoarchitectonic adventure

17

Cintia Belen Contreras[1], Galo J.A.A. Soler-Illia[1] and Omar Azzaroni[2]

[1]*Instituto de Nanosistemas, Escuela de Bio y Nanotecnologías, Universidad Nacional de San Martín, San Martín, Buenos Aires, Argentina* [2]*Instituto de Investigaciones Fisicoquímicas Teóricas y Aplicadas (INIFTA)—Departamento de Química—Facultad de Ciencias Exactas—Universidad Nacional de La Plata—CONICET, La Plata, Argentina*

17.1 Introduction—"soft chemistry" serving as a bridge toward nanoarchitected hybrid materials

Hybrid mesoporous architectures combining the properties of inorganic and organic materials constitute a remarkable category within the field of materials science [1−4]. Over the years, the blend of concepts from "sol−gel chemistry" and "soft matter," a concept often referred to as "*integrative chemistry*" [5,6], evolved as an attractive route toward the flexible realization of heteroarchitectures [7] and hierarchical nanosystems [8] with unprecedented control over functional tailoring [9,10].

One of the most attractive features of hybrid mesoporous materials relies on its the unique and thorough molecular control of their intrinsic topological and chemical characteristics, that is, composition, pore size, mesopore architecture, and morphology [11,12]. With the correct choice of building blocks and self-assembly conditions, it is possible to produce nanoarchitectured materials via sol−gel processes with precisely defined and tunable chemical functions incorporated into well-defined ordered mesostructured frameworks. With this in mind, the combination of supramolecular concepts with nanoscopic mesoporous solids has made great strides in the design of hybrid materials in which the harmony of noncovalent and covalent interactions leads to adjustable functions [13−15]. In addition, the increasing mastery in construction of functional nanoarchitected solids, by merging concepts and tools from self-assembly and "sol−gel" chemistry of inorganic precursors, has paved the way to materials exhibiting precise positioning of chemical functions on their surfaces or their inner spaces [4,9,16−19]. Exciting opportunities are revealed when we think in this manner. The synergism arising from the integration of molecular, supramolecular, and/or macromolecular functional units on the pore outlets of mesoporous scaffolds brings to bear a startling range of ideas related to the design of materials in which organic and inorganic entities play their roles side-by-side. For instance, nanostructured hybrid organic−inorganic nanocomposites present paramount

Materials Nanoarchitectonics. DOI: https://doi.org/10.1016/B978-0-323-99472-9.00001-8

advantages to facilitate integration and miniaturization of different devices, thus affording a direct connection between the inorganic and organic worlds. The ability to assemble and organize inorganic and organic components in a single material represents an exciting direction for developing novel multifunctional materials presenting a wide range of novel properties. For example, predefined functions hosted in an inorganic matrix can control the delivery of substances hosted in the mesoporous frameworks [20,21]. In this way, the chemically modified mesopores act as supramolecular-based nanoscopic gates able to control mass transport that can be opened and/or closed by specific chemical, physical, or biological signals.

We should bear in mind that nanoarchitected organic/inorganic hybrids are not simply physical mixtures of nanoscopic organic and inorganic building blocks. They can be defined as nanocomposites at the molecular scale, having at least one component, either the organic component or the inorganic component, with a characteristic length scale on the nanometer size. Then, more importantly, the properties of hybrid materials do not simply result from the sum of the individual contributions of their components, but also from the strong synergy created by an extensive hybrid interface. This explains why better control of the interface is not only leading to exciting developments in the field of nanoarchitected hybrid materials, but in the optimization of their useful properties [22].

It is clear that engineering the combination of dissimilar components at the nanometric and molecular level leads both to new challenges and opportunities for the development of novel and improved materials. For decades, the development methods and protocols for preparing hybrid nanoarchitected materials based on *"soft chemistry"* have attracted great interest both in academia and industry. These synthetic methods involve a *palette* of reactions carried out at room temperature in aqueous or organic solvents starting from molecular or nanoparticle precursors (alkoxides or metal salts, clusters, functional nanoparticles, monomers, amphiphiles, and so on). In the last few years, particular attention has been paid to synthetic strategies that permit precise location of inorganic and organic building blocks in well-defined "functional domains" that can be selectively addressed from the outside [1]. It is noteworthy that these "soft chemistry" conditions are identical to those in which many reactions of organic and organometallic chemistry, supramolecular chemistry, and polymer chemistry are performed. It is therefore possible with "soft chemistry" processes, originally developed by materials chemists, to produce materials in which the inorganic, organic, supramolecular, and polymer chemistry realms coexist harmonically.

The aim of this chapter is to present and discuss different strategies to synthesize nanoarchitected hybrid mesoporous materials furnished with specific functions. With this aim in mind, we have intended to select and discuss the main strategies and approaches, derived mainly from our own experience, rather than make an exhaustive report of methods, materials, and devices.

17.2 Design of hybrid organic—inorganic materials synthesized via sol—gel chemistry

17.2.1 Synthetic pathways

The mesostructured materials presenting ordered structures at the 2—50 nm scale are built through creating a nanocomposite with ordering at the mesoscale. There are essentially two general pathways to achieve these kinds of mesophases, which are a combination of sol—gel process, and the templating effect of supramolecular systems. Overall, a mesostructured phase is produced in a first stage, followed by template removal, which leads to the actual mesoporous material. The formation of this first mesostructured hybrid phase is critical in obtaining a final mesoporous material with tailored features, in a reproducible way. A complex coassembly of building blocks that will give rise to well-defined framework walls and pore template regions takes place during the precipitation or gelation of the systems. Control of this first step is essential in order to define the characteristic interaction lengths that control the nascent mesophase. The main driving forces toward obtaining organized templated mesophases have been presented in the literature, and the relevant thermodynamic and kinetic factors have been analyzed [1]. The interactions between the inorganic components and the organic template are among the most important thermodynamic drivers, and usually determine the feasibility of mesostructure formation, and its topology. These interactions are in turn determined by the composition of the initial systems, and the adequate size and hydrophilicity of the inorganic building blocks, in order to properly locate both kinds of building blocks in space. Regarding the kinetics, the cooperative formation of an organized hybrid mesostructure is the result of the delicate balance of phase separation/organization of the template and inorganic polymerization [4]. It has been proposed that processes linked to phase separation and organization at the hybrid interface between the inorganic building blocks and the template must be faster than the inorganic condensation that leads to "freezing" of a continuous matrix. Winning this "race toward order" leads to highly organized mesophases; if the inorganic condensation rate takes over the template self-assembly and the coassembly at the hybrid interface, only poorly ordered mesophases will result [23]. Recently, theoretical developments have helped to rationalize this complex self-assembly behavior, permitting the understanding of the organization processes as well as predicting the possibility of using preformed nanobuilding blocks with optimized size to achieve well-organized mesostructured and mesoporous systems [24,25].

There are globally four routes leading to obtaining mesostructured materials: (1) direct precipitation [4]; (2) true liquid crystal templating (TLCT) [26]; (3) evaporation-induced self-assembly (EISA) [27,28]; and (4) exotemplating [29], which are sketched in Fig. 17.1. We must stress that the use of these synthesis routes has afforded a wealth of different structures with a great diversity of pore size and topology, and an impressive variety of frameworks, ranging from oxides to phosphates, sulfides, metals, or polymers, to have been produced so far. Each of the synthesis routes represents a certain synthetic strategy and has been developed for a given type of framework—template combination. In addition, each route is suited to the final desired shaping of the material (i.e., powders, films, membranes, gels, etc.). Therefore, these two central aspects—*structural control* and *desired shaping—processing*—of the final material dictate the choice of a given synthetic route.

17.2.2 Structural control

The most relevant structural features of a mesoporous material are summarized in this chapter, along with the preferred methods employed to control them. Below we will briefly discuss these aspects by presenting selected classical examples. A detailed discussion of each of these parameters is out of the scope of this work, and readers are encouraged to refer to comprehensive reviews dealing with the synthesis parameters [30] and chemical strategies [31] leading to optimization of the synthesis.

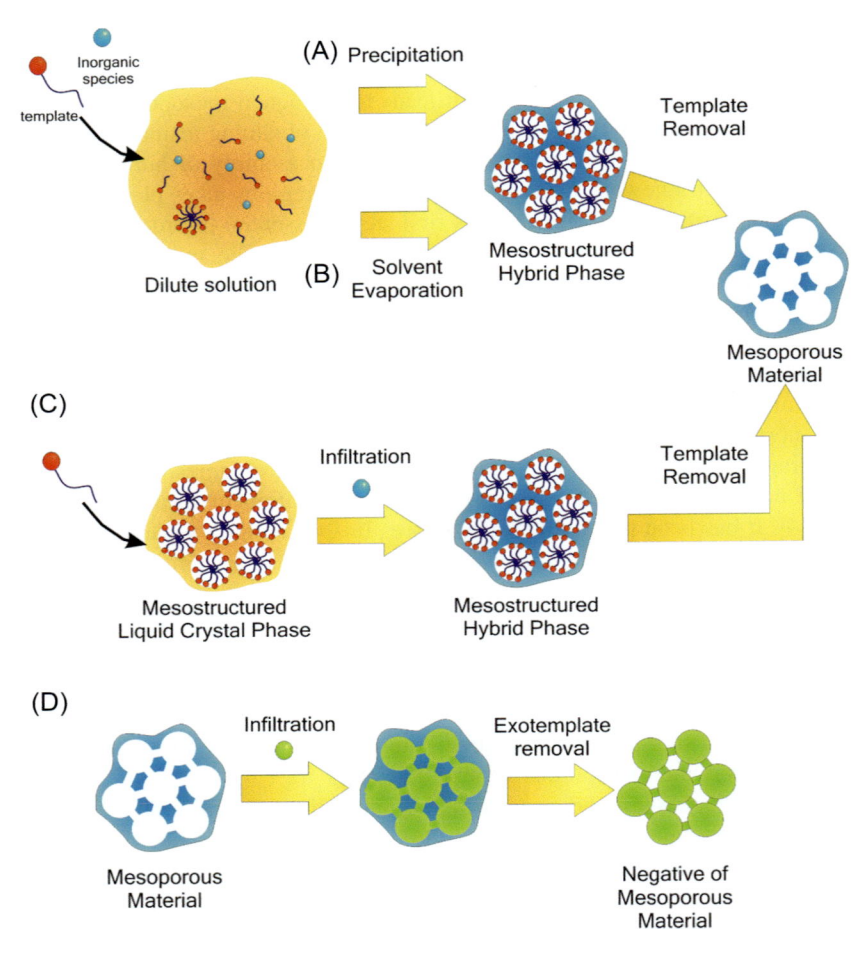

FIGURE 17.1

Scheme of the main synthesis routes to mesoporous materials: precipitation (A), true liquid crystal templating (TLCT) (B); evaporation-induces self-assembly (EISA) (C); and exotemplating (D).

Reproduced with permission from G.J.A.A. Soler-Illia, O. Azzaroni, Chem. Soc. Rev. 40 (2011) 1107.

17.2.2.1 *Pore topology*

Tremendous efforts are being dedicated to fabricating ordered mesoporous nanomaterials with controlled compositions, morphologies, mesostructures, and unique architectures, and to solve the mechanism of mesostructure formation and to explore their applications [32]. The *shape, spatial distribution*, and *interconnectivity of pores* constitute a central aspect, controlled by a variety of synthesis and processing variables. The mesostructures can be regulated by tuning the assembly of the soft micelles and the method is convenient, low cost, and thus used more often to synthesize mesoporous materials. In the soft-templating process, single micelles are the smallest units in the formation of ordered mesoporous. The ordered arrangement of single micelles and their assemblies show the way to the formation of mesostructures, as shown in Fig. 17.2: (a) the classical structure of the single micelle; (b) single micelle based nanostructure, obtained from precursors crosslinking on single micelles; (c—d) 1D and 2D mesoporous materials; (e) 3D mesoporous nanoparticles; (f) 3D core@shell mesoporous nanoparticles, gotten from the multilevel assembly of single micelles; and (g) asymmetric mesoporous nanoparticles, produced from the anisotropic assembly of single micelles.

A series of single micelle-based nano- and mesoarchitectures has been created by using different types of amphiphilic surfactants (ionic or nonionic) and block copolymers. Specifically, the self-assembly of block copolymers can occur both in the bulk and in solution. The bulk block copolymers with immiscible blocks can separate into microphases with various morphologies and the degree of microphase separation of diblock copolymers is determined by the segregation product and the level of complexity for their phase separation increases with the number of blocks [33]. The hydrophobic interaction of the hydrophobic blocks runs the aggregation of amphiphilic block copolymers chains. The formed aggregates are protected by hydrophilic corona in aqueous solutions, which reduce the contact of the hydrophobic block with water, thus minimizing the system energy. The aggregates mostly contain spherical or cylindrical micelles, bicontinuous structures, lamellae, vesicles, and their inverse structures, etc., as shown in Fig. 17.3. The general principle for the morphological transition can be simply explained by the classical cone—column geometric change mechanism, which is also applicable to microphase separation of block copolymer in the bulk. The packing parameter is affected by a number of factors, including the polymer composition and concentration, nature of the solvent, water content, addition of additives, interactions with the inorganic species, etc. [34]. The modifications of these influencing parameters allow flexible control over the morphology and dimension of block copolymers aggregates. Table 17.1 presents a quick overview of the most usual and reproducible block copolymers templates used in mesoporous materials production via the EISA method, along with their pore topology, and typical values of pore diameter.

On the other hand, the first reports on mesoporous materials used cationic cetyltrimethylammonium bromide (CTAB) as a supramolecular template. CTAB is a cationic template with a compact headgroup that leads to a variety of silica mesostructures: MCM-41 (2D hexagonal *p6m*) and MCM-48 (bicontinuous cubic *Ia3d*) and lamellar powders can be obtained by increasing the surfactant to silica ratio [4]. Film processing permits to obtain these mesophases as well as 3D hexagonal (*p63/mmc*), or micellar cubic (*Pm3n*) [28]. The shape of the template can be selected by the adequate choice of surfactants with different head-to-tail volume ratios, or hydrophilic—lipophilic balance (HLB) [51]. Gemini surfactant templates with a bulkier hydrophilic head with respect to CTAB lead to three-dimensional mesophases with higher curvature [52]. The nature of the framework precursors is also important, as the organized mesophases are formed by the coassembly of the template and the framework building blocks. For example, control of the hydrolysis and condensation of silica precursors is one of the key points that can direct to disordered or ordered structures in mesoporous thin films, due to the changes in the charge, size, hydrophilicity, and flexibility of the inorganic precursors that take place during aging of the precursor sols. These features control in turn the silica—template interactions that lead to an optimum assembly of both kinds of building blocks [23,53,54].

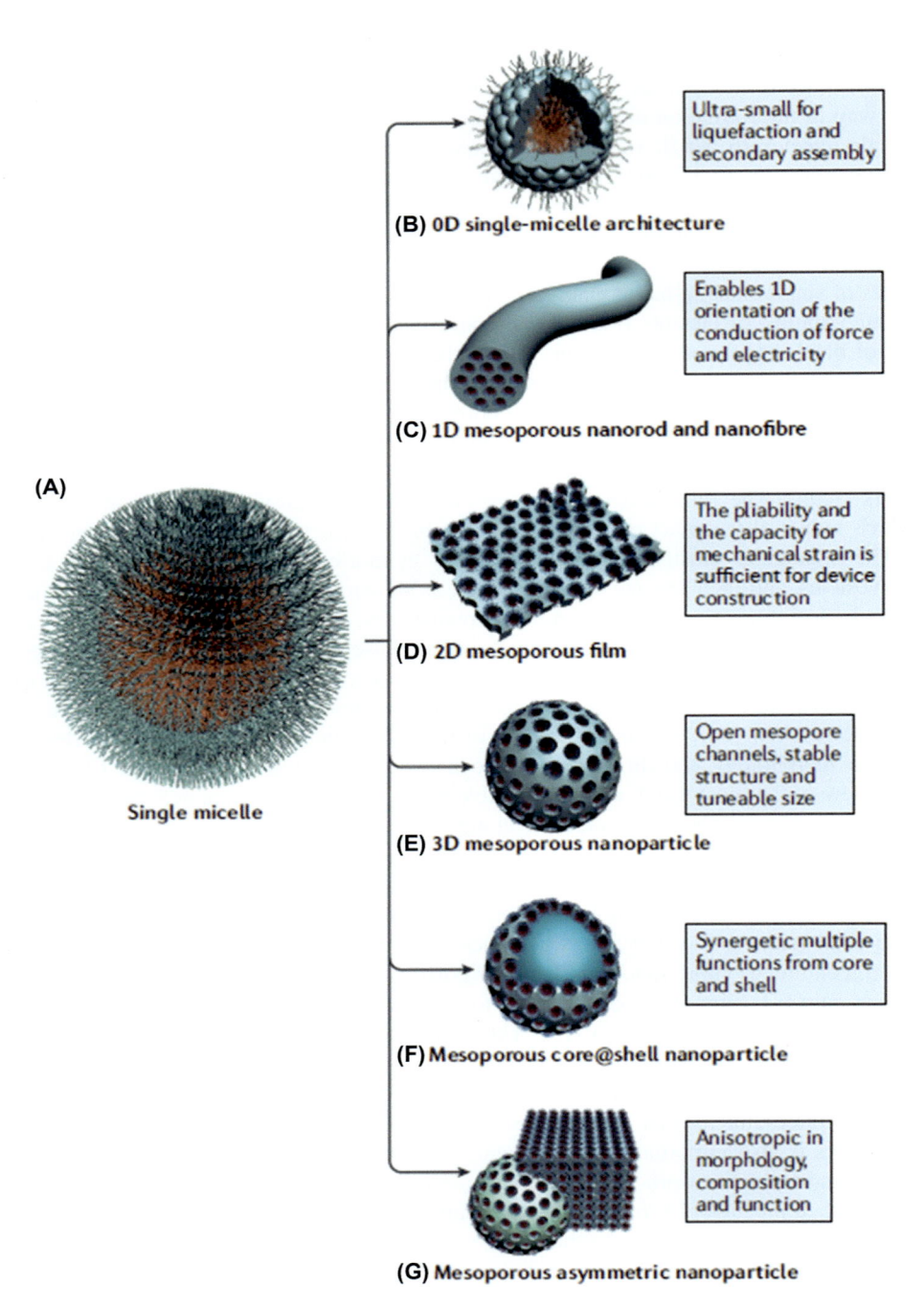

FIGURE 17.2

Single micelle directed synthesis of mesoporous materials with multilevel architectures.

Reproduced with permission from T. Zhao, A. Elzatahry, X. Li, D. Zhao, Nat. Rev. Mater. 2019 775.

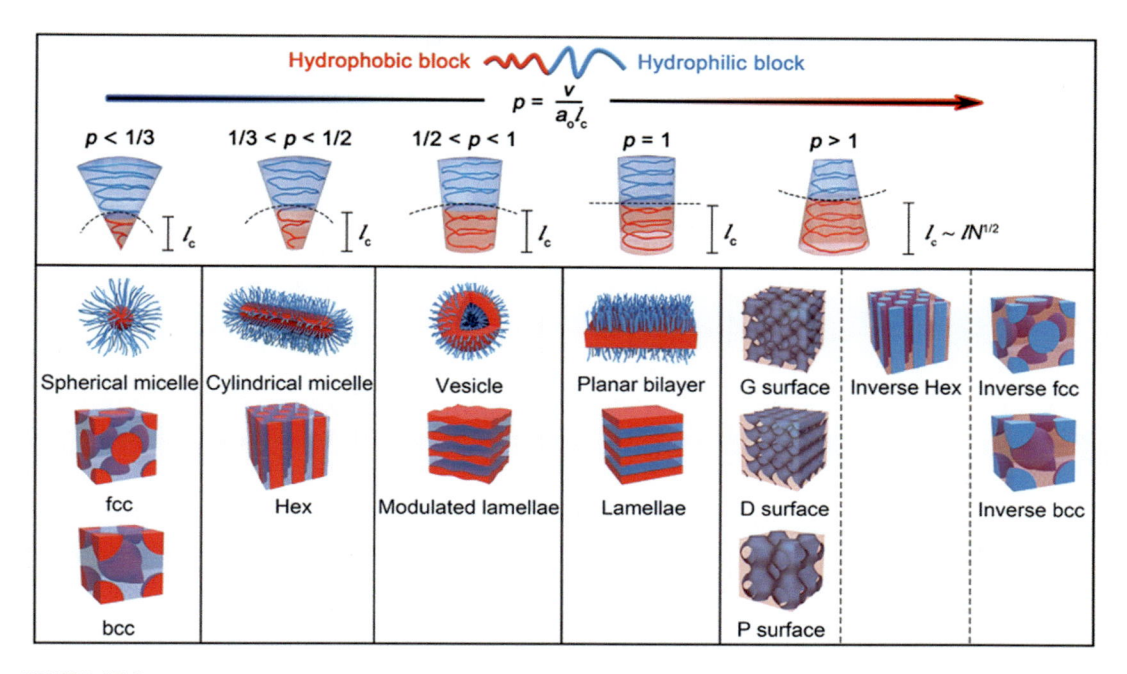

FIGURE 17.3

Schematic phase diagram of block copolymer self-assembly in solution, which presents the major morphologies.
Reproduced with permission from C. Li, Q. Li, Y.V. Kaneti, D. Hou, Y. Yamauchi, Y. Mai, Chem. Soc. Rev. 49 (2020) 4681–4736.

17.2.2.2 Pore size

The control of the pore size is essential to define the properties of mesoporous nanomaterials. Confinement-derived effects such as capillary condensation, size selectivity, or perm-selectivity can be regulated by adjusting the sizes and shapes of the mesopores, and the constrictions between them [55]. While pore sizes are in principle regulated externally through the choice of the template size, the measured pore diameter is also related to the pore topology and the surface composition (speciation, philicity, etc.); thus critical confinement-derived properties depend on understanding these features as a whole [56]. Micellar templates are uniform nanometric entities, composed by a well-defined aggregation number. Within a given mesostructure symmetry, pore size will mainly depend on the template size, and therefore selection of the templating molecule is essential to size tailoring. In the most common case of sol–gel derived microporosity, the micropore to mesopore ratio can be varied by thermal treatment of the mesoporous material, but also by submitting to carefully crafted postsynthetic treatments. The choice of the template also influences the presence of micropores, with nonionic templates and block copolymers being more prone to originate microporosity [57].

The first efforts to tailor pore size trusted in the use of ionic templates with different hydrophobic chain lengths, leading to small pore sizes (2–4 nm) and thin inorganic walls, such as the MCM families and derived materials [4]. The introduction of amphiphilic block copolymer (ABC) templates permitted a more flexible tailoring in the 5–20 nm pore diameter range, opening the way to

Table 17.1 Description of most usual block copolymers as templates.

Template	Pore structure	Symmetry	Pore size (nm)	Frame work	References
PS-*b*-PEO	Spherical	Fm3m	10.1−15.3 21.7	WO$_3$	[35]
		Primitive-cubic	16	TiO$_2$	[36]
	Cylindrical	p6mm	26−34	Carbon/silica	[37]
			17.4	Sn/carbon/silica	[38]
			14.5	In$_2$O$_3$	[39]
F127			2.6−3.8	Carbon	[40]
			3.3	N-doped carbon	[41]
			3.6−4.8	TiO$_2$	[42]
			6.2	ZrO$_2$	[31]
P123			2.6	TiO$_2$/silica	[43]
			9.8	TiO$_2$	[44]
			9.6	Black TiO$_2$	[45]
PI-*b*-PS-*b*-PEO	Bicontinuous	Ia3d	11−39	Carbon	[46]
			-	TiO$_2$/carbon	[47]
		I4$_1$32	31	Carbon	[48]
			31−35	Carbon	[46]
			42	CsTaWO$_6$	[49]
PMMA-*b*-PEO		Pn3m	72.8	Al$_2$O$_3$	[50]

the SBA 15 and related families of large-pore mesoporous nanomaterials. Precise tailoring of the template size can be achieved by regulating the polymerization degree of the hydrophilic or hydrophobic polymeric blocks. In addition, ABC are able to impart thicker walls, apart from being industrially available, hazard-free, and easy to remove from the mineral framework by thermal treatment or solvent extraction [58,59].

Commercially available soft templates including surfactants, and amphiphilic block copolymers have been intensively used to synthesize mesoporous nanomaterials with variable morphologies. For example, Pluronics F127 (EO106PO70EO106, where EO is ethylene oxide and PO is propylene oxide), an ABA-type block copolymer surfactant, creates spherical micelles with poly(propylene oxide) (PPO) cores and poly(ethylene oxide) (PEO) coronas (i.e., PPO@PEO) in an acid environment. Nevertheless, due to the limited choice of commercially available soft templates, the produced mesoporous materials usually show relatively small pore sizes (<10 nm) [60]. Micropores are usually present in mesoporous materials, as a consequence of the use of sol−gel soft methods. The incomplete condensation of the inorganic framework or the trapping of solvent or template molecules invariably leads to a texture in the subnanometer scale, which can be readily derived from nitrogen adsorption−desorption curves. Micropores can be desired, or an unwanted problem [61]. In the first case, methods have been used to produce zeolite-based materials presenting controlled micropores and mesopores, as a hierarchical ensemble [62]. In the most common case of sol−gel derived microporosity, the micropore to mesopore ratio can be varied by thermal treatment of the mesoporous material, but also by submitting to a careful postsynthetic treatment. The choice of the template also influences the presence of micropores, nonionic templates being more prone to originate microporosity [63]. Increase of surfactant concentration can lead to two different effects, depending on the template used. For ionic templates, a higher template molecule concentration will mostly result in more micelles; the interpore distances will shorten, but pore size will remain essentially constant, as well as the constrictions [64]. The case of nonionic templates that give rise to cage-like structures like SBA-16 is different, and strategies have been developed to independently tailor pore size and interpore constrictions. Some of the cubic mesophases (*Im3m* or *Ia3d* symmetry) could be considered as infinite periodical minimal surfaces [65]. Although in principle pore size can be arbitrarily tuned by the molecular weight of the porogen molecule, the use of surfactants as supramolecular templating agents presents a practical limit in the 10—20 nm pore diameter range.

Larger size, more complex molecules can segregate from solution, leading to inhomogeneous assembly with the inorganic building blocks and therefore to irregular pore formation. However, these phase separation processes can be harnessed in order to yield larger mesopores or macropores, or even hierarchical structures (as will be illustrated at the end of this section) [66]. In addition, larger molecules exhibit a wider conformational landscape, and tend to present slow assembly kinetics with the inorganic components. This results generally in systems where the organic molecule acts as a polymeric spacer rather than an assembled template, and results in poorly defined pores. Routes toward organized, larger pores imply therefore the use of the usual porogens with the addition of swelling agents such as trimethylbenzene (TMB) or similar molecules [59]. For example, the synthesis of large-pore ordered mesoporous carbon could be via two routes [37]: (1) increasing the molecular weight of hydrophobic segment of the tailor-made amphiphilic block copolymers; or (2) incorporating hydrophobic homo-polymer (h-PS) in the micelles of the block copolymers. In both cases, Step 1, coassembly of the block copolymer templates and precursor frame work; Step 2, calcination at a low temperature (450°C) to remove the templates; Step 3, carbonization at a high temperature ($> 800°C$) to obtain ordered mesoporous carbon (Fig. 17.4).

Therefore, various amphiphilic block copolymers with large molecular weights and special compositions have recently been explored, which have now become an important supplementary kind of soft template for mesoporous nanomaterials with unique framework compositions and functionalities [37]. These block copolymers can easily be synthesized via the well-established living polymerization methods such as reversible addition—fragmentation chain transfer polymerization (RAFT) and atom transfer radical polymerization (ATRP). Up until now, various PEO-based block copolymers, including PEO-*b*-polystyrene (PS), polyisobutylene (PIB), poly(methyl methacrylate) (PMMA), and poly(ethylene-*co*-butylene) (KLE), and poly(vinyl pyridine)-based copolymers, such as poly(2-vinyl pyridine)-*b*-polystyrene (P2VP-*b*-PS), were designed as templates to produce mesoporous materials. Among these tailor-made block copolymers, PEO-*b*-PS (or PMMA) copolymers are frequently studied since they can be easily synthesized. Their molecular weight can be simply controlled via the living radical polymerization process. Thus PEO-*b*-PS (or PMMA) block copolymers have intensively been used as a soft template for the controllable synthesis of mesoporous materials with unique functionality.

If the molecular weight of PS segment of the block copolymer increases, the hydrophobic volume of PEO-*b*-PS spherical micelles can be enlarged, producing a large pore size after selective removal of the templates. For example, PEO-*b*-PS copolymers with the same hydrophilic segments but different chain lengths of hydrophobic segments were used as the templates to produce mesoporous carbon and the pore size increased from 11.9 to 33.3 nm as the molecular weight of PS segment increased from 12 to 30.5 kg/mol [67].

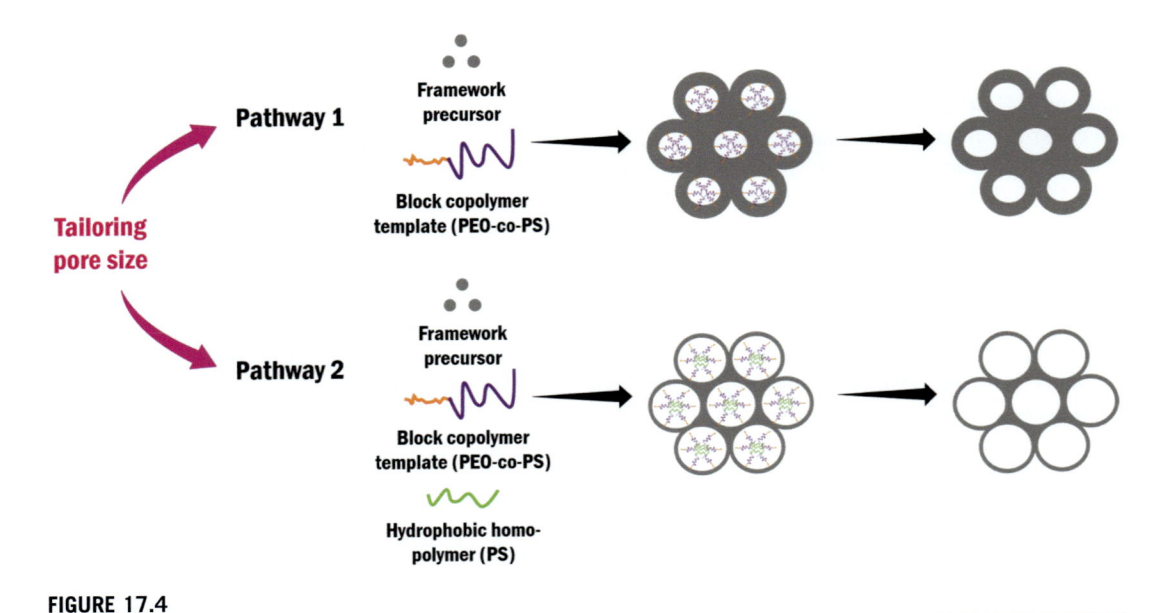

FIGURE 17.4

Synthesis of large-pore mesoporous carbon polymer and carbon via two routes.

On the other hand, PEO-*b*-PS (or PMMA) is usually dissolved in organic solvent due to its water-insolubility, thus for example in the synthesis of mesoporous silica via such an EISA method, it is usually difficult to completely remove the template via direct calcination or solvent extraction, resulting in inaccessible mesopores and an undetectable surface area. A considerable amount of ethanol can be produced during the hydrolysis of TEOS, which can weaken the interaction between silicate species and PEO segments and consequently cause PEO segments to retract from the silica walls. To deal with this problem, a hydrothermal recrystallization approach to create numerous "cracks" in the silica walls by micelle expansion was developed. The "cracks" in the as-made samples serve as micropore channels in the silica walls which allow air to enter for removal of the organic template via combustion [68]. After calcination in air, mesoporous silica with accessible large mesopores ($\sim$30.8 nm), high surface area (362 m^2/g), and large pore volume (0.66 cm^3/g) were finally obtained successfully.

Furthermore, EIAA method using PEO-*b*-PMMA has been employed for mesoporous silica particles synthesis. PEO-*b*-PMMA was selected as a typical water-insoluble template, which was first dissolved in THF/H$_2$O mixed solvent with volume ratio of 4:1 to form a clear and transparent solution in an open vessel [69]. When THF evaporates, PEO-*b*-PMMA molecules experience a microphase separation and form spherical micelles with PMMA cores and PEO shells. The silica oligomers are in the shells of uniform spherical micelles because of the hydrogen bonding with PEO segments. As THF further evaporates the composite micelles have a tendency to aggregate into a face-centered-cubic (*fcc*) closed-packing mesostructure to minimize the interface energy. Meanwhile, the crosslinking and condensation of silicate oligomers can fix the ordered mesostructure, and large crystal-like particles are obtained. As the PEO segments are inserted in the silica frameworks during the EIAA process, a large number of micropores can be produced in the pore walls during calcination. These micropores facilitate the contact between O_2 molecules and hydrophobic segments, resulting in complete decomposition of the hydrophobic segments. After removing the templates, the obtained mesoporous silica particles show unique crystal-like morphology with an *fcc* mesostructure, large pore size (~ 37.0 nm), large window size (~ 8.7 nm), high surface area (~ 508 m^2/g), and high pore volume (~ 1.46 cm^3/g).

The "20 nm diameter limitation" can be overcome by resorting to phase separation processes triggered by the combined use of sol−gel condensation and large-size polymeric templates. Spinodal phase separation followed by calcination has been used to achieve large mesopores, macropores, or multiscale structured materials in an inexpensive and reproducible way. These processes can take place in gelling matter, or in evaporating systems, such as xerogels or thin films. By this route, the condensation of inorganic or organic species, or their concentration after solvent evaporation, leads to a phase separation that acts as a porogenic process. In principle, the pore size and volume can be independently controlled by adequately choosing the concentrations of polymer and solvent, or using cosolvents [70].

The versatility of this process can be illustrated by the preparation of a variety of hierarchically porous oxide monoliths (silica, titania, zirconia, and mixed oxides) with controllable hierarchical pore architectures produced in one step. Monoliths are produced by exploiting a phase separation process triggered by the polymerization of furfuryl alcohol in acidic conditions, in the presence of inorganic precursors and Pluronics F127. The formation of poly(furfuryl alcohol) is concurrent with the sol − gel process, and the resulting colloidal dispersion acts as an in situ macropore template. The polymer template originates the macropores, whereas the block copolymer template assists the formation of mesopores, either by templating or by stabilizing the inorganic building blocks. Careful combination of furfuryl alcohol and F127 leads to control of the macropore and mesopore morphology [66].

Spinodal decomposition has also been adapted to the preparation of hierarchical meso−macroporous thin films. A one-pot synthesis route employs a combination of supramolecular templating with F127 and phase separation. Poly(propylene glycol) (PPG) is used as a pore enhancement agent in the presence of THF as a cosolvent. The balance between the PPG and the cosolvent is critical, in order to obtain larger mesopores through template swelling or phase separation. The presence of PPG leads to a phase separation, and therefore to a macro−mesoporous hierarchical mesostructure. When the cosolvent is added, the dissolution of PPG in the micelles is increased, leading to a more uniform population of intermediate size mesopores (15−40 nm diameter). An adequate balance of PPG and THF permits smooth control of the phase separation and micellar template swelling behavior, leading to bimodal pore systems with tunable pore diameters [71].

While phase separation techniques permit access to large pores and novel architectures, the large driving forces involved imply that the pore morphologies and architectures created by spinodal decomposition change greatly with small variations of the synthesis and processing variables (reagent ratio, solvent−cosolvent, curing temperature, drying conditions, processing temperature, etc.). Therefore, these synthetic routes are very sensitive to initial conditions and processing.

17.2.2.3 Nature of the inorganic framework

Since the beginning of this field, the production of mesoporous materials has been the most exhaustively explored topic. This is mostly due to the intrinsic interest in silica materials as potential substrates toward high surface area catalysts, and to the existence of numerous reliable synthesis routes toward the well-known MCM or SBA mesophases. Besides, the use of a variety of versatile silicon or organosilicon precursors opened the way to hybrid silica-based hybrid frameworks, containing dangling organic groups at the pore surface [3,56], or incorporated into the walls [72,73]. In addition, in recent years, a whole palette of mesoporous frameworks such as nonsilica oxides, sulfides, phosphates, carbons, or even microporous inorganic or organic frameworks has been developed [74]. This wide range of framework compositions is possible thanks to the use of different synthesis routes, as well as the thorough chemical control of several synthesis parameters [75].

An accurate control of the hydrolysis—condensation processes of silicon (IV) or high valence metals (Ti(IV), Zr(IV), Nb(V), Al(III)), permit production of pure and mixed mesoporous oxides in which framework composition is in principle controlled by the adequate mixture of the inorganic precursors (typically, chlorides, alkoxides, or acetylacetonates) that present similar high acidity. It has to be noted that these high-valence inorganic centers tend to yield amorphous or low-crystallinity mesostructured frameworks upon precipitation or related sol—gel soft processing. In the case of low-valent metals (such as the first transition series: M^{II}, M = Mn, Fe, Co, Ni, Cu, Zn, etc.), very few reliable reports of ordered mesoporous materials by direct synthesis exist. In this case, precipitation often leads to obtaining nonmesostructured crystalline basic salts. Wall thickness can be controlled by several parameters, the most usual is the template size, and the template:metal ratio that influences the intermicelle distance. The very nature of the template also influences the wall thickness: while ionic templates lead to thinner walls, nonionic surfactants (for example, presenting polymer-based hydrophilic blocks) give rise to thicker walls, due to a less defined inorganic—organic interface [1]. Notwithstanding, the use of nonionic templates often leads to micropores, due to the strong interactions between their hydrophilic heads and the uncondensed inorganic building blocks [76]. Aging at moderate temperature or under hydrothermal conditions enhances inorganic condensation and permits a better phase separation between the inorganic and organic components, leading to denser walls, and well-defined pores with less or no microporosity [1]. Control of the thermal treatment of the initially obtained mesostructured precursors is essential in order to keep the high surface area and well defined mesopores, while ensuring nanocrystalline walls, in the case of nonsilica. Low temperature treatment ($<250°C-300°C$) results in low porosity, due to partial template removal, and to poor mechanical and chemical stability, due to incomplete condensation of the inorganic frameworks. Temperatures higher than $300°C-350°C$ are necessary to completely remove the templates. At these temperatures, framework condensation is complete, in the case of oxides, therefore toughening the mesostructure. In this high temperature range, pore coarsening can begin to take place, resulting in loss of microporosity, and changes in the pore and neck shape. In the case of silica systems, evolution to higher temperatures ($600°C-700°C$) leads to a decrease in hydrophilicity due to dehydration of surface silanol groups.

Nonsilica oxides begin to undergo crystallization in the $300°C-400°C$ range. Extended growth of crystallites constitutes a problem, for it might lead to the total loss of surface area and porosity. Detailed studies carried out on mesoporous titania thin films showed that restricted crystallization of the anatase phase takes place through nucleation-growth processes in the confined environments of the mesopore walls. The mesopore system geometry determines the final shape and size of the anatase nanocrystals. Rapid anatase nucleation is followed by confined growth, the limits to which are set by the pore interfaces. Oriented growth and rearrangement occur because of these limitations, and the final pore and crystallite structures are intimately related to the low-temperature structures [77]. During thermal treatment, the surface area decreases, while the pore and neck sizes increase by coarsening. If the nucleation and growth processes can be controlled through gentle heating, a mesoporous, nanocrystalline, robust, and highly accessible titania framework can be achieved, which keeps the structure and topology of the original mesoporous system. Too high temperatures result in extended crystallization, and the original mesoporous structure can be lost. Therefore, an optimized thermal program is needed in order to take advantage of the concurrent processes of crystallization and pore coarsening for interesting properties such as photocatalysis [78]. Other strategies for obtaining mesoporous crystalline materials include controlled thermal treatment under an inert atmosphere, leading to partial template carbonization, or the use of templates that decompose at temperatures comparable to the crystallization threshold [79]. In both cases, the soft template serves as an organic or carbonaceous scaffold that holds the mesostructure in place during crystallization and sintering, in order to avoid extended crystallite growth that would lead to disruption of the pore structure.

A recent development in the synthesis of mesoporous materials involved the crosslinking of numerous functional groups like amines, strong acidic ionic liquids, sulfonates, triphenylphosphine, pyridine, sulfonates, and similar species onto porous polymers [80]. In situ crosslinking of the abovementioned groups with 1,4-bis(chloromethyl)benzene was done by using Lewis acid catalysts. No typical templates were used in this process. Such materials were noted to be hypercrosslinked porous polymers [81]. These systems have outstanding porosity with BET surface areas as high as $1520 \, m^2/g$ and the pore sizes are tunable from 4 to 130 nm. Such systems were shown to have acid, base, and metal sites available for catalysis reactions. Also with regard to porous polymers, other systems like porous metal sulfides have seen an upswing in the number of groups trying to make such systems. These porous metal sulfides are in general more difficult to make than porous metal oxides

17.2.2.4 *Control of pore surface*

In addition, the high surface area of mesoporous materials mostly contributes to their energetics. Surface energy, represented by an energy term dependent on the exposed area (dA) of the type $dG_{surf} = \gamma \, dA$ (where γ stands for the surface energy), controls thermodynamic and kinetics aspects (globally, stability and reactivity), that will greatly influence materials' properties, such as wettability, adsorption, or dissolution rate. Two important features are the pore specific surface area and the surface energy (γ). These parameters can be independently designed by adjusting synthesis and processing variables. The amount of surface area is generally controlled by the template choice, the $s = [surfactant]:[inorganic]$ ratio and the thermal treatment, which also affects pore size. The nature and reactivity of the pore surface can be tailored by numerous means: in the case of oxides, the density of M-OH groups (thus the hydrophilic character) can be controlled by exposure to solvents, and thermal treatment; high temperatures can lead to dehydroxylation, resulting in a hydrophobic surface. Surfaces with different philicity, and therefore different reactivity can incorporate a wealth of chemical species, such as inorganic cations or anions, or bifunctional molecules such as silanes, silazanes, diones, phosphates, carboxylates, etc. It must be pointed out that mesoporous materials manage to strongly adsorb species in solution, leading to partition and preconcentration of reagents in the pore networks. Molecules can remain just adsorbed, or react with the surface groups, leading to functionalized pore surfaces, which will present a modified behavior. The differential reactivity of the grafting group toward the surface species can be exploited in order to generate selective functionalization. Pore surface modification of mesoporous materials has vast implications in tailored materials for a wealth of applications such as adsorbents, catalysts, selective membranes, optical materials, biointerfaces, new electronics, etc. [82—85].

17.2.2.5 Pore contents

The pore filling of mesoporous materials with condensed vapors, liquids, molecules, polymers, or nanoparticles leads to extraordinary changes in their properties. Capillary condensation of vapors takes place after a critical pressure that is dependent on pore size and shape, and on γ, and can be explained by a modified Kelvin model. Nitrogen, argon, or krypton adsorption has been used as a major pore size characterization tool [86]. Besides, solvent condensation within mesopores leads to an important increase in material density, and subsequent loss of density contrast between walls and pores. Modification of pore size, shape, and surface nature change the threshold pressure (P_c) at which capillary condensation occurs. This parameter is roughly proportional to the inverse of the pore radius and directly proportional to the surface energy of the vapor at the material interface. Modification of pore volume also changes the electron density contrast between the empty and the filled porous structure. These effects have been advantageously used in mesoporous transparent thin films, in order to perform optical sensing [87]. The incorporation of functionalities can be reached in three pathways [52,56]: by subsequent attachment of organic components onto a pure matrix (grafting), by simultaneous reaction of condensable inorganic species and organic compounds (cocondensation by one-pot synthesis), and by the use of organic precursors that lead to periodic mesoporous material, as presented for silica-based mesoporous materials in Fig. 17.5.

In addition, repeated pore loading cycles permit inclusion of molecular compounds (assembly), beyond pore surface modification. In this way, nanosized entities can be produced within the pore systems, with properties different from the bulk. For example, a remarkable confinement effect has been reported for MCM-41 loaded with ibuprofen, which presents quasi-liquid-like behavior at ambient temperature, modifying its drug release performance. Inclusion of metal, semiconductor, or carbon nanoparticles (NP) within mesoporous materials has been studied, with particular interest in the catalysis or optics fields [88,89]. NP have been incorporated to the pore array either by capillary inclusion (limited to small nanoparticles in very accessible pore systems), or by in situ production. The second choice is the most frequently used, and soft chemical reduction, sequential precipitation, or electrochemical methods have been followed to load the mesoporous matrix with the desired NP [63]. One of the most important features is an excellent pore accessibility and interconnectivity. Repeated cycles composed of precursor uptake followed by precipitation or reduction steps must be performed to achieve the desired NP loading. Thus it is crucial to control the surface charge, in order to adjust the adsorption of the precursor species that will lead to the final nanocomposite material. It is important to stress that in order to harness these sorption processes, it is essential to take into account the real availability of the grafted organic group, which is modulated by their interactions with the inorganic framework, that is, the M-OH surface groups and their speciation [90].

17.3 Integration of functional molecular assemblies and supramolecular machineries into/onto mesostructured oxide supports

It is well-known that although a molecular monolayer is only a few nanometers thick it can completely change the properties of a surface. Molecular monolayers can be prepared by different strategies, such as the Langmuir–Blodgett technique or by direct chemisorption predefined

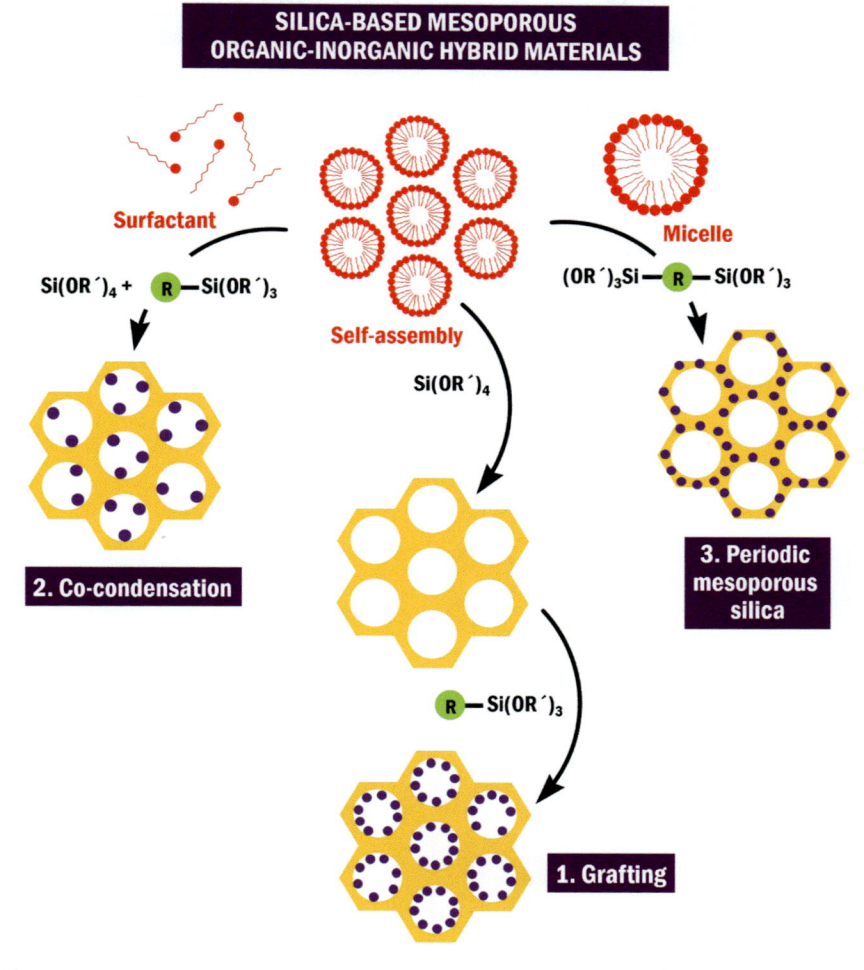

FIGURE 17.5

General synthetic pathway for silica based mesoporous organic—inorganic hybrid materials.

molecules. In the case of mesoporous oxide surfaces the chemisorption of functional silanes represents one of the most versatile strategies to integrate chemical units to the mesoporous platform [91]. Part of the appeal of this strategy stems from the stability of the monolayers, which is a vital issue in many applications. The stability is caused by partial in-plane crosslinkage of the molecules and possible covalent anchoring to the substrate

During recent decades, this strategy has been employed to create hybrid supramolecular nanoarchitectures that incorporate chemical entities which can act as a gate and allow controlled access to a certain site within the mesoporous environment. In other words, the derivatization of mesoporous films and particles with ad hoc silane monolayers facilitate the formation of heteroarchitectures that can control the entry/release of chemical species into or from mesoporous hosts.

For example, the assembly of photoactive units displaying photoisomerization processes on/in mesoporous silica leads to the formation of light-responsive nanostructured platforms. In this way, the decrease in the size of trans to cis isomers of azobenzene molecules attached to pore interiors (Fig. 17.6) offers a plausible strategy to regulate the transport of molecules through the mesopores.

As regards photoactive assemblies at mesoporous interfaces, Fujiwara et al. reported one of the first examples of a photocontrolled nanogated process in 3D hybrid scaffolds. Photoresponsive coumarin derivatives were grafted onto the pore outlets of mesoporous (MCM-41-type) solids with a pore diameter of approximately 2.5 nm [92]. Irradiation at >350 nm resulted in the photodimerization of the coumarin core and formation of the cyclobutane dimer, which closed the pores. The coumarin monomer could be regenerated and the pores reopened by photocleavage of the dimer by

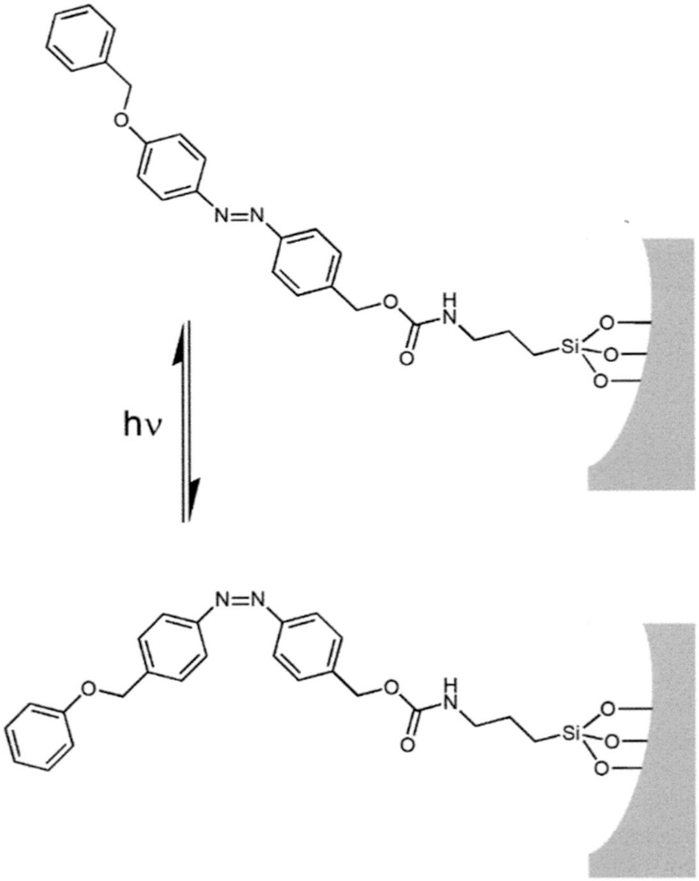

FIGURE 17.6

Azobenzene derivative tethered to the silica pore wall via the bifunctional placement strategy. The molecule is covalently linked to the silica pore wall while the hydrophobic nature of the rest of the molecule places that part in the organic region.

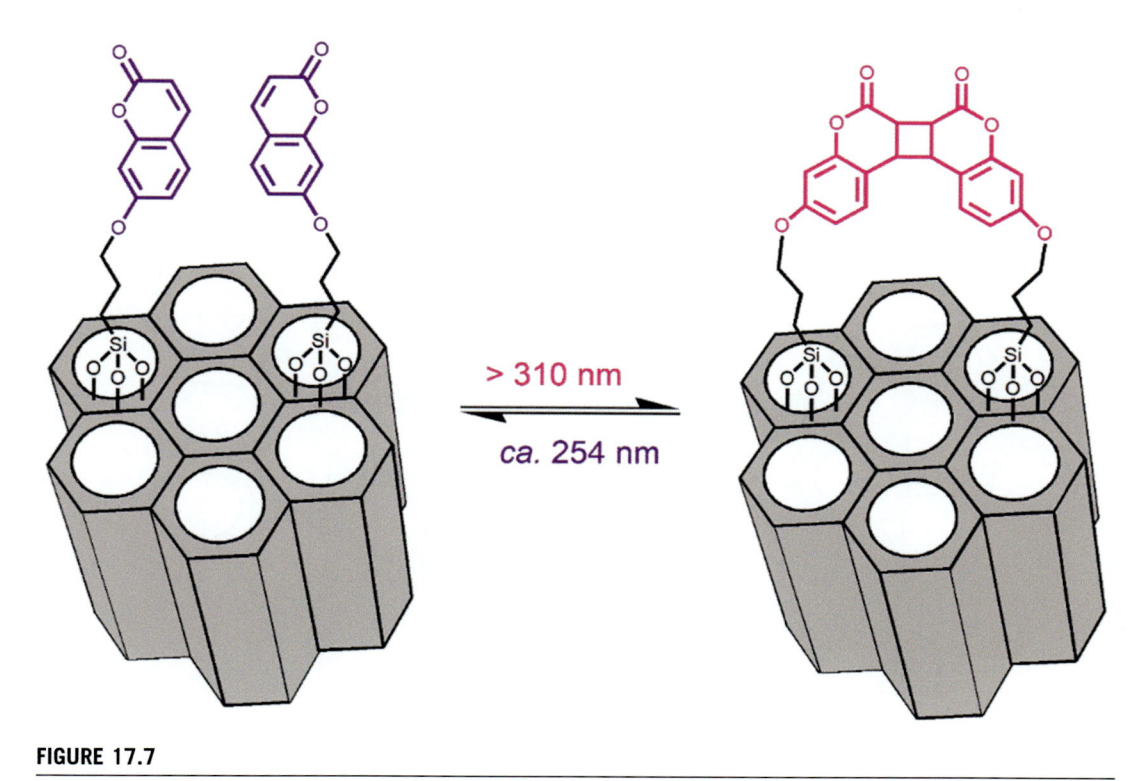

FIGURE 17.7

Opening and closing of coumarin-functionalized mesoporous materials.

using higher energy irradiation (250 nm, Fig. 17.7). This example shows how the use of a simple process (photodimerization) in combination with 3D mesoporous architectures allows regulation of a supramolecular function such as the release or uptake of a chemical species in a controlled way.

On the other hand, Martínez-Mañez and coworkers [93] reported the construction of gated hybrid systems that operate in aqueous solution and can be controlled ionically by pH modulation via the assembly of silane-terminated polyamines on mesoporous silica platforms. Fig. 17.8 shows a mesoporous silica scaffold with open pores that is functionalized with polyamines on the external surface. In these systems, the opening/closing of the mesopores arises from hydrogen-bonding interactions between less or unprotonated amines (open pores) and Coulombic repulsions between protonated amino groups (closed pores). At acidic pH values the amines are fully protonated, the gate is closed, and any access to the inner pores is precluded. In contrast, in the neutral pH region the amines are only partially protonated, the gate is open, and guest molecules can enter the pores. Interestingly, these platforms also exhibit an anion-controlled effect [94]. Under neutral pH conditions the gate is only open in the presence of small anions such as Cl^-, while bulky anions such as adenosine triphosphate (ATP) close the gate through formation of strong complexes with the amines at the pore outlets. Along these lines, Xiao and coworkers [95] reported a complementary system, formed by anchoring carboxylates in porous SBA-15 silica. In this case, the pores are

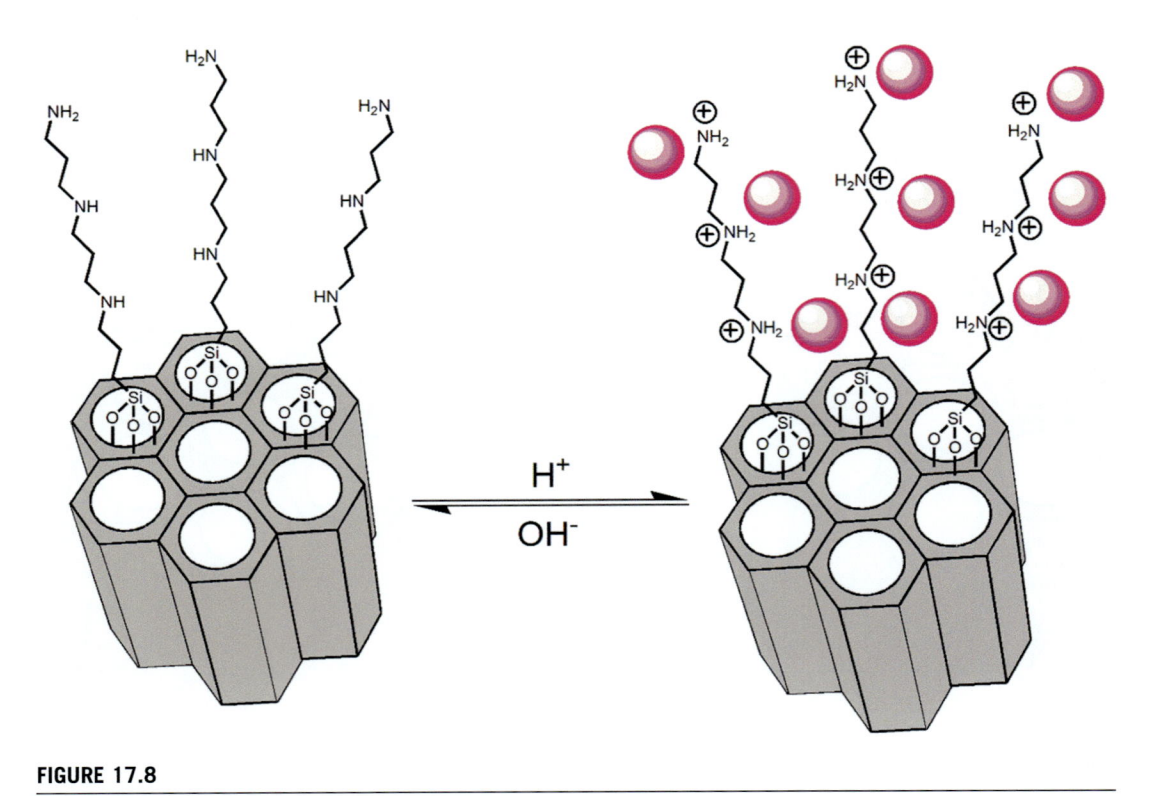

FIGURE 17.8

An ion-gated hybrid nanosystem in aqueous solution.

closed at neutral and basic pH values (the carboxylate state), whereas the pores remain open at acidic pH values (the carboxylic acid state).

On the other hand, the integration of supramolecular machineries into/onto mesoporous platforms gained increasing relevance as an appealing route to create "nanovalves" on account of the capabilities of these molecular machines to control functions at the nanoscale. The attachment/anchoring/assembly of supramolecular and molecular machines to surfaces has been achieved in a number of different classes of systems [15,20]. Stoddart, Zink, and coworkers proposed the construction of nanovalves by using movable molecular elements attached to mesoporous matrices with the ability to control the flow of molecules through the porous scaffold. Their approach was based on mechanically interlocked molecules that can be switched either chemically or electrochemically.

These authors reported the construction of redox-switchable and reversible molecular nanovalves (Fig. 17.9) employing bistable [2]rotaxanes as the controllable components [96]. The bistable [2] rotaxane consists of a 1,5-dioxynaphthalene (DNP) unit, a tetrathiafulvalene (TTF) unit, and two stoppers, as well as a cyclobis(paraquat-p-phenylene) (CBPQT^{4+}) ring that acts as the gate to control the passage of probe molecules into and out of nanopores. It was found that the CBPQT^{4+} ring of the bistable [2]rotaxane predominantly encircles the TTF unit rather than the DNP unit. Mesoporous MCM-41 particles were employed as a supporting platform and reservoir for the construction of

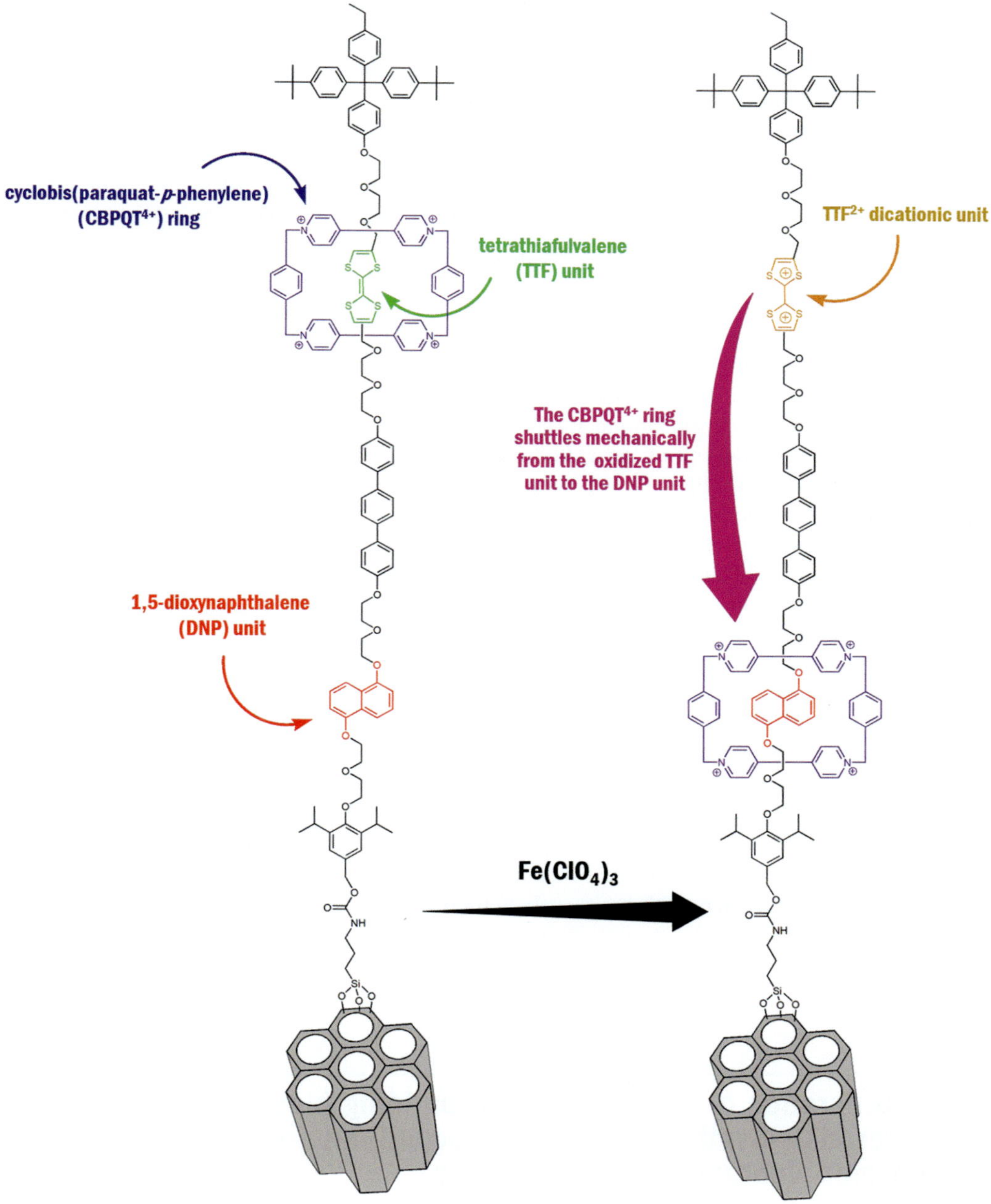

FIGURE 17.9

Graphical representation of the reversible, redox-controllable molecular nanovalves based on mesoporous silica particles, using a bistable [2]rotaxane that utilizes tetrathiafulvalene and DNP units as the electron-rich recognition sites that serve as stations for the encircling CBPQT^{4+} ring in the ground and oxidized states, respectively.

redox-controllable nanovalves. To this end, mesoporous silica particles MCM-41 were reacted with the bistable [2]rotaxane, thus forming the molecular nanovalves. When the TTF units in the bistable [2]rotaxanes are in their neutral state and encircled with the $CBPQT^{4+}$ rings, the nanovalves are open. However, the nanovalves can be closed by adding two equivalents of an oxidant, such as iron perchlorate hexahydrate, to oxidize the TTF unit on the rotaxane backbone. This process forces the $CBPQT^{4+}$ ring to shuttle mechanically from the oxidized TTF unit to the DNP unit, on account of the charge repulsion between the $CBPQT^{4+}$ ring and the oxidized TTF^{2+} dicationic unit, leading to closed nanovalves that completely block the entrance to the nanopores. Then, by adding an excess of ascorbic acid the oxidized TTF units are reduced to their neutral state, and the $CBPQT^{4+}$ rings move away from the openings of the nanopores, thus leading to the open configuration.

The same research groups reported the construction of hybrid mesoporous systems behaving like nanopistons, which are able to release encapsulated guest molecules in a controlled fashion under acidic conditions [97]. The mechanized mesoporous nanoparticles consisted of a monolayer of β-cyclodextrin (β-CD) rings positioned selectively around the entrance of the nanopores of the mesoporous nanoparticles. A rhodamine B/benzidine conjugate was prepared for use as the nanopistons for movement in and out of the cylindrical cavities provided by the β-CD rings on the surfaces of the nanoparticles (Fig. 17.10). Interestingly, the assembly/preparation process of the nanopiston system ensures that the β-CD rings are located only at the orifices of the nanopores on the surface of the nanoparticles, in which the β-CD rings become essentially an extension of the orifices of the nanopores. Thus the orifices of the nanopores are regulated by the uniform diameter of the β-CD rings. Since the β-CD rings on the surfaces of the nanoparticles can form complexes with a series of guest molecules in their hydrophobic cavities, molecular plugs can be designed and synthesized to respond to stimuli that trigger pH-, light-, or redox-operated drug delivery.

Supramolecular nanovalves sensitive to pH changes were prepared by assembling naphthalene-containing dialkylammonium stalks on mesoporous MCM-41 particles [98]. In these systems the mutual recognition motif between secondary dialkylammonium ions and dibenzo[24]crown-8 (DB24C8) is exploited as a gating mechanism. DB24C8 is a macrocyclic polyether, which is able to encircle dialkylammonium centers ($-CH_2NH_2^+CH_2-$), thus forming [2]pseudorotaxanes held together by multiple hydrogen bonds (Fig. 17.11) The DB24C8 rings recognize and self-assemble around the dialkylammonium ion stalks near the entrances of the nanopores to obtain closed nanovalves. Neutralization of the acidic dialkylammonium centers with bases such as triethylamine switches off the hydrogen bonds and causes dissociation of the DB24C8 rings from the [2]pseudorotaxanes. The field of nanomachinery incorporated into mesoporous materials is growing day by day, with exciting opportunities ahead [99].

17.4 Incorporation of macromolecular building blocks into mesoporous materials—Synthetic strategies toward functional hybrid polymer—inorganic mesostructures

17.4.1 Monomer impregnation/inclusion followed by polymerization

A straightforward strategy to incorporate macromolecular building blocks into mesoporous materials relies on the in situ polymerization of monomer-loaded mesopores. In a seminal work, Bein and

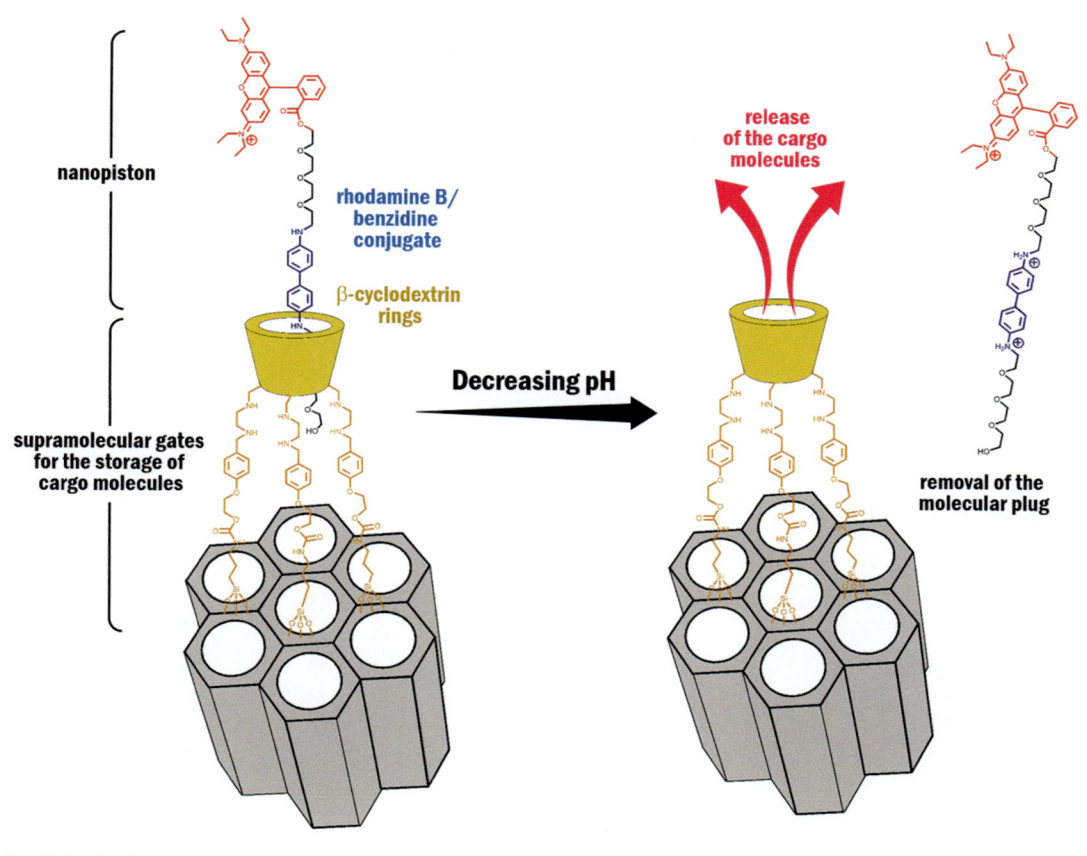

FIGURE 17.10

Schematic representation of nanopistons on mechanized silica nanoparticles.

collaborators developed a simple but effective protocol to build up functional macromolecular units within mesoporous frameworks [100,101]. Conducting filaments of polyaniline were prepared into 3-nm-wide mesochannels by adsorption of aniline vapor into the dehydrated host, followed by a reaction with peroxydisulfate, thus leading to encapsulated polyaniline macromolecular chains bearing several hundred aniline rings [102]. This approach based on the gas-phase incorporation of monomer units into the mesochannels was extended to different functional units leading to a plethora of hybrid platforms in which the polymeric building blocks were selectively confined within the pores [103]. For example, Bein et al. reported the synthesis of host—guest nanocomposites by the adsorption of methyl methacrylate (MMA) and its conversion to PMMA in the presence of benzoylperoxide within MCM-41 and MCM-48 mesochannels [104]. Nitrogen sorption isotherms confirmed the filling of the mesopores with the polymer while thermogravimetry evidenced that the polymer content of the composites increased with increasing pore volume. More important, scanning and transmission electron microscopy confirmed that polymer deposition did not occurred on the external host surface, thus evidencing that monomer preconcentration in the mesopores is a key

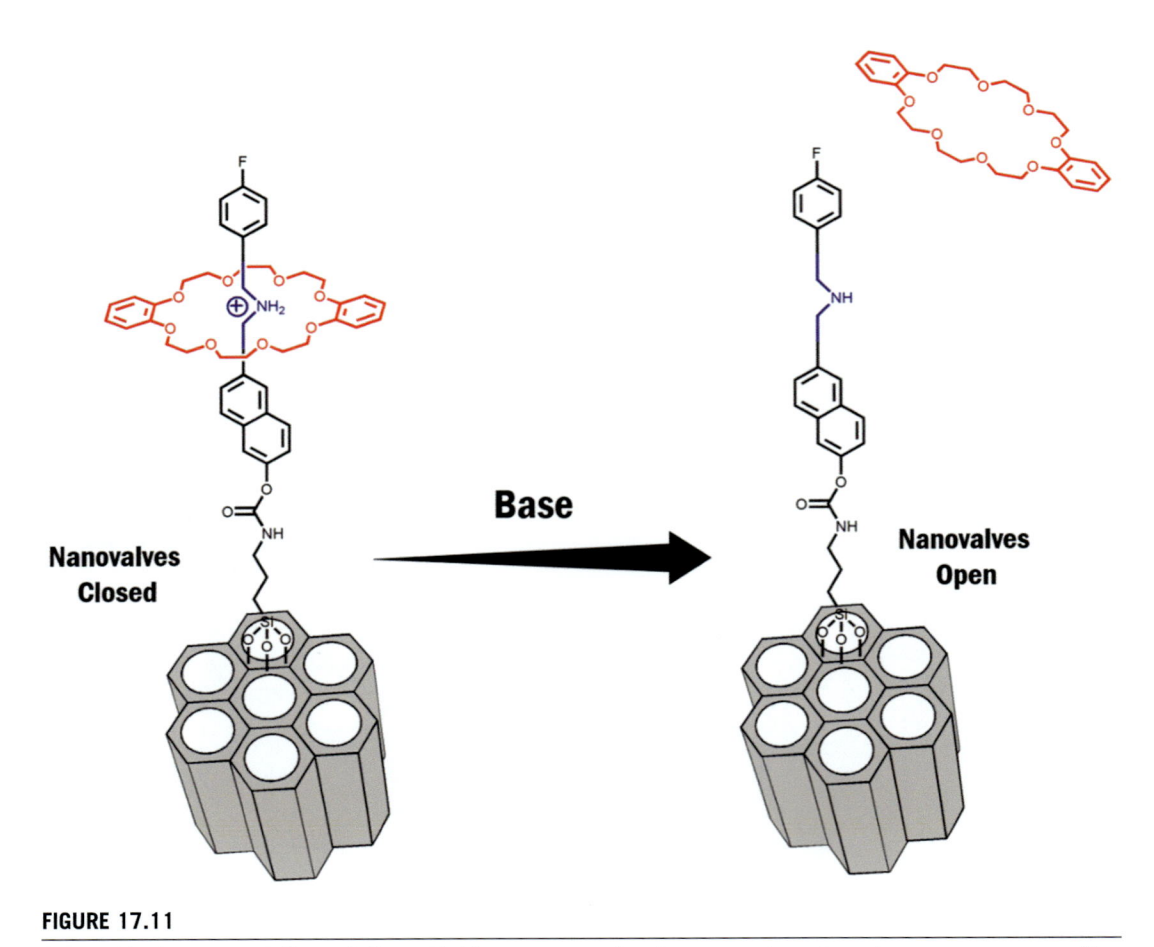

FIGURE 17.11

A graphical representation of the pH-driven supramolecular nanovalves based on mesoporous silica nanoparticles, using a DB24C8 crown ether and a dialkylammonium center-based pseudorotaxane as the gatekeeper. pH-controlled protonation/deprotonation closes or opens the gate, respectively, by virtue of "complexation" and "decomplexation."

step to attain spatial control over the polymerization reaction. Furthermore, polymers confined in the mesochannels did not show characteristic bulk behavior with respect to their glass transition temperature. The absence of a glass transition event in the composites prepared via gas-phase adsorption could be ascribed to nanoconfinement effects due to strong polymer–host interactions at the nanometer scale.

Hyeon et al. [105] exploited this methodology to create polypyrrole/poly(methyl methacrylate) coaxial nanocables through the sequential polymerization of methyl methacrylate and pyrrole monomers inside the channels of mesoporous SBA-15 silica, followed by the removal of the silica template SBA-15 (Fig. 17.12). The strategy consisted of incorporating methyl methacrylate (MMA)

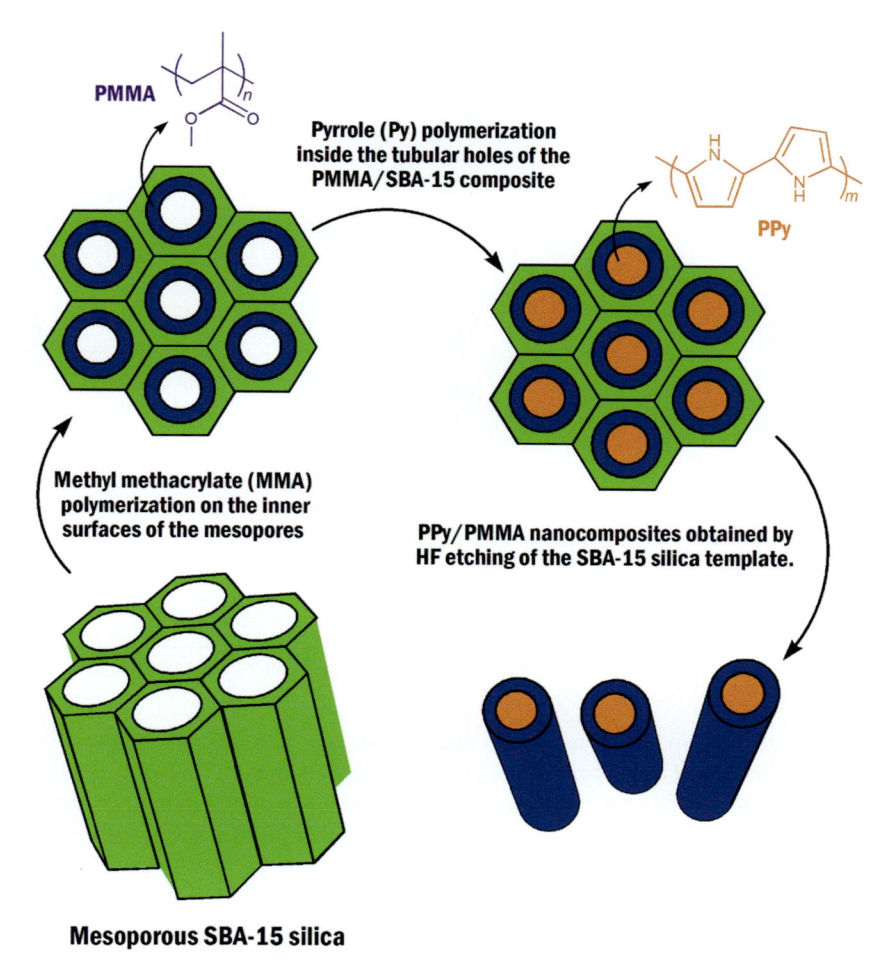

FIGURE 17.12

Sequential synthesis of polypyrrole-polymethacrylate composite architectures into the inner channels of mesoporous silica.

into the pores of SBA-15 silica by heating for 5 hours at 90°C under reduced pressure. Then, MMA was polymerized in the presence of benzoyl peroxide under an argon atmosphere at 70°C for 2 days and then 120°C for 2 hours, followed by evacuating in a vacuum oven at the same temperature for 18 hours. Next, pyrrole was loaded into the pores of the PMMA/SBA-15 composite using the same conditions as in the MMA incorporation, and then polymerized with 20 mL of 0.81 M aqueous $FeCl_3$ solution for 3 hours. The resulting solid was retrieved by filtration, followed by drying under vacuum at room temperature for 12 hours. To remove the silica template, the PPy/PMMA/SBA-15 composite was dispersed in an aqueous HF solution (48 wt.%) and stirred overnight.

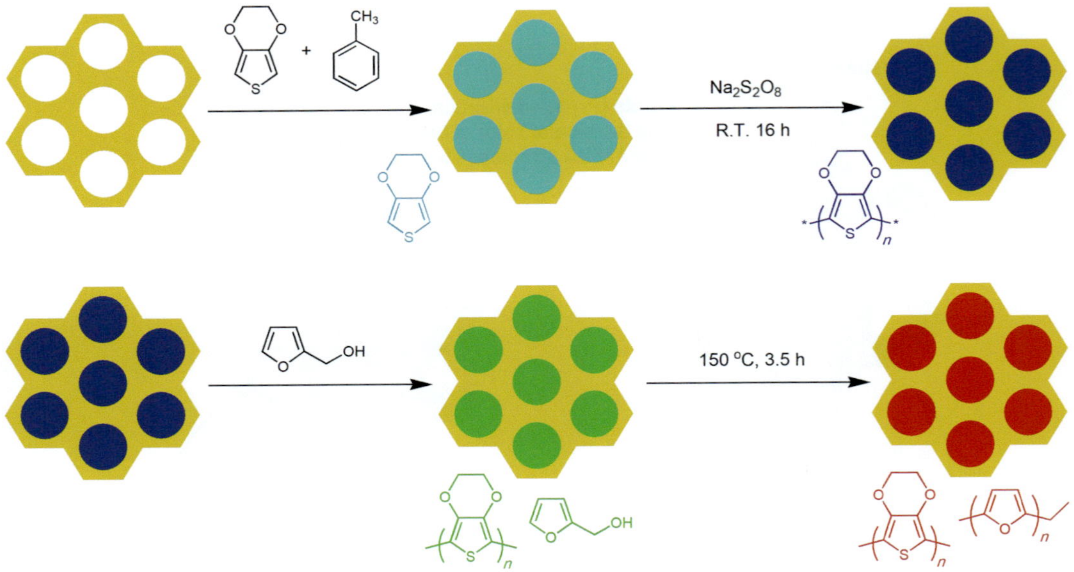

FIGURE 17.13

Sequential infiltration–polymerization steps necessary to accomplish the synthesis of PEDOT/PFA-mesoporous hybrids.

Sequential polymerization reactions within mesopores can also lead to nanoconfined polymer blends. Kelly et al. described the formation of poly(3,4-ethylenedioxythiophene) (PEDOT) blended with poly(furfuryl alcohol) (PFA) by a sequential infiltration–polymerization approach (Fig. 17.13) [106]. The PEDOT-modified mesoporous silica was prepared by filling the mesoporous silica with a toluenic solution of EDOT. The silica was previously heated to 120°C for 2 hours to remove adsorbed water and then cooled to room temperature. The EDOT solution is incorporated gradually to completely fill the pore volume of the silica by using a micropipette, and the mixture was agitated with a spatula for several minutes. The powder was transferred to a vial and dried at 60°C for a period of 1 hours, after which an aqueous solution of sodium persulfate was added to initiate the polymerization. Once the samples were purified, the polymerization of poly(furfuryl alcohol) was accomplished by drying in vacuo the PEDOT-modified samples at 85°C for 2 hours followed by the incorporation of furfuryl alcohol to completely fill the remaining pore volume of the mesoporous host. Finally, the monomer was polymerized by heating at 150°C for 3.5 hours. The filling of the mesopores and the polymer distribution within individual mesoporous particles were determined by a combination of energy-dispersive X-ray microanalysis, X-ray photoelectron spectroscopy, and nitrogen adsorption. The results suggest that when PEDOT is added to the silica host, followed by PFA, the phase separation of the two immiscible polymers is constrained by the dimensions of the silica mesopores, ensuring nanoscale contact between the two phases. The silica template can be removed by etching with 25% hydrofluoric acid, leaving behind a blended polymer microparticle. The etched microparticles exhibit macroporous morphologies different from that of pure PEDOT particles prepared by a similar route. The blended microparticles also appear to undergo limited phase separation provided that no evidence of polymer domain segregation was observed. Interestingly, when PFA is added to the host first, followed by PEDOT, the final composition of the blend is drastically altered. The reversal of the blending order results in a more amorphous, phase-separated material, thus demonstrating that nanoconfinement effects arising from the interaction of the polymers with the mesopore walls may dictate the factors that govern the structural reorganization of the constrained polymer blend.

The synthesis of nanocomposites by sequential chemical reactions within mesoporous has been also extended to the construction nanoconfined inorganic−hybrid architectures. Zhang et al. reported the fabrication of ZnO quantum dot/polythiophene (ZnO/PTh) into mesoporous silica (SBA-15) via a simple wet chemical two-step approach [107]. First, the SBA-15 was thermally treated at 120°C to remove the physically adsorbed water and then immersed in a thiophene solution (in methylene chloride). After sonication, the solution is slowly evaporated at 30°C over a period of 24 hours. This protocol enables the efficient incorporation of monomer units into the mesochannels. Then, the monomer-loaded mesoporous material is treated with an ethanolic solution containing H_2O_2 and $FeCl_3$ in order to initiate the polymerization of the confined monomers. Finally, the preparation of the ZnO/PTh/SBA-15 nanocomposites was accomplished by immersing the PTh/SBA-15 composites in an ethanolic solution of zinc acetate followed by treatment with an aqueous solution of Li(OH) to form ZnO quantum dots into the PTh-modified mesochannels. Photoresponse of ZnO/PTh/SBA-15 nanocomposites was studied with respect to its incident photon-to-collected electron conversion efficiency (IPCE) and morphology. The large increase in IPCE indicated that the ordered ZnO/PTh/ SBA-15 hybrid architecture has greatly improved the ability of charge collection and transportation. In addition, the presence of SBA-15 proved to be critical for controlling the interfacial morphology and hence enlarging the interfacial area of the inorganic−organic heterojunction. These results highlight the importance of hybrid nanostructured platforms as key enablers to photovoltaic cells provided that they are able to create more charge transfer junctions with high interfacial area.

Polymerization of mesopore-confined monomers can be also carried out by electrochemical (instead of chemical) means to produce, for example, mesoporous silica filled with polypyrrole [108]. In this case, prior to monomer adsorption, the mesoporous host (silica particles—SBA 15) required prolonged heating (300°C for 3 hours) to remove air and water in the channels. The host matrix and pyrrole monomer were separately placed in two glass tubes which were connected into a self-regulating system and kept at equilibrium under vacuum at room temperature for 24 hours. Thereafter, pyrrole molecules were driven into the channel leading to the pyrrole/SBA-15 nanocomposite. Then, pyrrole/SBA-15 nanocomposite was dispersed in water and the suspension was dropped on the surface of glassy carbon electrode. The pyrrole/SBA-15 modified electrode was subject to continuous cyclic electrode potential scans. Pyrrole molecules adsorbed in the channels of SBA-15 were electropolymerized and the PPy/SBA-15 modified electrode was obtained. The XRD, SEM, TEM, N_2 adsorption/desorption, and FT-IR studies confirmed that ordered mesostructure of SBA-15 remains unchanged after encapsulation and PPy is located in the channels of SBA-15.

Previous approaches were mostly based on the incorporation of monomer units via gas-phase adsorption. Even though this methodology has been proven successful by different authors [109], it requires careful preconditioning of the mesostructured host, that is, heating and vacuum, provided that air and moisture can seriously affect the incorporation of hydrophobic monomers into hydrophilic pores. The disadvantages of "gas-phase" inclusion methods can be overcome by "wet" strategies based on the use of monomer units displaying affinity to the pore walls [110]. Recently, Wolf et al. reported a new poly(*p*-phenylenevinylene) (PPV) composite material obtained through the incorporation of insoluble PPV polymer chains in the pores of monodisperse mesoporous silica spheres through an ion-exchange and in situ polymerization method [111]. The mesopores were templated by the alkylammonium surfactant leading to a scenario in which the interior of the mesopores contains alkylammonium—siloxide ion-pairs. Then, the surfactant-filled mesoporous silica spheres were refluxed in a methanolic solution of p-xylylenebis-(tetrahydrothiophenium chloride) to promote the ion exchange of the alkylammonium surfactant inside the pores for the doubly charged monomer. The basic siloxide sites were then able to deprotonate the monomer and cause it to undergo polymerization. Subsequent heating of the pPPV intermediate at 200°C in vacuo produces the fully conjugated PPV material confined in the mesoporous structure.

The use of "wet" chemical routes to incorporate the monomers and proceed with the polymerization reaction has been also extended to the use of free radical polymerization as a tool to build up mesopore-confined polymers. Ryoo et al. [112] exploited this straightforward route through radical polymerization of vinyl monomers inside mesoporous silica to prepare composite functional materials without altering the well-defined mesoporosity and locating the polymer entities selectively onto the silica mesopore walls. The experimental protocol demanded the incorporation of vinyl monomers, crosslinkers, and radical initiators onto the pore walls of mesoporous silica via wet impregnation method, followed by equilibration under reduced pressure in order to achieve uniform distribution (Fig. 17.14). Then, the monomers predominantly incorporated/adsorbed on the

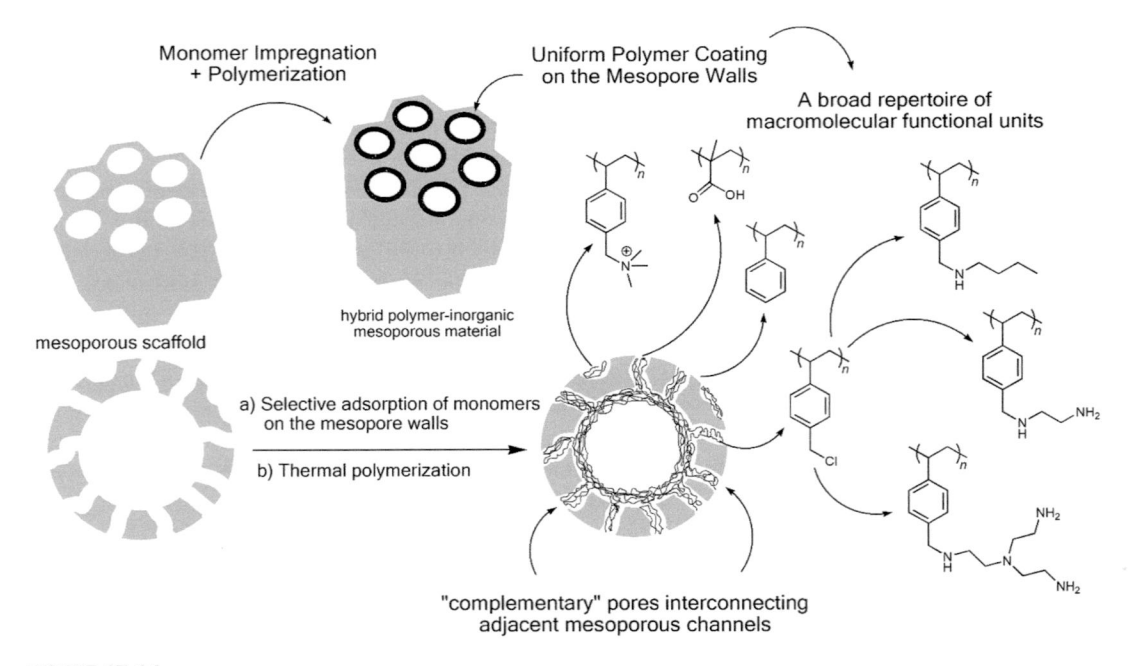

FIGURE 17.14

Synthetic strategy based on selective monomer adsorption followed by thermal polymerization for obtaining uniform functional polymer coatings on mesoporous frameworks.

mesopore walls were subsequently polymerized with heating to form a uniform polymer layer on the surface of the silica framework. Finally, the polymer–silica composite materials were washed with chloroform and ethanol in order to remove remaining monomers and loosely adsorbed polymers. N_2 adsorption isotherms and XRD characterization revealed that in situ polymerization of nanoconfined monomers led to uniform films on silica walls. This strategy enables a certain degree of control over the location of the confined polymers by tailoring the mesostructure of the silica framework as well as the polymerization conditions.

In this regard, it is worthwhile indicating that, even though free radical polymerization is the most versatile method for the polymerization of vinyl monomers, the propagation reaction is generally more difficult to control than ionic processes due to the irreversible termination of the growing polymer radicals through recombination and disproportionation reactions. However, diverse chemical and physical strategies have been devised to suppress such termination reactions. The former utilize transition metal complexes to stabilize growing polymer radicals via reversible interaction, while the latter employ restricted spaces such as micelles to provide an isolated reaction environment for each growing polymer radical. From the latter point of view, mesoporous materials displaying uniform arrays of nanoscopic channels may provide a confined but adequately large space for the synthesis of macromolecular building blocks. Along these lines, experimental work by Aida and coworkers [113] on the free radical polymerization of methyl methacrylate (MMA) within mesopores revealed the confined growth of high-molecular-weight polymer (PMMA) as well as the formation of long-living propagating polymer-radicals, as observed by electron paramagnetic resonance (EPR), is feasible. The molecular weight of PMMA within the mesopores could be controlled over a wide range by changing the monomer-to-initiator mole ratio. This indicates that mesopore-confined free radical polymerization enables the formation of a whole set of hybrid materials by simply choosing the adequate mesostructure and the desired monomer units [114]. Various commercial monomers are available, and their copolymerization can be a route to prepare multifunctional materials, including polyelectrolytes, hydrophobic, and thermoresponsive polymers. Hence, depending on the monomer, the resultant confined polymers can easily be postfunctionalized to incorporate diverse functional groups in high density, due to the open porous structure allowing facile access for the chemical reagent.

Kleitz et al. described the tailoring of mesoporous amine-functionalized polymer–silica composites by a two-step confined polymerization technique [115]. A functional vinyl monomer, chloromethylstyrene (CMS), was polymerized within the mesostructure leading to a uniform coating on the mesopore pore surface. In the second step, diverse amine-based moieties were attached to the polymer surface by nucleophilic substitution, generating a variety of nanoporous amino polymer–silica composites. This approach allows for a tuning of surface concentration of the organic groups either by varying polymer loading or by copolymerization of the CMS monomers with nonreactive monomers (styrene) as well as the facile incorporation of diverse types of amine groups, for example, secondary amines, diamines, linear or branched polyamines. These composite materials were shown to be active as catalysts in the Knoevenagel condensation reaction confirming their potential in liquid phase heterogeneous catalysis. Experimental evidence also reveals that the nature of the mesoporous framework and the polymerization conditions can lead to the appearance of blocking effects. The distribution and uniformity of the polymer layer along the channel walls depend on the content of polymer in the composite, which can be controlled by altering the polymerization conditions during the synthesis. At higher content, the polymer layer is formed heterogeneously in thickness along the channels, resulting in pore plugging and heterogeneous pore size, but without altering the ordered hexagonal mesostructure of the material [116]. The synthesis strategy using vinyl monomers can introduce remarkable advantages by incorporating organic moieties within the mesoporous silicas via the formation of a robust C−C bond rather than hydrolysis-susceptible siloxane bonds. However, polymers are not actually "grafted" to the pore walls but physically confined within the mesostructure. The presence of "complementary" pores interconnecting adjacent mesoporous channels is important for fixing the polymer phase provided that the macromolecular layer can acquire high stability through the formation of an interpenetrating network between the mesoporous framework and the polymers.

17.4.2 Functionalization of mesoporous materials with dendrimers and dendronized macromolecules

Dendrimers are macromolecular building blocks characterized by a regularly branched structure synthesized by step-by-step reactions. There are several distinctive features that make them attractive functional units for molecular design of mesopores, these are: accurate control over the whole molecular architecture, high level of monodispersity, and nanoscale dimensions. Synthesizing dendrons and other dendritic architectures within the mesoporous framework [117] introduces the possibility of generating hybrid materials displaying a high density of predesigned active sites that might be useful for designing catalytic platforms [118] or "active" membranes [119], just to name a few examples. Polyamidoamine (PAMAM) dendrimers up to the third generation were grown from amino-functionalized mesoporous silica via sequential Michael addition of methyl acrylate followed by amidation in the presence of ethylenediamine [120]. In-depth characterization of dendrimer-modified mesoporous silica using nitrogen adsorption, solid state NMR, FTIR, thermogravimetry, and chemical analysis revealed that dendrimer growth took place inside the channels with an average yield higher than 97%, all synthesis steps being included. The third generation was found to almost completely fill the pore system. These materials were then used as nanostructured supports to anchor rhodium species leading to rhodium-complexed polyamidoamine (PAMAM) dendrimers grown inside the channels of MCM-41. This was accomplished by phosphinomethylating and then complexing generations 0, 1, and 2 with rhodium. The Rh-PAMAM-G(0) and Rh-PAMAM-G(1) nanomaterials displayed very good activity in the hydroformylation of 1-octene (turnover frequency of 1800 h^{-1} at 70°C) and the catalyst could be recycled several times without loss in activity. These functional aspects make them excellent candidates for recyclable catalysts in olefin hydroformylation reactions. The use of dendritic mesoporous silicas is gaining increasing interest in catalysis. For example, by changing the chemical nature of dendritic networks (aliphatic or aromatic), it is possible to have predominantly base catalysts within a large range of well-defined controlled basicities. The hydrophobic nature of the dendritic mesostructures is advantageous in increasing reaction rates by altering local water concentration in the Knoevenagel condensation reaction [121]. Higher generation dendrimers were supposed to exhibit higher density of active sites, but instead they were found to be less effective than catalysts of lower generation. This can be attributed to steric crowding. Beyond generation 3 (G3), the mesoporous channels were essentially blocked and the dendrimer stopped growing. These data indicate that the pore system plays an important role in the synthesis of supported dendrimers of a desired generation.

Using larger pore (8.3 nm) SBA-15 silica with less-grafted propylamine, Acosta et al. synthesized up to G4 melamine-based dendrimers inside the support channels (Fig. 17.15) [117]. The porosity of the hybrid material can be modified either by using dendrimers of different generations,

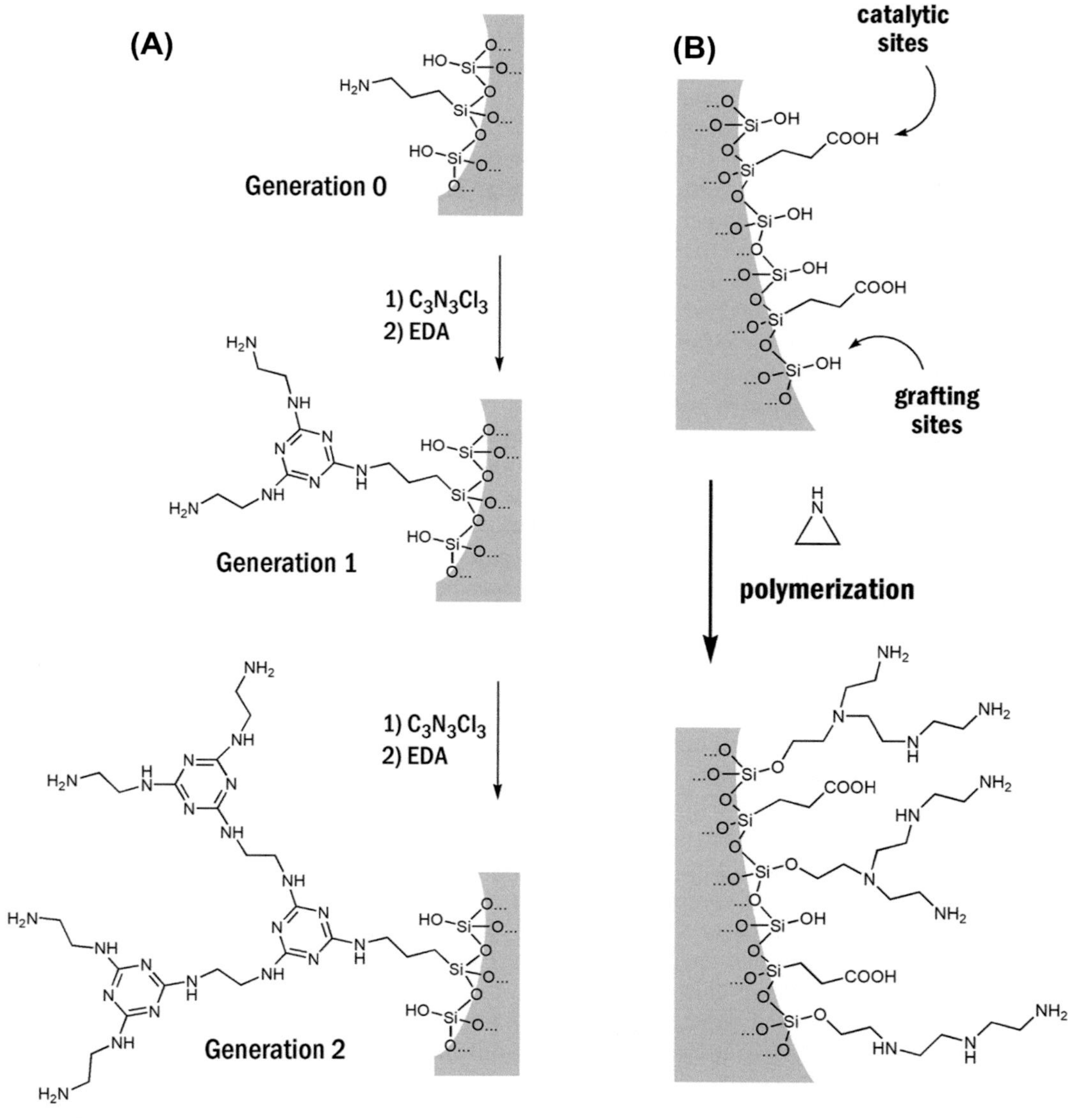

FIGURE 17.15

Synthesis of different generations of dendrimer-mesoporous hybrids through sequential reactions on the mesopore walls.

using different linkers in the dendrimer structure, or by controlling dendrimer loading. Porosimetry measurements indicated that the effective pore size of the hybrid and the total pore volume of the material can be controlled independently of one another. Copper sequestration was used as a probe to demonstrate that the terminal amino groups of the dendrimer are accessible and able to bind Cu (II) [122]. These melamine-based dendrimers [123] confined in the mesoporous structure have also shown interesting applications as CO_2 absorbers [124] as well as effective solid base catalysts for a diverse range of chemistries, which include Aldol chemistry and transesterification reactions. The dendron—modified mesoporous materials are active in both the nitroaldol (Henry) reaction and the transesterification of glyceryl tributyrate to afford methyl esters [125]. In both reactions it is observed that dendrons terminated with primary amines are more catalytically active than samples containing dendrons terminated with secondary amines. The first generation dendrons are the most active for both chemistries and larger pores displayed a higher activity that the smaller ones indicating the critical role of molecular transport and diffusion resistance in the catalytic functional properties of hybrid polymer—inorganic mesoporous films. The experimental results reported by Shantz and coworkers indicate that dendron catalysts are much more active and stable than simple amines attached to silica in the transesterification of triglycerides.

A different strategy to functionalize the mesoporous network with dendrimers relies on the implementation of a postsynthesis method using a novel amine dendritic precursor [126]. The synthesis of such building blocks was based on the introduction of a spacer unit with a terminal T-sylil function on some of the peripheral amine groups of poly(propyleneimine) dendrimers (Fig. 17.15). These precursors can react with silanol groups on the pore walls leading to the covalent modification of the mesoporous films with predesigned dendritic units. The postgrafting reaction should be carried out under anhydrous conditions in order to avoid self-condensation of dendritic precursors in the presence of water. This synthetic route represents an alternative to the preparation of amine-functionalized and dendrimer-functionalized mesoporous silica with a highly dense population of amine groups, avoiding several reaction steps of the iterative procedure usually followed for dendritic growth inside the channels of mesoporous materials. The use of different dendrimer generations and nominal degrees of surface functionalization can be exploited using simple experimental variables to finely tune the incorporation of amino group into the mesoporous framework. On the other hand, hyperbranched polymers prepared in a one-pot synthesis also constitute a valuable alternative to functionalize mesoporous frameworks with dendritic macromolecules. Linden and coworkers proposed a simple method based on hyperbranching polymerization for functionalizing mesoporous silica with high loading of amine groups [127]. These authors used an acid-catalyzed hyperbranching polymerization approach to produce reactive primary amino groups on the surface of mesoporous silica in the form of a surface-grown polyethyleneimine (PEI).

The polymer has been grown directly from the surface silanol groups utilizing a highly reactive, nonbulky monomer, aziridine. This technique can be implemented through the introduction of acetic acid into the mesoporous silica together with the monomer before polymerization or by using COOH—functionalized silica as the substrate for the polymerization in order to aid a complete surface functionalization. The surface polymerization of aziridine was straightforwardly performed with toluene as solvent, in which the SBA-15 substrate was suspended in the presence of catalytic amounts of acetic acid. The presence of the catalytically active COOH function on the pore walls enables a homogeneous growth of PEI. In this regard, due to the high amine loading, hyper-branched PEI-modified mesoporous silicas have shown interesting properties for CO_2 capture and could be extended to promising new materials for acid gas capture. The advantage of this hybrid material over previously reported adsorbents rests in its large CO_2 capacity and multicycle stability. The material was recycled by thermally desorbing the CO_2 from the surface with essentially no changes in capacity. Furthermore, the organic groups on the surface were stable in the temperature range between 25°C and 130°C due to the robust covalent attachment between the mesoporous support and the hyperbranched polymer [128].

17.4.3 Molecular assembly of polymerizable structure-directing agents

An attractive approach to build up macromolecular functional units within mesoporous materials is to utilize structure-directing agents bearing polymerizable groups [129]. Pioneering work of Brinker and coworkers described the use of polymerizable surfactants as both structure-directing agents and monomers in the evaporation-driven self-assembly of mesostructured materials (Fig. 17.16) [130]. Synthesis of polymerizable amphiphilic diacetylene molecules enabled the self-assembly of conductive, conjugated polymer/silica nanocomposites in thin-film forms suitable for integration into devices. The progressively increasing surfactant concentration drives self-assembly of diacetylene/silica surfactant micelles and their further organization into ordered, three-dimensional, liquid crystalline mesophases. Ultraviolet-light-initiated polymerization of the diacetylene units, accompanied by catalyst-promoted siloxane condensation, topochemically converts the

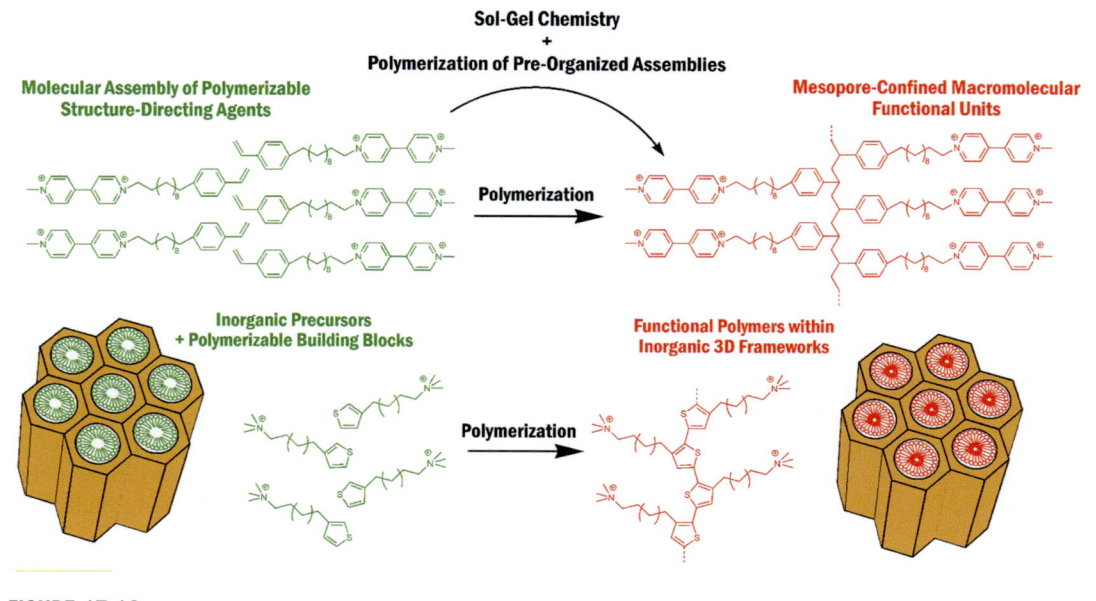

FIGURE 17.16

Scheme describing the formation of mesopore-confined macromolecular functional units through the molecular assembly and reaction of polymerizable structure-directing agents.

Adapted with permission from G.J.A.A. Soler-Illia, O. Azzaroni, Chem. Soc. Rev. 40 (2011) 1107.

colorless mesophase into the blue polydiacetylene/silica nanocomposite, preserving the highly ordered, self-assembled architecture. In a similar vein, responsive mesoporous silica was also synthesized through cooperative assembly of cetyltrimethylammonium bromide and silsesquioxanes containing a bridged diacetylenic group. The construction process involved the spontaneous organization of diacetylenic molecules around the surfactant liquid crystalline structure, forming a mesoscopically ordered composite with molecularly aligned diacetylenic units. Subsequent surfactant removal followed by topopolymerization gave rise to the responsive mesoporous silica embedded with polydiacetylene (PDA), a polymer that chromatically responds (e.g., blue to red) to a wide range of external stimuli [131].

Aida et al. [132] reported the synthesis and self-assembly of polypyrrole/silica nanocomposite material, templated by a pyrrole-containing surfactant, where polypyrrole domains segregated and insulated by one-dimensional silicate nanochannels were obtained after oxidative polymerization in the presence of $FeCl_3$. Along these lines, in recent years different groups have explored the use polymerizable surfactants as building blocks to create mesostructured functional materials displaying photoluminescent or electrochemical properties [133].

17.4.4 Complex micelles as structure-directing, functionalizing, and pore-generating agents

Recently, Gérardin and coworkers [134] introduced a method for the direct functionalization of mesoporous silica with strong polyacids of defined length, which can be homogeneously integrated in the mesopores. The proposed strategy relies on the use of polyion complex (PIC) assemblies as structure-directing agents, in order to incorporate predefined functionalities that are masked during the sol − gel process, thus preventing disruption of the mesostructuring process (Fig. 17.17). These authors have shown that PIC micelles, formed through the interaction of a weak polyacid-based double-hydrophilic block copolymer (DHBC) and an oppositely charged polyelectrolyte, are able to direct the structure of PIC−silica materials with exceptional control over textural and structural properties of the mesoporous materials [135,136]. In this particular method, the protocol involves the preservation of the DHBC in the PIC assembly in order confer strong acidic properties to the mesopores, using a strong polyacid-based DHBC containing a poly(styrene sulfonic acid) (PSS) block. This synthetic process operates under purely aqueous conditions and at room temperature, making it a fairly simple synthetic procedure. For instance, the resulting materials exhibited a very high volume density of -SO_3H functional groups as well as superprotonic conductivity, that is, $\sigma = 2.4 \times 10^{-2}$ S/cm at 363K (measured under 95% relative humidity conditions).

The versatility of the process can be easily exploited to control the ordered silica structure and to tune the pore properties of functionalized hybrid materials by choosing the adequate blocks in the polymeric assembly. Furthermore, this strategy could be explored to design Li- or Na-ion conducting materials.

17.4.5 Confined polymerization at "activated" mesoporous walls

Aida et al. reported the synthesis of crystalline polyethylene fibers (diameter 30−50 nm) by the polymerization of ethylene within mesoporous supports modified with titanocene and methyl-

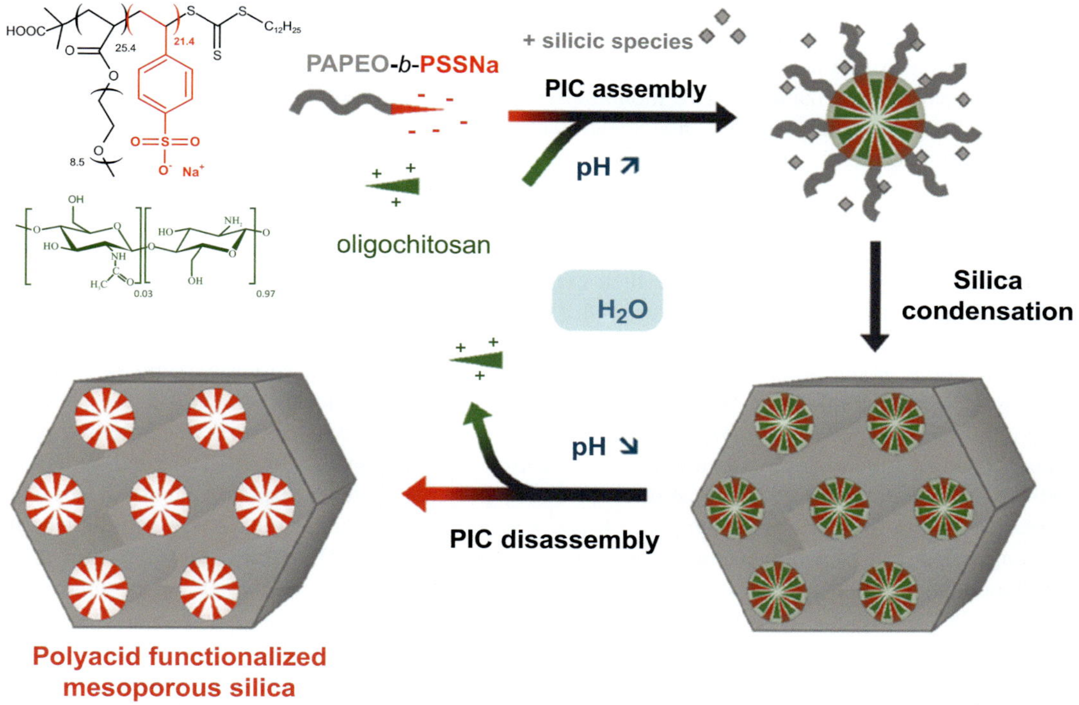

FIGURE 17.17

Illustration of the eco-designed pathway to form large-pore ordered silica functionalized with polyacid chains (mesoPIC-PSS).

Reproduced with permission from J. Richard, A. Phimphachanh, J. Schneider, S. Nandi, E. Laurent, P. Lacroix-Desmazes, et al., Chem. Mater. 34 (2022) 7828.

alumoxane (MAO) as a cocatalyst [137]. The polymerization of ethylene within this nanoconfined reactive environment gave a cocoon-like solid mass consisting of fibrous polyethylene (PE). Polarization microscopy of the PE fibers showed a clear birefringence in the fraying edge region, which suggests that the fibers are crystalline. These authors postulated that the formation of the crystalline PE fibers within the reactive mesopore can be explained by an "extrusion polymerization" mechanism similar to the biosynthesis of crystalline cellulose.

Polymer chains, formed at the activated titanocene sites within the individual mesopores, are extruded into the solvent phase and assembled to form extended-chain crystalline fibers. As the pore diameter of the mesoporous material (~ 3 nm) is much smaller than the lamellar length of ordinary PE crystals (~ 10 nm), the nascent polymer chains cannot fold within the narrow reaction channels of the support and therefore grow out of the porous framework before they assemble, resulting in the formation of extended-chain crystalline fibers. By using regularly arranged nanoscopic, one-dimensional polymerization reactors, the authors thus achieved oriented growth of polyethylene macromolecules that normally requires postprocessing steps. Chromium acetylacetonate

[Cr(acac)$_3$] complexes have been also grafted onto mesoporous materials resulting in a catalytic system that is able to polymerize ethylene under nanoconfinement at relatively low pressures [138]. This strategy also enables the formation of polymer blends with nanoscaled dispersion though space-confined polymerization of two monomers in mesoporous environments displaying dual catalytic sites. The duel catalysts offer independent active sites during the polymerization so as to generate the ultimate blends of the two polymers. Loading two catalytic systems within mesopores to make two polymers represents an attractive synthetic methodology to blend two polymers to nanoscale range through direct reaction without the need of a compatibilizer. The experimental protocol would simply rely on supporting the suitable catalysts/initiators onto the mesoporous framework and the dual catalytic system would be then exposed to two monomers simultaneously to generate polymer blends. Along these lines, Chan et al. [139] reported the blending of ethylene and syndiotactic polystyrene by pretreating the mesoporous support with the adequate catalysts. Ethylene homopolymer is made from a catalyst prepared by pretreating the mesoporous support with the cocatalyst, methylaluminoxane (MAO), prior to adding the metallocene precursor, zirconocene dichloride. On the other hand, syndiotactic polystyrene (sPS) is polymerized by pentamethylcyclopentadienyl titanium trimethoxide and MAO supported on mesoporous silicate. The combination of both catalytic systems led to the generation of binary polymer systems displaying unusual physical properties arising from nanoconfinement. PE and PS nanoblends demonstrated more significance in blending effectiveness when using space-confined polymerization. When PE and sPS are made simultaneously by dual catalysts within mesoporous channels, the crystallinity of PE is suppressed dramatically, as demonstrated by both DSC and XRD, while the physical blends usually merely show crystallinity reduction in proportion. Widenmeyer et al. applied surface organometallic chemistry to generate Sm(II) alkoxide, indenyl, and alkyl surface species via ligand exchange at mesoporous substrates [140]. Interestingly, Sm(II) grafting and subsequent ligand exchange did not markedly change the morphology and the microstructure of the samples. Such Sm(II)-modified organometallic−inorganic hybrid materials were able to initiate the graft polymerization of methyl methacrylate via a radical-initiated anionic coordination polymerization mechanism involving sterically unsaturated surface-confined samarium enolate moieties. The local environment of the Sm(II) surface centers (coordination sphere) may strongly affect their reactivity and, hence, the efficiency of MMA polymerization. Hybrid materials featuring the "smallest" ligands (methyl, methoxy) acted as the best initiators for graft polymerization. The polymer−inorganic nanocomposites revealed complete pore filling or blockage of the pore entrances as indicated by N$_2$ physisorption and scanning electron microscopy. The use of "activated" mesoporous walls to synthesize polymer chains under confinement has been also extended to the use of modified aluminum MCM-41 (Si/Al = 13, pore size = 3.2 nm) in which Ni(II) or Zn(II) species were introduced by ion exchange with aqueous salt solutions [141]. When these mesoporous aluminosilicates are evacuated to remove water and oxygen at 300°C (0.1 Torr, 24 hours), followed by exposure to alkyne vapor at 1 atm at various temperatures and time periods, an exothermic uptake of reagent produced dark brown or gray products. Combustion elemental analysis of the resulting solids revealed that approximately three-quarters of the pore volume can be filled with polymer. Although some polymer is formed at low temperatures with ethyne, significant pore loading is achieved above 150°C. Control experiments performing ethyne polymerization on unmodified Al/MCM- 41 (no Ni(II) or Zn(II) ions) did not reveal any evidence of polymer growth, thus indicating the requirement for Ni(II) or Zn(II), and the positive influence of porosity and confinement on ethyne polymerization. Zn(II)/MCM-41 treated with ethyne at 250°C for 24 hours generated the greatest polymer occupancy. Interestingly, Zn(II) ions have been reported not to catalyze such polymerization, furthermore their Lewis acidity (as has been suggested) cannot be the sole factor leading to ethyne polymerization, as many other M (II) or M(III) ions when incorporated into the mesoporous material failed to produce polymer. Hence, these experimental results evidence the emergence of interesting confinements effects that enhance the reactivity of Zn(II) ions, and consequently render them suitable species to promote the confined polymerization of ethyne. The rate and extent of polymerization within the mesoporous channels directly depend on the spatial distribution of catalytic sites as well as the diffusion rate of monomeric precursors and their local concentration near the catalytic sites [142]. Both factors can be adjusted by varying the pore size and by tailoring the catalytic activity of the surface through proper functionalization. To examine how these factors influence the nanoconfined polymerization of functional polymers, Pruski et al. studied the oxidative polymerization of 1,4-diethynylbenzene within two Cu^{2+}-activated mesoporous systems: silica and alumina materials (Cu-MCM and Cu-MAL, respectively) with different pore sizes and two different methods of Cu^{2+} surface functionalization.

One approach consisted of preparing the Cu^{2+}-incorporated MCM-41 silica material (Cu-MCM) by cocondensation using a Cu^{2+}-chelating molecule, *N*-[3-(trimethoxysilyl)propyl]ethylene-diamine, as the precursor, whilst the other approach consisted of impregnating the mesoporous alumina with Cu^{2+} ions. Experimental evidence provided by solid-state NMR and photophysical studies revealed that the preparation of the reactive mesopore environment has a strong influence on polymerization reaction. While the conjugated poly(phenylene butadiynylene) polymer (PPB) synthesized within the Cu-MAL substrates displayed characteristic features of polydyacetylene-type crosslinking and conformational heterogeneity in the three-dimensional arrangements of the polymeric chains. Conversely, the characterization of PPB polymer chains synthesized within the Cu-MCM provided sound evidence of the formation of isolated molecular wires (lack of crosslinking).

Spange et al. exploited the intrinsic reactivity of mesoporous walls to perform polymerization reactions under conditions of constricted geometry [143]. The cationic polymerization of suitable vinyl monomers can be initiated either by active protons derived from acidic surface groups or by immobilized cationic-active surface initiators. In particular, proton (H^+) surface initiation has been observed for aluminosilicates and for protic acids adsorbed on silica.

Pure silicate materials are usually unable to initiate directly the cationic surface polymerization of *N*-vinylcarbazole (NVC), styrene, or other vinyl monomers, even in the suspension of solvents which are established for cationic polymerization, for example, dichloromethane, toluene, or hexane. However, suitable initiators for cationic surface polymerization are halogenoarylmethanes which become cationically active on acidic surfaces. The mesopores are prone to preconcentrate species with strong affinity to the pore walls; hence, preferential polymerization in the mesochannels is feasible by keeping the concentration of the carbenium ion in the surrounding solution, or on the outer surface, as small as possible. Arylmethyl halides are essentially inactive in solution but are activated specifically on the inner surface of the mesopore. The synthesis of poly(vinyl ether)s or polyvinylcarbazole under the conditions of constricted geometry can be achieved by means of cationic host−guest polymerization of the corresponding monomers in the pores of MCM-41 (pore diameter 3.6 nm), MCM-48 (pore diameter 2.4 nm) and in nanoporous glasses (Gelsil with a pore diameter of 5 nm) with bis(4-methoxyphenyl)methyl chloride (BMCC) or triphenylmethyl chloride as the internal surface initiator.

In a similar way, the intrinsic acid–base properties of silica have been exploited to promote ring-opening polymerization (ROP) reactions. The functionalization of silica surfaces with (1,1′-ferrocenediyl) dimethylsilane, was first reported by Wrighton and coworkers [144] and utilizes the susceptibility of [1]ferrocenophanes to nucleophilic attack by the hydroxy groups on the surface of the silica wall. It has been found previously that reaction of [1]ferrocenophanes with nucleophiles can lead, depending on the reaction conditions, to oligomeric and high molecular mass species as well as stoichiometrically ring-opened species [145]. This methodology has subsequently been used to modify mesoporous silica, anchoring [1]silaferrocenophanes to the reactive –Si-OH sites present on the surface of the silica. Hence, the formation of grafted polymer chains initiated from the mesopore surface can also take place in the case of the ring-opening polymerization (ROP) of [1]silaferrocenophanes. In this context, Manners and coworkers [146] explored the ROP of [1] silaferrocenophanes inside mesoporous silica as a route to generate highly functional building blocks inside the channels (Fig. 17.18). Monomers were introduced into MCM-41 by vapor deposition through the combination of dehydrated MCM-41 and monomer in a Schlenk tube under vacuum to allow the monomer to sublime into the channels. NMR data of the composite materials revealed that, at low loadings, ring-opened monomeric species and oligomers are present. Differential scanning calorimetry studies indicated that fully loaded MCM-41 samples displayed a very broad exothermic transition, consistent with ring-opening polymerization in the 75°C–200°C temperature range. The broadening of the exothermic transition has been ascribed to the inhomogeneity of the trapping sites for the encapsulated polymer.

The chemical nature of the mesoporous walls can be a versatile tool to tune the reactivity of the confined environment in order to trigger polymerization processes under external stimuli. In this context, Stucky et al. have recently developed a route to produce polymer–inorganic mesoporous hybrids based on a semiconductor photopolymerization technique [147]. This method relies on an electric potential generated by the optical absorption by mesoporous semiconductors and the subsequent oxidation of monomer confined within the mesoporous structure. The semiconductor mesoporous scaffold acts as the electrode for electropolymerization, provided that the electrical driving force is supplied by photon energy rather than potentiostatic control. These authors explored the use of mesoporous TiO$_2$ as a photosensitizer and nanostructured template for creating hybrid TiO$_2$-polypyrrole materials. Photon absorption by an inorganic semiconductor can generate oxidative and reductive

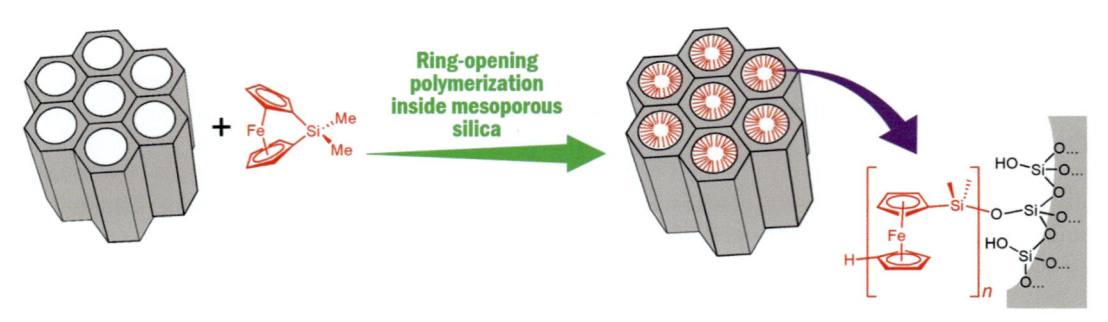

FIGURE 17.18

Simplified scheme describing the ring-opening polymerization of [1]silaferrocenophanes inside mesoporous silica as a route to generate functional building blocks inside the mesoporous channels.

equivalents to drive chemical reactions such as polymerization. Photopolymerization is initiated wherever photoexcited carriers in the inorganic sensitizer can oxidize or reduce a monomer. In this case, optical excitation of mesoporous TiO_2 was used to generate the electric potential necessary for triggering the oxidation and subsequent polymerization of the pyrrole monomer. This was simply achieved by immersing mesoporous TiO_2 in an aqueous solution containing pyrrole (0.2 M), sodium sulfate (0.1 M), and methyl viologen dichloride (10 mM) and subjected to UV illumination. The photopolymerization process was monitored by the quartz crystal microbalance, nitrogen sorption, and thermogravimetric techniques revealing that in situ generation of polypyrrole was observed to be self-limiting after approximately 20%−30% filling of the mesoporous TiO_2 network. This strategy introduces two interesting aspects that can have strong implications for the molecular design of functional polymer−inorganic mesoporous hybrids. First, the pore-confined macromolecules formed by this photoinduced polymerization technique should be in good electronic contact with the inorganic semiconductor phase because the polymerization reaction is locally initiated by charge transfer across the inorganic semiconductor−electrolyte interface. Second, the photopolymerization of the monomer can occur at any point on the mesoporous TiO_2 surface because monomer-containing solution and UV photons both penetrate the mesoporous network.

17.4.6 Infiltration of polymers into mesoporous frameworks

Incorporation of macromolecular building blocks into mesoporous materials via a polymer infiltration approach has received increasing attention. Seminal works from Tolbert and collaborators demonstrated that polymer infiltration from solution could be exploited as a straightforward route to confine polymer chains into nanoscale pores and tailor their topological characteristics. However, polymer infiltration into nansocale channels depends on the partitioning of the polymer from the solvent, and hence achieving a fine control over the polymer concentration, orientation, and uniformity in the corresponding nanocomposite is a nontrivial task. There are two important factors that require special attention. First, when the mesopore size is less that the radius of gyration of the solvated polymer, infiltration proceeds via a "reptation mechanism" (worm-like motion) that may require long processing times. Second, the polymer chains may suffer a significant loss of conformational entropy when they are confined in a mesopore, as a consequence spontaneous infiltration will greatly depend on the chemical interactions with the pore walls [148]. The infiltration of polyelectrolytes into mesoporous silica has been investigated in detail by Caruso and collaborators [149]. They studied the infiltration of poly(acrylic acid) (PAA) of different molecular weights (2−250 KDa), in amine-functionalized mesoporous silica particles with different pore sizes (4−40 nm). The surface charge of the nanopores and the charge density and conformation of PAA were tuned by changing the PAA solution conditions (e.g., pH and ionic strength) to which the particles were exposed. Thermogravimetric analysis and dynamic light scattering revealed that the extent of PAA infiltration strongly depends upon the relative sizes of the nanopores and the PAA molecules. Nanopores with diameters larger than 10 nm were capable of infiltrating a broader range of PAA molecular weights. The pH and ionic strength of the PAA solution govern the conformation of the macromolecules and the charge of the nanopores, and hence the ability of the macromolecules to infiltrate the mesoporous media. It was observed that the adsorption conditions play an important role in PAA infiltration. As the pH increases, the PAA charge density increases, and the polymer chains adopt a more rigid, linear conformation, resulting in a lower loading. When the pH of the PAA solution increased above 3, the amount of PAA loaded in the particles decreased due to the polymer chains adopting a more extended conformation. On the other hand, in the presence of salt, the degree of loading depends upon a balance between screening of the polyelectrolyte charge and screening of the particle surface charge. PAA loading decreases linearly with increasing salt concentration for PAA molecules below 100 KDa. The decrease in loading upon increasing salt concentration is even more pronounced in the case of low-molecular-weight PAA. This reduction in PAA loading with increasing ionic strength has been attributed to increased screening of the particle surface charge by salt ions, which weakens the electrostatic attraction between the PAA molecules and the mesopore walls. This indicates that besides polymer conformation electrostatic interactions between the polyelectrolyte and the pore walls also dominate to a great extent the infiltration process. Nanoconfinement within the mesopore may significantly influence the conformation of the adsorbed polyelectrolyte molecules, leading to the macromolecules exhibiting a more coiled conformation in nanopores than when adsorbed on planar surfaces. The nanoconfinement-driven conformational change may have profound effects on the functional features of the polyelectrolytes. For example, the preparation of diverse functional colloids involves the sequential assembly of interacting polymers within the nanopores of mesoporous particles. For the weak polyelectrolyte pair, PAA and poly(allylamine hydrochloride) (PAH), PAA is first deposited within the nanopores of amine-functionalized mesoporous particles, after which chemical or thermal cross-linking is performed to selectively form amide bonds between the carboxylic acid groups of the PAA and the primary amine groups grafted onto pore surface. Crosslinking is an important requirement in the preparation of these systems, which can be formed when two or more polyelectrolyte layers are deposited within the nanopores. In stark contrast, when the same PAA/PAH polyelectrolyte pair is deposited on nonporous particles, stable systems are produced without the need for crosslinking between the polyelectrolyte layers. The necessity for crosslinking in the construction of mesopore-confined PAA/PAH systems eloquently illustrates the importance of conformational and electrostatic changes taking place within the mesopore environment. Electrostatic interactions between infiltrated polyelectrolytes and the mesoporous walls have been exploited to build up an efficient pH-responsive carrier system. Xiao et al. constructed a mesoporous platform in which active molecules such as vancomycin can be stored and released from poly(dimethyldiallylammonium chloride) (PDDA)-loaded pores of SBA-15 by changing pH values at will. The amount of vancomycin stored in the mesopores is up to 36 wt.% at pH ~ 7. When the pH is at mild acidity, vancomycin is steadily released from the pores of SBA-15, thus resembling an "active" nanostructured framework that contains drug reservoirs and environment-sensitive pores, and the state of these pores (closed or opened) can be controlled by pH value. Polycations (PDDA) immobilized onto the anionic SBA-15 by electrostatic interactions were acting as "closed gates" for storage of drugs in the mesopores. Upon decreasing pH the ionized carboxylate species (COO−) are transformed into protonated groups (COOH) and polycations are separated from the surface of modified SBA-15, leading to opening of the gates for release of drugs from the mesopores [95].

The design of hybrid mesoporous materials incorporating polymeric assemblies through simple strategies [150] represents a very fertile research area offering major opportunities for controlling molecular transport through nanoporous interfaces [151,152]. Within this framework, Brunsen et al. [153] showed that the manipulation of the molecular transport properties of mesoporous silica thin films can by attained by the direct infiltration of polyelectrolytes into the inner environment of the 3D porous framework (Fig. 17.19). The infiltration/assembly of PAH alters the intrinsic cation-

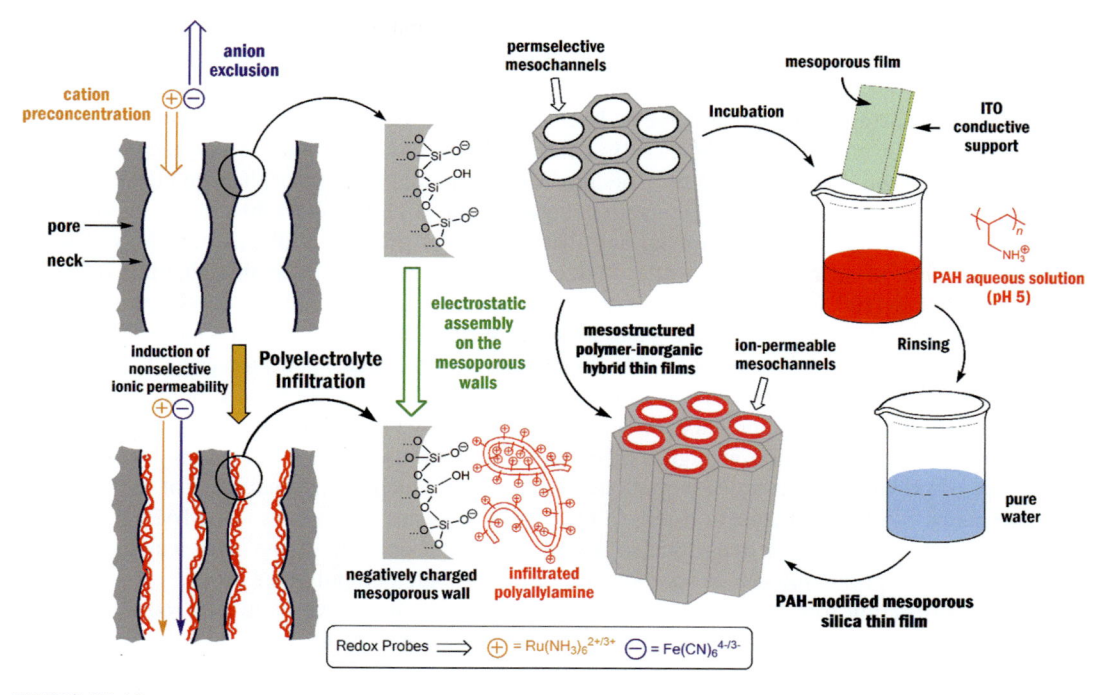

FIGURE 17.19

Schematic depiction of the infiltration of polyallylamine (PAH) inside the mesoporous silica film leading to the electrostatic assembly of the polyelectrolyte on the inner mesoporous walls.

Reproduced with permission from A. Brunsen, A. Calvo, F.J. Williams, G.J.A.A. Soler-Illia, O. Azzaroni, Langmuir 27 (2011) 4328.

permselective properties of mesoporous silica films, rendering them ion-permeable mesochannels and enabling the unrestricted diffusion of cationic and anionic species through the hybrid interfacial architecture. Contrary to what happens during the electrostatic assembly of PAH on planar silica films (quantitative charge reversal), the surface charge of the mesoporous walls is completely neutralized upon assembling the cationic PAH layer (i.e., no charge reversal occurs). These experimental results are a clear manifestation of the predominant role of nanoconfinement effects [154] in dictating the functional properties of polymer—inorganic hybrid nanomaterials.

On the other hand, alternative approaches involve the incorporation of uncharged hydrophobic semiconducting polymer into nanoporous materials by placing the materials in contact with a liquid solution of the polymer, either by spin-casting or direct immersion. Stucky and coworkers described a technique for incorporating the poly(3-hexyl thiophene) (RR P3HT) into titania mesopores by spin casting a film of the polymer on top of the titania film, and then heating at temperatures in the range of $100°C-200°C$ in order to fill 33% of the total volume with the semiconducting polymer. Conjugated polymer films were spin-cast on top of the mesoporous films, and then the samples were heated for various times and temperatures. Following the thermal treatment, the excess polymer was removed by rinsing the samples with toluene [155]. UV—Vis absorption measurements confirmed that the amount of polymer incorporated into the mesoporous sample remained constant after 5−10 minutes of solvent rinsing. It is worthwhile indicating that control experiments performed on similar polymers deposited on the top of bare ITO or glass dissolved very quickly during toluene rinsing. This observation reveals that polymer penetrates into the pores and remains there even if the film is rinsed in a good solvent. Even though the infiltration process implies a significant conformational entropy loss, the experimental evidence indicates that the polymer is remarkably stable within the titania mesopores. This has been attributed to the fact that the entropy loss is compensated by a strong enthalpic interaction between the highly polarizable main chain of the conjugated polymer and the polar titania, and that a chain of polymer will infiltrate only if some segments are able to adsorb on the walls of titania. In the case of this semiconducting polymer, upon heating from 100°C to 200°C its π-stacked structure evolves into a coiled configuration. Inside the pores coiled chains occupy a smaller titania surface area than rod-like chains, and consequently a greater number of polymer−wall contacts can be generated upon increasing the temperature. Interestingly, the absorption spectra of the infiltrated polymer chains were blue-shifted, thus suggesting that some chain segments remain locked in a coiled conformation and are unable to crystallize following infiltration at high temperatures. In a similar vein, Xi et al. also reported the critical role of polymer−pore wall interactions in defining the conformational state of the confined polymer chains. Studies performed on poly(p-phenylenevinylene) derivatives (DDMAPPV) bearing alkoxy side-chains of different lengths also revealed that the polarity of the pore wall has strong effects on the absorption/emission properties of confined conjugated polymers [156].

Polymer infiltration followed by chemical transformations can be exploited as a route to anchor presynthesized macromolecular building blocks into mesopores. Kruk et al. applied the "click chemistry" concept, that is, azide−alkyne cycloaddition, to covalently graft polymer chains to the surface of ordered mesoporous silica [157]. The surface of the silica was modified with aminopropyl groups that were converted to propargyl-bearing groups through a reaction with 4-pentynoyl chloride. The "clickable" mesopores were then reacted with azide-terminated polymers [158] including poly-(methyl methacrylate) (PMMA) and oligo(ethylene glycol). The combined infiltration and "grafting-to" strategy enabled the formation of covalently anchored uniform polymer films of thickness up to about 2 nm without any appreciable pore blocking, even for polymer loadings close to 25 wt.%. As expected, the infiltration and grafting process was affected by the molecular weight of the polymeric building blocks owing to an increasing steric hindrance in the case of infiltration of larger macromolecules. Similar experiments performed on higher-molecular-weight PMMA resulted in polymer loading of 18 wt.% and polymer film thickness of ~ 0.8 nm. Infiltration of macromolecular building blocks into mesoporous materials not only involves the use of linear polymers, mesopore-infiltrated dendrimers were also used to confer interesting functional properties to hybrid porous frameworks. Morán et al. reported a novel type of redox-active materials constituted of mesoporous silica hosts containing electroactive dendrimers within the ordered channels. The infiltrated redox-active functional units corresponded to poly(propyleneimine) dendrimers containing 4, 8, and 64 amidoferrocenyl moieties in MCM-41 mesoporous matrices (pore size ~ 3 nm) [159]. Prior to infiltrating the dendrimer, MCM-41 samples were treated with Me_2SiCl_2 in order to decrease the population of free Si-OH in the outer surface of the mesoporous samples provided that the silane anchoring reaction on external Si-OH groups occurs more rapidly than those located in the inner environment of the mesopore [160]. Incorporation of redox dendrimers (generations 1, 2, and 3) into the mesoporous material was achieved by refluxing a CH_2Cl_2 solution of the corresponding dendrimer in the presence of a pretreated sample of MCM-41 during

5 hours—12 hours. The integrity of the redox dendrimer within the mesopore was confirmed by IR and NMR spectroscopy, thus suggesting the formation of a stable dendrimer—matrix complex in which hydrogen bond interactions, CONH · · · OH—Si, may play a determinant role. X-ray diffraction and nitrogen adsorption isotherms confirmed the full occupancy of the channels by the smallest dendrimer, whereas less effective infiltration was observed in the case of bulkier dendrimers due to the emergence of pronounced steric hindrance (dendrimer loading for G1, G2, and G3 corresponded to 34, 15, and 8 wt.%, respectively). One remarkable feature of these hybrid materials is that the ferrocenyl units in the guest dendrimers are easily accessible to electrochemical oxidation as observed by cyclic voltammetry and differential pulse voltammetry experiments. Careful analysis of the electrochemical data revealed that the majority of G3 dendrimers were located on the outer part of the mesoporous matrix, whereas G2 dendrimers were located inside the MCM-41 channels as well as out of the mesoporous material. On the other hand, electrochemical data indicated that the whole population of G1 dendrimers was entirely located inside the mesopores. Electrochemical experiments showed that upon infiltrating the electroactive G1 dendrimers inside the mesopores, a more positive redox potential was obtained. This observation can be explained considering the decrease of effective electron density on the redox centers as a result of the binding of the amido-ferrocenyl moieties to the silanol groups inside the mesopores, rendering oxidation more difficult. In a similar vein, the infiltration of nanoparticle-loaded dendrimers into mesoporous materials has recently introduced a new strategy to prepare hybrid nanostructured materials displaying tailored catalytic properties [161].

Dendrimers represent versatile building blocks to synthesize very small metal nanoparticles (diameter ~ 1 nm) in an accurate and reproducible manner [162]. The globular architecture and chemical topology of dendrimers provide not only internal groups for nanoparticle growth upon reduction, but also a shell to prevent aggregation of the as-synthesized nanoparticles. Somorjai and coworkers report the synthesis of ~ 1 nm Rh and Pt nanoparticles in aqueous solution using PAMAM dendrimer templates and the subsequent loading onto SBA-15 mesoporous supports. The infiltration of the NP-loaded dendrimers was accomplished by sonicating the slurry formed by mixing the nanoparticle solution and the SBA-15. The infiltration process was performed in a solution with pH ~ 5. At this pH, the PAMAM dendrimer is positively charged, while the surface of SBA-15 silica is negatively charged (isoelectric point of silica is ~ 2). The strong electrostatic interactions between the NP-loaded dendrimers and the mesopore walls act as a driving force to fill the cavities the macromolecular functional units. Catalytic studies revealed that the hybrid NP-loaded mesostructured substrates were active for ethylene hydrogenation without any pretreatment. This was attributed to the 3D support provided by SBA-15 that may prevent dendrimer collapsing on the nanoparticle surface and blocking of their active sites. Catalytic activity was also demonstrated over the dendrimer encapsulated nanoparticles for the pyrrole hydrogenation reaction.

17.4.7 Nanostructured polymer—inorganic hybrids via surface-initiated polymerization in mesoporous hosts

Grafting of polymer chains onto mesoporous materials through the condensation of infiltrated end-functionalized polymers with reactive surface groups on the mesopore walls may lead low grafting densities because polymer chains have to diffuse against an increasing concentration. In this context, the "grafting from" approach, in which the polymer chains are grown from the initiator-modified mesopore wall [163], is expected to achieve high grafting densities provided that monomer diffusion into the reactive chain end is not significantly hindered by the growing polymer chains [164]. This type of surface-tethered polymeric assemblies is often referred to as "*polymer brushes*" [165−167]. Hybrid materials were synthesized by grafting polymer chains from the surface of ordered mesoporous silica (OMS) particles (mesopore diameter 9−14 nm), via surface-initiated atom transfer radical polymerization (SI-ATRP) [168] of methyl methacrylate or styrene [169]. A systematic study of the molar mass, molar mass distribution and chain-end structure of both the grafted chains grown from the mesoporous silica surface and the free chains produced in solution from an additional free initiator revealed the emergence of confinement effects on the results of the polymerization. The polymerizations of methyl methacrylate and styrene were perfectly controlled in the homogeneous medium via the ATRP mechanism. However, on the other hand, SEC and MALDI-TOF analyses of the growing polymer chains cleaved from the mesoporous particles confirmed the presence of dead chains of low molar mass resulting from termination via disproportionation.

Furthermore, a systematic study investigating the effect of varying the channel length revealed that the proportion of short dead chain decreased upon decreasing the average length of the cylindrical mesopore. This observation suggests a strong influence of the mass transport processes under nanoconfinement on the polymerization control. Kruk and Matyjaszewski [170] also used SI-ATRP polymerization to graft uniform layers of polyacrylonitrile (PAN), poly(2-(dimethylamino)ethyl methacrylate), and polystyrene on concave surfaces of cylindrical mesopores of diameter ~ 10 nm and spherical mesopores of diameter ~ 15 nm gradient (Fig. 17.20). In this work, the grafting process was optimized through the introduction of appropriate amounts of Cu(II) species that acted as a deactivator, allowing them to achieve better control over the nanoconfined polymerization reaction. Gas adsorption isotherms and gel permeation chromatography indicated that the SI-ATRP process resulted in polymer layers of controlled thickness (from several tenths of a nanometer to at least 2 nm) in which the macromolecular building blocks consisted of monodisperse polymer chains of controlled molecular weight. For example, in the case of PAN displaying grafting densities of 0.28 chains/nm^2, the degrees of polymerization (DP) ranged from DP 25 to 70, and the polydispersity indexes (PDI = Mw/Mn) were as low as 1.06−1.07. It is worthwhile to note that the addition of Cu(II) in the reaction mixture not only helps to promote a better control over the polymer film thickness but also mitigates pore blocking. If the surface-grafted polymer layer does not fill the pores completely, the tailored pores in the polymer-silica hybrids are accessible to the transport of different species, and the pore size distributions are similarly narrow as those of the corresponding silica supports [171]. The synthetic versatility and simplicity makes ATRP a powerful technique for designing well-ordered mesoporous organic−inorganic hybrid materials using widely available monomers, like N-isopropylacrylamide or glycidyl methacrylate, in order to attain highly functional nanostructured materials [172]. Recently, Cao and Kruk [173] demonstrated that atom transfer radical polymerization with activators regenerated by electron transfer (ARGET) can be implemented for grafting of polymer brushes from mesoporous supports. ARGET represents a major improvement in the versatility of ATRP, because it can be performed using a closed vial instead of using air-tight glassware (e.g.,: Schlenk tubes) and a vacuum line, and it involves low concentrations of copper catalyst (10−100 ppm vs 1000−10,000 ppm used in normal ATRP). The ARGET-based approach was illustrated on mesoporous SBA-15 silica (pore diameter 14 and 22 nm) modified

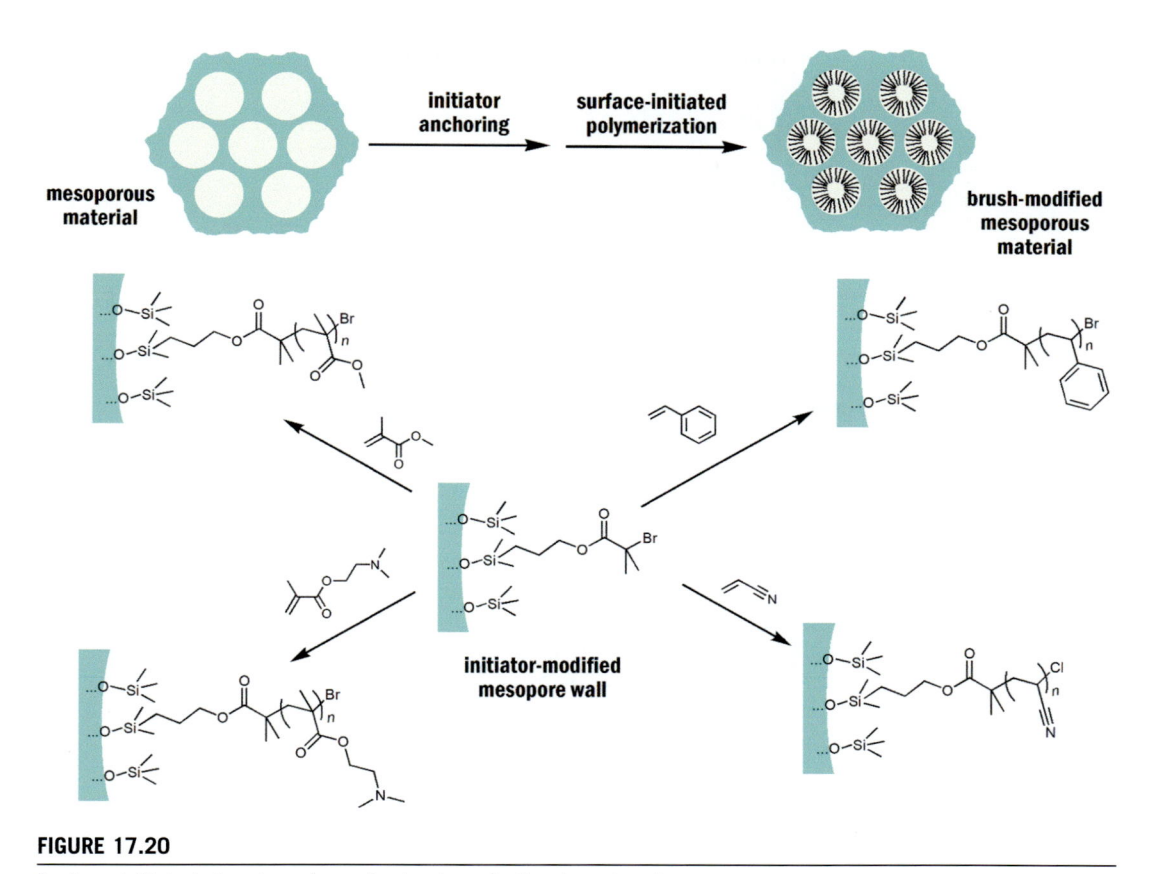

FIGURE 17.20

Surface-initiated atom transfer radical polymerization in ordered mesopores.

with poly(methyl methacrylate) (PMMA) and polystyrene (PS) brushes. Polymer loadings up to 36 wt.% and layer thicknesses of up to at least 2 nm were achieved in the polymerization process carried out in small vials without using a vacuum line. The polymer chains exhibited low polydispersity indexes (PDI $\sim 1.18-1.32$) for polymer loadings up to 29 wt.%, while a higher polydispersity (PDI ~ 2.1) was observed for higher loadings (48 wt.%).

Surface-initiated polymerization from mesoporous supports was further extended to N-carboxyanhydride (NCA) chemistry by Schantz and coworkers in order to create polypeptide-mesoporous hybrids (Fig. 17.21) [174]. Organic—inorganic nanocomposites were readily synthesized through the surface-initiated polymerization of N-carboxyanhydrides from amine-functionalized ordered mesoporous silica.

A combination of experimental techniques verified the formation of the peptide brushes, poly-Z-L-lysine (PZK) and poly-L-alanine (PA), and indicated that much of the polymer layer is formed within the silica mesopores. L-Lys(Z) peptide brushes can be synthesized and deprotected on the solid surface. L-Ala peptide brushes may also be synthesized, but mass spectrometry evidenced the appearance of nanoconfinement effects in polymer growth as the number of alanine units per brush

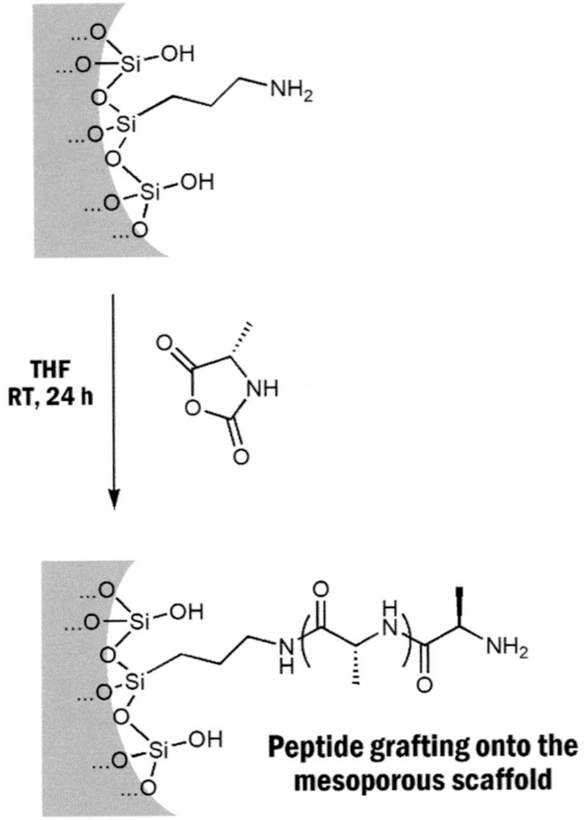

FIGURE 17.21

Scheme describing the grafting of peptides from the mesoporous silica surface.

appears dependent on the silica pore size/topology. Similar studies also indicate that the initial amine loading may strongly affect the filling of the mesoporous framework. In the PA–silica nanocomposites the porosity increases with decreasing amine loading. This has been attributed to an increase in spacing of polymers with equal or similar lengths inside the pores. These results show that NCA polymerization chemistry can be used to synthesize well-defined polypeptide-based composite materials in which the properties of the nanostructured hybrid and the grafted polymer can be tuned by altering the surface initiator loading, pore size, pore topology, and monomer identity.

The use of well-controlled nitroxide-mediated surface-initiated polymerization was also exploited as a route to build up polymer–mesoporous hybrids. Lenarda et al. reported the preparation and use of a TEMPO (2′,2′,6′,6′-tetramethyl-1′-piperidinyloxy)-based derivative, covalently tethered to the internal mesoporous walls of MCM-41 silica, to initiate the controlled radical polymerization of styrene (Fig. 17.22) [175]. The surface derivatization of the pore walls demanded simple sequential postgrafting steps to yield the desired initiator-modified mesoporous material.

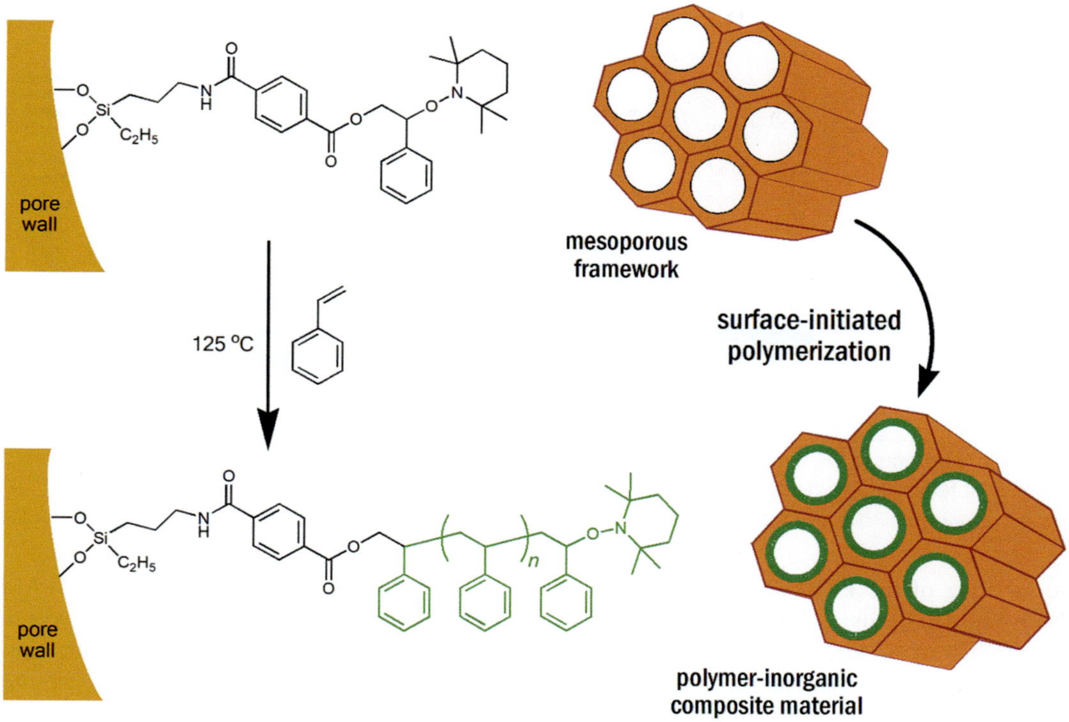

FIGURE 17.22

Controlled radical polymerization of polystyrene inside mesoporous frameworks by using TEMPO-modified pore walls.

The most important feature of these TEMPO-based initiators is the presence of a homolitically unstable alkoxyamine (C−ON) bond that permits, during polymerization, reversible homolysis of the covalent species, followed by monomer insertion and reversible recombination. The presence of inactive chain ends results in a dramatic reduction of the concentration of radical chain ends which, coupled with the inability of the nitroxide free radicals to initiate new chain growth, leads to a lower number of unwanted side reactions (termination, combination or disproportionation). This experimental route enabled the formation of polystyrene−mesoporous silica nanocomposites in which the styrene polymerization occurred inside the MCM-41 channels and the filling of the pores was tuned according to the polymerization conditions without affecting the structural and morphological features of the starting silica.

Surface-initiated polymerization was exploited as a strategy to build up a delivery system based on stimuli-responsive poly(N-isopropylacrylamide) (PNIPAM) brushes [176] synthesized inside a mesostructured matrix via atom transfer radical polymerization (ATRP). The control over drug release in response to the environmental temperature was investigated using ibuprofen (IBU) as a probe molecule. At low temperature, ibuprofen molecules drugs are confined in the pores owing to

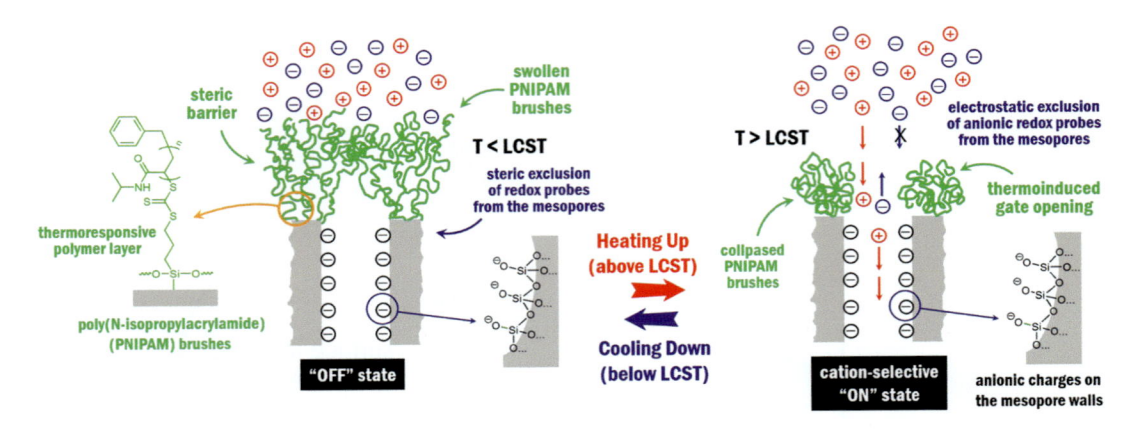

FIGURE 17.23

Schematic depiction of the ionic transport processes taking place in the hybrid polymer—inorganic interfacial assembly at temperatures below and above LCST.

Reproduced with permission from S. Schmidt, S. Alberti, P. Vana, G.J.A.A. Soler-Illia, O. Azzaroni, Chem. Eur. J. 2017, 23, 14500.

the swelling of the PNIPAM brushes and the formation of hydrogen bonding between PNIPAM chains and IBU provided that carboxylic acid groups of IBU could bind to both carbonyl oxygen and nitrogen of N-isopropylacrylamide monomer units via hydrogen bonding. Upon increasing temperature, the polymer chains become hydrophobic in the collapse of hydrogen bonds which in turn leads to the release of drug molecules from the pores. Consequently, this implies that the temperature-driven actuation of the PNIPAM chains can be used for controlled release of drugs [177,178]. In a similar fashion, Schmidt et al. [179] prepared mesoporous films derivatized with poly-(N-isopropylacrylamide) (PNIPAM) brushes via surface-initiated reversible addition—fragmentation chain transfer (SI-RAFT) polymerization (Fig. 17.23)

In this particular example, the combination of the thermoresponsive building blocks with mesoporous silica offers the possibility to create thermoactivated permselective interfaces using concerted functions inside and outside the mesoporous matrix. This singular feature arises from the synergistic combination of the electrostatic characteristics of the silica scaffold and thermocontrolled steric effects introduced by the capping brush layer. These unique transport properties of the hybrid interfacial are observed only in the presence of the cooperative interaction between the "gating" of the PNIPAM layer and the anion entry exclusion exerted by the mesoporous silica surface (Fig. 17.23). In other words, the interplay between the intrinsic acid—base properties of the mesoporous silica scaffold and the thermo-responsive characteristics of the PNIPAM layer leads to a functional assembly with advanced ionic transport properties. The overall system response in this case is a combination of an external stimulus (temperature change) and the intrinsic charge, which give as a result a behavior similar to an AND logical gate: only the appropriate charges will cross the barrier when the opening temperature is met. Other similar systems presenting a logical-like response using orthogonal functions have been created in the last years. Although the field is in its infancy, the impact of creating materials that can execute logical operations adapting to environment or crossed stimuli is enormous in several fields, from intelligent drug delivery to theranostics or remote sensing or actuating [180−182]. Recently, Andrieu-Brunsen and coworkers explored the use of surface-initiated photoiniferter polymerization as a strategy to functionalize mesoporous silica matrices (Fig. 17.24). One of the advantages of this strategy is its synthetic simplicity, requiring minimal reaction components and proceeding by the simple application of UV irradiation. In this way, surface-initiated polymerization activated by UV irradiation also permits spatial and temporal control over the surface functionalization to specify the location, the degree of functionalization as well as the possibility to create surface patterns with specific chemical groups. The researchers

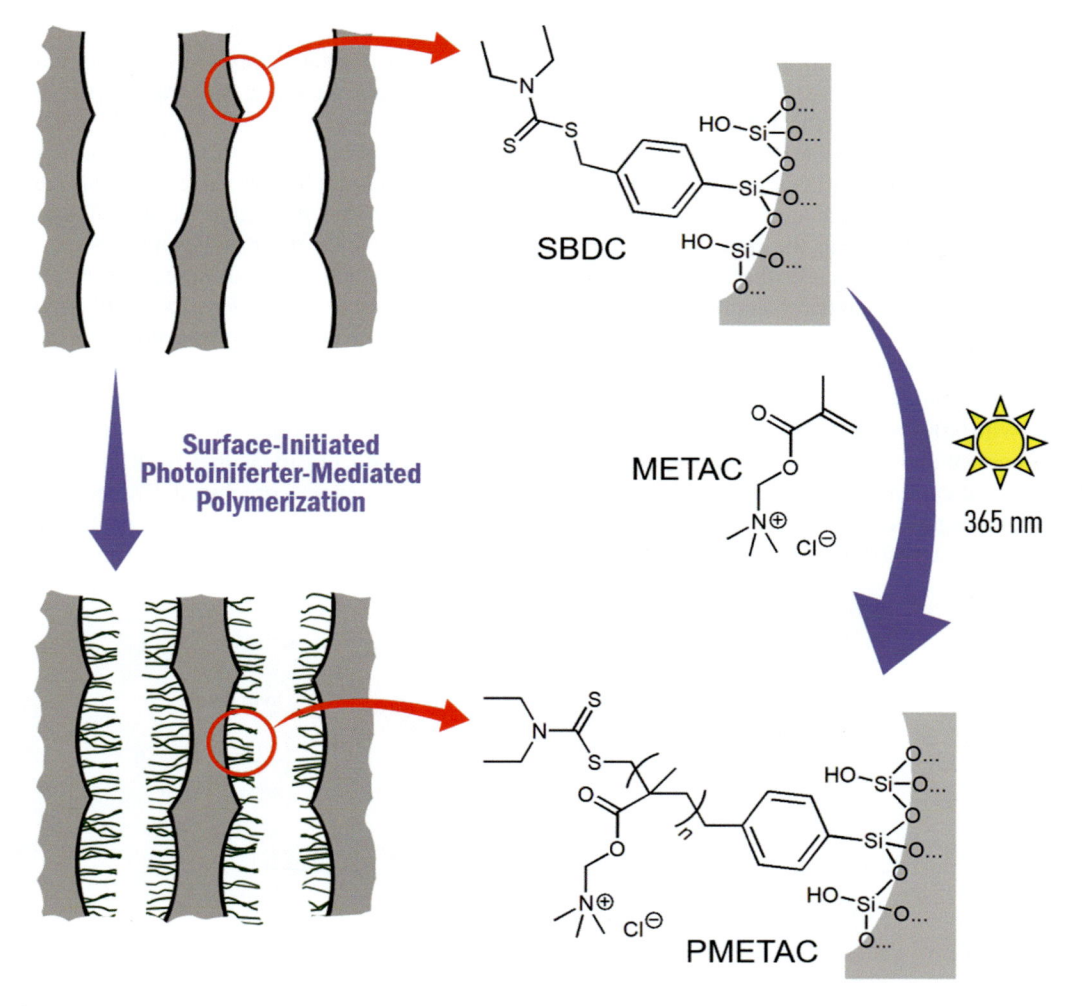

FIGURE 17.24

Schematic representation of the functionalization of mesoporous architectures following an UV-induced iniferter-initiated polymerization approach. SBDC: N,N-(diethylamino)dithiocarbamoyl-benzyl (trimethoxy) silane; METAC: [2-(methacryloyloxy)ethyl] trimethylammonium chloride.

reported the iniferter-initiated polymerization of carboxybetaine methacrylate, (CBMA), a zwitter-ionic monomer, from the interior and exterior surface of mesoporous silica substrates with pores ranging from 2 to 115 nm in diameter [183]. This strategy can also be implemented in a one-step approach through the incorporation of a photoiniferter, N,N-(diethylamino)dithiocarbamoyl-benzyl (trimethoxy)silane (SBDC), in a one-pot procedure through cocondensation with an inorganic precursor combining sol−gel chemistry and evaporation-induced self-assembly (EISA) [184]. This strategy facilitates the preparation of functional mesoporous thin films that are capable of undergoing polymerization upon exposure to UV light in the presence of a monomer. Interestingly, this approach has been also extended to the use of visible light to induce the polymer growth within the mesoporous matrix. By using gold−silver alloy nanoparticles as surface plasmon and thus nanoscale light source, it has been possible to functionalize mesoporous silica substrates with poly(dimethylaminoethyl methacrylate) layers [185].

The generation of charged macromolecular assemblies within nanoconfined mesoporous environments via surface-initiated polymerizations has gained increasing interest in different fields, such as molecular separation, dosing, or drug delivery. In particular, the molecular design of hybrid interfaces discriminating the transport of ionic species through which the passage of ions can be triggered or inhibited under the influence of an external stimulus has received increasing attention from the materials science community. Along these lines, Calvo et al. [186] described the creation of hybrid organic−inorganic assemblies displaying unique pH-dependent ionic transport properties originating from the combination of zwitterionic poly(methacryloyl-L-lysine) (PML) brushes and silica mesoporous matrices. In principle, the presence of pore wall-confined silanolate groups (Si-O^-), with pKa ~ 2, conferred permselective properties to the mesoporous silica thin films. Transport studies revealed that at pH significantly above the pKa of silica, that is, 8 and 5, the transport of cations was feasible whereas that free diffusion of anions was precluded. The mesopore walls are negatively charged and permselectively repel the transport of anions while at the same time allow the diffusion of cations (Fig. 17.25). However, under acidic conditions the transport of

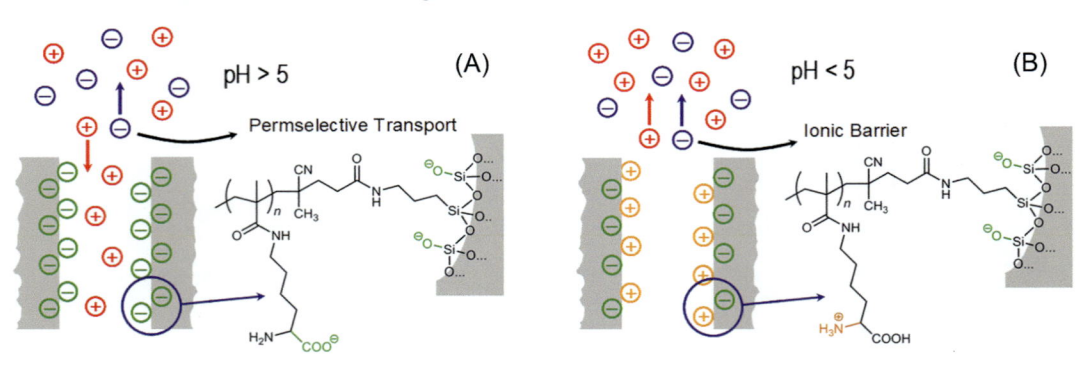

FIGURE 17.25

Schematic illustration of the ionic transport processes taking place in the mesoporous silica film modified with the zwitterionic polymer brushes at different pH values: (A) pH >5, permselective transport of cations and (B) pH <5, ionic barrier (exclusion of ionic species).

Reproduced with permission from A. Calvo, B. Yameen, F.J. Williams, G.J.A.A. Soler-Illia, O. Azzaroni. J. Am. Chem. Soc. 2009, 131, 10866.

both cationic and anionic species was completely inhibited (Fig. 17.25). The isoelectric point (pI) of the zwitterionic brush is ~5, and as such, it should be expected that at pH >5 the nanopore was negatively charged, that is, cation-permselective, and at pH <5 the same pore walls were positively charged, that is, anion-permselective. Intriguingly, PML brush-modified mesoporous film acted as an ionic barrier at pH <5 instead of behaving as an anion permselective membrane. The explanation for this particular behavior relies on the actual understanding of the physicochemical changes taking in the pore environment rather than merely analyzing the pH-induced changes occurring in the monomer units of the zwitterionic brush. In the pore walls the grafted polyzwitterionic chains coexist with silanol sites, which are negatively charged at pH >2. At pH >5 both the zwitterionic moieties and the SiO^- groups bear negative charges. As a result, the hybrid mesoporous film shows a remarkable cation-permselective behavior. Then, at pH <5 the zwitterionic monomers bear positive charges while the silanol groups are still negatively charged. This experimental scenario leads to the emergence of a zwitterionic, "bipolarly charged" mesopore in the $pI^{brush} > pH > pKa^{silica}$ range. In contrast to the typical Donnan exclusion phenomenon which refers to confined negative charges repelling anions and confined positive charges repelling cations, the confinement of both negative and positive charges leads to a very particular exclusion condition.

Within this framework, the modification of mesoporous silica thin films with phosphate-bearing polyprotic polymer brushes via surface-initiated polymerization of 2-(methacryloyloxy)ethyl phosphate (MEP) means to create hybrid interfacial architectures displaying arrays of stimuli responsive mesochannels (Fig. 17.26) [187]. In these hybrid platforms the ionic transport properties can be finely tuned in the presence of protons and calcium ions. In the case of protons, the electrostatic characteristics of the thermodynamically controlled environments arising from the multiple protonation states of phosphate groups are responsible for tuning the ionic transport of anionic and cationic species across the mesoporous framework over a wide range of pH. Increasing pH from 4 to 8 leads to a significant increase in (anion) permselectivity and (cation) preconcentration, thus reflecting the ability of the PMEP brush-modified mesopores to act as a selective "electrostatic nanovalve" precluding and boosting the anionic and cationic transport, respectively (Fig. 17.26). On the other hand, the complexation/chelation of phosphate groups with Ca^{2+} ions also allowed the generation of gate-like hybrid ensembles. These results demonstrate that the hybrid interface at pH 8 reversibly switches from low to high anionic conductance states depending on the formation of stable Ca^{2+}-phosphate complexes in the mesochannels. Hence, at high pH values, MEP-modified pores are strongly permselective precluding the transport of anionic probes; however, once Ca^{2+} is bound to the monomer units, the ionic mesochannel reaches the open state (Fig. 17.26).

Surface-initiated polymerization has also been employed to successfully integrate photoactive polymer brushes [188–191] on mesoporous inorganic materials. Among various types of stimuli that may be utilized to trigger actions, light-sensitivity is an attractive phenomenon for developing advanced hybrid materials capable of precise external modulation of chemical changes in mesoporous materials. Light-driven chemical changes therefore represent an appealing strategy; indeed, light can approach materials in a noncontacting manner also permitting various types of modulations in terms of intensity (flux) and energy (wavelength). In the case of photolabile "caged" compounds their chemical functionalities are blocked because of the presence of a covalently bound chromophore [192]. Light irradiation removes the cage and activates the protected functional groups. This is a particularly flexible approach, since a good number of photoremovable groups are

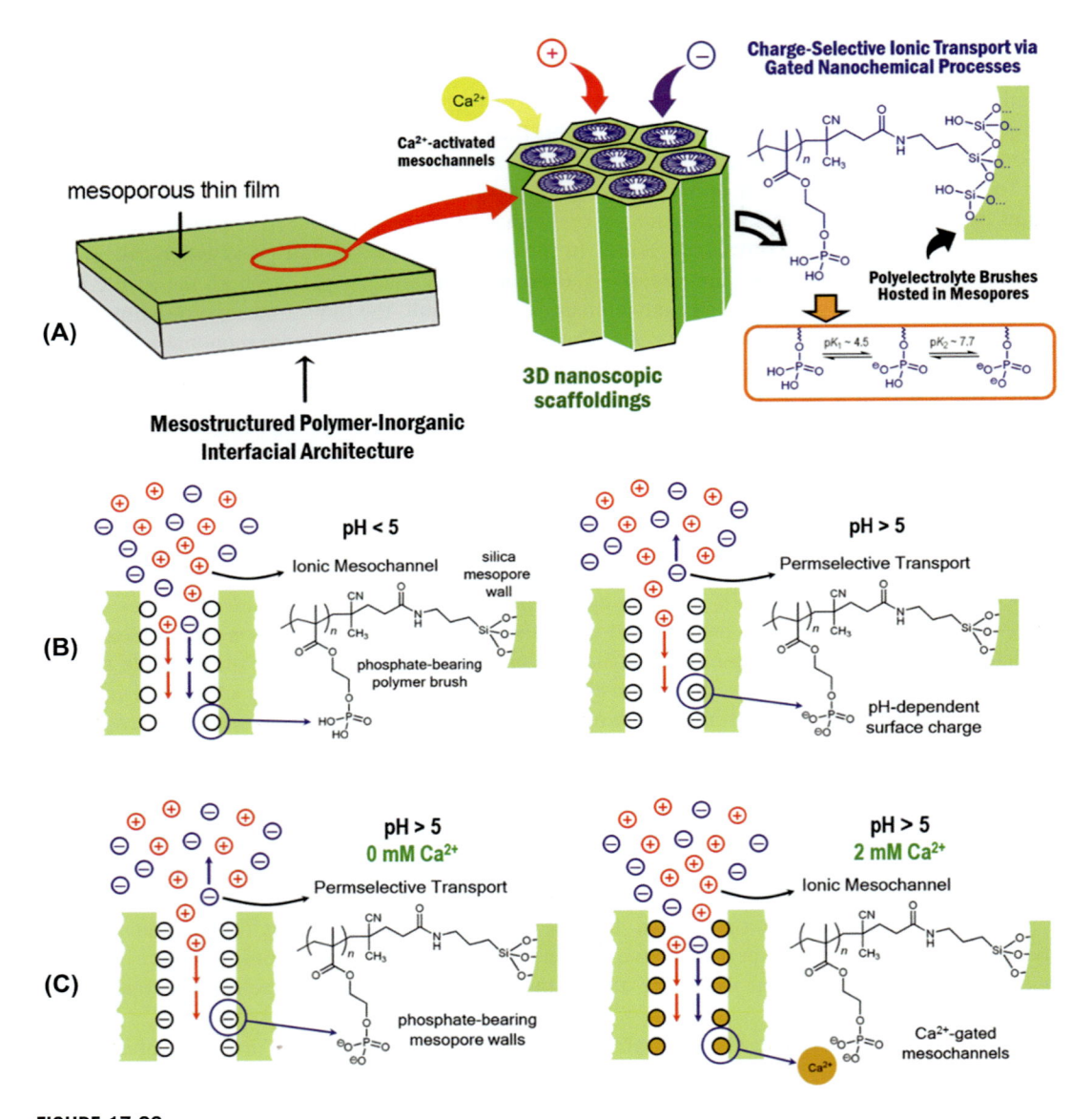

FIGURE 17.26

(A) Schematic depiction of the mesoporous film with poly/2-(methacryloyloxy)ethyl phosphate (PMEP) brushes. (B) Schematic depiction of the ionic transport processes taking place in the hybrid polymer—inorganic interfacial assembly at different pH's: (a) pH <5, ionic mesochannel (no exclusion of ionic species); (b) pH >5, permselective transport of cations. (C) Schematic depiction of the ionic transport processes taking place in the PMEP-modified mesopores at pH >5: (a) no added Ca^{2+} ions, permselective transport of cations; (b) 2 mM Ca^{2+}, chelation-induced formation of ionic mesochannels (no exclusion of ionic species).

Reproduced with permission from A. Brunsen, C. Díaz, L.I. Pietrasanta, B. Yameen, M. Ceolín, G.J.A.A. Soler-Illia, et al., Langmuir 28 (2012) 3583.

known that can be combined with the different organic functional groups and applied to generate materials displaying phototunable chemical states. In recent years different light responsive strategies have been applied as "triggers" for uncapping the pores and facilitating the molecular transport through mesoporous materials. In this sense, Brunsen et al. [193] described the integration of "caged" polymer brushes into mesoporous oxide thin films in order to create photoactive hybrid polymer—inorganic assemblies displaying light-activated gating and permselective transport of ionic species through 3D nanoscopic scaffoldings (Fig. 17.27).

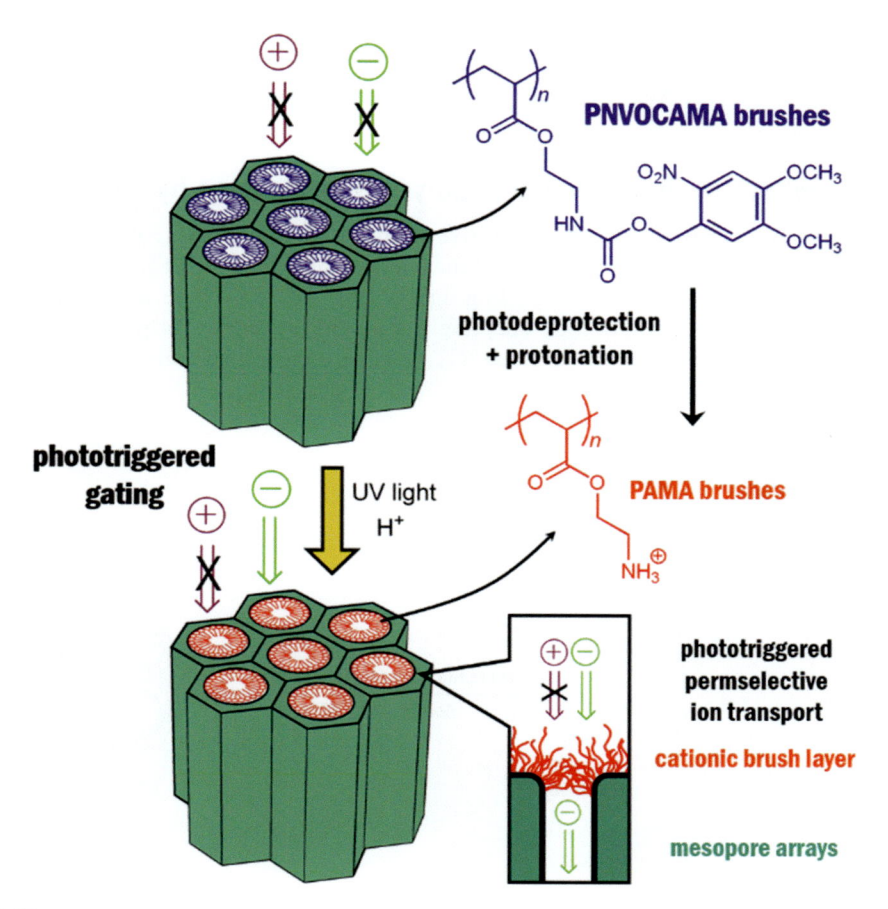

FIGURE 17.27

Schematic representation describing the modification of mesoporous silica films with photolabile polymer brushes and their subsequent "uncaging" to build up a permselective polycationic barrier in the outer region of the film.

Reproduced with permission from A. Brunsen, J. Cui, M. Ceolín, A. del Campo, G.J.A.A. Soler-Illia, O. Azzaroni, Chem. Comm. 48 (2012) 1422–1424.

Modification of the mesoporous film with photolabile polymer brushes was accomplished by reacting the surface amino groups with 2-bromoisobutyryl bromide in order to covalently anchor the polymerization initiators. The polymer brush growth took place via surface-initiated atom transfer radical polymerization of 2-[(4,5-dimethoxy-2-nitrobenzoxy) carbonyl] aminoethyl methacrylate (NVOCAMA) monomers in the presence of the adequate solvent and catalyst. This process led to the surface modification of the mesoporous film with covalently anchored PNVOCAMA brushes. Along these lines the authors were able to develop a hybrid mesoporous platform exhibiting light-activated permselectivity. In this scenario, the permselective ion transport is triggered when (upon irradiation) mesopores are "uncapped" and polymer brushes are massively switched between a "collapsed" hydrophobic state in which hydrated ions cannot enter the mesopores and a "swollen" hydrophilic state in which the cationic nature of the outer layer governs the transport through the vertically-oriented nanochannels (Fig. 17.27).

17.4.8 One-pot synthesis of functional polymer-mesoporous hybrids

Tolbert and coworkers synthesized optically active nanostructured composite materials by using an amphiphilic semiconducting polymer, a poly(phenylene ethynylene) (PPE), and a conventional ammonium surfactant as the structure-directing agents (Fig. 17.28). These hybrid materials were synthesized in basic aqueous solution by adapting a standard procedure for preparing mesoporous silica using cetyl trimethylammonium bromide (CTAB) as the surfactant and tetraethyl orthosilicate

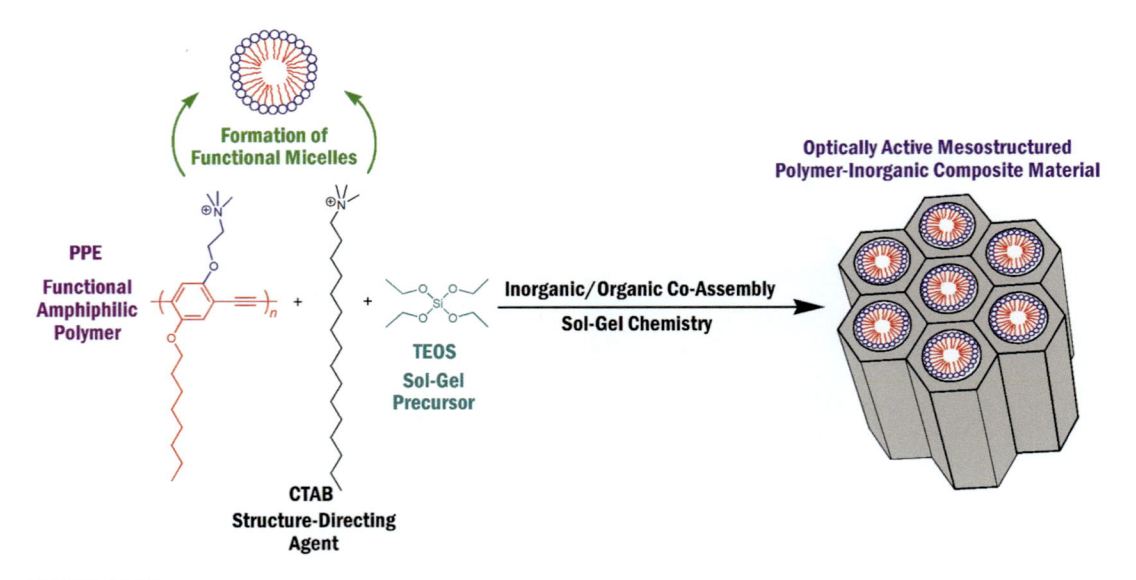

FIGURE 17.28

Synthesis of optically active nanostructured composite materials by using amphiphilic semiconducting polymers, a conventional surfactant as the structure-directing agents and tetraethyl orthosilicate as the silica source.

(TEOS) as the silica source. These authors replaced 1−5 wt.% of the CTAB in the formulation with PPE. The presence of the CTAB in addition to the PPE was necessary because the solubility of the polymer was insufficient to drive inorganic/organic coassembly. The ratio of CTAB/PPE is dictated by the solubility of the polymer in the aqueous solution. At the maximum PPE concentration (∼5% relative to CTAB) some polymer precipitation out of the solution might take place upon addition of TEOS, thus leading to the formation of composite materials with irreproducible amounts of polymer. However, solutions containing lower PPE concentrations (1%) yielded composite samples with consistent amounts of polymer incorporation (∼8%) with essentially no PPE left in solution after composite precipitation. The experimental protocol to prepare the sol−gel mixture indicates that PPE is added after the CTAB in order to facilitate the solubilization of the polymer. Thereafter, this mixture is stirred under mild heating (50°C) to completely dissolve the polymer. TEOS was then added, and the mixture was allowed to continue stirring for another 3 hours at room temperature. The resulting precipitate was then filtered and dried to achieve mesoporous silica incorporating semiconducting polymers into the 2D hexagonal architecture [194].

Frey et al. also reported the one-pot deposition of conjugated polymer-incorporated mesostructured metal oxide films, with control over charge and energy transfer [195]. The conjugated polymer dissolved in xylene is added dropwise into the polar precursor solution including the metal oxide precursor species and the block copolymer acting as structure directing agent (Fig. 17.29). The hydrophobicity and relatively low volatility of the xylene cosolvent drives the conjugated polymers into the hydrophobic domains of the self-organizing block copolymer mesophases that template the metal oxide scaffold.

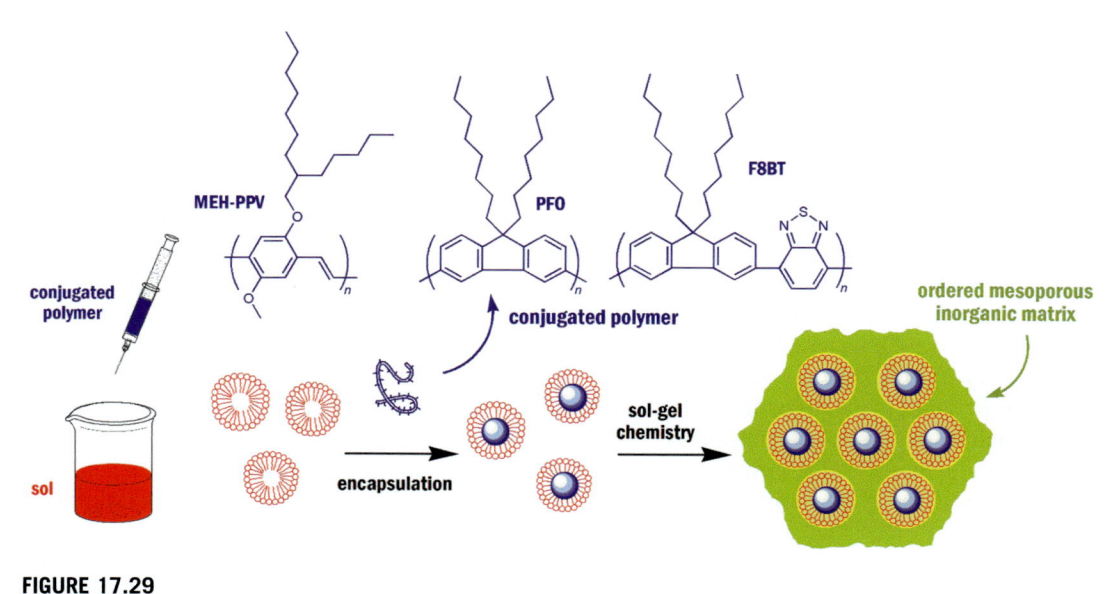

FIGURE 17.29

Schematic illustration of the one-pot deposition of conjugated polymer-incorporated mesostructured metal oxide films.

This experimental strategy enables the controlled incorporation of conjugated polymers into the organic domains of block copolymer−templated mesostructured metal oxides, thus leading to the deposition of conjugated polymer-incorporated 2D-hexagonal and 3D-cubic silica and titania films from aqueous solutions. Spatially locating the conjugated polymer guests in the organic domains of the hybrid mesostructure suppresses energy transfer between polymer chains in adjacent micelles. Hence, incorporation of red- and blue-emitting polymers in separate micelles resulted in simultaneous red and blue emission, that is, white light generation. The efficacy of the conjugated polymer-incorporated metal oxide films for optoelectronic devices was demonstrated by integrating the white-emitting 3D-cubic silica film, which supports carrier transport along the continuous through-film conjugated polymer pathways, as the active layer in white light emitting devices. The experimental protocol to prepare these highly functional films is quite straightforward. The conjugated polymers are first dissolved in xylene and then dropwise added into aqueous or aqueous/ethanolic solutions including a block copolymer surfactant and a metal oxide precursor species. Depending on surfactant concentration and type of metal-oxide precursor, conjugated polymer-incorporated 2D-hexagonal and 3D-cubic silica and titania films can be achieved through well-known evaporation-induced self-assembly processes. A similar strategy was also explored to build up mesostructured nanocomposite films by introducing presynthesized semiconducting polymers, such as blue-emitting poly(9,9-dioctylfluorenyl-2,7-diyl) (PFO), green-emitting poly(9,9 dioctylfluorenyl-2,7-diyl)-co-1,4-benzo-(2,19,3)-thiadiazole (F8BT), or red-emitting poly[2-methoxy-5(29-ethyl-hexyloxy)-1,4 phenylenevinylene] (MEHPPV), into a tetrahydrofuran (THF)−water homogeneous sol solution containing silica precursor species and a surface-active agent. Dovgolevsky et al. showed that depending on the concentration of the surface active agent, it was possible to prepare polymer−inorganic structured films with three different types of mesostructural order: (1) a 2D hexagonal mesophase silica with conjugated polymer guest species incorporated within the hydrophobic cylinders organized in domains aligned parallel to the substrate surface plane; (2) a lamellar mesophase silica with the layers oriented parallel to the substrate surface and the conjugated polymer guest species incorporated in the hydrophobic layers; or (3) an apparent intermediate phase consisting of a mixture of the hexagonal and lamellar phases in addition to worm-like aggregates with no appreciable orientational order. In particular, the continuous through-film conductive pathway provided by the intermediate phase has allowed the integration of ordered semiconducting polymer−silica nanocomposites into optoelectronic device architectures [196]. Along these lines, Zink and coworkers introduced an alternative method to locate macromolecular luminophores into mesostructured films through a one-pot procedure (Fig. 17.30). This method relies on the use of a water-soluble organic conducting polymer containing sulfonate groups, poly-(2,5-methoxy-propyloxysulfonate)phenylene vinylene (MPSPPV), that has a chain length of 1900 units. The polymer is dissolved in the initial sol and, as the film is formed, the organic backbone is incorporated in the organic region of the film while the sulfonate groups reside in the ionic interface region together with the positive alkylammonium headgroups of the surfactant.

The orientation of the polymers within the matrix was studied by fluorescence polarization. These studies revealed that polymer chains were preferentially oriented within the mesostructured matrix. This was explained by considering that most of the polymer chains are longer than the width of a cylindrical micelle and, consequently, the chains are forced to run parallel to the rods within which they are confined [197]. Sol−gel polymerization of tetraethoxysilane in the presence of amphiphilic phthalocyanine polymer has been reported to produce organic−inorganic composites

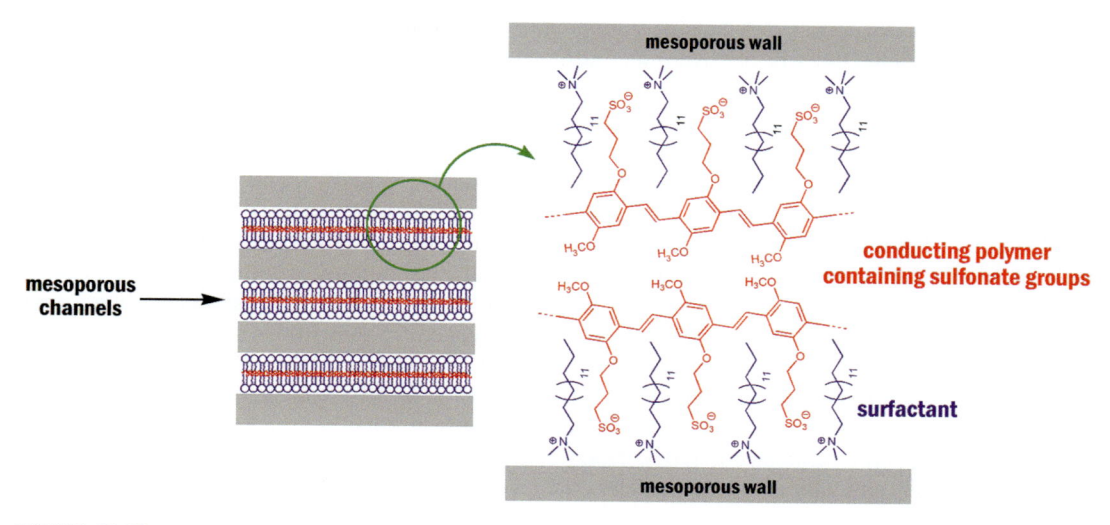

FIGURE 17.30

Scheme of the sol—gel film incorporating the conducting polymer within the mesoporous matrix. The figure depicts an ionic interfacial region in which the sulfonate groups of the functional polymer reside together with the positive alkylammonium headgroups of the surfactant.

with the rod-like phthalocaynine polymers incorporated within ordered hexagonal channels. The amphiphilic rod-like phthalocyanine polymer acted as a structure-directing agent for the sol—gel polymerization of an inorganic source. The inorganic framework served to isolate the rod-like conductive polymers which may lead to directional electronic conductivity through highly conductive phthalocyanine polymers within an ordered nanoscopic channel [198]. Finally, Fujiwara et al. reported the preparation of functional hybrid polymer—mesoporous silica composite materials by simply mixing hexadecyltrimethylammonium bromide, sulfonated polymers, such as Nafion or poly (sodium 4-styrenesulfonate), and TEOS in alkaline aqueous solution [199]. XRD characterization revealed that the crystallinity of hexagonal structure of composite materials was not affected by the incorporation of the polymers. However, when an excess amount of Nafion was mixed in the sol, a significant loss of the acid sites of Nafion was observed. This indicates that the sulfonated polymer might be incorporated in the wall framework of mesoporous silica matrix in close resemblance to a "framework polymer composite" of mesoporous silica.

17.5 Spatially-addressing macromolecular functional units on mesoporous supports—tailoring "inner" and "outer" chemistries in hybrid nanostructured assemblies

The bottom-up fabrication of (multi)functional materials requires a thorough control of the chemistry of the building blocks across multiple length scales. In a similar vein, molecular design of hybrid mesoporous materials demands new methodologies for the incorporation of functional units into mesoscale architectures in predetermined arrangements, thus enabling the formation of hierarchical materials in which *chemistry* and *topology* define the functional features of the composite material. Large efforts in contemporary materials science are being aimed at the processing of ordered mesoporous architectures, using low-cost and up-scalable strategies, in which functional macromolecular units are deliberately placed in specifically spatially separated regions of the mesostructure. The very possibility of providing new avenues to site-selectively incorporating functional building blocks into mesostructured platforms marks an important advance in the development of highly functional hybrid materials. Building up mesoporous materials in which shape, order, and spatial distribution of chemical groups control function and utility might be interesting for the creation of locally delimited nanoscopic reaction spaces in thin films or the construction of chemoresponsive mesostructured optical waveguides [93].

Benchmark examples of spatially controlled functionalization of mesostructured materials with macromolecular building blocks rely on the selective chemical derivatization of the inner and outer environment of the mesoporous framework [200]. Hong et al. [201] described the functionalization of the exterior surface of mesoporous silica particles via reversible addition−fragmentation chain transfer (RAFT) polymerization without affecting the mesoporous structure. Mesoporous silica particles with diameter of ~ 100 nm and mesopores ~ 2 nm in diameter were synthesized using cetyltrimethylammonium bromide (CTAB) as surfactant. Then the mesoporous material filled with CTAB was reacted with 5,6-epoxyhexyltriethoxysilane (EHTES) under reflux in toluene, forming EHTES-coated particles. Since the mesopores of the silica nanoparticles were filled with CTAB, EHTES was only grafted on the exterior surface of mesoporous particles. The presence of the surfactant in the mesopores precludes the functionalization on the inner mesoporous environment of the particle. After EHTES-coated particles were refluxed in a methanol solution of hydrochloric acid, the CTAB inside the mesopores was removed and the epoxyhexyl groups on the exterior surface of particle were converted into 5,6-dihydroxyhexyl units. This experimental protocol facilitates the unclogging of the mesopores, which in turn leads to a spatially-controlled derivatization of the outer surface with epoxy groups. Thereafter, the RAFT agent (S-1-dodecyl-S(α,α'-dimethyl-α''-acetic acid)-trithiocarbonate) was attached onto the mesoporous particle via esterification with the hydroxyl units on the surface catalyzed by DCC, forming mesoporous silica particles with RAFT agents on the exterior surface. Finally, pH-sensitive poly(acrylic acid) (PAA) chains were grown from the exterior surface of mesoporous particle via surface RAFT polymerization of acrylic acid using AIBN as the initiator. This strategy enabled the straightforward preparation core−shell mesoporous particles that can act as "active" nanocontainers for guest molecules provided that the pH-responsive PAA nanoshell can be reversibly opened and closed when triggered by pH change, thus resembling a nanovalve that regulates the loading and release of guest molecules from the mesoporous core. This strategy exploiting the synthetic versatility of RAFT polymerization and the structural features of mesoporous silica nanoparticles has been extended recently to diverse "responsive" core−shell systems [202].

In a similar vein, Schantz et al. demonstrated that amines and thiols can be selectively grafted to the exterior surface and within the pores of SBA-15 using a postsynthetic approach, in which external surface functionalization was achieved by reacting as-made SBA-15 with the desired organosilane prior to removing the template. The site-selective functionalization of the external mesoporous surface with amine groups was used for the grafting of large amounts of poly-Z-L-lysine in order to create novel hybrid materials [203].

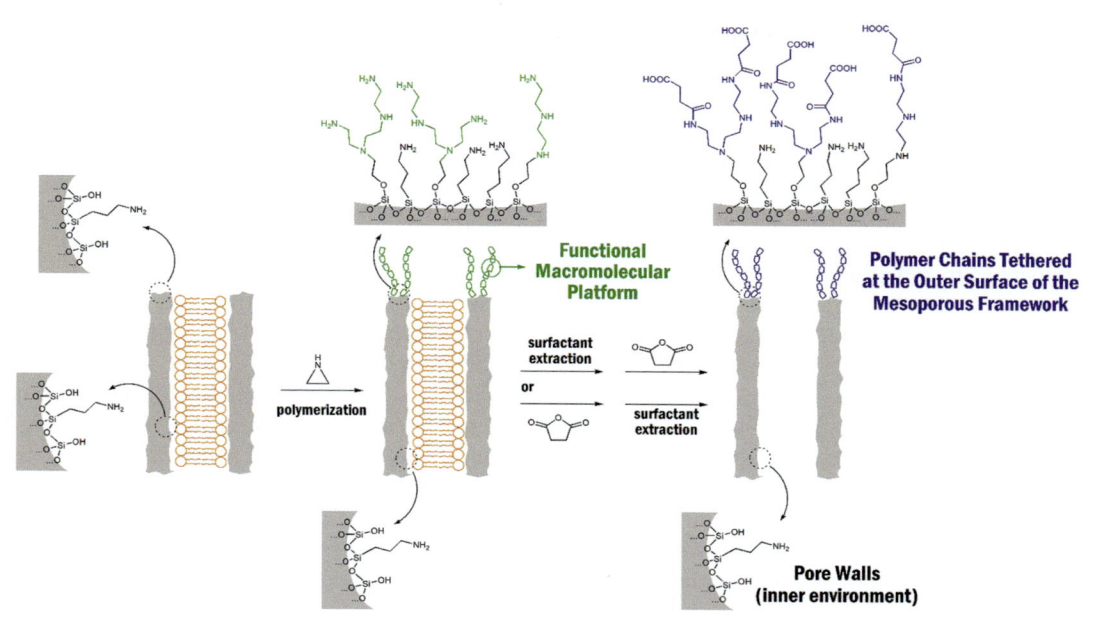

FIGURE 17.31

Schematic illustration showing the different steps involved in the functionalization of mesoporous silica with amino-terminated hyperbranched polymers preferentially located on the outer surface of the porous matrix and their subsequent conversion to carboxylic acid groups.

The spatially-controlled functionalization of mesoporous materials via preferential growth of hyperbranched poly(ethylene imine) (PEI) on the outer surface of the mesoporous silica was also demonstrated by Linden and coworkers [204]. As described above, the monomer aziridine was polymerized in the presence of mesoporous silica particles in which the surfactant used as structure-directing agent has not yet been extracted from the mesopores. This experimental strategy enables the facile incorporation of carboxylic acid functionalities preferentially to the outer part of the PEI layer by succinylation of the terminal amino groups. The relative concentration of carboxylic acid to amino groups in the PEI layer can be rationally controlled and, hence, a fine-tuning of the surface charge of the particles for a given application can be achieved. In addition, these authors demonstrated that the inner environment of the mesoporous silica core can be further functionalized with monolayer assemblies, for example, hexamethyldisilazane, even after growing the polymer shell from the outer surface (Fig. 17.31).

17.6 Conclusions

In this chapter, we have shown a broad variety of approaches to create hybrid nanoarchitected organic—inorganic mesostructured materials. From all the works presented here, it is clear that the combination of complementary properties from organic and inorganic worlds leads to integrated molecular systems exhibiting wide chemical flexibility, good stability, and excellent processability. As in many cases of organic/supramolecular/polymeric materials comprising many components, the functions and structures of hybrid organic–inorganic mesostructures are in general not just the sum of each component but can provide quite different, novel functional features from an emergent and collaborative behavior. We have provided a broad description of the current synthetic strategies as well as a discussion of the main protocols that are required to rationally design organic–inorganic mesoporous materials. The primary aim of this chapter was to discuss and describe the multiple approaches to functional nanoarchitected hybrids by combining ordered mesoporous materials and organic, supramolecular, and/or macromolecular building blocks. Central to this idea is the ability of developing orthogonal synthetic methods that permit reproducible location of functional domains in predetermined regions in space, either in bulk objects, colloids, or thin films. An adequate harnessing of functional location and their intercommunication will permit creation of autonomous nanosystems that sense their environment, adapt to external conditions, or are able to process signals in order to execute a preprogrammed task. Hopefully, this work will engender interest in acquiring a basic understanding and stimulate further explorations in nanoarchitectonics of hybrid mesostructured materials. In this context, the new horizons provided by hybrid organic–inorganic mesoporous materials appear very wide and the future offers the prospect of many developments in the adventure of creating novel matter, as chemists show an increased mastery in the construction of molecular and supramolecular assemblies in mesoporous environments, with responsive and adaptive behavior. This is an essential aspect as the conversion of organic/supramolecular/macromolecular functions, built-in within the inorganic framework, into macroscopic properties is a key element in materials nanoarchitectonics. These new tools are essential to rationally build the foundations of soft nanorobots and swarming matter [205]. With this in mind, we hope that this chapter can trigger a cascade of new, refreshing ideas in hybrid mesoporous materials as well as assist in the rational design of bioinspited functional materials based on nanoconfined organic, supramolecular, nanostructured or polymeric assemblies.

References

[1] G.J.A.A. Soler-Illia, O. Azzaroni, Chem. Soc. Rev. 40 (2011) 1107.
[2] L. Nicole, C. Laberty-Robert, L. Rozes, C. Sanchez, Nanoscale 6 (2014) 6267.
[3] D. Grosso, F. Ribot, C. Boissiere, C. Sanchez, Chem. Soc. Rev. 40 (2011) 829.
[4] G.J.A.A. Soler-Illia, C. Sanchez, B. Lebeau, J. Patarin, Chem. Rev. 102 (2002) 4093.
[5] M. Faustini, D. Grosso, C. Boissière, R. Backov, C. Sanchez, J. Sol–gel Sci. Technol. 70 (2014) 216.
[6] R. Backov, Soft Matter 2 (2006) 452.
[7] Q. Ji, M. Miyahara, J.P. Hill, S. Acharya, A. Vinu, S.B. Yoon, et al., J. Am. Chem. Soc. 130 (2008) 2376.
[8] K. Ariga, Q. Ji, J.P. Hill, A. Vinu, Soft Matter 5 (2009) 3562.
[9] K. Ariga, A. Vinu, Y. Yamauchi, Q. Ji, J.P. Hill, Bull. Chem. Soc. Jpn. 85 (2012) 1.
[10] K.C.-W. Wu, Y. Yamauchi, J. Mater. Chem. 22 (2012) 1251.
[11] M. Antonietti, G.A. Ozin, Chem. Euro. J. 10 (2004) 28.
[12] A. Vinu, T. Mori, K. Ariga, Sci. Tech. Adv. Mater. 7 (2006) 753.
[13] K. Ariga, Y.M. Lvov, K. Kawakami, Q. Ji, J.P. Hill, Adv. Drug. Deliv. Rev. 63 (2011) 762.

[14] A.B. Descalzo, R. Martínez-Máñez, F. Sancenón, K. Hoffmann, K. Rurack, Angew. Chem. Int. Ed. 45 (2006) 5924.

[15] S. Saha, K.C.-F. Leung, T.D. Nguyen, J.F. Stoddart, J.I. Zink, Adv. Funct. Mater. 17 (2007) 685.

[16] G.J.A.A. Soler-Illia, P.C. Angelomé, M.C. Fuertes, A. Calvo, A. Wolosiuk, A. Zelcer, et al., J. Sol−gel Sci. Technol. 57 (2011) 299.

[17] S. Mann, Nat. Mater. 8 (2009) 781.

[18] N.D. Petkovich, A. Stein, Chem. Soc. Rev. 42 (2013) 3721.

[19] A. Mehdi, C. Reye, R. Corriu, Chem. Soc. Rev. 40 (2011) 563.

[20] Z. Li, J.C. Barnes, A. Bosoy, J.F. Stoddart, J.I. Zink, Chem. Soc. Rev. 41 (2012) 2590.

[21] Q. He, J. Shi, J. Mater. Chem. 21 (2011) 5845.

[22] C. Sanchez, B. Julián, P. Belleville, M. Popall, J. Mater. Chem. 15 (2005) 3559.

[23] G.J.A.A. Soler-Illia, P. Innocenzi, Chem. Euro. J. 12 (2006) 4478.

[24] Q. Tang, P.C. Angelome, G.J.A.A. Soler-Illia, M. Müller, Phys. Chem. Chem. Phys. 19 (2017) 28249.

[25] N. Tarutani, Y. Tokudome, M. Jobbágy, G.J.A.A. Soler-Illia, Q. Tang, M. Müller, et al., Chem. Mater. 31 (2019) 322.

[26] G.S. Attard, J.C. Glyde, C.G. Göltner, Nature 378 (1995) 366−368.

[27] C.J. Brinker, Y. Lu, A. Sellinger, H. Fan, Adv. Mater. 11 (1999) 579.

[28] D. Grosso, F. Cagnol, G.J.A.A. Soler-Illia, E.L. Crepaldi, H. Amenitsch, A. Brunet-Bruneau, et al., Adv. Funct. Mater. 14 (2004).

[29] (a) F. Schüth, Angew. Chem. Int. Ed. 42 (2003) 3604.
(b) M. Tiemann, Chem. Mater. 20 (2008) 961.

[30] Y. Wan, D.Y. Zhao, Chem. Rev. 107 (2007) 2821.

[31] G.S. Armatas, G. Bilis, M. Louloudi, J. Mater. Chem. 21 (2011) 2997.

[32] T. Zhao, A. Elzatahry, X. Li, D. Zhao, Nat. Rev. Mater. (2019) 775.

[33] C. Li, Q. Li, Y.V. Kaneti, D. Hou, Y. Yamauchi, Y. Mai, Chem. Soc. Rev. 49 (2020) 4681−4736.

[34] G.J.A.A. Soler-Illia, E.L. Crepaldi, D. Grosso, C. Sanchez, Curr. Opin. Colloid Interface Sci. 8 (2003) 109.

[35] Y. Zhu, Y. Zhao, J. Ma, X. Cheng, J. Xie, P. Xu, et al., J. Am. Chem. Soc. 139 (2017) 10365.

[36] J. Zhang, Y. Deng, D. Gu, S. Wang, L. She, R. Che, et al., Adv. Energy Mater. 1 (2011) 241.

[37] J. Shim, J. Lee, Y. Ye, J. Hwang, S.-K. Kim, T.-H. Lim, et al., ACS Nano 6 (2012) 6870.

[38] J. Hwang, S.H. Woo, J. Shim, C. Jo, K.T. Lee, J. Lee, ACS Nano 7 (2013) 1036.

[39] Y. Ren, X. Zhou, W. Luo, P. Xu, Y. Zhu, X. Li, et al., Chem. Mater. 28 (2016) 7997.

[40] J. Fan, X. Ran, Y. Ren, C. Wang, J. Yang, W. Teng, et al., Langmuir 32 (2016) 9922.

[41] G.H. Wang, Z. Cao, D. Gu, N. Pfänder, A.C. Swertz, B. Spliethoff, et al., Angew. Chem., Int. Ed. 55 (2016) 8850.

[42] Y. Liu, R. Che, G. Chen, J. Fan, Z. Sun, Z. Wu, et al., Sci. Adv. 1 (2015) e1500166.

[43] W. Dong, Y. Yao, L. Li, Y. Sun, W. Hua, G. Zhuang, et al., Appl. Catal., B 217 (2017) 293.

[44] K. Zimny, T. Roques-Carmes, C. Carteret, M. Stebe, J. Blin, N. J. Chem. 116 (2012) 6585.

[45] W. Zhou, W. Li, J.-Q. Wang, Y. Qu, Y. Yang, Y. Xie, et al., J. Am. Chem. Soc. 136 (2014) 9280.

[46] J.G. Werner, T.N. Hoheisel, U. Wiesner, ACS Nano 8 (2013) 731.

[47] J.G. Werner, M.R. Scherer, U. Steiner, U. Wiesner, Nanoscale 6 (2014) 8736.

[48] J.G. Werner, S.S. Johnson, V. Vijay, U. Wiesner, Chem. Mater. 27 (2015) 3349.

[49] T. Weller, L. Deilmann, J. Timm, T.S. Dorr, P.A. Beaucage, A.S. Cherevan, et al., Nanoscale 10 (2018) 3225.

[50] Y. Liu, W. Teng, G. Chen, Z. Zhao, W. Zhang, B. Kong, et al., Chem. Sci. 9 (2018) 7705.

[51] J.N. Israelachvili, Intermolecular and Surface Forces: With, Applications to Colloidal and Biological Systems, 2nd edn., Elsevier, Amsterdam, 1992.

[52] Q. Huo, R. Leon, P.M. Petroff, G.D. Stucky, Science 268 (1995) 1324.

[53] (a) D. Grosso, F. Babonneau, P.-A. Albouy, H. Amenitsch, A.R. Balkenende, A. Brunet-Bruneau, et al., Chem. Mater. 14 (2002) 931.
     (b) X. Zhang, W. Wu, J. Wang, C. Liu, Thin Solid Films 515 (2007) 8376.

[54] P. Innocenzi, L. Malfatti, T. Kidchob, P. Falcaro, Chem. Mater. 21 (2009) 2555.

[55] C.B. Contreras, O. Azzaroni, G.J.A.A. Soler-Illia, Comprehensive Nanoscience and Nanotechnology, Second Edition, D.L. Andrews, R.H. Lipson and T. Nann (eds.), 2019, 1, chapter 16, 331.

[56] B. Coasne, A. Galarneau, R.J.M. Pellenq, F. Di Renzo, Chem. Soc. Rev. 42 (2013) 4141.

[57] R. Guillet-Nicolas, F. Bérubé, M. Thommes, M.T. Janicke, F. Kleitz, J. Phys. Chem. C. 121 (44) (2017) 24505.

[58] M. Templin, A. Franck, A. Du Chesne, H. Leist, Y. Zhang, R. Ulrich, et al., Science 278 (1997) 1795.

[59] D. Zhao, J. Feng, Q. Huo, N. Melosh, G.H. Fredrickson, B.F. Chmelka, et al., Science (1998) 548.

[60] J. Wei, Z. Sun, W. Luo, Y. Li, A.A. Elzatahry, A.M. Al-Enizi, et al., J. Am. Chem. Soc. 139 (5) (2017) 1706.

[61] R. Corriu, N. Trong Anh, Molecular Chemistry of Sol−gel Derived Nanomaterials, John Wiley and Sons, 2009. chapter 3, 27.

[62] K. Egeblad, C.H. Christensen, M. Kustova, C.H. Christensen, Chem. Mater. 20 (2008) 946.

[63] P.C. Angelomé, L.M. Liz-Marzán, J. Sol−gel Sci. Technol. 70 (2014) 180.

[64] G.J.A.A. Soler-Illia, A. Louis, C. Sanchez, Chem. Mater. 14 (2002) 750.

[65] (a) S.T. Hyde, Pure App Chem 64 (1992) 1617.
     (b) S.T. Hyde, Progress in Colloid and Polymer Science 82 (1990) 236.

[66] G.L. Drisko, A. Zelcer, V. Luca, R.A. Caruso, G.J.A.A. Soler-Illia, Chem. Mater. 22 (2010) 4379.

[67] L. Yan, D. Li, T. Yan, G. Chen, L. Shi, Z. An, et al., ACS Sustain. Chem. Eng. 6 (4) (2018) 5265.

[68] Y. Deng, T. Yu, Y. Wan, Y. Shi, Y. Meng, D. Gu, et al., J. Am. Chem. Soc. 129 (2007) 1690.

[69] J. Wei, H. Wang, Y. Deng, Z. Sun, L. Shi, B. Tu, et al., J. Am. Chem. Soc. 133 (2011) 20369.

[70] K. Nakanishi, N. Tanaka, Acc. Chem. Res. 40 (2007) 863.

[71] L. Malfatti, M.G. Bellino, P. Innocenzi, G.J.A.A. Soler-Illia, Chem. Mater. 21 (2009) 2763.

[72] S. Park, M. Santha Moorthy, C.S. Ha, NPG Asia Mater. 6 (2014) e96.

[73] P. Van Der Voort, D. Esquivel, E. De Canck, F. Goethals, I. Van Driessche, F.J. Romero-Salguero, Chem. Soc. Rev. 42 (2013) 3913.

[74] D. Gua, F. Schüth, Chem. Soc. Rev. 43 (2014) 313.

[75] B. Lebeau, A. Galarneaub, M. Linden, Chem. Soc. Rev. 42 (2013) 3661.

[76] (a) M. Impéror-Clerc, P. Davidson, A. Davidson, J. Am. Chem. Soc. 122 (2000) 11925.
     (b) N.A. Melosh, P. Lipic, F.S. Bates, G.D. Stucky, F. Wudl, G.H. Fredrickson, et al., Macromolecules 32 (1999) 4332.

[77] (a) B.L. Kirsch, E.K. Richman, A.E. Riley, S.H. Tolbert, J. Phys. Chem. B 108 (2004) 12698.
     (b) S.Y. Choi, M. Mamak, S. Speakman, N. Chopra, G.A. Ozin, Small 1 (2005) 226.

[78] Y. Sakatani, D. Grosso, L. Nicole, C. Boissière, G.J.A.A. Soler-Illia, C. Sanchez, J. Mater. Chem. 16 (2006) 77.

[79] D. Grosso, C. Boissière, B. Smarsly, T. Brezesinski, N. Pinna, P.A. Albouy, et al., Nat. Mater. 3 (2004) 787.

[80] S.L. Suib, Chem. Rec. 17 (12) (2017) 1169−1183.

[81] F. Liu, C. Liu, W. Kong, C. Qi, A. Zheng, S. Dai, Green. Chem. 18 (2016) 6536.

[82] A. Stein, B.J. Melde, R.C. Schroden, Adv. Mater. 12 (2000) 1403.

[83] J.-L. Shi, Z.-L. Hua, L.-X. Zhang, J. Mater. Chem. 14 (2004) 795.

[84] F. Hoffmann, M. Cornelius, J. Morell, M. Fröba, Angew. Chem. Int. Ed. 45 (2006) 3216.

[85] A. Mehdi, C. Reye, R. Corriu, Chem. Soc. Rev. 40 (2011) 573.

[86] S. Lowell, J.E. Shields, M.A. Thomas, M. Thommes, Characterization of Porous Solids and Powders: Surface Area, Pore Size and Density, Vol. 16, Springer Science & Business Media, 2006.

[87] S. Alberti, O. Azzaroni, G.J.A.A. Soler-Illia, Chem. Commun. 51 (2015) 6050.

[88] A. Calvo, M.C. Fuertes, B. Yameen, F.J. Williams, O. Azzaroni, G.J.A.A. Soler-Illia, Langmuir 26 (2010) 5559.

[89] M. Rafti, A. Brunsen, M.C. Fuertes, O. Azzaroni, G.J.A.A. Soler-Illia, ACS Appl. Mater. Interfaces 5 (2013) 8833.

[90] A. Calvo, P.C. Angelome, V.M. Sanchez, D.A. Scherlis, F.J. Williams, G.J.A.A. Soler-Illia, Chem. Mater. 20 (2008) 4661.

[91] S. Onclin, B.J. Ravoo, D.N. Reinhoudt, Angew. Chem. Int. Ed. 44 (2005) 6282.

[92] N.K. Mal, M. Fujiwara, Y. Tanaka, Nature 421 (2003) 350.

[93] R. Casasús, M.D. Marcos, R. Martínez-Máñez, J.V. Ros-Lis, J. Soto, L.A. Villaescusa, et al., J. Am. Chem. Soc. 126 (2004) 8612.

[94] J.V. Ros-Lis, B. Garcíaa-Acosta, D. Jiménez, R. Martínez-Máñez, F. Sancenón, J. Soto, et al., Chem. Soc. 126 (2004) 4064.

[95] Q. Yang, S. Wang, P. Fan, L. Wang, Y. Di, K. Lin, et al., Chem. Mater. 17 (2005) 5999.

[96] T.D. Nguyen, H.-R. Tseng, P.C. Celestre, A.H. Flood, Y. Liu, J.F. Stoddart, et al., PNAS 102 (2005) 10029.

[97] Y.-L. Zhao, Z. Li, S. Kabehie, Y.Y. Botros, J.F. Stoddart, J.I. Zink, J. Am. Chem. Soc. 132 (2010) 13016.

[98] T.D. Nguyen, K.C.-F. Leung, M. Liong, C.D. Pentecost, J.F. Stoddart, J.I. Zink, Org. Lett. 8 (2006) 3363.

[99] W. Chen, C.A. Glackin, M.A. Horwitz, J.I. Zink, Acc. Chem. Res. 52 (2019) 1531.

[100] T. Bein, P. Enzel, Angew. Chem. Int. Ed. Engl. 28 (1989) 1692.

[101] P. Enzel, T. Bein, J. Phys. Chem. 93 (1989) 6270.

[102] C.-G. Wu, T. Bein, Science 264 (1994) 1757.

[103] F.F. Fang, H. Jin Choi, W.S. Ahn, Compos. Sci. Technol. 69 (2009) 2088.

[104] K. Moller, T. Bein, R.X. Fischer, Chem. Mater. 10 (1998) 1841.

[105] J. Jang, B. Lim, J. Lee, T. Hyeon, Chem. Comm. (2001) 83.

[106] T.L. Kelly, K. Yano, M.O. Wolf, Langmuir 26 (2010) 421.

[107] B. Zhang, X. Chen, S. Ma, Y. Chen, J. Yang, M. Zhang, Nanotechnology 21 (2010) 065304.

[108] J. Dong, Y. Hu, J. Xu, X. Qu, C. Zhao, Electroanalysis 21 (2009) 1792.

[109] S.A. Johnson, D. Khushalani, N. Coombs, T.E. Mallouk, G.A. Ozin, J. Mater. Chem. 8 (1998) 13.

[110] (a) A.G. Pattantyus-Abraham, M.O. Wolf, Chem. Mater. 16 (2004) 2180.
(b) M.A. Álvaro, A. Corma, B. Ferrer, M.S. Galletero, H. García, E. Peris, Chem. Mater. 16 (2004) 2142.

[111] (a) T.L. Kelly, S.P.Y. Che, Y. Yamada, K. Yano, M.O. Wolf, Langmuir 24 (2008) 9809.
(b) T.L. Kelly, Y. Yamada, S.P.Y. Che, K. Yano, M.O. Wolf, Adv. Mater. 20 (2008) 2616.

[112] M. Choi, F. Kleitz, D. Liu, H.Y. Lee, W.-S. Ahn, R. Ryoo, J. Am. Chem. Soc. 127 (2005) 1924.

[113] S.M. Ng, S.-I. Ogino, T. Aida, K.A. Koyano, T. Tatsumi, Macromol. Rapid Commun. 18 (1997) 991.

[114] (a) D.-H. Choi, R. Ryoo, J. Mater. Chem. 20 (2010) 5544.
(b) M.T. Run, S.Z. Wu, D.Y. Zhang, G. Wu, Mater. Chem. Phys. 105 (2007) 341.
(c) M. Wainer, L. Marcoux, F. Kleitz, J. Mater. Sci. 44 (2009) 6538.

[115] R. Guillet-Nicolas, L. Marcoux, F. Kleitz, N. J. Chem. 34 (2010) 355.

[116] B.-S. Tian, C. Yang, J. Phys. Chem. C. 113 (2009) 4925.

[117] (a) S. Yoo, J.D. Lunn, S. González, J.A. Ristich, E.E. Simanek, D.F. Shantz, Chem. Mater. 18 (2006) 2935.
(b) D.M. Ford, E.E. Simanek, D.F. Shantz, Nanotechnology 16 (2000) S458.

[118] (a) P. Li, S. Kawi, J. Catalysis 257 (2008) 23.

(b) P. Li, S. Kawi, Cat. Today 131 (2008) 61.

[119] S. Yoo, S. Yeu, R.L. Sherman, E.E. Simanek, D.F. Shantz, D.M. Ford, J. Membr. Sci. 334 (2009) 16.

[120] (a) J.P.K. Reynhardt, Y. Yang, A. Sayari, H. Alper, Chem. Mater. 16 (2004) 4095.

(b) J.P.K. Reynhardt, Y. Yang, A. Sayari, H. Alper, Adv. Funct. Mater. 15 (2005) 1641.

[121] M.P. Kapoor, H. Kuroda, M. Yanagi, H. Nanbu, L.R. Juneja, Top. Catal. 52 (2009) 634.

[122] Y. Jiang, Q. Gao, H. Yu, Y. Chen, F. Deng, Microp. Mesop. Mater. 103 (2007) 316.

[123] E.J. Acosta, C.S. Carr, E.E. Simanek, D.F. Schantz, Adv. Mater. 16 (2004) 985.

[124] Z. Liang, B. Fadhel, C.J. Schneider, A.L. Chaffee, Microp. Mesop. Mater. 111 (2008) 536.

[125] Q. Wang, V. Varela Guerrero, A. Ghosh, S. Yeu, J.D. Lunn, D.F. Shantz, J. Cat. 269 (2010) 15.

[126] B. González, M. Colilla, C. López de Laorden, M. Vallet-Regí, J. Mater. Chem. 19 (2009) 9012.

[127] J.M. Rosenholm, A. Penninkangas, M. Lindén, Chem. Commun. (2006) 3909.

[128] (a) J.C. Hicks, J.H. Drese, D.J. Fauth, M.L. Gray, G. Qi, C.W. Jones, J. Am. Chem. Soc. 130 (2008) 2902.

(b) J.H. Drese, S. Choi, R.P. Lively, W.J. Koros, D.J. Fauth, M.L. Gray, et al., Adv. Funct. Mater. 19 (2009) 3821.

[129] (a) C.J. Bhongale, C.-H. Yang, C.-S. Hsu, Chem. Commun. 21 (2006) 2274.

(b) R. Guo, G. Li, W. Zhang, G. Shen, D. Shen, Chem Phy sChem 6 (2005) 2025.

[130] (a) Y. Lu, Y. Yang, A. Sellinger, M. Lu, J. Huang, H. Fan, et al., Nature 410 (2001) 913.

(b) Y. Yang, Y. Lu, M. Lu, J. Huang, R. Haddad, G. Xomeritakis, et al., J. Am. Chem. Soc. 125 (2003) 1269.

[131] H. Peng, J. Tang, L. Yang, J. Pang, H.S. Ashbaugh, C.J. Brinker, et al., J. Am. Chem. Soc. 128 (2006) 5304.

[132] M. Ikegame, K. Tajima, T. Aida, Angew. Chem. Int. Ed. 42 (2003) 2154.

[133] (a) G. Li, S. Bhosale, F. Li, Y. Zhang, R. Guo, H. Zhu, et al., Chem. Comm. (2004) 1760.

(b) G. Li, S. Bhosale, T. Wang, Y. Zhang, H. Zhu, J.-H. Fuhrhop, Angew. Chem. Int. Ed. 42 (2003) 3818.

(c) W. Zhang, J. Cui, C. Lin, Y. Wu, L. Ma, Y. Wen, et al., J. Mater. Chem. 19 (2009) 3962.

(d) Z. Yang, X. Kou, W. Ni, Z. Sun, L. Li, J. Wang, Chem. Mater. 19 (2007) 6222.

[134] J. Richard, A. Phimphachanh, J. Schneider, S. Nandi, E. Laurent, P. Lacroix-Desmazes, et al., Chem. Mater. 34 (2022) 7828.

[135] D. Houssein, J. Warnant, E. Molina, T. Cacciaguerra, C. Gérardin, N. Marcotte, Microporous Mesoporous Mater. 239 (2017) 244.

[136] E. Molina, M. Mathonnat, J. Richard, P. Lacroix-Desmazes, M. In, P. Dieudonné, et al., Beilstein J. Nanotechnol. 10 (2019) 144.

[137] (a) K. Kageyama, J.-i Tamazawa, T. Aida, Science 285 (1999) 2113.

(b) P. Lehmus, B. Rieger, Science 285 (1999) 2081.

[138] R.R. Rao, B.M. Weckhuysen, R.A. Schoonheydt, Chem. Commun. (1999) 44.

[139] S.-H. Chan, Y.-Y. Lin, C. Ting, Macromolecules 36 (2003) 8910.

[140] R. Anwander, I. Nagl, C. Zapilko, M. Widenmeyer, Tetrahedron 59 (2003) 10567.

[141] D.J. Cardin, S.P. Constantine, A. Gilbert, A.K. Lay, M. Alvaro, M.S. Galletero, et al., Chem. Soc. 123 (2001) 3141.

[142] V.S.-Y. Lin, D.R. Radu, M.-K. Han, W. Deng, S. Kuroki, B.H. Shanks, et al., J. Am. Chem. Soc. 124 (2002) 9040.

[143] (a) S. Spange, A. Graser, H. Muller, Y. Zimmermann, P. Rehak, C. Jager, et al., Chem. Mater. 13 (2001) 3698.

(b) S. Spange, A. Gräser, A. Huwe, F. Kremer, C. Tintemann, P. Behrens, Chem. Eur. J 7 (2001) 3722.

[144] A.B. Fischer, J.B. Kinney, R.H. Staley, M.S. Wrighton, J. Am. Chem. Soc. 101 (1979) 6501.

[145] S. O'Brien, J. Tudor, S. Barlow, M.J. Drewitt, S.J. Heyes, D. O'Hare, Chem. Comm. (1997) 641.

[146] (a) M.J. MacLachlan, M. Ginzburg, N. Coombs, N.P. Raju, J.E. Greedan, G.A. Ozin, et al., J. Am. Chem. Soc. 122 (2000) 3878.
(b) M.J. MacLachlan, P. Aroca, N. Coombs, I. Manners, G.A. Ozin, Adv. Mater. 10 (1998) 144.

[147] N.C. Strandwitz, Y. Nonoguchi, S.W. Boettcher, G.D. Stucky, Langmuir 26 (2010) 5319.

[148] G.D. Stucky, Nature, 410, 885.

[149] (a) A.S. Angelatos, Y. Wang, F. Caruso, Langmuir 24 (2008) 4224.
(b) Y. Wang, A.S. Angelatos, D.E. Dunstan, F. Caruso, Macromolecules 40 (2007) 7594.

[150] (a) G. Pérez-Mitta, L. Burr, J.S. Tuninetti, C. Trautmann, M.E. Toimil-Molares, O. Azzaroni, Nanoscale 8 (2016) 1470.
(b) B. Yameen, M. Ali, M. Álvarez, R. Neumann, W. Ensinger, W. Knoll, et al., Polymer Chemistry 1 (2010) 183.

[151] G. Pérez-Mitta, A. Albesa, F.M. Gilles, M.E. Toimil-Molares, C. Trautmann, O. Azzaroni, J. Phys. Chem. C. 121 (2017) 9070.

[152] G. Pérez-Mitta, W.A. Marmisollé, A.G. Albesa, M. Eugenia Toimil-Molares, C. Trautmann, O. Azzaroni, Small 14 (2018) 1702131.

[153] A. Brunsen, A. Calvo, F.J. Williams, G.J.A.A. Soler-Illia, O. Azzaroni, Langmuir 27 (2011) 4328.

[154] (a) F.M. Gilles, M. Tagliazucchi, O. Azzaroni, I. Szleifer, J. Phys. Chem. C 120 (2016) 4789.
(b) G. Laucirica, G. Pérez-Mitta, M.E. Toimil-Molares, C. Trautmann, W.A. Marmisollé, O. Azzaroni, J. Phys. Chem. C 123 (2019) 28997.

[155] K.M. Coakley, Y. Liu, M.D. McGehee, K.L. Frindell, G.D. Stucky, Adv. Funct. Mater. 13 (2003) 301.

[156] H. Xi, B. Wang, Y. Zhang, X. Qian, J. Yin, Z. Zhu, J. Phys. Chem. Solids 64 (2003) 2451.

[157] L. Huang, S. Dolai, K. Raja, M. Kruk, Langmuir 26 (2010) 2688.

[158] B. Yameen, M. Ali, M. Alvarez, R. Neumann, W. Ensinger, W. Knoll, et al., Polym. Chem. 1 (2010) 183.

[159] I. Díaz, B. García, B. Alonso, C.M. Casado, M. Morán, J. Losada, et al., Chem. Mater. 15 (2003) 1073.

[160] (a) D.S. Shephard, W. Zhou, T. Maschmeyer, J.M. Matters, C.L. Roper, S. Parsons, et al., Angew. Chem., Int. Ed. 37 (1998) 2719.
(b) T. Maschmeyer, R.D. Oldroy, G. Sankar, J.M. Thomas, I.J. Shannon, J.A. Klepetko, et al., Angew. Chem., Int. Ed. Engl. 36 (1997) 1639.

[161] W. Huang, J.N. Kuhn, C.-K. Tsung, Y. Zhang, S.E. Habas, P. Yang, et al., Nano Lett. 8 (2008) 2027.

[162] R.M. Crooks, M. Zhao, L. Sun, V. Chechik, L.K. Yeung, Acc. Chem. Res. 34 (2001) 181.

[163] A. Andrieu-Brunsen, S. Micoureau, M. Tagliazucchi, I. Szleifer, O. Azzaroni, G.J.A.A. Soler-Illia, Chem. Mater. 27 (2015) 808.

[164] (a) M. Save, G. Granvorka, J. Bernard, B. Charleux, C. Boissiére, D. Grosso, et al., Macromol. Rapid Commun. 27 (2006) 393.
(b) F. Audouin, H. Blas, P. Pasetto, P. Beaunier, C. Boissiére, C. Sánchez, et al., Macromol. Rapid Commun 29 (2008) 914.
(c) C. Li, J. Yang, P. Wang, J. Liu, Q. Yang, Micropor. Mesopor. Mater. 123 (2009) 228. K. Ikeda, M. Kida, et al., Polymer J., 2009, 41. 672.
(d) A. Martín, G. Morales, F. Martínez, R. van Grieken, L. Cao, M. Kruk, J. Mater. Chem. 20 (2010) 8026.

[165] O. Azzaroni, J. Polym. Sci. Part. A: Polym. Chem. 50 (2012) 3225.

[166] B. Yameen, A. Kaltbeitzel, G. Glasser, A. Langner, F. Muller, U. Gösele, et al., ACS Appl. Mater. Interfaces 2 (2010) 279.

[167] S.E. Moya, O. Azzaroni, T. Kelby, E. Donath, W.T.S. Huck, J. Phys. Chem. B 111 (2007) 7034.

[168] J. Pyun, K. Matyjaszewski, Chem. Mater. 13 (2001) 3436–3448.

[169] P. Pasetto, H. Blas, F. Audouin, C. Boissiére, C. Sanchez, M. Save, et al., Macromolecules 42 (2009) 5983.

[170] M. Kruk, B. Dufour, E.B. Celer, T. Kowalewski, M. Jaroniec, K. Matyjaszewski, J. Phys. Chem. B 109 (2005) 9216.

[171] M. Kruk, B. Dufour, E.B. Celer, T. Kowalewski, M. Jaroniec, K. Matyjaszewski, Macromolecules 41 (2008) 8584.

[172] (a) J. Moreno, D.C. Sherrington, Chem. Mater. 20 (2008) 4468.
(b) Q. Fu, G.V.R. Rao, L.K. Ista, Y. Wu, B.P. Andrewzjeski, L.A. Sklar, et al., Adv. Mater. 15 (2003) 1262.

[173] L. Cao, M. Kruk, Polym. Chem. 1 (2010) 97.

[174] J.D. Lunn, D.F. Shantz, Chem. Mater. 21 (2009) 3638.

[175] M. Lenarda, G. Chessa, E. Moretti, S. Polizzi, L. Storaro, A. Talon, J. Mater. Sci. 41 (2006) 6305.

[176] T.A. García, C.A. Gervasi, M.J. Rodríguez Presa, J.I. Otamendi, S.E. Moya, O. Azzaroni, J. Phys. Chem. C. 116 (2012) 13944.

[177] Z. Zhou, S. Zhu, D. Zhang, J. Mater. Chem. 17 (2007) 2428.

[178] Q. Fu, G.V. Rama Rao, T.L. Ward, Y. Lu, G.P. Lopez, Langmuir 23 (2007) 170.

[179] S. Schmidt, S. Alberti, P. Vana, G.J.A.A. Soler-Illia, O. Azzaroni, Chem. Eur. J. 23 (2017) 14500.

[180] S. Angelos, Y.-W. Yang, N.M. Khashab, J.F. Stoddart, J.I. Zink, J. Am. Chem. Soc. 131 (2009) 11344.

[181] W. Chen, C.-A. Cheng, D. Xiang, J.I. Zink, Nanoscale 13 (2021) 5497.

[182] A. Llopis-Lorente, B. De Luis-Fernández, A. García-Fernández, S. Jiménez-Falcao, M. Orzaez, F. Sancenón Galarza, et al., ACS Appl. Mater. & Interfaces 10 (2018) 26494.

[183] L. Silies, H. Didzoleit, C. Hess, B. Stühn, A. Andrieu-Brunsen, Chem. Mater. 27 (2015) 1971.

[184] J.C. Tom, R. Brilmayer, J. Schmidt, A. Andrieu-Brunsen, Polymers 9 (2017) 539.

[185] D. John, M. Stanzel, A. Andrieu-Brunsen, Adv. Funct. Mater. 31 (2021) 2009732.

[186] A. Calvo, B. Yameen, F.J. Williams, G.J.A.A. Soler-Illia, O. Azzaroni, J. Am. Chem. Soc. 131 (2009) 10866.

[187] A. Brunsen, C. Díaz, L.I. Pietrasanta, B. Yameen, M. Ceolín, G.J.A.A. Soler-Illia, et al., Langmuir 28 (2012) 3583.

[188] J. Cui, O. Azzaroni, A. del Campo, Macromol. Rapid Commun. 32 (2011) 1699.

[189] J. Cui, T.-H. Nguyen, M. Ceolín, R. Berger, O. Azzaroni, A. del Campo, Macromolecules 45 (2012) 3213.

[190] A.A. Brown, O. Azzaroni, W.T.S. Huck, Langmuir 25 (2009) 1744.

[191] A.A. Brown, O. Azzaroni, L.M. Fidalgo, W.T.S. Huck, Soft Matter 5 (2009) 2738.

[192] C.G. Bochet, J. Chem. Soc., Perkin Trans. 1 (2002) 125.

[193] A. Brunsen, J. Cui, M. Ceolín, A. del Campo, G.J.A.A. Soler-Illia, O. Azzaroni, Chem. Comm. 48 (2012) 1422–1424.

[194] A.P.-Z. Clark, K.-F. Shen, Y.F. Rubin, S.H. Tolbert, Nano Lett. 5 (2005) 1647.

[195] S. Kirmayer, S. Neyshtadt, A. Keller, D. Okopnik, G.L. Frey, Chem. Mater. 21 (2009) 4387.

[196] E. Dovgolevsky, S. Kirmayer, E. Lakin, Y. Yang, C.J. Brinker, G.L. Frey, J. Mater. Chem. 18 (2008) 423.

[197] R. Hernández, A.-Ch Franville, P. Minoofar, B. Dunn, J.I. Zink, J. Am. Chem. Soc. (2001) 123.

[198] M. Kimura, K. Wada, Y. Iwashima, K. Ohta, K. Hanabusa, H. Shirai, et al., Chem. Comm. 19 (2003) 2504.

[199] M. Fujiwara, K. Shiokawa, Y. Zhu, J. Mol. Cat. A: Chem. 264 (2007) 153.

[200] (a) J. Kecht, A. Schlossbauer, T. Bein, Chem. Mater. 20 (2008) 7207.
(b) C.-Y. Hong, X. Li, C.Y. Pan, Eur. Polym. J 43 (2007) 4114.
(c) X. Li, C.-Y. Hong, C.-Y. Pan, Polymer 51 (2010) 92.
(d) P.-W. Chung, R. Kumar, M. Pruski, V.S.-Y. Lin, Adv. Funct. Mater. 18 (2008) 1390.

[201] C.-Y. Hong, X. Li, C.-Y. Pan, J. Mater. Chem. 19 (2009) 5155.

[202]   (a) F. Roohi, M.M. Titirici, New J. Chem 32 (2008) 1409.
        (b) C.-Y. Hong, X. Li, C.-Y. Pan, J. Phys. Chem. C 112 (2008) 15320.
[203]   J.D. Lunn, D.F. Shantz, Chem. Comm. 46 (2010) 2926.
[204]   J.M. Rosenholm, A. Duchanoy, M. Lindén, Chem. Mater. 20 (2008) 1126.
[205]   A.C. Hortelao, C. Simó, M. Guix, S. Guallar-Garrido, E. Julián, D. Vilela, et al., Sci. Robot. 6 (2021).
        eabd2823.

Nanomaterials and catalysis

Tanna E.R. Fiuza[1,2], Danielle S. Gonçalves[1,3], Tathiana M. Kokumai[1], Karen A. Resende[1], Priscila Destro[1] and Daniela Zanchet[1]

[1]*Institute of Chemistry, University of Campinas (UNICAMP), Campinas, SP, Brazil* [2]*Brazilian Nanotechnology National Laboratory (LNNano), Brazilian Center for Research in Energy and Materials (CNPEM), Campinas, SP, Brazil* [3]*Karlsruhe Institute of Technology (KIT), Institute of Catalysis Research & Technology (IKFT), Baden, Württemberg, Germany*

18.1 Overview: nanoscience and catalysis

The extraordinary development of nanoscience in the last decades, strongly rooted in the understanding of fundamental aspects that govern the nanoscale properties, has opened a myriad of opportunities in several areas of research that have already impacted our daily lives. From new electronic, optical, and magnetic devices to biomedical applications, drug delivery, and intelligent materials, we could highlight many areas directly impacted by the advances in the design and understanding of nanomaterials. The catalysis field could not be different, taking advantage of the deeper comprehension of size-dependent properties of nanomaterials and strategies to design and probe their unique properties at the atomic scale. From the seminal works by studies on single-crystal surfaces, Fig. 18.1 summarizes a few examples of the opportunities to be explored to design more realistic model systems in catalysis inspired by the advances in nanoscience [1].

Catalysts are widely used in the chemical industry to increase the reaction rate by changing the energy landscape to transform reagents into the desired products (Fig. 18.2A) [2−5]. About 85%−90% of chemical products rely on a catalytic process, the great majority of them based on heterogeneous catalysts, in which reagents and products are in a different state than the catalysts. Heterogenous catalysts are usually solids (i.e., Al_2O_3, SiO_2, black carbon, ZrO_2, CeO_2, MgO) characterized by high surface areas that can act as catalysts by themselves or as supports to disperse a metallic phase (i.e., Pt, Pd, Rh, Ru, Ni, among others). Therefore, heterogeneous catalysts are indeed an important class of nanomaterials, and the size, shape, structure, and chemical composition directly affect the type and number of catalytic sites available for the reaction. Reducing the metallic particle size is essential to optimize the surface/volume ratio, exposing more catalytic sites and modulating the catalytic properties due to nanosized effects. The nanometric characteristic of the support can also play a crucial role, as well as the sites created at the metal−support interface.

In terms of performance, a catalyst must exhibit high activity (e.g., expressed in terms of reaction rate, that is, the converted amount of the reactant per volume of catalyst, mass, and time), selectivity (the amount of the reagent that is converted to the desired product), and long-term stability (the

Materials Nanoarchitectonics. DOI: https://doi.org/10.1016/B978-0-323-99472-9.00022-5

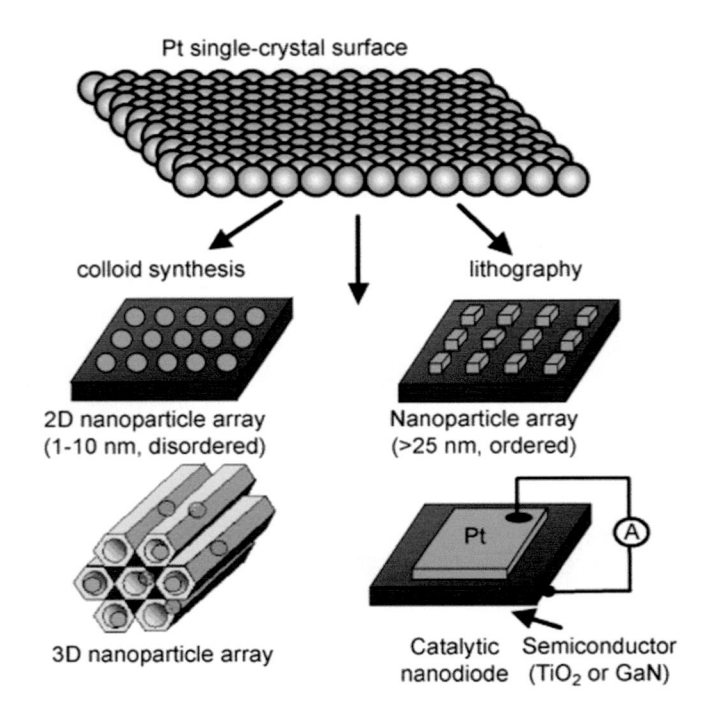

FIGURE 18.1

A few examples of the evolution from surface science to more realistic catalytic model systems based on the design of nanomaterials.

Reproduced with permission from [1].

lifetime of the catalyst), characteristics that depend on the number and nature of the catalytic sites. Moreover, it is highly desired that the catalyst be regenerated to recover its performance after several catalytic cycles. While maximizing the catalysts' activity was the main target in the past, the enhancement of catalysts' selectivity has been the challenge to overcome in the 21st century, motivated by the urgent environmental and economic needs to decrease the generation of residues and by-products. This challenge aligns with the shift in catalyst design from trial-and-error methods to rational approaches. In this aspect, nanoscience advances are a source of inspiration and opportunities.

In heterogeneous catalysts, chemical transformations take place at the surface. The catalytically active site, where the reagent adsorbs, chemical bounds are broken/formed, and product desorbs, can be, for example, a metal, an ion, a vacancy, or a metal—oxide interfacial site. The catalytic site depends not only on its chemical nature and oxidation state (i.e., $Pt^{+\delta}$) but can also depend on the local geometry. The reactions in which the rate depends on the local geometry, or equivalently, the rate normalized by the number of exposed sites depends on the particle diameter, were first named structure-sensitive reactions by Boudart in 1969 [6]. Examples of structure-sensitive reactions are hydrogenolysis reaction of alkanes, CH bond activation, and hydrogenation of CO [7]. Other reactions, however, are not size-dependent, that is, they are insensitive to the structure, such as

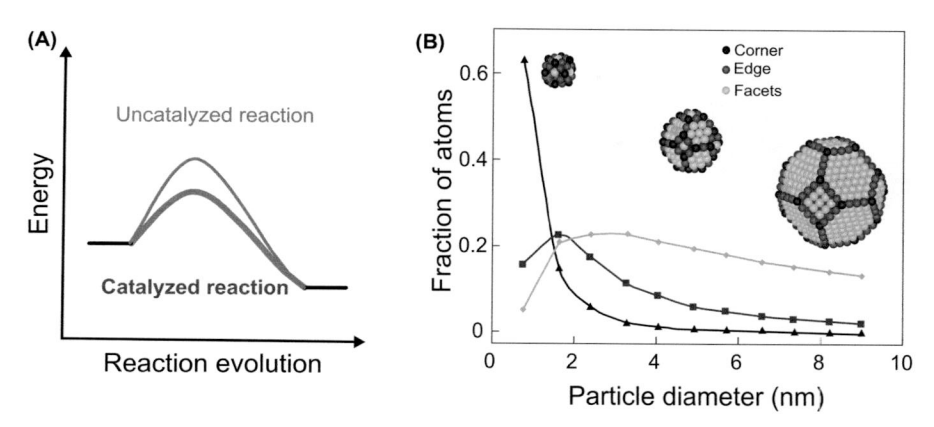

FIGURE 18.2

(A) Schematic representation of the potential energy diagram of a heterogeneous catalytic reaction, comparing the noncatalytic pathway (in red) with the catalytic one (in gray). (B) Scheme of the evolution of the fraction of atoms as a function of particle diameter according to their coordination: corner (black sphere), edge (dark blue sphere), and surface (light blue sphere) atoms.

Adapted from references [2–4].

hydrocarbon hydrogenation reactions. Whereas all catalytic processes are favored by increasing the surface/volume ratio of the catalysts, the structure-sensitive reactions are the most affected by the fine-tuning at the nanoscale.

Fig. 18.2B shows how the size of a nanoparticle impacts the contribution of different surface sites, represented by facets, edges, and corners, using a cuboctahedron as a model of an fcc (face-centered cubic) metallic nanoparticle [3,4]. In insensitive reactions, all sites show similar reactivity; in contrast, in structure-sensitive reactions, low coordination sites, that is, corners and edge/steps atoms, can present either higher or lower reactivity compared to atoms on the facets (or terraces) [7]. It becomes clear that by reducing the particles to the nanoscale, the ratio of the different sites varies (corner, edge, or terrace), and the particle size determines the catalytic performance in structure-sensitive reactions. The surface/volume ratio also increases by decreasing the particle size (as $1/R$; $R =$ radius of the particle), impacting the overall catalytic performance for reactions insensitive to the structure. Moreover, when the catalytically active nanoparticles are supported, unique interfacial sites can be formed at the nanoparticle–support interface, playing a crucial role in many reactions [8–12]. The total amount of interfacial sites is determined by the perimeter of the nanoparticle in contact with the support [8]. Finally, the electronic properties of the metallic phase are directly impacted by the particle size and interaction with the support.

Innumerous examples of the size-dependent performance of metal-supported catalysts can be found in the literature. For example, Co/oxides catalysts have been largely applied to steam reforming reactions and H_2 production [13–16]. While small particles are more easily oxidized depending on the reaction conditions, leading to catalyst deactivation, larger particles may favor carbon deposition and methane formation due to an unbalanced equilibrium between C–C and C–H cleavage and C oxidation steps. Fig. 18.3A shows the carbon deposition rate of CoCu-supported catalysts

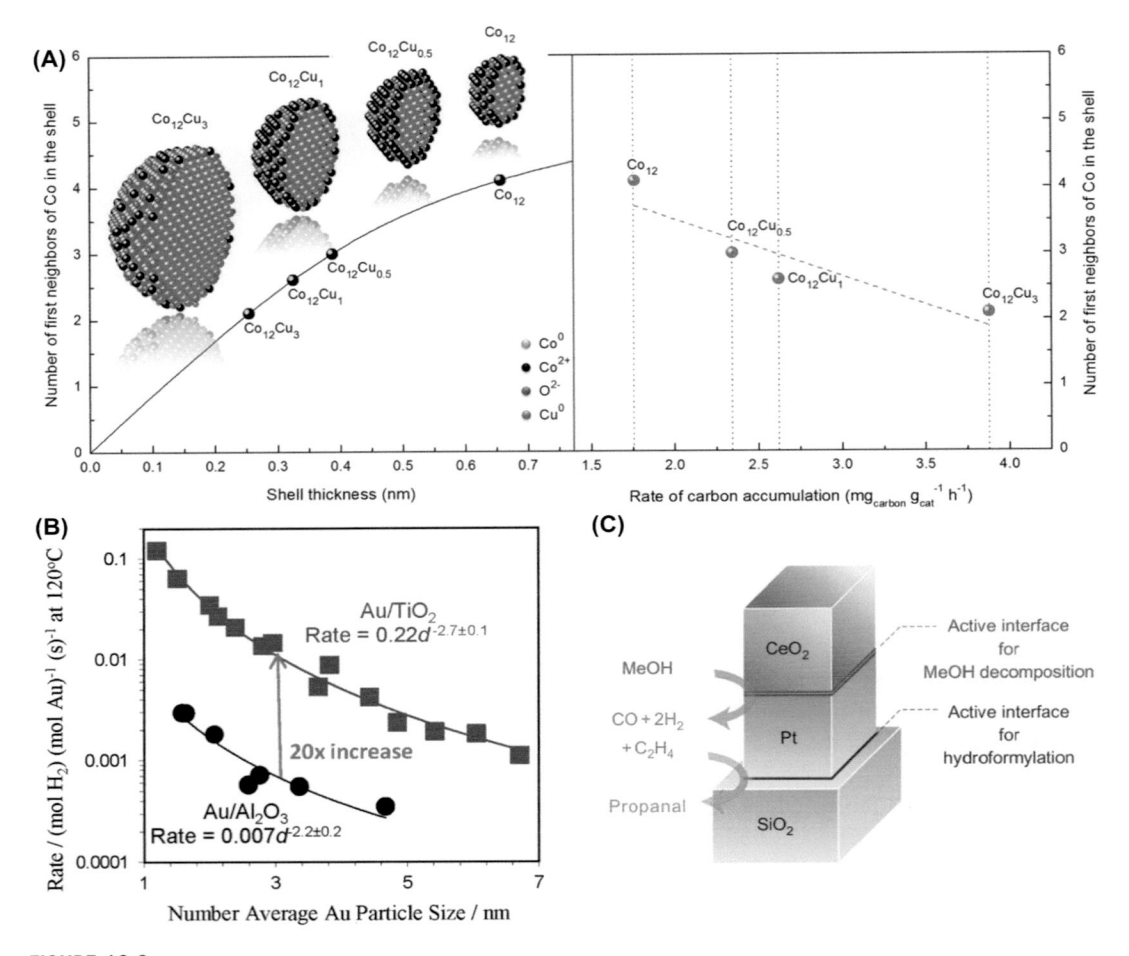

FIGURE 18.3

(A) The evolution of CoCu structure as a function of size and its impact on the carbon accumulation rate in steam reforming ethanol. (B) Rate per total mole of Au in WGS reaction as a function of Au particle size for Au/Al_2O_3 (●, black line) and Au/TiO_2 catalysts (■, blue line). The dependence of the rate with ~ d^{-2} (d = particle diameter) revealed the dependence with the particle perimeter (for surface, the dependence would be d^{-1} and for corners d^{-x}). (C) Illustration of the $CeO_2-Pt-SiO_2$ tandem catalyst for ethylene hydroformylation with MeOH over a tandem catalyst.

Reproduced with permission from (A) [17]; (B) [18]; (C) [19].

applied to the steam reforming of ethanol, showing the impact of the Cu on the number of Co−Co neighbors on the particle surface and the rate of carbon accumulation [17]. The relationship between the particle size and the reaction rate was also evaluated for Au-supported catalysts applied for water gas shift (WGS) reaction, an essential downstream reaction in H_2 production. In this work, by analyzing a large set of Au-supported catalysts, it becomes clear that the rate per mole of Au is higher for smaller nanoparticles, independent of the nature of the support (reducible or nonreducible), Fig. 18.3B [18]. It was possible to conclude that the dominant active sites for the WGS reaction in the Au-supported catalysts are the undercoordinated Au atoms (corner sites) and not the perimeter ones. An elegant work showing the role of the particle perimeter at the metal—oxide interface in the catalytic selectivity compared $Pt/CeO_2/SiO_2$ (one metal—oxide interface) with $CeO_2/Pt/SiO_2$ structure (two metal—oxide interfaces; tandem catalyst) for the formation of propanal from methanol in a cascade reaction [19]. While methanol decomposition forming CO and H_2 took place at the CeO_2—Pt interface, the conversion of CO, H_2, and ethylene hydroformylation occurred at the Pt—SiO_2 interface (Fig. 18.3C). By designing the interfacial sites, it was possible to significantly improve the catalytic behavior.

It is worth noting that the experimental distinction of the importance of each metallic site within a particle and the participation of metal-interface sites in the catalytic cycle is a very challenging task. Notably, in "real" samples characterized by a broad range of particle sizes, shapes, composition (in the case of bimetallic particles), and defects (in the case of oxides), it becomes even more complex [11]. Therefore, it becomes clear that optimizing the synthesis protocols to produce well-controlled (model) catalysts associated with advanced characterization tools with atomic resolution is pivotal to understanding the main aspects that can lead to more active, selective, and sustainable catalysts.

In this chapter, we chose to focus on wet chemistry methods such as colloidal synthesis, classified as a bottom-up approach, where small units carefully assemble, forming a defined nanostructure. Colloidal synthesis has gained much attention in the last decades due to its versatility, easy access, and low-cost apparatus. It has been explored to produce a myriad of nanoparticles with tunable size, shape, chemical composition, and structure. These preformed nanoparticles, typically 1—10 nm, can be used to prepare model metal-supported catalysts. Other methods, such as hydrothermal, sol—gel, and casting, have also been applied and examples will be given with regard to producing nanooxides of different sizes and shapes. The advances in nanoscience and heterogeneous catalyst development have been intrinsically linked to the ability to characterize the materials at the atomic scale and probe them under reaction conditions, and examples will be given. Finally, the understanding of AuCu-supported catalysts will be presented as a case study.

18.2 Design of model catalytic systems

The advances in the syntheses of nanomaterials have opened exciting avenues in the last decades to be explored with regard to model catalytic systems, as shown in Fig. 18.1. In heterogeneous catalysis, the impregnation method remains the most used method to produce metal-supported catalysts. In this case, the solution of a metal precursor is impregnated on a support, followed by thermal treatment (calcination and reduction) to form a dispersed metallic phase on the support surface. This classical method usually produces a metallic phase characterized by a broad particle size distribution, with the coexistence of single atoms and small clusters, leading to many different catalytic sites. Variations of the conventional impregnation methods, such as deposition—precipitation using the decrease of solubility of oxy/hydroxy metal species formed under basic pH, have been successfully applied in specific systems [20]. Although many catalysts prepared by impregnation have been empirically optimized by trial-and-error and successfully applied in many industrial reactions, a deep knowledge of the catalytic sites benefits the use of model systems [21,22]. Indeed, several strategies explored in the synthesis of nanomaterials have been successfully explored, such as atomic layer deposition [23], hydrothermal synthesis [24], sol—gel method [25], controlled surface reaction [26,27], and colloidal method [2,28].

This section will focus on the fundamental aspects of colloidal synthesis and examples of the fine-tuning of size, shape, chemical composition, and structure that can be achieved. Selected examples of nanooxides are presented. The use of these nanomaterials in catalytic applications is described.

18.2.1 Colloidal synthesis of metallic and hybrid nanoparticles targeting catalytic applications

The synthesis of well-defined nanoparticles is an underlying challenge and a requirement to explore the tunability of properties at the nanoscale fully. Typically, colloidal nanoparticles are formed in the liquid medium through the decomposition of organometallic/coordination complex precursors or the reduction of metal salts in the presence of stabilizers (or ligands) under controlled conditions (Fig. 18.4A) [29—40]. In the last 30 years, an enormous library of colloidal nanoparticles has been successfully synthesized by playing with concentration, precursors, reducing agents, ligands,

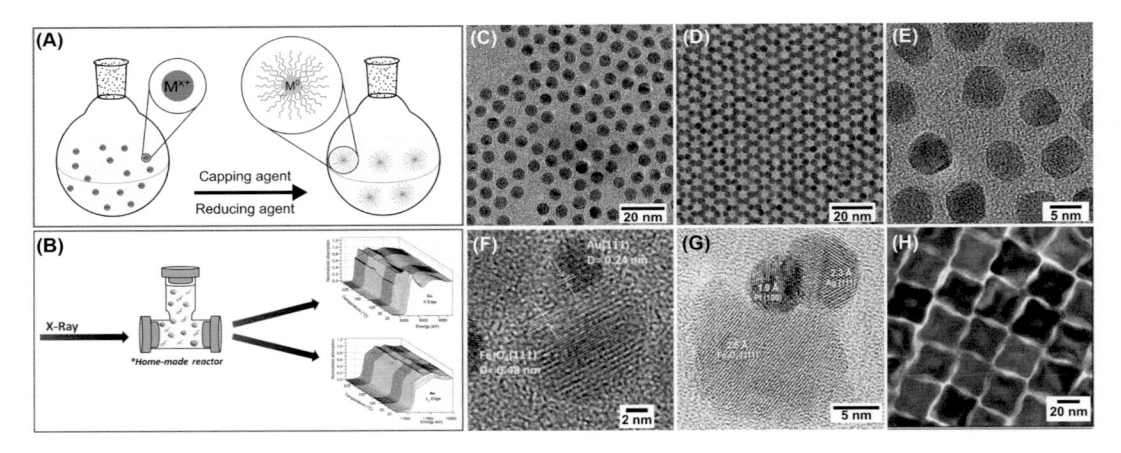

FIGURE 18.4

(A) Schematic representation of the colloidal synthesis, in which the metal M^{x+} precursor (red spheres) is reduced in the presence of capping agents, forming the colloidal metallic nanoparticles (yellow spheres) covered by organic ligands that hinder agglomeration [2]. (B) Apparatus to follow the Au nanoparticles' synthesis by X-ray absorption fine structure spectroscopy. (C—H) Example of nanoparticles obtained by colloidal methods. (C) 6 nm AuCu nanoalloy; scale bar 20 nm [2,30]. (D) Self-assembled 3 nm AuCu nanoalloy, scale bar 20 nm [2]. (E) 6 nm PtNi nanoalloy; scale bar 5 nm. (F) HRTEM of Au—Fe_xO_y dumbbells, highlighting the lattice spacing of Au (111) and Fe_xO_y (111), respectively; scale bar 5 nm. (G) HRTEM image of a representative Ag—Pt—Fe_3O_4 heterotrimer, highlighting the characteristic lattice spacings of Fe_3O_4 (311), Pt (100), and Ag (111); scale 5 nm. (H) Truncated octahedral FeO nanoparticles; scale bar 20 nm.

Adapted from (B) [29]; (F) [31]; (G) [32]; (H) [33].

solvents, injection time, and temperature, among other parameters. The development of special apparatus has allowed the in situ characterization of the nanoparticle growth process (Fig. 18.4B) [29,38]. Fig. 18.4C−H shows selected examples of different nanostructures obtained by colloidal synthesis.

According to classical theories of colloidal chemistry, the preparation of monodisperse nanoparticles (with a narrow size distribution) may be favored by a fast nucleation step followed by a controlled growth step. In the classical scheme proposed by LaMer (Fig. 18.5) [41], the precursors are initially introduced into the solution forming the monomers, in general, complexes or atomic aggregates linked to coordinated ligands/solvent molecules. The monomer concentration rapidly increases, and a nucleation burst occurs when it surpasses the critical supersaturation limit. The binding force or equilibrium constant determines the nucleation rate and nuclei formation concomitantly with the decrease of monomer concentration. Below the critical supersaturation limit, the nuclei and remaining monomers are consumed and organized to form nanoparticles. The nature of the precursors and ligands and the energy involved in the process determine its evolution. The key aspect of obtaining monodisperse nanoparticles is separating the nucleation and growth stages.

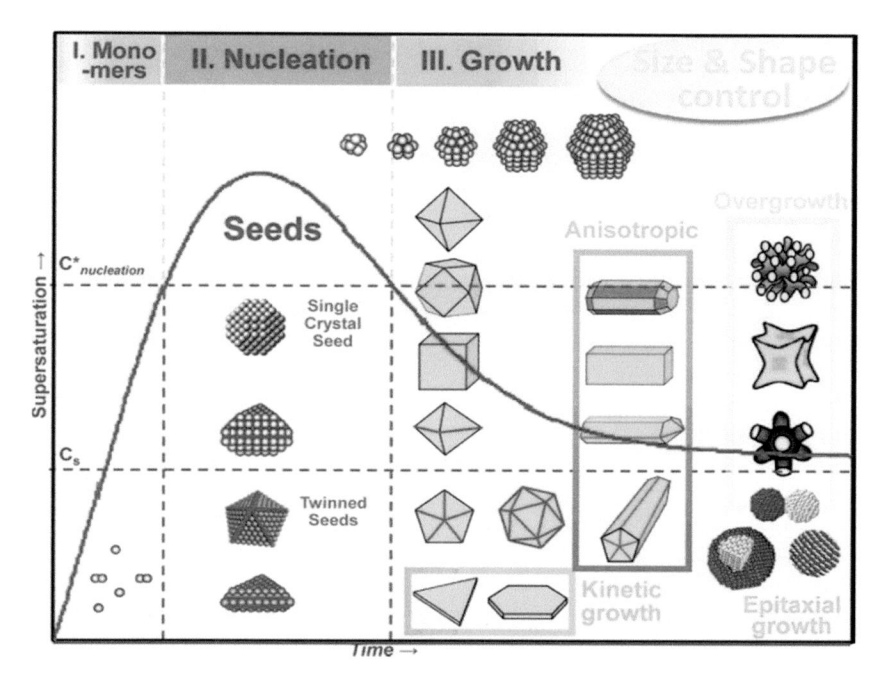

FIGURE 18.5

Schematic illustration of LaMer's proposed mechanism of particle formation, plotting the concentration of monomers against time, showing the different steps: initial stage, nucleation, and growth. The red curve shows the monomers' concentration variation as a time function. Examples of different seeds (nucleation stage) and final particles (growth stage) are shown.

Reproduced with permission from [41].

While the nucleation stage is crucial to determine the size distribution, the growth stage determines the final size and shape of the nanoparticles.

When looking at nanoscale effects, particle size is the primary parameter to control; the right combination of several synthetic parameters leads to size control [35,37,41]. Usually, stronger reducing agents, higher temperatures, and strong bind ligands favor the formation of smaller particles (<10 nm) with narrow size distribution (<10%). The metal precursor and capping ligands ratios are another parameter to be varied to tune the final particle size. Fig. 18.6A shows the changing of Ni nanoparticle size with the synthesis parameters [37]. In the case of larger particles, the seed method or multiple injections can favor particle growth and avoid the formation of new nuclei [42,43]. Pioneer works in the synthesis of colloidal particles and their use in catalysis were performed in the 1990s by El-Sayed and Somorjai groups [44,45] and many other examples

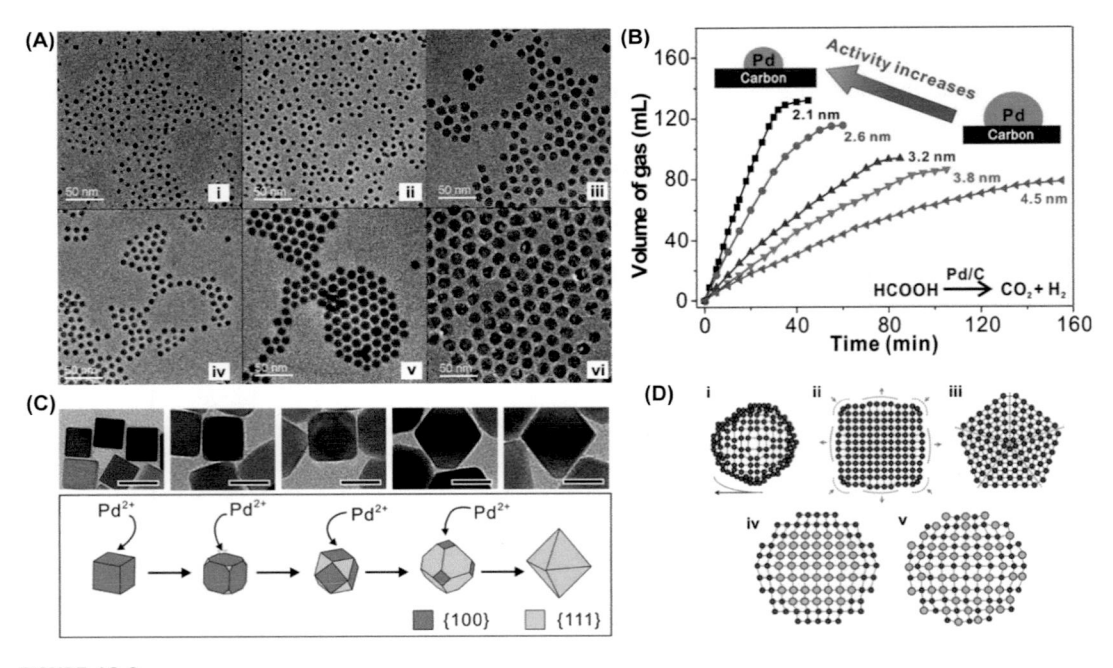

FIGURE 18.6

(A) TEM images of Ni nanoparticles with different sizes: i–vi: 4.8 ± 0.4, 7.5 ± 0.7, 11.7 ± 0.5, 8.8 ± 0.6, 12.8 ± 0.7, and 16.3 ± 0.4, respectively. (B) Plots of the volume of the generated gas ($CO_2 + H_2$) versus time from 5 mL solution of 0.6 M formic acid + 0.6 M sodium formate (additive) in the presence of 55 mg of Pd/C (2.3 wt.% Pd) catalysts with five different particle sizes at 25°C under ambient atmosphere, showing the increase of the catalytic activity with the smaller particle size. (C) TEM images of Pd polyhedrons grown on Pd cubic seeds using different amounts of Pd precursor (scale bar: 10 nm) and a scheme showing how continuous growth on the {100} planes led to the transformation of a Pd cube bound by {100} facets into an octahedron enclosed by {111} facets. (D) Illustration of the different sources of lattice strain: (i) surface relaxation due to size (compressive strain gradient), (ii) anisotropic strain due to shape, (iii) strain at grain boundaries due to twinning and unfilled volume, (iv) strain from epitaxy in core–shell structures, and (v) strain due to alloying.

Adapted from (A) [37]; (B) [42]; (C) [43]; (D) [44].

followed [8,37]. Regarding catalytic activity, using particles of different sizes allowed a clear probing of their impact on structure-sensitive reactions [13,46−48].

Besides size, morphology plays an important role in the properties of nanomaterials and catalysis. There is a direct correlation between the particle shape with the exposed crystallographic facets; the latter determines the local atomic coordination at the surface and, consequently, surface tension and reactivity. Several insightful theoretical and experimental studies on single crystals showed the impact of the crystal plane on catalytic performance. In colloidal synthesis, ligands, coordinated solvents, and ions that promote particle stability, avoiding particle aggregation, have also been explored to induce shape anisotropy during growth [49,50]. For instance, a seed-mediated approach was used to obtain polyhedral nanocrystals of Pd with controlled size, shape, and different proportions of {100} to {111} facets on the surface, including truncated cubes, cuboctahedrons, truncated octahedrons, and octahedrons (Fig. 18.6B) [51]. By varying the ratio of precursor to seed, the nanocrystals evolved from cubes into truncated cubes, cuboctahedrons, truncated octahedrons, and octahedrons due to faster growth along {100} directions relative to {111} directions. The control of the shape by the capping ligands is based on the differences in chemical affinity with the crystallographic facets; the facet with a slower growth rate dominates the surface. For example, citrate ions bind more strongly to {111} facets in Pd, leading to the preferential formation of octahedrons, icosahedrons, and decahedrons [52,53]. On the other hand, PVP/Br$^-$ preferentially binds to the {100} facets leading to the formation of nanocubes (Fig. 18.6C) [54,55]. By tuning the synthesis parameters, shapes such as cubes, octahedrons, tetrahedrons, decahedrons, icosahedrons, polyhedrons, and rods have been obtained with excellent uniformity. Somorjai and collaborators studied in detail hydrogenation and hydrogenolysis reactions and elegantly showed the correspondence of Pt (111) and (100) single crystals with nanoparticles with cubic and cuboctahedral shapes, respectively [56].

Both size decreases and shape modifications induce lattice strain on the surface of the nanoparticles (Fig. 18.6D). The reduction of nanoparticle size leads to an increase in surface energy and, as a result, boosts surface reactivity. In a clean surface nanoparticle, the smaller size causes more compression of atoms, reducing the surface area and the surface energy to compensate for the energy changes in the system [57]. However, considering most systems, including colloidal nanoparticles, other parameters have to be considered, such as capping ligands and impurities at the surface, interaction with support, and reagents, which affect this behavior. Changes in the Pt−Pt bond in Pt nanoparticles (1−3 nm) supported over Al_2O_3 as a function of particle size and adsorbates (H_2, He, CO) was detected; Pt−Pt bond distance of the ~1 nm Pt-in-He nanoparticles contracted about ~1.4%, related to the bulk value. Upon adsorption of H_2 or CO, the Pt−Pt bond distance relaxed towards the bulk Pt value (2.78 Å). By carefully engineering the surface lattice strain degree, molecules' and atoms' adsorption energies can be optimized for a particular chemical reaction [58,59].

Aiming to improve catalytic performance, the preparation of bimetallic nanoparticles has gained much attention. Combining two or more metals can produce unique properties targeting applications based on multifunctionalities. Colloidal nanoparticles with various morphologies/structures have been obtained, such as heterostructure, core−shell, chemically random alloys, intermetallics, and Janus (Fig. 18.7) [60]. The development of synthetic approaches to produce these complex nanoparticles has been one area that has significantly evolved in the last few years.

The composition can be controlled by the coreduction of two metals or chemical substitution to produce bimetallic nanoalloys. These nanoalloys can present a random substitution or form an intermetallic, that is, a chemically ordered alloy with a structure different from the parent metals [62,63]. Random fcc alloy PtFe nanoparticles could be prepared by the colloidal method with a

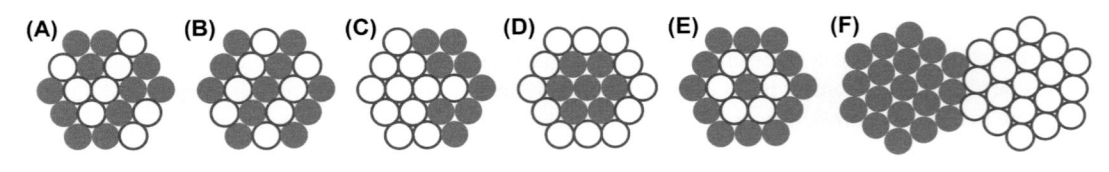

FIGURE 18.7

Examples of bimetallic particles: (A) random and (B) chemically ordered alloy, (C) subcluster with an interface, (D) core—shell, (E) onion-like, (F) heterostructure [61].

Adapted from references [60,61].

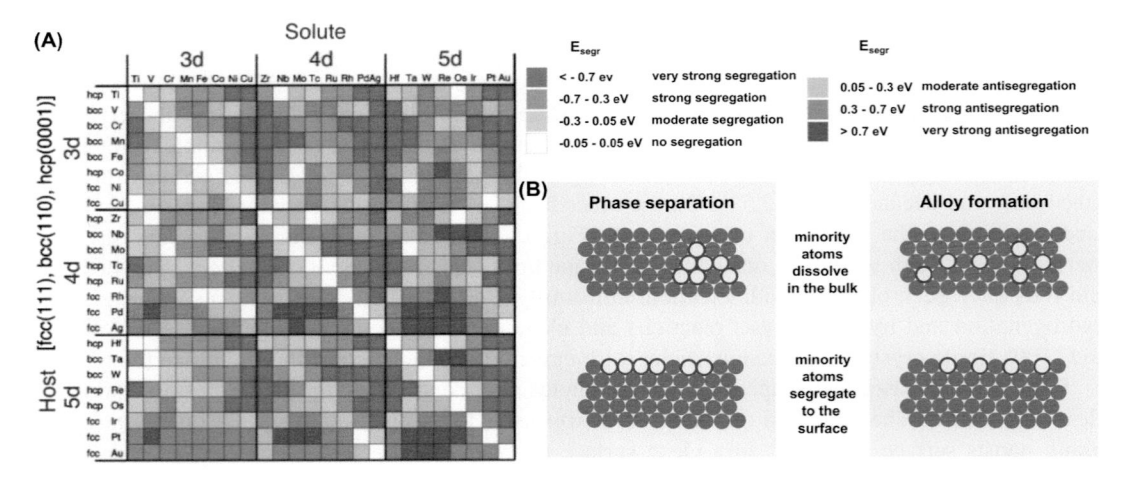

FIGURE 18.8

Surface segregation energies of a metal-transition impurity in closed-packed surfaces of transition metal hosts. From blue to red, the energies show the tendency of the impurity atom to form the alloy (blue) or segregate, forming a secondary phase (red). (B) Schematic representation of the segregation or dissolution of the impurity atoms discussed in (A). Both segregation and dissolution may occur inside the bulk or on the surface of the host structure.

Adapted from [67]; Based on reference [5].

narrow size distribution, tunable size, and composition [64]. An annealing induced the transformation to an ordered face-centered tetragonal (fct) phase, showing more appealing magnetic properties [65]. The occurrence of an ordered fct alloy is also known for other pairs of metals, such as AuCu, and depends on specific metal ratios, such as 1:3, 1:1, and 3:1 [30,66]. The alloy formation can be predicted theoretically or by the empirical rules of Hume—Rothery, according to the similarity of atomic radii, crystal structure, valence, and electronegativity [67,68]. Nevertheless, other factors can affect the atomic distribution in bimetallic nanoparticles, and segregation and enrichment of one metal related to the other on the surface can spontaneously occur. Fig. 18.8 presents the trend to segregate or form an alloy depending on the host and solute atom. Considering an AB bimetallic nanoparticle, the heterometallic A−B bonds must be stronger than the monometallic ones (A−A and B−B) to form the alloy. Also, the element with the lowest surface energy will segregate to the surface. The nanoparticles' surface energy is impacted by organic ligands and/or interaction with support; the metal that binds more strongly to the ligand and/or support surface will segregate.

Many studies in surface science have confirmed that the chemisorption properties of metal alloys differ significantly from those exhibited by pure metals due to modifications in their electronic and structural properties upon contact with different metal elements. A large amount of evidence accumulated on monometallic compounds indicates that the general properties of a metal center depend on three factors: (1) the nature of the metal, (2) the type of species bound to the metal, and (3) the structural geometry of the system. These factors should also influence the heteroatom binding properties in alloys and two mechanisms are behind this effect [68−71]. First, the local geometry of metal alloy is typically different from that of pure metals, known as the ensemble or geometric effect. For instance, the mean metal-to-metal bond length changes to accommodate the strain due to the different atomic sizes, modifying the overlap between some orbitals (Fig. 18.9A,B). Second, the formation of heteroatom bonds changes the electronic environment of the metal surface, known as the ligand effect (Fig. 18.9C). In general, metal alloys' electronic and structural properties are affected by both effects,

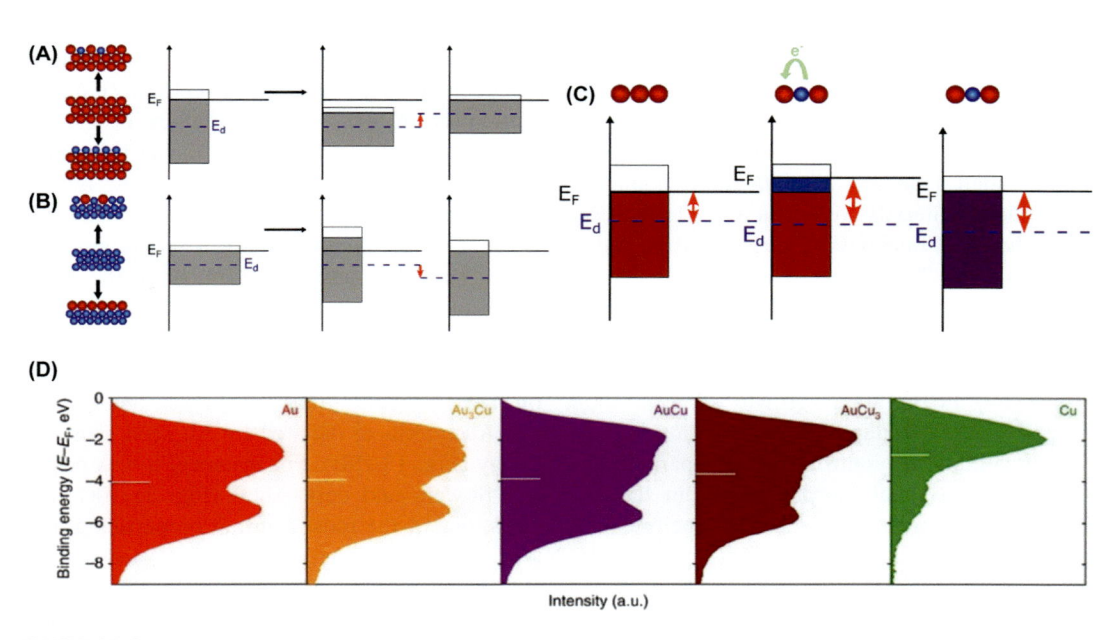

FIGURE 18.9

(A−B) Schematic representation of the effect of adding atoms with different sizes compared with the host structure on the d-band center. (C) Schematic representation of the ligand effect, in which the added atom affects the electronic structure. (A−C). (D) Surface valence band photoemission spectra of AuCu bimetallic nanoparticles and the monometallic counterparts. All the spectra are background corrected. The white bar indicates its center of gravity. For comparison, the upper limit of integration is fixed at −9.0 eV in binding energy.

Adapted from (A,C) [2,70]; (D) [71].

which are usually difficult to distinguish clearly. In addition, the size decrease in nanoparticles increases the percentage of low-coordinated atoms, which can enhance the geometry factor in nanoalloys. The surface energy becomes crucial, leading to unique dissolution/segregation patterns and interfaces not found in the corresponding bulk materials [68]. Metal alloys' chemical reactivity has been associated with shifting the d-band center closer to the Fermi level. Fig. 18.9D shows the valence band (VB) spectra of AuCu nanoparticles collected by XPS (X-ray photoelectron spectroscopy). It was found that the d-band widens, and its center is shifted downward (further away from the Fermi level) as the amount of copper in the alloy composition is increased. Therefore, combining two or more metals can provide an elegant way to finely adjust the electronic and structural properties that are usually unique. It is worth noting that metal−support interaction [30], pretreatments [72], alloy nature (random or chemically ordered) [73], and reaction conditions can deeply affect the alloy stability. It can impact the catalytic activity compared to the monometallic counterparts and the homogeneous alloy since the exposed sites are modified.

Unlike conventional alloys, a core−shell-like structure consists of an arbitrary metal (M) at the core coated by a second metal or oxide (N) layer and is denoted as M@N, Fig. 18.7D. Typically, the synthesis of core−shell structures involves seed-mediated synthesis, where preformed M metal nanoparticles are used as seeds for the nucleation of the second phase, N, which will form the shell around M [74]. The kinetic control of the synthesis conditions can favor the heteronucleation of the second phase on the seed surface against its homonucleation as a separate phase. The difference in the energy threshold between homogenous and heterogeneous nucleation depends on the wettability of the two phases. This parameter also impacts the formation of a core−shell structure or a more complex hybrid nanoparticle. The improvement in the synthesis protocols has led to the formation of core−shell nanoparticles with finely controlled shell thickness, of the order of an atomic monolayer. These materials are characterized by charge transfer and/or a pronounced stress effect according to the shell thickness. Charge transfer and strain can be quite significant for a one-monolayer thickness shell. The formation of imperfect shells, characterized by patches and islands of the shell, can be a consequence of the growth process or the matching of the two materials.

Complex hybrid particles such as dumbbells and yolk particles, among others, have also been synthesized [31,39,74−80]. One of the most studied heterostructures is the Au − Fe_3O_4 system, formed by the epitaxial growth of iron oxide on the Au seeds (heterogeneous nucleation) (Fig. 18.10A). The number of nucleation events at the Au interface, dictated by Au seed size and reaction conditions, controls the final particle morphology (Fig. 18.10B). For instance, it was possible to tune the final morphology from dumbbells to flowers and core−shell (Fig. 18.10C−E). Regarding the application of heterostructures in catalysis, one of the most exciting properties of the hybrid nanoparticles is to favor bifunctional mechanisms in which one specie is activated on the metallic sites (e.g., H_2), and the other is activated in the oxide phase (e.g., oxygenates). The epitaxial growth of the second phase creates a metal−oxide interface free of defects that can enhance charge transfer [31,78].

Independently of the nanoparticles' size, shape, or composition, the colloidal synthesis implies that the particles present capping ligands on the surface of the colloidal nanoparticles. Interesting examples can be found in the literature where the ligands have been exploited to produce a confined environment and tune the catalytic selectivity, mimicking an enzyme active site. However, these ligands may block partially (or totally) the surface sites for applications in gas-phase reactions. The removal of organic ligands can be performed in different ways, and the most common

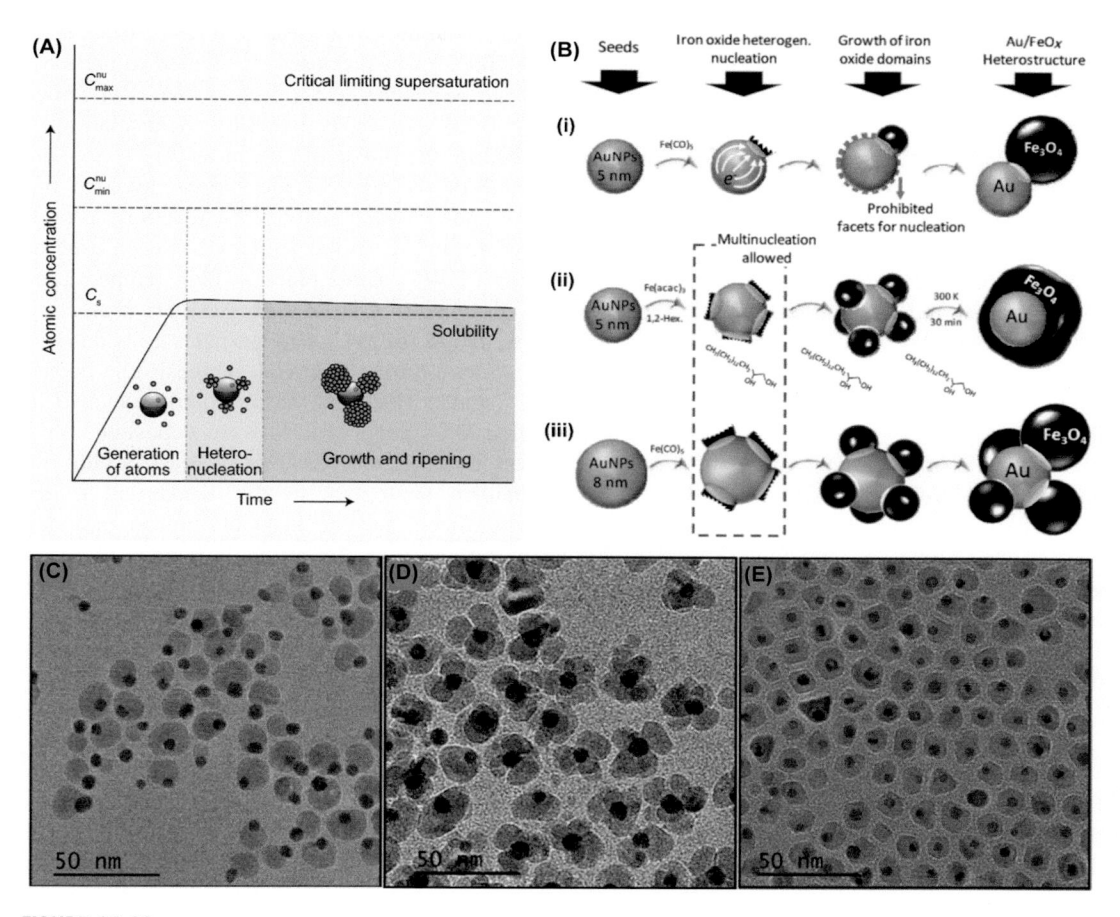

FIGURE 18.10

(A) LaMer's diagram for heterogeneous nucleation. Cs represents the saturated concentration of precursor atoms in the reaction solution at the reaction conditions. C_{numin} and C_{numax} indicate the bottom and top limit concentrations. (B) Schematic representation of Au$-$Fe$_3$O$_4$ heterostructures by the seed-assisted method. The different morphologies can be tuned by parameters such as seed size and solvents/capping ligands. (B$-$i) Formation of dumbbells (single Au/Fe$_3$O$_4$ interface). (B-ii) Multinucleation of Fe$_3$O$_4$ on the Au seed followed by forming a Fe$_3$O$_4$ shell on the Au core. (B-iii) Multinucleation of Fe$_3$O$_4$ on the Au seed forming Au/Fe$_3$O$_4$ flowers-like particles. (C$-$E) TEM images showing Au/Fe$_3$O$_4$ heterostructures with (C) dumbbells, (D) flower-like, and (E) core$-$shell morphologies. Scale bar: 50 nm.

Adapted from (A) [79]; (C$-$E) [81].

procedure, based on different protocols, involves a calcination step under airflow. Despite its efficiency, the thermal treatment may cause significant modifications to the nanoparticles, such as particle agglomeration, loss of shape anisotropy, and segregation. Thus different strategies have been developed to remove the ligands without heating. For instance, an extensive washing step, with the right choice of solvent, helps to clean the surface and facilitate the calcination step [13]. A fast activation protocol (<2 minutes) at high temperatures (>700°C) was proposed. It seemed to preserve the nanoparticles' characteristics, but this treatment has to be optimized for each system [81]. Other strategies have been explored, such as a Pt@SiO$_2$ core—shell catalyst prepared by encapsulating Pt colloidal nanoparticles in a porous SiO$_2$ shell or embedding the nanoparticles in an Al$_2$O$_3$ support produced by the sol—gel method [25,82]. The presence of the shell or the support avoided the growth of Pt nanoparticles during the thermal treatment and provided stability for reactions at high temperatures, such as CH$_4$ oxidation and CH$_4$ steam reforming. A similar strategy has been used to prepare WC$_x$ and Pt/WC$_x$ nanoparticles. The formation of WC$_x$ takes place at high temperatures (>800°C) by the carburation of WO$_x$ species in the presence of CH$_4$ and H$_2$ [83]. By protecting WO$_x$ nanoparticles with a SiO$_2$ shell, it was possible to obtain the WC$_x$ nanoparticles embedded in a sintered SiO$_2$ matrix (or Pt@WC$_x$ if a Pt precursor is added during the WO$_x$ synthesis and coprecipitated as oxy/hydroxy species inside the initial SiO$_2$ shell). The SiO$_2$ matrix could be etched in the presence of carbon black support on which the free WC$_x$ nanoparticles were deposited. As a final example of an elegant and elaborate work, we can highlight the synthesis of Pt@CoO hollow nanocrystals (yolk-shell structure) through a mechanism analogous to the Kinkerdall effect (Fig. 18.4I) [40]. Briefly, PtCo nanoparticles were synthesized, and the Pt@CoO hollow particles were formed under an oxidizing atmosphere due to the difference in the diffusion rates of the oxygen and cobalt atoms at the interface [40,84]. The CoO shell was porous and allowed the catalytic hydrogenation of ethene on the Pt surface [40].

The examples described in this section are just a glimpse of the numerous possibilities of colloidal metal nanoparticles in catalysis.

18.2.2 Nanooxides

Oxides are commonly used in heterogeneous catalysts as bulk catalysis or support; in both cases, the textural properties are important. In the case of bulk catalysts, the high surface area maximizes the exposed catalytic sites and controls the access to reagents and products; in the case of supported catalysts, it favors the dispersion of the metallic phase and can form unique catalytic sites at the metal—oxide interface. Although most of the common oxides used in catalysis are amorphous (i.e., SiO$_2$), low crystalline (γ-Al$_2$O$_3$), or have crystalline domains of nanometric sizes (i.e., Fe$_3$O$_4$), nanometric size effects in oxides are much less explored compared to metallic nanoparticles in heterogeneous catalysis. The oxide surface is usually characterized by Lewis acid sites (the metal cation) and basic sites (oxygen), hydroxy groups, oxygen vacancies, and defects that are highly dependent on the oxide nature and crystallographic plane. One way to classify the oxides is to consider their redox properties: nonreducible and reducible oxides [85,86]. The first class is usually used to disperse the metallic phase due to their low cost and earth abundance (i.e., SiO$_2$ and Al$_2$O$_3$). Although they do not participate directly in catalytic redox steps, it has been shown that the chemical nature of these oxides, particularly the hydroxy groups found on the surface, determines the dispersion and stability of the metallic phase. Nonreducible oxides are also broadly applied in acid—base catalysis, where the Lewis and Bronsted acid sites and oxygen basic sites impact its reactivity. Examples of these oxides are metal alkaline and alkaline earth oxides (e.g., CaO, MgO, ZrO$_2$, La$_2$O$_3$—basic catalysts), mixed oxides (e.g., SiO$_2$-Al$_2$O$_3$—acid catalysts), and zeolites (acid catalysts) [85]. On the other hand, reducible oxides are characterized by cations with variable valence bound to oxygen. The easy change in oxidation state occurs with the loss and gain of lattice oxygen (oxygen storage capability—OSC) that can participate in the reaction mechanism in addition to what is found in nonreducible oxides [86–88]. Transition metal oxides such as CeO_2, TiO_2, and FeO_x are some examples of this important class of materials. This section will focus mainly on reducible oxides.

The contrast in the chemical behavior of nonreducible and reducible oxides can be explained in terms of energy states. The 2p orbitals from oxygen comprise the highest occupied states from the VB. In contrast, the conduction band (CB) presents the lowest unoccupied states composed mainly of empty states from metal [89,90]. When a neutral O atom is removed from a nonreducible oxide, the electrons left in the structure cannot be accommodated in the empty states of metal since its states have high energy. The nonreducible oxides typically present high bandgap (E_g) values (i.e., SiO_2: $E_g = 8–10$ eV). On the other hand, the metals of reducible oxides present empty d orbitals composing the lowest empty states from CB, with a lower bandgap (<3 eV) (Fig. 18.11). The lower energy allows the accommodation of the excess electron in the empty levels of metal, and it is the cause of the change of the oxidation states of the metal (M^{n+} to $M^{(n-1)+}$) in these oxides [87,91].

The size and shape (and consequently the exposure of more reactive facets) are essential characteristics of oxide nanoparticles [92–95]. One of the most studied systems is CeO_2, which has a fluorite cubic structure. While irregular/polyhedra CeO_2 nanoparticles predominantly expose {111} facets (lowest energy), single-crystalline CeO_2 nanorods expose {110} and {100} facets and nanocubes {100} planes. The hydrothermal method has been successfully used to produce CeO_2 and other nanooxides. The size and shape can be controlled by the temperature, base concentration, and Ce precursor, specifically, the counterions of the Ce precursor [96–98]. While a low base concentration leads to nanopolyhedra, a much higher base concentration (>500 times increase) leads to nanorods or nanocubes; the first ones are favored at low temperatures (i.e., 100°C), the latter ones

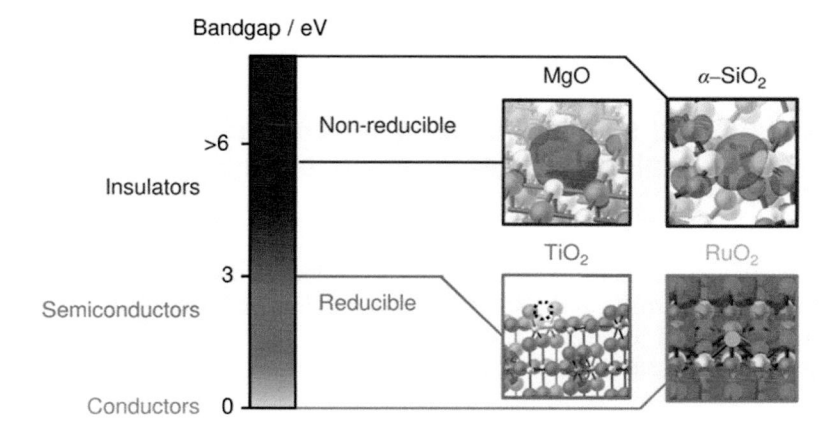

FIGURE 18.11

Schematic representation of the bandgap of reducible and nonreducible oxides.

Reproduced with permission from [89].

are favored at higher temperatures ($140°C-180°C$). It was also found that while Cl^- counterions favored the formation of nanorods, NO_3^- favored the formation of nanocubes [97]. By adding an appropriate amount of NO_3^- to the nanorods synthesis, it was possible to convert the nanorods into nanocubes.

It has been shown that the exposure of more reactive planes improved reactions such as the CO oxidation rate [99]. Catalytic studies of this reaction showed that bare CeO_2 nanorods are more active than cubes and octahedra and were associated with the effect of the exposed surface on the oxygen vacancy formation energy, nature, density of defects, and low coordination sites [99]. On the other hand, this trend can be changed in the case of metal-loaded CeO_2 catalysts. Pd/CeO_2 nanocubes were more active than rods or polyhedral, achieving 100% CO conversion below $100°C$ [100]. A better catalytic performance was also found in CO_2 methanation using Ru supported on CeO_2 nanocubes instead of nanorods and nanopolyhedra [101]. The results were correlated to the higher degree of reduction of CeO_2 in nanocubes in the presence of metal, increasing surface oxygen vacancies. Co/CeO_2 nanocubes also performed better than Co/CeO_2 nanorods in ethanol steam reforming [102]. In this case, the difference was associated with several factors: (1) higher Co dispersion on nanocubes, (2) higher oxygen mobility, (3) higher density of basic surface sites, and (4) easier reducibility. The works listed above also exemplify the key role of metal−oxide interfaces in redox catalytic reactions when dealing with nanometric oxides. Besides the interaction between the metallic phase and oxide support, which can lead to structural and electronic changes in the metallic phase, new catalytic sites can be formed at the interface and nearby regions. In addition, the presence of metallic species on the oxide surface stabilizes oxygen vacancies that would be commonly unstable depending on the reaction conditions [91].

Besides synthesizing nanooxides with different sizes and shapes, using templates to synthesize metal oxides is another exciting approach to producing nanostructured oxide supports with a high surface area to improve the metal−oxide interface [103,104]. For example, mesoporous oxides (Co_3O_4, NiO, MnO_2, Fe_2O_3, and CeO_2) could be produced using a KIT-6 template [105] and used to produce Pt-supported catalysts.

18.3 Challenges and selected state-of-art characterization techniques for nanomaterials and heterogeneous catalysts

In recent decades, the extraordinary development of nanomaterials was only possible due to the dissemination of characterization tools probing at the atomic scale and the continuous development of advanced instrumentation, culminating in a deep understanding of the structure−properties relationship and the impact of size reduction in several systems. Combining techniques is required to characterize nanomaterials and catalysts, from conventional and widely used materials science techniques to more sophisticated techniques using state-of-the-art microscopy techniques and synchrotron radiation. Table 18.1 lists the most common techniques to characterize nanomaterials and heterogeneous catalysts. To probe the impact of reaction conditions on the catalysts' properties, in situ and operando multitechniques instrumentation have been developed. In this section, we highlight a few selected advanced techniques to characterize nanomaterials with a particular interest in heterogeneous catalysts.

Table 18.1 Some of the most common techniques to characterize nanomaterials and heterogeneous catalysts.

N_2 physisorption	Textural properties, surface area, porous size distribution
Chemisorption (CO)	Metallic sites
Programmed temperature techniques (reduction, oxidation, desorption, reaction)	Reducibility, stability of adsorbates, reactivity
Infrared and Raman spectroscopy	Vibrational spectra, adsorbates
X-ray fluorescence, elemental analysis	Chemical composition
X-ray diffraction	Crystalline phase and domain size, unit cell, microstrain
X-ray absorption fine structure spectroscopy (XAFS, XANES, EXAFS)	Local symmetry of the absorbing atom, oxidation states, nearest neighbor coordination numbers, interatomic distances, disorder
X-ray photoemission spectroscopy	Surface atomic composition, oxidation state
Scanning electron microscopy (SEM)	Size, morphology, chemical composition.
Transmission electron microscopy (HRTEM, STEM, EDS, EELS, SAED, ePDF)	Size, morphology, lattice planes and atomic resolution, interface, defects, chemical composition, electronic states
Probe microscopies (STM, AFM)	Surface topology, roughness, adsorbates, electron density

Transmission electron microscopy (TEM) is one of the most essential tools for studying nanomaterials, providing information about size, shape, composition, crystalline structure, composition, and electronic state at the atomic scale [106−133]. The conventional TEM provides images suitable to characterize features typically >2 Å. But in the last years, the popularization of spherical aberration-corrected transmission electron microscopes (AC-TEM) was an inflection point and boosted the comprehension of nanomaterials, particularly supported catalysts. The significant improvement in the size of the electron probe (sub-angstrom) associated with the scanning TEM (STEM) mode and high-angle annular dark-field (HAADF) detector has allowed the clear identification of atomic details, such as the metal−support interface, and the detection of single atoms and small clusters dispersed on supports. The HAADF is a widely employed detection mode used to study many materials, including catalysts, because it provides qualitative information about the composition by the levels of contrast: heavier atoms appear brighter, with intensity (I) $\propto$ to Z^n, $(1.5 \leq n \leq 2.0$ and $Z =$ atomic number) [107]. Fig. 18.12A shows Pt single atoms spread on Fe_2O_3. Since $Z_{Pt} = 79$ and $Z_{Fe} = 27$, the Pt atoms are brighter and are easily visualized. AC-TEM also revealed the presence of highly dispersed CeO_x species on $AuCu/CeO_2−SiO_2$ catalysts, not detected by convention techniques whose amount depends on the CeO_2 loading [108].

In addition to image acquisition, TEM (especially in scanning mode—STEM) also provides quantitative chemical analysis by EDS (energy dispersive spectroscopy) and EELS (electron energy loss spectroscopy) [109−115]. Fig. 18.12B presents three different nanoparticles composed of Pt and Rh: $Pt_{0.5}Rh_{0.5}$ solid solution, Pt(core)−Rh(shell) nanoparticle, and Rh(core)−Pt(shell). This becomes clear by the EDS mapping the elemental distribution in each particle. On the other hand, EELS relies on the energy loss of the direct beam, originated by different processes of interaction with the material, such as excitation of plasmon modes and the photoelectron effect. The analysis of the absorption edges provides information about oxidation states and chemical bondings with atomic resolution [110−112]. For instance, EELS analysis can detect the presence of reduced Ti

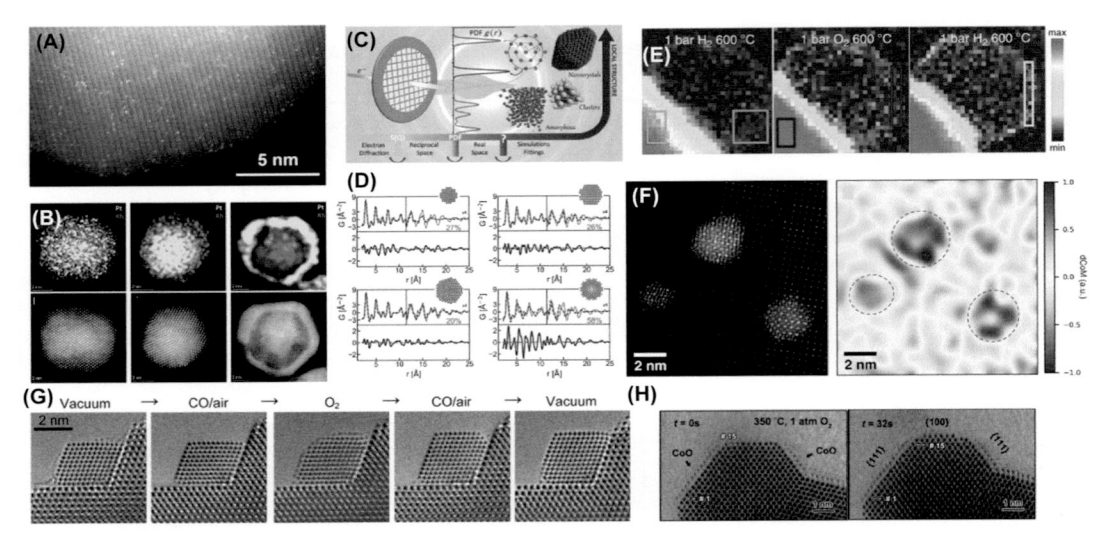

FIGURE 18.12

(A) High-resolution HAADF-STEM image of a $0.53Pt_1/Fe_2O_3$. The bright spots correspond to Pt single atoms spread on the support. (B) EDS elemental maps (top) and respective HAADF-STEM images (bottom) of a $Pt_{0.5}Rh_{0.5}$ solid solution nanoparticle, Pt(core)–Rh(shell) nanoparticle, and Rh(core)–Pt(shell) nanoparticle. Pt and Rh are mapped in green and red, respectively. Scale bar: 2 nm. (C) Schematic representation of a protocol to determine the local structure of nanomaterials using the pair distribution function from electron diffraction. (D) Comparison between measured (continuous line) and optimized ePDF (dashed line) for the four different model nanostructures: (A) fcc, (B) single twinned, (C) D_h, and (D) I_h. (E) EELS mapping (455–470 eV: Ti $L_{2,3}$ edge) of Pt nanoparticles supported on TiO_2 in H_2 at 600°C for 1 hour (left), after a switch to O_2 (middle), and back to H_2 (right). The increase of the signal intensity on the rounded particle indicates the presence of Ti on the Pt nanoparticle surface, corroborating the SMSI effect. (F) In the left, the HAADF image of an O-treated Au nanoparticle supported on $SrTiO_3$ taken before the 4D-STEM map is shown. On the right, the charge density map of the catalyst particles is shown, providing a direct correlation and the visualization of the atomic structure and charge distribution in heterogeneous catalysts. (G) Reversible change in a gold nanoparticle's morphology and surface structure in Au/CeO_2 under different environmental conditions. Scale bar: 2 nm. (H) Bright-field STEM images showing the {100} surface evolution of the oxidized Pt_3Co nanoparticle, with the appearance of a small step on the {100} surface after 32 seconds.

Adapted from [114]; (B) [115]; (C) [116]; (D) [117]; (E) [113]; (F) [118]; (G) [119]; (H) [120].

species on a metallic nanoparticle, proving the occurrence of the strong metal–support interaction (SMSI effect), in which the TiO_2 is reduced and migrates towards the metallic nanoparticle, forming a shell covering the nanoparticle surface (Fig. 18.12E) [113].

Electron diffraction is another essential tool in TEM analysis for determining crystalline structure. More recently, electron diffraction has been going far beyond identifying crystalline phases, allowing the investigation of nanocrystalline and disordered materials. The electron-based atomic pair distribution function (ePDF) has gained much attention for providing an exciting alternative to PDF obtained in synchrotron-light sources [116,121]. Although the data treatment is not trivial, the data acquisition is fast, using only a few μm of the sample (Fig. 18.12C). The ePDF has been used to understand different materials [122], including those related to catalysis, such as metallic nanoparticles [117,123] and metal oxides [124]. Fig. 18.12D presents the use of ePDF to analyze AuAg nanoparticles. More recently, another technique related to electron diffraction has gained much attention due to the possibility of analyzing different materials, including those sensitive to the focused electron beam from the STEM mode. The recent development of new detectors has boosted the use of STEM (particularly in AC-TEM) in many areas [125]. Direct detectors have gained attention in recent years since they allow recording of images with a higher signal-to-noise ratio (SNR) than conventional scintillator-coupled detectors. Thus it is now possible to collect data sets comprising the 2D diffraction patterns from each probe position in a 2D image ($2D \times 2D$). This technique is called "4D-STEM analysis." The 4D datasets are rich in information, including local structure, orientation, deformation, and electromagnetic fields [126,127]. 4D-STEM associated with DFT simulations was successfully used to reveal the charge transfer in a model catalyst composed of Au nanoparticles on $SrTiO_3$ support (Fig. 18.12F) [118].

In situ studies in TEM are challenging but important in catalytic applications since significant modifications can occur under different conditions. The development of specialized sample-holders adapted to conventional microscopes has allowed various in situ studies without a dedicated instrument. These include TEM image acquisition under heat, analysis in a liquid environment, under gas flow, or during electrochemical measurements [38,128,129]. The requirement to enclose the samples when using a liquid or gas environment, such as Si_3N_4 windows, hinders the atomic resolution. A more sophisticated alternative is environment TEM (ETEM). ETEM requires very sophisticated instrumentation, including a powerful differential pumping system to allow the introduction of gases close to the sample while preserving the high vacuum in the other parts of the microscope. In this case, it is possible to obtain detailed information about the effect of different reaction conditions on the surface of metallic nanoparticles [119,130,131]. In particular, evidence of surface changes and SMSI effect have been reported [119,130,132]. Fig. 18.12G shows the dynamic structural changes of Au nanostructures with different sizes supported on CeO_2(111) upon exposure to the vacuum and $CO + O_2$ atmosphere [133]. The reversibility of this process was also demonstrated (Fig. 18.12H) [119]. Fig. 18.12I shows the {100} surface evolution of the oxidized Pt_3Co nanoparticle: after 32 seconds, a small step raised on the {100} proves the highly dynamic evolution of the sample under an O_2 flow and heating [120]. These examples clearly show the impact of the reaction atmosphere in the exposed active sites and the relevance of in situ characterization in catalytic systems.

The advent of synchrotron light radiation was another milestone for materials science characterization. In catalysis, the design of reaction cells coupled to the intense synchrotron beam allowed in situ/in operando characterization of catalytic systems. In several cases, these analyses can reveal unique features of the catalyst that ex situ measurements could not detect once the catalytically active species are only formed under reaction conditions. Here, two of the most used techniques, X-ray diffraction (XRD) and X-ray absorption fine structure (XAFS) spectroscopy, are highlighted. With these two techniques, it is possible to follow the structural and electronic evolution of the catalyst under different reaction atmospheres, temperatures, and pressures and correlate these properties with the catalytic performance. The seminal work on Cu/ZnO catalysts for methanol synthesis showed the significant impact of the in situ characterization (Fig. 18.13) [134]. Besides combining both techniques in the same experiment, modern setups integrated them with other analytical

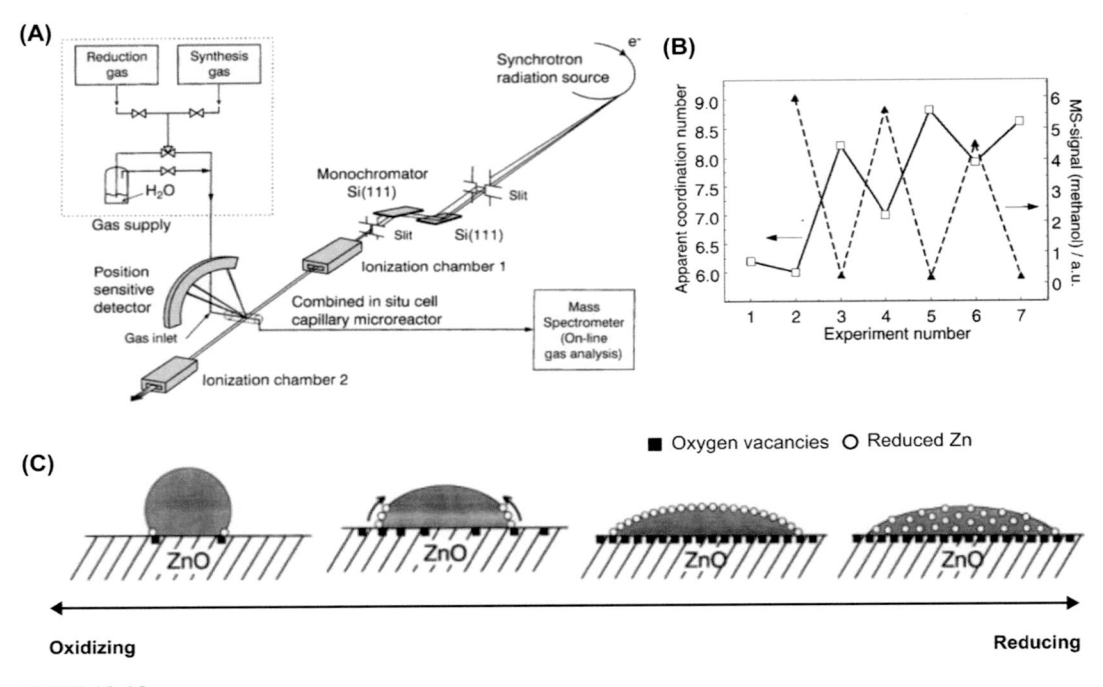

FIGURE 18.13

Seminal work by the Haldor Topsoe team on Cu/ZnO catalyst for methanol synthesis: (A) setup for in situ experiments, combining X-ray diffraction, X-ray absorption fine structure, and mass spectroscopy.
(B) Correlation of apparent coordination number and methanol production. Proposed model for the Cu/ ZnO catalyst Evolution of the Cu/ZnO system according to the atmosphere, from oxidizing to reducing: round Cu particles on ZnO, evolving to disk-like shape Cu particle, Zn—Cu alloy formation on the surface and finally brass alloy formation.

Adapted from reference [134].

techniques, such as Raman and infrared spectroscopies and gas analysis. Further examples of the broad application of synchrotron light in nanomaterials and catalysis and multitechnique approaches can be found elsewhere [135,136].

Although XRD can be performed in a laboratory diffractometer using a conventional X-ray tube as a source, nanomaterials are characterized by a lower SNR and broadening due to the small diffraction volume. Limited amounts of sample and diluted phases also provide limited information that can be obtained. In the case of in situ studies, the attenuation of the incident beam by the experimental cells or furnaces windows and the required time resolution to follow catalyst evolution under reaction conditions are significant barriers for low-intensity conventional X-ray sources. Several examples in the literature show that the high flux and brilliance of the synchrotron light sources have led the characterization of nanomaterials and catalysts to another level [137—141]. Changes in composition, segregation, formation of new phases, presence of microstrain, etc., can be revealed by the analysis of the XRD pattern.

XAFS analysis, including XANES (X-ray absorption near edge structure) and EXAFS (extended X-ray absorption fine structure), relies on energy scanning of the X-ray beam near an absorption edge. Although benchtop equipment became available, this technique mainly uses synchrotron radiation. Nowadays, the development of fast scanning monochromators associated with the high flux of synchrotron radiation allows obtaining a complete spectrum in a few seconds (quick EXAFS) [140,142,143]. It is a chemical-sensitive technique that provides information about the local symmetry of the absorbing atom, oxidation states, nearest neighbor coordination numbers, interatomic distances, and disorder [144]. XAFS does not require long-range atomic order like XRD, which is especially important for nanomaterials and catalysts. Since XRD and XAFS are complementary techniques, their combination gives information on crystalline and disordered species and their chemical nature.

Cu/ZnO catalysts are one of the most studied systems and have showcased the impact of in situ XRD and XAFS characterization (Fig. 18.13). The in situ analysis revealed the modification of Cu morphology depending on the reaction conditions [134]. During heating under a reductive atmosphere, the Cu–O contribution decreases until 200°C; a simultaneous increase of Cu–Cu contribution indicates the reduction of CuO forming metallic Cu. Under harsh reductive conditions (reduction at 600°C), CuZn alloy formation occurs. However, the formation alloy was not found under the typical reaction condition of methanol synthesis ($\sim$220°C) [134]. Nowadays, more sophisticated setups are available in the most important synchrotron laboratories in the world.

Even though XRD and XAFS are powerful tools for studying nanomaterials, they are not surface sensitive, a crucial aspect for several applications. In particular, the atomic chemical composition and electronic state can be different from the average values in nanomaterials since their high surface energy may favor, for example, charge transfer, oxidation, and atomic migration/segregation. In catalysis, understanding surface sites is critical since they are responsible for the activation of the species during the reaction. X-ray photoemission spectroscopy (XPS) is one of the most used surface-sensitive techniques, probing up to a few nanometers in depth by measuring the kinetic energy of photoelectrons ejected from the material's surface. However, XPS is an ultrahigh vacuum (UHV) technique due to the electrons' small inelastic mean free path, as it is mainly used ex situ [145]. This limitation is known as a pressure gap since the surface under reaction conditions can be quite different from those under UHV conditions. The development of ambient-pressure XPS (AP-XPS) was, therefore, a milestone in the study of nanomaterials and catalysis, although still not widely accessible [146–152]. Fig. 18.14A shows a schematic representation of an AP-XPS setup. For example, AP-XPS was used to follow the carbonaceous intermediates formed on Cu-based catalysts (Cu (111)), Cu/ZnO (000$\bar{1}$), CeO$_x$/Cu (111) and (Cu/CeO$_x$/TiO$_2$ (110)) under CO$_2$ conversion to CH$_3$OH [149]. It was possible to identify different intermediates (formate, carbonate, carboxylate) according to the catalyst, which was not formed when the same catalysts were measured under UHV. This result confirms that some adsorbed species only occur under reaction condition. It is possible to obtain different probe depths using different photon energies. This approach was used to estimate the percentage of Pt in each monolayer of PtCo nanoparticles of 4 nm under reducing conditions [150]. Fig. 18.14B shows the AP-XPS spectra and the schematic model corresponding to a seven-layer PtCo nanoparticle, revealing the access of reactants to Co is limited under CO$_2$ hydrogenation reaction conditions. Additionally, AP-XPS revealed significant changes in the surface composition of the PtCo nanoparticles under different gaseous atmospheres. The Pt-enriched surface under the H$_2$ atmosphere evolves to a composition containing a significant presence of Co when the atmosphere is CO and O$_2$ (Fig. 18.14C).

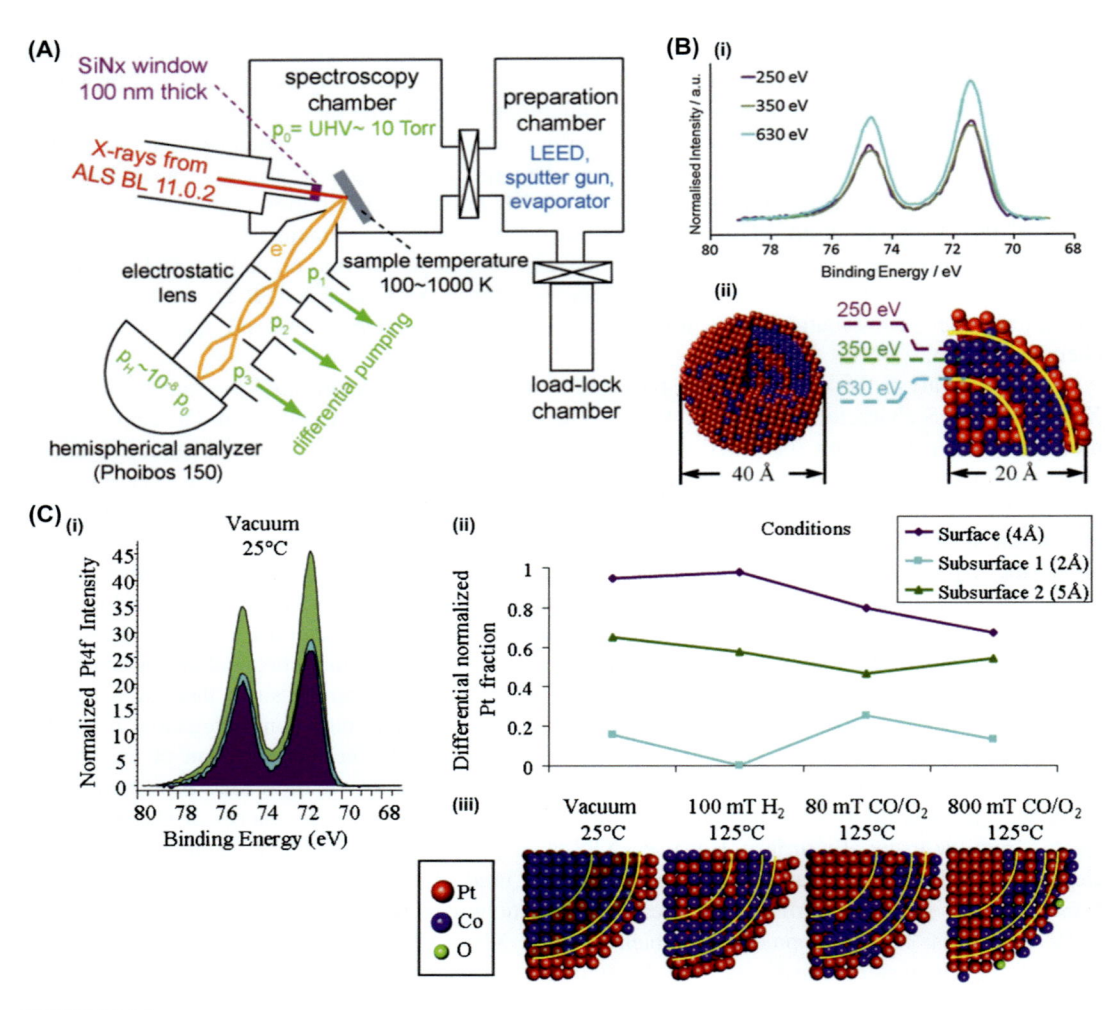

FIGURE 18.14

Schematic representation of the AP-XPS spectrometer at ALS beamline 11.0.2. (B-i) Ambient pressure Pt 4f XP Spectra of $Pt_{50}Co_{50}$, 4 nm nanoparticles during exposure to 0.1 torr of H_2 at three photon energies 250, 350, and 630 eV (probing depths: 0.48, 0.58, and 0.90 nm, respectively). (B-ii) Schematic model corresponding to a 7-layer nanoparticle containing the ratios of Co (blue) and Pt (red) atoms present (left) and a 2-D cross-section with probing depths marked (Co rich/Pt deficient region in between yellow guidelines) (right). (C) (C-i) Ambient pressure Pt 4f XP spectra of 4 nm CoPt alloy nanoparticles under vacuum and at 25°C at photoelectron energies of 250, 350, and 630 eV. (C-ii) Normalized atomic fraction of Pt corresponding to probing depths of around 4 Å, 6 Å, and 11 Å. The line colors correspond to the spectra in (a). (C-iii) Schematic representation showing the corresponding appearance of a cross-section of a 1/8th spherical cone through the nanoparticle. O. Yellow guidelines indicate the probing depths of each photon energy: 0–4 Å, 4–6 Å, and 6–11 Å.

Adapted from (A) [151]; (B) [150]; (C) [152].

While various characterization techniques provide information on the catalyst structure, gathering information about intermediate species on the catalyst surface during the reaction and the active sites over which reactant molecules are bound to form products is challenging. In this respect, infrared spectroscopy can fill the gap and provide information about the nature and strength of the bonds between adsorbed molecules and the catalyst, helping identify intermediate species and their adsorption sites [153]. Diffuse-reflectance infrared Fourier-transform spectroscopy (DRIFTS) has been helpful due to its easy operation. It can be associated with probe molecules, such as CO, CH_3OH, and CH_4, among others, and mass spectrometry to perform temperature-programmed desorption experiments, given information about the stability of the species formed on the surface and the fragmentation pattern as a function of temperature and atmosphere [153,154]. Such speciation can then be used to spot the catalytically active sites and address the importance and role of each site type in a reaction mechanism. Fig. 18.15A shows the DRIFTS study of the inverse ZrO_2/Cu catalyst at 220°C for 90 minutes under the typical CO_2 hydrogenation atmosphere (CO_2 (25%) + H_2 (75%), 8 mL/min) [155]. The reaction behavior of this catalyst was compared with the Cu/ZrO_2 catalyst under the same conditions. The authors found that the vibration peaks related to formate species are significantly weaker compared with the carbonates and methoxy species on the inverse ZrO_2/Cu catalyst. Also, the conversion of carbonates and formates into surface methoxy is

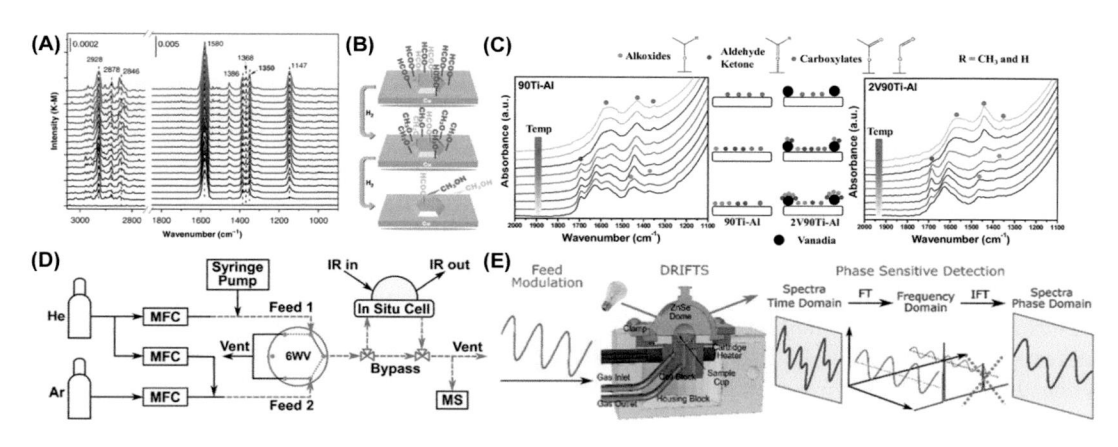

FIGURE 18.15

(A) In situ DRIFTS spectra of the CO_2 + H_2 reaction on ZrO_2/Cu-0.2 catalysts. The catalyst was exposed to a 75% H_2/25% CO_2 (8 mL/min) atmosphere at 220°C for 90 minutes (black to red lines). (B) Schematic representation of the reaction behaviors of the HCOO—Cu and HCOO—Zr intermediates on inverse ZrO_2/Cu catalyst. (A,B) (C) DRIFT spectra of surface species formed during propane adsorption on 90Ti-Al (left) and 2V90Ti-Al (right) catalysts in the temperature range of 293K—633K; in the middle, a schematic representation of the species formed on the catalysts' surface is shown. (D) ME-DRIFTS experimental setup for in situ experiments. MFC = mass flow controller, Ms = mass spectrometer, IR = infrared beam, and 4WV = 4-port two-position (dotted and solid lines) switching valve. (E) Schematic representation of the modulation excitation phase-sensitive detection-diffuse reflectance infrared Fourier transform spectroscopy (ME-PSD-DRIFTS) methodology.

Adapted from (A,B) [155]; (C) [156]; (D) [157]; (E) [158].

much faster when compared with the conventional Cu/ZrO_2 catalyst. Therefore, the activation of CO_2 and hydrogenation of the oxygenate intermediates is accelerated on the inverse catalysts. Fig. 18.15B shows a schematic representation of the proposed mechanism for the ZrO_2/Cu catalyst. Another nice example is the use of in situ DRIFTS to clarify the adsorption of alkanes on titania-doped catalysts (Fig. 18.15C), where the authors found different oxygenated species on the catalysts' surface and that the deposition of surface vanadia species suppresses the formation of precursors of CO_2, providing insights to further studies involving different alkanes [156].

Despite the rich information that DRIFTS can provide, identifying the true active species against spectator entities is challenging. A promising approach has been the use of transient experiments, which explore the perturbation of the catalytic system under a steady state to analyze the time evolution of adsorbed species under reaction conditions. For example, modulated excitation spectroscopy (MES) involves the periodic variation of an external parameter (pressure, temperature, reactant concentration). All species affected by those parameters will vary periodically with the same stimulation frequency, and the phase delay contains information about the kinetics of the process. Coupling MES with in situ DRIFTS (ME-DRIFTS) is a promising technique that allows the detection of species directly involved in the reaction with high sensitivity and selectivity [157−164]. Fig. 18.15D shows a scheme of the ME-DRIFTS experimental setup. Combined with phase-sensitive detection (PSD), it is possible to distinguish a low-intensity response from a high background signal, suppressing the signal coming from the spectator species as well as the background, thus evidencing the response of active entities. Fig. 18.15E presents a schematic representation of the ME-PSD-DRIFTS methodology [159,160]. This methodology has been successfully applied in several reactions by periodically varying the concentration of a reactant (c-MES), such as CO oxidation [161], WGS [162], and reverse WGS [163]. For example, the study of the reverse WGS reaction with a model $Pd/\gamma-Ga_2O_3$ catalyst and the details obtained of the reaction mechanism, identified Ga atoms as the sites for H_2 dissociation, forming Ga-H species that would hydrogenate the carbonate groups formed by CO_2 adsorption. The monodentate was identified as the more active intermediate among the formate species produced [164].

The ME-PSD methodology is relatively new, and its application in well-established reactions is essential to define the method's experiment conditions, limitations, and possibilities. Also, it is interesting to note that ME-DRIFTS can be coupled with other in situ/operando techniques. While the molecular information is obtained, the structural modifications during the reaction can be followed simultaneously by XRD and the electronic and local order changes can be probed by XAFS. Although these facilities are restricted to synchrotron light sources, the possibility of gathering so much information in one experiment has promoted the development of the required instrumentation [164].

18.4 Probing the catalyst complexity: the AuCu system

The AuCu system can be taken as a study case to discuss significant effects at the nanoscale and their impact on catalytic applications. Both Au and Cu elements arrange in an fcc lattice, and they are miscible and form chemically random alloys of any composition with the same fcc structure. Intermetallic phases can also be formed at specific Au:Cu ratios, the most common being 1:3 (fcc),

1:1 (fct), and 3:1 (fcc) Au:Cu ratios [165]. The differences in atomic radius between both elements and the difference in the molar heat of vaporization (related to the cohesion energy of the materials) are less than 15%, favoring the random substitution [166]. In perfect chemically random bulk alloys, the lattice parameters are linearly correlated to the composition, known as Vegard's law. Vegard's law has been extrapolated to the nanoscale and used to infer the chemical composition by XRD through the analysis of the lattice parameter. Commonly, small deviations from linearity occur, which is the case of the Au and Cu system [167]. It is worth noting, as previously discussed, that the Au:Cu ratio also directly impacts the ensemble and strain effects, changing the d-band center and its reactivity (see Fig. 18.9).

Colloidal synthesis has successfully produced AuCu nanoparticles in the $3-20$ nm size range and of variable composition [29,30,66,72,168]. Chemically ordered and random alloys have been prepared. Fig. 18.16 shows an overview of two colloidal synthesis methods to produce AuCu, characterization by TEM, XRD, XAFS (ex situ and in-situ), and proposed models of AuCu formation. In situ measurements during one-pot colloidal synthesis showed that Au seeds are first formed in the reaction medium, followed by Cu reduction and diffusion into the Au lattice forming the AuCu alloy. The final synthesis temperature was the main parameter that determined the Cu incorporation and the formation of chemically ordered alloy [29]. The one-pot approach has limitations, and a more versatile method to vary the final particle size uses preformed Au seeds (two-pot approach) instead of their in situ formation.

The surface energy becomes more important at the nanoscale, and bulk phase diagram modifications are expected. Theoretical and experimental studies indicate the Au tendency to segregate to the surface of bare AuCu nanoparticles (under vacuum). Nevertheless, the interaction of the nanoparticle surface with molecules (ligands, reagents) and support can profoundly affect the energy balance and change the atomic distribution within the nanoparticle. For example, under oxidizing conditions, the most oxyphilic metal segregates to the surface and oxidizes, in this case, forming CuO_x species [169]. The metal$-$support interaction can also work as a driving force to favor segregation at the interface. Theoretical calculations found that the CeO_2 induces the segregation of the Cu toward the AuCu$-CeO_2$ interface; a charge transfer from Cu to CeO_2 takes place, inducing the reduction of the Ce^{4+} to Ce^{3+} at the interface and the formation of CuO_x species by reverse oxygen spillover (Fig. 18.17) [170,171]. Finally, kinetic effects, determined by the synthesis conditions, can also change the trend of segregation [172,173].

The crucial impact of the support interaction and oxidizing and reducing reaction conditions has been nicely demonstrated for the AuCu system. Colloidal AuCu nanoparticles with average size of 14 nm were prepared, and the same batch of particles was used to prepare $AuCu/Al_2O_3$ and AuCu/SiO_2 catalysis [174]. The catalysts were submitted to mild calcination to remove the ligands, and the catalysts were tested in the CO oxidation reaction after oxidation/reduction pretreatments. The CO oxidation reaction ($CO_{(g)} + \frac{1}{2}O_{2(g)} \rightarrow CO_{2(g)}$) is an important reaction to convert CO to less nocive CO_2, and it has also been used as a probe reaction in redox catalysis. Fig. 18.18 shows the strategy used in this work and the main results. EDS-TEM mapping, in situ XRD, and XAFS analysis revealed the dealloy/realloy process that takes place under oxidizing/reducing conditions and the stabilization of Au nanoparticles and CuO_x species under reaction conditions with the formation of Au/CuO_x interface. The type of support directly impacts the formation of the Au/CuO_x species on the nanoparticle$-$support interface. On SiO_2, the Cu segregated from the AuCu alloy forms CuO_x species that remained near the Au nanoparticle. On the other hand, the CuO_x species

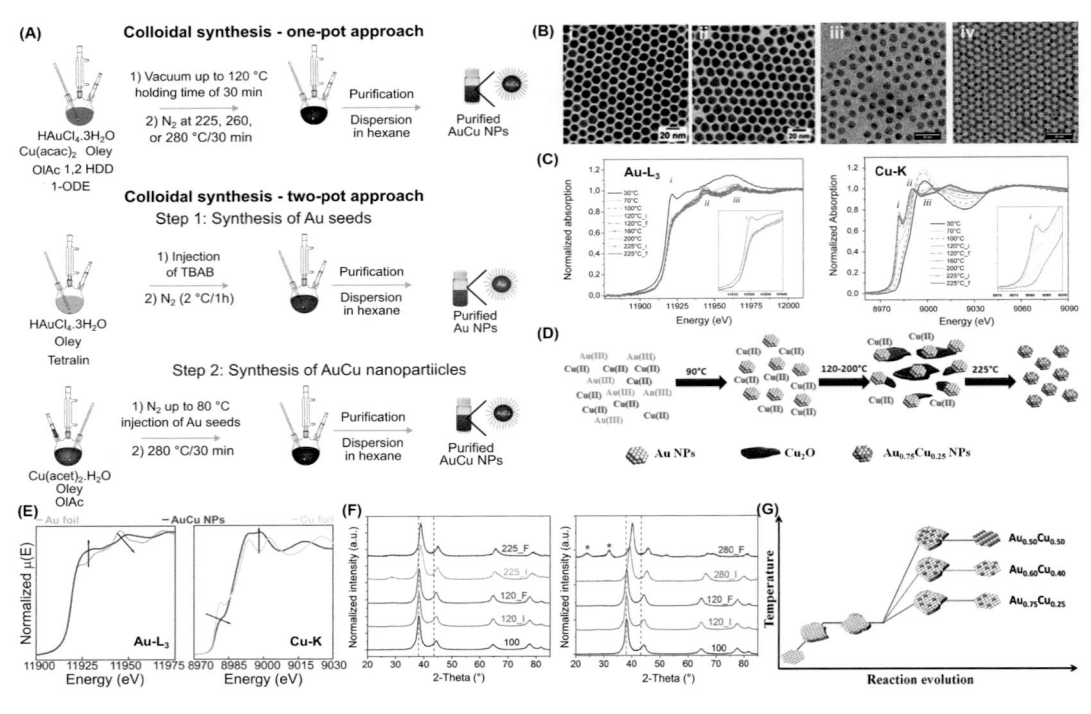

FIGURE 18.16

(A) Schematic representation of experimental protocols to synthesize AuCu nanoparticles by one-pot (top) and two-pot (bottom) approaches. (B) AuCu nanoparticles synthesized by the colloidal method: (B-i) one-pot at 225°C, (B-ii) one-pot at 280°C, (B-iii) two-pot using Au seeds with ∼6 nm, and (B-iv) two-pot using Au seeds with ∼3 nm. (C) In situ XANES spectra at Au-L_3 (left) and Cu-K (right) edges collected during the colloidal synthesis of AuCu nanoparticles by one-pot approach. (D) Schematic representation of the mechanism proposed for the AuCu formation based on the in situ XANES experiment. (E) XANES spectra at Au-L_3 and Cu-K edge from the AuCu nanoparticles shown in (B-iii), with apparent differences from the metallic foils due to alloy formation. (F) X-ray diffraction patterns from the AuCu nanoparticles synthesized by one-pot approach at 225°C and 280°C, corresponding to the nanoparticles shown in (B-i) and (B-ii), respectively. (G) Schematic representation of the different compositions and ordering of the AuCu nanoparticles synthesized by one-pot protocol depending on the final synthesis temperature, based on the XRD patterns shown in (F).

Adapted from (A,E) [30]; (B) [30,168]; (C,D) [29]; (F,G) [168].

supported on Al_2O_3 migrate away and spread on the support surface. The formation of an effective Au/CuO_x interface on SiO_2 led to a more active and stable catalyst for CO oxidation.

In the case of smaller AuCu nanoparticles (3.3 nm) and support with redox properties, such as CeO_2, the initial step to remove the ligands and pretreatments impacted not only the Au/CuO_x−CeO_2 interface but also the final nanoparticle size [175]. For these small particles supported on CeO_2, the activation in O_2 led to more extensive agglomeration and formation of larger particles

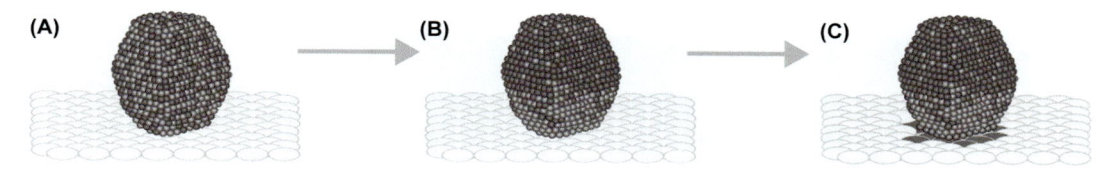

FIGURE 18.17

Schematic representation, based on theoretical calculations from reference [170], of (A) an AuCu nanoparticle in contact with a CeO_2 surface, (B) the Cu migration toward the AuCu–CeO_2 interface, exposing an Au-enriched surface and (C) reduction of Ce^{4+} to Ce^{3+} and Cu oxidation by reverse oxygen spillover at the interface. Au: purple spheres; Cu: green spheres; CeO_2: yellow; Ce^{4+}/Ce^{3+} near the interface: red spheres; and CuOx: blue spheres.

Adapted from [2].

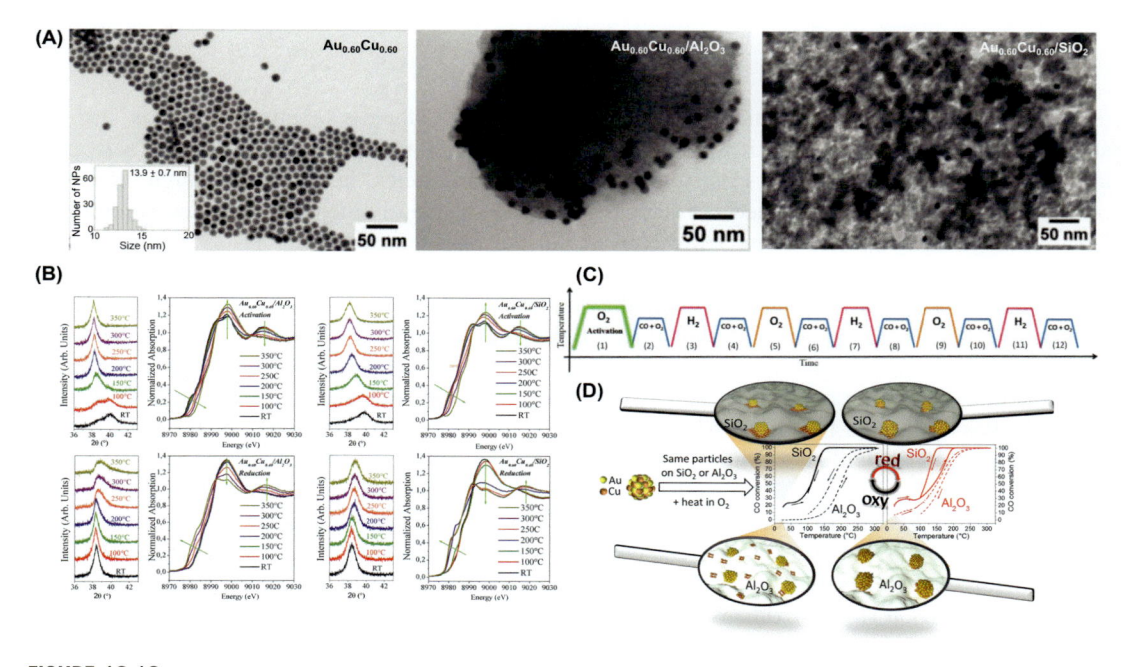

FIGURE 18.18

(A) $Au_{0.60}Cu_{0.60}$ colloidal nanoparticles and the corresponding $Au_{0.60}Cu_{0.60}/Al_2O_3$ and $Au_{0.60}Cu_{0.60}/SiO_2$ catalysts. Scale bar: 50 nm (B) in situ XRD and XAFS analysis of $Au_{0.60}Cu_{0.60}/Al_2O_3$ and $Au_{0.60}Cu_{0.60}/SiO_2$ catalysts under oxidizing and reducing conditions. (C) Catalytic test protocol. (D) The general overview of the performance of the catalysts in CO oxidation and the structural evolution under reaction conditions and pretreatments.

Adapted from reference [174].

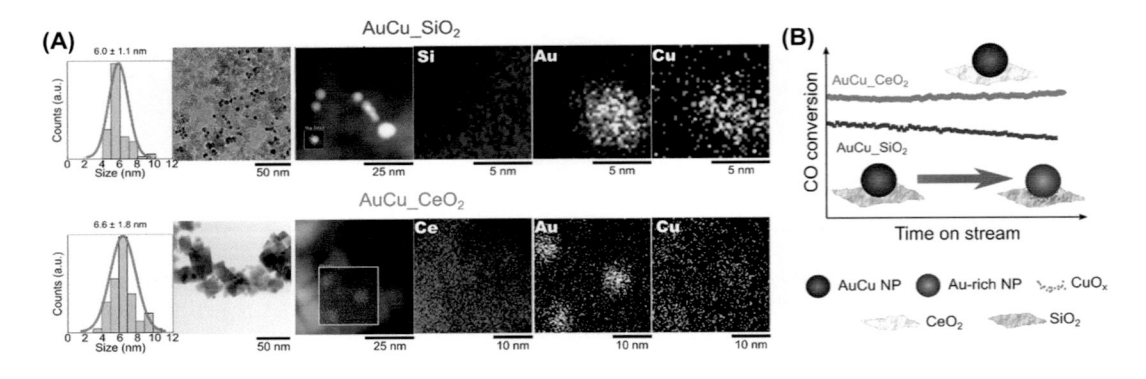

FIGURE 18.19

(A) TEM images of the AuCu/SiO₂ and AuCu/CeO₂ catalysts with the respective histogram of size distribution and EDS maps collected after calcination. (B) The performance of the catalysts in CO-PROX and the schematic representation of the evolution of the AuCu alloy on stream.

Adapted from reference [30].

(9.1 nm) compared to activation in H_2 (5.3 nm). Nevertheless, both conditions led to stable catalysts under the CO oxidation reaction. More interesting was the impact of the activation atmosphere on the CuO_x-CeO_2 species formed at the nanoparticles' interface. Although the nature of the CuO_x-CeO_2 species formed in each case could not be clearly identified, the activation in O_2 led to CuO_x-CeO_2 species being more active than the ones formed under H_2 (3.5 times higher CO_2 rate per Au exposed area in the first case).

CO oxidation reaction conditions favor the dealloy of the original AuCu nanoparticles and the formation of $Au-CuO_x$ species, as shown in the previous examples. A different picture, however, is found when the CO oxidation takes place in H_2-rich stream, known as CO-PROX (CO preferential oxidation) [30]. The analysis of AuCu/SiO₂ and AuCu/CeO₂ catalysts showed that the AuCu/CeO₂ catalysts resemble the AuCu/Al₂O₃ in terms of the formation and spread of the CuO_x phase on the support under oxidizing conditions (Fig. 18.19). The reducing atmosphere of CO-PROX promotes the partial realloy and stabilization of the AuCu nanoparticles in contact with the support. However, the analysis of the catalyst on the stream showed that the AuCu/SiO₂ catalyst was slowly deactivated, whereas the AuCu/CeO₂ catalyst was stable. It was possible to show that the CeO₂ support helps to stabilize the AuCu alloy under CO-PROX reaction conditions. In the case of the SiO₂ support, the Cu slowly oxidizes and segregates. These results showed that the AuCu alloy was more active for CO oxidation and the support that enhances the stability of this phase under reaction conditions leads to a more stable and active catalyst.

The fine-tuning of the impact of CeO₂ in the stability of the AuCu alloy nanoparticles under CO-PROX was evaluated by preparing CeO₂-SiO₂ supports with different nominal loadings (XCe, with $X = 4$, 13, and 19 wt.%) [108]. Fig. 18.20 shows the formation of dispersed CeOx species on the SiO₂ support, highly dispersed Ce atoms, and clusters in the 4Ce support. In situ XRD and XAFS showed the impact of the Ce loading in the extension of the AuCu realloy and the CuO_x-CeO_2 interaction under reducing conditions. The catalytic results showed that while the

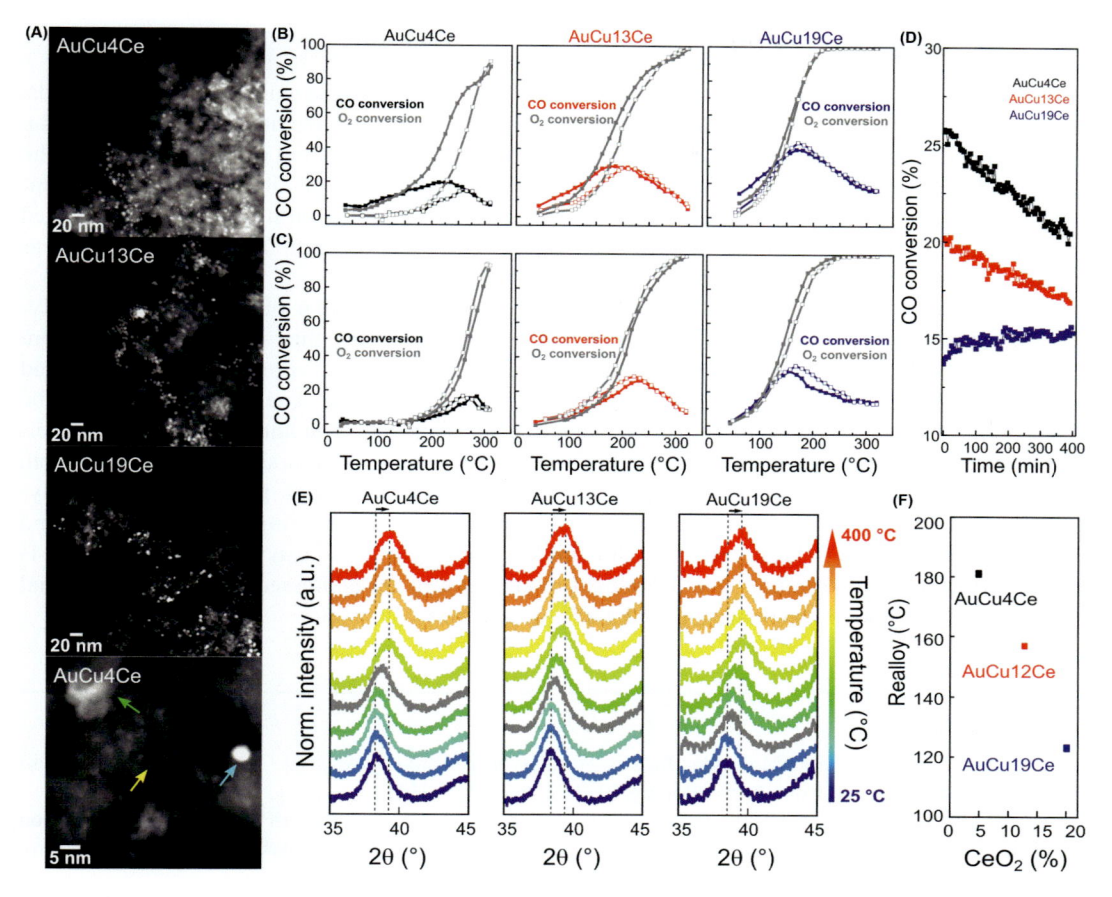

FIGURE 18.20

(A) HAADF-STEM images AuCu4Ce, AuCu13Ce, AuCu19Ce catalysts. Scale bar: 20 nm. At the bottom, a higher-resolution HAADF-STEM image of the AuCu4Ce catalyst is shown. Cyan, green, and yellow arrows indicate an AuCu nanoparticle, a larger CeO_2 particle, and the highly dispersed CeO_x species. Scale bar: 5 nm. (B—C) CO and O_2 conversion of (B) reduced and (C) oxidized AuCu4Ce, AuCu13Ce, and AuCu19Ce catalysts in CO-PROX reaction. (D) CO conversion of AuCuXCe catalysts in CO-PROX reaction at 170°C. (E) In situ X-ray diffraction (XRD) patterns of calcined AuCu4Ce, AuCu13Ce, and AuCu19Ce catalysts, respectively, during the reductive pretreatment (H_2/He 5% 50 mL/min); 5 °C min^{-1}. (F) Plot of the temperature in which the Cu starts to be reincorporated into the Au lattice as a function of the CeO_2 content, obtained from the in situ XRD patterns shown in (e).

Adapted from reference [108].

AuCu4Ce catalyst had the higher initial activity, it deactivated on the stream. The higher CeO_2 loading led to a more stable catalyst.

These results exemplify the large window of opportunities to tune the catalytic performance by the use of alloys and designing the metal—support interface.

18.5 Conclusions

In this chapter, we addressed the straight correlation between the nanoscience and catalysis fields. Strategies based on nanoscience to produce metal-supported catalysts have helped to bridge the gap between surface studies and applied catalysts and build a solid understating of these systems at the atomic scale. Selected works were presented to exemplify the exciting possibilities of tackling fundamental aspects of catalytic reactions. In this context, the use of colloidal nanoparticles associated with advanced characterization techniques has been a powerful approach. The AuCu system was described in more detail as a case study.

Going down in size, single-atom catalysts have become a new frontier of heterogeneous catalysis. Single-atom catalysis optimizes the use of metals by maximizing their dispersion. Still, more than that, it can lead to unique catalytic species due to the strong interaction with the support and reactants. Inspiration in both enzymatic and homogeneous catalysis has been revisited and can be used to guide strategies to design innovative heterogeneous catalysts. The knowledge already developed to manipulate the materials at the atomic and nanoscale can make a significant contribution. Multitechnique instrumentation using synchrotron radiation associated with in situ/operando analysis and AC-TEM and AP-XPS instrumentation has brought the understanding of nanomaterials and catalysts to another level. Their access is still limited due to the high costs and complexity but is expected to evolve in the future. The theoretical modeling was not addressed here, but it is crucial to building knowledge in nanomaterials and catalytic processes.

References

[1] G.A. Somorjai, J.Y. Park, Molecular factors of catalytic selectivity, Angew. Chem. Int. Ed. 47 (2008) 9212−9228. Available from: https://doi.org/10.1002/anie.200803181.

[2] T.E.R. Fiuza, The role of the support in the stabilization of the metallic phase in the preferential oxidation of CO (CO-PROX) reaction, Universidade Estadual de Campinas: Campinas, 2020. Available from: https://doi.org/10.47749/T/UNICAMP.2020.1129315.

[3] T.V.W. Janssens, B.S. Clausen, B. Hvolbæk, H. Falsig, C.H. Christensen, T. Bligaard, et al., Insights into the reactivity of supported Au nanoparticles: combining theory and experiments, Top. Catal. 44 (2007) 15−26. Available from: https://doi.org/10.1007/s11244-007-0335-3.

[4] C. Lentz, S.P. Jand, J. Melke, C. Roth, P. Kaghazchi, DRIFTS study of CO adsorption on Pt nanoparticles supported by DFT calculations, J. Mol. Catal. A Chem. 426 (2017) 1−9. Available from: https://doi.org/10.1016/j.molcata.2016.10.002.

[5] I. Chorkendorff, J.W. Niemantsverdriet, Concepts of Modern Catalysis and Kinetics, Wiley-VCH, 2007. ISBN 978-3-527-31672-4.

[6] M. Boudart, Catalysis by Supported Metals (1969) 153−166.

[7] R.A. Van Santen, Complementary structure sensitive and insensitive catalytic relationships, Acc. Chem. Res. 42 (2009) 57−66. Available from: https://doi.org/10.1021/ar800022m.

[8] M. Cargnello, V.V.T. Doan-Nguyen, T.R. Gordon, R.E. Diaz, E.A. Stach, R.J. Gorte, et al., Control of metal nanocrystal size reveals metal-support interface role for ceria catalysts, Science 341 (2013) 771−773. Available from: https://doi.org/10.1126/science.1240148.

[9] I. Ro, I.B. Aragao, Z.J. Brentzel, Y. Liu, K.R. Rivera-Dones, M.R. Ball, et al., Intrinsic activity of interfacial sites for Pt-Fe and Pt-Mo catalysts in the hydrogenation of carbonyl groups, Appl. Catal. B Environ. 231 (2018) 182−190. Available from: https://doi.org/10.1016/j.apcatb.2018.02.058.

[10] T. Li, F. Liu, Y. Tang, L. Li, S. Miao, Y. Su, et al., Maximizing the number of interfacial sites in single-atom catalysts for the highly selective, solvent-free oxidation of primary alcohols, Angew. Chem. Int. Ed. 57 (2018) 7795−7799. Available from: https://doi.org/10.1002/anie.201803272.

[11] I. Ro, J. Resasco, P. Christopher, Approaches for understanding and controlling interfacial effects in oxide-supported metal catalysts, ACS Catal. 8 (2018) 7368−7387. Available from: https://doi.org/10.1021/acscatal.8b02071.

[12] X.Y. Liu, A. Wang, T. Zhang, C.-Y. Mou, Catalysis by gold: new insights into the support effect, Nano Today 8 (2013) 403−416. Available from: https://doi.org/10.1016/j.nantod.2013.07.005.

[13] R.U. Ribeiro, J.W.C. Liberatori, H. Winnishofer, J.M.C. Bueno, D. Zanchet, Colloidal Co nanoparticles supported on SiO_2: synthesis, characterization and catalytic properties for steam reforming of ethanol, Appl. Catal. B Environ. 91 (2009) 670−678. Available from: https://doi.org/10.1016/j.apcatb.2009.07.009.

[14] C.N. Ávila-Neto, J.W.C. Liberatori, A.M. da Silva, D. Zanchet, C.E. Hori, F.B. Noronha, et al., Understanding the stability of co-supported catalysts during ethanol reforming as addressed by in situ temperature and spatial resolved XAFS analysis, J. Catal. 287 (2012) 124−137. Available from: https://doi.org/10.1016/j.jcat.2011.12.013.

[15] A.L.M. da Silva, J.P. den Breejen, L.V. Mattos, J.H. Bitter, K.P. de Jong, F.B. Noronha, Cobalt particle size effects on catalytic performance for ethanol steam reforming − smaller is better, J. Catal. 318 (2014) 67−74. Available from: https://doi.org/10.1016/j.jcat.2014.07.020.

[16] D. Zanchet, J.B.O. Santos, S. Damyanova, J.M.R. Gallo, J.M. C. Bueno, Toward understanding metal-catalyzed ethanol reforming, ACS Catal. 5 (2015) 3841−3863. Available from: https://doi.org/10.1021/cs5020755.

[17] C.N. Ávila-Neto, D. Zanchet, C.E. Hori, R.U. Ribeiro, J.M.C. Bueno, Interplay between particle size, composition, and structure of $MgAl_2O_4$-supported Co−Cu catalysts and their influence on carbon accumulation during steam reforming of ethanol, J. Catal. 307 (2013) 222−237. Available from: https://doi.org/10.1016/j.jcat.2013.07.025.

[18] M. Shekhar, J. Wang, W.-S. Lee, W.D. Williams, S.M. Kim, E.A. Stach, et al., Size and support effects for the water−gas shift catalysis over gold nanoparticles supported on model Al_2O_3 and TiO_2, J. Am. Chem. Soc. 134 (2012) 4700−4708. Available from: https://doi.org/10.1021/ja210083d.

[19] Y. Yamada, C.-K. Tsung, W. Huang, Z. Huo, S.E. Habas, T. Soejima, et al., Nanocrystal bilayer for tandem catalysis, Nat. Chem. 3 (2011) 372−376. Available from: https://doi.org/10.1038/nchem.1018.

[20] de Jong, K.P. Deposition precipitation, in: K.P. Jong de, (Ed.), Synthesis of Solid Catalysts; Wiley-VCH Verlag GmbH & Co. KGaA, Weinheim, Germany, pp. 111−134.

[21] J. Quinson, S. Neumann, T. Wannmacher, L. Kacenauskaite, M. Inaba, J. Bucher, et al., Colloids for catalysts: a concept for the preparation of superior catalysts of industrial relevance, Angew. Chemie Int. Ed. 57 (2018) 12338−12341. Available from: https://doi.org/10.1002/anie.201807450.

[22] P. Losch, W. Huang, E.D. Goodman, C.J. Wrasman, A. Holm, A.R. Riscoe, et al., Colloidal nanocrystals for heterogeneous catalysis, Nano Today 24 (2019) 15−47. Available from: https://doi.org/10.1016/j.nantod.2018.12.002.

[23] B.J. O'Neill, D.H.K. Jackson, J. Lee, C. Canlas, P.C. Stair, C.L. Marshall, et al., Catalyst Design with Atomic Layer Deposition, ACS Catal. 5 (2015) 1804−1825. Available from: https://doi.org/10.1021/cs501862h.

[24] Z. Kónya, V.F. Puntes, I. Kiricsi, J. Zhu, J.W. Ager, M.K. Ko, et al., Synthetic insertion of gold nanoparticles into mesoporous silica, Chem. Mater. 15 (2003) 1242−1248. Available from: https://doi.org/10.1021/cm020824a.

[25] P.J.S. Prieto, A.P. Ferreira, P.S. Haddad, D. Zanchet, J.M.C. Bueno, Designing Pt nanoparticles supported on CeO_2−Al_2O_3: synthesis, characterization and catalytic properties in the steam reforming and partial oxidation of methane, J. Catal. 276 (2010) 351−359. Available from: https://doi.org/10.1016/j.jcat.2010.09.025.

[26] A.C. Alba-Rubio, C. Sener, S.H. Hakim, T.M. Gostanian, J.A. Dumesic, Synthesis of supported RhMo and PtMo bimetallic catalysts by controlled surface reactions, ChemCatChem 7 (2015) 3881−3886. Available from: https://doi.org/10.1002/cctc.201500767.

[27] I.B. Aragao, I. Ro, Y. Liu, M. Ball, G.W. Huber, D. Zanchet, et al., Catalysts synthesized by selective deposition of Fe onto Pt for the water-gas shift reaction, Appl. Catal. B Environ. 222 (2018) 182−190. Available from: https://doi.org/10.1016/j.apcatb.2017.10.004.

[28] K. Na, Q. Zhang, G.A. Somorjai, Colloidal metal nanocatalysts: synthesis, characterization, and catalytic applications, J. Clust. Sci. 25 (2014) 83−114. Available from: https://doi.org/10.1007/s10876-013-0636-6.

[29] P. Destro, D.A. Cantaneo, D.M. Meira, G. dos Santos Honório, L.S. da Costa, J.M.C. Bueno, et al., Formation of bimetallic copper−gold alloy nanoparticles probed by in situ X-ray absorption fine structure spectroscopy, Eur. J. Inorg. Chem. 2018 (2018) 3770−3777. Available from: https://doi.org/10.1002/ejic.201800413.

[30] T.E.R. Fiuza, D. Zanchet, Supported AuCu alloy nanoparticles for the preferential oxidation of CO (CO-PROX, ACS Appl. Nano Mater. 3 (2020) 923−934. Available from: https://doi.org/10.1021/acsanm.9b02596.

[31] L.S. Costa, D. da; Zanchet, Pretreatment impact on the morphology and the catalytic performance of hybrid heterodimers nanoparticles applied to CO oxidation, Catal. Today 282 (2017) 151−158. Available from: https://doi.org/10.1016/j.cattod.2016.06.056.

[32] J.M. Hodges, A.J. Biacchi, R.E. Schaak, Ternary hybrid nanoparticle isomers: directing the nucleation of Ag on Pt−Fe$_3$O$_4$ using a solid-state protecting group, ACS Nano 8 (2014) 1047−1055. Available from: https://doi.org/10.1021/nn405943z.

[33] Y. Hou, Z. Xu, S. Sun, Controlled synthesis and chemical conversions of FeO nanoparticles, Angew. Chem. Int. Ed. 46 (2007) 6329−6332. Available from: https://doi.org/10.1002/anie.200701694.

[34] W. Baek, H. Chang, M.S. Bootharaju, J.H. Kim, S. Park, T. Hyeon, Recent advances and prospects in colloidal nanomaterials, JACS Au 1 (2021) 1849−1859. Available from: https://doi.org/10.1021/jacsau.1c00339.

[35] N.T.K. Thanh, N. Maclean, S. Mahiddine, Mechanisms of nucleation and growth of nanoparticles in solution, Chem. Rev. 114 (2014) 7610−7630. Available from: https://doi.org/10.1021/cr400544s.

[36] C.-J. Jia, F. Schüth, Colloidal metal nanoparticles as a component of designed catalyst, Phys. Chem. Chem. Phys. 13 (2011) 2457. Available from: https://doi.org/10.1039/c0cp02680h.

[37] H. Winnischofer, T.C.R. Rocha, W.C. Nunes, L.M. Socolovsky, M. Knobel, D. Zanchet, Chemical synthesis and structural characterization of highly disordered Ni colloidal nanoparticles, ACS Nano 2 (2008) 1313−1319. Available from: https://doi.org/10.1021/nn700152w.

[38] J.M. Yuk, J. Park, P. Ercius, K. Kim, D.J. Hellebusch, M.F. Crommie, et al., High-resolution EM of colloidal nanocrystal growth using graphene liquid cells, Science 336 (2012) 61−64. Available from: https://doi.org/10.1126/science.1217654.

[39] J.M. Hodges, J.R. Morse, J.L. Fenton, J.D. Ackerman, L.T. Alameda, R.E. Schaak, Insights into the seeded-growth synthesis of colloidal hybrid nanoparticles, Chem. Mater. 29 (2017) 106−119. Available from: https://doi.org/10.1021/acs.chemmater.6b02795.

[40] Y. Yin, R.M. Rioux, C.K. Erdonmez, S. Hughes, G.A. Somorjai, A.P. Alivisatos, Formation of hollow nanocrystals through the nanoscale kirkendall effect, Science 304 (2004) 711−714. Available from: https://doi.org/10.1126/science.1096566.

[41] K. An, G.A. Somorjai, Size and shape control of metal nanoparticles for reaction selectivity in catalysis, ChemCatChem 4 (2012) 1512−1524. Available from: https://doi.org/10.1002/cctc.201200229.

[42] N.R. Jana, L. Gearheart, C.J. Murphy, Seeding growth for size control of 5 − 40 Nm diameter gold nanoparticles, Langmuir 17 (2001) 6782−6786. Available from: https://doi.org/10.1021/la0104323.

[43] P. Dagtepe, V. Chikan, Quantized ostwald ripening of colloidal nanoparticles, J. Phys. Chem. C. 114 (2010) 16263−16269. Available from: https://doi.org/10.1021/jp105071a.

[44] T.S. Ahmadi, Z.L. Wang, T.C. Green, A. Henglein, M.A. El-Sayed, Shape-controlled synthesis of colloidal platinum nanoparticles, Science 272 (1996) 1924−1925. Available from: https://doi.org/10.1126/science.272.5270.1924.

[45] R.M. Rioux, H. Song, J.D. Hoefelmeyer, P. Yang, G.A. Somorjai, High-surface-area catalyst design: synthesis, characterization, and reaction studies of platinum nanoparticles in mesoporous SBA-15 silica, J. Phys. Chem. B 109 (2005) 2192−2202. Available from: https://doi.org/10.1021/jp048867x.

[46] J. Li, W. Chen, H. Zhao, X. Zheng, L. Wu, H. Pan, et al., Size-dependent catalytic activity over carbon-supported palladium nanoparticles in dehydrogenation of formic acid, J. Catal. 352 (2017) 371−381. Available from: https://doi.org/10.1016/j.jcat.2017.06.007.

[47] K.M. Bratlie, H. Lee, K. Komvopoulos, P. Yang, G.A. Somorjai, Platinum nanoparticle shape effects on benzene hydrogenation selectivity, Nano Lett. 7 (2007) 3097−3101. Available from: https://doi.org/10.1021/nl0716000.

[48] J.N. Kuhn, W. Huang, C.-K. Tsung, Y. Zhang, G.A. Somorjai, Structure sensitivity of carbon − nitrogen ring opening: impact of platinum particle size from below 1 to 5 Nm upon pyrrole hydrogenation product selectivity over monodisperse platinum nanoparticles loaded onto mesoporous silica, J. Am. Chem. Soc. 130 (2008) 14026−14027. Available from: https://doi.org/10.1021/ja805050c.

[49] T.C.R. Rocha, H. Winnischofer, D. Zanchet, Structural aspects of anisotropic metal nanoparticle growth: experiment and theory, Complex-Shaped Metal Nanoparticles, Wiley, 2012, pp. 215−238.

[50] H. Lee, C. Kim, S. Yang, J.W. Han, J. Kim, Shape-controlled nanocrystals for catalytic applications, Catal. Surv. Asia 16 (2012) 14−27. Available from: https://doi.org/10.1007/s10563-011-9130-z.

[51] M. Jin, H. Zhang, Z. Xie, Y. Xia, Palladium nanocrystals enclosed by {100} and {111} facets in controlled proportions and their catalytic activities for formic acid oxidation, Energy Environ. Sci. 5 (2012) 6352−6357. Available from: https://doi.org/10.1039/C2EE02866B.

[52] Y. Xia, Y. Xiong, B. Lim, S.E. Skrabalak, Shape-controlled synthesis of metal nanocrystals: simple chemistry meets complex physics? Angew. Chem. Int. Ed. 48 (2009) 60−103. Available from: https://doi.org/10.1002/anie.200802248.

[53] Y. Xiong, J.M. McLellan, Y. Yin, Y. Xia, Synthesis of palladium icosahedra with twinned structure by blocking oxidative etching with citric acid or citrate ions, Angew. Chem. Int. Ed. 46 (2007) 790−794. Available from: https://doi.org/10.1002/anie.200604032.

[54] B.T. Sneed, M.C. Golden, Y. Liu, H.K. Lee, I. Andoni, A.P. Young, et al., Promotion of the halide effect in the formation of shaped metal nanocrystals via a hybrid cationic, polymeric stabilizer: oCtahedra, cubes, and anisotropic growth, Surf. Sci. 648 (2016) 307−312. Available from: https://doi.org/10.1016/j.susc.2015.12.012.

[55] H. Zhang, M. Jin, Y. Xiong, B. Lim, Y. Xia, Shape-controlled synthesis of Pd nanocrystals and their catalytic applications, Acc. Chem. Res. 46 (2013) 1783−1794. Available from: https://doi.org/10.1021/ar300209w.

[56] C.J. Kliewer, C. Aliaga, M. Bieri, W. Huang, C.-K. Tsung, J.B. Wood, et al., Furan hydrogenation over Pt(111) and Pt(100) single-crystal surfaces and Pt nanoparticles from 1 to 7 Nm: a kinetic and sum frequency generation vibrational spectroscopy study, J. Am. Chem. Soc. 132 (2010) 13088−13095. Available from: https://doi.org/10.1021/ja105800z.

[57] B.T. Sneed, A.P. Young, C.-K. Tsung, Building up strain in colloidal metal nanoparticle catalysts, Nanoscale 7 (2015) 12248−12265. Available from: https://doi.org/10.1039/C5NR02529J.

[58] A. Khorshidi, J. Violet, J. Hashemi, A.A. Peterson, How strain can break the scaling relations of catalysis, Nat. Catal. 1 (2018) 263−268. Available from: https://doi.org/10.1038/s41929-018-0054-0.

[59] F. Liu, C. Wu, G. Yang, S. Yang, CO oxidation over strained Pt(100) surface: a DFT study, J. Phys. Chem. C. 119 (2015) 15500−15505. Available from: https://doi.org/10.1021/acs.jpcc.5b04511.

[60] K. An, S. Alayoglu, T. Ewers, G.A. Somorjai, Colloid chemistry of nanocatalysts: a molecular view, J. Colloid Interface Sci. 373 (2012) 1−13. Available from: https://doi.org/10.1016/j.jcis.2011.10.082.

[61] A. Zaleska-Medynska, M. Marchelek, M. Diak, E. Grabowska, Noble metal-based bimetallic nanoparticles: the effect of the structure on the optical, catalytic and photocatalytic properties, Adv. Colloid Interface Sci. 229 (2016) 80−107. Available from: https://doi.org/10.1016/j.cis.2015.12.008.

[62] G.D. Moon, S. Ko, Y. Min, J. Zeng, Y. Xia, U. Jeong, Chemical transformations of nanostructured materials, Nano Today 6 (2011) 186−203. Available from: https://doi.org/10.1016/j.nantod.2011.02.006.

[63] C.M. Cobley, Y. Xia, Engineering the properties of metal nanostructures via galvanic replacement reactions, Mater. Sci. Eng. R. Rep. 70 (2010) 44−62. Available from: https://doi.org/10.1016/j.mser.2010.06.002.

[64] S. Sun, C.B. Murray, D. Weller, L. Folks, A. Moser, Monodisperse FePt nanoparticles and ferromagnetic FePt nanocrystal superlattices, Science 287 (2000) 1989−1992. Available from: https://doi.org/10.1126/science.287.5460.1989.

[65] S. He, Y. Liu, H. Zhan, L. Guan, Direct thermal annealing synthesis of ordered Pt alloy nanoparticles coated with a thin N-doped carbon shell for the oxygen reduction reaction, ACS Catal. 11 (2021) 9355−9365. Available from: https://doi.org/10.1021/acscatal.1c02434.

[66] G. Guisbiers, S. Mejia-Rosales, S. Khanal, F. Ruiz-Zepeda, R.L. Whetten, M. José-Yacaman, Gold−copper nano-alloy, "tumbaga," in the era of nano: phase diagram and segregation, Nano Lett. 14 (2014) 6718−6726. Available from: https://doi.org/10.1021/nl503584q.

[67] A.V. Ruban, H.L. Skriver, J.K. Nørskov, Surface segregation energies in transition-metal alloys, Phys. Rev. B 59 (1999) 15990−16000. Available from: https://doi.org/10.1103/PhysRevB.59.15990.

[68] R. Ferrando, J. Jellinek, R.L. Johnston, Nanoalloys: from theory to applications of alloy clusters and nanoparticles, Chem. Rev. 108 (2008) 845−910. Available from: https://doi.org/10.1021/cr040090g.

[69] P. Liu, J.K. Nørskov, Ligand and ensemble effects in adsorption on alloy surfaces, Phys. Chem. Chem. Phys. 3 (2001) 3814−3818. Available from: https://doi.org/10.1039/b103525h.

[70] K. Jiang, H.-X. Zhang, S. Zou, W.-B. Cai, Electrocatalysis of formic acid on palladium and platinum surfaces: from fundamental mechanisms to fuel cell applications, Phys. Chem. Chem. Phys. 16 (2014) 20360−20376. Available from: https://doi.org/10.1039/C4CP03151B.

[71] D. Kim, J. Resasco, Y. Yu, A.M. Asiri, P. Yang, Synergistic geometric and electronic effects for electrochemical reduction of carbon dioxide using gold−copper bimetallic nanoparticles, Nat. Commun. 5 (2014) 4948. Available from: https://doi.org/10.1038/ncomms5948.

[72] P. Destro, Colloidal Nanoparticles for Heterogeneous Catalysis; Springer Theses; Springer International Publishing: Cham, 2019; ISBN 978-3-030−03549-5.

[73] W. Zhan, J. Wang, H. Wang, J. Zhang, X. Liu, P. Zhang, et al., Crystal structural effect of AuCu alloy nanoparticles on catalytic CO oxidation, J. Am. Chem. Soc. 139 (2017) 8846−8854. Available from: https://doi.org/10.1021/jacs.7b01784.

[74] L. Carbone, P.D. Cozzoli, Colloidal heterostructured nanocrystals: synthesis and growth mechanisms, Nano Today 5 (2010) 449−493. Available from: https://doi.org/10.1016/j.nantod.2010.08.006.

[75] C. Wang, H. Yin, S. Dai, S. Sun, A general approach to Noble metal − metal oxide dumbbell nanoparticles and their catalytic application for CO oxidation, Chem. Mater. 22 (2010) 3277−3282. Available from: https://doi.org/10.1021/cm100603r.

[76] H. Yu, M. Chen, P.M. Rice, S.X. Wang, R.L. White, S. Sun, Dumbbell-like bifunctional Au − Fe_3O4 nanoparticles, Nano Lett. 5 (2005) 379−382. Available from: https://doi.org/10.1021/nl047955q.

[77] Q. Huang, W. Li, Q. Lin, D. Pi, C. Hu, C. Shao, et al., A review of significant factors in the synthesis of hetero-structured dumbbell-like nanoparticles, Chin. J. Catal. 37 (2016) 681−691. Available from: https://doi.org/10.1016/S1872-2067(15)61069-5.

[78] S. Najafishirtari, T.M. Kokumai, S. Marras, P. Destro, M. Prato, A. Scarpellini, et al., Dumbbell-like $Au_{0.5}Cu_{0.5}@Fe_3O_4$ nanocrystals: synthesis, characterization, and catalytic activity in CO oxidation, ACS Appl. Mater. Interfaces 8 (2016) 28624−28632. Available from: https://doi.org/10.1021/acsami.6b09813.

[79] Y. Sun, Interfaced heterogeneous nanodimers, Natl. Sci. Rev. 2 (2015) 329−348. Available from: https://doi.org/10.1093/nsr/nwv037.

[80] P. Tancredi, L.S. da Costa, S. Calderon, O. Moscoso-Londoño, L.M. Socolovsky, P.J. Ferreira, et al., Exploring the synthesis conditions to control the morphology of gold-iron oxide heterostructures, Nano Res. 12 (2019) 1781−1788. Available from: https://doi.org/10.1007/s12274-019-2431-7.

[81] M. Cargnello, C. Chen, B.T. Diroll, V.V.T. Doan-Nguyen, R.J. Gorte, C.B. Murray, Efficient removal of organic ligands from supported nanocrystals by fast thermal annealing enables catalytic studies on well-defined active phases, J. Am. Chem. Soc. 137 (2015) 6906−6911. Available from: https://doi.org/10.1021/jacs.5b03333.

[82] S. Kim, S. Lee, W. Jung, Sintering resistance of $Pt@SiO_2$ core-shell catalyst, ChemCatChem 11 (2019) 4653−4659. Available from: https://doi.org/10.1002/cctc.201900934.

[83] S.T. Hunt, T. Nimmanwudipong, Y. Román-Leshkov, Engineering non-sintered, metal-terminated tungsten carbide nanoparticles for catalysis, Angew. Chem. Int. Ed. 53 (2014) 5131−5136. Available from: https://doi.org/10.1002/anie.201400294.

[84] W. Wang, M. Dahl, Y. Yin, Hollow nanocrystals through the nanoscale kirkendall effect, Chem. Mater. 25 (2013) 1179−1189. Available from: https://doi.org/10.1021/cm3030928.

[85] J.C. Védrine, Recent developments and prospectives of acid-base and redox catalytic processes by metal oxides, Appl. Catal. A Gen. 575 (2019) 170−179. Available from: https://doi.org/10.1016/j.apcata.2019.02.012.

[86] R.A. van Santen, I. Tranca, E.J.M. Hensen, Theory of surface chemistry and reactivity of reducible oxides, Catal. Today 244 (2015) 63−84. Available from: https://doi.org/10.1016/j.cattod.2014.07.009.

[87] M.V. Ganduglia-Pirovano, Oxygen defects at reducible oxide surfaces: the example of ceria and vanadia, Defects Oxide Surf. 58 (2015) 149−190. ISBN 978-3-319−14366-8.

[88] C.T. Campbell, Chemistry: oxygen vacancies and catalysis on ceria surfaces, Science 309 (2005) 713−714. Available from: https://doi.org/10.1126/science.1113955.

[89] N. Daelman, F.S. Hegner, M. Rellán-Piñeiro, M. Capdevila-Cortada, R. García-Muelas, N. López, Quasi-Degenerate States And Their Dynamics In Oxygen Deficient Reducible Metal Oxides, J. Chem. Phys. 152 (2020) 050901. Available from: https://doi.org/10.1063/1.5138484.

[90] A. Mehonic, A.J. Kenyon, in: J. Jupille, G. Thornton (Eds.), Defects at Oxide Surfaces, Vol. 58, Springer Series in Surface Sciences; Springer International Publishing, Cham, 2015. ISBN 978-3-319−14366-8.

[91] A. Ruiz Puigdollers, P. Schlexer, S. Tosoni, G. Pacchioni, Increasing oxide reducibility: the role of metal/oxide interfaces in the formation of oxygen vacancies, ACS Catal 7 (2017) 6493−6513. Available from: https://doi.org/10.1021/acscatal.7b01913.

[92] K. Zhou, X. Wang, X. Sun, Q. Peng, Y. Li, Enhanced catalytic activity of ceria nanorods from well-defined reactive crystal planes, J. Catal. 229 (2005) 206−212. Available from: https://doi.org/10.1016/j.jcat.2004.11.004.

[93] H.-W. Song, N.-Y. Kim, J. Park, J.-H. Ko, R.J. Hickey, Y.-H. Kim, et al., Shape-controlled syntheses of metal oxide nanoparticles by the introduction of rare-earth metals, Nanoscale 9 (2017) 2732−2738. Available from: https://doi.org/10.1039/C6NR07555J.

[94] S. Maya-Johnson, L. Gracia, E. Longo, J. Andres, E.R. Leite, Synthesis of cuboctahedral CeO_2 nanoclusters and their assembly into cuboid nanoparticles by oriented attachment, ChemNanoMat 3 (2017) 228−232. Available from: https://doi.org/10.1002/cnma.201700005.

[95] A.V. Nikam, B.L.V. Prasad, A.A. Kulkarni, Wet chemical synthesis of metal oxide nanoparticles: a review, CrystEngComm 20 (2018) 5091–5107. Available from: https://doi.org/10.1039/C8CE00487K.

[96] H.-X. Mai, L.-D. Sun, Y.-W. Zhang, R. Si, W. Feng, H.-P. Zhang, et al., Shape-selective synthesis and oxygen storage behavior of ceria nanopolyhedra, nanorods, and nanocubes, J. Phys. Chem. B 109 (2005) 24380–24385. Available from: https://doi.org/10.1021/jp055584b.

[97] Q. Wu, F. Zhang, P. Xiao, H. Tao, X. Wang, Z. Hu, et al., Great influence of anions for controllable synthesis of CeO_2 nanostructures: from nanorods to nanocubes, J. Phys. Chem. C. 112 (2008) 17076–17080. Available from: https://doi.org/10.1021/jp804140e.

[98] Z.-A. Qiao, Z. Wu, S. Dai, Shape-controlled ceria-based nanostructures for catalysis applications, ChemSusChem 6 (2013) 1821–1833. Available from: https://doi.org/10.1002/cssc.201300428.

[99] Z. Wu, M. Li, S.H. Overbury, On the structure dependence of CO oxidation over CeO_2 nanocrystals with well-defined surface planes, J. Catal. 285 (2012) 61–73. Available from: https://doi.org/10.1016/j.jcat.2011.09.011.

[100] G. Li, B. Wu, L. Li, Surface-structure effect of nano-crystalline CeO_2 support on low temperature CO oxidation, J. Mol. Catal. A Chem. 424 (2016) 304–310. Available from: https://doi.org/10.1016/j.molcata.2016.08.035.

[101] F. Wang, C. Li, X. Zhang, M. Wei, D.G. Evans, X. Duan, Catalytic behavior of supported Ru nanoparticles on the {1 0 0}, {1 1 0}, and {1 1 1} facet of CeO_2, J. Catal. 329 (2015) 177–186. Available from: https://doi.org/10.1016/j.jcat.2015.05.014.

[102] I.I. Soykal, B. Bayram, H. Sohn, P. Gawade, J.T. Miller, U.S. Ozkan, Ethanol steam reforming over Co/CeO_2 catalysts: investigation of the effect of ceria morphology, Appl. Catal. A Gen. 449 (2012) 47–58. Available from: https://doi.org/10.1016/j.apcata.2012.09.038.

[103] D.R. Carvalho, I.B. Aragao, D. Zanchet, $Pt-CeO_2$ catalysts synthesized by glucose assisted hydrothermal method: impact of calcination parameters on the structural properties and catalytic performance in PROX-CO, J. Nanosci. Nanotechnol. 18 (2018) 3405–3412. Available from: https://doi.org/10.1166/jnn.2018.14659.

[104] C. Bae, H. Yoo, S. Kim, K. Lee, J. Kim, M.M. Sung, et al., Template-directed synthesis of oxide nanotubes: fabrication, characterization, and applications, Chem. Mater. 20 (2008) 756–767. Available from: https://doi.org/10.1021/cm702138c.

[105] K. An, S. Alayoglu, N. Musselwhite, S. Plamthottam, G. Melaet, A.E. Lindeman, et al., Enhanced CO oxidation rates at the interface of mesoporous oxides and Pt nanoparticles, J. Am. Chem. Soc. 135 (2013) 16689–16696. Available from: https://doi.org/10.1021/ja4088743.

[106] D. Su, Advanced electron microscopy characterization of nanomaterials for catalysis, Green. Energy Env. 2 (2017) 70–83. Available from: https://doi.org/10.1016/j.gee.2017.02.001.

[107] A. Ponce, S. Mejía-Rosales, M. José-Yacamán, Scanning transmission electron microscopy methods for the analysis of nanoparticles, In *Nanoparticles in Biology and Medicine*, Humana Press, Totowa, NJ, 2012, pp. 453–471.

[108] T.E.R. Fiuza, S.D. Gonçalves, D. Zanchet, The impact of ceria loading on the $CuOx - CeO_2$ interaction and performance of $AuCu/CeO_2 - SiO_2$ catalysts in CO-PROX reaction, Eur. J. Inorg. Chem. 2021 (2021) 4222–4229. Available from: https://doi.org/10.1002/ejic.202100561.

[109] D.M. Koshy, G.R. Johnson, K.C. Bustillo, A.T. Bell, Scanning nanobeam diffraction and energy dispersive spectroscopy characterization of a model Mn-promoted Co/Al_2O_3 nanosphere catalyst for Fischer–Tropsch synthesis, ACS Catal. 10 (2020) 12071–12079. Available from: https://doi.org/10.1021/acscatal.0c02546.

[110] R.F. Egerton, M. Watanabe, Characterization of single-atom catalysts by EELS and EDX spectroscopy, Ultramicroscopy 193 (2018) 111–117. Available from: https://doi.org/10.1016/j.ultramic.2018.06.013.

[111] D.G. Stroppa, C.J. Dalmaschio, L. Houben, J. Barthel, L.A. Montoro, E.R. Leite, et al., Analysis of dopant atom distribution and quantification of oxygen vacancies on individual Gd-doped CeO_2 nanocrystals, Chem. − A Eur. J. 20 (2014) 6288−6293. Available from: https://doi.org/10.1002/chem.201400412.

[112] B. He, Y. Zhang, X. Liu, L. Chen, In-situ transmission electron microscope techniques for heterogeneous catalysis, ChemCatChem 12 (2020) 1853−1872. Available from: https://doi.org/10.1002/cctc.201902285.

[113] A. Beck, X. Huang, L. Artiglia, M. Zabilskiy, X. Wang, P. Rzepka, et al., The dynamics of overlayer formation on catalyst nanoparticles and strong metal-support interaction, Nat. Commun. 11 (2020) 3220. Available from: https://doi.org/10.1038/s41467-020-17070-2.

[114] S. Duan, R. Wang, J. Liu, Stability investigation of a high number density $Pt1/Fe_2O_3$ single-atom catalyst under different gas environments by HAADF-STEM, Nanotechnology 29 (2018) 204002. Available from: https://doi.org/10.1088/1361-6528/aab1d2.

[115] M. Jensen, B. Gonano, W. Kierulf-Vieira, P.J. Kooyman, A.O. Sjåstad, Innovative approach to controlled Pt−Rh bimetallic nanoparticle synthesis, RSC Adv. 12 (2022) 19717−19725. Available from: https://doi.org/10.1039/D2RA03373A.

[116] J.B. Souza Junior, G.R. Schleder, J. Bettini, I.C. Nogueira, A. Fazzio, E.R. Leite, Pair distribution function obtained from electron diffraction: an advanced real-space structural characterization tool, Matter 4 (2021) 441−460. Available from: https://doi.org/10.1016/j.matt.2020.10.025.

[117] L.M. Corrêa, M. Moreira, V. Rodrigues, D. Ugarte, Quantitative structural analysis of AuAg nanoparticles using a pair distribution function based on precession electron diffraction: implications for catalysis, ACS Appl. Nano Mater. 4 (2021) 12541−12551. Available from: https://doi.org/10.1021/acsanm.1c02978.

[118] M.J. Zachman, V. Fung, F. Polo-Garzon, S. Cao, J. Moon, Z. Huang, et al., Measuring and directing charge transfer in heterogenous catalysts, Nat. Commun. 13 (2022) 3253. Available from: https://doi.org/10.1038/s41467-022-30923-2.

[119] S. Takeda, Y. Kuwauchi, H. Yoshida, Environmental transmission electron microscopy for catalyst materials using a spherical aberration corrector, Ultramicroscopy 151 (2015) 178−190. Available from: https://doi.org/10.1016/j.ultramic.2014.11.017.

[120] S. Dai, Y. Hou, M. Onoue, S. Zhang, W. Gao, X. Yan, et al., Revealing surface elemental composition and dynamic processes involved in facet-dependent oxidation of Pt3Co nanoparticles via in situ transmission electron microscopy, Nano Lett. 17 (2017) 4683−4688. Available from: https://doi.org/10.1021/acs.nanolett.7b01325.

[121] T.L. Christiansen, S.R. Cooper, K.M.O. Jensen, There's no place like real-space: elucidating size-dependent atomic structure of nanomaterials using pair distribution function analysis, Nanoscale Adv. 2 (2020) 2234−2254. Available from: https://doi.org/10.1039/d0na00120a.

[122] J.B. Souza Junior, G.R. Schleder, F.M. Colombari, M.A. de Farias, J. Bettini, M. van Heel, et al., Pair distribution function from electron diffraction in cryogenic electron microscopy: revealing glassy water structure, J. Phys. Chem. Lett. 11 (2020) 1564−1569. Available from: https://doi.org/10.1021/acs.jpclett.0c00171.

[123] M.M. Hoque, S. Vergara, P.P. Das, D. Ugarte, U. Santiago, C. Kumara, et al., Structural analysis of ligand-protected smaller metallic nanocrystals by atomic pair distribution function under precession electron diffraction, J. Phys. Chem. C. 123 (2019) 19894−19902. Available from: https://doi.org/10.1021/acs.jpcc.9b02901.

[124] G.R. Schleder, G.M. Azevedo, I.C. Nogueira, Q.H.F. Rebelo, J. Bettini, A. Fazzio, et al., Decreasing nanocrystal structural disorder by ligand exchange: an experimental and theoretical analysis, J. Phys. Chem. Lett. 10 (2019) 1471−1476. Available from: https://doi.org/10.1021/acs.jpclett.9b00439.

[125] B.D.A. Levin, Direct detectors and their applications in electron microscopy for materials science, J. Phys. Mater. 4 (2021) 042005. Available from: https://doi.org/10.1088/2515-7639/ac0ff9.

[126] B.H. Savitzky, S.E. Zeltmann, L.A. Hughes, H.G. Brown, S. Zhao, P.M. Pelz, et al., Py4DSTEM: a software package for four-dimensional scanning transmission electron microscopy data analysis, Microsc. Microanal. 27 (2021) 712−743. Available from: https://doi.org/10.1017/S1431927621000477.

[127] R. Ritz, M. Huth, S. Ihle, M. Simson, H. Soltau, V. Migunov, et al., Imaging of electric fields with the PnCCD (S)TEM camera, European Microscopy Congress 2016: Proceedings, Wiley-VCH Verlag GmbH & Co. KGaA, Weinheim, Germany, 2016, pp. 376−377.

[128] M.A.L. Cordeiro, P.A. Crozier, E.R. Leite, Anisotropic nanocrystal dissolution observation by in situ transmission electron microscopy, Nano Lett. 12 (2012) 5708−5713. Available from: https://doi.org/10.1021/nl3029398.

[129] J.E. Evans, K.L. Jungjohann, N.D. Browning, I. Arslan, Controlled growth of nanoparticles from solution with in situ liquid transmission electron microscopy, Nano Lett. 11 (2011) 2809−2813. Available from: https://doi.org/10.1021/nl201166k.

[130] P.A. Crozier, R. Wang, R. Sharma, In situ environmental TEM studies of dynamic changes in cerium-based oxides nanoparticles during redox processes, Ultramicroscopy 108 (2008) 1432−1440. Available from: https://doi.org/10.1016/j.ultramic.2008.05.015.

[131] N. Ta, J. Liu, (Jimmy), S. Chenna, P.A. Crozier, Y. Li, et al., Stabilized gold nanoparticles on ceria nanorods by strong interfacial anchoring, J. Am. Chem. Soc. 134 (2012) 20585−20588. Available from: https://doi.org/10.1021/ja310341j.

[132] H. Yoshida, H. Omote, S. Takeda, Oxidation and reduction processes of platinum nanoparticles observed at the atomic scale by environmental transmission electron microscopy, Nanoscale 6 (2014) 13113−13118. Available from: https://doi.org/10.1039/C4NR04352A.

[133] Y. He, J.-C. Liu, L. Luo, Y.-G. Wang, J. Zhu, Y. Du, et al., Size-dependent dynamic structures of supported gold nanoparticles in CO oxidation reaction condition, Proc. Natl. Acad. Sci. 115 (2018) 7700−7705. Available from: https://doi.org/10.1073/pnas.1800262115.

[134] J.-D. Grunwaldt, A. Molenbroek, N.-Y. Topsøe, H. Topsøe, B. Clausen, In situ investigations of structural changes in Cu/ZnO catalysts, J. Catal. 194 (2000) 452−460. Available from: https://doi.org/10.1006/jcat.2000.2930.

[135] N.S. Marinkovic, S.N. Ehrlich, P. Northrup, Y. Chu, A.I. Frenkel, Synchrotron catalysis consortium (SCC) at NSLS-II: dedicated beamline facilities for in situ and operando characterization of catalysts, Synchrotron Radiat. N. 33 (2020) 4−9. Available from: https://doi.org/10.1080/08940886.2020.1701367.

[136] S.R. Bare, A. Boubnov, J. Hong, A.S. Hoffman, The consortium for operando and advanced catalyst characterization via electronic spectroscopy and structure (Co-ACCESS) at stanford synchrotron radiation lightsource (SSRL), Synchrotron Radiat. N. 33 (2020) 15−19. Available from: https://doi.org/10.1080/08940886.2020.1701369.

[137] L. Mino, E. Borfecchia, J. Segura-Ruiz, C. Giannini, G. Martinez-Criado, C. Lamberti, Materials characterization by synchrotron X-ray microprobes and nanoprobes, Rev. Mod. Phys. 90 (2018) 025007. Available from: https://doi.org/10.1103/RevModPhys.90.025007.

[138] A.I. Frenkel, Q. Wang, N. Marinkovic, J.G. Chen, L. Barrio, R. Si, et al., Combining X-ray absorption and X-ray diffraction techniques for in situ studies of chemical transformations in heterogeneous catalysis: advantages and limitations, J. Phys. Chem. C. 115 (2011) 17884−17890. Available from: https://doi.org/10.1021/jp205204e.

[139] V. Middelkoop, A. Vamvakeros, D. de Wit, S.D.M. Jacques, S. Danaci, C. Jacquot, et al., 3D Printed Ni/Al$_2$O$_3$ based catalysts for CO$_2$ methanation - a comparative and operando XRD-CT study, J. CO$_2$ Util. 33 (2019) 478−487. Available from: https://doi.org/10.1016/j.jcou.2019.07.013.

[140] M.A. Newton, Operando catalysis using synchrotron methods, Catal. Struct. React. 3 (2017) 2−4. Available from: https://doi.org/10.1080/2055074X.2017.1281605.

[141] O. Martin, C. Mondelli, A. Cervellino, D. Ferri, D. Curulla-Ferré, J. Pérez-Ramírez, Operando synchrotron X-ray powder diffraction and modulated-excitation infrared spectroscopy elucidate the CO_2 promotion on a commercial methanol synthesis catalyst, Angew. Chem. Int. Ed. 55 (2016) 11031−11036. Available from: https://doi.org/10.1002/anie.201603204.

[142] C. La Fontaine, S. Belin, L. Barthe, O. Roudenko, V. Briois, ROCK: a beamline tailored for catalysis and energy-related materials from Ms time resolution to Mm spatial resolution, Synchrotron Radiat. N. 33 (2020) 20−25. Available from: https://doi.org/10.1080/08940886.2020.1701372.

[143] L. Plais, C. Lancelot, C. Lamonier, E. Payen, V. Briois, First in situ temperature quantification of CoMoS species upon gas sulfidation enabled by new insight on cobalt sulfide formation, Catal. Today 377 (2021) 114−126. Available from: https://doi.org/10.1016/j.cattod.2020.06.065.

[144] S. Calvin, XAFS for Everyone, CRC Press, 2013. ISBN 9781439878644.

[145] A.M. Venezia, X-ray photoelectron spectroscopy (XPS) for catalysts characterization, Catal. Today 77 (2003) 359−370. Available from: https://doi.org/10.1016/S0920-5861(02)00380-2.

[146] D.E. Starr, H. Bluhm, Z. Liu, A. Knop-Gericke, M. Hävecker, Application of ambient-pressure X-ray photoelectron spectroscopy for the in-situ investigation of heterogeneous catalytic reactions, In-situ Characterization of Heterogeneous Catalysts, John Wiley & Sons, Inc, Hoboken, NJ, USA, 2013, pp. 315−343.

[147] F.(Feng) Tao, Design of an in-house ambient pressure AP-XPS using a bench-top X-ray source and the surface chemistry of ceria under reaction conditions, Chem. Commun. 48 (2012) 3812. Available from: https://doi.org/10.1039/c2cc17715c.

[148] J. Knudsen, J.N. Andersen, J. Schnadt, A versatile instrument for ambient pressure X-ray photoelectron spectroscopy: the lund cell approach, Surf. Sci. 646 (2016) 160−169. Available from: https://doi.org/10.1016/j.susc.2015.10.038.

[149] J. Graciani, K. Mudiyanselage, F. Xu, A.E. Baber, J. Evans, S.D. Senanayake, et al., Highly active copper-ceria and copper-ceria-titania catalysts for methanol synthesis from CO_2, Science 345 (2014) 546−550. Available from: https://doi.org/10.1126/science.1253057.

[150] S. Alayoglu, S.K. Beaumont, F. Zheng, V.V. Pushkarev, H. Zheng, V. Iablokov, et al., CO_2 hydrogenation studies on Co and CoPt bimetallic nanoparticles under reaction conditions using TEM, XPS and NEXAFS, Top. Catal. 54 (2011) 778−785. Available from: https://doi.org/10.1007/s11244-011-9695-9.

[151] S. Yamamoto, H. Bluhm, K. Andersson, G. Ketteler, H. Ogasawara, M. Salmeron, et al., In situ X-ray photoelectron spectroscopy studies of water on metals and oxides at ambient conditions, J. Phys. Condens. Matter 20 (2008) 184025. Available from: https://doi.org/10.1088/0953-8984/20/18/184025.

[152] F. Zheng, S. Alayoglu, V.V. Pushkarev, S.K. Beaumont, C. Specht, F. Aksoy, et al., In situ study of oxidation states and structure of 4nm CoPt bimetallic nanoparticles during CO oxidation using X-ray spectroscopies in comparison with reaction turnover frequency, Catal. Today 182 (2012) 54−59. Available from: https://doi.org/10.1016/j.cattod.2011.10.009.

[153] E. Stavitski, Infrared spectroscopy on powder catalysts, In-situ Characterization of Heterogeneous Catalysts, John Wiley & Sons, Inc, Hoboken, NJ, USA, 2013, pp. 241−265.

[154] F. Zaera, New advances in the use of infrared absorption spectroscopy for the characterization of heterogeneous catalytic reactions, Chem. Soc. Rev. 43 (2014) 7624−7663. Available from: https://doi.org/10.1039/C3CS60374A.

[155] C. Wu, L. Lin, J. Liu, J. Zhang, F. Zhang, T. Zhou, et al., Inverse ZrO_2/Cu as a highly efficient methanol synthesis catalyst from CO_2 hydrogenation, Nat. Commun. 11 (2020) 5767. Available from: https://doi.org/10.1038/s41467-020-19634-8.

[156] D. Shee, G. Deo, In situ DRIFT studies of alkane adsorption on vanadia supported titania-doped catalysts, Catal. Today 325 (2019) 25−32. Available from: https://doi.org/10.1016/j.cattod.2018.06.003.

[157] B.S. Patil, P.D. Srinivasan, E. Atchison, H. Zhu, J.J. Bravo-Suárez, Design, modelling, and application of a low void-volume in situ diffuse reflectance spectroscopic reaction cell for transient catalytic studies, React. Chem. Eng. 4 (2019) 667−678. Available from: https://doi.org/10.1039/C8RE00302E.

[158] P.D. Srinivasan, B.S. Patil, H. Zhu, J.J. Bravo-Suárez, Application of modulation excitation-phase sensitive detection-DRIFTS for in situ/operando characterization of heterogeneous catalysts, React. Chem. Eng. 4 (2019) 862−883. Available from: https://doi.org/10.1039/C9RE00011A.

[159] A. Urakawa, T. Bürgi, A. Baiker, Sensitivity enhancement and dynamic behavior analysis by modulation excitation spectroscopy: principle and application in heterogeneous catalysis, Chem. Eng. Sci. 63 (2008) 4902−4909. Available from: https://doi.org/10.1016/j.ces.2007.06.009.

[160] P. Müller, I. Hermans, Applications of modulation excitation spectroscopy in heterogeneous catalysis, Ind. Eng. Chem. Res. 56 (2017) 1123−1136. Available from: https://doi.org/10.1021/acs.iecr.6b04855.

[161] E. Del Río, S.E. Collins, A. Aguirre, X. Chen, J.J. Delgado, J.J. Calvino, et al., Reversible deactivation of a $Au/Ce_{0.62}Zr_{0.38}O_2$ catalyst in CO oxidation: a systematic study of CO_2-triggered carbonate inhibition, J. Catal. 316 (2014) 210−218. Available from: https://doi.org/10.1016/j.jcat.2014.05.016.

[162] J. Vecchietti, A. Bonivardi, W. Xu, D. Stacchiola, J.J. Delgado, M. Calatayud, et al., Understanding the role of oxygen vacancies in the water gas shift reaction on ceria-supported platinum catalysts, ACS Catal. 4 (2014) 2088−2096. Available from: https://doi.org/10.1021/cs500323u.

[163] A. Aguirre, S.E. Collins, Selective detection of reaction intermediates using concentration-modulation excitation DRIFT spectroscopy, Catal. Today 205 (2013) 34−40. Available from: https://doi.org/10.1016/j.cattod.2012.08.020.

[164] J. Nilsson, P.-A. Carlsson, N.M. Martin, E.C. Adams, G. Agostini, H. Grönbeck, et al., Methane oxidation over Pd/Al_2O_3 under rich/lean cycling followed by operando XAFS and modulation excitation spectroscopy, J. Catal. 356 (2017) 237−245. Available from: https://doi.org/10.1016/j.jcat.2017.10.018.

[165] Z. Xu, E. Lai, Y. Shao-Horn, K. Hamad-Schifferli, Compositional dependence of the stability of AuCu alloy nanoparticles, Chem. Commun. 48 (2012) 5626. Available from: https://doi.org/10.1039/c2cc31576a.

[166] R. Ferro, A. Saccone, Elements of alloying behaviour systematics, Intermetallic Chemistry, Elsevier, 2008.

[167] H. Okamoto, D.J. Chakrabarti, D.E. Laughlin, T.B. Massalski, The Au − Cu (gold-copper) system, J. Phase Equilibria 8 (1987) 454. Available from: https://doi.org/10.1007/BF02893155.

[168] P. Destro, M. Colombo, M. Prato, R. Brescia, L. Manna, D. Zanchet, $Au_{1−x}Cu_x$ colloidal nanoparticles synthesized via a one-pot approach: understanding the temperature effect on the Au:Cu ratio, RSC Adv. 6 (2016) 22213−22221. Available from: https://doi.org/10.1039/C6RA02027E.

[169] A. Wang, X.Y. Liu, C.-Y. Mou, T. Zhang, Understanding the synergistic effects of gold bimetallic catalysts, J. Catal. 308 (2013) 258−271. Available from: https://doi.org/10.1016/j.jcat.2013.08.023.

[170] S. Chen, L. Li, W. Hu, X. Huang, Q. Li, Y. Xu, et al., Anchoring high-concentration oxygen vacancies at interfaces of $CeO_{2−x}/Cu$ toward enhanced activity for preferential CO oxidation, ACS Appl. Mater. Interfaces 7 (2015) 22999−23007. Available from: https://doi.org/10.1021/acsami.5b06302.

[171] L. Zhang, H.Y. Kim, G. Henkelman, CO oxidation at the Au−Cu interface of bimetallic nanoclusters supported on CeO_2 (111), J. Phys. Chem. Lett. 4 (2013) 2943−2947. Available from: https://doi.org/10.1021/jz401524d.

[172] I. Zegkinoglou, L. Pielsticker, Z.-K. Han, N.J. Divins, D. Kordus, Y.-T. Chen, et al., Surface segregation in CuNi nanoparticle catalysts during CO_2 hydrogenation: the role of CO in the reactant mixture, J. Phys. Chem. C. 123 (2019) 8421−8428. Available from: https://doi.org/10.1021/acs.jpcc.8b09912.

[173] M. Nolan, Charge transfer and formation of reduced Ce^{3+} upon adsorption of metal atoms at the ceria (110) surface, J. Chem. Phys. 136 (2012) 134703. Available from: https://doi.org/10.1063/1.3697485.

[174] P. Destro, T.M. Kokumai, A. Scarpellini, L. Pasquale, L. Manna, M. Colombo, et al., The crucial role of the support in the transformations of bimetallic nanoparticles and catalytic performance, ACS Catal. 8 (2018) 1031−1037. Available from: https://doi.org/10.1021/acscatal.7b03685.

[175] T.E.R. Fiuza, D.S. Gonçalves, I.F. Gomes, D. Zanchet, CeO_2-supported Au and AuCu catalysts for CO oxidation: impact of activation protocol and residual chlorine on the active sites, Catal. Today (2020). Available from: https://doi.org/10.1016/j.cattod.2020.07.034.

Design of supramolecular chemosensor arrays and their applications to optical chips

19

Yui Sasaki and Tsuyoshi Minami

Institute of Industrial Science, The University of Tokyo, Tokyo, Japan

19.1 Introduction

From the viewpoint of food analysis, environmental monitoring, and diagnosis, simultaneous detection of multiple analytes (i.e., chemical species and ions) in real samples is in high demand to obtain accurate results [1,2]. To date, such nonvisible analytes in real samples have been assessed by utilizing instrumental methods including spectrophotometer, voltammetry, mass spectrometer, and high-performance liquid chromatography. Although those instrumental technologies have achieved accurate real-sample analysis, the requirement of large equipment and trained personnel, the time-consuming measurements, and the complicated preprocessing of samples could be a concern for the realization of on-site analysis. Therefore, the size-miniaturization of the analytical devices is desirable.

According to the definition in the official journal of the International Union of Pure and Applied Chemistry, a term of a chemical sensor is described as "a device that transforms chemical information, ranging from the concentration of a specific sample component to total composition analysis, into an analytically useful signal." [3] In other words, chemical sensors can be referred to as systemized devices consisting of receptors to detect analytes and transducers to transfer chemical information upon detecting analytes. The types of transducers are classified into optical, electrochemical, electric, mass sensitive, magnetic, thermometric, and other devices for physical stimuli such as radiation. Among them, optical responses are promising candidates as simple output signals because chemical information based on analyte detection can be easily recognized by the naked eye [4]. In chemical sensors, the receptors are generally classified into artificial and biogenic materials, of which the design defines the ability of chemical sensors [3]. In this chapter, strategically designed receptors based on host−guest chemistry are defined as artificial materials [5−13]. Detection mechanisms of chemical sensing using artificial receptors or biogenic materials (i.e., enzymes and antibodies) rely on noncovalent interactions between receptor units and analytes. Notably, biogenic materials are superior to artificial receptors for selective detection because of ideal complementarity against analytes, which are typical lock-and-key models [14,15]. On the other hand, artificial receptors are superior to biogenic materials for multidetection because of the inherent cross-reactivity [16,17]. Such multisensing systems, which are referred to as sensor arrays [18−20] could capture multiple analytes simultaneously and perform powerful analyses using

Materials Nanoarchitectonics. DOI: https://doi.org/10.1016/B978-0-323-99472-9.00014-6

computational techniques (i.e., pattern recognition methods) [17,20−24]. Given the chemical and physical robustness of synthetic materials, artificial receptors can be potentially employed as feasible detection portions of chemical sensors in real-world scenarios [25]. The smallest sized-chemical sensors utilizing artificial receptors are chemosensors [26−29], which can visualize molecular recognition information through optical responses (i.e., color or fluorescence changes) by capturing analytes (Fig. 19.1). Therefore, chemosensor arrays combined with pattern recognition techniques are promising molecular sensor platforms for on-site analysis. Indeed, the chemosensor arrays have successfully classified and quantified analytes in real samples (biomarkers, food samples, drinks, hazardous, etc.), whereas the sensing system of the solution-state chemosensor arrays using spectrophotometers to record spectral changes could become bottlenecks for on-site analysis [25]. Hence, solid-state chemosensor arrays and easy-to-record portable systems are required to achieve on-site real-sample detection.

For mass production of chemical sensor chips (devices), printing technologies are potent methods and have been widely employed in organic electronics [30−32] and analytical chemistry [33−36] to this date. The printing methods (inkjet, screen, gravure printing, etc.) can be applied to various types of flexible substrates such as plastic films and paper, enabling the fabrication of disposal and low-cost chemical sensors. In addition, those microfabrication techniques to align the orientation of patterns have contributed to not only circuits of organic electronics but also paper-based analytical devices. Paper possesses separation ability due to its capillary action and high absorbability, and has been used as a sensor substrate of microfluidic devices. The printable paper-based chemical sensor devices are capable of being manufactured as disposable devices that can detect analytes in small volumes of samples. Moreover, the employment of easy-to-record apparatuses

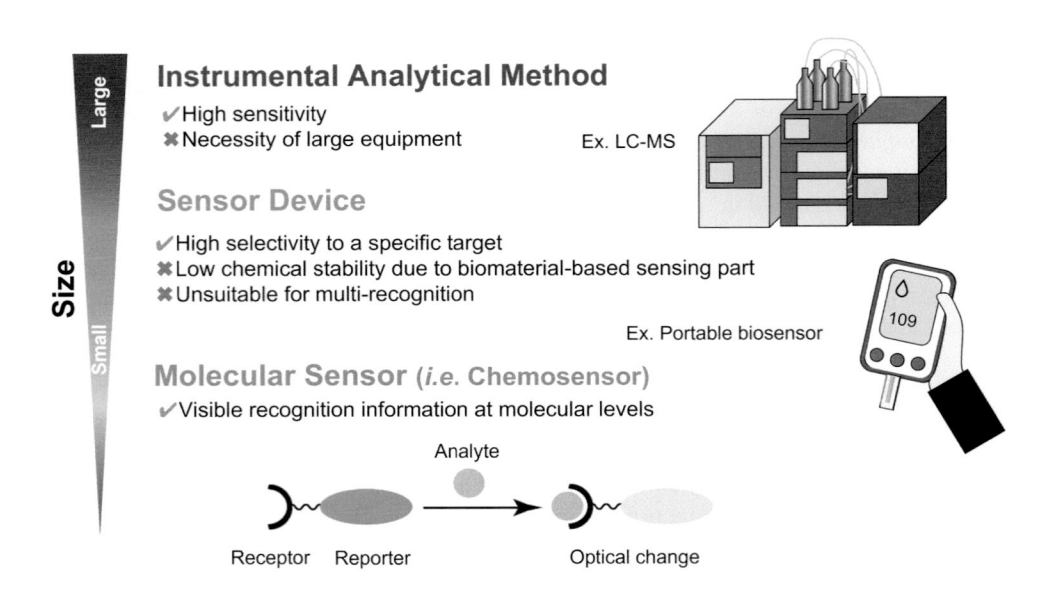

FIGURE 19.1

Conceptual illustration of the size-miniaturization of analytical devices.

such as smartphones, digital cameras, flatbed scanners, etc. could facilitate the success of on-site detection of analytes [37,38] by the combination of the printable paper-based chemical sensors. Furthermore, unique patterns of paper-based chemical sensors such as barcodes [39] and QR codes [40−42] were developed for rapid sensing by the mobile-type-recordable method, which successfully displayed the optical changes of the chemical sensors on the surface of the paper by the analyte detection. However, both qualitative and quantitative detections of multiple analytes in real samples by the abovementioned sensor devices are still challenging because of the insufficient dataset in the conventional design of paper-based sensor devices. Thus this chapter focuses on the approaches for establishing paper-based chemosensor arrays for on-site analysis based on supramolecular chemistry.

19.2 General introduction of chemosensors

19.2.1 General introduction of chemosensor arrays

Pattern recognition in social life that deals with huge amounts of personal information is extensively utilized to save and manage privacy data enabling the association of individuals. Biometrics is an established technology to automatically identify individuals based on biological features and/or behavioral characteristics [43], which has been applied to criminal investigations since the 1980s. Such technology has been employed for plant maintenance since the 1990s and personal information management since the 2000s in association with an exponential progress of informatization. According to the general classification, the biometric modality means the features for the identification should be universality, uniqueness, and permanence, which are categorized into biological features (fingerprint, face, iris, hand, etc.) and behavioral characteristics (voice, signature, etc.). Among them, a fingerprint made of ridge and valley lines is a representative feature utilized in biometrics owing to the permanent minutiae. The types of the unique pattern on a finger surface are loop, arch, and whorl, and the inherent minutiae range from differences in the position and the direction of ridge ending, bifurcation, delta, core, etc. on the pattern [44]. Therefore, the fingerprint contains the various pattern of features on the finger's surface, resulting in information-rich data for identification and verification in biometrics applications.

The concept of chemosensor arrays attributes to the molecular recognition in the mammalian olfactory system that shows an ideal cross-reactive response for the discrimination of abundant odorant molecules [45]. The chemosensors aligned on the array show various colorimetric and/or fluorescent responses by capturing analytes, allowing chemical sensing at molecular levels. The optical responses on the array upon detecting analytes correspond to the types of analyte species and their concentrations, whereby the characteristics of the optical responses containing $\Delta \lambda$ (i.e., changes in wavelength), Δ absorbance, or Δ fluorescence intensities are referred to as "minutiae of chemosensors" in an inset data for pattern recognition. The minutiae obtained by optical properties of the chemosensors rely on the chemosensor designs by tuning the chemical structures of receptor units, reporter units, or both units, which contribute to preparing the fingerprint-like response pattern [24]. After designing appropriate chemosensors, pattern recognition is performed through the following three steps: data collection, data analysis, and the visualization of datasets (Fig. 19.2). The chemosensors in solution are pipetted into plastic-type 96, 384, or 1536 well-microtiter plates

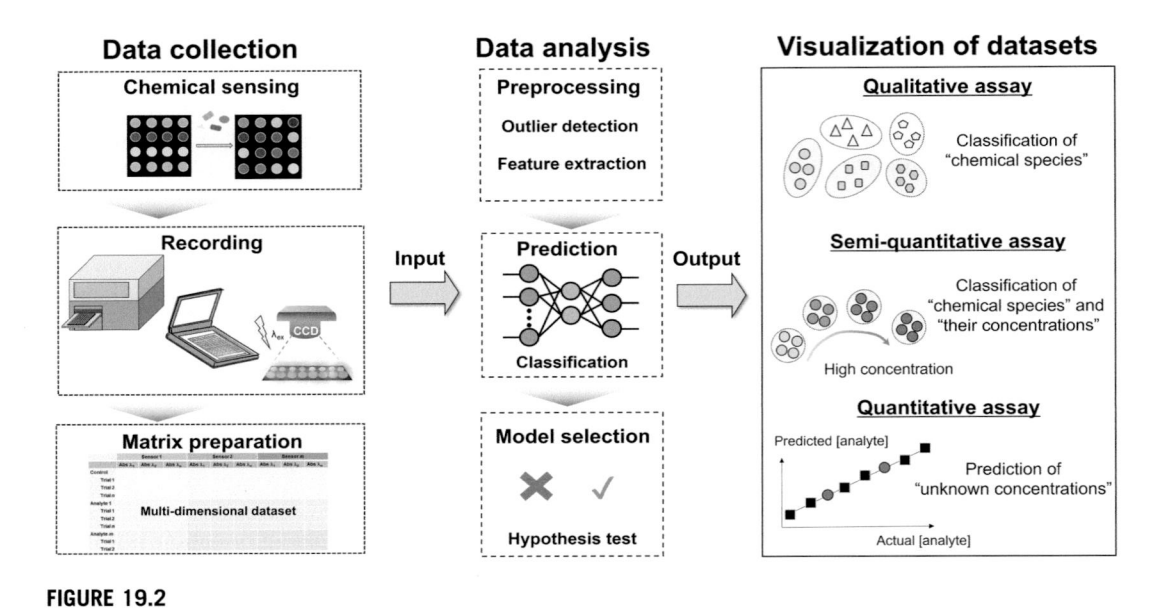

FIGURE 19.2

General scheme to obtain analytical results using a chemosensor array.

for chemical sensing. In the data collection, the optical responses of the chemosensor arrays to analytes are recorded as absorbance or fluorescence intensity at corresponding wavelengths using spectrophotometers, followed by the preparation of a dataset. The acquired dataset consists of variables on the x-axis and observation with repetitions on the y-axis, resulting in the multidimensional matrix (Fig. 19.3). In contrast to the abovementioned solution-state chemosensor arrays, the observed optical responses of solid-state chemosensor arrays are recorded as digital images by using portable apparatuses such as flatbed scanners, digital cameras, and smartphones. Therefore, the color intensities originating from the analyte detection are extracted for preparing the dataset through an imaging analysis process. With analytical software such as ImageJ, Python, and MATLAB®, the acquired images are split into each color channel (red, green, blue (RGB), grayscale, Y (luma component), C_b (blue-difference chroma components), and C_r (red-difference chroma components), etc.), followed by the extraction of the color intensities from each split image [46]. Thus, the dataset for the solid-state chemosensor arrays is constructed by the color intensities at each color channel (Fig. 19.3). Next, the preprocessing of the multidimensional dataset for highly accurate analysis is performed by using statistical methods. For example, Student's t-test is one of the parametric methods, which is utilized for removing outliers from the dataset. Moreover, analysis of variance (ANOVA) classified in the same methods are applied to visualize the contribution of variables. As a result, noise factors identified by ANOVA are statistically eliminated from the initially obtained datasets. By using the treated datasets referred to as inset datasets, pattern recognition is performed for the classification and quantification of analytes. The purpose of pattern recognition is generally classified into qualitative and semiquantitative assay (i.e., classification of types of analytes and their concentrations), and quantitative assay (i.e., prediction of unknown

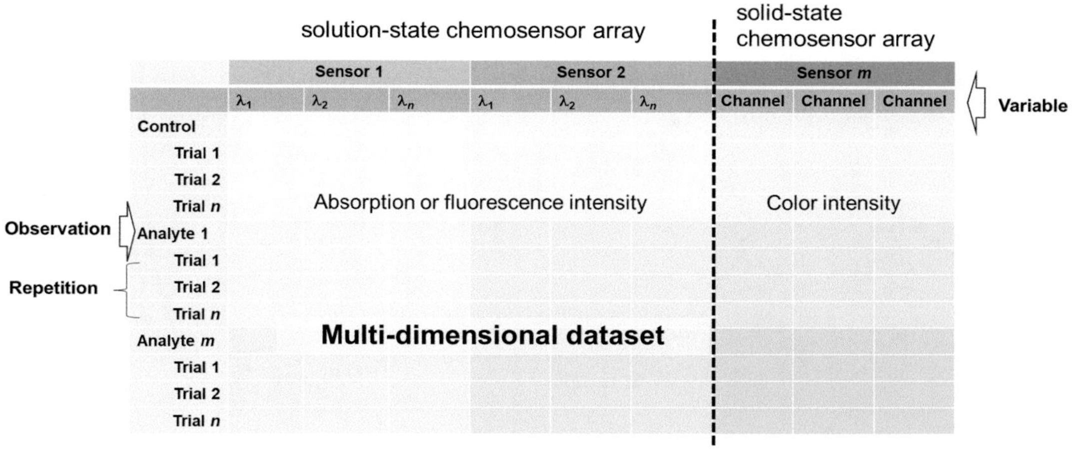

FIGURE 19.3

Schematic illustration of the multidimensional dataset for pattern recognition.

concentrations in mixtures of analytes). More importantly, results of qualitative and/or semiquantitative assay could provide a correlation between the position of distributed clusters and their response space, which enables not only visualization of a magnitude of optical responses but also group categorization based on types of analytes. The discrimination power of the chemosensor array could be evaluated by hypothesis-free and hypothesis-driven strategies with chemometrics techniques. The hypothesis-free strategy could provide us with a trend of features of the dataset, which enables evaluating the quality of the inset data using an unsupervised method [e.g., principal component analysis (PCA) and hierarchical clustering analysis] [21]. In contrast, the hypothesis-driven strategy is utilized to assess the quality of detectability of the chemosensors for pattern recognition with cross-validation, which is applied by supervised methods [e.g., linear discriminant analysis (LDA), support vector machine (SVM), and artificial neural network] [17]. Although such powerful computational analytical techniques can be applied to pattern recognition, the quality of chemical sensing relies on the design of chemosensors. Hence, the next section focuses on the strategy of chemosensor designs to prepare information-rich inset data for pattern recognition.

19.2.2 Strategy of general chemosensor array designs

Herein, the strategy of chemosensor array designs to acquire the minutiae for the fingerprint-like response pattern is based on tuning the chemosensors of (1) receptor units, (2) reporter units, or (3) both units. Based on dye chemistry, the optical properties of dye and fluorescent probes are generally tuned by the introduction of the donor and acceptor systems, the substituent effects, etc. On the other hand, the optical profiles of chemosensors are tuned by noncovalent interactions with analytes, which rely on the binding affinities of the receptor units. An example of the strategy employing different receptors with the same reporter units is a fluorescent chemosensor array for proteins using conjugated polymers based on water-soluble poly(p-phenyleneethynylene) (PPE) [47]

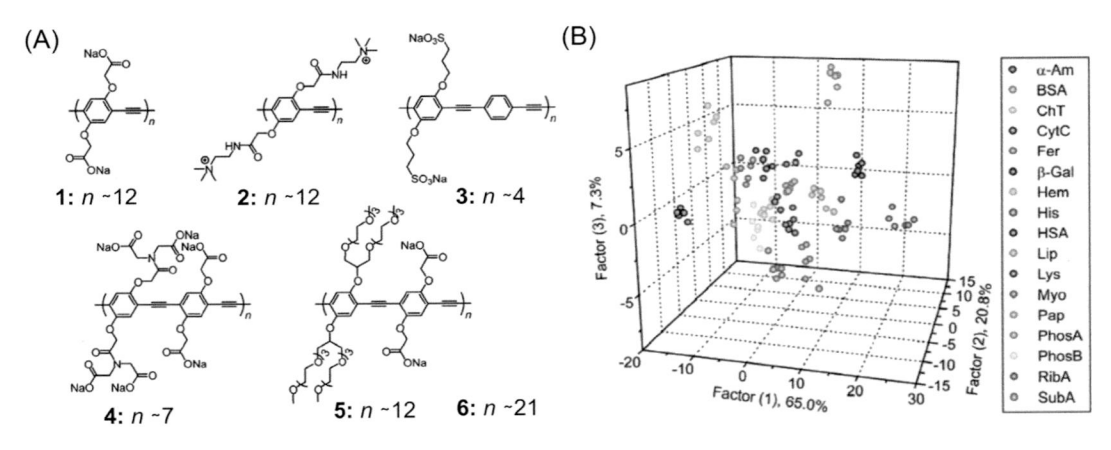

FIGURE 19.4

(A) Chemical structures of poly(p-phenyleneethynylene) derivatives (**1–6**) for pattern recognition of proteins.
(B) Result of the discrimination of 17 proteins by linear discriminant analysis.

Reproduced with permission from O.R. Miranda, C.C. You, R. Phillips, I.B. Kim, P.S. Ghosh, U.H.F. Bunz et al., Array-based sensing of proteins using conjugated polymers, J. Am. Chem. Soc. 129 (2007) 9856–9857. Copyright 2007 American Chemical Society.

derivatives (**1**: $n \sim 12$, **2**: $n \sim 12$, **3**: $n \sim 4$, **4**: $n \sim 7$, **5**: $n \sim 12$, and **6**: $n \sim 21$) reported by Bunz and Rotello et al. (Fig. 19.4A) [48]. The conjugated polymer-type chemosensors possess different binding affinities owing to the unique designs of the positive or negative charged units introduced into neutral pendants, allowing electrostatic interactions with the target proteins [i.e., a-amylase (a-Am), bovine serum albumin (BSA), a-chymotrypsin (ChT), cytochrome c (Cytc), ferritin (Fer), β-galactosidase (β-Gal), hemoglobin (Hem), histone (His), human serum albumin (HSA), lipase (Lip), lysozyme (Lys), myoglobin (Myo), papain (Pap), acid phosphatase (PhosA), alkaline phosphatase (PhosB), ribonuclease A (RibA), and subtilisin A (SubA)]. In general, obtaining information-rich datasets based on multiple optical properties is limited by the strategy of employing different receptors with the same reporter units, which attributes to the difficulties of providing color variations against a single reporter unit. In this regard, each selected proteins possess different basis properties including molecular weights, cofactors (e.g., metal ions), isoelectric points, and UV absorption, implying that a synergy effect of their unique features could be minutiae in the fluorescence fingerprint-like response pattern for discrimination of all proteins (Fig. 19.4B).

An example of array design based on the strategy of employing different reporters with the same receptor units for carboxylate drugs is a dual read-out-type chemosensor array reported by Anzenbacher et al. [49]. The chemosensor array was designed based on a donor and acceptor system using octametylcalix[4]pyrrole skeleton [50,51] (**7–13**) for carboxylate drugs, which was applied to urinalysis as the challenging application in the fields of host–guest chemistry. The changes in the optical properties of the synthesized chemosensors depended on partial intramolecular charge transfer (ICT) [52] triggered by anion recognition through hydrogen bonds. Moreover, a turn-on-type C_3-symmetric chemosensor (**14**) was further added to the library for preparing information-rich optical response patterns (Fig. 19.5A). In contrast to the strategy of employing the

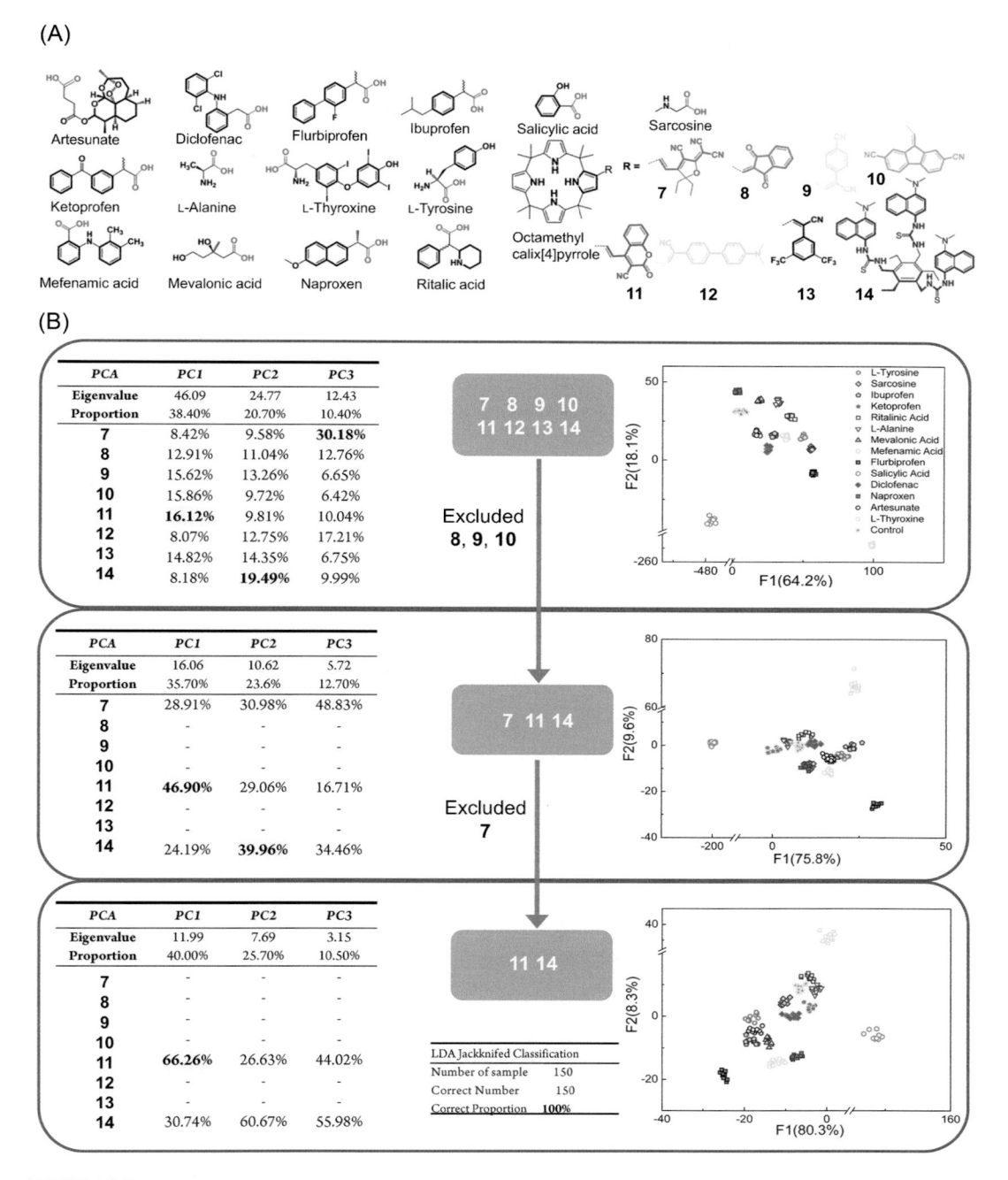

FIGURE 19.5

(A) Chemical structures of chemosensors (**7–14**) for the discrimination of target carboxylates. (B) Results of the statistical validation with principal component analysis and linear discriminant analysis against the classification of 14 carboxylates.

different receptor units, the introduction of different reporter units could provide variables as optical responses. Indeed, the strategically designed chemosensors displayed a variety of optical response patterns stemming from complemental anion recognition, while all chemosensors did not equally contribute to pattern recognition. The statistical validation with PCA and LDA based on the simultaneous discrimination of 14 carboxylates (i.e., artesunate, diclofenac, flurbiprofen, ibuprofen, ketoprofen, L-alanine, L-thyroxine, L-tyrosine, mefenamic acid, mevalonic acid, naproxen, ritalinic acid, salicylic acid, and sarcosine) clarified the contribution of each chemosensor and the effect of exclusion of numbers of chemosensors from the library (i.e., decrease of response space) (Fig. 19.5B).

The next example for the preparation of an information-rich dataset is a fluorescent chemosensor library based on the combination of the different receptors and reporters reported by Hong et al. [53]. A fluorescence library for the detection of metal ions (i.e., Na^+, Mg^{2+}, K^+, Ca^{2+}, Mn^{2+}, Fe^{3+}, Fe^{2+}, Co^{2+}, Ni^{2+}, Cu^{2+}, Cu^+, Zn^{2+}, Ag^+, Cd^{2+}, Hg^{2+}, Cr^{3+}, and Pb^{2+}) and biological anions [i.e. pyrophosphate (PPi), deoxyadenosine triphosphate (dATP), deoxycytidine triphosphate (dCTP), deoxyguanosine triphosphate (dGTP), deoxythymidine triphosphate (dTTP), adenosine triphosphate (ATP), adenosine diphosphate (ADP), and adenosine monophosphate (AMP)] included 35 chemosensor candidates (seven moieties for metal ion recognition × five fluorophore units) (**15–49**) (Fig. 19.6). Certainly, the versatile combination of different receptors and reporters can offer the information-rich dataset to discriminate various types of analytes simultaneously, whereas the synthesis burden against low contributed chemosensors could prevent the establishment of analytical tools in real-world scenarios. Thus an approach to realizing the sophisticated design of chemosensor arrays is summarized in the next section.

19.2.3 "Zero" synthetic supramolecular chemosensor arrays

From the viewpoint of analytical chemistry, the final goal of sensor fabrication is "zero" synthesis which means the preparation of chemosensor arrays without any organic synthesis in laboratories [46,54–60]. Focusing on molecular self-assembly [61,62], that is the principle of zero-synthesis, its origin is attributes to the biological molecular recognition system. The molecular recognition system has played a crucial role in four billion years of process of molecular evolution [63], which is obvious as shown in an assembled structure of DNA formed by base pairs. The helical structure of DNA revealed in 1953 [64] is a spontaneously self-organized complex and endows the capability of encoding genome information. The discovery of such attractive functionalities of self-assembled structures in biological systems should be a *Eureka* in science history, which further increased our interest owing to the beautiful shape of the biological architecture based on a complete golden ratio. Indeed, supramolecular materials inspired by an artistic aspect of such functional architectures in biological systems have been created [65–68], and the endeavor of investigating unique characteristics of artificial material led us on a journey toward the understanding of molecular recognition in *Mother Nature*. The unique function of the artificial materials can be tuned by self-assembly and self-disassembly, which are fundamental roles in supramolecular chemistry and could be key to the establishment of the concept for zero-synthesis supramolecular chemosensor arrays.

The pioneering work of the molecular self-assembled chemosensor without organic synthesis was proposed by Anslyn et al., who employed an intermolecular complex for citrate detection in beverages using a C_3-symmetric receptor (**50**) and an indicator (**51**) (Fig. 19.7) [69,70]. The self-

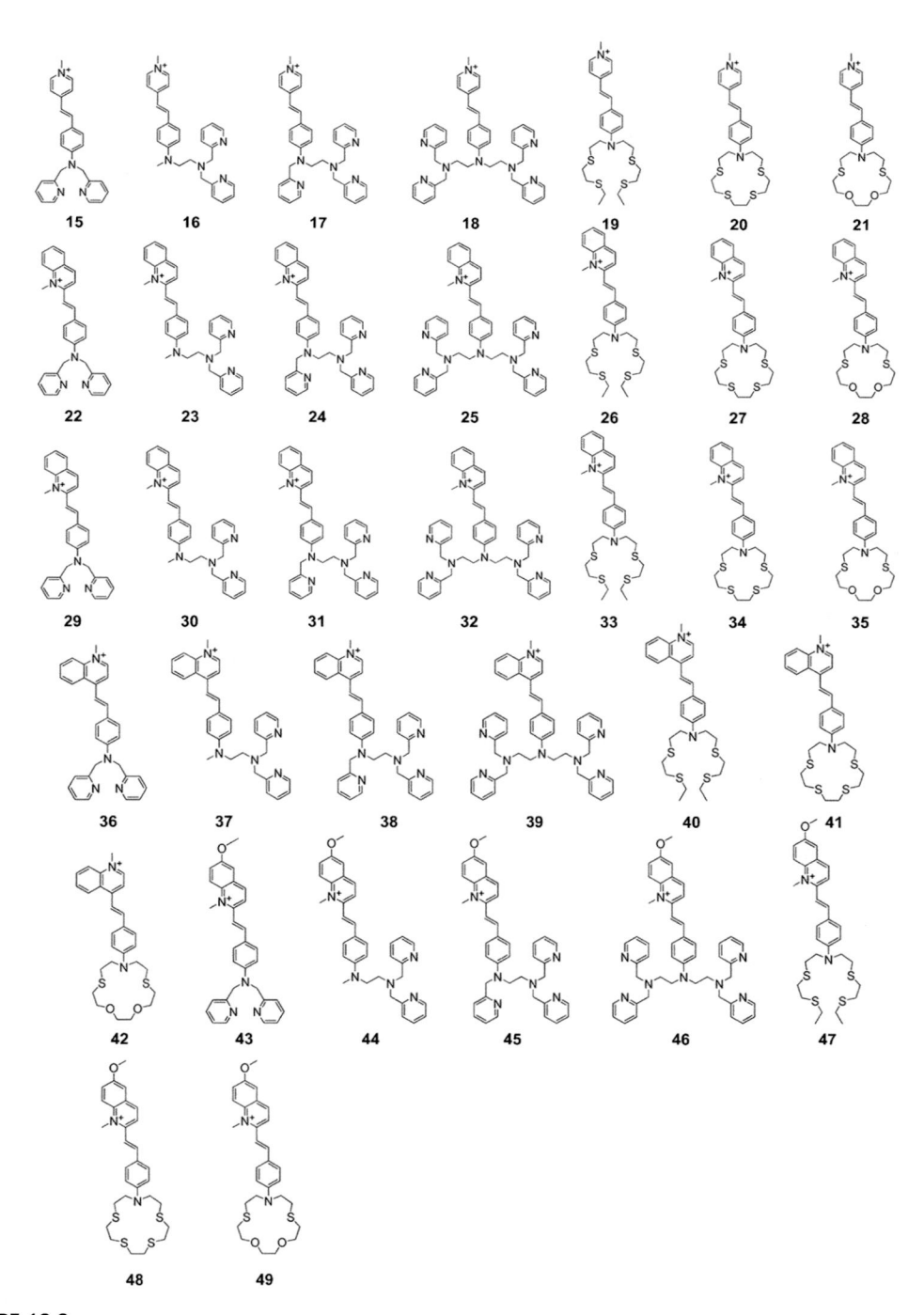

FIGURE 19.6

Chemical structures of the fluorescent chemosensor library (**15–49**) for the detection of cations and anions.

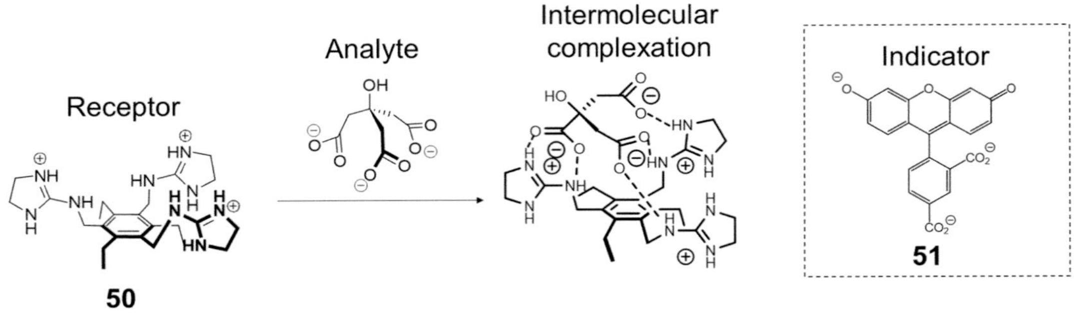

FIGURE 19.7

The C_3-symmetric receptor (**50**) and the indicator (**51**) for the detection of citrate based on the competitive assay.

Reproduced with permission from S.L. Wiskur, H. Ait-Haddou, J.J. Lavigne, E.V. Anslyn, Teaching Old Indicators New Tricks, Acc. Chem. Res. 34 (2001) 963–972.

assembly and self-disassembly of the intermolecular complex are manipulated by the difference in the binding affinities between the complex of receptor – indicator and the complex of receptor – analyte, which is called indicator displacement assay (IDA) [71–73]. This section focuses on the approaches of zero-synthetic supramolecular chemosensor arrays endowed with noncovalent reversible interaction into optical sensing systems to obtain an information-rich dataset for pattern recognition.

An example of the zero-synthetic supramolecular chemosensor array reported by Minami et al. [58] is a turn-on-type fluorescent sensing system for the discrimination of 14 monosaccharides including phosphorylated saccharides [i.e., glucose-6-phosphate (G6P), fructose-6-phosphate (F6P), glucose (Glc), fructose (Fru), galactose (Gal), mannose (Man), arabinose (Ara), xylose (Xyl), rhamnose (Rha), talose (Tal), ribose (Rib), fucose (Fuc), N-acetyl-glucosamine (NAcGlc), and N-acetyl-galactosamine (NAcGal)], which is fabricated by two catechol fluorophores [i.e., esculetin (**52**) and 4-methylesculetin (**53**)] and a receptor (i.e., 3-nitrophenylboronic acid (**54**) [58,74–76]). Based on the IDA principle, the fluorophores display light greenish-blue or light blue fluorescence in water, while turn-off phenomena (i.e., fluorescent quenching) arise from photoinduced electron transfer (PeT) [77], being caused by the formation of a boronate ester with the receptor (**54**). The key point of this self-assembled system is the fluorescent on–off signal manipulated by the role in the nitrophenyl group [78] of the receptor (**54**) as a quencher unit. Upon adding target saccharides, turn-on phenomena (i.e., fluorescent recovery) occurs according to the competitive assay between the complex of the receptor – catechol fluorophore and the complex of the receptor – saccharide (Fig. 19.8A). It goes without saying not only the chemosensor arrays were fabricated by just mixing the commercially available materials, but also the obtained optical contrast in the fluorescent self-assembled system was significant as unique minutiae in the dataset even in the small numbers of building blocks. The acquired fingerprint-like fluorescence pattern was applied to the qualitative detection of target saccharides and achieved 100% correct classification with LDA (Fig. 19.8B). Moreover, the simultaneous quantitative detection of Glc, G6P, and F6P was demonstrated by using

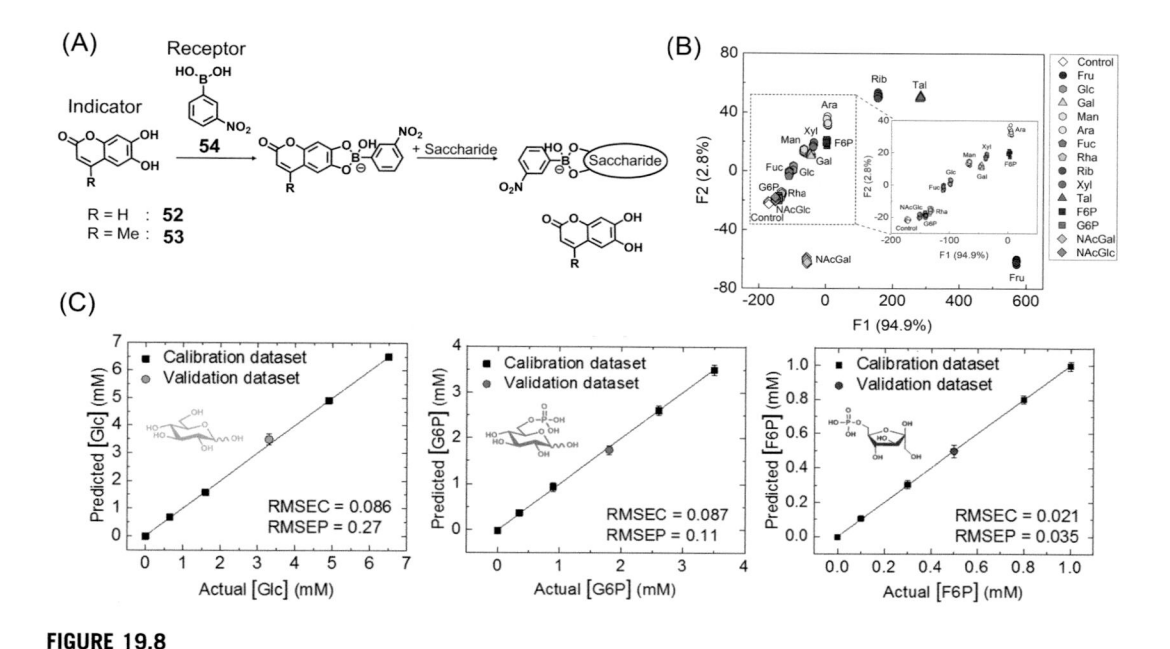

FIGURE 19.8

(A) Schematic illustration for the fluorescent detection mechanism of saccharides based on the indicator displacement assay using two catechol fluorophores (**52** and **53**) and the receptor (**54**). (B) Result of the qualitative assay against 14 saccharides by linear discriminant analysis. (C) Regression analysis against Glc, G6P, and F6P by support vector machine.

Reproduced with permission from Y. Sasaki, É. Leclerc, V. Hamedpour, R. Kubota, S. Takizawa, Y. Sakai, et al., Simplest Chemosensor Array for Phosphorylated Saccharides, Anal. Chem. 91 (2019) 15570–15576. Copyright 2019 American Chemical Society.

SVM toward the monitoring of the pseudo glycolysis pathway, resulting in an accurate prediction of the saccharides and their concentrations (Fig. 19.8C). Furthermore, the zero-synthetic fluorescent supramolecular chemosensor array performed indirect monitoring of the glycolytic activity of human-induced pluripotent stem (hiPS) cells. The demonstration was the first example of achieving the discrimination of phosphorylated saccharides simultaneously using a chemosensor array [79–81].

Minami et al. also proposed the noncovalent reversible interaction-based supramolecular recognition system using an "optical manipulator" to obtain optical contrast for an information-rich dataset [82]. They designed a self-assembled colorimetric chemosensor array consisting of three catechol dyes [i.e., alizarin red S (**55**), bromopyrogallol red (**56**), and pyrogallol red (**57**)] and an optical manipulator (**54**) for the simultaneous detection of 11 metal ions (i.e., Ni^{2+}, Cu^{2+}, Zn^{2+}, Cd^{2+}, Co^{2+}, Fe^{2+}, Hg^{2+}, Al^{3+}, Pb^{2+}, Ga^{3+}, and Ca^{2+}). The catechol dye and the optical manipulator spontaneously form a boronate ester accompanied by a blue shift with an isosbestic point, while the color recovery of the catechol dye occurs in the presence of target metal ions according to the competitive assay (Fig. 19.9A). With the LDA, the self-assembled chemosensor array

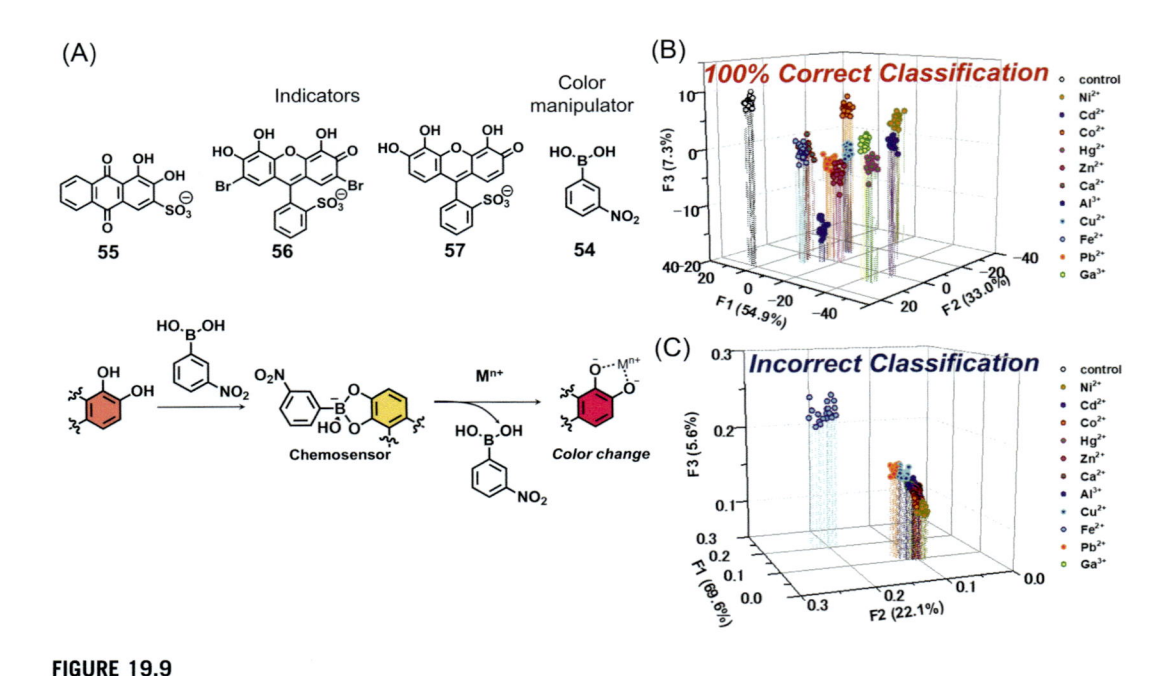

FIGURE 19.9

(A) Schematic illustration for the colorimetric detection mechanism of metal ions based on the competitive assay using 3 catechol dyes (**55** − **57**) and the manipulator (**54**). Result of the qualitative assay against 11 metal ions in (B) the presence or (C) the absence of **54**.

Reproduced with permission from Y. Sasaki, T. Minamiki, S. Tokito, T. Minami, A molecular self-assembled colourimetric chemosensor array for simultaneous detection of metal ions in water, Chem. Commun. 53 (2017) 6561−6564. Copyright 2017 Royal Society of Chemistry.

performed the discrimination of target metal ions with high accuracy (Fig. 19.9B). In the proposed sensing system, the color changes in the catechol dyes depend on the equilibrium between the complex of the catechol dye − manipulator and the complex of the catechol dye − metal ion. Namely, the equilibrium of the complexation of the catechol dye and the manipulator shifts to the superior complexation of the catechol dyes and target metal ions, resulting in larger wavelength shifts ($\Delta \lambda$) of the catechol dyes in the presence of the metal ions than without using the optical manipulator. As a proof-of-concept, the discrimination of 11 metal ions was performed by the dye array, meaning without using the optical manipulator. The LDA result suggested the unsuccessful discrimination of all metal ions, which was probably due to the insufficient inset dataset. This clarified not only the role of the optical manipulator but also the power of supramolecular interactions for pattern recognition (Fig. 19.9C). Overall, the proposed chemosensor array based on coordination chemistry [83] and dynamic covalent chemistry [84] that enables strategic tuning of optical properties of the self-assembled and self-disassembled structures would become a "genuine" supramolecular chemosensor array.

19.3 **Realization of chemosensor arrays for real sample analysis**

Paper is important in intellectual development over the long history of human civilization [85], being used to record information because of its light weight, high accessibility, and easy manufacture. To date, such paper materials that are familiar to our daily life have been widely applied to portable and disposable chemical sensors. The paper-based chemical sensors can easily visualize molecular recognition information through optical responses, suggesting that the paper devices can be referred to as portable devices which encode chemical information. As aforementioned, the potential of paper for on-site analytical devices has been clarified by the combination of unique designs of the sensor devices and easy-to-record tools, while simultaneous multiple detections have not been fully achieved by the conventional paper-based chemical sensors. Therefore, Section 19.3 shows the strategy for pattern recognition by using the chemosensor arrays embedded in the paper and their applications for real-sample detection by imaging analysis.

19.3.1 **Approaches to paper-based chemosensor arrays**

An example of the paper-based chemosensor array is a fluorescent detection of carboxylate anions reported by Xu and Bonizzoni [86]. The target carboxylate anions ubiquitously play roles in a biological system, and the detection of them is important in the fields of diagnosis and environmental assessments. To this end, they employed polycationic amine-terminated poly(amidoamine) (PAMAM) dendrimers [87], enabling interaction with target carboxylate analogs (i.e., citrate, isocitrate, malate, tricarballylate, maleate, oxaloacetate, lactate, tartrate, and glycolate) through noncovalent bonding. In this regard, the PAMAM dendrimer (i.e., fifth generation) consisting of 1,2-diaminoethane core modified by 128 surface amino groups (**58**) can easily form complexes with a small sized-fluorophore [i.e., calcein (**59**)], while the fluorophore is disassembled from the ensemble of the PAMAM dendrimer upon the addition of the target anions based on the IDA (Fig. 19.10A). Therefore, the IDA-based sensing system by the self-assembled complex of the PAMAM dendrimer and the fluorophore (**59**) was applied to the fluorescent detection of carboxylate anions on paper substrates. In this assay, a platform of the paper-based chemosensor array was designed based on a format of a standard 96-well microtiter plate, and the sensing ability was assessed from the standpoints of different types of paper (Fig. 19.10B). In the first selection of paper, transparency films possessed high loading capacity, whereby the overspreading of solution could be avoided. Although the fluorophore (**59**) embedded with the paper substrate displayed an optical response under wet conditions, the fluorescent emission derived from **59** was suppressed by the scattering effect on the surface of the substrate under dry conditions. In the case of a TLC plate, the optical response stemming from the analyte detection on the substrate was minimized because of its lower loading capacity. On the other hand, common office printer paper possessing a large loading capacity could be utilized as the solid-state support for chemical sensors owing to its hydrophobic surface allowing the avoidance of ink bleed, while the suppression of optical responses according to analyte detection could be observed on the office printer paper because of the paper originally being treated by fluorophores. Indeed, the IDA-based chemosensor array on paper substrate exhibited the failure for accurate discrimination of target oxyanions. In contrast, filter paper and chromatography paper [34,88], which do not originally contain fluorophores, have also been

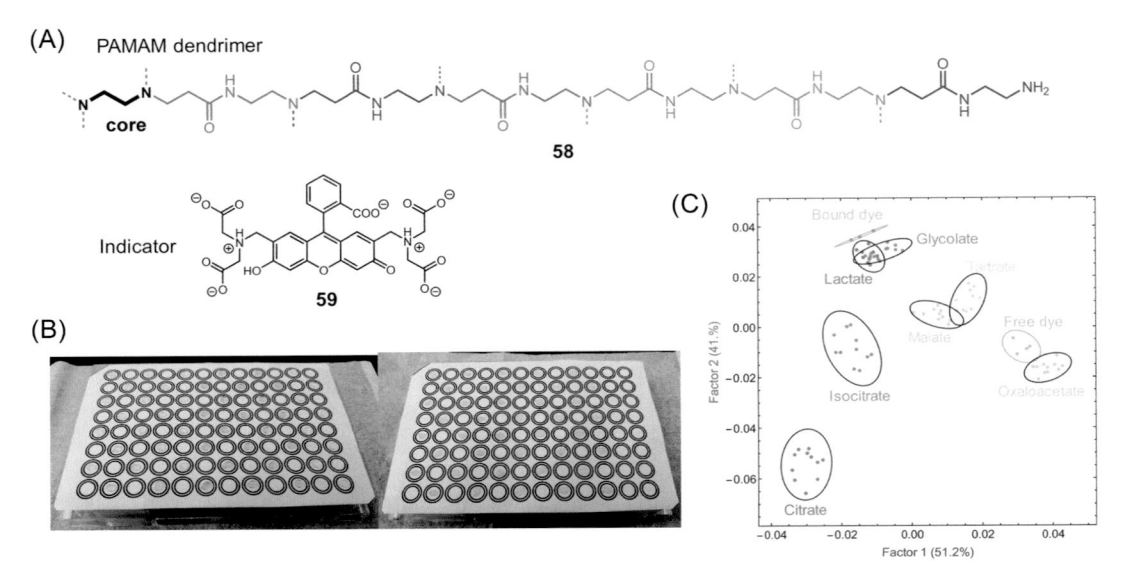

FIGURE 19.10

(A) Chemical structures of the fifth-generation PAMAM dendrimer derivative (**58**) and calcein (**59**). (B) Photograph of the manufactured 96-well paper-based chemosensor array. (C) Result of the qualitative assay against seven carboxylate anions and reference samples (i.e., bound dye and free dye) by linear discriminant analysis. Ellipsoids indicate 95% confidence.

Reproduced from Y. Xu, M. Bonizzoni, Disposable paper strips for carboxylate discrimination, Analyst 145 (2020) 3505–3516 with permission from The Royal Society of Chemistry.

applied to the solid-state fluorescent chemosensor array and showed appropriate discriminability by LDA (Fig. 19.10C). Overall, the canonical score plots based on qualitative oxyanion detection revealed that the detectability of the 96-well microtiter paper-based chemosensor array depended on the inherent properties of solid-state support materials.

As shown above, the advantages of paper-based chemosensor arrays are printability, low cost, disposability, low sample consumption, etc., while some drawbacks such as a backscattering effect to cause noise and an uneven structural surface of the paper to affect repeatability are still of concern. Caglayan et al. compared the detectability of synthesized fluorescent chemosensors (**60** − **63**) on a 96-well microtiter plate and a solid-support material (i.e., chromatography paper) against doping drugs and analogs (i.e., meldonium, carnitine, arginine, urea, acetylcholine, hydrochlorothiazide, modafinil, and aspirin) (Fig. 19.11A) [89]. Meldonium—classified as a cardioprotective and antiischemic drug—is prohibited by the World Anti-Doping Agency (WADA) [90], and thus its accurate detection is desirable. In addition, the selective detection of meldonium is required considering the structural similarity with biomarkers in human samples (e.g., acetylcholine, arginine, and carnitine) [91]. Given the high demand for easy-to-use doping test kits for athletes, they decided to develop a fluorescent chemosensor array comprising four ethidium bromide derivatives [92] for pattern recognition of the doping drugs in an artificial urine sample. In this study, the

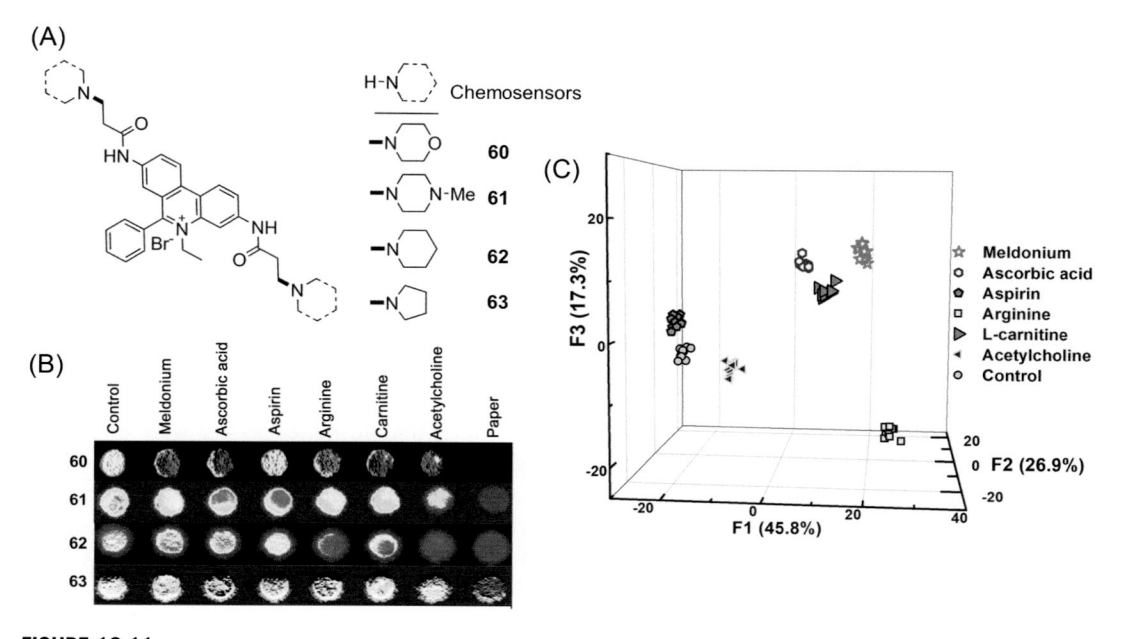

FIGURE 19.11

(A) Chemical structures of fluorescent chemosensors (**60–63**) for the discrimination of meldonium and its derivatives. (B) Captured digital image of fluorescence responses of chemosensors on the paper substrate in the presence of analytes. (C) Result of the qualitative assay against drug analogs by linear discriminant analysis.

Reproduced from E. Yalcin, C. Erkmen, T. Taskin-Tok, M.G. Caglayan, Fluorescence chemosensing of meldonium using a cross-reactive sensor array, Analyst 145 (2020) 3345–3352 with permission from The Royal Society of Chemistry.

discrimination power of the fluorescent chemosensor array was evaluated in both solution and solid states. The optical responses of the fluorescent chemosensors in the artificial urine sample were recorded by using a spectrophotometer, which contributed to the discrimination of seven clusters (i.e., six analytes and control) by LDA. Moreover, the same fluorescent chemosensors were also applied to solid-state sensing on a paper microzone plate and those optical responses under UV light irradiations at 254 or 365 nm were rapidly captured using a smartphone along with four different optical filters including band-pass and long-pass. As shown in Fig. 19.11B, the acquired digital image showed various fluorescence responses depending on the combination of the chemosensors and the targets. The captured images were subsequently treated using open-source software (i.e., Fiji) for imaging analysis, followed by qualitative analysis with LDA. Indeed, the canonical score plot by LDA distributed all clusters with 100% correct classification, which indicated the feasibility of paper-based doping analysis combined with the smartphone (Fig. 19.11C). However, the simple instrumental setup containing a smartphone, color filters, and a UV lamp was unfortunately insufficient to obtain acceptable analytical results in the semiquantitative and quantitative assays because of the inadequate dataset by the simplification of the recording system. Therefore, further approaches to obtain the information-rich dataset for quantitative detection by utilizing paper-based chemosensor arrays and easy-to-record apparatuses were required in the early stage of pattern recognition for on-site analysis. The described examples of paper-based chemosensor arrays have revealed the potential of arrays as a sensor platform for qualitative detection of multiple analytes, whereas quantitative detection is still challenging in the sensing system by conventional paper-based chemical sensors.

19.3.2 Paper-based zero-synthetic supramolecular chemosensor arrays

As the next attempt for real-sample analysis by the paper-based chemosensor arrays, Minami et al. focused on the optimization of the fabrication conditions to acquire the information-rich dataset using the printed chemosensor arrays [46]. To this end, an experimental design (e.g., a central composite design) [93] was employed to evaluate the contribution of parameters for device fabrication. In this assay, office printing methods including a wax printer and an inkjet printer were employed to easily fabricate 96-well-microtiter paper-based printed chemosensor array devices. The optimization based on the central composite design with 108 experiments was performed to obtain appropriate sensor designs (e.g., well length, width, and shape) and printing conditions (e.g., printing cycles and sample volumes). According to the optimization results, a square shape with 5×5 mm^2 was decided for the printing pattern to decrease the coffee-ring effect of inks on filter paper. The optimized chemosensor array device was subsequently applied to saccharide discrimination in a soft drink based on imaging analysis and pattern recognition. The colorimetric detection mechanism for the target saccharides relied on the IDA system utilizing off-the-shelf materials with four catechol dyes [i.e., **55 − 57**, and pyrocatechol violet (**64**)] and the receptor (**54**) (Fig. 19.12A). The zero-synthetic supramolecular chemosensors were utilized as color inks to simultaneously fabricate the uniform multiple sensing sites on paper. As shown in Fig. 19.12B, the printed chemosensors on paper displayed the colorful response pattern by adding target saccharides, followed by recording with a flatbed scanner. Subsequently, imaging analysis was proceeded by using an in-house-developed imaging algorithm constructed with MATLAB to extract color intensities at seven color channels (i.e., RGB, gray, and YC$_b$C$_r$). The extracted color intensities for each channel showing the fingerprint-like response pattern clarified the contribution of seven color channels against 12 analytes including mono- and disaccharides [i.e., Glc, Fru, Rha, Fuc, Xyl, Rib, Gal, arabinose (Ara), N-acetyl-glucosamine (Ace), sucrose (Suc), maltose (Mal), and lactose (Lac)] (Fig. 19.12C). With the information-rich dataset obtained by the abovementioned protocol, the qualitative assay against 12 saccharides was demonstrated by employing LDA. Remarkably, the LDA canonical score plot exhibited not only the successful distinguishment of 13 clusters (12 saccharides and control) but also the grouping of monosaccharides as Group 1 and disaccharides as Group 2 with 100% correct classification (Fig. 19.12D). In this regard, the color intensity extracted from the digital images correlated to the difference in the chemical structures of the target saccharides, which contributed to the grouping of monosaccharides and disaccharides. Especially, the position of Fru was far from the cluster of control, while the cluster of Ace was close to control, indicating that the binding affinities of the supramolecular chemosensors to target saccharides corresponded to the response space in the LDA canonical score plot. Moreover, the paper-based chemosensor array was applied to a semiquantitative assay against Fru in the presence of an interferent saccharide (i.e., Glc). The distributed clusters depending on the increase of Fru concentrations suggested the discrimination

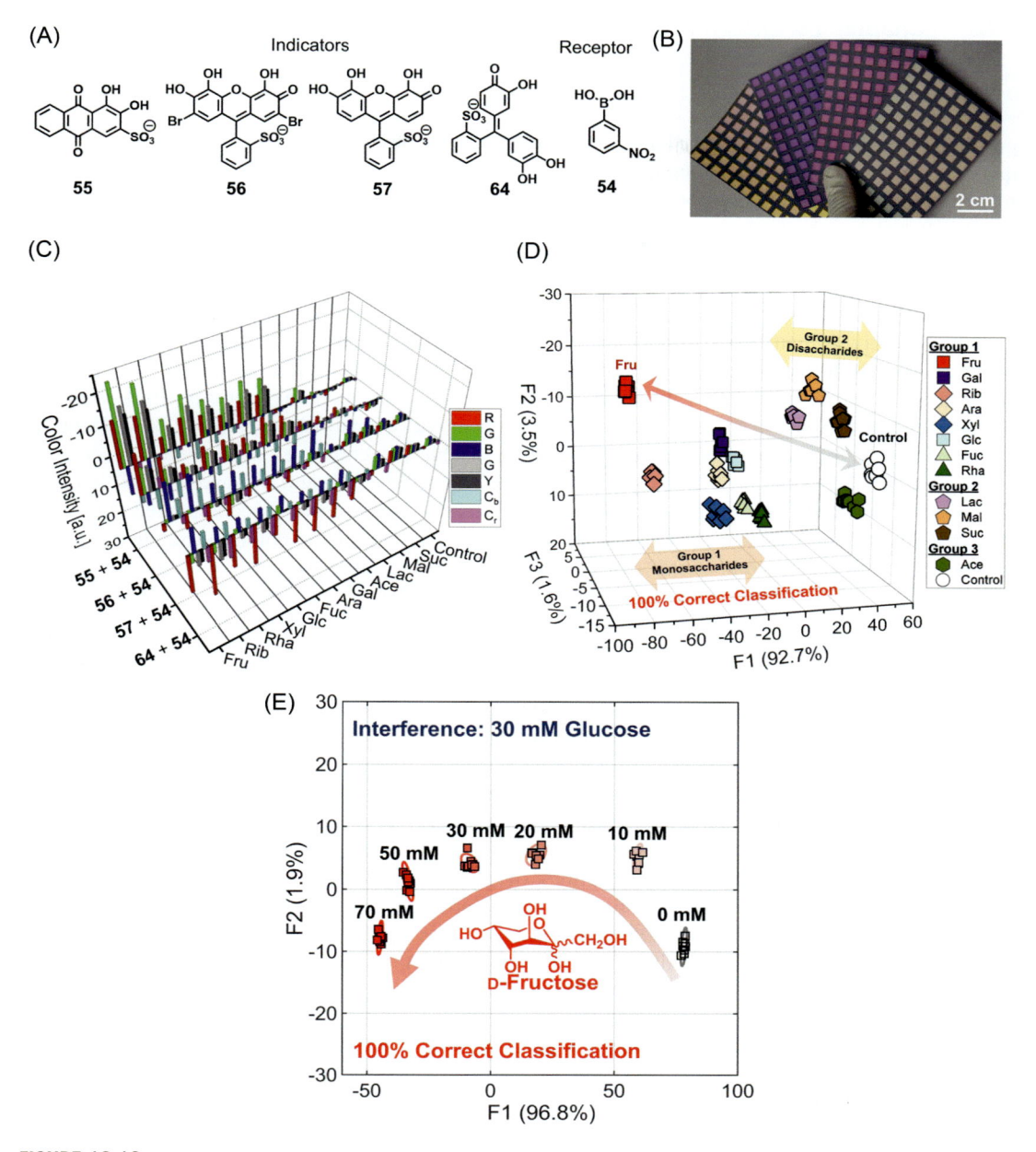

FIGURE 19.12

(A) Chemical structures of catechol dyes (**55–57**, **64**) and the receptor (**54**) for the discrimination of saccharides. (B) Captured digital image of the 96-well-microtiter paper-based printed chemosensor array. (C) Colorimetric fingerprint-like response pattern of chemosensors on the paper substrate in the presence of 12 saccharides. (D) Result of the qualitative detection of saccharides by linear discriminant analysis (LDA). (E) Result of the semiquantitative assay against Fru in presence of Glc by LDA. Ellipsoids indicate 95% confidence.

power of the manufactured paper-based chemosensor array device (Fig. 19.12E). Furthermore, the quantitative assay of Fru in a commercial soft drink was performed by using SVM. The result of regression analysis with low root mean square errors for calibration (RMSEC) and prediction (RMESP) values clarified the high accuracy of the developed 96-well-microtiter paper-based printed chemosensor array. Moreover, a demonstrated spike test achieved high recovery rates (99%–105%), and the limit of detection of the paper-based chemosensor was estimated to be 2.5×10^{-4} M.

Subsequently, Minami et al. applied a zero-synthetic supramolecular chemosensor array to phosphonate sensing on the printed paper-based device [59]. The quantification of commercial herbicide glyphosate is in high demand because of its potential carcinogenic activity [94], whereas portable chemical sensors for glyphosate have not been fully established [95,96]. To develop an easy-to-detect glyphosate system, a commercially available catechol dye-based chemosensor array on disposable filter paper was designed. The catechol dyes (**55 − 57**, and **64**) and a metal ion (i.e., Zn^{2+}) spontaneously form coordination binding-based chemosensors accompanied by color changes. Upon the addition of target oxyanions, the disassembly of the colorimetric chemosensors occurs according to the competitive assay among the catechol dye, the Zn^{2+} ion, and glyphosate (Fig. 19.13A) [97]. The colorimetric chemosensor solutions were independently printed on filter paper, resulting in the portable printed paper-based sensor array for high-throughput quantification of glyphosate. The printed colorimetric chemosensor array showed gradual color changes with an increase of glyphosate concentrations (Fig. 19.13B). The visible optical changes on the paper-based chemosensor array were rapidly captured by using the office flatbed scanner and imaging analysis constructed the inset data for pattern recognition. In Fig. 19.13C, each cluster was distributed by LDA in accordance with the increase in glyphosate concentrations, indicating the success of semiquantitative detection of glyphosate with 100% accuracy. Furthermore, the usability of the facile chemosensor array device was accessed by regression analysis in both tap water and commercial herbicide. The prediction dataset was correctly distributed on the calibration line built by SVM, which means that the unknown concentrations of glyphosate were predicted with high accuracy (Fig. 19.13D). In addition, the applicability of the chemosensor array device in real-world scenarios was investigated. Indeed, the limit of detection value was estimated at 16 mg/L, which is lower than the health advisory value per day for children (20 mg/L).

As a further challenging attempt toward high-throughput food analysis, Minami et al. proposed a printed 384-well microtiter paper-based fluorescent chemosensor array for food analysis [60]. The selected targets included a monosaccharide group (i.e., Fru, Rib, Xyl, Glc, Ara, Gal, and Man) and a disaccharide group (i.e., Suc and Mal) and a sulfur-containing amino acid group [i.e., homocysteine (HCys), L-cysteine (Cys), glutathione reduced form (GSH), and glutathione oxidized form (GSSH)], which are essential taste components and markers for monitoring food freshness. To achieve not only the simultaneous group categorization (i.e., monosaccharides, disaccharides, and sulfur-containing amino acids) but also discrimination of 13 types of analytes, the chemosensor array requires selectivity against group categories while having cross-reactivity against analytes in the categorized groups. Thus the catechol fluorophores (i.e., **53** and **55**) and two optical manipulators (i.e., **54** [58,74−76] for saccharides and the Zn^{2+} ion [98−101] for sulfur-containing amino acids) were employed as the building blocks, offering a zero-synthetic supramolecular chemosensor array based on coordination chemistry and dynamic covalent chemistry (Fig. 19.14A). More importantly, the inherent properties of fluorophores with a cumarin skeleton (i.e., **53**) [98−101] and a

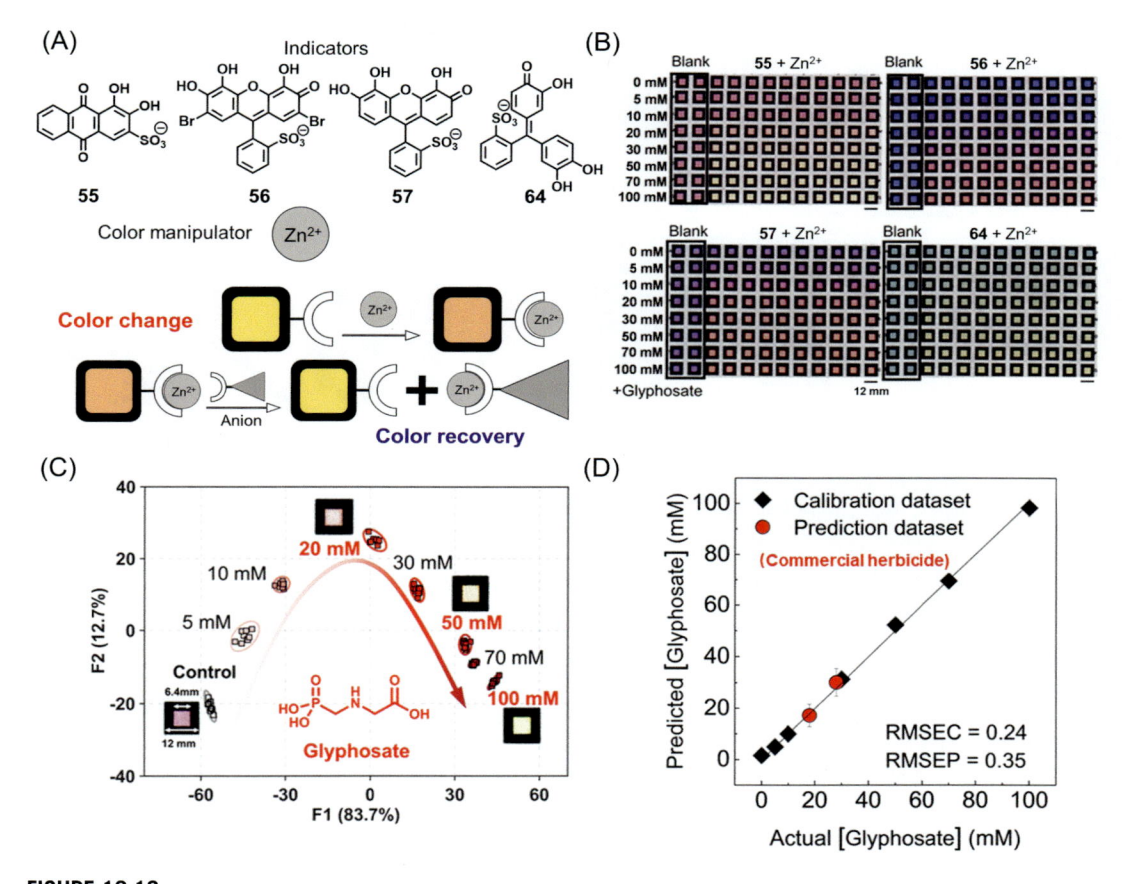

FIGURE 19.13

(A) Chemical structures of catechol dyes (**55**–**57**, **64**) and the schematic illustration of the colorimetric phosphonate detection mechanism based on the competitive assay. (B) Captured digital image of recorded color responses of the chemosensors on the paper substrate in the presence of glyphosate. (C) Result of the semiquantitative detection of glyphosate by linear discriminant analysis. Ellipsoids indicate 95% confidence. (D) Regression analysis of glyphosate in commercial herbicide by support vector machine.

Reproduced with permission from Z. Zhang, V. Hamedpour, X. Lyu, Y. Sasaki, T. Minami, A printed paper-based anion sensor array for multi-analyte classification: on-site quantification of glyphosate, ChemPlusChem 86 (2021) 798–802. Copyright 2021 John Wiley & Sons.

hydroxyanthraquinone skeleton (i.e., **55**) [102–106] can provide different fluorescent profiles including turn-on and turn-off responses upon complexation with the color manipulators. Therefore, the optical profiles of the fluorescent supramolecular chemosensors show the fingerprint-like response pattern upon the addition of different types of analytes. In contrast to the conventional 96-well microtiter plate-based chemosensor arrays, the proposed 384-well-type paper device could incorporate varied chemical information because of the increased well numbers. In addition, the

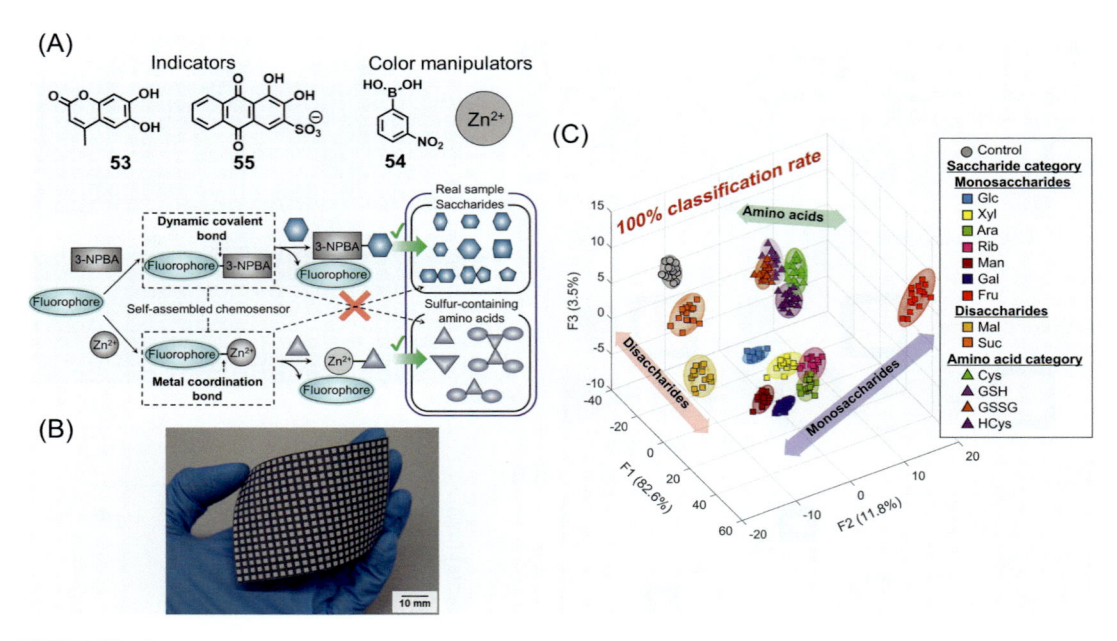

FIGURE 19.14

(A) Chemical structures of catechol fluorophores (**53** and **55**) and the manipulator (**54**) and schematic illustration of the fluorescent detection mechanism based on the competitive assay that endowed the selectivity against group categories (i.e., discrimination of saccharides and sulfur-containing amino acids) while the cross-reactivity against analytes in the categorized groups (i.e., discrimination of 13 analytes). (B) The captured digital image of the 384 well-microtiter paper-based printed chemosensor array. (C) Result of the qualitative detection of saccharides and sulfur-containing amino acids by linear discriminant analysis. Ellipsoids indicate 95% confidence.

Reproduced with permission from X. Lyu, Y. Sasaki, K. Ohshiro, W. Tang, Y. Yuan, T. Minami, Printed 384-well microtiter plate on paper for fluorescent chemosensor arrays in food analysis, Chem. Asian J. 17 (2022) e202200479. Copyright 2022 John Wiley & Sons.

miniaturized well size allows the decrease in sample volumes ($1\,\mu L/4\,mm^2$ of each well) (Fig. 19.14B). The obtained inset data was applied to the simultaneous qualitative detection of nine saccharides and four sulfur-containing amino acids, resulting in 100% classification along with group categorization of monosaccharides, disaccharides, and sulfur-containing amino acids based on imaging analysis and pattern recognition (Fig. 19.14C). The spike and recovery test for a diluted freshly made tomato juice was finally performed by using the proposed supramolecular chemosensor array device, and the recovery rates represented 91%—114% for Fru and 92%—127% for GSH, respectively.

Taken together, the strategy combined the approach to maximize optical properties of chemosensor arrays based on supramolecular chemistry and the appropriate techniques for optimization and fabrication of paper devices could open a new avenue for the realization of portable paper-based chemical sensors enabling simultaneous qualitative and quantitative assay against various types of analytes.

19.4 Conclusion and perspective

This chapter summarized the general introduction of chemosensor arrays and actual approaches to obtain information-rich datasets for pattern recognition. Inspired by biometrics, the design of chemosensor arrays offers unique optical profiles as minutiae in the inset data. Therefore, Section 19.2 described the representative examples of chemosensor arrays designed by the strategy based on tuning the chemosensors of receptor units, reporter units, or both units. Next, the concept of zero-synthetic supramolecular chemosensor arrays, which is attributed to sophisticated biological molecular recognition systems, was summarized as the idea of practical sensor designs for the realization of easy-to-prepare chemical sensors. The described designs of chemosensor arrays based on supramolecular chemistry enabled tuning of the optical properties of the self-assembled and self-disassembled structures by adding analytes. Indeed, the combination of off-the-shelf materials as the building blocks for supramolecular chemosensors contributed to not only the minimization of synthetic burden but also to obtaining optical contrasts referred to as minutiae of chemosensors. To date, the feasibility of chemosensor arrays was maximized by the combination of solid-state substrates and portable recording apparatuses, whereas simultaneous quantitative detection has been still concerned. Hence, the optimization and fabrication of paper-based chemosensor arrays for the quantification of multiple analytes in real samples were summarized in Section 19.3. By employing printing methods in the device fabrication process, the chemosensor solutions can be utilized as printing inks. In addition, the appropriate building blocks for zero-synthetic supramolecular chemosensor arrays played significant roles in displaying optical properties depending on the types of analytes and their concentrations on paper, allowing high detectability. As shown, the proposed chemosensor arrays showed applicability to on-site real sample analysis, while further improvement of their sensitivity comparable with conventional spectrophotometers would be required to discriminate the optical response of chemical sensing from noise scattering. In addition, the development of software enabling simultaneous data correction and analysis would be also necessary to facilitate the establishment of portable chemical sensor devices.

In conclusion, the authors believe the endeavor of the realization of chemical sensors for those who are not familiar with chemistry will lead us on a journey toward the maximization of supramolecular materials.

References

[1] V. Gubala, L.F. Harris, A.J. Ricco, M.X. Tan, D.E. Williams, Point of care diagnostics: status and future, Anal. Chem. 84 (2012) 487−515.

[2] S.R. Corrie, J.W. Coffey, J. Islam, K.A. Markey, M.A.F. Kendall, Blood, sweat, and tears: developing clinically relevant protein biosensors for integrated body fluid analysis, Analyst 140 (2015) 4350−4364.

[3] A. Hulanicki, S. Glab, F. Ingman, Chemical sensors: definitions and classification, Pure Appl. Chem. 63 (1991) 1247−1250.

[4] A.W. Czarnik, Fluorescent chemosensors for ion and molecule recognition, American Chemical Society, Washington, DC, 1993.

[5] C.J. Pedersen, Cyclic polyethers and their complexes with metal salts, J. Am. Chem. Soc. 89 (1967) 7017−7036.

[6]   S.C. Peacock, D.J. Cram, High chiral recognition in α-amino-acid and -ester complexation, J. Chem. Soc., Chem. Commun. (1976) 282−284.

[7]   J.-M. Lehn, Supramolecular chemistry: concepts and perspectives, Wiley-VCH, New York, 1995.

[8]   L. Pu, Fluorescence of organic molecules in chiral recognition, Chem. Rev. 104 (2004) 1687−1716.

[9]   R. Joseph, C.P. Rao, Ion and molecular recognition by lower rim 1,3-di-conjugates of calix[4]arene as receptors, Chem. Rev. 111 (2011) 4658−4702.

[10]   J.B. Wittenberg, L. Isaacs, Complementarity and preorganization, supramolecular chemistry, John Wiley & Sons, Inc, New York, 2012.

[11]   X. Sun, T.D. James, Glucose sensing in supramolecular chemistry, Chem. Rev. 115 (2015) 8001−8037.

[12]   P.A. Gale, E.N.W. Howe, X. Wu, Anion receptor chemistry, Chem 1 (2016) 351−422.

[13]   A.P. Davis, Biomimetic carbohydrate recognition, Chem. Soc. Rev. 49 (2020) 2531−2545.

[14]   J.-P. Behr, The lock-and-key principle: the state of the art−100 years on, Wiley-VCH, New York, 1994.

[15]   W.J. Peveler, M. Yazdani, V.M. Rotello, Selectivity and specificity: pros and cons in sensing, ACS Sens. 1 (2016) 1282−1285.

[16]   J.W. Gardner, Detection of vapours and odours from a multisensor array using pattern recognition Part 1. Principal component and cluster analysis, Sens. Actuators B Chem. 4 (1991) 109−115.

[17]   P. Anzenbacher, Jr., P. Lubal, P. Buček, M.A. Palacios, M.E. Kozelkova, A practical approach to optical cross-reactive sensor arrays, Chem. Soc. Rev. 39 (2010) 3954−3979.

[18]   J.J. Lavigne, S. Savoy, M.B. Clevenger, J.E. Ritchie, B. McDoniel, S.-J. Yoo, et al., Solution-based analysis of multiple analytes by a sensor array: toward the development of an "electronic tongue", J. Am. Chem. Soc. 120 (1998) 6429−6430.

[19]   K.J. Albert, N.S. Lewis, C.L. Schauer, G.A. Sotzing, S.E. Stitzel, T.P. Vaid, et al., Cross-reactive chemical sensor arrays, Chem. Rev. 100 (2000) 2595−2626.

[20]   N.A. Rakow, K.S. Suslick, A colorimetric sensor array for odour visualization, Nature 406 (2000) 710−713.

[21]   P.C. Jurs, G.A. Bakken, H.E. McClelland, Computational methods for the analysis of chemical sensor array data from volatile analytes, Chem. Rev. 100 (2000) 2649−2678.

[22]   D.G. Smith, I.L. Topolnicki, V.E. Zwicker, K.A. Jolliffe, E.J. New, Fluorescent sensing arrays for cations and anions, Analyst 142 (2017) 3549−3563.

[23]   Y. Geng, W.J. Peveler, V.M. Rotello, Array-based "chemical nose" sensing in diagnostics and drug discovery, Angew. Chem. Int. Ed. 58 (2019) 5190−5200.

[24]   Y. Sasaki, R. Kubota, T. Minami, Molecular self-assembled chemosensors and their arrays, Coord. Chem. Rev. 429 (2021) 213607.

[25]   Y. Sasaki, X. Lyu, W. Tang, H. Wu, T. Minami, Supramolecular optical sensor arrays for on-site analytical devices, J. Photochem. Photobiol. C: Photochem. Rev. 51 (2022) 100475.

[26]   B. Wang, E.V. Anslyn, Chemosensors: principles, strategies, and applications, John Wiley & Sons, Inc, Hoboken, 2011.

[27]   J. Wu, B. Kwon, W. Liu, E.V. Anslyn, P. Wang, J.S. Kim, Chromogenic/fluorogenic ensemble chemosensing systems, Chem. Rev. 115 (2015) 7893−7943.

[28]   L. You, D. Zha, E.V. Anslyn, Recent advances in supramolecular analytical chemistry using optical sensing, Chem. Rev. 115 (2015) 7840−7892.

[29]   C. Guo, A.C. Sedgwick, T. Hirao, J.L. Sessler, Supramolecular fluorescent sensors: an historical overview and update, Coord. Chem. Rev. 427 (2021) 213560.

[30]   Y. Diao, L. Shaw, Z. Bao, S.C.B. Mannsfeld, Morphology control strategies for solution-processed organic semiconductor thin films, Energy Environ. Sci. 7 (2014) 2145−2159.

[31]   Y. Khan, A. Thielens, S. Muin, J. Ting, C. Baumbauer, A.C. Arias, A new frontier of printed electronics: flexible hybrid electronics, Adv. Mater. 32 (2020) 1905279.

[32] Y. Sasaki, X. Lyu, W. Tang, H. Wu, T. Minami, Polythiophene-based chemical sensors: toward on-site supramolecular analytical devices, Bull. Chem. Soc. Jpn. 94 (2021) 2613−2622.

[33] A.W. Martinez, S.T. Phillips, G.M. Whitesides, E. Carrilho, Diagnostics for the developing world: microfluidic paper-based analytical devices, Anal. Chem. 82 (2010) 3−10.

[34] K. Yamada, T.G. Henares, K. Suzuki, D. Citterio, Paper-based inkjet-printed microfluidic analytical devices, Angew. Chem. Int. Ed. 54 (2015) 5294−5310.

[35] Y. Xia, J. Si, Z. Li, Fabrication techniques for microfluidic paper-based analytical devices and their applications for biological testing: a review, Biosens. Bioelectron. 77 (2016) 774−789.

[36] Y. Sasaki, X. Lyu, Q. Zhou, T. Minami, Indicator displacement assay-based chemosensor arrays for saccharides using off-the-shelf materials toward simultaneous on-site detection on paper, Chem. Lett. 50 (2021) 987−995.

[37] Y. Lu, Z. Shi, Q. Liu, Smartphone-based biosensors for portable food evaluation, Curr. Opin. Food Sci. 28 (2019) 74−81.

[38] Y.-H. Shin, M.T. Gutierrez-Wing, J.-W. Choi, Review—recent progress in portable fluorescence sensors, J. Electrochem. Soc. 168 (2021) 017502.

[39] L. Guan, J. Tian, R. Cao, M. Li, Z. Cai, W. Shen, Barcode-like paper sensor for smartphone diagnostics: an application of blood typing, Anal. Chem. 86 (2014) 11362−11367.

[40] X. Ji, W. Chen, L. Long, F. Huang, J.L. Sessler, Double layer 3D codes: fluorescent supramolecular polymeric gels allowing direct recognition of the chloride anion using a smart phone, Chem. Sci. 9 (2018) 7746−7752.

[41] A. Katoh, K. Maejima, Y. Hiruta, D. Citterio, All-printed semiquantitative paper-based analytical devices relying on QR code array readout, Analyst 145 (2020) 6071−6078.

[42] J.A.M. Conrado, R. Sequinel, B.C. Dias, M. Silvestre, A.D. Batista, J.Fd.S. Petruci, Chemical QR code: a simple and disposable paper-based optoelectronic nose for the identification of olive oil odor, Food Chem. 350 (2021) 129243.

[43] A.K. Jain, R. Bolle, S. Pankanti, in: Biometrics: personal identification in networked society, Springer US, Boston, MA, 1996.

[44] A. Farina, Z.M. Kovács-Vajna, A. Leone, Fingerprint minutiae extraction from skeletonized binary images, Pattern Recognit. 32 (1999) 877−889.

[45] S. DeMaria, J. Ngai, The cell biology of smell, J. Cell Biol. 191 (2010) 443−452.

[46] X. Lyu, V. Hamedpour, Y. Sasaki, Z. Zhang, T. Minami, 96-Well microtiter plate made of paper: a printed chemosensor array for quantitative detection of saccharides, Anal. Chem. 93 (2021) 1179−1184.

[47] I.-B. Kim, A. Dunkhorst, J. Gilbert, U.H.F. Bunz, Sensing of lead ions by a carboxylate-substituted PPE: multivalency effects, Macromolecules 38 (2005) 4560−4562.

[48] O.R. Miranda, C.C. You, R. Phillips, I.B. Kim, P.S. Ghosh, U.H.F. Bunz, et al., Array-based sensing of proteins using conjugated polymers, J. Am. Chem. Soc. 129 (2007) 9856−9857.

[49] Y. Liu, T. Minami, R. Nishiyabu, Z. Wang, P. Anzenbacher, Jr., Sensing of carboxylate drugs in urine by a supramolecular sensor array, J. Am. Chem. Soc. 135 (2013) 7705−7712.

[50] P. Anzenbacher, Jr., K. Jursíková, J.L. Sessler, Second generation calixpyrrole anion sensors, J. Am. Chem. Soc. 122 (2000) 9350−9351.

[51] D.S. Kim, J.L. Sessler, Calix[4]pyrroles: versatile molecular containers with ion transport, recognition, and molecular switching functions, Chem. Soc. Rev. 44 (2015) 532−546.

[52] Z.R. Grabowski, K. Rotkiewicz, W. Rettig, Structural changes accompanying intramolecular electron transfer: focus on twisted intramolecular charge-transfer states and structures, Chem. Rev. 103 (2003) 3899−4032.

[53] H.-W. Rhee, S.W. Lee, J.-S. Lee, Y.-T. Chang, J.-I. Hong, Focused fluorescent probe library for metal cations and biological anions, ACS Comb. Sci. 15 (2013) 483−490.

[54] H. Miyaji, J.L. Sessler, Off-the-shelf colorimetric anion sensors, Angew. Chem. Int. Ed. 40 (2001) 154−157.

[55] A.M. Mallet, Y. Liu, M. Bonizzoni, An off-the-shelf sensing system for physiological phosphates, Chem. Commun. 50 (2014) 5003−5006.

[56] E.G. Shcherbakova, T. Minami, V. Brega, T.D. James, P. Anzenbacher Jr., Determination of enantiomeric excess in amine derivatives with molecular self-assemblies, Angew. Chem. Int. Ed. 54 (2015) 7130−7133.

[57] Y. Huang, P. Cheng, C. Tan, Visual artificial tongue for identification of various metal ions in mixtures and real water samples: a colorimetric sensor array using off-the-shelf dyes, RSC Adv. 9 (2019) 27583−27587.

[58] Y. Sasaki, É. Leclerc, V. Hamedpour, R. Kubota, S. Takizawa, Y. Sakai, et al., Simplest chemosensor array for phosphorylated saccharides, Anal. Chem. 91 (2019) 15570−15576.

[59] Z. Zhang, V. Hamedpour, X. Lyu, Y. Sasaki, T. Minami, A printed paper-based anion sensor array for multi-analyte classification: on-site quantification of glyphosate, ChemPlusChem 86 (2021) 798−802.

[60] X. Lyu, Y. Sasaki, K. Ohshiro, W. Tang, Y. Yuan, T. Minami, Printed 384-well microtiter plate on paper for fluorescent chemosensor arrays in food analysis, Chem. Asian J. 17 (2022) e202200479.

[61] G.M. Whitesides, J.P. Mathias, C.T. Seto, Molecular self-assembly and nanochemistry: a chemical strategy for the synthesis of nanostructures, Science 254 (1991) 1312−1319.

[62] K. Ariga, T. Kunitake, Molecular self-assembly, in: Supramolecular Chemistry—Fundamentals and Applications: Advanced Textbook, Springer Berlin Heidelberg, Berlin, Heidelberg, 2006.

[63] M. Kimura, T. Ohta, On some principles governing molecular evolution, Proc. Natl. Acad. Sci. U.S.A. 71 (1974) 2848−2852.

[64] J.D. Watson, F.H.C. Crick, Molecular structure of nucleic acids: a structure for deoxyribose nucleic acid, Nature 171 (1953) 737−738.

[65] V. Balzani, A. Credi, F.M. Raymo, J.F. Stoddart, Artificial molecular machines, Angew. Chem. Int. Ed. 39 (2000) 3348−3391.

[66] J.-P. Sauvage, From chemical topology to molecular machines (Nobel lecture), Angew. Chem. Int. Ed. 56 (2017) 11080−11093.

[67] S. Kassem, T. van Leeuwen, A.S. Lubbe, M.R. Wilson, B.L. Feringa, D.A. Leigh, Artificial molecular motors, Chem. Soc. Rev. 46 (2017) 2592−2621.

[68] K. Ariga, T. Mori, T. Kitao, T. Uemura, Supramolecular chiral nanoarchitectonics, Adv. Mater. 32 (2020) 1905657.

[69] S.L. Wiskur, H. Ait-Haddou, J.J. Lavigne, E.V. Anslyn, Teaching old indicators new tricks, Acc. Chem. Res. 34 (2001) 963−972.

[70] A. Metzger, E.V. Anslyn, A chemosensor for citrate in beverages, Angew. Chem. Int. Ed. 37 (1998) 649−652.

[71] E.A. Katayev, Y.A. Ustynyuk, J.L. Sessler, Receptors for tetrahedral oxyanions, Coord. Chem. Rev. 250 (2006) 3004−3037.

[72] M. Kitamura, S.H. Shabbir, E.V. Anslyn, Guidelines for pattern recognition using differential receptors and indicator displacement assays, J. Org. Chem. 74 (2009) 4479−4489.

[73] A.C. Sedgwick, J.T. Brewster, T. Wu, X. Feng, S.D. Bull, X. Qian, et al., Indicator displacement assays (IDAs): the past, present and future, Chem. Soc. Rev. 50 (2021) 9−38.

[74] Y. Kubo, T. Ishida, A. Kobayashi, T.D. James, Fluorescent alizarin−phenylboronic acid ensembles: design of self-organized molecular sensors for metal ions and anions, J. Mater. Chem. 15 (2005) 2889−2895.

[75] Y. Kubo, A. Kobayashi, T. Ishida, Y. Misawa, T.D. James, Detection of anions using a fluorescent alizarin−phenylboronic acid ensemble, Chem. Commun. (2005) 2846−2848.

[76] Y. Sasaki, Z. Zhang, T. Minami, A saccharide chemosensor array developed based on an indicator displacement assay using a combination of commercially available reagents, Front. Chem. 7 (2019) 49.

[77] A.P. de Silva, H.Q.N. Gunaratne, T. Gunnlaugsson, A.J.M. Huxley, C.P. McCoy, J.T. Rademacher, et al., Signaling recognition events with fluorescent sensors and switches, Chem. Rev. 97 (1997) 1515−1566.

[78] T. Ueno, Y. Urano, H. Kojima, T. Nagano, Mechanism-based molecular design of highly selective fluorescence probes for nitrative stress, J. Am. Chem. Soc. 128 (2006) 10640−10641.

[79] T. Imada, H. Kijima, M. Takeuchi, S. Shinkai, Selective binding of glucose-6-phosphate, 3,4-dihydroxyphenylalanine (DOPA) and their analogs with a boronic-acid-appended metalloporphyrin, Tetrahedron 52 (1996) 2817−2826.

[80] L.A. Cabell, M.-K. Monahan, E.V. Anslyn, A competition assay for determining glucose-6-phosphate concentration with a tris-boronic acid receptor, Tetrahedron Lett. 40 (1999) 7753−7756.

[81] S. Horie, Y. Kubo, Fluorescence-based indicator displacement assay for phosphosugar detection using zinc(II) dipicolylamine-appended phenylboronic acid, Chem. Lett. 38 (2009) 616−617.

[82] Y. Sasaki, T. Minamiki, S. Tokito, T. Minami, A molecular self-assembled colourimetric chemosensor array for simultaneous detection of metal ions in water, Chem. Commun. 53 (2017) 6561−6564.

[83] G.F. Swiegers, T.J. Malefetse, New self-assembled structural motifs in coordination chemistry, Chem. Rev. 100 (2000) 3483−3538.

[84] S.J. Rowan, S.J. Cantrill, G.R.L. Cousins, J.K.M. Sanders, J.F. Stoddart, Dynamic covalent chemistry, Angew. Chem. Int. Ed. 41 (2002) 898−952.

[85] V.W. von Hagen, Paper and civilization, Sci. Mon. 57 (1943) 301−314.

[86] Y. Xu, M. Bonizzoni, Disposable paper strips for carboxylate discrimination, Analyst 145 (2020) 3505−3516.

[87] A.M. Jolly, M. Bonizzoni, Intermolecular forces driving encapsulation of small molecules by PAMAM dendrimers in water, Macromolecules 47 (2014) 6281−6288.

[88] Y. He, Y. Wu, J.-Z. Fu, W.-B. Wu, Fabrication of paper-based microfluidic analysis devices: a review, RSC Adv. 5 (2015) 78109−78127.

[89] E. Yalcin, C. Erkmen, T. Taskin-Tok, M.G. Caglayan, Fluorescence chemosensing of meldonium using a cross-reactive sensor array, Analyst 145 (2020) 3345−3352.

[90] C. Görgens, S. Guddat, J. Dib, H. Geyer, W. Schänzer, M. Thevis, Mildronate (Meldonium) in professional sports − monitoring doping control urine samples using hydrophilic interaction liquid chromatography − high resolution/high accuracy mass spectrometry, Drug. Test. Anal. 7 (2015) 973−979.

[91] A.A. Azaryan, A.Z. Temerdashev, E.V. Dmitrieva, Determination of Meldonium in human urine by HPLC with tandem mass spectrometric detection, J. Anal. Chem. 72 (2017) 1057−1060.

[92] E. Yalçın, H. Duyar, H. Ihmels, Z. Seferoğlu, Spectroscopic studies on the interactions of 5-ethyl-6-phenyl-3,8-bis((3-aminoalkyl)propanamido)phenanthridin-5-ium derivatives with G-quadruplex DNA, Spectrochim. Acta A Mol. Biomol. Spectrosc. 196 (2018) 432−438.

[93] V. Hamedpour, P. Oliveri, C. Malegori, T. Minami, Development of a morphological color image processing algorithm for paper-based analytical devices, Sens. Actuators B Chem. 322 (2020) 128571.

[94] K. Crump, The potential effects of recall bias and selection bias on the epidemiological evidence for the carcinogenicity of glyphosate, Risk Anal. 40 (2020) 696−704.

[95] T. Minami, Y. Liu, A. Akdeniz, P. Koutnik, N.A. Esipenko, R. Nishiyabu, et al., Intramolecular indicator displacement assay for anions: supramolecular sensor for glyphosate, J. Am. Chem. Soc. 136 (2014) 11396−11401.

[96] Y. Liu, M. Bonizzoni, A supramolecular sensing array for qualitative and quantitative analysis of organophosphates in water, J. Am. Chem. Soc. 136 (2014) 14223−14229.

[97] V. Hamedpour, Y. Sasaki, Z. Zhang, R. Kubota, T. Minami, Simple colorimetric chemosensor array for oxyanions: quantitative assay for herbicide glyphosate, Anal. Chem. 91 (2019) 13627−13632.

[98] J. Wang, H.-B. Liu, Z. Tong, C.-S. Ha, Fluorescent/luminescent detection of natural amino acids by organometallic systems, Coord. Chem. Rev. 303 (2015) 139−184.

[99] R. Kaushik, P. Kumar, A. Ghosh, N. Gupta, D. Kaur, S. Arora, et al., Alizarin red S−zinc(II) fluorescent ensemble for selective detection of hydrogen sulphide and assay with an H$_2$S donor, RSC Adv. 5 (2015) 79309−79316.

[100] X. Lyu, W. Tang, Y. Sasaki, J. Zhao, T. Zheng, Y. Tian, et al., Toward food freshness monitoring: coordination binding−based colorimetric sensor array for sulfur-containing amino acids, Front. Chem. 9 (2021) 685783.

[101] Y. Sasaki, X. Lyu, R. Kubota, S. Takizawa, T. Minami, Easy-to-prepare mini-chemosensor array for simultaneous detection of cysteine and glutathione derivatives, ACS Appl. Bio Mater. 4 (2021) 2113−2119.

[102] G. Springsteen, B. Wang, Alizarin Red S. as a general optical reporter for studying the binding of boronic acids with carbohydrates, Chem. Commun. (2001) 1608−1609.

[103] H. Kunkely, A. Vogler, Fluorescence of alizarin complexone and its metal complexes, Inorg. Chem. Commun. 10 (2007) 355−357.

[104] S. Say-Liang-Fat, J.-P. Cornard, Al(III) complexation by alizarin studied by electronic spectroscopy and quantum chemical calculations, Polyhedron 30 (2011) 2326−2332.

[105] J.W. Tomsho, S.J. Benkovic, Elucidation of the mechanism of the reaction between phenylboronic acid and a model diol, Alizarin Red. S, J. Org. Chem 77 (2012) 2098−2106.

[106] X. Sun, M.L. Odyniec, A.C. Sedgwick, K. Lacina, S. Xu, T. Qiang, et al., Reaction-based indicator displacement assay (RIA) for the colorimetric and fluorometric detection of hydrogen peroxide, Org. Chem. Front. 4 (2017) 1058−1062.

3D graphene fabrication and application for energy storage systems

20

Yuta Nishina[1] and Rizwan Khan[2]

[1]*Research Core for Interdisciplinary Sciences, Okayama University, Okayama, Japan* [2]*Department of Electrical Engineering, Kwangwoon University, Seoul, Republic of Korea*

20.1 Introduction

Carbon-based materials have attracted considerable attention in electrochemical applications due to their abundance, stability, and relatively environment-friendly characteristics. Carbon materials have been widely used as electrode materials [1]. Some of the promising properties of carbon-based electrodes include stabilities under electrochemical conditions and wide potential windows for long-term redox cycles [2]. Carbon allotropes can exist in various forms, such as 0-D fullerenes, 1D carbon nanotubes (CNTs), 2D graphene, 3D graphite, diamond, and amorphous carbons. Each carbon allotrope has different physical and electrochemical properties (Fig. 20.1) [3]. Among them, graphene emerged as an attractive candidate for a variety of applications due to its unique structure and properties [4]. Since the first discovery of graphene by Novoselov and Geim (2010 Nobel Prize winners in physics), extensive research has been done on this newly discovered carbon material [5], owing to its excellent electrical conductivity (15,000 cm^2/Vs), large surface area (2630 m^2/g), excellent thermal conductivity (between 1500 and 2500 W/mK), high mechanical strength (Young's modulus of 1 TPa and intrinsic tensile strength of 130 GPa), and low density [6].

2D graphene has been widely investigated in different fields such as energy applications (secondary batteries, capacitors and fuel cell) and biomedical applications (sensors and imaging etc.) [7,8]. However, although 2D graphene has presented excellent performance in different applications, two serious problems should be addressed: (1) graphene sheets (GSs) are easily aggregated; and (2) graphene sheets lack pores ranging from micro- to macro-sized [9]. A 3D graphene-based framework with high porosity overcomes these two problems. Therefore, researchers have focused on the development of various 3D graphene-based structure to enhance their performance for catering to their potential application in electronic devices and biomedical application (Fig. 20.2) [10]. Different methods have been applied for the synthesis of large-scale 3D graphene structure. Vickery et al. synthesized 3D graphene for the first time through a freeze-drying method in 2009 [11]. Since then, various methods have been reported for the synthesis of 3D graphene, such as hydrogels, aerogels, sponges, organogels, foams, sugar blowing, and an electrochemical approach [12]. For all the structures of 3D graphene, the aim is to achieve high performance that can be utilized in supercapacitors (SCs) and batteries [13]. For SCs, 3D graphene with large surface area and porous structure has shown excellent

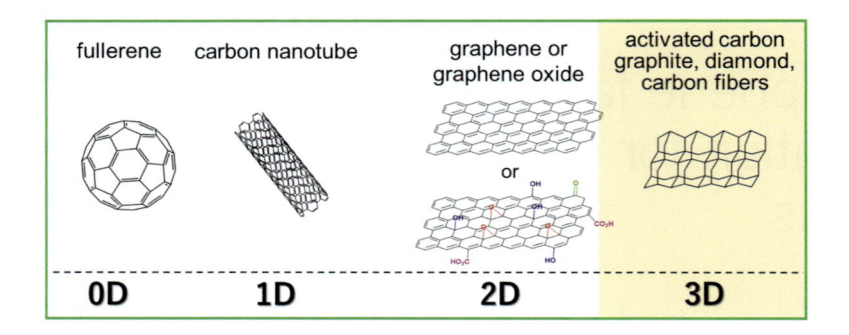

FIGURE 20.1

Carbon-based materials with various dimensions.

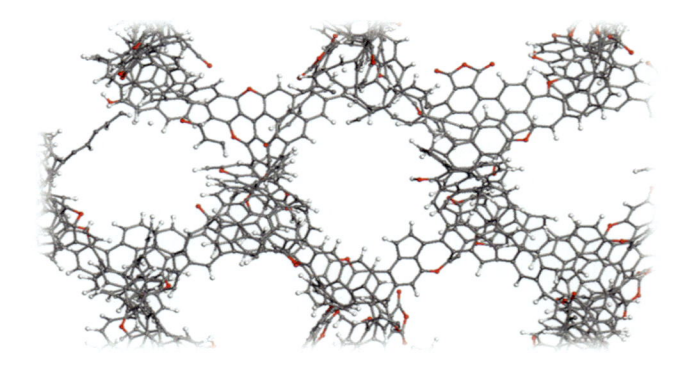

FIGURE 20.2

Schematic representation of 3D graphene.

properties in enhancing power density and capacity without compromising energy density, as well as good stability due to the well-defined structures that facilitate fast ion diffusion and transportation [14,15]. For lithium-ion batteries (LIBs), 3D graphene with optimized pores size hinders the capacity fade issues and enhances the capacity [16]. In this chapter, we first introduce some synthesis methods of 3D graphene. After this, the application of 3D graphene-based materials for SCs and LIBs will be discussed. Finally, a summary will be provided to show the usefulness of 3D graphene materials.

20.2 Fabrication of 3D graphene-based nanomaterials

There has been extensive research in the synthesis and fabrication of 3D graphene-based nanomaterials, especially related to the control of their structure and properties for different applications. This chapter mainly focuses on the synthesis and fabrication of 3D graphene structures. The fabrication of a 3D graphene-based structure may be divided into two main types: (1) template-assisted method, and (2) template-free chemical method.

20.2.1 Template-assisted method

The template-assisted method is further divided into two main types: (1) template-assisted chemical vapor deposition (CVD) method, and (2) template-assisted chemical method. The detail of each is given below.

20.2.1.1 Template-assisted chemical vapor deposition method

The CVD method is widely used for the synthesis of different types of carbon materials such as graphene, CNTs, and fullerene. Cheng et al. first reported the synthesis of flexible three-dimensional interconnected graphene using the CVD method [17]. The basic steps involved using this method are: (1) the formation of a graphene layer on a 3D foam-like structure; (2) polymerization; and (3) etching, which result in the formation of a 3D foam-like interconnected graphene structure. In general, Cu, ZnO, aluminum oxide, and Ni foam have been used as scaffolds and catalysts in the preparation of 3D nanostructures. The formed 3D scaffolds are coated with polymer. Finally, the template (Cu, Ni foam, ZnO, Al_2O_3, etc.) and polymer coating is dissolved in an appropriate solvent to obtain the 3D graphene network. Through the choice of proper scaffolds (Ni or Cu foam), the structures and pore size can be controlled in the resulting 3D graphene films [18]. The structural and surface characterization of graphene foam is shown in Fig. 20.3. The template-assisted CVD technique has been widely used for the fabrication of 3D graphene with excellent properties. However, the structural stability of these 3D graphenes needs further improvement.

20.2.1.2 Template-assisted chemical method

Graphene derivatives, such as graphene oxides (GO) and reduced graphene oxides (RGO), are used as starting materials. First, the functional graphene precursor is deposited either on metal foam or polymer beads. After this, a chemical or electrochemical process is employed to remove the functional groups on graphene. Finally, metal foams or polymer beads are dissolved in the suitable solutions to achieve a 3D graphene structure (Fig. 20.4A). The resulting materials show a well-defined and porous structure (Fig. 20.4B). Typically, Cu foam, Ni foam, MnO_2, $CaCO_3$, and polystyrene beads have been employed as scaffolds [19]. The template-assisted chemical method has been widely used to obtain 3D graphene structures for different applications. However, this method involves multiple steps and requires additional chemicals to achieve a 3D graphene structure.

20.2.2 Self-assembly or template-free chemical method

The template-free chemical method or self-assembly method has been considered to be a promising strategy for the fabrication 3D nanostructures. GO can act as a building block for self- assembly into 3D macrostructures such as aerogels, papers, hydrogels, hybrid, and sponge structures. This is because GO sheets possess an abundance of oxygenated functional groups on their surfaces, which are hydrophilic in nature, and the repulsive forces between these groups favor the formation of a well-ordered nanostructure. This method is further classified into two types based on the interaction between the graphene sheets (physical interaction and covalent interactions). Most of these 3D-based solids are produced through the van der Waals interactions of GO building blocks. Following chemical reduction or the gelation process, drying via a special process is essential to obtain 3D nanostructures. However, the flexibility of these 3D structures is poor without the addition of some

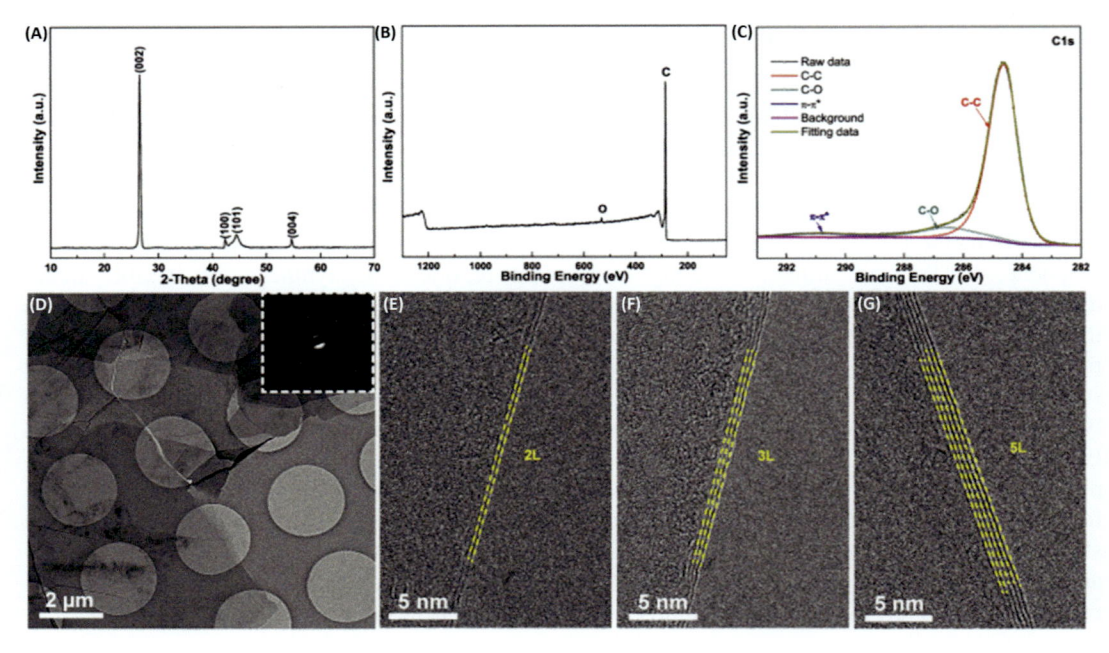

FIGURE 20.3

Structural and spectroscopic characterization of graphene foam. (A) XRD pattern of the graphene foam. XPS survey scan (B) and C1s XPS spectrum (C) of the foam. (D) TEM image of the thin graphene sheets in the graphene foam. (E—G) High-resolution TEM images of the graphene sheets with different numbers of layers (E: 2 layers; F: 3 layers; G: 5 layers).

Reproduce from M. Huang, C. Wang, L. Quan, T.H.-Y. Nguyen, H. Zhang, Y. Jiang, et al., CVD growth of porous graphene foam in film form, Matter. 3 (2020) 487–497 with permission from Elsevier, copy right 2020.

organics or polymers. To tackle these problems, the covalent interaction between graphene sheets was proposed and efforts have been made with various types of network linkers such as resorcinol, glutaraldehyde, DNA, polyallylamine, etc. (Fig. 20.5A) [20]. The surface morphology of such 3D interconnected graphene sheets are shown in Fig. 20.5B. Other methods for fabrication have also been reported such as electrochemical approaches, freeze-drying technique, and sugar blowing.

20.2.3 Electrochemical approaches

Several research groups have synthesized 3D graphene base composites through electrochemical method (Fig. 20.6A) [21]. The electrochemical method either employs the electrochemical deposition method or the electrochemical leavening method. In the deposition method, the GO suspension first is electrochemically reduced and then directly deposited on the surface of the electrode. The structure and morphology of the 3D graphene can be precisely controlled through tuning the deposition rate and time (Fig. 20.6B). In the electrochemical leavening method, current is applied to exfoliate the graphite papers.

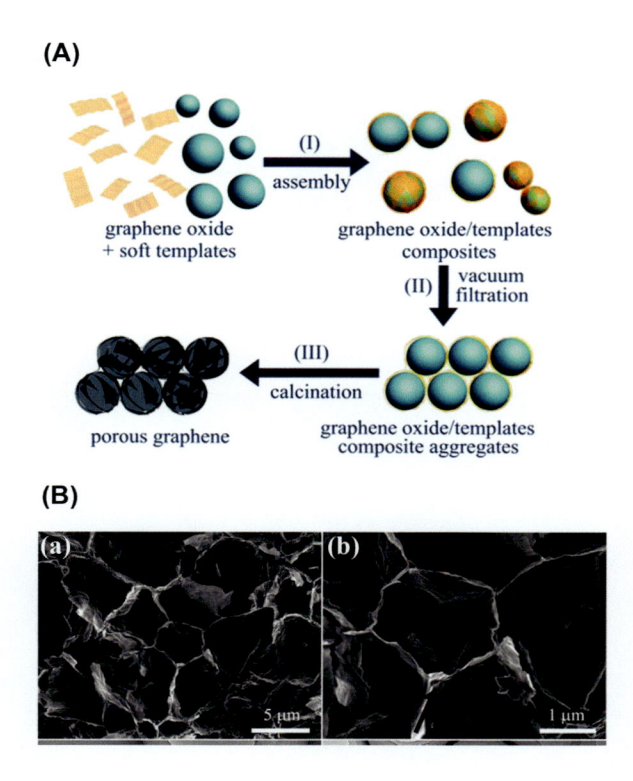

FIGURE 20.4

(A) Synthesis of porous graphene foam, and (B) SEM images in low (a) and high (b) resolution.

Reproduce from X. Huang, B. Sun, D. Su, D. Zhao, G. Wang, Soft-template synthesis of 3D porous graphene foams with tunable architectures for lithium–O2 batteries and oil adsorption applications, J. Mater. Chem. A. 2 (2014) 7973–7979 with permission from Elsevier, copy right 2014.

20.2.4 Freezing technique

The freezing technique is widely used for large-scale production of 3D graphene nanostructures. The freezing method is simpler and more cost-effective than the self-assembly approach. Liu et al. employed the freezing technique followed by thermal reduction to obtain a flexible 3D graphene papers type structure (Fig. 20.7A and B) [22]. A 3D graphene structure can be also produced on a large scale from low-cost graphene paper through the combination of a modified Hummer method and freezing technique. The as-prepared 3D nanostructure can be combined with polymer, organic molecules, or metal nanoparticles for energy storage applications.

20.2.5 Sugar blowing method

The sugar blowing method uses NH_4Cl and glucose as a nitrogen and carbon source, respectively. In this method a large amount of bubbles are produced during the heating of NH_4Cl and glucose

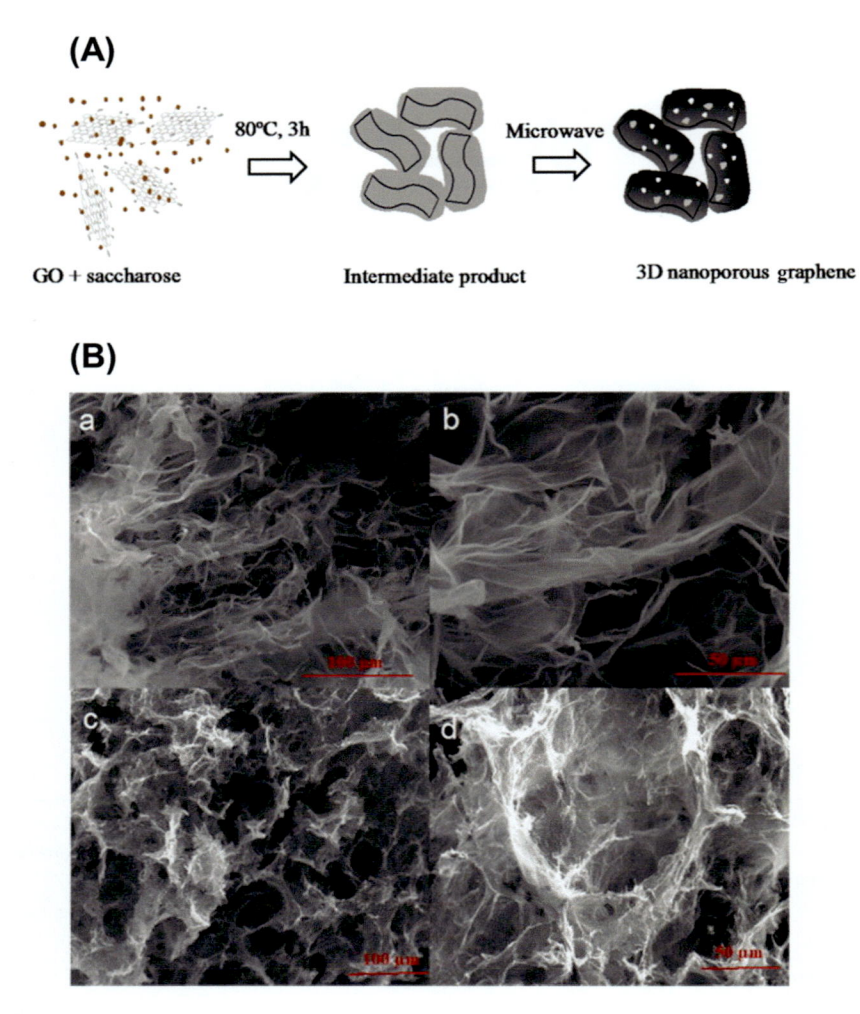

FIGURE 20.5

(A) Synthesis of 3D nanoporous graphene, (B) SEM image of GO (a, b) and 3D porous graphene (c, d).

Reproduce from T.T. Vu, T.C. Hoang, T.H.L. Vu, T.S. Huynh, T.V. La, Template-free fabrication strategies for 3D nanoporous Graphene in desalination applications, Arab. J. Chem. 14 (2021) 103088 with permission from Elsevier, copy right 2021.

(Fig. 20.8A) [23]. After this 3D interconnected and porous graphitic films are progressively generated after high-temperature annealing (Fig. 20.8B−D); high temperature (e.g., >1300°C) is needed in this method. The heating rate is an important parameter for controlling the 3D nanostructure to obtain a high surface area (Fig. 20.8E).

(A)

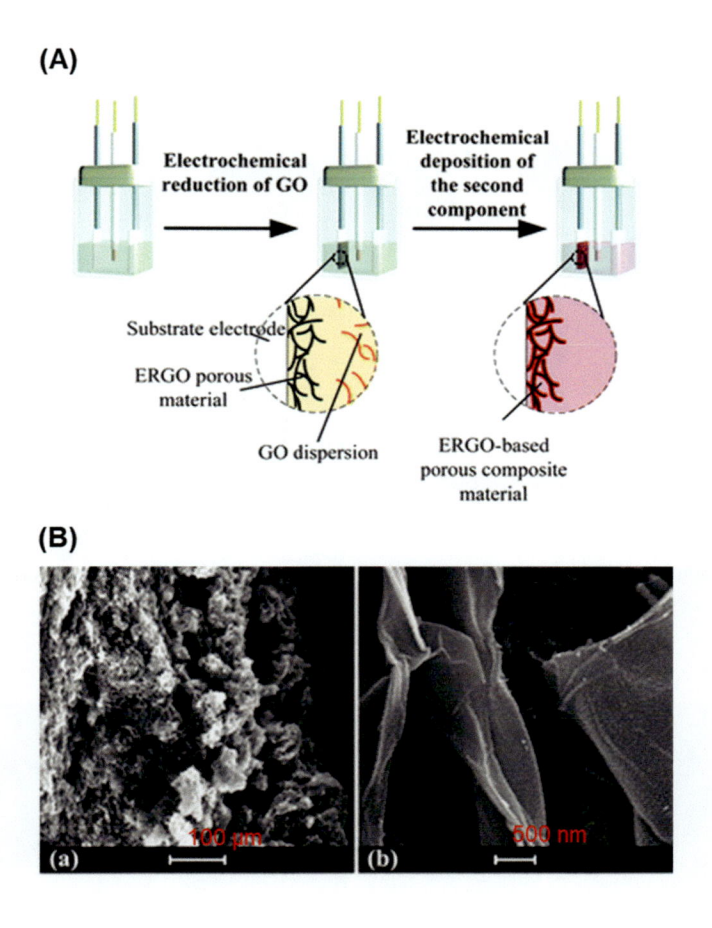

(B)

FIGURE 20.6

(A) Electrochemical approach for the synthesis of 3D graphene. (B) SEM images of the 3D graphene—polyaniline composite materials synthesized with the method of chronoamperometry at 0.65 V; (a) 100 μm, (b) 500 nm.

Reproduce from K. Chen, L. Chen, Y. Chen, H. Bai, L. Li, Three-dimensional porous graphene-based composite materials: electrochemical synthesis and application, J. Mater. Chem. 22 (2012) 20968–20976 with permission from royal chemical society, copy right 2012.

20.3 Applications for energy storage systems

20.3.1 Application of 3D graphene for supercapacitors

SCs are attractive for high-power devices because of their rapid charge—discharge performances. High-performance SCs have large specific capacitances, high-power capabilities, and ultralong cycle lives [24]. SCs are classified into two types: electrical double-layer capacitors (EDLCs) and pseudocapacitors. EDLCs store energy using an electrostatic mechanism at electrode/electrolyte

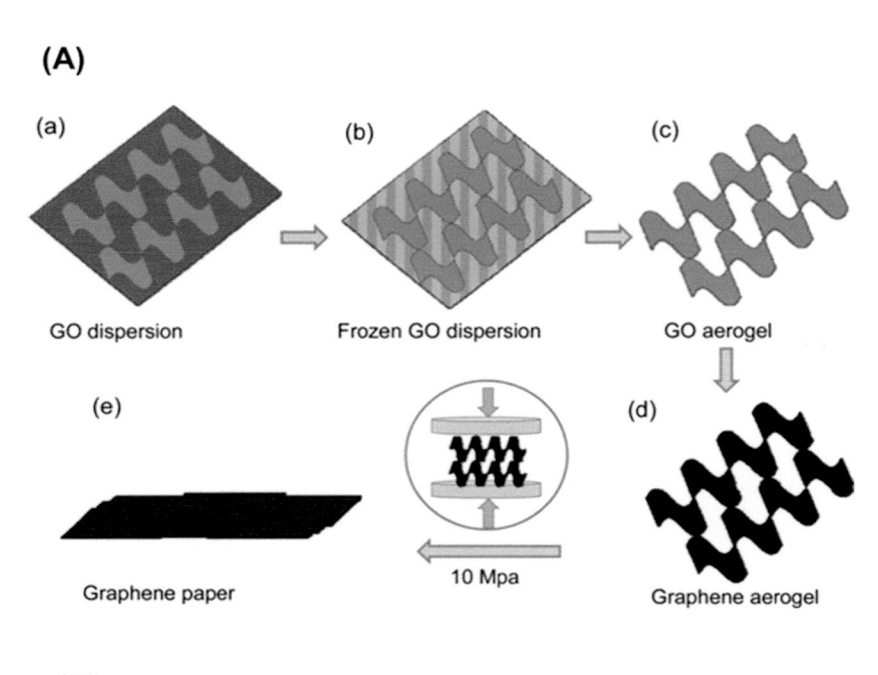

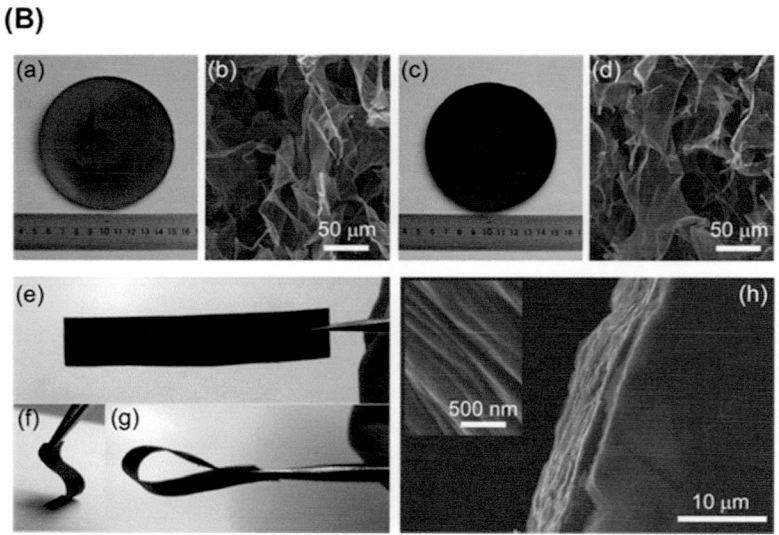

FIGURE 20.7

(A) Synthesis of graphene paper. (a) GO dispersion in water (b) dispersion frozen at −50°C. (c) GO aerogel achieved by freeze drying, (d) Graphene aerogel obtained by treating (c) at 200°C in air. (e) Mechanical pressing of the graphene aerogel to form graphene paper; (B) Digital image (a) and SEM image (b) of a GO aerogel. Digital image (c) and SEM image (d) of a graphene aerogel. Digital images (e − g) and SEM images (h) of graphene paper.

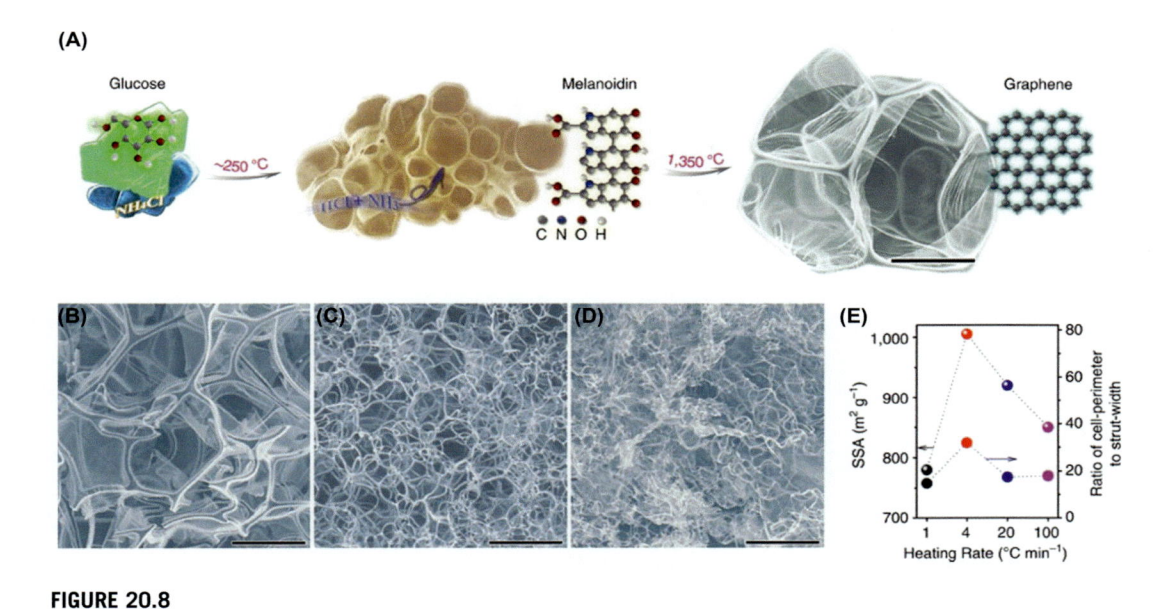

FIGURE 20.8

(A) Synthesis of 3D graphene nanostructure through sugar blowing method; (B−D) SEM images of the 3D graphene nanostructure formed by the heating rate of 1, 20, or 100°C min^{-1}, respectively. Scale bars: 200 μm, and (E) changes of specific surface area versus heating rates.

Reproduce from X. Wang, Y. Zhang, C. Zhi, X. Wang, D. Tang, Y. Xu, et al., Three-dimensional strutted graphene grown by substrate-free sugar blowing for high-power-density supercapacitors, Nat. Commun. 4 (2013) 2905, copy right 2013.

interfaces; in contrast, pseudocapacitance is based on the redox reactions at electrode surfaces (Fig. 20.9) [25]. The electrode materials for SCs should possess a porous structure and a large surface area, which are important to allow the fast transportation of ions at the interface of electrolyte and electrode materials. Among the expected applications of 3D graphene materials, SCs have attracted much attention [26]. Graphene-based nanomaterials have been extensively investigated in SCs applications due to their unique properties [27]. The structure and morphology of the graphene-based nanomaterials are playing an important role to enhance the performance of SCs. Especially, the development of 3D graphene nanostructure has altered this situation and led to new advances for fabricating high-performance SCs due to their light weight, high porosity, and conductivity [28]. Therefore, 3D graphene and it composites are being actively investigated with regard to high-performance SCs.

20.3.1.1 Porous 3D graphene nanostructure

Significant effort has been made for the preparation of well-defined porous 3D graphene nanostructures and investigated as electrode materials for SCs. For example, a 3D graphene aerogel structure with hierarchical pore system was reported that demonstrated a capacitance of 226 F/g after 5000 cycles (Table 20.1). The high capacitance was achieved due to the synergistic effect between meso- and macropores. The macropores shorten the diffusion distances from the external electrolyte to the

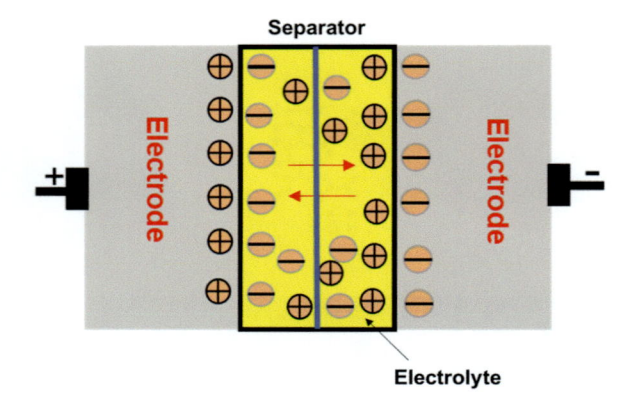

FIGURE 20.9

Schematic representation of the supercapacitors.

Table 20.1 Porous 3D graphene nanostructures for supercapacitors.

Structure	Preparation method	Specific capacitance	References
3D graphene framework	Hard template	113 F/g	[29]
3D porous graphene sheet	Soft template	245 F/g	[30]
	Template free	331 F/g	[31]
	Hydrothermal method	298 F/g	[32]
	Microwave-Ar plasma	203 F/g	[33]
	Gas foaming	231 F/g	[34]
	Gas exploitation	426 F/g	[35]
3D porous graphene hydrogel/aerogel	Chemical vapor deposition	38.2 F/cm^3	[36]
	Hydrothermal method	196 F/g	[37]
	Heating and mixing	226 F/g	[38]
3D graphene ball	Soft template	390 F/g	[39]
3D graphene monolith	Soft template	217 F/g	[40]
3D graphene network	Hydrothermal method	325 F/g	[41]
3D graphene aerogel	Freeze-casting method	554 F/g	[28]
3D graphene foam	Electrophoretic deposition	62.2 mF/cm^2	[42]
3D functional graphene	Amino acid treatment	295 F/g	[43]
3D functional graphene	Electrostatic self-assembly	378 F/g	[44]
Porous graphene oxide	Microwave	568 F/g	[45]

interior surfaces. The mesopores and micropores can enhance ion transport and charge storage. Similarly, a bicontinuous 3D graphene porous structure was constructed using the CVD method. The resulting composite materials presented a high capacitance of 245 F/g and a capacitance retention of 96% after 6000 cycles (Table 20.1). It was shown that graphene with a porous network showed higher electrochemical performance than a compact counterpart due to its larger active surface area. In the follow-up studies, the capacitance was further enhanced by incorporating pseudo-capacitance materials into the 3D graphene architecture.

The combination of 3D graphene with other active materials, including conductive polymers, metal, metal oxides, and hydroxides, is an efficient way to improve the electron transfer rates of nanocomposites, which enhance the electrochemical performance of SCs.

20.3.1.2 Composite of 3D graphene and metals for supercapacitors

A variety of nanocomposites have been synthesized by combining 3D graphene with metal oxides, and then investigated as electrode materials for SCs. For example, 3D graphene$-MnO_2$ nanocomposite was fabricated by a self-assembly approach. The prepared composite material showed a specific capacitance of 242 F/g compared to 137 F/g of pure 3D graphene (Table 20.2). Similarly, 3D graphene was combined with metal hydroxide and applied as an electrode material for high-performance SCs. The nanocomposite of 3D graphene with $Ni(OH)_2$ was prepared, which showed a specific capacitance of 1632 F/g (Table 20.2). Furthermore, ternary 3D graphene-based composites have been fabricated as electrode materials in SCs to achieve high electrochemical performance. For example, $Ni-Al-CNT-RGO$ ternary composite was applied as an electrode material, and it displayed a capacitance of 1562 F/g at a current rate of 5 mA/cm^2 (Table 20.2). The high capacitance was achieved due to the synergistic effect among the four components. The high electrochemical performance in these composites is attributed to high porosity, fast diffusion of ions, and high conductivity. The literature survey showed that the composite of metals oxide or hydroxide with 3D graphene is ideal for high-performance SCs.

20.3.1.3 Composite of 3D graphene with conductive polymer

Functional graphene structures have gained more interest as high-performance electrode materials for SCs due to their robustness, flexibility, and high porosity. In particular, the composites of 3D graphene with conductive polymer are, promising for high-performance SCs. Polymers offer several

Table 20.2 Composites of 3D graphene with metal oxides and metal hydroxides for supercapacitors.

Structure	Preparation method	Specific capacitance	References
3D graphene-MnO_2	Self-assembly	242 F/g	[46]
3D $NiCo_2S_4$/Ni graphene	In situ grown	1740 mF/cm^2	[47]
Graphene/Ni/Mn	Hydrothermal carbonization	1648 F/g	[48]
Graphene/$Ni(OH)_2$	Solvothermal-induced self-assembly	1632 F/g	[49]
$Ni-Al-CNT-RGO$	Solvothermal	1562 F/g	[50]
3D $CoNi_2S_4$/graphene	Ultrasonication-hydrothermal	1141 F/g	[51]
Graphene/Ni/Co	Microwave plasma-CVD	765 F/cm^3	[52]
SnS/S-doped graphene	Nonhydrolytic thermal annealing	642 F/g	[53]
3D porous graphene/$Ni(OH)_2$	Vacuum filtration self-assembly	2321 F/cm^3	[54]
3D graphene/MnO	Flash reduction process	1706 F/g	[55]
3D graphene/CoO_3	Hydrothermal method	688 F/g	[56]

Table 20.3 Composites of 3D graphene with conductive polymer for supercapacitors.

Structure	Preparation method	Specific capacitance	References
N-graphene/PPy	Electropolymerization	196 mF/cm^2	[57]
Graphene/Ni/PPy	Electropolymerization	350 F/g	[58]
N-graphene/PANI	In situ polymerization	521 F/g	[59]
3D graphene/PANI	Electrodeposition	751 F/g	[60]
3D graphene/PANI	In situ polymerization	800 F/cm^3	[61]
3D graphene/PANI	Hydrothermal method	1182 F/g	[62]
CNT/graphene/PANI	Hydrothermal method	568 F/cm^3	[63]

characteristic properties such as precise control of redox potentials, a variety of counterions, and flexible or even bendable electrodes. For example, the composite of polypyrrole (PPy) and 3D graphene was synthesized and employed as a flexible electrode material for SCs. The prepared material showed high specific capacitance of 350 F/g at a current density of 1.5 A/g (Table 20.3). Among other conductive polymers, PPy has been extensively investigated as an electrode for high-performance SCs due to its high stability and improved charge storage ability, which not only enhances the capacitance but also improves the flexibility of the 3D nanostructure-based electrode. Besides PPy, polyaniline (PANI), another widely studied conductive polymer, was combined with 3D graphene and applied as electrode material for SCs. PANI possesses excellent conductivity, high theoretical capacitance, fast redox reaction, and has a simple synthesis process. For example, a 3D graphene/PANI nanocomposite array was fabricated as an electrode material and it displayed an energy density of 126 W/hkg and a power density of 7622 W/kg. The specific capacitance of the resulting 3D network structure was 751 F/g (Table 20.3). Also, another type of 3D graphene/PANI hybrid nanostructure was reported that showed a high volumetric capacitance of over 800 F/cm^3. The composites of conductive polymer and 3D graphene showed high electrochemical performance due to excellent surface area, high porosity, and better conductivity.

20.3.1.4 Chemical doping method

Chemical doping of 3D graphene has been widely investigated to obtain high-performance SCs. Various heteroatoms have been doped in 3D graphene structures and high capacitance has been obtained. For example, a nitrogen-doped graphene framework was applied as a SC, which showed a high specific capacitance of 326 F/g (Table 20.4). Similarly, 3D nitrogen-doped graphene was synthesized through a hydrothermal method using NH_3BF_3 as both a source of nitrogen and boron and also a reducing agent. The resulting material was applied as an electrode material for SCs, which exhibited a capacitance of 239 F/g at 0.2 A/g (Table 20.4). Doping of 3D graphene with heteroatoms improves wettability, electrical conductivity, and allows a fast transport rate of ions in solution.

20.3.2 3D graphene for lithium ion batteries

Lithium ion batteries (LIBs) are the most widely used rechargeable batteries with the advantages of low toxicity, high capacity, and high working voltage (Fig. 20.10) [75]. With such useful properties, LIBs have emerged as one of the key power sources in the electronic markets, although LIBs are still

Table 20.4 Doping of 3D graphene with heteroatoms and their application in supercapacitors.

Structure	Preparation method	Specific capacitance	References
3D N-graphene	Hydrothermal	326 F/g	[64]
3D N, B-codoped graphene	Hydrothermal	239 F/g	[65]
3D N-doped graphene	Hydrothermal	114 F/g	[66]
3D S, N-codoped graphene	Heating and mixing	343 F/g	[67]
3D N, P-codoped graphene	Hydrothermal	453 F/g	[68]
3D N-doped graphene	Heating and mixing	405 F/g	[69]
3D N-doped graphene	In situ polymerization	521 F/g	[59]
3D S, N-codoped graphene	Hydrothermal method	89 F/g	[70]
N,P,F-codoped graphene	Hydrothermal and freeze-drying	319 F/g	[71]
3D S, N-codoped graphene	Hydrothermal method	399 F/g	[72]
3D S-doped graphene	Laser direct writing method	22 mF cm^2	[73]
3D O-doped graphene	Heating and mixing	173 F/g	[74]

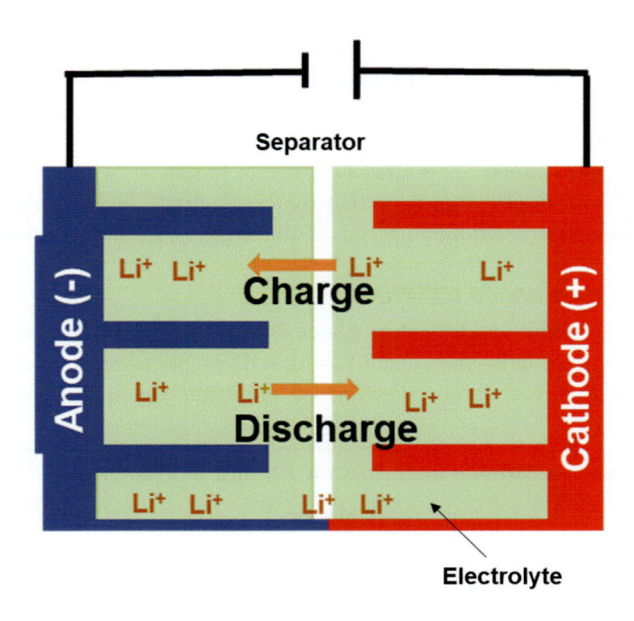

FIGURE 20.10

Schematic representation of the lithium ion batteries.

not ideal candidate for large-volume and high-power applications [76]. Recently, researchers have been working on improving the electrochemical performance of LIBs by applying graphene and its composite as electrode materials. Graphene has been applied either as a cathode support or anode material for LIBs [77]. Unfortunately, there are two shortcomings of graphene-based electrodes for LIBs applications. First, graphene suffers from agglomeration and restacking, as a result decreasing the performance of LIBs. Secondly, the large irreversible capacity makes it less suitable for practical application. To overcome these problems, 3D graphene-based materials have been developed and applied as electrodes in LIBs. 3D graphene with porous structure has been constructed and combined with active components (such as conducting polymer, transition metal oxides, silicon, tin, and metal sulfide) to improve the electrochemical performance of LIBs [78].

20.3.2.1 3D graphene as cathode materials

Transition metal oxides with different structures are widely used as cathode materials for Li-ion batteries. The metal oxides-based cathode material showed high electrochemical performance. However, these materials show poor cycling stability. The electrochemical performance was improved by combining $LiFePO_4$ with carbonaceous materials. In particular, 3D graphene-based materials enhance the conductivity and facilitate fast kinetics for Li-ion insertion/extraction. For example, ultrathin graphite foam was prepared by the CVD method on Ni foam and then $LiFePO_4$ was drop-casted on it. The resulting electrode materials exhibited a capacity of 70 mA/hg with 99.9% columbic efficiency (Table 20.5). Similarly, $LiFePO_4$ nanoplatelets was wrapped in a 3D nitrogen-doped graphene porous structure. The resulting electrode materials showed high electrochemical performance. The porous 3D structure allows fast transportation of ions and improves the conductivity. Nitrogen doping improves the wettability and electronic conductivity. 3D graphene was fabricated by applying silica spheres as a template, and then the composite of 3D graphene with Li_2FeSiO_4/C was prepared by a sol−gel method. The material showed high rate capability and good cycling stability compared to the composite of 2D graphene with Li_2FeSiO_4/C. The high electrochemical performance of 3D graphene/Li_2FeSiO_4/C could be attributed to the 3D porous structure of graphene, which allows fast transportation of ions (Table 20.5). These results suggested that the conductivity and electrochemical performance of metals or metals oxide-based cathode materials is improved by combination with 3D graphene.

20.3.2.2 3D graphene as anode materials

In order to replace the graphite anode, 3D graphene has been directly applied as an anode material to achieve high capacity. However, the capacity of the 3D graphene structure is significantly loss in the first cycle due to the irreversible reaction of graphene functional groups with Li-ion. To

Table 20.5 3D graphene as cathode materials for lithium ion batteries.

Structure	Preparation method	Capacity	References
Ultrathin graphite foam	CVD method using Ni foam	158 mA/hg	[79]
3D N-doped graphene	Hydrothermal method	155 mA/hg	[80]
3D graphene/Li_2FeSiO_4 /C	Sol−gel method	313 mA/hg	[81]
RGO	Thermal annealing	170 mA/hg	[82]
	Hydrothermal reduction	172 mA/hg	[83]
	Chemical reduction	58 mA/hg	[84]
Graphene hydrogel	Hydrothermal reduction	52 mA/hg	[85]
Graphene grass	Hydrothermal reduction	63 mA/hg	[86]
Porous graphene macroform	Hydrothermal reduction	66 mA/hg	[87]
3D crosslinked graphene	Hydrothermal method	210 mA/hg	[88]

overcome this problem, 3D graphene has been combined with other high-capacity active materials such as Li-alloy or transition metal oxides-based anode materials. For example, 3D graphene nanocomposites with TiO_2 were prepared by a one-step hydrothermal method. TiO_2 nanocrystals were grown in situ onto a graphene surface using glucose as a linker. This mesoporous structure demonstrated a high reversible capacity, excellent Li storage, and high rate capability. 3D graphene was decorated with a honeycomb-like MoS_2 structure and applied as anode materials for Li-ion batteries. The resulting nanocomposite material showed a high discharge capacity of 1235 mA/hg due to the large surface area, high conductivity, and macroporous structure of 3D graphene (Table 20.6). Group IV elements, such as Si, Ge, and Sn could be used as high-performance anode materials for Li-ion batteries. These materials show 10 times higher capacity than graphite. However, these materials show poor cycling stability due to a large change in volume during charging/discharging. This problem could be overcome by incorporating 3D graphene into their structure. For example, the composite of Si and 3D graphene was prepared and applied as anode materials for Li-ion batteries. Si nanoparticles were embedded in a graphene structure. The resulting nanocomposite exhibited a capacity of 983 mA/g in the first cycle and retained a capacity of 370 mA/hg after 100 cycles (Table 20.6).

20.3.2.3 3D graphene interconnected sheets in lithium ion batteries

In this method, the active materials are intercalated in space between the graphene sheets forming electrode materials. For example, H-Fe^3O^4 was encapsulated between the graphene sheets to form 3D conductive network and applied as electrode materials in LIBs [101]. The resulting composite materials showed a high capacity of 1555 mA/hg. The high capacity is attributed to the porous 3D network of graphene sheets which allows fast intercalation of electron and lithium ion into the pores. Also, the composite of Fe_3O_4 and graphene were fabricated as electrode materials, which presented a high capacity of ~ 850 mA/hg at 150 mA/g [102]. Similar performance improvement has also been achieved in other 3D graphene framework supported metal oxides, such as graphene/

Table 20.6 3D graphene as anode materials for lithium ion batteries.

Structure	Preparation method	Capacity	References
3D graphene/MoS_2	Hydrothermal-self assembly	1235 mA/hg	[89]
3D graphene/Ge sponge	Freeze drying-thermal reduction	1258 mA/hg	[90]
3D graphene/Si	Mixing and heating	370 mA/hg	[91]
3D graphene sheet/Si/Au	Vacuum filtration and hot pressing	1520 mA/hg	[92]
3D graphene/Sn	Plasma enhanced-CVD	1037 mA/hg	[93]
3D graphene/Si nanowire	Electrostatic self-assembly	1335 mA/hg	[94]
3D graphene/SnO2	Solvothermal-self assembly	905 mA/hg	[95]
N-doped graphene/Fe_3O_4	Hydrothermal method	1130 mA/hg	[96]
3 VS4/graphene	Hydrothermal method	551 mA/hg	[97]
3D SiOC@C/RGO	Hydrothermal method	676 mA/hg	[98]
S,N-codoped graphene/Si	Hydrothermal method	941 mA/hg	[99]
Cu_3P/N-RGO	Heating-up method	705 mA/hg	[100]

Co_3O_4 [103], graphene/MnO_2 [104], and graphene/Si [105]. This type of approach enhances the surface area, porosity, and electrical conductivity of 3D graphene-based electrode materials.

20.3.2.4 3D bilayer graphene framework in lithium ion batteries

In this method, the layered active materials are deposited on the surface of graphene sheet forming 3D integrated electrode. Such an assembly was prepared by solvothermal/hydrothermal synthesis, sputtering, and atomic layer deposition. For example, a 3D porous graphene-framework MoS_2 electrode was assembled by a simple hydrothermal method. The prepared material showed a high capacity of 1200 mA/hg, and capacity retention of 100% after 3000 cycles [106]. Similarly, Fe_3O_4 nanoparticles were deposited on a graphene foam and presented a better capacity of 785 mA/hg [107]. 3D graphene networks have also been applied as substrate for cathode materials of LIBs, however, there is no contribution of graphene to the overall capacitance in a high-voltage window. Impressively, by applying the high-voltage window electrolyte ($LiFePO_4$), graphene accounted for the total mass of the electrode. The specific capacity of the graphene/ LFP cathode (98 mA/hg) was almost 23% higher than that of the aluminum/LFP cathode and 170% higher than that of the nickel-foam/LFP cathode. Furthermore, the capacity retention of the electrode was 98% after 500 cycles [108]. Compared to the traditional materials, 3D graphene showed high surface area, better electrical conductivity, a well-defined porous structure, and good interconnectivity.

20.3.2.5 3D graphene arrays network in lithium ion batteries

In this type of structure, the nanoarrays of active materials are assembled with 3D graphene networks to obtain high-performance electrodes. 3D array nanostructures possess a high surface area, allow easy passage for electrolyte, and provide a short diffusion path for Li ions. For example, graphene-supported $Li_4Ti_5O_{12}$ composite structure was frst synthesized by a hydrothermal method. The resulting composite showed a specific capacity of 160 mA/hg, and showed capacity retention of 100% after 500 cycles [108]. An MoS_2 array nanostructure was fabricated on graphene, forming 3D honeycomb-like electrodes, which showed specific capacity of 1235 mA/hg and capacity retention of 85.8% after 60 cycles, higher than the bulk MoS_2 materials [89]. All these results indicate that 3D graphene-based nanostructures are promising electrode materials for stable and high-performance Li-ion batteries.

20.4 Conclusion

This chapter has summarized the fabrication of 3D graphene and its application in SCs and LIBs in the last few years. Different fabrication methods of 3D graphene were discussed, such as template-assisted, self-assembly, electrochemical approaches, freezing, and sugar blowing. Depending on the synthesis process, various kinds of 3D graphene structure can be prepared. The application of 3D graphene and their composites for SCs and LIBs were discussed in detail. The literature survey suggests that the unique porous and well-defined structures of 3D graphenes are promising for application in advanced energy storage systems. The electrochemical performance of the 3D graphene-based electrode can be improved by producing composites with metals, metals oxides, and polymers, and by doping with heteroatoms.

References

[1] R. Khan, Y. Nishina, Covalent functionalization of carbon materials with redox-active organic molecules for energy storage, Nanoscale 13 (2021) 36−50. Available from: https://doi.org/10.1039/D0NR07500K.

[2] L.L. Zhang, X.S. Zhao, Carbon-based materials as supercapacitor electrodes, Chem. Soc. Rev. 38 (2009) 2520−2531. Available from: https://doi.org/10.1039/B813846J.

[3] H. Badenhorst, A review of the application of carbon materials in solar thermal energy storage, Sol. Energy 192 (2019) 35−68. Available from: https://doi.org/10.1016/j.solener.2018.01.062.

[4] Y. Yang, A.M. Asiri, Z. Tang, D. Du, Y. Lin, Graphene based materials for biomedical applications, Mater. Today 16 (2013) 365−373. Available from: https://doi.org/10.1016/j.mattod.2013.09.004.

[5] K.S. Novoselov, A.K. Geim, S.V. Morozov, D. Jiang, Y. Zhang, S.V. Dubonos, et al., Electric field effect in atomically thin carbon films, Science 306 (2004) 666−669. Available from: https://doi.org/10.1126/science.1102896.

[6] V. Singh, D. Joung, L. Zhai, S. Das, S.I. Khondaker, S. Seal, Graphene based materials: Past, present and future, Prog. Mater. Sci 56 (2011) 1178−1271. Available from: https://doi.org/10.1016/j.pmatsci.2011.03.003.

[7] J. Zhu, D. Yang, Z. Yin, Q. Yan, H. Zhang, Graphene and graphene-based materials for energy storage applications, Small 10 (2014) 3480−3498. Available from: https://doi.org/10.1002/smll.201303202.

[8] B. Fadeel, C. Bussy, S. Merino, E. Vázquez, E. Flahaut, F. Mouchet, et al., Safety assessment of graphene-based materials: focus on human health and the environment, ACS Nano 12 (2018) 10582−10620. Available from: https://doi.org/10.1021/acsnano.8b04758.

[9] M. Wang, Y. Niu, J. Zhou, H. Wen, Z. Zhang, D. Luo, et al., The dispersion and aggregation of graphene oxide in aqueous media, Nanoscale 8 (2016) 14587−14592. Available from: https://doi.org/10.1039/C6NR03503E.

[10] Z. Sun, S. Fang, Y.H. Hu, 3D Graphene materials: from understanding to design and synthesis control, Chem. Rev. 120 (2020) 10336−10453. Available from: https://doi.org/10.1021/acs.chemrev.0c00083.

[11] J.L. Vickery, A.J. Patil, S. Mann, Fabrication of graphene−polymer nanocomposites with higher-order three-dimensional architectures, Adv. Mater 21 (2009) 2180−2184. Available from: https://doi.org/10.1002/adma.200803606.

[12] M. Ding, C. Li, Recent advances in simple preparation of 3D graphene aerogels based on 2D graphene materials, Front. Chem 10 (2022). Available from: https://doi.org/10.3389/fchem.2022.815463. Available from: https://www.frontiersin.org/article/. accessed March 19, 2022.

[13] A.R. Thiruppathi, B. Sidhureddy, E. Boateng, D.V. Soldatov, A. Chen, Synthesis and electrochemical study of three-dimensional graphene-based nanomaterials for energy applications, Nanomaterials (Basel) 10 (2020) 1295. Available from: https://doi.org/10.3390/nano10071295.

[14] A.H. Reaz, S. Saha, C.K. Roy, M.M. Hosen, T.S. Shuvo, M.M. Islam, et al., Performance improvement of supercapacitor materials with crushed 3D structured graphene, J. Electrochem. Soc 169 (2022) 010521. Available from: https://doi.org/10.1149/1945-7111/ac4930.

[15] Y. Ping, Y. Gong, Q. Fu, C. Pan, Preparation of three-dimensional graphene foam for high performance supercapacitors, Prog. Nat. Sci.: Mater. Int 27 (2017) 177−181. Available from: https://doi.org/10.1016/j.pnsc.2017.03.005.

[16] D. Sui, M. Chang, Z. Peng, C. Li, X. He, Y. Yang, et al., Graphene-based cathode materials for lithium-ion capacitors: a review, Nanomaterials 11 (2021) 2771. Available from: https://doi.org/10.3390/nano11102771.

[17] Z. Chen, W. Ren, L. Gao, B. Liu, S. Pei, H.-M. Cheng, Three-dimensional flexible and conductive interconnected graphene networks grown by chemical vapour deposition, Nat. Mater 10 (2011) 424−428. Available from: https://doi.org/10.1038/nmat3001.

[18] M. Huang, C. Wang, L. Quan, T.H.-Y. Nguyen, H. Zhang, Y. Jiang, et al., CVD growth of porous graphene foam in film form, Matter 3 (2020) 487–497. Available from: https://doi.org/10.1016/j.matt.2020.06.012.

[19] X. Huang, B. Sun, D. Su, D. Zhao, G. Wang, Soft-template synthesis of 3D porous graphene foams with tunable architectures for lithium−O2 batteries and oil adsorption applications, J. Mater. Chem. A 2 (2014) 7973–7979. Available from: https://doi.org/10.1039/C4TA00829D.

[20] T.T. Vu, T.C. Hoang, T.H.L. Vu, T.S. Huynh, T.V. La, Template-free fabrication strategies for 3D nanoporous Graphene in desalination applications, Arab. J. Chem 14 (2021) 103088. Available from: https://doi.org/10.1016/j.arabjc.2021.103088.

[21] K. Chen, L. Chen, Y. Chen, H. Bai, L. Li, Three-dimensional porous graphene-based composite materials: electrochemical synthesis and application, J. Mater. Chem 22 (2012) 20968–20976. Available from: https://doi.org/10.1039/C2JM34816K.

[22] F. Liu, T.S. Seo, A controllable self-assembly method for large-scale synthesis of graphene sponges and free-standing graphene films, Adv. Funct. Mater 20 (2010) 1930–1936. Available from: https://doi.org/10.1002/adfm.201000287.

[23] X. Wang, Y. Zhang, C. Zhi, X. Wang, D. Tang, Y. Xu, et al., Three-dimensional strutted graphene grown by substrate-free sugar blowing for high-power-density supercapacitors, Nat. Commun 4 (2013) 2905. Available from: https://doi.org/10.1038/ncomms3905.

[24] K.K. Patel, T. Singhal, V. Pandey, T.P. Sumangala, M.S. Sreekanth, Evolution and recent developments of high performance electrode material for supercapacitors: a review, J. Energy Storage 44 (2021) 103366. Available from: https://doi.org/10.1016/j.est.2021.103366.

[25] A. Borenstein, O. Hanna, R. Attias, S. Luski, T. Brousse, D. Aurbach, Carbon-based composite materials for supercapacitor electrodes: a review, J. Mater. Chem. A 5 (2017) 12653–12672. Available from: https://doi.org/10.1039/C7TA00863E.

[26] L.L. Zhang, R. Zhou, X.S. Zhao, Graphene-based materials as supercapacitor electrodes, J. Mater. Chem 20 (2010) 5983–5992. Available from: https://doi.org/10.1039/C000417K.

[27] Q. Ke, J. Wang, Graphene-based materials for supercapacitor electrodes − a review, J. Materiomics 2 (2016) 37–54. Available from: https://doi.org/10.1016/j.jmat.2016.01.001.

[28] B.B. Sahoo, N. Kumar, H.S. Panda, B. Panigrahy, N.K. Sahoo, A. Soam, et al., Self-assembled 3D graphene-based aerogel with Au nanoparticles as high-performance supercapacitor electrode, J. Energy Storage 43 (2021) 103157. Available from: https://doi.org/10.1016/j.est.2021.103157.

[29] L. Wu, W. Li, P. Li, S. Liao, S. Qiu, M. Chen, et al., Powder, paper and foam of few-layer graphene prepared in high yield by electrochemical intercalation exfoliation of expanded graphite, Small 10 (2014) 1421–1429. Available from: https://doi.org/10.1002/smll.201302730.

[30] J.-C. Yoon, J.-S. Lee, S.-I. Kim, K.-H. Kim, J.-H. Jang, Three-dimensional graphene nano-networks with high quality and mass production capability via precursor-assisted chemical vapor deposition, Sci. Rep 3 (2013) 1788. Available from: https://doi.org/10.1038/srep01788.

[31] M. Yu, Y. Huang, C. Li, Y. Zeng, W. Wang, Y. Li, et al., Building three-dimensional graphene frameworks for energy storage and catalysis, Adv. Funct. Mater 25 (2015) 324–330. Available from: https://doi.org/10.1002/adfm.201402964.

[32] Y. Xu, Z. Lin, X. Zhong, X. Huang, N.O. Weiss, Y. Huang, et al., Holey graphene frameworks for highly efficient capacitive energy storage, Nat. Commun 5 (2014) 4554. Available from: https://doi.org/10.1038/ncomms5554.

[33] T. Odedairo, J. Ma, Y. Gu, W. Zhou, J. Jin, X.S. Zhao, et al., A new approach to nanoporous graphene sheets via rapid microwave-induced plasma for energy applications, Nanotechnology 25 (2014) 495604. Available from: https://doi.org/10.1088/0957-4484/25/49/495604.

[34] J. Hao, Y. Liao, Y. Zhong, D. Shu, C. He, S. Guo, et al., Three-dimensional graphene layers prepared by a gas-foaming method for supercapacitor applications, Carbon 94 (2015) 879−887. Available from: https://doi.org/10.1016/j.carbon.2015.07.069.

[35] Y. Zhao, S. Huang, M. Xia, S. Rehman, S. Mu, Z. Kou, et al., N-P-O co-doped high performance 3D graphene prepared through red phosphorous-assisted "cutting-thin" technique: a universal synthesis and multifunctional applications, Nano Energy 28 (2016) 346−355. Available from: https://doi.org/10.1016/j.nanoen.2016.08.053.

[36] K. Qin, J. Kang, J. Li, E. Liu, C. Shi, Z. Zhang, et al., Continuously hierarchical nanoporous graphene film for flexible solid-state supercapacitors with excellent performance, Nano Energy 24 (2016) 158−164. Available from: https://doi.org/10.1016/j.nanoen.2016.04.019.

[37] Y. Xu, Z. Lin, X. Huang, Y. Liu, Y. Huang, X. Duan, Flexible solid-state supercapacitors based on three-dimensional graphene hydrogel films, ACS Nano 7 (2013) 4042−4049. Available from: https://doi.org/10.1021/nn4000836.

[38] Z.-S. Wu, Y. Sun, Y.-Z. Tan, S. Yang, X. Feng, K. Müllen, Three-dimensional graphene-based macro- and mesoporous frameworks for high-performance electrochemical capacitive energy storage, J. Am. Chem. Soc 134 (2012) 19532−19535. Available from: https://doi.org/10.1021/ja308676h.

[39] J.Y. Lee, K.-H. Lee, Y.J. Kim, J.S. Ha, S.-S. Lee, J.G. Son, Sea-urchin-inspired 3D crumpled graphene balls using simultaneous etching and reduction process for high-density capacitive energy storage, Adv. Funct. Mater 25 (2015) 3606−3614. Available from: https://doi.org/10.1002/adfm.201404507.

[40] M. Kota, X. Yu, S.-H. Yeon, H.-W. Cheong, H.S. Park, Ice-templated three dimensional nitrogen doped graphene for enhanced supercapacitor performance, J. Power Sources 303 (2016) 372−378. Available from: https://doi.org/10.1016/j.jpowsour.2015.11.006.

[41] S. Fan, L. Wei, X. Liu, W. Ma, C. Lou, J. Wang, et al., High-density oxygen-enriched graphene hydrogels for symmetric supercapacitors with ultrahigh gravimetric and volumetric performance, Int. J. Hydrog. Energy 46 (2021) 39969−39982. Available from: https://doi.org/10.1016/j.ijhydene.2021.09.227.

[42] J. Wang, C. Xu, D. Zhang, A three-dimensional electrode fabricated by electrophoretic deposition of graphene on nickel foam for structural supercapacitors, N. J. Chem 45 (2021) 18567−18574. Available from: https://doi.org/10.1039/D1NJ02815D.

[43] M. Haghshenas, M. Mazloum-Ardakani, L. Amiri-Zirtol, F. Sabaghian, Arginine-functionalized graphene oxide for green and high-performance symmetric supercapacitors, Int. J. Hydrog. Energy 46 (2021) 30219−30229. Available from: https://doi.org/10.1016/j.ijhydene.2021.06.170.

[44] N. An, Z. Guo, J. Xin, Y. He, K. Xie, D. Sun, et al., Hierarchical porous covalent organic framework/graphene aerogel electrode for high-performance supercapacitors, J. Mater. Chem. A 9 (2021) 16824−16833. Available from: https://doi.org/10.1039/D1TA04313G.

[45] J. Ma, Y. Yamamoto, C. Su, S. Badhulika, C. Fukuhara, C.Y. Kong, One-pot microwave-assisted synthesis of porous reduced graphene oxide as an electrode material for high capacitance supercapacitor, Electrochim. Acta 386 (2021) 138439. Available from: https://doi.org/10.1016/j.electacta.2021.138439.

[46] S. Wu, W. Chen, L. Yan, Fabrication of a 3D MnO_2/graphene hydrogel for high-performance asymmetric supercapacitors, J. Mater. Chem. A 2 (2014) 2765−2772. Available from: https://doi.org/10.1039/C3TA14387B.

[47] H. Wan, J. Liu, Y. Ruan, L. Lv, L. Peng, X. Ji, et al., Hierarchical configuration of $NiCo_2S_4$ nanotube@Ni−Mn layered double hydroxide arrays/three-dimensional graphene sponge as electrode materials for high-capacitance supercapacitors, ACS Appl. Mater. Interfaces. 7 (2015) 15840−15847. Available from: https://doi.org/10.1021/acsami.5b03042.

[48] H. Chen, X. Chang, D. Chen, J. Liu, P. Liu, Y. Xue, et al., Graphene-karst cave flower-like Ni−Mn layered double oxides nanoarrays with energy storage electrode, Electrochim. Acta 220 (2016) 36−46. Available from: https://doi.org/10.1016/j.electacta.2016.10.019.

[49] R. Wang, A. Jayakumar, C. Xu, J.-M. Lee, Ni(OH)2 nanoflowers/graphene hydrogels: a new assembly for supercapacitors, ACS Sustain. Chem. Eng 4 (2016) 3736−3742. Available from: https://doi.org/10.1021/acssuschemeng.6b00362.

[50] W. Yang, Z. Gao, J. Wang, J. Ma, M. Zhang, L. Liu, Solvothermal one-step synthesis of Ni−Al layered double hydroxide/carbon nanotube/reduced graphene oxide sheet ternary nanocomposite with ultrahigh capacitance for supercapacitors, ACS Appl. Mater. Interfaces. 5 (2013) 5443−5454. Available from: https://doi.org/10.1021/am4003843.

[51] J. Shen, J. Wu, L. Pei, M.-T.F. Rodrigues, Z. Zhang, F. Zhang, et al., CoNi2S4-graphene-2D-MoSe2 as an advanced electrode material for supercapacitors, Adv. Energy Mater 6 (2016) 1600341. Available from: https://doi.org/10.1002/aenm.201600341.

[52] G. Xiong, P. He, D. Wang, Q. Zhang, T. Chen, T.S. Fisher, Hierarchical Ni−Co hydroxide petals on mechanically robust graphene petal foam for high-energy asymmetric supercapacitors, Adv. Funct. Mater 26 (2016) 5460−5470. Available from: https://doi.org/10.1002/adfm.201600879.

[53] C. Liu, S. Zhao, Y. Lu, Y. Chang, D. Xu, Q. Wang, et al., 3D porous nanoarchitectures derived from SnS/S-doped graphene hybrid nanosheets for flexible all-solid-state supercapacitors, Small 13 (2017) 1603494. Available from: https://doi.org/10.1002/smll.201603494.

[54] B. Hou, X. Jin, L. Jiang, Y. Li, C. Qiu, D. Han, et al., Flexible porous graphene/nickel hydroxide composite films with 3D ion transport channels for high volumetric performance asymmetric super-capacitor, Appl. Surf. Sci 569 (2021) 151036. Available from: https://doi.org/10.1016/j.apsusc.2021.151036.

[55] H. Zhang, D. Yang, T. Ma, H. Lin, B. Jia, Flash-induced ultrafast production of graphene/MnO with extraordinary supercapacitance, Small Methods 5 (2021) e2100225. Available from: https://doi.org/10.1002/smtd.202100225.

[56] S. Jadhav, R.S. Kalubarme, N. Suzuki, C. Terashima, B. Kale, S.W. Gosavi, et al., Probing electrochemical charge storage of 3D porous hierarchical cobalt oxide decorated rGO in ultra-high-performance supercapaci-tor, Surf. Coat. Technol 419 (2021) 127287. Available from: https://doi.org/10.1016/j.surfcoat.2021.127287.

[57] X. Yang, A. Liu, Y. Zhao, H. Lu, Y. Zhang, W. Wei, et al., Three-dimensional macroporous polypyrrole-derived graphene electrode prepared by the hydrogen bubble dynamic template for supercapacitors and metal-free catalysts, ACS Appl. Mater. Interfaces. 7 (2015) 23731−23740. Available from: https://doi.org/10.1021/acsami.5b07982.

[58] Z. Zhang, K. Chi, F. Xiao, S. Wang, Advanced solid-state asymmetric supercapacitors based on 3D gra-phene/MnO2 and graphene/polypyrrole hybrid architectures, J. Mater. Chem. A 3 (2015) 12828−12835. Available from: https://doi.org/10.1039/C5TA02685G.

[59] Y. Wang, S. Tang, S. Vongehr, J. Ali Syed, X. Wang, X. Meng, High-performance flexible solid-state carbon cloth supercapacitors based on highly processible N-graphene doped polyacrylic acid/polyaniline composites, Sci. Rep 6 (2016) 12883. Available from: https://doi.org/10.1038/srep12883.

[60] M. Yu, Y. Ma, J. Liu, S. Li, Polyaniline nanocone arrays synthesized on three-dimensional graphene net-work by electrodeposition for supercapacitor electrodes, Carbon 87 (2015) 98−105. Available from: https://doi.org/10.1016/j.carbon.2015.02.017.

[61] Y. Xu, Y. Tao, X. Zheng, H. Ma, J. Luo, F. Kang, et al., A metal-free supercapacitor electrode material with a record high volumetric capacitance over 800 F cm − 3, Adv. Mater 27 (2015) 8082−8087. Available from: https://doi.org/10.1002/adma.201504151.

[62] L. Zhang, D. Huang, N. Hu, C. Yang, M. Li, H. Wei, et al., Three-dimensional structures of graphene/polyaniline hybrid films constructed by steamed water for high-performance supercapacitors, J. Power Sources 342 (2017) 1−8. Available from: https://doi.org/10.1016/j.jpowsour.2016.11.068.

[63] D. Wu, C. Yu, W. Zhong, Bioinspired strengthening and toughening of carbon nanotube@polyaniline/graphene film using electroactive biomass as glue for flexible supercapacitors with high rate

performance and volumetric capacitance, and low-temperature tolerance, J. Mater. Chem. A 9 (2021) 18356−18368. Available from: https://doi.org/10.1039/D1TA05729D.

[64] L. Sun, L. Wang, C. Tian, T. Tan, Y. Xie, K. Shi, et al., Nitrogen-doped graphene with high nitrogen level via a one-step hydrothermal reaction of graphene oxide with urea for superior capacitive energy storage, RSC Adv 2 (2012) 4498−4506. Available from: https://doi.org/10.1039/C2RA01367C.

[65] Z.-S. Wu, A. Winter, L. Chen, Y. Sun, A. Turchanin, X. Feng, et al., Three-dimensional nitrogen and boron co-doped graphene for high-performance all-solid-state supercapacitors, Adv. Mater 24 (2012) 5130−5135. Available from: https://doi.org/10.1002/adma.201201948.

[66] P. Chen, J.-J. Yang, S.-S. Li, Z. Wang, T.-Y. Xiao, Y.-H. Qian, et al., Hydrothermal synthesis of macroscopic nitrogen-doped graphene hydrogels for ultrafast supercapacitor, Nano Energy 2 (2013) 249−256. Available from: https://doi.org/10.1016/j.nanoen.2012.09.003.

[67] X. Chen, X. Chen, X. Xu, Z. Yang, Z. Liu, L. Zhang, et al., Sulfur-doped porous reduced graphene oxide hollow nanosphere frameworks as metal-free electrocatalysts for oxygen reduction reaction and as supercapacitor electrode materials, Nanoscale 6 (2014) 13740−13747. Available from: https://doi.org/10.1039/C4NR04783D.

[68] J. Li, X. Yun, Z. Hu, L. Xi, N. Li, H. Tang, et al., Three-dimensional nitrogen and phosphorus co-doped carbon quantum dots/reduced graphene oxide composite aerogels with a hierarchical porous structure as superior electrode materials for supercapacitors, J. Mater. Chem. A 7 (2019) 26311−26325. Available from: https://doi.org/10.1039/C9TA08151H.

[69] N.A. Elessawy, J. El Nady, W. Wazeer, A.B. Kashyout, Development of high-performance supercapacitor based on a novel controllable green synthesis for 3D nitrogen doped graphene, Sci. Rep 9 (2019) 1129. Available from: https://doi.org/10.1038/s41598-018-37369-x.

[70] X. Liu, Z. Lu, X. Huang, J. Bai, C. Li, C. Tu, et al., Self-assembled S,N co-doped reduced graphene oxide/MXene aerogel for both symmetric liquid- and all-solid-state supercapacitors, J. Power Sources 516 (2021) 230682. Available from: https://doi.org/10.1016/j.jpowsour.2021.230682.

[71] A. Kumar, C.-S. Tan, N. Kumar, P. Singh, Y. Sharma, J. Leu, et al., Pentafluoropyridine functionalized novel heteroatom-doped with hierarchical porous 3D cross-linked graphene for supercapacitor applications, RSC Adv 11 (2021) 26892−26907. Available from: https://doi.org/10.1039/D1RA03911C.

[72] Q. Liu, L. Zhang, H. Chen, J. Jin, N. Wang, Y. Wang, et al., Sulfur and nitrogen co-doped three-dimensional graphene aerogels for high-performance supercapacitors: a head to head vertical bicyclic molecule both as pillaring agent and dopant, Appl. Surf. Sci 565 (2021) 150453. Available from: https://doi.org/10.1016/j.apsusc.2021.150453.

[73] X. Sun, X. Liu, F. Li, Sulfur-doped laser-induced graphene derived from polyethersulfone and lignin hybrid for all-solid-state supercapacitor, Appl. Surf. Sci 551 (2021) 149438. Available from: https://doi.org/10.1016/j.apsusc.2021.149438.

[74] Z. Li, B. Li, L. Du, W. Wang, X. Liao, H. Yu, et al., Three-dimensional oxygen-doped porous graphene: sodium chloride-template preparation, structural characterization and supercapacitor performances, Chin. J. Chem. Eng 40 (2021) 304−314. Available from: https://doi.org/10.1016/j.cjche.2020.11.042.

[75] N. Nitta, F. Wu, J.T. Lee, G. Yushin, Li-ion battery materials: present and future, Mater. Today 18 (2015) 252−264. Available from: https://doi.org/10.1016/j.mattod.2014.10.040.

[76] J. Michael, E.Kendrick Lain, Understanding the limitations of lithium ion batteries at high rates, J. Power Sources 493 (2021) 229690. Available from: https://doi.org/10.1016/j.jpowsour.2021.229690.

[77] H.-H. Chang, T.-H. Ho, Y.-S. Su, Graphene-enhanced battery components in rechargeable lithium-ion and lithium metal batteries, C 7 (2021) 65. Available from: https://doi.org/10.3390/c7030065.

[78] Z. Chen, H. Li, R. Tian, H. Duan, Y. Guo, Y. Chen, et al., Three dimensional graphene aerogels as binder-less, freestanding, elastic and high-performance electrodes for lithium-ion batteries, Sci. Rep 6 (2016) 27365. Available from: https://doi.org/10.1038/srep27365.

[79] H. Ji, L. Zhang, M.T. Pettes, H. Li, S. Chen, L. Shi, et al., Ultrathin graphite foam: a three-dimensional conductive network for battery electrodes, Nano Lett 12 (2012) 2446−2451. Available from: https://doi.org/10.1021/nl300528p.

[80] B. Wang, W.A. Abdulla, D. Wang, X.S. Zhao, A three-dimensional porous $LiFePO_4$ cathode material modified with a nitrogen-doped graphene aerogel for high-power lithium ion batteries, Energy Environ. Sci. 8 (2015) 869−875. Available from: https://doi.org/10.1039/C4EE03825H.

[81] H. Zhu, X. Wu, L. Zan, Y. Zhang, Three-dimensional macroporous graphene−Li_2FeSiO_4 composite as cathode material for lithium-ion batteries with superior electrochemical performances, ACS Appl. Mater. Interfaces. 6 (2014) 11724−11733. Available from: https://doi.org/10.1021/am502408m.

[82] D.P. Dubal, P. Gomez-Romero, All nanocarbon Li-Ion capacitor with high energy and high power density, Mater. Today Energy 8 (2018) 109−117. Available from: https://doi.org/10.1016/j.mtener.2018.03.005.

[83] F. Tu, S. Liu, T. Wu, G. Jin, C. Pan, Porous graphene as cathode material for lithium ion capacitor with high electrochemical performance, Powder Technol 253 (2014) 580−583. Available from: https://doi.org/10.1016/j.powtec.2013.12.008.

[84] V. Aravindan, D. Mhamane, W.C. Ling, S. Ogale, S. Madhavi, Nonaqueous lithium-ion capacitors with high energy densities using trigol-reduced graphene oxide nanosheets as cathode-active material, ChemSusChem 6 (2013) 2240−2244. Available from: https://doi.org/10.1002/cssc.201300465.

[85] H. Wang, C. Guan, X. Wang, H.J. Fan, A. High, Energy and power Li-Ion capacitor based on a TiO_2 nanobelt array anode and a graphene hydrogel cathode, Small 11 (2015) 1470−1477. Available from: https://doi.org/10.1002/smll.201402620.

[86] H. Li, L. Shen, J. Wang, S. Fang, Y. Zhang, H. Dou, et al., Three-dimensionally ordered porous TiNb2O7 nanotubes: a superior anode material for next generation hybrid supercapacitors, J. Mater. Chem. A 3 (2015) 16785−16790. Available from: https://doi.org/10.1039/C5TA02929E.

[87] L. Ye, Q. Liang, Y. Lei, X. Yu, C. Han, W. Shen, et al., A high performance Li-ion capacitor constructed with $Li_4Ti_5O_{12}$/C hybrid and porous graphene macroform, J. Power Sources 282 (2015) 174−178. Available from: https://doi.org/10.1016/j.jpowsour.2015.02.028.

[88] D. Sui, L. Xu, H. Zhang, Z. Sun, B. Kan, Y. Ma, et al., A 3D cross-linked graphene-based honeycomb carbon composite with excellent confinement effect of organic cathode material for lithium-ion batteries, Carbon 157 (2020) 656−662. Available from: https://doi.org/10.1016/j.carbon.2019.10.106.

[89] J. Wang, J. Liu, D. Chao, J. Yan, J. Lin, Z.X. Shen, Self-assembly of honeycomb-like MoS2 nanoarchitectures anchored into graphene foam for enhanced lithium-ion storage, Adv. Mater 26 (2014) 7162−7169. Available from: https://doi.org/10.1002/adma.201402728.

[90] J. Qin, X. Wang, M. Cao, C. Hu, Germanium quantum dots embedded in N-doping graphene matrix with sponge-like architecture for enhanced performance in lithium-ion batteries, Chem. − A Eur. J 20 (2014) 9675−9682. Available from: https://doi.org/10.1002/chem.201402151.

[91] J. Ji, H. Ji, L.L. Zhang, X. Zhao, X. Bai, X. Fan, et al., Graphene-encapsulated si on ultrathin-graphite foam as anode for high capacity lithium-ion batteries, Adv. Mater 25 (2013) 4673−4677. Available from: https://doi.org/10.1002/adma.201301530.

[92] H.-J. Kim, S.E. Lee, J. Lee, J.-Y. Jung, E.-S. Lee, J.-H. Choi, et al., Gold-coated silicon nanowire−graphene core−shell composite film as a polymer binder-free anode for rechargeable lithium-ion batteries, Phys. E: Low-Dimensional Syst. Nanostruct 61 (2014) 204−209. Available from: https://doi.org/10.1016/j.physe.2014.03.030.

[93] N. Li, H. Song, H. Cui, C. Wang, Sn@graphene grown on vertically aligned graphene for high-capacity, high-rate, and long-life lithium storage, Nano Energy 3 (2014) 102−112. Available from: https://doi.org/10.1016/j.nanoen.2013.10.014.

[94] Y. Zhu, W. Liu, X. Zhang, J. He, J. Chen, Y. Wang, et al., Directing silicon−graphene self-assembly as a core/shell anode for high-performance lithium-ion batteries, Langmuir 29 (2013) 744−749. Available from: https://doi.org/10.1021/la304371d.

[95] R. Wang, C. Xu, J. Sun, L. Gao, H. Yao, Solvothermal-induced 3D macroscopic SnO_2/nitrogen-doped graphene aerogels for high capacity and long-life lithium storage, ACS Appl. Mater. Interfaces. 6 (2014) 3427−3436. Available from: https://doi.org/10.1021/am405557c.

[96] Y. Chang, J. Li, B. Wang, H. Luo, H. He, Q. Song, et al., Synthesis of 3D nitrogen-doped graphene/Fe_3O_4 by a metal ion induced self-assembly process for high-performance Li-ion batteries, J. Mater. Chem. A 1 (2013) 14658−14665. Available from: https://doi.org/10.1039/C3TA13370B.

[97] B. Liu, X. Ren, J. Yin, K. Zhu, J. Yan, K. Ye, et al., Anchored graphene aerogel as a conductive agent-free electrode for high-performance lithium-ion batteries, ACS Appl. Energy Mater 5 (2022) 567−574. Available from: https://doi.org/10.1021/acsaem.1c03083.

[98] Self-assembled homogeneous SiOC@C/graphene with three-dimensional lamellar structure enabling improved capacity and rate performances for lithium ion storage - ScienceDirect, (n.d.). https://www.sciencedirect.com/science/article/pii/S0008622321010022 (accessed April 12, 2022).

[99] S. Wang, F. Zhang, W.-J. Yu, H. Tong, F. Liu, C. Wang, S/N-co-doped GNRs-wrapped Si nanoparticles composite as an anode material for Li-ion batteries, Ionics (2022). Available from: https://doi.org/10.1007/s11581-022-04527-1.

[100] C. Guo, K. Pan, Y. Xie, L. Li, Monodispersed copper phosphide nanocrystals in situ grown in a nitrogen-doped reduced graphene oxide matrix and their superior performance as the anode for lithium-ion batteries, Inorg. Chem. Front. (2022). Available from: https://doi.org/10.1039/D1QI01456K.

[101] R. Wang, C. Xu, J. Sun, L. Gao, C. Lin, Flexible free-standing hollow Fe_3O_4/graphene hybrid films for lithium-ion batteries, J. Mater. Chem. A 1 (2013) 1794−1800. Available from: https://doi.org/10.1039/C2TA00753C.

[102] W. Wei, S. Yang, H. Zhou, I. Lieberwirth, X. Feng, K. Müllen, 3D graphene foams cross-linked with pre-encapsulated Fe_3O_4 nanospheres for enhanced lithium storage, Adv. Mater 25 (2013) 2909−2914. Available from: https://doi.org/10.1002/adma.201300445.

[103] R. Wang, C. Xu, J. Sun, Y. Liu, L. Gao, C. Lin, Free-standing and binder-free lithium-ion electrodes based on robust layered assembly of graphene and Co_3O_4 nanosheets, Nanoscale 5 (2013) 6960−6967. Available from: https://doi.org/10.1039/C3NR01392H.

[104] A. Yu, H.W. Park, A. Davies, D.C. Higgins, Z. Chen, X. Xiao, Free-standing layer-by-layer hybrid thin film of graphene-MnO_2 nanotube as anode for lithium ion batteries, J. Phys. Chem. Lett. 2 (2011) 1855−1860. Available from: https://doi.org/10.1021/jz200836h.

[105] B. Wang, X. Li, X. Zhang, B. Luo, M. Jin, M. Liang, et al., Adaptable silicon−carbon nanocables sandwiched between reduced graphene oxide sheets as lithium ion battery anodes, ACS Nano 7 (2013) 1437−1445. Available from: https://doi.org/10.1021/nn3052023.

[106] Y. Gong, S. Yang, Z. Liu, L. Ma, R. Vajtai, P.M. Ajayan, Graphene-network-backboned architectures for high-performance lithium storage, Adv. Mater 25 (2013) 3979−3984. Available from: https://doi.org/10.1002/adma.201301051.

[107] J. Luo, J. Liu, Z. Zeng, C.F. Ng, L. Ma, H. Zhang, et al., Three-dimensional graphene foam supported Fe_3O_4 lithium battery anodes with long cycle life and high rate capability, Nano Lett 13 (2013) 6136−6143. Available from: https://doi.org/10.1021/nl403461n.

[108] N. Li, Z. Chen, W. Ren, F. Li, H.-M. Cheng, Flexible graphene-based lithium ion batteries with ultra-fast charge and discharge rates, Proc. Natl Acad. Sci. U S A 109 (2012) 17360−17365. Available from: https://doi.org/10.1073/pnas.1210072109.

Index

Printed in the United States
by Baker & Taylor Publisher Services